Mechanical and Electrical Systems for Construction Managers

AMERICAN TECHNICAL PUBLISHERS
ORLAND PARK, IL 60467-5756

American Technical Publishers, Inc., Editorial Staff

Editor in Chief:
Jonathan F. Gosse
Vice President—Production:
Peter A. Zurlis
Art Manager:
James M. Clarke
Copy Editor:
Catherine A. Mini
Cover Design:
Jennifer M. Hines
Illustration/Layout:
Jennifer M. Hines
William J. Sinclair
Lauren A. Gately

2 3 4 5 6 7 8 9 – 10 – 9 8 7 6 5 4 3 2 1

Printed in the United States of America

ISBN 978-0-8269-9360-1

Acknowledgments

The author and publisher are grateful to the following companies, organizations, and individuals for providing photographs, information, and technical assistance:

Mechanical-Electrical Academic Consortium

Advantage Drills Inc.
Air Conditioning Contractors of America
ALCO Controls Division Emerson Electric Company
Ansonics, Inc.
A.O. Smith Corporation
ASHRAE Handbook - Fundamentals
Bacharach, Inc.
Badger Meter, Inc.
Baldor Electric Co.
Broan-NuTone LLC
Bussman Mfg., a McGraw-Edison Co. Division
Carrier Corporation
Cherne Industries, Inc.
Cleaver-Brooks
Cooper Wiring Devices
Copper Development Association
Crane Valves
Danfoss Drives
DOE/NREL, Ben Kroposki
DOE/NREL, Warren Gretz
Dresser Instruments
Ductsox
Dwyer Instruments Inc.
Eaton Corp., Cutler-Hammer Products
Eljer Plumbingware, Inc.
Elkay Mfg. Co.
Fluke Corporation
The Foxboro Company
General Electric Company
Greenheck
Greenlee Textron Inc.
Honeywell Chemicals
Honeywell, Inc.
Ideal Industries, Inc.
ITT Bell & Gossett
Jackson Systems, LLC
Jenkins Bros.
Jun-Air USA, Inc.
Klein Tools, Inc.
Kohler Co.
Lau, a Division of Tonkins Industries
Lennox Industries Inc.
Leviton Manufacturing Co., Inc.
Lunkenheimer Co.
McDonnell-Miller
McQuay International
Milwaukee Electric Tool Corp.
Moen, Inc.
Mueller Co.
MTE Corporation
National Wood Flooring Association
NIBCO, Inc.
Plan Ahead, Inc.
Ranco Inc.
Reed Manufacturing Co.
Ridge Tool Company
Riello Burners
Rockwell Automation, Allen-Bradley Company, Inc.
Ruud Lighting, Inc.
Sarco Co., Inc.
Siemens
Sloan Valve Company
Sporlan Valve Company
The Stanley Works
Sterling
Superior Boiler Works, Inc.
The Trane Company
TSI Incorporated
Uponor Wirsbo
Victaulic Company of America
The Wadsworth Electric Mtg. Co., Inc.
Watts Regulator Company
York International Corp.

Contents

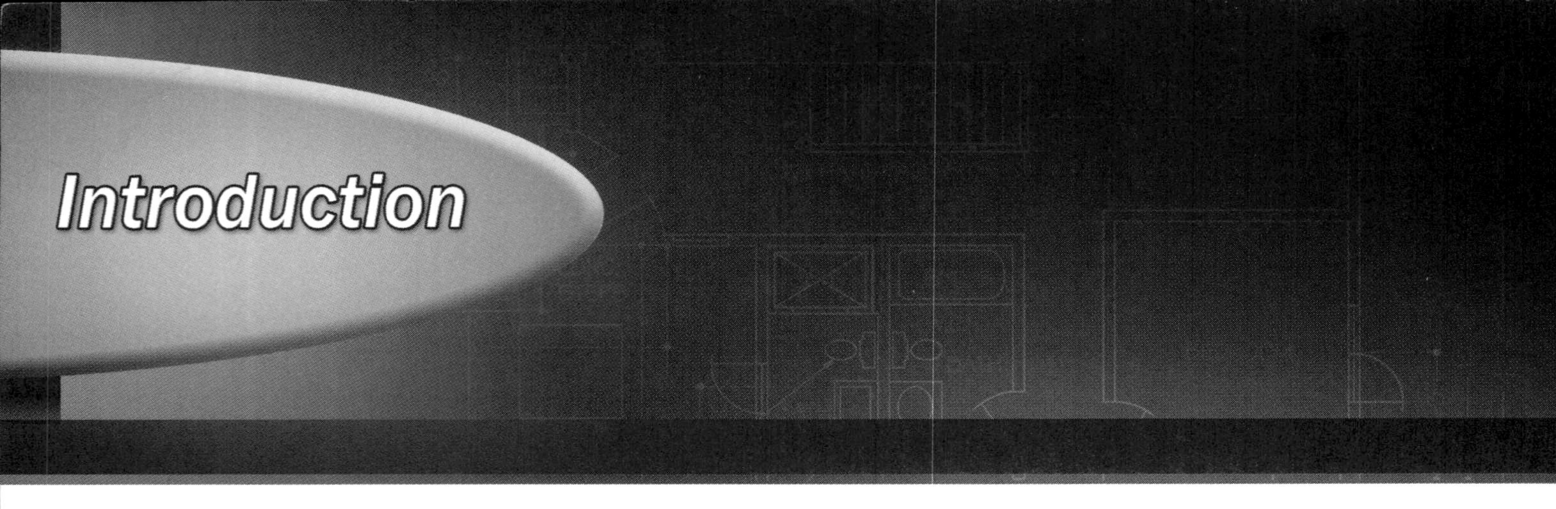

Introduction

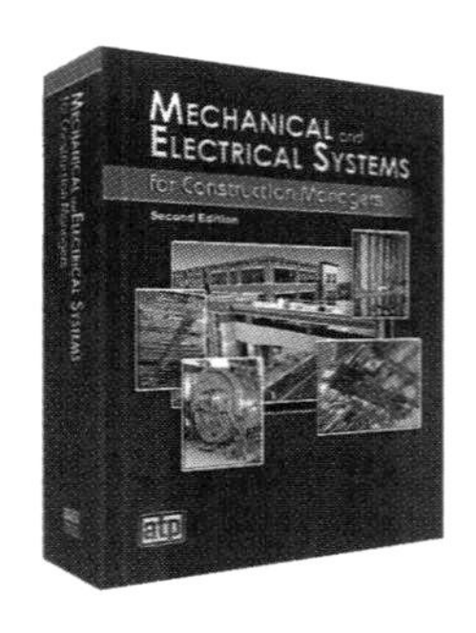

Today's construction managers require a thorough understanding of all aspects of a wide range of mechanical, electrical, and plumbing systems. *Mechanical and Electrical Systems for Construction Managers,* 2nd Edition, provides a comprehensive overview of mechanical and electrical systems as required by general contractors, construction managers, and supervisors. Information on mechanical and electrical systems is presented with concise text and comprehensive illustrations.

This edition contains the latest information on building automation systems, including control devices, signals, and logic. Each chapter includes an introduction that provides an overview of key content found in the chapter. Detailed illustrations clearly convey concepts, equipment, and practices. Tech Facts provide informative technical information related to key chapter content. The text also includes application photographs and illustrations that enhance the text.

Each chapter concludes with Review Questions that test for comprehension of the content covered. The Appendix contains useful tables and charts used in the trade. An extensive Glossary provides easy-to-find definitions for key terms. The comprehensive Index uses cross-references so information can be found quickly and easily.

Mechanical and Electrical Systems for Construction Managers Workbook is designed to reinforce information presented in *Mechanical and Electrical Systems for Construction Managers.* Each chapter in the workbook covers information from the corresponding chapter in the textbook. The textbook may be used as a reference to complete the questions in the workbook. The appropriate textbook pages should be carefully reviewed before completing the review questions. When studying the textbook, particular attention should be paid to illustrations, examples, and italicized terms.

The question types used in the workbook include true-false, multiple choice, completion, and matching. For true-false questions, circle T if the statement is true or F if the statement is false. For multiple choice and matching questions, write the letter of the correct answer in the answer blank next to the question. For completion questions, write the correct answer(s) in the answer blank next to the question.

Mechanical and Electrical Systems for Construction Managers and *Mechanical and Electrical Systems for Construction Managers Workbook* are two of many high-quality training products available from American Technical Publishers, Inc. To obtain more information about related training products, visit the American Tech website at www.go2atp.com.

The Publisher

Features

***Application Photos** enhance text and illustrations*

***Review Questions Test** for comprehension of content covered*

***Chapter Introductions** provide overview of key content found in chapter*

***Detailed illustrations** clearly convey concepts, equipment, and practices*

***Tech Facts** provide informative technical information related to key chapter content*

Fluke Corporation

Handheld scope meters are used when signal measurements must be taken in the field.

Review Questions

1. What is the difference between an analog display and a digital display?
2. What causes ghost voltages?
3. Why might a voltage indicator be used?
4. Describe the procedure for using a voltage tester.
5. Why should personal protective equipment be worn when using a DMM?
6. How can an in-line ammeter be used to measure current within a circuit?
7. What can be gained by measuring resistance in a circuit?
8. Define speed.
9. Why is a scope sometimes more useful than a DMM when examining voltage?
10. List four safety rules for using test instruments.

MECHANICAL AND ELECTRICAL SYSTEMS FOR CONSTRUCTION MANAGERS

24

Electrical Plans and Connections

An electrical plan is a drawing and list that indicates the electrical devices to be used and the electrical wiring methods. A good electrical plan also indicates the location of all major electrical devices and connections. Electrical connections must be mechanically and electrically secure. Electrical connections require supplies such as solder, tape, wire nuts, and wire markers.

ELECTRICAL PRINTS

Pictorial Drawings

Electrical Layouts

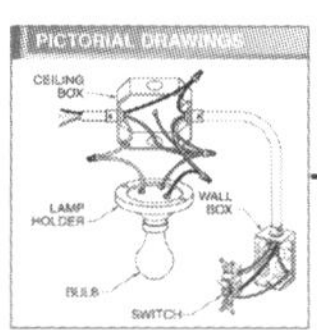

Figure 24-1. Pictorial drawings show the physical details of electrical components as seen by the eye.

RESIDENTIAL ELECTRICAL CIRCUITS

INDOOR AIR QUALITY

Health Effects

TSI Inc.

Supply air ductwork that is not properly maintained may affect indoor air quality.

INDOOR AIR QUALITY GUIDELINES

Figure 8-8. Energy conservation measures reduce the amount of outside air used in a building, which contributes to the buildup of indoor air contaminants.

Today's buildings must meet the needs of current and future occupants. Construction managers must be well-versed in the methods of building efficiency. Mechanical and electrical systems must be taken into account when planning the efficiency of a building beyond basic construction. Mechanical systems include both heating, ventilation, and air conditioning (HVAC) and plumbing systems.

1 Mechanical and Electrical Systems

SCOPE

Today's building construction is considerably more complex than it has ever been. In the past, not all buildings were designed with adequate ventilation, electrical support, or water supply and removal systems in place. Constant progress has been made in terms of designing and includes systems for everyday comfort, lighting, or water supply within a building.

HVAC History

Individuals have attempted to improve the comfort of their living environment for thousands of years. Early control of comfort levels in a living environment was done by hand. For example, an individual may have controlled the temperature in a living environment by adding fuel to a fire or allowing the fire to die down.

Credit is often given to the Romans for having created the first central heating systems. They used empty spaces underneath floors to guide hot air from a furnace. As structures grew in size and contained a greater number of rooms, hot air from a central fire was regulated manually by opening or closing a diffuser (damper) through the use of an adjusting pulley. By the 18th century, water-based systems were designed and used to circulate building heat.

Early attempts to produce air conditioning were made by Dr. John Gorrie in the mid-1800s. Dr. Gorrie developed an ice machine and had hopes of expanding the technology to cool buildings. Unfortunately, he was never able to get the financing for this work and air conditioning was not practically applied until 1902. In that year, an electrical air conditioning system was invented. The system was mainly used in workplaces. By the 1950s, air conditioning systems became more popular for residential use. **See Figure 1-1.** Today, almost every residential and commercial building has systems in place for HVAC.

Figure 1-1. Today, heating and cooling systems can be combined into one combination unit.

Plumbing History

The plumbing trade has an old and interesting history. There is evidence of sanitation and plumbing skill from ancient times, such as water-conveying aqueducts constructed of terra cotta and brick, and holes in the ground lined with tile and used as bathtubs. Early Egyptian, Grecian, and Roman rulers promoted sanitary facilities. An individual who worked in the sanitary field in ancient Rome was called a "plumbarius," taken from the Latin word "plumbum," which means lead. The term "plumbarius" was appropriate since the work consisted of shaping lead. Until recently, lead was commonly used for water supply piping and wastewater removal.

In American history, plumbing developments began as early as 1652. Boston developed the first city water system in the colonies, primarily for firefighting and domestic use. In 1857, Julius Adams developed the framework upon which modern sewage disposal was based. Adams developed designs and guidelines that made modern-day sanitary engineering possible. After the Civil War, plumbing systems were improved slowly but steadily. Patents were issued on traps and methods of ventilation. Public water supply and sewage disposal systems became more common, and plumbing became a necessity.

Around 1900, the first prototype of a siphonic washdown water closet was developed. The original siphonic washdown water closet sometimes failed to provide the necessary siphonic action and allowed the waste to overflow. Up to 1900, few urban homes provided more than a hydrant and a slop hopper for waste disposal. After the turn of the century, hoppers and washdown water closets, sinks, and bathtubs were provided inside a building.

From 1929 to 1954, sales of plumbing products and heating equipment increased from $498 million to $2.33 billion, reflecting a desire for sanitary plumbing facilities in homes and offices. Scientific methods began to play a role in constructing plumbing systems. Fixture traps were ventilated, and hot and cold running water was introduced. The design of the siphonic washdown water closet was improved during this period, and states began developing legislation for the control of sanitation. Modern manufacturing methods provided materials and equipment that could be scientifically incorporated into a plumbing system. **See Figure 1-2.** Plumbing systems are now an integral part of the process of creating a new building or home.

Figure 1-2. Pathways and piping for water supply and wastewater removal must be included in the design of a new building.

Electrical History

In 600 BC, the Greeks discovered a substance, amber (fossilized sap), that when rubbed against wool, caused other substances to be attracted to it. The word electricity is derived from the Greek word for amber. Today, this attraction is referred to as static electricity. Static electricity is one of two basic forms of electricity. The other form of electricity is generated.

Evidence has been found that basic forms of electricity were produced as early as 250 BC. Between the 16th and 18th centuries, advances in electricity were made across Europe and in Japan. In the mid-eighteenth century, Benjamin Franklin invented the lightning rod and discovered the relationship between positive and negative electricity.

In the early 1900s, electricity came into wide distribution in many cities. This new power source was adapted to provide automatic building environment control. Electricity allowed increased control and required less time and attention than a manual control system. Early electric controls were large, bulky, unreliable, inaccurate by modern standards, and dangerous, partly because early electrical power quality and distribution were poor.

Since that time, standards have improved greatly and electricity is the most widely used form of energy. **See Figure 1-3.** Electricity is used to provide energy for traditional electrical applications such as lighting, heating, cooling, cooking, communication, and transportation.

Figure 1-3. Electrical power is often produced and then distributed across a wide region. Transformers may be used to facilitate this distribution.

Tech Facts

Electricity is rated in terms of voltage. A volt is defined as electric potential difference. As a unit, the volt derives its name from the physicist Alessandro Volta, who invented the first chemical battery.

BUILDING DESIGN PLANNING

Space required and allotted for a building should be one of the first considerations in planning a new project. Whether the building will use a low-rise, high-rise, or skyscraper layout will determine the types of systems needed to sustain any occupants using the building. As the overall building structure and exterior design become more elaborate, the storage space needs for the systems will increase, and thus the cost will increase as well.

Space Utilization

The space a building occupies will be influenced by many factors. Current federal, state, and local building codes must be followed prior to the start of and during the building of a new project. Building shape and height will be influenced by the intended use and the approximate number of future occupants. Buildings constructed to hold multiple businesses, for instance, may require more floors than traditional low-rise buildings can offer.

Building Efficiency Factors

Some states may offer incentives to contractors to utilize energy-efficient systems and practices on new construction. Incentives may be provided for reducing lighting loads, installing high-efficiency AC units and windows, and using solar methods of heating.

Energy-efficient building measures often require a higher initial budget. However, in complying with state codes, benefits are seen in incentives offered by gas and electric companies and in long-term gas and electric bill savings.

DOE/NREL, Ben Kroposki

Energy-saving measures must be taken into account when designing new construction. The photovoltaic cells on this building provide shade and reduce electricity needs.

COST FACTORS

Historically, initial cost planning for new construction included building materials, tools, and labor. New construction will be affected not only by the materials and number of workers needed, but also by the requirements of mechanical and electrical systems. The more extensive an electrical wiring system is, the more cost added to the project. Similar considerations must be given to plumbing and HVAC. The added systems in the building will often require the construction of storage space.

Roughly 45% of a new construction budget will be attributed to these systems. A project will often incur plumbing and HVAC costs of 30%. Another 15% of the cost comes from electrical needs.

BUILDING CHECKLIST

HVAC, plumbing, and electrical systems will be required on any new project. The scope of the systems needed should be identified early in the process. Planning for these systems early will make it easier to budget the proper amount of money, space, and labor.

HVAC Systems

HVAC systems can include different types of heating systems and air conditioning systems. Heating systems are often split into forced-air or steam, and hydronic heating.

Plumbing Systems

Plumbing systems will mainly include piping for sanitary, vent, and storm water drainage. Water supply needs will affect piping. Basic fixtures and appliances must also be considered.

Electrical Systems

Electrical systems will include circuits, switches, and transformers. Electrical systems entail more than lighting needs and power for machinery. The span of electrical systems will likely include fire alarm and building security systems as well.

Review Questions

1. When was the air conditioner invented?
2. From what is the word “electricity” derived?
3. What is the origin of the title “plumbarius”?
4. How can a new project benefit from the installation of energy-efficient systems?
5. What percentage of a construction budget is generally set aside for HVAC, plumbing, and electricity?

There are a variety of plumbing materials that are used for plumbing systems. Pipe and fitting materials can be classified as plastic, copper, cast iron soil pipe, or steel. Valves are used to regulate fluid flow within a plumbing system. Water meters measure and indicate water usage for a plumbing system water supply. Local plumbing codes establish which plumbing materials may or may not be used for a plumbing system.

2 Plumbing Materials

The potable water supply, sanitary drainage and vent, and storm water drainage systems are constructed using pipe, fittings, valves, and meters. *Pipe* is cylindrical tube used for conveying potable water, wastewater, water-borne waste, and air from one location to another. A *fitting* is a device fastened to the ends of pipes to make connections between individual pipes. A *valve* is a fitting used to regulate fluid flow within a system. A *meter* is a device used to measure and indicate fluid flow, such as a gas or water meter.

Plumbing pipe and fitting materials are classified into four groups:

- plastic
- copper
- cast iron soil pipe
- steel

Local plumbing codes specify the type of piping material that may be used for each plumbing system. Local plumbing codes are based on local conditions such as soil types, ground conditions, local rainfall, and frost or freezing conditions. Consult the local plumbing code to ensure that only code-approved material is being used for plumbing systems.

"Size" is the general term commonly used when referring to pipe diameter, and should not be taken as the actual outside diameter of the pipe. For example, the actual outside diameters of 4″ size copper tube and cast iron soil pipe are 4⅛″ and 4⅜″, respectively.

PLASTIC PIPE AND FITTINGS

Plastics are a family of synthetic materials manufactured from petroleum-based products and chemicals, including oil, natural gas, and coal. During the manufacturing process, raw materials are converted into resins, which are classified into two general types—thermosetting and thermoplastic. A *thermosetting resin* is plastic resin that cannot be remelted after it is formed and cured in its final shape. A *thermoplastic resin* is a plastic resin that can be heated and reformed repeatedly with little or no degradation in physical characteristics. Thermoplastics are commonly used for plumbing pipe and fittings.

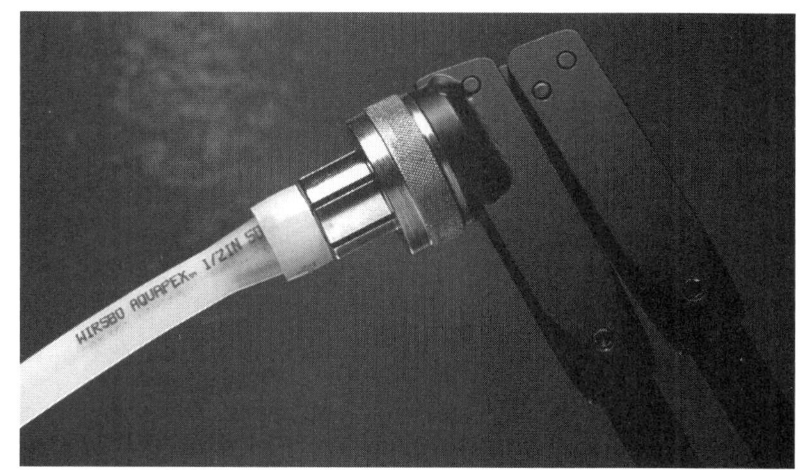

Uponor Wirsbo

Some cross-linked polyethylene (PEX) tubing is expanded to accept a fitting and then returns to its original shape to secure the fitting in position..

Plastic pipe is manufactured by the extrusion process in which plastic resins are heated, softened, and forced through a circular die, which forms the cylindrical pipe shape. **See Figure 2-1.** Plastic pipe fittings are manufactured by the injection molding process in which plastic resins are heated, softened, and forced into a cool cavity that is shaped like the desired fitting.

Plastic pipe and fittings are lightweight, inexpensive, and easily joined. In addition, plastic pipe and fittings:

- are resistant to most household chemicals, acids, and other corrosive liquids
- have smooth interior walls to ensure flow of contents and minimize sludge and slime buildup
- have low thermal conductivity, which allows liquids being conveyed to maintain more uniform temperatures
- do not decay and resist bacteria growth that could cause offensive odors
- have good flexibility, which allows long pipe runs with a minimum number of joints
- do not conduct electricity and are not subject to galvanic or electrolytic corrosion

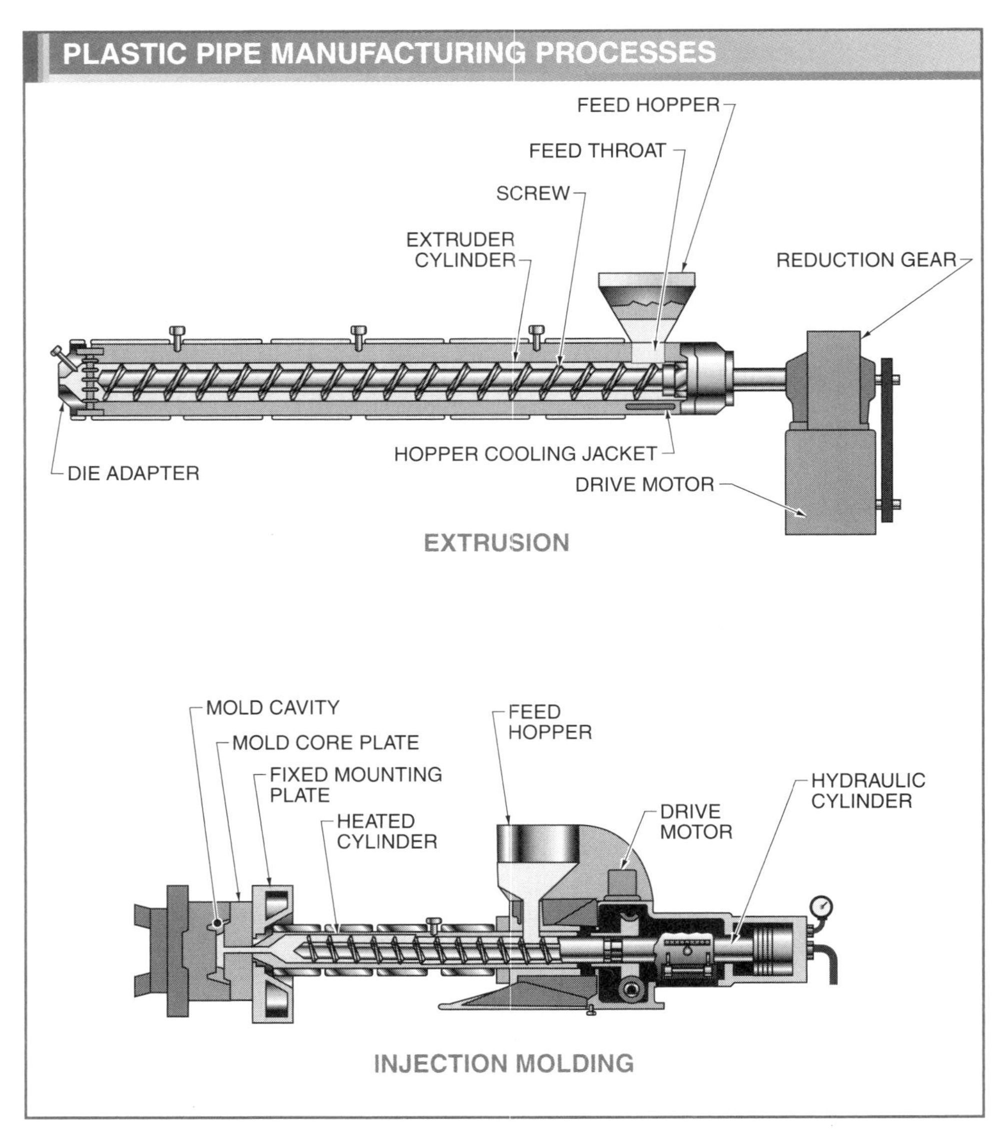

Figure 2-1. Plastic pipe and fittings are manufactured using the extrusion or the injection molding process.

Plastic pipe and fittings have low heat resistance and give off harmful vapors when burned. In addition, plastic pipe and fittings have a very high rate of expansion and contraction when heated and cooled. In some applications, an expansion loop must be made to compensate for the expansion and contraction of plastic tubing. **See Figure 2-2.**

An *expansion loop* is a loop in plastic tubing and provides an area for the tubing to expand and contract without stressing. Always refer to manufacturer recommendations regarding expansion loop specifications. Plastic pipe and fittings are flexible and require hangers and supports spaced at closer intervals than metallic pipe. When compared to metallic pipe, plastic pipe and fittings have a low crush resistance.

A variety of plastic piping materials are available. **See Figure 2-3.** Plastic piping materials used for plumbing or plumbing-related systems include:

- acrylonitrile-butadiene-styrene (ABS)
- polyvinyl chloride (PVC)
- chlorinated polyvinyl chloride (CPVC)
- cross-linked polyethylene (PEX)

In addition, polyethylene (PE), polybutylene (PB), and polypropylene (PP) are permitted for plumbing applications in some jurisdictions. Always consult the local plumbing code regarding materials that can be used in the local jurisdiction.

Plastic pipe is available in a variety of colors for easy identification. The applications generally represented by the various colors are:

- gas distribution—yellow, or black with yellow stripes
- water distribution—black, light blue, white, clear, or gray
- sewer and mains—green, white, black, or gray
- drainage, waste, and vent piping—black, or white
- hot and cold water distribution—tan, red, white, blue, silver, or clear
- fire sprinklers—orange
- industrial process piping—dark gray if PVC; light gray if CPVC

Plastic pipe is marked so that it can be readily identified even if it is cut into short pieces. Most standards require plastic pipe to include the manufacturer name or trademark; recognized standardization authority; pipe size; resin type; "DWV" for drainage, waste, and venting applications; schedule number; and laboratory seal or mark attesting to appropriateness for potable water usage if the pipe is to be used for potable water applications. **See Figure 2-4.**

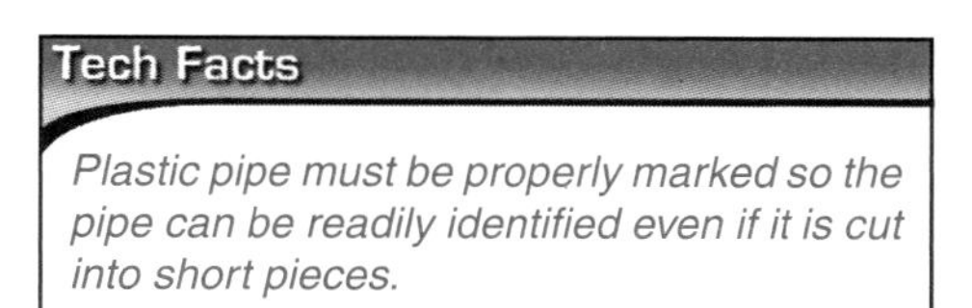

Plastic pipe must be properly marked so the pipe can be readily identified even if it is cut into short pieces.

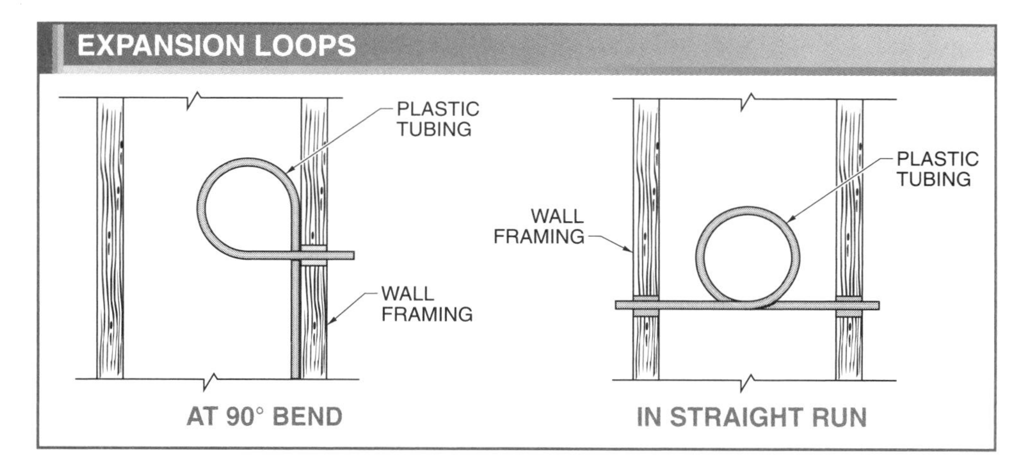

Figure 2-2. An expansion loop in PEX plastic tubing compensates for expansion and contraction of the tubing due to temperature changes.

PLASTIC PIPING MATERIALS							
Applications	Piping Material						
	ABS	PVC	CPVC	PEX	PE	PB	PP
Drain, Waste, and Vent	•	•					•
Hot and Cold Water Distribution			•	•		•	•
Outside Sewers and Mains	•	•					
Subsoil Septic Fields		•			•		
Fire Sprinkler Piping			•			•	
Water Piping	•	•	•	•	•	•	

Figure 2-3. A variety of plastic piping materials are used for plumbing or plumbing-related applications.

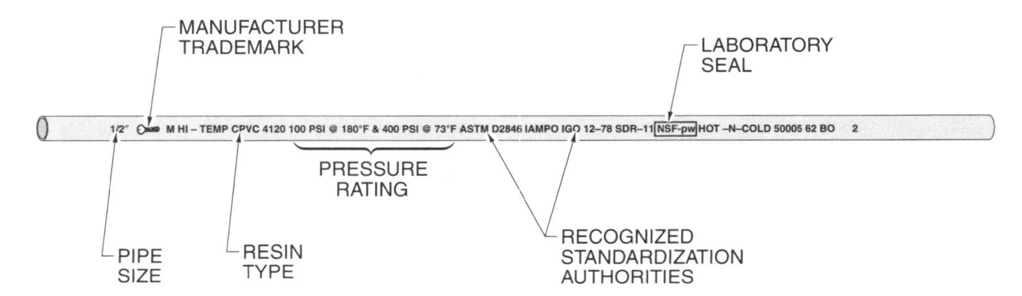

Figure 2-4. Plastic pipe is marked with manufacturer name or trademark, recognized standardization authority, pipe size, resin type, schedule number, and laboratory seal or mark.

Acrylonitrile-Butadiene-Styrene (ABS) Pipe and Fittings

Acrylonitrile-butadiene-styrene (ABS) pipe and fittings, or Schedule 40 ABS DWV, is black plastic pipe and fittings used for sanitary drainage and vent piping, and aboveground and underground storm water drainage. ABS pipe and fittings are easier and less expensive to install than metal pipe, and take less time to rough-in than metal DWV products. In addition, ABS pipe and fittings have an operational temperature range of –40°F to 180°F.

A variety of ABS DWV fittings are available for plumbing and plumbing-related applications. **See Figure 2-5.** ABS DWV pipe is made with a cellular core to reduce the amount of ABS resin needed to manufacture the pipe, which reduces the cost of the pipe. Installing ABS DWV pipe is a one-step process; a priming step is not needed, and therefore the installation cost of the pipe is reduced. ABS DWV pipe and fittings are available in sizes ranging from 1¼″ to 6″, and in 10′ and 20′ pipe lengths.

Polyvinyl Chloride (PVC) Pipe and Fittings

Polyvinyl chloride pipe and fittings are the most commonly used type in the plumbing industry. *Polyvinyl chloride (PVC) pipe and fittings* are plastic pipe and fittings used for sanitary drainage and vent piping, aboveground and underground storm water drainage, water mains, and water service lines. PVC DWV pipe and fittings are joined by solvent cementing. PVC DWV pipe and fittings are white colored, and fittings are similar in shape to ABS DWV fittings. PVC Schedule 80 pipe and fittings, commonly used for industrial pressure applications, are dark gray in color.

Tech Facts

PVC pipe and fittings are combustible; however, they are difficult to ignite and will not continue to burn without an outside heat source.

Figure 2-5. ABS DWV fittings are used for sanitary drainage and vent piping.

PVC pipe is manufactured by the extrusion process in a variety of sizes ranging from 1¼″–6″, and in 10′ and 20′ pipe lengths. PVC pipe up to 16″ size is available for use in underground drainage piping. PVC pipe and fittings have outstanding physical properties, including excellent corrosion and chemical resistance. PVC pipe and fittings must not be used to store and/or convey compressed air or other compressed gases.

Even though ABS and PVC pipe and fittings have similar applications and are similar in appearance, ABS and PVC must not be interchanged in piping systems. Some plumbing codes permit ABS pipe and PVC pipe and fittings to be joined when connecting a building drain to a building sewer if the proper transition couplings (mission couplings) and adapters are used.

Since ABS and PVC pipe and fittings have similar applications, the material/installation cost and local plumbing code restrictions must be considered when selecting the proper material for a job. While PVC pipe and fittings may be less expensive than ABS pipe and fittings, a two-step process is required to join PVC pipe and fittings, and only one step is required to join ABS pipe and fittings; therefore the overall cost of PVC pipe and fittings may be greater since the installed cost is higher.

The maximum developed length of rigid plastic drainage, waste, and vent piping, including ABS and PVC DWV pipe, is 35′. When ABS or PVC DWV pipe is installed in a building, the pipe must be protected at penetrations of fire-rated walls, floors, and ceilings using a firestop, such as caulk, foam, or a restricting collar. **See Figure 2-6.**

Figure 2-6. A firestop, such as fire-resistant caulk, protects ABS DWV pipe at penetrations of fire-rated walls, floors, and ceilings.

Tech Facts

PVC and ABS pipe should not be used for drainage applications where the temperature of the waste is over 140°F.

Chlorinated Polyvinyl Chloride (CPVC) Pipe and Fittings

Chlorinated polyvinyl chloride pipe and fittings are commonly used for hot and cold water distribution systems. *Chlorinated polyvinyl chloride (CPVC) pipe and fittings* are cream-colored thermoplastic materials specially formulated to withstand higher temperatures than other plastics, and are used in potable water distribution, corrosive industrial fluid handling, and fire suppression systems. **See Figure 2-7.** Chlorinated polyvinyl chloride (CPVC) plastic hot- and cold-water distribution systems, are typically rated for 180°F at 100 psi of pressure. CPVC pipe is joined by solvent cementing.

CPVC pipe is available in sizes ranging from ½″ to 12″. CPVC pipe for plumbing systems is manufactured using the extrusion process in sizes ranging from ½″ to 2″ copper tube size (CTS). Industrial CPVC pipe is manufactured using the extrusion process in sizes ranging from ¼″ to 12″ in Schedule 40 and Schedule 80 wall thickness. CPVC pipe is available in 10′ lengths. Plastic pipe is available as Schedule 40 or 80 pipe. Schedule 40 plastic pipe is standard weight pipe, and the walls of Schedule 80 pipe are approximately one-third thicker than Schedule 40 pipe.

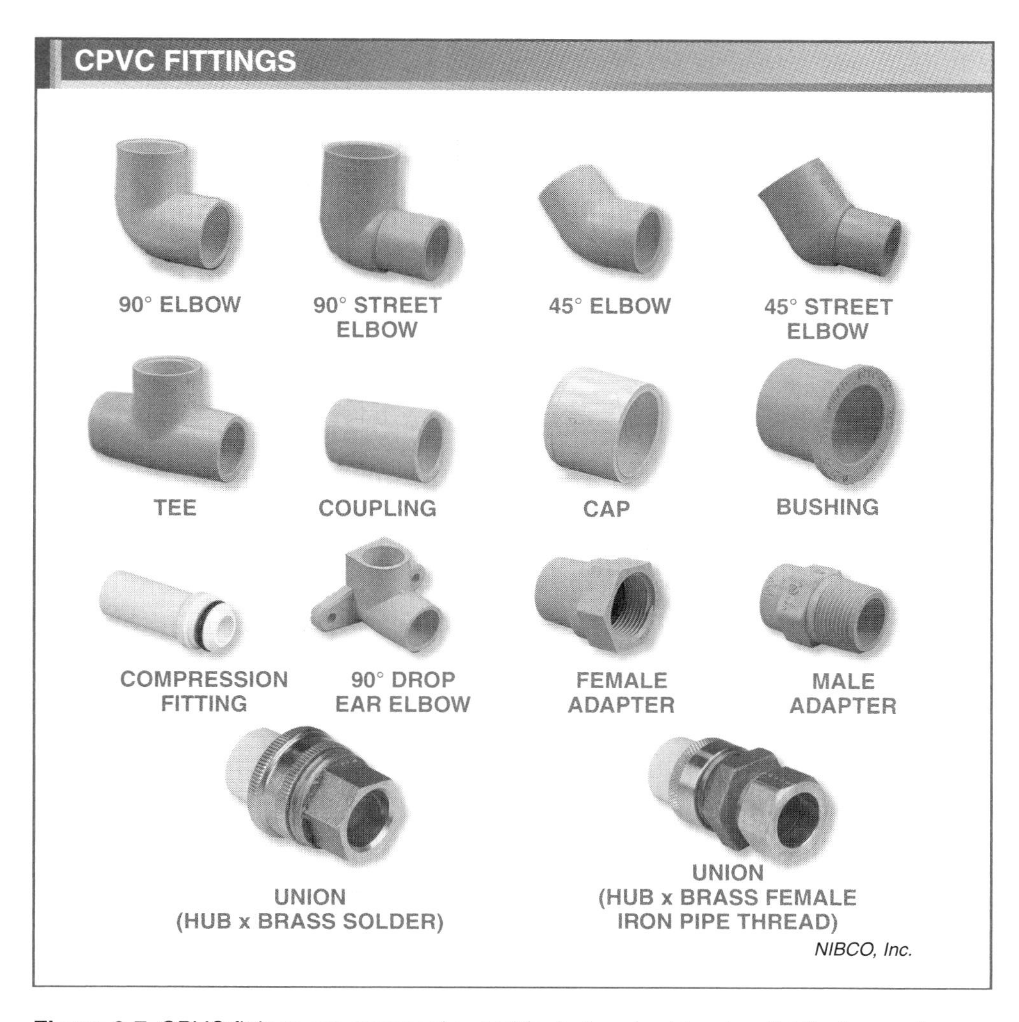

Figure 2-7. CPVC fittings are commonly used for hot and cold water distribution systems.

Cross-Linked Polyethylene (PEX) Pipe and Fittings

Cross-linked polyethylene (PEX) is a thermosetting plastic made from medium- or high-density cross-linkable polyethylene and is used for water service piping and cold and hot water distribution piping. PEX tubing and fittings are available in sizes ranging from ¼″ to 2″. **See Figure 2-8.** Tubing is available in straight lengths of 20′, and coils of 100′, 300′, 400′, 500′, and 1000′.

PEX offers many advantages over other plumbing materials, such as copper, including:

- faster installation
- chemical and corrosion resistance
- superior strength
- high-temperature and high-pressure resistance

Various manufacturing processes are used for PEX including the Engel, silane, and radiation processes. In each of the manufacturing processes, molecular chains of PEX resin are linked to create a plastic material that is durable within a wide range of temperatures and pressures. The Engel process is performed prior to extrusion of the PEX tubing, while the silane and radiation processes are performed during or after the tubing is extruded. The *Engel process,* or peroxide process, is a PEX manufacturing process in which peroxides (heat-activated chemicals) release molecules for cross-linking. The Engel process provides the most precise control of cross-linking, resulting in a more uniform product than produced with the other processes. The *silane process* is a PEX manufacturing process in which silane molecules are bonded to polyethylene molecules during the extrusion process, resulting in greater manufacturing efficiency and productivity. The *radiation process* is a PEX manufacturing process in which polyethylene is subjected to high-energy electrons to form the cross-linked bond.

Tech Facts

PEX tubing must be kept out of direct sunlight, which causes it to break down. A person installing PEX tubing and fittings must be trained by a representative of the pipe and fitting manufacturer. Tubing and fittings must be from the same manufacturer.

PEX tubing manufactured using the Engel process has shape or thermal memory, in which the tubing returns to its original shape after being deformed. When installing fittings, the tubing is expanded using an expander tool and the fitting is inserted into the tubing. The tubing returns to its original shape, securing the fitting in position. Crimping rings are not required for PEX tubing manufactured with the Engel process, but may be used with certain types of fittings. PEX tubing manufactured using the silane or radiation processes does not have shape or thermal memory. Therefore, fittings must be secured into position using crimping rings. The ribbed end of the fitting is placed in the tubing, and a crimping tool is used to compress the crimping ring.

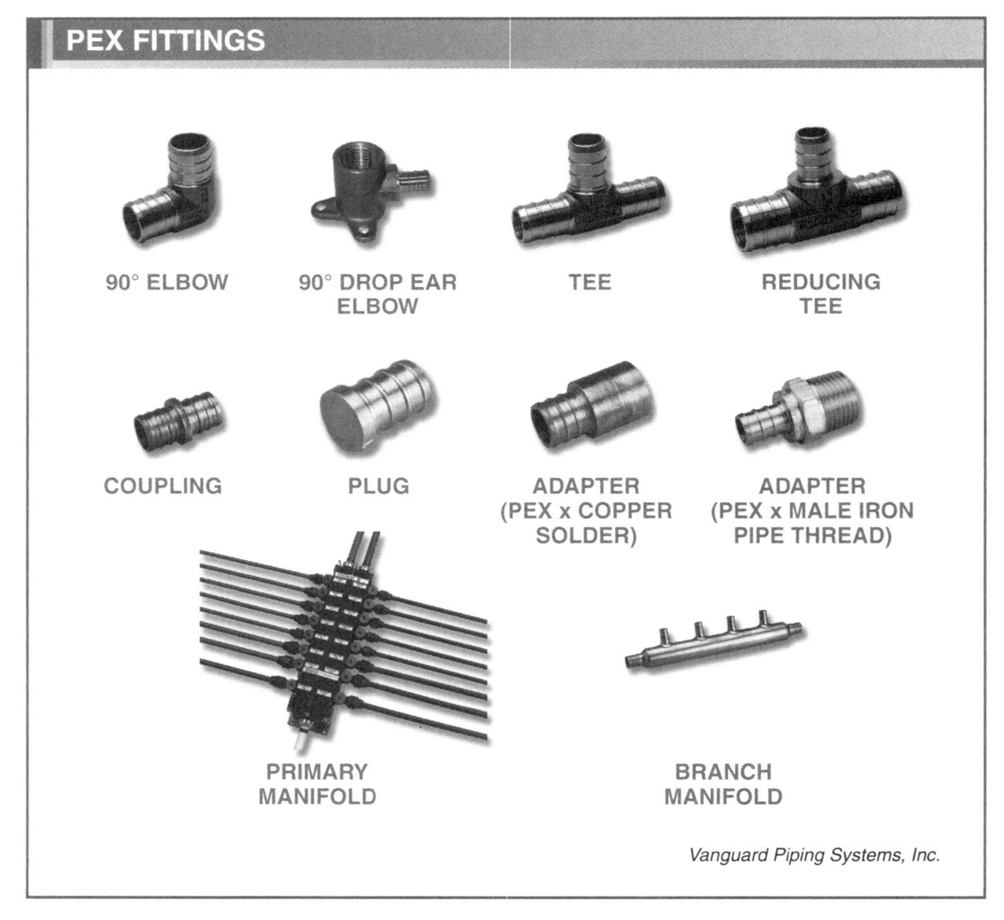

Figure 2-8. PEX fittings are used for water service piping and cold and hot water distribution piping.

PEX can be installed using conventional plumbing design in which water distribution pipes and fixture branches convey water to fixtures. As an alternative, PEX can be piped as a home run. A *home run* is a plumbing design in which centrally located manifolds distribute water to each fixture with dedicated hot and cold water lines. Even though additional PEX tubing is required to pipe a home run, labor is saved since connections are only required at the manifold and point-of-use valve. A home run reduces the pressure drop and temperature fluctuations in a plumbing system.

COPPER TUBE AND FITTINGS

Since the late 1940s, over 14 billion pounds of copper tube has been manufactured in the United States for plumbing and plumbing-related applications. Copper tube, or tubing, is used for water supply and distribution piping, fire suppression systems, and heating, ventilating, and air conditioning (HVAC) applications. Prior to the development of plastic and no-hub soil pipe, copper was also used for sanitary waste and vent piping in residential construction.

Copper tube is manufactured using either the extrusion or piercing process. In the extrusion process, a copper billet is heated and forced through a die and over a mandrel. The clearance between the mandrel and die determines the thickness of the tube wall. In the piercing process, one end of a heated cylindrical copper billet is fed between horizontal rotating rolls. The rolls force the heated billet onto a piercing plug, creating a void in the middle of the billet. The extruded or pierced tubes are cold-drawn to smaller sizes by pulling the tube through a die.

Copper tube is manufactured in four different wall thicknesses or types—K, L, M, and DWV. Type K copper tube has the thickest wall, followed by type L, type M, and type DWV with the thinnest wall. For any given diameter of copper tube, the outside diameter (OD) of all copper tube types is the same, which is ⅛″ larger than the nominal or standard size. The inside diameter of copper tube is determined by the wall thickness. For example, the outside diameter of ½″ type K and L copper tube is ⅝″ (.625″). The inside diameters of ½″ type K and L copper tube are .527″ and .545″, respectively. **See Figure 2-9.**

Copper tube is available as either drawn or annealed. *Drawn copper tube,* or hard copper, is copper tube that is pulled through a single die or series of dies to achieve a desired diameter. Drawn copper tube is available in nominal sizes up to 8″ and in 20′ straight lengths. *Annealed copper tube,* or soft copper, is drawn copper tube that is heated to a specific temperature and cooled at a predetermined rate to impart desired strength and hardness characteristics. Annealed copper tube is available in 20′ straight lengths and in coils ranging from 40′ to 100′ in length. **See Figure 2-10.**

Tech Facts

Because of the high cost of labor, the copper press fitting method was developed to reduce copper tubing installation time. In the copper press fitting method, use of soldering is eliminated, which reduces preparation time.

Reed Manufacturing Co.

A hinged pipe cutter is used to cut steel, plastic, or copper pipe in place.

SIZES AND WEIGHTS OF COPPER TUBE													
Nominal Sizes*	Outside Diameter*	Inside Diameter*				Wall Thickness*				Weight†			
	Types K-L-M-DWV	Type K	Type L	Type M	Type DWV	Type K	Type L	Type M	Type DWV	Type K	Type L	Type M	Type DWV
¼	.375	.305	.315	.325	—	.035	.030	.025	—	.145	.126	.106	—
⅜	.500	.402	.430	.450	—	.049	.035	.025	—	.269	.198	.145	—
½	.625	.527	.545	.569	—	.049	.040	.028	—	.344	.285	.204	—
⅝	.750	.652	.666	.690	—	.049	.042	.030	—	.418	.362	.263	—
¾	.875	.745	.785	.811	—	.065	.045	.032	—	.641	.455	.328	—
1	1.125	.995	1.025	1.055	—	.065	.050	.035	—	.839	.655	.465	—
1¼	1.375	1.245	1.265	1.291	1.295	.065	.055	.042	.040	1.04	.884	.682	.650
1½	1.625	1.481	1.505	1.527	1.541	.072	.060	.049	.042	1.36	1.14	.940	.809
2	2.125	1.959	1.985	2.009	2.041	.083	.070	.058	.042	2.06	1.75	1.46	1.07
2½	2.625	2.435	2.465	2.495	—	.095	.080	.065	—	2.93	2.48	2.03	—
3	3.125	2.907	2.945	2.981	3.035	.109	.090	.072	.045	4.00	3.33	2.68	1.69
3½	3.625	3.385	3.425	3.459	—	.120	.100	.083	—	5.12	4.29	3.58	—
4	4.125	3.857	3.905	3.935	4.009	.134	.110	.095	.058	6.51	5.38	4.66	2.87
5	5.125	4.805	4.875	4.907	4.981	.160	.125	.109	.072	9.67	7.61	6.66	4.43
6	6.125	5.741	5.845	5.881	5.959	.192	.140	.122	.083	13.9	10.2	8.92	6.10
8	8.125	7.583	7.725	7.785	7.907	.271	.200	.170	.109	25.9	19.3	16.5	10.6
10	10.125	9.449	9.625	9.701	—	.338	.250	.212	—	40.3	30.1	25.5	—
12	12.125	11.315	11.565	11.617	—	.405	.280	.254	—	57.8	40.4	36.7	—

* in in.
† in lb/ft

Figure 2-9. For any given diameter of copper tube, the outside diameter is the same and the inside diameter is determined by the wall thickness.

During manufacture, drawn and annealed copper tube are permanently stamped every 18″ with the tube type, name or trademark of the manufacturer, and the country of origin. In addition, drawn copper tube is identified with a colored stripe and lettering. The colors used to identify the drawn copper tube are:

- green—type K
- blue—type L
- red—type M
- yellow—type DWV

Tech Facts

Copper tube fittings are joined to copper tube using soldered, brazed, press fit, rolled groove, flared, or compression joints. The fittings can be made as cast copper alloy (copper mixed with other metals), cast bronze (copper mixed with tin), or wrought copper (99.9% pure copper).

Copper Tube Fittings

Cast copper alloy, cast bronze, or wrought copper tube fittings are used to connect lengths of copper tube. Cast copper alloy fittings are an alloy (combination) of various metals, including copper, tin, and zinc, and are cast in sand molds. Cast bronze fittings, which are an alloy of copper and tin, are manufactured in a manner similar to cast copper alloy fittings. Wrought copper fittings are made from commercially pure copper (99.9% pure), and are formed by a hammering process. Cast copper and bronze fittings have a rough surface texture and wrought copper fittings have a smooth and shiny finish. Installation and material costs for cast and wrought fittings are approximately equal. However, some fittings are not available as both wrought and cast fittings.

COPPER TUBE SIZES AND LENGTHS				
Tube Type	Drawn Tube Color Code	Lengths†		
		Nominal Sizes*	Drawn	Annealed
K	Green	**Straight lengths**		
		¼–8	20	20
		10	18	18
		12	12	12
		Coils		
		¼–1	–	60
			–	100
		1¼–1½	–	60
		2	–	40
			–	45
L	Blue	**Straight lengths**		
		¼–8	20	20
		12	18	18
		Coils		
		¼–1	–	60
			–	100
		1¼–1½	–	60
		2	–	40
			–	45
M	Red	**Straight lengths**		
		¼–12	20	–
DWV	Yellow	**Straight lengths**		
		1¼–8	20	–

* in in.
† in ft

Figure 2-10. Drawn and annealed copper tube is available in a wide range of diameters and lengths.

Copper tube fittings are joined to copper tube using soldering or brazed, press fit, rolled groove, flared, or compression joints. Solder joint fittings are connected to copper tube by soldering or brazing. Press fit and Rolled groove copper fittings are available in wrought copper and cast bronze. Flared joint copper fittings are a mechanical joint fitting manufactured only in cast bronze.

Solder Joint Fittings. Copper solder joint fittings are available in pressure or DWV patterns. Solder joint pressure fittings are used for aboveground water supply applications with types K, L, and M drawn copper tube. **See Figure 2-11**.

Solder joint DWV fittings are used for drain, waste, and venting applications with type DWV copper tube. Pressure fittings have a deeper solder socket than DWV fittings since pressurized water flows through the tube and fittings. Change-in-direction DWV fittings, such as 90° elbows or tees, have a larger radius (sweep) than pressure fittings to prevent stoppage. **See Figure 2-12.** Solder joint fittings are available in ⅛″–12″ sizes. Brazed joint fittings are the same as solder joint fittings, but the brazing process is slightly different from the soldering process.

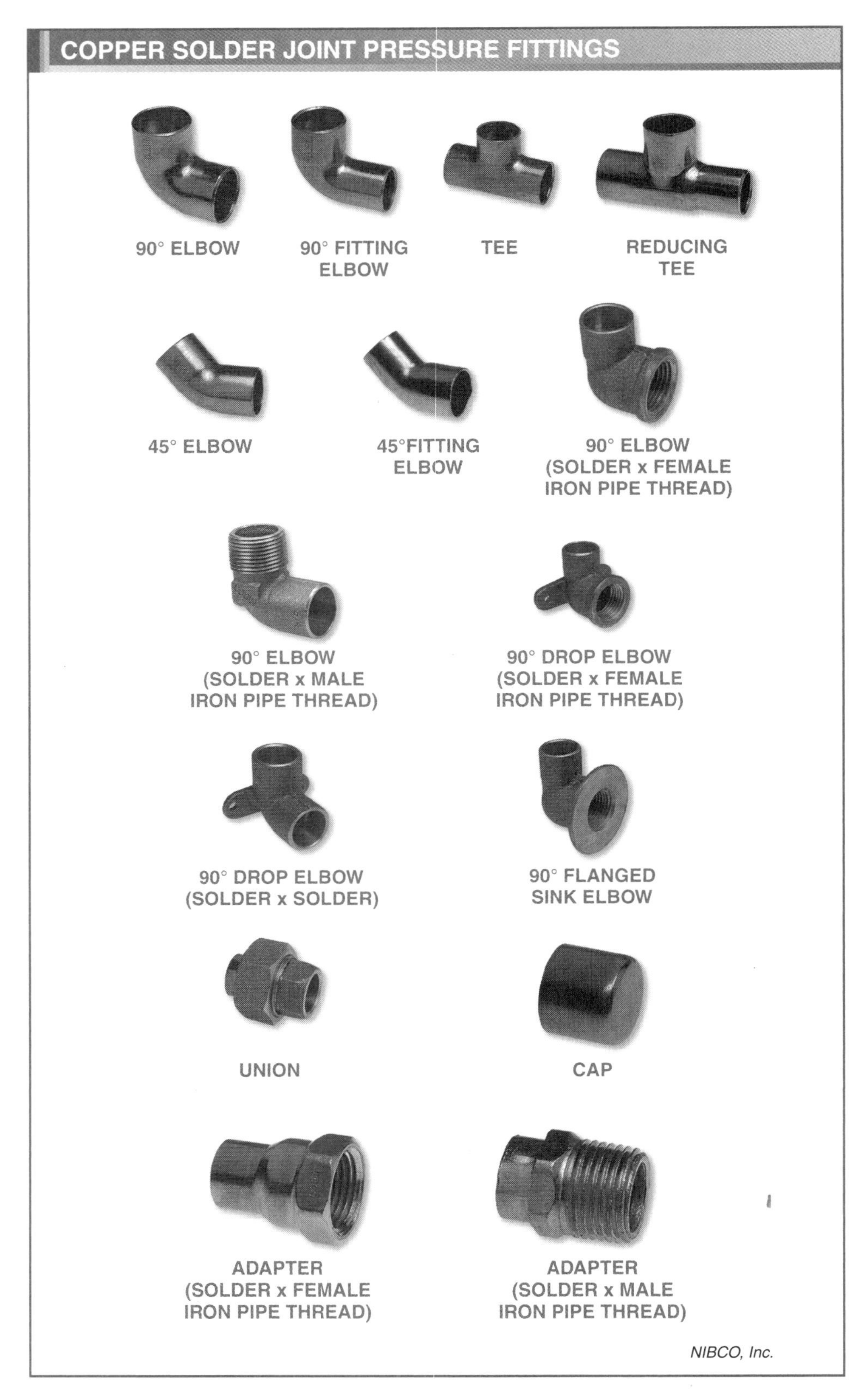

Figure 2-11. Copper solder joint pressure fittings are used for aboveground water supply applications with types K, L, and M drawn copper tube.

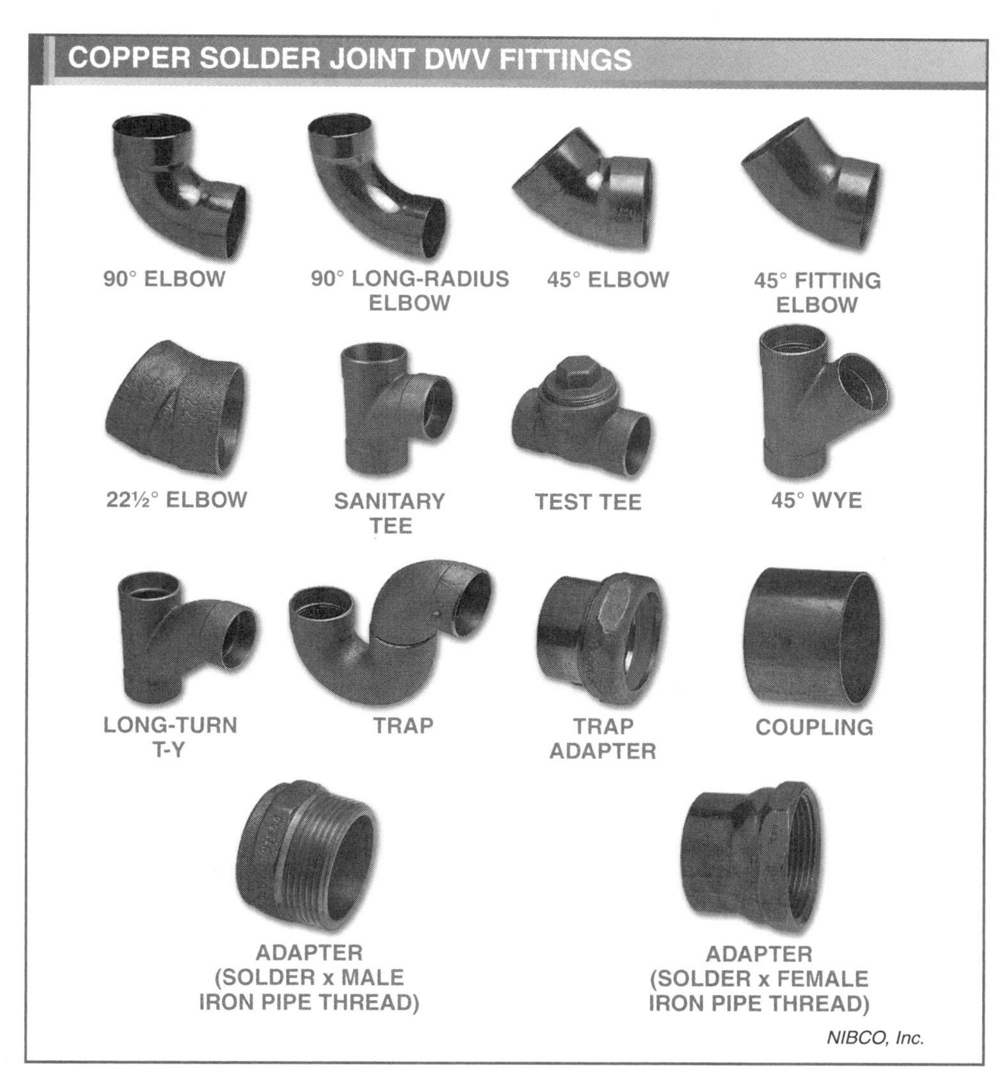

Figure 2-12. Copper solder joint DWV fittings are used for drain, waste, and venting applications with type L, M, or DWV copper tube.

Copper Press Fittings. Copper press fittings are manufactured with an ethylene propylene diene monomer (EPDM) O-ring seal contained within a recess in the fitting socket. The fitting is joined to the tube using a pressing tool that mechanically presses the fitting onto the pipe. Copper press fittings are manufactured in ½″ to 4″ sizes for aboveground water supply and distribution piping. Although copper press fittings are more expensive than solder joint fittings, the installation time is much less, resulting in lower installation cost. **See Figure 2-13.**

Rolled Groove Joint Fittings. Copper rolled groove joint fittings are used for aboveground potable water supply applications. Sizes of rolled groove joint fittings range from 2″ to 6″ and are available in types L and M drawn copper tube. **See Figure 2-14.** Copper rolled groove joint water mains are easier and faster to install than soldered copper piping.

Tech Facts

Copper solder joint pressure fittings have a deeper solder socket than DWV fittings since pressurized water flows through the tube and fittings.

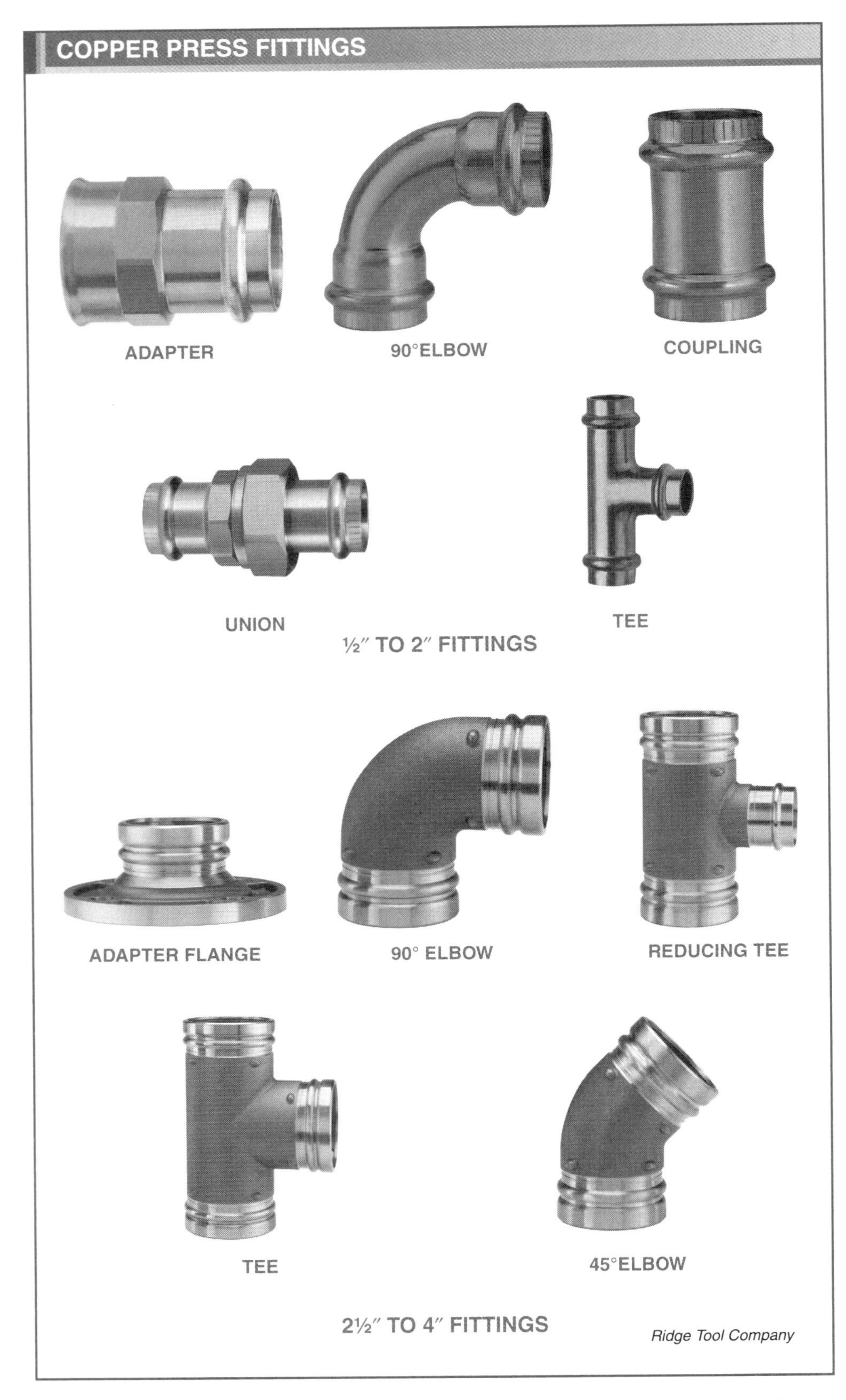

Figure 2-13. Copper press fittings are used for aboveground potable water supply piping.

COPPER ROLLED GROOVE JOINT FITTINGS

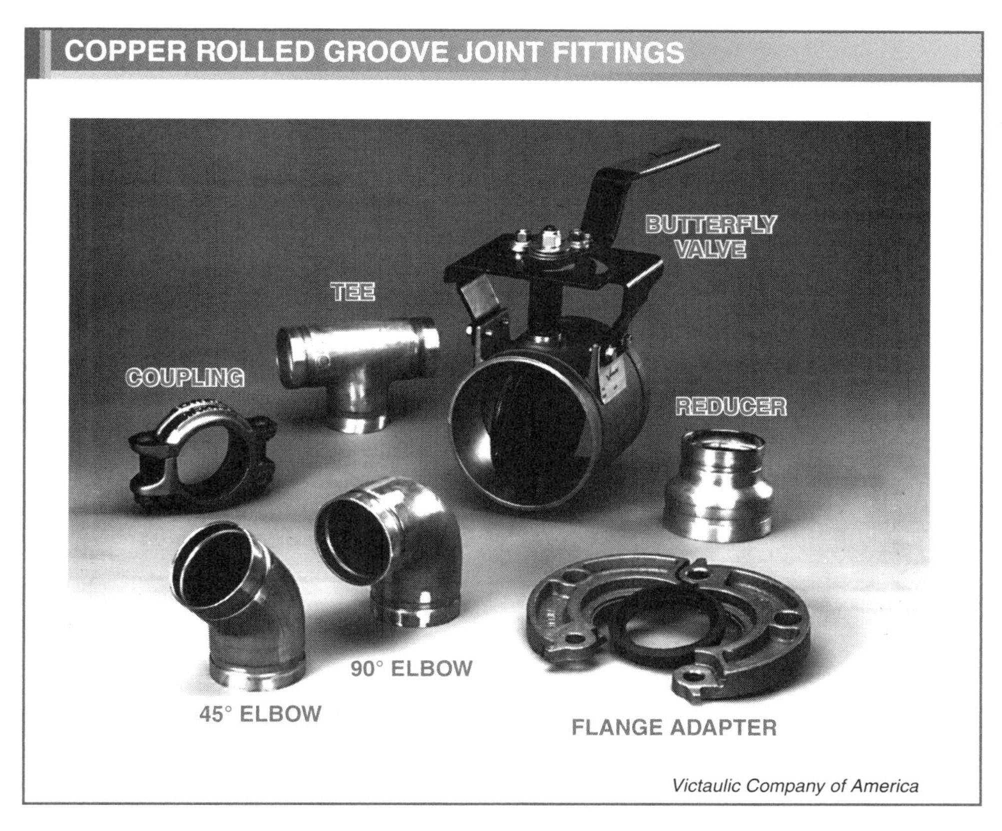

Victaulic Company of America

Figure 2-14. Copper rolled groove joint fittings are used for aboveground potable water supply applications.

Flared Joint Fittings. Flared joint fittings are used with types K, L, and M annealed copper tube, and are typically used for underground water service applications. Flared joint fittings cannot be used with drawn copper tube since drawn copper tube is subject to splitting when flared. Flared joints can be easily disassembled by loosening the tube nut. Flared joint fittings are available in ⅜″–3″ sizes. **See Figure 2-15.**

Compression Joint Fittings. Compression joint fittings are used on copper tube to make connections that may need to be disassembled. Compression joint fittings are used to install fixture supply tubes and fixture shutoff valves, and are sometimes installed in locations where it is difficult or unsafe to solder copper tube. Compression joint fittings are used with types K, L, and M annealed copper tube for aboveground applications.

CAST IRON SOIL PIPE AND FITTINGS

Cast iron soil pipe and fittings are manufactured from gray cast iron, which is strong and corrosion-resistant due to its metallurgical structure. Cast iron soil pipe is centrifugally cast to ensure uniform wall thickness, straight length, and a smooth inner wall. Molten gray iron is poured into a spinning pipe mold where centrifugal force propels the molten iron against the mold walls, causing it to solidify and form the pipe walls. As the gray iron solidifies, large graphite flakes form within the pipe walls to inhibit corrosion. When the pipe has cooled, it is coated with bituminous petroleum asphalt to prevent corrosion during storage and use, and to improve its appearance.

Cast iron soil pipe and fittings are leakproof, nonabsorbent, corrosion-resistant, and are easily cut and joined with the proper tools. Cast iron soil pipe and fittings provide a quiet plumbing system since cast iron does not transmit the sound of water draining through pipe. However, cast iron soil pipe is heavy, has a low tensile strength, and may crack and break if not handled properly.

Centrally located manifolds in a PEX home run distribute water to individual fixtures with dedicated hot and cold water lines.

Cast iron soil pipe and fittings are available in no-hub and bell-and-spigot patterns. No-hub and bell-and-spigot cast iron soil pipe and fittings are primarily used for aboveground and underground sanitary drainage, vent, and storm water drainage piping. No-hub cast iron soil pipe and fittings are primarily used for aboveground applications, and bell-and-spigot pipe and fittings are typically used for underground applications.

No-Hub Cast Iron Soil Pipe and Fittings

No-hub, or hubless, cast iron soil pipe and fittings provide a faster means of joining pipe and fittings than bell-and-spigot cast iron soil pipe. No-hub cast iron soil pipe and fittings are joined using a mechanical coupling consisting of a neoprene sleeve and a stainless steel band equipped with screw clamps. Ends of the pipe and fittings are inserted into the sleeve and stainless steel band and properly aligned. The screw clamps are tightened to secure the sleeve and band in position. No-hub cast iron soil pipe and fittings are available in sizes ranging from 1½″ to 15″, and in 10′ lengths. **See Figure 2-16.**

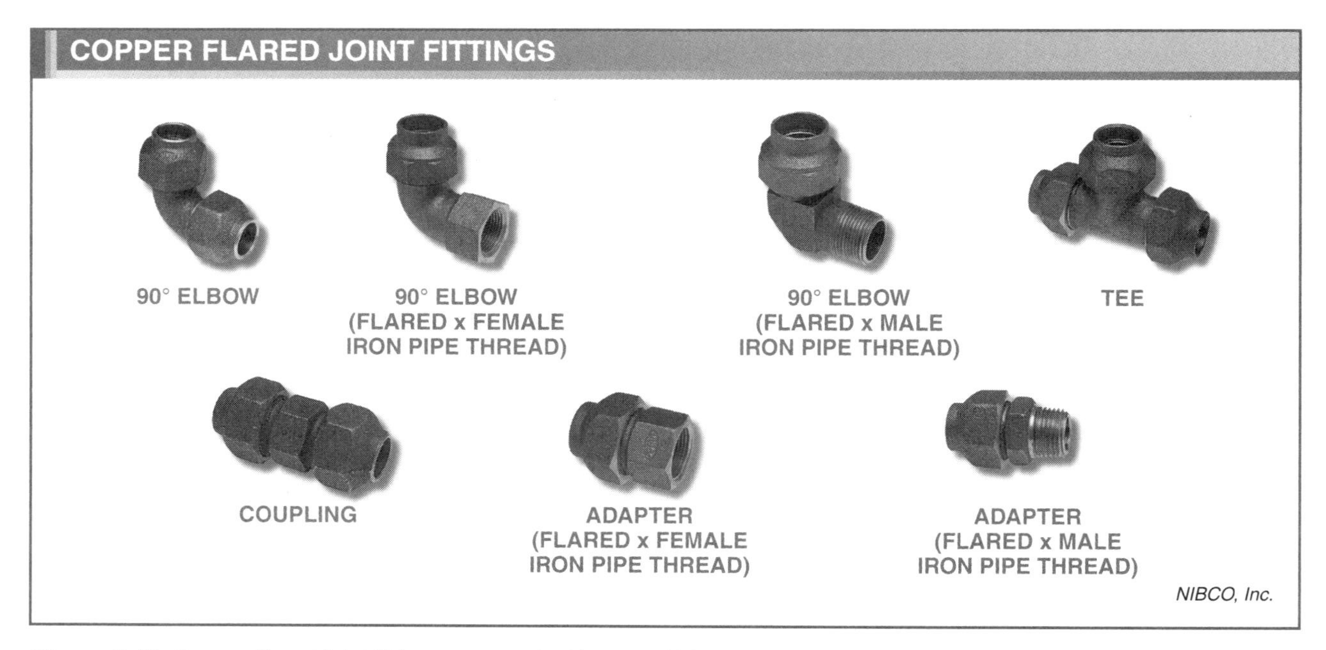

Figure 2-15. Copper flared joint fittings are used with types K, L, and M annealed copper tube, and are typically used for underground water service applications.

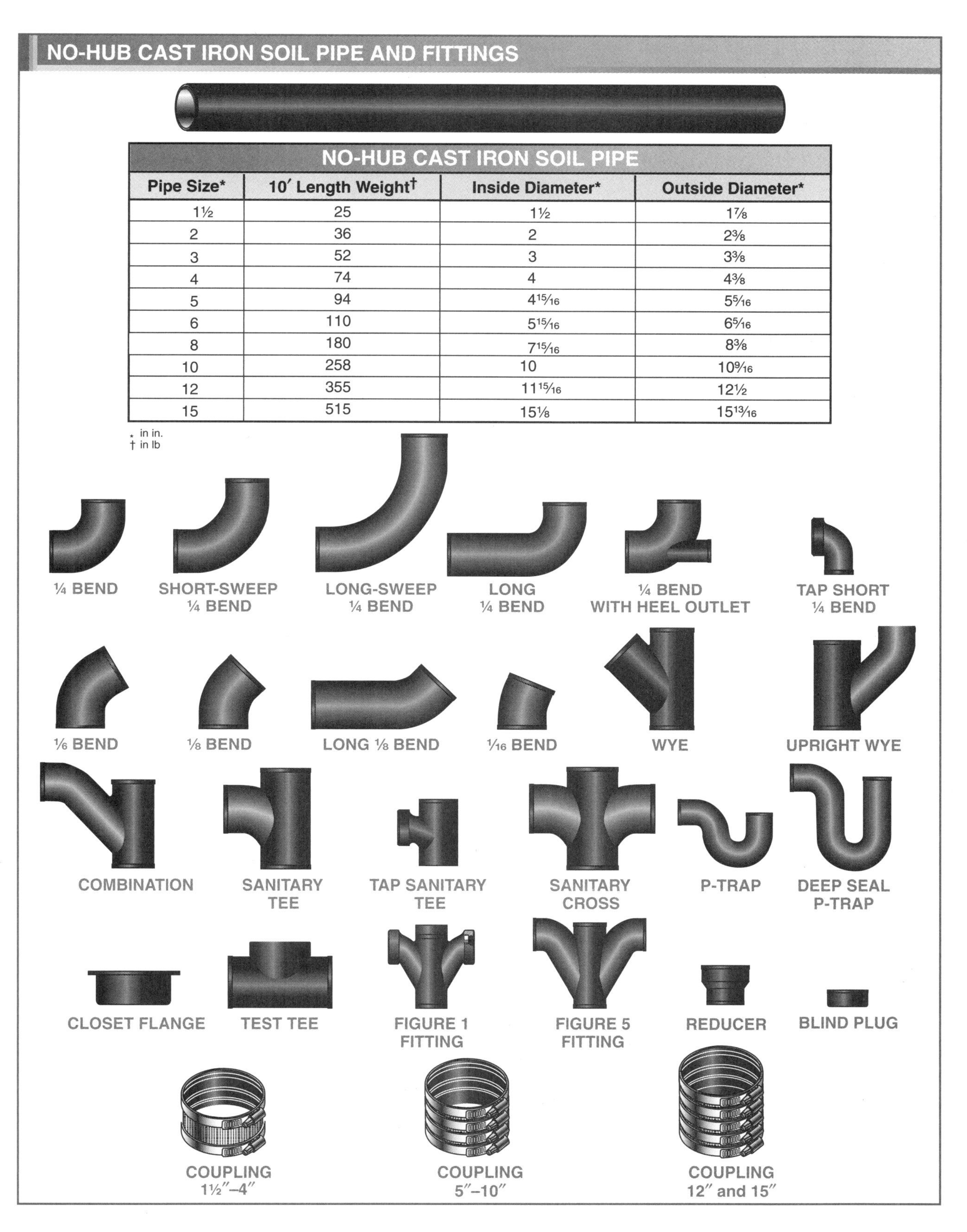

NO-HUB CAST IRON SOIL PIPE AND FITTINGS

NO-HUB CAST IRON SOIL PIPE			
Pipe Size*	**10′ Length Weight†**	**Inside Diameter***	**Outside Diameter***
1½	25	1½	1⅞
2	36	2	2⅜
3	52	3	3⅜
4	74	4	4⅜
5	94	$4\frac{15}{16}$	$5\frac{5}{16}$
6	110	$5\frac{15}{16}$	$6\frac{5}{16}$
8	180	$7\frac{15}{16}$	8⅜
10	258	10	$10\frac{9}{16}$
12	355	$11\frac{15}{16}$	12½
15	515	15⅛	$15\frac{13}{16}$

* in in.
† in lb

Figure 2-16. No-hub cast iron soil pipe and fittings are primarily used for aboveground applications.

Bell-and-Spigot Cast Iron Soil Pipe and Fittings

Bell-and-spigot cast iron soil pipe and fittings have a bell, or hub, at one end of the pipe and fittings. The spigot, or plain end, of another piece of pipe is inserted into the bell to join them. The space between the bell and spigot is sealed with a preformed rubber compression gasket. Single-hub bell-and-spigot cast iron soil pipe is available in 3½′, 5′, and 10′ lengths. Double-hub pipe (pipe with 2 hubs on each end) is available in 30″ lengths. Double-hub pipe minimizes waste that results when cutting shorter pieces from longer lengths of pipe. **See Figure 2-17.**

Bell-and-spigot cast iron soil pipe and fittings are available in a wall thickness designated as service weight in sizes ranging from 2″ to 15″. Service weight pipe and fittings are marked with the letters "SV." At one time, extra-heavy bell-and-spigot cast iron soil pipe and fittings, marked with XH, were manufactured. However, due to developments in plumbing materials, XH pipe and fittings are no longer available. XH pipe and fittings were larger and heavier than SV pipe and fittings and cannot be interchanged with them.

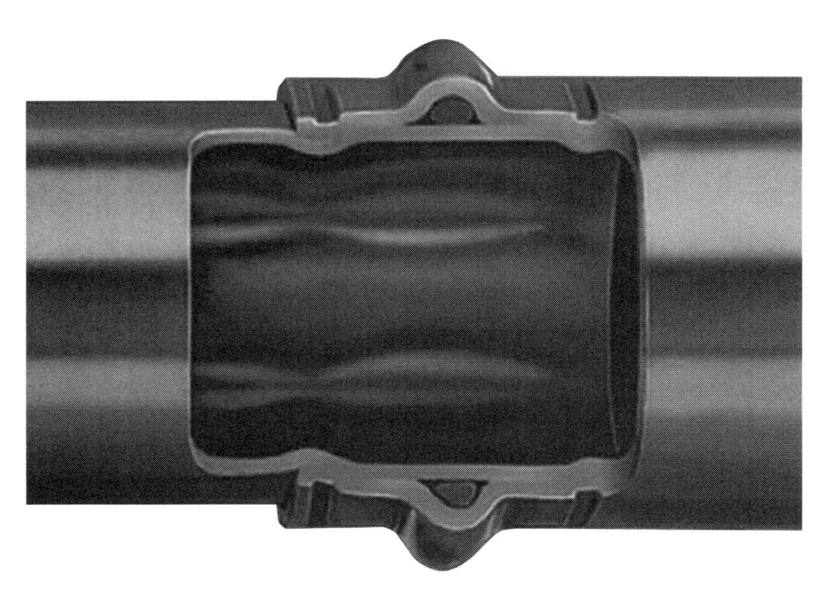

Copper press fittings are mechanically pressed onto copper pipe with a pressing tool that presses the fitting and pipe together on each side of the EDPM O-ring, and also presses the O-ring to the pipe to form a watertight seal.

Tech Facts

"Cast iron" is a common term used to describe many types of iron. The term "cast" identifies the method used to produce the finished product but does not describe the material itself. The casting process involves pouring molten iron into a mold.

Iron is also identified by color or physical properties. Cast iron soil pipe and fire hydrants are manufactured from gray iron. Gray iron is the most common type of cast iron. "Gray" refers to the color of the iron after it has cooled. Ductile iron is cast iron manufactured by adding magnesium to molten gray iron.

STEEL PIPE AND FITTINGS

Steel pipe and fittings are used for water distribution, sanitary waste and vent, storm water drainage, and gas piping systems. Steel pipe is relatively inexpensive, strong, and rugged, and is not easily damaged by rough handling. However, due to their weight and installation cost, steel pipe and fittings have been replaced by less expensive materials for many applications. Steel pipe is available in nominal pipe sizes ranging from ⅛″ to 12″, in several different wall thicknesses. *Nominal pipe size,* or iron pipe size (IPS), is the approximate inside diameter of steel pipe.

Steel pipe is manufactured from mild carbon steel as welded or seamless pipe. *Welded pipe,* or butt-welded or continuous weld pipe, is steel pipe manufactured by drawing flat steel strips through a die to form a cylindrical shape and then electric butt-welding the seam to create a leakproof joint. Welded pipe is available in 21′ lengths. *Seamless pipe* is steel pipe made by piercing a solid cylindrical steel billet with a series of mandrels while passing the billet through rollers. Seamless pipe is typically available in 21′ lengths, although random lengths ranging from 16′ to 48′ are available.

Steel pipe is coated after it is manufactured to protect it against corrosion. *Black pipe* is steel pipe that is coated with varnish to protect it against corrosion. *Galvanized pipe* is steel pipe that is cleaned and dipped into a hot (870°F) molten zinc bath to create a protective coating. Galvanized pipe is typically used for potable water piping, such as water distribution systems.

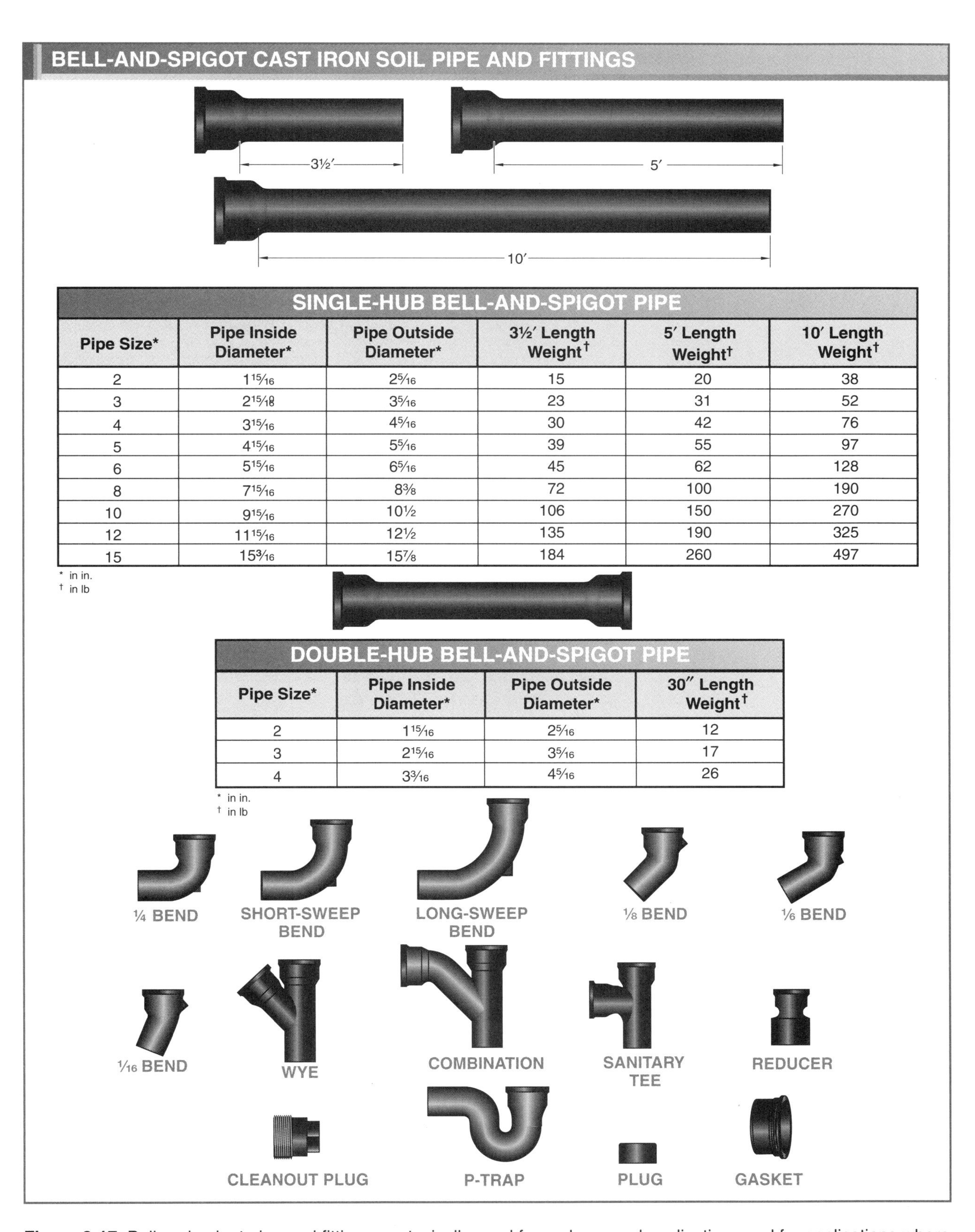

SINGLE-HUB BELL-AND-SPIGOT PIPE

Pipe Size*	Pipe Inside Diameter*	Pipe Outside Diameter*	3½′ Length Weight†	5′ Length Weight†	10′ Length Weight†
2	1 15/16	2 5/16	15	20	38
3	2 15/16	3 5/16	23	31	52
4	3 15/16	4 5/16	30	42	76
5	4 15/16	5 5/16	39	55	97
6	5 15/16	6 5/16	45	62	128
8	7 15/16	8 3/8	72	100	190
10	9 15/16	10 1/2	106	150	270
12	11 15/16	12 1/2	135	190	325
15	15 3/16	15 7/8	184	260	497

* in in.
† in lb

DOUBLE-HUB BELL-AND-SPIGOT PIPE

Pipe Size*	Pipe Inside Diameter*	Pipe Outside Diameter*	30″ Length Weight†
2	1 15/16	2 5/16	12
3	2 15/16	3 5/16	17
4	3 3/16	4 5/16	26

* in in.
† in lb

Figure 2-17. Bell-and-spigot pipe and fittings are typically used for underground applications and for applications where long, straight pipe runs are required.

The steel pipe is joined using threaded or grooved joints. Threaded and coupling pipe is used for threaded joints. *Threaded and coupled (T and C) pipe* is steel pipe used for threaded joints that has threads on both ends of the pipe and a coupling on one end. *Grooved end pipe* is steel pipe used for grooved joints. Grooved end pipe is steel pipe used for grooved joints that has grooves on both ends. Plain-end pipe is also available. *Plain-end pipe* is steel pipe that is not threaded or grooved on the ends. Threads or grooves are cut or rolled into the pipe ends to join the pipe with fittings. Steel pipe is available as Schedule 40 or 80 pipe. Schedule 40 steel pipe is standard weight pipe, and the walls of Schedule 80 pipe are approximately one-third thicker than Schedule 40 pipe.

Even though steel pipe cuts and threads easily, threaded joints are typically only used when specified since installation of the pipe, including threading and joining pipe, is relatively time-consuming and expensive. When steel pipe is specified, nipples are used to reduce the amount of cutting and threading required for threaded joints. A *nipple* is a short piece of pipe, typically less than 12″ in length, with threads on each end. A *close nipple* is a nipple that is threaded its entire length. A *shoulder nipple* is a nipple that is threaded on the ends and has a short portion of unthreaded pipe in the middle. Nipples up to 6″ in length are available in ½″ increments. Nipples 7″–12″ in length are available in 1″ increments.

Malleable Iron Threaded Fittings

Malleable iron threaded fittings join and change direction of black and galvanized pipe. **See Figure 2-18.** Malleable iron threaded fittings are available with a varnish or galvanized finish, and are commonly used for natural gas piping in ½″–2″ sizes. Malleable iron threaded fittings are available as Class 150 or 300 fittings. Class 150 fittings are standard weight fittings, while Class 300 fittings are extra-heavy fittings.

Malleable iron threaded fittings are made from gray cast iron in a sand mold and are heat treated by controlling the cooling rate over a 72-hour period. Heat treating changes the grain structure of the iron, making it a tough, elastic material. Malleable iron threaded fittings should not be heated or welded since the grain structure of the fittings will change, resulting in a loss of the toughness and elasticity properties.

IDENTIFYING PIPE FITTINGS

Pipe fittings are available in many standard configurations, including elbows, tees, wyes, and crosses. Specialty pipe fittings are also available, but typically are special ordered when needed for a job. Some pipe fittings, such as tees and wyes, are identified by the letter of the alphabet that they resemble, while other fittings, such as elbows and crosses, are identified by their general shape.

Most pipe fittings have at least two openings; elbows (ells) have two openings, tees and wyes have three openings, and crosses have four openings. The openings can be the same size or different sizes. A *straight fitting* is a pipe fitting in which all openings are the same dimension. A straight fitting is referred to by its nominal size and configuration, such as ½″ 90° ell or ¾″ tee. A *reducing fitting* is a pipe fitting in which the dimension of at least one opening is smaller than other openings, such as a ½″ × ¼″ 90° ell or ¾″ × ½″ tee.

When identifying reducing fittings, the sizes of the inlet and outlet of the run are listed first, followed by the size of other outlets. **See Figure 2-19.** For reducing elbows and couplings, the largest opening of the run is named first, for example, ¾″ × ½″ ell. For tees, wyes, sanitary tees, and combination wye and ⅛ bends, the largest opening of the run is named first, followed by the smaller run, then the side outlet; for example, ¾″ × ½″ × ½″ tee. For crosses with the same diameter run openings and two side openings of the same size, the run size is named first followed by the side outlet size; for example, 3″ × 2″ cross.

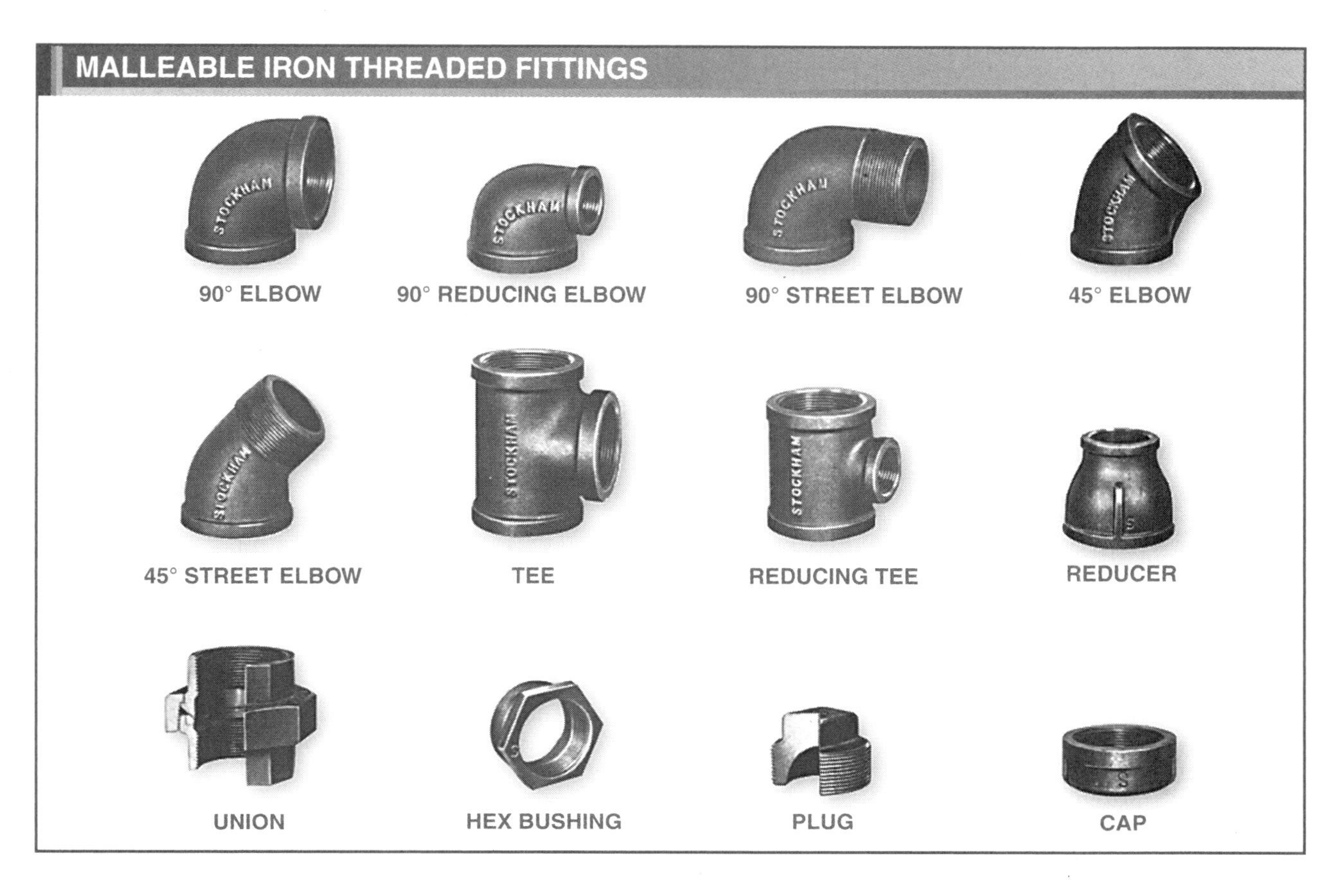

Figure 2-18. Malleable iron threaded fittings are available with either a black (varnish) or galvanized finish.

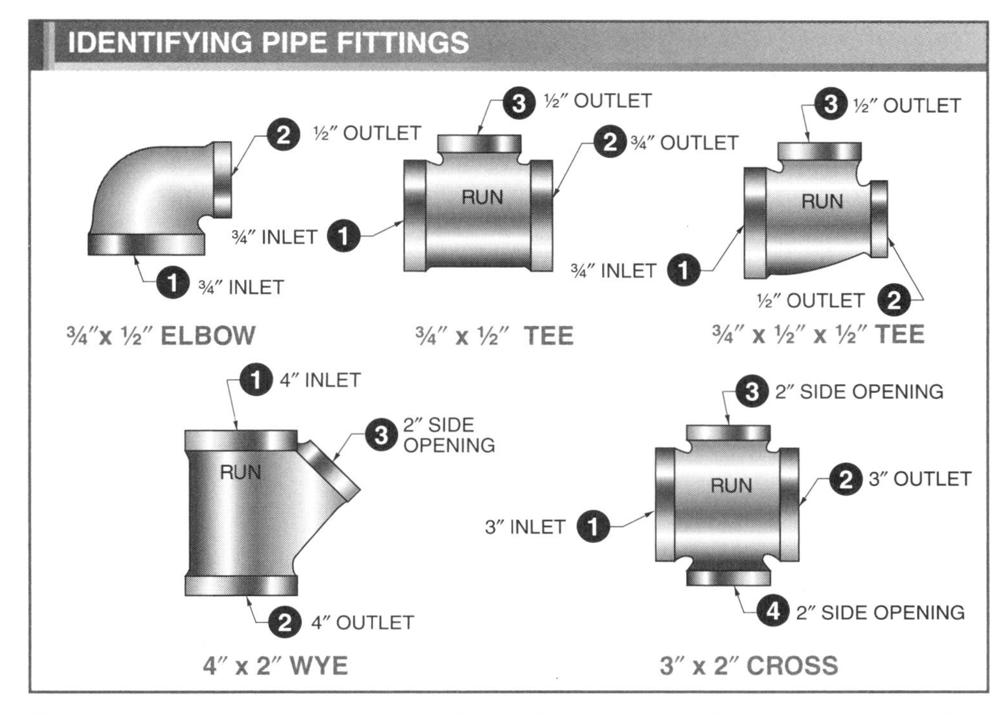

Figure 2-19. Reducing fittings are identified by listing the size of the inlet and outlet of the run first, followed by the size of other outlets.

PLUMBING VALVES

A *valve* is a fitting used to regulate fluid flow within a system. Valves are used to turn the fluid (water or other liquid or gas) flow on and off, or to regulate the direction, pressure, and/or temperature of a fluid within the system. Several types of valves are used in plumbing and plumbing-related systems including:

- gate valves
- globe valves
- compression stop valves
- stop-and-waste valves
- sillcocks
- boiler drains
- core cocks
- ball valves
- butterfly valves
- check valves
- backwater valves
- pressure-reducing valves
- relief valves

Tech Facts

Leaking valves and faucets account for a large volume of wasted water. A valve or faucet leaking 60 drops per minute will waste approximately 2300 gal. of water per year. Leaking valves and faucets should be repaired as soon as possible to conserve water and to avoid excessive water and sewer charges.

Valve bodies are available as cast bronze or cast iron. In general, valve bodies for 2″ and smaller valves are manufactured from cast bronze with bronze internal components. Valve bodies for 2½″ and larger valves are typically manufactured from cast iron and have bronze internal components.

Valves turn fluid flow on and off, or regulate the direction, pressure, and/or temperature of a fluid. A *full-way valve,* or shutoff valve, is a valve designed to be used in its fully open or fully closed position. The disk and seat of full-way valves, such as gate valves, may be damaged if the valve is used to throttle fluid flow. In general, full-way valves installed on water supply piping are the same size as the pipe on which they are installed. Full-way valves are installed on the water supply for multifamily dwellings so repairs can be made in one apartment without interrupting water service to other apartments, and in commercial buildings so the water supply can be shut off for individual rooms or fixtures without interrupting water service to other rooms or sections of the buildings. Full-way valves should be placed in locations that can be easily accessed when repairs are required. A *control valve,* or throttling valve, is a valve designed to control fluid flow rate by partially opening or closing the valve. Throttling valves, such as compression stops or globe valves, are installed on fixture supply pipes for individual fixtures.

Rated valves may be specified for certain applications. A *rated valve* is a valve that meets or exceeds engineering criteria for the normal pressure range of the fluids contained within the system it is controlling. The pressure rating for a rated valve is marked in raised letters on the side of the valve body. Most rated valves are marked 125# SWP or 200# WOG, indicating the valve is rated for use at a maximum of 125 lb steam working pressure (SWP) or 200 lb water, oil, or gas (WOG) pressure.

Gate Valves

A *gate valve* is a valve that has an internal gate that slides over an opening through which water flows. Gate valves are full-way valves used to regulate fluid flow. In a gate valve, a threaded stem raises and lowers a wedge-shaped disk (gate), which fits against a valve seat within the valve body. Gate valves are typically installed in piping systems where they remain completely open or completely closed most of the time, such as gate valves on each side of a water meter. When the wedge-shaped disk is retracted from the seat, a gate valve permits a straight and unrestricted fluid flow through the valve.

Gate valves have a split- or solid-wedge disk. A *split-wedge disk gate valve* is a gate valve in which a two-piece bronze wedge fits against the seat to restrict fluid flow. The parallel halves of a split-wedge disk valve are forced outward onto the valve seat by the closing pressure of the valve, providing a good seal even if scale is trapped on one of the valve seats. Split-wedge disk gate valves should be installed with the valve stem in the vertical position. A *solid-wedge disk gate valve* is a gate valve in which a one-piece solid bronze wedge fits against the valve seat to restrict fluid flow. A solid-wedge disk gate valve can be installed with the valve stem in any position.

Three stem and screw configurations are available for gate valves—rising stem-outside stem and yoke, rising stem-inside screw, and nonrising stem-inside screw. **See Figure 2-20.** A *rising stem-outside stem and yoke (OS&Y) gate valve* is a gate valve in which the threaded stem of the valve rises as the valve is opened. When the handwheel is turned, the stem rises as the yoke bushing engages the stem threads, and provides a visual indication of whether the valve is open. A *rising stem-inside screw gate valve* is a gate valve in which the unthreaded stem and handwheel rise as the valve is opened to indicate the position of the wedge disk. Adequate clearance must be provided above the valve, however, since the stem and handwheel rise during operation. A *nonrising stem-inside screw gate valve* is a gate valve in which neither the handwheel nor the stem rises when the valve is opened. Nonrising stem-inside screw gate valves are used where there is inadequate clearance for the operation of a rising stem valve. However, since the stem and handwheel do not rise during operation, there is no way to visually check the open or closed position of a nonrising stem valve.

Globe Valves

A *globe valve* is a valve that has a disk (globe) that rises or lowers over a seat through which water flows. **See Figure 2-21.** Due to the internal water passage configuration, fluid flowing through the valve changes direction several times, resulting in turbulence, resistance to the fluid flow, and a pressure drop in the system. Globe valves are recommended on installations requiring frequent operation, throttling, and/or a positive shutoff when closed, including plumbing fixture supply pipes. Globe valves must be installed with the flow direction arrow pointing in the downstream direction.

The circular disks, or washers, within the valve body are replaceable composition disks made for use with different fluids, including hot water, cold water, and chemicals. Globe valves have rising stems, and can only be installed in applications where adequate clearance is provided.

Tech Facts

An angle supply valve used to control the water supply to individual plumbing fixtures is a type of globe valve. Common plumbing fixtures that use angle supply valves include water closets, lavatories, and kitchen sinks. Angle supply valves found in residential applications typically have a chrome finish since they are visible in the residence.

Angle supply valves are used to control the flow of water into water closets. When repairing or replacing a water closet, the angle supply valve is closed.

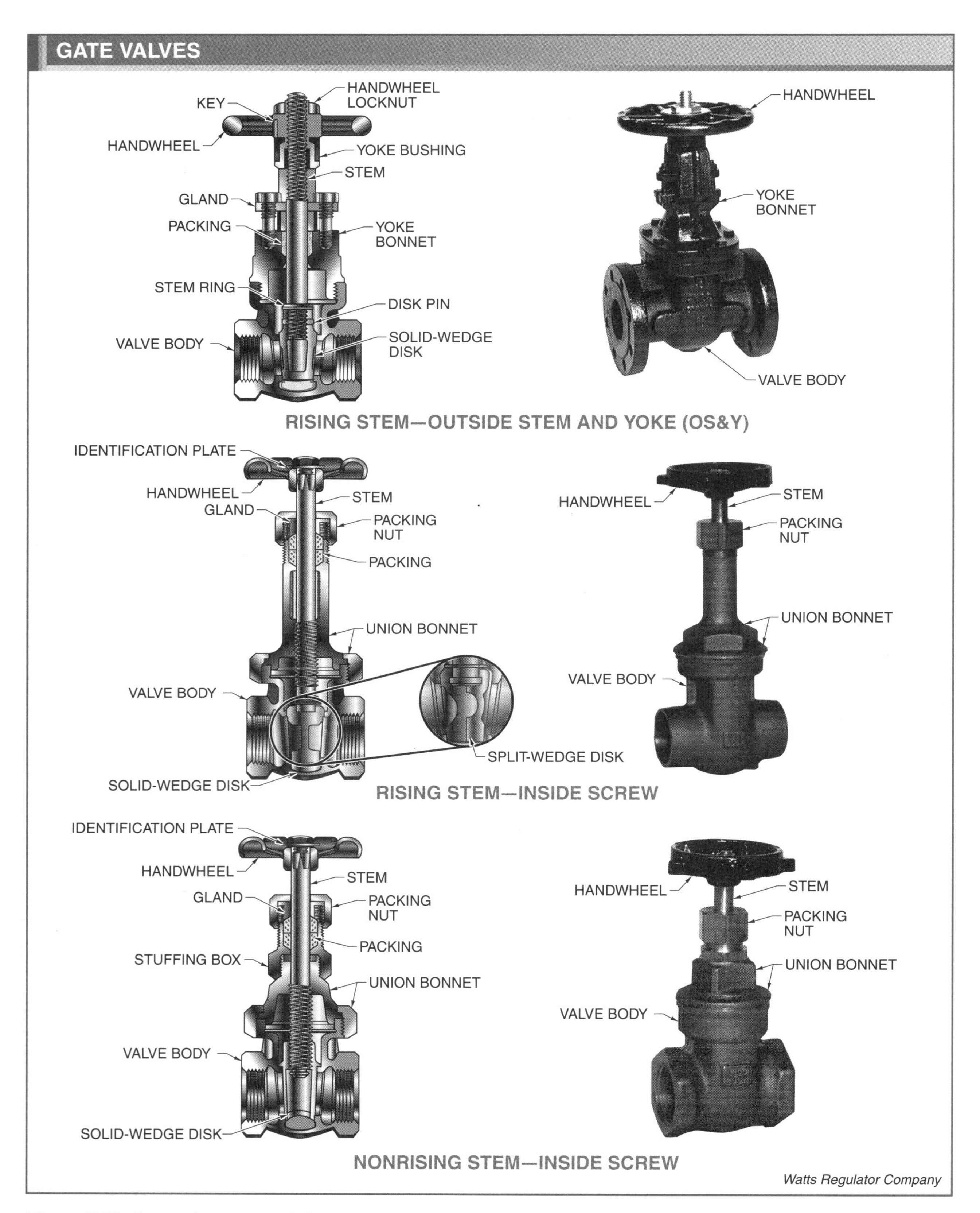

Figure 2-20. Gate valves are available in rising stem—outside stem and yoke (OS&Y), rising stem—inside screw, and nonrising stem—inside screw types.

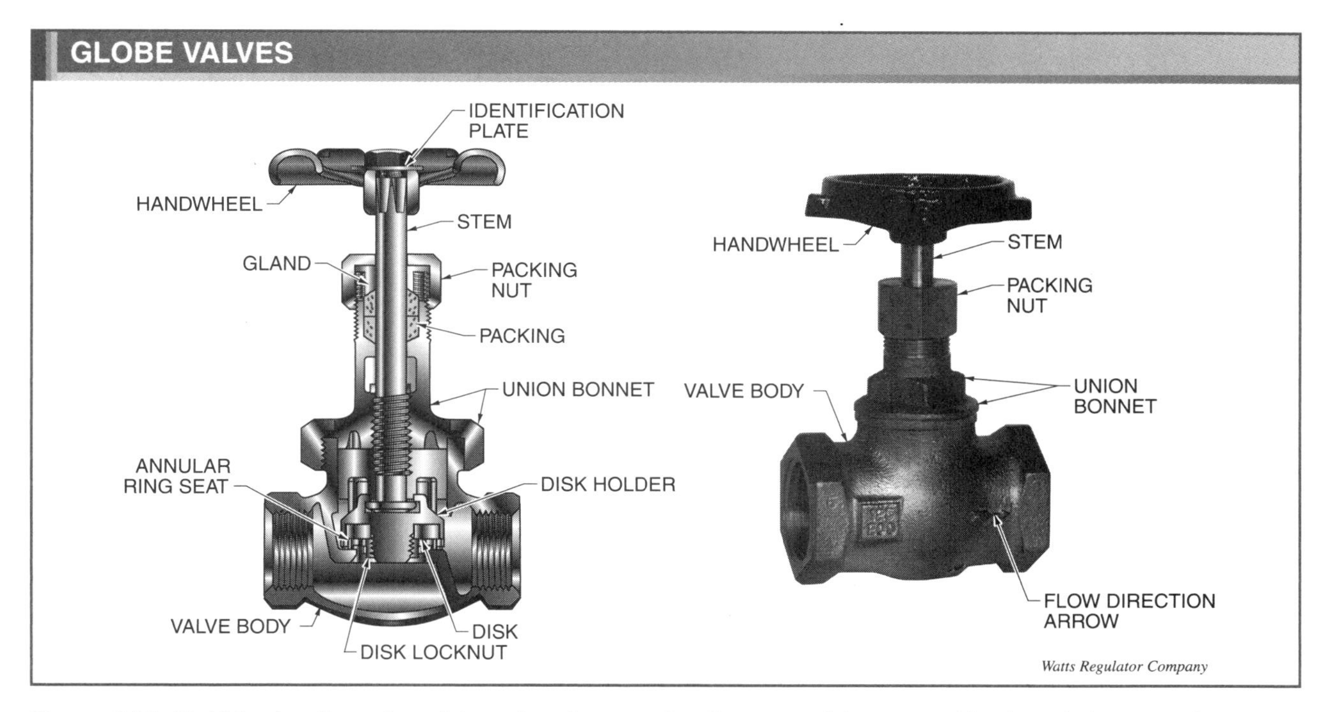

Figure 2-21. Fluid flowing through a globe valve changes direction several times, resulting in turbulence, resistance to fluid flow, and a pressure drop.

Globe valves are rated or non-rated, depending on the application. A *rated globe valve* is a globe valve that has a full-size valve seat opening. For example, a 1″ rated globe valve has a 1″ diameter seat opening. A *non-rated globe valve,* or compression stop, is a globe valve in which the valve seat diameter is less than the stated size of the valve. A *stop-and-waste valve* is a non-rated globe valve with a side port in the valve body, which is used to drain fluid from the outlet side of the valve. **See Figure 2-22.** Stop-and-waste valves are used to control water flow to fixtures, such as sillcocks, that are subject to freezing. Water is drained from the valve body and downstream piping to prevent freezing.

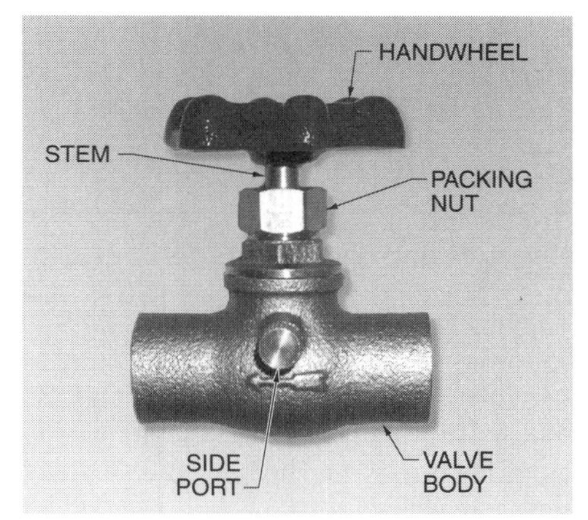

Watts Regulator Company

Figure 2-22. Water is drained from the side port of a stop-and-waste valve.

Angle Valves. An *angle valve* is a globe valve in which the inlet and outlet are at 90° to each other. **See Figure 2-23.** Angle valves are commonly used in place of a globe valve and 90° elbow since angle valves provide less resistance to water flow and reduce the number of joints required, thus reducing installation time.

Tech Facts

Most globe valves are marked with a direction arrow to indicate flow. Typically, a raised arrow is molded into the casting of the valve to indicate direction. Globe valves must be installed with this arrow pointing in the downstream direction.

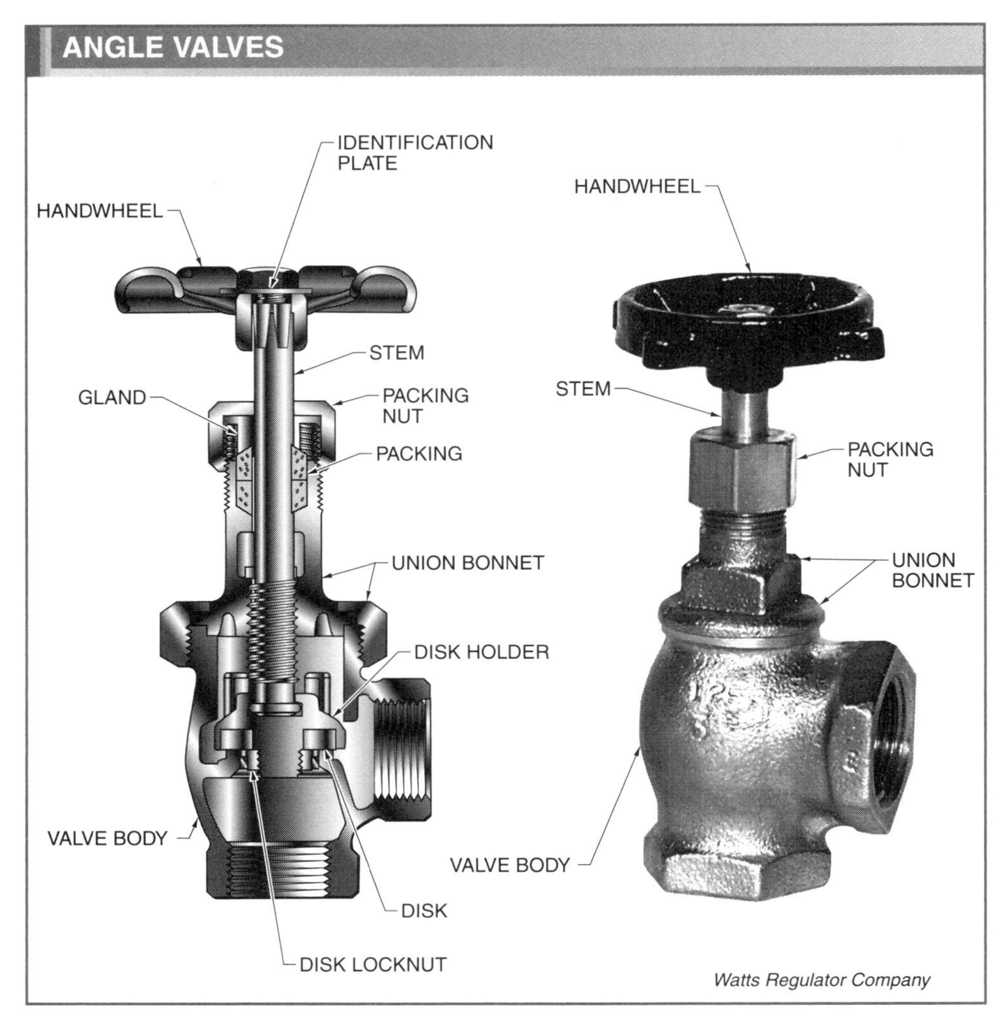

Figure 2-23. The inlet and outlet of an angle valve are at 90° to each other.

Sillcocks and Boiler Drains. Sillcocks and boiler drains are angle compression stops. A *sillcock,* or hose bibb, is a valve with integral external threads installed on the exterior of a building for attachment of a garden hose. Frost-free and standard sillcocks are available. **See Figure 2-24.**

Frost-free sillcocks are installed in areas subject to cold weather conditions. The valve of a frost-free sillcock, located within a sillcock valve housing, is inside the building to prevent the sillcock from freezing during cold weather. A frost-free sillcock must be installed with a slight pitch toward the exterior of the building to allow the valve to drain properly. Standard sillcocks do not provide the freeze resistance of frost-free sillcocks, but are installed in the same manner as a frost-free sillcock. A full-way valve should be installed inside the building to shut off the sillcock water supply if repairs need to be made.

Tech Facts

Globe valves allow water to flow by raising or lowering the disk contained within the valve body. The valve is opened or closed by turning the handwheel to change the disk position. Turning the handwheel counterclockwise moves the disk away from the valve seat to open the valve. Turning the handwheel clockwise moves the disk toward the valve seat to close the valve. Sediment or debris accumulating on the valve seat or disk face may prevent a globe valve from completely closing.

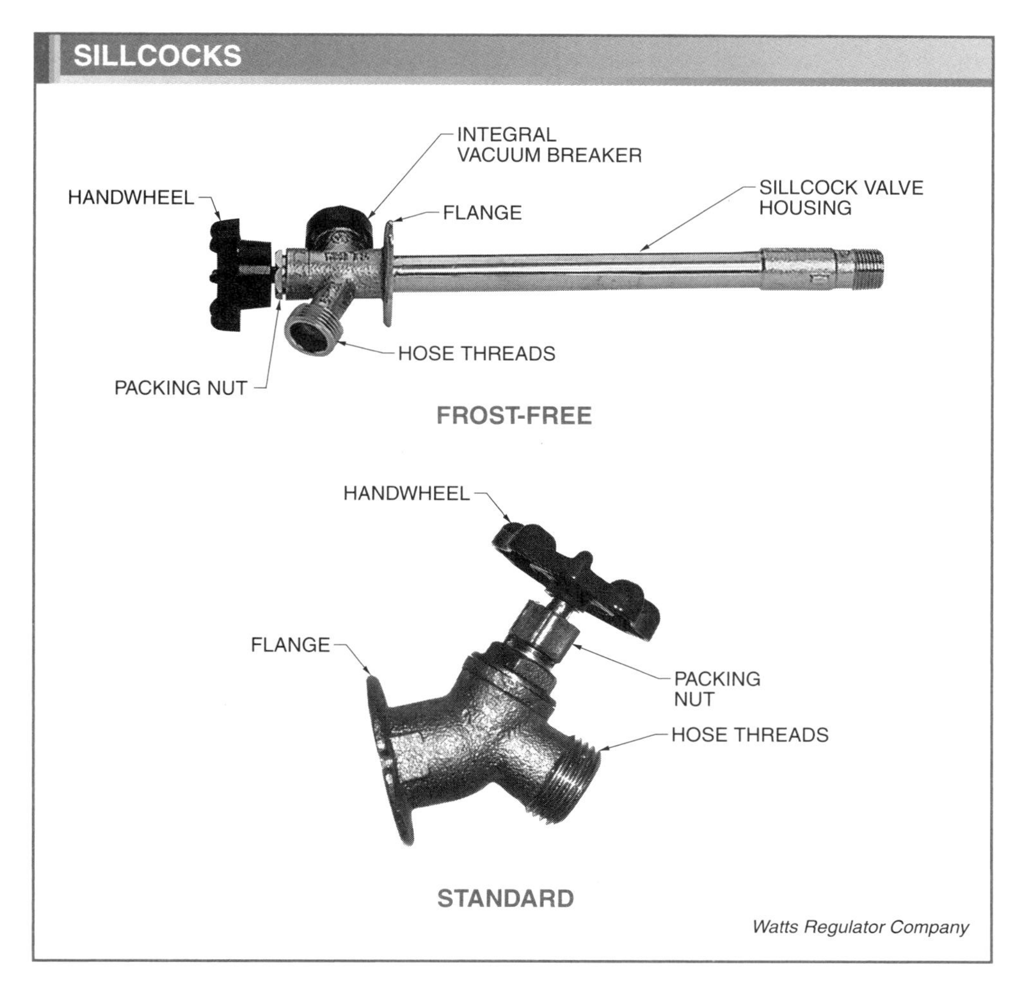

Figure 2-24. Frost-free and standard sillcocks provide a connection for a garden hose.

Valves with hose threads, including sillcocks, must be installed with a vacuum breaker to prevent back siphonage. **See Figure 2-25.** A *vacuum breaker* is a backflow prevention device that consists of a spring-loaded check valve that seals against an atmospheric outlet when water is turned on. Vacuum breakers are an integral part of a sillcock or may be attached to existing sillcocks.

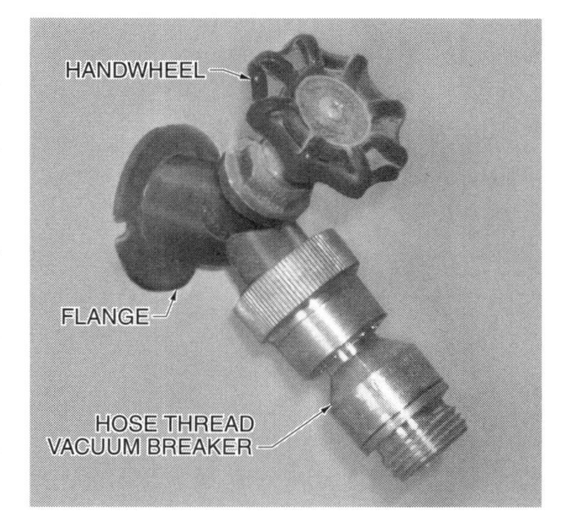

Figure 2-25. Vacuum breakers must be installed on valves with external hose threads.

A *boiler drain* is a valve with hose threads that is installed on a tank, such as a water heater, to drain and/or flush the tank. Sillcocks and boiler drains are similar, except that a sillcock has a mounting flange while a boiler drain threads directly into the tank. **See Figure 2-26.**

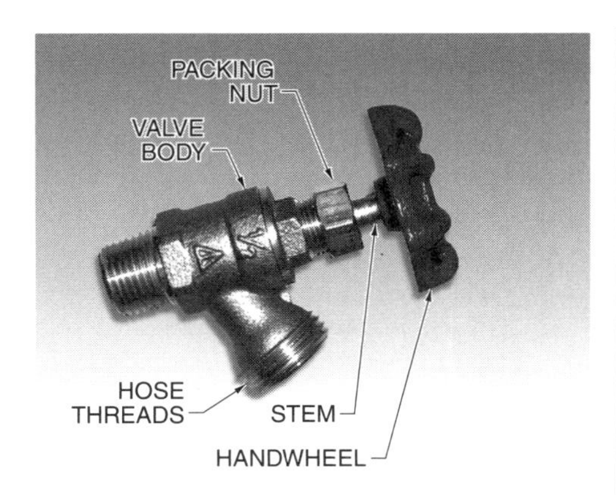

Figure 2-26. Boiler drains are installed on tanks to drain and/or flush the tanks.

Core Cocks

A *core cock,* or plug valve, is a valve through which water or gas flow is controlled by a circular core or plug that fits closely in a machined seat. As the core is turned, water flows through a port in the core. Core cocks, like gate valves, are full-way valves. However, opening or closing a core cock requires only a 90° turn of the valve handle or key.

Corporation, curb, and gas cocks are common core cocks. **See Figure 2-27.** A *corporation cock* is a core cock placed on the water main to which the water service of the building is connected. A *curb cock* is a core cock installed on the water service to turn on or off the potable water flow to a building. A *gas cock* is a valve for controlling gas flow to a gas appliance.

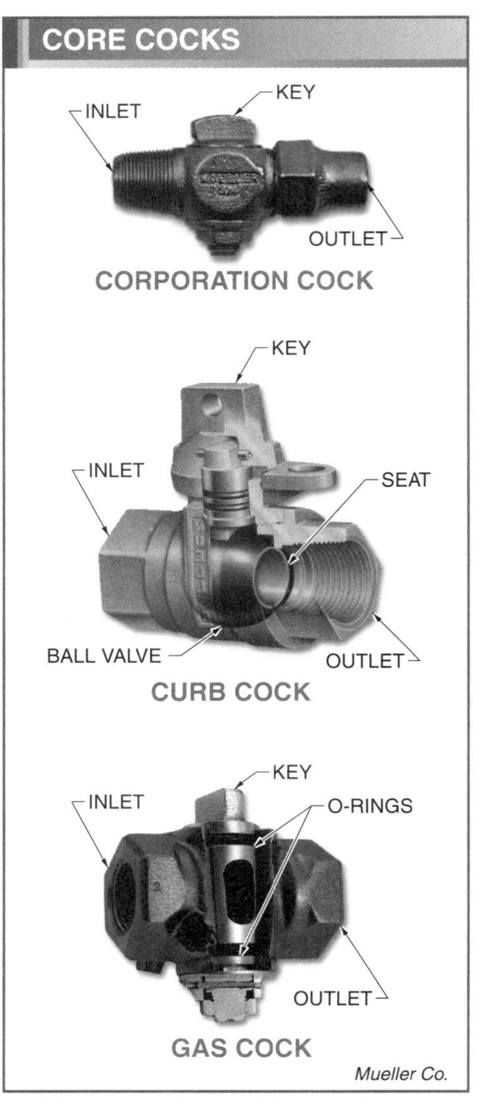

Figure 2-27. A core cock requires a 90° turn of the valve handle or key to completely open or close the valve.

Ball Valves

A *ball valve* is a valve in which fluid flow is controlled by a ball that fits tightly against a resilient (pliable) seat in the valve body. As the ball is turned, water flows through a port in the ball. **See Figure 2-28.** A ball valve requires only a 90° rotation of the handle to open or close the valve. Ball valves are full-way valves that can be used for throttling fluid flow, and are commonly installed on water supply piping (instead of gate valves or globe valves), and on low-pressure gas piping instead of a lever-handle gas cock.

Butterfly Valves

A *butterfly valve* is a valve used to control fluid flow; it consists of a rotating disk that seats against a resilient material within the valve body. The disk is attached to a shaft that controls the rotation of the disk. **See Figure 2-29.** A butterfly valve is fully opened or closed by a 90° rotation of the handle, and is a full-way valve that can be used for throttling fluid flow. The main advantage of a butterfly valve is its thin body, allowing the valve to be installed where other valves cannot fit.

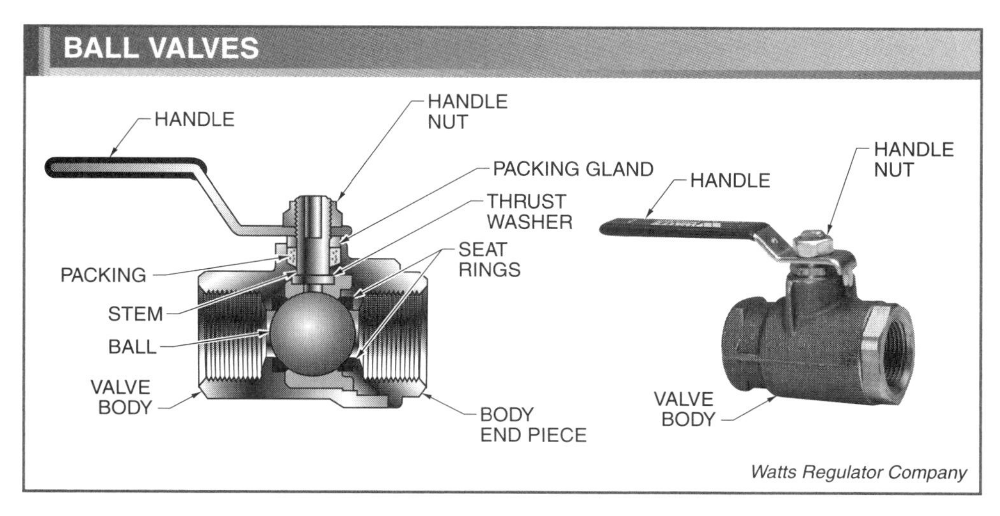

Figure 2-28. A ball valve controls fluid flow by a ball that fits tightly against a resilient seat in the valve body.

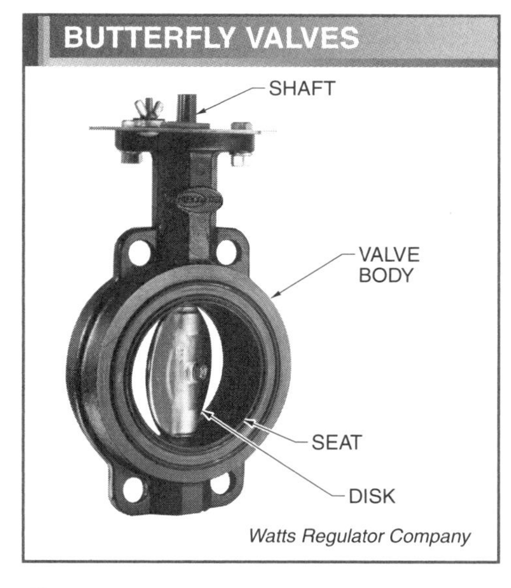

Figure 2-29. A butterfly valve controls fluid flow by a rotating disk that seats against a resilient material within the valve body.

Check Valves

A *check valve* is a valve that allows flow in only one direction. Check valves react automatically to changes in the pressure of the fluid flowing through the valve, and close when pressure changes occur.

Check valves are available as swing check and lift check valves. A *swing check valve* is a check valve in which backflow is prevented through the use of a hinged disk within the valve body. **See Figure 2-30.** In its normal operating condition, fluid flows straight through the valve and holds open the hinged disk. When backflow occurs, the hinged disk swings down into position. Swing check valves provide little resistance to fluid flow, and are commonly used with gate valves in installations where fluids are moving at a low velocity and in which there is seldom a change in fluid flow direction.

A *lift check valve* is a check valve in which backflow is prevented through the use of a disk that moves vertically within the valve body. **See Figure 2-31.** In its normal operating condition, fluid pressure forces the disk from its seat, allowing fluid to flow. When backflow occurs, the disk drops onto its seat, preventing backflow from occurring. Lift check valves have a high resistance to fluid flow due to the fluid passageway within the valve body. Lift check valves are commonly used with globe and angle valves in installations where frequent changes in flow direction occur.

Backwater Valves. A *backwater valve* is a check valve used to prevent the backflow of sewage into a building. Ball-type backwater valves and swing check backwater valves can be integral parts of floor drains to prevent flooding caused by stoppages in the building drain or building sewer. **See Figure 2-32.** Swing check backwater valves can also be installed on building drain branches serving basement plumbing fixtures in areas where the sewer main is subject to stoppage or flooding. A *ball-type backwater valve* is a backwater valve in which backflow is prevented through the use of a ball enclosed within the valve body. A *swing check backwater valve* is a backwater valve in which backflow is prevented through the use of a hinged disk or flapper.

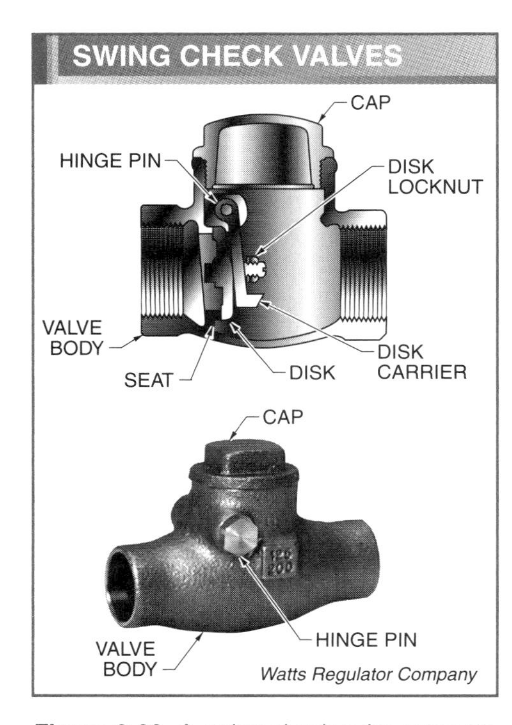

Figure 2-30. A swing check valve prevents backflow through the use of a hinged disk within the valve body.

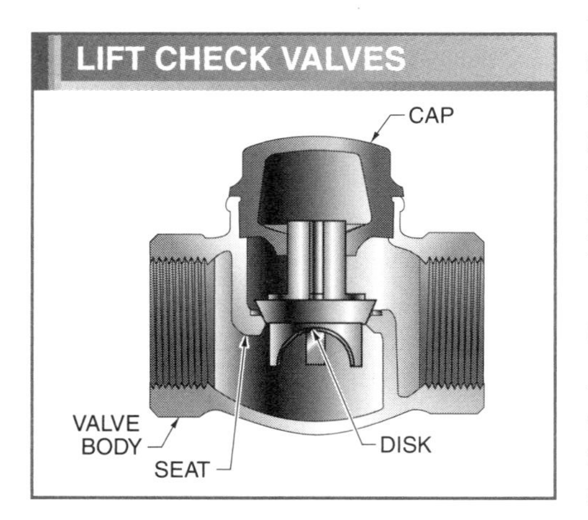

Figure 2-31. A lift check valve prevents backflow by using a disk that moves vertically within the valve body.

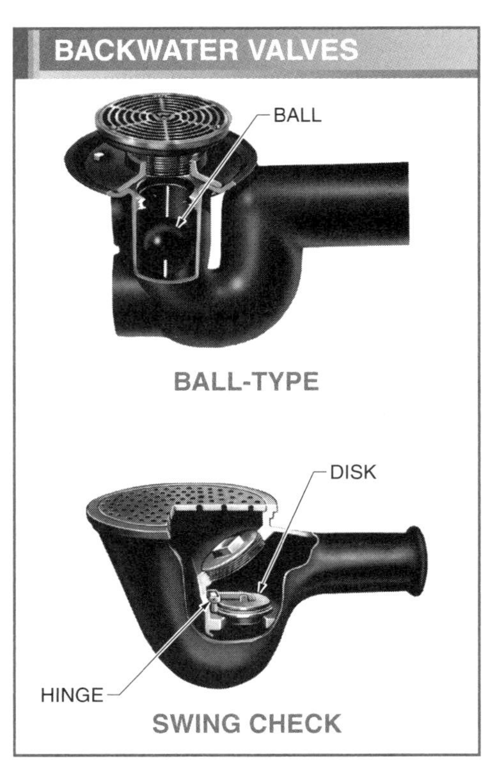

Figure 2-32. Backwater valves prevent the backflow of sewage into a building.

Pressure-Reducing Valves

A *pressure-reducing valve,* or pressure-regulating valve, is an automatic device used to convert high and/or fluctuating inlet water pressure to a lower or constant outlet pressure. Pressure-reducing valves are installed near the water meter on a building water service to reduce excessive water main pressure. Pressure-reducing valves have an adjustment screw to adjust the outlet pressure. A strainer must be installed with a pressure-reducing valve to prevent dirt and debris from entering the valve mechanism; many pressure-reducing valves contain an integral strainer. **See Figure 2-33.**

Relief Valves

A *relief valve* is a safety device that is activated to open when pressure and/or temperature in a closed plumbing system exceeds safe operating limits. When the pressure and/or temperature returns to safe operating levels, the relief valve returns to the closed position. Relief valves protect water heaters, hot water storage tanks, and boiler tanks from overheating and possible explosion.

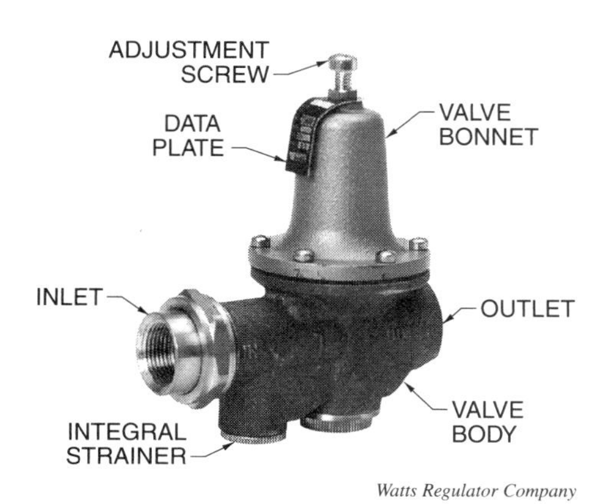

Figure 2-33. A pressure-reducing valve converts high and/or fluctuating inlet water pressure to a lower and constant outlet pressure.

Relief valves are used to relieve temperature and pressure, or for pressure relief only. **See Figure 2-34.** A *temperature and pressure (T&P) relief valve* is a safety device used to protect against excessive temperature and/or pressure in a water heater. A *pressure relief valve* is a safety device used to automatically lower excessive pressure in a closed plumbing system. Pressure relief valves are installed on some types of commercial and industrial water heating equipment.

Tech Facts

A temperature and pressure (T&P) relief valve must be installed in a water heater to protect against excessive temperature and/or pressure.

WATER METERS

Water meters measure and indicate water usage for a building so that the building owner can be charged for the amount of water used. A *water meter* is a device used to measure, in cubic feet or gallons, the amount of water flowing through the water service. A water meter is installed at the end of the water service pipe—either directly inside or outside of the building walls, in accordance with local plumbing codes. A full-way valve, such as a gate or ball valve, is installed on each side of the water meter. Three types of water meters are disc, turbine, and compound meters.

Disc Water Meters

A *disc water meter,* or displacement meter, is a water meter used to measure water flow through small water services. Disc water meters, which are accurate devices for measuring small volumes of water, are available in ⅝″–2″ sizes.

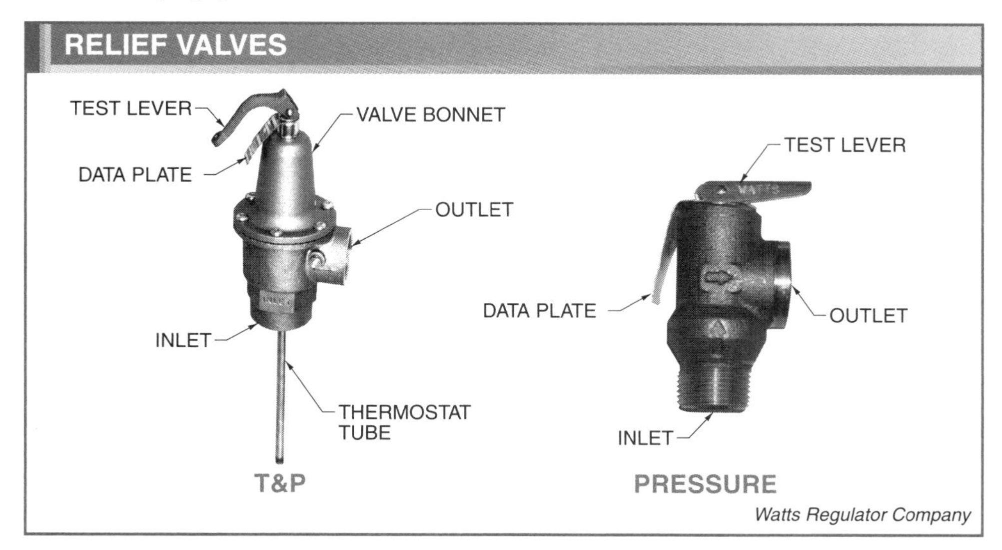

Figure 2-34. Relief valves relieve temperature and pressure, or pressure only. T&P relief valves are typically installed on residential water heaters and pressure relief valves are installed on boiler tanks.

During operation of a disc water meter, water enters the meter and flows into the measuring chamber in which the nutating disc is fitted. As the water passes through the measuring chamber, the disc rotates in a wobbling (nutating) motion and physically displaces all of the water from the measuring chamber. Disc rotations are transmitted through the gear train to the recording dial of the sealed register. **See Figure 2-35.**

Turbine Water Meters

A *turbine water meter* is a water meter used to measure large and constant volumes of water in buildings. During operation of a turbine water meter, water strikes the blades of the rotor (turbine), causing it to turn. The rotating motion is transferred by means of a magnetic coupling to a vertical spindle and then to the gears in the register. **See Figure 2-36.** Turbine water meters are available in 1½″–20″ sizes.

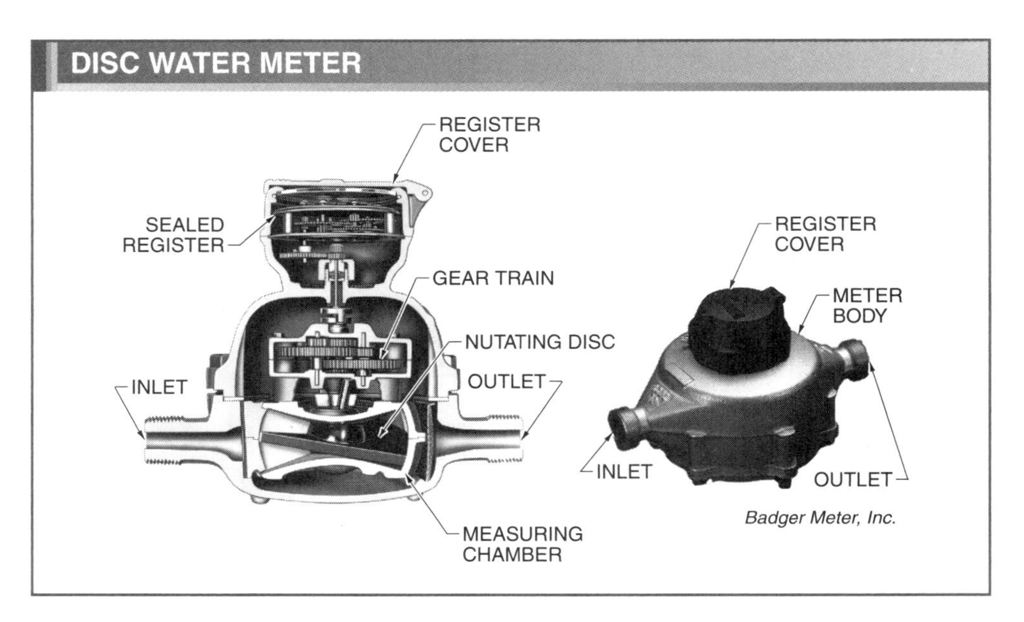

Figure 2-35. A disc water meter is used to measure water flow through small water services.

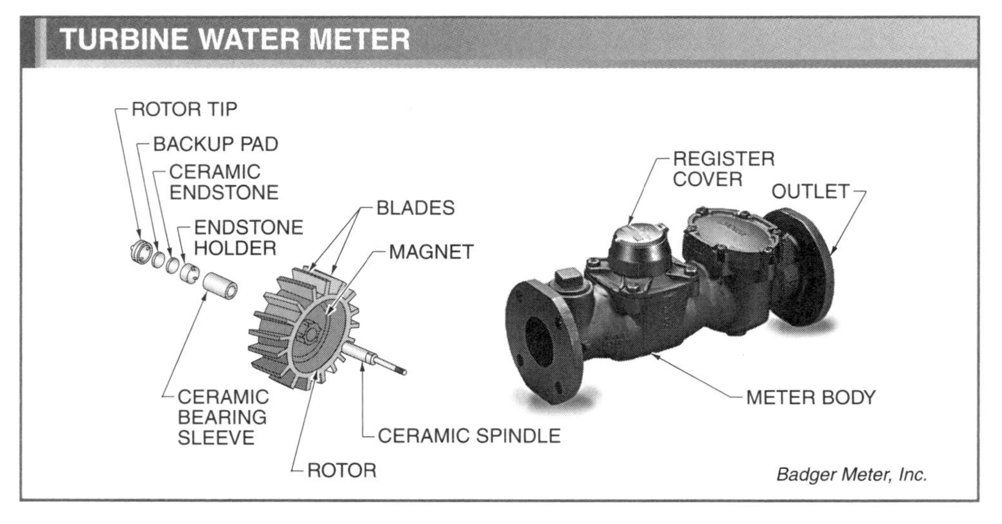

Figure 2-36. A turbine water meter is used to measure large and constant volumes of water in buildings.

Compound Water Meters

A *compound water meter* is a water meter that combines a disc and turbine meter, and is used in buildings in which there is a large fluctuation of water flow, such as an office building, which has large water usage during business hours and little water usage during evening hours and weekends. Compound water meters are available in 2″–10″ sizes.

When water enters a compound water meter during a period of little water usage, the disc portion of the meter measures the water volume since the heavy-duty valve prevents the water from entering the turbine portion of the meter. During periods of large water usage, the water forces the heavy-duty valve open, which in turn closes water flow to the disc portion of the meter and forces the water into the turbine portion of the meter where the volume is measured. **See Figure 2-37.**

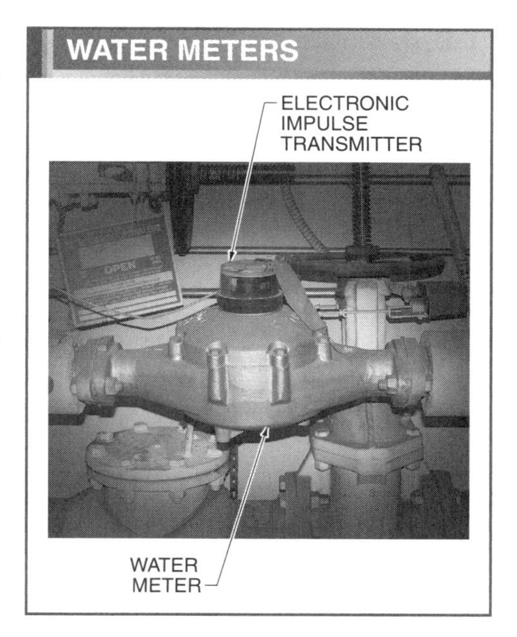

Figure 2-38. Outside registers and electronic impulse transmitters allow water meters to be read without entering the building.

Outside Meter Reading

Water meters are available with outside registers and electronic radio transmitters to eliminate entering the building to read the meter. **See Figure 2-38.** Outside registers receive an electronic signal transmitted from the meter to the register via a two-conductor wire. A meter reader must visually read each register and record the reading for billing.

Electronic radio transmitters read the amount of water passing through the meter and store this data. The meter reader drives down the street in a vehicle containing a radio that signals the electronic radio transmitter to send the meter reading to a data-recording computer in the vehicle. This system eliminates the need for a meter reader to walk up to each building.

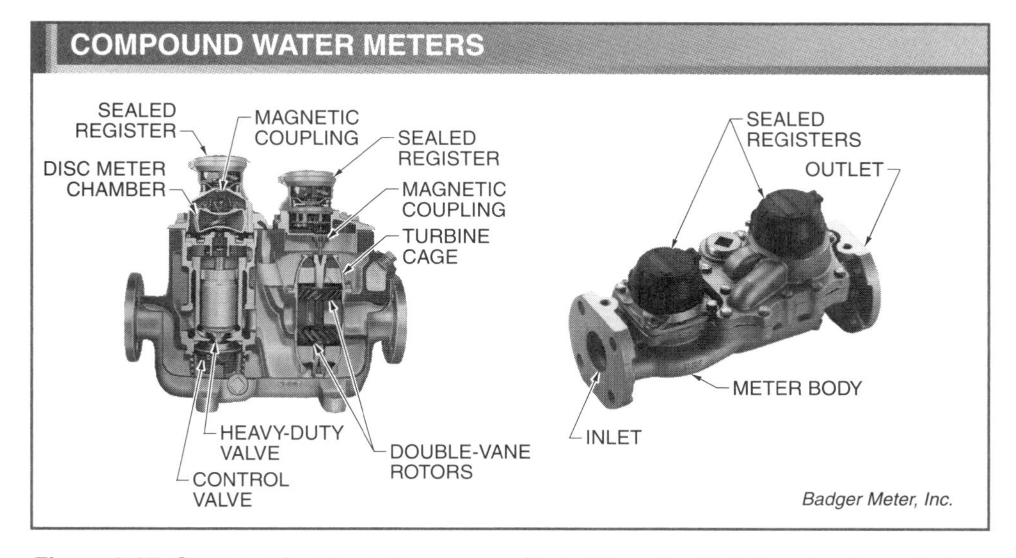

Figure 2-37. Compound water meters are used in buildings in which there is a large fluctuation of water flow.

Outside Registers

Outside registers allow a water meter to be read without personnel entering the building, and are available for disc, turbine, and compound water meters. An electrical or electronic signal is transmitted from a water meter to the register to indicate the proper meter reading. A self-contained generator, mounted on a water meter, is connected to the outdoor register using a two-conductor wire. The disc spindle of the meter rotates as water flows through the meter. The rotating motion is transferred to a spring-biased, six-pole magnet in the generator using a reduction gear train and an escape gear. When the escape gear is released, the biasing springs return the magnet to its original position. This action produces a low-voltage pulse (approximately 8 V) in coils located near the magnet. The pulse is transmitted over the wire to a solenoid in the outdoor register. The outdoor register advances one digit for every pulse received.

Review Questions

1. List an application for the following pipe and fittings:
   A. ABS and PVC pipe and fittings
   B. CPVC pipe and fittings
   C. PEX tubing and insert fittings
   D. DWV copper pipe and fittings
   E. Types L and M copper pipe and solder joint pressure fittings
   F. Types K and L annealed copper pipe and flare fittings
   G. Types L and M drawn copper with grooved joint fittings
   H. bell-and-spigot cast iron soil pipe
   I. no-hub cast iron soil pipe
   J. steel pipe with malleable iron fittings
2. List an application for the following valves:
   A. gate valve
   B. butterfly valve
   C. globe valve
   D. check valve
   E. angle valve
   F. backwater valve
   G. core cock
   H. pressure-reducing valve
   I. ball valve
   J. relief valve
   K. stop-and-waste valve
3. List the three types of water meters. Which type would be used in a one-family home? In a large office building?
4. What is an advantage of an outside water meter register?

The purpose of sanitary drainage is to move wastewater and waterborne waste from plumbing fixtures and appliances to sanitary sewers and away from buildings. Vent piping is used to pressure-balance the drainage piping, which protects fixture trap seals. Stormwater drainage piping conveys rainwater and other precipitation to the storm sewer or place of disposal.

3 Sanitary Drainage, Vent, and Stormwater Drainage Piping

Sanitary drainage piping conveys wastewater and waterborne waste from the plumbing fixtures and appliances to the sanitary sewer. Vent piping provides circulation of air to or from a sanitary drainage system and also provides air circulation within the sanitary drainage piping to protect trap seals from siphonage or back pressure. Stormwater drainage piping conveys rainwater or other precipitation to the storm sewer or other place of disposal. The ability to design and install sanitary drainage, vent, and stormwater drainage systems is essential for the following reasons:

- On many small jobs, especially new construction and remodeling projects, a plumber will have to size the piping systems for the entire job since piping drawings will not be included as part of the prints.
- On large jobs, even if the plumbing systems are shown in the prints, a plumber is responsible for ensuring that the systems are sized and installed within the requirements of the local plumbing code.

A plumber sizes the sanitary drainage, vent, and stormwater drainage systems of a building based on the following:

- number and type of plumbing fixtures and appliances, floor drains, and roof drains
- locations of the plumbing fixtures and appliances, floor drains, and roof drains
- locations of the water supply, sanitary drainage, vent, and stormwater piping
- pipe and fittings available
- locations of conflict with structural features of the building or with other trades whose systems must also be installed in the building

SANITARY DRAINAGE PIPING

Soil pipes convey discharge containing fecal matter from water closets or similar fixtures, with or without the discharge of other fixtures, to the building drain or building sewer. Waste pipes convey only liquid waste that is free from fecal matter. Since waste pipes do not convey solid or semisolid materials, waste pipes are typically sized smaller than soil pipes in a building.

Horizontal and vertical pipes are used to construct sanitary drainage, vent, and stormwater drainage piping. A *horizontal pipe* is any pipe or fitting that makes an angle of less than 45° with the horizontal plane. A *vertical pipe* is any pipe or fitting that makes an angle of 45° or less with the vertical plane. **See Figure 3-1.**

Drainage Fixture Unit System

Sanitary drainage piping must be properly sized before it can be installed. Sanitary drainage piping is sized according to the drainage fixture unit system. A *drainage fixture unit (dfu)* is a measure of the probable discharge of wastewater and waterborne waste into the drainage system formulated in tests conducted by the International Association of Plumbing and Mechanical Officials, which developed the Uniform Plumbing Code. Standard plumbing fixtures were individually tested and the amount of liquid waste that could be discharged through their waste outlets in a given time interval was carefully measured. It was found that a lavatory, which is one of the smaller plumbing fixtures, would discharge approximately 7½ gal. of water in one min through its waste outlet. Since 7½ gal. equals approximately 1 cu ft of water, 7½ gal. was established as the basis of the drainage fixture unit system, that is:

1 drainage fixture unit (dfu) = 7½ gal. per min (gpm)

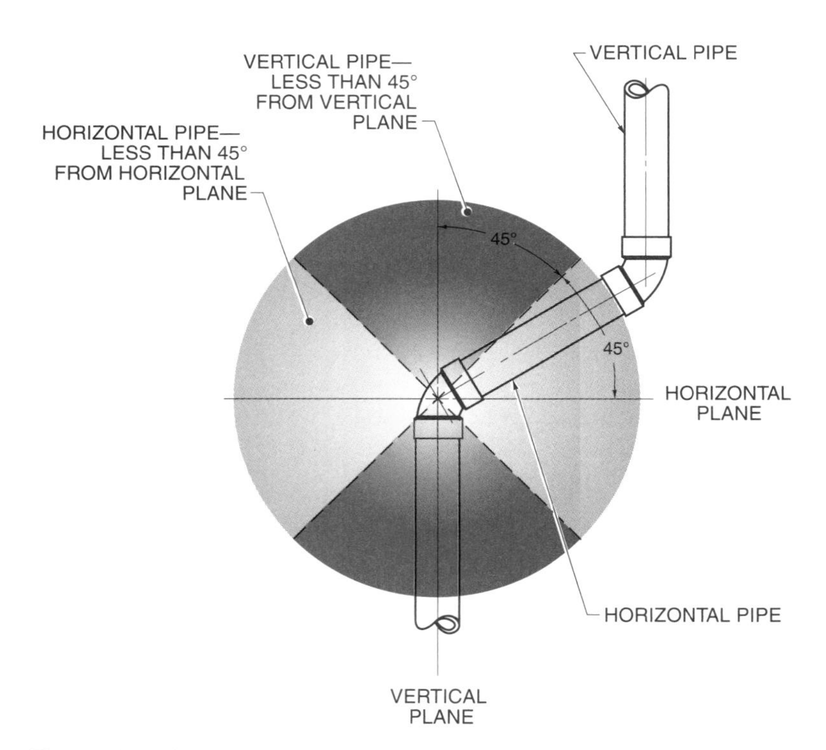

Figure 3-1. Horizontal and vertical pipes are used to construct sanitary drainage piping, vent piping, and storm water drainage piping.

Wastewater discharge rates of plumbing fixtures have been established based on 1 dfu being equal to 7½ gal. of water per minute of waste discharge. Drainage fixture unit values for various plumbing fixtures are typically found in a plumbing code, along with minimum fixture trap and drain size. **See Figure 3-2.** The minimum fixture trap and drain size is the smallest size pipe into which the fixture may drain. This size must be used even if another sizing table indicates that a smaller size pipe would convey an equal number of drainage fixture units of waste. Always refer to the plumbing code adopted in your particular area for accepted drainage fixture unit values and minimum size trap and drain sizes.

Sizing the Building Drain and the Building Sewer

Properly sized horizontal drainage pipes can discharge up to a certain number of drainage fixture units of waste without subjecting the plumbing system to plus or minus pressure. Changes in direction, materials, grades, and other factors affect the discharge capacities of drainage pipes and were considered in the design of sizing tables. **See Figure 3-3.**

Discharge capacities for different sizes of horizontal drainage pipes, including building sewers, building drains, and building drain branches from stacks, are typically found in plumbing codes. Based on sizing tables, a plumber can establish the total discharge of all the fixtures in a building in drainage fixture units and select a drain size to serve the demand. **See Figure 3-4.**

For example, a plumbing installation consists of 30 water closets, 28 lavatories, four drinking fountains, three wall-hung urinals with 2″ traps, four 2″ floor drains, and two service sinks. The total of the drainage fixture unit values of all the fixtures is 238 dfu:

No.			dfu		Total
30	Water closets	x	6	=	180
28	Lavatories	x	1	=	28
4	Drinking fountains	x	1	=	4
3	Urinals	x	4	=	12
4	2″ Floor drains	x	2	=	8
2	Service sinks	x	3	=	6
			Total dfu	=	238

A 5″ building drain and building sewer sloped at ¼″ per foot is required.

In addition to the volume of wastewater and waterborne waste that is conveyed by a horizontal drainage pipe, the following other factors must be taken into consideration:

- No water closet is permitted to discharge its waste into a horizontal drainage pipe less than 3″ in diameter.
- No more than two water closets may drain into a 3″ horizontal drainage pipe.
- A building drain should not be less than 4″ in diameter if it receives the discharge of three or more water closets.
- Building sewers must be at least 4″ in diameter.

DRAINAGE FIXTURE UNIT VALUES FOR COMMON PLUMBING FIXTURES		
Type of Fixture	Drainage Fixture Unit Value*	Minimum Fixture Trap and Drain Size†
Clothes washer (domestic use)	2	1½
Bathtub with or without shower	2	1½
Bidet	2	1½
Drinking fountain	1	1¼
Dishwasher, domestic	2	1½
Dishwaster, commerical	4	2
FLOOR DRAIN:		
2″ waste	2	2
3″ waste	3	3
4″ waste	4	4
Lavatory	1	1¼
Laundry tray (one or two compartment)	2	1½
Shower stall, domestic	2	1½
SINK:		
Combination, sink and tray (with disposal unit)	3	1½
Combination, sink and tray (with one trap)	2	1½
Domestic	2	1½
Domestic, with disposal unit	2	1½
Service	3	2
Soda fountain	2	1½
Commercial, flat rim, bar, or counter	3	1½
Wash, circular or multiple (per set of faucets)	2	1½
URINAL:		
Wall-hung, 2″ trap	4	2
Wall-hung, 1½″ trap	2	1½
Trough (per 6′ section)	2	1½
Water closet	6	3
Unlisted Fixture or Trap Size:		
1¼″	1	
1½″	2	
2″	3	
2½″	4	
3″	5	
4″	6	

* in dfu

† in in.

Figure 3-2. Pipe sizes for sanitary drainage and vent piping are based on drainage fixture unit values.

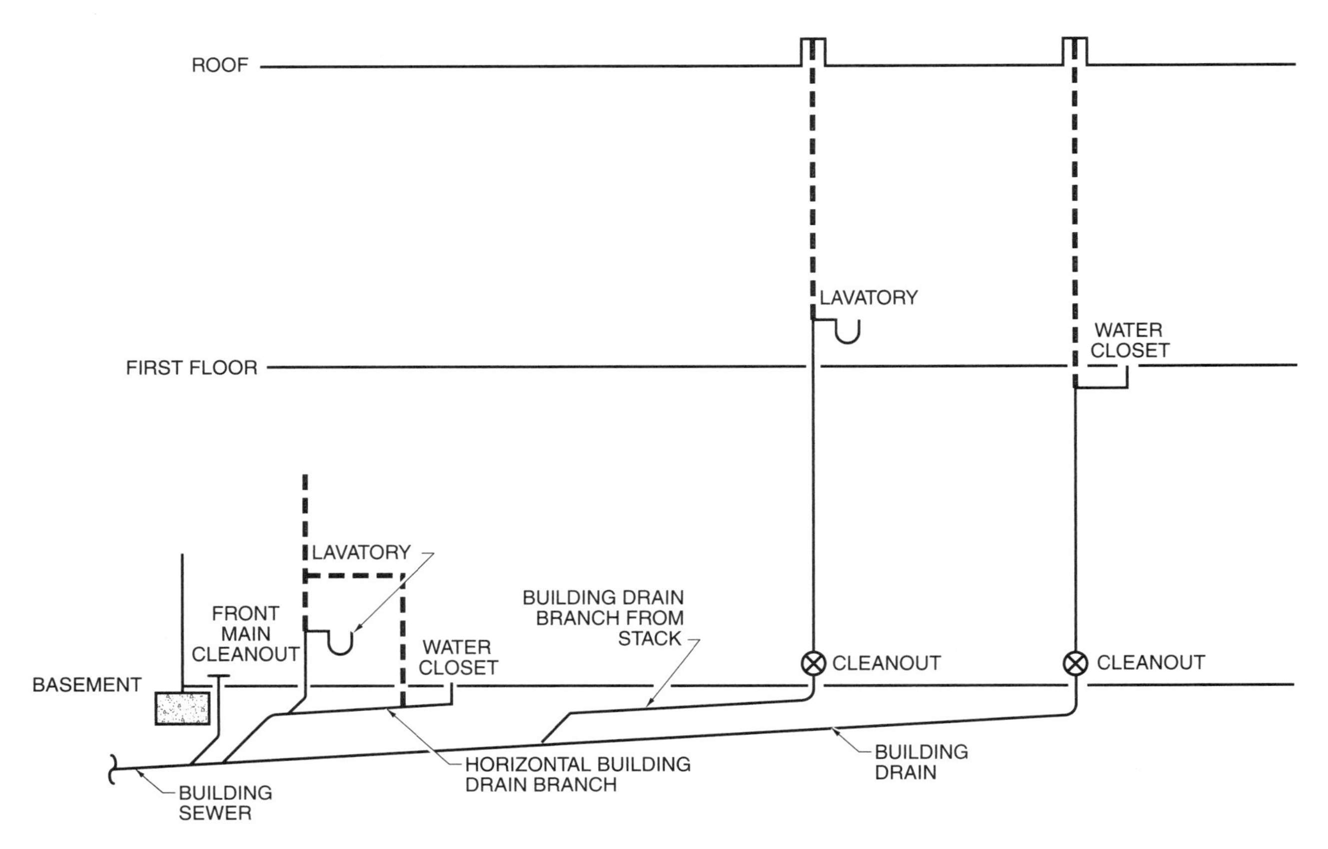

Figure 3-3. Properly sized horizontal drainage pipes can discharge a certain number of drainage fixture units of waste without subjecting the plumbing system to plus or minus pressure.

MAXIMUM LOADS* FOR BUILDING SEWER, BUILDING DRAIN, AND BUILDING DRAIN BRANCHES FROM STACKS†				
Drain Diameter‡	Slope§			
	1/16	1/8	1/4	1/2
1¼	—	—	—	—
1½	—	—	—	—
2	—	—	21	26
2½	—	—	24	31
3‖	—	36#	42#	50#
4	—	180	216	250
5	—	390	480	575
6	—	700	840	1000
8	1400	1600	1920	2300
10	2500	2900	3500	4200
12	3900	4600	5600	6700
14	7000	8300	10,000	12,000

* in dfu
† every building drain that receives the discharge of three or more water closets shall not be less than 4″ in diameter
‡ in in.
§ in in/ft
‖ no water closet shall discharge into a drain less than 3″ in diameter
\# not over two water closets

Figure 3-4. The sizes of building sewer, building drain, and building drain branches from stacks are based on the slope of the pipe and the potential dfu discharge.

Properly sized horizontal drainage pipe should be approximately one-third full of sewage to ensure proper scouring action. The efficiency of horizontal drainage pipe does not increase if the pipe is a size larger than is necessary. Scouring action is vastly reduced by increasing the drainage pipe size. With a larger pipe, solids are carried along the bottom of the pipe and, because water flow within the larger pipe is shallow and slow, the solids become separated from the water and remain in the drainage piping. Eventually, lack of scouring action may result in stoppage of the drainage or branch piping, and often the entire building drain is affected. On the other hand, a drain that is too small is overtaxed by flow and may result in siphonage, back pressure, and basement flooding.

Sizing Horizontal Branch Drains

Horizontal branch drains are another type of horizontal drainage piping. A *horizontal branch drain* is drainage pipe extending horizontally from a soil or waste stack or building drain, with or without vertical sections or branches. A horizontal branch drain receives the discharge from one or more fixture drains on the same floor as the horizontal branch and conveys it to the soil or waste stack or to the building drain. Most plumbing codes require all underground drainage pipes to be at least 2″ in diameter. Similar to building sewers and building drains, horizontal branch drain sizes are determined using a sizing table. **See Figure 3-5.**

For example, per the Maximum Loads for Horizontal Branch Drains table, a 3″ horizontal branch drain is required to serve two water closets, two lavatories, two bathtubs, and two domestic kitchen sinks.

No.			dfu		Total
2	Water closets	x	6	=	12
2	Lavatories	x	1	=	2
2	Bathtubs	x	2	=	4
2	Kitchen sinks	x	2	=	4
		Total dfu		=	22

MAXIMUM LOADS* FOR HORIZONTAL BRANCH DRAINS†

Drain Diameter‡	Horizontal Branch Drain*— ¼ in/ft
1¼	1
1½	3
2	6
2½	12
3§	32‖
4	160
5	360
6	620
8	—
10	—
12	—
15	—

* in dfu
includes horizontal branches of building drain
† every building drain that receives the discharge of three or more water closets shall not be less than 4″ in diameter
‡ in in.
§ no water closets shall discharge into drain less than 3″ in diameter
‖ not over two water closets

Figure 3-5. The size of horizontal branch drains is based on the grade of the drain.

Sizing Soil and Waste Stacks

After the horizontal building sewer and building drain pipe are sized, vertical pipes or stacks that empty into the horizontal drainage pipes must be sized. A stack is any vertical line of soil, waste, or vent piping extending through one or more stories. Wastewater does not flow down a stack in slugs of wastewater separated by pockets of air; rather, it flows down the stack in a sheet around the inside walls of the pipe. In addition, wastewater and waterborne waste does not continue to accelerate as it falls down a tall stack. Instead, the wastewater and waterborne waste reaches its maximum velocity after falling approximately two stories or floor levels.

When sizing soil and waste stacks, branch intervals must be considered. A *branch interval (BI)* is a vertical length of stack at least 8′ high within which the horizontal branches from one story or floor of the building are connected to the stack. In general, one branch interval equals one floor of plumbing fixture drains.

Depending on the location of the plumbing fixtures, some branch intervals extend more than one story of building height. **See Figure 3-6.**

Soil and waste pipes must be properly supported.

The Maximum Loads for Soil and Waste Stacks table, similar to other sizing tables, is based on the drainage fixture unit method. **See Figure 3-7.** The Maximum Loads for Soil and Waste Stacks table combines three tables into one and lists the maximum number of drainage fixture units that may empty into:

- stacks of not more than three stories, with a maximum of three branch intervals
- stacks of more than three stories or three branch intervals
- a stack on any one story or branch interval

Stacks that are 2″ and larger in diameter and that are over three stories or branch intervals high have a greater drainage fixture unit capacity than shorter stacks. Other considerations when sizing stacks are:

- No water closets are permitted to drain into a stack less than 3″ in diameter.
- No more than two water closets may drain into a 3″ stack in any one story or branch interval.

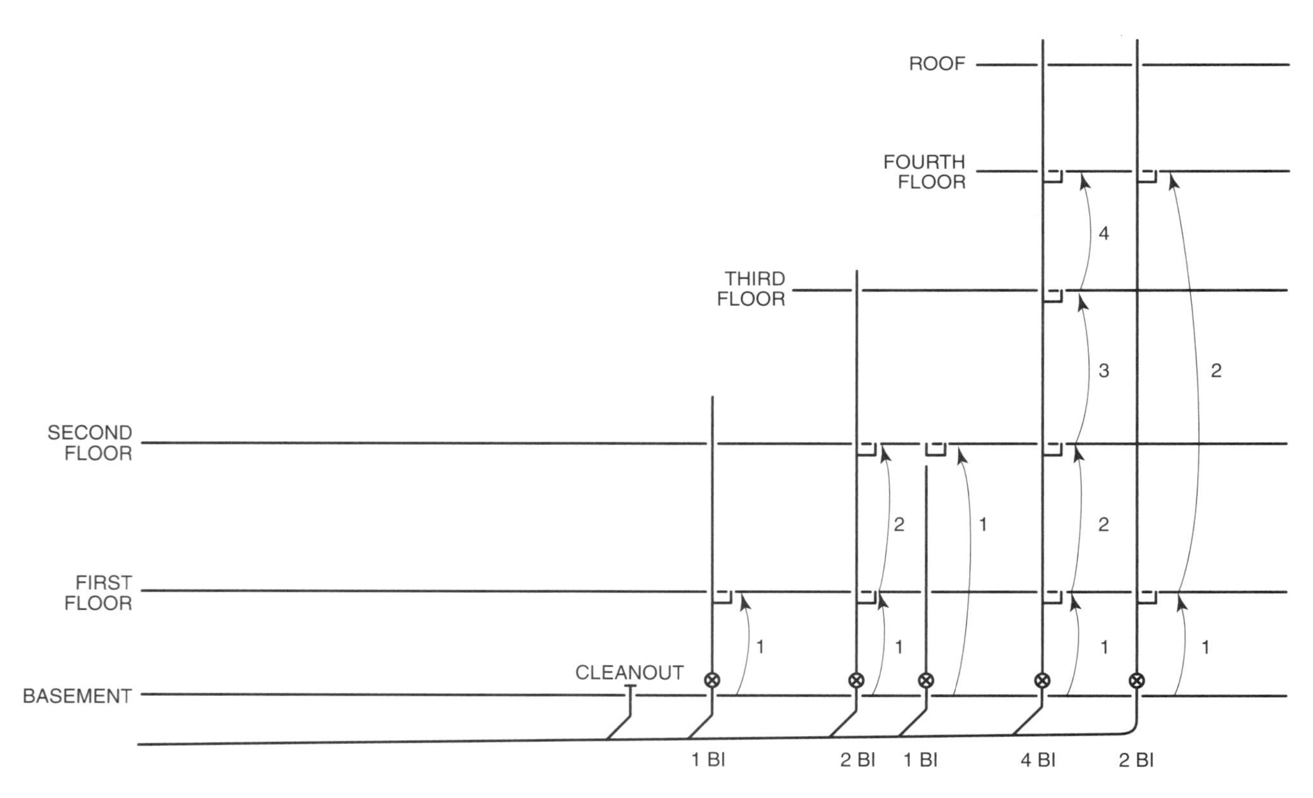

Figure 3-6. A branch interval equals one floor of plumbing fixtures but may exceed one story depending on the location of the fixtures.

MAXIMUM LOADS FOR SOIL AND WASTE STACKS[*]			
Stack Diameter[†]	Stacks Not More than Three Stories or Branch Intervals (BI)	Stacks More than Three Stories or Branch Intervals (BI)	Total at One Story or Branch Interval (BI)
1¼[‡]	2	2	1
1½[‡]	4	4	2
2[‡]	9	18	6
2½[‡]	20	42	9
3	36[§]	72[§]	24[‖]
4	240	500	90
5	540	1100	200
6	960	1900	350
8	—	3600	600
10	—	5600	1000
12	—	8400	1500

[*] in dfu
[†] in in.
[‡] no water closets permitted on a stack less than 3″ in diameter
[§] not over six water closets permitted, and not over six branch intervals on a 3″ soil stack
[‖] not over two water closets permitted

Figure 3-7. Soil and waste stack diameter is based on the potential waste discharge and the branch intervals of the building.

- No more than six water closets may drain into a 3″ stack, and 3″ stacks may not have more than 6 BI of waste emptying into them.
- No soil or waste stack is permitted to be smaller than the largest horizontal branch connected to it.
- Any building in which plumbing is installed must have at least one 3″ or larger diameter stack vent or vent stack extended full size through the roof. The full-size stack should be located farthest from where the building drain leaves the building.

Stack sizing varies depending on the floor on which the fixtures are installed in a building. For example, a 1½″ diameter stack is required to serve a building with two domestic kitchen sinks with 1½″ traps and drains, rated at 2 dfu each, and located on different floors of the building. **See Figure 3-8.**

However, if the same sinks and drains are installed on the same level of the building, a 2″ stack is required because only 2 dfu may drain into a 1½″ pipe on any one story or branch interval. **See Figure 3-9.**

Offset Stacks. In some situations, a stack cannot continue vertically for its entire length due to structural obstructions or stack location. A stack must be offset if it cannot continue vertically over its entire length. An *offset* is a combination of elbows or bends that brings one section of the pipe out of line but into a line parallel with the other section.

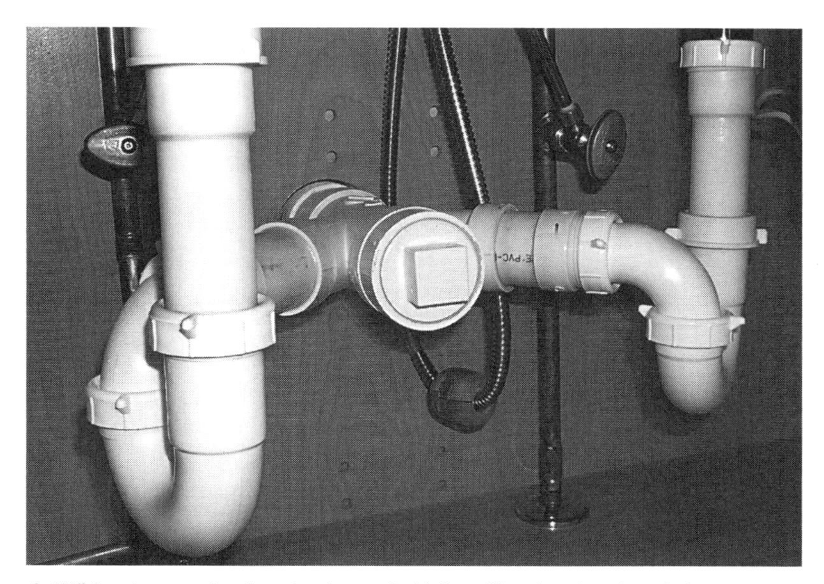

A 1½″ lavatory waste pipe may be vented into a 2″ water closet vent pipe.

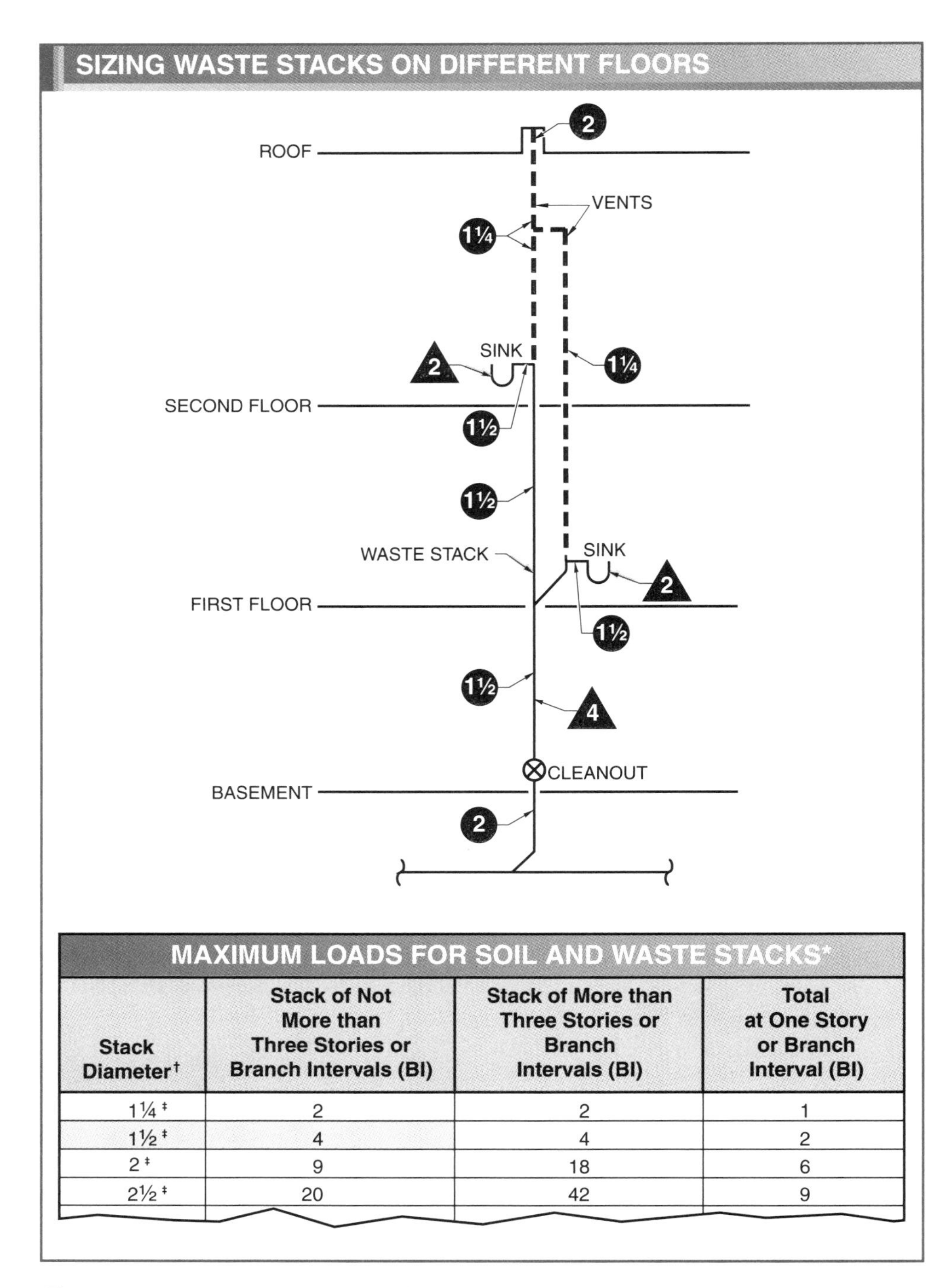

MAXIMUM LOADS FOR SOIL AND WASTE STACKS*

Stack Diameter†	Stack of Not More than Three Stories or Branch Intervals (BI)	Stack of More than Three Stories or Branch Intervals (BI)	Total at One Story or Branch Interval (BI)
1¼‡	2	2	1
1½‡	4	4	2
2‡	9	18	6
2½‡	20	42	9

Figure 3-8. When two sinks with 1½″ traps and drains rated at 2 dfu each are located on different floors of a building, a 1½″ diameter stack is required.

Fittings such as ⅛ bends, ¼ bends, and short-sweep ⅛ and ¼ bends are commonly used to create offsets, depending on the joist depth or space available above the ceiling where drainage pipe is installed.

A stack that is offset 45° or less from the vertical plane is sized as though it is a straight vertical stack. **See Figure 3-10.** However, if a stack is offset more than 45° from the vertical plane, the procedure for sizing the stack is:

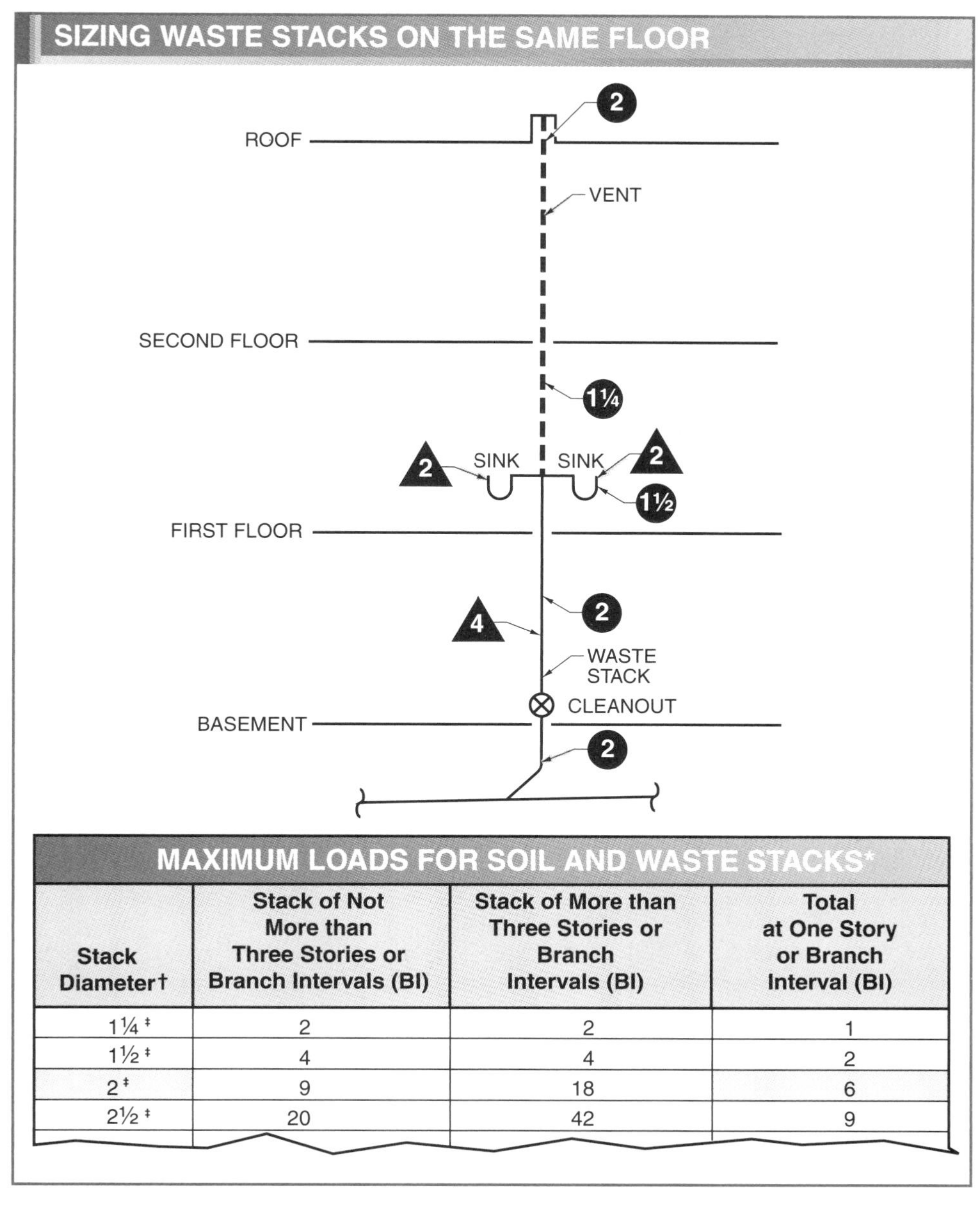

MAXIMUM LOADS FOR SOIL AND WASTE STACKS*

Stack Diameter†	Stack of Not More than Three Stories or Branch Intervals (BI)	Stack of More than Three Stories or Branch Intervals (BI)	Total at One Story or Branch Interval (BI)
1¼ ‡	2	2	1
1½ ‡	4	4	2
2 ‡	9	18	6
2½ ‡	20	42	9

Figure 3-9. When two sinks with 1½″ traps and drains rated at 2 dfu each are located on the same floor, a 2″ diameter stack is required.

1. The portion of a stack above the offset is sized the same as a straight stack (based on the total number of drainage fixture units above the offset). **See Figure 3-11.**
2. The offset portion of the stack is sized like a horizontal building drain branch.
3. The portion of the stack below the offset is sized at least as large as the offset.

Tech Facts

In order to determine sanitary drainage piping size, drainage fixture units (dfus) were developed by the American Bureau of Standards. Drainage fixture units are based on the discharge of lavatories in gallons per minute. A lavatory is the smallest fixture typically found in a home and all other fixtures are compared to it.

In buildings of five or more stories with offset stacks or a building with a horizontal building drain branch from the stack, fixtures should not be installed on the floor in which the offset occurs if waste from fixtures four or more stories above the offset is discharged into the stack. Fixtures in these buildings that drain into the horizontal portion of an offset stack or horizontal building drain branch are subject to trap seal loss from back pressure. Trap seal loss due to back pressure is prevented if fixture drains near the base of the stack or stack offset connect to the horizontal pipe at least 8′ from the offset (measured vertically or horizontally). **See Figure 3-12.** The drain may also connect back into the vertical portion of the stack 2′ below the offset. Stacks that are offset above the highest fixture drain connected to the stack are not affected by trap seal loss from back pressure because of adequate ventilation and do not need to be connected using the 8′ or 2′ guidelines. Offsets in the upper vertical portion of a stack are sized the same as stack vents.

When sizing offset stacks, the portion above the offset is sized first, followed by the offset, and then the portion below the offset.

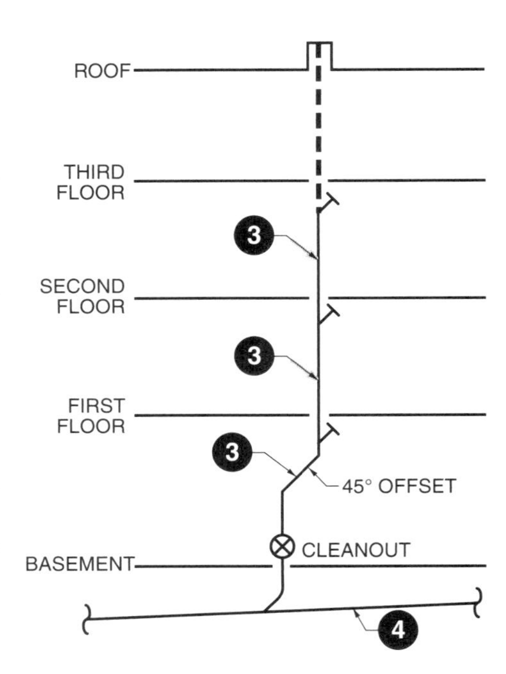

Figure 3-10. A stack that is offset 45° or less from the vertical plane is sized as a straight vertical stack.

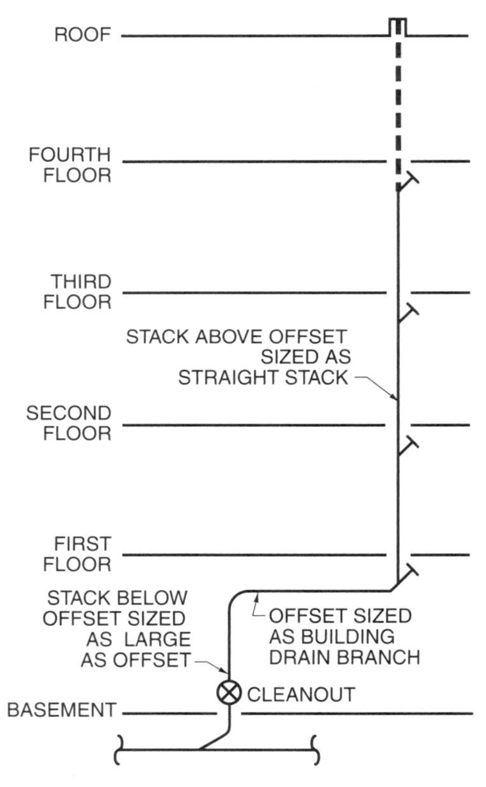

Figure 3-11. A stack offset more than 45° from the vertical plane is sized in three parts—stack above the offset, offset, and stack below the offset.

An offset is used for soil or waste stacks if the stacks cannot continue vertically over their entire length.

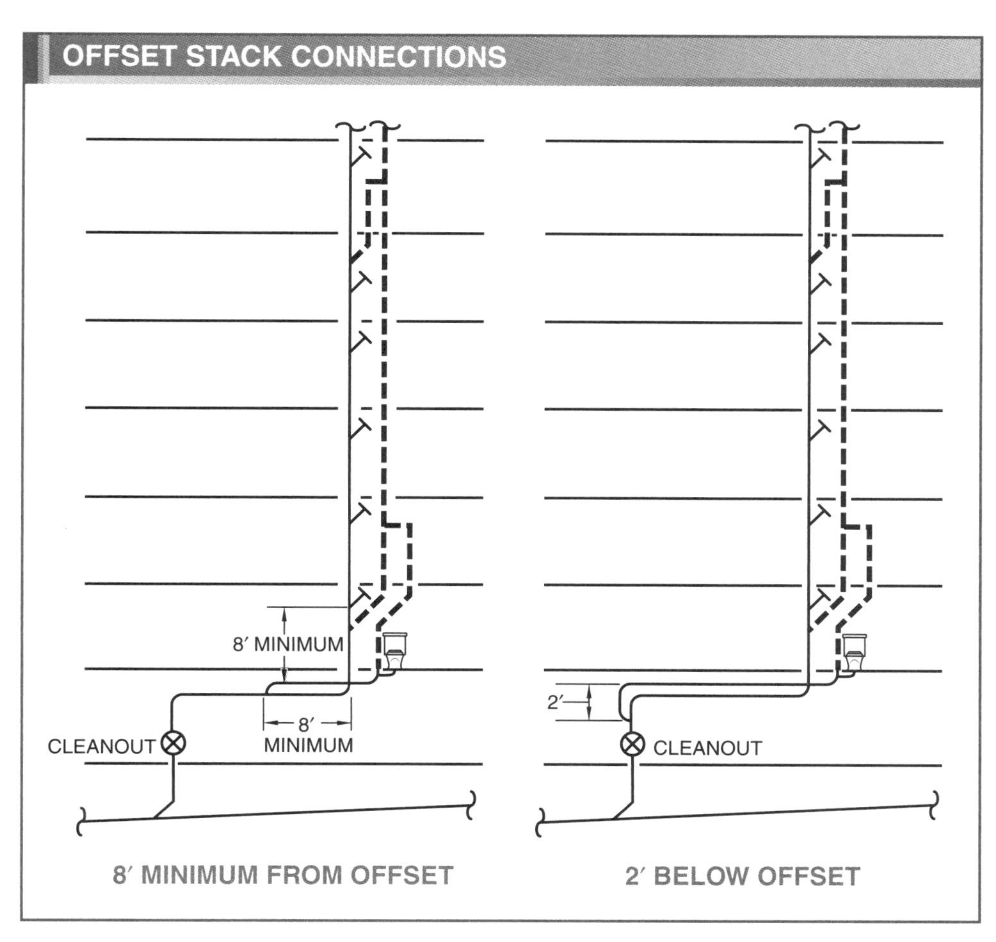

Figure 3-12. Trap seal loss due to back pressure is prevented when fixture drains near the base of the stack or stack offset are properly connected.

For example, assume that 11 dfu of waste empty into the stack on each floor for a total of 44 dfu (4 × 11 = 44) on the entire stack. **See Figure 3-13.** A 3″ stack is large enough for the section of the stack above the offset since the stack is more than three stories. The size of the offset section of the stack is based on the slope and number of drainage fixture units. For example, if the slope is ¼″ per foot with 44 dfu draining into the stack, a 4′ pipe must be used. A 4″ pipe is used for the section of the stack below the offset since it must be at least as large as the offset.

DRAINAGE PIPING INSTALLATION

Several items must be considered when installing drainage piping, including the grade or pitch of horizontal drainage piping, changes in direction of drainage piping, and location of cleanouts in the drainage piping system. Properly installed drainage systems provide adequate drainage for wastewater and waterborne waste, allowing the system to be self-scouring and free from stoppage. Cleanouts provide access to the drainage piping system if a stoppage occurs.

Tech Facts

Horizontal drainage piping must be installed so that the piping falls toward the waste disposal location. The typical grade for properly installed horizontal drainage piping should be ¼″ fall per foot of pipe.

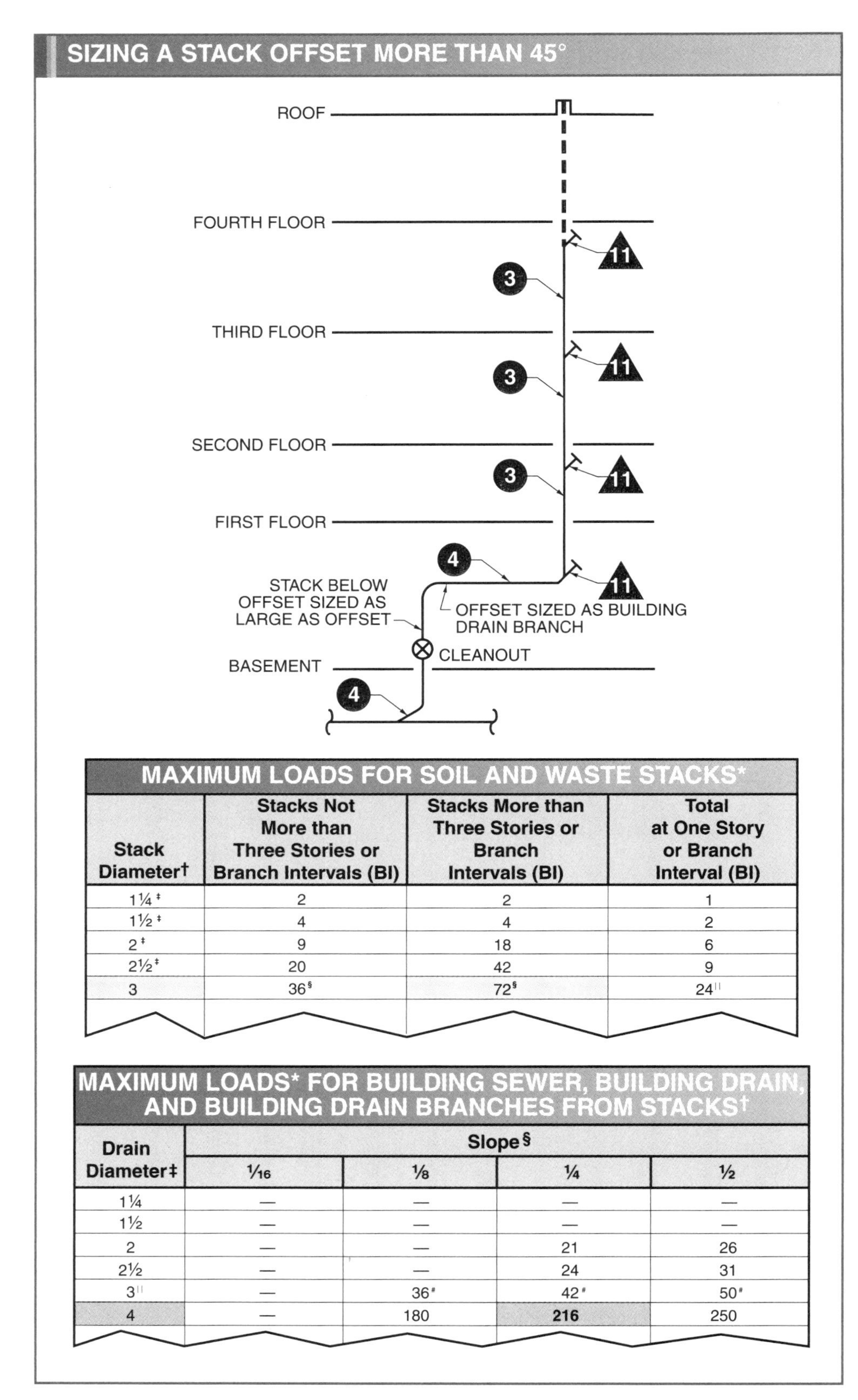

MAXIMUM LOADS FOR SOIL AND WASTE STACKS*

Stack Diameter†	Stacks Not More than Three Stories or Branch Intervals (BI)	Stacks More than Three Stories or Branch Intervals (BI)	Total at One Story or Branch Interval (BI)
1¼ ‡	2	2	1
1½ ‡	4	4	2
2 ‡	9	18	6
2½ ‡	20	42	9
3	36 §	72 §	24 ‖

MAXIMUM LOADS* FOR BUILDING SEWER, BUILDING DRAIN, AND BUILDING DRAIN BRANCHES FROM STACKS†

Drain Diameter‡	Slope§			
	1⁄16	⅛	¼	½
1¼	—	—	—	—
1½	—	—	—	—
2	—	—	21	26
2½	—	—	24	31
3 ‖	—	36 #	42 #	50 #
4	—	180	**216**	250

Figure 3-13. When sizing offset stacks, the portion above the offset is sized first, followed by the offset, and then the portion below the offset.

Horizontal Drainage Piping Grade

When horizontal drainage piping size is determined, the grade at which the pipe is installed must be calculated. *Grade,* or pitch, is the slope of a horizontal run of pipe and is expressed as a fractional inch per foot length of pipe; for example, ¼″ per foot.

Horizontal drainage piping is graded ¼″ per foot of run to provide adequate drainage. **See Figure 3-14.** Horizontal pipes graded at ¼″ per foot allow wastewater and waterborne waste to achieve the necessary velocity and discharge capacity so that the pipes can scour themselves and function properly without producing plus or minus pressures in the plumbing system. However, because of basement floor depth and inadequate sanitary sewer main depth, the building drain and/or building sewer grade may be less than ¼″ per foot. In addition, a long sewer may require less grade because the accumulated or total pitch results in a deep building drain outlet. When laying pipe with a slight grade (1/16″–⅛″), a leveling instrument should be used. A leveling instrument, such as a transit level or laser transit level, ensures a consistent grade for pipe along its entire length and pipe that is free of sagging. A sagging pipe results in sections of the pipe being filled with water.

A pitch greater than ¼″ per foot provides greater velocity and discharge capacity of wastewater and waterborne waste in soil and waste pipes, and it may also decrease the depth of waste necessary to provide self-scouring action. A greater pitch might also generate minus pressure if the drain were filled to capacity.

Changes in Direction

Changes in direction of pipes, or turns, are common in sanitary drainage piping. *Change in direction* refers to the various turns that may be required in drainage piping. Changes in direction in drainage piping must be made by the appropriate use of fittings such as Ys, long- or short-sweep ¼ bends, ⅙, ⅛, or 1/16 bends, or by a combination of these fittings.

Care must be taken in selecting the proper fittings for changes in direction in drainage piping to prevent waste stoppages. **See Figure 3-15.** Long-sweep fittings lessen the probability of waste stoppage. Short-sweep ¼ bends may be used for changes of direction from a horizontal plane to the vertical plane. Long-sweep ¼ bends, two ⅛ bends, or a combination wye and ⅛ bend may be used where the change of direction is

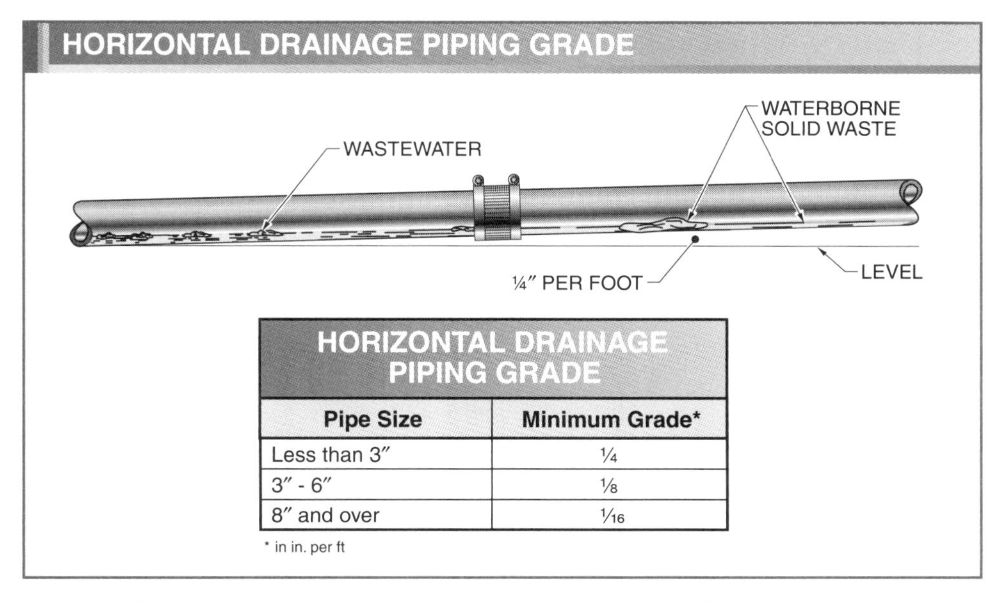

HORIZONTAL DRAINAGE PIPING GRADE	
Pipe Size	**Minimum Grade***
Less than 3″	¼
3″ - 6″	⅛
8″ and over	1/16

* in in. per ft

Figure 3-14. Horizontal drainage piping must be properly graded to ensure adequate drainage and self-scouring action inside the pipe.

from the vertical plane to the horizontal plane or from the horizontal plane to the horizontal plane. **See Figure 3-16.** Fittings that are not illustrated but have the equivalent sweep of the changes in direction fittings can be used for the same type of changes in direction.

Tech Facts

Properly sized horizontal drainage pipe should be approximately one-third full of sewage to ensure proper scouring action. Too much sewage will not allow solid waste to flow through the drainage pipe.

SELECTING CHANGES IN DIRECTION

Changes In Direction from	Size	Fittings
Vertical to Horizontal	Less Than 3″	Long-Sweep ¼ Bend or Long-Turn T-Y
Vertical to Horizontal	3″ and Larger	Long-Sweep ¼ Bend, Two ⅛ Bends, Combination, or Wye and ⅛ Bend
Horizontal to Vertical	All Sizes	Short-Sweep ¼ Bends or Long-Turn 90° Drainage Elbow
Horizontal to Horizontal	All Sizes	Long- or Short-Sweep ¼ Bend or Extra Long-Turn 90° Drainage Elbow

Figure 3-15. Various turns may be required for drainage piping. It is important that the appropriate fittings are used.

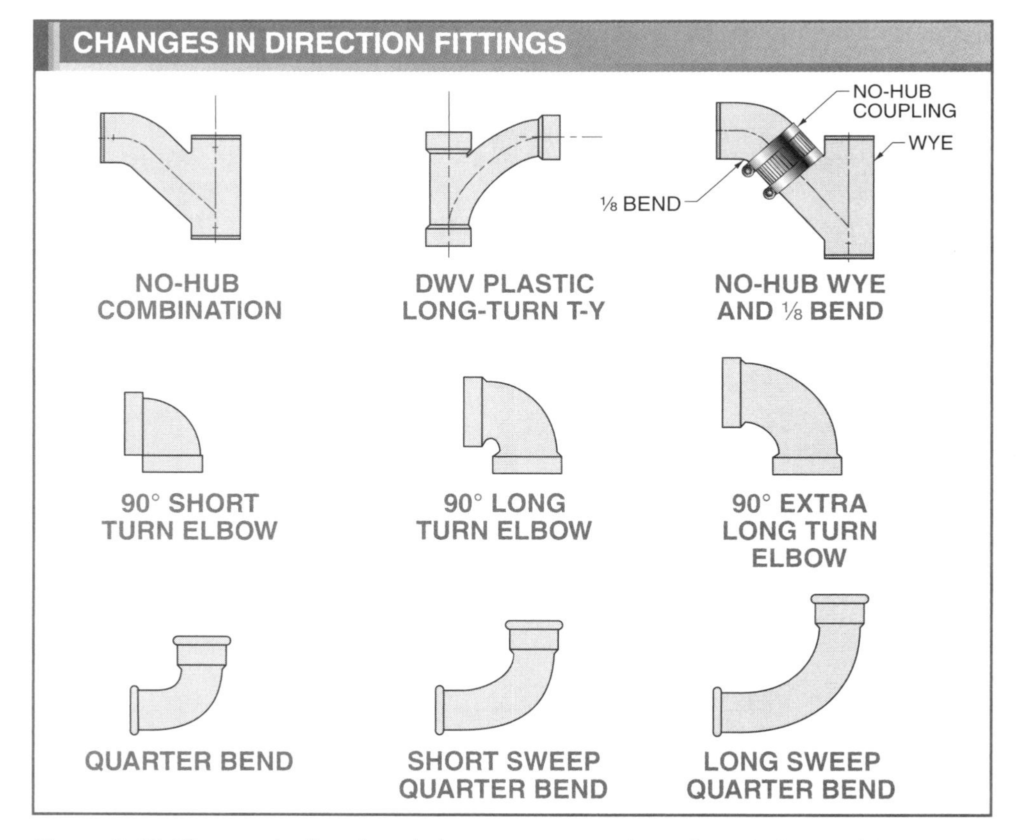

Figure 3-16. Changes in direction of pipes are common in sanitary drainage piping.

Cleanouts

A *cleanout* is a fitting with a removable cap or plug installed in sanitary drainage piping to allow access to the piping for removing stoppages and cleaning the interior of the pipe. Sanitary drainage piping must be equipped with an adequate number of properly sized and placed cleanouts to allow easy access to the system if stoppage occurs.

Cleanouts consist of tee or wye fittings installed in a drainage line and the unused opening capped with the appropriate removable cap or plug. **See Figure 3-17.** In general, a cleanout is the same size as the drainage pipe it serves (up to 4″ maximum). A 4″ cleanout is typically the largest size cleanout installed in a sanitary drainage system regardless of the size of the drain line it serves. Cleanouts should be provided in the following locations:

- front main cleanout—at the outside wall of the building at the connection of the building sewer and building drain
- stack base cleanout—at the base of all vertical soil or waste stacks
- at all 90° changes in direction
- at the upper terminal of all horizontal branch drains
- every 50′ on 3″ and smaller horizontal drainage pipe; every 100′ on 4″ and larger horizontal drainage pipe

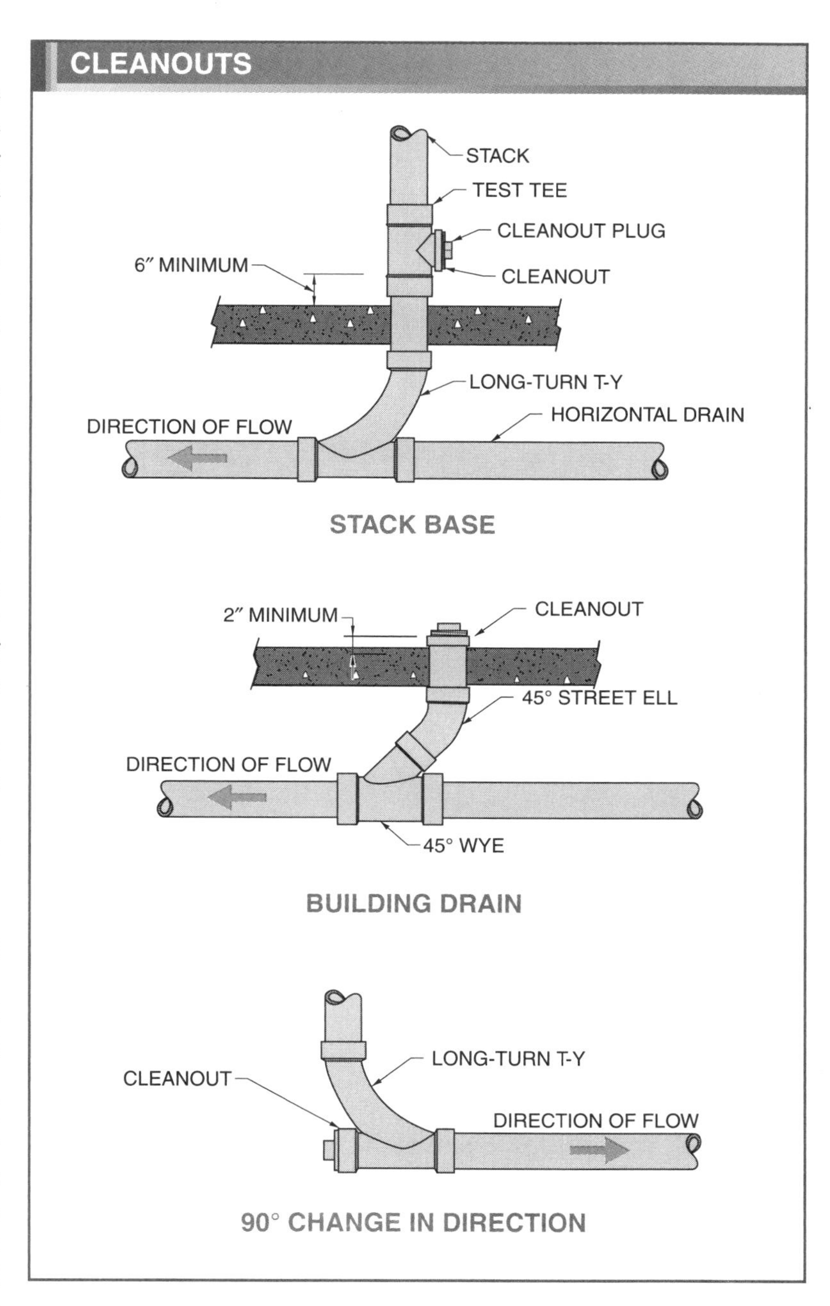

Figure 3-17. Cleanouts provide access to the piping for removing stoppage and cleaning the interior of the pipe.

A front main cleanout is always placed at the outside wall of the building and may be placed either inside or outside of the building. **See Figure 3-18.** A front main cleanout should be a full wye fitting placed in the direction of flow of the drain and should extend a minimum of 2″ above the finished floor or grade level so the cleanout opening cannot be used for a drain. The cleanout should be placed flush with the floor if the cleanout is in a traffic area. Stack base cleanouts should be located at least 6″ above the floor for easy access and to prevent their use as floor drains. Cleanouts at the upper terminals of horizontal branch drains can be eliminated if there is a plumbing fixture trap or a plumbing fixture with an integral trap that can be easily removed and used for a cleanout.

Cleanouts must be easily accessible. If cleanouts are located in walls or ceilings, access panels are constructed to provide access. Cleanouts for underground piping must be closely watched during backfilling operations so the cleanouts remain intact and are not buried or damaged.

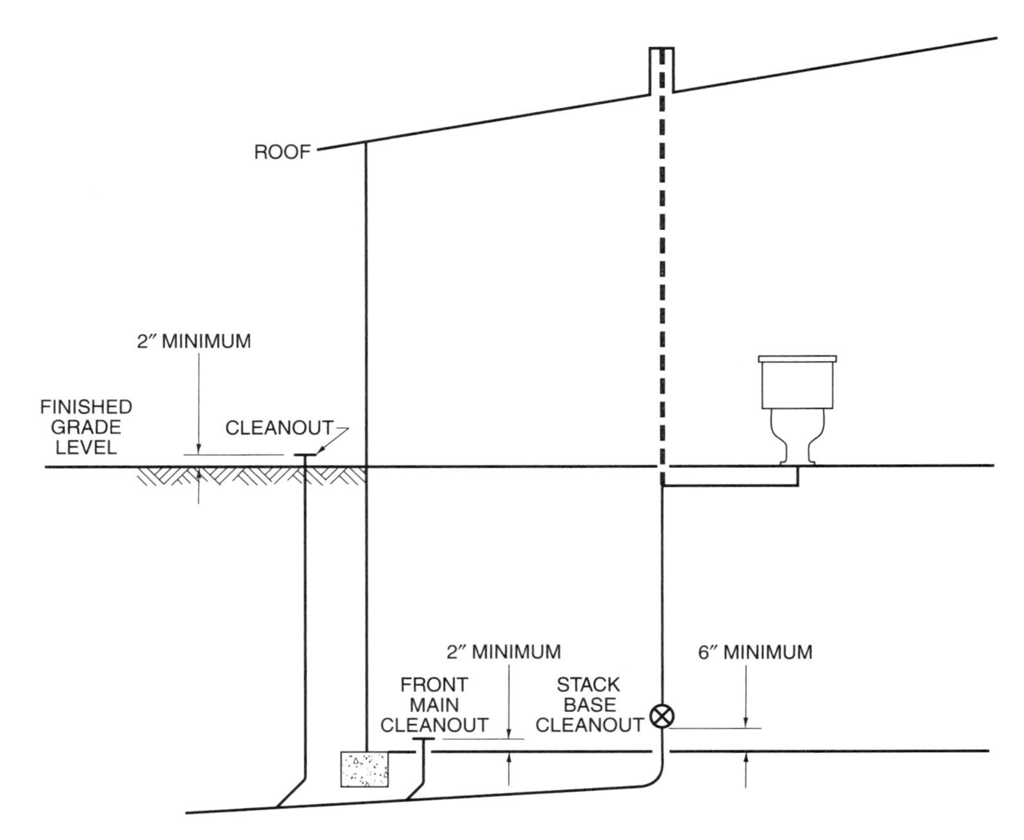

Figure 3-18. A front main cleanout is placed at the outside wall of a building and may be placed either inside or outside of the building.

SANITARY DRAINAGE PIPING VENTING

Vent pipes are extensions of drain pipes and are installed when the drain pipes are installed. Vent pipes ventilate building sanitary drainage piping and prevent trap siphonage and back pressure. Since every plumbing fixture has a water seal trap, every trap must be vented to protect its seal. A properly installed vent system provides air circulation within sanitary drainage piping and ensures that trap seals will not be subject to a pressure of more than 1″ of water column. Air circulation within the drainage system allows wastewater to flow freely through the pipes and also removes objectionable gases from the sanitary drainage system.

Trap siphonage and back pressure are a result of pressure differences between the drainage system and the atmosphere. The *atmosphere* is a blanket of gases that surrounds the earth, is approximately 100 miles thick, and contains about 21% oxygen, 78% nitrogen, and 1% other gases. Approximately one-half the total weight of the atmosphere is below 18,000 feet. A 1 sq in. column of air that extends the height of the atmosphere exerts 14.7 pounds of pressure (psi) on the earth's surface at sea level. Atmospheric pressure increases below sea level since the volume of air in the 1 sq in. column is greater. Atmospheric pressure decreases above sea level since the total volume of air in the 1 sq in. column is reduced at higher elevations.

Compressibility is a common property of the gases that compose the atmosphere. Air can be compressed, and when it is, it develops a pressure greater than its normal sea level pressure (14.7 psi). *Minus pressure* is a pressure less than 1 atmosphere (14.7 psi). *Plus pressure* is a pressure greater than one atmosphere.

Trap Seal Loss

Trap seal loss is a common problem in drainage piping. Trap seal loss is attributed directly to inadequate trap venting and the subsequent minus and plus pressures in the drainage system. During siphonage, the trap contents are forced into the waste piping of the drainage system by the atmospheric pressure on the fixture side of the trap seal. During back pressure, the trap contents are forced to the fixture side of the trap. Back pressure occurs when the pressure within drainage piping is greater than atmospheric pressure on the fixture side.

Retarded Flow in the Drainage System

Retarded flow in a drainage system is the result of improper atmospheric conditions, insufficient venting, or improper fitting installation. Due to the compressibility of the gases in air, the air may be compressed to exceed atmospheric pressure. If a drainage system is not properly vented, water flow in a soil pipe compresses the air ahead of it, developing pressure greater than atmospheric pressure. Increased pressure causes retarded flow in a vertical stack and, therefore, affects the discharge capacity of its branches.

A partial vacuum may also develop in the drainage system, affecting the discharge capacity of the system. A partial vacuum is the result of atmospheric pressure on the flow side of the waste resisting the movement of the wastewater because a minus pressure develops on the opposite side. A partial vacuum indicates a lack of proper relief ventilation, partial closure of a vent pipe terminal, or an excessively long vent pipe.

Removal of Objectionable Gases

In addition to being foul-smelling, sewer gases contain chemical elements that, when combined with moist air, create acids that corrode vent pipes. Hydrogen, which is present in all acids, is found in large quantities in a sanitary drainage system. Hydrogen may be found in its free state, but it is usually combined with other elements to create compounds such as hydrogen sulfide. Hydrogen sulfide in a sanitary drainage system is objectionable because it absorbs additional oxygen from moisture in the drainage system, creating sulfuric acid, which is a highly corrosive acid.

Horizontal drainage piping is designed to flow about one-third full. The air space along the top two-thirds of the pipe allows passage of sewer gases, especially in building sewer and building drain pipes. The air space allows passage of sewer gases along the top of the pipe, up the stack, and into the atmosphere.

Venting Methods

Various venting methods can be applied to a plumbing installation and will primarily be based on the manner in which the plumbing fixtures are located and grouped. A completed vent piping system combines several methods including stack vents, vent stacks, individual vents, common vents, branch vents, and wet vents.

At one time, all fixture traps were individually vented. Based on experimentation and the introduction of different soil and waste pipe relief vents, individual trap venting was found to be unnecessary in most situations and very costly. Even though individual trap venting virtually eliminates trap seal loss, other methods of venting can be used more economically.

Several venting methods are combined to create an effective and efficient vent piping system. The various methods are classified as:

- Venting methods used to ventilate soil and waste pipes. A *stack vent* is the extension of a soil or waste stack above the highest horizontal drain connected to the stack. A *vent stack* is a pipe provided specifically to prevent trap siphonage and back pressure. Stack vents, vent stacks, and various relief vents are classified according to the purpose they serve. A *relief vent* is a vent whose primary purpose is to provide additional air circulation between drainage and vent systems or to serve as an auxiliary vent on specially designed systems. A *yoke vent* is a vent pipe connecting upward from a soil or waste stack to a vent stack to prevent pressure differences in the stacks. Relief and yoke vents serve the fixture trap only in an indirect way and maintain atmospheric pressure in the sanitary waste system.

- Venting methods used to protect trap seals against back pressure and siphonage. An *individual vent* is a vent pipe that vents a fixture trap and connects with the vent system above the fixture or terminates in the open air. A *back vent* is a vent pipe that connects to a waste pipe on the sewer side of its trap to prevent siphonage. A *common vent* is a vent pipe that connects at the junction of two fixture drains and serves as a vent for both fixture drains. A *wet vent* is a portion of a vent pipe through which liquid waste flows.

Vent Grades. The vent pipe must be graded slightly back toward a soil or waste pipe so that water cannot accumulate in it.

Sizing Vent Pipes

Vent pipes are sized based on the drainage fixture units connected to the vent pipe and the developed length of vent pipe. *Developed length* is the length of vent pipe measured along the centerline of the pipe and fittings. **See Figure 3-19.** Developed length must be considered when sizing vent piping because friction between air in motion within vent pipes and the interior surface of the pipe reduces the flow and the volume of air moving through the vent pipe. Pipe smaller than 1¼″ in diameter is unsuitable for venting. Although air flow within a smaller pipe may be adequate, the vent can easily become plugged.

Stack Vents. A stack vent is the extension of a soil or waste stack above the highest horizontal drain connected to the stack. **See Figure 3-20.** Stack vents admit air to the plumbing system and provide an outlet for sewer gases. A stack vent is usually the terminal for other vent pipes, such as individual and group fixture vents and vent stacks.

Vents are sized according to the Size and Length of Individual, Branch, Circuit, and Stack Vents table. Stack vent sizing is based on the total number of drainage fixture units draining into the waste portion of the stack and the developed length of the vent. **See Figure 3-21.** Even though the Size and Length of Individual, Branch, Circuit, and Stack Vents table indicates that 2½″ pipe may be used to vent drainage systems, 2½″ pipe is never installed since no drain, waste, or vent fittings are made for use with 2½″ pipe.

For example, a 3″ stack vent is required for a group of fixtures consisting of six stalled on the second floor of a 100′ tall building. Following is the procedure to determine the drainage fixture unit value of this group of fixtures:

No.			dfu		Total
6	Water closets	x	6	=	36
4	Lavatories	x	1	=	4
3	Urinals	x	4	=	12
2	Showers	x	2	=	4
			Total dfu	=	56

Assuming that each floor is 10′ high, the stack vent has a developed length of approximately 90′ from the first floor ceiling to the top of the stack (100′ – 10′ = 90′). Based on the Size and Length of Individual, Branch, Circuit, and Stack Vents table, a maximum of 72 dfu may be vented with either a 2″ pipe with a developed length of 50′, a 2½″ pipe with a developed length of 80′, or a 3″ pipe with a developed length of 400′. A 3″ stack vent is required since the developed length of the vent is 90′.

Vent Stacks and Main Vents. A vent stack provides air circulation to and from the drainage system. Vent stacks are also main vents. A *main vent* is the principal artery of the vent system to which vent branches may be connected. Vent stacks or main vents are required in buildings with individual vents, relief vents, or branch vents on three or more branch intervals. Vent stacks or main vents are also terminals for smaller individual and group fixture vents.

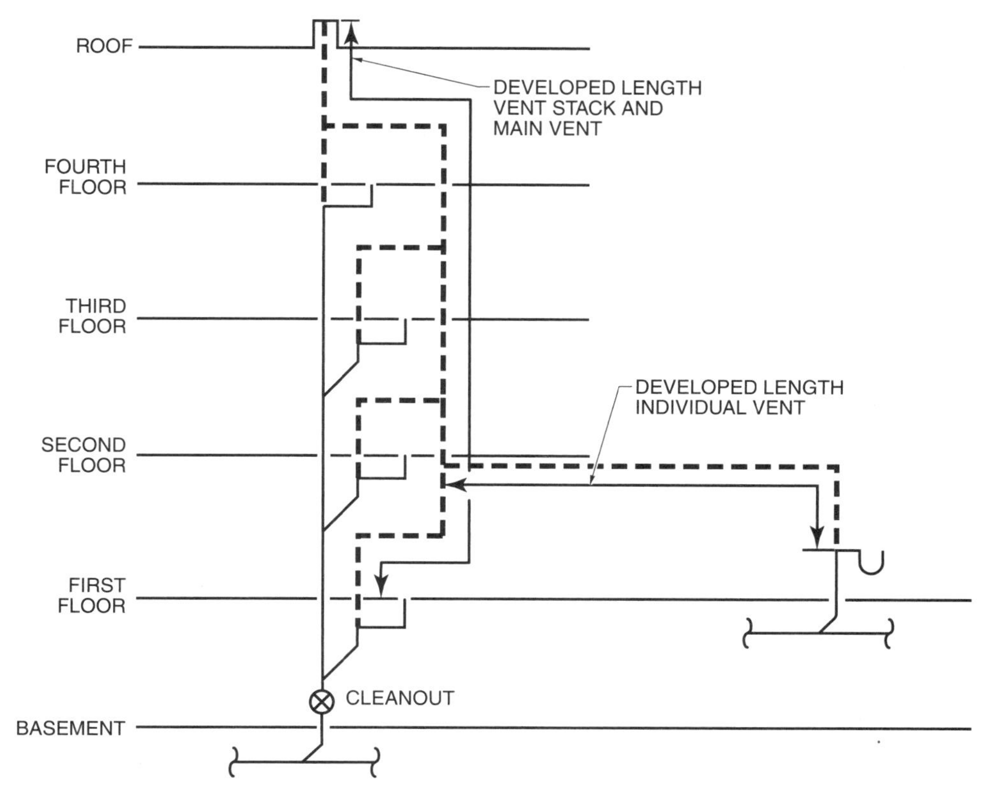

Figure 3-19. Developed length of vent pipe is measured along the centerline of the pipe and fittings.

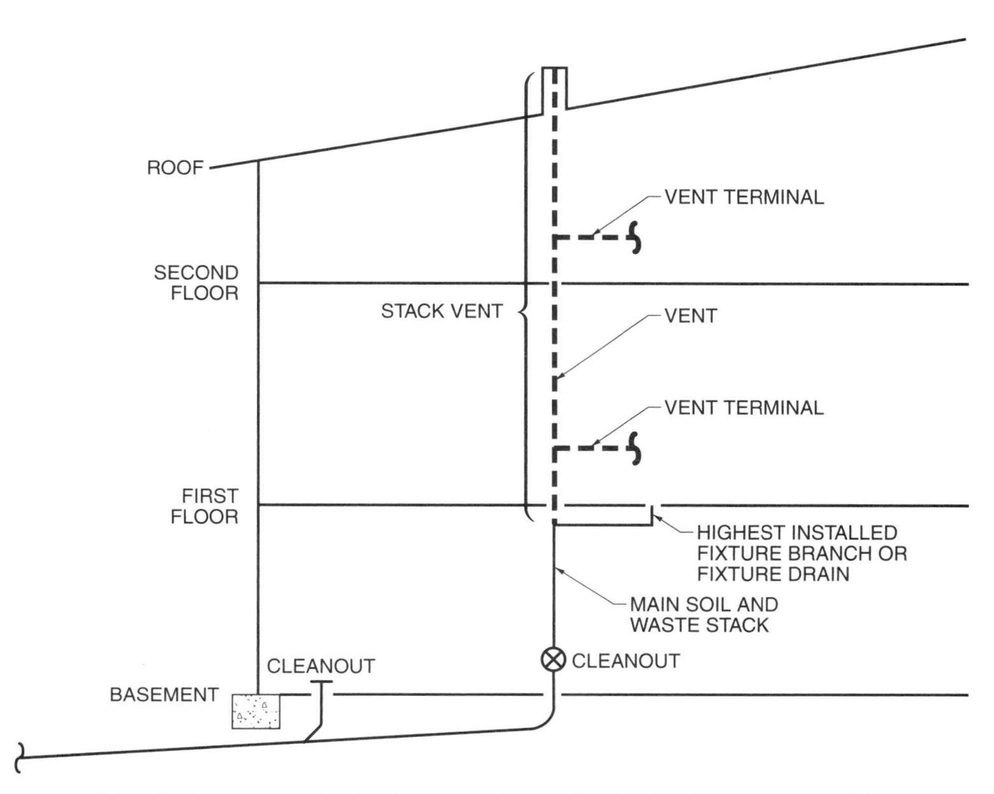

Figure 3-20. A stack vent extends above the highest horizontal drain connected to a soil or waste stack.

SIZE AND LENGTH OF INDIVIDUAL, BRANCH, CIRCUIT, AND STACK VENTS								
Fixture Units Connected*	**Vent Diameter†**							
	1¼	**1½‡**	**2**	**2½**	**3**	**4**	**5**	**6**
	Maximum Developed Length of Vent§							
2	50	‖						
4	40	200	‖					
8	#	150	250					
10		100	200	‖				
24		50	150	400	‖			
42		30	100	300	500			
72		#	50	80	400			
240			#	50	200	‖		
500				#	180	700	‖	
1100					50	200	700	

* in dfu
† in in.
‡ except 6 dfu fixtures
§ in ft
‖ unlimited length
not permitted

Figure 3-21. Stack vent size is based on the total number of drainage fixture units draining into the waste portion of a stack and the developed length of the vent.

Vent branches are connected to a main vent prior to the main vent exiting the roof of a building.

The vent stack is usually located close to the soil pipe stack, but its actual location depends on the building construction. **See Figure 3-22.** A vent stack is installed when the soil pipe stack is installed. Openings are left at the correct height and proper floors to accommodate fixture trap vents. A vent stack or main vent begins at the base of the soil stack and relieves back pressure that might occur at this location. A vent stack is connected to the waste stack at the lower end with a wye and ⅛ bend and connects with the stack vent (uppermost portion of the soil stack). However, if the vent stack or main vent is not run within a few feet of the soil stack, the vent may continue through the roof separately.

In a multistory building of more than five branch intervals, the waste stack and the vent stack must be reconnected with a yoke vent every five branch intervals, counting from the top interval down. A yoke vent connects upward from a soil or waste stack to a vent stack to prevent pressure differences in the stacks. A yoke vent is a type of relief vent that provides additional air circulation between drainage and vent systems. The yoke vent connection is the same size as the vent stack to which it connects.

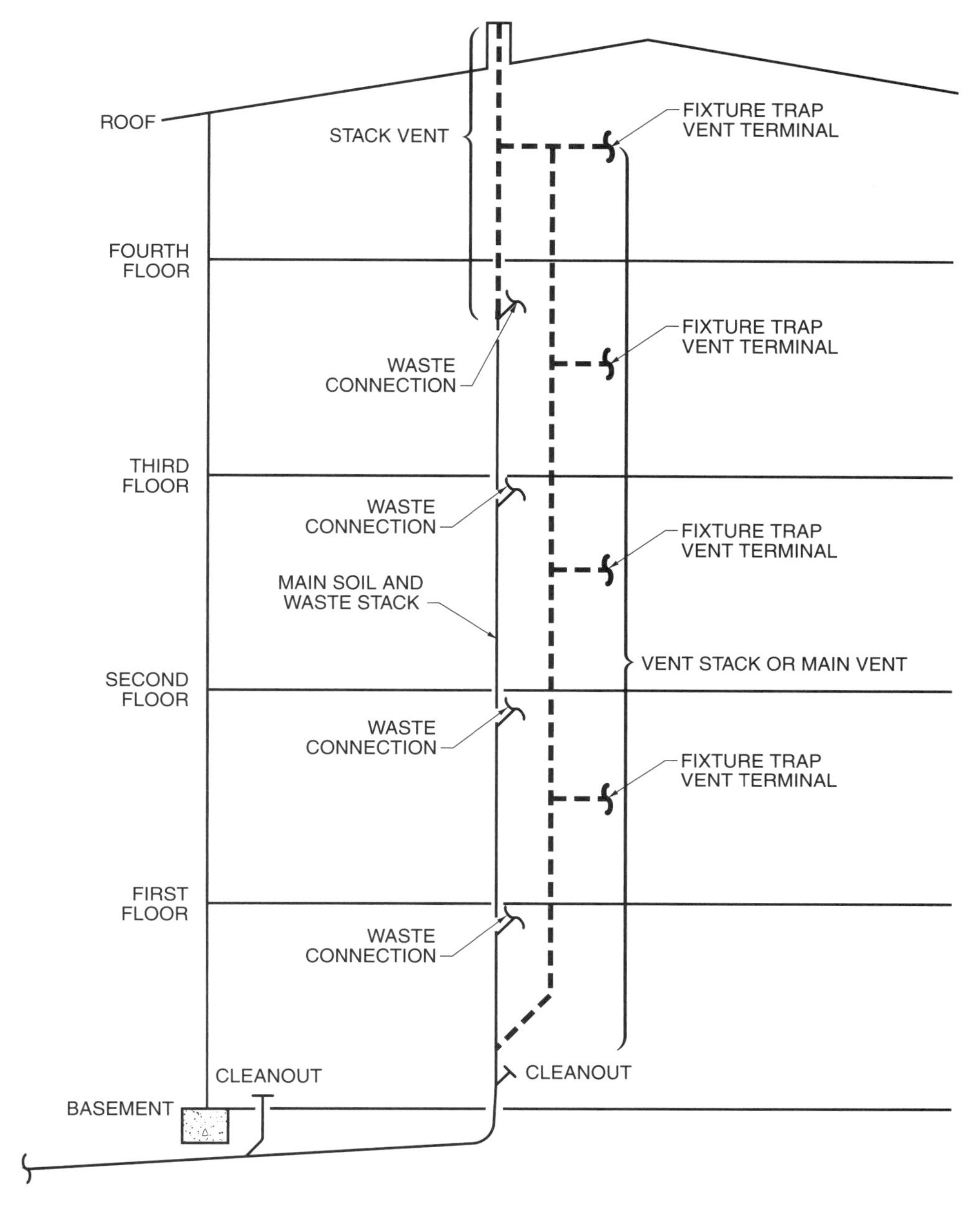

Figure 3-22. The vent and soil stacks are installed at the same time and are usually located close to one another.

A yoke vent is connected to a soil stack with a wye and ⅛ bend. When connected to a waste stack, a yoke vent is connected to the waste stack below the horizontal fixture branch drain for that floor and to the vent stack at least 3′ above the floor level. **See Figure 3-23.** Sufficient space must be allowed between the soil or waste stack and the vent stack so that the connection between the yoke vent and soil or waste stack can be easily made.

Vent stack and main vent sizing is based on the size of the soil or waste stack, number of drainage fixture units connected to the stack, and the developed length of the vent. **See Figure 3-24.** For example, a 3″ vent stack (90′ developed length) is required in an 8-story apartment building when a 4″ size soil stack is used to serve 88 dfu. Based on the Size and Lengths of Vent Stacks table, a 4″ stack with 240 dfu or less connected to it requires a 3″ vent stack.

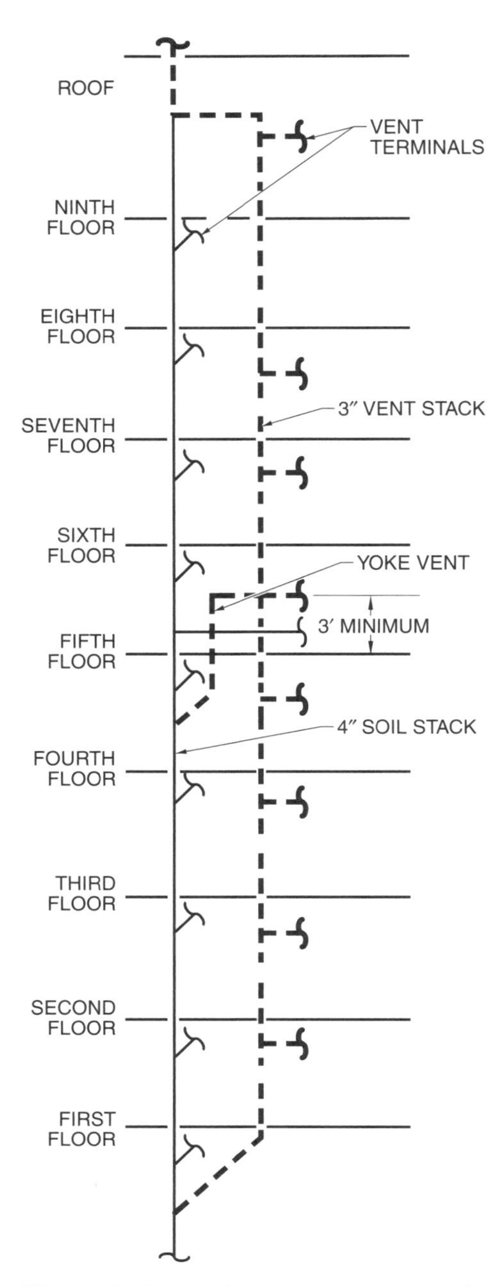

Figure 3-23. A yoke vent is connected to the waste stack below the horizontal fixture branch drain for that floor and to the vent stack at least 3′ above the floor.

Stack Terminals. Stack vent and vent stack terminals are extended through the roof to vent sewer gas to the outside air. Vent terminals typically extend at least 1′ through the roof to prevent rainwater on the roof from draining into the terminals and to prevent objects on the roof from falling into and blocking the vent opening. Since sewer gas is foul-smelling, a vent terminal should be located at least 10′ from windows, doors, or other ventilation openings. If vent terminals must be located closer than 10′, extend the pipe at least 2′ above the opening. If a vent terminal passes through a roof that serves another purpose, such as a sun deck, the vent should terminate at least 7′ above the roof.

Roof terminals for vents must be sealed to the roof surface to prevent rainwater, snow, and other moisture around the pipe from leaking into the building. **See Figure 3-25.** Roof jackets or roof flanges are installed around the vent terminals to make the roof watertight. Some roof jackets are adjustable for different roof pitches and stack height variations caused by expansion and contraction of the stack or settling of the building.

In colder climates, unwanted closure of the vent terminal due to frost is a common problem and can cause trap seal loss. Air within a plumbing system is usually close to the moisture saturation point. Condensation occurs when humid air is emitted through the stack terminal and freezes rapidly. One method of preventing frost closure of a vent terminal is to increase the size of the terminal before it passes through the roof. In cold climates, vent pipes passing through the roof should be at least 2″ pipe. Another method of preventing frost closure is the use of a roof jacket that allows 1″ of free air space around the vent terminal.

Individual Vents. An individual vent, or back vent, vents an individual fixture trap and terminates into a branch vent, stack vent, vent stack, or the open air. Individual vents are referred to as back vents because they are commonly installed directly in back of the fixtures they serve in a plumbing system.

Tech Facts

Installing individual fixture vents is the easiest way to eliminate trap seal loss, since there is no negative or positive pressure from the other fixture traps.

SIZE AND LENGTHS OF VENT STACKS												
Soil or Waste Stack Size*	Fixture Units Connected†	Diameter Vent*										
		1¼	1½	2	2½	3	4	5	6	8	10	12
		Maximum Developed Length of Vent‡										
1¼	2	50										
1½	4	40	200									
2	9		100	200								
2	18		50	150								
2½	42		30	100	300							
3	72			50	80	400						
4	240			40	70	250						
4	500				50	180	700					
5	540					150	600					
5	1100					50	200	700				
6	1900						50	200	700			
8	2200							150	500			
8	3600							60	250	800		
10	3800								200	600		
10	5600								60	250	800	
12	6000									200	600	
12	8400									100	300	900
15	10,500									50	200	600
15	50,000										75	180

* in in.
† in dfu
‡ in ft

Figure 3-24. Vent stack and main vent sizing is based on the size of soil or waste stack, number of drainage fixture units connected to the stack, and the developed length of the vent.

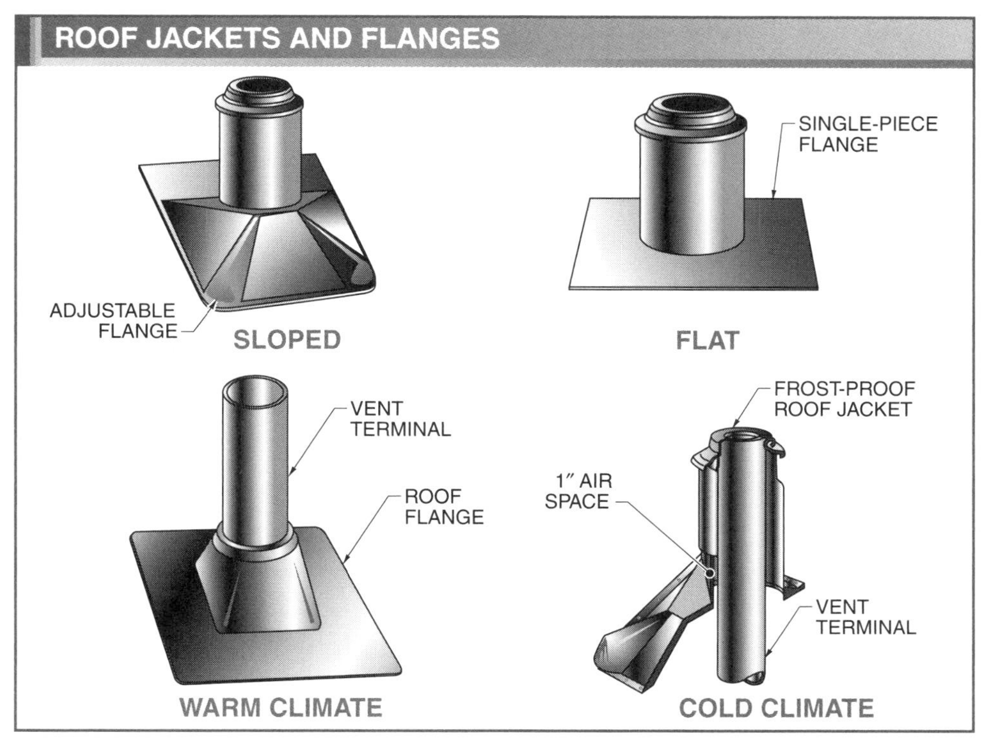

Figure 3-25. Roof jackets or roof flanges are installed around vent terminals to prevent rainwater from leaking into the building.

An individual vent is the most practical method of venting a fixture trap. Trap seal loss is virtually eliminated when an individual vent is used since the plumbing system is pressure-balanced at every fixture trap. **See Figure 3-26.** The most common individual vent is a continuous vent. A *continuous vent* is a vertical vent that is a continuation of the drain to which it connects.

When venting wall-hung or cabinet-set fixtures, such as sinks, lavatories, and drinking fountains, the fixture trap discharges into the side opening of a sanitary drainage tee. The top opening of the drainage tee is the connection for an individual vent. The bottom opening of the drainage tee connects to the waste or soil stack.

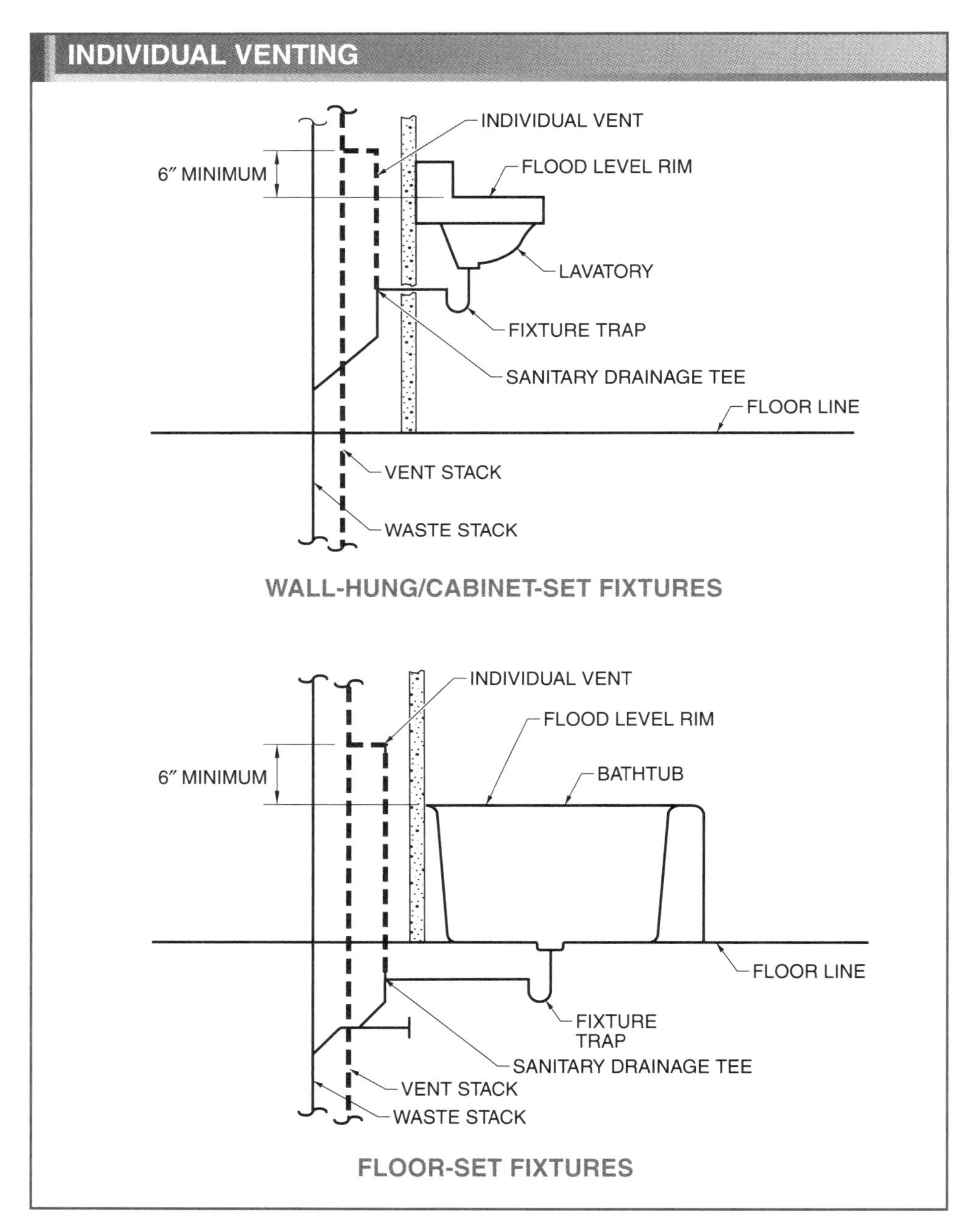

Figure 3-26. Individual venting eliminates trap seal loss because the plumbing system is relieved of minus and plus pressure at every fixture trap.

The maximum distance between the fixture trap and its vent is based on the fixture drain size. **See Figure 3-27.** Individual vents and other fixture trap vents should be connected as close to the trap as possible, usually to the fixture drain pipe directly below and in back of the fixture. Dirt, rust, or other foreign material in the vent drops into the waste line and is carried away by the discharge of the fixture, thus maintaining a clear vent. If an individual vent is to be reconnected to the vent stack, it should be connected at least 6″ above the flood level rim of the fixture served by the vent so that the vent will not serve as a waste line for the fixture if the fixture waste pipe is clogged. A *flood level rim* is the top edge of a fixture from which water overflows.

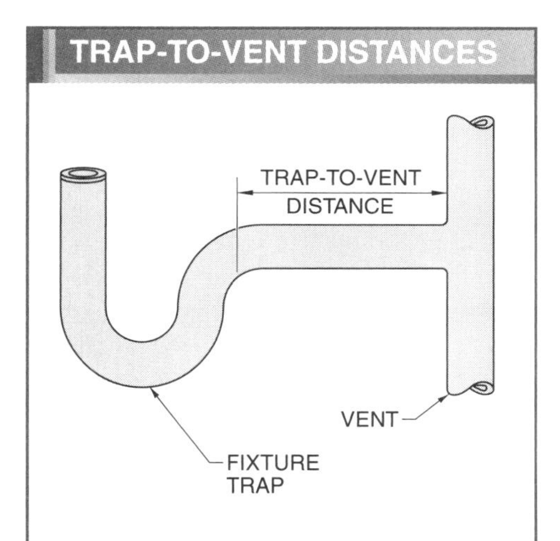

FIXTURE TRAP-TO-VENT DISTANCE*

Fixture Drain Size†	Trap-to-Vent Distance
1¼	2′-6″
1½	3′-6″
2	5′
3	6′
4	10′

* developed length between trap of water closet or similar fixture and its vent shall not exceed 4′

† in in.

Figure 3-27. Trap-to-vent distance is based on the fixture drain size.

Individual vents can also be installed for floor-set fixtures, such as urinals, shower baths, and bathtubs. The fixture trap discharges into the side opening of a sanitary drainage tee, with the top opening serving as an individual vent connection. The vent stack connection must be at least 6″ above the flood level rim of the fixture served by the vent. Floor-set water closets in a basement must be properly vented to provide adequate ventilation and air circulation. **See Figure 3-28.** Flat venting of water closets should be avoided since stoppage of the fixture drain can result in waste backing into the vent and plugging it.

An individual vent is sized based on the drainage fixture unit value of each fixture and the developed length of the individual vent to the vent stack, stack vent, or roof terminal. An individual fixture vent should never be smaller than one-half the fixture drain diameter.

For example, based on a 50′ developed length of the vent pipe, the following are individual vent sizes for common fixtures: Based on the Size and Length of Individual, Branch, Circuit, and Stack Vents table, a 1½″ vent pipe may be required for a fixture (such as a water closet) discharging 6 dfu. However, plumbing codes do not permit water closets to be vented by 1½″ vent pipes and require 2″ vent pipes for all water closet vents.

Fixture Type	Individual Vent Minimum Size	dfu
Lavatories	1¼″	1
Drinking fountain	1¼″	1
Domestic sink	1¼″	2
Domestic shower stall	1¼″	2
Bathtub	1¼″	2
Laundry tray	1¼″	2
Service sink	1½″	3
Water closet	2″	6

Tech Facts

Do not discharge a water closet into an unvented waste pipe. Fixtures that discharge their waste into a waste pipe directly above a water closet must be vented to ensure the trap seal remains intact.

SIZE AND LENGTH OF INDIVIDUAL, BRANCH, CIRCUIT, AND STACK VENTS

Fixture Units Connected*	Vent Diameter†							
	1¼	1½‡	2	2½	3	4	5	6
	Maximum Developed Length of Vent§							
2	50	‖						
4	40	200	‖					
8	#	150	250					
10		100	200	‖				
24		50	150	400	‖			
42		30	100	300	500			
72		#	50	80	400			
240			#	50	200	‖		
500				#	180	700	‖	
100					50	200	700	

* in dfu
† in in.
‡ except 6 dfu fixtures
§ in ft
‖ unlimited length
\# not permitted

For ease of identification, drainage and vent piping in large commercial structures may be labeled.

Common Vents. A common vent, or unit vent, connects two fixture drains and serves as a vent for both fixture drains. The fixtures discharge waste into a sanitary cross, tapped cross, or figure 1 fitting. The top opening of the cross or fitting is used for a common vent connection, which is completed in the same manner as an individual vent. **See Figure 3-29.** Common venting is used on two similar fixtures with the same vertical drain heights, which are installed on opposite sides of a partition, and is common for fixture traps serving apartment and hotel bathrooms. Apartment and hotel bathrooms are usually located back to back and the waste, vent, and water pipes are run in a common wall between the rooms. Common vent design and principles are similar to individual venting.

Pipe size for a common vent is determined similarly to an individual vent except that the pipe size must be adequate to vent the total drainage fixture unit value of both fixtures. For example, a 1½″ pipe is required as a common vent for two lavatory traps. Each lavatory is rated at 1 dfu, and the developed length of the vent is 60′. The total of the drainage fixture unit values is 2 dfu. Since 50′ is the maximum developed length of a 1¼″ vent pipe that vents 2 dfu, a 1½″ vent is required. The same procedure is applied to other fixtures and the vent size increases as the drainage fixture unit value of fixtures becomes greater.

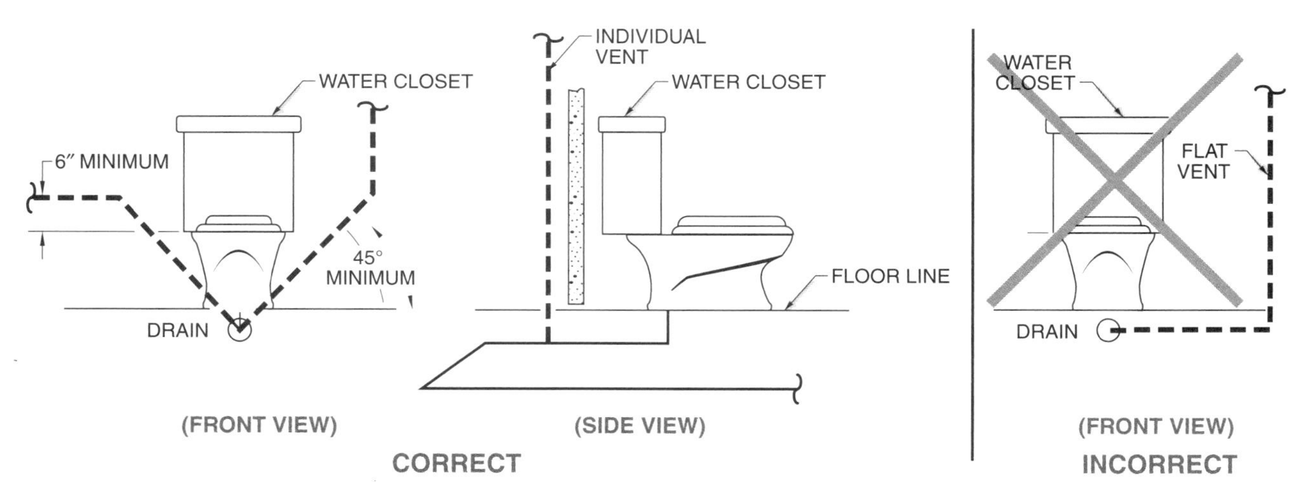

Figure 3-28. Floor-set water closets in a basement must be properly vented to provide adequate ventilation and air circulation.

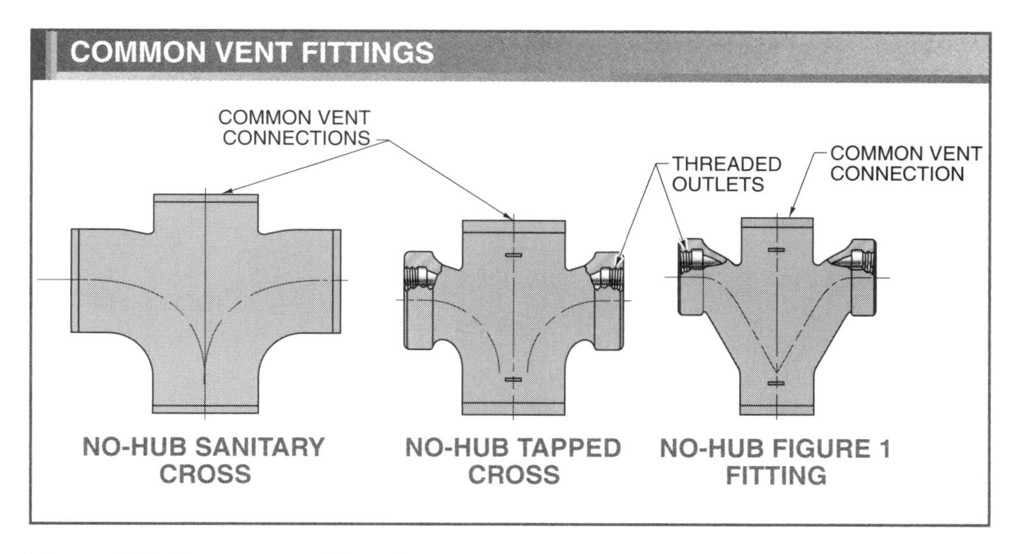

Figure 3-29. Common vent fittings include no-hub sanitary and tapped crosses and no-hub figure 1 fittings.

Common venting is used for lavatory, bathtub, and water closet installations in back-to-back rooms. **See Figure 3-30.** However, it may be inconvenient to common vent bathtubs, showers, and other floor-set fixtures due to space limitations. When water closets are the highest fixtures on a stack, they may be common vented to that stack. In this installation, the stack vent also serves as a common vent. **See Figure 3-31.**

Branch Vents. A *branch vent* is a vent pipe connecting two or more individual vents with a stack vent or vent stack. Branch vents commonly run horizontally behind a group of fixtures. When a branch vent runs horizontally and ties several individual fixture vents together, the branch vent must be at least 6″ above the flood level rim of the highest fixture connected to the vent. Branch vent size is based on the number of drainage fixture units connected to it and the developed length of the branch vent from its vent stack or stack vent to the farthest fixture drain served by the branch vent.

For example, what size branch vent is required for a bathroom containing six water closets, four lavatories, four wall-hung urinals with 2″ traps, and one service sink with a developed length of 76′ to the nearest vent stack?

No.			dfu		Total
6	Water closets	x	6	=	36
4	Lavatories	x	1	=	4
4	Urinals	x	4	=	16
1	Service sink	x	3	=	3
			Total dfu	=	59

SIZE AND LENGTH OF INDIVIDUAL, BRANCH, CIRCUIT, AND STACK VENTS

Fixture Units Connected*	Vent Diameter†						
	1¼	1½‡	2	2½	3	4	5
	Maximum Developed Length of Vent§						
2	50	∥					
4	40	200	∥				
8	#	150	250				
10		100	200	∥			
24		50	150	400	∥		
42		30	100	300	500		
72		#	50	80	400		
240			#	50	200	∥	

Based on the Size and Length of Individual, Branch, Circuit, and Stack Vents table, a 2½″ branch vent is required.

Wet Vents. A *wet vent* is the portion of a vent pipe through which liquid waste flows. Wet venting is a venting method commonly used for small groups of residential bathroom fixtures. Wet venting is a permitted practice in some code jurisdictions.

However, always consult the plumbing code that is in effect in the area where the plumbing work is being performed. Some plumbing authorities believe wet venting is an effective method for maintaining balanced pressure to prevent fixture trap seal loss and that fixture discharge into the line scours the vent, keeping it clean. Some plumbing authorities believe that vent pipe used for wet venting will become fouled rapidly and its diameter greatly reduced, thus resulting in stoppage and trap seal loss.

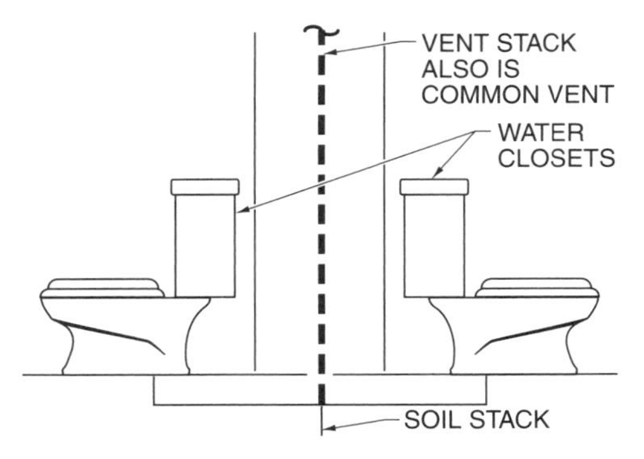

Figure 3-31. When water closets are the highest fixtures on a stack, they may be common vented to that stack.

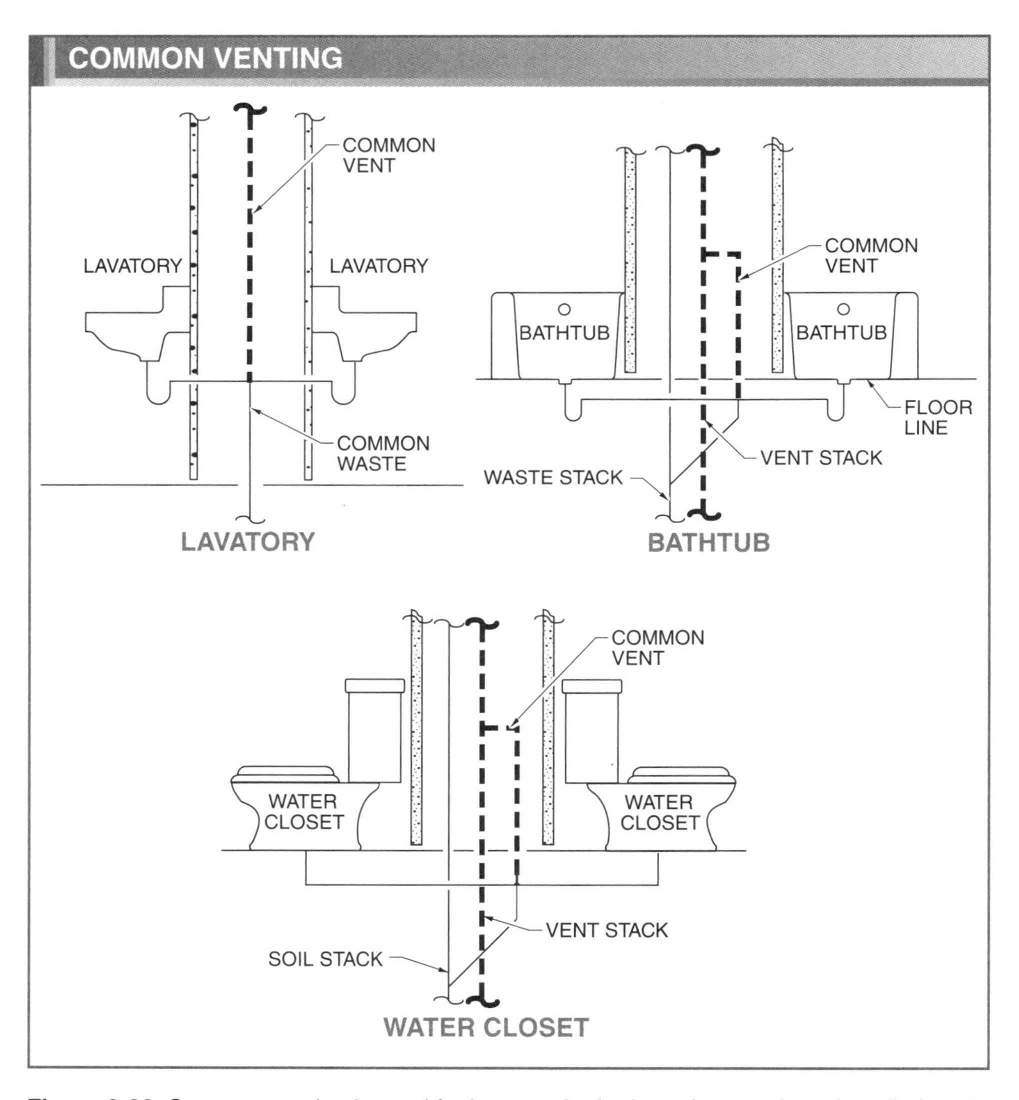

Figure 3-30. Common venting is used for lavatory, bathtub, and water closet installations in back-to-back rooms.

A basic wet vent is a common vent with two fixtures whose horizontal drain openings are at different heights and installed back to back. In this installation, the vertical drain is sized one pipe size larger than the upper fixture drain, but in no case is it smaller than the lower fixture drain. See **Figure 3-32.** For example, a 1½″ vertical drain is required when the lavatory waste is above the kitchen sink waste. The wet vent of the vertical drain pipe would be 1½″ pipe instead of the 1¼″ that would normally be required for a lavatory drain. However, if the kitchen sink drain is above the lavatory drain, the wet vent portion is sized as 2″ pipe instead of 1½″ pipe, which is the normal size for a kitchen sink drain. As a general rule, not more than 1 dfu may drain into a 1½″ wet vent or more than 4 dfu into a 2″ wet vent.

Tech Facts

In 1874, an unknown American plumber solved the problem of venting by balancing air pressure within a plumbing system with outside atmospheric pressure. This prevented siphonage and blowout of trap seals.

A typical wet vent installation is a residential bathroom. **See Figure 3-33.** In this installation, the vertical portion of the lavatory waste is the wet vent for the bathtub drain. In this case, the lavatory waste is increased to 1½″ pipe, but the lavatory vent, which vents 3 dfu, remains 1¼″ pipe. Another wet vent application is an installation in which the kitchen sink and lavatory are back to back and the waste pipe for these two fixtures is the wet vent for the bathtub. **See Figure 3-34.** In this installation, the wet vent portion of the vertical drain must be 2″ pipe because it drains 3 dfu, and the branch vent to the stack is 1½″ pipe because it vents 5 dfu (lavatory [1 dfu] + kitchen sink [2 dfu] + bathtub [2 dfu] = 5 dfu).

Wet vents are often used in residential basement bathroom installations. **See Figure 3-35.** In this installation, the lavatory drain is a wet vent for the water closet. The size of the wet vent portion of the pipe is not increased because the 2″ pipe required for a water closet vent is more than one pipe size larger than a lavatory waste, which is usually 1¼″.

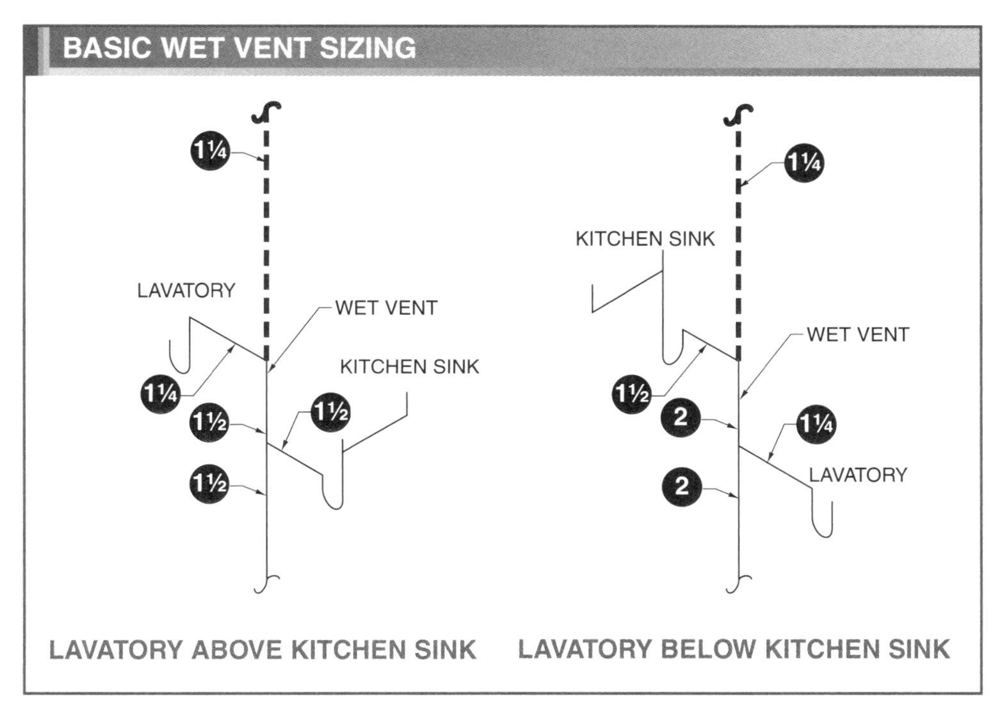

Figure 3-32. A basic wet vent is a common vent with two fixtures whose horizontal drain openings are at different heights and installed back to back.

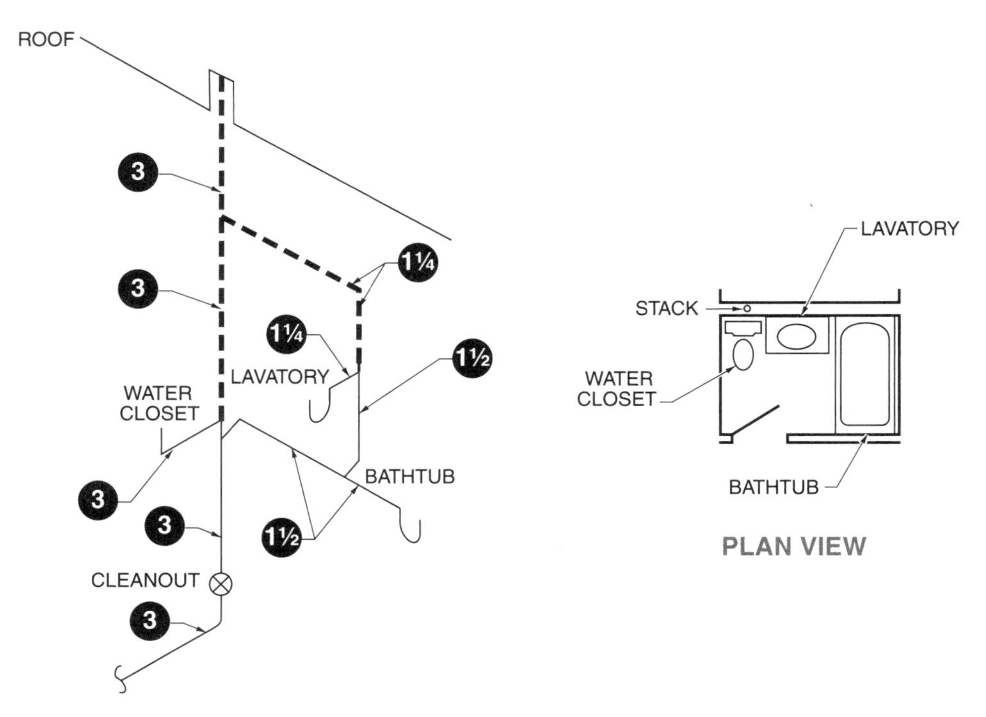

Figure 3-33. The vertical portion of the lavatory waste is the wet vent for the bathtub drain.

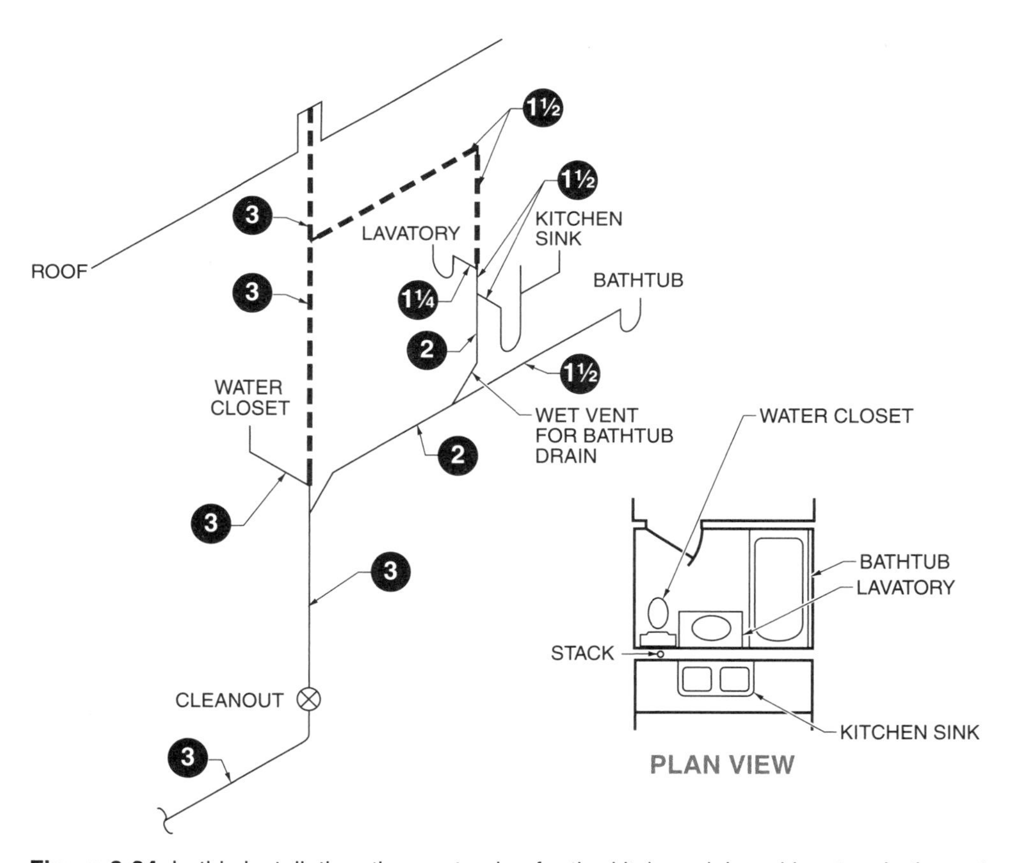

Figure 3-34. In this installation, the waste pipe for the kitchen sink and lavatory is the wet vent for the bathtub.

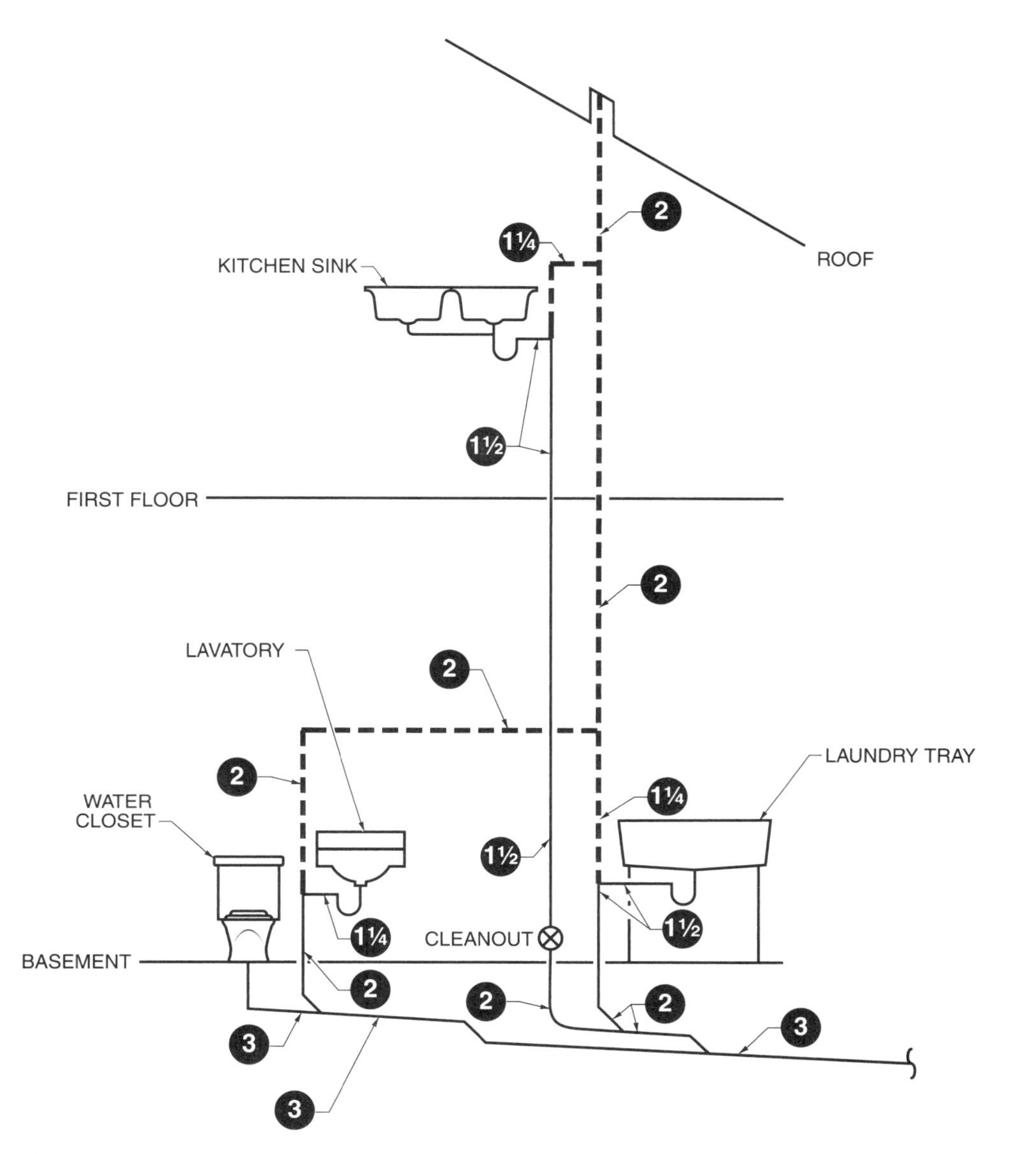

Figure 3-35. When the lavatory drain is the wet vent for the water closet, it is not necessary to increase the wet vent size.

Another type of wet vent is a stack group. A *stack group* is a group of fixtures located next to a stack so vents may be reduced to a minimum using the proper fittings. **See Figure 3-36.** A stack group is commonly used in one-story homes or when the bathroom is located on the top floor of a building. When a stack group is sized, the individual fixture drains are their normal size, but the stack vent is the same size as the soil stack. If the individual fixture trap arms exceed the maximum distance cited in the Fixture Trap-to-Vent Distance table, the fixtures must be revented.

STORMWATER DRAINAGE PRINCIPLES

The stormwater drainage system conveys rainwater and other precipitation to the storm sewer or other place of disposal. At one time, surface and rainwater was discharged into the building drain, which also served the plumbing fixtures of the building. This practice was considered satisfactory because the public sewer drained into a river, lake, or natural drainage basin, and the discharge of a large volume of clear wastewater did not create a problem. However, as the population grew, the need for sewage treatment became more apparent. Currently, most municipalities maintain sewage disposal plants where the sewage treatment process consists of liquefying suspended organic materials. Rainwater passing through a disposal plant affects the treatment process and is therefore separated from the waterborne organic waste before it enters the treatment plant. Rainwater is relatively clean and can be discharged into a natural drainage terminal, such as a drainage basin, without negatively affecting the ecology. Drainage basins are a common practice for large industrial plants located in areas away from municipalities.

Tech Facts

In a properly installed plumbing system, every fixture contains a trap and every trap contains a vent.

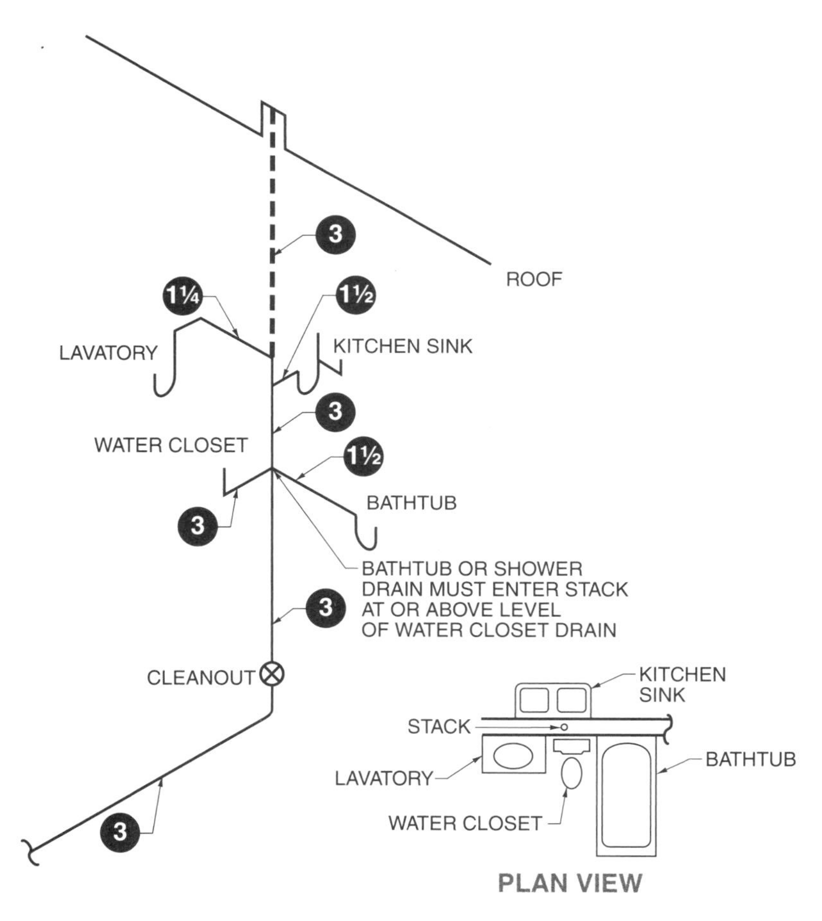

Figure 3-36. A stack group is located next to a stack so vents may be kept to a minimum.

In most municipalities private buildings are required to have drainage systems that connect to the municipality storm sewer. Discharging rainwater into storm sewers eliminates the discharge of rainwater leaders into gutters and over sidewalks where it may become a pedestrian hazard.

Stormwater drainage pipes are installed at the same time as sanitary drainage and vent pipes. In some structures, such as in tall buildings, stormwater and sanitary drainage pipes are in the same trench and run parallel to one another in pipe shafts. Stormwater drainage piping shares several common features with sanitary drainage piping:

- Changes in direction in stormwater drains are made with the same fittings as those required for sanitary drainage piping.
- Stormwater drain pipes are required to have cleanouts in the same locations as sanitary drainage pipes.
- Horizontal storm drains are graded at least ¼″ per foot. However, excessive grade in stormwater drainage piping is not a concern since stormwater is relatively free of solid material.

Building Storm Drain Installation

Building storm drains convey rainwater from roof drains through a rainwater leader and into a storm sewer. If a storm sewer is not available, rainwater is piped to a drainage basin, such as a pond. The building storm drain is placed under the basement floor if there is an adequate elevation difference between the building storm drain and storm sewer main, or it is suspended from the basement ceiling if the elevation difference is inadequate.

Roof Drains. A roof drain receives rainwater collecting on a roof surface and discharges it into a rainwater leader. Roof drain bodies are available for a variety of applications. **See Figure 3-37.** Most roof drains have no-hub outlet connections.

When drains are installed, flashing is installed around the drain body and sealed to the roof to form a watertight connection.

Refer to the drain manufacturer recommendations for flashing installation details. A cast iron, cast aluminum, or plastic strainer basket, which extends several inches above the roof, is attached to the drain to prevent stones, leaves, and other debris from entering the stormwater system. A flat drain cover is installed if the roof is used as a sun deck or another application.

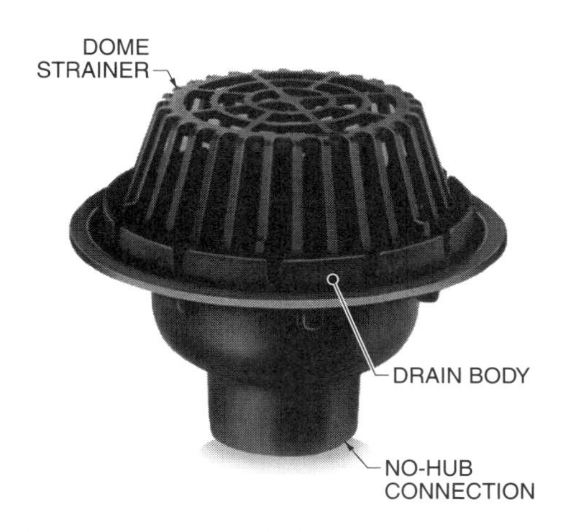

Figure 3-37. Roof drains receive rainwater collecting on a roof surface.

Storm Drain Traps. In most applications, storm drain traps are not required when a storm drainage system connects to municipality storm sewer mains. Storm drain traps should be used when a roof drain body is located within 10′ of a door, window, or any other opening to a building. **See Figure 3-38.**

Rainwater leaders usually run along a column or are placed in vertical shafts constructed specifically for pipe. Leaders typically extend vertically from the base fitting through the floors of the building and terminate below the roof where they connect to a roof drain.

When planning the location of a building storm drain beneath a basement floor, the building storm drain and all branches are designed so that they will not cross over main runs of the building sanitary drain. If it is necessary to suspend the storm drain from the basement ceiling, the drain should not conflict with heating mains, ventilation ducts, windows, or beams.

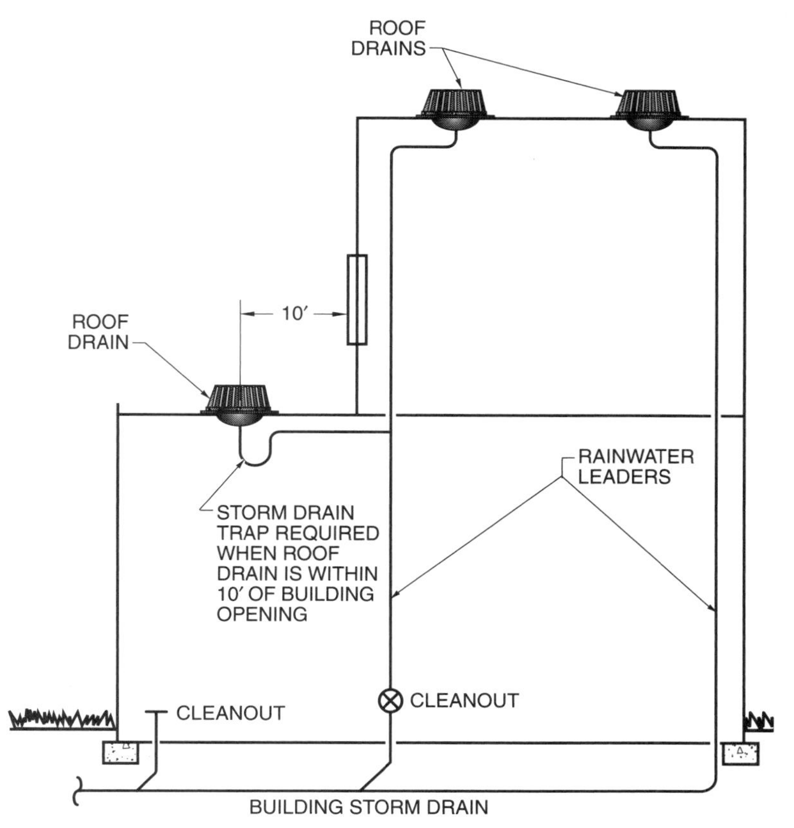

Figure 3-38. Storm drain traps should be used when drain bodies or conductors are located within 10′ of a door, window, or other opening.

Sizing Building Storm Drains and Rainwater Leaders

Projected roof area is the primary factor in determining size of building storm drains and rainwater leaders. The *projected roof area* is the area (in sq ft) of a portion of the roof drained by a particular pipe. In addition to the projected roof area, the size of the building storm drains is based on the horizontal storm drain slope. **See Figure 3-39.** A greater slope allows rainwater to move quickly and efficiently through the system, allowing more rainwater to be conveyed. Based on the Horizontal Storm Drain Size table and using a slope of ⅛″ per foot, 4″ branch drain pipes are used to drain areas up to 1880 sq ft, a 5″ pipe drains areas up to 3340 sq ft, a 6″ pipe drains areas up to 5350 sq ft, and an 8″ pipe drains areas up to 11,500 sq ft. **See Figure 3-40.**

HORIZONTAL STORM DRAIN SIZE*			
Drain Diameter†	Maximum Projected Roof Area For Drains of Various Slopes‡		
	⅛ in/ft Slope	¼ in/ft Slope	½ in/ft Slope
3	822	1160	1644
4	1880	2650	3760
5	3340	4720	6680
6	5350	7550	10,700
8	11,500	16,300	23,000
10	20,700	29,200	41,400
12	33,300	47,000	66,600
15	59,500	84,000	119,000

* based on maximum rate of rainfall of 4″ per hr. If maximum rate of rainfall is more or less than 4″ per hr, adjust the projected roof area by multipying by 4 and dividing by the maximum rate of rainfall (in in. per hr)

† in in.

‡ in sq ft

Figure 3-39. Projected roof area and horizontal storm drain slope determine the size of horizontal storm drains.

All vertical rainwater leaders drain 1500 sq ft of projected roof area and require 3″ pipe. **See Figure 3-41.** With a ⅛″ per foot slope, it is not uncommon for horizontal pipe to be larger than the leader into which it drains. The underground pipe is reduced on the vertical rise. **See Figure 3-42.**

RAINWATER LEADER SIZE*	
Leader Size†	Maximum Projected Roof Area‡
2	720
2½	1300
3	2200
4	4600
5	8650
6	13,500
8	29,000

* based on maximum rate of rainfall of 4″ per hr. If maximum rate of rainfall is more or less than 4″ per hr, adjust the projected roof area by multiplying by 4 and dividing by the maximum rate of rainfall (in in. per hr).

† in in.

‡ in sq ft

Figure 3-41. Rainwater leader size is based on the maximum projected roof area.

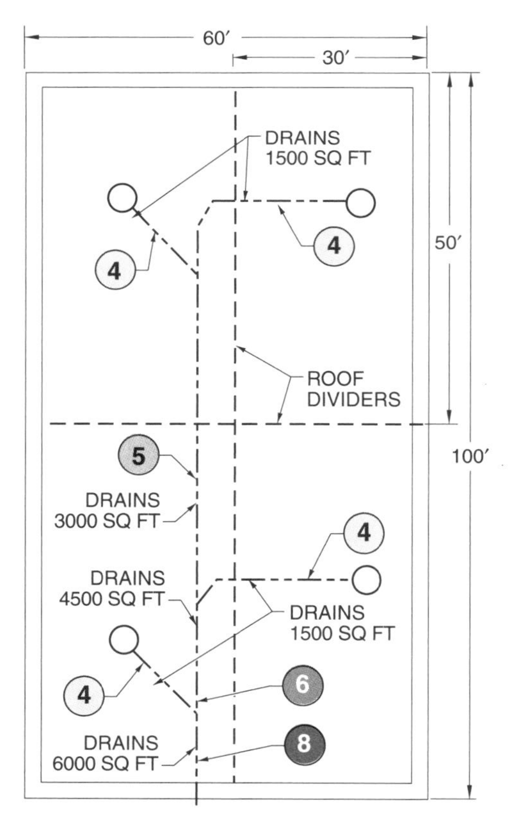

Figure 3-40. Projected roof area is used to determine the size of building storm drains and rainwater leaders.

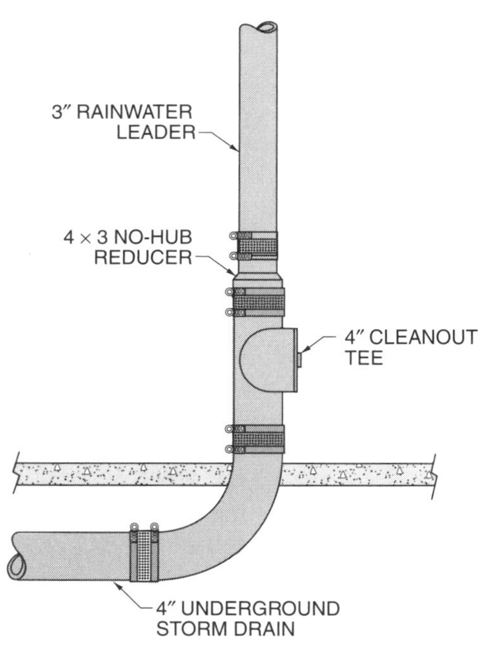

Figure 3-42. A reducer is used to transition from a rainwater leader to a storm drain.

Review Questions

1. What is a plumber's responsibility for proper and safe installation of sanitary drainage facilities in a building?
2. Define the term "drainage fixture unit." How is this unit determined as a standard for calculating the sizes of drainage piping?
3. What is the minimum size of building sewer pipe permitted by most plumbing codes?
4. From the tables in the text, determine the individual dfu values and the horizontal branch drain pipe size required to serve all of the following fixtures:

   Three water closets

   Four lavatories

   One bathtub

   Two kitchen sinks

   Two laundry trays
5. Define the term "stack." Distinguish between a soil stack and a stack vent.
6. At what point does a stack change from a soil stack to a stack vent?
7. List as many approved fittings or combinations of fittings as possible for making the following changes of direction in drainage piping:

   A. Horizontal turns

   B. Horizontal to vertical turns

   C. Vertical to horizontal turns

   D. Combination turns with cleanouts
8. What are the minimum sizes for a soil stack and building drain for a residence? What is the minimum size for a drain branch under a basement floor?
9. List two reasons for venting a plumbing system.
10. Define "stack group."
11. What is a "trap arm"?
12. Define "continuous waste and vent."
13. What is the purpose of a yoke vent?
14. Why do some plumbing codes require that a vent stack be enlarged to at least 2″ before it passes through a building roof?
15. On what basis is the size of the stormwater drainage system calculated?

Sanitary drainage and vent piping systems vary in size. The size of a piping system is dependent on the total number of drainage fixture units (dfus). The dfu value is determined by the type and number of fixtures attached to the piping system.

4 Sizing Sanitary Drainage and Vent Piping

Sanitary drainage and vent piping systems are constructed using a variety of piping methods. Drainage and venting principles are consistent regardless of the type of building, whether residential or commercial structures. Drainage and venting loads are greater in most commercial structures that have a large number of occupants and more plumbing fixtures, making plumbing system design more challenging.

ONE-STORY, ONE-FAMILY DWELLING

Plumbing codes define sanitary drainage and vent piping requirements for residential and commercial structures. Minimum requirements are the absolute minimum level of piping that can be installed in a building. For some construction projects, minimum requirements are specified in the prints and specifications. For other construction projects of a similar nature, additional piping, such as an individual vent, is installed to ensure a higher level of protection against trap siphonage and back pressure.

A typical one-story, one-family dwelling may consist of a bathroom (with water closet, lavatory, and bathtub) and kitchen sink on the first floor, and a laundry tray and floor drain with 2″ trap in the basement. **See Figure 4-1.** The dfu values for the fixtures in the dwelling are based on the Drainage Fixture Unit Values for Common Plumbing Fixtures table. The dfu values for the fixtures in the example one-story, one-family dwelling are:

No.			dfu		Total
1	Water closet	x	6	=	6
1	Lavatory	x	1	=	1
1	Bathtub	x	2	=	2
1	Kitchen sink	x	2	=	2
1	Laundry tray	x	2	=	2
1	Floor drain	x	2	=	2
			Total dfu	=	15

Minimum Requirements

Based on the drainage fixture unit values of the fixtures, the minimum sizes of the building sewer, building drain, and building drain branches are determined. The building drain, which is 3″ pipe sloped at ¼″ per foot, is increased to 4″ pipe for the building sewer, with the building sewer terminating at the sanitary sewer main or septic tank. A front main cleanout is provided on the inside of the foundation wall at the base of the stack. The front main cleanout extends 2″ minimum above the floor to prevent a homeowner or occupant from removing the cleanout plug and using the cleanout as a floor drain.

A 2″ building drain branch serves the laundry tray and the basement floor drain. The 2″ pipe is reduced to 1½″ pipe above the basement floor, and a cross fitting is installed to provide an opening for the laundry tray waste and a cleanout. The 1½″ pipe is reduced to 1¼″ at the top of the cross to vent the laundry tray trap. The 1¼″ vent pipe connects to the 3″ stack at a point 6″ above the flood level rim of the kitchen sink (3′-6″ above the first-floor elevation).

A stack group is used to plumb the bathroom of the one-story dwelling, representing the minimum requirement for a waste and vent pipe installation. These minimum requirements may not be permitted in areas with strict plumbing codes since the bathtub, lavatory, and kitchen sink are not individually vented.

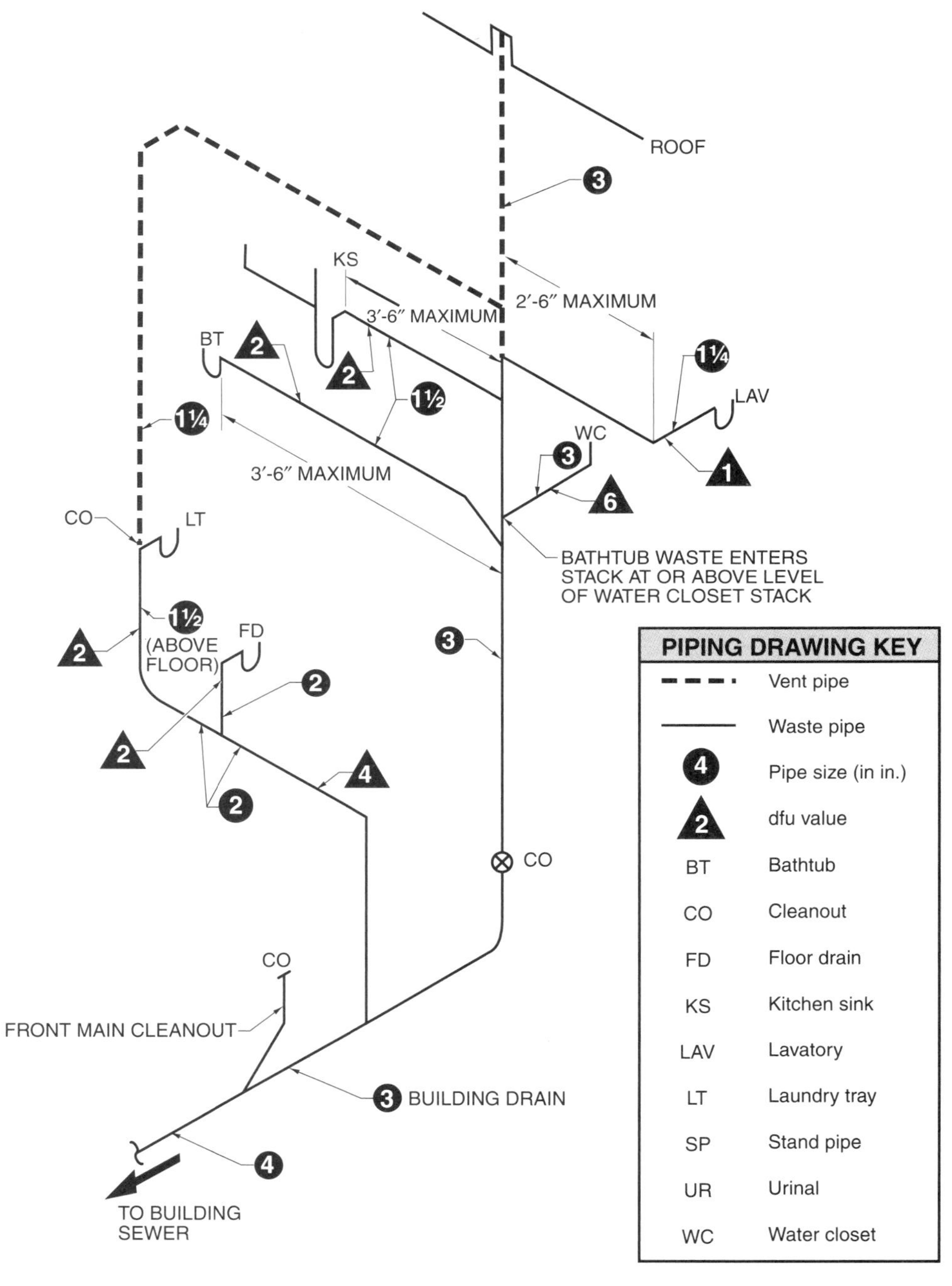

Figure 4-1. A typical one-story, one-family dwelling consists of a bathroom with water closet, lavatory, and bathtub, kitchen sink, laundry tray, and floor drain.

However, for most plumbing codes, individually venting the bathtub, lavatory, and kitchen sink is not necessary, provided that the distance between the trap and the stack does not exceed 2′-6″ for a 1¼″ lavatory trap or 3′-6″ for 1½″ bathtub and kitchen sink traps. The bathtub drain enters the stack at or above the water closet opening.

Individually Vented Fixtures

In some jurisdictions, additional venting may be necessary to satisfy plumbing code requirements. **See Figure 4-2.** All fixtures except the floor drain are vented. The building drain from the stack is 4″ pipe with a 4″ cleanout at the inside of the foundation wall. A 2″ branch pipe serves the floor drain, laundry tray, and stand pipe.

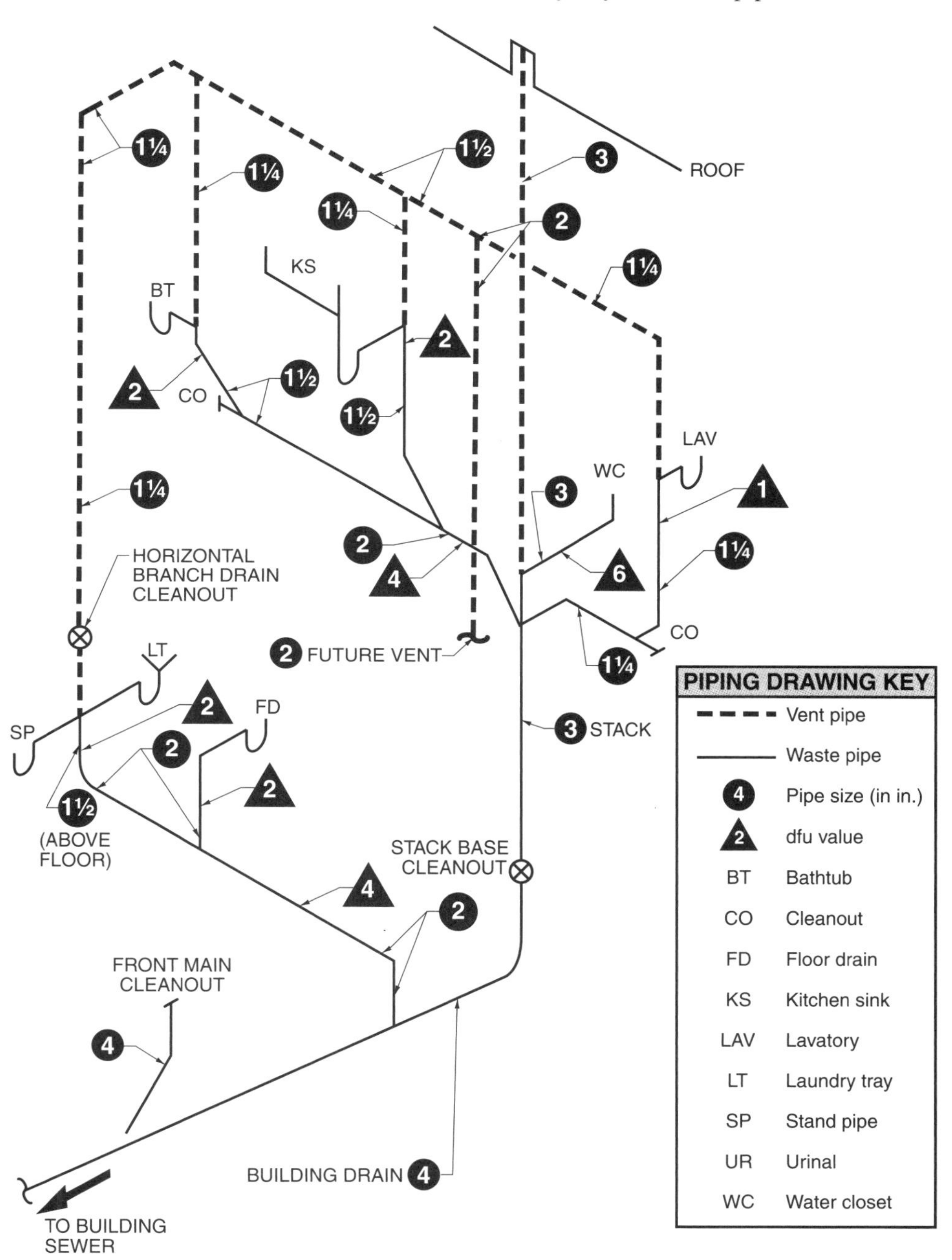

Figure 4-2. Individually vented fixture traps provide greater protection from trap siphonage and back pressure.

All fixtures draining into the stack have a continuous waste and vent. The horizontal branch drain that serves the lavatory is 1¼″ pipe. A 2″ branch pipe serves the kitchen sink and bathtub, and a 1½″ pipe serves the bathtub fixture trap.

The horizontal vent connected to the stack on the right side that serves the lavatory drain is 1¼″ pipe. The vent on the other side is 2″ pipe to the first tee, which is a future vent. A future vent is provided so there is a proper size vent opening if a basement bathroom is installed in the future. After the future vent opening, the vent is reduced to 1½″ pipe to serve the kitchen sink and bathtub vents. After the bathtub vent, the vent pipe is reduced to 1¼″ for the basement laundry tray and standpipe vents.

A front main cleanout is installed on the building drain inside the foundation wall, a stack base cleanout is installed at the base of the 3″ soil stack, and individual cleanouts are installed at the ends of the horizontal branches draining into the stack. A cleanout is also installed on the vent pipe above the laundry tray and standpipe to serve the horizontal branch drain.

TWO-STORY, ONE-FAMILY DWELLING

Two-story, one-family dwellings typically include plumbing fixtures on both floors as well as in a basement. The first floor may have a water closet, lavatory, and kitchen sink, and the second floor has a water closet, lavatory, and bathtub. A laundry tray and floor drain are located in the basement. **See Figure 4-3.** Based on the DFU Values for Common Plumbing Fixtures table, the dfu values for the fixtures are:

No.			dfu		Total
2	Water closets	x	6	=	12
2	Lavatories	x	1	=	2
1	Bathtub	x	2	=	2
1	Kitchen sink	x	2	=	2
1	Laundry tray	x	2	=	2
1	Floor drain	x	2	=	2
			Total dfu	=	22

Tech Facts

Some plumbing codes do not require individual venting of the bathtub, lavatory, and kitchen sink in residential construction provided that the trap-to-vent stack distance is not excessive.

Individually Vented Fixtures

The building drain from the stack is 4″ pipe with a 4″ cleanout on the inside of the foundation wall. A 2″ branch pipe serves the floor drain and laundry tray. All fixtures draining into the stack have a continuous waste and vent. The kitchen sink is located where it is impractical to drain it into the stack; instead, it drains into a building drain branch that serves the laundry tray and floor drain.

A building drain branch receives wastes from the laundry tray and floor drain in the basement and the kitchen sink on the first floor. The building drain branch is 2″ pipe below the basement floor and is reduced to 1½″ above the floor for the laundry tray and kitchen sink wastes. The laundry tray vent is 1¼″ pipe, which continues up to the first floor where it is connected to the 1¼″ kitchen sink vent at a point 6″ above the kitchen sink flood level rim (approximately 3′-6″ above the kitchen floor). The 1¼″ vent pipe continues upward to a point immediately below the roof, where it is increased to 2″ pipe before passing through the roof.

The first-floor water closet is set off to the side of the 3″ stack. The water closet drains into a 3″ sanitary tee and a wye in the stack. The water closet may be connected using a vented closet tee or cross. A *vented closet tee* and *vented closet cross* are specially designed fittings for offsetting a water closet on a lower floor from the stack to properly vent the water closet. **See Figure 4-4.** Vented closet tees and crosses are available with side openings to accept the wastes of other fixtures on the same floor. Waste from the lavatory on the first floor is drained by 1¼″ pipe into either a wye below the closet bend or into the side opening of the vented closet tee.

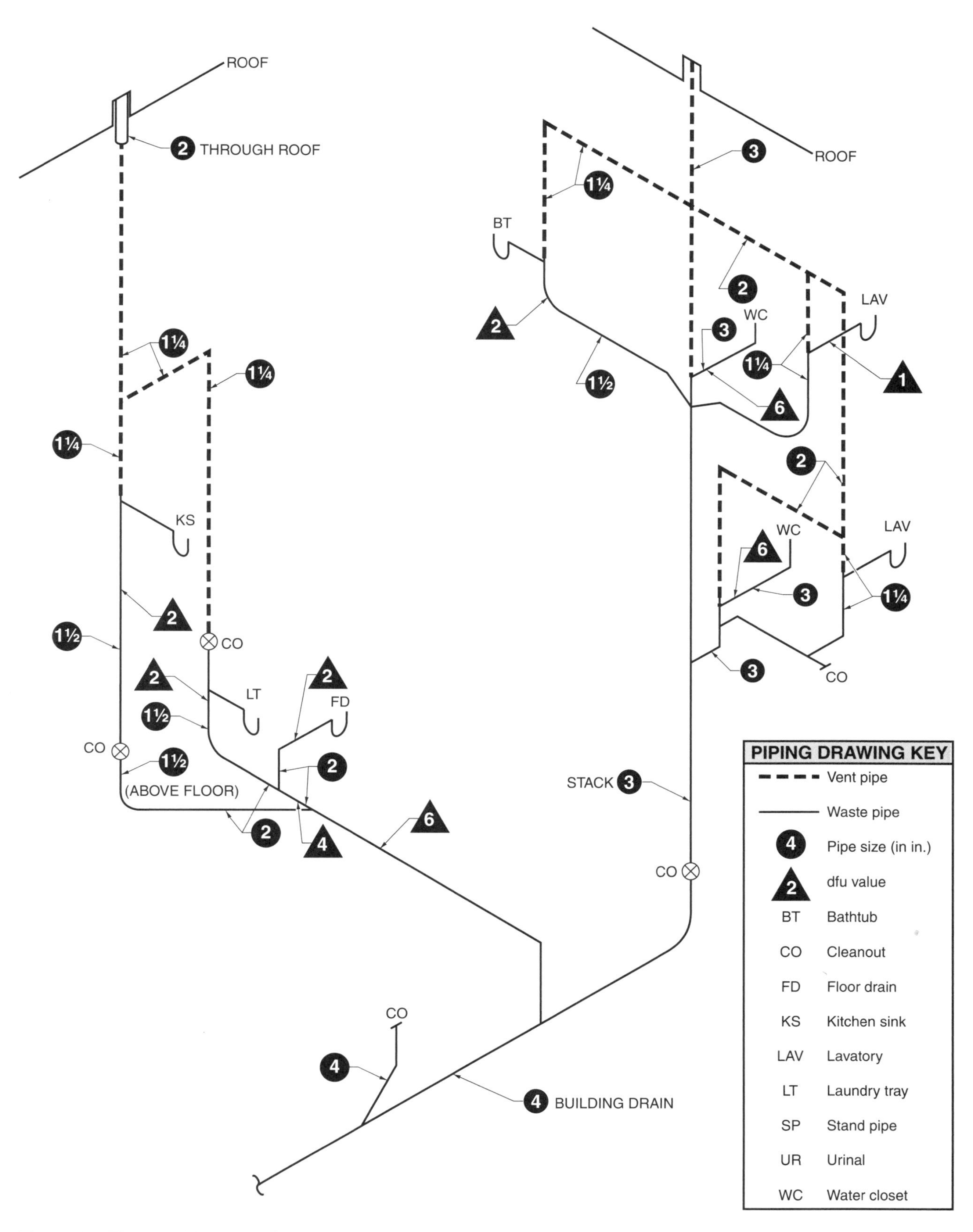

Figure 4-3. The plumbing plan of a typical two-story residence shows individual venting of all fixtures.

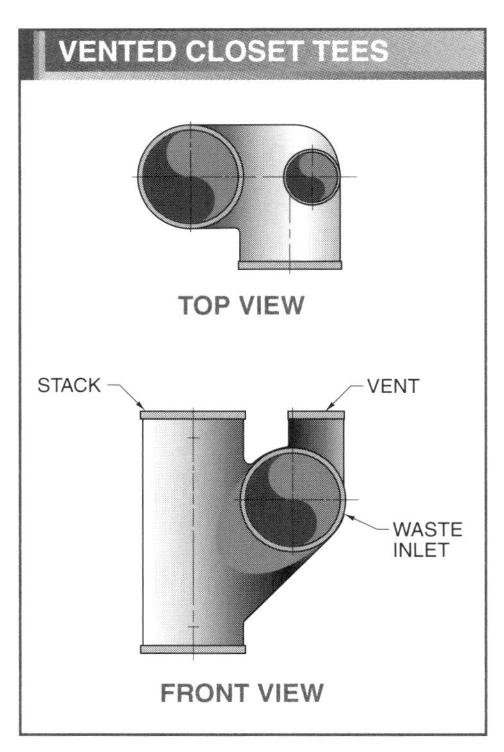

Figure 4-4. A no-hub vented closet tee offsets a water closet on a lower floor from the stack to properly vent the water closet.

The first-floor 2″ water closet vent receives the 1¼″ lavatory vent and continues to the second floor. On the second floor, the 2″ vent is connected to the 1¼″ vent from the second-floor lavatory and ties back into the stack at least 6″ above the flood level rim of the lavatory (approximately 3′ above the second floor). The second-floor lavatory waste discharges into one side of a double wye placed in the stack below the closet bend, and the 1½″ bathtub waste discharges into the other side. The bathtub vent is 1¼″ pipe and ties back into the stack.

Cleanouts are installed on the building drain on the inside of the foundation wall, at the base of the 1½″ kitchen sink waste stack, on the laundry tray vent stack, and on the horizontal branch serving the first-floor lavatory waste. Cleanouts are not included on the horizontal branches serving the second-floor bathtub and lavatory because there would be no access to the cleanouts in a residence.

DUPLEX RESIDENCE

A duplex residence contains similar plumbing fixtures on each floor or area of the duplex. **See Figure 4-5.** Each floor of the duplex consists of a kitchen with a sink, and a bathroom with a water closet, lavatory, and bathtub. The basement consists of a bathroom with a water closet and a lavatory, two laundry trays, and a floor drain. Based on the DFU Values for Common Plumbing Fixtures table, the dfu values for the fixtures are:

No.			dfu		Total
3	Water closets	x	6	=	18
3	Lavatories	x	1	=	3
2	Bathtubs	x	2	=	4
2	Kitchen sinks	x	2	=	4
2	Laundry trays	x	2	=	4
1	Floor drain	x	2	=	2
		Total dfu		=	35

The lower end of the building drain is 4″. The main stack, which is installed in the partition directly behind the water closet, is 3″ pipe and has bathroom branch connections on the first and second floors.

The basement water closet is vented with a 2″ wet vent, and the 1¼″ basement lavatory waste is connected to the 2″ vent. The 2″ vent extends from the basement bathroom to the first floor, where it is connected to vents from first-floor fixtures. The laundry trays are connected to a 1½″ common waste pipe and are common vented with a 1¼″ pipe that is connected to the vent portion of the kitchen sink stacks.

A 1½″ pipe is used for the bathtub waste of the first-floor bathroom. The 1½″ waste pipe also receives the lavatory waste and discharges it into the stack at or below the closet bend. The bathtub is wet vented through the lavatory waste, which is increased to 1½″ pipe. The 1¼″ lavatory and bathtub vent connects to the 2″ vent for the first-floor water closet, and this vent extends to the second floor. The second-floor bathroom waste pipe installation is the same as the first-floor bathroom installation. The 2″ vent from the basement and first-floor fixtures connects to the 3″ stack on the second floor.

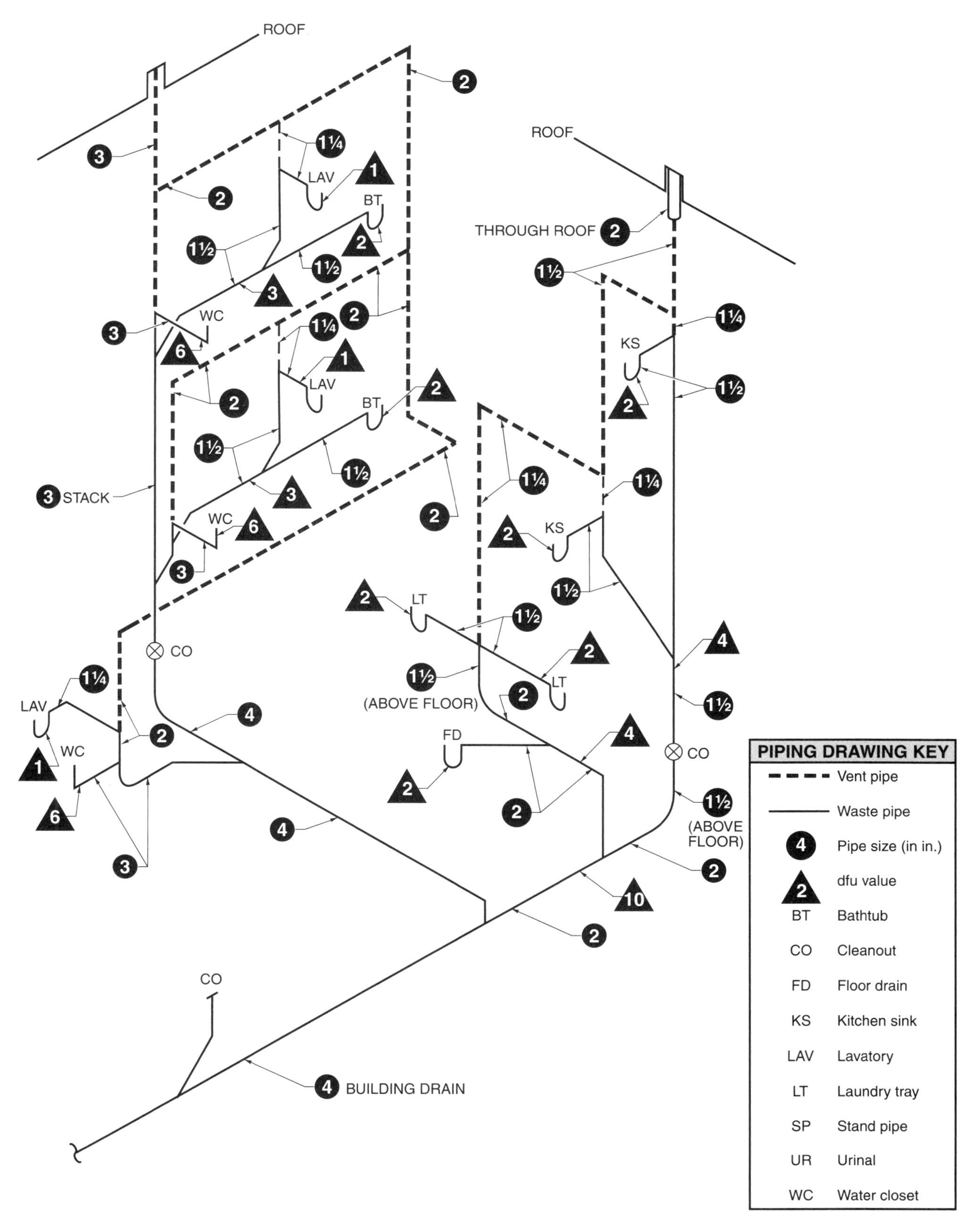

Figure 4-5. A duplex residence typically includes similar plumbing fixtures on each floor of the duplex.

Both kitchen sinks discharge waste into the same 1½″ waste pipe, and each kitchen sink is individually vented with 1¼″ pipe. At the tee where the laundry tray vent connects to the first-floor kitchen sink vent, the vent is increased to 1½″ pipe. This 1½″ pipe continues to the second floor to connect to the 1¼″ vent from the second-floor kitchen sink. The vent then continues as 1½″ pipe to a point below the roof, where it is increased to 2″ pipe before passing through the roof.

The underground pipe serving the kitchen sink, laundry trays, and floor drain is 2″ pipe because it is a building drain branch from a stack. As a branch from a stack, it is sized based on the Maximum Loads for Building Sewer, Building Drain, and Building Drain Branches from Stacks table, and it may drain 21 dfu at ¼″ slope. If this 2″ pipe was considered a horizontal fixture branch, it would be sized from the Maximum Loads for Horizontal Branch Drains table, and could only drain 6 dfu.

Individual vents may be installed to allow the installation to conform to stricter plumbing codes. Individual vents would need to be installed for the basement lavatory and for the bathtubs on the first and second floors.

MULTIFAMILY DWELLING

Multifamily dwellings, such as apartments, have greater demands placed on the plumbing system than a one-family residence. For design efficiency and ease of plumbing system installation, common fixtures are installed back-to-back with one another. When designing plumbing systems for multifamily dwellings, the systems may be broken down into individual soil or waste stacks, such as a stack serving the bathrooms and another stack serving the kitchens. Each unit of a common multifamily dwelling might contain a kitchen sink and a bathroom, which includes a bathtub, lavatory, and water closet. Based on the DFU Values for Common Plumbing Fixtures table, the dfu values for the fixtures in a four-unit multifamily dwelling are:

No.			dfu		Total
4	Water closets	x	6	=	24
4	Lavatories	x	1	=	4
4	Bathtubs	x	2	=	8
4	Kitchen sinks	x	2	=	8
			Total dfu	=	44

Bathroom Stack Minimum Requirements

One of the stacks in a multifamily dwelling serves the bathrooms for the four units. **See Figure 4-6.** The waste portion of the soil stack is 3″ pipe, since only two water closets discharge their waste at one story or branch interval. The vent portion of the soil stack can be reduced to 2″ pipe through the roof if there is another full-size stack in the building.

The second-floor water closets empty into a 3″ sanitary cross in the stack. The first-floor water closets drain into a 3″ sanitary cross, which then discharges waste into a 3″ wye in the stack, or the water closets discharge their waste into a vented closet cross.

The 1½″ bathtub waste pipes on each floor discharge their waste into a common 2″ waste pipe. The waste then drains into a wye placed below the water closet waste fitting or into the side opening of a vented closet cross. A 2″ common lavatory waste is installed, since 2 dfu are discharged into the waste pipe on each floor and the pipe is also a wet vent for the bathtubs. A maximum of 1 dfu may drain into a 1½″ wet vent, and a maximum of 4 dfu may drain into a 2″ wet vent.

A 2″ vent pipe for the first-floor water closet connects to the 1½″ vent serving the lavatories and bathtubs. The 2″ vent continues to the second floor to connect to the 1½″ lavatory and then connects to the 3″ stack vent.

Bathroom Stack with Individually Vented Fixtures

Individually vented fixtures virtually eliminate siphonage and back pressure in a drainage system. **See Figure 4-7.** In this installation, the back-to-back lavatory waste pipe is reduced to 1½″ pipe with a 1¼″ common vent for the two lavatories. The bathtubs are vented with a 1¼″ common vent pipe, while discharging their waste into 1½″ waste pipes. The remainder of the waste and vent piping and sizing is the same as the waste and vent pipe sizing in Figure 4-6.

Tech Facts

Vents balance the air pressure on each side of a trap seal, which protects the trap seal from siphonage. Vents also allow sewer gas to rise into the open air without exposing the occupants of a building to the gas. Sewer gas may cause health problems to occupants if it is not properly ventilated.

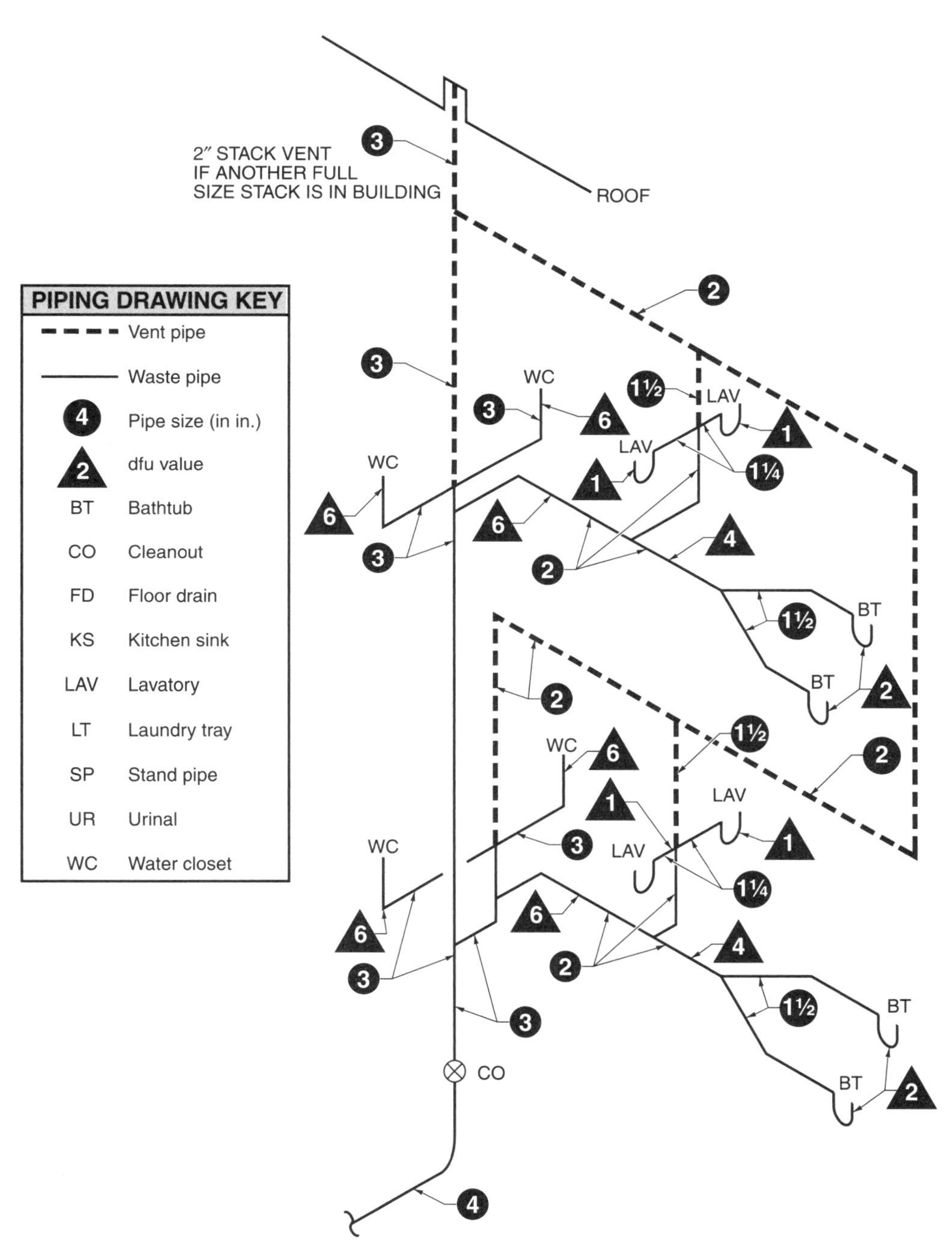

Figure 4-6. Back-to-back bathrooms are common in multifamily dwellings such as an apartment.

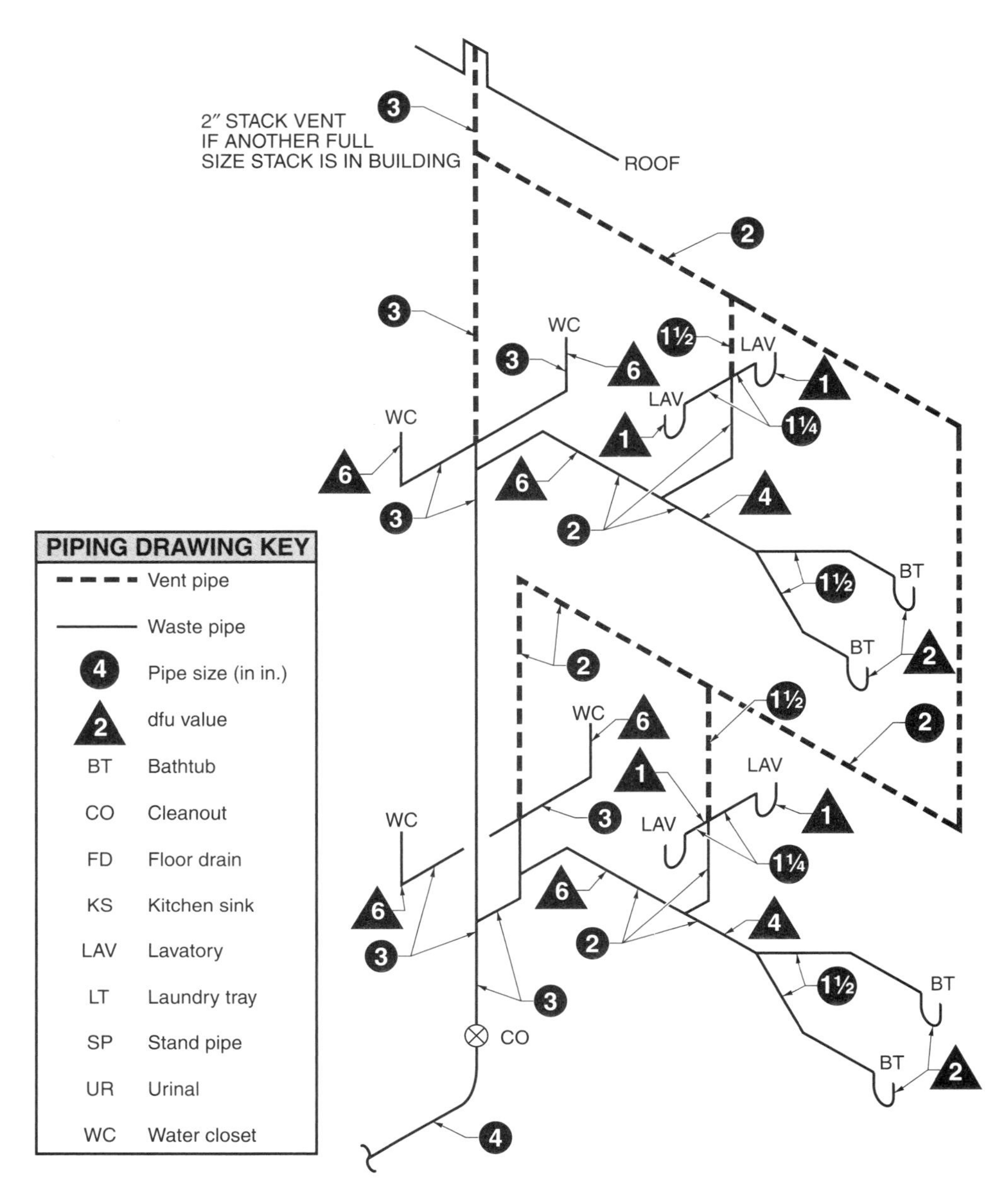

Figure 4-7. Fixtures can be individually vented in multifamily dwellings with back-to-back bathrooms.

Kitchen Sink Waste Stack

A kitchen waste stack may be installed in a multifamily dwelling to convey kitchen waste. **See Figure 4-8.** A 2″ pipe is used for the lowest portion of the waste stack because it receives 8 dfu of discharge (4 kitchen sinks × 2 dfu = 8 dfu). The portions of the stack that receive only the waste from first-floor kitchen sinks and only the waste from the second-floor kitchen sinks are sized as 2″ pipe. Per the Maximum Size for Soil and Waste Stacks table, a 1½″ waste stack cannot be used, since only 2 dfu are permitted to drain into a 1½″waste stack on any one floor or branch interval.

A 1¼″ pipe is used to vent the kitchen sinks. The 1¼″ pipe extends to the second floor, where it is connected to the stack vent above the second-floor kitchen sinks. The 1¼″ vent from the second-floor kitchen sinks is increased to 1½″ at a tee and continues to a point below the roof, where it is increased to 2″ pipe before passing through the roof.

MULTISTORY BUILDING BATHROOM STACK

Some multistory buildings, such as hotels, require only a bathroom stack. Bathroom waste and vent pipe connections are individually vented, similar to bathroom waste and vent piping in a residence. The bathtub trap is wet vented through the lavatory waste pipe. **See Figure 4-9.**

A 4″ soil stack is required for the waste from the bathrooms, since there are more than six stories or branch intervals of bathrooms in the building. Based on the Size and Length of Vent Stacks table, a 3″ vent stack is required. The procedure for determining the vent stack size is:

1. A total of 72 dfu discharge into the waste stack.

No.			dfu		Total
8	Water closets	x	6	=	48
8	Lavatories	x	1	=	8
8	Bathtubs	x	2	=	16
			Total dfu	=	72

2. The developed length of the stack is 80′ (8 floors × 10′ floor-to-ceiling height).
3. In the Size and Length of Vent Stacks table, locate the 4″ waste stack in the left column for the 240 dfu discharge. Since the developed length of the stack is 80′, a 3″ vent diameter is required to vent the installation.

The 3″ vent stack is connected to the waste stack below the first-floor bathroom. The vent stack also has a 3″ yoke vent, which is installed between the third and fourth floors. The vent stack continues to the roof where it is either reconnected to the 4″ soil stack or extended through the roof.

TWO-STORY INDUSTRIAL BUILDING BATHROOM PIPING

In public and some private buildings, gender-specific bathroom facilities must be provided. A women's bathroom typically includes water closets and lavatories, while a men's bathroom includes water closets, lavatories, and urinals. **See Figure 4-10.** The men's bathroom, located on the first floor, contains two water closets, two wall-hung urinals with 2″ traps, and two lavatories that are located on the same wall. The women's bathroom, located on the second floor, contains four water closets and four lavatories. The women's bathroom fixtures are located on the same wall as the men's bathroom fixtures.

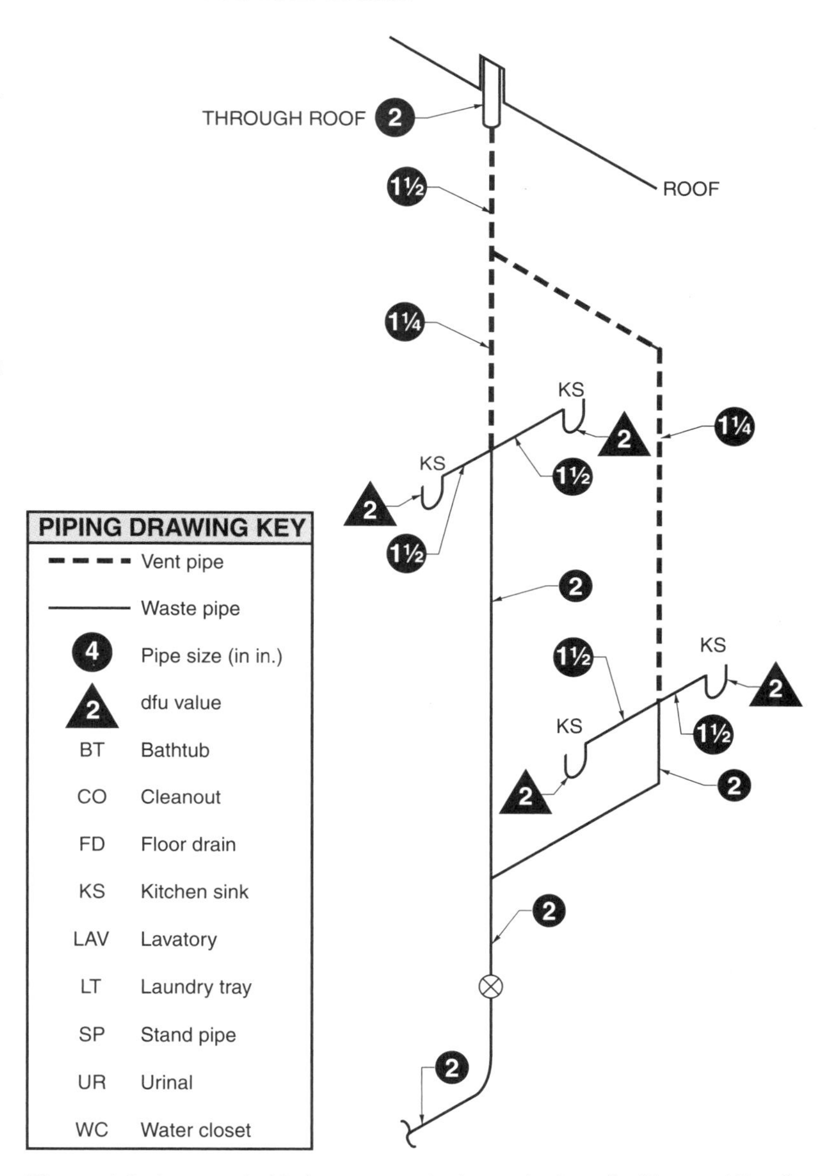

Figure 4-8. A separate kitchen waste stack may be installed in a multifamily dwelling to convey kitchen waste.

The first-floor water closets are connected to a 4″ horizontal branch pipe, which is connected to the 4″ stack. The horizontal branch pipe is sized per the

Maximum Loads for Horizontal Branch Drains table. The other end of the horizontal branch pipe is reduced to 2½″pipe to receive the first-floor urinal and lavatory wastes:

No.			dfu		Total
2	Wall-hung urinals	x	4	=	8
2	Lavatories	x	1	=	2
		Total dfu			10

However, piping sized with 2½″ pipe would be constructed with 3″ pipe since 2½″ pipe and fittings are not available.

At the first urinal, the horizontal waste pipe is reduced to 2″ pipe. After the second urinal, the waste pipe is reduced to 1½″ pipe for the portion that receives the waste from both lavatories, and then is reduced again to 1¼″ pipe for the last lavatory waste.

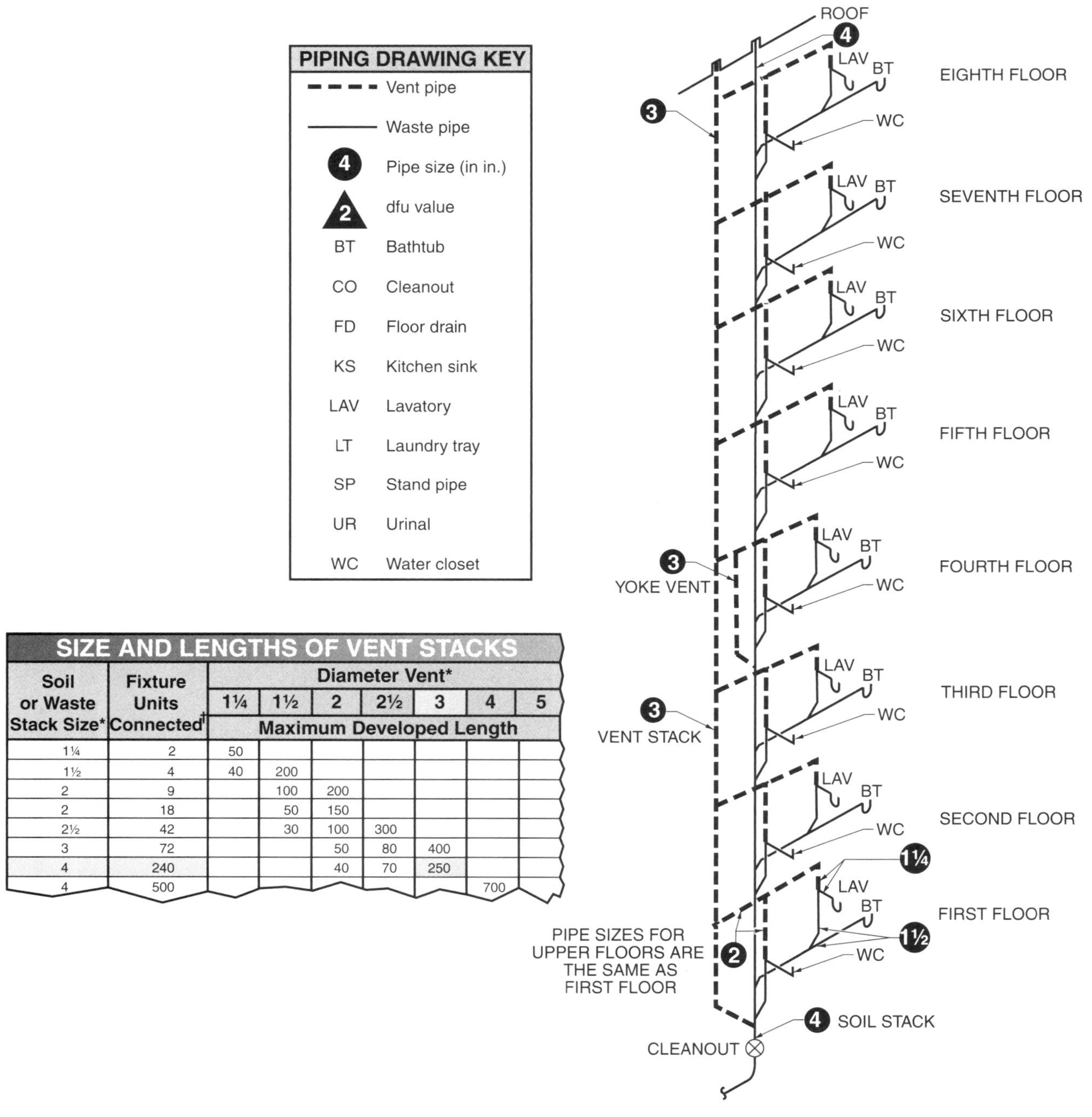

SIZE AND LENGTHS OF VENT STACKS

Soil or Waste Stack Size*	Fixture Units Connected†	Diameter Vent*						
		1¼	1½	2	2½	3	4	5
		Maximum Developed Length						
1¼	2	50						
1½	4	40	200					
2	9		100	200				
2	18		50	150				
2½	42		30	100	300			
3	72			50	80	400		
4	240			40	70	250		
4	500						700	

Figure 4-9. The bathtubs in a multistory bathroom stack may be wet vented through the lavatory waste pipe.

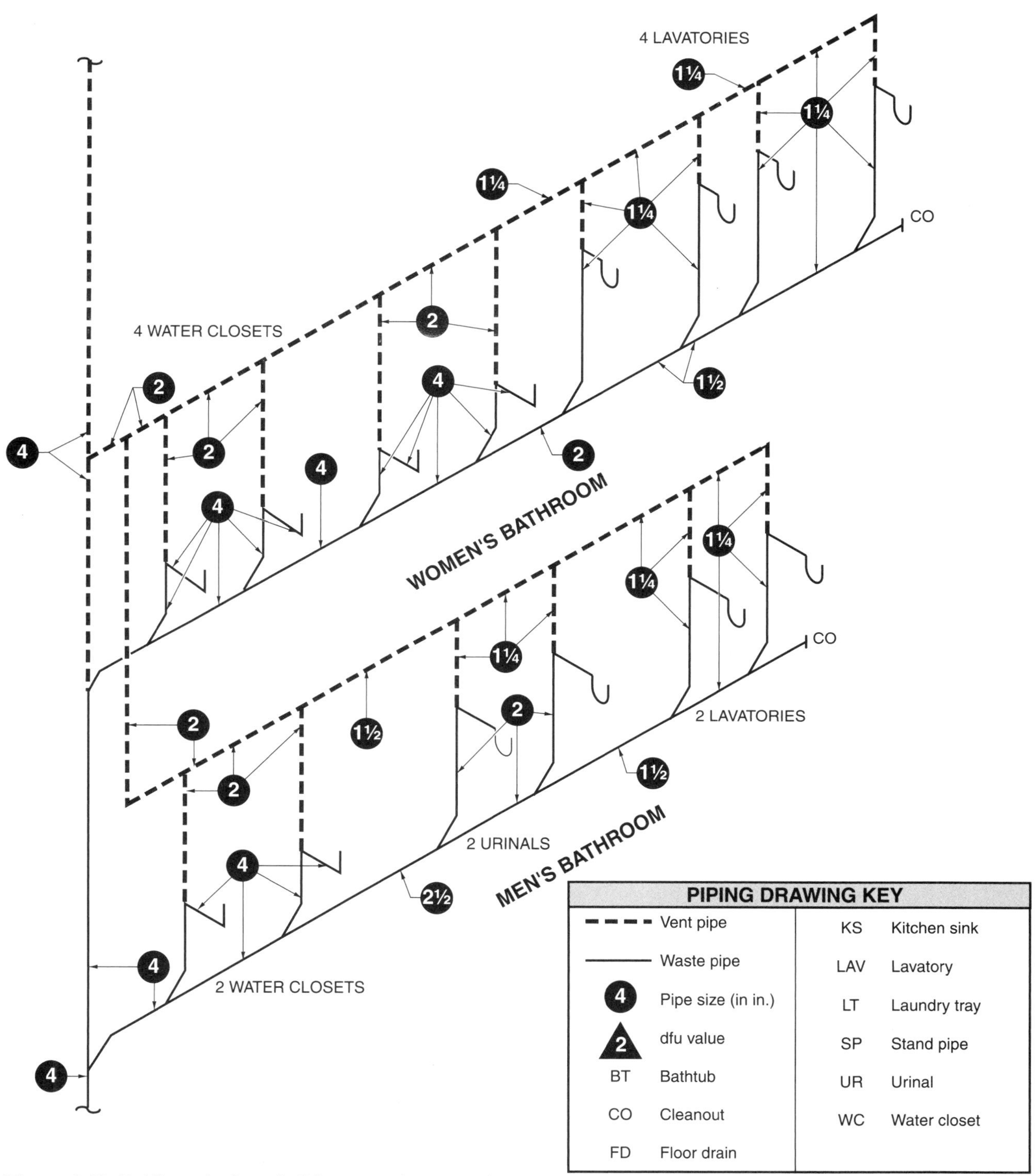

Figure 4-10. Public and private buildings may have gender-specific facilities.

The urinals and lavatories are individually vented with 1¼″ pipe. The 1¼″ individual fixture vents are connected to a 1¼″ horizontal vent pipe, which is increased to 1½″ pipe after the second urinal vent. The horizontal vent pipe receives the 2″ individual vents from the water closets and is then extended to the second floor, where the 2″ vent pipe is connected to the second-floor horizontal vent pipe.

The second-floor water closets discharge their waste into a 4″ horizontal branch pipe, which is connected to the 4″ stack. The other end of the horizontal branch is reduced to 2″ pipe to receive the waste from four lavatories. After the first lavatory waste, the horizontal branch pipe is reduced to 1½″ pipe for the waste from the next two lavatories, and then is reduced again to 1¼″ pipe for the last lavatory waste.

The lavatories on the second floor are individually vented with 1¼″ pipes, which are connected to a 1¼″ horizontal vent pipe. The horizontal vent pipe is increased to 2″ pipe where the 2″ individual vent of the first water closet is connected to it. The 2″ horizontal vent pipe continues toward the 4″ stack vent, and is connected to the 2″ individual vents from the remaining water closets. A 2″ tee is provided for the connection for the 2″ vent from the first-floor bathroom, before the vent is connected to the stack vent.

An aerial lift may be used to position plumbers above previously placed items or built-in furniture when making repairs or installing new piping.

TWO-STORY OFFICE BUILDING BATHROOM PIPING

Buildings with a larger occupancy require a larger number of fixtures to provide adequate facilities for the occupants. If possible, the men's and women's bathrooms are located back-to-back on each floor so the waste and vent piping can be economically installed. **See Figure 4-11.** Each of the men's bathrooms contains four wall-hung water closets, two urinals with 2″ traps, three lavatories, and a 2″ floor drain between the urinals. Each of the women's bathrooms contains five wall-hung water closets, four lavatories, and a 2″ floor drain between the first and second water closet stalls. The wall-hung water closets in both bathrooms are back-to-back on a common wall, while the other fixtures are on the opposite bathroom walls. Wall-hung water closet supporting chair carrier fittings are available in a variety of styles to accommodate most installations. **See Figure 4-12.**

Water closets on both floors are connected to a 4″ horizontal branch constructed with water closet supporting chair carrier fittings. The back-to-back water closets are common vented with 2″ pipe, and the remaining water closet in the women's bathroom also has a 2″ vent. The first-floor water closet vents are connected to a 2″ horizontal vent that extends to the ceiling, where the vent receives a 1¼″ vent from the women's bathroom lavatories and the 1¼″ vent from the men's bathroom urinals and lavatories. The 2″ vent continues to the second floor, where it connects to the 4″ stack vent.

The first-floor women's bathroom lavatories are individual 1¼″ waste pipes. The horizontal waste pipe, which receives the individual 1¼″ lavatory wastes, is increased to 1½″ pipe at the second lavatory waste pipe, and then to 2″ pipe at the fourth lavatory waste pipe. The 2″ waste pipe extends toward the stack, where it receives the 2″ waste pipe from the women's bathroom floor drain before it discharges into a double wye in the 4″ stack.

The horizontal waste pipe for the first-floor men's bathroom is 1¼″ pipe for the first lavatory and 1½″ pipe for the second and third lavatories. The 1½″ horizontal waste pipe is increased to 2″ for the first urinal and to 2½″ for the second urinal. The 2½″ waste pipe extends toward the stack, receiving the waste from the 2″ floor drain under the urinals. The waste pipe then discharges into the double wye in the 4″ stack.

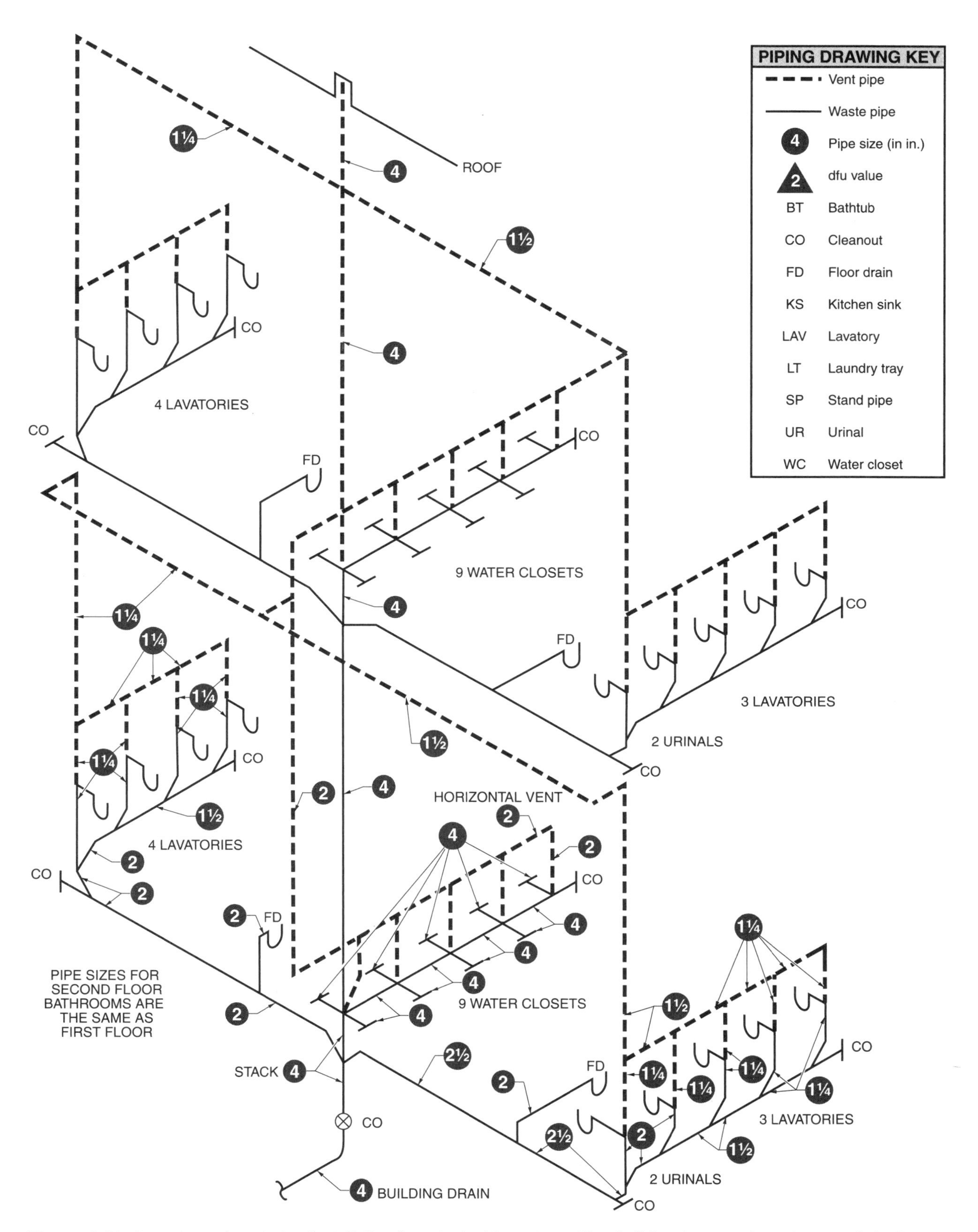

Figure 4-11. A waste and vent pipe installation for a typical two-story office building has men's and women's bathrooms back-to-back on each floor.

The individual vents for the first-floor men's bathroom lavatories and urinals is 1¼″ pipe. The horizontal vent pipe for the lavatory vents is 1¼″, but is increased to 1½″ pipe at the first urinal connection. The 1½″ vent continues upward, where it is connected to a 2″ vent from the water closets. The first-floor women's bathroom lavatory vents are 1¼″ pipe. The primary differences in piping between the first- and second-floor bathrooms is that the 2″ horizontal vent pipe for the water closets, the 1¼″ vent from the women's bathroom lavatories, and the 1½″ vent from the men's bathroom lavatories and urinals on the first floor are connected directly to the 4″ stack vent.

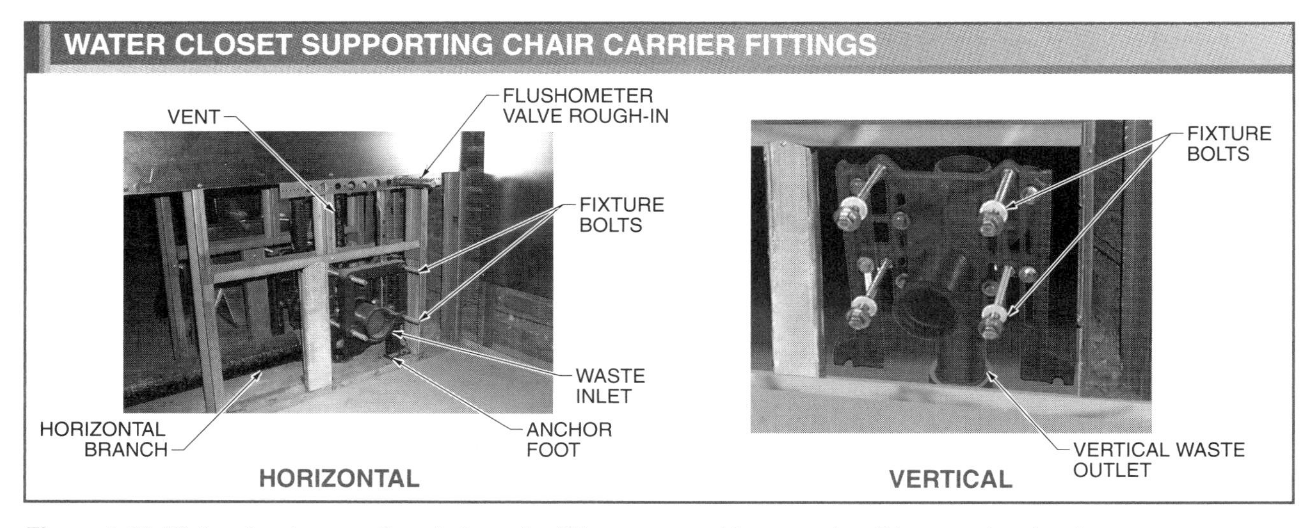

Figure 4-12. Water closet supporting chair carrier fittings are used to support wall-hung water closets.

Review Questions

1. Give the abbreviations or symbols for:
   Waste pipe
   Cleanout
   Vent pipe
   Floor drain
   Size of pipe
   dfu value
   Stand pipe
   Laundry tray
2. Define "stack group."
3. What is the purpose of a vented closet tee?
4. Why are common fixtures installed back-to-back with one another in multifamily dwellings?
5. What is the advantage of individually vented fixtures?

There are many different types and sizes of buildings that require water to be conveyed to plumbing fixtures within the buildings. It is important to properly size the water supply piping in the buildings based on the available pressure, fixture demand, and height of the particular building. An adequate amount of water must be supplied to the plumbing fixtures and appliances to ensure that they function properly.

5 Sizing Water Supply Piping

The building water supply system must be sized properly so that the pipes can convey an adequate supply of water to the plumbing fixtures and appliances to allow them to function properly.

The five factors that determine the size of water supply piping are:

- available water pressure
- fixture demand
- length of piping
- height of building
- flow pressure needed at top floor

AVAILABLE WATER PRESSURE

The available pressure is the water pressure in the street water main or other supply source. Common street water main pressure is 45 psi–60 psi. Water pressure within a building must never be allowed to exceed 80 psi. If the street water main pressure exceeds 80 psi, a pressure-reducing valve must be installed on the water service at the point where the pipe enters the building. **See Figure 5-1.** A *pressure-reducing valve* is an automatic device used to convert high and/or fluctuating inlet water pressure to a lower or constant outlet pressure. When street water main pressure fluctuates widely throughout the day, the water supply system of the building must be designed on the basis of the minimum pressure available.

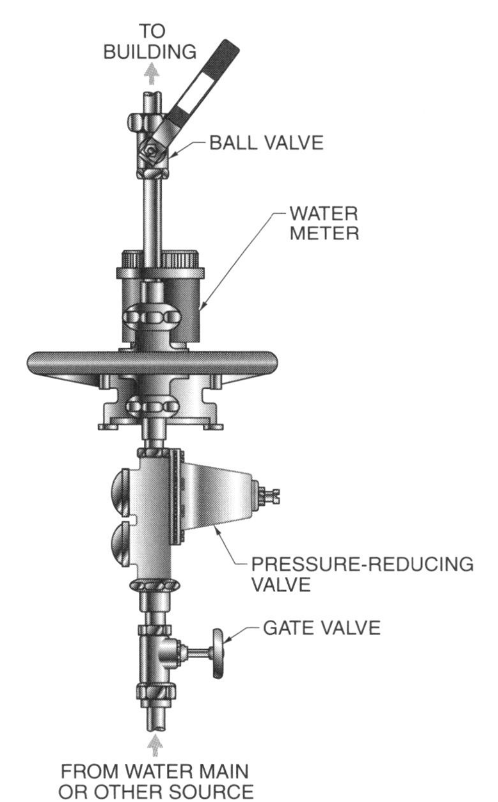

Figure 5-1. A pressure-reducing valve converts high and/or fluctuating inlet water pressure to a lower or constant outlet pressure.

FIXTURE DEMAND

Each plumbing fixture served by the water supply system has a specified flow rate. *Flow rate* is the volume of water used by a fixture in a given amount of time. Flow rate is measured in gallons per minute (gpm). **See Figure 5-2.** The total demand on a water supply system if all fixtures are used simultaneously is determined by adding the minimum flow rates for all the plumbing fixtures within a building. However, all plumbing fixtures in a building are rarely used simultaneously.

MINIMUM FLOW RATES FOR COMMON PLUMBING FIXTURES	
Type of Fixture	**Flow Rate***
Standard lavatory faucet	2.0
Self-closing basin faucet	2.5
Sink faucet, ⅜″	4.5
Sink faucet, ½″	4.5
Bathtub faucet	6.0
Laundry tray faucet, ½″	5.0
Shower	5.0
Ball cock for water closet	3.0
Flushometer valve for water closet	15–35
Flushometer valve for urinal	15.0
Drinking fountain	0.75
Sillcock or wall hydrant	5.0

* in gpm

Figure 5-2. Minimum flow rates for fixtures and appliances must be considered when designing water supply systems.

A system was developed to estimate the total demand on a water supply system, based on the reasonable assumption that plumbing fixtures are not all used simultaneously. A *water supply fixture unit (wsfu)* is a measure of the estimated water demand of a plumbing fixture. A plumbing fixture, such as a sink, water closet, or lavatory, is assigned a wsfu value based on the:

- fixture flow rate when the fixture is used
- average time water is actually flowing when a fixture is being used
- frequency that the fixture is used

For example, a domestic dishwasher for private use has a 2 wsfu demand and a dishwasher for public use, such as in a restaurant, has a 4 wsfu demand since a domestic dishwasher does not use as much water and is not used as frequently as a dishwasher in a restaurant.

The type of dwelling in which the fixture is installed must also be considered when a wsfu value is assigned. Fixtures installed in private or private-use dwellings are not used as frequently as public or public-use fixtures. In plumbing fixture classification, private or private-use fixtures are installed in residences, apartments, individual hotel bathrooms, and similar installations where fixtures are intended for use of one family or individual. Public or public-use fixtures are installed in areas where the general public is admitted, such as schools or public restrooms in retail stores, and other installations where fixture use is generally unrestricted.

TYPE AND LENGTH OF PIPING

Water in street water mains is typically under a pressure of 45 psi–60 psi, which is adequate to serve a moderate-size plumbing installation. Water in a water supply pipe connected to a street water main is under the same pressure as the main while water within the pipe is at rest. However, when water is suddenly drawn from the pipe, such as when a fixture is used, a decided pressure drop results at the outlet. *Pressure loss due to friction* is the pressure variation resulting from friction within the pipe between the street water main and the water supply outlet where the water is being used.

Pressure loss due to friction occurs in all water supply piping due to the resistance resulting from water contacting the interior pipe surface and resistance between water molecules. Galvanized pipe, copper tube, and PVC, CPVC, and PEX plastic pipe and tubing are used to convey water to fixtures. Each material has a different resistance to water flow within the tube, pipe, and tubing. **See Figure 5-3.**

Flow resistance also occurs as water passes through valves, fittings, and changes in direction. Pressure loss due to friction increases as more pipe, fittings, valves, and other devices are installed in the water supply system. **See Figure 5-4.** As pressure loss within water supply piping increases, the discharge capacity of fixture supply pipes decreases. When designing a water supply piping system, pressure loss due to friction must be considered so that adequate water pressure is available at the outlet and proper plumbing fixture and appliance operation is ensured.

Length of piping is the only factor that can be affected by a plumber when installing a plumbing system. Larger pipe sizes are used to reduce the amount of friction within the pipe and fittings. In addition, ball valves are used to decrease the pressure loss due to friction since their design allows water to flow in a straight line through the valves.

HEIGHT OF BUILDING

Water pressure in a water supply system is decreased due to the height to which the water must flow. A column of water loses .434 psi of pressure for every 1′ of elevation or head. If the highest outlet of a water supply system is 50′ above the water main or other water source, the pressure loss due to head is 21.7 lb (.434 × 50). Pressure loss due to head is subtracted from the available pressure to determine whether there is adequate pressure to raise the water to the required height within the building.

PRESSURE LOSS DUE TO FRICTION IN TYPE M COPPER TUBE

	Pressure Loss per 100′ of Tube† Standard Type M Tube Size‡							
Flow Rate*	**3/8**	**1/2**	**3/4**	**1**	**1 1/4**	**1 1/2**	**2**	**2 1/2**
1	2.5	0.8	0.2					
2	8.5	2.8	0.5	0.2				
3	17.3	5.7	1.0	0.3	0.1			
4	28.6	9.4	1.8	0.5	0.2			
5	42.2	13.8	2.6	0.7	0.3	0.1		
10		**46.6**§	8.6	2.5	0.9	0.4	0.1	
15			17.6	5.0	1.9	0.9	0.2	
20			**29.1**§	8.4	3.2	1.4	0.4	0.1
25				12.3	4.7	2.1	0.6	0.2

Copper Development Association

PRESSURE LOSS DUE TO FRICTION IN PEX TUBING

	Pressure Loss per 100′ of Tubing†			
Flow Rate*	**3/8**	**1/2**	**5/8**	**3/4**
1	6.8	1.6	0.6	0.3
2	23.4	5.4	2.0	1.04
3			4.2	2.15
4				

* in gpm

† in psi

‡ in in.

§ flow velocity > 10 ft/sec

Figure 5-3. Each water supply piping material has a different resistance to water flow within the tube, pipe, and tubing.

FRICTION ALLOWANCE FOR FITTINGS AND VALVES*

Fitting Size†	Equivalent Tube Length‡								
	Standard Elbow		90° Tee		Valve				
	90°	45°	Side Branch	Straight Run	Coupling	Ball	Gate	Butterfly	Check
3/8	.5	—	1.5	—	—	—	—	—	1.5
1/2	1	.5	2	—	—	—	—	—	2
5/8	1.5	.5	2	—	—	—	—	—	2.5
3/4	2	.5	3	—	—	—	—	—	3
1	2.5	1	4.5	—	—	.5	—	—	4.5
1 1/4	3	1	5.5	.5	.5	.5	—	—	5.5
1 1/2	4	1.5	7	.5	.5	.5	—	—	6.5
2	5.5	2	9	.5	.5	.5	.5	7.5	9
2 1/2	7	2.5	12	.5	.5	—	1	10	11.5
3	9	3.5	15	1	1	—	1.5	15.5	14.5
3 1/2	9	3.5	14	1	1	—	2	—	12.5
4	12.5	5	21	1	1	—	2	16	18.5
5	16	6	27	1.5	1.5	—	3	11.5	23.5
6	19	7	34	2	2	—	3.5	13.5	26.5
8	29	11	50	3	3	—	5	12.5	39

* Allowances are for streamlined soldered fittings and recessed threaded fittings. Double the allowances for standard threaded fittings.
† in in.
‡ in ft

Figure 5-4. Flow resistance occurs as water passes through valves, fittings, and changes in direction of piping.

FLOW PRESSURE NEEDED AT TOP FLOOR

Every plumbing fixture requires a minimum flow pressure to function properly. *Flow pressure,* or working water pressure, is the water pressure in the water supply pipe near an outlet, such as a faucet, and is measured while the outlet is wide open and flowing. Flow pressure ranges from 8 psi for faucets and tank-type water closets to 25 psi for certain models of flushometer valves. **See Figure 5-5.**

Insufficient flow pressure results in inadequate amounts of water flowing through the pipe and being delivered to the fixture. Inadequate flow pressure prevents water closets from flushing properly and results in inadequate flow rates from faucets. Flow pressure is determined by subtracting the pressure loss due to friction and pressure loss due to head from the available pressure:

Flow Pressure = Available Pressure – (Pressure Loss due to Friction + Pressure Loss due to Head)

MINIMUM FLOW PRESSURES FOR PLUMBING FIXTURES

Type of Fixture	Flow Pressure*
Standard lavatory faucet	8
Self-closing basin faucet	8
Sink faucet, 3/8″	8
Sink faucet, 1/2″	8
Bathtub faucet	8
Laundry tray faucet, 1/2″	8
Shower	8
Ball cock for water closet	8
Flushometer valve for water closet	15–25
Flushometer valve for urinal	15
Drinking fountain	15
Sillcock or wall hydrant	10

* in psi

Figure 5-5. Flow pressure is the water pressure in the water supply pipe near an outlet and is measured while the outlet is wide open and flowing.

For example, a water supply must be provided to water closets with flushometer valves on the sixth floor of a commercial structure. Available water pressure is

50 psi. The water supply is piped using 2″ copper tube, which has a pressure loss due to friction of 2.5 psi per 200′ of tube. The flushometer valves are located 65′ above the water main; therefore, the pressure loss due to head is 28.2 psi (65′ × .432 = 28.2 psi). Determine whether adequate flow pressure is available for a flushometer valve requiring 40 psi.

Flow Pressure = *Available Pressure* – (*Pressure Loss due to Friction* + *Pressure Loss due to Head*)

where

Flow Pressure = flow pressure (in psi)

Available Pressure = available pressure (in psi)

Pressure Loss due to Friction = pressure loss due to friction (in psi)

Pressure Loss due to Head = pressure loss due to height (in psi)

Flow Pressure = 50 psi – (2.5 psi + 28.2 psi)

Flow Pressure = 50 – 30.7

Flow Pressure = 19.3 psi; not enough flow pressure available for the flushometer valve

SIZING WATER SUPPLY PIPING

Plumbing systems for residential structures are generally designed by plumbers. Similar to sizing drainage, waste, and vent piping, proper design of a water supply system for larger and more complex plumbing projects, such as public buildings, involves a larger number of details and calculations. Therefore, plumbing systems for large projects are typically designed by mechanical engineers.

> **Tech Facts**
>
> *Ground water and surface water constitute the sources of water used for residential, commercial, and industrial applications. Ground water is water that seeps or flows downward from the Earth's surface and saturates subsurface soil or rock. Surface water is water on the Earth's surface, such as water in streams, rivers, lakes, or reservoirs.*

As a general rule, and based on the available pressure, demand, type and length of piping, height of the building, and flow pressure needed at the top floor, the minimum sizes for water supply piping are:

- For ¾″ pipe:
  - minimum size of water service for any building from street main to water meter
  - minimum size of water supply pipe (first section of water distribution piping within building)
  - minimum size of fixture supply or fixture branch pipe to a sillcock
  - minimum size of fixture supply or fixture branch pipe to water heater
  - minimum size of hot water supply pipe on outlet side of water heater
- For ½″ pipe:
  - minimum size of concealed water supply piping
  - minimum size for three fixtures or less in the same bathroom of a house. **See Figure 5-6.**
  - minimum size of fixture branch pipe, except individual fixture branch pipe for specified fixtures and appliances. **See Figure 5-7.**
- For ⅜″ or ¼″ pipe:
  - fixture supply pipe, if it is not concealed within building partitions or is over 30″ long

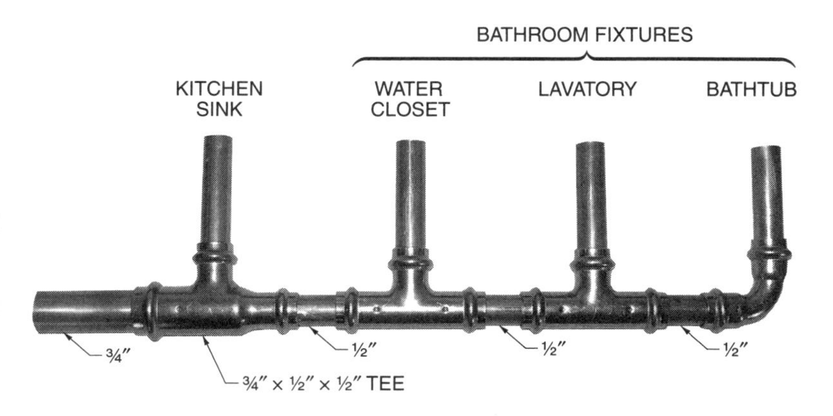

Figure 5-6. The fixture supply pipe for three fixtures or fewer in the same bathroom of a house is a minimum of ½″ size.

MINIMUM SIZES OF FIXTURE BRANCH PIPE	
Type of Fixture	Nominal Pipe Size*
Clothes washer (domestic)	½
Bathtub	½
Dishwasher (domestic)	½
Drinking fountain	½
Kitchen sink (domestic)	½
Kitchen sink (commercial)	¾
Lavatory	½
Laundry tray	½
Shower (single-head)	½
Sillcock or wall hydrant	¾
Sink (service)	½
Sink (flushing rim)	¾
Urinal (flushometer valve)	¾
Water closet (flush tank)	½
Water closet (flushometer valve)	1

* in in.

Figure 5-7. The minimum size of fixture branch pipe is ½″.

One-Story, One-Family Dwelling

Plumbers generally design plumbing systems for residential structures. **See Figure 5-8.** In a typical one-story, one-family dwelling, a laundry tray, clothes washer, and water heater may be located in the basement. A kitchen sink, water closet, lavatory, and bathtub are installed on the first floor. Two sillcocks are also installed—one on the left side of the house and one at the rear.

The water service for most one-family dwellings is ¾″ size. A ¾″ gate valve is installed on the water service as the pipe enters the building. A water meter of the appropriate size is also installed—usually a disk water meter in a residential structure since a relatively small water demand is required in a residence. A ¾″ ball valve is installed on the outlet side of the water meter. Valves are placed on each side of the water meter so that water flow can be shut off if the water meter must be removed for repairs.

Water distribution piping, such as mains, fixture branches, and fixture supply pipe, conveys water from the water service to the point of use. A main is the principal pipe artery for a water supply, and fixture branches extend from the main to fixture supply pipes. In a typical one-family dwelling, a ¾″ main is installed, with fixture branch and fixture supply pipe reduced to ½″ size. In the example one-family dwelling, a ¾″ main extends from the outlet-side ball valve to a ¾″ tee and provides cold water to the sillcock on the left side of the house. A ¾″ sillcock fixture branch pipe extends from the tee to the sillcock. A ¾″ ball valve is installed before the branch pipe passes through the dwelling wall so that water flow can be shut off if the sillcock must be repaired or replaced. In addition, the valve can be shut off during winter months in cold climates so that water within the sillcock body can be drained to avoid freezing.

The cold water main continues as ¾″ pipe from the sillcock fixture branch tee to another ¾″ tee supplying water to the water heater fixture branch pipe. The ¾″ water heater fixture branch pipe is fitted with a ¾″ ball valve before it connects to the water heater. The valve is shut off when the water heater is drained or if the water heater must be repaired or replaced.

The cold water main continues from the water heater fixture branch pipe as ¾″ pipe to a ¾″ tee supplying water to the fixture branch pipes for the first-floor and basement fixtures. The cold water main extends from the tee as ¾″ pipe to a ball valve and the sillcock at the rear of the house.

A short section of pipe extends from the ¾″ tee to a ½″ × ½″ × ¾″ tee that supplies cold water to first-floor and basement fixtures. The fixture branches and fixture supply pipes are sized as ½″ pipe for the water closet, lavatory, and bathtub in the bathroom. In addition, the fixture branches and fixture supply pipes are sized as ½″ pipe for the kitchen sink, and the laundry tray and clothes washer in the basement.

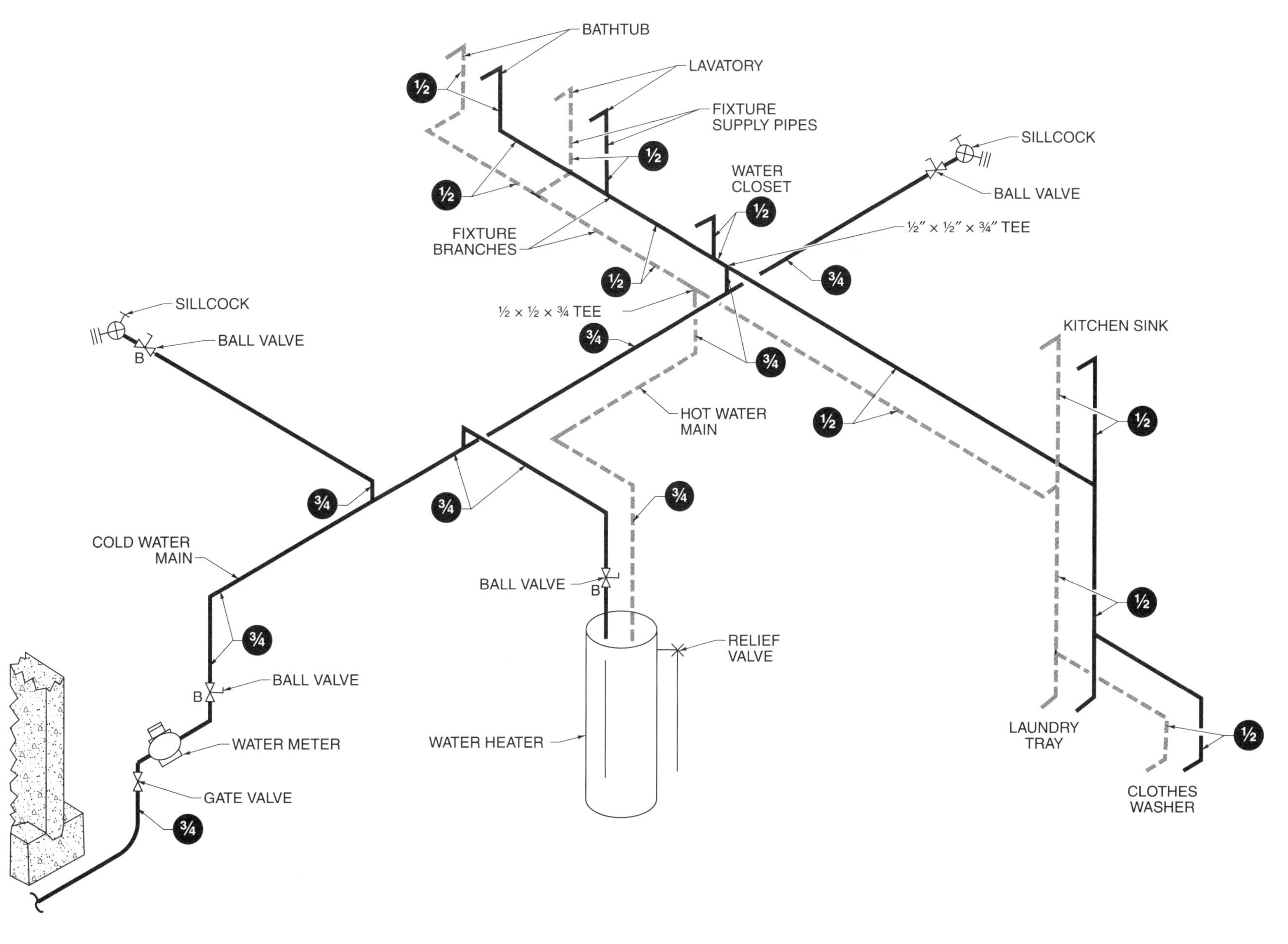

Figure 5-8. A typical one-story, one-family dwelling contains a laundry tray, clothes washer, and water heater in the basement.

Hot water pipe is sized as ¾″ pipe at the outlet of the water heater. The ¾″ hot water main continues toward the rear of the residence to convey hot water to the first-floor and basement fixtures. A ½″ × ½″ × ¾″ tee is installed to reduce pipe size to ½″ for the first-floor and basement fixture branches serving the lavatory, bathtub, kitchen sink, laundry tray, and clothes washer.

Sizing Water Supply Piping in Larger Installations

Water supply piping sizing for larger installations is based on the available water pressure, fixture demand, type and length of piping, height of the building, and flow pressure needed at the top floor. The *water supply fixture unit sizing method* is a common method of properly sizing water supply piping for buildings that require a 2″ or smaller water service and in which the distribution piping does not exceed 2½″ in size. Water supply piping sizing for installations requiring larger pipe sizes should be completed by a mechanical engineer. The procedure for sizing water supply piping is:

1. Determine the total demand on the water supply piping. When determining the total cold water supply fixture unit (cwsfu) demand, either step 1A or step 1B will be used.

   A. For buildings with water closets that are not fitted with flushometer valves, total the cwsfu demand for the entire building by adding the wsfu demand of individual fixtures. **See Figure 5-9.**

   B. For buildings with water closets fitted with flushometer valves, total the cwsfu demand for the entire building in wsfu from the Water Supply Fixture Units table and assign the following values to each flushometer valve beginning with the most remote flushometer valve on each branch:

   For the first flushometer valve........... 40 wsfu
   For the second flushometer valve..... 30 wsfu
   For the third flushometer valve.......... 20 wsfu
   For the fourth flushometer valve........ 15 wsfu
   For the fifth flushometer valve (and all additional flushometer valves).. 10 wsfu

   The flushometer valve supply pipe may never be smaller in size than the size of the valve inlet.

   C. Total the hot water supply fixture unit (hwsfu) demand for the building using the values from the Water Supply Fixture Units table and assign this value to the water heater.

2. Determine the developed length of piping from the water meter to the most remote outlet if the water pressure at the meter is known, or from the water main to the most remote outlet if the water pressure at the meter is unknown.
3. Determine pressure loss due to height by identifying the difference in head (elevation) between the water meter (or other water supply source) and highest water supply outlet and multiplying the difference by 0.5 psi.
4. Determine the working water pressure by subtracting the pressure loss due to height from the available water pressure.
5. Select the pressure range from the Water Supply Pipe and Water Meter Sizes for Water Supply Systems table that includes the pressure value determined in step 4. **See Figure 5-10.**

An angle supply valve provides a transition between a fixture branch and water closet fixture supply pipe.

WATER SUPPLY FIXTURE UNITS		
Type of Fixture	Fixture Units*	
	Private Use	Public Use
Clothes washer (each pair of faucets)	2	4
Bathtub with or without shower	2	4
Drinking fountain (each head)	—	1
Dishwasher	2	4
Lavatory	1	2
Laundry tray (each pair of faucets)	2	4
Shower (each head)	2	4
Sillcock	3	5
SINK:		
Standard kitchen	2	4
Bar	1	2
Washup (each set of faucets)	—	2
Washup (circular spray)	—	4
Service	—	4
URINAL:		
Stall	—	5
Wall-hung	—	5
Trough (per 6′ section)	—	3
Water closet (flush tank)	3	5
Water closet (flushometer valve)†	—	10
Water supply outlets for unused items are computed at their maximum demand, but in no case less than:		
3/8″	1	2
1/2″	2	4
3/4″	3	6
1″	6	10

* in wsfu
† Branches and mains serving water closets or similar flushometer valves may be sized from this table when the following values are assigned to each flushometer valve, beginning with the most remote valve on each branch.

For the first flushometer valve .... 40 fixture units
For the second flushometer valve .... 30 fixture units
For the third flushometer valve ....20 fixture units
For the fourth flushometer valve .... 15 fixture units
For the fifth flushometer valve (and all additional flushometer valves)...........10 fixture units

Figure 5-9. The cwsfu demand for an entire building is determined by adding the wsfu demand of individual fixtures if the building does not include water closets fitted with flushometer valves.

6. Select the Maximum Allowable Length column that equals or exceeds the total developed length determined in step 2. Once this column has been selected, the water meter, water service, and fixture branches within the building are sized from this column.
7. Identify the value in the Maximum Allowable Length column that equals or exceeds the cwsfu demand determined in step 1A or 1B, depending on whether the water closets are fitted with flushometer valves. Move to the left in that row to select the water meter and water service size.
8. Next, size the water distribution main from the adjacent Main and Branches column in the Water Supply and Water Meter Sizes for Water Supply Systems table. The cold water main is the largest size water supply pipe in the building.
9. Size each piece of cold water supply pipe starting with the most remote fixture from the water meter and working back toward the meter.

WATER SUPPLY AND WATER METER SIZES FOR WATER SUPPLY SYSTEMS*											
Water Meter and Water Service†	Main and Branches†	Maximum Allowable Length‡									
		40	60	80	100	150	200	250	300	400	500
Pressure Range—30 psi to 45 psi											
¾	½	6	5	4	4	3	2	—	—	—	—
¾	¾	18	16	14	12	9	6	—	—	—	—
¾	1	29	25	23	21	17	15	13	12	10	9
1	1	36	31	27	25	20	17	15	13	12	10
1	1¼	54	47	42	38	32	28	25	23	19	17
1½	1¼	90	68	57	48	38	32	28	25	21	19
1½	1½	151	124	105	91	70	57	49	45	36	31
2	1½	210	162	132	110	80	64	53	46	38	32
1½	2	220	205	190	176	155	138	127	120	105	96
2	2	372	329	292	265	217	185	164	147	124	107
2	2½	445	418	390	370	330	300	280	265	240	220
Pressure Range—46 psi to 60 psi											
¾	½	9	8	7	6	5	4	3	2	—	—
¾	¾	27	23	19	17	14	11	9	8	6	5
¾	1	44	40	36	33	28	23	21	19	17	14
1	1	60	47	41	36	30	25	23	20	18	15
1	1¼	102	87	76	67	52	44	39	36	30	27
1½	1¼	168	130	106	89	66	52	44	39	33	29
1½	1½	270	225	193	167	128	105	90	68	62	52
2	1½	360	290	242	204	150	117	98	84	67	55
1½	2	380	360	340	318	272	240	220	198	170	146
2	2	570	510	470	430	368	318	280	250	205	173
2	2½	680	640	610	580	535	500	470	440	400	365
Pressure Range—Over 60 psi											
¾	½	11	9	8	7	6	5	4	3	2	—
¾	¾	34	28	24	22	17	13	11	10	8	—
¾	1	63	53	47	42	35	30	27	24	21	18
1	1	87	66	55	48	38	32	29	26	22	19
1	1¼	140	126	108	96	74	62	53	47	39	34
1½	1¼	237	183	150	127	93	74	62	54	43	37
1½	1½	366	311	273	240	186	154	130	113	88	73
2	1½	490	395	333	275	220	170	142	122	98	82
1½	2	380§	380§	380§	380§	370	335	305	282	244	212
2	2	690§	670	610	560	478	420	375	340	288	245
2	2½	690§	690§	690§	690§	690§	650	610	570	510	460

* in wsfu
† in in.
‡ in ft
§ maximum allowable load on meter

Figure 5-10. Water meter, water service, main, and branch pipe sizes are based on the available pressure, length of water supply piping, and wsfu demand.

10. At the point where the cold water supply to the water heater is provided, size the cold water supply to the water heater and the hot water supply out of the water heater using the hwsfu demand calculated in step 1C.

11. At the point where the cold water supply to the water heater branches from the cold water main, add the hwsfu demand to the cwsfu demand and size the supply pipe accordingly. If the total wsfu demand requires the cold water main size to exceed the size of the building supply pipe, the size of the cold water main is not increased.
12. Size the hot water supply piping by using the same column of the Water Supply Pipe and Water Meter Sizes for Water Supply Systems table starting at the hot water supply outlet that is farthest from the water heater and size back toward the heater.

Sizing Water Supply Piping in a Four-Unit, Multifamily Dwelling. Multifamily dwellings, such as apartments, place greater demands on the water supply than a one-family residence. Each unit of a common multifamily dwelling might contain a kitchen sink, lavatory, water closet with flush tank, and bathtub. The water heater and a laundry room with one laundry tray and two clothes washers may be located in the basement of the building. For the application example, the conditions assumed are:

- water pressure at water meter = 60 psi
- developed length of the cold water piping from the water meter to the most remote cold water outlet = 60′
- head (elevation) difference between water meter and highest water supply outlet = 12′

The procedure for sizing the water supply piping for a common four-unit, multifamily dwelling is:

1. Determine the total demand on the water supply piping in cwsfu and hwsfu. **See Figure 5-11.** Step 1B is omitted since the building is not equipped with flushometer valves.
   A. Since water closets in the multifamily dwelling are not fitted with flushometer valves, the cwsfu demand for the entire building is determined by adding the wsfu demand of the individual fixtures. The cwsfu demand for the entire building is determined by referencing the Private Use column of the Water Supply Fixture Units table. Each apartment includes:

No.		cwsfu
1	Kitchen sink	2
1	Water closet	3
1	Lavatory	1
1	Bathtub	2
Total cwsfu demand for each apartment =		8

Total cwsfu demand for four apartments = *Total cwsfu for each apartment* × *No. of apartments*

Total cwsfu demand for four apartments = 8 cwsfu / apartment × 4 apartments

Total cwsfu demand for four apartments = 32 cwsfu

The cwsfu demand for common fixtures, such as sillcocks outside the building and fixtures in the laundry room, is calculated as:

No.		cwsfu
2	Sillcocks (3 cwsfu each)	6
1	Laundry tray (2 cwsfu)	2
2	Clothes washers (2 cwsfu each)	4
Total cwsfu demand for common fixtures =		12

WATER SUPPLY FIXTURE UNITS

Type of Fixture	Fixture Units* Private Use
Clothes washer (each pair of faucets)	2
Bathtub with or without shower	2
Drinking fountain (each head)	—
Dishwasher	2
Lavatory	1
Laundry tray (each pair of faucets)	2
Shower (each head)	2
Sillcock	3
SINK: Standard kitchen	2
Bar	1
Washup (each set of faucets)	—
Washup (circular spray)	—
Service	—
URINAL: Stall	—
Wall-hung	—
Trough (per 6′ section)	—
Water closet (flush tank)	3

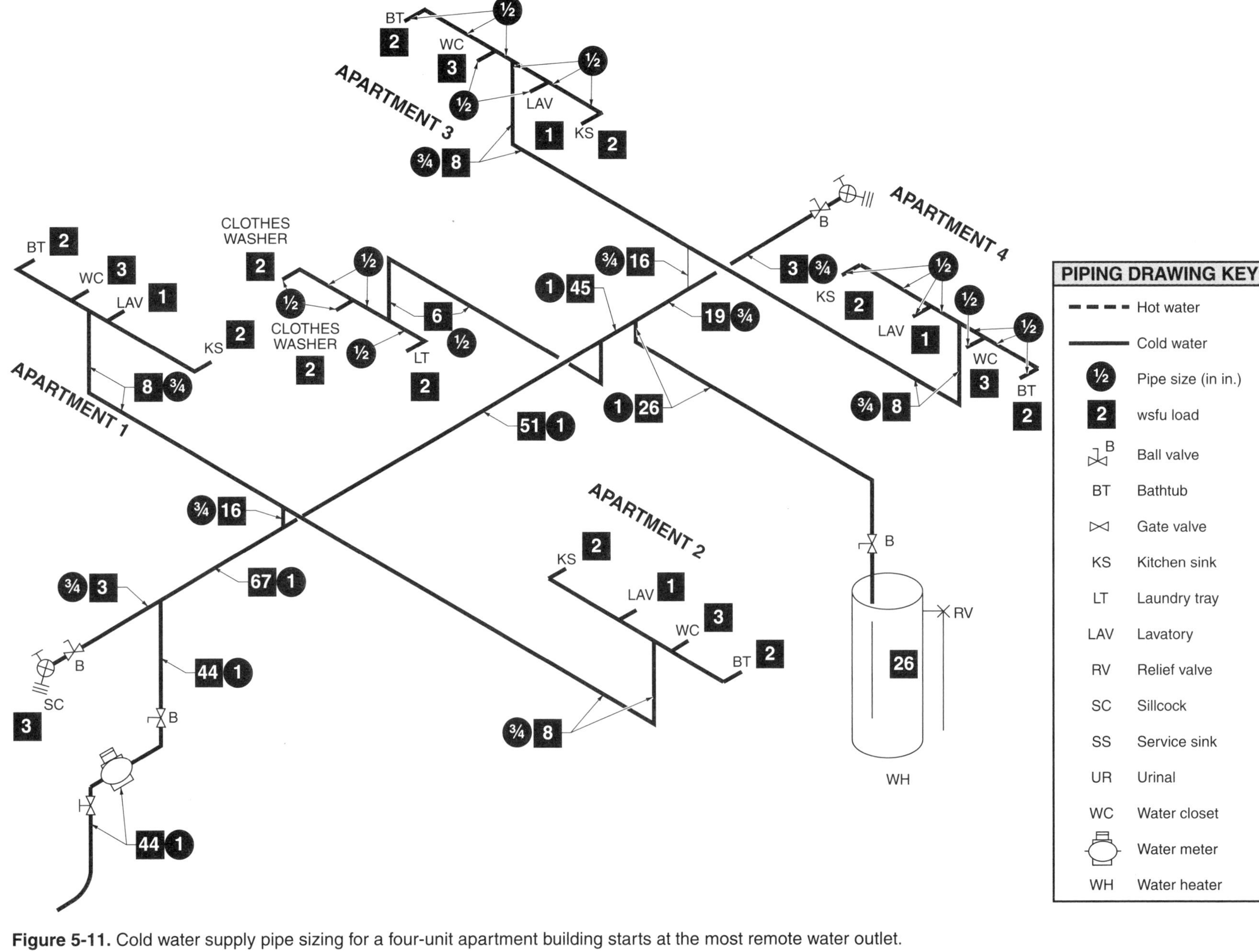

Figure 5-11. Cold water supply pipe sizing for a four-unit apartment building starts at the most remote water outlet.

The cwsfu demand for the entire building is calculated as:

Total cwsfu demand for four apartments	= 32
Total cwsfu demand for common fixtures	= 12
Total cwsfu demand for entire building	= 44

C. The hwsfu demand for the entire building is determined by referencing the Private Use column of the Water Supply Fixture Units table. Each apartment includes:

No.		hwsfu
1	Kitchen sink	2
1	Lavatory	1
1	Bathtub	2
	Total hwsfu demand each apartment	= 5

Total hwsfu demand for four apartments = Total hwsfu demand for each apartment × No. of apartments

Total hwsfu demand for four apartments = 5 hwsfu / apartment × 4 apartments

Total hwsfu demand for four apartments = 20 hwsfu

The hwsfu demand for common fixtures, such as the laundry tray and clothes washer, is calculated as:

No.		hwsfu
1	Laundry tray (2 hwsfu)	2
2	Clothes washers (2 hwsfu each)	4
	Total hwsfu demand for common fixtures	= 6
	Total hwsfu demand for four apartments	= 20
	Total hwsfu demand for entire building	= 26

This demand of 26 hwsfu is assigned to the water heater.

2. Developed length of the application example = 60′.
3. Determine pressure loss due to height.

   Pressure Loss due to Height = Head × .5

   where

   Pressure Loss due to Height = pressure (in psi)

   Head = head (in ft)

   .5 = constant

   Pressure Loss due to Height = 12′ × .5

   Pressure Loss due to Height = 6 psi
4. Determine the working water pressure by subtracting the pressure loss due to height from the available water pressure.

   Working Water Pressure = Available Water Pressure – Pressure Loss due to Height

   where

   Working Water Pressure = pressure range (in psi)

   Available Water Pressure = available water pressure (in psi)

   Pressure Loss due to Height = pressure loss (in psi)

   Working Water Pressure = 60 psi – 6 psi

   Working Water Pressure = 54 psi
5. A building with 54 psi working water pressure is sized from the 46 psi–60 psi pressure range of the Water Supply and Water Meter Sizes for Water Supply Systems table.
6. The water supply pipe for the entire building is sized from the 60′ column of the Water Supply and Water Meter Sizes for Water Supply Systems table.
7. The 44 cwsfu value calculated in step 1A falls between the 40 and 47 wsfu values in the Water Supply and Water Meter Sizes for Water Supply Systems table, and is rounded up the next highest value. The water meter and water service size of 1″ pipe is determined by moving left to the appropriate column.

Water Meter and Water Service†	Main and Branches†	Maxim	
		40°	60°
Pressure Range—46 psi to 60 psi			
¾	½	9	8
¾	¾	27	23
¾	1	44	40
1	1	60	47
1	1¼	102	87
1½	1¼	168	130
1½	1½	270	225
2	1½	360	290
1½	2	380	360
2	2	570	510
2	2½	680	640

8. The water supply main size is determined to be 1″ from the adjacent Main and Branches column in the Water Supply Pipe and Water Meter Sizes for Water Supply Systems table.
9. Cold water supply pipe is sized beginning with the most remote fixture from the water meter and working back toward the meter. In the application example, the fixtures in apartment 3 are farthest from the water meter.

Although there is only 8 cwsfu demand on the cold water fixture branch pipe conveying water to apartment 3, the pipe is sized as ¾″ because it serves four fixtures. However, ½″ fixture supply pipe is installed for the individual fixtures in apartment 3. Cold water supply pipe sizing for the other apartments is identical to apartment 3.

Cold water main size does not increase where fixture branch pipe from apartment 3 joins the fixture branch pipe from apartment 4 since a ¾″ pipe serves a 23 cwsfu demand. The same sizing also applies to the point where the fixture branch pipes join for apartments 1 and 2.

Since a ¾″ pipe serves 23 cwsfu, the cold water main continues as ¾″ pipe to the sillcock at the rear of the building. The sillcock must be supplied by a ¾″ branch pipe, as sized from the Minimum Sizes of Fixture Water Branch Pipe table. A ½″ pipe will properly serve the 6 cwsfu demand of the laundry tray and two clothes washers located in the laundry room.

10. The water heater demand is 26 hwsfu and requires 1″ cold water supply pipe and a 1″ hot water supply out of the water heater.
11. At the laundry room fixture branch, the cold water supply fixture unit demand increases to 51 cwsfu. The Water Supply Pipe and Water Meter Sizes for Water Supply Systems table indicates that the cold water main should increase to 1¼″ pipe at this point. Even though the total wsfu demand requires the cold water main size to exceed the size of the building supply pipe, the cold water main size is not increased in size. Therefore, the cold water main extending back to the water meter is sized as 1″ pipe.
12. The hot water supply piping is sized using the same column of the Water Supply Pipe and Water Meter Sizes for Water Supply Systems table as was used to size the cold water supply piping. The sizing of the hot water supply piping starts at the hot water supply outlet that is farthest from the water heater—apartment 1, in the application example—and is sized back toward the heater. **See Figure 5-12.** The three fixtures in this apartment that require a hot water supply—kitchen sink, lavatory, and bathtub—are properly served by ½″ fixture branch and fixture supply pipes. The hot water piping for the other three apartments is sized similarly using ½″ pipe.

At the point where the hot water supply to apartments 1 and 2 joins the hot water main, a ½″ × ½″ × ¾″ tee is used to transition from ½″ to ¾″ pipe. A ¾″ hot water main is large enough to supply the combined 16 hwsfu demand of apartments 1 and 2 and the laundry room. The 6 hwsfu demand of the laundry room fixtures is properly served by ½″ size fixture branch and fixture supply pipes.

Fixture branch pipe size is increased to 1′ at the point where the hot water main for apartments 1 and 2 and the laundry room joins the supply to apartments 3 and 4. A 1″ fixture branch pipe extends back to the outlet of the water heater.

Sizing Water Supply Pipe in a Public Building. Public buildings place a large demand on the water supply and must be sized properly to ensure adequate water pressure for fixtures at remote locations. Each rest room of a public building may contain a battery of similar fixtures, such as flushometer valve water closets, as well as fixtures for custodians and employee lunch areas. For the application example, the conditions assumed are:

- Available pressure at street water main = 50 psi.

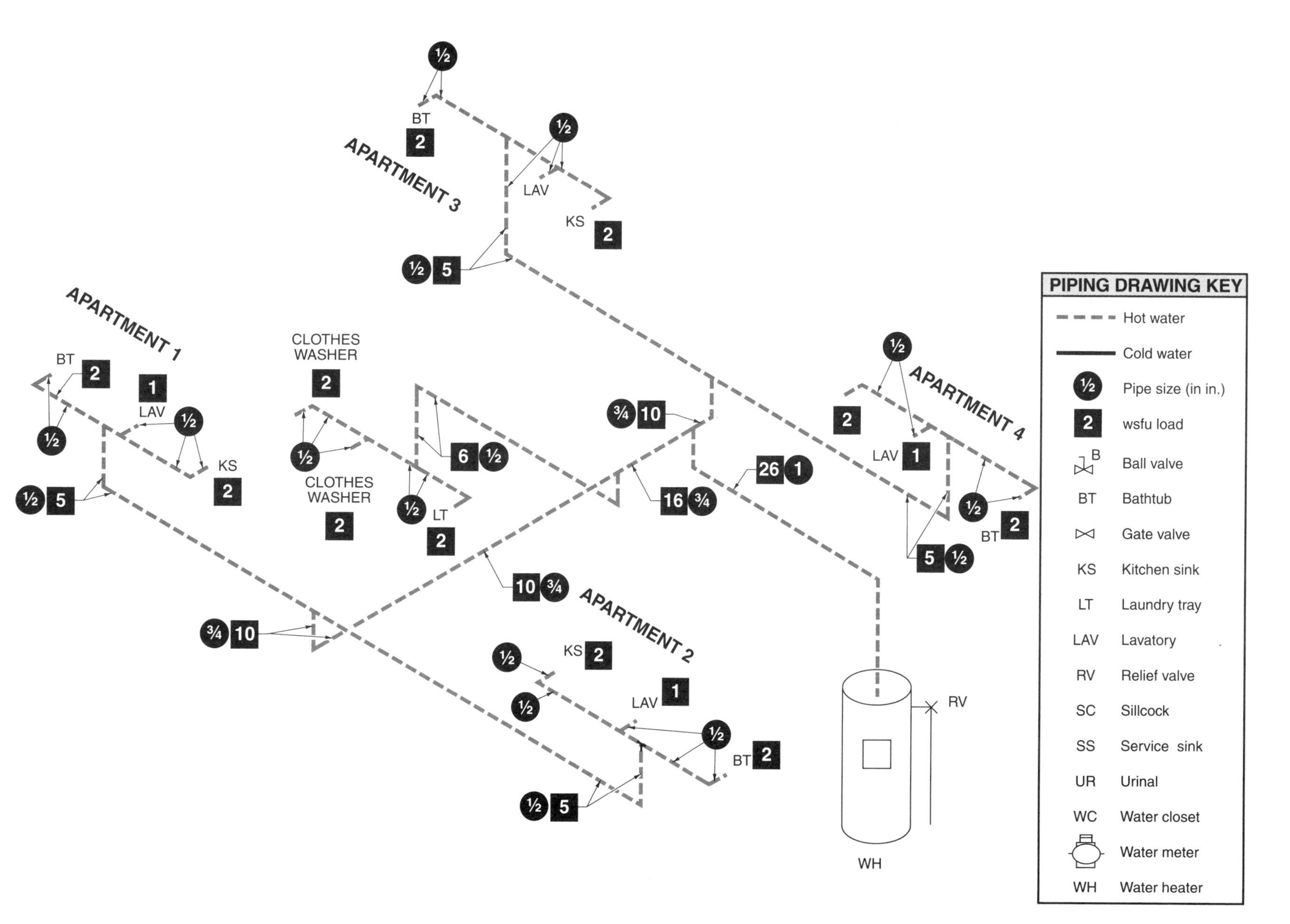

Figure 5-12. Hot water supply pipe sizing for a four-unit apartment building starts at the hot water supply outlet that is farthest from the water heater.

- Developed length of the cold water supply piping from street water main to most remote cold water outlet = 200′.
- Head (elevation) difference is insignificant because the application example is a one-story building.
- All water closets are installed with flushometer valves.
- wsfu demand of stall-type urinals in the men's rest room = 5 wsfu each.

The procedure for sizing the water supply piping for a public building is:

1. Determine the total demand on the water supply piping in cwsfu and hwsfu. **See Figure 5-13.**

   B. Since water closets in the public building are fitted with flushometer valves, the cwsfu demand for the entire building is determined by adding the wfsu demand of the individual fixtures and assigning the appropriate values to the flushometer valves. The cwsfu demand for the entire building is determined by referencing the Public Use column of the Water Supply Fixture Units table and totaling the cwsfu demand.

Women's rest room:

No.		cwsfu
1	First water closet	40
1	Second water closet	30
1	Third water closet	20
1	Fourth water closet	15
1	Fifth water closet	10
1	Sixth water closet	10
4	Lavatories (2 cwsfu each)	8
	Total cwsfu demand for women's rest room	= 133

Men's rest room:

No.		cwsfu
1	First water closet	40
1	Second water closet	30
1	Third water closet	20
1	Fourth water closet	15
2	Urinals (5 cwsfu each)	10
2	Lavatories (2 cwsfu each)	4
	Total cwsfu demand for men's rest room	= 119

Unisex rest room:

No.		cwsfu
1	First water closet	40
1	Lavatory	2
	Total cwsfu demand for unisex rest room	= 42

The cwsfu demand for common fixtures, such as a service sink, kitchen sink, and sillcock, is calculated as:

WATER SUPPLY FIXTURE UNITS

Type of Fixture	Fixture Units*	
	Private Use	Public Use
Dishwasher	2	4
Lavatory	1	2
Sillcock	3	5
SINK:		
Standard kitchen	2	4
Bar	1	2
Washup (each set of faucets)	—	2
Washup (circular spray)	—	4
Service	—	4
URINAL:		
Stall	—	5

* in wsfu
† Branches and mains serving water closets or similar flushometer valves may be sized from this table when the following values are assigned to each flushometer valve, beginning with the most remote valve on each branch.
For the first flushometer valve .... 40 fixture units
For the second flushometer valve .... 30 fixture units
For the third flushometer valve .... 20 fixture units
For the fourth flushometer valve .... 15 fixture units
For the fifth flushometer valve (and all additional flushometer valves) .... 10 fixture units

No.		cwsfu
1	Service sink	4
1	Kitchen sink	4
1	Sillcock	5
Total cwsfu demand for common fixtures		= 13

The cwsfu demand for the entire building is calculated as:

Total cwsfu demand for women's rest room	= 133
Total cwsfu demand for men's rest room	= 119
Total cwsfu demand for unisex rest room	= 42
Total cwsfu demand for common fixtures	= 13
Total cwsfu demand for entire building	= 307

C. The hwsfu demand for the entire building is determined by referencing the Public Use column of the Water Supply Fixture Unit table.

No.		hwsfu
7	Lavatories (2 hwsfu each)	14
1	Kitchen sink	4
1	Service sink	4
	Total hwsfu demand =	22

WATER SUPPLY AND WATER METER SIZES FOR WATER SUPPLY SYSTEMS*

Water Meter and Water Service†	Main and Branches†	Max 40°	200	250
Pressure Range—46 psi to 60 psi				
¾	½	9	4	3
¾	¾	27	11	9
¾	1	44	23	21
1	1	60	25	23
1	1¼	102	44	39
1½	1¼	168	52	44
1½	1½	270	105	90
2	1½	360	117	98
1½	2	380	240	220
2	2	570	318	280
2	2½	680	500	470

* in wsfu
† in in.

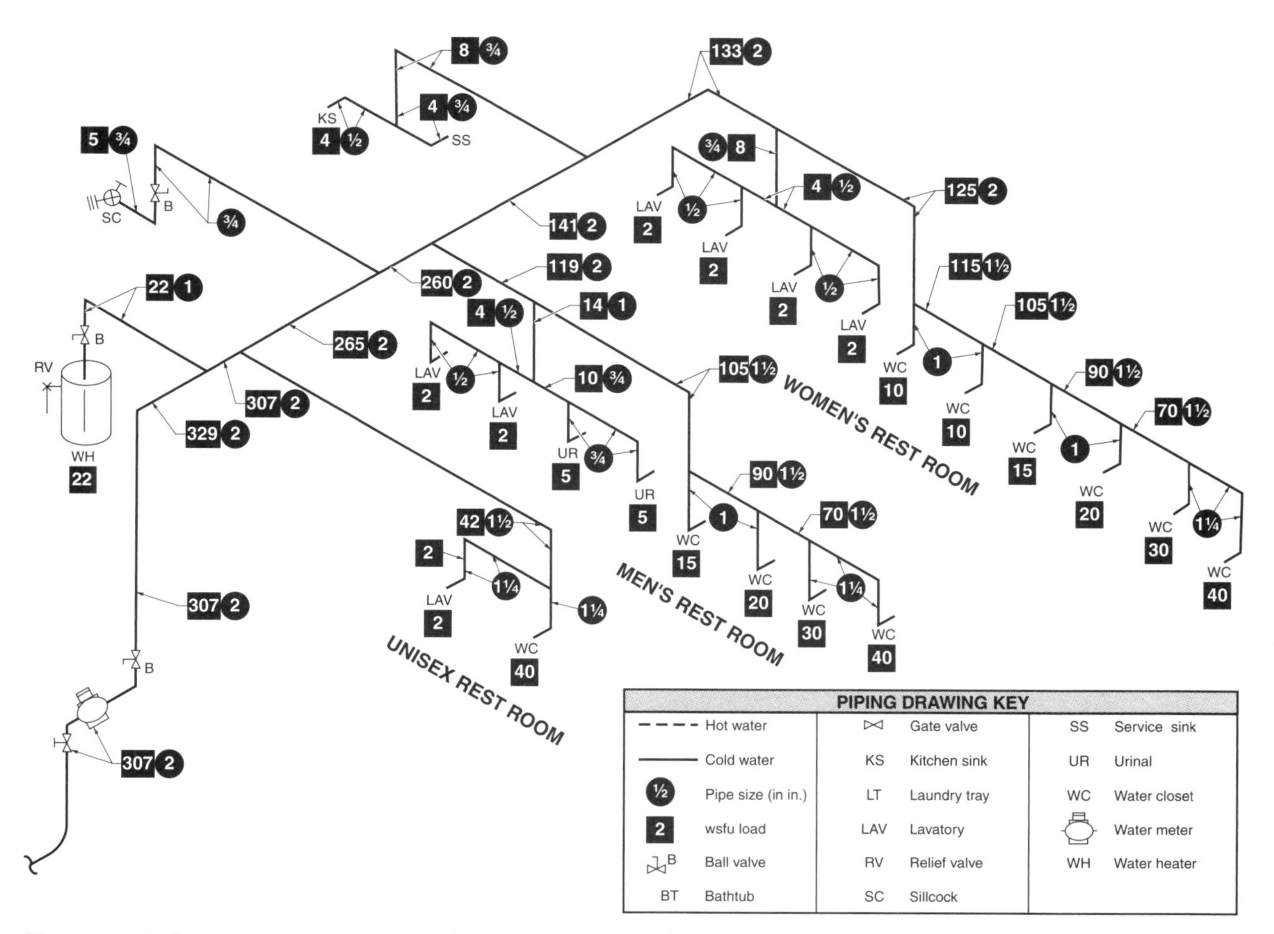

Figure 5-13. Cold water supply pipe sizing for a public building starts with determining the total demand on the water supply piping.

2. The developed length of the application example = 200′.
3. Pressure loss due to height is negligible in the application example since the public building is a one-story structure and the elevation difference is insignificant.
4. Determine the working water pressure by subtracting the pressure loss due to height from the available water pressure.

   Working Water Pressure = *Available Water Pressure* – *Pressure Loss due to Height*

   where

   Working Water Pressure = pressure range (in psi)

   Available Water Pressure = available water pressure (in psi)

   Pressure Loss due to Height = pressure loss (in psi)

   Working Water Pressure = 50 psi – 0 psi

   Working Water Pressure = 50 psi
5. A building with 50 psi of working water pressure is sized from the 46 psi–60 psi pressure range of the Water Supply and Water Meter Sizes for Water Supply Systems table.
6. Since the developed length for the application example is 200′, the water supply pipe for the entire building is sized from the 200′ column of the Water Supply and Water Meter Sizes for Water Supply Systems table.
7. The water meter and water service pipe are 2″.
8. The water distribution main size is also determined to be 2″.
9. Cold water pipe is sized beginning with the fixture farthest from the water meter, which is the water closet on the far right in the women's rest room. The water closet, rated at 40 cwsfu demand, must be supplied with 1¼″ fixture branch and fixture supply pipe. The second water closet in the group, rated at 30 cwsfu demand, is also supplied with 1¼″ fixture supply pipe. Although the fixture branch pipe serving these two water closets is 1¼″ size, the fixture supply pipe to the flushometer valves would be sized as 1″ pipe. A 1¼″ fixture branch pipe supplying water to the two remote water closets serves a 70 cwsfu demand.

The next three water closets in the women's rest room, rated at 20, 15, and 10 cwsfu, are supplied with 1″ fixture supply pipes. The final water closet in the women's rest room, rated at 10 cwsfu, is supplied with a 1″ fixture supply pipe even though the sizing table indicates a ¾″ pipe can supply this demand. This pipe must be sized as 1″ since the supply pipe for a flushometer valve may never be smaller in size than the size of the valve inlet. The Minimum Sizes of Fixture Branch Pipe table indicates that a water closet with a flushometer valve has a 1″ pipe inlet.

A 1½″ fixture branch pipe supplies the third, fourth, and fifth water closets in the women's rest room. At the point where the final water closet fixture branch pipe joins the cold water main, the demand increases to 125 cwsfu and 2″ pipe is used. A 2″ cold water main is installed back to the water meter.

The fixture supply pipe for the women's rest room lavatories, rated at 2 cwsfu each, is ½″ size. The 8 cwsfu demand of all lavatories is supplied with a ¾″ fixture branch pipe that extends from the 2″ cold water main.

The 8 cwsfu demand of the kitchen sink and service sink is supplied with a ¾″ fixture branch pipe. The fixture branch pipe reduces to ½″ size for the fixture supply pipes.

The water supply pipe sizing in the men's rest room starts at the most remote outlet, which is the water closet rated at 40 cwsfu. The water closet is supplied with 1¼″ fixture branch and fixture supply pipes. The second water closet, rated at 30 cwsfu, is supplied with the 1¼″ fixture supply pipe. However, the fixture supply pipe to the flushometer valves would be 1″ pipe. The fixture branch pipe supplying the first two water closets, which serves 70 cwsfu demand, is 1½″ pipe. The third and fourth water closets in the men's rest room, rated at 20 and 15 cwsfu, respectively, are supplied with 1″ fixture supply pipes and 1½″ fixture branch pipe. The fixture branch pipe supplying water to the water closets is sized as 1½″ pipe since the total demand is 105 cwsfu, which is the maximum demand that can be supplied with a 1½″ pipe.

The men's rest room urinals are each supplied with ¾″ fixture supply and fixture branch pipes. The fixture branch pipe serves 10 cwsfu. The lavatories in the men's rest room with a combined demand of 4 cwsfu are supplied with ½″ fixture branch and fixture supply pipes. At the point where the ½″ lavatory fixture branch joins the ¾″ urinal fixture branch pipe, the demand increases to 14 cwsfu and the pipe increases to 1″ size. At the point where the 1″ fixture branch pipe serving the lavatories and urinals joins the pipe serving the four water closets in the men's rest room, the demand increases from 105 cwsfu to 119 cwsfu, and the cold water supply increases to 2″ size.

The sillcock is served with a ¾″ fixture branch pipe.

Water supply sizing in the unisex rest room begins at the most remote water supply outlet, which is the water closet. The water closet, rated at 40 cwsfu, requires a 1¼″ fixture supply pipe. The lavatory in the unisex rest room, rated at 2 cwsfu, is served with a ½″ fixture supply pipe that extends from 1¼″ fixture supply pipe. A 1¼″ fixture branch pipe is adequate to serve the water closet and lavatory, rated at 42 cwsfu.

10. The water heater must serve a demand of 22 hwsfu (as calculated in step 1C), and is supplied with a 1″ fixture supply pipe.
11. At the point where the water heater fixture supply pipe joins the cold water main, the demand on the main increases to 329 cwsfu. However, it is not necessary to increase the cold water main size above the 2″ size originally selected, even though a larger size is indicated.
12. The hot water supply pipe sizing begins at the fixtures farthest from the water heater, which are the women's rest room lavatories. **See Figure 5-14.** The lavatories are served by ½″ fixture supply and fixture branch pipes. The fixture branch pipe size is increased to ¾″ with a ½″ × ½″× ¾″ tee at the point where the hot water supplies for the lavatories join together, rated at 8 hwsfu demand.

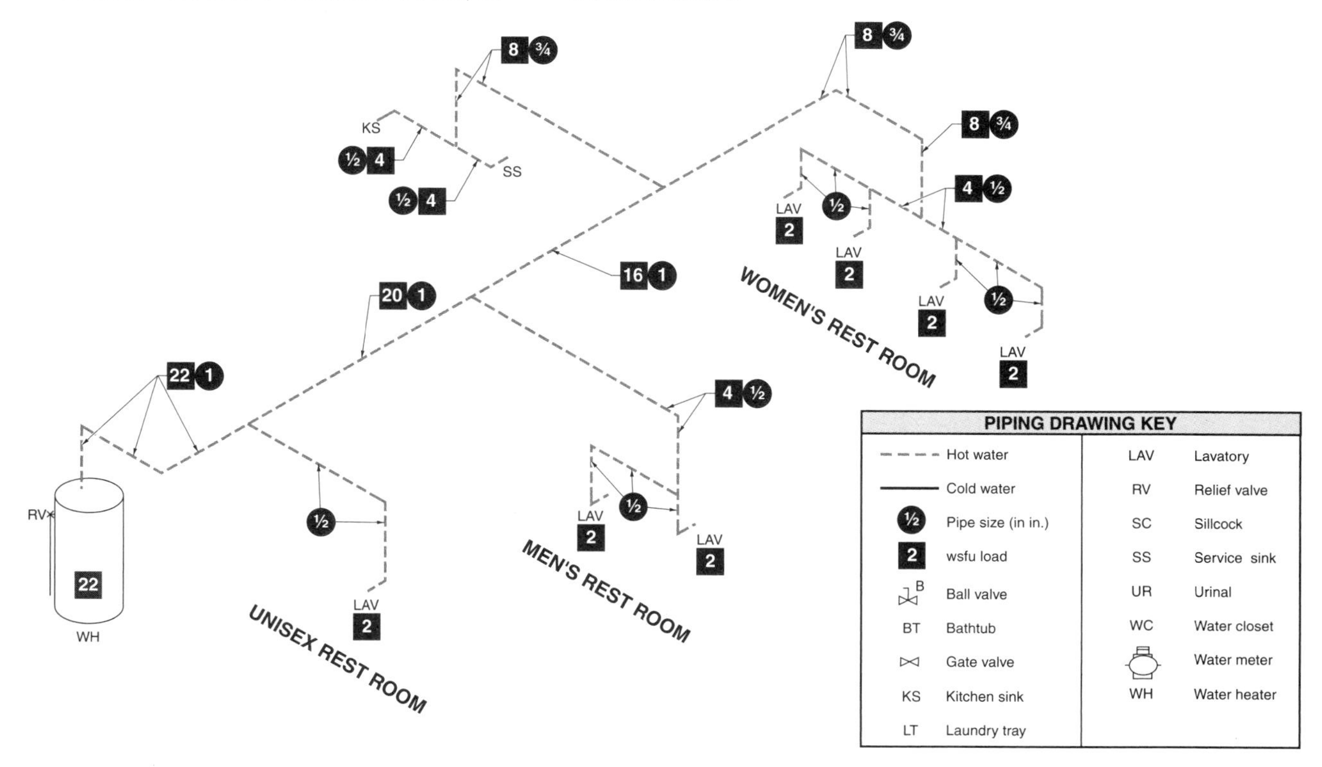

Figure 5-14. The hot water supply pipe sizing for a public building begins at the fixture farthest from the water heater.

The kitchen and service sinks, rated at 4 hwsfu each, are sized for ½″ fixture supply pipe. The combined total of 8 hwsfu requires a ¾″ fixture branch pipe. At the point where the hot water fixture branch pipe for the women's rest room and the fixture branch pipe serving the kitchen and service sinks join, the demand increases to 16 hwsfu and the hot water main increases to 1″ size. The 1″ hot water main continues back to the water heater. The men's rest room lavatories are supplied with ½″ fixture branch and fixture supply pipes that extend from the 1″ hot water main. The lavatory in the unisex rest room is also supplied with a ½″ fixture branch pipe.

Review Questions

1. Define "flow rate" as it applies to a building water supply.
2. How is the total demand determined for a building water supply system?
3. Define "available pressure."
4. Why is the water supply fixture unit system used to determine pipe sizes in buildings?
5. How does pressure loss due to friction affect the sizing of pipe in a building water supply system?
6. Where do you start when sizing a water supply for a building? Why?
7. Sketch the water supply piping of your home or apartment building and size it.
8. Sketch the water supply piping in the men's and women's rest rooms of a public building and size it.

Plumbing fixtures and appliances are connected to the water supply piping so that water can be distributed to all of the fixtures throughout a building. Also, drainage piping is connected to all fixtures in order to remove waterborne waste from the building.

6 Plumbing Fixtures and Appliances

The majority of the piping installed by plumbers serves one of two uses: it either supplies water to a plumbing fixture or appliance, or it drains the waterborne waste from a fixture or appliance after the fixture has been used.

A *plumbing fixture* is a receptacle or device that is connected permanently or temporarily to the water distribution system, demands a supply of potable water, and discharges the waste directly or indirectly into the sanitary drainage system. Common fixtures include lavatories, water closets, and bathtubs.

An *appliance* is a plumbing fixture that performs a special function and is controlled and/or energized by motors, heating elements, or pressure- or temperature-sensing elements. Common appliances include water heaters and water softeners. Control of potable water and wastewater flow is accomplished through the use of fixture trim.

Fixture trim is the water supply and drainage fittings installed on a fixture or appliance to control water flow into a fixture and wastewater flow from the fixture to the sanitary drainage system.

Plumbing fixtures are manufactured from durable, corrosion-resistant, and nonabsorbent materials, and have smooth surfaces that can be easily cleaned. Materials used to manufacture plumbing fixtures are vitreous china, marble, enameled cast iron, enameled steel, stainless steel, fiberglass, and plastic resins. To stand up to the hard use they will receive, plumbing fixtures and appliances must be functional, well built, and attractively designed. Plumbing fixtures and appliances are available in a variety of styles and colors to coordinate with the room color scheme.

In 1992, the Americans with Disabilities Act (ADA) was enacted to provide greater accessibility to public and commercial buildings for people with physical disabilities. The ADA applies to all new public and commercial buildings. While the ADA does not specifically apply to residential construction, a variety of ADA-compliant residential fixtures and appliances are available to accommodate people with physical disabilities. ADA-compliant fixtures and appliances provide users with additional maneuvering room and special features such as grab bars. ADA-compliant fixtures and appliances must be positioned at the proper height, and may require specialized fittings or trim to operate properly.

WATER CLOSETS

A *water closet,* or toilet, is a water-flushed plumbing fixture that receives human liquid and solid waste in a water-containing receptacle and, upon flushing, conveys the waste to a soil pipe. Most water closets are manufactured from vitreous china although some institutional water closets are made from stainless steel. *Vitreous china* is a ceramic compound fired at high temperature to form a nonporous material with a ceramic glaze fused to the surface. Vitreous china does not absorb odors and is easily cleaned with a soft cloth. Water closets are rated at 6 dfu, and require a minimum of 3″ waste and 2″ vent pipes. Water closets equipped with a flush tank require a ½″ water supply, while water closets equipped with flushometer valves require a 1″ water supply. Water closets are floor-set or wall-hung. **See Figure 6-1.** A *floor-set water closet* is a water closet installed directly on the floor, and is common in residential construction. A *wall-hung water closet* is a water closet suspended from the wall on a supporting chair carrier, and is used in commercial construction.

Water closet bowls are elongated or round front (also called plain bowl). See **Figure 6-2.** Elongated bowls measure approximately 14″ wide by 18½″ long, measured from the seat hinge holes to the front outside rim edge. Round front water closet bowls measure approximately 14″ wide by 16½″ long measured from the seat hinge holes to the front outside rim edge. Round front water closet bowls are installed in residential construction, while elongated water closet bowls are installed in commercial and public buildings.

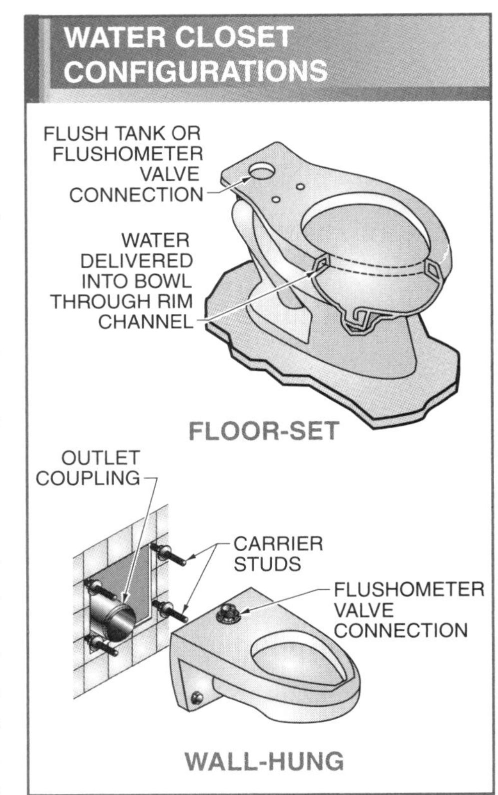

Figure 6-1. Water closets are floor-set or wall-hung.

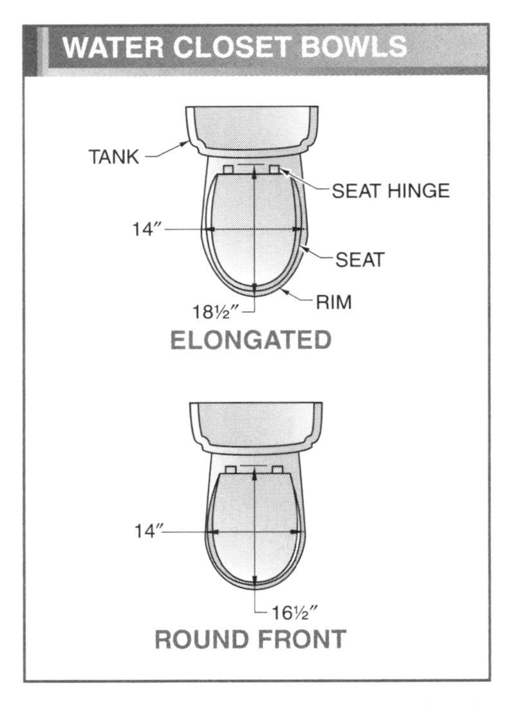

Figure 6-2. Elongated water closet bowls measure approximately 14″ × 18½″ and round bowls measure approximately 14″ × 16½″.

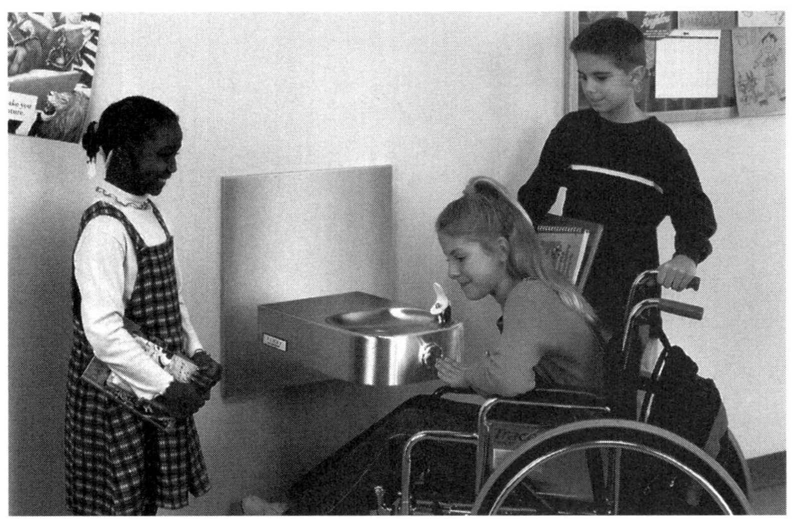

Elkay Mfg. Co.

An ADA-compliant water cooler is designed and installed to provide users with physical disabilities room to maneuver.

Water Closet Bowl Operating Principles

Water closet quality is based on the efficiency with which it flushes, or eliminates waste from the bowl. The siphonic flushing action of a water closet is produced by a decrease in pressure on the outlet side of the trap caused by water passing through the outlet passageway when the water closet is flushed.

Depending on the style of water closet, the water spot, trap seal, and passageway configuration vary. **See Figure 6-3.** The *water spot,* or water surface, is the surface area of the water in the water closet bowl when a flush is completed, and is usually measured in inches of width by length. The *trap seal* is the height of water in a water closet bowl, which prevents sewer gases from entering the living area. Trap seal is measured while the water is at rest, and is measured from the top of the dam or water spot to the inlet of the passageway. The *passageway* is the channel that connects a water closet bowl to the outlet. Since the passageway configuration is not circular, the passageway diameter is measured by the largest diameter ball it will pass, and is generally ⅛″ larger than the maximum size ball that can pass through it.

The passageway outlet diameter of a water closet trap is smaller than the passageway inlet diameter. The trap is constructed of short offsets and turns to slow the flow of water through it and build a head of water in the water closet bowl. When a water closet is not being used, equal pressure is present on each side of the trap. When a water closet is flushed, water passing through the passageway eliminates air in the passageway outlet, producing a partial vacuum. Atmospheric pressure and the head of water on the inlet side of the trap force waste from the fixture.

A variety of water closet designs are available, including siphon jet, gravity-fed, and blowout. A *siphon jet water closet* is a water closet with a siphonic passageway at the rear of the bowl and an integral flush rim and jet. A gravity-fed water closet, or rim jet water closet, is a water closet that uses the gravity from water falling from the water closet tank to begin the flushing action. A *blowout water closet* is a water closet with a nonsiphonic passageway at the rear of the bowl and an integral flush rim and jet. Blowout water closets are used in commercial construction with flushometer valves. The water spot and passageway diameter of siphon jet, gravity-fed, and blowout water closets vary with the design and manufacturer.

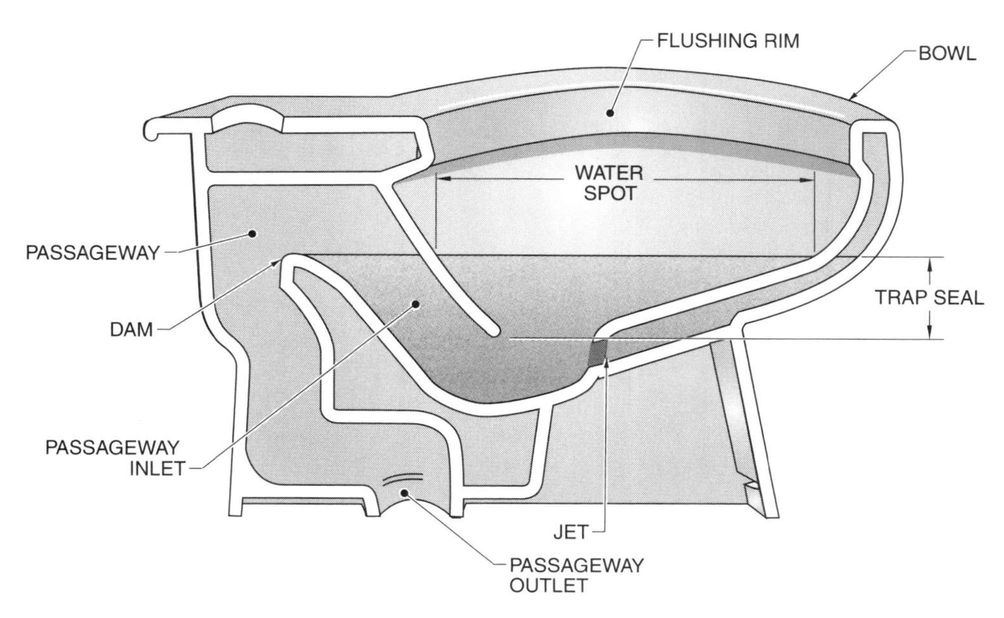

Figure 6-3. The water spot, trap seal, and passageway configuration of a water closet varies depending on the style of the fixture.

Siphon Jet Water Closets. The flushing action of wall-hung and floor-set siphon jet water closets provides the siphonage necessary to properly remove waste from the water closet bowl. A *jet* is an orifice (opening) at the base of the bowl that directs water into the passageway inlet to help create the siphonic flushing action. **See Figure 6-4.** The flushing action of a siphon jet water closet consists of these steps:

1. Water enters the water closet bowl through the rim holes and jet. The water level rises in the bowl and begins to flow over the dam.
2. Water continues to enter the passageway and flows steadily over the dam. Water flows to the rear wall of the passageway, and is then redirected to the opposite wall, creating a water curtain across the passageway and preventing air from entering the passageway.
3. The passageway water level begins to rise. As the water level in the passageway outlet rises, air from the passageway mixes with water exiting the closet.
4. The passageway fills and a siphon action is created, flushing the bowl.

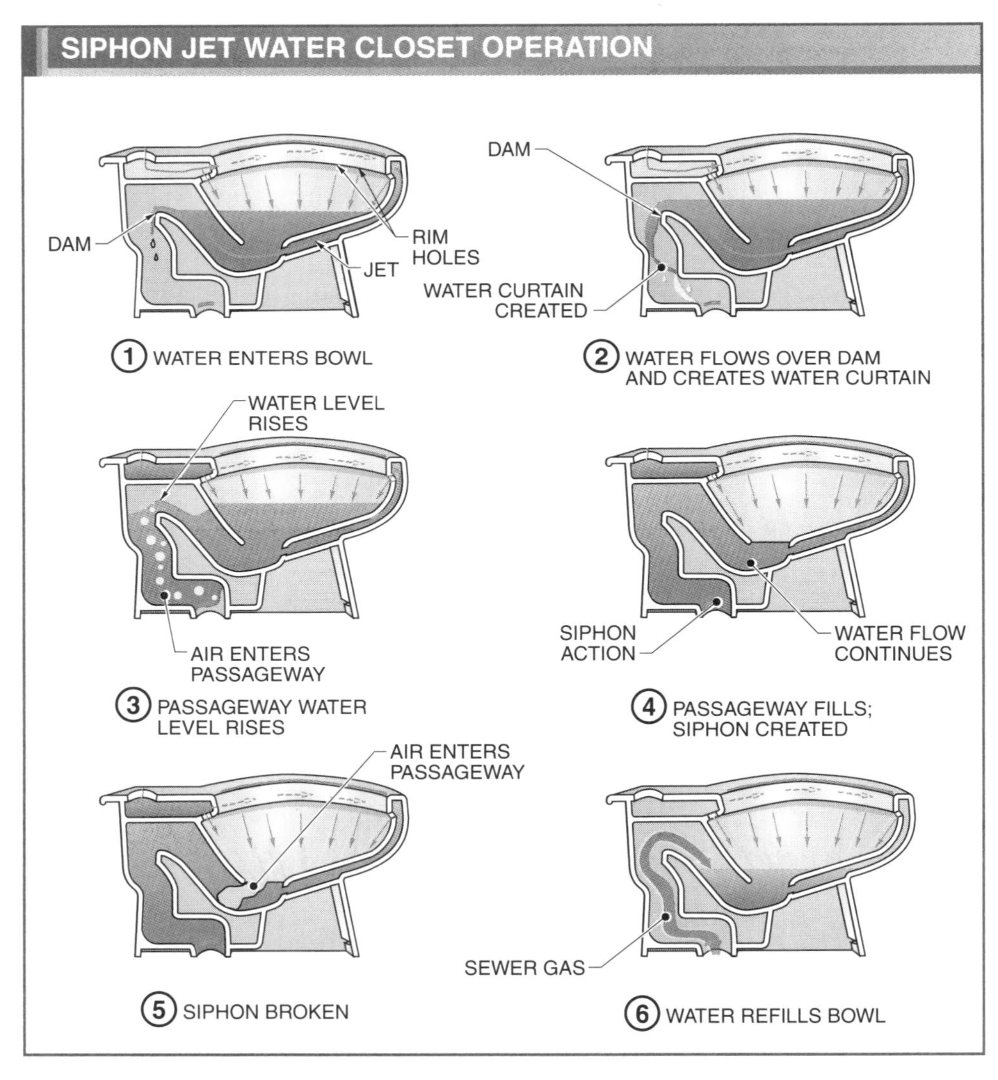

Figure 6-4. The flushing action of a siphon jet water closet provides the siphonage necessary to properly remove waste from the water closet bowl.

5. The siphon is broken and the flushing action stops as the water level in the bowl drops enough for air to be introduced into the passageway, and the flush stops.
6. Water continues to flow through the rim holes to refill the water closet bowl to prevent sewer gases from entering the living space.

Gravity-Fed Water Closets. A gravity-fed water closet is a floor-set fixture that uses the gravity of falling water from the water closet tank to begin the flushing action. **See Figure 6-5.** The flushing action of a gravity-fed water closet consists of these steps:

1. Water is released into the water closet bowl when the trip lever is pushed.
2. Water enters the water closet bowl through rim holes, raising the water level and scouring the bowl.
3. Water flows over the dam, and a siphonic action is created due to the weight of the water and gravity.
4. Wastewater and waterborne waste are carried out of the bowl and into the soil pipe.

Blowout Water Closets. A blowout water closet is primarily used for installations such as healthcare facilities where a large volume of waste is anticipated. The flushing action of a blowout water closet requires a large water jet directed into the water closet passageway inlet. **See Figure 6-6.** Blowout water closets can flush large amounts of toilet paper and waste without becoming plugged. However, blowout water closets require 3.5 gallons of water to flush properly. The flushing action of a blowout water closet consists of these steps:

1. Water is initially released into the flush rim of the water closet to scour the bowl when the flushometer valve is activated. Additional water is directed to the priming jet.
2. The priming jet force draws the contents of the bowl into the passageway inlet and forces (blows) the contents into the passageway outlet to flush the bowl.
3. Water continues to flow into the water closet bowl through the flush rim to refill the bowl and provide protection against sewer gases entering the living area.

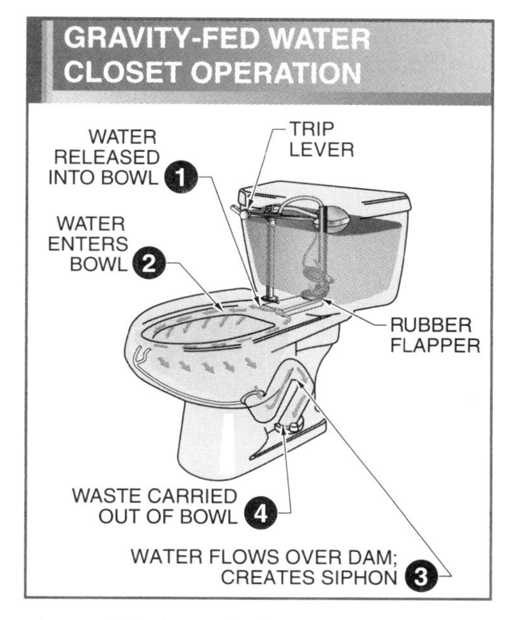

Figure 6-5. A gravity-fed water closet uses the gravity of falling water from the water closet tank to begin the flushing action.

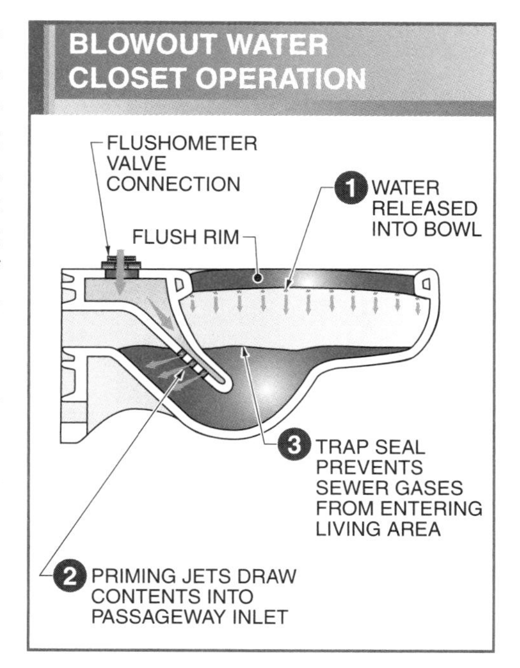

Figure 6-6. The flushing action of a blowout water closet uses a large water jet directed into the water closet passageway inlet.

Wall-hung blowout water closets are anchored to the supporting chair carrier with three bolts; siphon jet water closet bowls have four bolts.

Water Closet Flush Devices

The Energy Policy Act of 1992 mandated that the maximum volume of water used for flushing water closets, except blowout water closets, is 1.6 gallons per flush (gpf). Flush devices, such as flush tanks (also called toilet tanks), pressure tanks, and flushometer valves, are designed to deliver the proper amount of water to the water closet bowl and provide the necessary scouring action to clean the bowl.

Flush Tank Operation. Flush tanks are usually used on residential water closets. A *flush tank* is a reservoir that retains a supply of water used to flush one water closet. Flush tank operation is quiet, and requires only low water pressure and a small volume of water for proper operation. A flush tank can be supplied with ½″ pipe, which lowers installation cost. Most residential water closets are two-piece construction, consisting of the water closet bowl and flush tank. However, some water closets are one-piece construction with an integral flush tank within the body of the water closet.

Electronic flushometer valves use a sensor to activate the flush cycle.

A water closet flush tank consists of either a float valve or a ball cock assembly connected to the water supply and a flush valve assembly including an overflow tube, valve seat, and a rubber flapper attached to a trip lever with a lift chain or plastic strap.

A ball cock assembly consists of a seat faced with a soft rubber compression washer. The rubber washer, which is fastened to the float valve plunger, moves up and down over the seat. The plunger is connected to a brass float rod, which has the brass or plastic float threaded onto the end. In a float valve assembly the float is a moving part of the assembly body. **See Figure 6-7.**

To flush the water closet, the trip lever on the outside of the flush tank is pressed, which raises the rubber flapper from the flush valve seat. The water in the flush tank quickly flows by gravity into the water closet bowl, creating a siphonic action in the bowl. The flapper floats back down onto its seat as the water level decreases and is sealed against the seat by the action of the water passing into the closet bowl. When the flapper has reseated, the flush is stopped and the water closet tank refills with water.

The flush valve assembly includes an overflow tube, which bypasses the flush valve seat. The overflow tube allows any ball cock or water control assembly leakage to enter the closet bowl rather than overflow the tank. Water from the refill tube is piped into the overflow tube. Since water closets are flushed by siphonic action that removes most of the water from the trap, the refill tube replaces the trap seal to prevent sewer gases from passing through the water closet trap. Water flow through the refill tube stops when the float reaches the proper height.

Pressure Tank Operation. A *pressure flush water closet* is a water closet in which water is stored in a 1.6 gallon pressure tank before it is discharged to the fixture at a high velocity. The pressure tank is mounted within the water closet tank, and requires a ½″ water supply pipe at a minimum water supply pressure of 25 psi to operate properly. **See Figure 6-8.** The operation of a pressure flush water closet consists of these steps:

1. Water enters the pressure tank under normal working water pressure and compresses air in the tank.

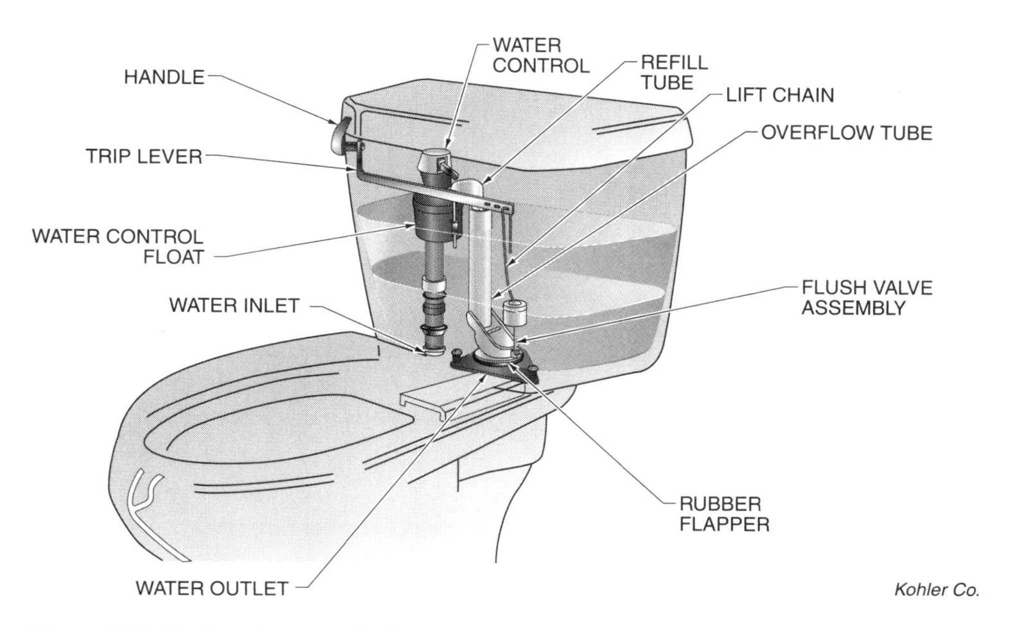

Figure 6-7. Flush tanks are typically used on residential water closets.

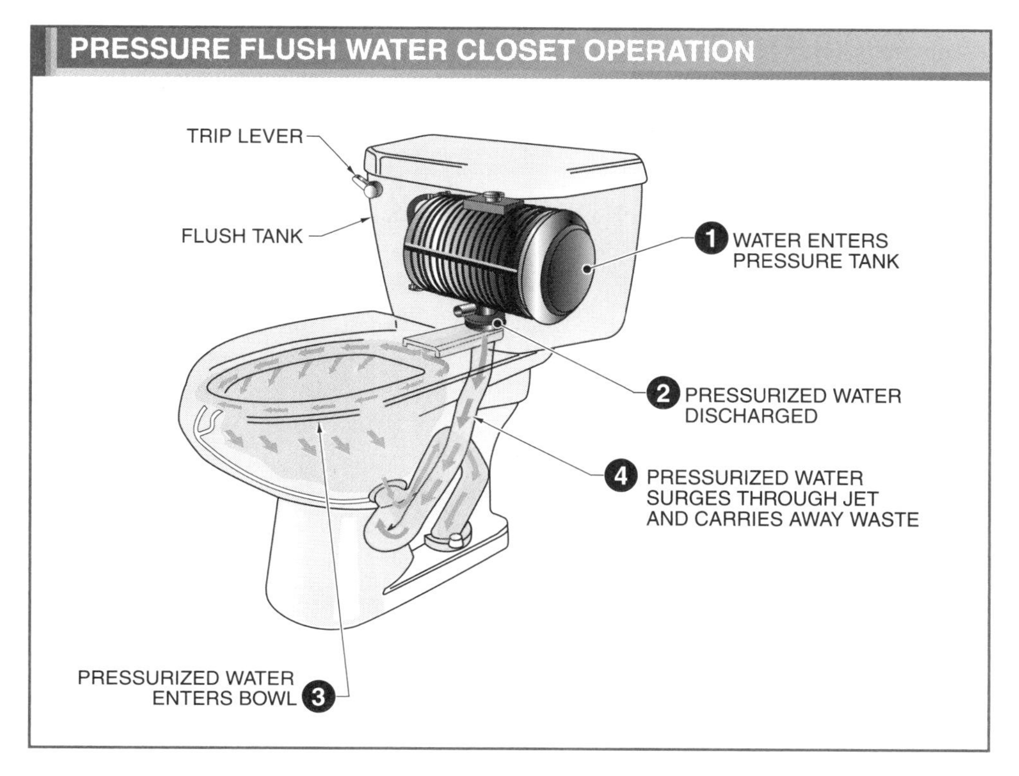

Figure 6-8. A pressure flush water closet is a water closet in which water is stored in a 1.6 gal. pressure tank before it is discharged to the fixture at a high velocity.

2. The trip lever is pushed and pressurized water is discharged to the water closet bowl and water jet.
3. Pressurized water enters the bowl through the rim holes, providing the necessary scouring action to clean the bowl.
4. Pressurized water surges through the jet, creating a siphonic action in the fixture passageway that carries the waste through the trap passageway and into the soil pipe. Water continues to flow into the water closet bowl through the flush rim to refill the bowl and provide protection against sewer gases entering the living area.

Flushometer Valves

A *flushometer valve* is a flush device actuated by direct water pressure to supply a fixed quantity of water for flushing purposes. It is commonly used to flush water closets, urinals, and some types of hospital fixtures. A flushometer valve can be operated in 6- to 10-second intervals. A flushometer valve has a small number of working parts and is very easy to maintain. Water closet flushometer valves are available in 1.6 gpf models for new construction and 3.5 gpf models for blowout water closets and for repair of water closets installed before 1992. Water closet flushometer valves must be supplied with a 1″ pipe.

Diaphragm Flushometer Valves. Diaphragm flushometer valves are the most common flushometer valves. A *diaphragm flushometer valve* is a flushometer valve in which a segmented diaphragm within the valve body controls the flushing water by equalizing pressure on both sides of the diaphragm.

When the flushometer valve is not in use, the valve is divided by a diaphragm into upper and lower chambers with equal water pressure on both sides. Greater surface area on top of the diaphragm holds the diaphragm onto its seat. **See Figure 6-9.** Handle movement in any direction pushes the plunger, which tilts the relief valve and allows water to escape from the upper chamber into the lower chamber. Water pressure in the lower chamber increases and raises the relief valve, disk, diaphragm, and guide, allowing water to flow through the valve outlet and through the flush tube to flush the fixture.

A small amount of water flows through the bypass port in the diaphragm while the valve is operating, gradually refilling the upper chamber and equalizing the pressure again. The diaphragm returns to its seat and closes the valve as the upper chamber fills. If the bypass port becomes plugged with dirt or other debris, a flushometer valve will continue to flush until the dirt or debris is cleared since no water can enter the upper chamber to equalize the pressure and stop the flush.

Water Closet Seats

Water closet seats are installed above the flush rim of water closets to support a user. Water closet seats are manufactured from solid plastic or pressed wood fibers and are coated with a smooth, nonabsorbent finish. Water closet seats are available in a variety of colors with closed or open fronts. Closed front water closet seats are used for residential applications and open front seats are used for commercial and public applications.

ADA-Compliant Water Closet Installation

The top of an ADA-compliant water closet seat must be between 17″ and 19″ above finished floor level. **See Figure 6-10.** ADA-compliant floor-set water closets are manufactured with the closet bowl 1½″ to 2″ taller than standard water closets. Wall-hung water closet supporting chair carrier fittings are adjusted to raise the waste outlet piping to 7½″ to 8″ above the floor, depending on the water closet manufacturer. Water closet flush tanks and flushometer valves are interchangeable between standard and ADA-compliant closets.

An ADA-compliant water closet must be positioned in the rest room stall or bathroom to provide adequate clear floor space to allow a wheelchair-bound user to maneuver within the stall or bathroom and to permit side transfer of the user from a wheelchair to the water closet. Grab bars are installed at the back wall and close wall side of the water closet to aid the user in transferring from a wheelchair to the fixture. Detail drawings on the job specifications should be referred to for the proper positioning of grab bars for use with a water closet.

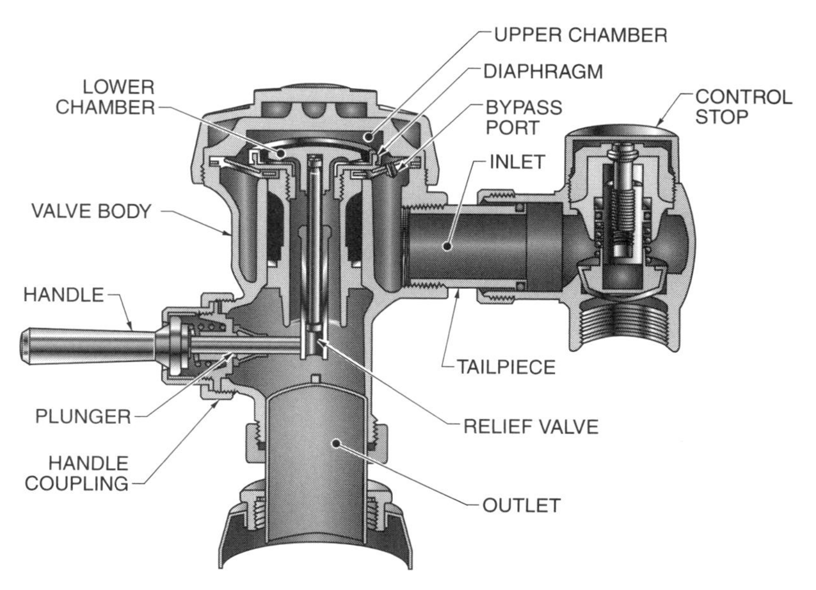

Figure 6-9. A diaphragm flushometer valve contains a relief valve diaphragm within the valve body that discharges water to a fixture by equalizing pressure on both sides of the valve.

Tech Facts

Iron in water can produce stains on plumbing fixtures when the amount of iron is .3 parts per million or greater.

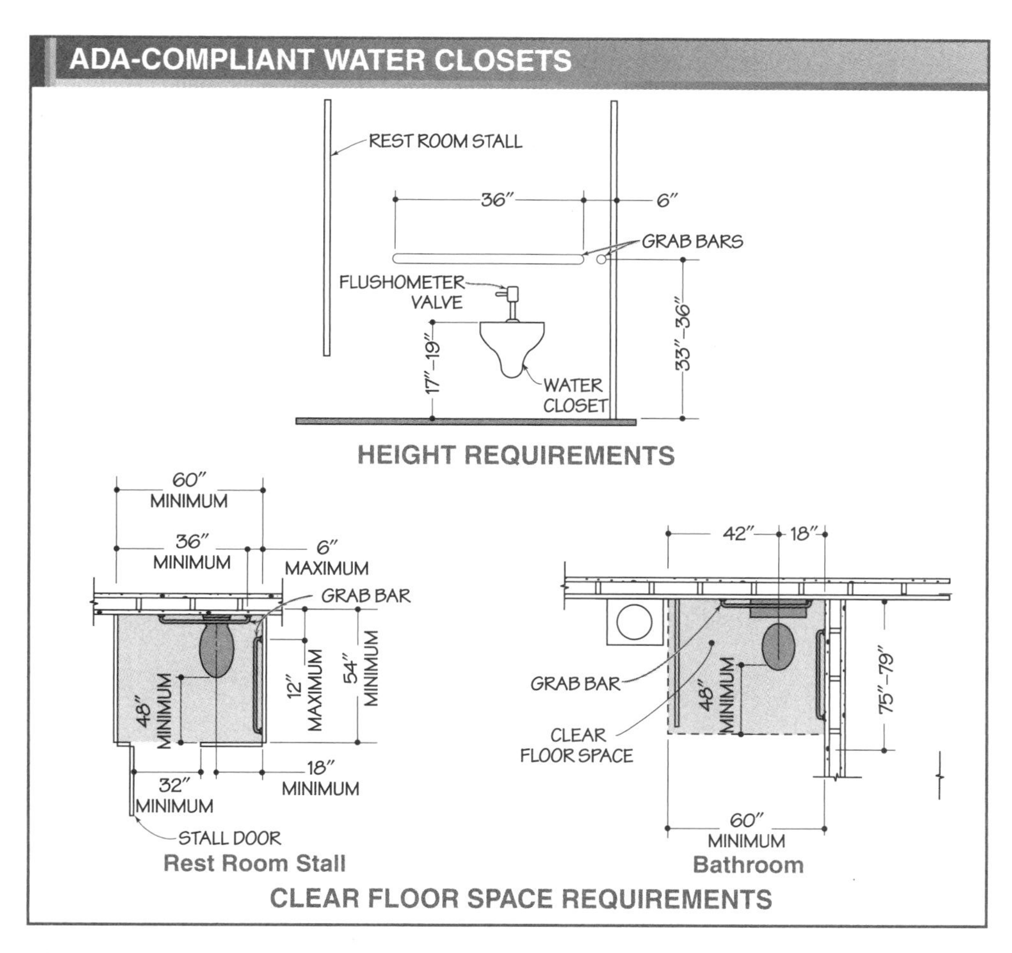

Figure 6-10. Water closet height, adequate clear floor space, and grab bars are required for ADA-compliant water closets.

URINALS

A *urinal* is a plumbing fixture that receives only liquid body waste and conveys the waste through a trap seal into a sanitary drainage system. Urinals are manufactured from smooth, nonabsorbent materials such as vitreous china, enameled cast iron, or stainless steel and are available in various flushing actions including washout, siphon jet, and blowout.

Washout Urinals

Washout urinals are the most common type of urinals installed. A *washout urinal* is a urinal in which the liquid waste is washed from a trap rather than flushed. Water enters the top of a washout urinal and flows into the fixture bowl through openings in the top rim. The water spreads across the back of the urinal and flows by gravity through the trap into the sanitary drainage system. A washout urinal has a restricted opening over the trap inlet consisting of a stainless steel strainer, china grate, or small openings cast into the fixture to prevent debris from entering the trap and plugging the fixture drain.

Washout urinals are available in stall, wall-hung, or trough models. **See Figure 6-11.** A stall urinal provides access to all users regardless of their height. Stall urinals are installed with the lip slightly below the floor level to provide proper drainage for the area adjacent to the fixtures. A strainer, which is placed over the waste outlet of the urinal, is connected to a separate P-trap installed below floor level with a preformed rubber gasket. Stall urinals are rated at 3 dfu, and require a 2″ waste pipe and P-trap and a 1¼″ vent pipe.

Wall-hung urinals have integral or exposed P-traps. Integral-P-trap wall-hung washout urinals have a 2″ P-trap within the body of the urinal. The waste outlet is covered with a strainer, or several small holes are cast into the bottom of the fixture. Wall-hung washout urinals with integral P-traps are rated at 3 dfu, and require 2″ waste and 1¼″ vent pipes. Exposed-trap wall-hung washout urinals are rated at 2 dfu, and have a 1½″ P-trap and waste pipe and 1¼″ vent pipe.

Trough urinals are manufactured from enameled cast iron and stainless steel and are available in 3′, 4′, 5′, and 6′ lengths. Trough urinals are commonly installed in sports facilities and other buildings where there are large crowds. Water enters a trough urinal through a perforated flush pipe attached to the upper back edge of the fixture. Waste flows out through a strainer located in the middle of the urinal and into an exposed 1½″ P-trap. Trough urinals are rated at 2 dfu per 6′ section and require a 1¼″ vent pipe.

Siphon Jet Urinals

A *siphon jet urinal* is a wall-hung urinal in which the trap seal is forced from the trap through a large opening at the trap inlet. A jet at the base of the bowl directs water into the trap inlet to create the siphonic flushing action. Siphon jet urinals have a large opening at the bottom of the urinal over the trap inlet. **See Figure 6-12.** Siphon jet urinals are rated at 3 dfu, have an integral 2″ trap, and require 2″ waste and 1¼″ vent pipes.

Blowout Urinals

A *blowout urinal* is a wall-hung urinal with a nonsiphonic passageway at the rear of the bowl and an integral flush rim and jet. **See Figure 6-13.** Blowout urinals are flushed with a large jet of water that blows the waste out of the fixture trap. Blowout urinals have a large opening to the trap inlet to accommodate nonliquid waste. Blowout urinals have noisy flushing action, and are installed in rest rooms when noise is not a concern, such as in sports complexes. Blowout urinals are rated at 3 dfu, have integral 2″ P-traps, and require 2″ waste and 1¼″ vent pipes. Blowout urinals require a 1″ water supply to provide the volume of water necessary to force waste from the trap.

Tech Facts

Waterless urinals were developed to conserve water and promote cleanliness.

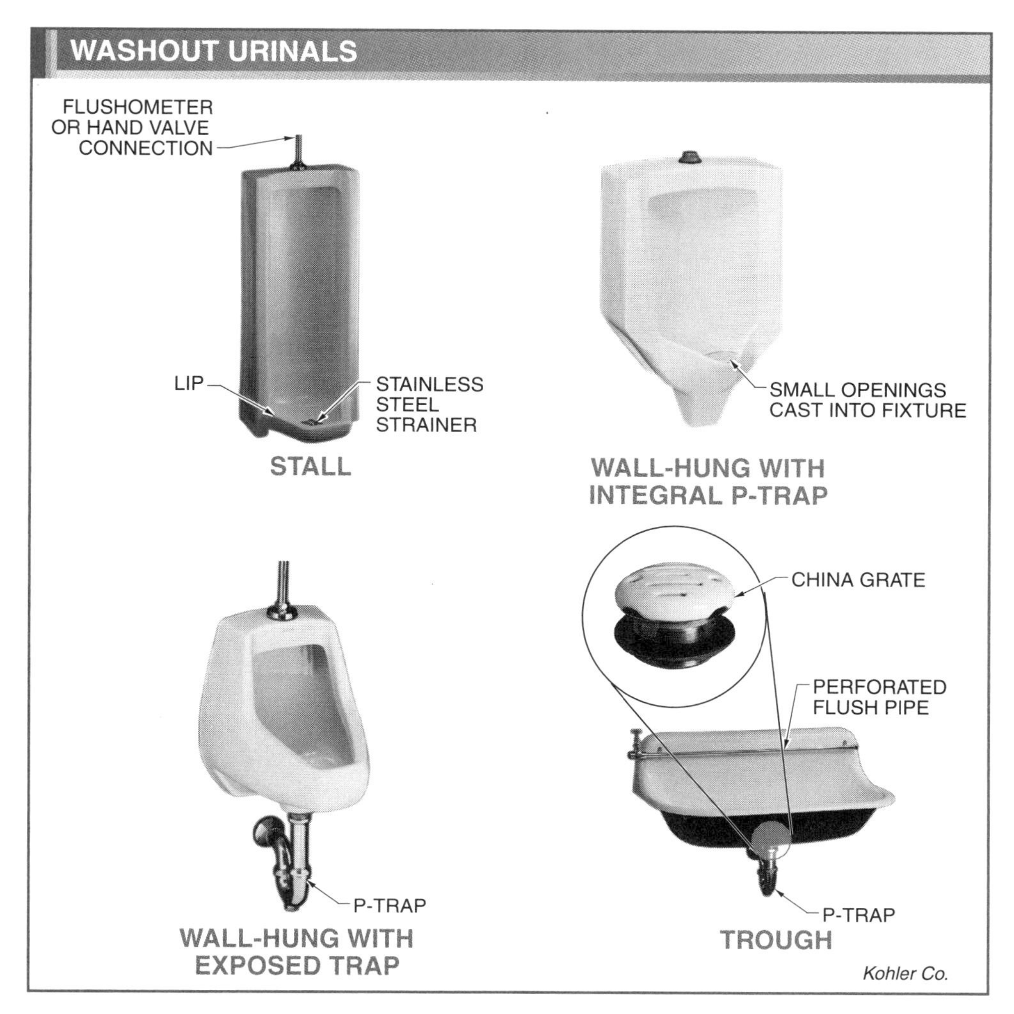

Figure 6-11. Washout urinals are available in stall, wall-hung, or trough models.

Waterless Urinals

To conserve water, lower maintenance and repair bills, and create an odor-free restroom, waterless urinals have been developed. Waterless urinals do not have a flush valve or water supply pipe connected to the fixture. A waterless urinal contains a removable cartridge with a floating liquid sealant in the drain opening at the bottom of the fixture. Liquid waste passes through a floating liquid sealant, which closes to seal off odors, and into the drainage piping. The removable cartridge also provides the trap seal as the fixture itself does not have an integral trap. The cartridge and sealant fluid must be periodically removed and replaced by the building custodial staff. **See Figure 6-14.**

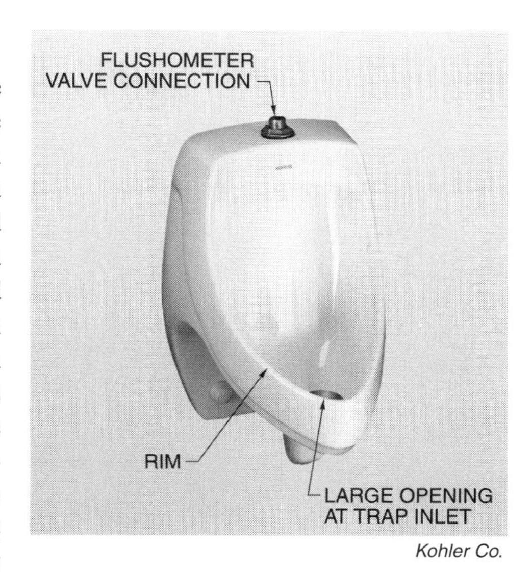

Figure 6-12. A siphon jet urinal has a large opening over the trap inlet.

Tech Facts

Soft abrasive cleaners should be used to clean vitreous china fixtures and appliances. Avoid strong abrasive cleaners as they will scratch and dull surfaces.

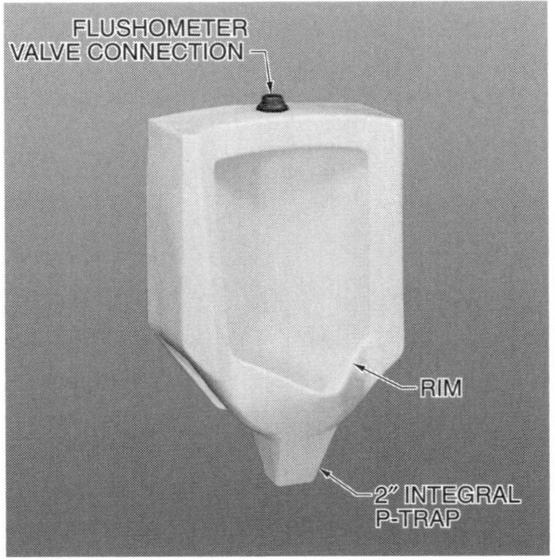

Kohler Co.

Figure 6-13. Blowout urinals are flushed with a large jet of water that blows the waste out of the fixture trap.

ADA-Compliant Urinals

Standard wall-hung urinals are used as ADA-compliant urinals, but the height of the fixture is lowered to accommodate the user. The top of the flush rim of an ADA-compliant wall-hung urinal must be no more than 17″ above the finished floor level and the flushometer valve handle must be a maximum of 44″ above finished floor level. Floor-set urinals are ADA-compliant if the flushometer valve is a maximum of 44″ above finished floor level. ADA-compliant urinals must have at least 30″ of clear floor space to provide adequate access to the fixture. **See Figure 6-15.**

Urinal Flush Devices

The Energy Policy Act of 1992 mandated that the maximum volume of water used for flushing urinals is 1.0 gallons per flush (gpf). Urinals are flushed with manual or electric sensor flushometer valves to deliver water to the fixture for waste removal. Urinal flushometer valves require a minimum of ¾″ water supply pipe.

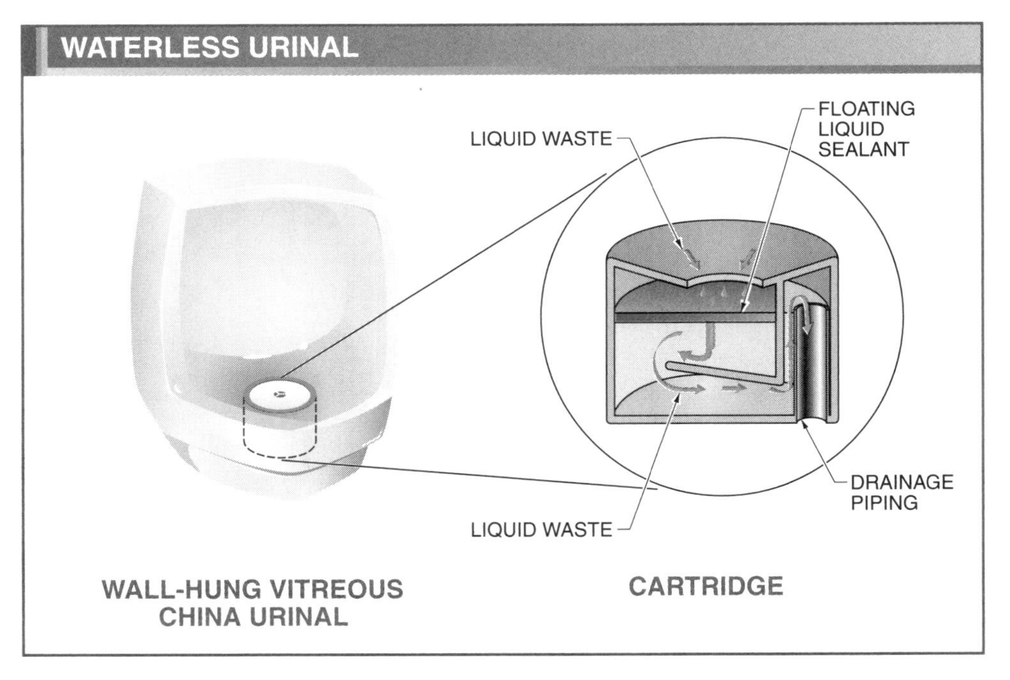

Figure 6-14. A waterless urinal contains a cartridge in the bottom of the fixture.

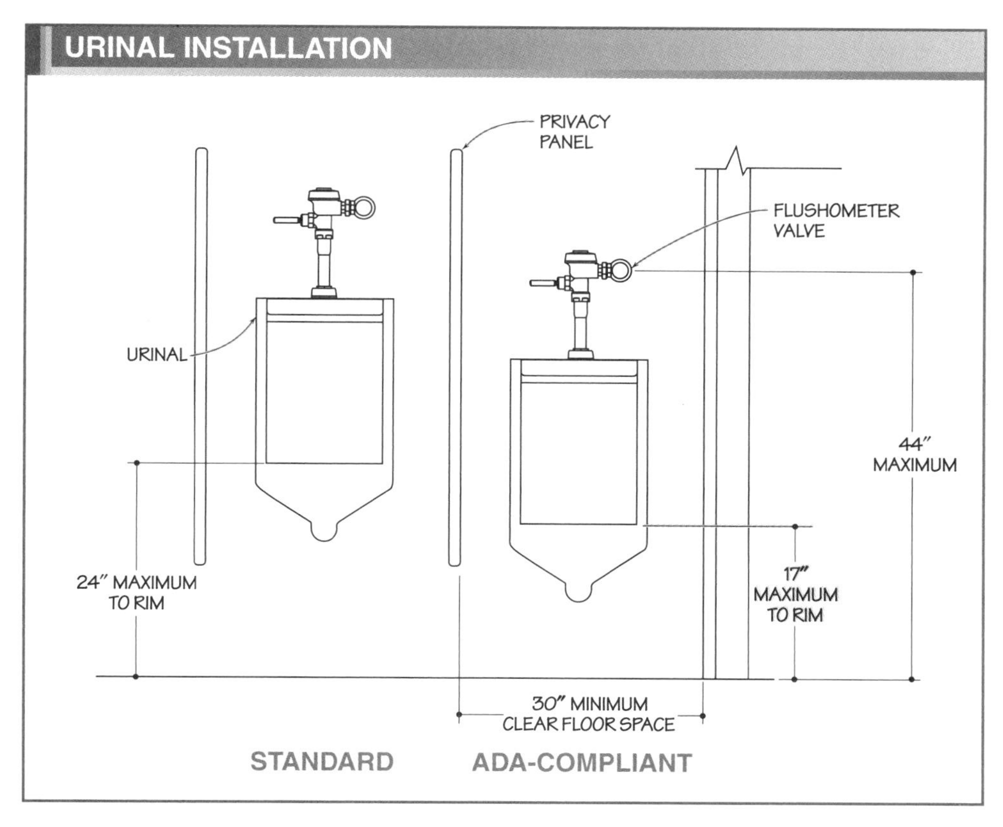

Figure 6-15. Urinals that are ADA-compliant are installed with their front lip 7″ lower than standard urinals.

The flushing of urinals in public restrooms presents a problem. The user of the urinal does not always flush the fixture after use. The electronic flushometer was developed to automatically flush the urinal after each use. **See Figure 6-16.** Electronic flushometer valves sense the presence of a user with a small infrared light beam which is constantly emitted from the valve. **See Figure 6-17.** When a user enters the effective range of the light beam, the beam is reflected into the sensor and transformed into a low-voltage electrical signal to activate the circuit. The circuit continues in a "hold" mode while the user remains within the effective range. As the user steps away from the fixture, another electrical signal energizes the fixture control solenoid that activates the flush cycle. After flushing, the circuit automatically resets and is ready for the next user. Electronic flushometer valves may be either battery-powered or direct wired to low-voltage current.

Some urinals contain electronic sensor-controlled solenoid valves within the fixture body that discharge the flushing water through an air gap into the flush rim. These valves are battery-powered or direct wired to low-voltage current. Urinals that use this type of flushing system use only ½ gal. of water per flush and are supplied with a ½″ water pipe. For ease of cleaning and to prevent vandalism, the electronic sensor is the only external component of an integral flushometer valve.

LAVATORIES

A *lavatory,* or washbasin, is a plumbing fixture used to wash the hands and face. Lavatories are installed in bathrooms and rest rooms, are rated at 1 dfu, and require a 1¼″ P-trap, waste pipe, and vent pipe. Lavatory water supply pipe is a minimum of ¼″ size. Lavatories are available in a wide variety of sizes, colors, and shapes, and are made from vitreous china, enameled cast iron, marble, glass, stainless steel, or acrylic plastics. Lavatories are available in wall-hung, pedestal, and countertop models. **See Figure 6-18.**

Tech Facts

Waterless urinals are becoming a popular alternative to fixture valve urinals. Unlike fixture valve urinals that must be connected to the water supply, waterless urinals are only connected to a drain line.

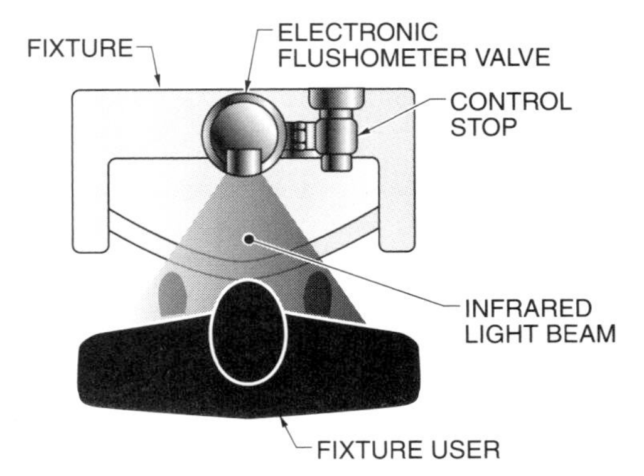

Figure 6-17. An electronic flushometer valve uses a small infrared light beam to sense the presence of a user.

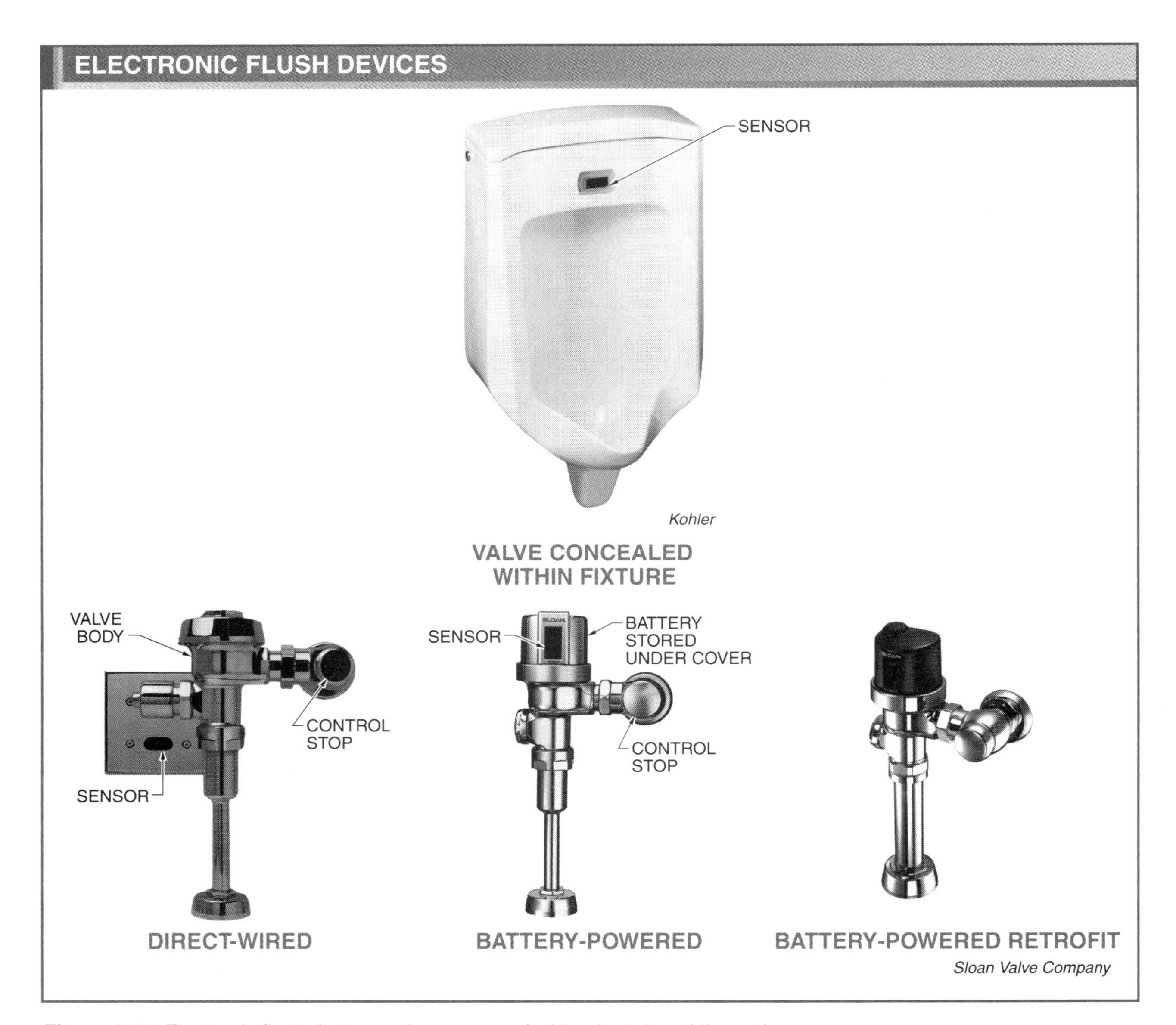

Figure 6-16. Electronic flush devices valves are used with urinals in public washrooms.

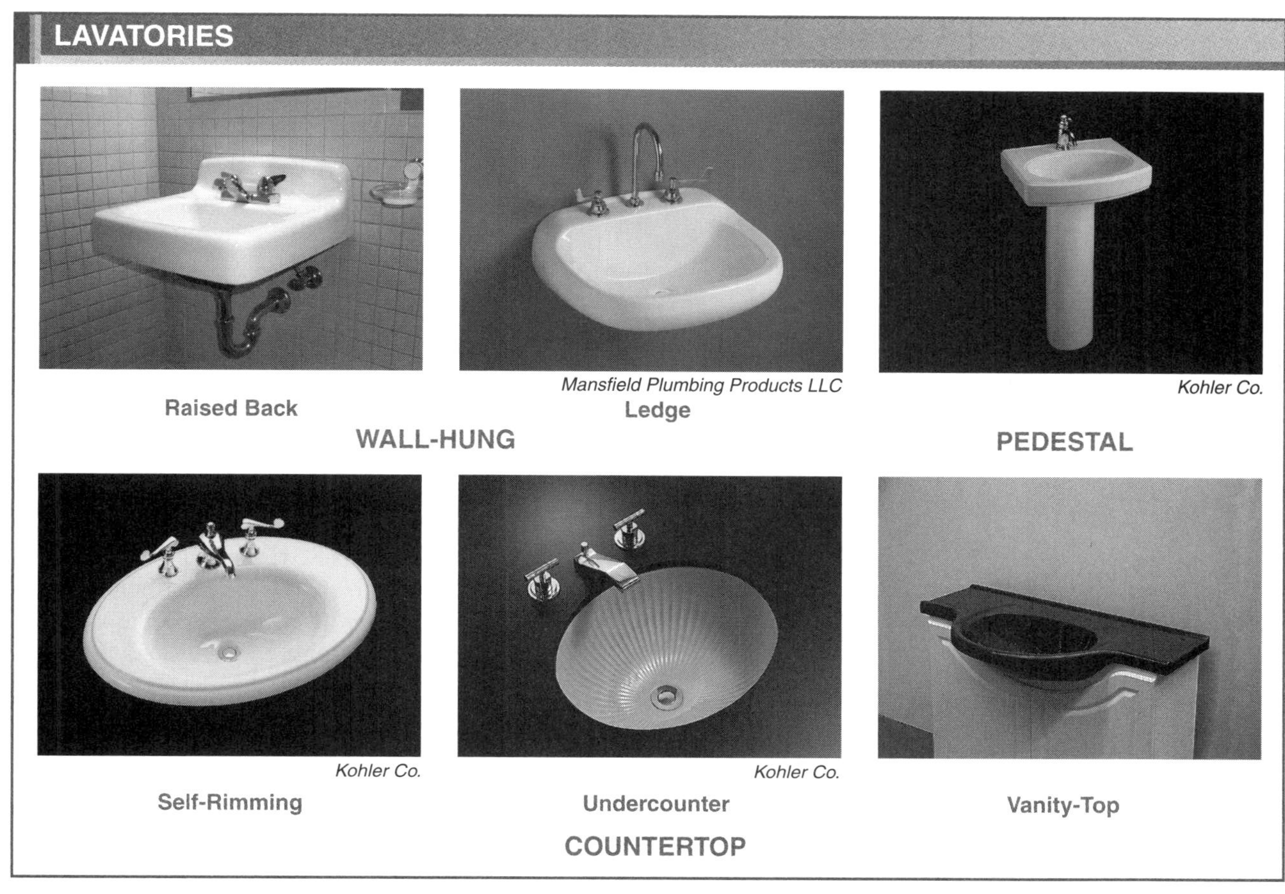

Figure 6-18. Lavatories are available in wall-hung, pedestal, and countertop models.

A *wall-hung lavatory* is a lavatory supported by a stamped steel or cast iron bracket fastened to a backing board installed between wall framing members when the fixture is roughed in. The flood level rim of a wall-hung lavatory is typically set at 31″ above the finished floor level to permit the fixture user to stand slightly bent over the lavatory to allow washing water to drain from the arms back into the basin. Wall-hung lavatories are available in raised-back and ledge models. Raised-back, wall-hung lavatories support themselves better than ledge lavatories since there is more bearing surface on the back of the fixtures.

A *pedestal lavatory* is a two-piece lavatory with the washbasin resting directly on a pedestal base. The washbasin section is attached to the wall with a stamped steel bracket. A backing board is installed immediately behind the washbasin between the studs to provide proper support for the lavatory when the fixture is roughed in. Even though the washbasin is attached to the wall, the pedestal prevents downward movement of the basin.

A *countertop lavatory,* or vanity lavatory, is a lavatory installed in an opening of a bathroom or rest room cabinet or countertop, or resting on a cabinet frame. Several types of countertop lavatories are available including self-rimming, under-counter, and vanity-top lavatories.

A *self-rimming lavatory* is a lavatory in which the bowl is placed in an opening in the countertop, and the edge of the fixture rests directly on top of the countertop. A bead of adhesive caulk is placed under the lavatory edge prior to installation and another bead of caulk is placed around the edge of the rim after installation to prevent water from flowing between the countertop and lavatory. An *under-counter lavatory* is a lavatory attached to the underside of a countertop using rim clamps or other proprietary hardware. A bead of adhesive caulk is placed along the upper surface of the lavatory edge prior to installation so water does not leak into the cabinet. A *vanity-top lavatory* is a one-piece washbasin and countertop installed on top of a bathroom or rest room cabinet or supported from wall framing members. Vanity-top lavatories are attached to bathroom or rest room cabinets using adhesive caulk or the hardware supplied with the fixture.

ADA-Compliant Lavatories

Standard lavatories can be used as ADA-compliant lavatories if the proper clearances are allowed for users. A minimum of 30″ by 48″ of clear floor space must be provided for a lavatory to allow unobstructed forward approach by a user. The clear floor space extends a maximum of 19″ under the lavatory. The flood level rim of ADA-compliant lavatories must be less than 34″ above the finished floor level. **See Figure 6-19.** A minimum of 29″ must be provided below a lavatory rim to provide knee clearance. In addition, a 6″ × 9″ minimum space must be provided for toe clearance. For ADA compliance, countertop lavatories in public buildings are installed with the countertop a maximum of 34″ above finished floor level. The bottom of a countertop apron cannot extend lower than 29″ from finished floor level.

ADA-compliant lavatories are available for installation in public and commercial buildings. **See Figure 6-20.** A wall-hung, ADA-compliant lavatory is mounted on a concealed carrier fitting so the flood level rim is positioned a maximum of 34″ above finished floor level. ADA-compliant lavatories are fitted with wrist blade handles or single-handle faucets. A *wrist blade handle* is an ADA-compliant faucet handle that does not require tight grasping by the hand or twisting of the wrist to properly operate the faucet. In public buildings, at least one lavatory is designated as an ADA-compliant lavatory, depending on the occupancy of the building. P-traps of ADA-compliant lavatories may be offset back from the centerline of the fixtures to provide an unobstructed approach by the user. Because the water supply piping and drainage piping can become hot, they are wrapped or covered to prevent user injury.

Lavatory Trim

Lavatory trim includes a faucet that controls the cold and hot water supply and the lavatory drain fitting through which water drains from the lavatory. Faucets and drain fittings are available separately or as combination fittings, which include a lavatory faucet and a pop-up drain assembly.

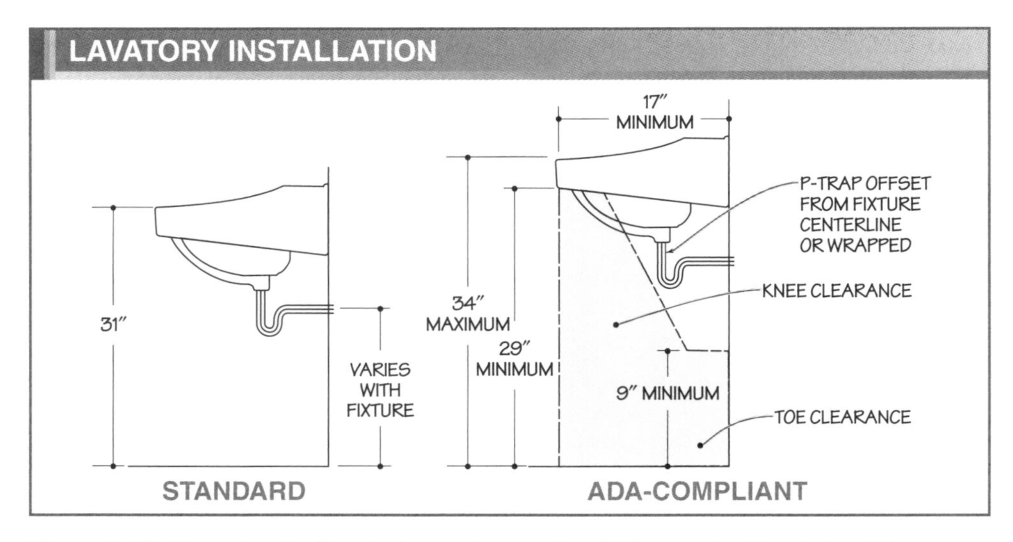

Figure 6-19. Many standard lavatories can be used as ADA-compliant lavatories if the proper clearances are allowed for users.

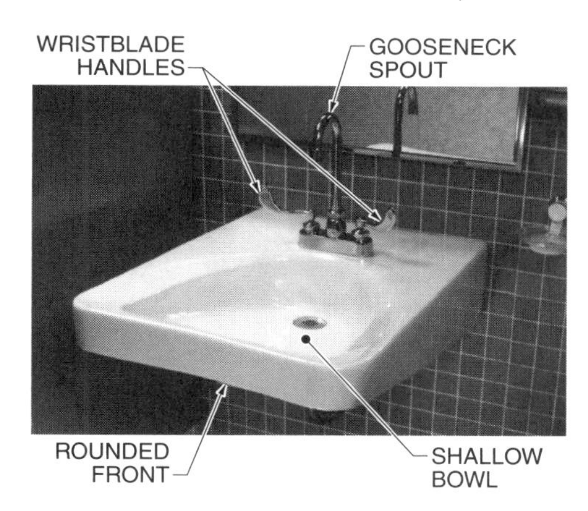

Figure 6-20. ADA-compliant lavatories are mounted on a concealed carrier fitting so the flood level rim is positioned a maximum of 34″ above finished floor level.

Lavatory faucets are classified as compression faucets or port control faucets. A *compression faucet* is a faucet in which the flow of water is shut off by means of a washer that is forced down (compressed) onto its seat, as in a globe valve. Most lavatory compression faucets are combination faucets in which the cold and hot water compression valves are joined in one faucet body and provided with a common mixer spout. Combination faucets allow the user to mix hot and cold water to the desired temperature prior to the water being delivered to the lavatory basin.

A *port control faucet,* or single-handle faucet, is a single-handle noncompression faucet that contains ports for the hot and cold water supply and a cartridge or ceramic disc that opens and closes the ports as the faucet handle is moved or rotated. **See Figure 6-21.** The cartridge or disc opens or closes one or both ports in the faucet body to provide water at the desired temperature and volume. In a cartridge-type port control faucet, a cam connected to the faucet handle controls water flow around a rubber sleeve within the cartridge. When the faucet is in the OFF position, the cam is fully inserted into the rubber sleeve and securely seals the sleeve against the hot and cold water ports. When the faucet is turned on to deliver both hot and cold water, the cam is withdrawn from the sleeve, allowing the sleeve to flex and water to flow. When the faucet is turned on to deliver only hot or cold water, the cam is withdrawn and the bevel is aligned with one of the ports by turning the faucet handle. The sleeve flexes, allowing water to flow.

Ceramic disc-type port control faucets control water flow through the use of two ceramic discs that slide back and forth across each other in a watertight seal. Since the ceramic disks are very hard, they are unaffected by temperature extremes, corrosion, or hard water conditions. When the faucet is in the OFF position, the upper disk is positioned so the hot and cold water supply ports and the outlet port are covered. When the faucet is turned on to deliver both hot and cold water, the upper disk is rotated to allow water to flow through the hot and cold water supply ports and outlet port before exiting the faucet spout. When the faucet is turned on to deliver only hot or cold water, the upper disk is rotated so that only one of the supply ports is open, allowing only one water temperature to flow to the spout.

Kohler Co.

A port control faucet is a single-handle faucet that contains a cartridge or disc that opens or closes the water supply ports in the faucet body to provide water at the desired temperature and volume.

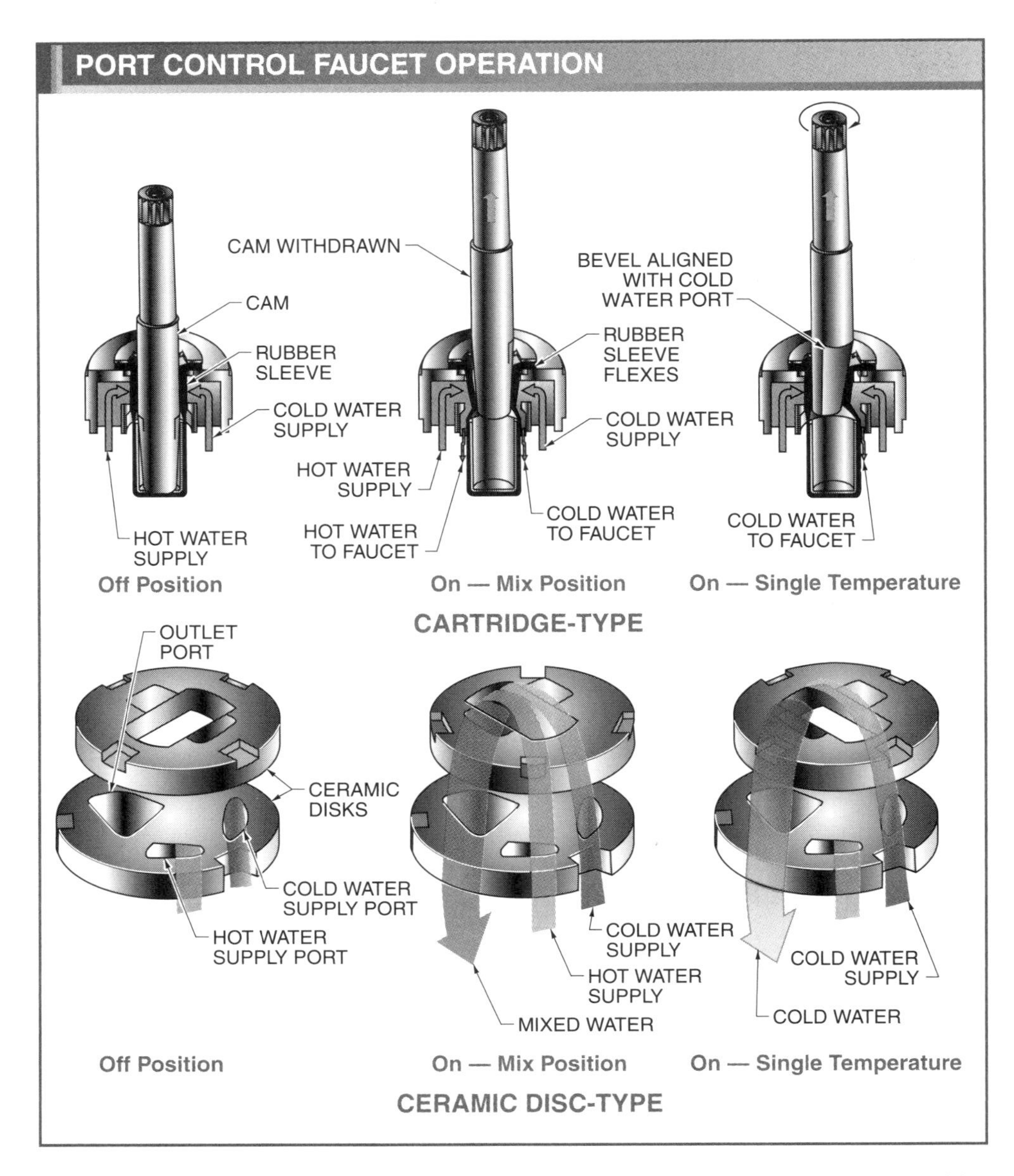

Figure 6-21. A port control faucet contains ports for the hot and cold water supply and a cartridge or ceramic disc that opens and closes the ports as the faucet handle is moved or rotated.

Lavatory drain fittings are installed in the bottom of a lavatory bowl. A *pop-up waste fitting* is a lavatory drain fitting that consists of a brass waste outlet into which a sliding metal or plastic stopper is fitted. A lever passes out the side of the drain fitting and is connected to a lift rod on top of the lavatory. The rod is lifted to lower the stopper and allow the lavatory to fill, or depressed to raise the stopper and drain the lavatory. Lavatory drain fittings are also available with a perforated grate over the drain opening for use in public buildings.

Combination fittings include centerset, concealed, and port control faucets. **See Figure 6-22.** A *centerset faucet* is a combination lavatory fitting that consists of two faucet handles, spout, and pop-up waste fitting lift rod mounted on a raised base. Centerset faucets are installed on lavatories with faucet mounting holes on 4″ centers. Water is supplied to the hot and cold water faucets and is mixed in the valve body under the raised base before being delivered to the spout. A *concealed faucet,* or bottom-mounted faucet, is a combination lavatory fitting that consists of one or two faucet handles and spout mounted above the lavatory or countertop and pop-up waste fitting lift rod, with the faucet bodies below the fixture. Concealed faucets are used on lavatories where the faucet mounting holes are drilled on 8″ or 12″ centers. A port control faucet is a combination lavatory fitting that consists of a single-handle faucet, spout, and pop-up waste fitting lift rod mounted on a raised base. Hot and cold water is supplied to the faucet through separate supply lines and mixed within the faucet body before being delivered to the spout.

Electronic Faucets

Electronic faucets are installed on lavatories in public restrooms to conserve water and for hygiene. An electronic faucet contains a solenoid valve that is activated when the user places their hands under the spout. Tempered (mixed hot and cold) water will flow from the faucet until the hands are removed or until the faucet reaches its automatic time-out setting. **See Figure 6-23.** The control module for an electronic faucet is mounted below the lavatory and may be either battery-powered or plugged into an electric receptacle provided for the faucet transformer. Electronic faucets are ADA-compliant for use on lavatories.

BATHTUBS

A *bathtub* is a plumbing fixture used to bathe the entire body. Bathtubs are manufactured from enameled cast iron, enameled pressed steel, fiberglass, or acrylic plastics, and are available in a variety of sizes, shapes, and colors. Bathtubs are rated at 2 dfu, and require 1½″ P-trap and waste pipe and 1¼″ vent pipe. Bathtubs require a minimum of ½″ water supply pipe.

Bathtubs are right-hand or left-hand, depending on the waste opening location. A *right-hand bathtub* is a bathtub with the drain on the right end as a person faces the tub. A *left-hand bathtub* is a bathtub with the drain on the left end as a person faces the tub. Bathtubs have a built-in pitch to provide proper drainage of the fixture yet allow the flood level rim to remain level.

Tech Facts

Water temperatures over 125°F can cause either severe burns instantly or death from scalding. Children, senior citizens, and people with physical disabilities are most susceptible to scalding. The water heater temperature should be adjusted to avoid burns and scalding.

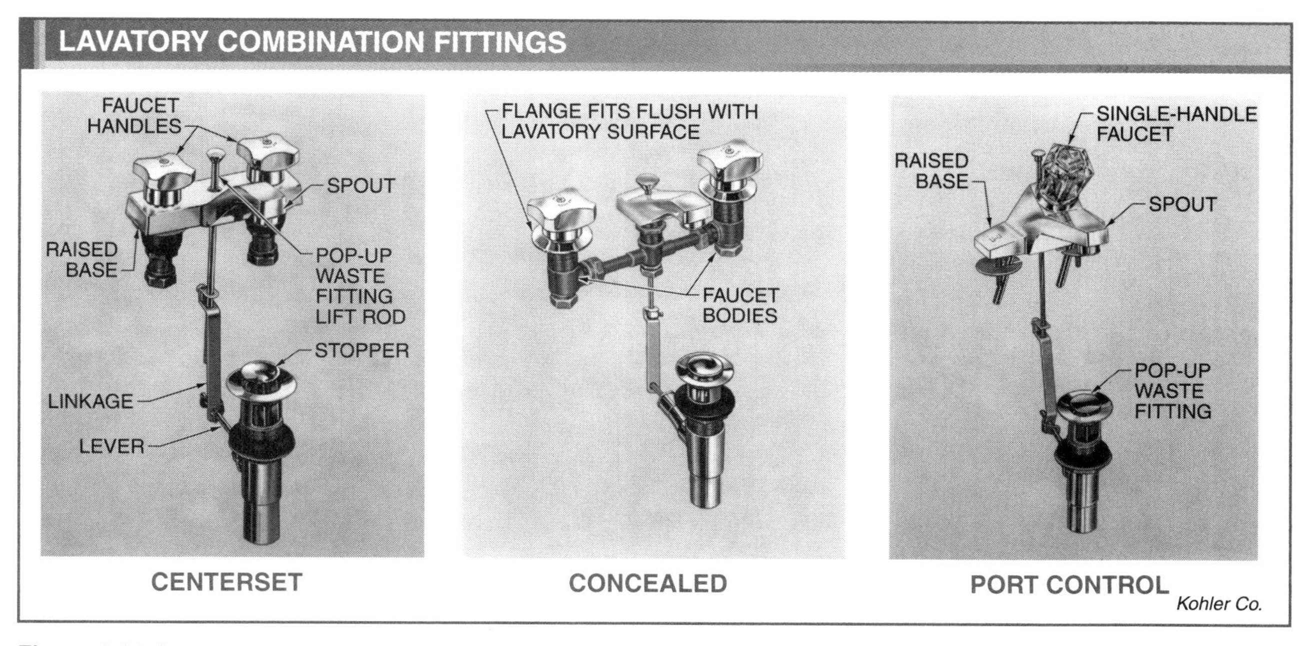

Figure 6-22. Lavatory combination fittings include centerset, concealed, and port control faucets.

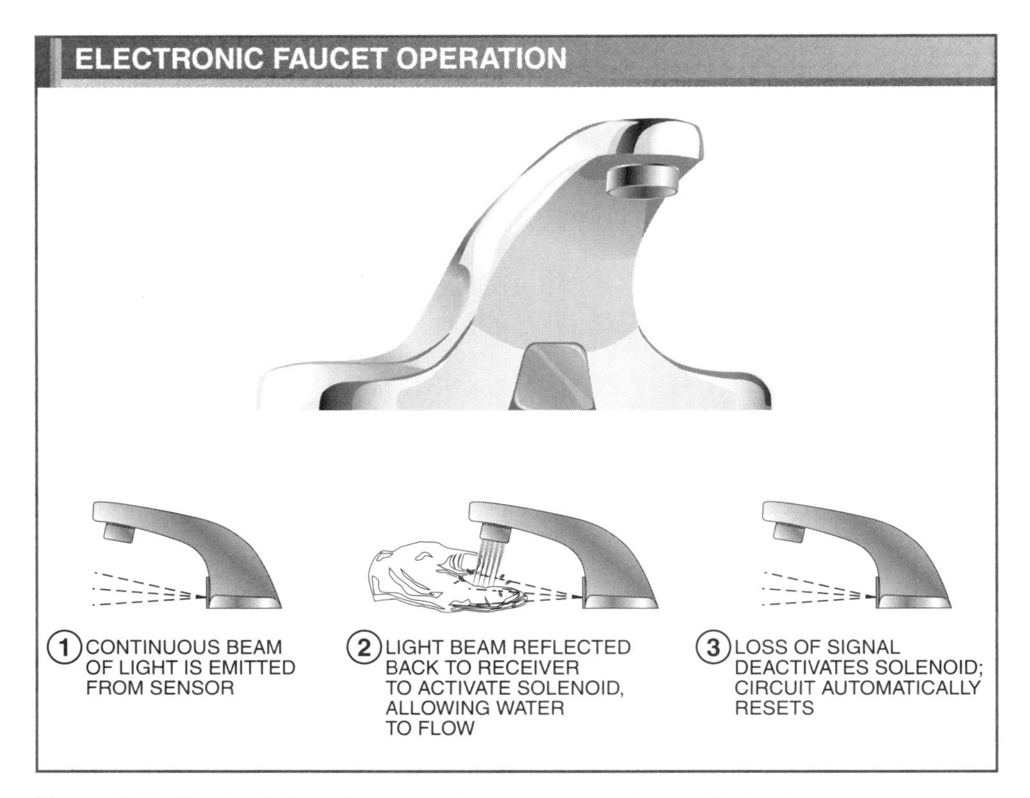

Figure 6-23. Electronic faucets are used to conserve water and for hygiene.

Bathtubs are available in a variety of designs including recessed, freestanding, and drop-in. **See Figure 6-24.** A *recessed bathtub,* or built-in bathtub, is a bathtub permanently attached or built into the walls and floor of the bathroom. Recessed baths are supported by backing attached to the studs on the back wall of the tub, and the front apron rests on the floor.

Some recessed bathtubs are integral bath and shower combination assemblies. Integral bath and shower combination assemblies are common in residential construction since the shower walls eliminate the need for ceramic tile or other water-resistant wall coverings in the bathtub area.

A *freestanding bathtub* is a bathtub supported by legs and is not permanently attached to the bathroom walls or floor. A *drop-in bathtub* is a bathtub that is installed in an enclosure that supports the fixture. Drop-in bathtubs provide greater water capacity than standard bathtubs.

ADA-Compliant Bathtubs

ADA-compliant bathtubs are equipped with a seat and grab bars to assist the user in entering and exiting the fixture. The faucet handles are positioned at the foot of the bathtub along the outside edge to provide easy user access to the controls. The showerhead is equipped with a hose so that it can be adjusted vertically on an adjustment bar or removed for hand use.

When an ADA-compliant bathtub seat is in the fixture, a minimum of 30″ by 60″ or 48″ by 60″ of clear floor space is required, depending on the direction of tub access. **See Figure 6-25.** When the seat is positioned at the head of the bathtub, a 30″ by 75″ minimum clear floor space is required. Grab bars are required for ADA-compliant bathtubs at the foot and head ends and side wall of a bathtub when the seat is in the fixture. Grab bars are required on the foot end and side wall of a bathtub when the seat is positioned at the head of the fixture.

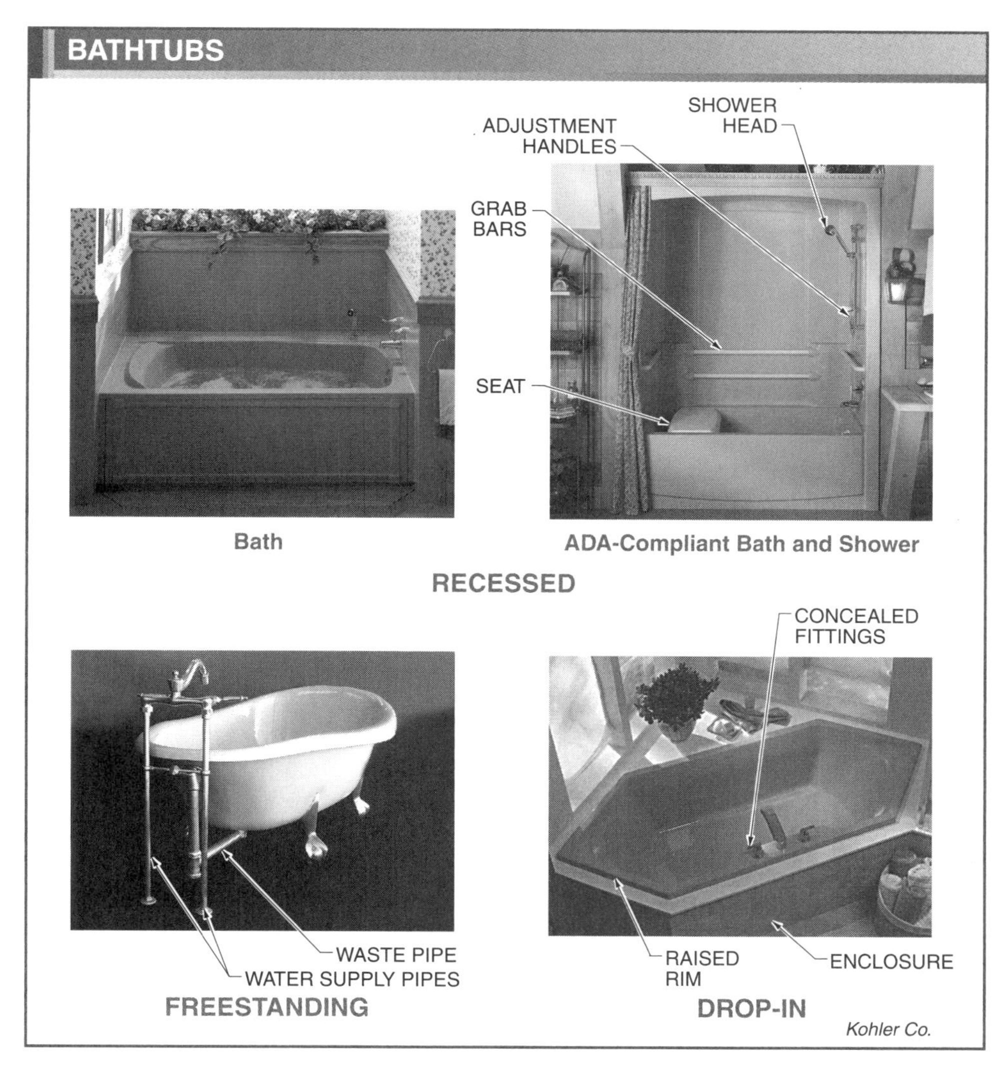

Figure 6-24. A bathtub is a plumbing fixture used to bathe the entire body and is available in recessed, freestanding, and drop-in designs.

Whirlpool Bathtubs

A *whirlpool bathtub* is a plumbing fixture equipped with water and air circulation equipment and used to bathe and massage the entire body. **See Figure 6-26.** An electric pump assembly enclosed below the whirlpool bathtub circulates water and air through hydrojets positioned along the sides and ends of the fixture. Whirlpool bathtubs are manufactured from enameled cast iron, fiberglass, or acrylic plastic, and are available in a variety of sizes, shapes, and colors. Whirlpool bathtubs, like built-in bathtubs, are rated at 2 dfu and require a minimum of 1½″ waste and 1¼″ vent pipes. Since whirlpool tubs usually contain more water than built-in bathtubs, many whirlpool bathtubs are installed with a 2″ waste pipe to allow the fixture to drain faster.

Whirlpool bathtubs are right-hand or left-hand, depending on the waste opening location, and have a built-in pitch to provide proper drainage of the fixture yet allow the flood level rim to remain level. Whirlpool bathtubs are available in recessed and drop-in designs. For larger whirlpool bathtubs, additional framing support will be required below the fixture to support the combined weight of the bathtub, the water it contains, and the user of the tub.

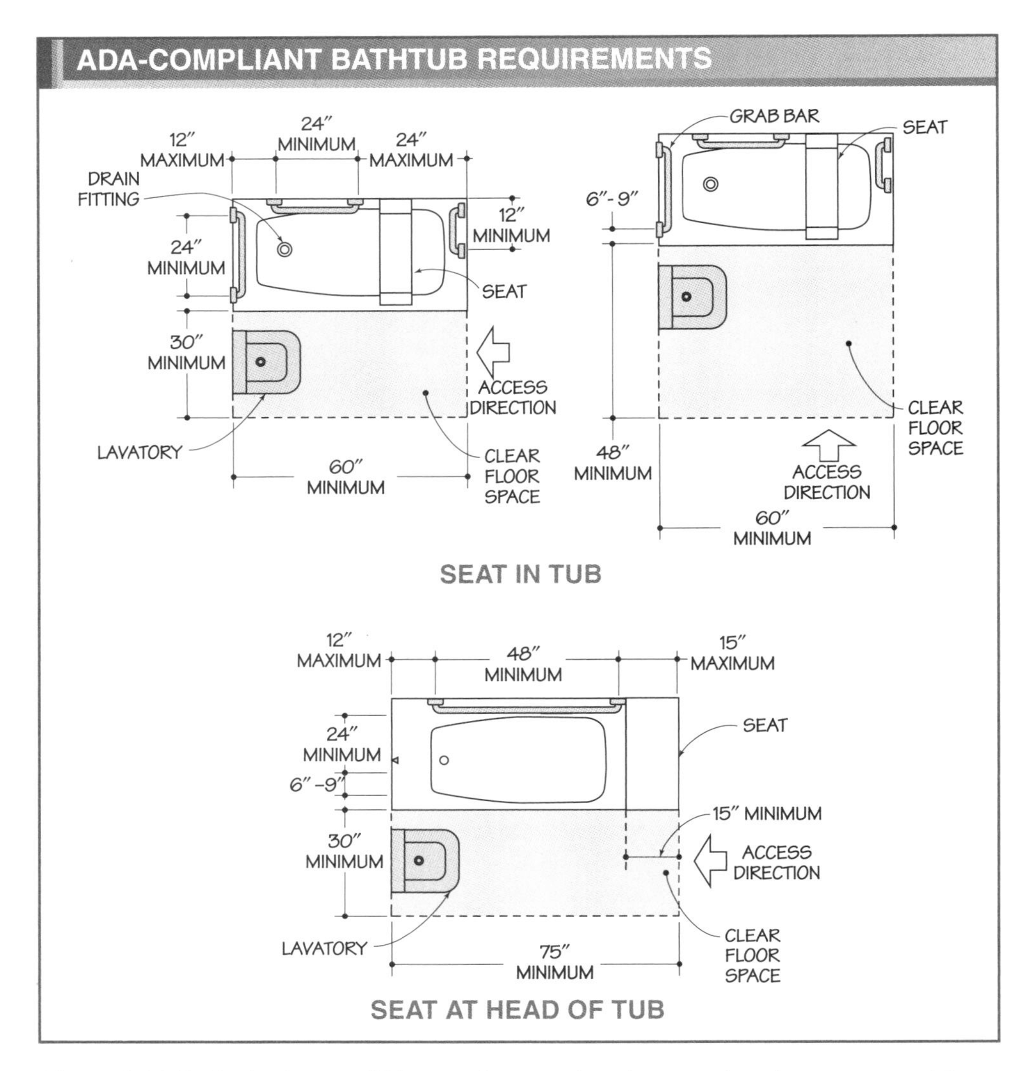

Figure 6-25. Bathtubs that are ADA-compliant require adequate clear floor space, grab bars, and a seat.

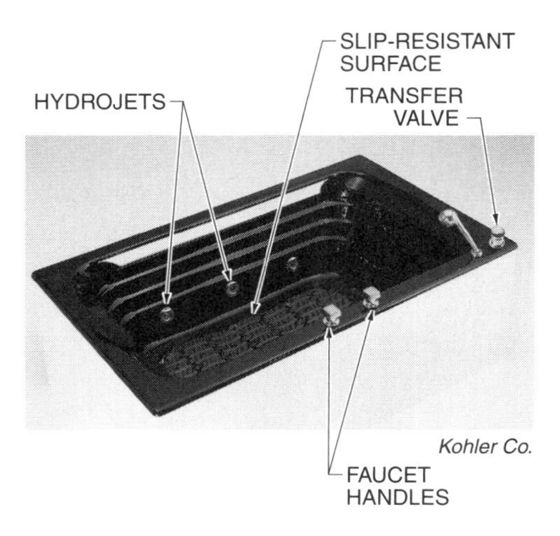

Figure 6-26. Whirlpool bathtubs are equipped with hydrojets that provide massaging action.

Bathtub Faucets

Water is supplied to standard and whirlpool bathtubs through overrim bathtub fittings. An *overrim bathtub fitting* is a bathtub water supply fitting that consists of a faucet assembly and mixing spout. Overrim bathtub fittings are installed in the foot of a bathtub with the spout outlet above the flood level rim of the fixture to prevent a cross-connection. Overrim bathtub fitting faucets are compression faucets or port control faucets. **See Figure 6-27.**

In many installations, a bathtub is used as a bathtub and shower, requiring a combination bath and shower fitting to divert water flow from the faucet spout to the showerhead. The faucet spout of a combination bath and shower fitting contains a diverter attached to an external knob which is raised to divert water flow from the bathtub spout to the showerhead.

Bathtub Drain Fittings

A *combination waste and overflow fitting* is a bathtub drain fitting that is an outlet for bathtub waste and allows excess water to drain from the fixture so that it does not overflow onto the bathroom floor. **See Figure 6-28.** Combination waste and overflow fittings are manufactured from plastic or brass tubing.

A *lift waste fitting* is a combination waste and overflow fitting in which a lifting mechanism within the overflow tube is connected by a lever to the stopper in the waste shoe (bathtub drain outlet). When the control handle is turned in one direction, the stopper raises and allows water to drain from the bathtub. When the control handle is turned in the opposite direction, the stopper drops into the drain and allows the bathtub to fill with water. All the parts of a lift waste fitting are accessible through the tub overflow and the waste outlet.

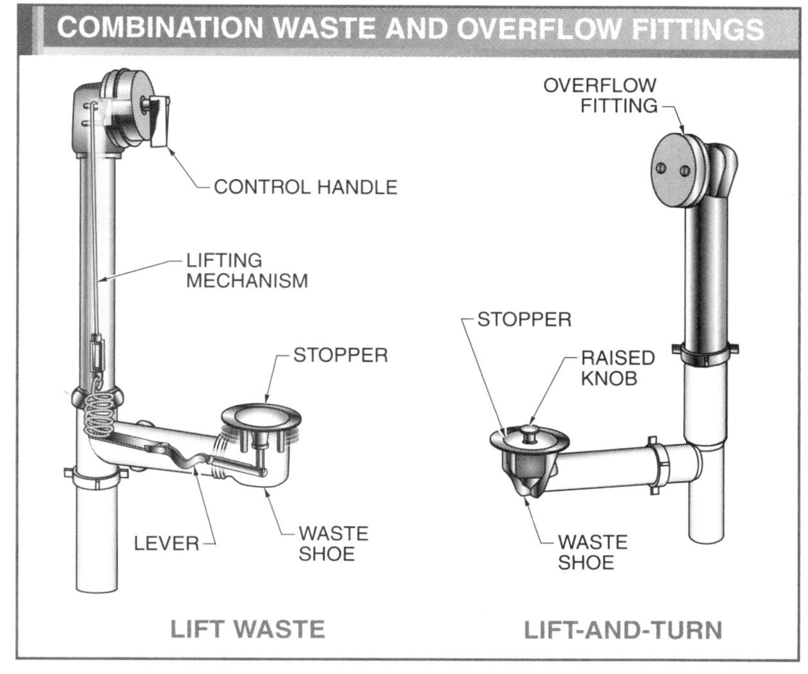

Figure 6-28. Combination waste and overflow fittings allow excess water to drain from a fixture so that it does not overflow onto the bathroom floor.

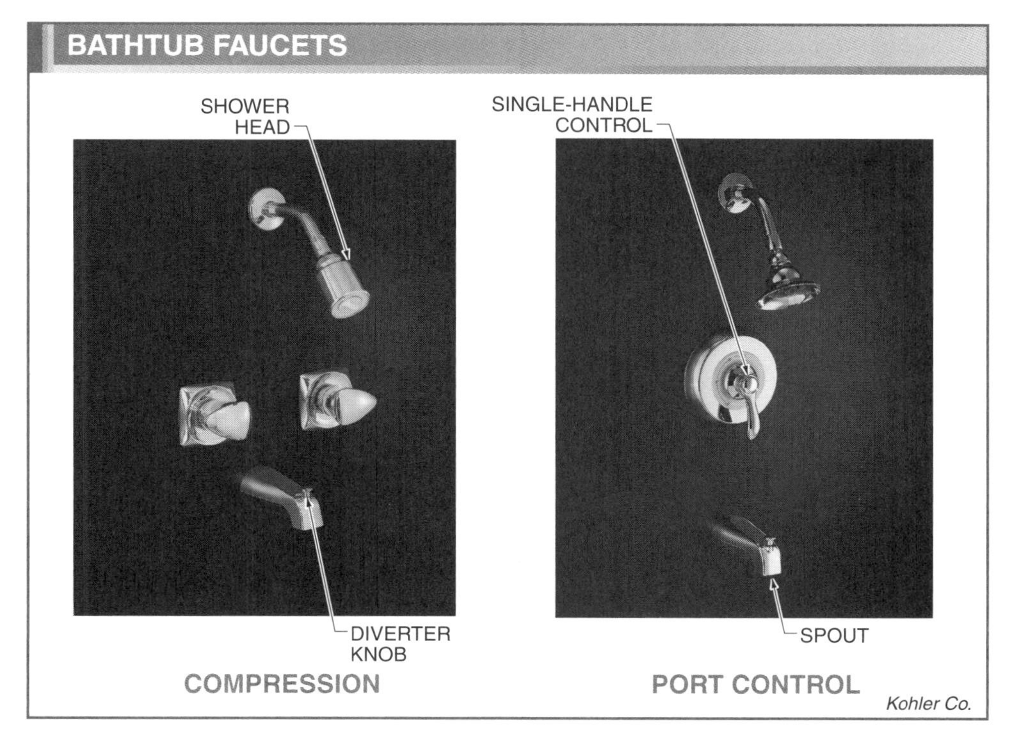

Figure 6-27. Overrim bathtub fittings consist of a faucet assembly and mixing spout.

A *lift-and-turn waste fitting* is a combination waste and overflow fitting consisting of a stopper with a raised knob at the top that is raised and rotated to allow the fitting to remain in the drain position. The stopper is rotated and lowered down into the waste shoe to allow the bathtub to fill with water.

SHOWERS

A *shower,* or shower bath, is a plumbing fixture that discharges water from above a person who is bathing. Showers are rated at 2 dfu, and require 1½″ P-trap and waste pipe and 1¼″ vent pipes. However, most showers are supplied with a 2″ waste outlet to allow the fixture to drain faster. Showers require ½″ hot and cold water supply pipes. Showers vary in size and complexity from single-showerhead installations used in residential construction to a series of showerheads for shower rooms in commercial and public buildings, such as schools and fitness facilities.

A shower enclosure is constructed with one-piece or multipiece prefabricated assemblies. **See Figure 6-29.** One-piece assemblies are only used in new construction since they are difficult to fit through door openings and slide into position. One-piece fiberglass or acrylic assemblies provide a complete shower enclosure and eliminate potential leaks at the joints between the walls and shower base that may occur with multipiece assemblies. Shower doors equipped with tempered glass are installed along the shower entrance to prevent water from splashing out of the shower.

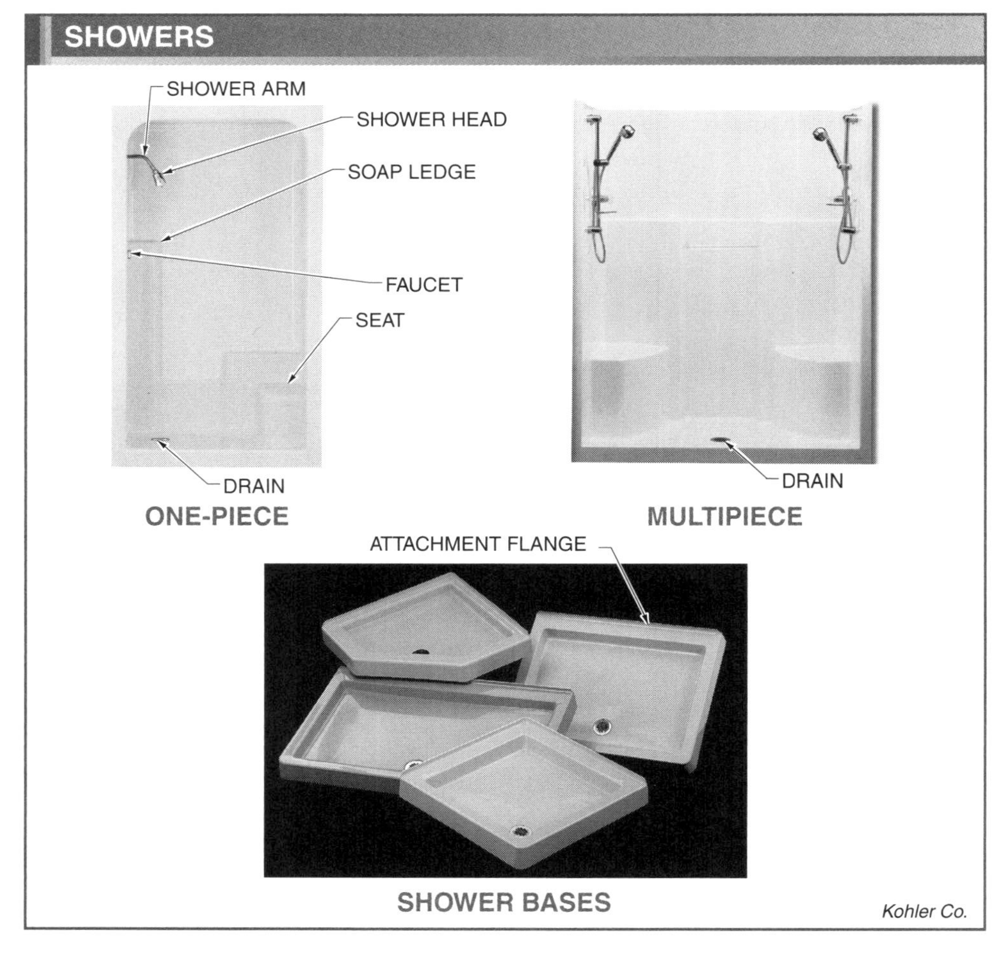

Figure 6-29. Shower enclosures are constructed with one-piece or multipiece prefabricated assemblies.

Multipiece assemblies consist of a shower base and wall panels and must be assembled on the job site per manufacturer instructions. Proper installation procedures must be followed to ensure a leakproof enclosure.

Ceramic tile shower baths may be constructed in almost any shape and size to fit the available space within the bathroom. For a ceramic tile shower, the plumber installs either a fiberglass or acrylic shower base or a waterproof membrane fastened to a drain body. The tile setter completes the construction of the shower enclosure over water-resistant cement wallboard.

A shower base must be properly installed to ensure a sound and leakproof enclosure. Manufacturer instructions vary and should be referred to prior to installation of a shower base. Many fiberglass and acrylic shower bases require a mortar base. The drain pipe must be positioned before the base is applied to the subfloor. Mortar is spread and the shower base is positioned. In ceramic tile installations, a rubber membrane is applied directly below the mortar base to provide a leakproof surface, and tile is applied over the membrane to complete the shower base installation. The drain strainer is installed and secured to the membrane to ensure a leakproof joint.

ADA-Compliant Showers

ADA-compliant showers include a seat, grab bars, and an adjustable shower-head attached to a hose, and provide for easy user access. **See Figure 6-30.** An ADA-compliant shower seat is an integral feature of a shower enclosure or is separately installed 18″ above the shower floor. Grab bars are installed 33″ to 36″ above the shower floor. Control faucets are located approximately 36″ to 38″ above the shower floor. Large ADA-compliant showers provide wheelchair access into the stall for bathing.

Tech Facts

The process of heating water releases gases, such as chlorine, oxygen, hydrogen sulfide, and carbon dioxide, which are present in water.

Shower Water Supply Valves

Compression or port control faucets are installed in showers to control the volume and temperature of water flowing from the showerhead. **See Figure 6-31.** Plumbing codes require an integral temperature-control device in port control valves to maintain a safe water temperature and protect against a sudden change in water temperature. A *pressure-balancing valve* is an integral temperature-control device that prevents surges of hot and cold water through the use of a sensitive diaphragm within the device. Pressure-balancing valves are adjusted to the desired temperature with a screw stop located behind the trim plate, and maintain a constant water temperature within ±5°F. A *thermostatic valve* is an integral temperature-control device that regulates the flow of hot and cold water through the use of a thermostat within the device. A thermostat within the valve allows water up to a maximum predetermined temperature to pass through the valve. Water temperature is adjusted using a screw located behind the shower trim plate.

Shower Drains

Shower waste is discharged into a strainer fitting or floor drain in the bottom of the shower. The drain fitting is connected to a pipe extending up from a 2″ P-trap below the floor with a rubber compression gasket.

BIDETS

A *bidet* is a plumbing fixture used to bathe the external genitals and posterior parts of the body and also to provide relief for certain health conditions. Bidets consist of a low-set basin supplied with hot and cold water and a drain equipped with a pop-up waste fitting. **See Figure 6-32.** Bidets are rated at 2 dfu, and require 1½″ P-trap and waste pipe and 1¼″ vent pipe. Bidets require a minimum of ¼″ water supply pipe.

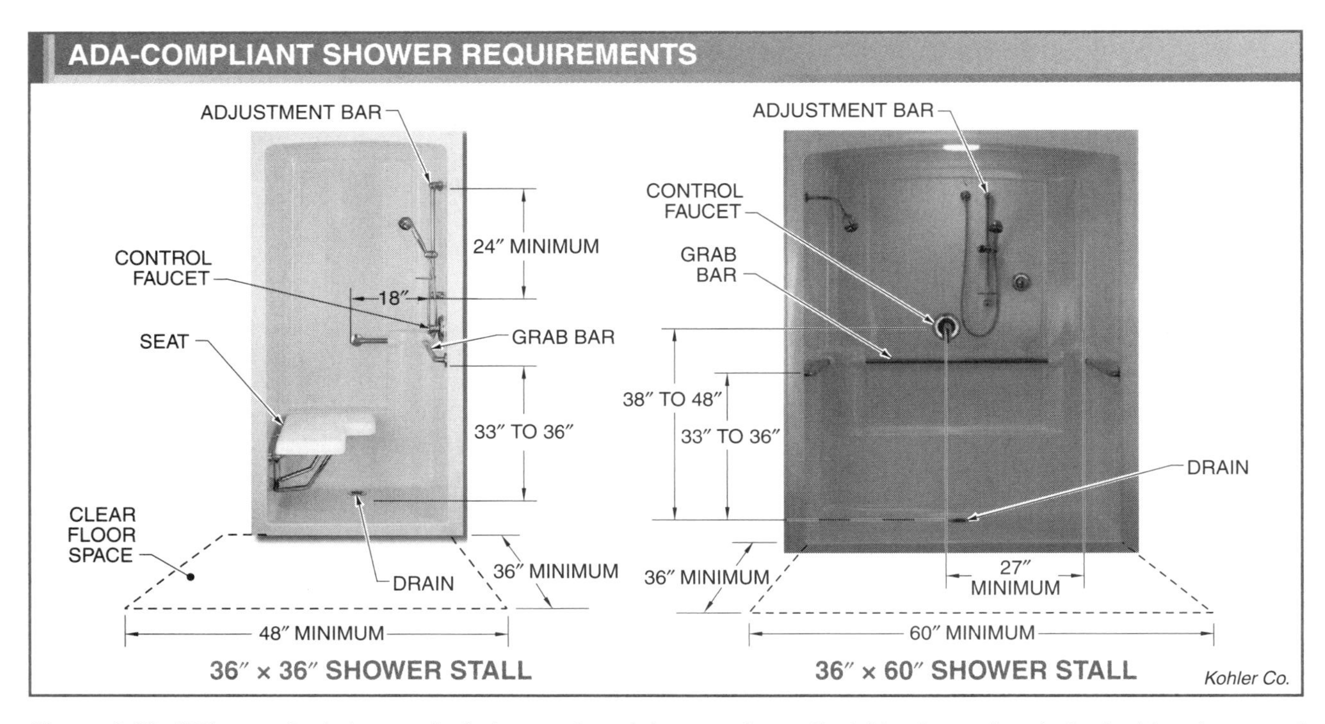

Figure 6-30. ADA-compliant showers include a seat, grab bars, and an adjustable shower head attached to a hose, and provide for easy user access.

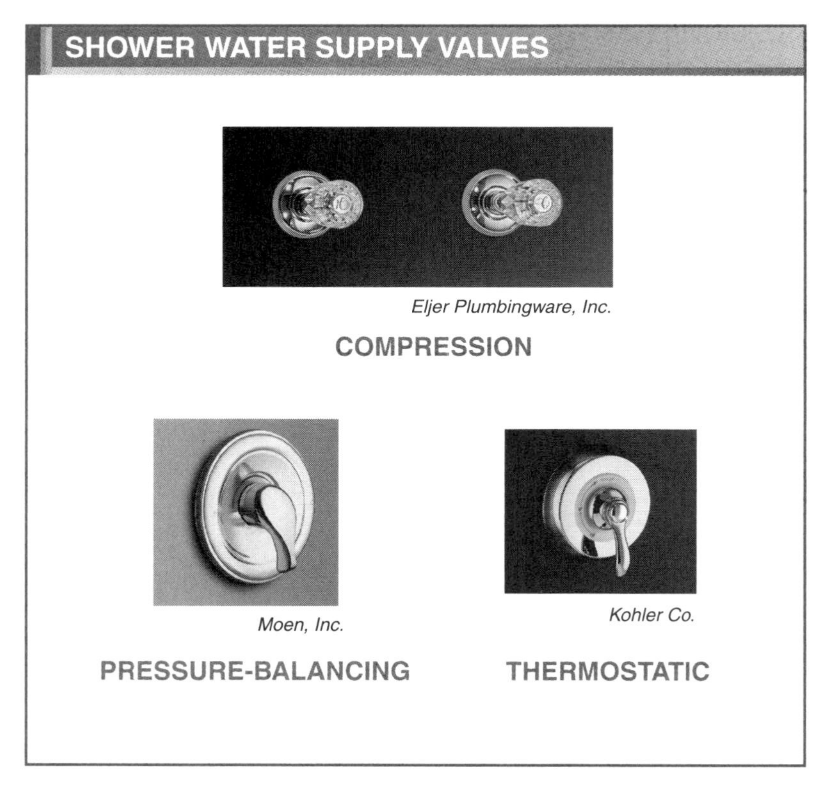

Figure 6-31. Compression and port control faucets with integral temperature control devices control the volume and temperature of water flowing from the shower head.

Figure 6-32. Bidets consist of a low-set basin supplied with hot and cold water and a drain equipped with a pop-up waste fitting.

A user sits astride the fixture and faces the hot and cold water faucet handles to wash the genital area. Water flow and temperature are controlled by faucets mounted on the back of the bidet; water enters the fixture through either a vertical spray in the center of the bowl or a flush rim that maintains bowl cleanliness.

KITCHEN SINKS

A kitchen sink is a shallow, flat-bottomed plumbing fixture that is used to clean dishes and prepare food. Kitchen sinks are available in a variety of shapes and sizes, and are manufactured from enameled cast iron, enameled pressed steel, stainless steel, and acrylic. Kitchen sinks are rated at 2 dfu, and require a 1½″ P-trap and waste pipe and 1¼″ vent pipe. Kitchen sinks require a minimum of ½″ water supply pipe. A kitchen sink has up to three basins, and is available in self-rimming and under-counter models. **See Figure 6-33.**

A *self-rimming kitchen sink* is a kitchen sink in which the bowl is placed in an opening in the countertop and the edge of the fixture rests directly on top of the countertop. A bead of caulk is placed under the sink edge prior to installation, and it is secured to the countertop with rim clamps.

An *undercounter kitchen sink* is a kitchen sink attached to the underside of a countertop using rim clamps or other proprietary hardware. A bead of caulk is placed along the upper surface of the edge prior to installation so water does not leak into the cabinet. Some acrylic under-counter sinks are bonded to the acrylic countertop with epoxy when it is fabricated.

ADA-Compliant Kitchen Sinks

ADA-compliant kitchen sinks feature a shallow bowl with the drain opening offset to the back, a single-handle faucet, and wrapped or shielded water supply and drainage piping. **See Figure 6-34.** The maximum distance from the finished floor level to the flood level rim of an ADA-compliant kitchen sink is 34″. The kitchen sink and countertop must be supported by the wall and adjoining cabinets.

Kitchen Sink Faucets

Kitchen sink faucets are available with swing spouts so a single spout can supply water to more than one basin. **See Figure 6-35.** Compression or port control faucets provide water flow and temperature control. Some spouts are connected to the water supply with a flexible hose and can be removed from their bases for use as a sprayer. A separate flexible hose and sprayer attachment are also installed with kitchen sinks. Water flow from the sink spout is diverted to the sprayer when the lever on the sprayer is depressed.

Figure 6-33. Kitchen sinks are available in self-rimming, metal-framed, and under-counter models.

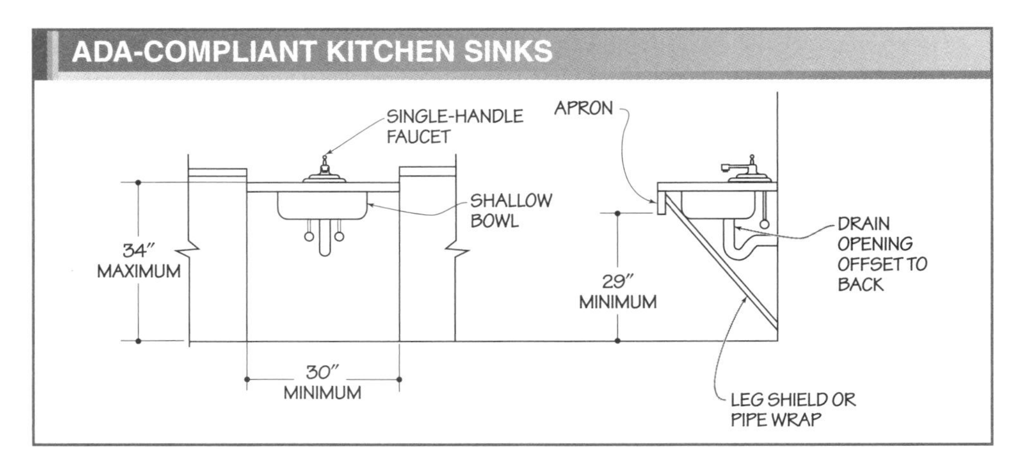

Figure 6-34. ADA-compliant kitchen sinks feature a shallow bowl with the drain opening offset to the back, a single-handle faucet, and wrapped or shielded water supply and drainage piping.

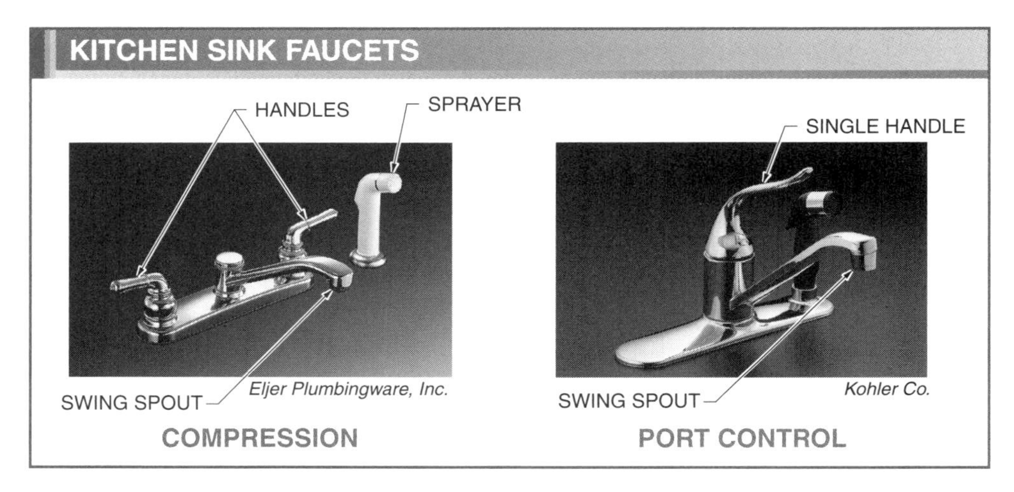

Figure 6-35. Kitchen sink faucets have swing spouts so a single spout can supply water to more than one basin.

Kohler

Lavatory faucets can have air gaps of several inches above the flood level rim to prevent back siphonage.

Kitchen Sink Drain Fittings

Kitchen sinks require a basket strainer for each basin to prevent solid food particles from entering and clogging the drainage piping. A *basket strainer* is a drain fitting installed in a kitchen sink that consists of a strainer body fitted with a fixed strainer and a removable basket with a rubber stopper. A bead of caulk or plumber's putty is placed around the drain opening of the sink basin. The strainer body is placed in the drain opening of the sink and a locknut and gasket below the sink secure the body in position. A flanged tailpiece connects the kitchen sink drain to the waste pipe. The removable basket, or crumb cup, is placed in the drain body and the knob is rotated until the stopper properly seats in the strainer body to retain water.

Continuous waste fittings are used when multibasin kitchen sinks are installed. **See Figure 6-36.** A *continuous waste fitting* is a drainage fitting that consists of a section of horizontal drainage pipe and sanitary tee and is used to convey waste from a kitchen sink drain to a common P-trap. Waste from the P-trap is conveyed to the drainage piping.

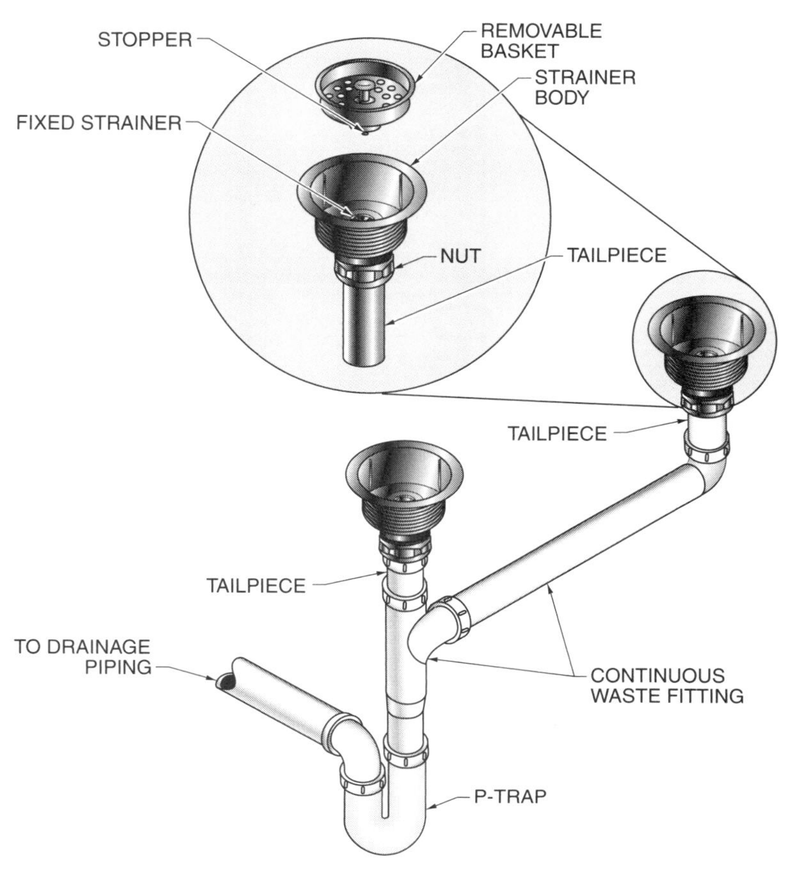

Figure 6-36. Continuous waste fittings consist of a section of horizontal drainage pipe and sanitary tee, and are used to convey waste from a kitchen sink drain to a common P-trap.

FOOD WASTE DISPOSERS

A *food waste disposer,* or garbage disposal, is an electric appliance supplied with water from the kitchen sink faucet, and that grinds food waste into pulp and discharges the pulp into the drainage system. Domestic food waste disposers are rated at 2 dfu; however, the drainage fixture unit value is not added to the drainage fixture unit load of the kitchen sink drainage piping, since the disposer waste is conveyed through the kitchen sink drain.

A domestic food waste disposer is mounted below a kitchen sink basin and replaces the basket strainer assembly. **See Figure 6-37.** Plumber's putty is placed around the drain opening of the sink basin. A sink flange is placed in the opening and is secured in place with a gasket and backer ring on the bottom of the sink. A mounting ring is then installed at the bottom of the flange and mounting screws are inserted and tightened in the mounting ring to provide a sound mounting surface for the food waste disposer. The food waste disposer is secured to the mounting ring and the electrical and waste connections are then made.

A food waste disposer installed in a multibasin kitchen sink discharges its waste into a continuous waste fitting. **See Figure 6-38.** A baffle tee joining the food waste disposer waste pipe and waste pipe for the other basin contains an internal baffle to divert the disposer waste downward and prevent it from backing up into the connected basin.

The mounting ring of a food waste disposer is installed at the bottom of the kitchen sink flange, and the mounting screws are tightened to provide a sound mounting surface for the disposer.

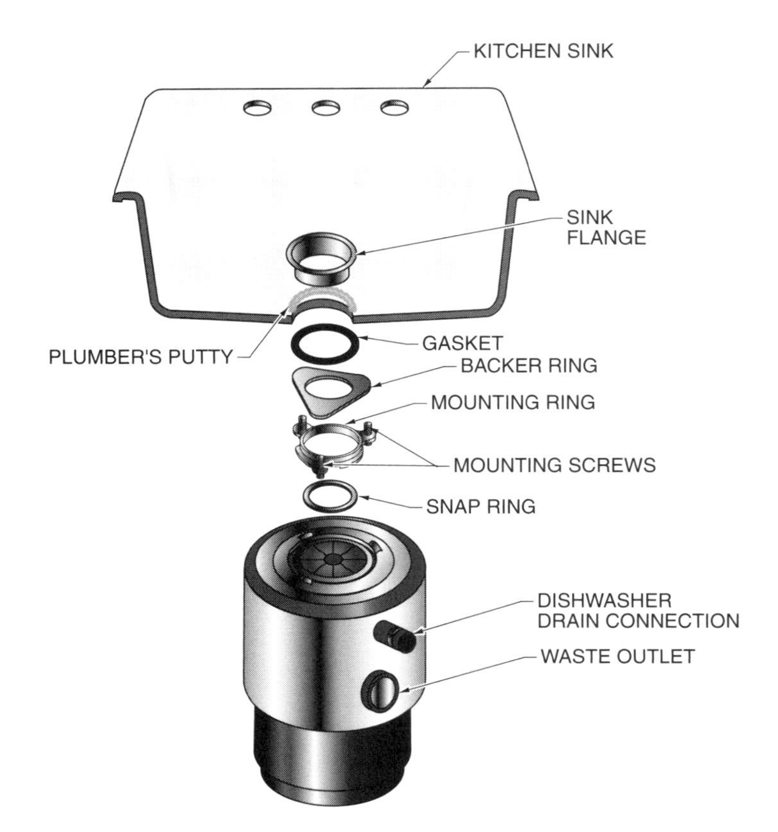

Figure 6-37. A food waste disposer grinds food waste into pulp and discharges the pulp into the drainage system.

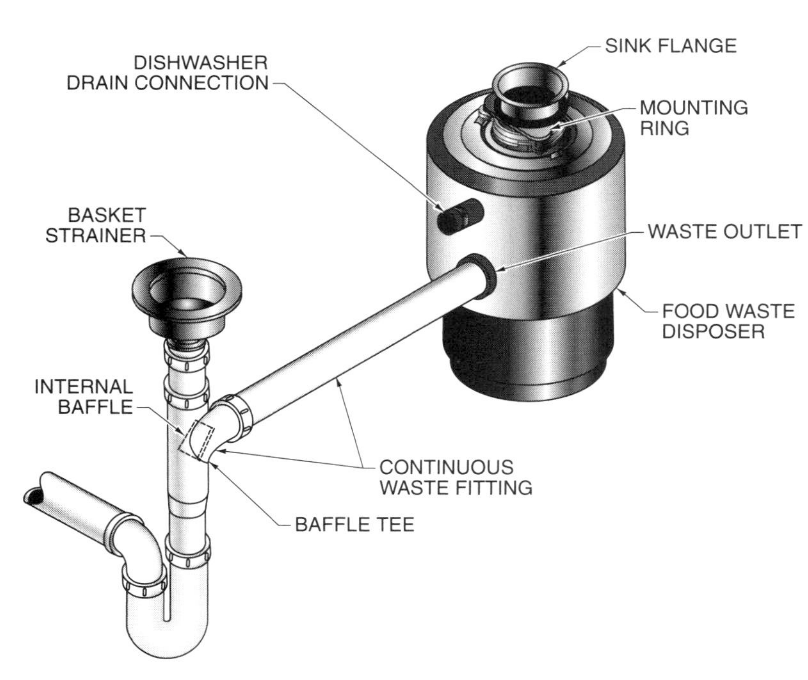

Figure 6-38. Food waste disposers installed in a multi-basin kitchen sink discharge waste into a continuous waste fitting.

Tech Facts

Domestic dishwashers are rated at 2 dfu. The waste is pumped out of the appliance and into the kitchen sink waste pipe or food waste disposer tailpiece.

DOMESTIC DISHWASHERS

A *dishwasher* is an electric plumbing appliance used to wash dishes. A domestic dishwasher is rated at 2 dfu. Domestic dishwasher waste is pumped out of the appliance into the kitchen sink waste pipe or food waste disposer tailpiece. The waste outlet of a domestic dishwasher with pumped-out waste is connected to a dishwasher drain connection on the food waste disposer or to a dishwasher tailpiece below the basket strainer using a rubber hose. A high loop must be installed in the rubber hose between the dishwasher and drain connection to allow the dishwasher to drain properly and to prevent backflow of waste from the kitchen sink and/or food waste disposer into the dishwasher. **See Figure 6-39.**

Some plumbing codes require installing an air gap fitting in discharge piping from domestic dishwashers to prevent backflow of waste from a kitchen sink into the appliance. A dishwasher air gap fitting is usually installed in one of the holes provided on the back ledge of the kitchen sink. The dishwasher discharge hose extends from the appliance to the air gap fitting and then to the kitchen sink drain or food waste disposer.

LAUNDRY TRAYS

A *laundry tray,* or laundry tub, is a plumbing fixture used for prewashing clothes that is commonly installed in a residential laundry room. Laundry trays are supplied with hot and cold water and a drain connection. A laundry tray may also receive waste from the clothes washer and store the water from the appliance if it is equipped with a water-reuse cycle. Laundry trays are rated at 2 dfu, and require a 1½″ P-trap and waste pipe, 1¼″ vent pipe, and ½″ hot and cold water supply piping.

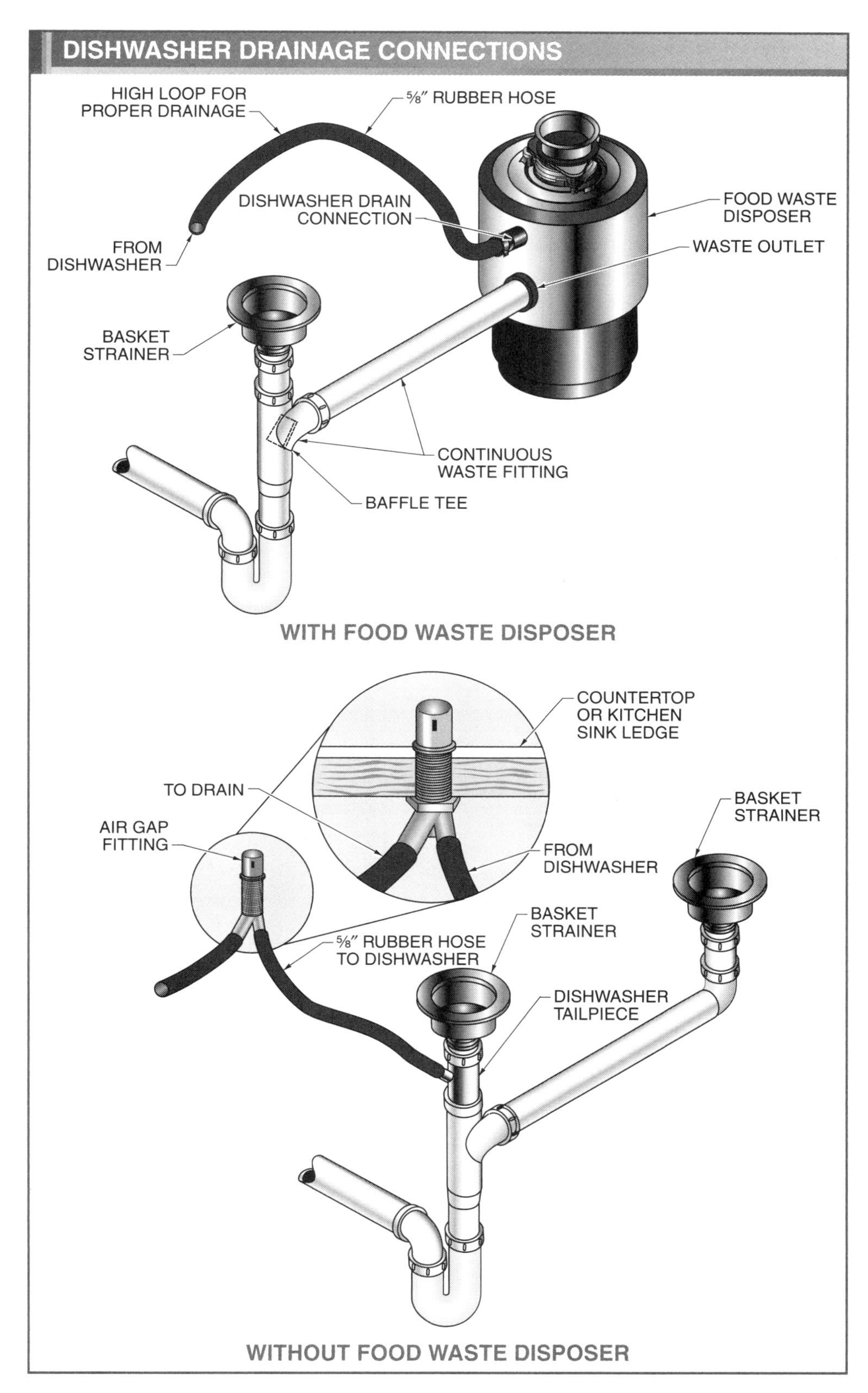

Figure 6-39. The waste outlet of a domestic dishwasher with pumped-out waste is connected to a dishwasher drain connection on the food waste disposer or to a dishwasher tailpiece below the basket strainer using a rubber hose.

Laundry trays are manufactured from fiberglass or plastic and are available in single-compartment or double-compartment models. **See Figure 6-40.** Laundry trays are wall-hung or floor-set. Wall-hung laundry trays are installed on mounting brackets supplied with the fixture, while floor-set models rest on plastic or steel legs.

Laundry Tray Faucets

Laundry tray faucets are two-handle compression faucets with swing spouts that are mounted on the back ledge of the fixture. Laundry tray faucets are supplied with water from above when the fixture is installed in the basement or lower level of a structure, or from below the laundry tray when the fixture is installed on upper levels. **See Figure 6-41.** Different styles of laundry tray faucets are used for each application.

Laundry Tray Drain Fittings

Laundry tray wastewater flows through a strainer fitting which uses a rubber stopper to retain water in the fixture. Strainer fittings are used in each laundry tray compartment to prevent lint and other debris from clogging the drainage piping. A cast iron or plastic laundry tray drainage fitting is installed for a double-compartment laundry tray to join the waste lines of both compartments before it is connected to a 1½″ P-trap.

CLOTHES WASHER OUTLET BOXES

Clothes washer outlet boxes are installed in residences in which the clothes washer is located on an upper level. A *clothes washer outlet box* is a plastic enclosure that accommodates water supply and waste connections for a clothes washer. **See Figure 6-42.** Outlet boxes are provided with or without valves preinstalled in the fixture. Knockouts are provided for the drainage piping and also for the valves if they are not preinstalled in the outlet box. Outlet boxes are installed by attaching the mounting brackets to studs when roughing in the plumbing for other fixtures. The outlet box is positioned with the faceplate flush with the finished wall surface. Although washing machine outlet boxes are rated at 2 dfu, they require a 2″ P-trap and waste pipe in order to accommodate the volume of water pumped from a clothes washer. The box requires a minimum of a 1¼″ vent and ½″ hot and cold water supply piping.

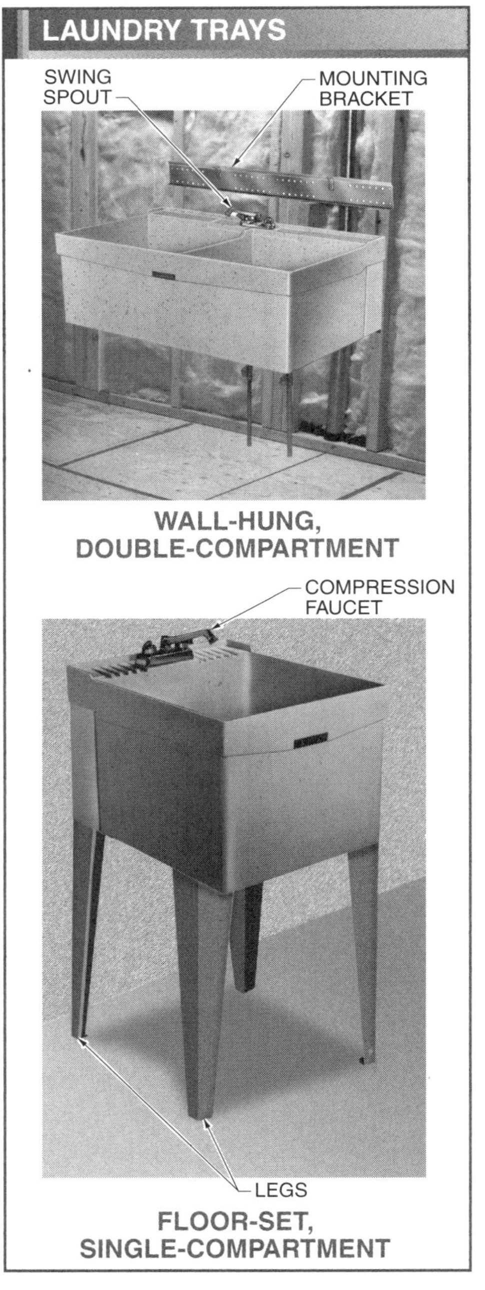

Figure 6-40. Wall-hung and floor-set laundry trays are manufactured from fiberglass or plastic and are available in single-compartment or double-compartment models.

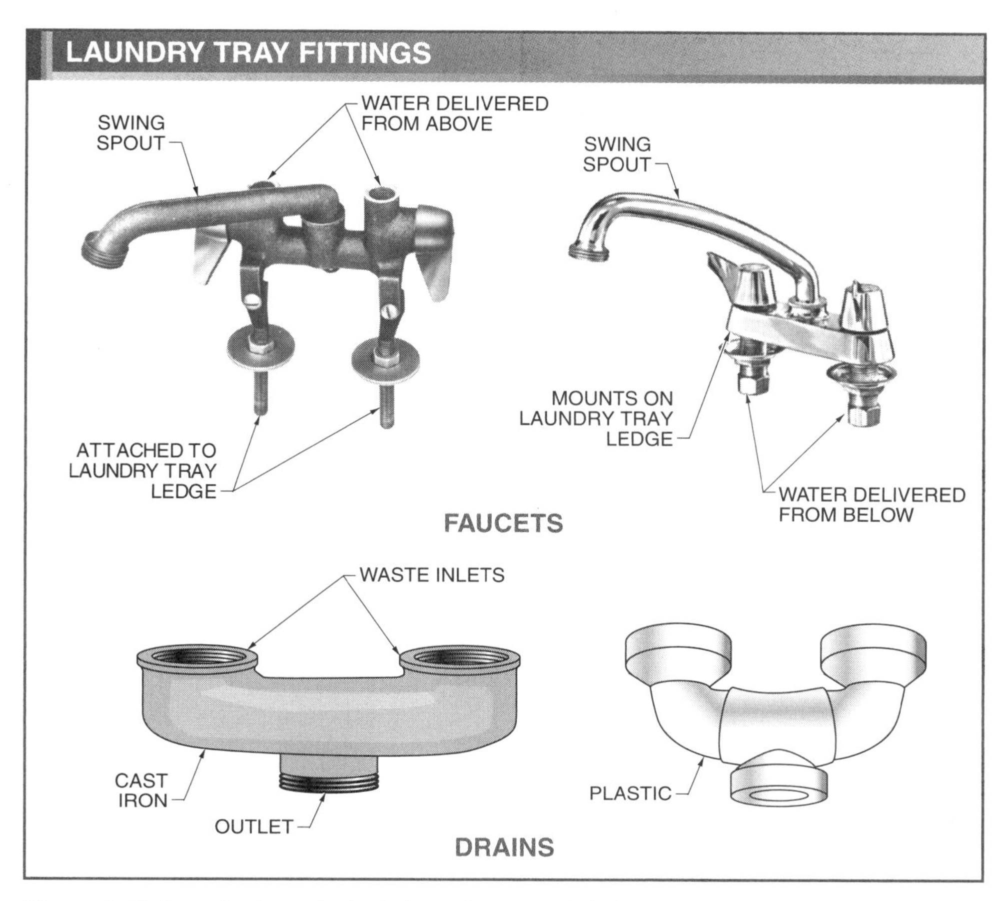

Figure 6-41. Laundry tray trim includes a faucet and drain.

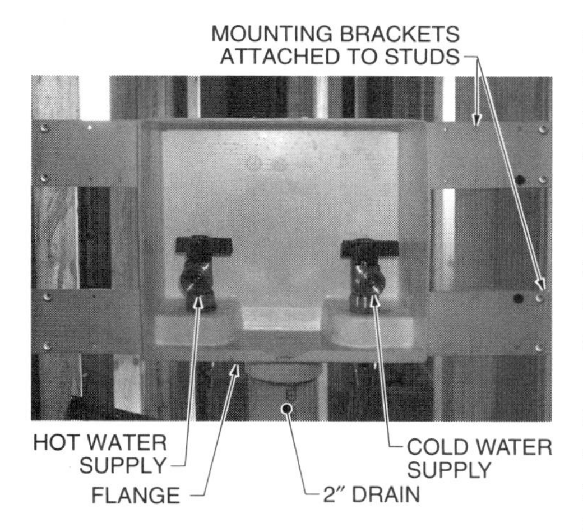

Figure 6-42. Clothes washer outlet boxes accommodate water supply and waste connections for a clothes washer.

FLOOR DRAINS

A *floor drain* is a cast iron or plastic plumbing fixture set flush with the finished floor and used to receive water drained from the floor and convey it to the drainage system. In addition, floor drains are used in locations where there is a possibility of overflow, leakage, and/or spillage of liquid waste onto the floor, such as residential laundry and utility rooms, public rest rooms, custodial closets, shower room entrances and exits, building entryways, and garages.

The dfu rating of a floor drain depends on the outlet size of the drain. A 2″ floor drain is rated at 2 dfu, a 3″ floor drain is rated at 3 dfu, and a 4″ floor drain is rated at 4 dfu. The dfu rating of a floor drain is added to the sanitary drainage piping dfu rating into which the floor drain discharges its waste; the dfu rating is not added to the vent piping. Some plumbing codes do not require individual vents for floor drains if they are installed within 25 feet of a vented drainage pipe. Plumbing codes require that floor drains used as part of a fixture, such as a shower drain, or for any receptor with a built-up threshold and a water supply, be equipped with a common seal P-trap that is vented according to the Trap-to-Vent Distance table.

Floor Drain Types

A variety of floor drains are available, including floor drains with and without integral P-traps and dry pan floor drains. **See Figure 6-43.** Floor drains with integral P-traps are installed below ground level in buildings. Floor drains without integral traps are typically installed in upper floor levels and require separate P-traps. A dry pan floor drain is installed in areas of buildings subject to freezing, such as entryways and garages. Water collected by dry pan floor drains is piped to a trap located in a heated area of the building.

Floor Drain Minimum Requirements

Plumbing codes have minimum requirements for floor drains to ensure adequate drainage capacity and to prevent backflow, including:

- 2″ minimum outlet size
- Removable grate or strainer
- Combined free area of the holes in a grate or strainer equal to or greater than drain outlet size
- Deep-seal P-trap with 3″ minimum trap seal
- Basement floor drains equipped with a backwater valve. **See Figure 6-44.**

Tech Facts

Floor drains that contain a ball-type backwater valve and cleanout are typically used in basements. The ball-type backwater valve is used to prevent wastewater from rising through the drain and into the living area.

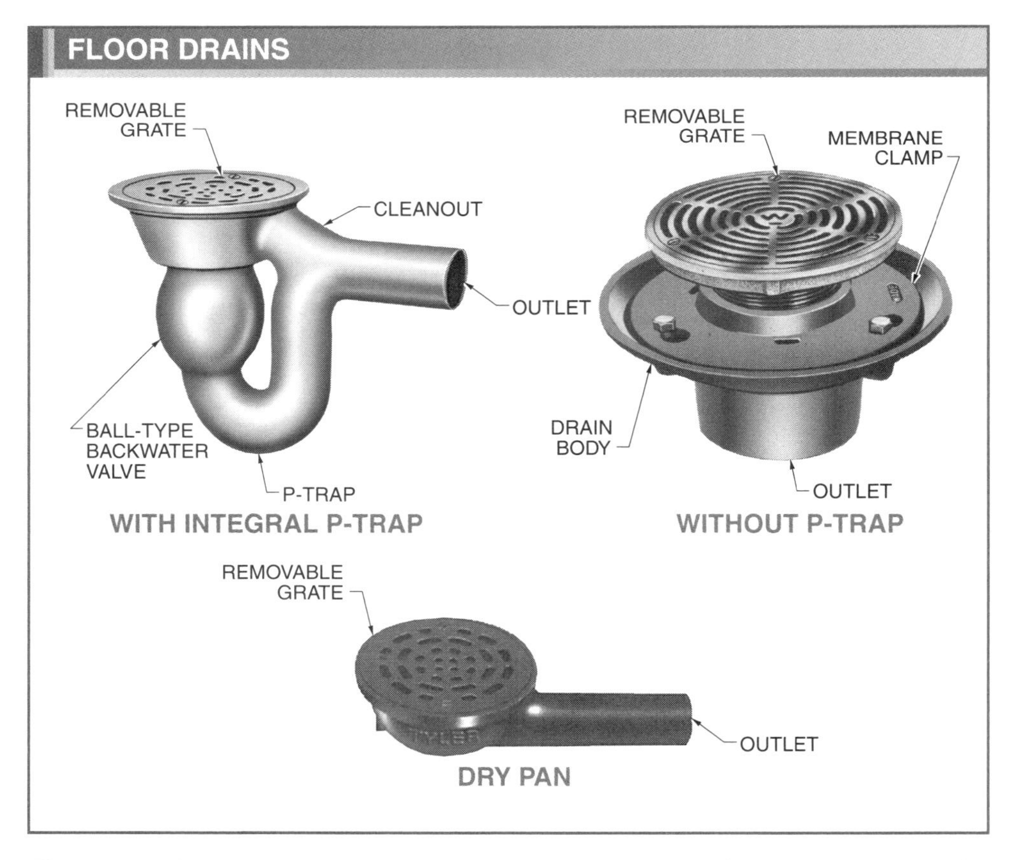

Figure 6-43. Cast iron or plastic floor drains are set flush with the finished floor and receive water drained from the floor.

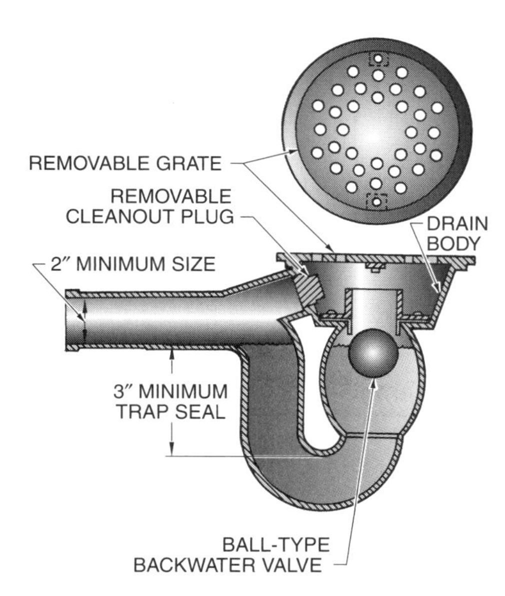

Figure 6-44. Basement or below-grade floor drains must be equipped with backwater valves.

Floor Sinks

A *floor sink* is a floor drain installed in commercial kitchens and food markets to indirectly receive waste from food preparation and storage equipment and fixtures. **See Figure 6-45.** Floor sinks are indirect waste receptors, and receive waste from equipment and fixtures that are not directly connected to the drainage system. An indirect waste receptor is installed so that if the drainage piping becomes clogged, waste backs up onto the floor rather than into food preparation and storage equipment and fixtures. Floor sinks are manufactured from enameled cast iron and stainless steel, and are available with 2″, 3″, and 4″ outlet sizes. A dome strainer is usually installed over the waste outlet of a floor sink to catch large food scraps, and a removable grate covers the fixture at floor level. Various grate sizes and shapes accommodate drain outlets from different equipment piped to the floor sink.

Floor sink dfu ratings vary with the drain outlet size—a 2″ floor sink is rated at 2 dfu, a 3″ floor sink is rated at 3 dfu, and a 4″ floor sink is rated at 4 dfu. Floor sinks must be properly vented; underground floor sinks require a minimum 2″ vent.

DRINKING FOUNTAINS AND WATER COOLERS

A *drinking fountain* is a wall-hung plumbing fixture that delivers a stream of drinking water through a nozzle at an upward angle to permit the user of the fixture to conveniently drink from the fountain. **See Figure 6-46.** Drinking fountains are manufactured from vitreous china, enameled cast iron, and stainless steel. A *water cooler* is a wall-hung plumbing appliance that incorporates an electric cooling unit into a drinking fountain to provide cooled drinking water at a desired temperature. The cooling unit reduces the temperature of the drinking water to a desired temperature before delivering it to the nozzle. The ideal temperature for drinking water is 50°F. Drinking fountains and water coolers are rated at 1 dfu, and require a 1¼″ P-trap, waste pipe, and vent pipe. Drinking fountains and water coolers require a minimum of ½″ water supply pipe.

The installation height of drinking fountains and water coolers varies depending on where the fixtures are installed. Standard drinking fountain and water cooler heights range from 36″ to 44″ from the finished floor level to the nozzle. The nozzles of drinking fountains and water coolers installed in elementary schools are 30″ above finished floor level. ADA-compliant drinking fountains and water coolers are installed with the nozzles a maximum of 36″ above finished floor level. Bilevel drinking fountains and water coolers are available to accommodate varying user heights. Bilevel drinking fountains share a common cooling unit, water supply, and drain.

Drinking Fountain and Water Cooler Sanitary Features

Drinking fountains and water coolers are installed in public areas such as hallways and gymnasiums. Plumbing codes require drinking fountains and water coolers to be manufactured with the following:

- A bowl constructed of a nonabsorbent material
- A mouth protector provided over the nozzle

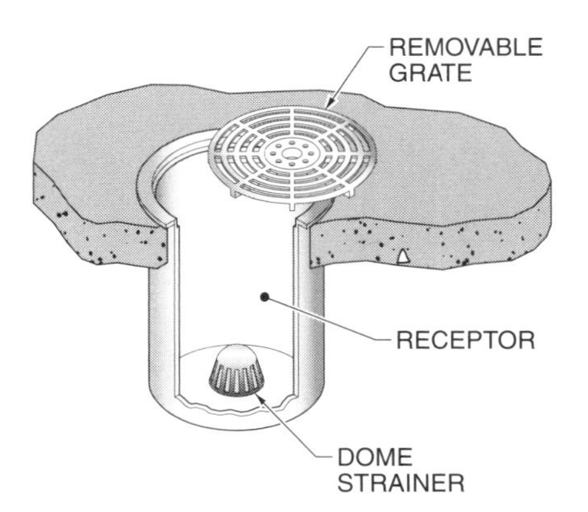

Figure 6-45. Floor sinks are installed in commercial kitchens and food markets to receive waste from food preparation and storage fixtures and equipment.

- The drinking water delivered at an angle so water cannot fall back onto the nozzle
- A nozzle located above the flood level rim
- The water supply separate from the waste pipe

Plumbing codes prohibit the installation of drinking fountains and water coolers in rest rooms for sanitary reasons.

Drinking fountain and water cooler water supply fittings consist of a self-closing compression stop valve and a stream-regulating valve attached to the fountain nozzle. A self-closing compression stop valve reduces the amount of wasted drinking water. A strainer fitting is installed at the drain opening, connected to a 1¼″ P-trap.

ADA-Compliant Drinking Fountains

ADA-compliant drinking fountains or water coolers are installed with the nozzle a maximum of 36″ from the finished floor level. In addition, there must be a minimum of 27″ of knee clearance and 9″ of toe clearance at the base of the fixtures to accommodate wheelchair-bound users. **See Figure 6-47.** The amount of clear floor space required varies with the installation. Drinking fountains or water coolers recessed in an alcove require a minimum of 30″ to 48″ of clear floor space. Built-in or freestanding drinking fountains and water coolers require a minimum of 48″ by 30″ of clear floor space in front of the fountain.

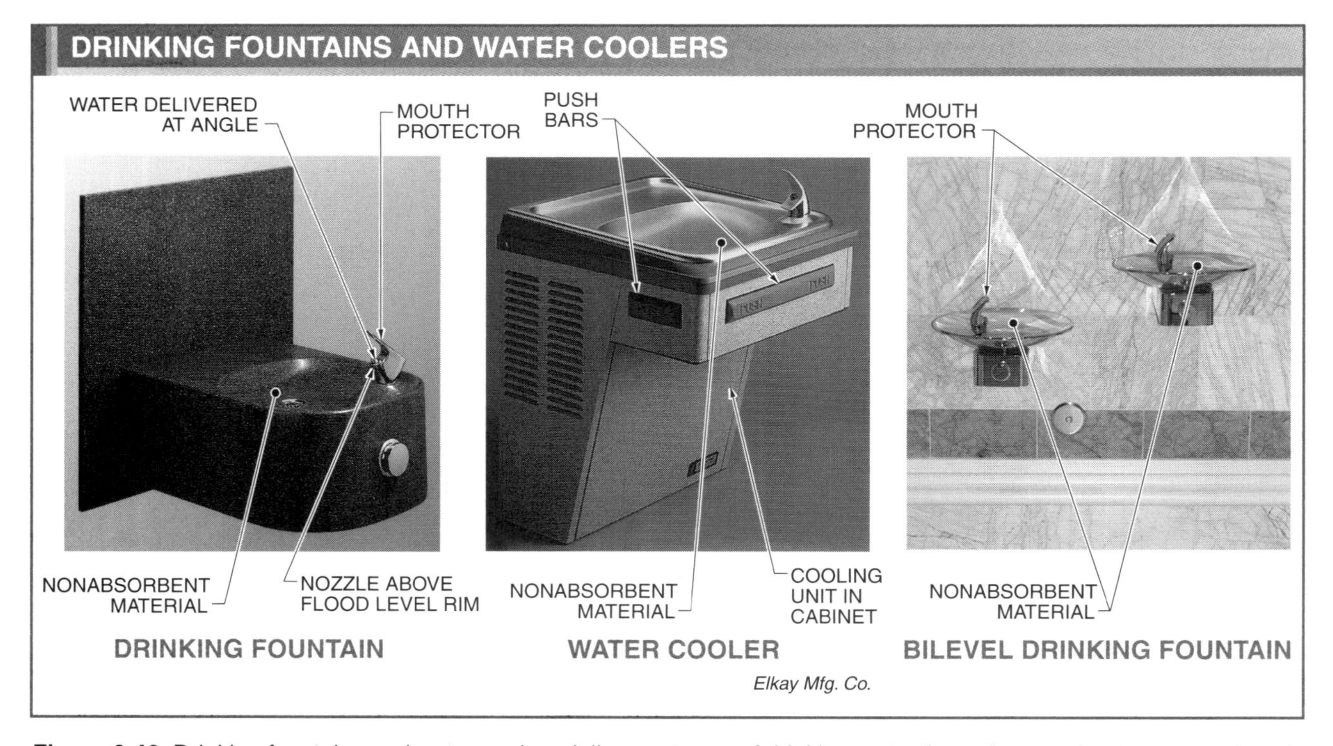

Figure 6-46. Drinking fountains and water coolers deliver a stream of drinking water through a nozzle at an upward angle to allow a user to conveniently drink from the fixture.

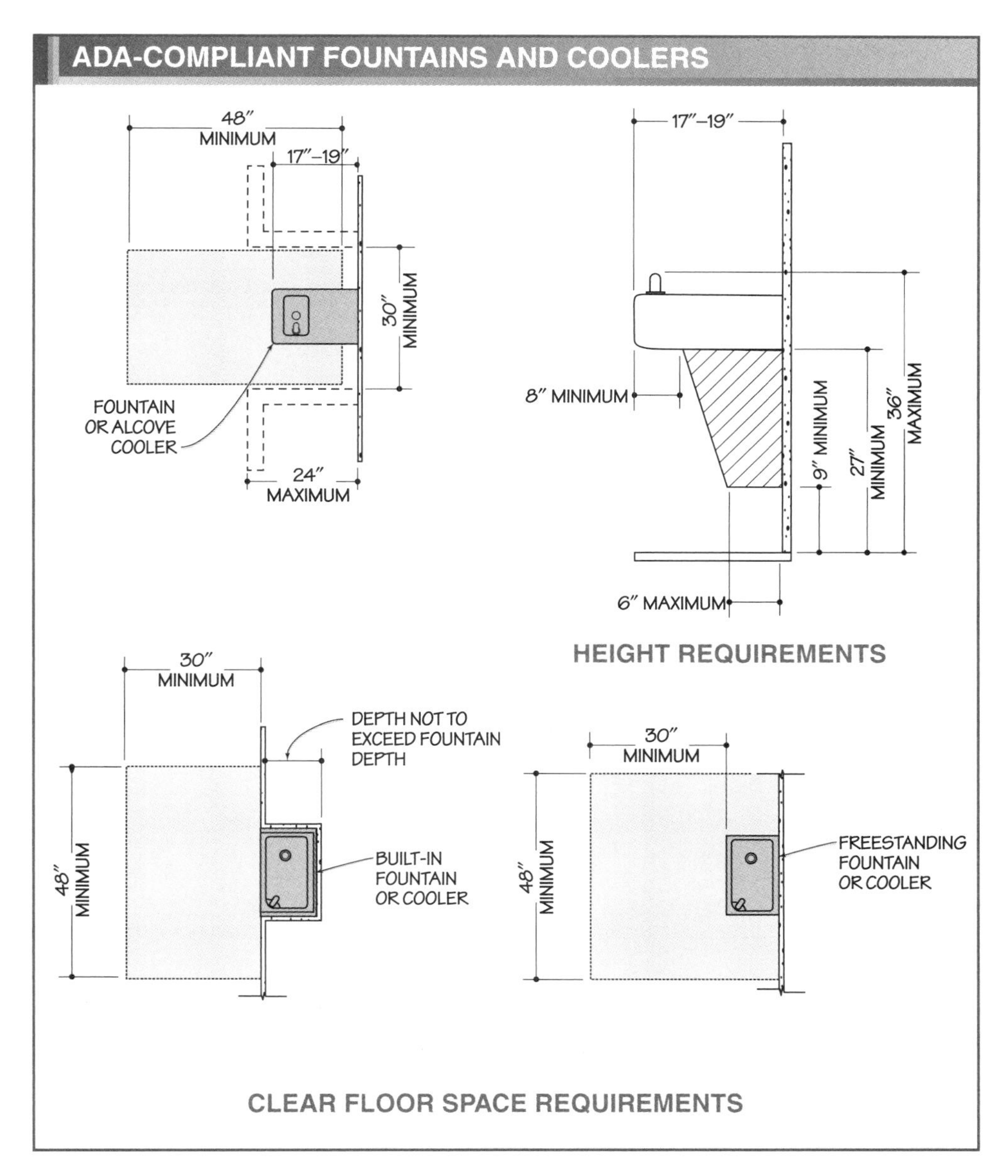

Figure 6-47. Drinking fountains and water coolers that are ADA-compliant are installed with their nozzles a maximum of 36″ from the finished floor level and require a minimum knee clearance of 27″ and toe clearance of 9″ at the base of the fixtures.

SERVICE SINKS AND MOP BASINS

Service sinks and mop basins are installed in custodian closets and building maintenance areas for use by custodial and maintenance personnel. A *service sink,* or slop sink, is a high-back sink with a deep basin used for filling and emptying scrub pails, rinsing mops, and disposing of cleaning water. Service sinks are manufactured of enameled cast iron and are supplied with a common seal P-trap. **See Figure 6-48.** A *mop basin,* or mop receptor, is a floor-set basin used for cleaning and other maintenance tasks. Mop basins are manufactured of fiberglass or enameled cast iron and are available in a variety of sizes. Service sinks and mop basins are rated at 3 dfu, and require a 2″ P-trap and waste pipe and 1½″ vent pipe.

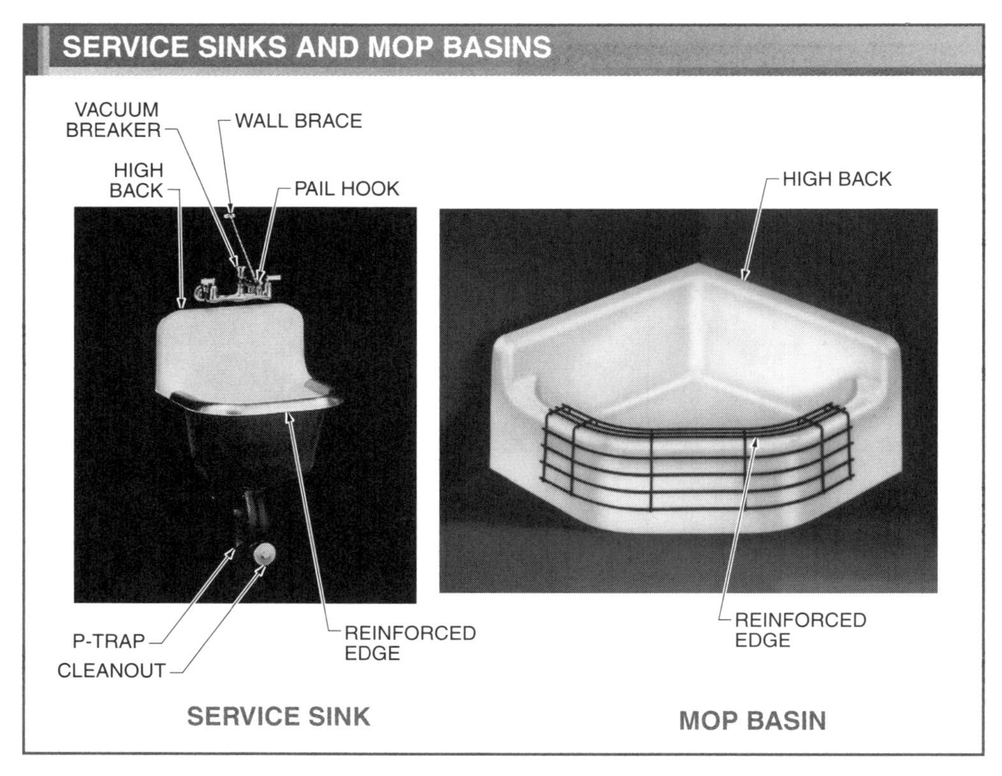

Figure 6-48. Service sinks and mop basins are installed in custodian closets and building maintenance areas for use by custodial and maintenance personnel.

Service Sink and Mop Basin Water Supply Fittings

A compression faucet with hose threads and a pail hook is usually installed for service sinks and mop basins. Service sink and mop basin faucets are equipped with a wall brace to support the weight of a bucket of water. Since the faucet is equipped with hose threads cast onto the spout, the fixture requires an integral vacuum breaker or a hose thread vacuum breaker. Service sinks and mop basins are provided with either ½″ or ¾″ hot and cold water supply pipes.

Service Sink and Mop Basin Drain Fittings

Service sinks are provided with a strainer fitting for the waste outlet and an exposed P-trap. While not required by plumbing codes, most P-traps have an integral cleanout to clear stoppages in the drainage piping. Mop basins are provided with a strainer fitting that is connected to a P-trap installed below the floor level.

WATER SOFTENERS

Potable water contains varying amounts of natural minerals such as calcium, magnesium, and sodium. These minerals dissolve in water as it passes through underground limestone deposits. Small amounts of calcium and magnesium are acceptable for potable water, but larger amounts can result in mineral deposits on skin, clothing, piping, and other surfaces that come into contact with the water. *Hard water* is potable water that contains excessive amounts of calcium and magnesium.

Water hardness is measured in either grains per gallon or parts per million (ppm). One grain is equal to .002 ounces of calcium or magnesium per U.S. gallon of water. Most water supplies contain from 3 grains per gallon to 50 grains per gallon. (By comparison, a common aspirin tablet contains 5 grains of aspirin. Therefore, water 5 grains hard would contain one aspirin tablet of hardness minerals per U.S. gallon of water.)

One part per million equals 1 lb of calcium or magnesium per 1 million pounds of water. One grain per U.S. gallon equals 17.1 parts per million.

Some effects of hard water include the following:

- Calcium and magnesium curdle soaps and some detergents, reducing the amount of lathering.
- Curdled soap clogs clothing fabric, dulling colored fabrics and graying white fabrics.
- Curdled soap sticks to skin after bathing and leaves a dull film on hair.
- Dishes, glasses, and silverware are water spotted and eventually may become etched by the minerals.
- A scum or ring is left in bathtubs, lavatories, and sinks.
- When hard water is heated, calcium tends to precipitate out (or drop out) of the water. This calcium deposit builds up a scale in the flush rims of water closets and urinals as well as in hot water pipes, water heaters, pots and pans, coffee makers, boilers, and radiators, reducing their efficiency.

Water Softening Process

Water softeners remove minerals present in potable water. When minerals are dissolved in water, they break down into two or more components known as ions. *Ions* are individual atoms or groups of atoms that carry an electrical charge. In any stable liquid, such as water, the number of positively charged ions is equal to the number of negatively charged ions. Dissolved calcium and magnesium are positively charged ions and result in an imbalance in the number of negatively and positively charged ions in the solution. Therefore, positively charged "hard" calcium and magnesium ions must be exchanged for negatively charged "soft" sodium ions. A *water softener* is a plumbing appliance that removes dissolved minerals, such as calcium and magnesium, from water by an ion-exchange process.

Zeolite is a synthetic resin bead used as the ion-exchange medium in the ion-exchange process. Zeolites remove mineral ions from solution and replace them with sodium (salt) ions through an interaction that occurs on the surface of the granular zeolite particles in the water softener. **See Figure 6-49.** The operation of the ion-exchange process is as follows:

1. Zeolite, which is precharged with sodium ions, is contained within the water softener tank.
2. During operation, untreated water flows into the softener and the zeolite removes calcium and magnesium ions from the water and replaces them with sodium ions. Treated "softened" water exits the water softener outlet.
3. The zeolite surface becomes saturated with mineral ions and is depleted of sodium ions, gradually slowing the ion-exchange process.
4. The zeolite is regenerated by stripping the calcium and magnesium ions from the zeolite and recharging the zeolite by running a brine (concentrated sodium chloride) solution through the water softener.

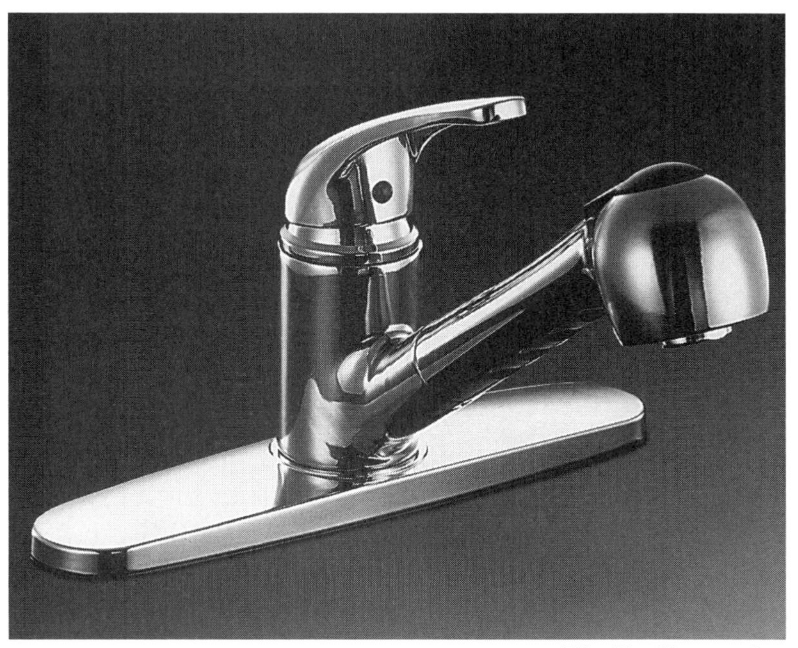

Eljer Plumbingware, Inc.

Kitchen sink faucets may incorporate a pull-out sprayer into the spout.

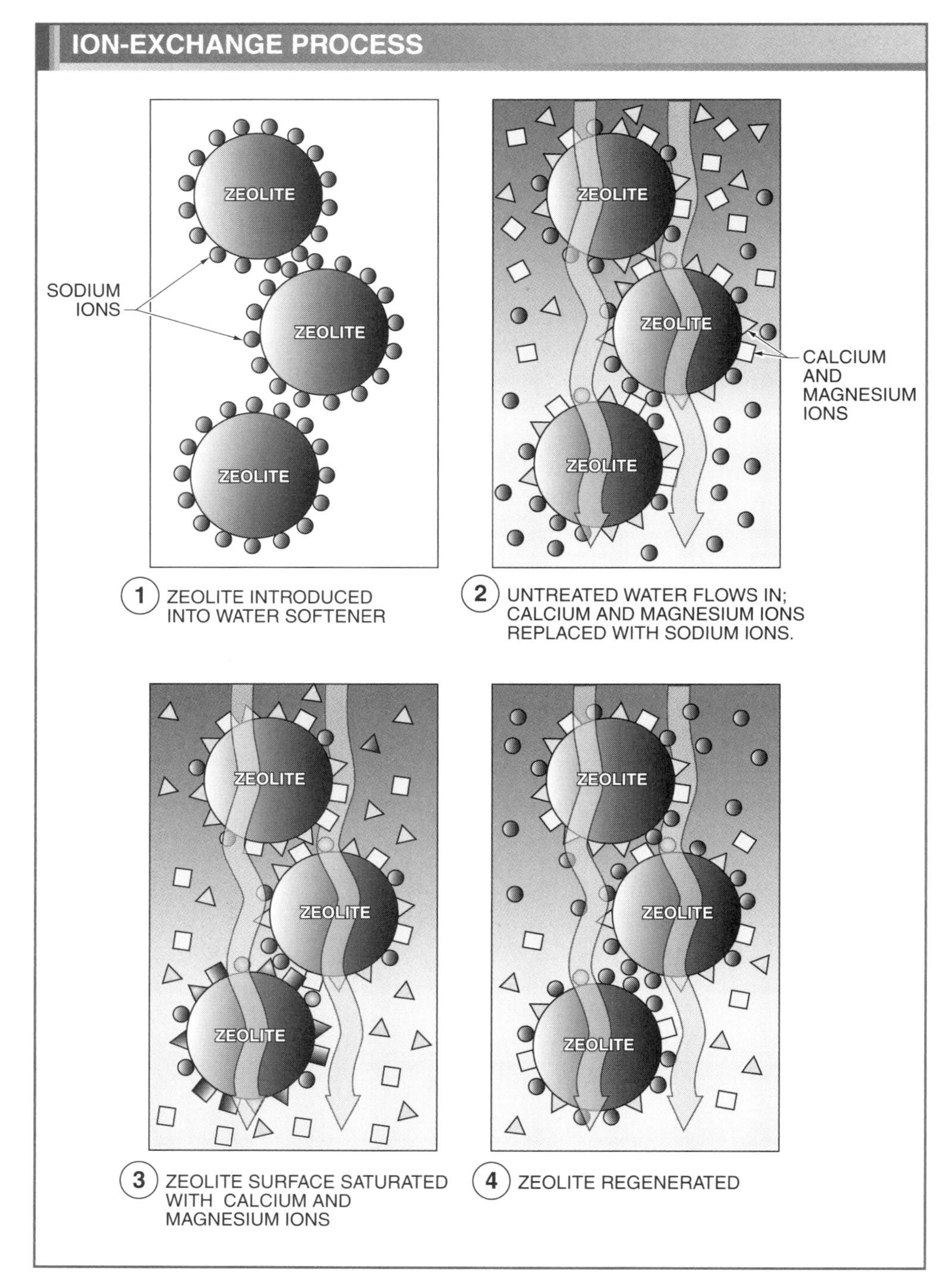

Figure 6-49. Water softeners use an ion-exchange process to remove calcium and magnesium ions from water.

The brine solution removes calcium and magnesium ions from the zeolite surface and forces sodium ions onto the same surface. The calcium and magnesium mineral waste and excess salt solution are flushed down the drain. The frequency with which a softener must be regenerated depends on two factors—hardness of incoming water and amount of water used.

Automatic Water Softeners

Automatic water softeners are installed for residential applications since the only maintenance required by a homeowner is the addition of salt to the brine tank. An automatic water softener is equipped with a programmable demand timer control to automate the service (softening) and regeneration cycles of the appliance.

Two types of residential automatic water softeners are cabinet and two-tank softeners. **See Figure 6-50.** A *cabinet automatic water softener* is a water softener that has a mineral tank within the brine tank. It is the most common automatic water softener since it consumes less floor space than a two-tank model. A *two-tank automatic water softener* is a water softener that has separate mineral (zeolite) and brine tanks. Both types of water softener operate similarly. A mineral tank stores the zeolite, which exchanges sodium ions for calcium and magnesium ions in the water. The brine tank stores the sodium (salt) that is dissolved in water to make the brine solution in the brine well. The brine well is a vessel that contains a brine valve assembly used to draw the brine solution into the mineral tank to regenerate the zeolite. A multiple-port valve assembly diverts the water and brine solutions in and out of the mineral tank in proper sequence for operation and regeneration cycles. A programmable demand timer control automates the operation of the multiple-port valve assembly. The demand timer control measures the amount of water used and regenerates the water softener when the softening capacity of the zeolite has been reached.

Automatic Water Softener Installation. An automatic water softener is piped into a residential water supply and provides soft water to all the plumbing fixtures and appliances except sillcocks and kitchen sink cold water faucets. Soft water is not supplied to the kitchen sink since the sodium content of soft water may adversely affect individuals with certain health conditions.

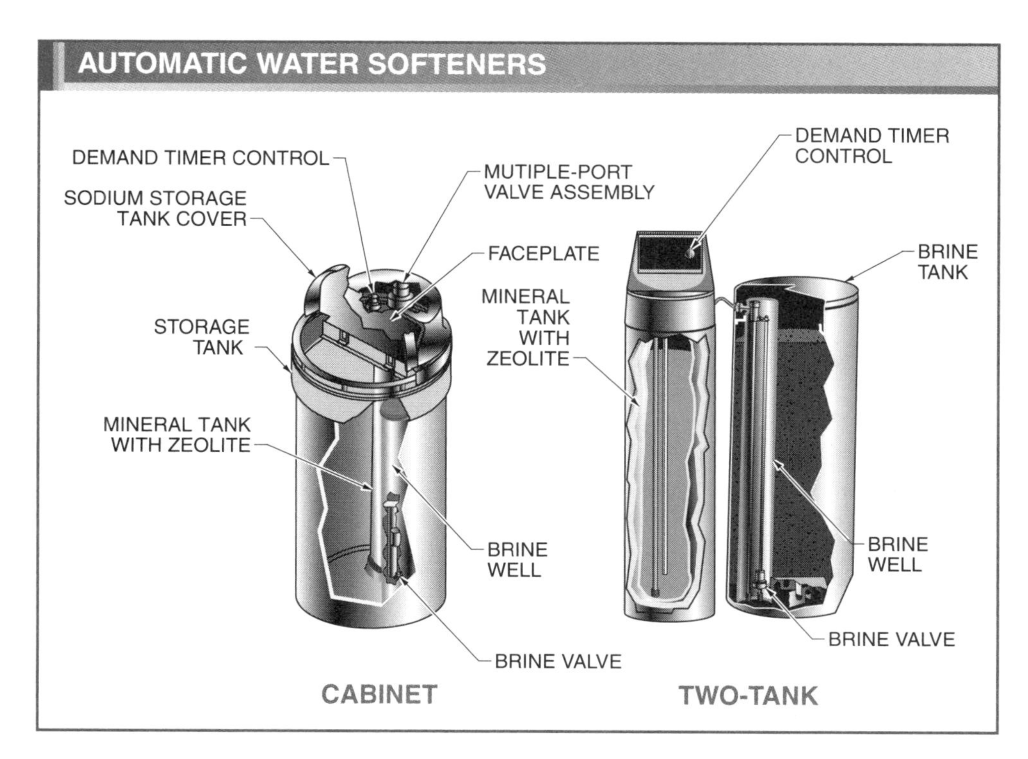

Figure 6-50. Automatic water softeners are available in cabinet and two-tank models.

A bypass valve is provided on water softener piping so maintenance can be performed on the water softener without interrupting the water supply to the remainder of the residence. The bypass valve can be obtained when the water softener is purchased, or a bypass valve can be constructed with three ball valves, two tees, and two short lengths of pipe. **See Figure 6-51.**

Valve drain and overflow drain hoses must be provided for automatic water softeners. A valve drain hose extends from the water softener to a floor drain and conveys the discharge during regeneration from the appliance to the floor drain. An overflow drain hose extends from the brine tank to a floor drain and conveys overflow from the water softener.

When installing an automatic water softener, the time of day and the water hardness number are programmed into the demand timer control. The water hardness number is obtained from the municipal water department or by analysis of a water sample from the residence. The regeneration cycle typically occurs between midnight and 5:00 AM, when water demand is low.

WATER FILTERS

In addition to calcium and magnesium, varying amounts of other undesirable elements are also contained in the water supply. A *water filter* is a plumbing appliance that removes sand, sediment, chlorine, lead, and other undesirable elements from water and protects water heaters and other fixtures and appliances from collecting residue. **See Figure 6-52.**

Water filters are installed at the water service or at the point of water use. Household water filters installed at the water service remove sand, sediment, and other undesirable debris before the water enters the water softener or water heater. Different filter elements are available, depending on the undesirable elements being removed from the water. Some water filter elements remove sediment or rust, while other elements reduce the chlorine levels present in the water supply. Filter elements for household water filters must be replaced every 2 months to 6 months, depending on the levels of undesirable elements in the water.

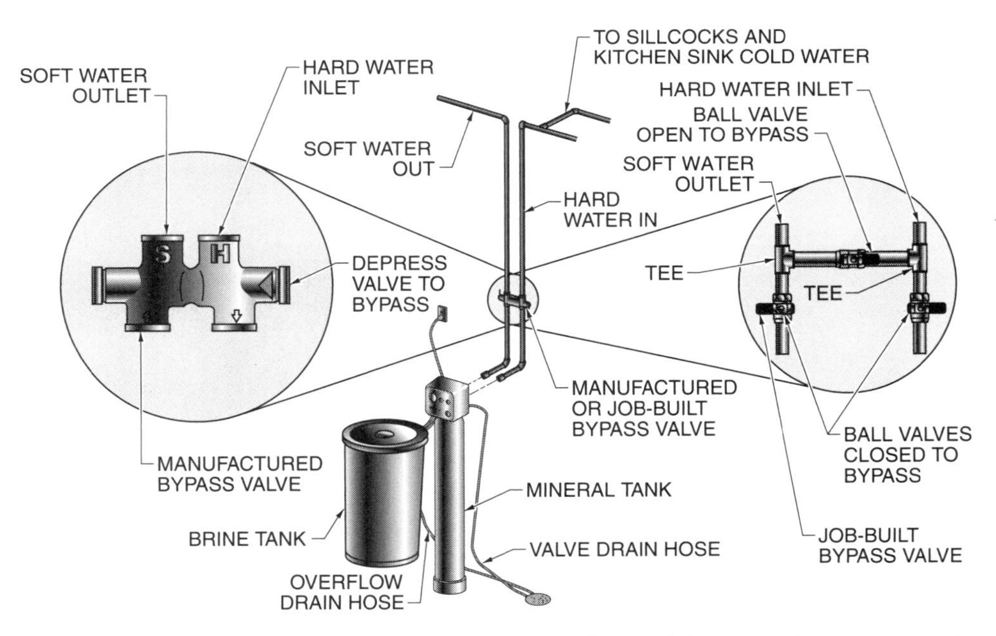

Figure 6-51. A bypass valve is installed on water softener piping so maintenance can be performed on the water softener without interrupting the water supply for the remainder of the building.

Elkay Mfg. Co.

Figure 6-52. Water filters remove undesirable elements from water.

Point-of-use filters utilize carbon filters or the reverse osmosis process to eliminate undesirable elements resulting in water odors or poor-tasting water. The carbon filtration process removes lead, chlorine, volatile organic compounds (VOC), and sediment from drinking and cooking water, depending on the water filter element installed in the water filter. Carbon filters are installed in the water line servicing a refrigerator with an integral icemaker and water dispenser, or are installed under a kitchen sink cabinet. Some carbon filter elements are designed to remove lead, rust, and other sediments. Other filter elements remove chemicals such as herbicides and pesticides that may be present in water due to farm run-off.

A reverse osmosis filter removes up to 99% of the undesirable elements in a water supply including lead, mercury, copper, pesticides, herbicides, nitrates, and tannins. Reverse osmosis filters are commonly installed under the kitchen sink cabinet and provide filtered water to a second faucet mounted on the kitchen sink.

WATER HEATERS

A *water heater* is a plumbing appliance used to heat water for purposes other than heating a structure. For most residential applications, a gas or electric water heater is installed near the water supply inlet and supplies hot water to plumbing fixtures throughout the residence. A hot water system is a closed system that is subjected to extreme pressure and temperature. Water heater components such as the jacket, storage tank, and top cover are manufactured from strong and durable materials to resist the pressure and temperature produced by a water heater.

Gas Water Heaters

A *gas water heater* is a water heater that utilizes heat produced by the combustion of natural or liquefied petroleum gas (LPG) to heat cold water contained within a storage tank. **See Figure 6-53.** Cold water entering a gas water heater is heated as it moves through the storage tank. A painted steel jacket and top cover enclose a storage tank and the insulation that prevents heat loss from the water in the storage tank. A glass lining in the storage tank protects the steel tank from the corrosive effects of hot water. Natural gas or LPG is ignited in the main burner at the bottom of the storage tank, and the heated gases flow through the flue inside the tank to heat the water. The flue is fitted with a baffle to retard the flow of hot gases to provide maximum heat transfer. The gases exit the water heater through a draft hood and are vented to the outside of the building.

The top of a gas water heater storage tank has three openings. The cold water inlet is connected to the cold water supply pipe. The inlet is attached to a dip tube inside the storage tank. The hot water outlet is connected to the supply pipe that conveys hot water to fixtures in the building. A magnesium anode rod is incorporated into the hot water outlet fitting of some water heaters to prevent electrolytic deterioration of the water heater. A relief valve opening is provided at the top of the storage tank for the installation of a T&P relief valve. The side of the storage tank has two openings, one of which accommodates the water heater control and the other is for the tank drain valve. The drain valve opening is located as close to the tank bottom as possible to provide an outlet for water and sediment when draining the water heater.

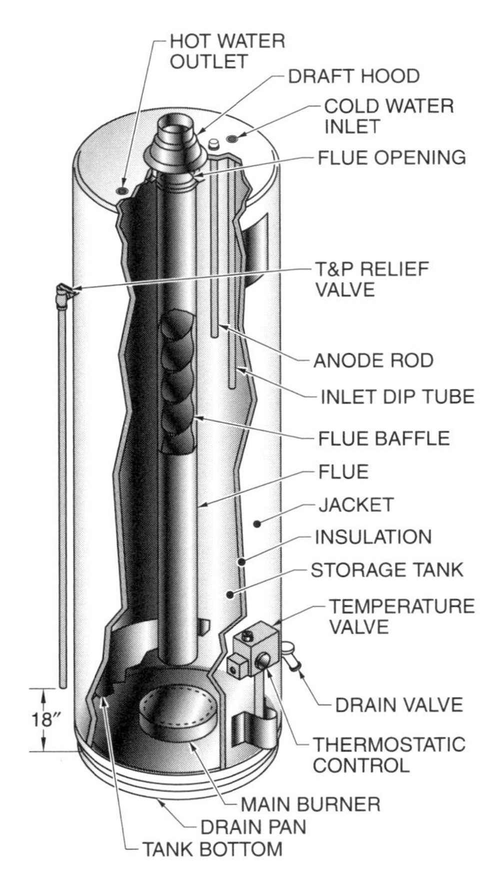

Figure 6-53. Gas water heaters utilize heat produced by the combustion of natural gas or LPG to heat cold water contained within a storage tank.

A dip tube extends from the cold water inlet at the top of the storage tank toward the bottom of the tank. The dip tube conveys incoming cold water through the stored hot water and discharges the cold water at the bottom of the storage tank. Cold water is not allowed to mix with the heated water and is delivered to the tank bottom where it is heated by the main burner. A dip tube has a small antisiphon orifice near the top end of the tube to prevent hot water in the tank from being siphoned out during an interruption in the cold water supply.

A thermostatic control consists of a sensing element immersed in the water in the storage tank and a temperature control dial used to set the water heater to the desired temperature. When hot water is drawn from the storage tank, cold water replaces it in the storage tank. The element senses the temperature decrease and allows gas to flow to the main burner to heat the water. The thermostatic control shuts off gas flow when the water reaches the desired temperature.

The main burner is the primary combustion device of a gas water heater; it burns natural gas or LPG to generate the heat needed to operate the water heater. A safety pilot ignites the gas at the main burner when the thermostat turns on the gas supply. If the safety pilot is extinguished or if the safety pilot flame becomes too small to ignite the burner, the gas supply to the water heater is automatically shut off by the thermocouple. A thermocouple is a small electric generator made of two different metals that are firmly joined and is used as a safety device for gas-powered appliances. A safety pilot generates a small amount of electricity as the flame heats the thermocouple and holds open the safety shutoff gas valve. If the pilot flame becomes too small or if it is extinguished, the thermocouple then does not produce enough electricity, and a spring closes the gas valve.

Electric Water Heaters

An *electric water heater* is a water heater that utilizes heat produced by the flow of electricity through a resistance wire contained in the heating elements to heat cold water contained within the storage tank. **See Figure 6-54.** Cold water entering an electric water heater is heated as it moves through the storage tank and exits the water heater as hot water. A painted steel jacket and top cover enclose a storage tank surrounded by insulation; the insulation prevents heat loss from the water in the storage tank. A glass lining in the storage tank protects the steel tank from the corrosive effects of hot water.

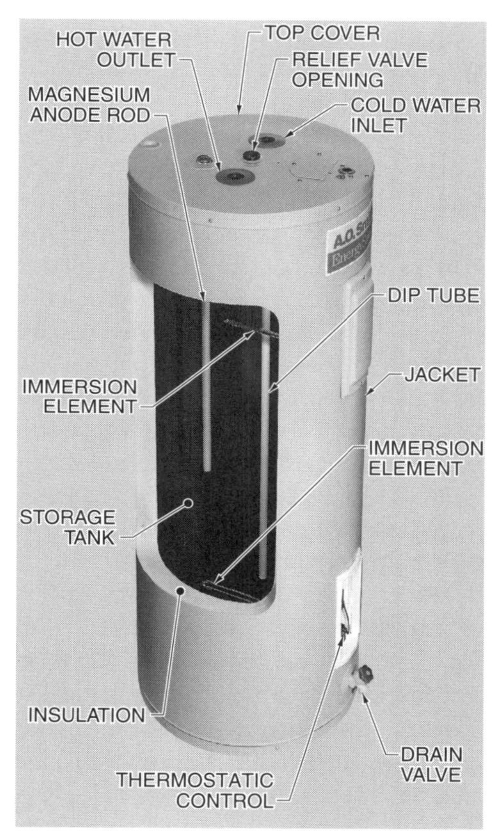

A.O. Smith Corporation

Figure 6-54. Electric water heaters utilize heat produced by the flow of electricity through a resistance wire contained in the heating elements to heat cold water contained within the storage tank.

The top of a storage tank has three openings—cold water inlet, hot water outlet, and relief valve opening. The cold water inlet is attached to a dip tube inside the storage tank. The hot water outlet is connected to the supply pipe that conveys hot water to fixtures in the building. A magnesium anode rod is incorporated into the hot water outlet fitting of some water heaters to prevent electrolytic deterioration of the water heater. A relief valve opening is provided at the top of the storage tank for the installation of a T&P relief valve.

The side of a storage tank has an opening for each of the immersion elements and an opening for a drain valve. The drain valve opening is located as close to the tank bottom as possible.

An *immersion element* is an electric heating device that is inserted into the storage tank of an electric water heater and makes direct contact with the water to provide fast and efficient heat transfer to the water. Each immersion element has a separate automatic thermostat. Electric water heaters must be filled with water whenever the immersion elements are operated. Dry firing (operating the elements without water) will cause the elements to burn out since there is no water to carry the heat away from the element.

A dip tube extends from the cold water inlet at the top of the storage tank toward the bottom of the tank. The dip tube conveys incoming cold water through the stored hot water and discharges the cold water at the bottom of the storage tank. The dip tube has a small antisiphon orifice near the top end of the tube to prevent hot water in the tank from being siphoned out during an interruption in the cold water supply.

An automatic thermostat is the primary device that starts and stops the flow of electricity to the immersion elements, and may be adjusted to the desired water temperature. The thermostat in an electric water heater senses the outside surface temperature of the storage tank. Some electric water heaters use a thermostat that has a remote sensing bulb located in an immersion element.

Electric water heaters have a high limit control that stops the flow of electricity to the immersion element when the tank surface adjacent to the device reaches a predetermined temperature. A high limit control protects against excessive water temperature caused by a defective thermostat or grounded immersion element.

Relief Valves

A water heater can become dangerous if high temperature or high pressure develops within the storage tank. High temperature may develop in a water heater storage tank due to water heater control failure. If the control fails to turn off the heat source, the water in the storage tank will be heated to excessive and unsafe temperatures.

Gauge Pressure (psi)	Boiling Point (°F)
0	212
30	274
60	307
90	331

In an open container at sea level, water boils and vaporizes (turns to steam) at 212°F. When water is placed in a closed vessel under pressure and heated, as in a water heater, the boiling point of water increases in direct proportion to the pressure within the vessel.

Superheated water is water under pressure that is heated above 212°F without becoming steam. When the pressure within a container of superheated water is suddenly released by a break in a pipe or a rupture of a water heater tank, superheated water immediately converts to steam and expands 1700 times. The rapid expansion of water converting to steam can result in an explosion of the water heater.

Excessive pressure can develop within a water heater storage tank from high water service pressure, water hammer, or thermal expansion. High water service pressure can be reduced to an acceptable working pressure using a pressure-reducing valve on the water service.

Water hammer is a water supply system defect in which a loud noise is created when a quick-closing valve, such as a clothes washer valve or ball valve, is suddenly closed. A momentary pressure increase is created in the water supply system, forcing water in the pipes to abruptly hit the ends of long, straight runs of pipe. Water hammer is controlled by the installation of a water hammer arrestor. A *water hammer arrestor,* or shock absorber, is a device installed on water supply pipe near the fixture with the quick-closing valve to control the effects of water hammer. **See Figure 6-55.** A water hammer arrestor contains a bellows that compensates for additional pressure in the water supply. When water is flowing properly, the bellows is collapsed. When water pressure increases, the bellows absorbs the water pressure force by expanding.

Thermal expansion occurs when a liquid, such as water, is heated. Water expands approximately 2.5% in volume for every 100°F of temperature increase. If water cannot expand within the water supply system, the pressure increase could result in damage to the system.

Warning: High temperature and excessive pressure within a water heater are potentially dangerous situations and should be handled properly to avoid damage or injury. When a water heater storage tank ruptures due to high pressure, a stream of water is released from the crack until the pressure is relieved. However, if the bottom of a water heater storage tank ruptures from high temperature, the water heater blows upward, possibly through the roof of the building.

T&P relief valves are quick-opening valves installed in water heaters to prevent high temperature and/or high pressure from developing within water heater storage tanks. A temperature and pressure (T&P) relief valve is an automatic self-closing safety valve installed in the opening in a water heater that protects against the development of high temperature and/or high pressure within the storage tank. **See Figure 6-56.**

T&P relief valves are selected on the basis of temperature, pressure, and relief capacity. The temperature relief portion of a T&P relief valve must open before the water temperature reaches the atmospheric boiling point. T&P relief valves for domestic water heaters are set to open at 210°F. The pressure relief portion of the valve must open before the pressure within the system exceeds the working pressure of the lowest pressure-rated component in the system. T&P relief valves are set to open at a minimum of 125 psi. When the temperature of water in a water heater storage tank is excessive, the T&P relief valve opens and releases a large volume of water. When excessive pressure is generated within the storage tank, the valve opens and a small amount of water drains from the drain. A T&P relief valve must be able to release high-temperature water at a rate equal to or faster than the rate at which the water heater can generate it. The data plate or label on a T&P relief valve must contain information about the temperature and pressure setting of the valve, the relief capacity of the valve, and the recognized standardization authority such as ASME International or AGA (American Gas Association).

The minimum inlet and outlet size of a T&P relief valve is ¾″. A full-size drain, or drip line, must extend from a T&P relief valve to within 18″ of the floor or to a safe place of disposal. The end of the drain cannot have threads or a valve installed to prevent closing the drain. The extension thermostat of a T&P relief valve is installed in a storage tank opening or the hot water outlet with the temperature-sensing element in the top 6″ of the storage tank.

The test lever of a T&P relief valve is used to verify the proper operation of the valve. T&P relief valves should be tested once a year by manually opening the valve. Water should flow from the valve if the T&P valve is operating properly. If water does not flow from the valve, it must be replaced immediately.

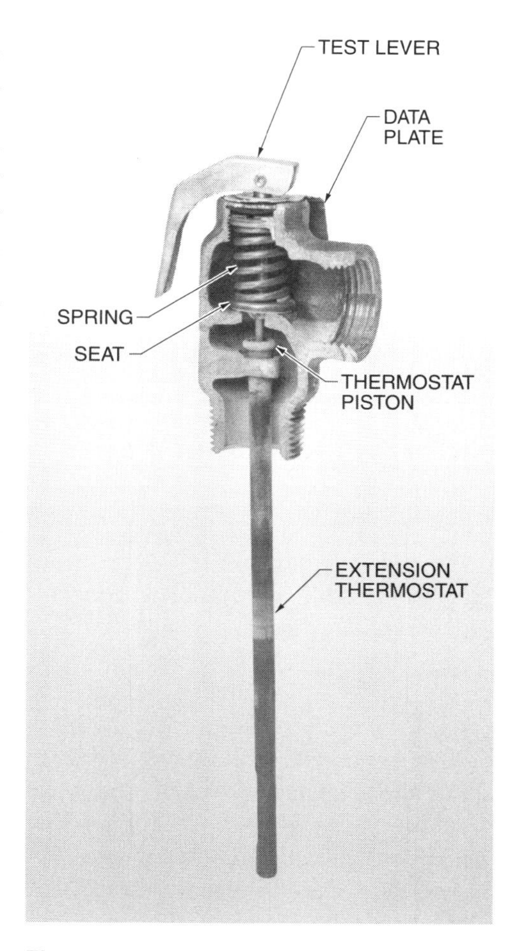

Figure 6-56. A temperature and pressure (T&P) relief valve is installed in the opening in a water heater and protects against the development of high temperature and/or high pressure within the storage tank.

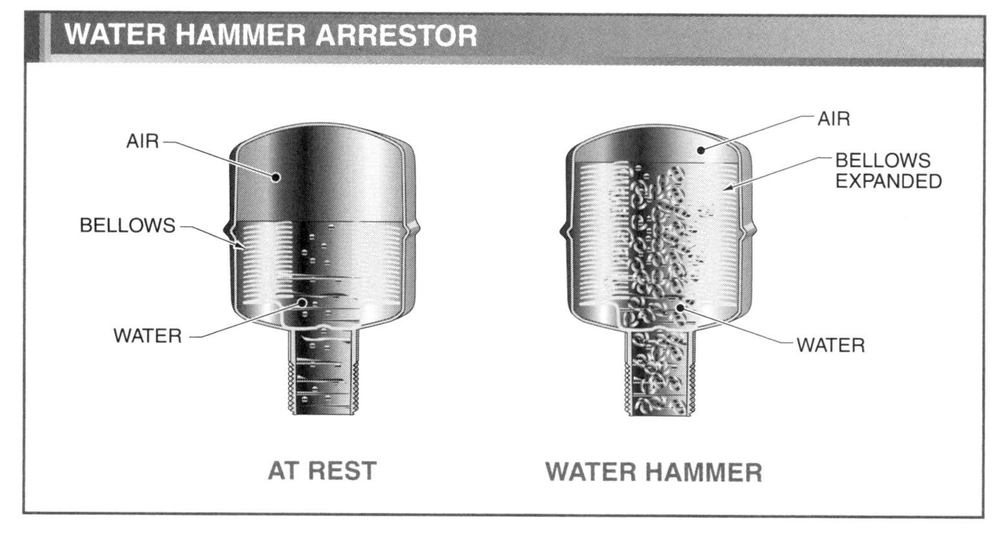

Figure 6-55. A water hammer arrestor is installed on water supply pipe near the fixture with the quick-closing valve to control the effects of water hammer.

PLUMBING FIXTURE AND APPLIANCE INSTALLATION

Plumbing fixtures and appliances must be handled carefully during installation. Proper plumbing fixture and appliance installation occurs in two phases—rough-in and finishing. *Rough-in,* or rough plumbing, is the installation of all parts of a plumbing system that can be completed prior to the installation of the fixtures. Rough-in includes installation of the following:

- Drainage piping
- Vent piping
- Water supply piping
- Any necessary fixture supports

Plumbing fixture and appliance manufacturers provide a rough-in drawing for plumbing fixtures and appliances to aid in the installation of the fixture or appliance. A rough-in drawing provides information regarding the proper placement of the drainage, vent, and water supply piping so they fit the fixture or appliance when it is installed. In addition, rough-in drawings provide dimensions of each fixture and information regarding necessary supports or backing.

Finishing is the final installation, or setting, of plumbing fixtures and appliances. With the exception of bathtubs, shower bases, and showers, plumbing fixtures and appliances are installed after the rooms in which they are to be installed have been decorated and are ready to be occupied. Each plumbing fixture or appliance has unique installation procedures. However, the procedure typically involved in finishing plumbing fixtures and appliances is as follows:

1. Uncrate fixture or appliance and its trim and inspect for damage and/or defects.
2. Check measurements for the waste and water supply piping with the rough-in drawing.
3. Attach fixture wall or floor supports.
4. Attach fixture trim, if necessary.
5. Hang or set the fixture.
6. Align, level, and/or plumb the fixture.
7. Secure fixture fasteners.
8. Connect water supply and waste piping to the fixture or appliance.
9. Purge water supply piping to relieve air and remove any dirt in the piping.
10. Test the water supply and waste connections.
11. Clean and inspect the fixture.
12. Caulk the fixture where it meets a wall and/or floor to make a watertight connection.

Review Questions

1. Define "plumbing fixture" and "plumbing appliance," and identify the major difference between fixtures and appliances.
2. List five materials from which plumbing fixtures are made.
3. Describe the flushing action of a siphon jet water closet.
4. Describe the action of a flush tank mechanism.
5. Explain the operation of a flushometer valve.
6. What is the standard drain and vent size for a water closet?
7. Describe the operation of a port control faucet.
8. Describe two methods for trapping a double compartment kitchen sink.
9. List two ways to connect the dishwasher waste to the drainage piping.
10. Which plumbing fixture may not require a vent?
11. Define indirect waste receptor and identify the fixture for which it may be required.
12. List five sanitary requirements for a drinking fountain or water cooler.
13. Describe the ion-exchange process.

Plumbing systems need to be tested and inspected to ensure that the installed plumbing system is properly installed according to the local municipality's plumbing code. An improperly installed plumbing system will not be safe and may contain leaks. It is the responsibility of the plumber to test the plumbing system.

7 Testing and Inspecting Plumbing Systems

Plumbing systems are subjected to appropriate tests and inspections as the various phases of plumbing installation are completed to ensure the systems are safe, properly installed, and free from leaks. Testing plumbing systems is the responsibility of the plumber. Plumbing system inspection is performed by a plumbing inspector or building official. The testing of plumbing systems is enforced by local municipalities.

The entire plumbing system must be tested and inspected to certify that systems meet plumbing code requirements. In general, plumbing systems are tested and inspected in the sequence in which they are installed:

1. Sanitary drainage and vent piping
2. Stormwater drainage system
3. Potable water supply and distribution system
4. Plumbing fixture installation

For some installations, several plumbing systems can be tested and inspected simultaneously, for example when the building sewer and storm sewer are installed in the same trench. Larger plumbing installations, such as sanitary drainage and vent piping in multistory buildings, are tested and inspected in sections to identify and correct defects so construction performed by other trades can proceed in a timely manner.

PLUMBING SYSTEM TESTS

Plumbers use the following tests to ensure that the plumbing systems of a building are free from leaks:

- Air test
- Water test
- Final air test

Local plumbing codes specify the plumbing system test recognized in that state or municipality.

Air Test

An *air test* is a plumbing system test in which inlets and outlets to the system are sealed and air is forced into the system until a uniform air pressure of 5 psi is reached and maintained for 15 min without additional air being added to the system.

To perform an air test, the following items are needed:

- Devices to seal the openings in the system
- Test gauge assembly
- Hand pump or air compressor
- Soap solution to check for leaks (usually in a spray bottle)

A hand pump or portable air compressor is used to force air into the plumbing system, and is typically used when testing smaller installations, such as a one-family dwelling. **See Figure 7-1.** For larger installations, such as commercial or public buildings, an engine-driven air compressor is used as the compressed air source. An adequate supply of air hose is required to convey air from the air compressor to the test opening connection.

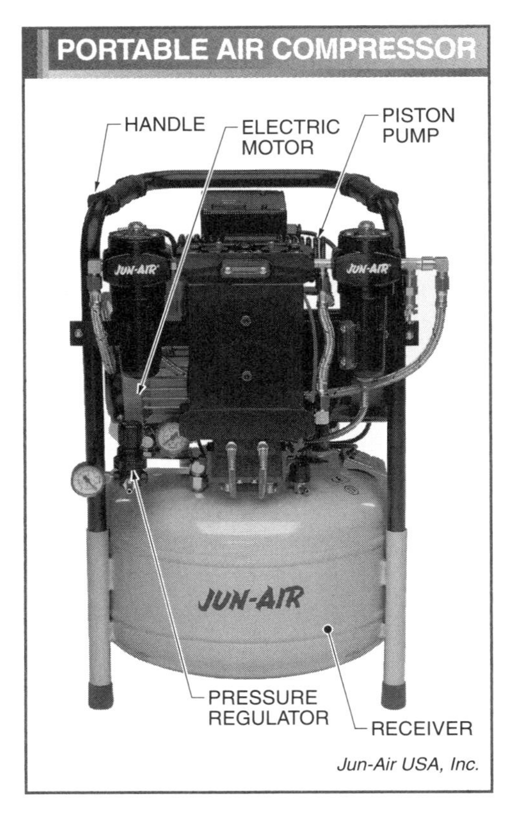

Figure 7-1. Portable air compressors are used to supply air for the test.

Several types of devices are used to seal or plug openings in plumbing systems when performing an air test, including test plugs and caps and pipe plugs and caps. An *inflatable test plug* is an inflatable rubber device inserted into the plumbing system to seal openings during an air or water test. Deflated test plugs are properly positioned within the pipe and inflated with a hand pump or air compressor to seal an opening. **See Figure 7-2.** Inflatable test plugs are available in ¾″ to 18″ pipe sizes. A *mechanical test plug* is a test device inserted into the end of a pipe or other opening and secured in position by tightening a hex or wing nut. The O-ring of a mechanical test plug is inserted into the pipe or opening and the hex or wing nut is tightened, resulting in the O-ring expanding and sealing the opening or pipe. Mechanical test plugs are available in ⅜″ to 12″ pipe sizes.

Warning: Personal injury or property damage may occur if a test plug fails when a plumbing system is under pressure. Always wear a protective helmet (hard hat) and eye protection when performing air and water tests using test plugs and do not stand directly in front of an outlet where a test plug is installed. Attach an inflation extension hose to a test plug so the plug can be inflated or deflated without working directly in front of the test plug outlet. Do not remove a test plug from a pipe or other opening until testing pressure has been released from the plumbing system.

A *test cap* is a reinforced rubber cap installed on the outside of an opening and secured in position with a stainless steel hose clamp. **See Figure 7-3.** Since a test cap is visible when properly installed, there is less possibility of the test cap being left in place after testing and inspection. Test caps are available in 1½″ to 8″ pipe sizes.

Pipe plugs and caps are installed on pipe ends using solder, solvent cement, or threads, depending on the type of piping material being sealed. A *pipe plug* is a test device installed on the inside of the pipe end. A *pipe cap* is a test device installed on the outside of the pipe end. Copper pipe caps are soldered to the pipe ends. Plastic pipe plugs and caps are solvent-cemented to the pipe ends, and are removed by cutting off the pipe end and plug or cap. Malleable iron plugs and caps have male and female threads, respectively, and are removed from the pipe end using a pipe wrench.

A *test gauge assembly* is a test device used to measure pressure within waste and vent, water, gas, air, or other piping systems to ensure that the system is maintaining the proper pressure. In addition to measuring air pressure within a plumbing system, the test gauge assembly also permits air to be admitted to the system. Job-built test gauge assemblies are constructed of a ball valve, reducing tee, two nipples, and a test gauge. **See Figure 7-4.**

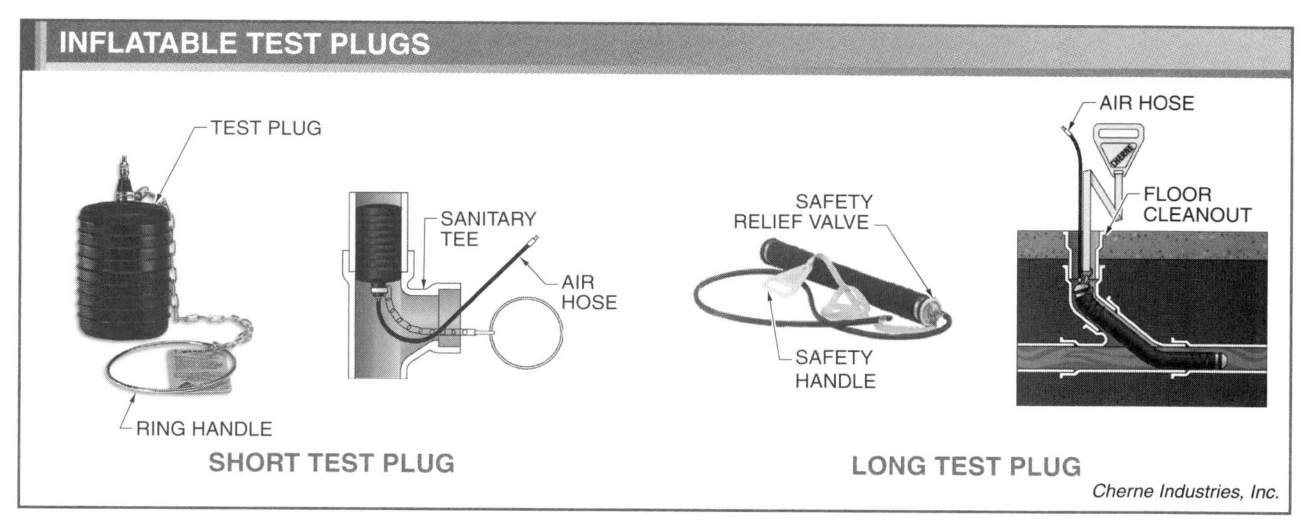

Figure 7-2. Inflatable test plugs must be properly positioned before conducting an air or water test.

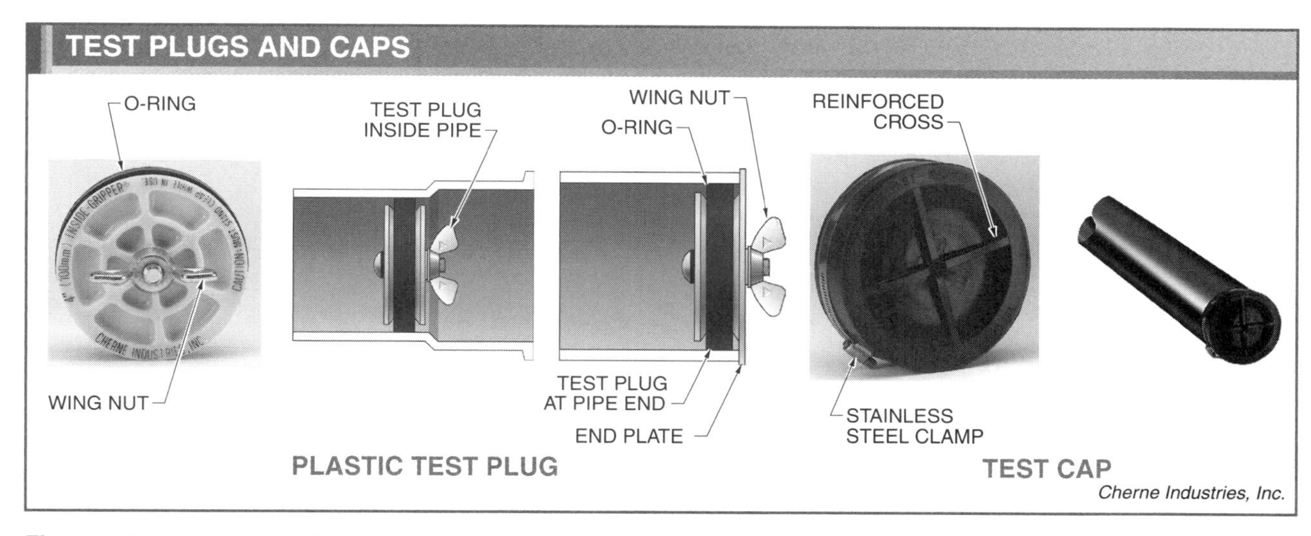

Figure 7-3. Mechanical test plugs and caps are installed on fixture drains.

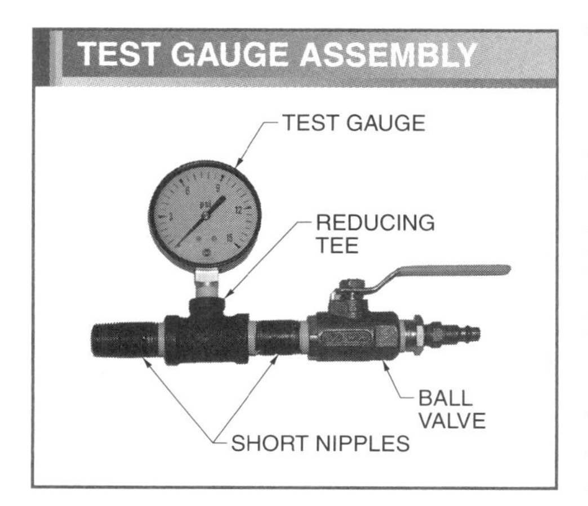

Figure 7-4. A test gauge assembly for an air test can be job-built.

Water Test

A *water test* is a plumbing system test in which pipe openings are sealed with plugs or caps and the pipe is filled with water to provide a specified amount of pressure to the plumbing system to determine the tightness of the system. For a water test, the plumbing system must be filled with water to a point equal to a 10′ head. For some installations, an extension pipe may be temporarily installed to provide a 10′ head. A water hose, and test and pipe plugs or caps are required to perform a water test of a plumbing system.

Water tests are typically performed on DWV systems since it is less time-consuming to perform a water test than an air test. After pipe openings are sealed, water is introduced into the system through the highest vent opening. The water level must remain constant for 15 min prior to inspection. If the water level is maintained, no leaks are present in the system. If the water level decreases, there are leaks in the system that must be repaired before approval.

Plumbers prefer water tests since it is easier to locate leaking joints and defects in piping due to the presence of water at the leak or defect. However, a water test cannot be performed during cold weather conditions where the temperature may fall below 32°F and water may freeze in the pipes.

Final Air Test

A final air test (also called a manometer test) is a test of the plumbing fixtures and their connections to the drainage system. The traps are filled, the vent stacks and building drain are plugged, and a manometer tube is inserted into a plumbing fixture (usually a water closet). Air is introduced to the system until a pressure of 1″ of water column is reached. This pressure must be maintained for the period of the test. **See Figure 7-5.**

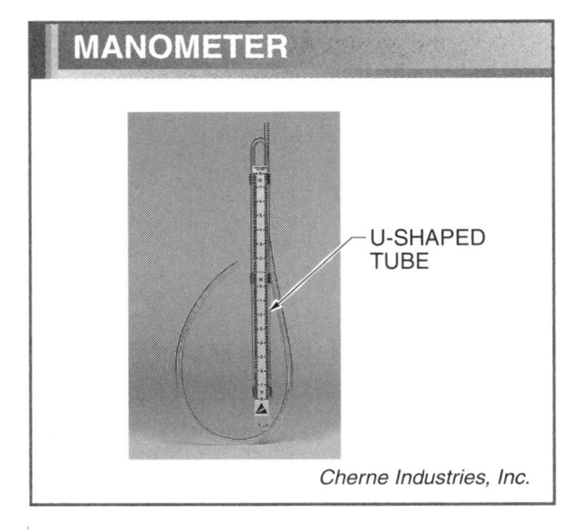

Figure 7-5. A manometer is used to verify a 1″ water column for a final air test.

SANITARY DRAINAGE AND VENT PIPING TESTS

Underground sanitary drainage and vent piping is tested and inspected before it is covered with earth and backfilled. Aboveground piping is tested and inspected prior to being covered with drywall or other wallcovering to ensure that it does not leak. Air or water tests are performed on sanitary drainage and vent piping.

Sanitary Drainage and Vent Piping Air Tests

The following steps are required to air test sanitary drainage and vent piping:

1. Plug the piping at the front main cleanout with an inflatable rubber test plug. **See Figure 7-6.**
2. Plug the vent openings in the roof and the fixture drains with the appropriate mechanical test plug, test cap, or inflatable test plug. **See Figure 7-7.**
3. If the building piping is being tested in sections, isolate the section being tested with short test plugs inserted in the appropriate test tees. **See Figure 7-8.**
4. Attach the test gauge assembly to the system; this is usually done at a lavatory or sink drain opening. **See Figure 7-9.**
5. Connect the hose from the air compressor to the test gauge and force air into the system to 5 psi.
6. Watch the gauge for a short period. If the pressure drops, the leak(s) must be found and repaired. Large leaks may be located by the sound of air escaping from the leak. Small leaks will have to be located by spraying the pipe joints with a soap solution. It may be necessary to even spray the pipe and fittings since they are sometimes defective.
7. Continue the test procedure until the system holds 5 psi of air for 15 min.
8. After the pipe has been inspected, carefully release the air in the system and remove all test caps and plugs from the piping.

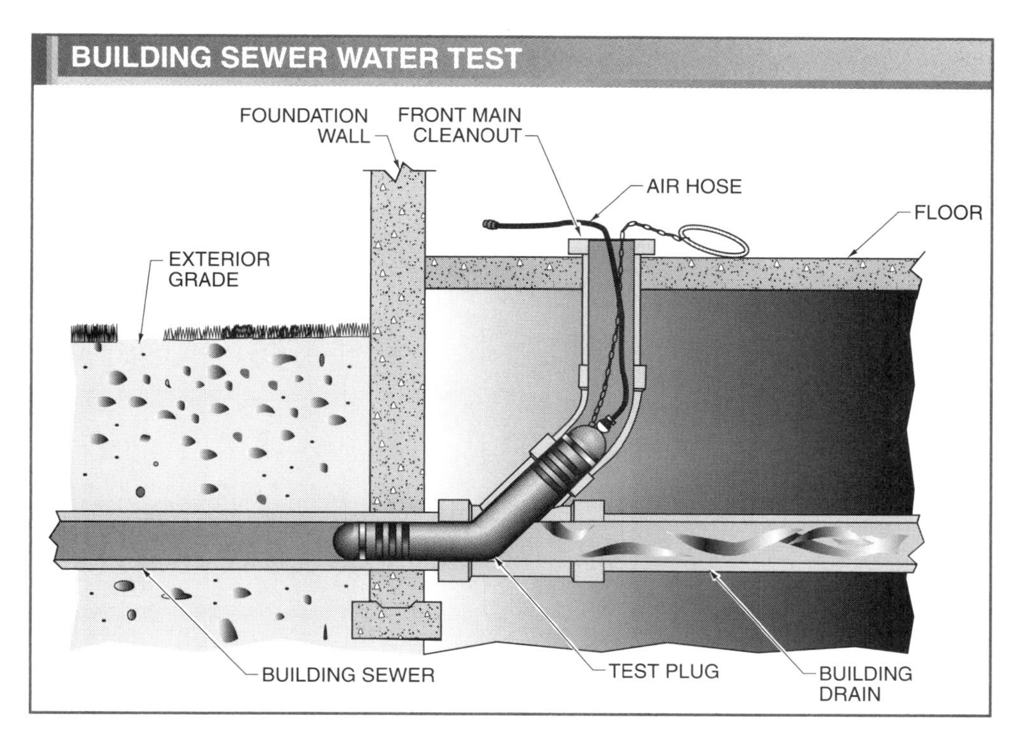

Figure 7-6. A test plug is placed at the wye beneath a front main cleanout when performing an air or water test of the building sewer.

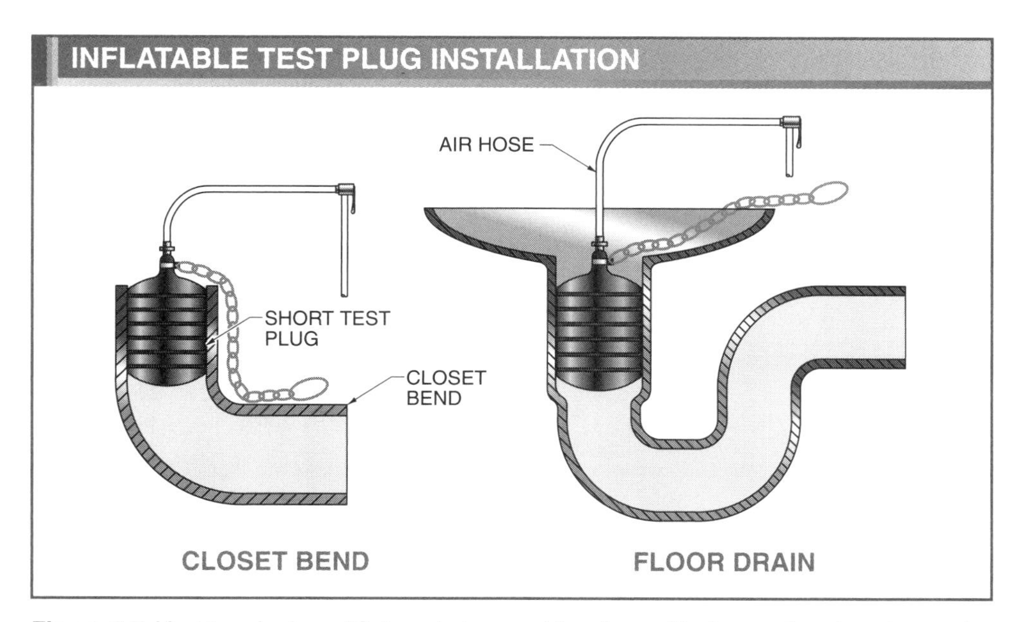

Figure 7-7. Vent terminals and fixture drains must be plugged before performing air or water tests.

Tech Facts

Plumbers use different plumbing system tests to make sure that the plumbing systems of a building are free from leaks. Plumbing system inspection is performed by a plumbing inspector or building official.

Sanitary Drainage and Vent Piping Water Test

The following steps are required to water test sanitary drainage and vent piping:

1. Plug the piping at the front main cleanout with an inflatable rubber test plug.

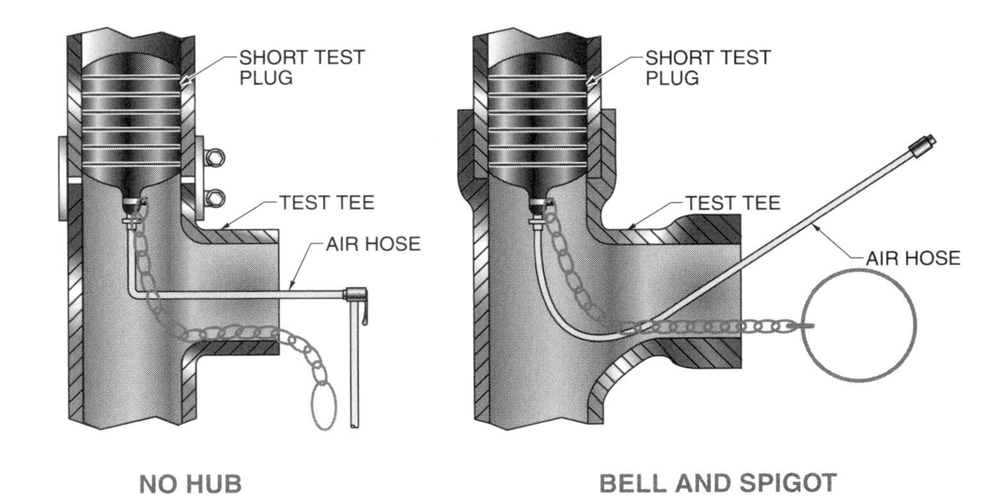

Figure 7-8. Inflatable test plugs are inserted into test tees to isolate sections of piping.

Figure 7-9. When performing an air test, the 5 psi reading on a test gauge assembly must be maintained for 15 min.

2. Plug all the vent openings in the roof except one, and plug the fixture drains with the appropriate mechanical test plug, test cap, or inflatable test plug.
3. If the building piping is being tested in sections, isolate the section being tested with inflatable sealing plugs inserted in the appropriate test tees. **See Figure 7-10.**
4. In a small building or a house, introduce water into the system through the unplugged roof vent opening until the pipe overflows.
5. In a multistory building, isolate the building into sections of about five floors and fill the piping in this section so that all the piping being tested is subjected to at least a 10′ head of water. The water level must be clearly visible.
6. Watch the water level in the pipe for a short time. If the water level in the pipe drops, inspect the piping for water leak(s) and repair them.
7. Continue the test procedure until the system holds a 10′ head for 15 min without the addition of water.
8. Upon completion of a water test, the water must be removed from the system. In homes and small buildings, the air is simply released from the inflatable rubber test plug placed in the front main cleanout, allowing the water to drain into the building sewer. **See Figure 7-11.** In tall buildings where the pipe is tested in sections, the inflatable sealing plug is deflated and the water drains away.
9. Remove all test caps and plugs from the piping.

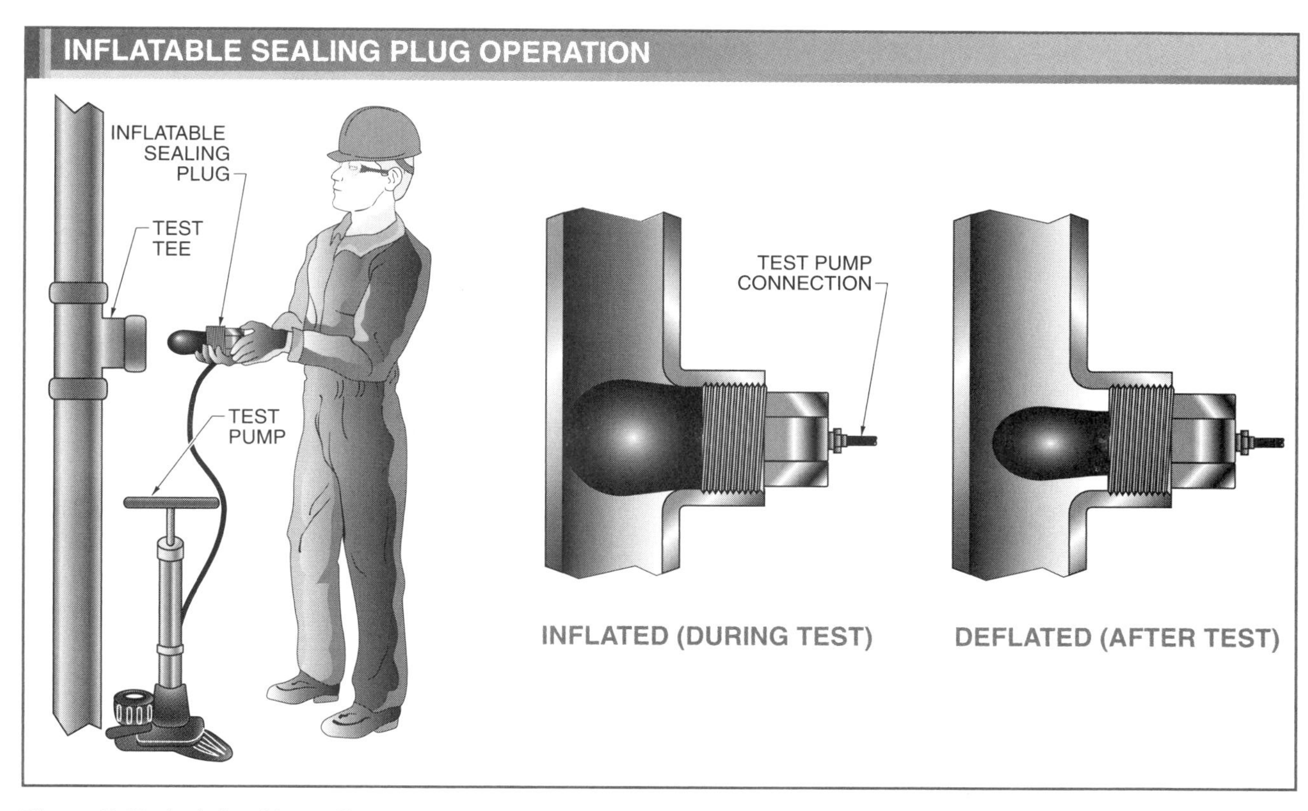

Figure 7-10. An inflatable sealing test plug is used to isolate piping for a water test.

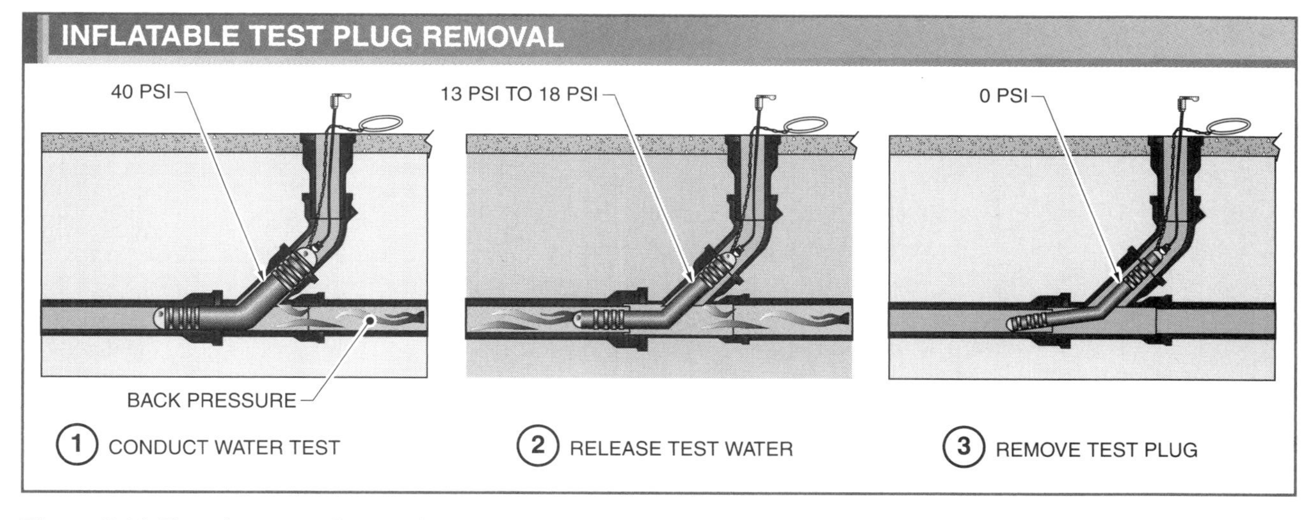

Figure 7-11. Test plugs must be carefully removed upon completion of a water test.

STORMWATER DRAINAGE PIPING TESTS

Stormwater drainage piping located within the building, including the building storm drain and rainwater leaders, is tested with an air test or water test in a manner similar to the sanitary drainage and vent piping. Exterior rainwater leaders that do not connect to the storm sewer system are not required by plumbing codes to be tested.

POTABLE WATER SUPPLY AND DISTRIBUTION PIPING AIR TESTS

An air test of the potable water supply and distribution piping is conducted in a manner similar to an air test on the sanitary drainage and vent piping. The differences that exist between an air test performed on the potable water supply and distribution piping and on the sanitary drainage and vent piping are the following:

- The pressure at which the potable water supply and distribution piping is tested is considerably larger. Potable water supply and distribution piping is typically tested at 1½ times the working water pressure or 150 psi, whichever is greater.
- The test duration for a potable water supply and distribution piping air test is 12 hr to 24 hr.
- Since the pressure and duration of a potable water supply and distribution piping air test are greater than the pressure and duration of sanitary drainage and vent piping air tests, openings in the water piping are sealed with the appropriate pipe caps and plugs and by closing the valves at the water meter and branch piping. If the test gauge assembly must remain on the piping system overnight, the assembly is sealed with a pipe cap or plug and the ball valve handle is removed to prevent tampering with the test.

Leaks in potable water supply and distribution piping are located with a soapy water solution or with an ultrasonic testing device.

Tech Facts

CPVC is a cream-colored thermoplastic material used in potable water distribution, industrial corrosive fluid handling, and fire suppression systems. CPVC solvent-cemented joints for ½" to 1" pipe should cure a minimum of 2 hr above 60°F before testing, with additional curing time required for larger-diameter pipe and cooler temperatures. CPVC piping is hydrostatically tested at line pressure (150 psi maximum) for 1 hr. After testing, the piping system should be flushed a minimum of 10 min.

Ultrasonic Leak Detection

Since leaks in potable water supply and distribution piping are usually small, they are difficult to locate with a soapy water solution. An ultrasonic leak detector is commonly used to locate small leaks in the system. An *ultrasonic leak detector* is a sensitive microphone probe and amplifier that is used to locate leaks in water supply and distribution piping. **See Figure 7-12.** An ultrasonic leak detector operates in the ultrasonic range of 35,000 hertz (Hz) or cycles per second to 45,000 Hz, which is the sound frequency generated by pressurized air as it escapes from pipes. The sound frequency is picked up by the microphone probe and converted into sound that can be heard by the person operating the leak detector.

When using an ultrasonic leak detector, a plumber places the headphones over the ears, turns on the amplifier, and passes the microphone probe along the pipe being tested. As the test probe approaches the leak, the sound generated by the leak becomes louder. In some cases, the amplified sound may be heard upon entering the room.

FINAL AIR TESTS

A *final air test* is a test of the plumbing fixtures and their connections to the sanitary drainage system. A final air test is performed after all plumbing fixtures are set and their traps are filled with water. Fixture connections are tested by plugging the stack openings on the roof and the building drain where it leaves the building, and introducing air into the system to achieve a pressure of a 1″ water column. The 1″ water column must remain constant for the duration of the inspection—usually 15 min—without the introduction of additional air.

A manometer is used to verify the 1″ water column for a final air test. A manometer is a clear U-shaped tube partially filled with water; it is used to measure pressure within a closed system. **See Figure 7-13.** The procedure for performing a final air test is as follows:

Figure 7-12. An ultrasonic leak detector is a sensitive microphone probe and amplifier that is used to locate leaks in water supply and distribution piping.

1. Fill all fixture and floor drain traps with water.
2. Plug all roof stack openings.
3. Plug the building drain at the front main cleanout using a test plug.
4. Insert the manometer hose through a water closet trap seal and remove any water from the hose by blowing into one end.
5. Fill the manometer with water to the zero inch mark on the manometer ruler and place the manometer on a stable surface.
6. Attach the hose extending from the water closet to the manometer.
7. Insert another hose through the water closet trap seal and blow air into the system until a 1″ pressure difference is shown and maintained on the manometer. Only a small amount of air is required to be introduced into the system to produce the required pressure difference.

During a final air test, the piping is pressurized to 1″ water column. **See Figure 7-14.** Since some plumbing fixture traps contain only a 2″ water seal, excessive pressure may blow out the traps and void the final air test. During the duration of a final air test, large volumes of water cannot be added to the system quickly without blowing out fixture traps. For example, water closets or urinals cannot be flushed since large volumes of water are added to the system during the flushing process. Final air tests should be performed when other tradesworkers are not working in the building. If it is not possible to work alone in the building, turn off all flushometer valves and empty all water closet flush tanks during a final air test to prevent flushing of the fixtures. If an air test is performed over the course of more than one day, the test plug in the front main cleanout opening must be removed. Before removing the plug from the front main cleanout, remove plugs from roof vent openings to prevent siphonage of fixture trap seals.

Although an acceptable air or water test of the roughed-in plumbing typically ensures an acceptable final air test, many things can affect the system between the times when the initial tests and final air test are performed. Careless backfilling and compaction, or the use of heavy equipment over underground piping, may cause broken piping and leaking pipe joints. Aboveground piping can be broken at the floor line by careless workers. Drywall screws, nails, and other fasteners can pierce piping within the wall cavities.

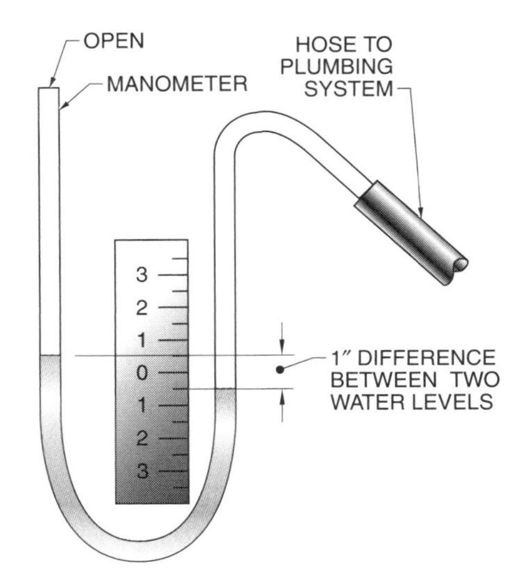

Figure 7-13. For a final air test, a pressure difference of 1″ water column must be maintained.

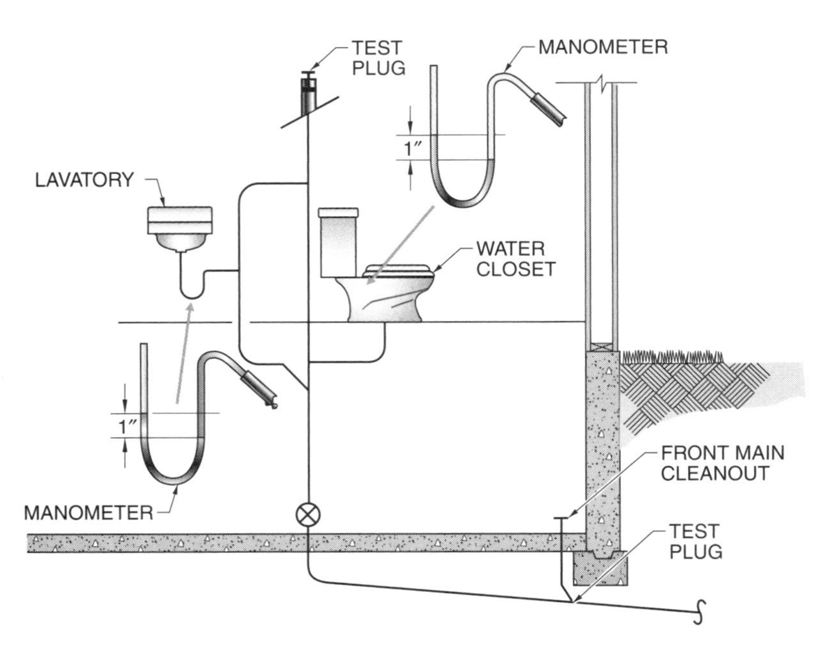

Figure 7-14. The DWV system, including all fixture traps, must be tested with a pressure equal to a 1″ water column.

If the manometer does not maintain the 1″ water column, the procedure for locating and repairing the leak(s) is as follows:

1. Check water closet bowls for proper setting and connection. Verify that the appropriate thickness of wax ring was used to set the bowl or the proper amount of bowl-setting compound was used during installation. Inspect for internal leaks in the water closet bowl by listening for air being released in the closet bowl, or by using a soapy water solution or odorant.
2. Inspect all exposed fixture traps, and repair any that are leaking.
3. Verify that underground piping is watertight by flooding the piping with water. Water is introduced into the underground piping at the stack cleanout farthest from the front main cleanout until the water level is flush with the cleanout opening. Observe the water level for 15 min to 30 min to see if the level drops. If the water level is maintained, the underground piping is watertight. However, if the water level drops, a leak is present which must be located and repaired.
4. To isolate a leak to a single stack, plug each stack cleanout with a test plug and apply the manometer test to each stack.
5. If a leak cannot be easily located, smoke or odorant leak detection tests may be conducted to determine the leak location.

SMOKE LEAK DETECTION

Leaks in the sanitary drainage and vent piping system can also be located using a smoke test. A smoke test is performed by filling fixture traps with water, plugging the building drain at the front main cleanout, and introducing a thick, odorous smoke into the system. When smoke appears at stack openings in the roof, the openings are closed and a pressure of 1″ water column is applied to the system and maintained for 15 min.

A smoke chamber is used to introduce smoke into the piping system. A *smoke chamber* is a device constructed of a short section of large-diameter pipe and two reducer fittings in which a smoke cartridge is placed. **See Figure 7-15.** The smoke cartridge is ignited and inserted into the pipe chamber, and the chamber is closed. Compressed air is applied to one end of the smoke chamber to distribute smoke throughout the piping system. Leaks are identified by locating smoke escaping from the piping.

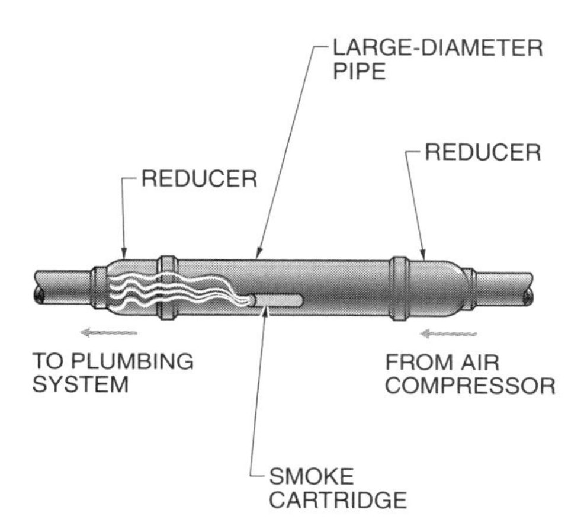

Figure 7-15. A smoke chamber is constructed of a short section of large-diameter pipe and two reducer fittings. A smoke cartridge is placed inside the device to produce thick, odorous smoke.

Since a smoke test relies on the smoke odor to locate leaks, care must be taken not to get the odor on plumbers' clothing. Typically, one plumber is responsible for the smoke chamber and igniting the smoke cartridge and another plumber is responsible for entering the building to locate the leaks. If possible, a smoke chamber should be located either outside the building or in a room without plumbing fixtures.

Odorant Leak Detection

Leaks in the sanitary drainage and vent piping system can also be located using an odorant test. To perform an odorant test, fixture traps are filled with water and the building drain is plugged at the front main cleanout. Two ounces of oil of peppermint or oil of wintergreen are poured into each stack and the building drain. **See Figure 7-16.** The stack and drain openings are closed and a pressure of 1″ water column is applied to the piping. If leaks are present, a sweet smell will be detected in the area of a leak.

When performing an odorant test, care must be taken to avoid spilling odorant on the plumber or plumber's clothing. Typically, one plumber is responsible for introducing the odorant into the piping and another plumber is responsible for entering the building to locate the leaks.

Tech Facts

The plumbing system tests recognized in a state or municipality are specified by local plumbing codes. Some tests may or may not be recognized by a municipality, depending on which plumbing code the municipality uses.

PLUMBING TEST PROCEDURES

A plumbing contractor must furnish the equipment, material, power, and labor necessary for plumbing tests and inspections. Piping must be exposed until it has been tested and inspected. Plumbing inspectors typically require that concealed or covered untested piping be exposed prior to conducting air or water tests. A common procedure for performing plumbing tests is the following:

1. Assemble testing apparatus.
2. Seal all openings.
3. Apply the test.
4. Check for leaks.
5. Fix defects.
6. Call for an inspection and test.
7. Assist the inspector at time of inspection and test.
8. Remove testing apparatus upon completion of test and inspection.

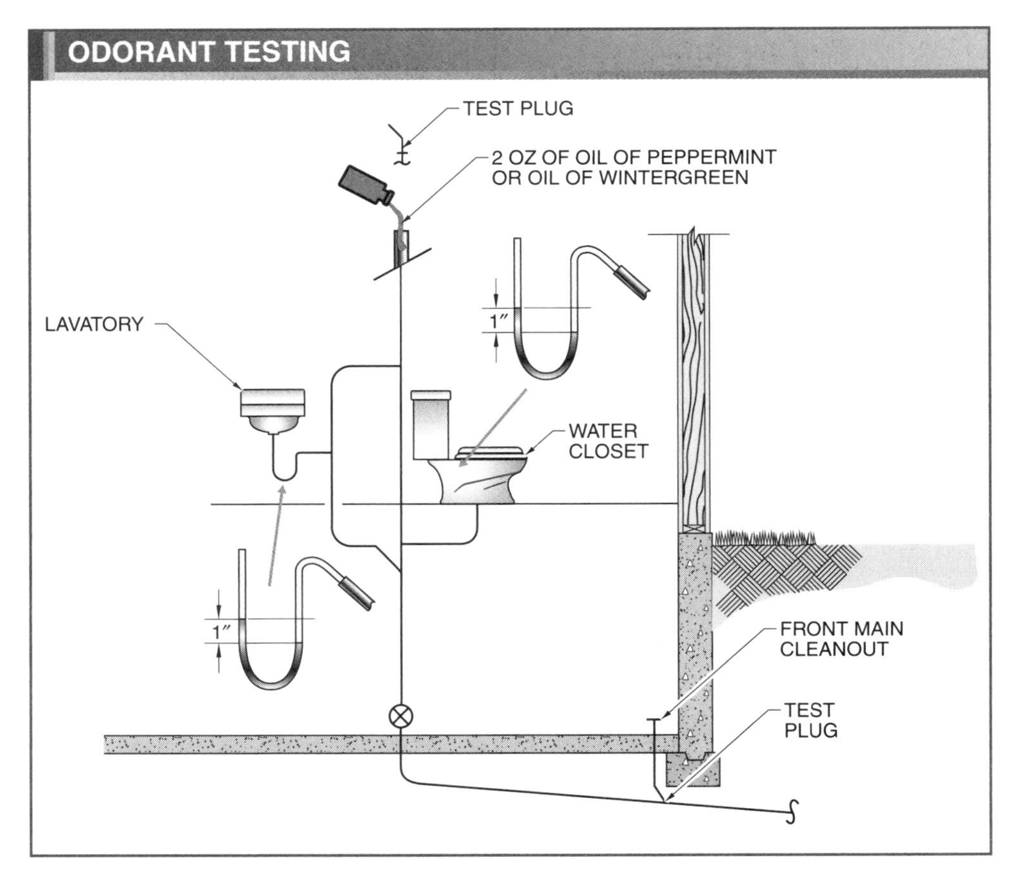

Figure 7-16. An odorant is used to locate leaks during a final air test.

Assemble Testing Apparatus

Plumbing tests involve subjecting plumbing systems to pressure. Verify that the appropriate test plugs and caps and pipe plugs and caps are available for sealing openings within the system being tested.

Assemble the proper equipment for applying pressure to the plumbing system including an air compressor or hand pump for an air test, or a potable water supply to fill piping when performing a water test. In addition to a pressure source, the proper hose to route compressed air or water from the source to the system is required.

A test gauge assembly and manometer are required for plumbing tests to ensure that the appropriate pressure is maintained within the system. Ensure that the test gauge and manometer are properly calibrated.

A means of locating leaks in a plumbing system, such as a spraybottle of soapy water solution, is also required when performing air tests. A smoke chamber and smoke cartridge or an odorant may be necessary for final air tests.

Seal All Openings

Plumbing tests are pressure tests in which pressure must be contained within the system by sealing the openings with test plugs or caps and/or pipe plugs or caps. In a final air test, fixture traps are filled with water to seal the trap.

Apply Tests

Plumbing tests are first performed by a plumber without an inspector being present. Leaking pipe joints and defective materials should be located and identified.

Fix Defects

After locating and identifying leaking pipe joints and defective materials, corrective action must be taken. Leaking joints must be repaired; cracked and defective pipe and fittings must be removed and replaced. Plumbing fixtures that will not pass a final air test must also be removed and stored carefully so that they can be returned to the manufacturer.

Call for an Inspection and Test

After a plumber is satisfied that a plumbing system meets test requirements, a plumbing inspector should be notified that the work is ready for inspection. Plumbing inspectors typically require a minimum notice of one working day before an inspection so they can schedule the inspection.

Tech Facts

A plumber is required to furnish all of the equipment necessary to conduct plumbing system tests. The local plumbing code will specify which type or types of tests must be conducted. A plumber may complete a practice test of the plumbing system before the official test with the plumbing inspector. The practice test will allow repairs to be made.

Assist Inspector During Test

Plumbing system tests are performed when the plumbing inspector arrives at the job site to test and inspect the work. Plumbing inspectors may require a plumber to remove a test plug or cap to ensure that pressure has reached all parts of the system under test. A plumber should escort a plumbing inspector around the job for a visual inspection of the work to verify that it meets plumbing code requirements.

Remove Testing Apparatus

Upon completion of the test and inspection, all testing apparatus must be removed. Test plugs and caps—especially plugs and caps placed on roof vent terminals—are removed and any water or air used to perform the plumbing test is drained from the system. Test nipples and pipe caps typically remain in water supply and sanitary drainage openings after testing to prevent the openings from being covered when finished walls are erected.

Review Questions

1. When should the inspection of the different parts of the plumbing system be performed?
2. Why is the drainage piping of a multistory building tested in sections?
3. List the pipes to be plugged and the equipment to be used to make a water test of an entire drainage system.
4. Which system or parts of the system for stormwater drainage must be tested? Which parts do not need to be tested?
5. Describe the test typically performed on the water supply and distribution system of a building before final inspection.
6. How much pressure is applied to the potable water supply and distribution system when performing a test? How long must the pressure be maintained?
7. Describe the final air test procedure.
8. What should be done with pipe, fittings, and fixtures that fail in a plumbing test?
9. Why must the test plugs and caps be removed from the roof vent terminals when a final air test is completed?

Heating, ventilating, and air conditioning (HVAC) systems used in residential, commercial, and industrial buildings provide comfort to occupants. A feeling of comfort results when temperature, humidity, circulation, filtration, and ventilation of the air are controlled. HVAC systems are designed to fulfill comfort requirements with maximum efficiency.

8 Comfort

HVAC systems contain mechanical, pneumatic, electrical, electronic, and chemical components. HVAC system components include blowers, compressors, valves, dampers, actuators, fuels, refrigerants, and controls. Precautions must be taken and safety rules followed when working with HVAC systems.

COMFORT

Comfort is the condition that occurs when a person cannot sense a difference between themselves and the surrounding air. Comfort occurs when no differences exist or undesirable conditions have been corrected. The five requirements for comfort are proper temperature, humidity, circulation, filtration, and ventilation. **See Figure 8-1.** *Discomfort* is the condition that occurs when a person can sense a difference between themselves and the surrounding air.

Temperature

Controlling the temperature of the human body is an important physiological function. *Physiological functions* are the natural physical and chemical functions of an organism. Physiological systems regulate body temperature to maintain comfort. Normal body temperature is 98.6°F. The body has natural heating and cooling systems to maintain this temperature. These systems control heat output by responding to the conditions of the air according to the internal temperature of the body. The body responds by controlling blood flow at the surface of the skin, radiating heat from body surfaces, or using evaporation of perspiration from skin. When the body is clothed, the body's temperature control system provides comfort at an air temperature of approximately 75°F. If the air temperature varies much above or below 75°F, the body begins to feel uncomfortably warm or uncomfortably cool.

Temperature control in building spaces is provided by warm air from heating equipment or cool air from air conditioning equipment. Heating equipment supplies the proper amount of heat in cold weather to offset heat loss from a building. Heating equipment is rated in Btu per hour (Btu/hr). A *Btu (British thermal unit)* is the amount of heat required to raise 1 lb of water 1°F. Air conditioning equipment supplies the proper amount of cooling in hot weather to offset heat gain to building spaces. Air conditioning equipment is rated in Btu per hour or ton of cooling. A *ton of cooling* is the amount of heat required to melt a ton of ice (2000 lb) over a 24-hour period. One ton of cooling equals 12,000 Btu/hr or 288,000 Btu per 24-hour period ($12,000 \times 24 = 288,000$). **See Figure 8-2.**

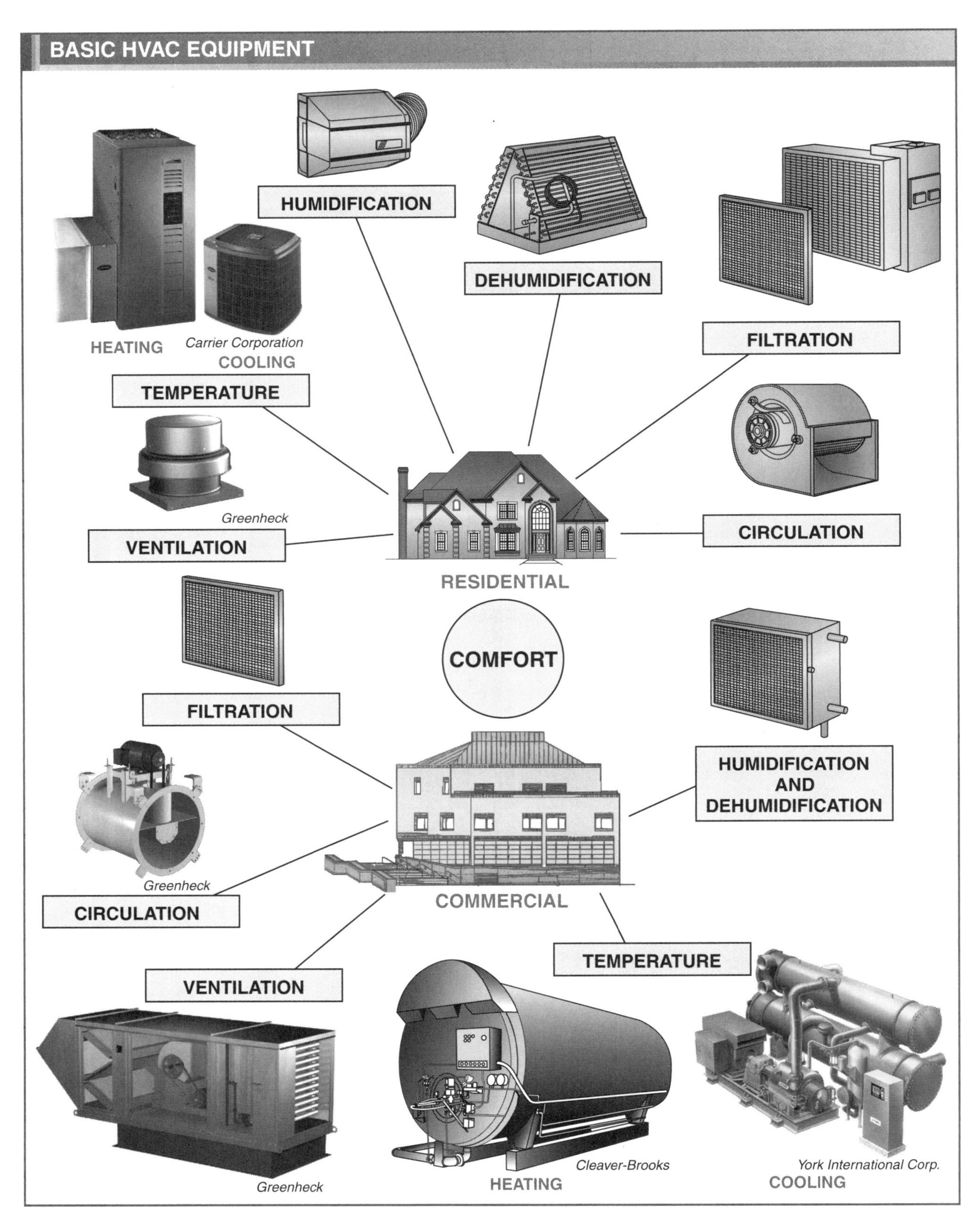

Figure 8-1. HVAC equipment works together in residential or commercial systems to provide the proper temperature, humidity, circulation, filtration, and ventilation required for comfort.

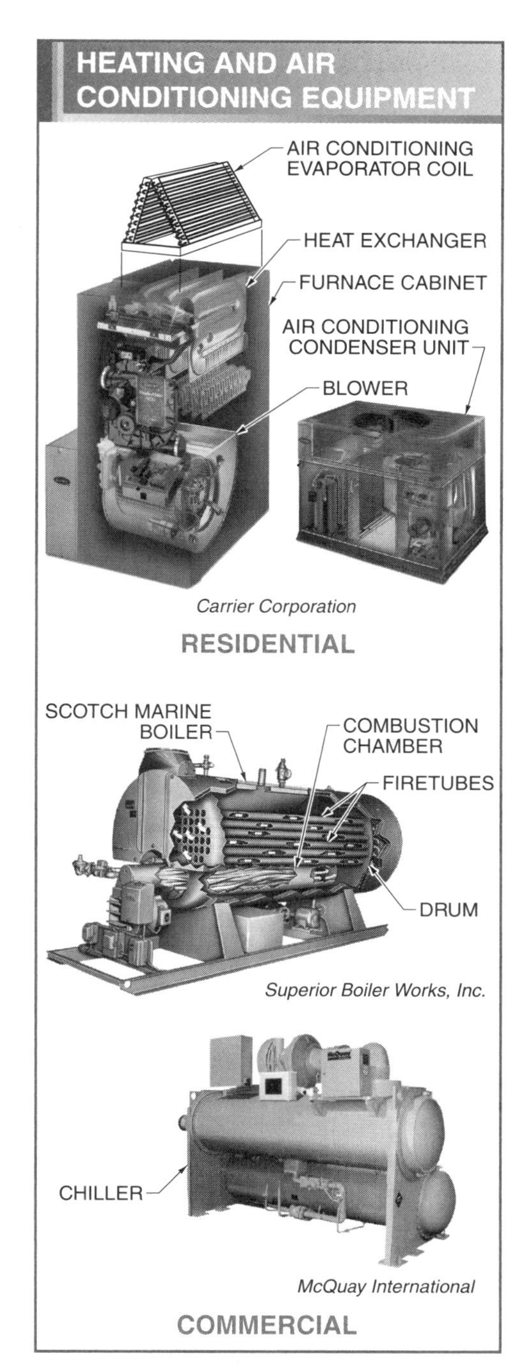

Figure 8-2. Heating and air conditioning equipment controls the temperature in building spaces.

The operation of heating and air conditioning equipment is a function of temperature control in the HVAC system. For maximum comfort, temperature control equipment maintains air temperature in a building within 1°F or 2°F of the temperature necessary for comfort.

Humidity

Humidity is the amount of moisture (water vapor) in air. Humidity is always present in air. A low humidity level indicates dry air that contains little moisture. A high humidity level indicates damp air that contains a significant amount of moisture.

Relative Humidity. *Relative humidity* is the amount of moisture in air compared to the amount of moisture the air would hold at the same temperature if it were saturated (full of water). Relative humidity is always expressed as a percentage. For example, air at 50% relative humidity holds one-half of the amount of moisture it would hold at the same temperature if it were saturated. Air at 60% relative humidity holds 60% of the amount of moisture it would hold at the same temperature if it were saturated. The amount of moisture required to saturate the air changes as the dry bulb temperature changes. In addition, the relative humidity and the capacity to hold moisture increases as the dry bulb temperature increases. Humidity is important in determining comfort.

Humidity affects comfort because it determines how slowly or rapidly perspiration evaporates from the body. Evaporation of perspiration cools the body. The higher the relative humidity, the slower the evaporation rate. The lower the relative humidity, the faster the evaporation rate. For example, with no temperature change and an increase in relative humidity, a person feels warmer because of the slower evaporation rate. With no temperature change and a decrease in relative humidity, a person feels cooler because of the faster evaporation rate.

Comfort is usually attained at normal cooling and heating temperatures with a relative humidity of about 50%, which is 50% saturated. If the humidity is too low, a higher air temperature is required to feel comfortable. If the humidity is too high, a lower air temperature is required for the same feeling of comfort. In some cases the humidity in a building may be too low or too high for comfort. Humidity level is controlled by humidifiers and dehumidifiers. **See Figure 8-3.**

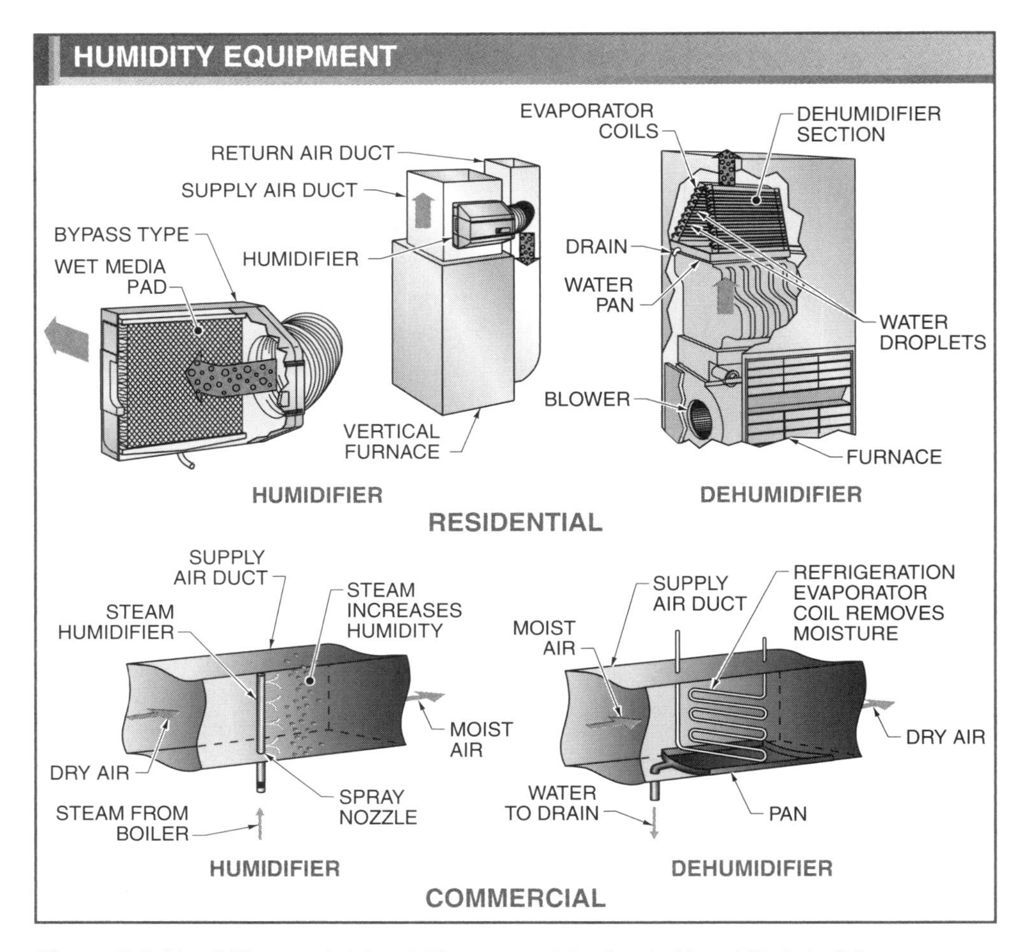

Figure 8-3. Humidifiers and dehumidifiers control the level of humidity in building spaces.

A *humidifier* is a device that adds moisture to air by causing water to evaporate into the air. In cold climates and dry climates, the humidity in a building may be too low for comfort. For example, in cold climate areas such as Chicago in the winter, the relative humidity is low. In dry climate areas such as southern Arizona, relative humidity is also low. In these cases, humidifiers are used in buildings to add moisture to the air and to maintain a comfortable humidity level.

A *dehumidifier* is a device that removes moisture from air by causing moisture to condense. *Condensation* is the formation of liquid (condensate) as gas or vapor cools below its dew point. *Dew point* is the temperature below which moisture begins to condense. In buildings with swimming pools or a large number of potted plants, the humidity level may be too high for comfort.

Tech Facts

Certain HVAC applications may call for a humidity transmitter. A humidity transmitter measures the amount of moisture in the air compared to the amount of moisture the air could hold if it were saturated. Room and duct humidity transmitters are the most common. Humidity transmitters are easily affected by different factors including corrosion and dirt and are inaccurate at humidity extremes. For these reasons, transmitters should be used as relative guides to measuring humidity levels and not as absolute measures.

Circulation

Circulation is the movement of air. Air in a building must be circulated continuously to provide maximum comfort. A total comfort system always includes gently circulating air that is clean and fresh. In a building where there is improper circulation, air rapidly becomes stagnant and uncomfortable. *Stagnant air* is air that contains an excess of impurities such as CO_2 and lacks the oxygen required to provide comfort. When air is improperly circulated, temperature stratification occurs. *Temperature stratification* is the variation of air temperature in a building space that occurs when warm air rises to the ceiling and cold air drops to the floor. Temperature stratification in a building space causes discomfort. **See Figure 8-4.**

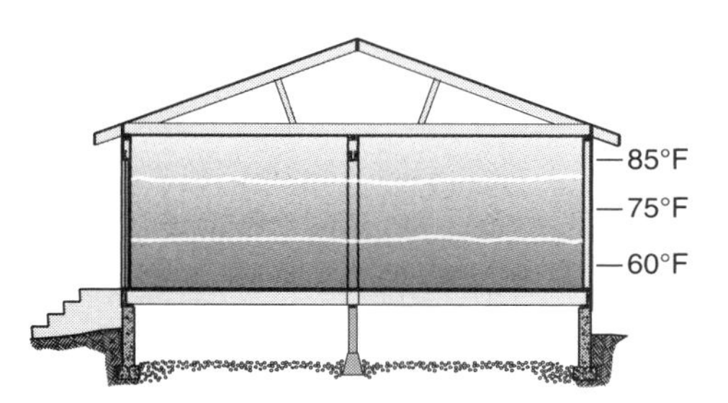

Figure 8-4. Temperature stratification occurs when there is improper circulation in a building space.

Air velocity is the speed at which air moves from one point to another. Air velocity is measured in feet per minute (fpm). If air is not moving, heat cannot be carried away from skin surfaces. If heat is carried to the surface of the skin but is not carried away by the air, the body will overheat. Air circulation is important because it helps cool the body by causing perspiration to evaporate. An increase in air velocity increases the rate of evaporation of perspiration from the skin, which causes a person to feel cool. A decrease in air velocity or no air velocity reduces the evaporation rate, which causes a person to feel warm. Air movement is used to supply clean, fresh air and to remove stagnant air. Air movement of 40 fpm is considered ideal.

Supply air ductwork and registers are used to distribute air to different building spaces. Supply air ductwork is sized and located for maximum efficiency. The type, location, and size of registers determine the amount of supply air introduced into each building space and the distribution pattern of the air. Return air ductwork and grills are used to move air from building spaces back to the blower. Return air ductwork and grills, like supply air ductwork and registers, are sized and located for maximum efficiency. **See Figure 8-5.**

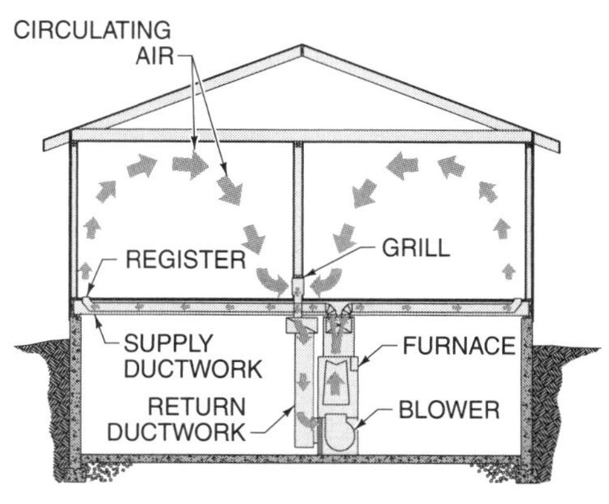

Figure 8-5. Supply and return air ductwork are sized and located to provide efficient flow of air through building spaces.

Filtration

Filtration is the process of removing particles and contaminants from air that circulates through an air distribution system. Circulated air in a building must be clean and includes both return air that is recirculated and fresh air that is used for ventilation. The air in a forced-air heating system is automatically cleaned by filters that are placed in the air ductwork where the air enters the heating or cooling equipment. **See Figure 8-6.** The primary consideration in air filter selection is the level of clean air desired. Filters are commercially available in low-, medium-, and high-efficiency levels.

Ventilation

Ventilation is the process that occurs when outdoor air is brought into a building. Ventilation and air circulation are necessary for comfort inside a building. Offensive odors and vapors are created in closed building spaces. The best method for keeping the air in building spaces comfortable is to dilute the contaminated air. This is accomplished by introducing fresh air from outdoors into recirculated (filtered) air.

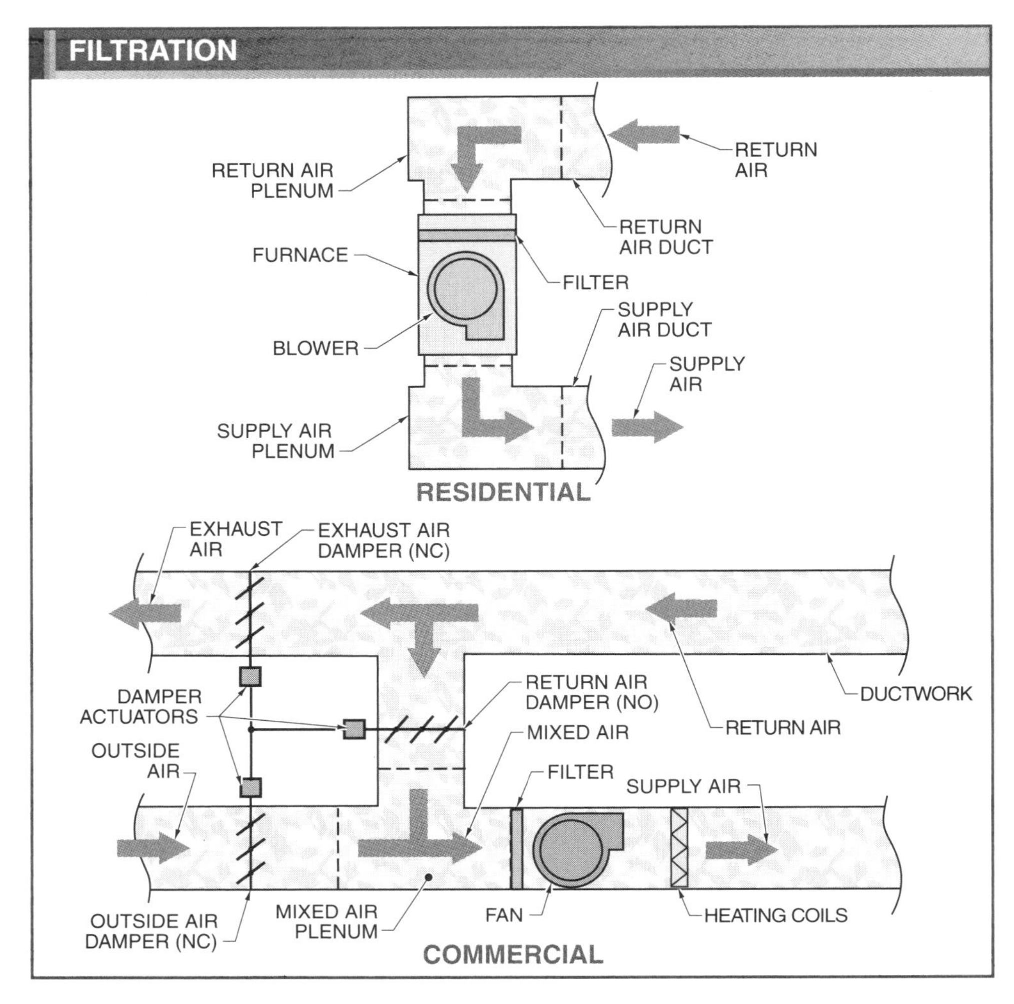

Figure 8-6. Circulated air is cleaned by filters in heating and cooling equipment or filters placed in the return air ductwork near the equipment.

Some odors that originate in buildings are simply offensive, while others may be dangerous. Offensive odors are eliminated by bringing fresh air into a system from outdoors and introducing it into the return air portion of the HVAC system. **See Figure 8-7.** The amount of ventilation necessary for comfort is determined by the number of occupants and the kind of activity taking place inside the building. If odors or contaminants originating in a building are offensive or potentially toxic, the air in the building may require continuous exhaust with no recirculation. The building would require 100% makeup air for ventilation. *Makeup air* is air that is used to replace air that is lost to exhaust.

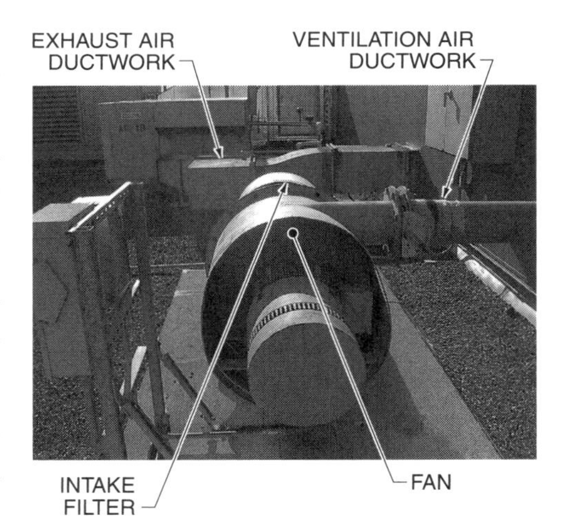

Figure 8-7. The best method of keeping the air in building spaces comfortable is to dilute the contaminated air by introducing fresh air from outdoors into the recirculated (filtered) air.

Tech Facts

Mold is one of the biggest contributors to poor air quality. Mold thrives in buildings that have poor airflow, high humidity, or water leaks.

INDOOR AIR QUALITY

Because many people spend much time in buildings with heating, cooling, and ventilation systems, indoor air quality (IAQ) has become a growing concern. The concerns have increased since energy conservation measures were instituted in buildings during the 1970s as a result of the oil embargo. Energy conservation measures reduce the amount of outside air used in a building, which contributes to the buildup of indoor air contaminants. In general, indoor air quality (IAQ) is the quality of the air in building spaces. **See Figure 8-8.** Effects of poor indoor air quality include fatigue, dizziness, headaches, and irritation of the eyes, nose, and throat. Common effects are similar to those from colds or other viral diseases, so it is often difficult to determine if the symptoms are a result of exposure to indoor air pollution. If the symptoms decrease or go away when the person is away from the building space and return when the person is in a building space, an effort must be made to identify the source. Some effects are made worse by an inadequate supply of outside (ventilating) air. Some effects are from the building heating, cooling, or humidifying equipment itself. **See Figure 8-9.**

Health Effects

Two terms that are associated with the health effects of buildings are "building related illness" and "sick building syndrome." A building related illness is an illness that is traced to a building space and attributed to a definite cause. These include Legionnaires' disease, asthma, hypersensitivity pneumonitis, and humidifier fever. These diseases pose serious risks, but can be treated. Sick building syndrome is an illness that can be traced to a building space but not to a specific cause. Building personnel may complain of dry or burning mucous membranes in the nose, eyes, and throat; sneezing; stuffy or runny nose; fatigue; headache; dizziness; nausea; irritability; and forgetfulness. Other factors may contribute to the problem such as poor lighting, noise, vibration, thermal discomfort, or psychological stress.

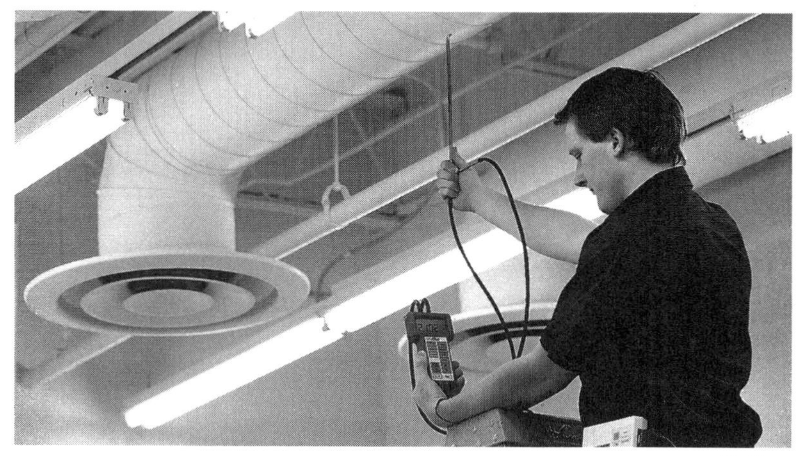

TSI Inc.

Supply air ductwork that is not properly maintained may affect indoor air quality.

INDOOR AIR QUALITY GUIDELINES		
Temperature*	**Humidity†**	**Carbon Dioxide‡**
Winter — 70 to 74 Summer — 72 to 76	Min. 30 Max. 60	Well ventilated — 600 to 1000 Average — 800
Combustable Gases	**Formaldehyde‡**	**Hydrogen Sulfide‡**
Max. 10% LEL (lower explosive limit)	Max. —0.1	Max. 8 hr — 10 Max. 15 min — 15

* in °F
† in percent
‡ in parts per million (ppm)

Figure 8-8. Energy conservation measures reduce the amount of outside air used in a building, which contributes to the buildup of indoor air contaminants.

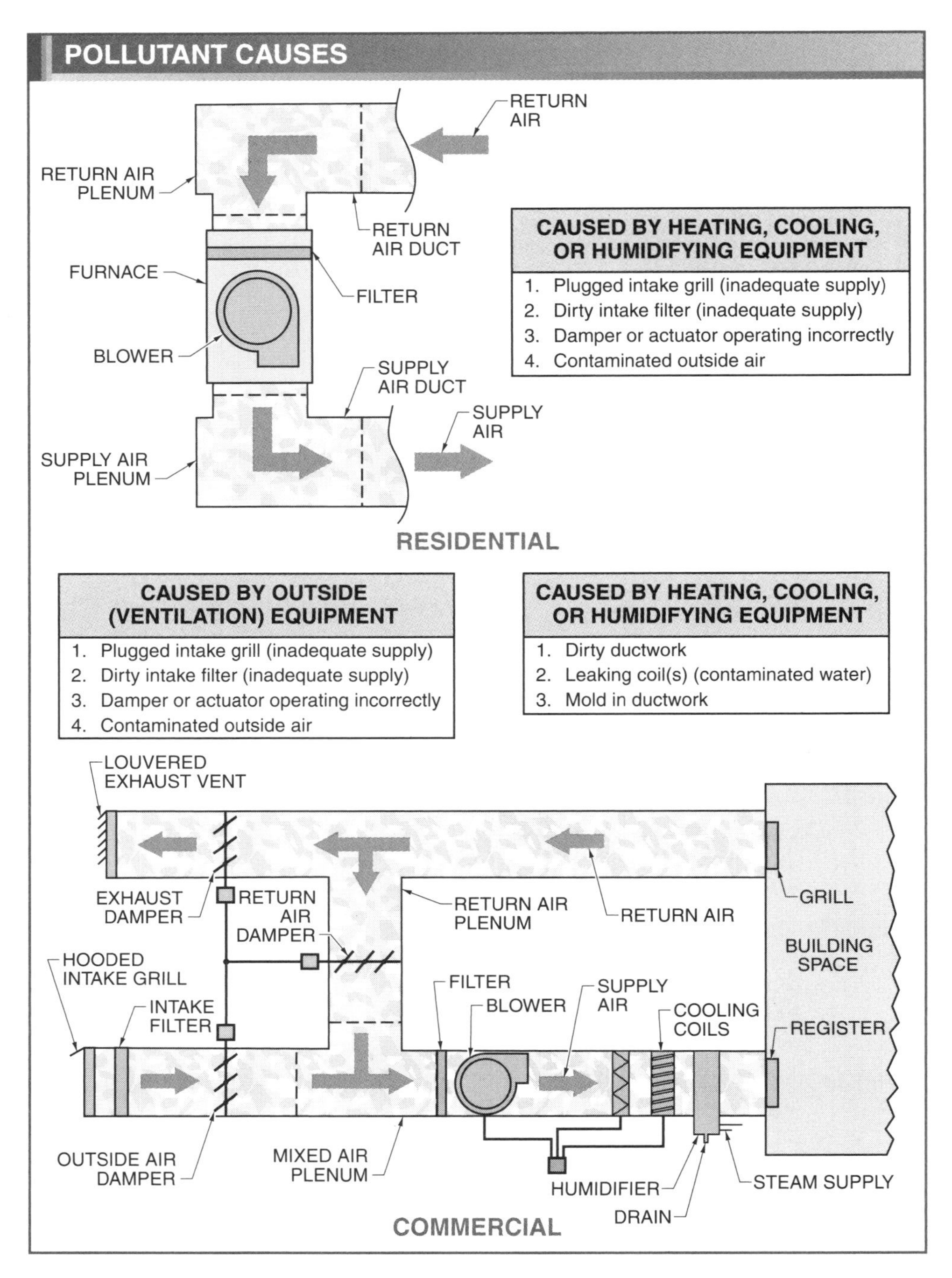

Figure 8-9. In some building spaces there may be one pollutant, while other spaces have a combination of pollutants.

Outbreaks of illness among many personnel in a single building may occur in some cases, and in others health symptoms show up only in individual workers. According to some World Health Organization experts, up to 30 percent of new or remodeled residential and commercial buildings have unusually high rates of health and comfort complaints that are related to indoor air quality. **See Figure 8-10.**

There are three main reasons for poor indoor air quality in building spaces:

BUILDING SPACE CONTAMINANTS		
Health and Comfort Complaints	**Sources**	**Pollutant**
Headache, fatigue, queasiness, poor vision, poor concentration, heart pains, and possibly death	Malfunctioning gas appliances, fireplaces, kerosene heaters, grills, running cars in attached garages, contaminated outside air	Carbon monoxide
Difficult breathing, lung damage, respiratory infections	Malfunctioning gas appliances, fireplaces, kerosene heaters, grills, running cars in attached garages	Nitrogen dioxide
Eye, nose, and throat irritation, cancer of the bladder, stomach, skin, and lungs	Tobacco smoke, fireplaces, kerosene heaters, grills, self-cleaning ovens, hobbies, contaminated outside air	Polycyclic aromatic hydrocarbons
Eye, nose, and throat irritation, emphysema, bronchitis, allergies, asthma, respiratory and ear infections	Tobacco smoke, fireplaces, kerosene heaters, grills, incense burning, hobbies, contaminated outside air	Particles

Figure 8-10. According to some World Health Organization experts, up to 30 percent of residential and commercial buildings have unusually high rates of health and comfort complaints that are related to indoor air quality.

- The presence of indoor air pollution sources.
- Poorly designed, maintained, or operated ventilation systems.
- Uses of the building that were unanticipated or poorly planned for when the building was designed or renovated.

Indoor Air Pollution Sources

Probably the most important factor influencing indoor air quality is the presence of pollutant sources. Environmental tobacco smoke was the largest source of pollution in building spaces, but with the advent of local, state, and federal laws that limit or abolish smoking indoors, the potential effects of tobacco smoke have been dramatically reduced.

Another pollutant source that has been dramatically reduced by regulations is asbestos. Asbestos was used in hundreds of products such as pipe and furnace insulation, fireproofing, flooring, millboard, textured paints, and many other materials.

Formaldehyde is a source of pollution that is found in the glues in furniture, in carpeting, and in a host of other building materials and equipment. Volatile organic compounds that are considered pollutants are found in many products from construction materials, carpets, furnishings, cleaning materials, air fresheners, paints, adhesives, and copying machines to company processes and activities.

Biological contaminants from dirty ventilation systems or water-damaged walls, ceilings, and carpets add to the sources of pollutants found in all building spaces. Biological contaminants include hair, skin, animal dander, mold, and mildew. Pesticides used for pest management in building spaces have the potential to be a very serious pollutant source if not used correctly.

Ventilation Systems

The ventilation systems in buildings are designed to be operated during heating and cooling demand times, or on their own to bring fresh outdoor air into building spaces. Poorly designed, installed, operated, or maintained ventilation systems contribute to indoor air problems in several ways. Problems may arise when, in an effort to save energy, ventilation systems are not used to bring in adequate amounts of outdoor air. Inadequate ventilation also occurs if air supply registers and air return grills within each room are blocked or placed in such a way that proper air circulation does not occur in the building space. Improperly located outdoor air intakes can also bring in air contaminated with automobile or truck exhaust, combustion emissions from furnaces and boilers, fumes from waster containers, or air from lowlands or compost piles. Ventilation systems can be a source of indoor pollution themselves by spreading biological contaminants that have multiplied in cooling towers, humidifiers, dehumidifiers, air conditioners, filters, or the inside surfaces of building ductwork.

Use of a Building

Buildings originally designed for one purpose may end up being converted to use for other purposes. When building renovations are not planned and executed properly, the room partitions and ventilation system can contribute to indoor air quality problems by restricting air circulation or by not providing an adequate supply of outdoor air. The ventilation system is the most overlooked portion of a renovation project because it is hidden above the ceiling or in wall cavities.

Control of Pollutants

The control of all indoor air pollutants in a building is nearly impossible, but there are some strategies that will control most pollutants so that they will not affect building occupants. Common strategies are elimination, source control, dilution, and preventive maintenance.

The best way to control indoor air pollution in a building space is to eliminate the product or material of contamination from the building. An example occurred after the Environmental Protection Agency (EPA) completed a major assessment of the respiratory health risks of environmental tobacco smoke in 1992, and smoking was banned from public buildings in some states. The elimination of tobacco smoke from public buildings dramatically improved the indoor air quality in affected buildings.

By controlling the source of the pollution, pollution emissions can be reduced. Some sources, like those that contain asbestos, can be sealed or enclosed; others, like combustion equipment or appliances, can be adjusted to decrease the amount of emissions. Another way of controlling pollutions at the source is to vent the emissions to a location where they do not create a problem. For example, by placing a fume hood over a printing press, the emissions from the printing process can be vented outside.

The approach of controlling indoor air pollution by dilution is a common approach used for many years. The low cost of energy before the oil embargo of the 1970s allowed occupied spaces to be ventilated with large amounts of fresh air. After the oil embargo the price of energy dramatically increased. One method used to try to control the escalating cost of energy was to decrease the fresh air supplied to building spaces for ventilation. Reducing the amount of fresh air used created a large indoor air quality problem because pollutants were allowed to build up in building spaces. To counteract the problem, the American Society of Heating, Refrigerating, and Air-Conditioning Engineers, Inc. (ASHRAE) recommended that the ventilation requirements be increased in all residential, commercial, and industrial buildings. The recommendations were adopted into regulations in the late 1970s in most municipalities. The normal building space requirement for fresh outside air (ventilation) is 20 cubic feet per minute per person in the space. The

McQuay International

Chillers are used to remove heat from water that circulates through a building for cooling purposes.

increased ventilation requirement did not eliminate indoor air pollution problems but it does tend to eliminate the buildup of pollutants and keep the pollutants at acceptable levels. **See Figure 8-11.**

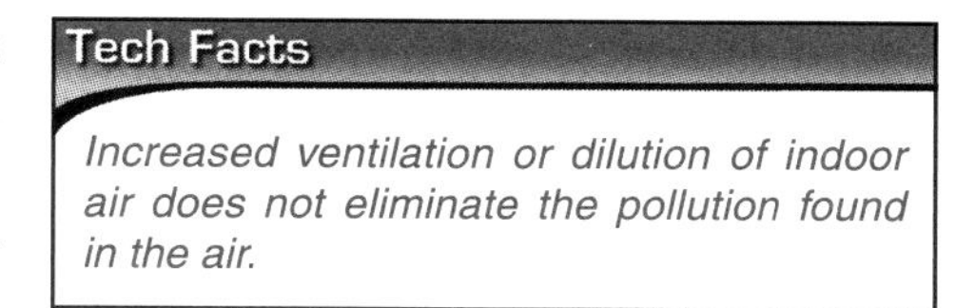

Increased ventilation or dilution of indoor air does not eliminate the pollution found in the air.

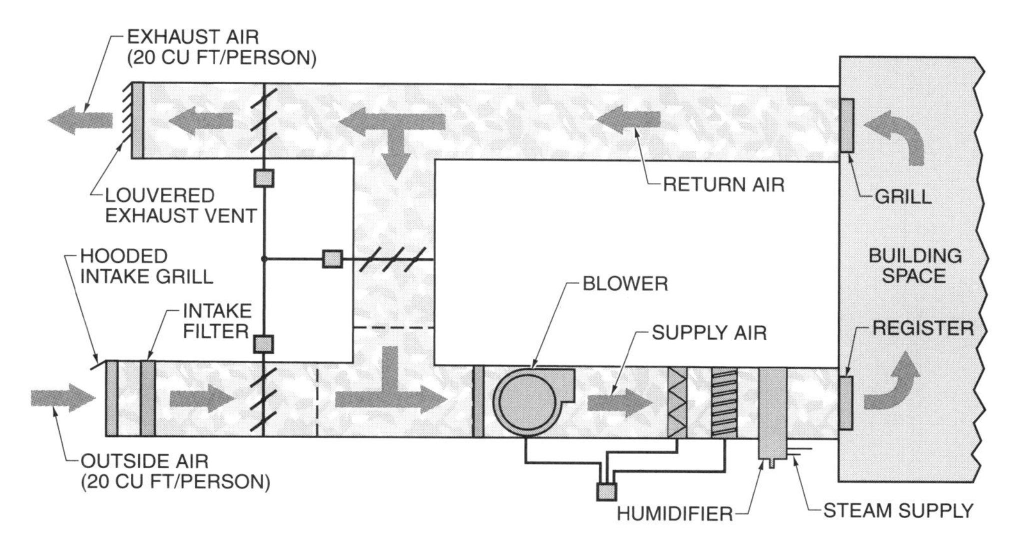

Figure 8-11. Normal building space requirement for fresh outside air (ventilation) is 20 cubic feet per minute per person in the space.

Review Questions

1. Define *comfort* as it relates to heating and air conditioning systems.
2. List the five requirements for comfort.
3. At what temperature do most individuals feel comfortable?
4. Define *humidity* as it relates to building spaces.
5. Define *relative humidity*.
6. At what relative humidity do most individuals feel comfortable?
7. What piece of equipment is used to add humidity to building spaces?
8. Define *condensation*.
9. At what velocity should air in a building space be moving for most individuals to feel comfortable?
10. How do humidifiers add moisture to air?
11. Define *temperature stratification* and explain what causes it.
12. In a comfort system, what are supply ductwork and registers used for?
13. In a comfort system, what are return air ductwork and grills used for?
14. What device is used to clean the air as it circulates in a heating or air conditioning system?
15. What is used to dilute contaminated air?

16. Explain what is meant by indoor air quality (IAQ) and list possible indoor air quality complaints.
17. Describe two types of negative health effects related to building spaces.
18. List common indoor air pollution sources.
19. Describe how indoor air pollutants can be controlled.
20. List three reasons for poor indoor air quality.

9 Psychrometrics

Psychrometrics is the branch of physics that describes the properties of air and the relationships between them. Properties of air are the temperature, humidity, enthalpy, and volume. Psychrometric charts are used to show the relationships between the various properties of air in any condition. When air is conditioned, one or more of the properties of air change. When one property changes, the others are affected. If any two properties of air are known, the others can be found by using a psychrometric chart.

PROPERTIES OF AIR

Atmospheric air is the mixture of dry air, moisture, and particles. *Dry air* is the elements that make up atmospheric air with the moisture and particles removed. *Moist air* is the mixture of dry air and moisture. The *properties of air* are the characteristics of air, which are temperature, humidity, enthalpy, and volume. The properties of air determine the condition of the air, which is related to comfort.

Temperature

Temperature is the measurement of the intensity of heat. Temperature is measured with a dry bulb thermometer. A *dry bulb thermometer* is a thermometer that measures dry bulb temperature. *Dry bulb temperature* (*db*) is the measurement of sensible heat. Temperature controls in HVAC systems turn equipment ON and OFF in response to changes in sensible heat. Dry bulb temperature is expressed in degrees Fahrenheit and degrees Celsius.

Humidity

Humidity is moisture (water vapor) in air The volume of moisture in the air compared to the total volume of air is small. Humidity is expressed as either relative humidity or humidity ratio.

Relative Humidity. *Relative humidity* (*rh*) is the amount of moisture in the air compared to the amount it would hold if the air were saturated. Relative humidity is always expressed as a percentage. For example, air that has a relative humidity of 55% holds 55% of the moisture it can hold at the same temperature when it is saturated.

Humidity Ratio. *Humidity ratio* (W) is the ratio of the mass (weight) of the moisture in a quantity of air to the mass of the air and moisture together. Humidity ratio indicates the actual amount of moisture found in the air. Humidity ratio is expressed in grains (gr) of moisture per pound of dry air (gr/lb) or in pounds of moisture per pound of dry air (lb/lb). A *grain* is the unit of measure that equals $\frac{1}{7000}$ lb. For example, 1 lb of air contains 78 gr or .0111 lb of moisture ($1 \div 7000 \times 78 = .0111$ lb).

Humidity represents latent heat. Latent heat is heat identified by a change of state and no temperature change. Therefore, latent heat cannot be measured with a thermometer.

A certain quantity of heat is required to change water (liquid) into water vapor (gas). The heat required to make this change is latent heat. Because a certain amount of latent heat is required to evaporate a certain amount of water to water vapor, the amount of moisture (water vapor) in the air represents the amount of latent heat. **See Figure 9-1.** Heat must be removed from the air for the moisture to condense back to water.

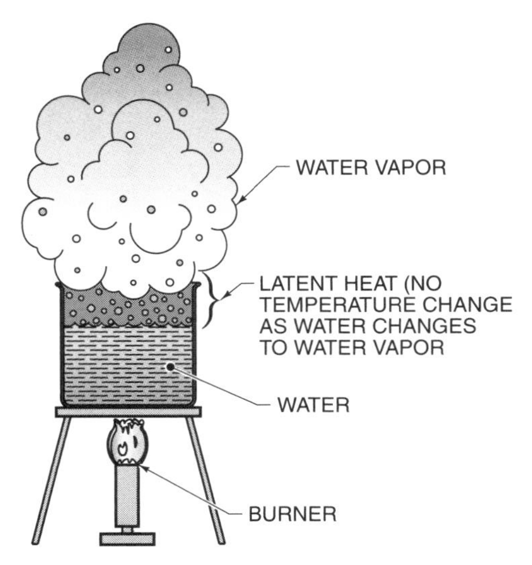

Figure 9-1. Latent heat is identified by the amount of water vapor in the air.

Wet Bulb Temperature. *Wet bulb temperature (wb)* is a measurement of the amount of moisture in the air. Wet bulb temperature is measured with a wet bulb thermometer. A *wet bulb thermometer* is a thermometer that has a small cotton sock placed over the bulb. The end of the sock is wetted with distilled water. The sock draws water up from the sock end, which keeps the sock and the bulb of the thermometer wet.

A wet bulb thermometer measures humidity by measuring the temperature of the air when the bulb of the thermometer is wet. The bulb is cooled as water evaporates from the sock. Because of this cooling effect, the thermometer reads lower than it would without the sock. The cooling effect is a function of the rate of evaporation of water from the sock. The rate of evaporation is a function of the amount of moisture in the air. The drier the air, the quicker the evaporation. Wet bulb temperature readings indicate the amount of moisture in the air. **See Figure 9-2.**

Dew Point. *Dew point (dp)* is the temperature below which moisture in the air begins to condense. Dew point varies with the dry bulb temperature and the amount of humidity in the air. At dew point, air is saturated with moisture and the dry bulb and wet bulb temperatures are the same. Dew point is also known as saturation temperature. On a psychrometric chart, the dew point values are the same as the saturation temperature values.

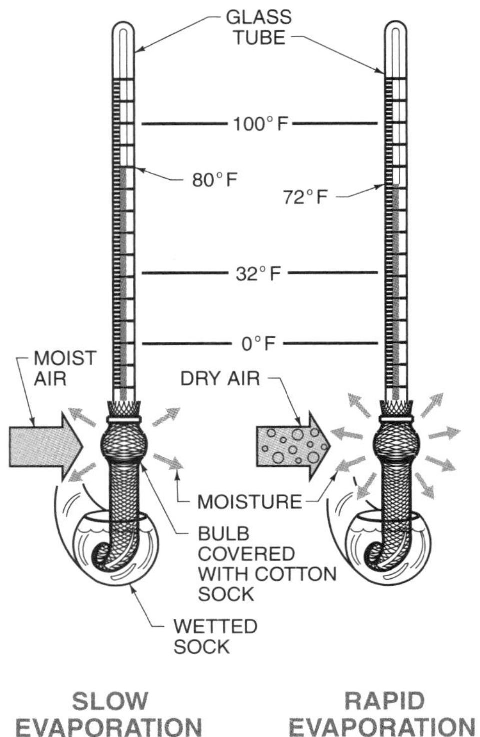

Figure 9-2. Wet bulb temperature readings indicate the amount of moisture in the air. The cotton sock serves as a wick, which keeps the bulb of the thermometer wet.

Humidity Measurement Equipment. A *psychrometer* is an instrument used for measuring humidity that consists of a dry bulb thermometer and a wet bulb thermometer mounted on a common base. A psychrometer measures humidity by comparing the temperature readings on the dry bulb and wet bulb thermometers. *Wet bulb depression* is the difference between the wet bulb and dry bulb temperature readings and is directly related to the amount of moisture in the air. A *sling psychrometer* is an instrument used for measuring humidity that consists of a wet bulb and a dry bulb thermometer mounted on a base. The base is mounted on a handle so it can be rotated rapidly. **See Figure 9-3.** Air flows over the bulbs of the thermometers when the sling psychrometer is rotated. As the sling psychrometer is rotated, the water on the wet bulb thermometer evaporates. The sling psychrometer is rotated in the air until the temperature reading on each thermometer stabilizes. The readings are taken immediately. Charts or graphs must then be used to find the relative humidity from the two temperature readings. **See Appendix.**

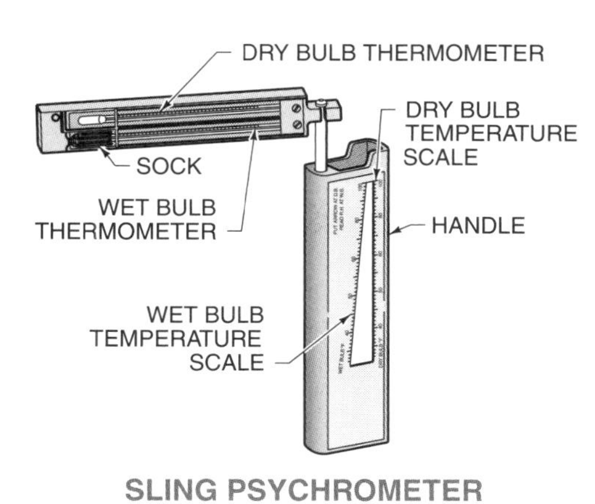

Figure 9-3. Sling psychrometers measure wet and dry bulb temperature simultaneously. Charts or graphs are required to determine relative humidity.

Humidity can also be measured with a hygrometer. A *hygrometer* is any instrument used for measuring humidity. A psychrometer is one kind of hygrometer. A *dimensional change hygrometer* is a hygrometer that operates on the principle that some materials absorb moisture and change size and shape depending on the amount of moisture in the air. A material in the hygrometer is exposed to the humidity in the air. As the material expands or contracts based on the moisture absorbed or evaporated, a linkage moves an indicator according to the motion of the material. Materials used in dimensional change hygrometers include hair, wood, and plastic.

Electrical impedance hygrometers measure humidity electronically. An *electrical impedance hygrometer* is a hygrometer based on the principle that the electrical conductivity of a substance changes as the amount of moisture in the air changes. Sensors on the hygrometer are covered with a salt-base substance such as lithium chloride. The electrical conductivity of the substance changes with the amount of moisture in the air. The hygrometer senses the amount of electricity conducted and gives a reading based on that amount. Most hygrometers give direct percentage readings of humidity (relative humidity). **See Figure 9-4.**

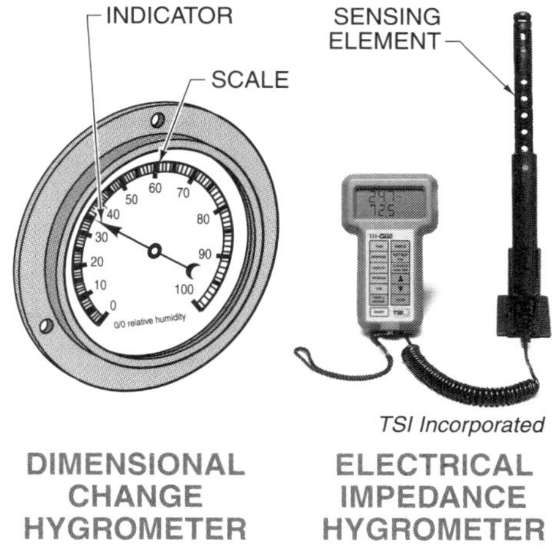

Figure 9-4. Hygrometers measure and provide readings of relative humidity. Hygrometers are based on the principle that the characteristics of certain material change with the amount of moisture in the air.

Enthalpy

Any material that has a dry bulb temperature above absolute zero contains heat. *Enthalpy* (*h*) is the total heat contained in a substance measured from a baseline of 32°F, and is the sum of sensible heat and latent heat. Enthalpy of air is expressed in Btu per pound of moist air. The enthalpy of air at different conditions is found on a psychrometric chart.

Volume

Air, like most substances, expands when heated and contracts when cooled. *Volume* (*V*) is the amount of space occupied by a three-dimensional figure. It is expressed in cubic units such as cubic inches or cubic feet. *Specific volume* (v) is the volume of a substance per unit of the substance. The specific volume of air is expressed in cubic feet per pound at a given temperature. Because air expands and contracts at different temperatures and moisture content, air volume is always varying.

Standard Conditions. The properties of air are related so that a change in one of the properties causes a change in the other three. Properties of air are compared at standard conditions. *Standard conditions* are values used as a reference for comparing properties of air at different elevations and pressures. One pound of dry air and its associated moisture at standard conditions has a pressure of 29.92″ Hg (14.7 psia), temperature of 68°F, volume of 13.33 cu ft/lb, and density of .0753 lb/cu ft. Standard conditions are used when comparing the relationships between the different properties of air such as on a psychrometric chart.

Relationships between Properties

The properties of moist air are related in such a way that a change in one brings about changes in one or more of the others. The two properties used most often for identifying specific conditions of the air are temperature and humidity.

Some HVAC systems are equipped with humidifiers or dehumidifiers to ensure that the humidity (the amount of water in the air) inside a building is in the comfort zone.

A change in any property of the air affects other properties. Both dry bulb and wet bulb temperatures must be taken into account when considering the effect of changes in temperature on the other properties of air. Together, dry bulb and wet bulb temperatures affect relative humidity, humidity ratio, dew point, enthalpy, and volume.

A change in wet bulb temperature changes the humidity ratio and the relative humidity of the air. A change in wet bulb temperature indicates that moisture has been added to or removed from the air. Wet bulb temperature, humidity ratio, and relative humidity are all directly related to the amount of moisture in the air. A change in one property changes the others.

A change in both dry bulb and wet bulb temperatures affects enthalpy, which is total heat content. A change in dry bulb temperature indicates a change in sensible heat. A change in wet bulb temperature indicates a change in latent heat.

A change in dry bulb temperature affects the specific volume of the air. If the dry bulb temperature increases, the specific volume increases. If the dry bulb temperature decreases, the specific volume will decrease.

When considering the effect of changes in humidity on other properties of air, the humidity ratio must be considered. As the humidity ratio increases, the latent heat content of the air also increases. As the humidity ratio decreases, the latent heat content of the air also decreases.

PSYCHROMETRIC CHART

A *psychrometric chart* is a chart that defines the condition of the air for any given property. **See Figure 9-5.** Each of the properties of air is shown on the chart. The properties of air for any condition can be determined by using a psychrometric chart. Psychrometric charts are available for standard conditions and special conditions such as higher- or lower-than-normal pressures or higher- or lower-than-normal temperatures.

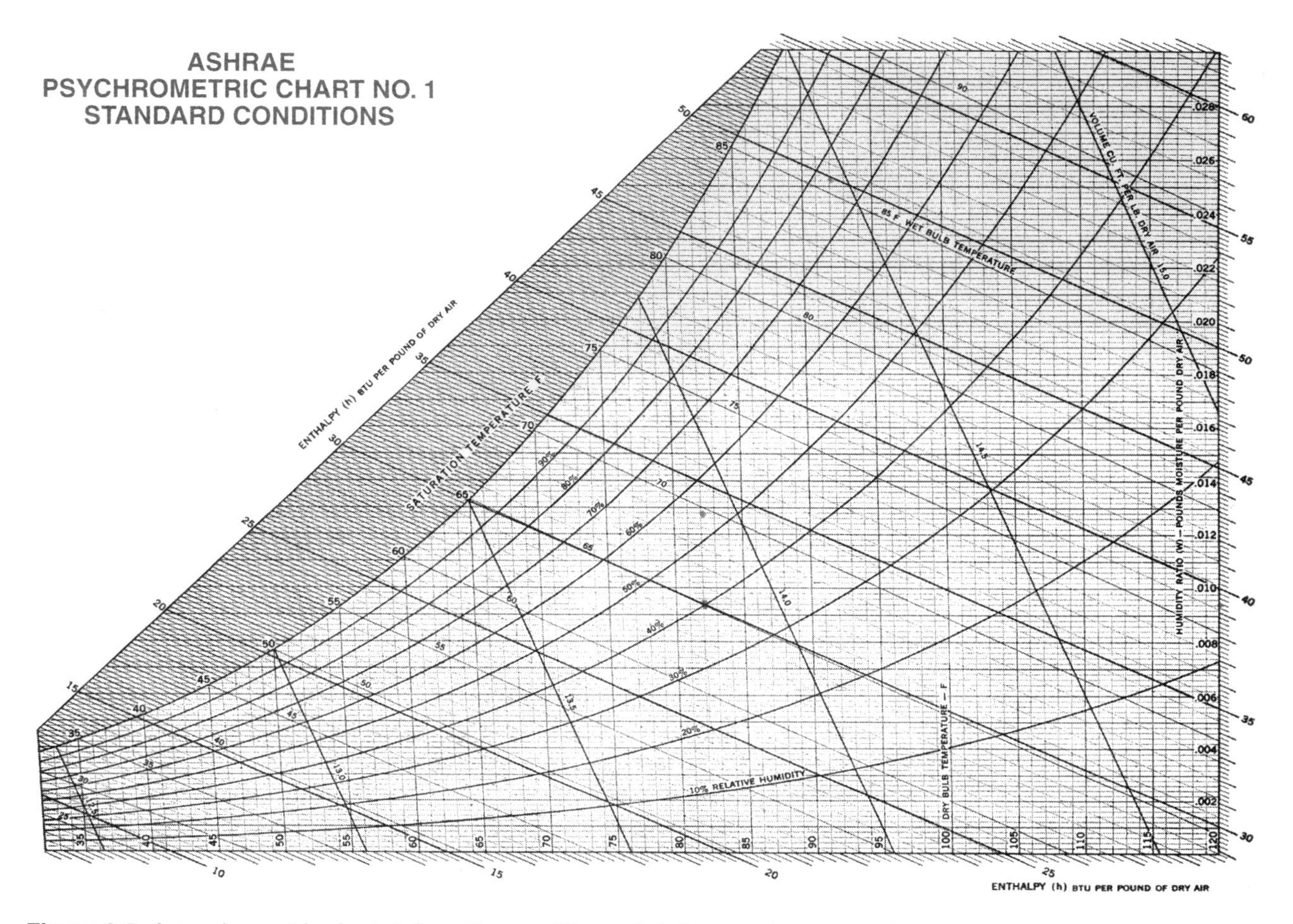

Figure 9-5. A psychrometric chart defines the conditions of air for any given property.

Using the Psychrometric Chart

The properties of air found on a psychrometric chart are dry bulb temperature, relative humidity, humidity ratio, wet bulb temperature, dew point, enthalpy, and specific volume. The properties of the air define the condition of the air. **See Figure 9-6. See Appendix.**

The dry bulb temperature scale is found along the bottom of the chart. Vertical lines on the chart identify the dry bulb temperature of the air at a given point. The dry bulb temperature scale begins at the left side of the chart and increases to the right. On a chart for air at standard conditions, the dry bulb temperature scale begins at about 35°F and extends to about 120°F.

Relative humidity of air is found on the curving lines that run from the bottom left side of the chart up to the right side. Relative humidity is expressed as the percentage of saturation of the air by moisture. The scale begins at 10% at the bottom right of the chart and increases up to the saturation line, which is 100%.

The humidity ratio scale is found on the right side of the chart. The humidity ratio for air is found on the horizontal lines that run across the chart. The humidity ratio scale begins at the bottom of the chart and extends to the top. On a chart for standard conditions, humidity ratio begins at 0 lb of moisture per pound of dry air and increases to about .030 lb of moisture per pound of dry air.

Tech Facts

Buildings often require the use of humidifiers and dehumidifiers to maintain humidity within the comfort zone. A psychrometric chart can be used to determine the correct temperature and humidity needed for comfort.

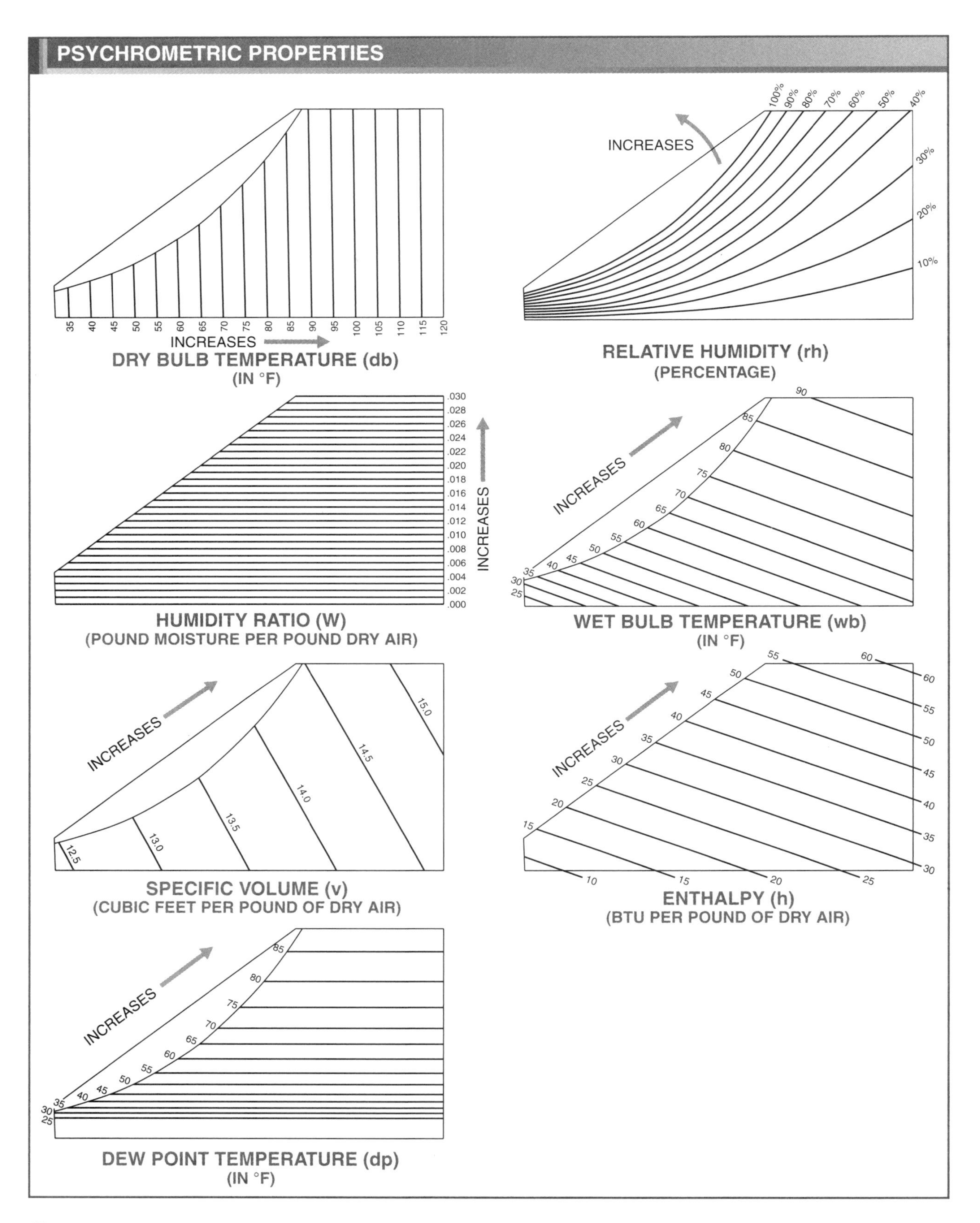

Figure 9-6. If two properties of air are known, the remaining properties can be found using a psychrometric chart.

The wet bulb temperature and dew point scales are found along the curve of the chart. The scale begins at the bottom left and increases as the curve extends up to the right. On a chart for standard conditions, the scale begins at about 25°F and increases to about 90°F. Lines related to wet bulb temperature run from the scale on the curve of the chart diagonally down to the right at about a 35° angle. Any point along one of the wet bulb temperature lines defines the amount of moisture in the air at that point. Lines related to dew point run horizontally from the scale on the curve to the right side of the chart. The scale is the same as that used for wet bulb temperature. The *saturation line* is the curve where the wet bulb temperature and dew point scales begin. Air at any point on this line is saturated with moisture.

The enthalpy scale is located above the saturation line on the curve of the chart. The numbers on the enthalpy scale almost coincide with lines that extend from the wet bulb temperature lines on the chart. On some charts an enthalpy scale runs below the dry bulb temperature scale, which runs across the bottom of the chart. Enthalpy is found by extending a line through a known point approximately parallel to the wet bulb temperature lines and beyond the saturation line to the enthalpy scale. On a chart for standard conditions, enthalpy begins at about 10 Btu/lb of air at the bottom of the scale and increases to about 60 Btu/lb of air.

Specific volume of air is found on the lines that run at a steep angle from the saturation line to the bottom of the chart. The specific volume scale is shown on the specific volume lines or on an extension of the lines at the bottom of the chart. On a chart for standard conditions, specific volume begins at about 12.5 cu ft/lb of air and increases to about 15.0 cu ft/lb of air.

If any two properties of air are known, the others can be found by locating the point defined by those properties. This is done by applying the procedure:

1. To find the properties of air at 90°F and 40% relative humidity, first locate the 90°F dry bulb temperature line on the scale across the bottom of the chart. Follow this line upward until it crosses the 40% relative humidity line. The point where the two lines intersect defines the condition of the air. This point is used to find the other properties of the air. **See Figure 9-7.**

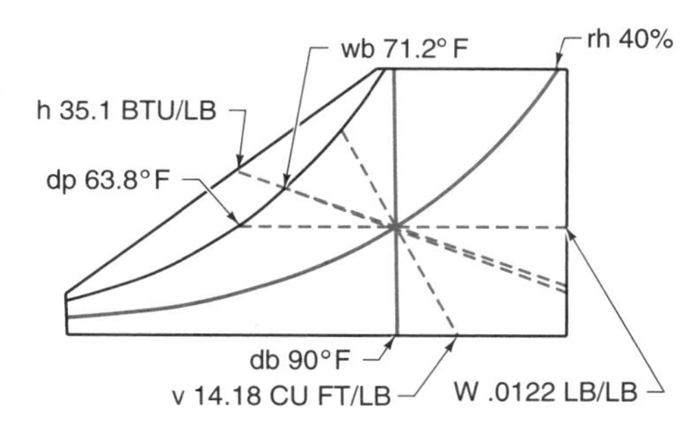

Figure 9-7. All properties of air are found by the intersection of the lines that represent any two properties.

2. To find humidity ratio at this point, follow the horizontal line that runs through the point to the humidity ratio scale on the right side of the chart. The humidity ratio is .0122 lb of moisture per pound of dry air.
3. To find wet bulb temperature, follow the wet bulb temperature line that runs through the point to the wet bulb temperature scale on the curve of the chart. The wet bulb temperature is 71.2°F.
4. To find dew point, follow the horizontal line that runs close to or through the point to the dew point scale on the curve. The dew point is 63.8°F.
5. To find enthalpy, extend the wet bulb temperature line for the point through the curve to the enthalpy scale. The enthalpy at saturation for the point is 35.1 Btu/lb.
6. To find specific volume, locate the specific volume lines on both sides of the point and approximate the specific volume for the point. The specific volume is about 14.18 cu ft/lb.

Example: Finding Properties of Air at a Specific Point

The air in a building has a dry bulb temperature of 75°F and 50% relative humidity.

Find the wet bulb temperature, humidity ratio, dew point, enthalpy, and specific volume of the air.

1. Locate the 75°F dry bulb temperature line on the scale on the bottom of the chart. Follow this line upward until it crosses the 50% relative humidity line. The point where the two intersect is the point that defines the condition of the air at the two given properties. **See Figure 9-8.**

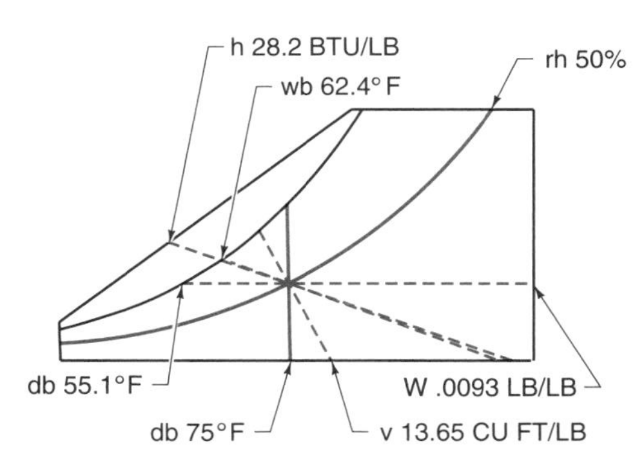

Figure 9-8. Properties of air at 75°F and 50% relative humidity are found at the intersection of the 75°F dry bulb temperature line and the 50% relative humidity line.

2. Find humidity ratio by following the horizontal line that runs through the point to the humidity ratio scale on the right side of the chart. The humidity ratio is .0093 lb of moisture per pound of dry air.
3. Find wet bulb temperature by following the wet bulb temperature line that runs through the point to the wet bulb temperature scale on the curve of the chart. The wet bulb temperature is 62.4°F.
4. Find dew point by following the horizontal line that runs closest to or through the point to the dew point scale on the curve. The dew point temperature is 55.1°F.
5. Find enthalpy by extending the wet bulb temperature line for the point through the curve to the enthalpy scale. The enthalpy at saturation for the point is 28.2 Btu/lb.
6. Find specific volume by locating the specific volume lines on both sides of the point and approximating the specific volume for the point. The specific volume for the point is approximately 13.65 cu ft/lb.

Applying the Psychrometric Chart

Engineers and air conditioning technicians use psychrometric charts to define properties of moist air during processes that change the properties of air. Heating, cooling, humidification, and ventilation are the most common processes that change the properties of air.

Heating. When moist air is heated, sensible heat is added and there is no change in humidity. If only sensible heat changes, the properties of heated air are determined by horizontal movement between points on the chart. The horizontal movement is the difference between the initial condition and final condition of the air. *Initial condition* is the point that represents the properties of air before it goes through a process. *Final condition* is the point that represents the properties of air after it goes through a process. The point that represents the final condition of the air when sensible heat is added is located directly to the right of the original point on the same humidity ratio line.

The final condition of the air when sensible heat is added is found by applying the procedure:

1. The point that represents the initial condition of the air is located at the intersection of the dry bulb temperature and relative humidity lines on the chart. If air has a dry bulb temperature of 60°F and 40% relative humidity, the intersection of these two lines represents the initial condition of the air. **See Figure 9-9.**
2. The point that represents the final condition of the air is located directly to the right of the initial point on the same humidity ratio line where it intersects the final dry bulb temperature line. If sensible heat is added to the air raising it to a temperature of 100°F, the intersection of the humidity ratio line from the initial point and the 100°F dry bulb temperature line represents the final condition of the air.

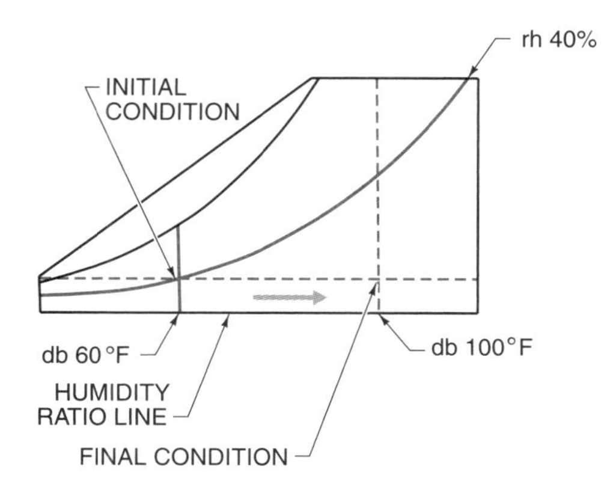

Figure 9-9. The point that represents the condition of air when sensible heat is added is located on the same humidity ratio line at the intersection with the final temperature.

Example: Finding Points—Heated Air

Air enters a furnace at a dry bulb temperature of 75°F and 50% relative humidity and leaves the furnace at a dry bulb temperature of 94°F. Find the point that represents the condition of the air as it leaves the furnace.

1. Locate the point at the intersection of the 75°F dry bulb temperature line and the 50% relative humidity line. This point represents the initial condition of the air. **See Figure 9-10.**

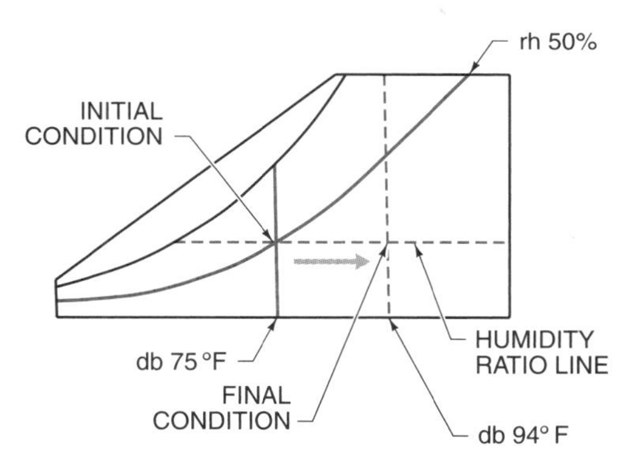

Figure 9-10. To find the point that represents the condition of the heated air, follow the same humidity ratio line to the 94°F dry bulb temperature line.

2. Locate the intersection (to the right) of the humidity ratio line of the initial point with the 94°F dry bulb temperature line. This point represents the final condition of the air.

Cooling. Psychrometric charts are often used to find the total cooling that occurs when air passes through an air conditioning coil. The properties of cooled air are found by locating the initial condition and final condition of the air on the chart and identifying the appropriate properties of the air at each point. The differences between the two conditions indicate the changes made to the air during the process.

Total cooling is found by applying the procedure:

1. The points that represent the initial condition and final condition of the air are located at the intersection of the lines that identify the two given properties. If the air has an initial dry bulb temperature of 80°F and an initial wet bulb temperature of 71°F, the intersection of the two lines (point 1) represents the initial condition of the air. **See Figure 9-11.** If the air has a final dry bulb temperature of 60°F and a final wet bulb temperature of 50°F, the intersection of the two lines (point 2) represents the final condition of the air.

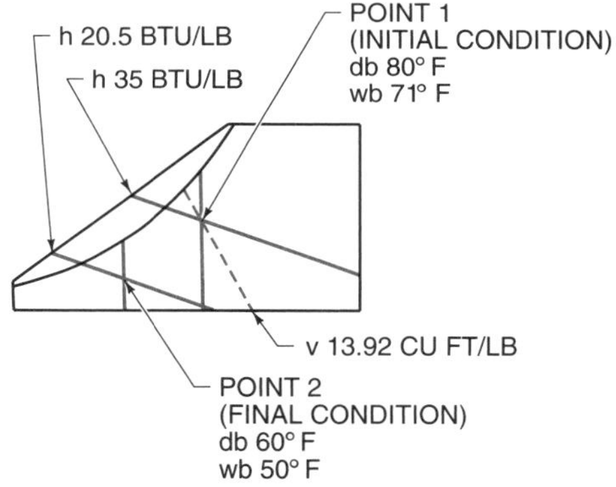

Figure 9-11. Properties of cooled air are found by locating the initial and final enthalpy and initial specific volume of the air on the chart.

2. Enthalpy of the air at the initial condition and final condition is found by using points 1 and 2 and the enthalpy scale on the chart. The enthalpy of the air at the initial condition, point 1, is 35 Btu/lb. The enthalpy at the final condition, point 2, is 20.5 Btu/lb.

3. The amount of heat removed from the air is found by subtracting the enthalpy of the air at the final condition from the enthalpy of the air at the initial condition.

   $Heat\ removed = h_{point\ 1} - h_{point\ 2}$

   $Heat\ removed = 35 - 20.5$

   $Heat\ removed =$ **14.5 Btu/lb**

4. Specific volume of the air at the initial condition is found on the chart by locating the specific volume lines on both sides of point 1 and approximating the specific volume for that point. Specific volume at the initial condition is 13.92 cu ft/lb.

5. To find total cooling, first find the weight of the air being cooled per minute (rate of cooling). If the quantity of air being cooled is 11,000 cfm (cubic feet per minute), the weight of the air being cooled per minute is found by dividing 11,000 cfm by the specific volume.

   $$Rate\ of\ cooling = \frac{Quantity\ of\ air}{v}$$

   $$Rate\ of\ cooling = \frac{11{,}000}{13.92}$$

   $Rate\ of\ cooling =$ **790.23 lb/min**

6. Total cooling is found by multiplying the rate of cooling by the amount of heat removed from the air. Because the capacity of cooling equipment is usually rated in Btu per hour, the result is multiplied by 60.

   $Total\ cooling = Rate\ of\ cooling \times Heat\ removed$

   $Total\ cooling = 790.23 \times 14.5 \times 60$

   $Total\ cooling =$ **687,500.1 Btu/hr**

Example: Finding Points—Total Cooling

A cooling coil cools 12,500 cfm of air. The air enters the coil at a dry bulb temperature of 74°F and a wet bulb temperature of 63.2°F. The air leaves the coil at a dry bulb temperature of 55°F and a wet bulb temperature of 52.5°F. Find the total cooling as the air passes through the cooling coil.

1. Locate point 1 at the intersection of the 74°F dry bulb temperature line and the 63.2°F wet bulb temperature line. Locate point 2 at the intersection of the 55°F dry bulb temperature line and the 52.5°F wet bulb temperature line. **See Figure 9-12.**

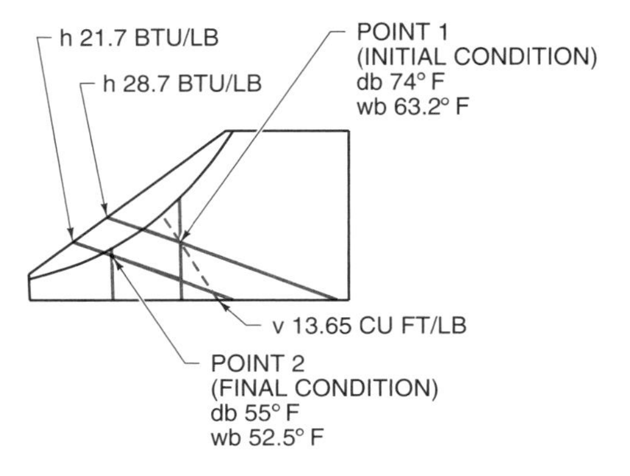

Figure 9-12. Total cooling is the amount of heat removed from air as the air passes through an air conditioning coil.

2. Locate the enthalpy at each point. The enthalpy at point 1 is 28.7 Btu/lb. The enthalpy at point 2 is 21.7 Btu/lb.

3. Subtract the enthalpy of the air leaving the coil from the enthalpy of the air entering the coil. This is the amount of heat removed from the air.

   $Heat\ removed = h_{point\ 1} - h_{point\ 2}$

   $Heat\ removed = 28.7 - 21.7$

   $Heat\ removed =$ **7.0 Btu/lb**

4. Find the specific volume of the air at initial condition (from the chart). The specific volume is 13.65 cu ft/lb.

5. Find the rate of cooling.

   $$Rate\ of\ cooling = \frac{Quantity\ of\ air}{v}$$

   $$Rate\ of\ cooling = \frac{12{,}500}{13.65}$$

   $Rate\ of\ cooling =$ **915.75 lb/min**

6. Find the total cooling.

   $Total\ cooling = Rate\ of\ cooling \times Heat\ removed$

   $Total\ cooling = 915.75 \times 7.0 \times 60$

   $Total\ cooling =$ **384,615 Btu/hr**

Humidification. The humidity in a building is often too low for comfort or for a process. The quantity of water that must be added to air to increase the relative humidity is found by applying the procedure:

1. The points that represent the initial condition and final condition of the air are located at the intersection of the lines that identify the two given properties. **See Figure 9-13.** If air has a dry bulb temperature of 68°F and a relative humidity of 45%, the point that represents the initial condition of the air is found at the intersection of the two lines (point 1). To raise the relative humidity to 57%, a certain quantity of water must be added to the air. The point that represents the desired final condition of the air (point 2) is found at the intersection of the 57% relative humidity line and the same dry bulb temperature line (68°F).
2. Humidity ratio at these two points is found by following the horizontal lines that run through the points to the scale on the right side of the chart. The humidity ratio at point 1 is .0066 lb/lb and the humidity ratio at point 2 is .0083 lb/lb.
3. The amount of water that must be added to each pound of air to increase the relative humidity of the air from point 1 to point 2 is found by subtracting the humidity ratio of point 2 from the humidity ratio of point 1.

   Humidity ratio difference = $W_{point\ 2} - W_{point\ 1}$

   Humidity ratio difference = $.0083 - .0066$

   Humidity ratio difference = **.0017 lb/lb**
4. To find the amount of water to be added to the air in a building space, first determine the volume of air in the space. If the dimensions of the space are 60′ × 50′ × 10′, the volume of the space is found by applying the formula:

   $V = l \times w \times h$

   $V = 60 \times 50 \times 10$

   $V =$ **30,000 cu ft**
5. The total weight of the air in a building space is found by dividing the volume of the air by the specific volume of the air at point 1 (initial condition). Specific volume at this point is 13.44 cu ft/lb.

   $$Total\ weight\ of\ air = \frac{V}{v_{point\ 1}}$$

   $$Total\ weight\ of\ air = \frac{30,000}{13.44}$$

   Total weight of air = **2232.14 lb**
6. The weight of the water to be added is found by multiplying the total weight of the air by the humidity ratio difference, which is .0017 lb of water per pound of air.

   Weight of water to be added = *Total weight of air* × *Humidity ratio difference*

   Weight of water to be added = $2232.14 \times .0017$

   Weight of water to be added = **3.79 lb**

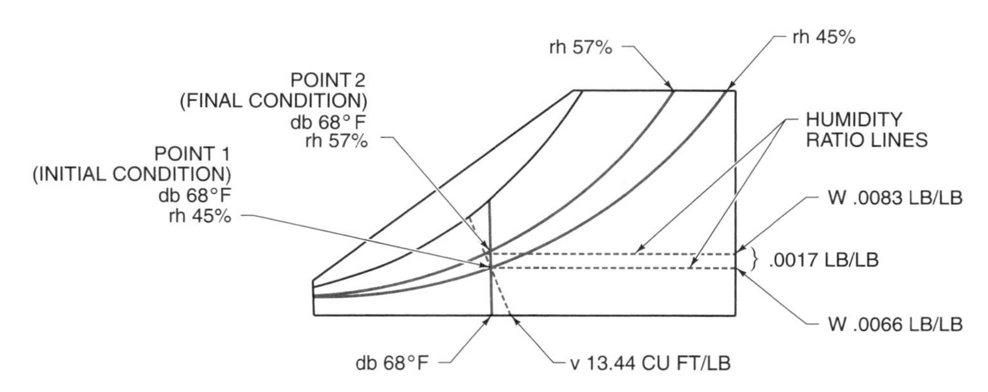

Figure 9-13. Properties of moist air are determined by finding the initial and final humidity ratio and the initial specific volume of the air on the chart.

7. To find the number of gallons of water to be added to the air, divide the weight of the water by 8.345, which is the weight of 1 gal. of water.

$$\textit{Gallons of water} = \frac{\textit{Weight of water to be added}}{8.345}$$

$$\textit{Gallons of water} = \frac{3.79}{8.345}$$

$$\textit{Gallons of water} = \mathbf{.454\ gal.}$$

Example: Finding Points—Humidification

The air in a building has a dry bulb temperature of 72°F and 40% relative humidity. The building is 45′ long, 37′ wide, and has a 10′ high ceiling. Find the quantity of water required to raise the relative humidity to 55%.

1. Locate point 1 at the intersection of the 72°F dry bulb temperature line and the 40% relative humidity line. Locate point 2 at the intersection of the 72°F dry bulb temperature line and the 55% relative humidity line. **See Figure 9-14.**
2. Locate the humidity ratio of the air at each point. Humidity ratio at point 1 is .0066 lb of moisture per pound of air and the humidity ratio at point 2 is .0092 lb of moisture per pound of air.
3. Find the humidity ratio difference.

$$\textit{Humidity ratio difference} = W_{point\ 2} - W_{point\ 1}$$

$$\textit{Humidity ratio difference} = .0092 - .0066$$

$$\textit{Humidity ratio difference} = \mathbf{.0026\ lb/lb}$$

4. Find the volume of a building space.

$$V = l \times w \times h$$

$$V = 45 \times 37 \times 10$$

$$V = \mathbf{16{,}650\ cu\ ft}$$

5. Find the total weight of the air in a space.

$$\textit{Total weight of air} = \frac{V}{v_{point\ 1}}$$

$$\textit{Total weight of air} = \frac{16{,}650}{13.55}$$

$$\textit{Total weight of air} = \mathbf{1228.78\ lb}$$

6. Find the weight of water to be added.

$$\textit{Weight of water to be added} = \textit{Total weight of air} \times \textit{Humidity ratio difference}$$

$$\textit{Weight of water to be added} = 1228.78 \times .0026$$

$$\textit{Weight of water to be added} = \mathbf{3.19\ lb}$$

7. Find the number of gallons of water.

$$\textit{Gallons of water} = \frac{\textit{Weight of water to be added}}{8.345}$$

$$\textit{Gallons of water} = \frac{3.19}{8.345}$$

$$\textit{Gallons of water} = \mathbf{.382\ gal.}$$

Air Mixtures. Air conditioning systems use ventilation to mix makeup air from outdoors with return air from building spaces. The condition of the air mixture is found by applying the procedure:

1. The point that represents the condition of the makeup air is located at the intersection of the two given properties. If makeup air is at a dry bulb temperature

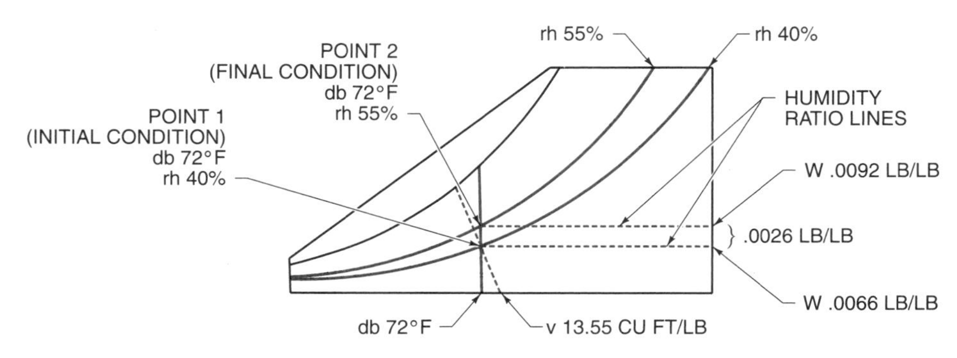

Figure 9-14. The amount of water to be added to the air is the amount that will change the condition of the air to the desired humidity.

of 60°F and 65% relative humidity, the intersection of these two lines represents the condition of the makeup air. **See Figure 9-15.**

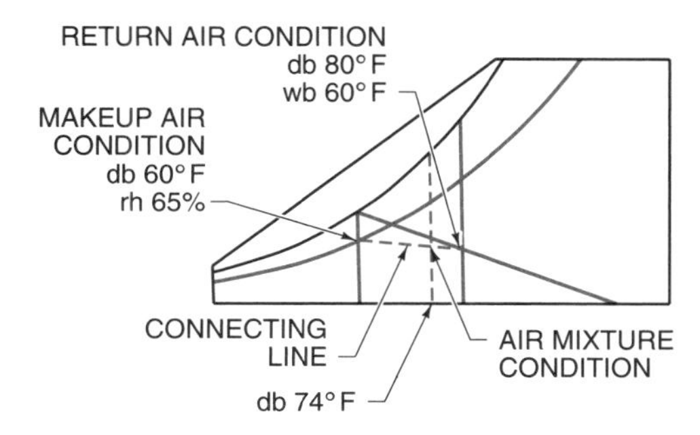

Figure 9-15. The properties of an air mixture are found by locating a point on a line drawn between the makeup air condition and the return air condition.

2. The point that represents the condition of the return air is located at the intersection of the two given properties. If return air is at 80°F dry bulb temperature and 60°F wet bulb temperature, the intersection of these two lines represents the condition of the return air.
3. Draw a line on the chart that connects the point that represents the condition of the makeup air and the point that represents the condition of the return air (connecting line).
4. Multiply the percentage of makeup air and return air by the corresponding dry bulb temperature. The resulting values are the temperature ratio of the makeup air and return air. If air consists of 30% makeup air and 70% return air, the temperature ratio of the makeup air and return air is determined by applying the formula:

   Temperature ratio = *db* × *Percentage of air*

   For makeup air:

   Temperature ratio = *db* × *Percentage of air*

   Temperature ratio = 60 × .3

   Temperature ratio = **18°F**

   and

   For return air:

   Temperature ratio = *db* × *Percentage of air*

   Temperature ratio = 80 × .7

   Temperature ratio = **56°F**
5. Find the dry bulb temperature of the air mixture by adding the temperature ratios.

   Air mixture db = *Makeup air temperature ratio* + *Return air temperature ratio*

   Air mixture db = 18 + 56

   Air mixture db = **74°F**
6. The point that represents the condition of the air mixture is found at the intersection of the 74°F dry bulb temperature line and the connecting line. This point represents the condition of the makeup air and the return air mixture.

Example: Finding Points—Air Mixture

Twenty percent makeup air at 90°F dry bulb temperature and 70% relative humidity is mixed with 80% return air at 75°F dry bulb temperature and 65°F wet bulb temperature. Find the final condition of the air mixture.

1. Locate the point that represents the condition of the makeup air at the intersection of the 90°F dry bulb temperature line and the 70% relative humidity line. **See Figure 9-16.**
2. Locate the point that represents the condition of the return air at the intersection of the 75°F dry bulb temperature line and the 65°F wet bulb temperature line.

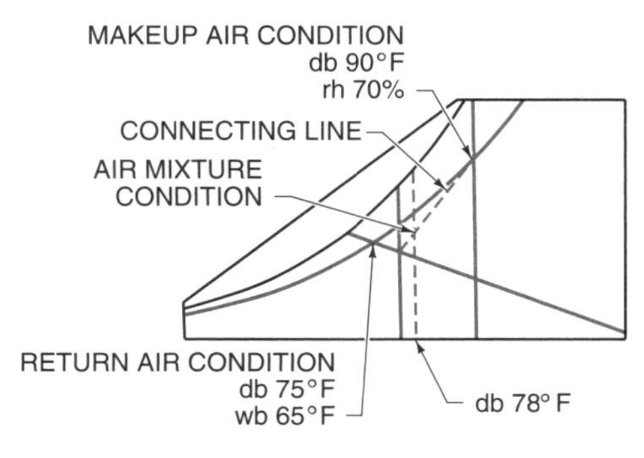

Figure 9-16. The condition of the air mixture is located at the intersection of the 78°F dry bulb line and the connecting line.

3. Draw a line on the chart that connects the point that represents the condition of the makeup air and the point that represents the condition of the return air (connecting line).
4. Find temperature ratio for makeup air and return air.

   For makeup air:

   Temperature ratio = *db* × *Percentage of air*

   Temperature ratio = 90 × .2

   Temperature ratio = **18°F**

   For return air:

   Temperature ratio = *db* × *Percentage of air*

   Temperature ratio = 75 × .8

   Temperature ratio = **60°F**
5. Find the dry bulb temperature of the air mixture.

   Air mixture db = *Makeup air temperature ratio* + *Return air temperature ratio*

   Air mixture db = 18 + 60

   Air mixture db = **78°F**
6. The point that represents the condition of the air mixture is found at the intersection of the 78°F dry bulb temperature line and the connecting line. This point represents the condition of the makeup air and the return air mixture.

Review Questions

1. Define temperature.
2. List the four properties of air.
3. List and explain the two ways humidity is expressed.
4. What does wet bulb temperature indicate?
5. Define dry bulb temperature.
6. Define moist air.
7. What kind of heat does moisture in the air represent? Explain.
8. What is the relative humidity of air that holds 35% of the moisture it can hold at a given pressure and temperature?
9. How does a wet bulb thermometer work?
10. Define hygrometer. List and describe two kinds.
11. What effect does an increase in dry bulb temperature have on the specific volume of air?
12. List the seven properties of air identified on a psychrometric chart.
13. What property of air identifies the total heat content of air?
14. Where is the dry bulb temperature scale found on a psychrometric chart?
15. What three properties of air are identified by the scale that runs along the curve of a psychrometric chart?
16. How is specific volume found on a psychrometric chart?
17. Where is the enthalpy scale found on a psychrometric chart?
18. What is the wet bulb temperature of the air when the dry bulb temperature is 56°F and the relative humidity is 60%?
19. What is the relative humidity when the dry bulb temperature is 93°F and the wet bulb temperature is 76°F?
20. How much water (in gallons) must be added to the air of a space 80′ × 40′ × 10′, at 82°F dry bulb temperature and 65°F wet bulb temperature to change it to 55% relative humidity at the same dry bulb temperature?

Forced-air heating systems consist of a heating unit, distribution system, and controls. Supply air is heated by a furnace and distributed to building spaces by a blower through supply air ductwork. Air is returned to the furnace through return air ductwork. Forced-air heating systems are used in small and medium-size buildings where the ductwork is not long. Some special-purpose heating units do not require ductwork. The air in a special-purpose heating unit is heated and distributed within a specific building space.

10 Forced-Air Heating Systems

FORCED-AIR HEATING SYSTEMS

A *forced-air heating system* is a heating system that uses air to carry heat. The air is heated and distributed through a building to control the temperature in the building spaces. The major parts of a forced-air heating system are the heating unit, distribution system, and controls. Different types of forced-air heating systems are used for different applications. The kind of system used for a particular application depends on the amount of heat required and the physical layout of the building. The two main types of forced-air heating systems are the central forced-air and modular forced-air heating system.

Central Forced-Air Heating Systems

A *central forced-air heating system* is a forced-air heating system that uses a centrally located furnace to produce heat for a building. Supply air ductwork runs from the heating unit to the building spaces. Return air ductwork runs from the building spaces back to the heating unit.

Modular Forced-Air Heating Systems

A *modular forced-air heating system* is a forced-air heating system that uses more than one heating unit to produce heat for a building. Each heating unit produces heat for a module or zone of the building. A *zone* is a specific section of a building that requires separate temperature control. Heating units in modular forced-air heating systems are usually located on roofs.

FURNACES

The furnace is the central element in a forced-air heating system. A *furnace* is a self-contained heating unit that includes a blower, burner(s) and heat exchanger or electric heating elements, and controls. **See Figure 10-1.** Furnaces are available in different sizes and styles for a variety of applications. Forced-air furnaces are categorized by direction of airflow, fuel or energy used, dimensions, and heating capacity.

Furnaces are categorized based on the direction of airflow out of the furnace. The three most common styles of furnaces are upflow, horizontal, and downflow (counterflow).

An *upflow furnace* is a furnace in which heated air flows upward as it leaves the furnace. Return air enters through or near the bottom of the upflow furnace and exits out of the top. **See Figure 10-2.** A *horizontal furnace* is a furnace in which heated air flows horizontally as it leaves the furnace. Return air enters horizontally at one end of the furnace, and supply air exits horizontally at the other end. Horizontal furnaces are used where headroom is limited, such as in attics or crawl spaces of buildings. A *downflow furnace* is a furnace in which heated air flows downward as it leaves the furnace. Return air enters through the top of the furnace and supply air exits out the bottom. Downflow furnaces are used where the supply ductwork is located below the furnace, such as when the furnace is located on one floor and the ductwork is installed in the ceiling space of the floor or basement below, where the ductwork is located in a concrete slab floor, or where the ductwork is under the floor in a crawl space.

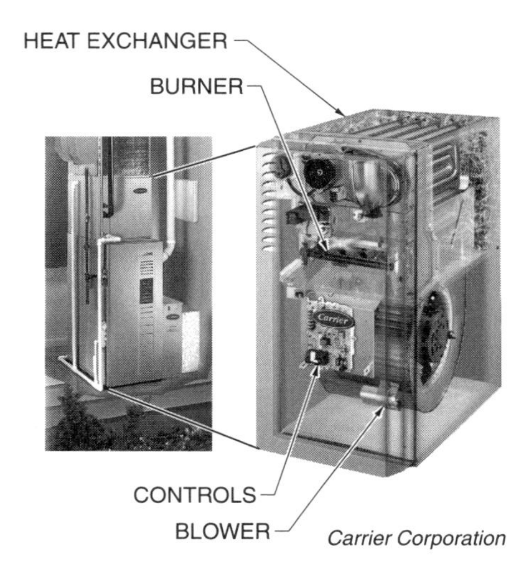

Figure 10-1. A furnace is the central element in a forced-air heating system.

Furnaces produce heat by either combustion or electrical energy. The heat in a combustion furnace is produced by burning fuel, which may be coal, wood, fuel oil, or gas fuel. Many furnaces such as coal-burning and wood-burning furnaces are made for burning a specific kind of fuel, but some furnaces can burn any solid fuel. Fuel oil-burning furnaces usually burn Grade No. 2 fuel oil. Gas fuel-burning furnaces burn natural or LP gas. Some gas fuel-burning furnaces can burn both kinds of gas fuel. Natural gas-burning furnaces are used in areas where natural gas pipelines are in place and the fuel is available. LP gas-burning furnaces are used where the fuel must be transported to the point of use in tanks. The differences between natural gas- and LP gas-burning furnaces are the size of the orifices (openings) on the burner(s) and the pressure of the gas fuel.

Electric furnaces produce heat as electricity flows through resistance heating elements. Air is heated as it passes by the hot elements. National safety codes must be followed when installing any kind of furnace. Check jurisdictional codes for further regulations.

Manufacturer's specification sheets are used to find the characteristics of a furnace, which are dimensions, input rating, output rating, and efficiency. **See Figure 10-3.** The

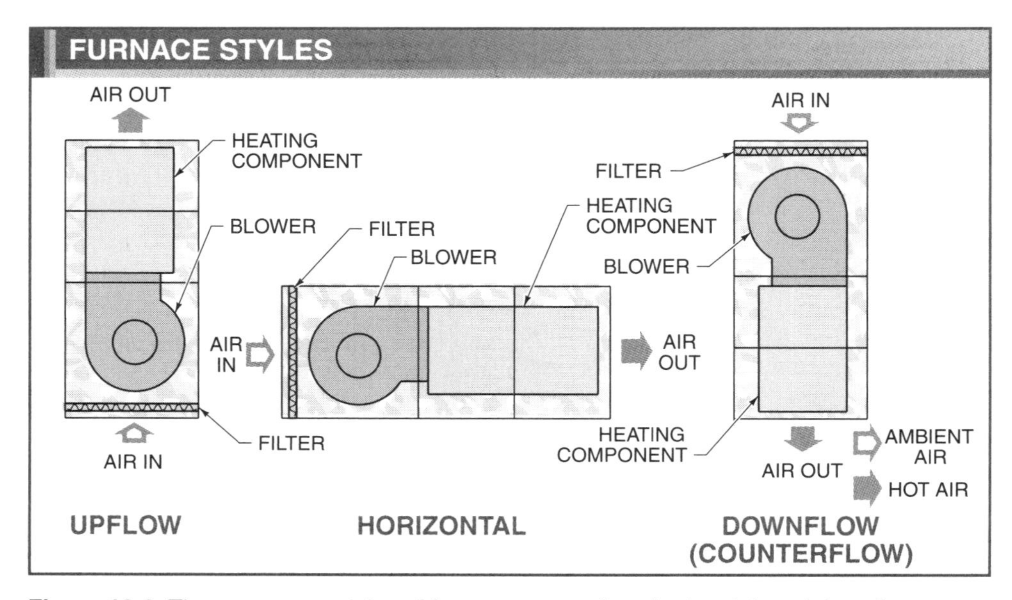

Figure 10-2. Three common styles of furnaces are upflow, horizontal, and downflow.

UPFLOW GAS FURNACE CHARACTERISTICS (STANDING PILOT)							
Model Number	Dimensions*			Input Rating†	Output Rating†	Shipping Weight‡	AFUE§
	D	W	H				
005-AX-10	28.5	14	46	40,000	30,000	100	75
050-AX-6	28.5	18	46	50,000	38,000	120	76
070-AX-5	28.5	20	46	75,000	58,000	120	77
300-AX-6	28.5	24	46	132,000	105,000	205	80
400-AX-0	28.5	24	46	150,000	127,000	255	85
58MSA060	28.5	17.5	39.8	55,000	35,000	171	90
58MTA100	28.5	21	39.8	100,000	75,000	234	93
58MUP120	28.5	24	39.8	120,000	97,000	270	96.6

* in in.
† in Btu/hr
‡ in lb
§ Annual fuel utilization efficiency (in percent)

Figure 10-3. Furnace characteristics are found on manufacturer specification sheets.

dimensions of a furnace depend on the size and arrangement of the components in the furnace. *Input rating* is the heat produced in Btu/hr per unit of fuel burned. Input rating is found by multiplying the heating value of the fuel by the flow rate. Heating value is the amount of heat produced per unit of fuel in Btu per hour. The heating value of a fuel depends on the chemical makeup of the fuel. *Flow rate* is the rate at which a furnace burns fuel. A furnace produces a given amount of heat for each unit of fuel burned. Input rating is found by applying the formula:

$IR = HV \times Q$

where

IR = input rating (in Btu/hr)

HV = heating value (in Btu/hr)

Q = flow rate (in cu ft/hr)

Example: Finding Input Rating

A natural gas-fired furnace burns 100 cu ft of gas fuel per hour. Find the input rating of the furnace. *Note:* The heating value for natural gas is 1000 Btu/cu ft.

$IR = HV \times Q$

$IR = 1000 \times 100$

IR = **100,000 Btu/hr**

Output rating (heating capacity) is the actual heat output produced by a heater in Btu/hr after heat losses from draft. Output rating is found by multiplying the input rating by the efficiency rating of the furnace, which is provided by the manufacturer. *Efficiency rating* is the comparison of the furnace input rating with the output rating. Efficiency rating is the evaluation of how well a furnace burns fuel. Output rating is found by applying the formula:

$OR = IR \times ER$

where

OR = output rating (in Btu/hr)

IR = input rating (in Btu/hr)

ER = efficiency rating (in percent)

Example: Finding Output Rating

A furnace with an input rating of 100,000 Btu/hr has an efficiency rating of 80%. Find the output rating for the unit.

$OR = IR \times ER$

$OR = 100{,}000 \times .80$

OR = **80,000 Btu/hr**

Efficiency for a typical furnace is about 80%. Twenty percent of the heat produced rises up the flue. This heat produces draft in the stack, which is necessary for proper furnace operation. *Draft* is the movement of air across a fire and through a heat exchanger. Because modern condensing furnaces allow less heat for producing a draft, these furnaces operate with higher efficiency ratings.

Combustion furnace components include a cabinet, blower, burner(s), heat exchanger, and filter. Electric furnace components include a cabinet, blower, resistance heating element(s) in place of the burner(s) and heat exchanger, and filter. These components are arranged differently, but all of these components are found in most furnaces. **See Figure 10-4.**

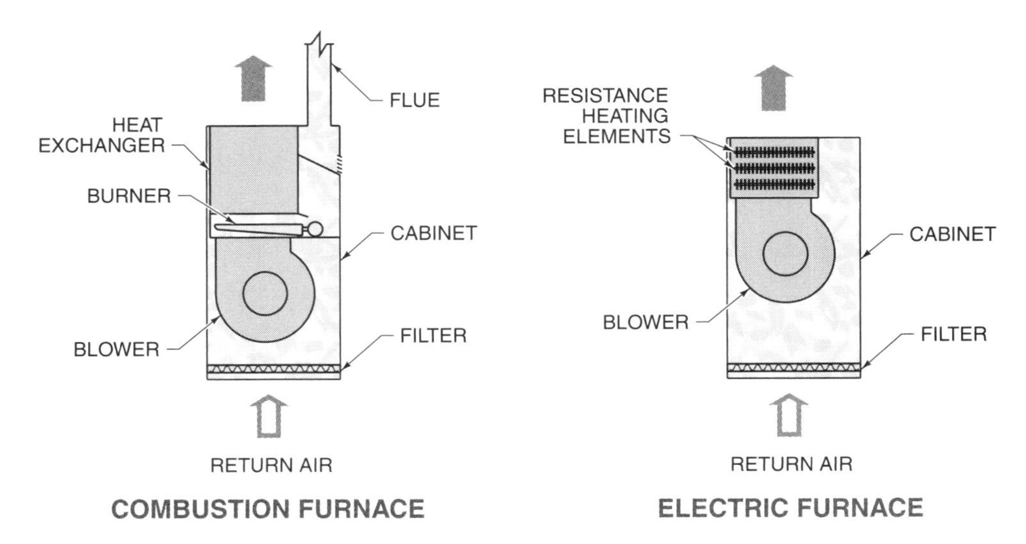

Figure 10-4. Combustion and electricity are used to produce heat for furnaces.

Cabinets

A *cabinet* is a sheet metal enclosure that completely covers and provides support for the components of a furnace. Cabinets for small furnaces are made of heavy-gauge sheet metal which provides support. Larger furnaces have a frame within the sheet metal cabinet, which supports heavier components and adds strength to the furnace.

Furnace cabinets completely enclose the other components except where the return air and supply air duct connections are made. Louvered access panels on the front of the cabinet allow combustion airflow to the burner(s) and enclose the burner vestibule, controls, and blower compartment. The *burner vestibule* is the area where the burner(s) and controls are located. **See Figure 10-5.**

Blowers

A *blower* is a mechanical device that consists of moving blades or vanes that force air through a venturi. A *venturi* is a restriction that causes increased pressure as air moves through it. The blower in a forced-air heating system is the component that moves air through the heat exchanger or the resistance heating element and through the ductwork to building spaces.

A blower may be a part of the furnace or a separate part within a forced-air heating system. Blowers are available in different sizes for different applications. Blowers used in forced-air systems include propeller fans, centrifugal blowers, and axial-flow blowers. **See Figure 10-6.**

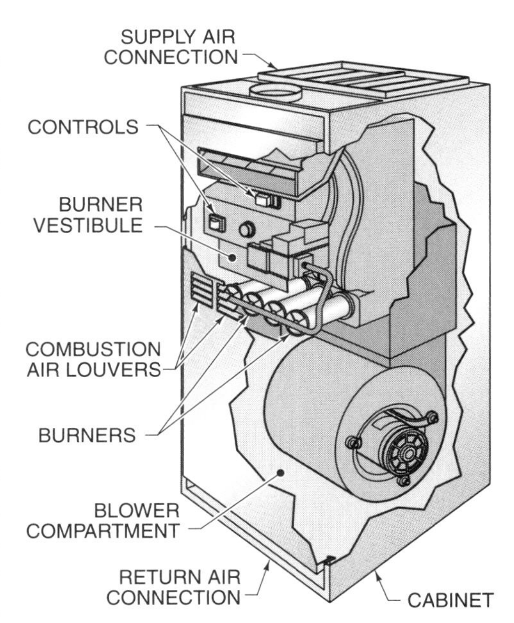

Figure 10-5. The burner vestibule is the area where the burner(s) and controls are located. Louvered access panels on the cabinet allow combustion air to flow to the burners.

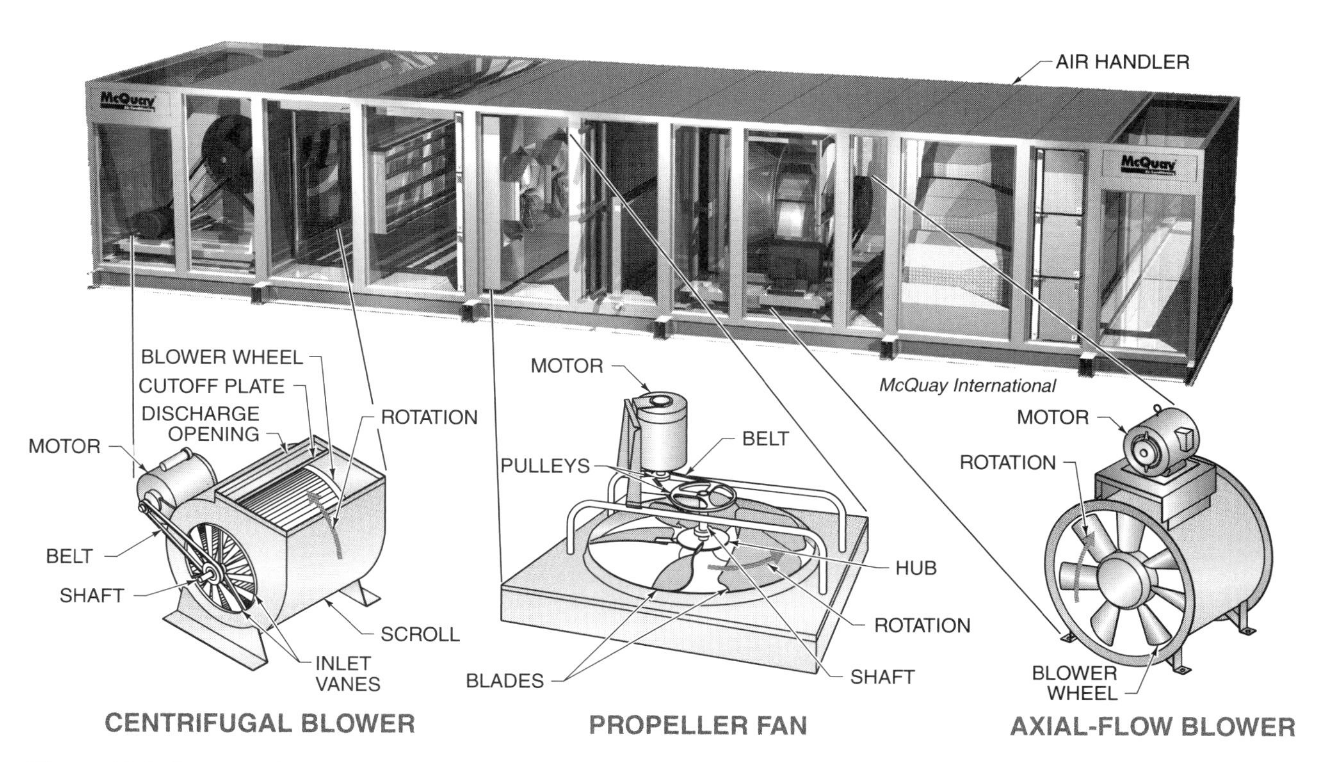

Figure 10-6. Propeller fans, centrifugal blowers, and axial-flow blowers are the three types of blowers used to move air in forced-air systems.

Propeller Fans. A *propeller fan* is a mechanical device that consists of blades mounted on a central hub. The hub may be mounted directly on the shaft of a motor or may be turned by a motor with a pulley and belt arrangement. The angle at which the blades are mounted moves air through the opening. Propeller fans are used mostly in applications where there is no ductwork and outdoors on air conditioning condensers.

Centrifugal Blowers. A *centrifugal blower* consists of a scroll, blower wheel, shaft, and inlet vanes. The *scroll* is a sheet metal enclosure that surrounds the blower wheel. The *blower wheel* is a sheet metal cylinder with curved vanes along its perimeter. The blower wheel rotates on the blower shaft. *Inlet vanes* are adjustable dampers that control the airflow to the blower. Return air is drawn into the air inlet and through the inlet vanes to the center of the wheel. Air passes through the vanes of the blower wheel and is thrown off by centrifugal force through the discharge opening in the scroll. *Centrifugal force* is the force that pulls a body outward when it is spinning around a center.

Axial Flow Blowers. An *axial flow blower* is a blower that contains a blower wheel, which works like a turbine wheel. The blower wheel is mounted on a shaft with its axis parallel to the airflow. The wheel turns at high speed. The angle of the blades moves air by compressive and centrifugal force. *Compressive force* is the force that squeezes air together. Axial flow blowers are used in medium- to high-pressure forced-air heating systems.

Blower motors are electric motors that provide the mechanical power for turning blower wheels. A *blower drive* is the connection from an electric motor to a blower wheel, which is a motor-to-wheel connection. Two kinds of motor-to-wheel connections used with blowers are belt drive and direct drive systems.

A *belt drive system* is a motor-to-wheel connection that has a blower motor mounted on the scroll. The blower motor is connected to the blower wheel through a belt and sheave arrangement. A *sheave* is a pulley, which is a grooved wheel. One or more sheaves are mounted on the motor shaft and the blower wheel shaft. V belts are closed-looped belts made of rubber, nylon, polyester, or rayon. They are used to transmit power from the motor shaft to the blower wheel shaft. **See Figure 10-7.**

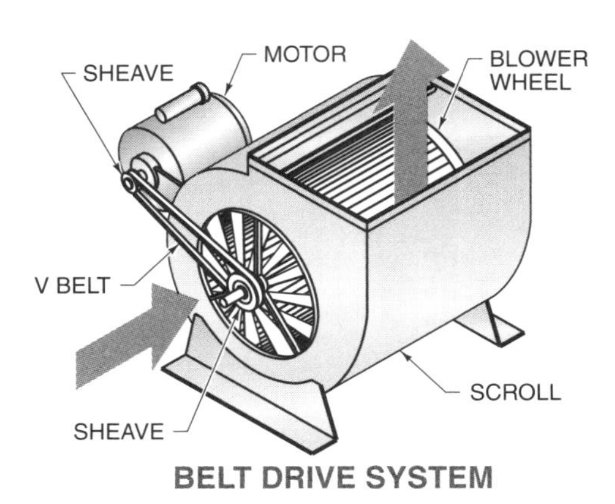

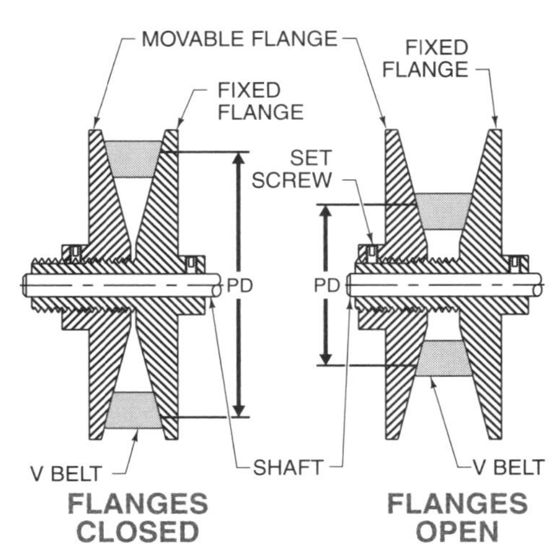

Figure 10-7. A variable-speed belt drive system is a motor-to-wheel connection that has a blower motor connected to the blower wheel through a belt and adjustable sheave arrangement.

The speed of the blower wheel, in revolutions per minute (rpm), determines the volume of air that will flow through a blower. The speed of the blower wheel is indirectly proportional to the diameters of the sheaves used in the system. In some belt drive systems, an adjustable blower sheave makes blower wheel speed adjustments. The speed of a belt drive blower is found by applying the formula:

$$N_b = \frac{N_m \times PD_m}{PD_b}$$

where

N_b = speed of blower wheel (in rpm)

N_m = speed of motor (in rpm)

PD_m = diameter of motor sheave (in inches)

PD_b = diameter of blower sheave (in inches)

Example: Finding Belt Drive Blower Speed

A belt drive blower has a motor speed of 1725 rpm, a 3″ motor sheave, and a 7″ blower sheave. Find the speed of the blower.

$$N_b = \frac{N_m \times PD_m}{PD_b}$$

$$N_b = \frac{1725 \times 3}{7}$$

$$N_b = \frac{5175}{7}$$

$$N_b = \mathbf{739.29\ rpm}$$

A *direct drive system* is a motor-to-wheel connection that has a blower wheel mounted directly on the motor shaft. The blower wheel turns as the motor turns the shaft. The motor is mounted in the center of the air inlet of the blower. A direct drive blower is normally connected to a multispeed motor. Changing the speed (rpm) of the motor changes the speed of the blower. Large blowers use four- or five-speed motors and small blowers use two- or three-speed motors. **See Figure 10-8.**

The performance characteristics of a blower are horsepower (HP), speed, volume, and static pressure. These characteristics are found on blower performance charts. *Static pressure* is pressure that acts through weight only with no motion. Static pressure is expressed in inches of water column. Blower performance charts illustrate the performance characteristics of a particular type and size of blower in cubic feet per minute, static pressure, revolutions per minute, and horsepower. For example, a blower with a 1 HP motor has a speed of 1733 rpm and moves 1350 cfm of air to produce 2.5″ WC static pressure. **See Figure 10-9.**

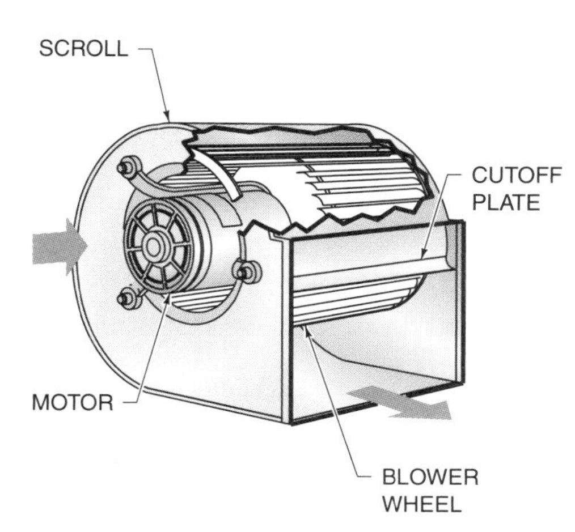

Figure 10-8. A direct drive system is a motor-to-wheel connection that has a blower wheel mounted directly on the motor shaft. Changing the speed of the blower is accomplished by changing the speed of the motor.

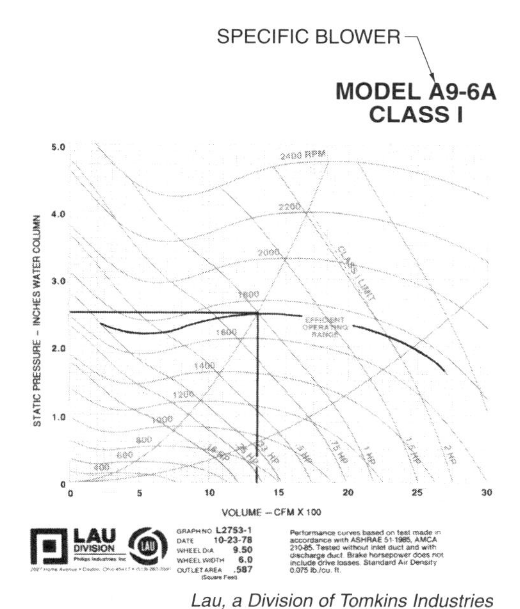

Figure 10-9. Blower performance charts show performance characteristics, which are horsepower, speed, volume, and static pressure.

Burners

A *burner* is the heat-producing component of a combustion furnace. Some furnaces use one large burner while others use many smaller burners. A burner mixes air and fuel to provide a combustible mixture, supplies the air-fuel mixture to the burner face where combustion takes place, and meters fuel to maintain a controlled firing rate. Atmospheric burners, power burners, and low excess air burners are three types of burners.

Atmospheric Burners. An *atmospheric burner* is a burner that uses ambient air supplied at normal atmospheric air pressure for combustion air. *Ambient air* is unconditioned atmospheric air. Atmospheric burners burn either gas fuel or fuel oil to produce heat.

A *gas fuel-fired atmospheric burner* is a burner that mixes ambient air with a gas fuel to create a flame. The gas is directed through a manifold to burner tubes. A *manifold* is a pipe that has outlets for connecting other pipes. A *burner tube* is a tube that has an opening on one end and burner ports located along the top. A spud (fitting) that contains a small orifice is located in the outlets of the manifold. An *orifice* is a precisely sized hole through which gas fuel flows. The gas fuel flows from the manifold, through the spud and orifice, and into the burner tubes. **See Figure 10-10.**

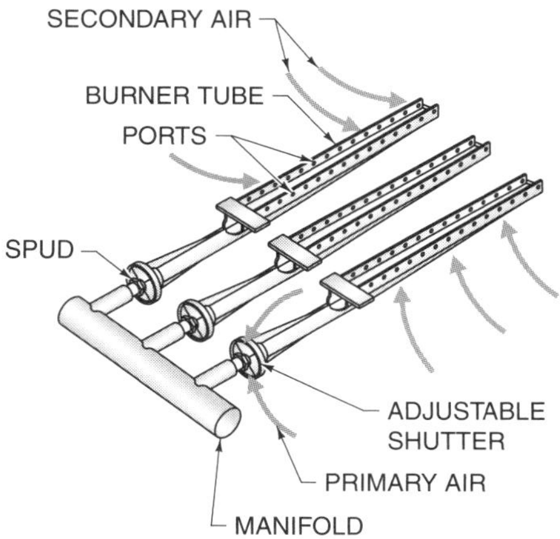

Figure 10-10. Gas fuel flows from the manifold through the spud into the burner tubes in a gas fuel-fired atmospheric burner. Primary air is drawn into the burner through the adjustable shutter. Secondary air is drawn into the flame at the burner ports.

A burner tube has an adjustable shutter, which is the primary air inlet. The gas fuel enters the burner tube at a high velocity, which draws air in through the adjustable shutter. Primary air mixes with the gas and produces a combustible air-fuel mixture.

The air-fuel mixture is ignited at the burner ports with a pilot light or an electric spark. Secondary air is drawn into the flame at the burner ports.

A *pilot burner* is a small burner located near the burner tubes. The pilot burner produces a pilot light, which is a small standing flame. The pilot light ignites the air-fuel mixture when the gas fuel valve opens. The pilot light may be a standing flame (one that burns constantly) or may be lighted electrically on each call for heat.

An *electric spark igniter* is a device that produces an electric spark. The electric spark is used to ignite either a pilot burner or main burner. On a call for heat, an electrode is energized near the pilot burner or burner face. When the flame is established, the electrode igniter is de-energized until the next call for heat. Pilot burners and electric spark igniters are controlled by combustion safety control systems. **See Figure 10-11.**

A *hot surface igniter* uses a small piece of silicon carbide that glows when electric current passes through it. The current is usually 120 VAC. The silicon carbide material is very durable, lasts a long time, and gives few problems.

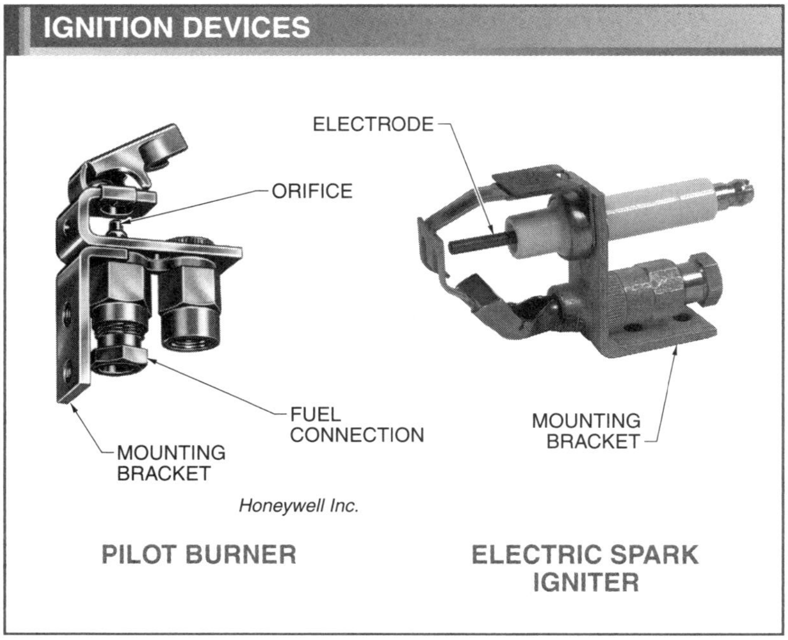

A gas fuel valve controls the flow of fuel to a gas fuel-fired atmospheric burner. A *gas fuel valve* is a 100% shutoff safety valve that controls the flow of fuel to the main burner and the pilot burner. If combustion does not occur on a call for heat, the valve will close and will not allow gas fuel to flow to the main burner nor to the pilot burner. A combination valve is a modern valve that has a built-in fuel pressure regulator, which regulates the pressure of the gas fuel that enters the burner. **See Figure 10-12.**

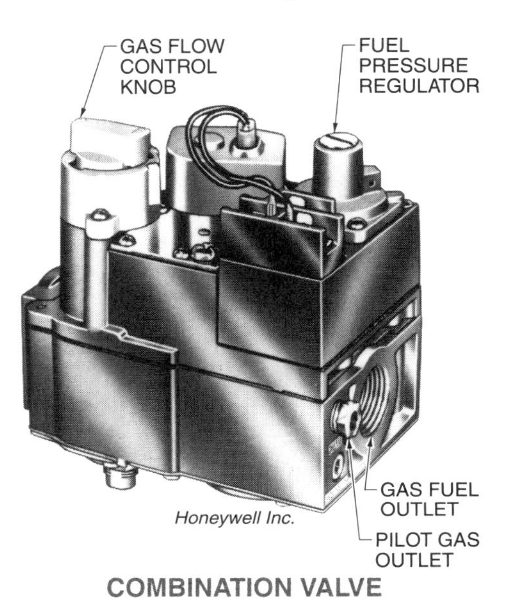

Figure 10-12. A combination valve controls the flow of gas fuel to the main burner and the pilot burner.

IGNITION DEVICES

ELECTRODE
ORIFICE
FUEL CONNECTION
MOUNTING BRACKET
MOUNTING BRACKET
Honeywell Inc.
PILOT BURNER
ELECTRIC SPARK IGNITER

Figure 10-11. The air-fuel mixture in a gas fuel-fired atmospheric burner is ignited at the burner ports by a pilot burner or an electric spark igniter.

The heat output of an atmospheric burner depends on the size of the burner and orifice and the pressure of the fuel. Most natural gas burners are designed to operate with pressure of 3.5″ WC. LP gas burners are designed to operate with pressure of 10.5″ WC. The holes in the orifice are sized to provide the amount of fuel that will produce the rated heat output for the particular burner.

Power Burners. A *power burner* is a burner that uses a fan or blower to supply and control combustion air. Air and fuel are introduced under pressure at the burner face. Two basic kinds of power burners are gas fuel and fuel oil power burners.

A *gas fuel power burner* is a power burner that uses natural or LP gas and has a fan or blower on the outside of the combustion chamber. The blower provides combustion air, which is blown into the combustion chamber under pressure through a perforated bulkhead. **See Figure 10-13.** The proper air-fuel mixture is attained by using a properly sized orifice in the gas fuel line and by adjusting the combustion air blower. An electric spark igniter starts combustion. Gas fuel power burners are used in commercial gas fuel furnaces with an output greater than 240,000 Btu/hr.

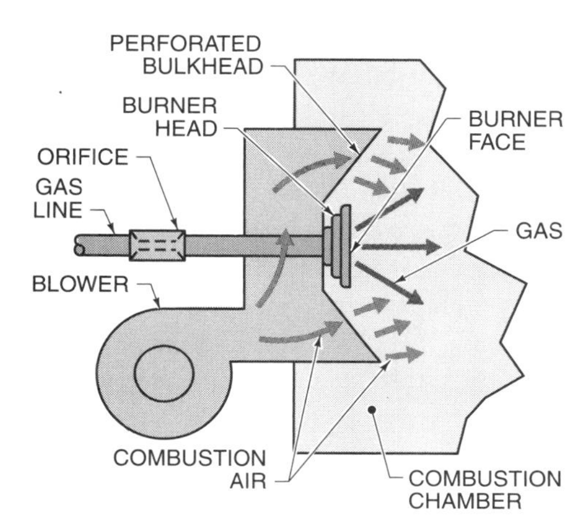

Figure 10-13. Combustion air in a gas fuel power burner is provided by a fan or blower.

A *fuel oil power burner* is a burner that atomizes fuel oil. A fuel oil power burner consists of a fuel oil pump, burner assembly, and combustion air blower. **See Figure 10-14.** The fuel oil pump draws oil from a storage tank and pumps it through the burner assembly at about 100 psi. The burner assembly consists of a tube that holds a fuel oil line, nozzle, and electrodes. Combustion air is forced through the tube by the combustion air blower. The fuel oil line carries oil from the pump to the nozzle. The nozzle releases atomized fuel oil, which mixes with combustion air at the burner face. The electrodes produce an electric spark that ignites the air-fuel mixture. The heating capacity of a fuel oil power burner depends on the fuel oil pressure and size of the nozzle. Nozzles are sized in gallons per hour (gph) of fuel oil flow.

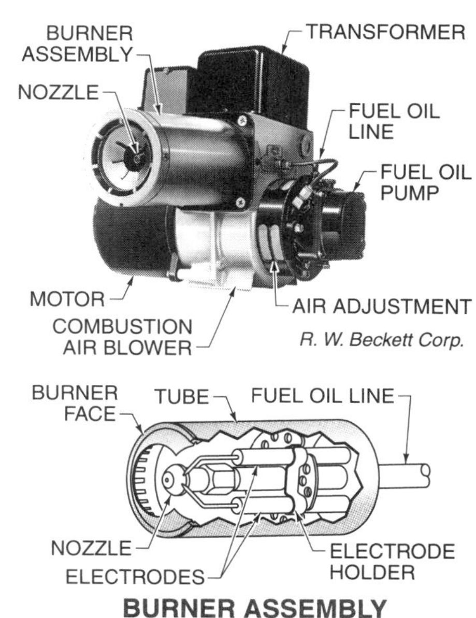

Figure 10-14. The oil in a fuel oil power burner is atomized as it flows through the nozzle. Combustion air mixes with the oil at the burner face.

Low Excess Air Burners. Low excess air burners are used in many new furnace designs. A *low excess air burner* is a burner that uses only the amount of air necessary for complete combustion. This ensures that the combustion will be as efficient as possible. Pulse burners are low excess air burners.

A *pulse burner* is a low excess air burner that introduces the air-fuel mixture to the burner face in small amounts or pulses. **See Figure 10-15.** The pulses of the mixture are controlled by the pressure in the firing chamber. A pulse burner is more efficient than a conventional burner because no excess air is introduced. This ensures a more complete combustion, which produces cleaner products of combustion. Draft over the fire is produced within the burner so that no heat rises up the stack. Most pulse burners are more than 90% efficient.

A *resistance heating element* is an electric heating element that consists of a grid of electrical resistance wires that are attached to a support frame with ceramic insulators. Nichrome, a nickel-chromium alloy, is used for electrical resistance wire. Nichrome wire resists the flow of electricity and becomes red-hot when connected to an electrical circuit. Resistance heating elements are installed in a furnace so that air is heated as it passes through the furnace and flows over the hot elements. The electrical circuit is protected against overcurrent and overheating by fuses and temperature-sensing limit switches. **See Figure 10-16.**

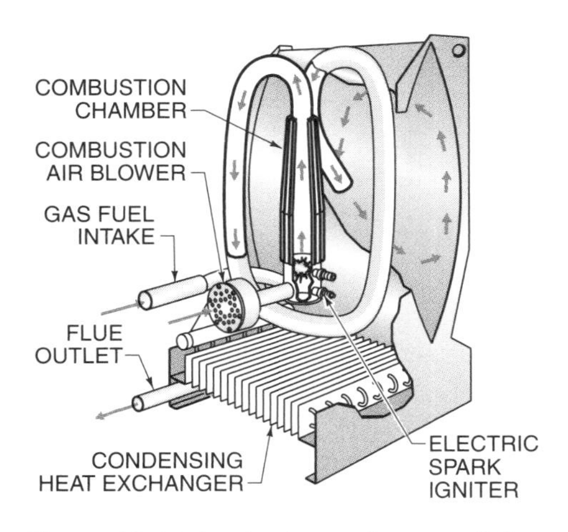

Figure 10-15. A pulse burner is more efficient than conventional burners because it uses low excess air.

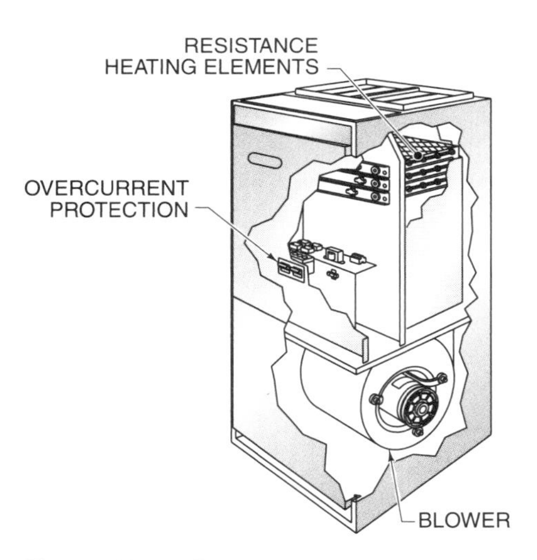

Figure 10-16. Resistance heating elements provide heat for an electric furnace.

Heat Exchangers

A *heat exchanger* is any device that transfers heat from one substance to another without allowing the substances to mix. The heat exchanger in a forced-air furnace transfers heat from the hot products of combustion to cool air. The products of combustion often contain toxic chemical compounds. Heat exchangers are corrugated and curved to allow movement but are strong enough to prevent deformation. Heat exchangers are often coated with corrosion-resistant ceramic glazing.

The two types of heat exchangers used in conventional forced-air furnaces are the clam shell heat exchanger and drum heat exchanger. The clam shell heat exchanger is usually used with multiple-burner atmospheric burners. The drum heat exchanger is used with fuel oil-fired atmospheric burners and power burners.

Clam Shell Heat Exchangers. A *clam shell heat exchanger* is a heat exchanger that has multiple clam-shaped sections. The sections are made of medium-gauge sheet metal or cast iron. Each section consists of two clam-shaped pieces of metal that are placed edge-to-edge and then welded together to produce an airtight fit.

Combustion takes place at the burner openings, which are located at the bottom of each section. A burner is installed inside each burner opening. When the burner is ignited, hot products of combustion rise through the inside of the sections of the heat exchanger and out the flue openings at the top of each section. **See Figure 10-17.** Air is blown between the sections by a blower. Heat transfer occurs because of the temperature difference between the hot products of combustion and the cool air.

Hot flue gas that leaves the flue openings is collected in the draft diverter before the gas rises up the flue. A *draft diverter* is a box made of sheet metal that runs the width of the heat exchanger. The draft diverter is open across the bottom, which allows dilution air to mix with the flue gas as the flue gas leaves the heat exchanger. *Dilution air* is atmospheric air that mixes with, dilutes, and cools the products of combustion. A draft diverter assures a constant draft and eliminates downdrafts in the flue that would affect burner operation. **See Figure 10-18.**

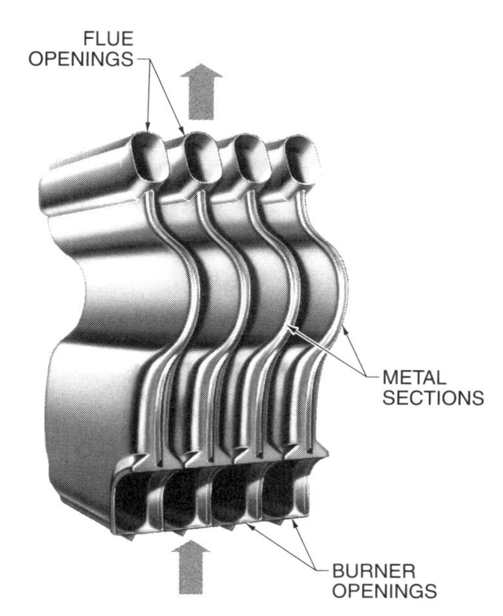

Figure 10-17. Products of combustion flow through the clam-shaped sections of a clam shell heat exchanger.

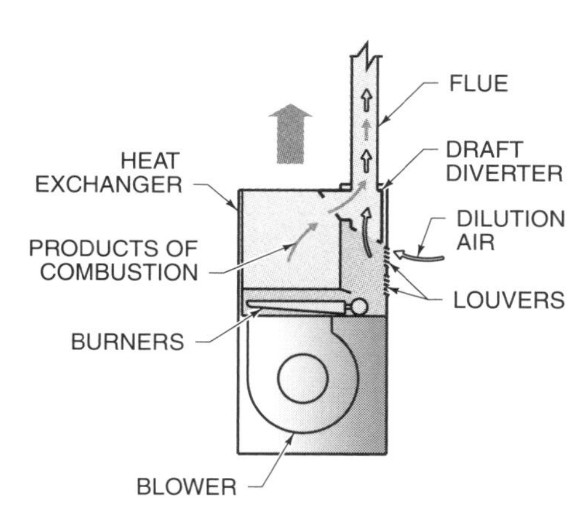

Figure 10-18. A draft diverter eliminates downdrafts that affect burner operation.

Drum Heat Exchangers. A *drum heat exchanger* is a round drum or tube that is located on a combustion chamber to make the products of combustion flow through it. **See Figure 10-19.** The *combustion chamber* is the area in a heating unit where combustion takes place. Combustion chambers are designed to retain heat, which helps ignite the fuel.

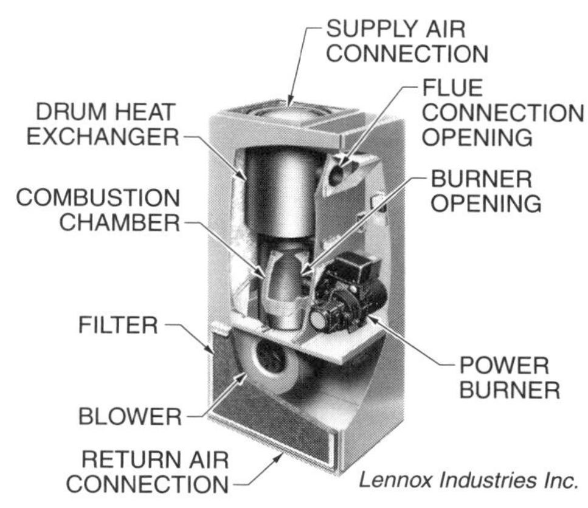

Figure 10-19. Products of combustion flow from the combustion chamber through a drum heat exchanger in a furnace with a power burner.

As the hot products of combustion flow through a drum heat exchanger, air is blown around and across the outside surface of the heat exchanger. Heat is transferred through the walls of the heat exchanger because of the temperature difference between the hot products of combustion inside and the cool air outside.

A drum heat exchanger used with a power burner is connected directly to the flue. A *damper* is a device that controls airflow. A barometric damper is installed in the flue above the drum heat exchanger. A *barometric damper* is a metal plate positioned in an opening in the flue so that atmospheric pressure can control the airflow through the combustion chamber and flue. The plate is balanced so atmospheric pressure can open and close the damper to allow dilution air to enter the flue to control draft. **See Figure 10-20.**

High-efficiency furnaces use condensing heat exchangers. A *condensing heat exchanger* is a heat exchanger that reduces the temperature of the flue gas below the dew point temperature of the heat exchanger. A condensing heat exchanger removes the latent heat of vaporization from the products of combustion and uses it to heat air. Removing this heat causes the moisture produced during combustion to condense inside the heat exchanger. A high-efficiency furnace must be provided with a drain to remove the condensate. **See Figure 10-21.**

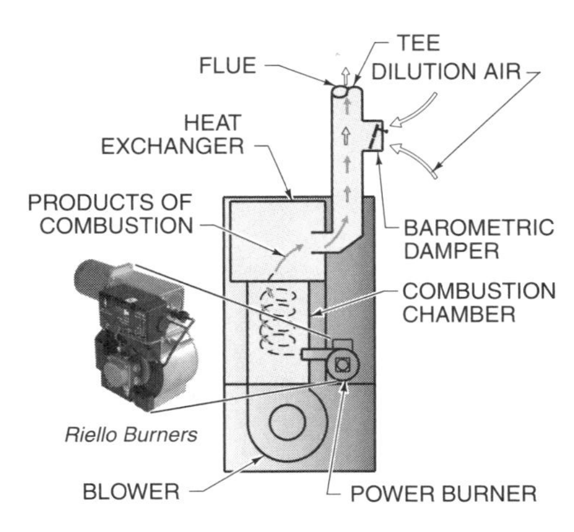

Figure 10-20. A barometric damper controls the draft of a furnace with a power burner by allowing dilution air to enter the flue.

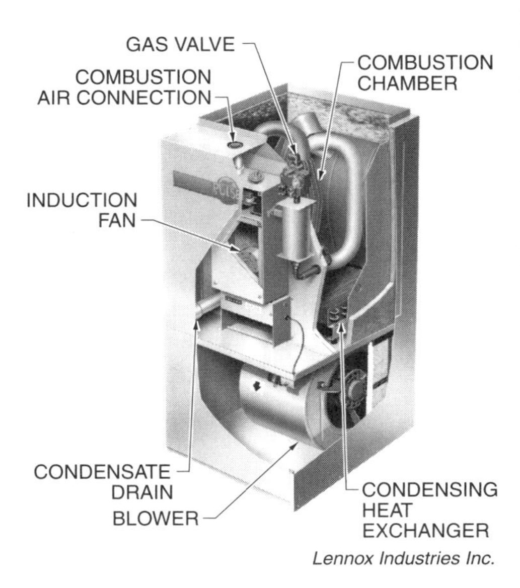

Figure 10-21. A high-efficiency furnace removes the latent heat of vaporization from the products of combustion.

High-efficiency furnaces reduce the temperature of the flue gas to the dew point temperature of the condensing heat exchanger. The lower temperature provides less energy to carry the flue gas up the flue and out the stack. A combustion air blower or draft inducer is required to provide positive pressure in the flue. *Positive pressure* is pressure greater than atmospheric pressure. A *combustion air blower* is a blower used to provide combustion air at a positive pressure at the burner face. A *draft inducer* is a blower installed in the flue pipe to provide positive pressure in the flue, which carries the products of combustion up the stack. A smaller flue such as a 2½″ diameter plastic pipe can be used with a draft inducer.

Filters

A *filter medium* is any porous material that removes particles from a moving fluid. The filter media is the part of a filter that separates particles from air. Filters clean return air before the air enters a furnace. All forced-air systems should contain filters. Some furnaces have racks in the blower compartment for standard filters. If there is no filter rack in the furnace, a rack must be built in the return duct at the duct connection to the furnace.

Airborne particulate matter may create an environment inside a building that is unhealthy for occupants. Indoor air quality (IAQ) is a designation of the contaminants present in the air. Airborne particulate matter includes dirt particles, spores, pollen, leaves, etc. The size of airborne particles is measured in microns. A micron is a unit of measure equal to .000039″.

Furnaces used in residential buildings are supplied with low-efficiency filters. *Low-efficiency filters* are filters that contain filter media made of fiberglass or other fibrous material. The fibers are treated with oil to help them hold dust and dirt. These filters are 1″ to 2″ thick and are mounted in a light cardboard frame. Low-efficiency filters remove about 40% of large airborne particles such as dust and dirt and should be disposed of when they accumulate dirt. Low-efficiency filters are used for normal residential filtering applications. **See Figure 10-22.**

Medium-efficiency filters are filters that contain filter media made of dense fibrous mats or filter paper. These filters are used in applications such as office buildings that require more filtering efficiency than low-efficiency filters provide. Medium-efficiency filters remove from 40% to 80% of common-size particulate matter.

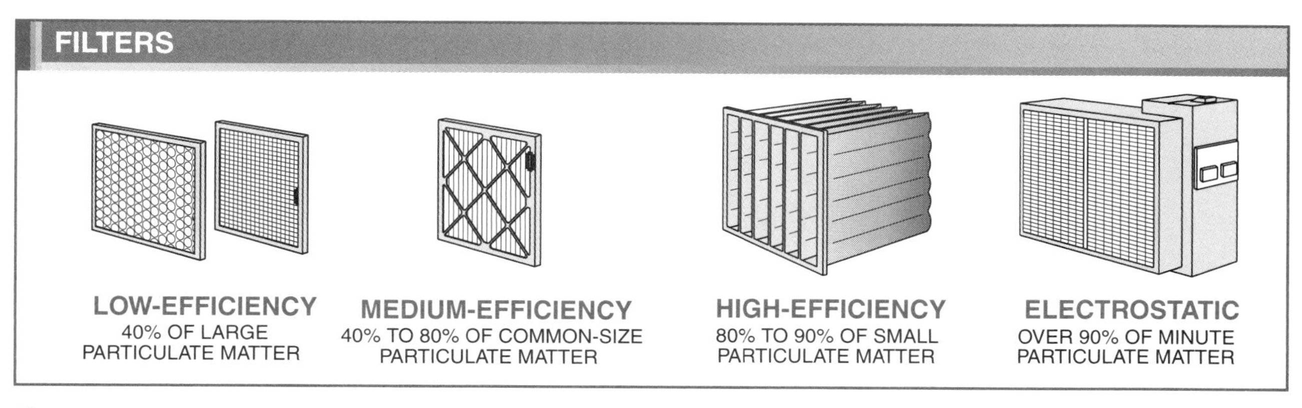

Figure 10-22. Filters remove particulate matter from the air. The kind of filter used depends on air quality requirements.

High-efficiency filters are filters that contain filter media made of large bags of filter paper. The bag shape increases the surface area of the filter, which reduces the velocity of the air through the filter media. The reduced velocity increases the filtering efficiency. Bag filters are installed in racks or frames with prefilters. *Prefilters* are filters installed ahead of bag filters in the airstream to filter large particulate matter. Bag filters are used in hospitals, laboratories, or electronic component production facilities where a high degree of filtration is required. Filtering efficiency with bag filters is 80% to 90% of small particulate matter.

The highest degree of filtering efficiency is attained with an electrostatic filter. *Electrostatic filters* are devices that clean the air as the air passes through electrically charged plates and collector cells. Electrostatic filters remove smaller particulate matter from the air than other filter types. Filtering efficiency with electrostatic filters is about 90% to 99% of particulate matter as small as bacteria. Prefilters are usually installed ahead of electrostatic filters to remove the large particulate matter.

DISTRIBUTION SYSTEMS

An air distribution system is the supply air ductwork, registers, return air ductwork, and grills that are used to circulate air through a building. The distribution system directs heated supply air from the furnace to the building spaces that require heat and returns air from building spaces to the furnace. **See Figure 10-23.**

Ductwork

Air distribution systems are categorized by the layout of the ductwork. Three basic layouts are perimeter loop, radial, and trunk and branch.

Perimeter Loop Systems. A *perimeter loop system* consists of a single loop of ductwork with feeder branches that supply air to the loop. Perimeter loop systems are used in special limited situations. The supply plenum is located in the center and the branches extend outward from it. A *supply plenum* is a sealed sheet metal chamber that connects the furnace supply air opening to the supply ductwork. **See Figure 10-24.**

Radial Systems. A *radial system* consists of branches that run out radially from the supply plenum of a furnace. Radial systems are used where ductwork can be run in a crawl space, attic, or duct chase. A *duct chase* is a special space provided in a building for installing ductwork.

Tech Facts

A heating system may contain a boiler or electric heating elements to provide heat for building spaces. In addition, the system may contain an air-handling unit and ductwork to move the air to different building spaces, as well as terminal air-handling units to transfer heat to the air.

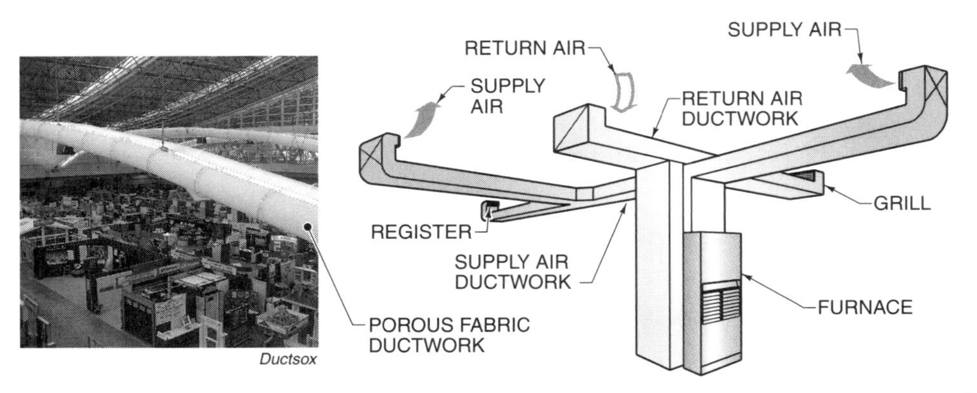

Figure 10-23. An air distribution system consists of supply air ductwork, registers, return air ductwork, and grills.

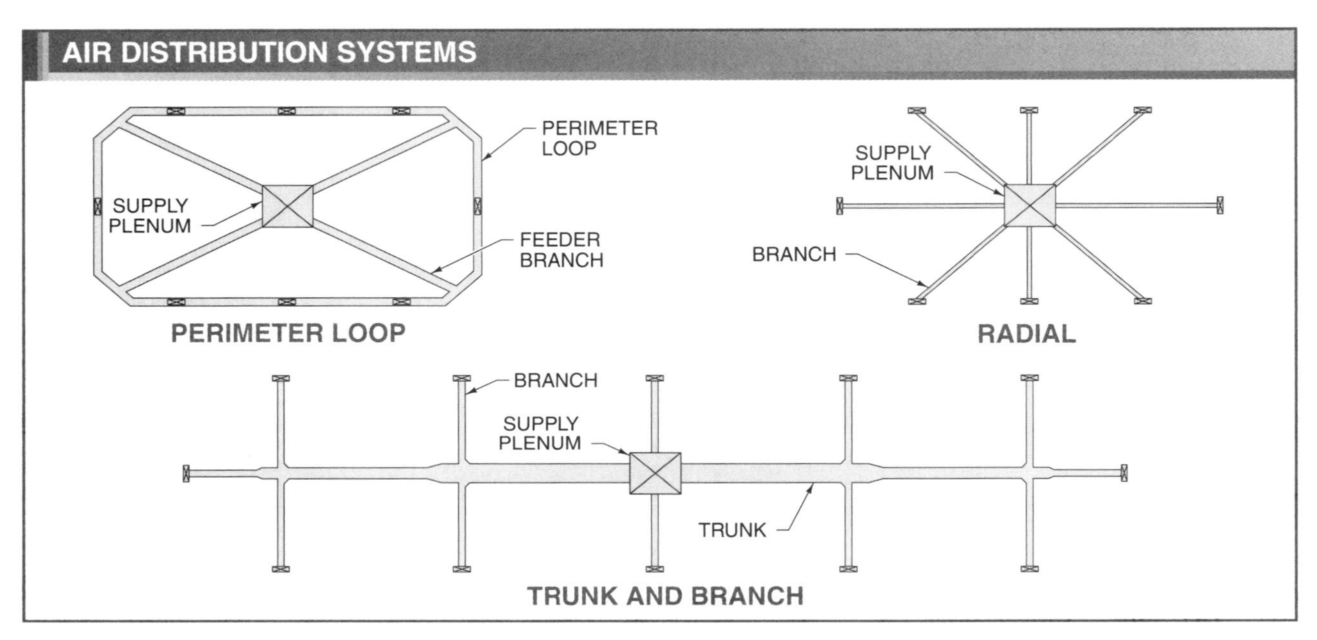

Figure 10-24. Air distribution systems are categorized by the layout of the ductwork.

Trunk and Branch Systems. A *trunk and branch system* consists of one or more trunks that run out from the supply plenum of a furnace. A *trunk* is a main supply duct that extends from the supply plenum. Branches extend from the trunks to each register. A trunk and branch system is installed with the trunks running parallel with beams or support members of a building and the branches running at right angles to the trunks. Trunk and branch systems make it possible to keep ductwork close to building support members. In many cases the ductwork is run in joist spaces and then covered by the building finish materials.

Flexible and Rigid Ductwork. Ductwork may be either flexible or rigid. Flexible ductwork makes installations and retrofits much easier, especially when space is at a premium. Many terminal units in commercial buildings use flexible ductwork for their final connections. A bonus is that flexible ductwork absorbs vibrations and pulsations that otherwise might cause noise. A disadvantage is that the pressure drop through flexible ductwork is higher than with other materials. Today, residential and commercial buildings use sheet metal or ductboard materials for rigid ductwork. **See Figure 10-25.**

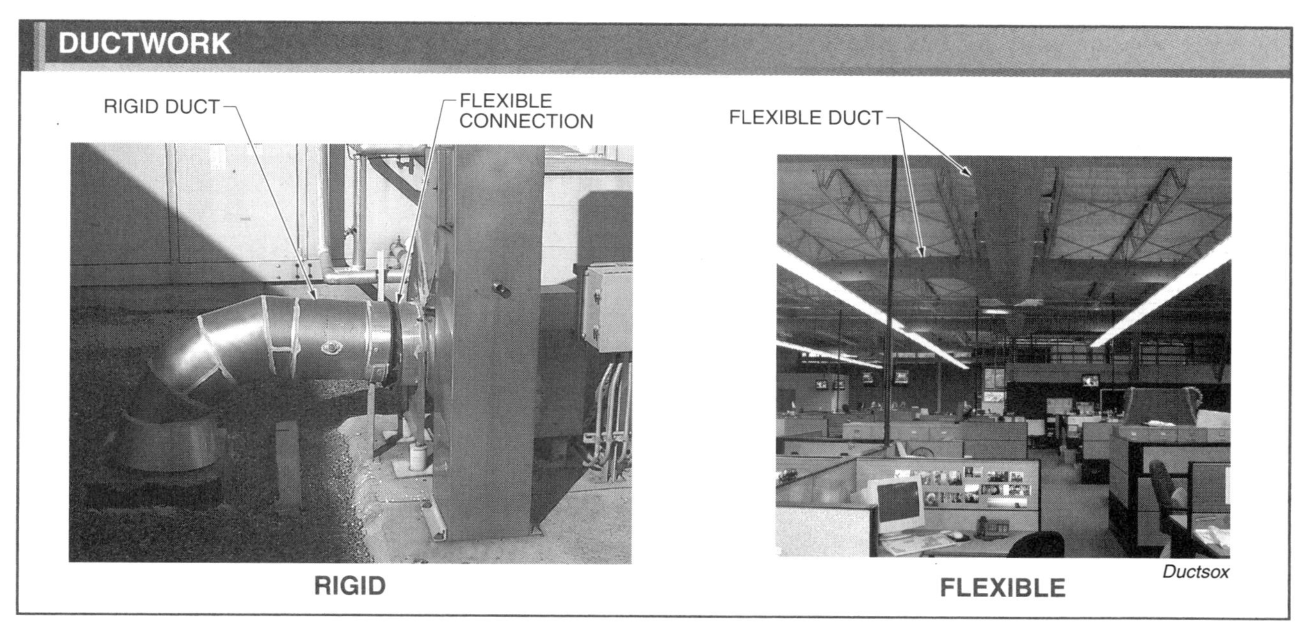

Figure 10-25. Ductwork may be rigid or flexible depending on installation requirements.

Registers and Grills

A *register* is the device that covers the opening of the supply air ductwork. A register consists of a panel with vanes, a frame for mounting, and a damper for controlling airflow. **See Figure 10-26.** The airflow pattern from the register is controlled by the vanes on the panel. Registers throw air straight out from the register and/or spread air out evenly in all directions. Registers are sized according to the quantity and velocity of air required in each building space. Registers are located for efficient distribution of air to each building space.

The airflow pattern determines the area of influence of the air from the register. The *area of influence* is the area from the front of the register to a point where the air velocity drops below 50 fpm. Air velocity above 50 fpm produces an uncomfortable draft. **See Figure 10-27.**

Tech Facts

The amount of ventilation to be supplied to an occupied space is based on the standards set by the American Society of Heating, Refrigerating, and Air Conditioning Engineers in the standard ANSI/ASHRAE 62.1, Ventilation for Acceptable Indoor Air Quality.

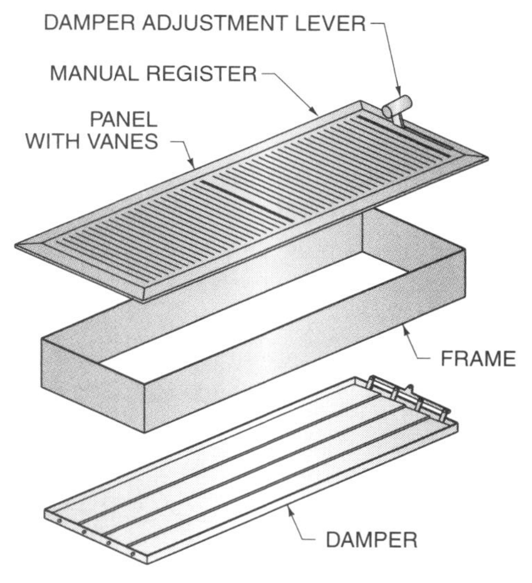

Figure 10-26. Registers consist of a panel with vanes, a frame for mounting, and a damper for controlling airflow.

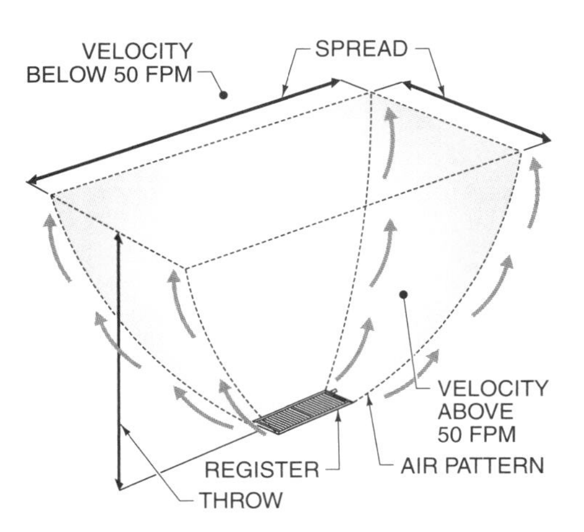

Figure 10-27. Registers throw air straight out from the front of the register and/or spread the air out evenly in all directions.

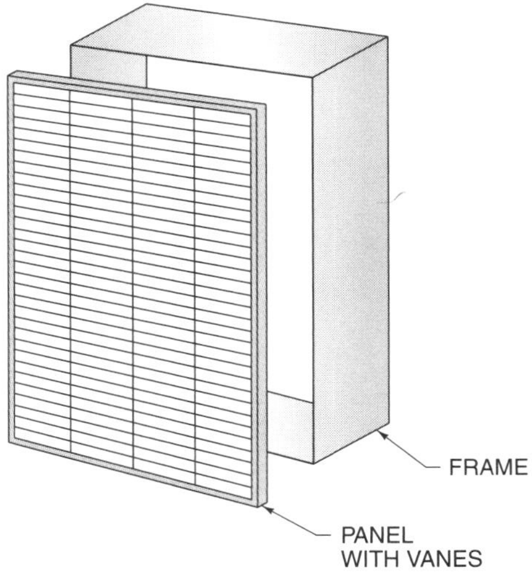

Figure 10-28. Grills consist of a panel with vanes and a frame that holds the panel in place.

A *grill* is the device that covers the opening of the return air ductwork. A grill consists of a decorative panel with vanes and frame that holds the panel in place. The vanes are arranged to block the view to the ductwork. Grills are sized and located to return the supply air back to the furnace. Grills are located on ceilings, walls, or floors, depending on the ductwork system and the location of the ductwork in the building frame. Buildings have fewer grills than registers because grills are larger and more centrally located than registers. It is also common for grills to be located high in the spaces being conditioned, which prevents temperature stratification of air in the rooms. **See Figure 10-28.**

Ductwork may be wrapped in insulation to decrease hot and cold air transfer to unoccupied building spaces.

CONTROLS

Controls operate the individual components of a furnace. Controls are installed on a furnace by the manufacturer or an air conditioning technician to maintain safe, efficient furnace operation. Power, operating, safety, and combustion safety controls are the different kinds of controls.

Power Controls

Power controls control the flow of electricity to a furnace. Power controls are installed in the electrical circuit between the power source and the furnace. Power controls include disconnects, fuses, and circuit breakers. Power controls should be installed by a licensed electrician per Articles of the NEC®.

Disconnects. A *disconnect* is a manual switch that shuts OFF the current to electrical equipment. When a disconnect is open, no electric current flows. When a disconnect is closed, electric current flows. At least two disconnects are installed in the electrical power circuit to a furnace. One disconnect is located in the electrical service panel where the circuit originates. The other disconnect is located on or near the furnace. **See Figure 10-29.** Most heating systems use manual disconnects. A *manual disconnect* is a protective metal box that contains fuses or circuit breakers and the disconnect. A manual disconnect has a handle that extends outside the box. The handle allows manual opening or closing of the disconnect without opening the box.

A *conductor* is a material that has very little resistance which permits electrons to move through it easily. Conductors in an electrical power circuit are sized to carry the electricity required to operate the equipment. Fuses or circuit breakers are placed in the electrical power circuit to protect the conductors from excessive current flow.

Fuses. A *fuse* is an electric overcurrent protection device used to limit the rate of current flow in a circuit. A fuse will blow (burn out) if an overcurrent condition occurs, which breaks the circuit and shuts OFF the current flow before the circuit is damaged. **See Figure 10-30.** Large heating systems use cartridge fuses to protect electrical circuits. A *cartridge fuse* is a snap-in type electrical safety device that contains a wire that is designed to carry a specific amount of current. If an overcurrent condition occurs, the fuse wire will melt and the circuit will open.

Circuit Breakers. A *circuit breaker* is an overcurrent protection device with a mechanical mechanism that may manually or automatically open a circuit when an overload condition or short circuit occurs. Circuit breakers that are used as disconnects have a switch that can be used to manually open and close the circuit. Circuit breakers are used for combination disconnects and fusing in small heating systems.

Operating Controls

Operating controls are controls that cycle equipment ON or OFF. Operating controls include transformers, thermostats, blower controls, relays, contactors, magnetic starters, and solenoids.

Transformers. A *transformer* is an electric device that uses electromagnetism to change (step-up or step-down) AC voltage from one level to another. A transformer consists of a primary coil and a secondary coil. Each coil is a wire that is wound on a metal core. **See Figure 10-31.** The primary side of the transformer is connected to an electrical power source. The secondary side of the transformer is connected to an electrical load.

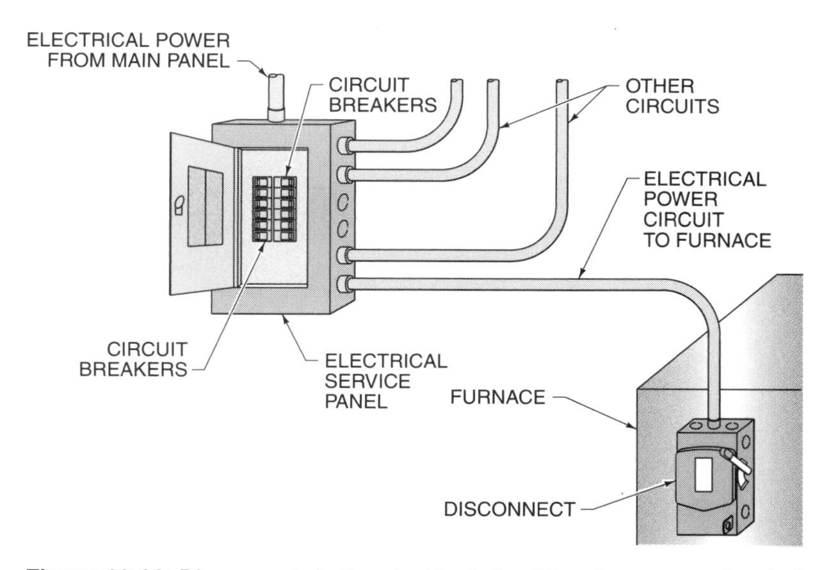

Figure 10-29. Disconnects in the electrical circuit to a furnace are located at the electrical service panel and on or near the furnace.

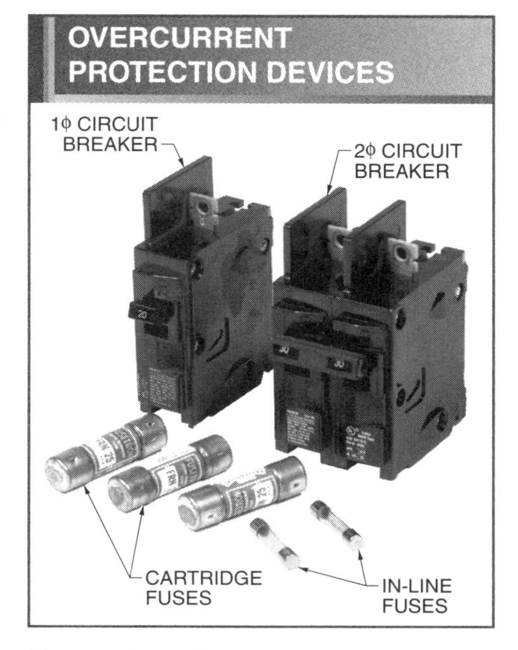

Figure 10-30. Fuses and circuit breakers protect the wiring and components in an electrical circuit from excessive current flow.

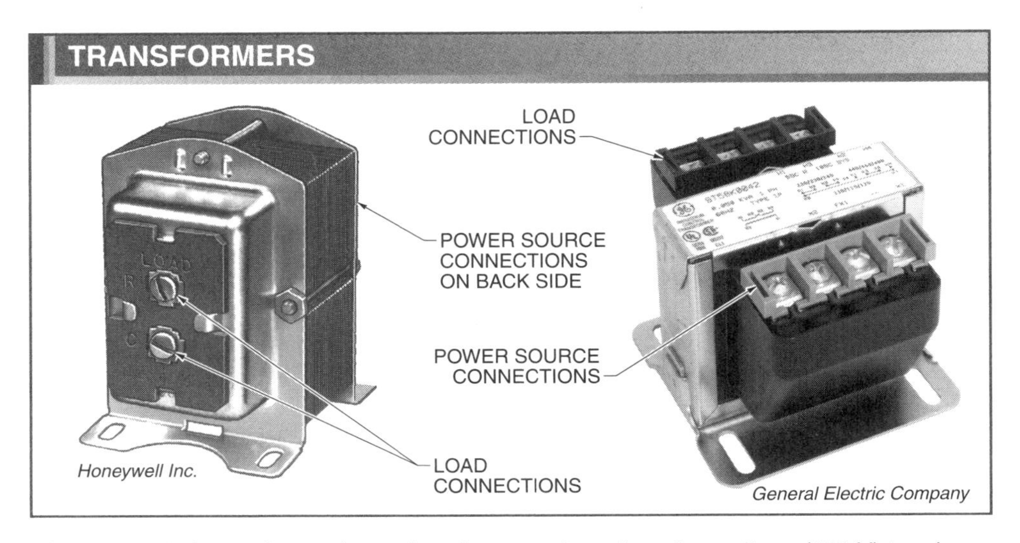

Figure 10-31. A transformer is used to change voltage from line voltage (120 V) to a lower voltage (24 V) for control systems.

When current flows through the primary coil, an electromagnetic field is produced. The electromagnetic field causes an electric current in the secondary coil. If the coil on the primary side has more windings than the coil on the secondary side, the voltage on the secondary side is less than the voltage on the primary side. This is a step-down transformer. If the coil on the secondary side has more windings than the coil on the primary side, the voltage on the secondary side is greater than the voltage on the primary side. This is a step-up transformer. Most control circuit transformers are step-down transformers with a secondary voltage of 24 V. These step-down transformers are installed at the factory as part of the furnace control package. Step-up transformers are used for the ignition of fuel. Step-up transformers increase the secondary voltage to 10,000 V to 15,000 V.

Thermostats. A *thermostat* is a temperature-actuated electric switch. A thermostat operates and controls the burner(s) or heating elements in a heating unit. When the temperature at the thermostat falls below a setpoint, the thermostat closes a switch, which completes the electrical circuit. When the temperature at the thermostat rises above the setpoint, the thermostat opens the switch, which opens the electrical circuit. **See Figure 10-32.**

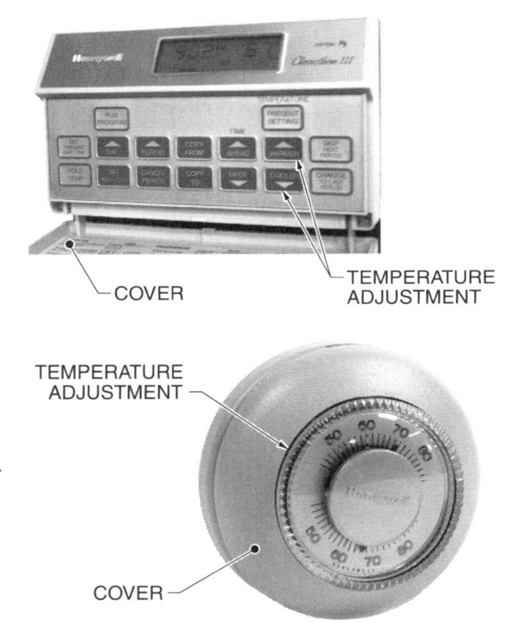

Figure 10-32. A thermostat is a temperature-actuated electric switch that controls and operates the burner(s) or heating elements in a heating unit.

Thermostats may be low-voltage or line voltage. Low-voltage thermostats carry 24 V and are more sensitive to temperature change than line voltage thermostats. Line voltage thermostats carry 120 V or 240 V. The contacts and the temperature-sensing element in line voltage thermostats must be large enough to handle the high voltage.

The operating parts of a thermostat are the sensor, switch, and setpoint adjustor. The sensor in a thermostat may be a bimetal element, remote bulb, or electronic circuit. A *bimetal element* is a sensor that consists of two different kinds of metal that are bonded together into a strip or a coil. The metals expand at different rates when heated. When a bimetal element is heated, the element bends away from the metal with the greater rate of expansion. When the bimetal element is cooled, the element bends toward the metal with the greater rate of expansion. The sensitivity and temperature range of the bimetal element increases when the element is coiled. The movement of the element actuates a switch that controls the flow of electric current. **See Figure 10-33.**

A *remote bulb* is a sensor that consists of a small refrigerant-filled metal bulb connected to the thermostat by a thin tube. The refrigerant in the metal bulb is the sensor in a remote bulb thermostat. The refrigerant in the metal bulb vaporizes or condenses in response to temperature changes. Pressure exerted by the refrigerant vapor is transmitted to a bellows element in the thermostat. The pressure change expands or contracts the bellows element, which actuates the switch that controls the flow of electric current.

Tech Facts

Thermostats are an integral part of an HVAC control system. Controlled devices, such as dampers, fans, and heating and cooling coils, are operated by electronic or pneumatic controllers that receive electronic or pneumatic signals from the thermostats.

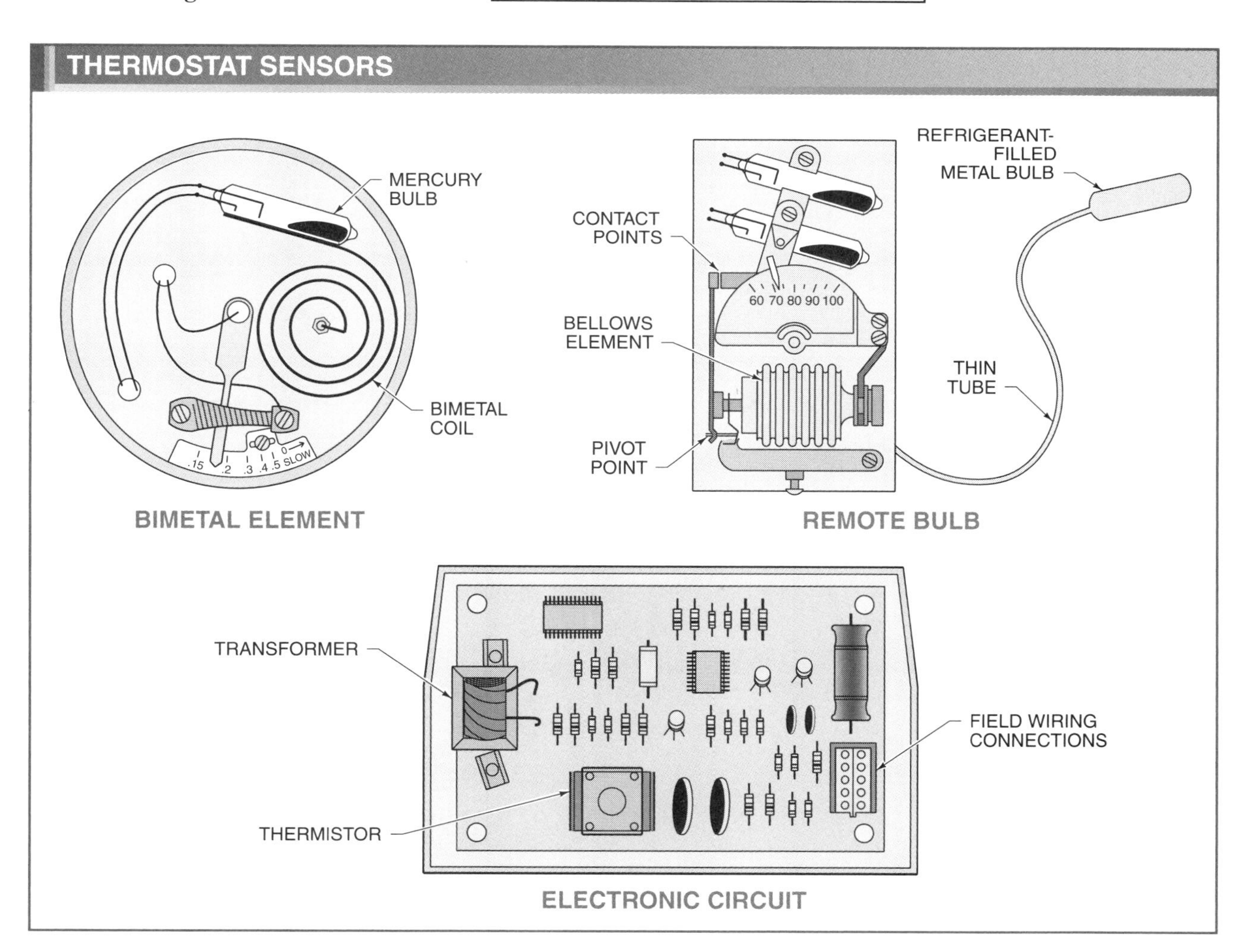

Figure 10-33. The sensor in a thermostat may be a bimetal element, remote bulb, or electronic circuit.

Electronic sensors are electronic devices that sense temperature changes. A *thermistor* is an electronic device that changes resistance in response to a temperature change. Thermistors are direct- or reverse-acting. The electrical resistance of a direct-acting thermistor increases with a temperature increase. The electrical resistance of a reverse-acting thermistor decreases with a temperature increase.

Open contact and mercury bulb are the two types of switches used in thermostats. Open contact switches may be sealed in glass to protect the contacts from dirt and oxide buildup. The switch mechanism opens or closes the contacts quickly, with positive action, to prevent electrical arcing (sparks) and burning of the contacts, which causes oxide buildup. The contacts consist of a material that readily conducts electricity. Open contact switches are often actuated by a bimetal element but may be actuated by any mechanical motion. **See Figure 10-34.**

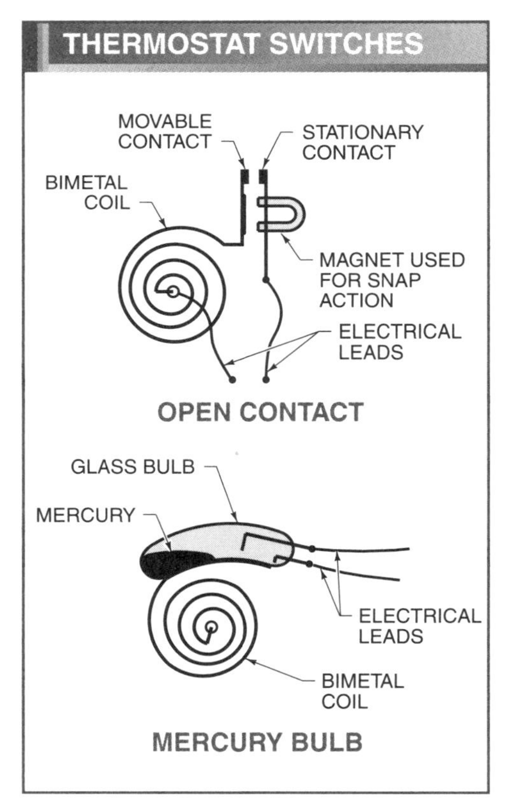

Figure 10-34. Open contact and mercury bulb switches are used in thermostats.

Mercury bulb switches are closed glass bulbs that contain a small quantity of mercury. The bulb is mounted on a bimetal coil. Exposed electrical leads are embedded in the bulb. When the bimetal coil expands or contracts, the glass bulb tips one way or the other. When the mercury moves to the end that holds the electrical leads, contact is made and the circuit is closed. When the mercury—an electrical conductor—moves to the other end, contact is broken and the circuit is open.

Setpoint temperature is the temperature at which the switch in a thermostat opens and closes. Most thermostats are built so that the setpoint can be changed manually. The *setpoint adjustor* is a lever or dial that indicates the desired temperature on an exposed scale. **See Figure 10-35.**

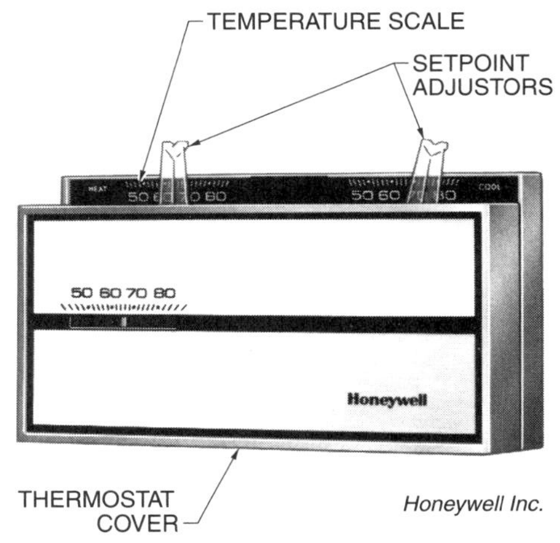

Figure 10-35. The setpoint adjustor indicates the temperature at which the switch opens or closes.

Setpoint temperature on a remote bulb thermostat is adjusted on the linkage between the remote bulb and the switch. Moving the adjustor shortens or extends the linkage. The movement increases or decreases the temperature at which the switch opens or closes.

When a thermostat senses the setpoint temperature and shuts OFF the burner(s), the heat that remains in the furnace will raise the temperature in a building space above the setpoint. The switch turns the burner(s) ON when the temperature drops approximately two degrees below the setpoint and shuts the burner(s) OFF when the temperature rises approximately two degrees above the setpoint. The *differential* is the difference between the temperature at which the switch in the thermostat turns the burner(s) ON and the temperature at which the thermostat turns the burner(s) OFF. The differential is necessary to prevent rapid cycling of the burner(s). Because of the differential, the actual temperature in a building space usually rises slightly above the setpoint during an ON cycle and drops slightly below the setpoint during an OFF cycle. A heat anticipator is used to prevent the temperature from rising above the setpoint.

A *heat anticipator* is a small heating element that is located inside a thermostat. The heat anticipator is wired with the thermostat contacts. When the thermostat calls for heat, the heat anticipator produces heat. A heat anticipator improves temperature control in a building space by providing enough heat to turn the thermostat OFF before the room temperature increases above the setpoint. When enough heat is produced, the heat anticipator is disconnected from the circuit and does not provide any false heat. **See Figure 10-36.**

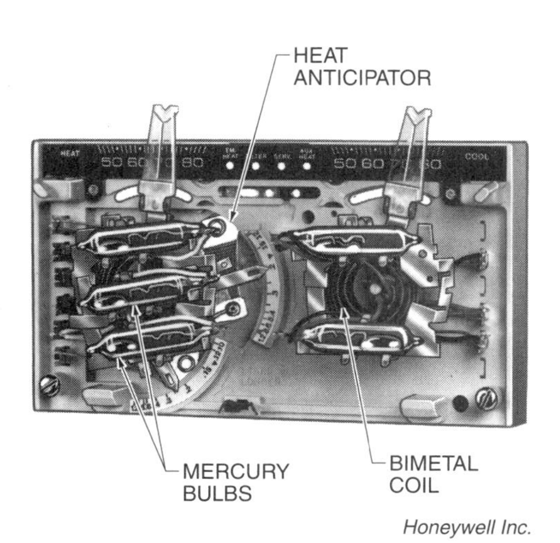

Figure 10-36. A heat anticipator improves temperature control in a building space.

Programmable and Setback Thermostats. Low voltage bimetal and remote bulb thermostats have been replaced in many applications by electronic thermostats. An electronic thermostat has an electronic thermistor element which is used to sense temperature. In most cases the electronic thermostat is more accurate than older electromechanical types. This accuracy leads to greater comfort in building spaces. The electronic circuitry also permits other features to be available. These features include fault and alarm indication with a light emitting diode (LED), and setback functions.

Thermostat setback is the reduction in heating setpoint at night when occupants are asleep or the space is unoccupied. There is also a setup feature available for cooling systems that will cause the setpoint to increase when the space is unoccupied. Both of these strategies save significant amounts of energy and equipment run time. These thermostats are often called programmable because the setpoint and time schedule parameters can be adjusted or programmed into the controller electronics. This is almost always performed using a keypad and display on the thermostat. **See Figure 10-37.**

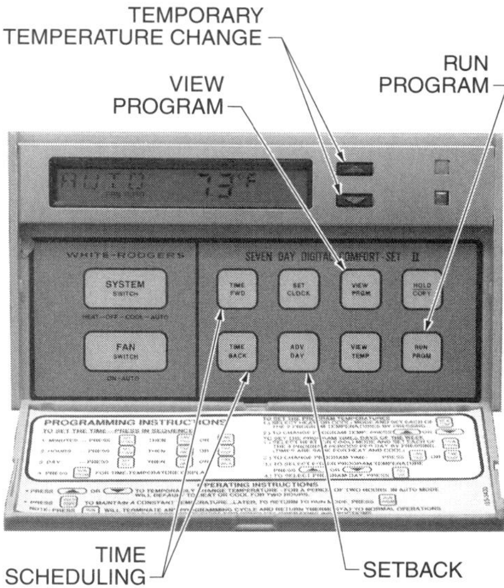

Figure 10-37. Low-voltage bimetal and remote bulb thermostats have been replaced in many applications by electronic thermostats.

Blower Controls. A *blower control* is a temperature-actuated switch that controls the blower motor of a furnace. A blower control consists of a bimetal element that operates an electric switch. The electric switch closes when the temperature increases and opens when the temperature decreases.

A blower control is installed in a furnace so the bimetal element is located in the air stream of the furnace near the heat exchanger. When the furnace burner ignites, the air near the heat exchanger is heated. The temperature change closes the electric switch, which turns the blower motor ON. When the furnace burner shuts OFF, the air around the heat exchanger cools. This temperature change opens the electric switch, which turns the blower motor OFF. **See Figure 10-38.**

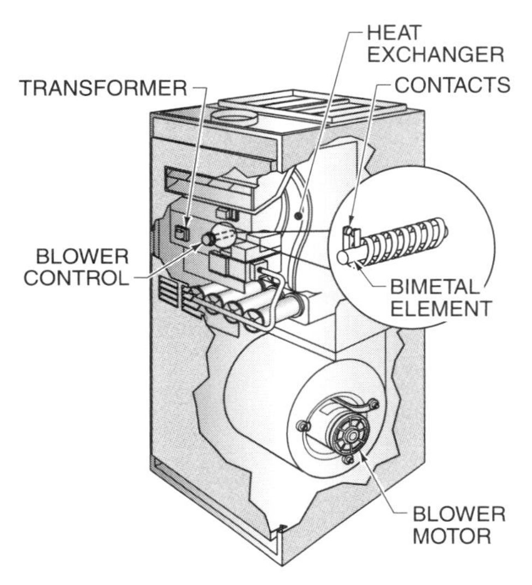

Figure 10-38. A blower control is located so that the bimetal element is near the heat exchanger.

To save energy and help ensure comfort, multispeed fans are being used instead of single-speed fans. In a multispeed fan system, the fan has more than one set speed. In this system, when only a small amount of heat is needed the fan runs at low set speed, but if the demand for heat increases the fan speed increases to a higher set speed.

Variable Frequency Drives. In addition to multispeed fans, large fan motors may use a variable frequency drive (VFD). **See Figure 10-39.** A VFD is located in the power supply between the power distribution and the fan motor. A VFD can change the fan speed from 0 rpm to maximum rpm. This provides a great amount of adjustability in the amount of air delivered to a space. VFDs are reliable and save significant amounts of fan energy if applied and installed properly. VFDs must be installed and serviced by trained factory personnel.

Relays. A *relay* is an electric device that controls the flow of electric current in one circuit with another circuit. Relays are often used in control circuits where a low-voltage circuit controls a line voltage circuit. The different types of relays used in heating system controls are electromechanical relays, contactors, and magnetic starters.

An *electromechanical relay* is an electric device that uses a magnetic coil to open or close one or more sets of contacts. One contact is fixed, and the other contact is located on a movable arm that is controlled by the coil. The moving contact is held in the initial position by a spring. When the coil is energized, the contacts open or close depending on the action desired. **See Figure 10-40.**

Relays are identified by poles and by the position of the contacts when the control circuit is de-energized. *Poles* are the number of load circuits that contacts control at one time. SP stands for single-pole circuit. DP stands for double-pole circuit. The position of the contacts is identified as normally open (NO) or normally closed (NC). *Throws* are the number of different closed contact positions per pole, which is the number of circuits that each individual pole controls. Circuits are identified as single-throw (ST) circuits or double-throw (DT) circuits. Relays are available for many combinations of poles, but the contact positions are generally limited to NO, NC, or a combination of the two.

A *contactor* is a heavy-duty relay that can be rebuilt. A contactor has a coil and contacts that are designed to operate with high electric current, which is required to run large electric motors. Contactors may be used for controlling compressors or large motors in a system.

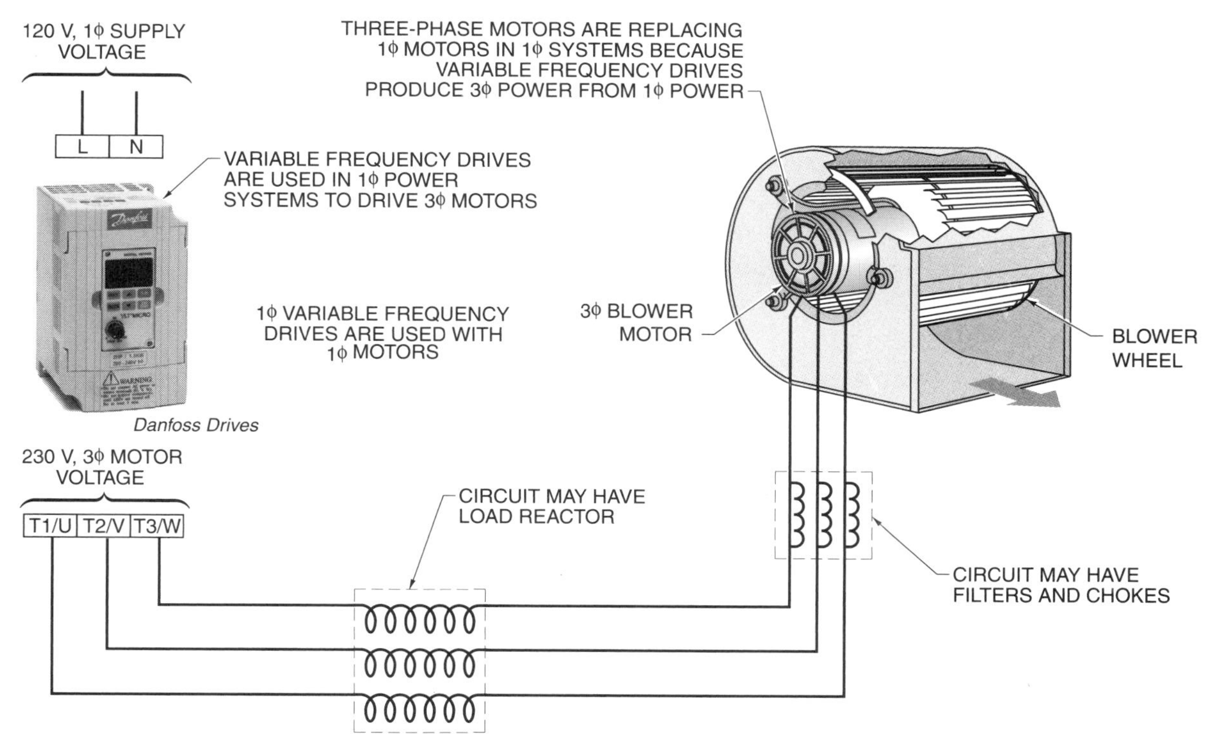

Figure 10-39. In addition to multispeed fans, large fan motors use variable frequency drives (VFDs).

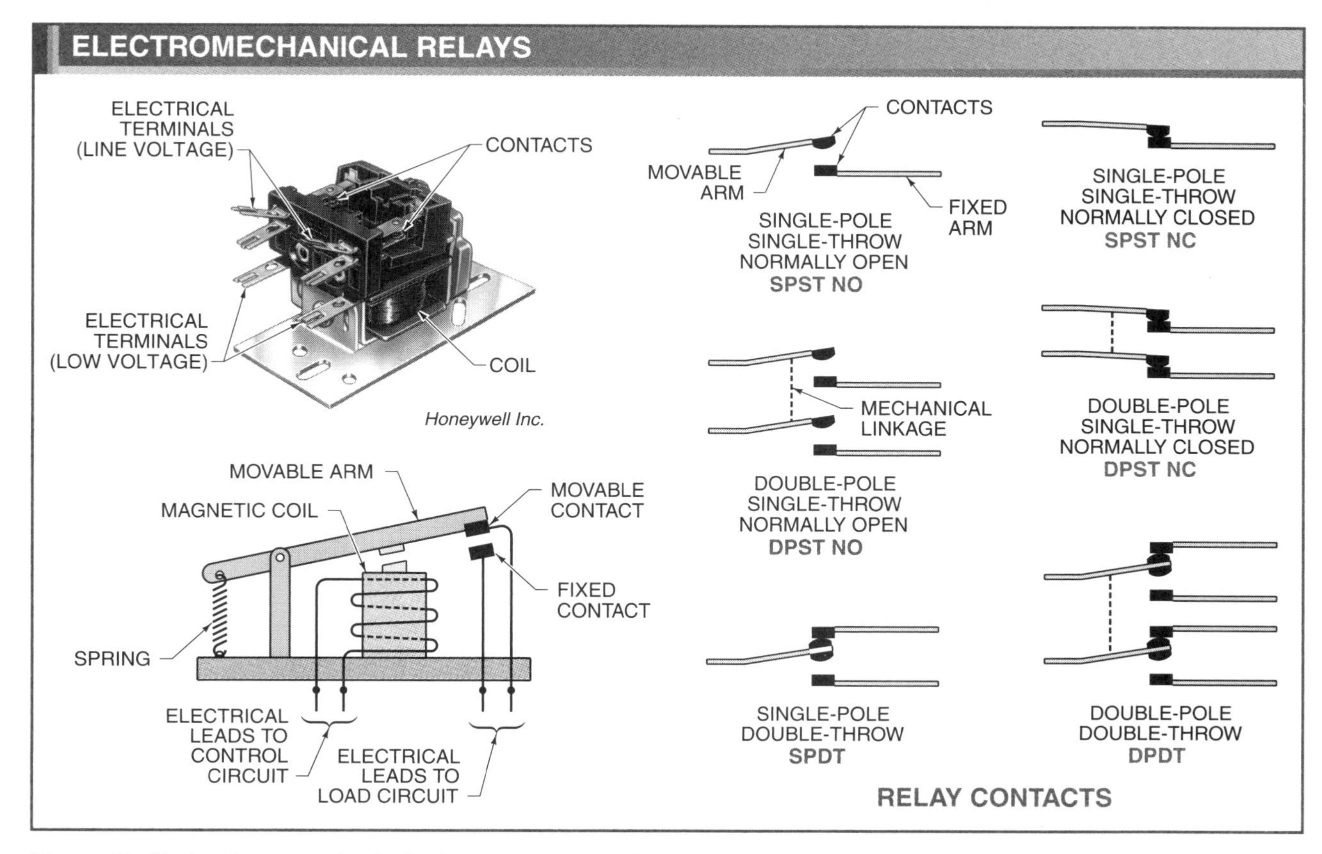

Figure 10-40. An electromechanical relay uses a magnetic coil to open or close a set of contacts.

A *magnetic starter* is a contactor with an overload relay (contact) added to it. An *overload relay* is an electric switch that protects a motor against overheating and mechanical overloading. An overload relay opens the circuit to a motor if excessive electric current or heat is present. Magnetic starters are used in motor circuits when the motors do not have internal overloads. **See Figure 10-41.**

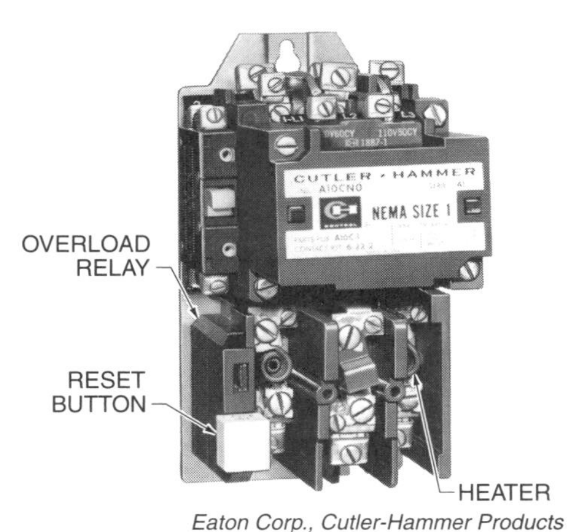

Eaton Corp., Cutler-Hammer Products

Figure 10-41. A magnetic starter is a contactor that has an overload relay added to it.

When a motor starts, it draws a tremendous momentary inrush of current that is normally five, or even six to eight, times normal running current. Fuses or circuit breakers (overcurrent protection) must be designed to handle the momentary inrush of current without opening the circuit when the motor starts. Overload protection must open the motor circuit when current increases while the motor is running. Electric motors are damaged when higher than nameplate current is sent to the motor without blowing the fuses or tripping the circuit breakers.

Motors overheat when they are overloaded. *Overload* occurs when a motor is connected to an excessive load. For example, a ½ HP motor is overloaded when connected to a ¾ HP load. When a motor is overloaded, it draws more electric current than it is designed to carry. Overload causes motors to overheat and breaks down wiring insulation in the motor.

A *bimetal overload relay* contains a set of contacts that are actuated by a bimetal element. If the temperature around the bimetal element rises because of excessive current flow, the element opens the contacts, which shuts the motor OFF. When the temperature drops, the element closes the contacts, which turns the motor ON. **See Figure 10-42.**

Safety Controls

Safety controls are controls that monitor the operation of a furnace. If a furnace causes a hazard to personnel or equipment, safety controls will shut the furnace OFF.

Limit Switches. A *limit switch* is an electric switch that shuts a furnace OFF if the furnace overheats. If the furnace temperature rises above a safe temperature, the limit switch shuts the burner(s) or electric heating element OFF. A bimetal element senses the temperature of the air around the switch and opens the electric switch if the temperature rises above a setpoint. A faulty fuel valve, broken blower belt, or faulty blower motor could cause a furnace to overheat. Most limit switches are automatic electric switches which reset automatically when the temperature drops below the setpoint. Manual-reset limit switches are used for some applications. **See Figure 10-43.**

Combustion Safety Controls

Combustion safety controls are safety controls that shut down the burner(s) if a malfunction occurs. Combustion safety controls monitor firing to make sure that ignition occurs and that the flame remains ON during a call for heat. Some combustion safety controls reset automatically after a shutdown, but others must be reset manually. Manual-reset safety controls require that the burner(s) be checked before the furnace is reignited. Combustion safety controls include stack switches, pilot safety controls, flame rods, and flame surveillance controls.

Tech Facts

Limit switches typically open when the temperature of the furnace reaches approximately 200°F. When the limit switch opens, power to the gas valve is shut off and the burners are turned off.

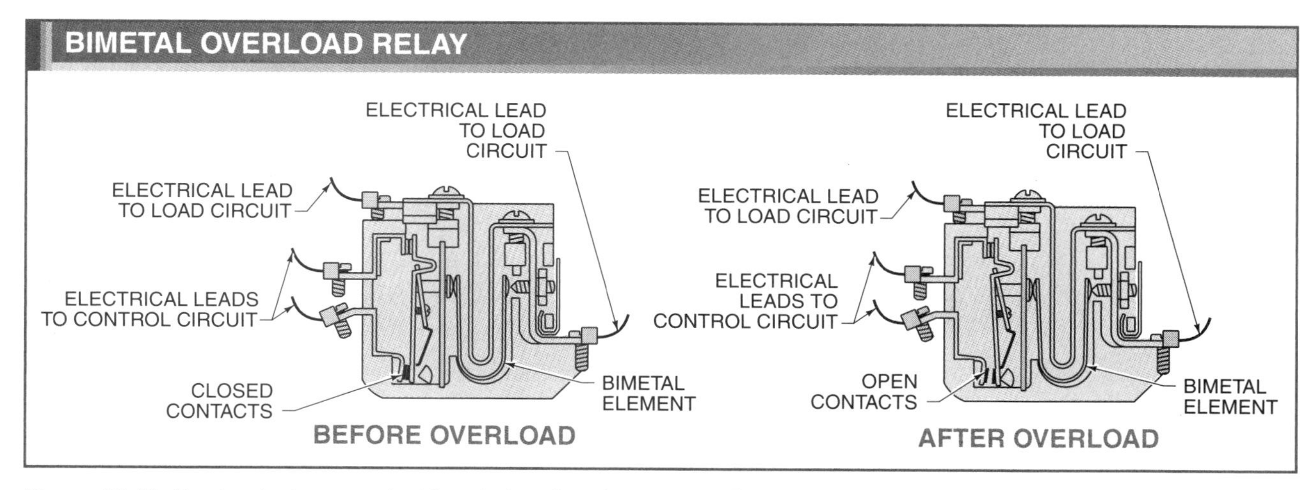

Figure 10-42. Overload relays are electric switches that shut a motor OFF if it overheats.

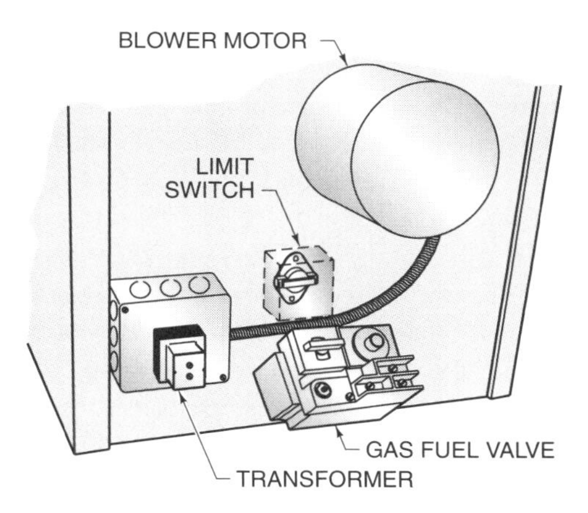

Figure 10-43. Limit switches shut a furnace OFF if the furnace overheats.

Stack Switches. A *stack switch* is a mechanical combustion safety control device that contains a bimetal element that senses flue-gas temperature and converts it to mechanical motion. A stack switch is installed in the flue near the furnace. On a call for heat, current flows through a safety switch heater in the burner control circuit. The safety switch heater is wired to a set of normally closed (NC) contacts. The current flow through the NC contacts allows the fuel valve to remain open.

If the flue-gas temperature does not rise in approximately 90 sec, the safety switch heater opens the NC contacts. If ignition does occur, the bimetal element expands. This moves a metal rod that closes a set of NO contacts (normally open contacts) in the burner control circuit. This allows the fuel valve to remain open. **See Figure 10-44.** A mechanical linkage connects the bimetal element and the contacts in the control box. The contacts open or close as the bimetal element moves.

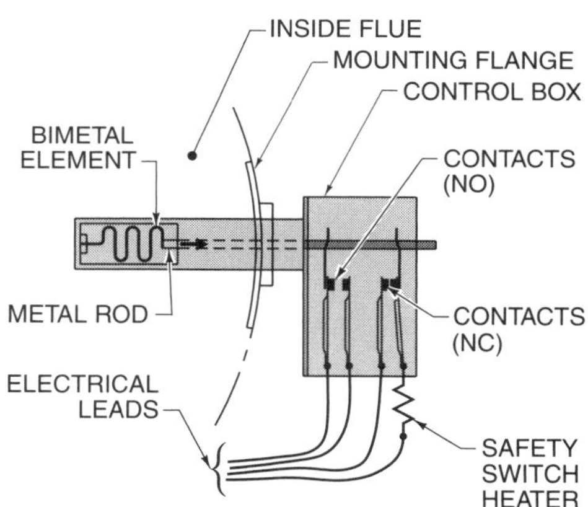

Figure 10-44. A stack switch is a combustion safety control device that converts temperature to mechanical motion.

Pilot Safety Controls. A *pilot safety control* is a safety control that determines if the pilot light is burning. Burners that use a pilot light for initial combustion have pilot safety controls. A pilot safety control is an electric combustion safety control that contains a thermocouple. A *thermocouple* is a pair of electrical wires (usually constantan and copper) that have different current-carrying characteristics and that are welded together at one end (hot junction). The thermocouple is installed in the pilot flame, which heats the hot junction. When the hot junction is heated, a low-voltage electric signal is produced. The free ends of the wires (cold junction) are connected to an electromagnetic coil in a safety valve ahead of the gas fuel valve or in a special valve within the gas fuel valve.

When the pilot light is burning, the low-voltage electric signal generated by the thermocouple produces a magnetic field in the coil. The magnetic field holds the safety valve open. If the pilot light goes out, there is no magnetic field and the safety valve closes. A pilot safety control monitors the pilot light in a burner and does not allow the gas fuel valve to open or fuel oil burner to ignite unless the pilot flame is established. Pilot safety controls are used mainly on gas fuel-burning equipment that has a standing pilot.

Flame Rods. A flame rod is an electronic combustion safety control used on large commercial furnaces. A *flame rod* is an electronic combustion safety control that uses a flame to conduct electricity. As a furnace is firing, a control device sends out a low-voltage electric signal to a metal rod located in the flame. If a flame is established on a call for heat, the safety circuit is closed by the flame. If a flame is not established within a reasonable length of time, the furnace shuts OFF. **See Figure 10-45.**

Flame Surveillance Controls. A *flame surveillance control* is an electronic combustion safety control that consists of a light-sensitive device that detects flame. Cadmium sulfide is a light-sensitive substance used in flame surveillance controls. The resistance of cadmium sulfide to electric current depends on the intensity of the light that strikes the material. When a cadmium sulfide cell (cad cell) is exposed to light, the resistance to the flow of electricity through the cell is low. When the cad cell is in darkness, the resistance through the cell is high. Electrical leads are connected to each side of the cell. Current flow through the cell is monitored to determine if the cell detects light.

The cad cell is mounted on a burner in such a way that it is in direct line of sight with the flame from the burner. If the burner ignites and a flame is established on a call for heat, the resistance through the cell is low. Low resistance allows the electric signal to pass to the control center, which actuates the furnace. If the burner does not ignite, the resistance through the cell is high, which prevents the electric signal from reaching the control center. After a reasonable time, the control center shuts the furnace OFF. Cad cell combustion safety control devices are used on furnaces that contain fuel oil burners. **See Figure 10-46.**

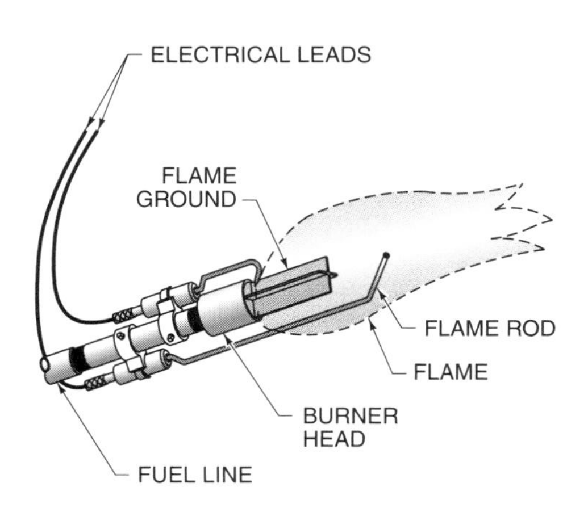

Figure 10-45. A flame rod uses the burner flame as an electrical conductor.

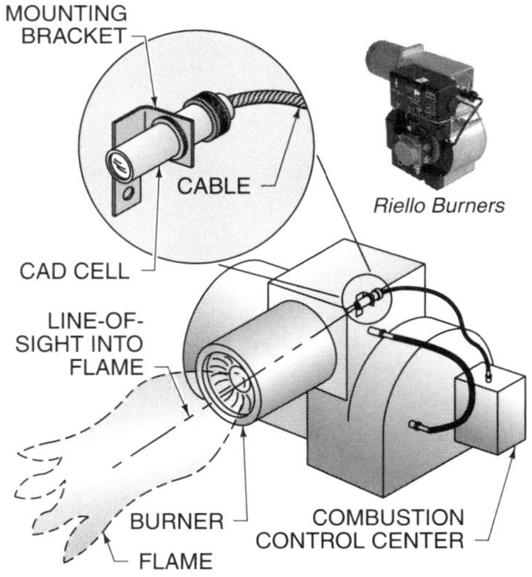

Figure 10-46. A cad cell is a combustion safety control device that detects flame.

SPECIAL PURPOSE HEATING UNITS

Special purpose heating units are furnaces built for special applications. These applications include large open spaces, spaces with a common air supply but not a central heating unit, spaces that use 100% makeup air, and spaces that do not require heat but have people that need heat.

Unit Heaters

A *unit heater* is a self-contained heating unit that is not connected to ductwork. Unit heaters burn gas fuel, use resistance heating elements, or have hot water or steam coils. Unit heaters are used as space heaters. **See Figure 10-47.** A blower in the unit heater draws air in through the back of the unit and heated air is blown out through louvers in the front. A unit heater is normally suspended from the ceiling or from the roof frame of a building. A fuel line, control circuit, and electric conductors are run to the heater.

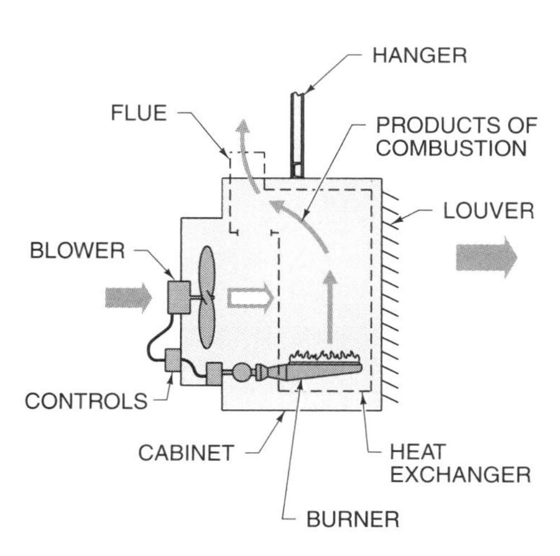

Figure 10-47. A unit heater is a self-contained heating unit that is not connected to ductwork.

Duct Heaters

A *duct heater* is a unit heater that is installed in a duct and supplied with air from a remote blower. **See Figure 10-48.** The blower for a duct heater is usually controlled by a central control system. Each duct heater has an air-proving limit switch that shuts the unit OFF if no air is moving in the duct. Duct heaters usually use resistance heating elements but may be heated by hot water or steam. Duct heaters are used in large systems that have a central blower for reheat in heating and ventilation systems. *Reheat* is heat that is supplied at the point of use while air supply comes from a central location.

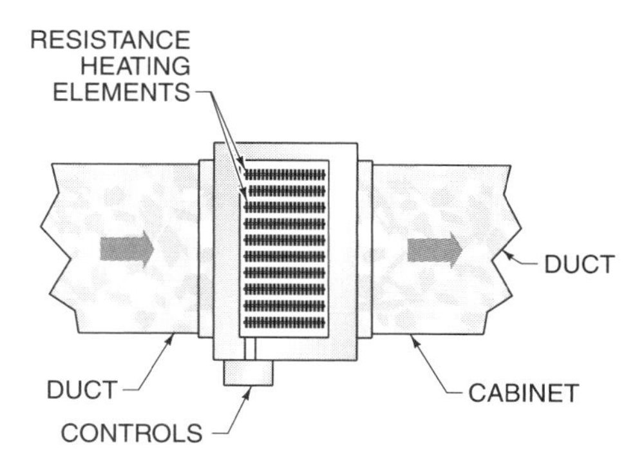

Figure 10-48. A duct heater is located inside ductwork and uses air from a remote blower.

Direct-Fired Heaters

A *direct-fired heater* is a unit heater that does not have a heat exchanger. The blower blows air directly through the combustion chamber and out the supply end of the unit. Direct-fired heaters are used in applications where 100% makeup air is required for ventilation and 100% of the air is constantly exhausted. **See Figure 10-49.**

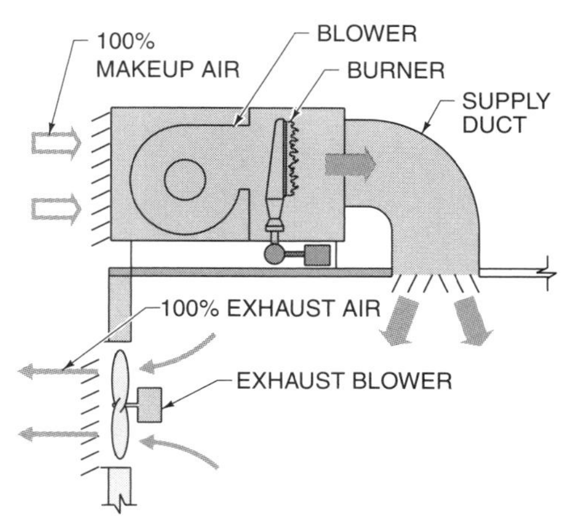

Figure 10-49. A direct-fired heater is a unit heater that does not have a heat exchanger.

Lau, a Division of Tomkins Industries

Blowers are used to create positive and negative pressures that move conditioned air through a building.

Infrared Radiant Heaters

An *infrared radiant heater* is a heating unit that heats by radiation only. An infrared radiant heater has a heat source such as a gas fuel flame, resistance heating elements, or hot water coil that heats a surface to a temperature high enough to radiate energy. **See Figure 10-50.** Infrared radiant heaters do not heat the air. Radiant energy is converted to heat when the radiant energy waves strike an opaque object.

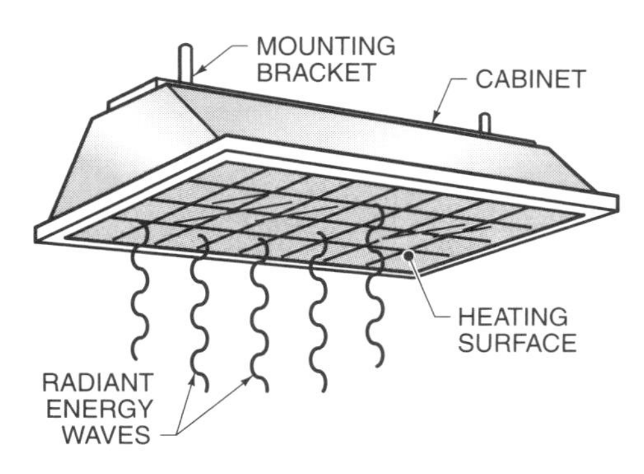

Figure 10-50. The surface of an infrared radiant heater is at a temperature high enough to radiate energy waves.

Rooftop Units

A *rooftop unit* is an air-cooled packaged unit that is located on the roof of a building. **See Figure 10-51.** These units also commonly provide cooling as well. The mounting of the unit on the roof creates valuable free space inside the building and may make maintenance access easier. However, this ease of access may be offset by the technician's exposure to the elements during service and the possibility of having to carry heavy tools and equipment up to the roof when performing service.

Wall Units

Wall units are packaged units that provide heat to a building space by using a fan and heat exchanger mounted in a metal enclosure that is mounted against an outside wall. **See Figure 10-52.** These units are commonly used in commercial buildings such as hotels and schools.

TROUBLESHOOTING AND SERVICING FORCED-AIR HEATING SYSTEMS

Proper tools, supplies, and materials are required to troubleshoot and service forced-air heating systems. Basic questions asked when troubleshooting include:

- Are the building spaces are too hot or too cold?
- How long has the problem existed?
- Has the unit ever worked properly?
- Is there a pattern to the problem such as occurring at a certain time of day?
- Has the unit recently been serviced or changed?

Tools and Supplies

To properly service or troubleshoot forced-air heating systems all safety equipment must be available and in good order. Tools should include a set of standard hand tools such as screwdrivers and wrenches. Specialized tools such as an electronic thermometer, digital multimeter, manometer, flue gas analyzer, and fine particle analyzer may also be needed. Units such as electronic ignition systems may require specialized analyzers and procedures. **See Figure 10-53.**

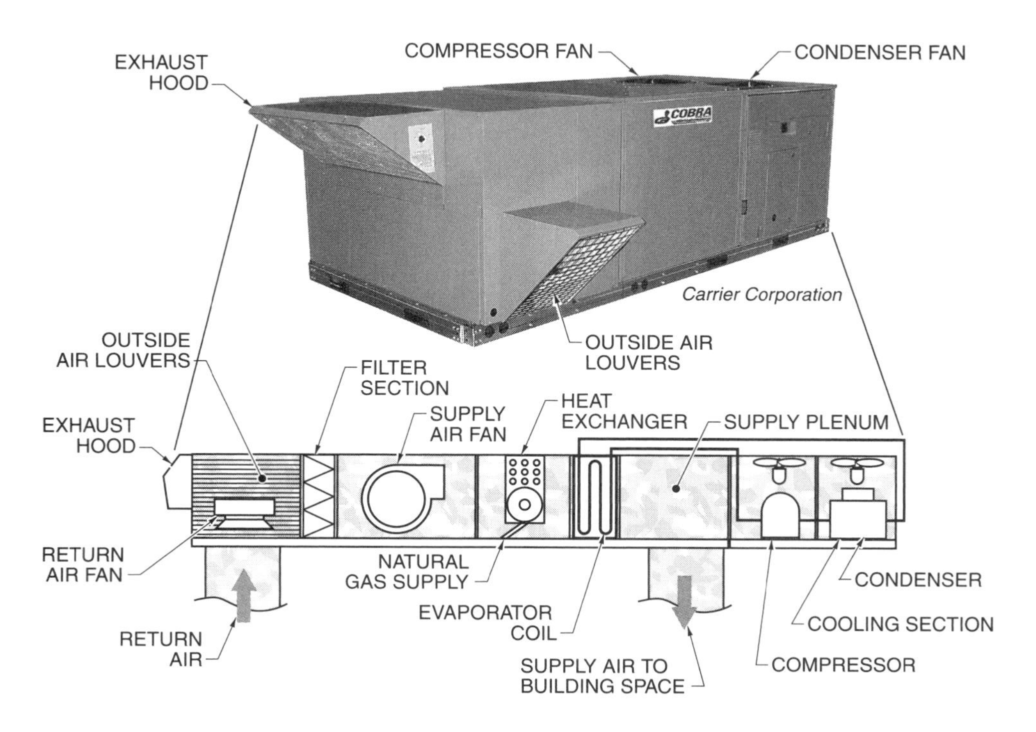

Figure 10-51. Rooftop packaged units provide heat and commonly provide cooling to a building space but are installed on the roof of a building.

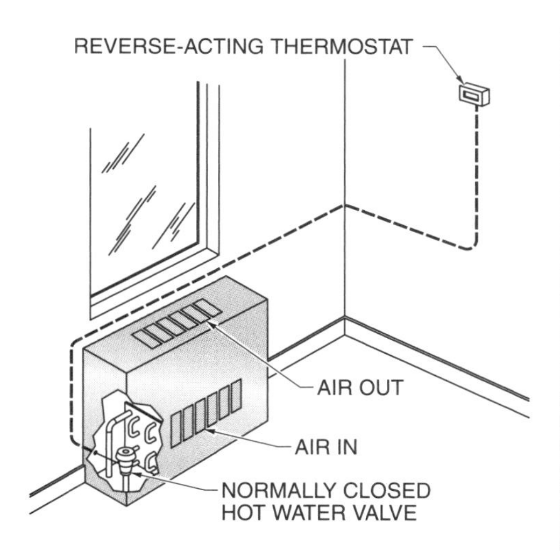

Figure 10-52. Wall units are used in commercial buildings to provide heat to a building space by using a fan and heat exchanger mounted in a metal enclosure that is against an outside wall.

Tech Facts

Special-purpose heating units are typically used in large open spaces and spaces with a common air supply, such as a warehouse or gymnasium.

Figure 10-53. Tools such as an electronic thermometer, digital multimeter, manometer, flue gas analyzer, and fine particle analyzer may be needed along with other specialized tools to service forced-air heating systems.

Manufacturer's specialized training may be required to service proprietary units. When servicing any unit, all applicable safety procedures must be followed, along with specific recommendations by the manufacturer. Any service and troubleshooting literature available from the manufacturer should be obtained and used as well. Individual model units may require specific procedures. Consumable supplies typically include filters, electrical tape, lubricant, and fasteners.

Tech Facts

When using a wrench, do not use a pushing motion. A pulling motion should always be used to help protect the hand and knuckles from injury.

Installation, Servicing, and Troubleshooting Procedures

When troubleshooting forced-air heating systems, the usual complaint from the customer is that they are too hot or too cold. Before checking specific parts of the HVAC system, there are some basic questions that may be asked such as how long the problem has existed, whether the equipment has ever worked properly, whether there have been any recent changes to the building, etc. For the power supply, the incoming line voltage should be at the proper level. The step-down transformer should be providing 24 VAC. The power supply, including transformer, is checked using a DMM. **See Figure 10-54.**

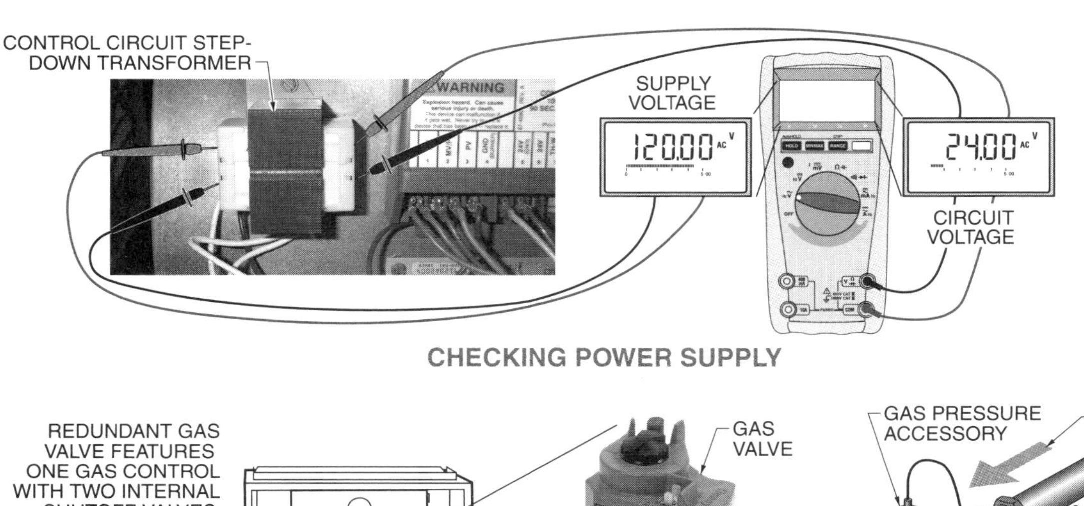

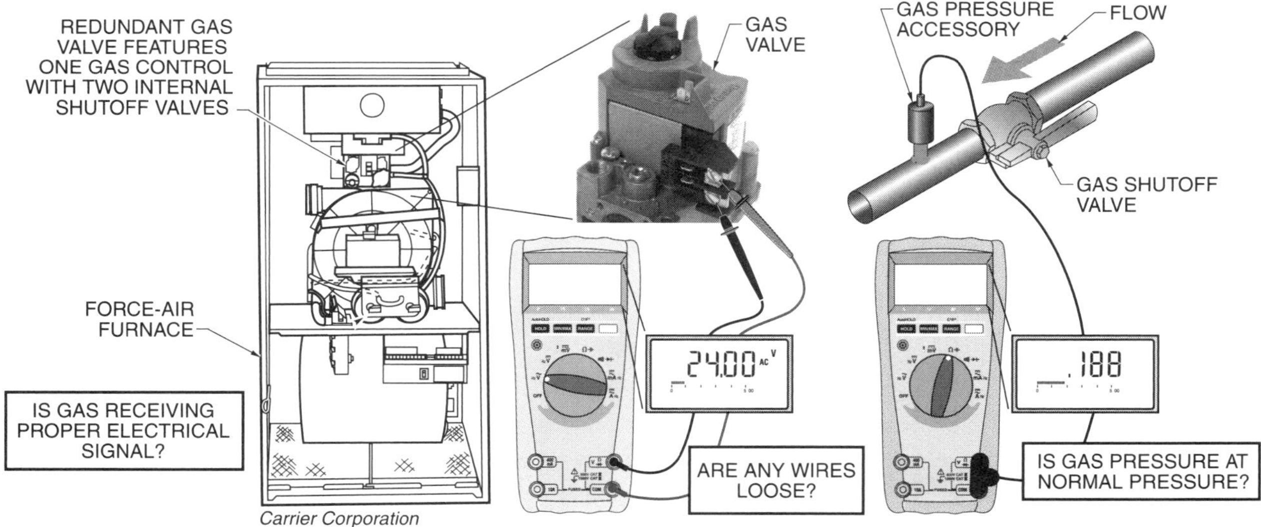

Figure 10-54. When troubleshooting a forced-air heating system, the power supply, including the transformer and gas valve, must be checked for proper operation.

The thermostat should be checked for damage. The setpoint and anticipator should be set properly on the thermostat. The thermostat may have a knob, slider, or pointer that is used for adjustment. The gas valve or heating device should be working properly. DMMs are normally used to check these devices. The gas pressure should be at normal pressure (about .125 psi to .25 psi or 3.5″ WC to 6.9″ WC). Wires are checked for any loose connections. External problems such as unit sizing or space reconfiguration should be checked as well. **See Figure 10-55.**

Routine Service Procedures. All manufacturers have routine preventive maintenance procedures available for their products. These procedures usually involve routine cleaning, filter replacement, and possible lubrication of the fan and fan motor bearings. Service intervals may be monthly or based on the number of hours the equipment has operated.

FORCED-AIR HEATING TROUBLESHOOTING MATRIX

Problem	Possible Cause	Possible Repair
STANDING PILOT SYSTEMS		
1. No heat – Burner does not light	A. Disconnect open	Close disconnect
	B. Fuse or circuit breaker open	Determine cause – replace fuse, reset circuit breaker
	C. Bad wiring	Repair connection or replace bad wiring
	D. Bad transformer	Determine cause and replace transformer
	E. Open pilot safety circuit	Replace pilot unit
	F. Pilot not lit	Light pilot
	G. Open limit circuit	Check and possibly replace limit
2. Pilot will not stay lit	A. Bad thermocouple	Check and replace
	B. Pilot flame aimed wrong	Change direction of flame
	C. Wrong gas valve for thermocouple	Change thermocouple
3. Burner lights but fan does not run	A. Bad fan motor	Replace fan motor
	B. Bad fan relay	Replace fan relay
	C. Bad fan switch	Replace fan switch
	D. Bad wiring or connections in fan circuit	Repair connections and/or replace wiring
4. Not enough heat	A. Undersized furnace	Add insulation to building space or change to larger furnace
	B. Burner cycles due to low airflow	Find obstruction to increase air flow
	C. Low gas pressure	Check and adjust gas pressure
	D. Leaking supply or return ducts	Repair air ducts
ELECTRONIC IGNITION SYSTEM		
1. No heat – burner not lit	A. See above for standing pilot	
	B. Gas valve may be OFF	Check manufacturer's troubleshooting matrix – Turn gas valve ON
2. No heat – fan is running	A. Dirty or bad flame sensor	Clean or replace flame sensor
	B. Bad circuit board	Check and change circuit board

Figure 10-55. Troubleshooting matrices are helpful when troubleshooting forced-air heating systems and can be found at manufacturer web sites.

Review Questions

1. List and describe the two main forced-air heating systems.
2. List the three different furnaces as categorized by the direction of airflow.
3. List and describe the basic parts of a furnace.
4. Why are louvers located on the access panels of a furnace cabinet?
5. List and describe the three different blowers used in forced-air heating systems.
6. Why are centrifugal blowers used most commonly in low- to medium-pressure forced-air heating systems?
7. List and describe the two drive systems used on blowers in forced-air heating systems.
8. Describe the process of changing the speed of the blower wheel of a belt drive blower.
9. How is the speed of the blower wheel of a direct drive blower changed?
10. List and describe the three functions of a burner.
11. What is the main difference between an atmospheric burner and a power burner?
12. List and describe the two most common heat exchangers.
13. What is the main function of a heat exchanger in a furnace?
14. List the four general categories of filters and the filtering efficiency of each.
15. What are the components of an air distribution system?
16. List and describe three duct layout systems used in forced-air systems.
17. When the temperature rises above setpoint in a forced-air heating system, does the switch in a thermostat open or close? Explain.
18. List three types of temperature sensors used often in thermostats.
19. How would a blower control operate in the furnace of a forced-air heating system on a call for heat?
20. Describe the magnetic action of a relay.
21. How do overload relays protect motors against overheating conditions? Explain.
22. List and describe four combustion safety controls.
23. List and describe four special-purpose heating units.
24. List some common tools and supplies needed to properly service or troubleshoot forced-air heating systems.

Steam and hydronic heating systems use steam or water to carry heat. Steam and hydronic heating systems heat water in a boiler at a central location and distribute the steam or heated water through supply piping to building spaces. The steam or heated water then flows through terminal devices that transfer the heat from the steam or water into the air of the building space. The steam or water returns to the boiler through return piping.

11 Steam and Hydronic Heating Systems

STEAM AND HYDRONIC HEATING SYSTEMS

Most commercial and some residential buildings use steam or hot water systems to supply the required heat. Boilers located in central mechanical rooms supply the steam or hot water. Water carries one Btu per degree change per pound of water while steam carries 970 Btu for every pound of steam that condenses back into a liquid. A heating unit that uses hot water to heat air allows easier control of the air temperature than a steam system. Steam systems rely on the difference in pressure to move the steam (vapor) from the boiler through piping to individual heating units where the steam gives up the heat as it condenses back into a liquid, and then the condensate is pumped back to the boiler through return piping. In a hot water heating system the heated water is distributed by a circulating pump through supply piping to heating units and returned to the boiler in return piping.

A common practice is to use a combination of steam and hot water systems to furnish all the heating and processing needs of a commercial building. In this type of system a boiler is used to produce the steam required by building processes. Some or all of the steam needed to heat the building is piped to a heat exchanger to heat water. The hot water is then circulated through the heating units by supply and return piping to heat building space air.

STEAM HEATING SYSTEMS

A *steam heating system* is a heating system that uses steam to carry heat from the point of generation to the point of use. Steam heating systems are used because steam carries more Btu per pound of steam than water, allowing steam to be used where there is a limited amount of room for the distribution system. Another advantage to steam heating systems is that the steam does not need to be pumped to the location of use. Steam is a vapor that flows any time there is a difference in pressure from one area (boiler) to another area (heating unit). The boiler is located in a central plant and the steam is distributed to individual buildings of campus complexes or utility districts. **See Figure 11-1.**

In individual buildings, the boiler is normally located on the lowest floor, usually in a basement mechanical room. One of the deciding factors in locating the boiler is where is the air handing equipment located. If the air handing equipment is located on several floors throughout the building, the boiler is more likely to be located in the basement because the steam will flow up through the floors through the supply piping without the assistance of a pump, and the condensed liquid (condensate) can flow back to the boiler by gravity through the return piping. If all of the air handing equipment is located on a single floor, the boiler would most likely be located near that equipment to minimize the length of piping runs and heat loss.

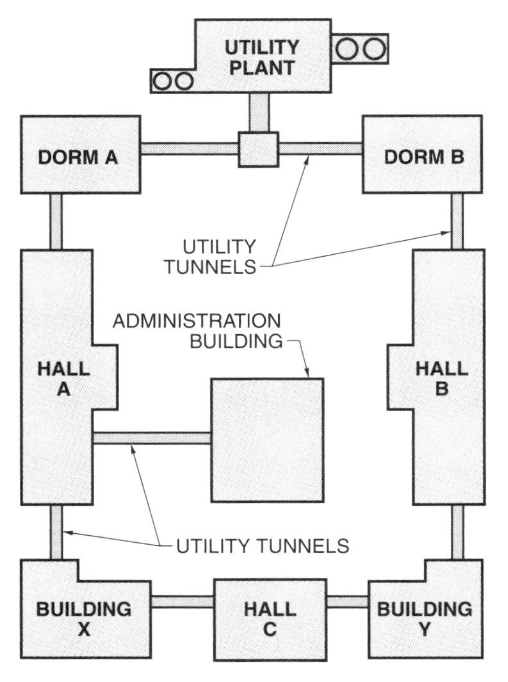

Figure 11-1. The boiler is located in a central plant and the steam is distributed to individual buildings of campus complexes or utility districts.

Where boilers are used to supply steam to a campus complex of buildings, the central plant is in a single building somewhere in the campus complex. The steam is distributed to each building in the complex through underground piping or through tunnels that may supply additional utilities to the buildings. The condensate is collected in each of the buildings and pumped back to the building housing the boiler or boilers.

In some large municipalities there are utility districts established to supply all or most of the steam needs of the municipality district. Buildings that utilize supplied steam are hospitals, office buildings, and shopping malls. The steam is piped to the individual buildings using underground tunnels, similar to those of the campus complex. The steam is metered at the entrance of the building for billing purposes. In most of these systems the condensate is not returned to the central boiler plant because of the expense involved in pumping it back to the plant and the possibility of the condensate being contaminated.

Steam heating systems consist of a boiler, fittings, accessories, a steam supply system, heat exchangers, terminal devices, controls, and the condensate return system. **See Figure 11-2.** A boiler heats water to above the boiling point to produce steam. The fittings on a boiler and steam system provide safe and efficient operation of the boiler. Accessories are the required systems and components to support the operation of a boiler such as the fuel supply system and the feed water system. The *steam supply system* is the piping and support systems that provide a path for steam flow from the boiler to the point of use. Heat exchangers transfer the heat from the steam to a medium (normally air) to heat a building space. Controls are used to regulate the temperature of the medium being used, for example the temperature of air discharged from a heating coil. The *condensate return system* is a system used to return condensate back to the boiler after all the useful heat has been removed.

HYDRONIC HEATING SYSTEMS

A *hydronic heating system* is a heating system that uses water or other fluid to carry heat from the point of generation to the point of use. Hydronic heating systems are used where heating equipment is located far from building spaces that require heat. Hydronic heating systems are also used where a central heating plant heats several buildings. Hydronic heating systems consist of a boiler, fittings, accessories, hot water supply system, circulating pump, terminal devices, controls, and hot water return system. **See Figure 11-3.**

A hydronic heating system uses a boiler to heat water and store water. The fittings maintain the safe and efficient operation of the boiler. Accessories are required to support the operation of the boiler and associated systems. The piping system distributes the hot water to building spaces. Terminal devices transfer the heat from the hot water to the air in the building spaces that require heat. A circulating pump moves the water from the boiler through the supply piping system to the terminal devices and back to the boiler. Controls regulate the temperature of the water throughout the hydronic heating system and the temperature of the air in the building spaces. Controls operate the boiler and circulating pump, which produce and distribute hot water to the building spaces.

Tech Facts

Hydronic heating systems use hot water boilers to heat water to approximately 200°F. The heated water that is used to heat building spaces returns to the boiler at approximately 160°F through return piping to be reheated and sent through the heating system again. Hydronic heating systems heat room air rather than ducted air.

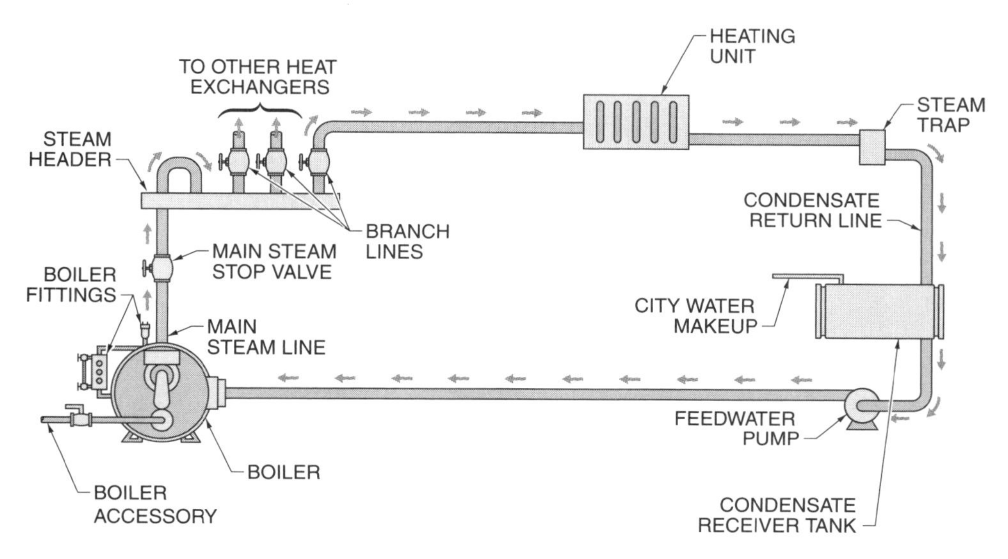

Figure 11-2. Steam heating systems consist of a boiler, fittings, accessories, steam supply system, heat exchangers, terminal devices, controls, and the condensate return system.

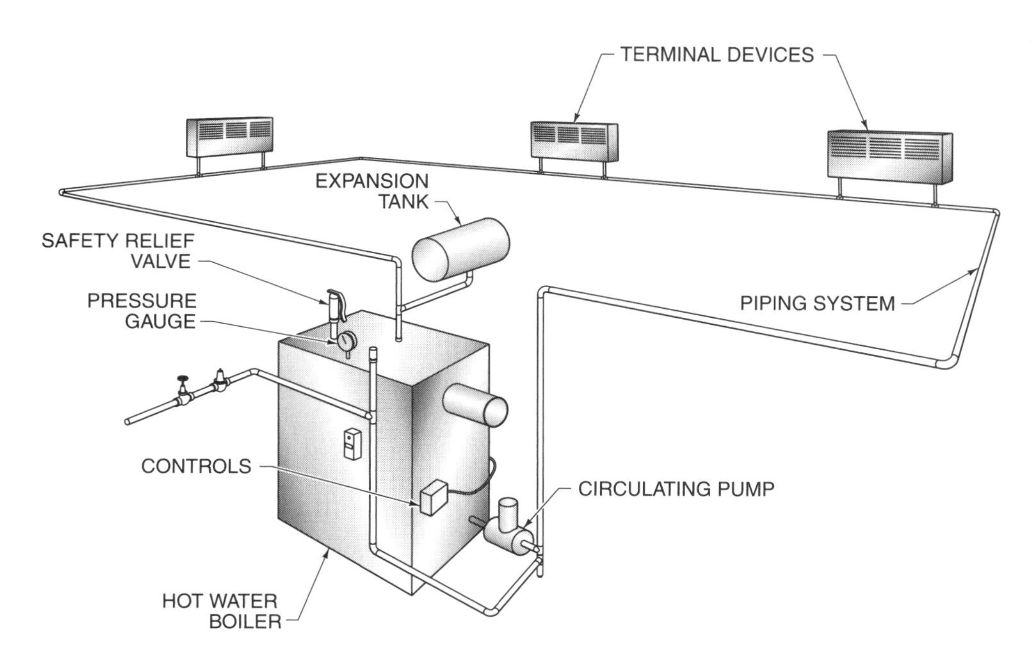

Figure 11-3. Hydronic heating systems consist of a boiler, fittings, accessories, hot water supply system, circulating pump, terminal devices, controls, and hot water return system.

Boilers

The boiler is the central element of a steam or hydronic heating system. A *boiler* is a pressure vessel that safely and efficiently transfers heat to water. A *pressure vessel* is a tank or container that operates at a pressure greater than atmospheric pressure. Boilers use steam or hot water to carry heat. The heat is transferred from the steam or hot water to the air in building spaces. Hot water boilers and piping systems are completely filled with water. The steam boiler and piping system are filled with steam. Boilers are classified by working pressure, temperature, method of heat production, and materials of construction. ASME International classifies boilers into low-pressure hot water heating boilers, hot water supply boilers, power hot water (high temperature) boilers, low-pressure steam heating boilers, power steam boilers, and small power boilers.

Low-Pressure Hot Water Heating Boilers. A *low-pressure hot water heating boiler* is a boiler in which water is heated for the purpose of supplying heat at pressures not to exceed 160 psi and temperatures not to exceed 250°F.

Hot Water Supply Boilers. A *hot water supply boiler* is a boiler having a volume exceeding 120 gal., a heat input exceeding 200,000 Btu/hr, or an operating temperature exceeding 200°F that provides hot water for uses other than heating. Hot water supply boilers are found in residential buildings and in small low-rise commercial buildings because of the limited pressure, temperature, and Btu capabilities of such boilers.

Power Hot Water (High Temperature) Boilers. A *power hot water (high temperature) boiler* is a boiler used for heating water or liquid to a pressure exceeding 160 psi or to a temperature exceeding 250°F. Because of the higher pressures and temperatures it produces, this type of boiler is utilized in large high-rise commercial buildings.

Low-Pressure Steam Heating Boilers. A *low-pressure steam heating boiler* is a boiler operated at pressures not to exceed 15 psi of steam. This type of boiler is used for heating purposes only and is normally used in residential and commercial buildings where steam is not needed for any other purpose.

Power Steam Boilers. A *power steam boiler* is a boiler in which steam or other vapor is used at pressures exceeding 15 psi. Power steam boilers are used in commercial buildings where steam is utilized for other processes as well as heating. Power steam boilers are also used in the utility industry for the production of electricity.

Small Power Boilers. A *small power boiler* is a boiler with pressures exceeding 15 psi but not exceeding 100 psi and having less than 440,000 Btu/hr heat input. A small power boiler is a special classification used where steam is required for a process but pressure and heating capacity need to be limited.

Heat Production

Boilers burn fuel or use electricity to create steam or to heat water. Combustion boilers burn fuel such as natural gas, fuel oil, or coal to produce heat. Natural gas- and fuel oil-fired boilers are used in many buildings. Coal boilers are normally used in large industrial plants. Electric boilers use resistance heating elements to produce heat. Electric boilers are used mainly in residential or small commercial buildings.

Combustion Boilers. Boilers that produce heat by combustion are made of steel or cast iron. *Steel boilers* are welded together to form a single unit that has a specific output rating. Steel boilers have removable manhole covers for maintenance and service. *Cast iron (sectional) boilers* are made of hollow sections that are bolted together. Openings between the sections provide for water and flue-gas connections. Hot water flows inside the sections and hot products of combustion flow outside the sections. Heat is transferred from the products of combustion through the cast iron to the water. The number of sections in a cast iron boiler determines the output rating of the boiler. Adding or removing sections to the boiler achieves a required output rating. **See Figure 11-4.**

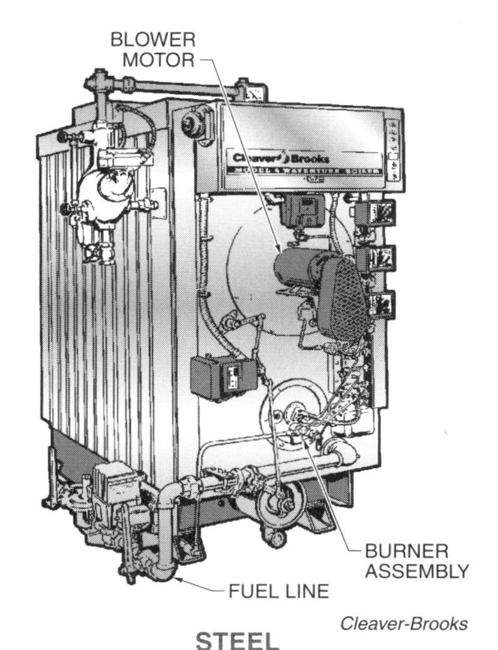

CLEAN OUT DOORS
PRESSURE GAUGE
CAST IRON SECTIONS
SECTION BEING REMOVED
OPENING FOR GAS FUEL OR FUEL OIL BURNER
CAST IRON (SECTIONAL)

Figure 11-4. Steel boilers are usually welded together to form a single unit with a specific output rating. Cast iron (sectional) boilers consist of hollow sections that are bolted together, and by adding or removing sections to the boiler a required output rating is achieved.

A *firetube boiler* heats water that surrounds the fire tubes as the hot gases of combustion pass through the tubes. The combustion chamber is located below or at one end of the boiler. A breeching connected to the flue stack collects the hot products of combustion and directs the gases up the stack. **See Figure 11-5.**

A *watertube boiler* heats water as hot products of combustion pass over the outside surfaces of the water tubes. Water fills the boiler and the water tubes. The water tubes run to each end of the boiler through the firebox (combustion chamber). The combustion chamber is located below or at one end of the boiler. A breeching connected to the flue stack collects the hot products of combustion.

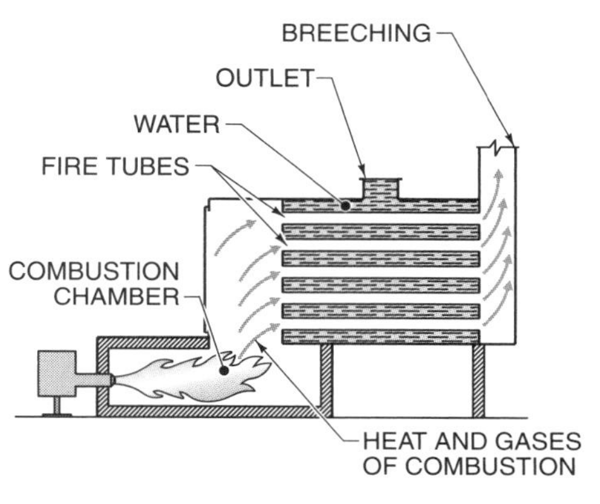

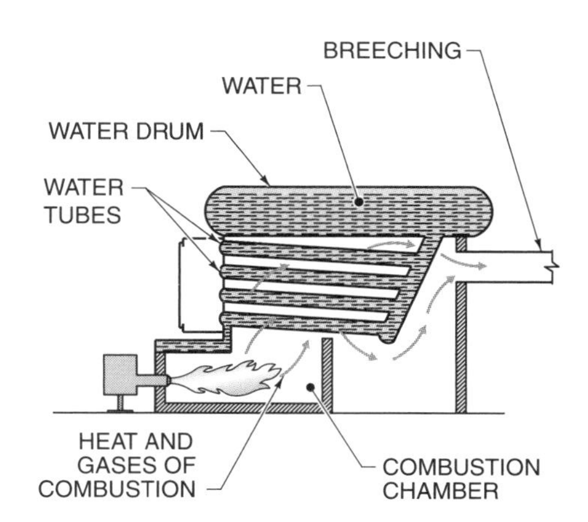

Figure 11-5. Fire tubes hold the hot products of combustion, which heat the water that surrounds the fire tubes. Water tubes hold water heated by the hot products of combustion, which surround the water tubes.

Electric Boilers. An *electric boiler* is a boiler that uses heat produced by resistance heating elements to produce steam or heat water; electric boilers are classified separately from combustion boilers. The resistance heating elements used in boilers are immersion heaters. *Immersion heaters* are copper rods enclosed in insulated waterproof tubes that are installed so that water surrounds the heaters. No heat exchanger is required because heat is transferred directly from the immersion heater to the water. **See Figure 11-6.**

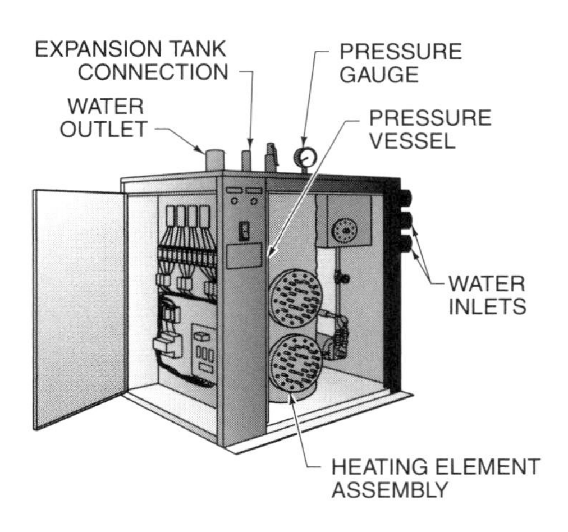

Figure 11-6. An electric boiler is a boiler that uses heat produced by resistance heating elements to produce steam or heat water directly.

Input rating is the heat produced in Btu/hr per unit of fuel burned. *Output rating* is the actual heat output produced by a heater in Btu/hr after heat losses from draft. The output rating is normally 75% to 80% of the input rating for combustion boilers. *Gross unit output* is the heat output of a boiler when it is fired continuously. *Net unit output* is gross boiler output multiplied by a percentage of loss because of pickup. *Pickup* is additional heat that is needed to warm the water in a hydronic heating system once a period of off-time, such as night, has passed. The pickup factor of a boiler can be as high as 30% to 50%. Electric boilers have output ratings of 100% because all the energy used to produce heat is transferred to the water.

The amount of heat produced by a combustion boiler depends on the size of the boiler and the kind of fuel used. Gas fuel- and fuel oil-fired boilers have output ratings from 100,000 Btu/hr for residential applications to millions of Btu/hr for commercial applications. Coal-fired boilers are normally used for industrial applications and have output ratings of hundreds of millions of Btu/hr. The amount of heat produced by an electric boiler depends on the size of the boiler and the voltage used. Electric boilers are rated in kilowatts per hour (kW/h). One kilowatt per hour equals 3414 Btu.

FITTINGS

Fittings are components directly attached to the boiler and boiler devices that are required for the operation of the boiler. *Accessories* are components not directly attached to the boiler. **See Figure 11-7.** The major differences between the two systems are the temperature and pressure ratings.

The fittings on steam boilers include safety valves, steam pressure gauges, water columns, bottom blowdown valves, surface blowdown valves, fusible plugs, vent valves, and pressure controls. Hot water boiler fittings are similar to those of steam boilers. In place of a safety valve a safety relief valve is used and instead of a steam pressure gauge a pressure-temperature gauge is used. Since a hot water boiler is completely filled with water, there is no need for water columns, surface blowdown valves, or fusible plugs. In place of the pressure controls, Aquastats™ are utilized because the boiler is used to maintain water temperature.

Safety Valves

A *safety valve* is a steam boiler fitting (valve) that prevents the boiler from exceeding a maximum allowable working pressure (MAWP). The safety valve is the most important valve on a steam boiler. If pressure in the steam boiler exceeds the safety valve setting, the safety valve opens to relieve the excess pressure. Failure of the safety valve can lead to boiler explosion. The safety valve must be located at the highest part of the steam side of the boiler and connected directly to the boiler shell. There must not be any valves located between the safety valve and the boiler or between the safety valve and the point of valve discharge.

Safety valves are designed to open rapidly when the pressure in the boiler exceeds a setpoint. The safety valve stays open until sufficient steam is released and there is a predetermined drop in pressure inside the boiler. When the desired decrease in pressure is achieved the safety valve closes quickly. ASME International code requires at least one safety valve to be mounted on each boiler. If the heating surface of the boiler exceeds 500 sq ft it must have two or more safety valves.

The capacity of a safety valve is the amount of steam, in pounds per hour (lb/hr), that the safety valve is capable of venting at rated pressure. The safety valve capacity is listed on the data plate attached to the safety valve. **See Figure 11-8.**

Safety Relief Valves. A *safety relief valve* is a valve on a hot water boiler that prevents excessive pressure from building up during a malfunction. A safety relief valve is mounted on the top of the boiler with no other valves between it and the boiler. If the pressure in the boiler exceeds the maximum allowable working pressure (MAWP), the safety relief valve opens and discharges hot water to an appropriate drain in the boiler room. ASME International code requires that hot water boilers have at least one safety relief valve. The hot water boiler safety relief valve is rated in Btu relieved per hour. It consists of an inlet, outlet, body, sealing diaphragm, spring, and try lever. **See Figure 11-9.**

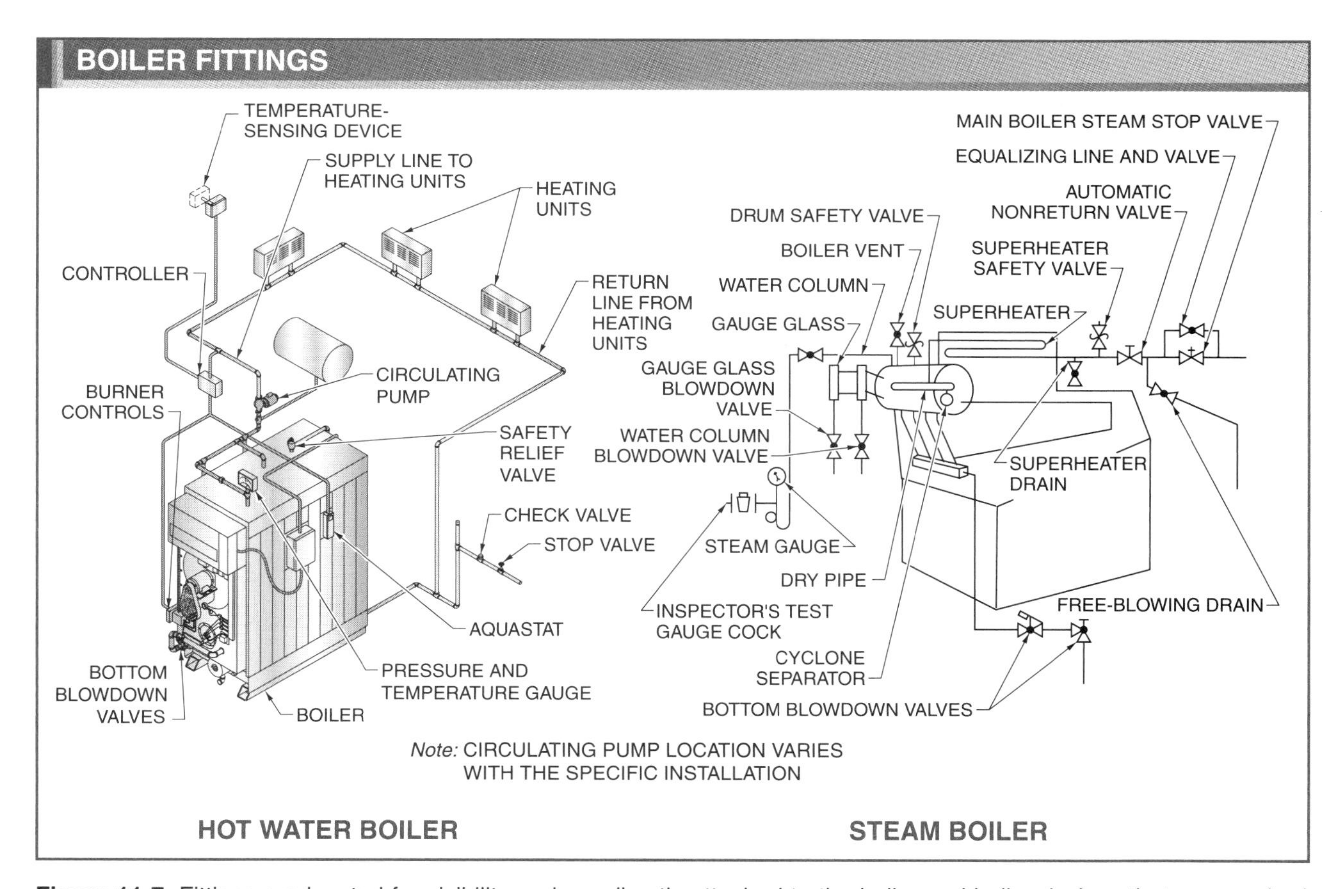

Figure 11-7. Fittings are located for visibility and are directly attached to the boiler and boiler devices that are required for the operation of the boiler.

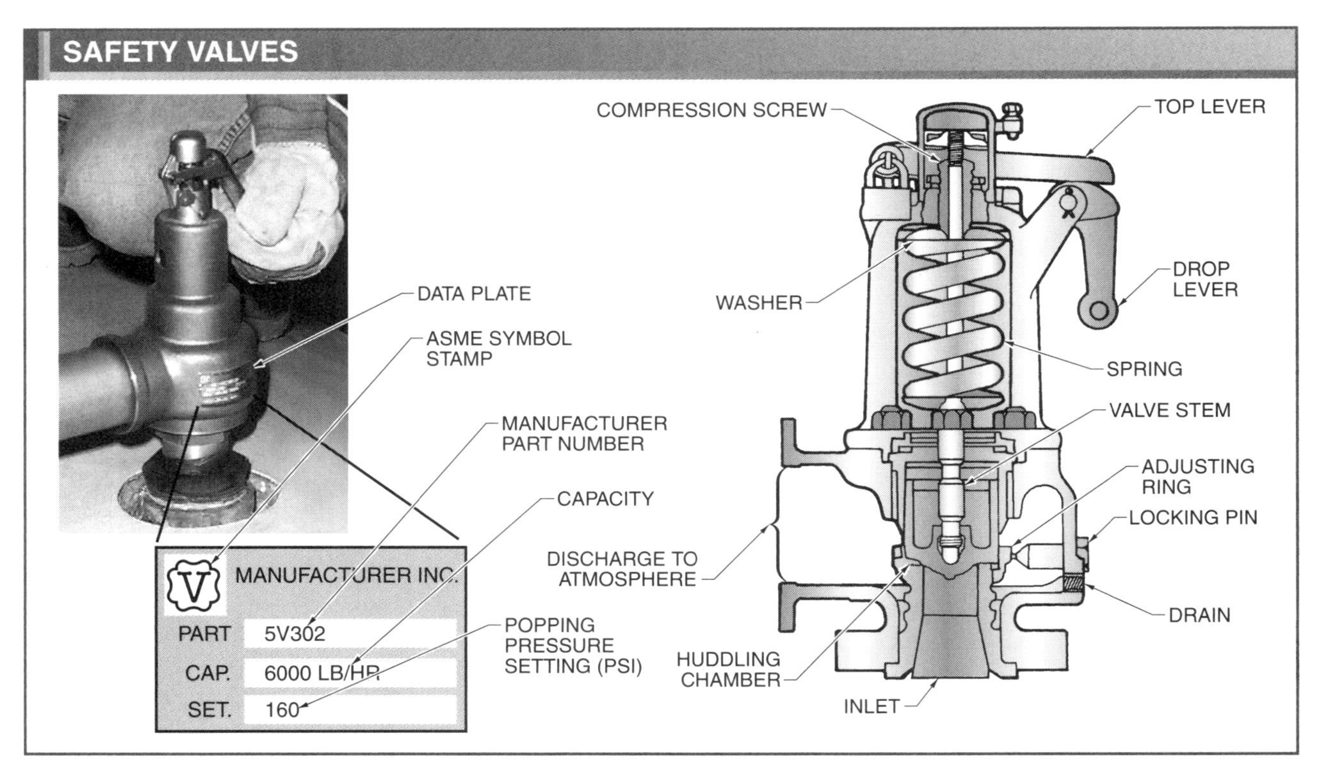

Figure 11-8. The capacity of a safety valve is the amount of steam in pounds per hour (lb/hr) that the safety valve is capable of venting at rated pressure.

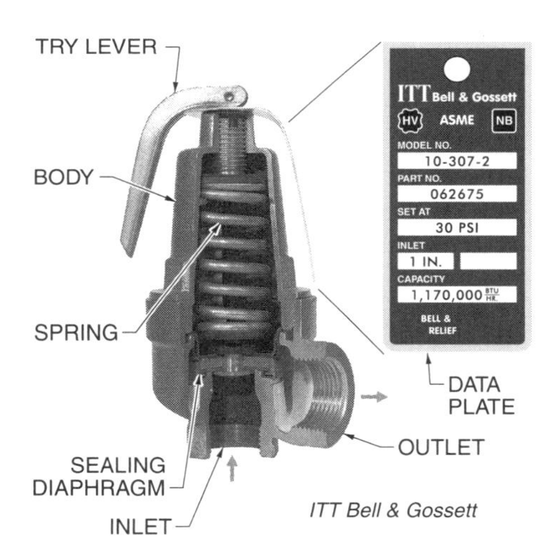

Figure 11-9. A safety relief valve is a valve on a hot water boiler that prevents excessive pressure from building up during a malfunction and is rated in Btu relieved per hour (Btu/hr).

Steam Pressure Gauges

A *steam pressure gauge* is a boiler fitting (gauge) that displays the amount of pressure in pounds per square inch (psi) inside a steam boiler. A steam pressure gauge allows the operator to monitor steam pressure changes in the boiler and maintain steam pressure within the safe operating range. A steam pressure gauge is connected to the highest part of the steam side of the boiler and must be positioned to allow easy viewing by the boiler operator. On many steam boilers a compound gauge is used. A *compound gauge* is a pressure gauge that indicates vacuum in inches of mercury (Hg) and pressure in pounds per square inch (psi) on the same gauge. **See Figure 11-10.**

Pressure-Temperature Gauges

A pressure-temperature gauge is used on hot water boilers to ensure safe and efficient boiler operation. A *pressure-temperature gauge* is a gauge that measures the pressure and temperature of the water at the point on the boiler where the gauge is located. A pressure-temperature gauge is installed at the top and front of the boiler so the gauge can be seen easily. **See Figure 11-11.**

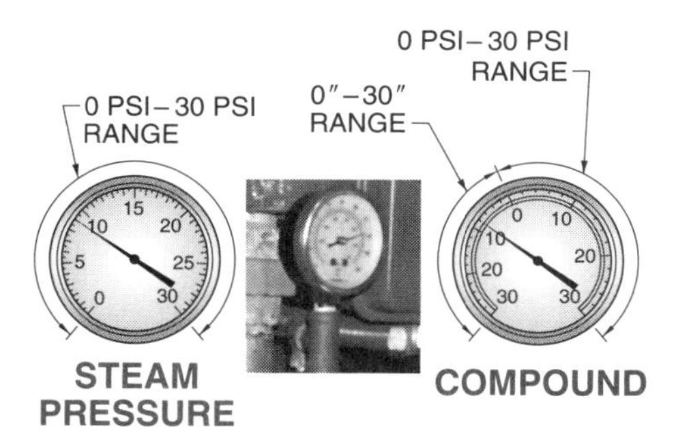

Figure 11-10. Pressure gauges indicate the amount of pressure (psi) in a boiler.

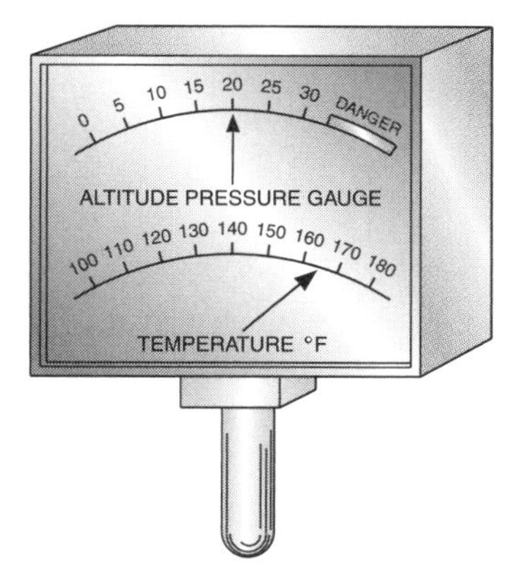

Figure 11-11. A pressure-temperature gauge measures the pressure and temperature of boiler water at the point on the boiler where the gauge is located.

Water Columns

A *water column* is a steam boiler fitting that reduces the movement of boiler water in the gauge glass to provide an accurate reading of the water level. A *gauge glass* is a tubular glass column that indicates the water level in the boiler. It is fitted with isolation valves to facilitate repair and replacement of the gauge glass. The ASME International *Boiler and Pressure Vessel Code* does not require a water column on all boilers. However, most steam boilers have a water column. Some small boilers that do not encounter rapid changes in the load may have the gauge glass mounted directly on the boiler shell.

All steam boilers must have two methods of determining the water level in the boiler. The gauge glass is the first and easiest method for determining boiler water level. A second method for determining boiler water level is try cocks. *Try cocks* are valves located on the water column that are used to determine the boiler water level if the gauge glass is not functional. **See Figure 11-12.**

To indicate proper boiler water level (approximately one-half of the gauge glass), steam and water must be discharged from the middle try cock only. Water discharged from the top try cock indicates a high water condition in the boiler. Steam discharged from the bottom try cock indicates a low water condition in the boiler. Both high water and low water conditions lead to serious problems in the boiler or steam piping system. Too high a water level in the boiler leads to water escaping from the boiler into the steam piping and causing water hammer. *Water hammer* is a banging noise caused by water in steam lines moving rapidly and hitting obstructions such as elbows and valves. Water hammer can lead to ruptures in the steam piping. Too low a water level in the boiler allows some of the heating surface of the boiler to be exposed. Without the water present to dissipate the heat, the heating surface can become overheated and become weak. The weakening of the heating surface leads to total failure of the area and boiler meltdown occurs.

Hydronic heating systems may need testing, such as performing a try lever test by opening the try lever for 5 sec to 10 sec and allowing the valve disc to snap closed.

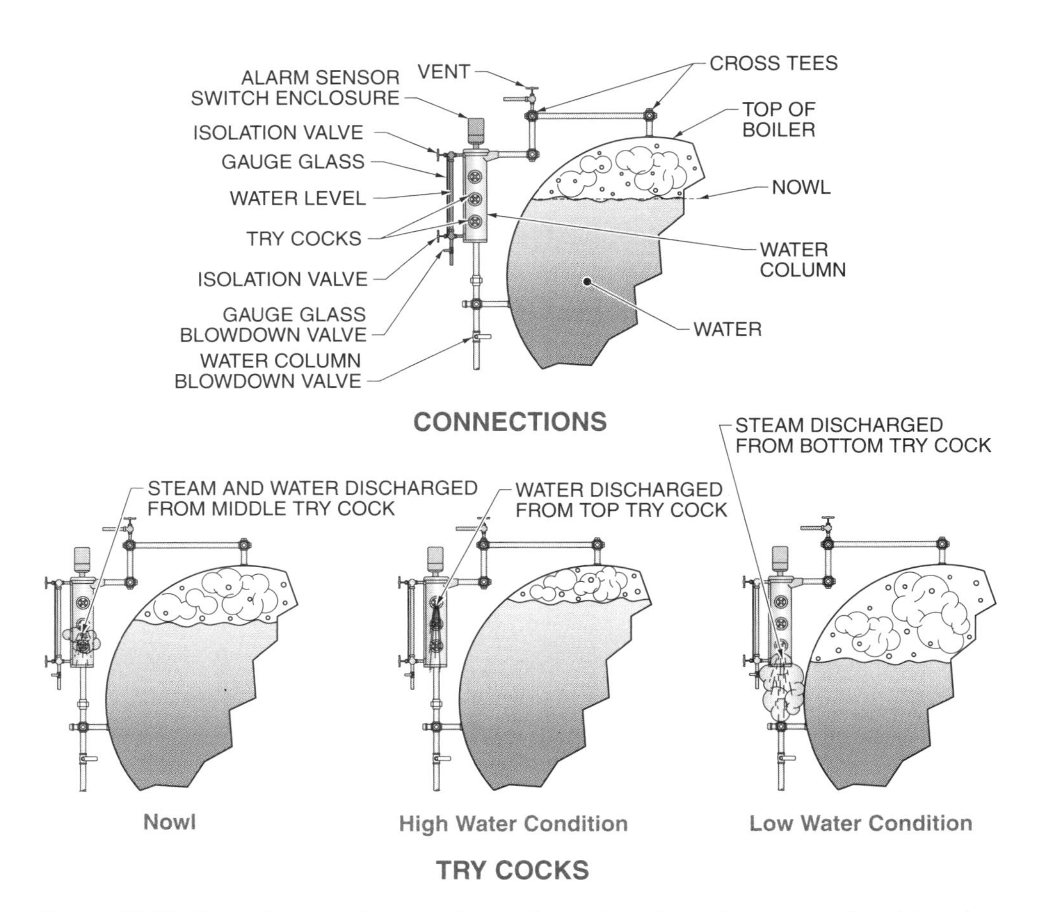

Figure 11-12. Try cocks are valves located on the water column that are used to determine the boiler water level if the gauge glass is not functional.

Bottom Blowdown Valves

A *bottom blowdown valve* is a valve located on the blowdown line located at the lowest point of the waterside of a boiler, used to release water from the bottom of the boiler. All boilers have bottom blowdown valves. **See Figure 11-13.** The size, capacity, and type of boiler determine the number and type of valves allowed by the ASME International code. Steam boilers normally have two valves on the bottom blowdown line of the boiler. One valve is a lever (quick-opening) valve and the other valve is a screw (slow-opening) valve. Some manufacturers place two slow-opening valves on the bottom blowdown line. The quick-opening valve is always located closest to the boiler shell. Bottom blowdown is performed for water level control, sludge and sediment removal, chemical concentration control, or dumping the boiler.

Surface Blowdown Valves

A *surface blowdown valve* is a valve located on the surface blowdown line used to release water and surface impurities from the steam boiler. **See Figure 11-14.** The surface blowdown line piping is located at the normal operating water level (NOWL) of the boiler. Surface impurities increase the surface tension by causing a film, which prevents steam bubbles from breaking through the surface of the water. Increased water surface tension can lead to foaming, causing a rapid fluctuation in the water level. Foaming causes a false water level reading.

Tech Facts

Steam boilers must have at least one safety valve for each 500 sq ft of heating surface according to the ASME code.

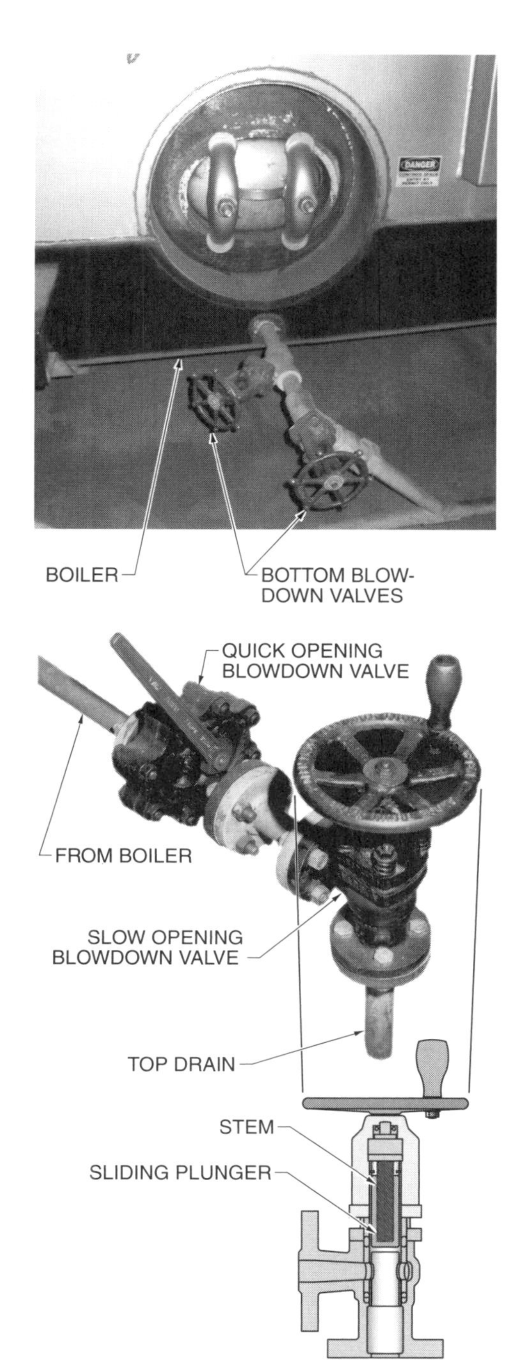

Figure 11-13. Bottom blowdown valves drain boiler water and sludge from the bottom of a boiler.

Fusible Plugs

A *fusible plug* is a warning device used to indicate extreme overheating of the boiler from a low water condition. A fusible plug is generally used on coal-fired steam boilers and serves as a last warning to the boiler operator of a low water condition. Overheating of the boiler can cause boiler tubes to melt. The fusible plug is a brass or bronze plug with a tapered hollow core filled with a tin alloy. The core melts at approximately 450°F. If the water is not covering the fusible plug to dissipate the heat, the core will melt from the high temperature of the gasses of combustion. Once the core melts, the steam blows through the tapered hole, making a whistling noise and warning the operator of the low water condition. **See Figure 11-15.**

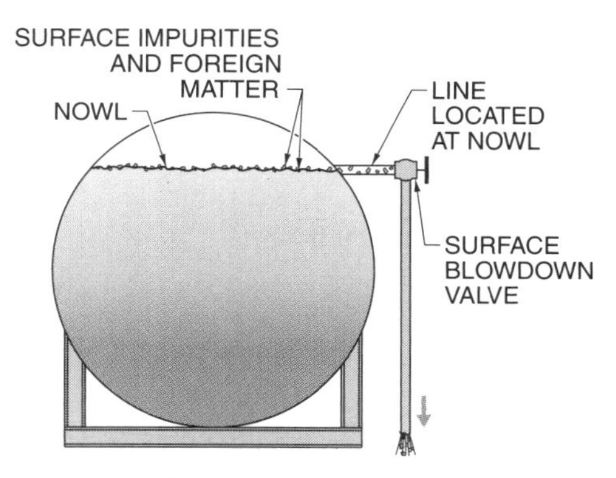

Figure 11-14. The surface blowdown valve controls the amount of water and surface impurities released from the boiler at the NOWL.

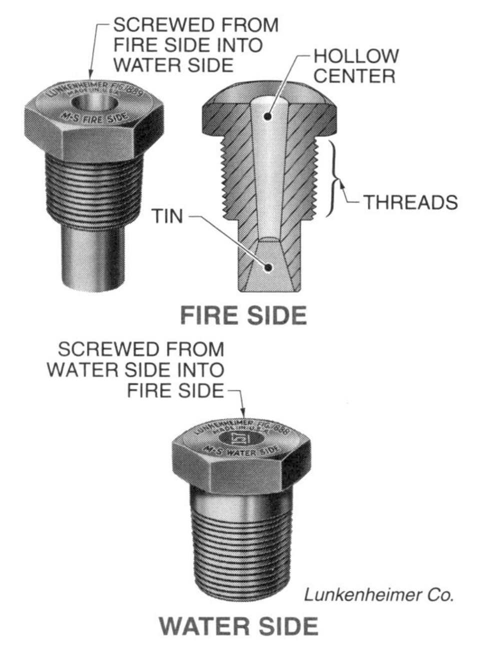

Figure 11-15. During an extreme condition, the tin in a fusible plug melts, allowing steam to exit and alert the boiler operator with a whistle.

Vent Valves

A *vent valve (boiler vent or air cock)* is a boiler fitting used to vent air from inside the boiler. The vent valve is opened when filling the boiler with water to prevent air from being trapped inside the boiler. The vent valve may be installed on the boiler vent line (½″ or ¾″ piping) that is connected to the highest part of the boiler. Not all boilers are equipped with a vent valve.

When the water is drained from a steam boiler for servicing, the vent valve must be opened to prevent a vacuum from forming. The vent valve is also used to prevent a dangerous vacuum from forming inside a steam boiler when it is taken out of service.

Pressure Controls

A *pressure control* is a switch on a steam boiler that starts the burner, controls the firing rate, or shuts down the burner based on the steam pressure in the boiler. A pressure control regulates the amount and frequency of firing in a burner. Like a steam pressure gauge, the pressure control must be protected from live steam with a siphon. The pressure control must be installed in a true vertical position to function properly. If the pressure control is mounted incorrectly, expansion and contraction of the siphon may cause the pressure control to move, resulting in improper pressure control of the boiler. **See Figure 11-16.**

There are normally three pressure controls used to control the pressure in a steam boiler, an operating pressure control, a modulating pressure control, and a high-pressure limit control. The operating control turns the burner on at a specified steam pressure and off at the boiler's desired operating pressure. Most of the higher-pressure steam boilers utilize a modulating pressure control that controls the firing rate of the burner, normally between 30% and 100%, depending on the steam load. The high-pressure limit control is required on a steam boiler as a safety device in the event of a failure of the operating pressure control. The high-pressure limit control is a manually reset type of control.

Aquastats™

An *Aquastat*™ is a device that controls the starting, stopping, and firing rate of the burner by sensing the temperature of the water in a hot water boiler. An Aquastat™ consists of a remote bulb sensor that senses changes in boiler water temperature and acts like a switch. The electric circuit is energized or de-energized to turn the burner ON or OFF depending on the temperature setpoint. Most hot water boilers are equipped with a second Aquastat™ set at a higher temperature limit to provide a safety backup to the primary Aquastat™, which is normally a manual reset type.

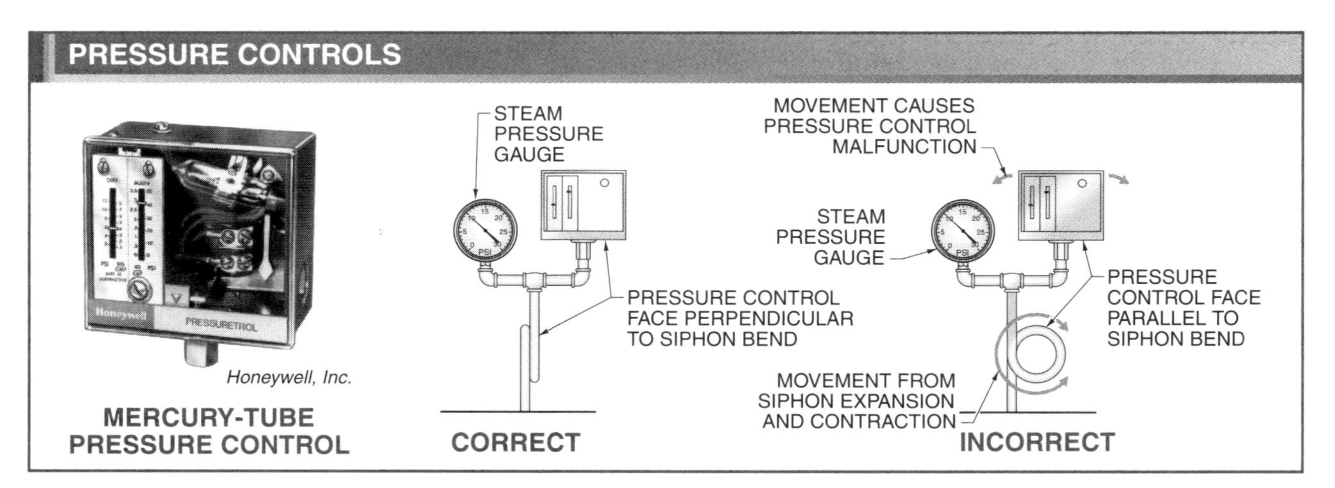

Figure 11-16. Pressure controls are switches on steam boilers that start the burner, control the firing rate, or shut down the burner based on the steam pressure in the boiler.

On large hot water boilers there may be a modulating Aquastat™ to control the boiler's firing rate to limit the number of times the boiler cycles ON and OFF. **See Figure 11-17.**

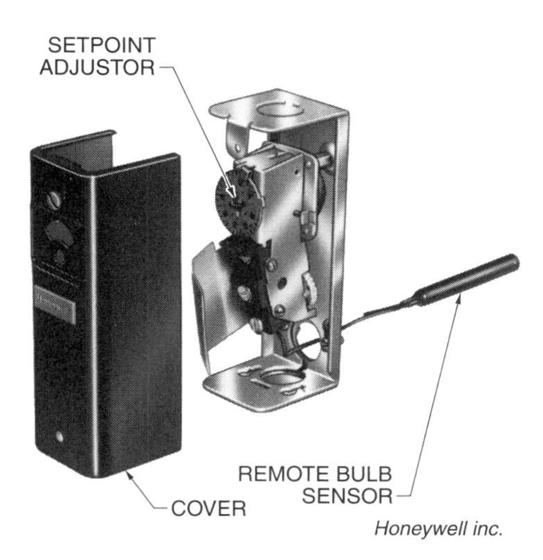

Figure 11-17. An Aquastat™ senses the temperature of boiler water and cycles the burners as necessary to maintain the boiler water temperature at the setpoint.

STEAM BOILER HEATING SYSTEM ACCESSORIES

The feedwater, steam, fuel, and draft systems of a steam boiler require many accessories that must be controlled for maximum safety and efficiency. The complexity of these systems varies depending on the size and type of boiler.

Feedwater Systems

A *feedwater system* is a system that supplies the proper amount of water to the boiler. The feedwater system requires accessories to supply the correct amount of water in the proper condition to the boiler. A feedwater system accessory is equipment that is not directly attached to the boiler and that controls the quantity, pressure, and/or temperature of water supplied to the boiler. Maintaining the correct water level in the boiler is critical for safety and efficiency. Feedwater is treated and regulated automatically to meet the demand for steam. Valves are installed in feedwater lines to permit access for maintenance and repair. The feedwater system and accessories must be capable of supplying water to the boiler in all circumstances. **See Figure 11-18.**

A steam heating system uses heat from the steam boiler to provide comfort in the building spaces. Water in the boiler is heated and turned to steam. Steam leaves the boiler through the main steam line (boiler outlet) where it enters the main steam header. From the main steam header, branch lines direct the steam to the heating units. Heat is released to the building spaces as steam travels through the heating units. Steam in the heating units cools and turns into condensate. The condensate is separated from the steam by steam traps that allow condensate, but not steam, to pass. The condensate is directed through the condensate return line to the condensate return tank. The feedwater pump pumps the condensate and/or makeup water back to the boiler through check valves and stop valves on the feedwater line. The feedwater enters the boiler and is turned to steam to repeat the process.

Feedwater Valves. A *feedwater valve* is a valve that controls the flow of makeup water into a boiler to compensate for losses. The makeup water piping has a check valve and stop valve. A *check valve* is a valve that allows flow in only one direction. Check valves control the direction of water flow. Many different kinds of check valves are available. The ball, poppet, or disk in a check valve opens when fluid pushes against it in one direction but closes when the direction of flow reverses. **See Figure 11-19.**

Gate valves and globe valves are used as stop valves. A *gate valve* is a valve that has an internal gate that slides over the opening through which water flows. A gate valve is a two-position valve that is fully open or fully closed and is not designed to throttle or regulate the flow of water. When the gate is open, the valve allows full flow. When the gate is closed, no water flows. Gate valves have little resistance to water flow when open and are sealed tight when closed. **See Figure 11-20.**

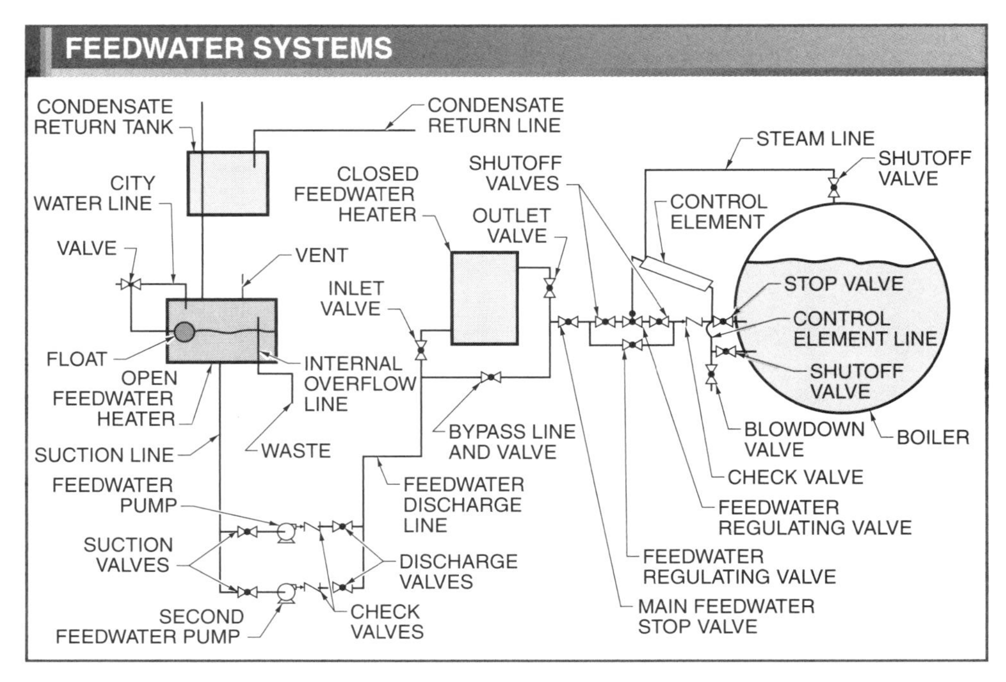

Figure 11-18. The feedwater system must supply sufficient water at the proper temperature and pressure to maintain a normal operating water level (NOWL).

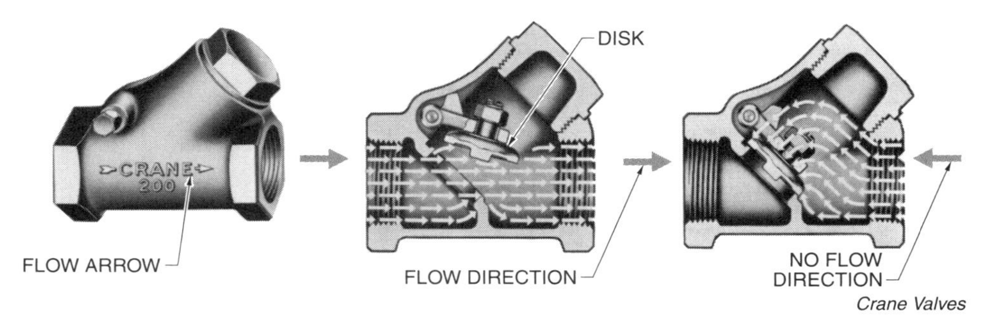

Figure 11-19. Check valves allow boiler water to flow in only one direction.

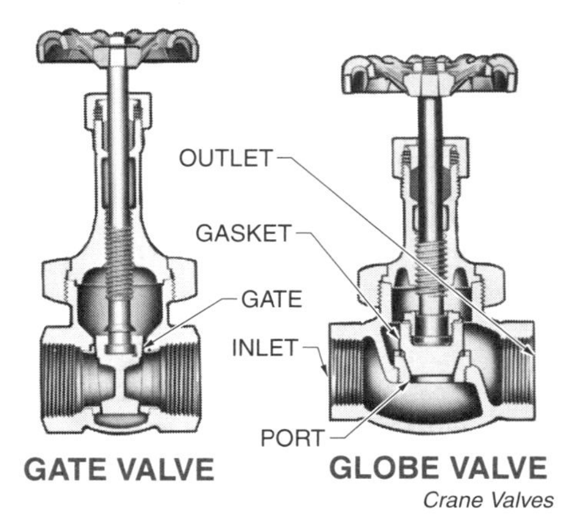

Figure 11-20. Gate valves are used to completely stop boiler water flow. Globe valves regulate boiler water flow but are sometimes used to stop boiler water flow.

A *globe valve* is a valve that has a disk (globe) that rises or lowers over a seat through which water flows. The seat is at right angles to the flow of water. The water must make two 90° turns to exit the valve. A globe valve is an infinite position valve that is used to regulate water flow. The resistance and pressure drop through the valve are controlled by the amount the valve is opened.

A *stop valve* is a valve that stops water flow. Stop valves are located anywhere in a steam heating system where water may have to be completely shut OFF. Gate valves are often used as stop valves because gate valves are either open or closed. Globe valves may also be used as stop valves.

Condensate Return Tanks. A *condensate return tank* is a steam boiler accessory that collects condensate returned from heating units. Condensate that is not contaminated is pumped back to heat feedwater or is used by the boiler. **See Figure 11-21.** The condensate return tank and pump are installed at the lowest point in the building where steam is used. Condensate is directed to the condensate return tank through condensate return lines. The condensate return tank is vented to atmosphere to prevent pressure from building up in the condensate return system. The condensate return tank is equipped with an automatic makeup water feeder to maintain a minimum water level in the tank. The feedwater (a combination of condensate and makeup water) is then pumped to the boiler on a demand for more water by the boiler.

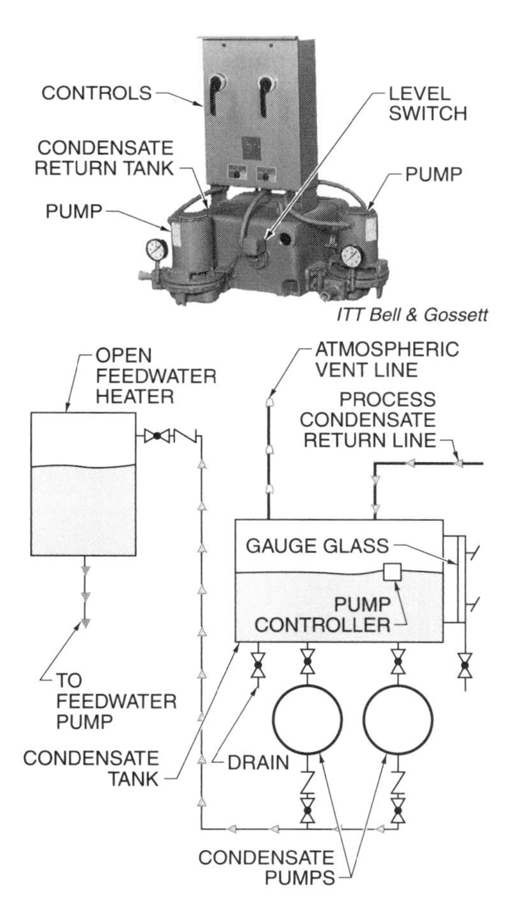

Figure 11-21. The condensate return tank collects condensate from the terminal heating units to be returned back to the boiler.

Feedwater Pumps. A *feedwater pump* is a steam boiler accessory used to pump feedwater into the boiler. Water flow produced by the feedwater pump must overcome boiler water pressure and system resistance to maintain the normal operating water level (NOWL) in the boiler. Feedwater pumps vary in size depending on system requirements. Feedwater pumps must have adequate capacity to maintain the required boiler water level under all operating conditions.

The most common feedwater pump is the centrifugal feedwater pump driven by an electric motor. A *centrifugal feedwater pump* is a feedwater pump that converts centrifugal force to flow for transporting feedwater. *Centrifugal force* is the force that pulls a body outward when it is spinning around a center. The feedwater is thrown out of the impeller against the pump housing. The pump housing directs feedwater to piping connected to the boiler. The discharge flow rate of a centrifugal feedwater pump is dependent on impeller size and speed. The pressure rating of a centrifugal feedwater pump is dependent on the number of stages. **See Figure 11-22.**

Feedwater Heaters. A *feedwater heater* is a steam boiler accessory that heats feedwater before the feedwater enters the boiler. Feedwater heater systems are required on some boilers to collect and remove oxygen and other gasses that may cause corrosion. Maintaining the proper feedwater temperature also reduces the amount of scale-causing material that settles on the heating surfaces of the boiler drum.

Feedwater Regulators. A *feedwater regulator* is a steam boiler accessory that maintains the NOWL in the boiler. The most common way to accomplish this on steam boilers used for heating purposes is by controlling the feedwater pump using a float feedwater regulator. **See Figure 11-23.** A bowl containing a float is mounted at the NOWL. As the water level drops in the boiler, the float in the regulator drops, activating a switch that turns the feedwater pump ON. When the water level is restored to its normal level, the float deactivates the switch, turning the feedwater pump OFF.

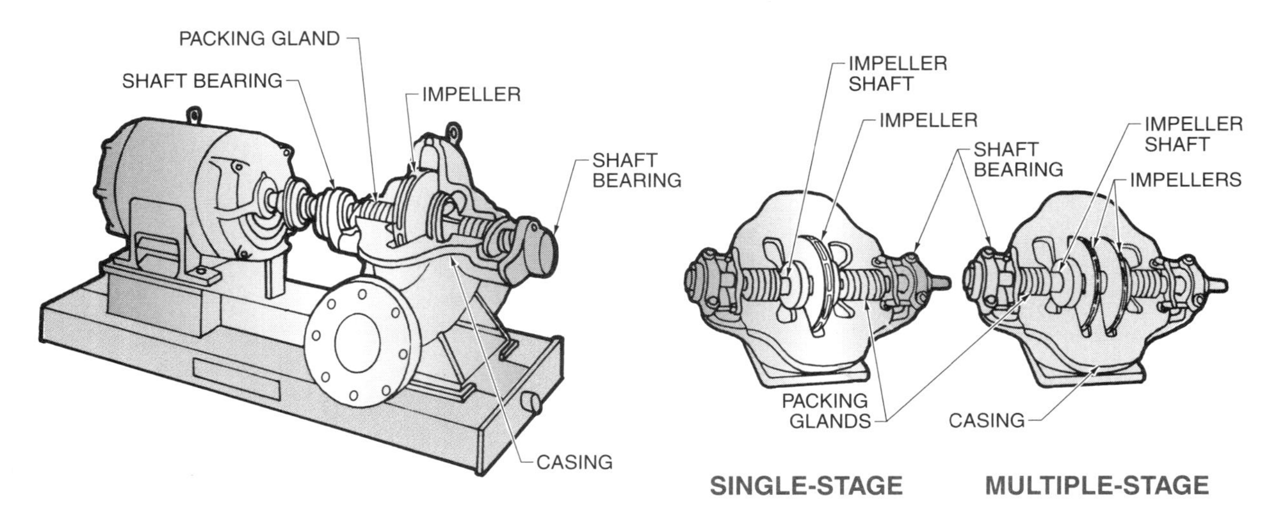

Figure 11-22. The discharge flow rate of a centrifugal feedwater pump is dependent on impeller size and speed, with the pressure rating of the pump being dependent on the number of stages.

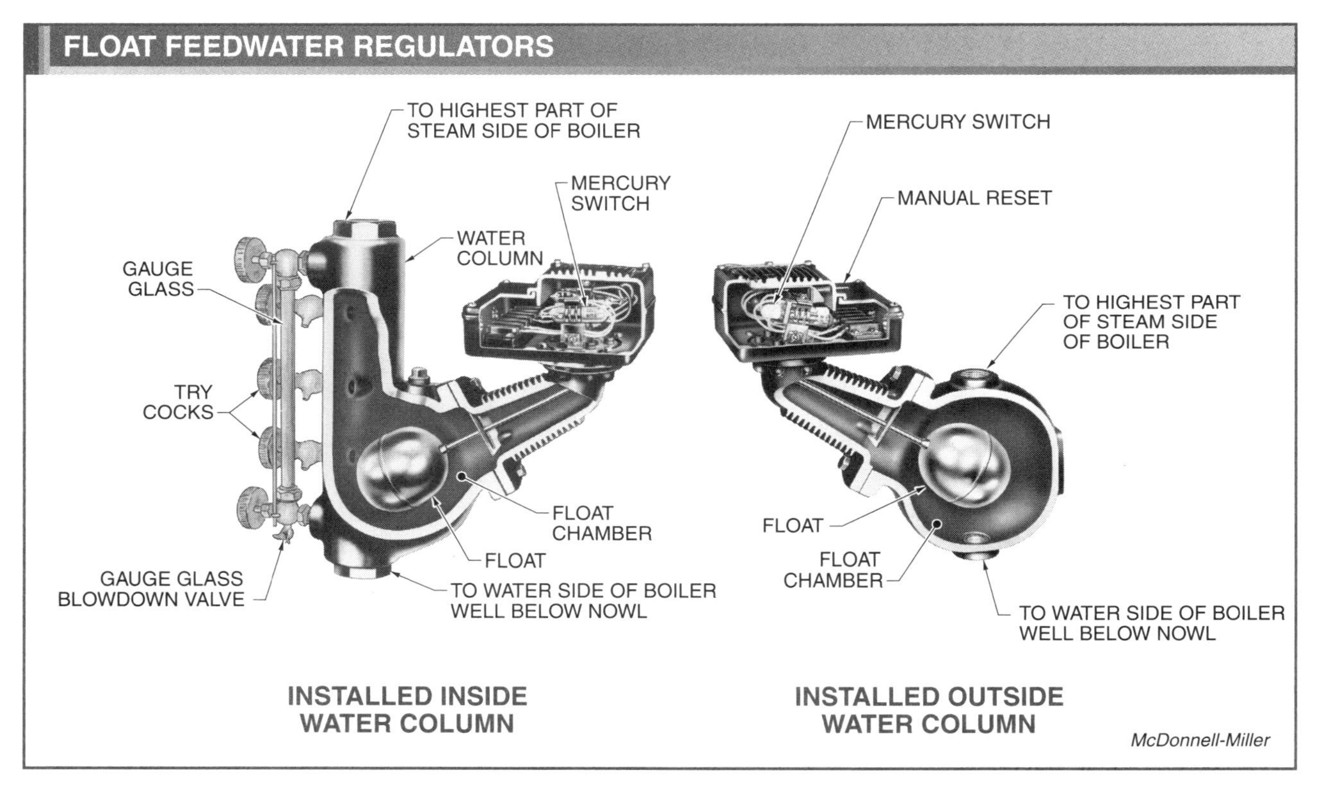

Figure 11-23. Float feedwater regulators can be installed attached to or separated from the water column, depending on their design.

Low Water Fuel Cutoffs. A *low water fuel cutoff* is a steam boiler fitting that shuts the burner OFF in the event of a low water condition in the boiler. An automatic makeup feedwater system can malfunction or there could be an interruption in the water supply of the facility. A loss of water can lead to the burning out of tubes and/or a boiler explosion. Depending on the application, the low water fuel cutoff is normally integrated into the water column on the boiler. **See Figure 11-24.**

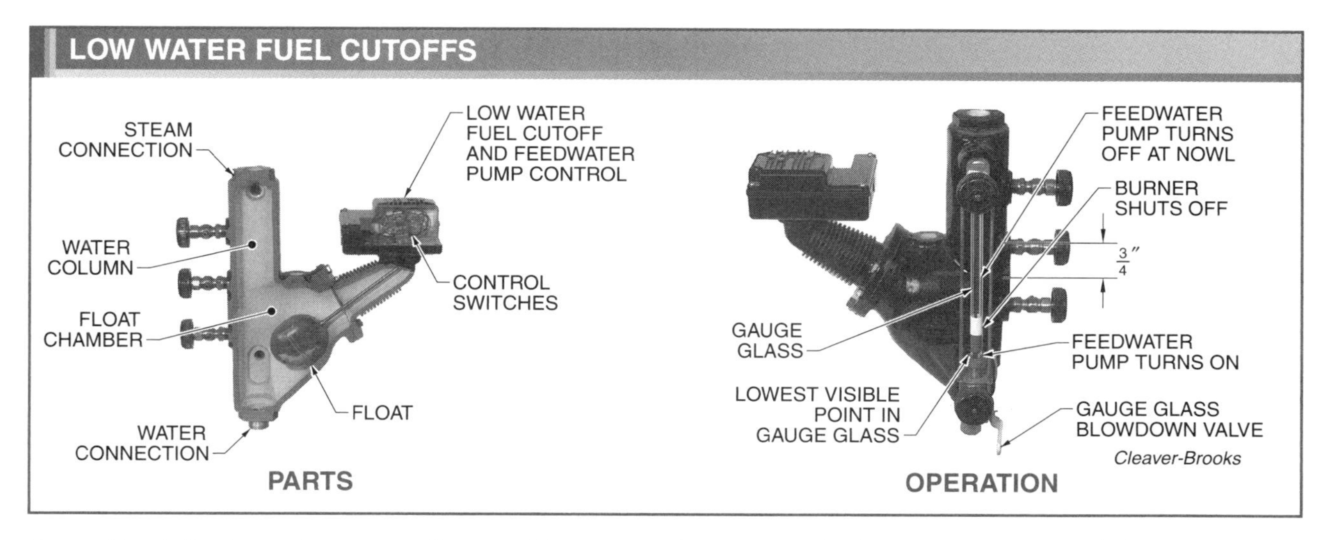

Figure 11-24. The low water fuel cutoff shuts off the burner when water in the boiler drops to an unsafe level.

The low water fuel cutoff is located slightly below the NOWL. The top line connects to the highest part of the steam side of the boiler. The bottom line connects to the water side well below the NOWL. A blowdown line and valve are used to keep the float chamber free of sludge and sediment. The primary function of a low water fuel cutoff is to de-energize the burner circuit and shut down the burner if the water level in the boiler drops below a safe operating level.

Steam Systems

A *steam system* is a system that collects, controls, and distributes the steam produced in the boiler. Steam system accessories are used to control steam and commonly include steam valves, steam headers, heating units, steam traps, and steam strainers.

Steam Valves. Steam generated in the boiler is piped to various locations to be used for heating or other applications. Different valves are used to regulate the flow of steam. Steam from the boiler is directed by the steam header and controlled by the main steam stop valve. A *main steam stop valve* is a valve used to place the boiler on line or take the boiler off line. It is usually an os&y valve. **See Figure 11-25.** An *os&y valve* is a gate valve that indicates whether it is open or closed by the position of the stem. An os&y valve allows steam to flow when the gate is lifted to the open position and the stem is up. It stops steam flow when the gate is lowered to close the valve and the stem is in. An os&y valve is always used either in the full open position or full closed position to prevent damage to the valve. When wide open, steam flows unrestricted through the valve.

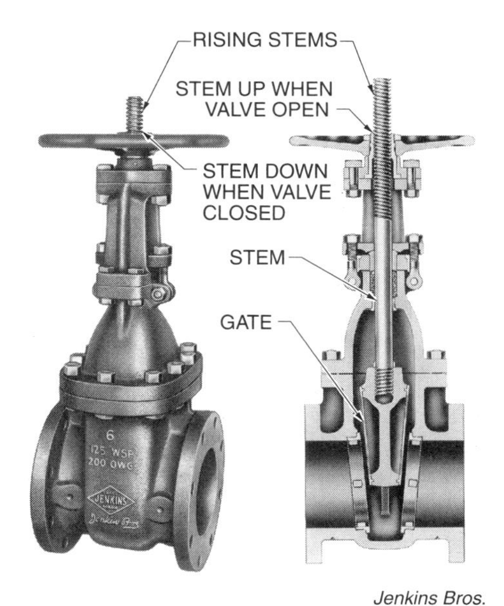

Figure 11-25. An os&y valve is used as the main steam stop valve for a boiler and does not restrict the flow of steam when open.

Steam Headers. A *steam header* is a distribution pipe that supplies steam to branch lines. Steam headers are constructed of pipe that is strong enough to withstand the pressure and velocity of steam that passes from the boiler to the branch lines. Steam headers and branch lines must be supported to prevent damage from strain. In addition, the heating and cooling of steam lines causes expansion and contraction during normal usage. Different methods are used to compensate for the expansion and contraction that takes place. **See Figure 11-26.**

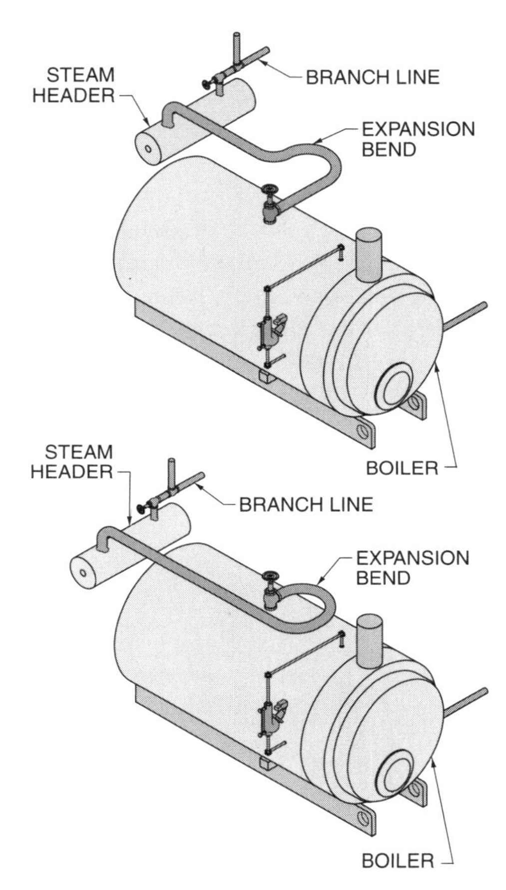

Figure 11-26. Expansion bends in steam lines allow movement created by expansion and contraction without causing damage to the steam header or branch lines.

Heating Units. A *heating unit* is a heat exchanger such as a radiator or coil in which heat that was transported by steam is transferred to the air in a building space. Heating units are located in building spaces for maximum efficiency and are designed to maximize heat transfer from the steam to the building spaces. When the heating unit releases the heat, a steam trap removes air and condensate from the remaining steam.

Steam Traps. A *steam trap* is a steam boiler accessory that removes air and condensate from steam lines and heating units. Air must be removed from the steam lines to maintain the supply of steam for maximum efficiency. Steam is permitted to bypass a steam trap but condensate is diverted to the condensate tank. **See Figure 11-27.** Steam traps operate automatically and increase the overall steam system efficiency by removing air and condensate without the loss of steam. Steam temperature drops slightly as it travels from the boiler through the steam header and from the steam header to the heating units. Condensate that is formed must be removed.

Steam Strainers. A *steam strainer* is a steam system accessory that is located in steam lines before the steam trap to remove foreign matter from the steam that could cause a steam trap malfunction. Steam trap function is greatly affected by steam strainer function. Dirt, rust, scale, and/or other particles in steam lines can lead to steam trap failure. Small particles of foreign matter clog orifices in the steam trap and prevents the proper discharge of condensate. **See Figure 11-28.**

Fuel Systems

The *fuel system* is a system that provides fuel for combustion to produce the necessary heat in the boiler. Steam boiler fuel system accessories are used to store, supply, and control the fuel supplied to the boiler. In the combustion process, fuel mixed with air is burned to produce the heat necessary to operate the boiler. The heat from the gases of combustion is transferred throughout the boiler for maximum efficiency. **See Figure 11-29.** Fuels normally used in boilers include fuel oil, natural gas, and coal. In some cases electricity is used to produce steam in a steam heating system. The type of fuel used is determined by the design of the boiler, the price and availability of the fuel, and compliance with air pollution regulations. The type of fuel used determines the fuel system accessories.

Fuel oil systems include all the devices required to safely and efficiently operate the fuel oil burner. Fuel oil accessories clean, control the temperature, and regulate the pressure of fuel oil. **See Figure 11-30.**

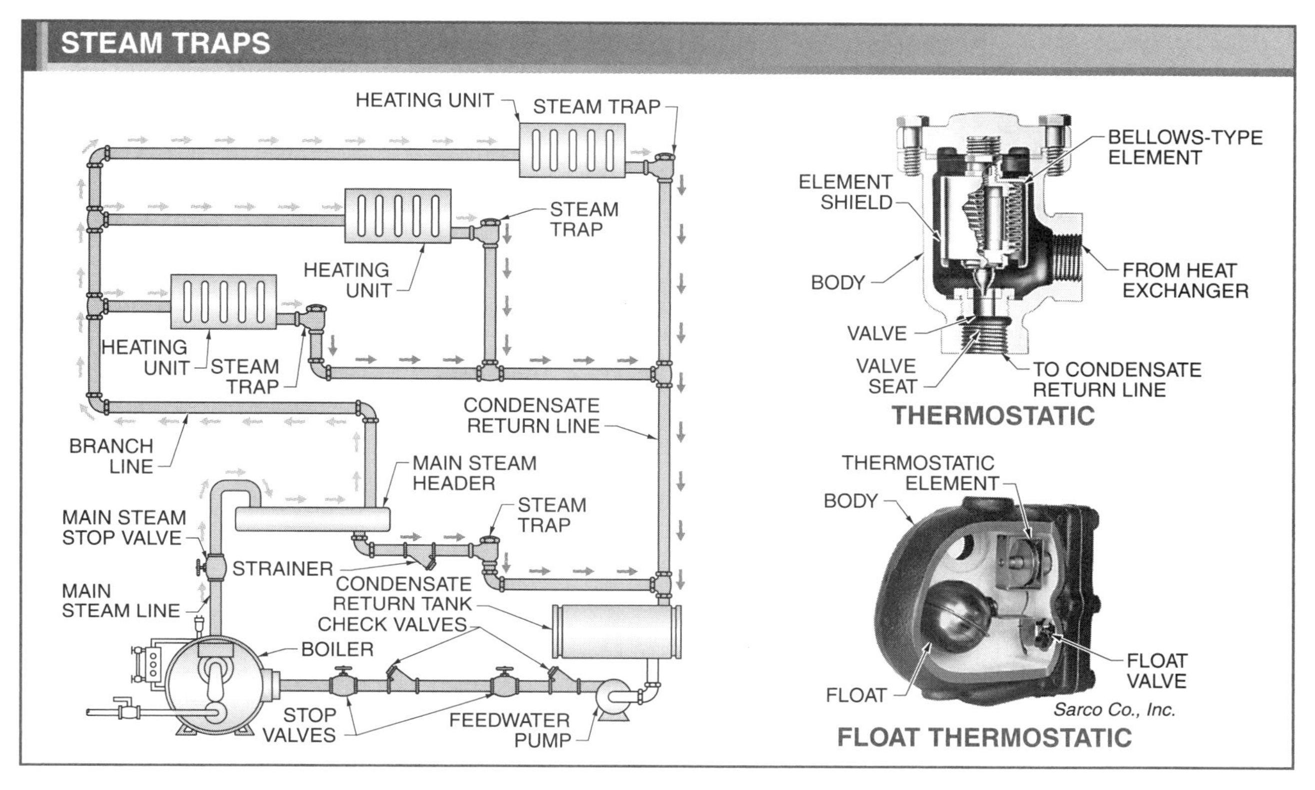

Figure 11-27. Steam traps operate automatically and are located throughout the steam system after each device where steam is used or heat is released.

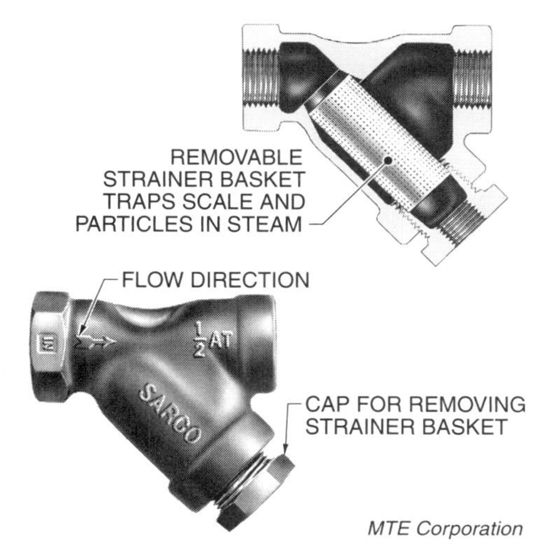

Figure 11-28. Steam strainers located before steam traps must be routinely cleaned to ensure efficient removal of foreign matter from steam.

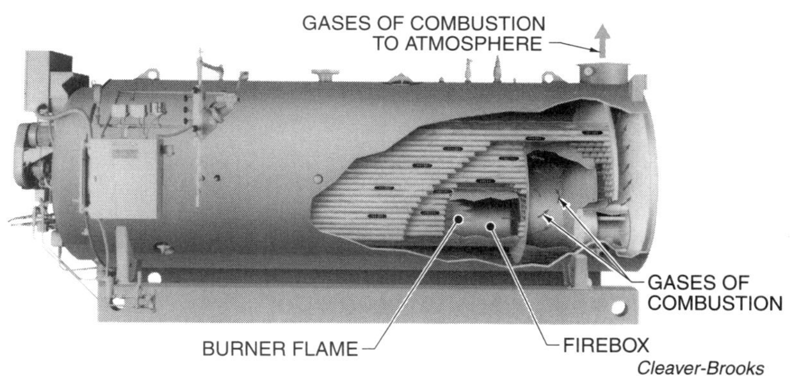

Figure 11-29. The fuel system prepares fuel for combustion for maximum heat transfer efficiency.

Tech Facts

Steam traps are used to remove condensate from steam lines and heating units. In addition, steam traps remove air from the steam lines.

Fuel oil purity varies, and equipment to remove foreign matter is required to prevent buildup in the lines that restrict flow to the burner, causing inefficient combustion. Fuel oil accessories for a heavy fuel oil system require additional equipment for heating of the fuel oil for proper burner operation. Fuel oil accessories include storage tanks, pumps, strainers, and heaters.

There are two systems in use when natural gas is used as a fuel for steam boilers, a low-pressure gas system and a high-pressure gas system. A *low-pressure gas system* is a system in which gas is mixed with air at approximately atmospheric pressure. **See Figure 11-31.** A *high-pressure gas system* is a system in which gas is mixed at a pressure of up to

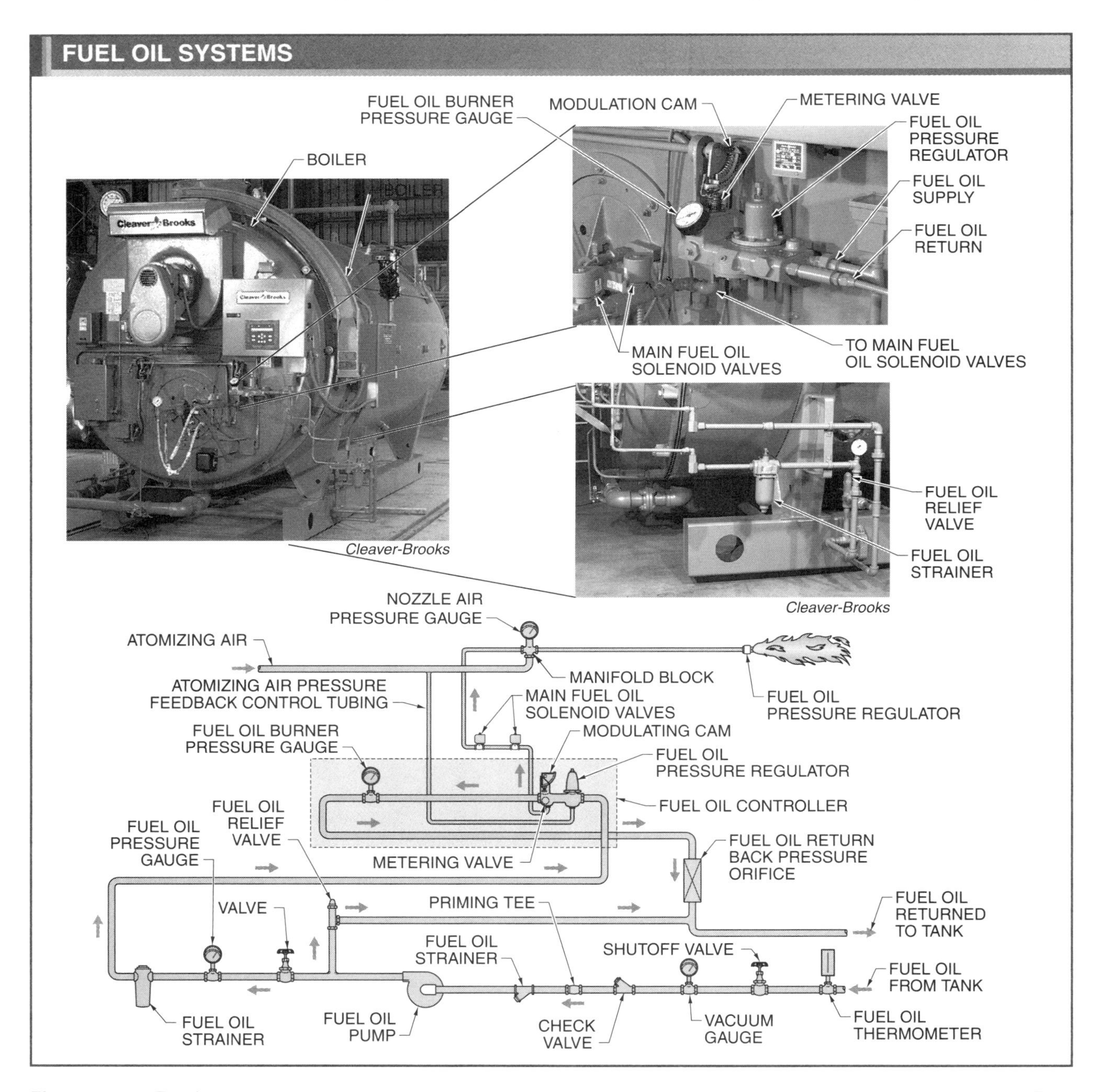

Figure 11-30. The fuel oil system includes all the devices required to safely and efficiently operate the fuel oil burner. Fuel oil accessories clean, control the temperature, and regulate the pressure of fuel oil.

5 psi with air. The main components of a gas fuel system include shutoff cocks, gas pressure regulators, main gas valves, butterfly valves, low and high gas pressure switches, and vent valves.

Tech Facts

Combination burners reduce fuel costs by switching to a secondary fuel to obtain off-peak rates.

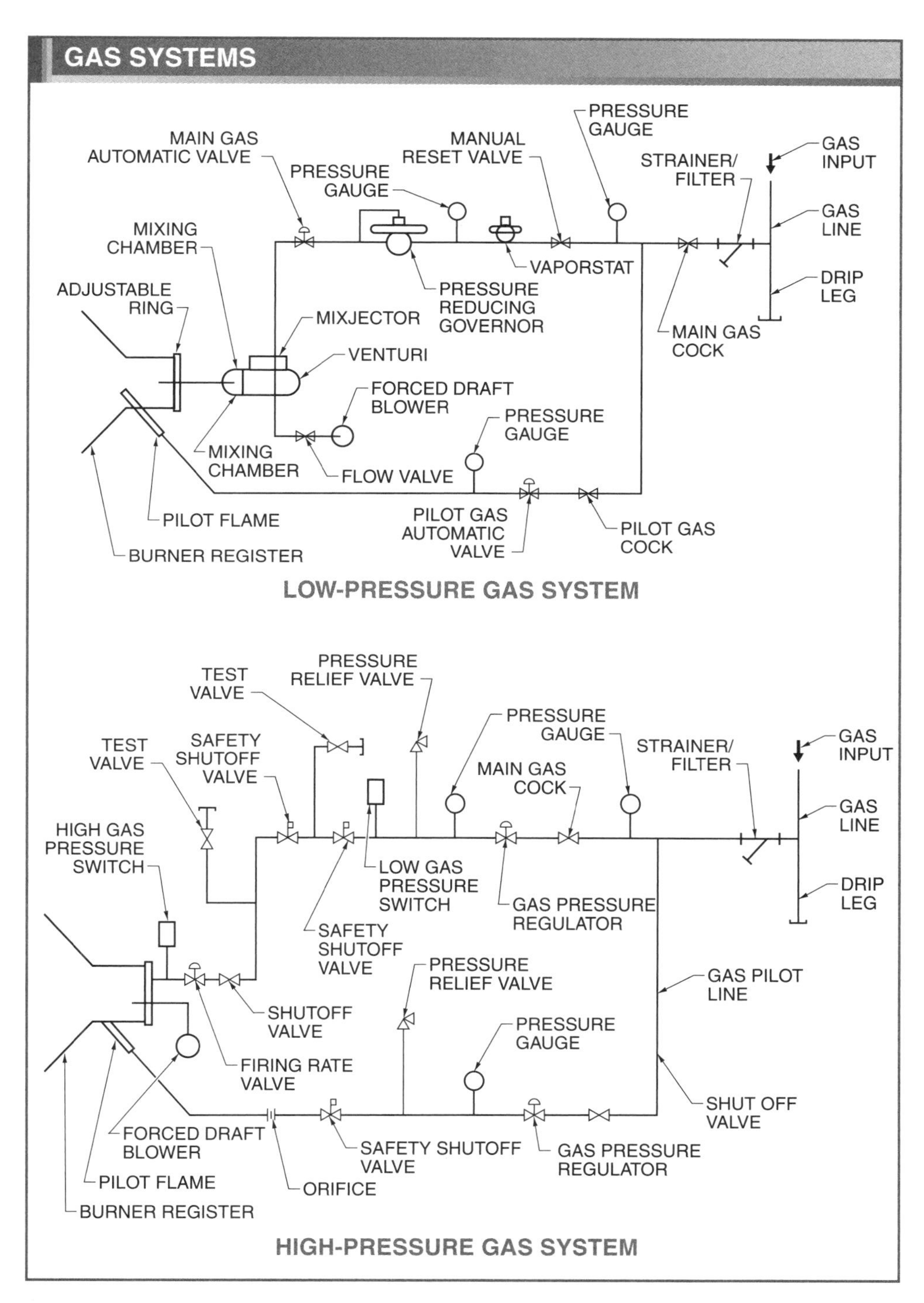

Figure 11-31. In a low-pressure gas system, the rate of combustion is controlled by the amount of air supplied to the venturi. In a high-pressure gas system, the rate of combustion is controlled by the gas pressure regulator, the safety shutoff valves, and the firing rate valve.

Pulverized coal used for firing steam boilers is fed into the boiler by stokers or by hand firing. *Pulverized coal* is coal that is ground to a fine powder of the consistency of talcum powder and then blown into the boiler where it is burned in suspension similar to natural gas or fuel oil. A *stoker* is a mechanical device for feeding coal to the burner at a constant rate. Efficient coal stokers have replaced hand firing in all commercial applications. Coal was commonly used to fire steam heating boilers, but fuel oil and natural gas have become more popular due to price, availability, and pollution standards. In the very large boiler plants used in the production of electricity, coal is used most extensively.

Fuel Oil Storage Tanks. In a fuel oil system, fuel oil is stored in a fuel oil tank. The location and installation of the fuel oil tank must comply with Environmental Protection Agency (EPA) regulations. Some of the fuel oil that is pumped from the tank goes to the burner, where it is burned. The remaining fuel oil is directed through the fuel oil pressure regulator and returned back to the tank.

A flame scanner is a safety device that shuts down the burner any time there is a flame failure.

Fuel Oil Pumps. Fuel oil pumps may be direct drive or belt drive. **See Figure 11-32.** Fuel oil pumps are commonly internal gear pumps. An *internal gear pump* is a positive-displacement pump that consists of a small drive gear mounted inside a large internal ring gear. The gears rotate in the same direction. A crescent seal separates the low- and high-pressure areas of the pump. Fuel oil is trapped between the teeth of the gears as the gears rotate and is discharged from the pump as the gears remesh at the discharge side.

Fuel Oil Strainers. A *fuel oil strainer* removes foreign matter from a fuel oil system. Fuel oil strainers use a fine mesh screen to trap foreign matter. On heavy fuel oil systems, strainers are located on the discharge line of the fuel oil heater and on the suction line before the fuel oil pump. On a light fuel oil system, a strainer is normally located on the discharge line of the pump. The type of fuel oil used determines how coarse the mesh is for the screen, and whether it is a single-basket strainer or a duplex strainer. **See Figure 11-33.**

Fuel Oil Heaters. A *fuel oil heater* is a boiler component used to heat cold heavy fuel oil to assist in pumping and allow for efficient burning. Both steam and electric fuel oil heaters are used. A steam boiler normally uses a steam fuel oil heater. A steam fuel oil heater consists of a tube-and-shell heat exchanger with the steam being on the inside of the tubes. An electric fuel oil heater is used on small boilers and hot water boilers, and as a backup on large steam boilers that have a steam fuel oil heater.

Fuel Oil Burners. A *fuel oil burner* is a boiler component that provides atomized fuel oil to the boiler. Fuel oil burners atomize fuel oil to increase mixability with combustion air for maximum combustion efficiency. Fuel oil burners vary depending on the methods used to atomize the fuel oil and the type of fuel oil used. Fuel oil burners commonly used include air atomizing, steam atomizing, pressure atomizing, and rotary cup burners. **See Figure 11-34.**

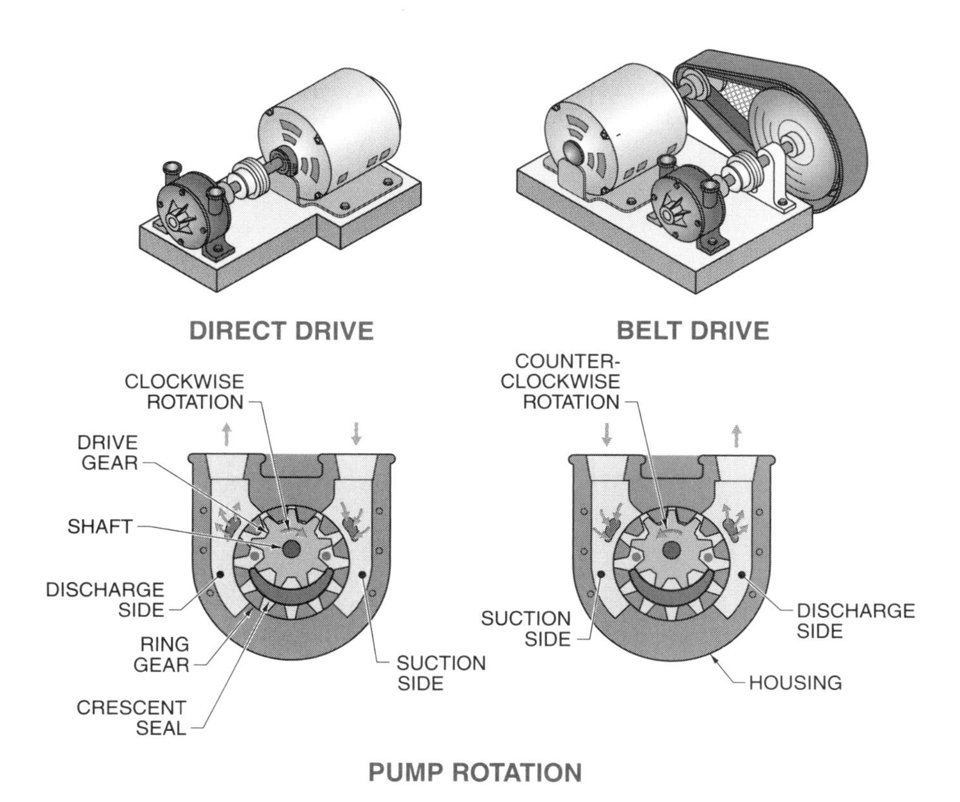

Figure 11-32. Fuel oil pumps provide the flow required in the fuel oil system for the transport of fuel from the fuel oil tank to the burner.

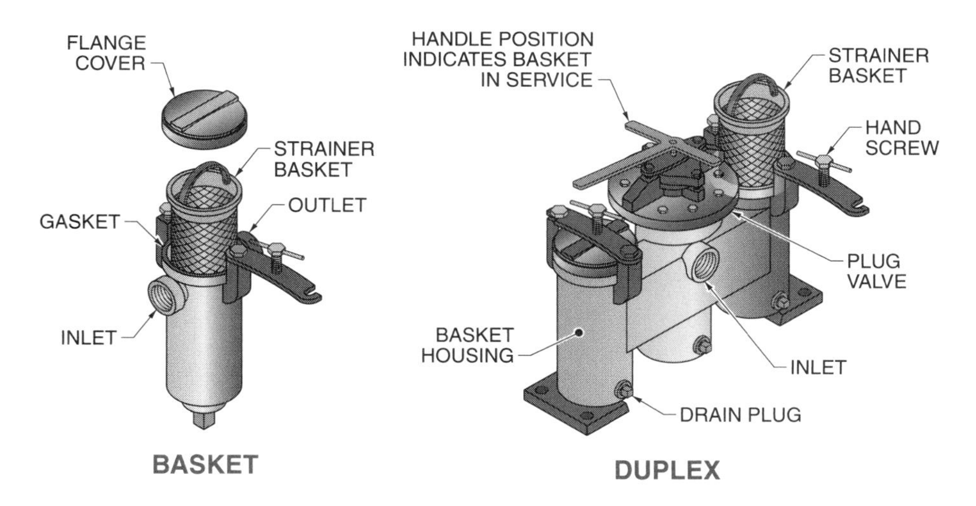

Figure 11-33. Fuel oil strainers are located between the fuel oil tank and the fuel oil pump to remove foreign matter and protect the pump and other fuel oil system devices.

Gas Shutoff Cocks. A *gas shutoff cock* is a special valve used on natural gas fuel systems to manually isolate the fuel system or parts of the system. Gas cocks are normally located before the gas fuel train, after the main gas valve, and on the pilot light gas line. These valves are normally a quick opening, quick closing valve that utilizes special packing and grease to prevent leaks around the valve stem.

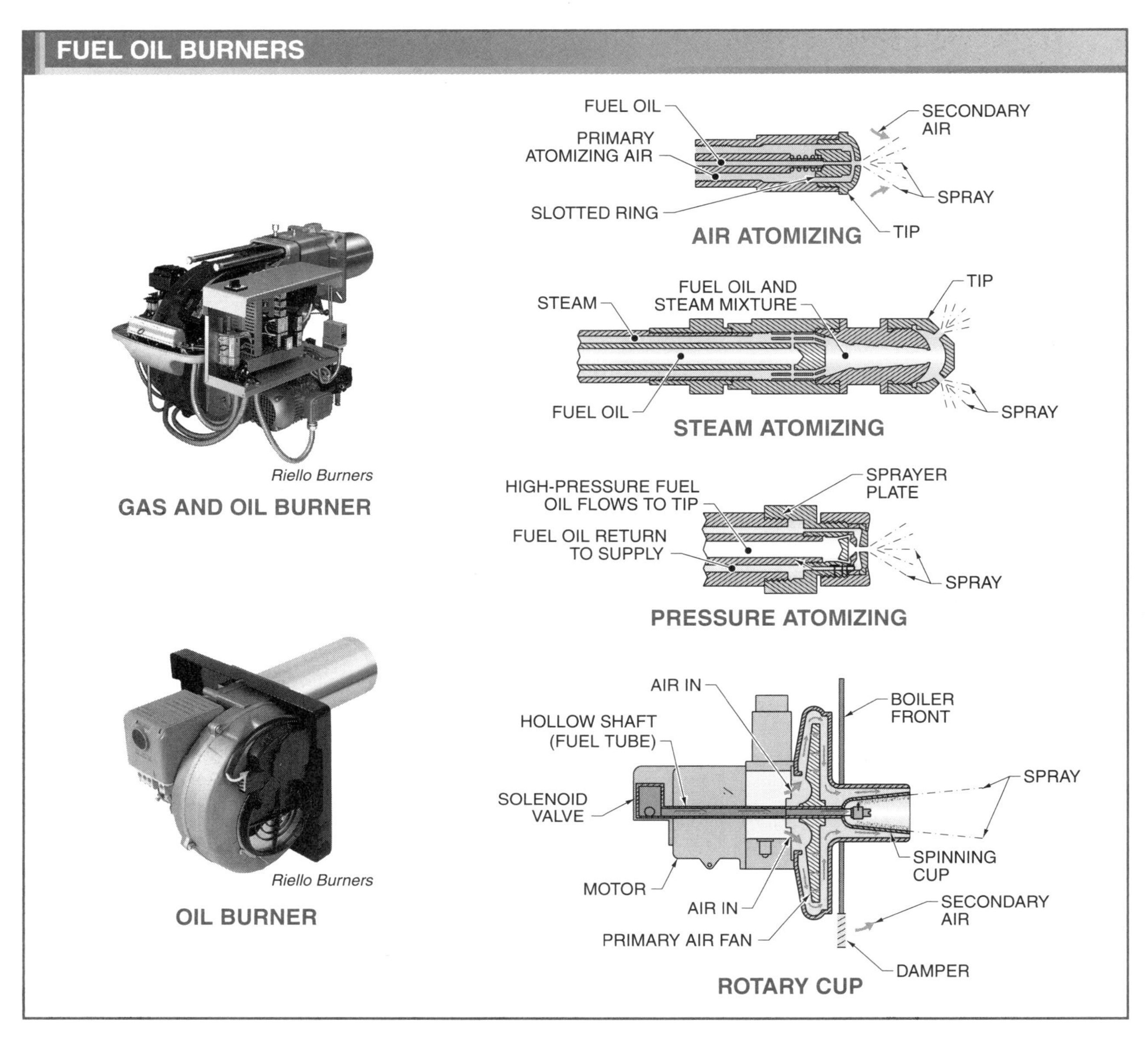

Figure 11-34. Fuel oil burners atomize fuel oil to increase mixability with combustion air for maximum combustion efficiency.

Gas Pressure Regulators. A *gas pressure regulator* reduces the pressure from the utility to the gas pressure required for the burner. The gas pressure regulator maintains the required gas pressure at the burner through all firing rates.

Main Gas Valves. A *main gas valve* is an electrically operated valve used to either allow gas to flow to the burner or to stop the flow of gas to the burner. On small gas-fired boilers, the main gas valve is an electrically operated solenoid valve. Electrically operated solenoid valves are also used to control gas flow for pilot light systems. On large boilers the main gas valves are hydraulic-electric valves. Hydraulic-electric valves have a small electric motor that turns a small hydraulic pump that pumps oil from a reservoir in the valve actuator into a cylinder. As the oil enters the cylinder the piston in the cylinder works against the closing spring pressure to open the valve.

Gas Butterfly Valves. A *gas butterfly valve* is a metering valve that controls the volume of gas sent to a burner. The butterfly valve (flow valve) is normally connected by linkage to a damper that regulates the volume of air to the burner so that the fuel-air ratio remains at the desired mix to burn the fuel completely with the minimal amount of air.

Low and High Gas Pressure Switches. Low- and high-pressure gas pressure switches are switches used to monitor the gas pressure in the gas line. A *low-pressure gas switch* is a normally open switch that is held closed by gas pressure and will open if the gas pressure is too low. A *high-pressure gas switch* is a normally closed switch that remains closed and opens if the gas pressure is too high.

Gas Vent Valves. *A gas vent valve* is a valve used as a safety device on high-pressure gas systems that use two main gas valves and/or two pilot light valves. Gas vent valves are electric solenoid valves that are opened when the main gas valves are closed. If the main gas valve leaks, the gas would be vented outside of the building to atmosphere. When the main gas valve opens, the vent valve closes, and the boiler will operate as normal.

Boiler Combustion Safety Controls. The *boiler combustion safety control* is a control that is similar to the combustion safety controls used on forced-air furnaces. Boiler combustion safety controls monitor and control the burner start-up sequence, the main flame during operation, and the shutdown sequence. These controls include a programmer that controls the burner operation. A *programmer* is a control device that functions as the mastermind of the burner control system to control the firing cycle of a boiler.

The programmer monitors the safety circuit and the flame safeguard circuit. The safety circuit includes all of the temperature switches, pressure switches, and water level switches to ensure all conditions are normal. The flame safeguard circuit monitors the presence or absence of a flame for the pilot light and the main flame.

Draft Systems

A *draft system* is a boiler system that regulates the flow of air into and from the boiler burner. For fuel to burn efficiently, the proper amount of oxygen must be provided. The proper amount of primary air and secondary air must be provided for complete fuel combustion and to discharge the gases of combustion. **See Figure 11-35.**

Figure 11-35. A draft system regulates the flow of air into and out of the burner and discharges the gases of combustion from the firebox into the atmosphere.

During operation of the boiler, there is a flow of air into the fuel, a flow of hot gases of combustion to the heating surfaces, and a flow of gases of combustion to the atmosphere. Each of these processes has resistance to flow from obstructions such as tubes, tube sheets, firebox walls, dampers, and baffles. To overcome the resistance to flow, either natural draft or mechanical draft is employed. The type used depends on the size and capacity of the boiler. Mechanical draft includes forced draft, induced draft, and balanced draft.

Natural Draft. *Natural draft* is draft produced by natural action resulting from temperature and density differences between atmospheric air and the gases of combustion. **See Figure 11-36.** Air for combustion is heated, expands, and becomes lighter. Colder, heavier air pushes in to replace the hot gases of combustion. A difference in pressure is created as the cold air for combustion replaces the hot gases of combustion going up the chimney. By directing the flow of air or hot gases to the chimney, the amount of draft can be increased or decreased. The maximum amount of natural draft can be increased with the height of the chimney. Controlling the temperature of the gases of combustion can also regulate natural draft. The more a gas is heated, the faster it rises. Boilers that employ natural draft are limited in size because of the limitations on chimney height.

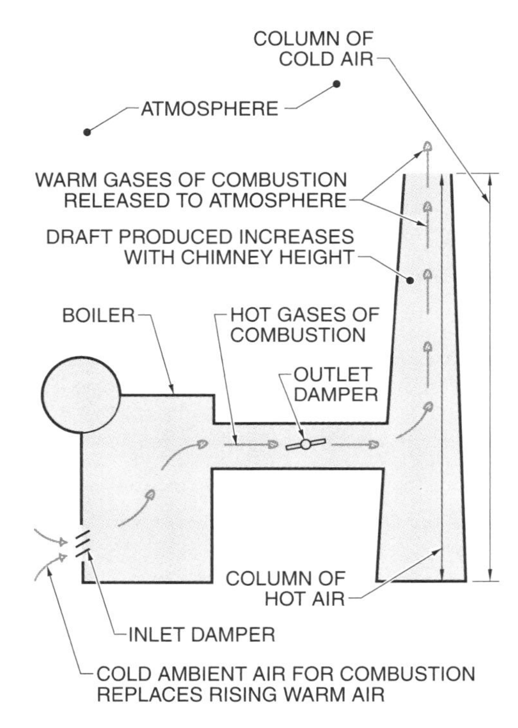

Figure 11-36. The height of the chimney affects the amount of natural draft produced.

Forced Draft. *Forced draft* is mechanical draft created by air pushed through the firebox by blowers located on the front of the boiler. Forced draft air is forced through the atomized fuel in a boiler, so the fire burns hotter because more air is available for combustion. This creates a pressure inside the firebox in comparison to ambient air. If a leak appeared in a firebox wall, the poisonous gases of combustion could leak into the boiler room, creating a hazard to the boiler operator(s). **See Figure 11-37.**

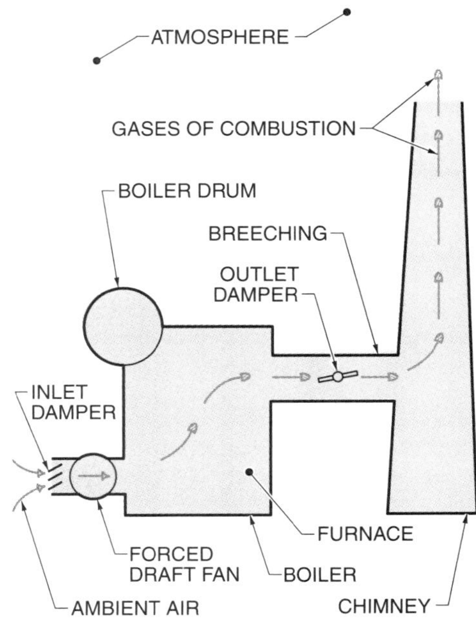

Figure 11-37. Forced draft is mechanical draft that pushes air through the firebox by blowers located on the front of the boiler, allowing the fire to burn hotter because more air is available for combustion.

Induced Draft. *Induced draft* is a mechanical draft of air pulled through the boiler firebox by a blower located in the breaching after the boiler. The blower pulls air out of the boiler and pushes the air up the chimney. A vacuum is created in the firebox, causing ambient air to take the place of the air that has been forced up the chimney. If there were a leak in the firebox of a boiler using induced draft, air from the boiler room would be allowed to enter the firebox uncontrolled, cooling the gases of combustion and decreasing boiler efficiency. **See Figure 11-38.**

Balanced Draft. *Balanced draft* is mechanical draft that uses a combination of forced draft and induced draft. This type of draft is normally utilized on large utility boilers that have a high resistance to the flow of the gases of combustion. There is both a forced draft fan and an induced draft fan. The forced draft fan pulls ambient air in and pushes it through the firebox, and the induced draft fan pulls air out of the firebox and pushes it up the chimney. The air pressure in the firebox is slightly higher than the ambient air pressure. A leak in the firebox wall would allow a slight amount of air into the firebox. **See Figure 11-39.**

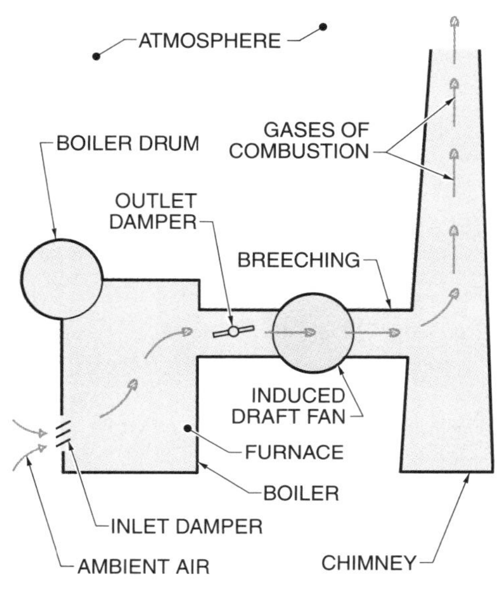

Figure 11-38. Induced draft produces a vacuum in the firebox and breeching to create airflow through the firebox.

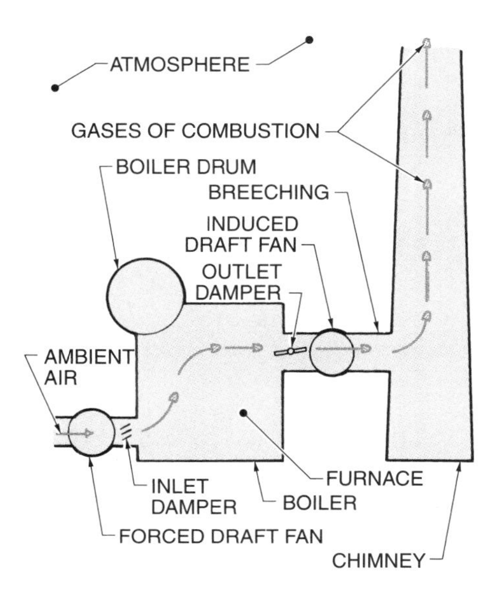

Figure 11-39. Balanced draft allows for more precise control of combustion for maximum efficiency in large boilers.

ACCESSORIES FOR HOT WATER BOILERS AND HYDRONIC HEATING SYSTEMS

Some of the accessories for hot water boilers and hydronic heating systems are the same as those for steam boilers. The fuel system and draft system are the same as those for small steam boilers. Large hot water boilers may utilize large, sophisticated fuel and draft systems to accommodate the different types of fuel and the large volumes of fuel consumed. Hot water boilers do not have feedwater or steam systems. The water heated by the boiler is constantly circulated throughout the entire heating system. Accessories used in a hydronic heating system include circulating pumps, expansion tanks, air vents, pressure-reducing valves, and flow control valves.

Circulating Pumps

A *circulating pump* is a pump that moves water from the boiler through the piping system, through terminal devices, and then through return piping of a hydronic heating system. A circulating pump may be connected on either the supply or return piping but is normally connected close to the boiler on the return piping. The number and size of the circulating pumps is determined by the volume of water to be moved and the resistance to flow in the system. Hydronic heating systems that have many piping loops that feed different zones of a building have a pump located on each loop.

Most circulating pumps that are used on hydronic heating systems are centrifugal pumps. A *centrifugal pump* is a pump that has a rotating impeller inside a cast iron or steel housing. An *impeller* is a plate with blades that radiate from a central hub (eye). The water inlet is an opening in the pump housing that is at the center of the impeller. The water outlet is an opening in the pump housing that is located on the outer perimeter of the pump housing (volute). **See Figure 11-40.**

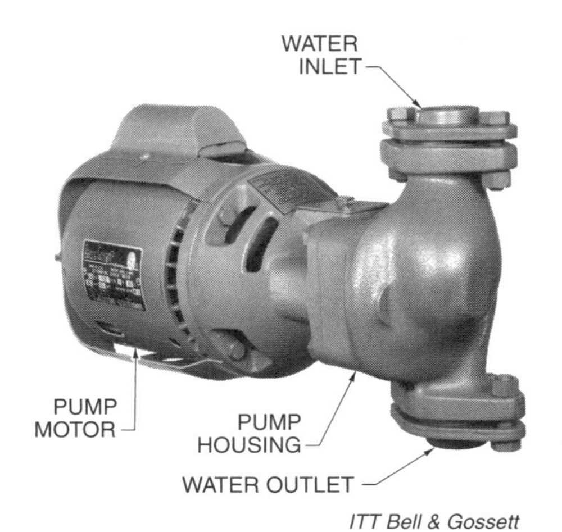

Figure 11-40. A centrifugal pump contains a rotating impeller inside a cast iron or steel housing.

Return piping is connected to a centrifugal pump at the water inlet. Supply piping is connected to a centrifugal pump at the water outlet. The impeller of a circulating pump rotates at speeds from 1150 rpm to 3550 rpm. As the impeller rotates, a low-pressure area is created at the eye of the impeller. The impeller throws water out of the water outlet by centrifugal force.

Circulating pumps are normally driven by electric motors. The motor is often part of the pump housing on a circulating pump that has a pumping rate of 0 gpm to 100 gpm. The motor on a circulating pump that has a pumping rate above 100 gpm is normally connected by an in-line connection (coupling) on the shaft of the motor. Pipe-mounted circulating pumps are small pumps that are installed in the piping of a hydronic system. Floor-mounted circulating pumps are large pumps that must have the piping run to the pump. **See Figure 11-41.**

Expansion Tanks

All components, piping, and terminal devices of hydronic heating systems operate while full of water. Water expands as it is heated. An *expansion tank* is a vessel that allows the water in a hydronic heating system to expand without raising the system water pressure to dangerous levels. An expansion tank is partially filled with air in the normal operating condition. As water in the system expands, the water flows into the tank, compressing the air. The air in the tank acts as a cushion. **See Figure 11-42.**

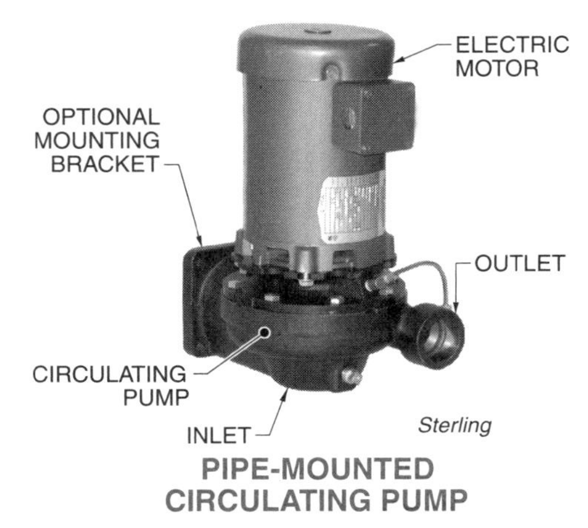

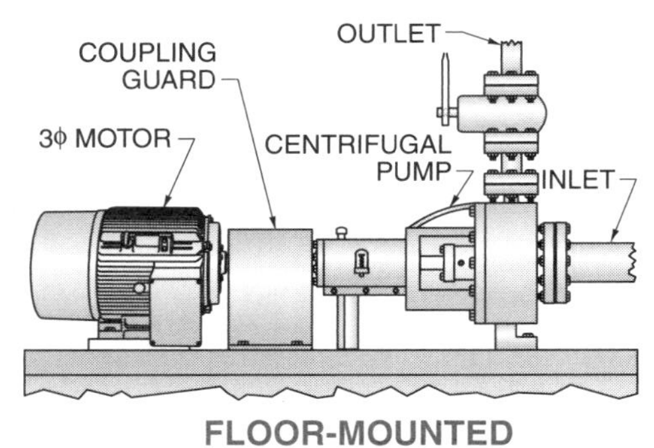

Figure 11-41. Pipe mounted circulating pumps are small pumps (0 gpm to 100 gpm) that are installed in the piping of a hydronic system. Floor mounted circulating pumps are large pumps (over 100 gpm) that must have the system piping run to the pump.

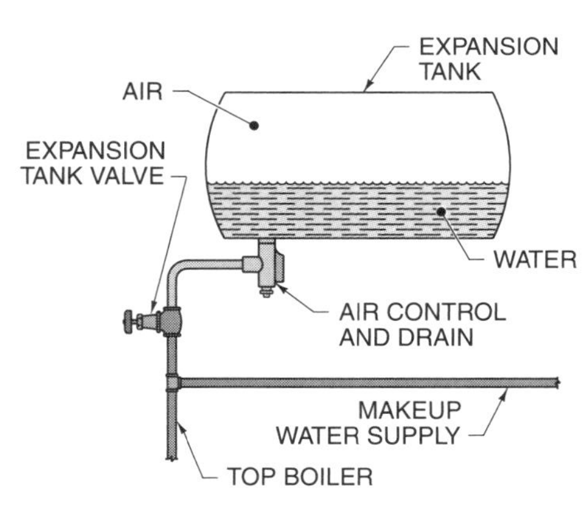

Figure 11-42. An expansion tank is a device that is partially filled with air and acts as a cushion as water in the system expands.

Air Vents

An *air vent* is an automatic or manual device that vents air from a hydronic heating system without allowing water to escape or air to enter the system. Air vents are installed at the highest parts of the piping system where the air rises as it comes out of solution with the water. As an air-water solution is heated, the air separates from the solution and rises because air is lighter than water. This air is then vented from the hydronic heating system through air vents.

Pressure-Reducing Valves

A *pressure-reducing valve* is an automatic device used to convert high and/or fluctuating inlet water pressure to a lower or constant outlet pressure. Pressure-reducing valves reduce the pressure of makeup water to between 12 psi to 18 psi so that the water can be used in a boiler. A pressure-reducing valve is located in the makeup water piping that leads to a boiler. **See Figure 11-43.**

Flow Control Valves

A *flow control valve* is a valve that regulates the flow (gpm) of water in a hydronic heating system. A flow control valve may be manual or automatic. A *manual flow control valve* is a globe valve that manually controls the flow of water in a hydronic heating system. An *automatic flow control valve* is a check valve that opens or closes automatically. When a circulating pump is ON, the flow control valve opens to allow water flow. When the circulating pump is OFF, the flow control valve closes and prevents water from circulating by natural circulation. A flow control valve is located on the outlet piping of a boiler. **See Figure 11-44.**

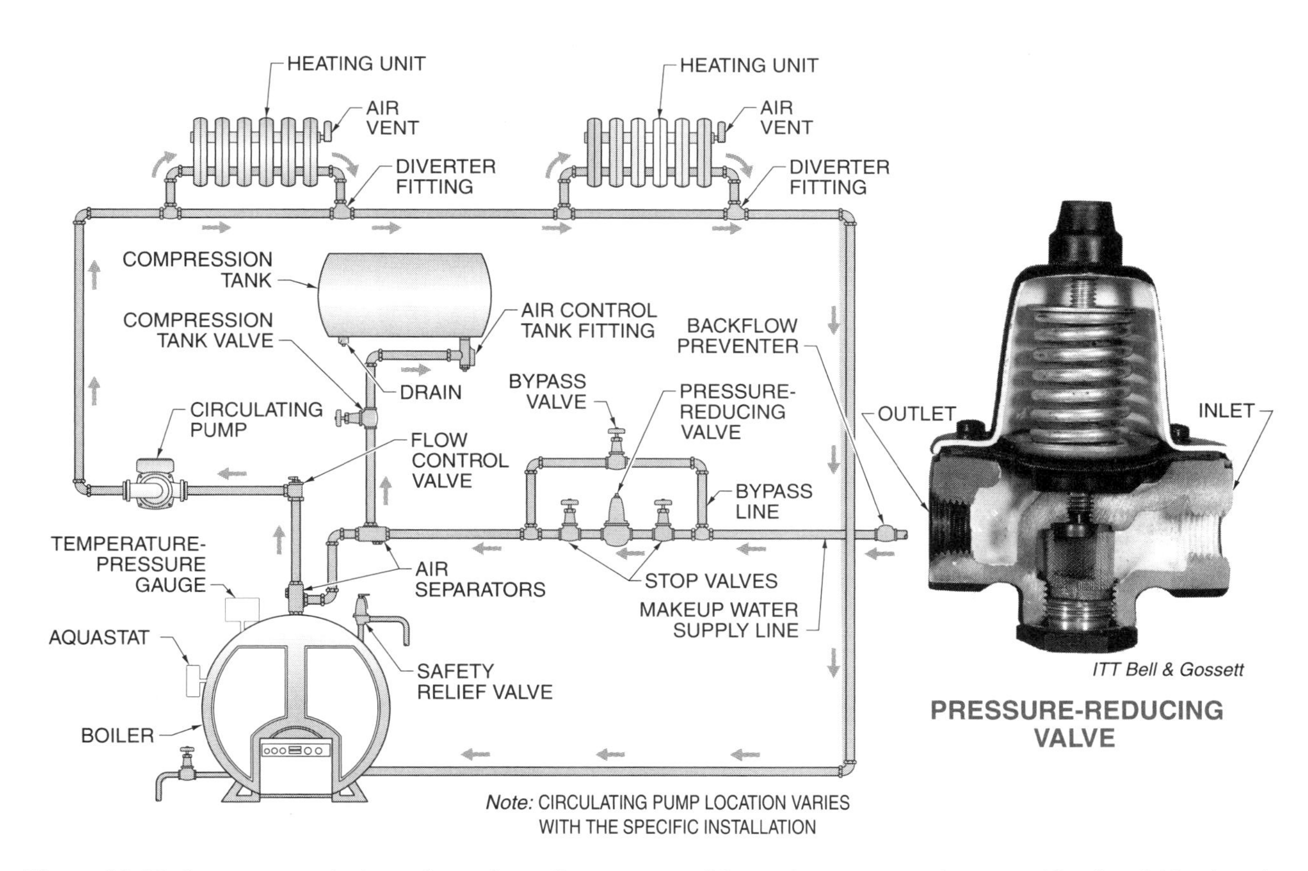

Figure 11-43. A pressure-reducing valve reduces the pressure of the makeup water to between 12 psi and 18 psi so that the boiler can use the water, or it controls the pressure of the hot water in the heating system.

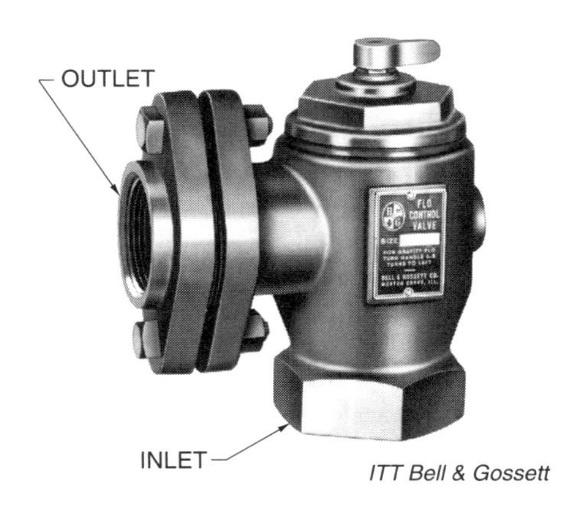

Figure 11-44. Flow control valves regulate the flow (amount) of water in a hydronic heating system.

PIPING SYSTEMS

Piping systems in hydronic heating systems are used to distribute heated water from a boiler into the building and return the water to the boiler. Piping systems vary depending on the degree of temperature control desired and the specific application. In most systems the pipe is made of steel or cast iron. In small systems the pipe may be copper. Piping systems used for hydronic heating systems include one-, two-, three-, and four-pipe systems. Because of variations in buildings, each piping system is designed for a specific kind of building.

One-Pipe Piping Systems

One-pipe piping systems are used in small- to medium-size buildings. The two kinds of one-pipe systems are series and primary-secondary systems.

Series. A *one-pipe series piping system* is a piping system that circulates water through one pipe, looped through each terminal device. As the water flows through the system, a temperature drop occurs at each terminal device.

The inlet temperature at each terminal device is lower than the inlet temperature at the preceding terminal device. The inlet temperature of the last terminal device in the system is lower than the inlet temperature of any other terminal device. A one-pipe series piping system is an economical system to install and gives good temperature control when used in small buildings. **See Figure 11-45.**

Primary-Secondary. A *one-pipe primary-secondary piping system* is a piping system that circulates boiler water through a primary loop. The terminal devices are connected in parallel to the primary loop by a secondary loop. Each terminal device has an individual secondary loop. The secondary loop can be connected to the primary loop by a diverter fitting.

A *diverter fitting* is a tee that meters the water flow. Water flow is adjusted for the gallons per minute required by the terminal device from the secondary loop. The return water from each terminal device returns to the primary loop so the temperature drop of each device affects the next terminal device in the system, but the flow rate in each terminal device is controlled within its own secondary loop. Flow control allows the heat output of the terminal device to be controlled.

Two-Pipe Piping Systems

Two-pipe systems are used in medium- to large-size residential and commercial buildings. Two-pipe systems consist of a supply pipe that carries hot water from the boiler to terminal devices and a return pipe that carries water back to the boiler. Two-pipe systems include direct-return and reverse-return systems. **See Figure 11-46.**

Direct-Return. A *two-pipe direct-return piping system* is a piping system that circulates the supply water in the opposite direction of the circulation of the return water. One side of a terminal device is connected to the supply piping and the other side is connected to the return piping. Water in the return piping flows from each terminal device directly back to the boiler through the return piping. A direct-return piping system is difficult to balance because the water that flows through the final terminal device must be pumped back through the whole piping system to the boiler.

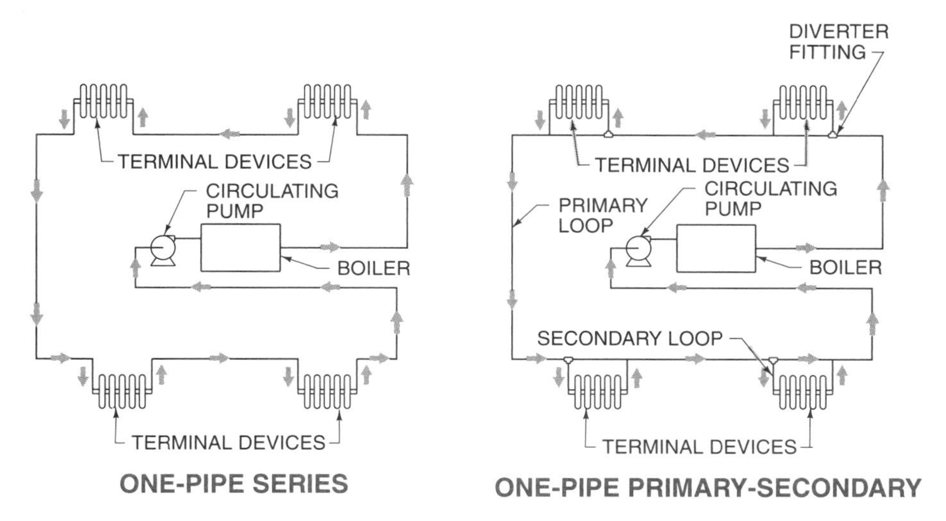

Figure 11-45. One pipe is used to supply and return boiler water in one-pipe hydronic piping systems.

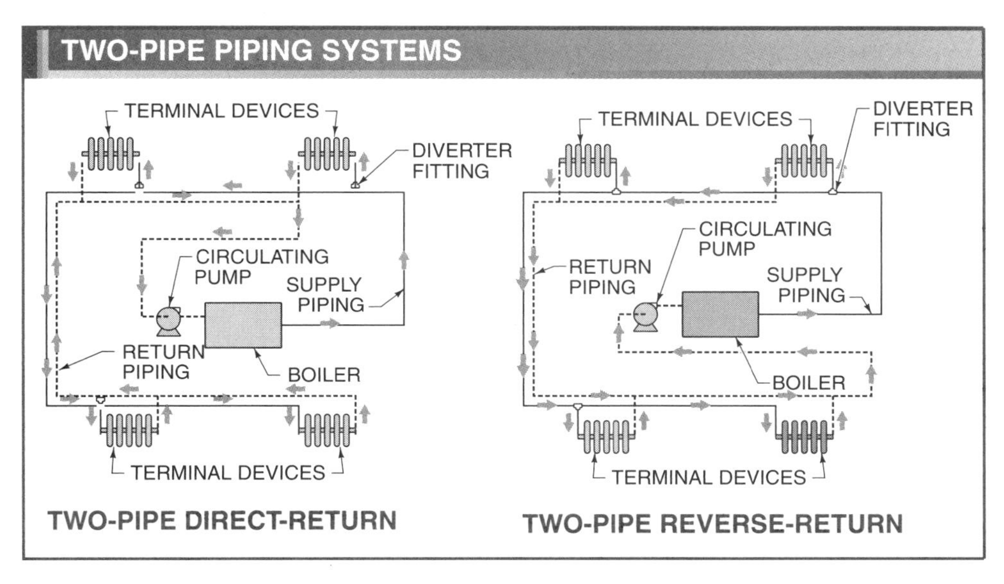

Figure 11-46. Two separate pipes are used to supply and return boiler water in two-pipe hydronic piping systems.

Balancing is the adjustment of the resistance to the flow of water through the piping loops in a system. Balancing assures the proper amount of water flow through each terminal device for the amount of heat required from each terminal device.

Reverse-Return. A *two-pipe reverse-return piping system* is a piping system that circulates return water in the same direction as supply water. The same flow direction makes the system easy to balance because the distance around the piping system is the same for the water flowing through each terminal device.

Three-Pipe Piping Systems

A three-pipe piping system has two supply pipes and one return pipe. A three-pipe piping system is used when different parts of a system require heating and cooling simultaneously. One supply pipe is connected to the boiler and the other supply pipe is connected to the chiller. **See Figure 11-47.**

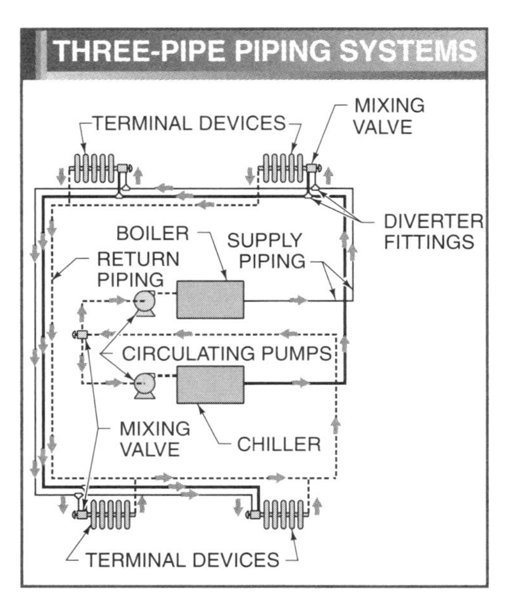

Figure 11-47. A three-pipe hydronic piping system uses two supply pipes and one return pipe.

Both supply pipes are connected to each terminal device. The return piping is a common line that runs from each terminal device back to the boiler and chiller. Mixing valves regulate the flow of water from the two supply pipes to the terminal devices in the system. A *mixing valve* is a three-way valve that has two inlets and one outlet and is used to mix two water supplies into one desired flow to the terminal device. On a call for heat, water from the boiler supply pipe flows into the terminal device. On a call for cooling, water from the chiller supply pipe flows into the terminal device. The water in the common return line then flows to a three-way diverting valve to be directed back to the boiler and/or chiller. A *diverting valve* is a three-way valve that has one inlet with two outlets to divert the water flow to the boiler and/or chiller, depending on building needs.

Three-pipe piping systems provide good control of air temperature but are more expensive to install than two-pipe piping systems. Three-pipe piping systems are also more expensive to operate because the hot and cold return water are mixed.

Four-Pipe Piping Systems

A four-pipe piping system separates heating and cooling. The terminal devices are connected to both heating pipes and cooling pipes. Water flow to the terminal devices is controlled by valves. Heating and cooling can be achieved at any terminal device when desired. Four-pipe piping systems are expensive to install but provide excellent control of air temperature and are more economical to operate than three-pipe piping systems. **See Figure 11-48.**

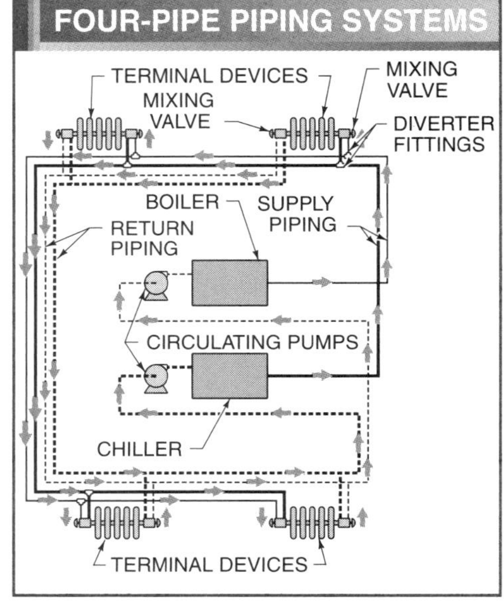

Figure 11-48. A four-pipe hydronic piping system uses supply and return heating piping and supply and return cooling piping.

TERMINAL DEVICES

A *terminal device* is a heating system component that transfers heat from the hot water in a hydronic heating system to the air in building spaces. A supply pipe runs from the boiler to each terminal device and a return pipe leaves each terminal device and runs back to the boiler. Radiant heating and convection heating terminal devices are the two types of terminal devices.

Radiant Heating

Thermal radiation is heat transfer that occurs as radiant energy waves transport energy from one object to another. Radiant heating systems are based on thermal radiation. *Radiant heating* is heating that occurs when a surface is heated and the surface gives off heat in the form of radiant energy waves.

The *radiant surface* is the heated surface from which radiant energy waves are generated. The radiant energy waves are absorbed only by opaque objects in direct line of sight of the radiant surface and are not absorbed by the air. Opaque objects absorb the radiant energy waves and become heated. As the objects are heated, they give up some of the heat to the air around them, which raises the temperature of the air.

A *hydronic radiant heater* is a heater that has a radiant surface that is heated by hot water to a temperature high enough to radiate energy. The three hydronic radiant heaters are surface radiation systems, radiant panels, and radiators. **See Figure 11-49.**

Surface Radiation Systems. A *surface radiation system* is a heating system that uses the interior surfaces of a room as radiant surfaces. A piping coil or grid is built into the construction material of the ceiling, walls, or floor of a room. Insulation is installed behind the piping and a thin layer of surfacing material is installed over the insulation.

The radiant surface is heated by hot water that flows through the pipes. When the temperature of the radiant surface is higher than the temperature of objects in the room, heat flows from the radiant surface to the objects. The amount of radiant heat produced is regulated by controlling the temperature of the radiant surface.

Radiant Panels. A *radiant panel* is a factory-built panel with a radiant surface that is heated by a piping coil or grid built into it. Radiant panels are individual heaters that are not an integral part of a building. Radiant panels are enclosed in an insulated cabinet. Radiant panels are used for spot heating. *Spot heating* provides heat for a specific building space. Buildings are not heated by spot heating from radiant panels.

Radiators. *Radiators* are terminal devices that consist of hollow metal coils or tubes. Hot water passes through the coils or tubes. Heat from the hot water is transferred to the tubes, which transfer the heat to the air. The air in a building space is heated by radiation and convection when using radiators.

Tech Facts

Fin-tube convectors are terminal devices that contain a section of finned tube, ¾" to 2" in diameter, that is used to transfer heat.

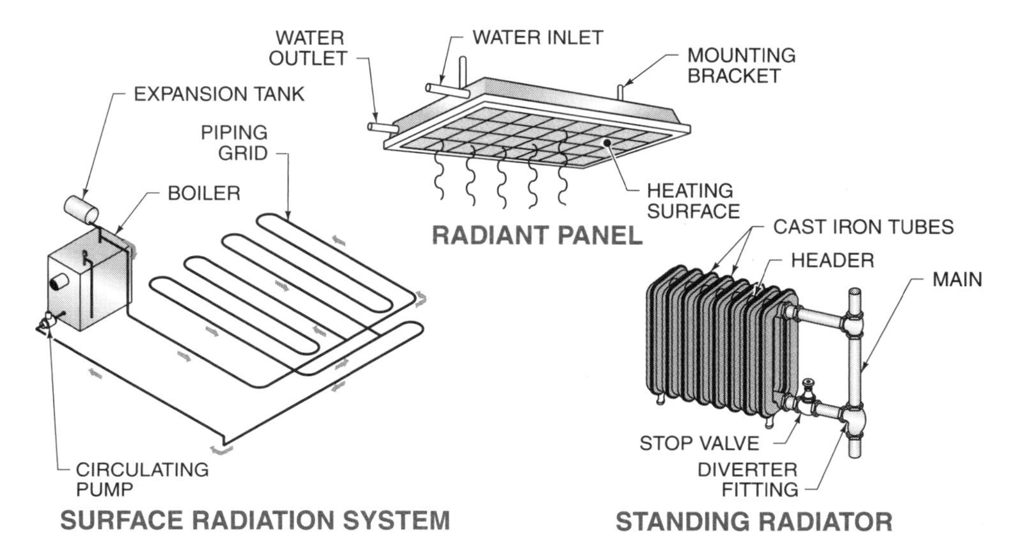

Figure 11-49. Radiant heating devices use hot water to heat a surface to a temperature high enough for heat to flow from the surface.

A *standing radiator* is a section of vertical tubes that are connected with headers at each end. Hot water from a boiler flows through the tubes. Heat is transferred to the air by radiation and convection when air flows over the tubes. Standing radiators are usually made of cast iron. Because the units are unattractive, baseboard radiators are typically used. A *baseboard radiator* is a section of horizontal tubes that are connected with headers at each end, and that can be concealed.

Convection Heating

Convection heating is based on heat transfer by convection. *Convection heating* is heat transfer that occurs when currents circulate between warm and cool regions of a fluid. The two kinds of convection heaters are cabinet convectors and baseboard convectors. **See Figure 11-50.**

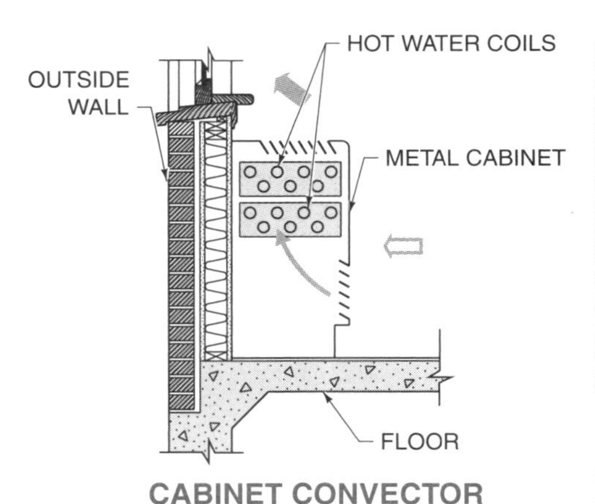

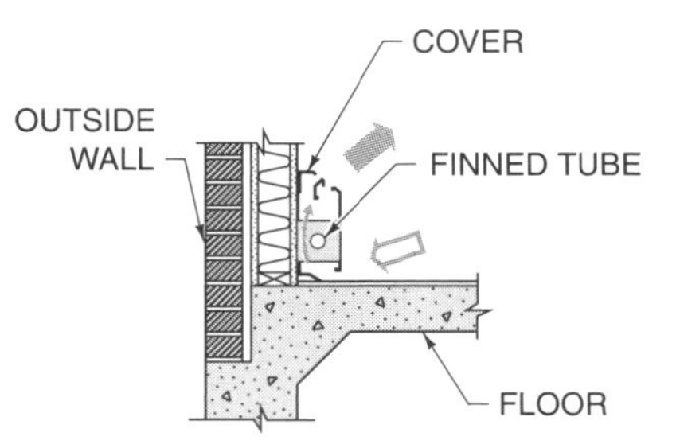

Figure 11-50. Convection heating devices contain hot water coils that heat air as the air flows over the hot water coils.

Cabinet Convectors. A *cabinet convector* is a convection heater that is enclosed in a cabinet. The hot water coil or tubes and the front of the heater are covered by a sheet metal cabinet. The cabinet has a grill in the bottom for cool air to enter and a grill on the top for warm air to exit. Cabinet convectors can be assembled in sections to produce the amount of heat required to heat a building space. Cabinet convectors are installed under windows along the outside walls of a room.

Baseboard Convectors. A *baseboard convector* is a convection heater that is enclosed in a low cabinet that fits along a baseboard. Baseboard convectors are smaller than cabinet convectors and use a section of finned tube to transfer heat. A *finned tube* is a copper pipe with aluminum fins. The fins provide a large heat transfer surface area. Baseboard convectors are installed at floor level along outside walls. **See Figure 11-51.**

Figure 11-51. Baseboard convectors are smaller than cabinet convectors and use a section of finned tube to transfer heat.

Forced Convection Heating

Forced convection heating creates convective currents by mechanically moving air past a hot water coil. Forced convection heaters use a blower or fan to move air. The blower draws air in the bottom or back of the heater, moves it past the hot water coil, and distributes the air to the building space. All forced convection heaters have a hot water coil, blower, controls, and a sheet metal cabinet. The three kinds of forced convection heaters used in hydronic heating systems are unit heaters, cabinet heaters, and unit ventilators. **See Figure 11-52.**

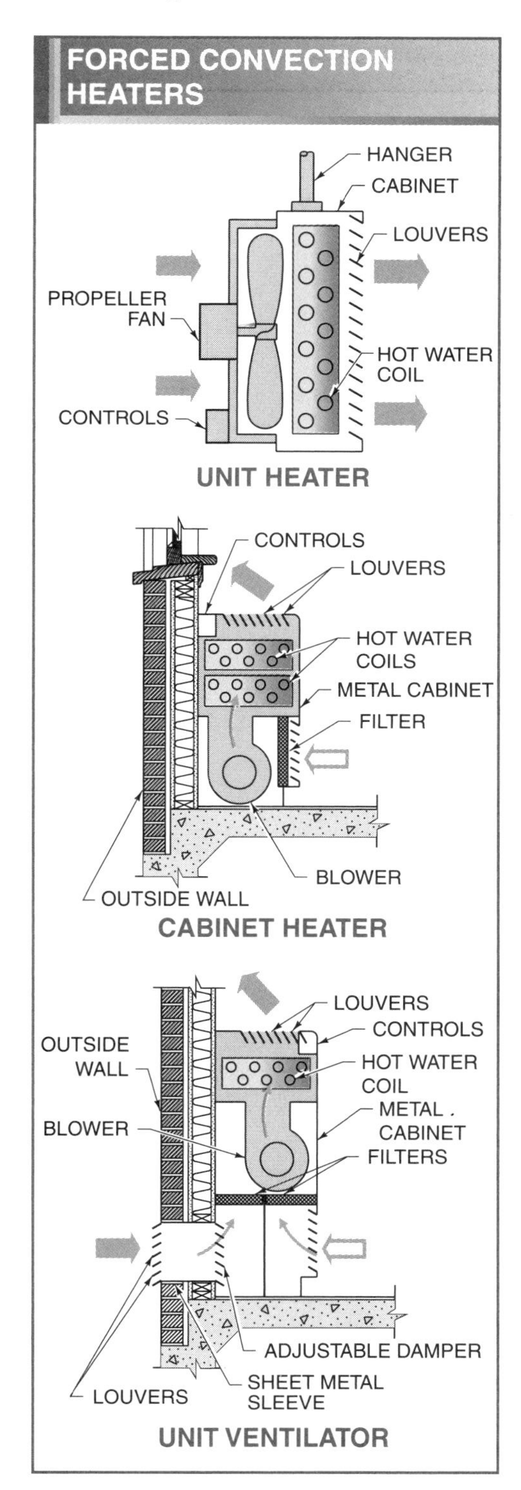

Figure 11-52. Forced convection heaters contain blowers that move air.

Unit Heaters. A *unit heater* is a self-contained heating unit that is not connected to ductwork. Unit heaters are normally used as space heaters. Unit heaters are suspended from the ceiling framework of a building space. Return air is drawn in the back of the unit by the fan. The air is blown across the hot water coil and out the front of the unit through louvers. Unit heaters are made in different sizes and have different heating capacities. The heat output of a unit is controlled by regulating the flow of water and the speed of the fan.

Cabinet Heaters. A *cabinet heater* is a forced convection heater that has a blower, hot water coils, filter, and controls in one cabinet. The blower creates air flow across the hot water coils. Cabinet heaters differ from cabinet convectors in that a blower moves the air and a filter cleans the air. The blower adds energy to the air so that directional louvers can be used on the top or front of the heater to distribute the air more efficiently in the building space. The size of a cabinet heater depends on the heat output required. Cabinet heaters are installed along the outside wall of a room under a window.

Unit Ventilators. A *unit ventilator* is a forced convection heater that has a blower, hot water coil, filters, controls, and a cabinet that has an opening for outdoor ventilation. The blower creates an air flow of mixed outside air and return air (mixed air) across the hot water coil. Unit ventilators differ from cabinet heaters in that they have an outside opening to bring in air from outdoors for ventilation. The cabinet fits on an outside wall so that the connection to the outdoors can be made with an opening in the back of the cabinet for an outdoor air inlet. A sheet metal sleeve extends from the inlet through the wall. A louver covers the opening to the outdoors. The outdoor air inlet has an adjustable damper on the inside for regulating the amount of air admitted to the unit for ventilation. Unit ventilators are often used in buildings that have a large number of people and a need for constant ventilation.

Terminal Device Controls

Terminal device controls control the temperature of the air in a building space by regulating the air and water flow through the terminal device. Circulating pumps and zone valves regulate water flow and blowers regulate air flow. Circulating pumps are normally switched ON or OFF by a signal from a thermostat, which produces the flow of hot water in a hydronic piping system. A circulating pump control provides hot water for a complete hydronic heating system or secondary piping loop in a system.

Zone Valves. A *zone valve* is a valve that regulates the flow of water in a zone or terminal device of a building. A *zone* is a specific section of a building that requires separate temperature control. Zone valves may be manual or automatic. A *manual zone valve* is a valve that is set by hand to regulate the flow of water in a piping loop. An *automatic zone valve* is a valve that is opened or closed automatically by a valve motor, electric solenoid coil, or pneumatic actuator. Automatic zone valve actuators are driven by an electric or electronic control system or by a pneumatic control system, with the control system controlled by a zone thermostat. **See Figure 11-53.**

Figure 11-53. Zone valves regulate the flow of water in a control zone or terminal device of a building.

Valves used for controlling water flow are either digital or modulating valves. A *digital valve* is a two-position (ON/OFF) valve. The valve is either completely open or completely closed. A *modulating valve* is an infinite position valve. The valve may be completely open, completely closed, or at any intermediate position in response to the control signal the valve receives.

Low-Temperature Limit Controls. A low-temperature limit control (freeze stat) is used on terminal devices that use outdoor air for ventilation. A *low-temperature limit control* is a temperature-actuated electric switch that may energize a damper motor to shut the damper and open the water valve if the ventilation air temperature drops below a setpoint (35°F). The terminal unit fan will continue to run. In other applications, the low-temperature limit control shuts the terminal unit down, completely closes the damper, turns the fan OFF, and opens the water valve. Freeze-protection controls are required when outdoor air temperature drops below 32°F, because water in the coil of the terminal unit may freeze. **See Figure 11-54.**

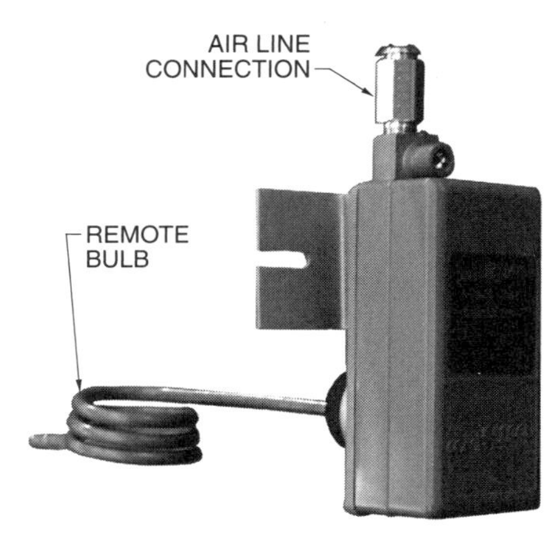

Figure 11-54. A low-temperature limit control (freeze stat) is used on terminal devices that use outdoor air for ventilation to protect the devices from freezing when outdoor air temperature drops below 32°F, because water in the coil of the terminal unit may freeze.

Thermostats. Thermostats used with terminal devices regulate water and air flow by controlling the circulating pump(s), zone valves, and terminal device blowers. A thermostat is used to control digital valves (two-position). On a call for heat, the thermostat opens the valve. When the temperature reaches the thermostat setpoint, the thermostat closes the valve.

A proportional thermostat is used to control modulating valves. A *proportional thermostat* is a thermostat that contains a potentiometer that sends out an electric signal that varies as the temperature varies. A *potentiometer* is a variable-resistance electric device that divides voltage proportionally between two circuits. An actuator receives the signal and converts it to mechanical action in the modulating valve.

Programmable Thermostats. A *programmable thermostat* is a stand-alone thermostat that can be programmed by the building occupants. **See Figure 11-55.** The standard features of a programmable thermostat are multiple day and night setpoints for heating and cooling, deadband settings, and override capabilities. The ability to program setpoints into the thermostat for the times throughout the day or night the space is occupied and a separate programmed setpoint for the times when the space is unoccupied, conserves energy.

Deadband settings are another energy conservation idea that is standard on most programmable thermostats. The building occupants are able to set the individual setpoints for the desired temperatures but the program built into the thermostat will not allow the heating setpoints to overlap the cooling setpoints in order to avoid heating and cooling at the same time. In many cases there is a minimum of a 2°F differential between the heating setpoint and the cooling setpoint. *Deadband* is the setpoint differential when no mechanical heating or cooling is allowed because the temperature is between the two setpoints.

Blowers. The blowers in terminal devices are controlled by relays, which cycle the blower in the terminal device ON or OFF. The relays are the same as those used in forced-air heating systems. A signal from the thermostat closes the relay contacts which supply electricity to the blower motor. Thermostats control the blower in a unit heater or unit ventilator. Thermostats used to control terminal devices may be located inside the cabinet of the unit that they control or located remotely in the space being heated. **See Figure 11-56.**

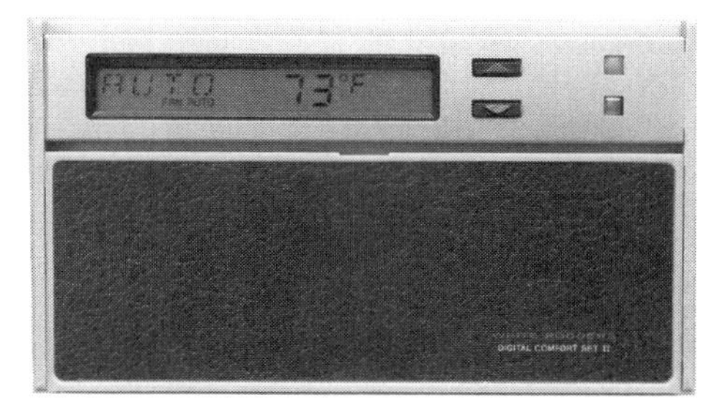

ACCESS DOOR CLOSED

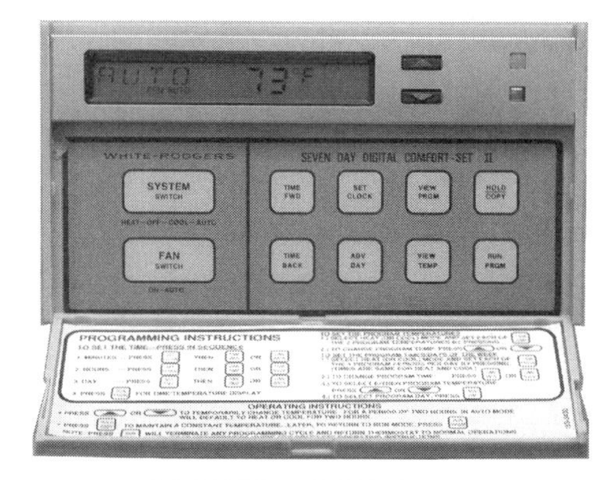

ACCESS DOOR OPEN

Figure 11-55. The standard features of a programmable thermostat are multiple day and night setpoints for heating and cooling, deadband settings, and override capabilities.

Water Strainers. A *water strainer* is a device in a hydronic heating system that removes particles from the water inside the piping system. If not removed, the particles can plug small openings (orifices) in controls and coils. The particles can also lodge under the seat of a valve, causing the valve to leak when closed. Some particles when suspended in water become very abrasive, causing valves and pump impellers to wear prematurely. The strainers are normally located before pumps and control valves in the system to protect the devices.

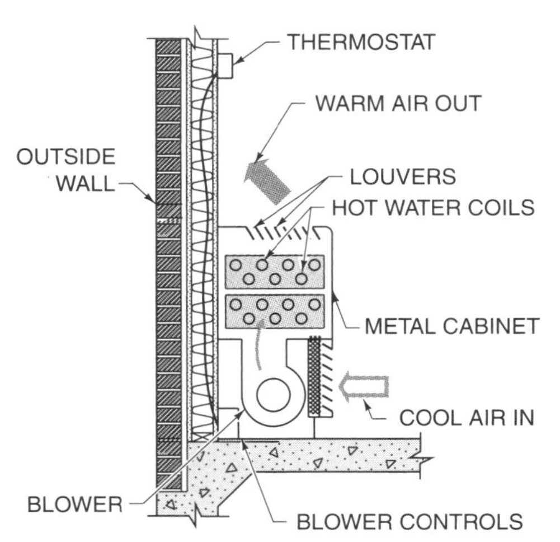

Figure 11-56. The blowers in terminal devices are controlled by thermostat-controlled relays.

TROUBLESHOOTING AND SERVICING STEAM AND HYDRONIC HEATING SYSTEMS

To troubleshoot and service steam and hydronic heating systems, individuals must be familiar with the complete system. Many times a faulty component causes problems in other components of the system or makes the components appear to be faulty. Individuals must learn to troubleshoot at the component level. Troubleshooting to the component level requires good electrical skills.

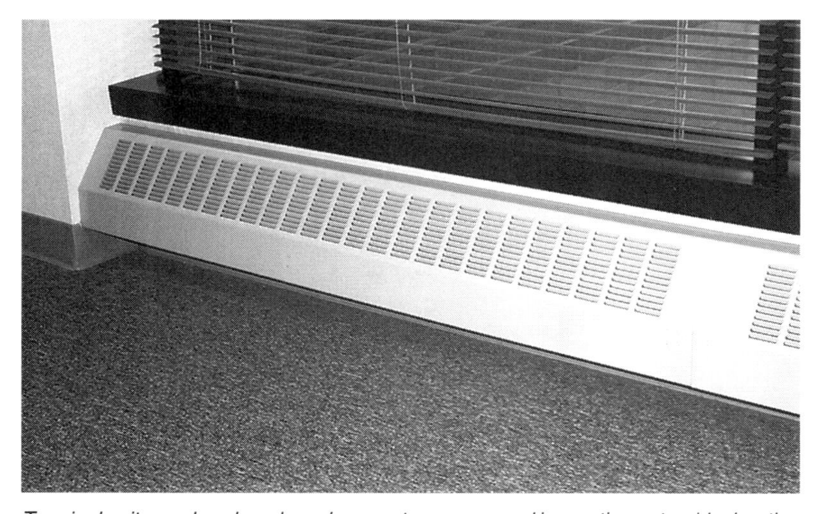

Terminal units, such as baseboard convectors, may need inspection or troubleshooting if there is a problem with comfort in a building space.

Tools and Supplies

Instruments for taking readings of voltage, amperage, resistance, temperature, water pressure, and airflow are required with troubleshooting steam and hydronic heating systems. Digital multimeters can be used to test for voltage, amperage, and resistance. If amperage is to be tested, a separate clamp-on amp-probe accessory is used with the multimeter. With the appropriate accessories, temperature can also be sensed with a multimeter. To check pressure signals, gauges or gauge sets are used.

To measure pressure in a pipe, pressure gauges are normally used. To measure temperature of air flowing out of terminal units a thermometer is used. To measure the airflow to or from heating units an anemometer is used with the appropriate accessories. Knowing how to use and read these instruments correctly and being sure to certify the instruments' accuracy ensure accurate readings. **See Figure 11-57.**

Routine Service Procedures

Local, state, and federal regulations require periodic inspections of all boilers by an authorized inspector. Most insurance companies also require periodic inspections to qualify for insurance. Periodic inspections are usually scheduled when the boiler is normally down in the summer or off season. This scheduled downtime is used to accomplish maintenance and replacement or repairs that cannot be done when the boiler is operating. While periodic inspection pertains primarily to the waterside and fireside of the boiler, it is an excellent opportunity for the servicing of all components of the boiler and accessories including piping, valves, pumps, gaskets, and refractory. Comprehensive cleaning, painting, and replacement of expendable items should be planned for this time period along with the coordination of any extensive repairs or replacements.

Prior to placing a boiler into service after initial installation, after extensive repairs, or after the boiler has been down for an extended period, a complete inspection should be made of all controls, connecting piping, wiring, and all fasteners such as nuts, bolts, and setscrews to ensure no damage or misadjustment has occurred. A well-planned preventive maintenance program avoids unnecessary downtime or costly repairs and promotes safety. It is recommended by most boiler manufacturers that a boiler room log or record be maintained, recording daily, weekly, and monthly maintenance activities. **See Figure 11-58.** A log is valuable in recognizing anomalies and obtaining length of service from equipment.

Tech Facts

Hydronic heating systems use less water by volume than air distribution systems to distribute the same amount of heat, since water has a higher specific heat and is more dense than air.

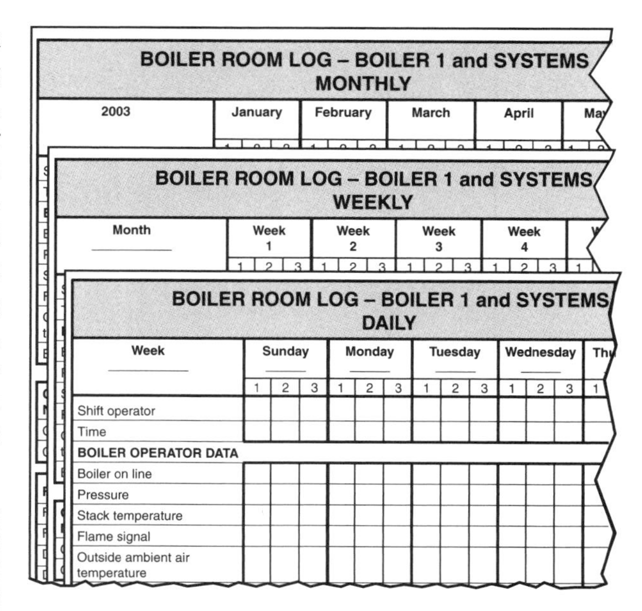

BOILER ROOM LOG – BOILER 1 and SYSTEMS
MONTHLY

2003	January	February	March	April	Ma

BOILER ROOM LOG – BOILER 1 and SYSTEMS
WEEKLY

Month	Week 1	Week 2	Week 3	Week 4

BOILER ROOM LOG – BOILER 1 and SYSTEMS
DAILY

Week	Sunday			Monday			Tuesday			Wednesday			Th
	1	2	3	1	2	3	1	2	3	1	2	3	1
Shift operator													
Time													
BOILER OPERATOR DATA													
Boiler on line													
Pressure													
Stack temperature													
Flame signal													
Outside ambient air temperature													

Figure 11-58. Most boiler manufacturers recommend that a boiler room log or record be maintained, recording daily, weekly, and monthly maintenance activities.

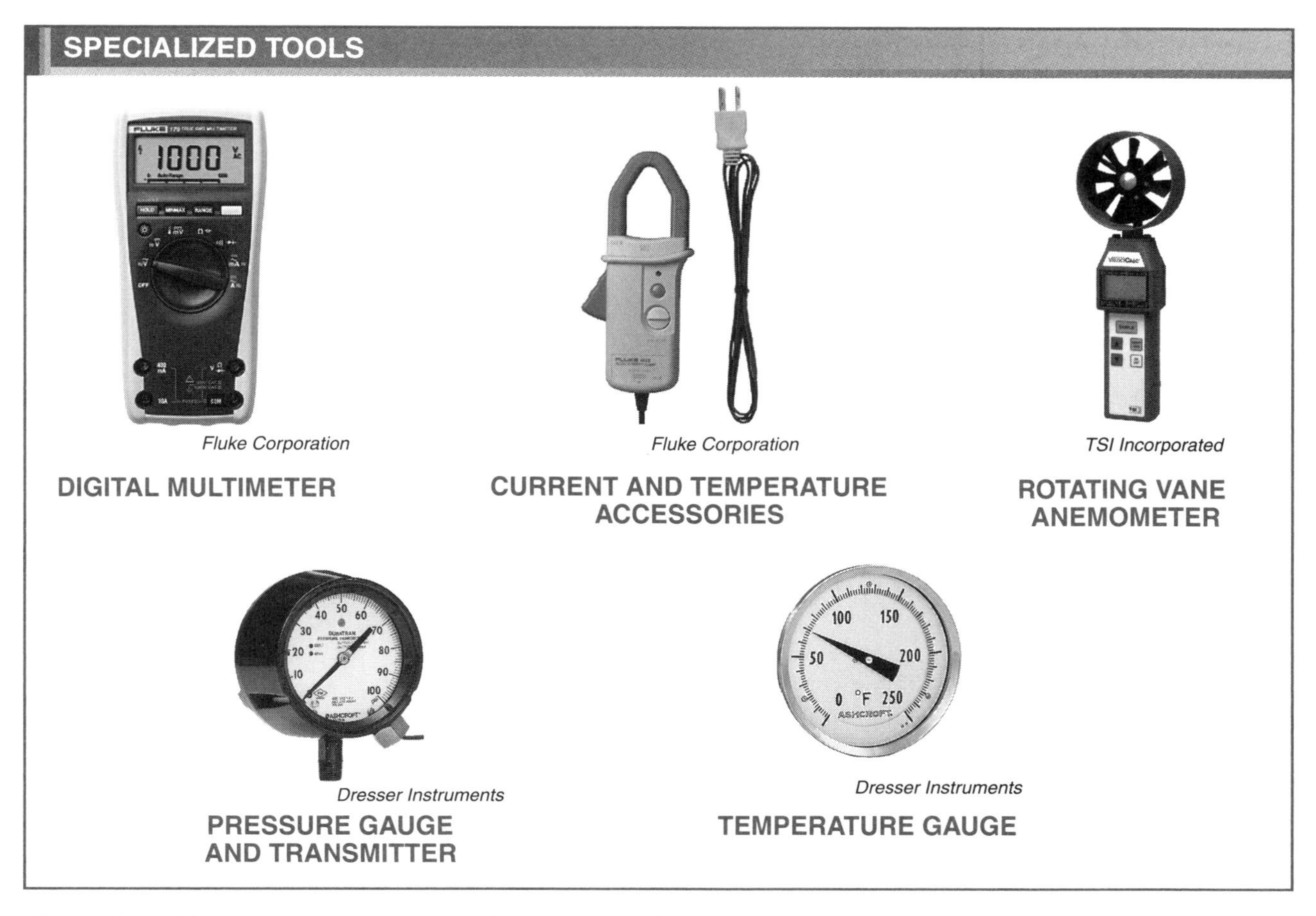

Figure 11-57. Test instruments must be read correctly and their accuracy certified.

Review Questions

1. What are the advantages of a steam heating system?
2. List common components of a steam heating system.
3. What is a hydronic heating system?
4. List the four ways boilers are classified.
5. What are the two major types of combustion boilers? Describe their operation.
6. What is the difference between gross unit output and net unit output when finding the amount of heat produced by a boiler?
7. List and describe eight types of boiler fittings.
8. In what units do steam pressure gauges display pressure?
9. What is water hammer? Describe possible effects of water hammer.
10. Why is bottom blowdown performed?
11. List two types of valves that are used as stop valves.
12. Describe the function of a condensate return tank.
13. List and describe five steam system accessories used to control steam.
14. What are six major components of a gas fuel system?
15. List and describe three types of mechanical draft.
16. What is the function of an expansion tank?
17. How do direct-return and reverse-return piping systems differ?
18. How does radiant heating heat building spaces? List three types of radiant heating systems.
19. What four components are present in all forced convection heaters?
20. What is the function of a zone valve?
21. List three advantages programmable thermostats have over non-programmable thermostats.

Refrigeration is the process of moving heat from an area where it is undesirable to an area where it is not objectionable. Refrigeration is based on a law of physics that states that matter gains or losses heat when it changes state. The two main types of refrigeration processes are mechanical compression and absorption. Mechanical compression refrigeration uses mechanical equipment to produce refrigeration. Absorption refrigeration uses the absorption of one chemical by another chemical and heat transfer to produce refrigeration.

12 Refrigeration Principles

REFRIGERATION PRINCIPLES

Refrigeration is the process of moving heat from an area where it is undesirable to an area where the heat is not objectionable. According to the second law of thermodynamics, heat always flows from a material at a high temperature to a material at a low temperature.

A *refrigeration system* is a closed system that controls the pressure and temperature of a refrigerant to regulate the absorption and rejection of heat by the refrigerant. One side of a refrigeration system decreases the temperature and pressure on the refrigerant, which causes the refrigerant to absorb heat from the medium (air or water) in the system. The air or water is cooled when heat is absorbed from the air or water. The air or water is then used for cooling building spaces. The other side of a refrigeration system increases the temperature and pressure on the refrigerant, which causes the refrigerant to reject heat to the air or water in the system. The air or water is heated and used for heating or the heat is exhausted to the atmosphere.

Refrigeration applications include commercial and industrial refrigeration and air conditioning. A commercial or industrial refrigeration system uses mechanical equipment to produce a refrigeration effect for applications other than human comfort. An air conditioning system produces a refrigeration effect to maintain comfort within a building space.

MECHANICAL COMPRESSION REFRIGERATION

Mechanical compression refrigeration is a refrigeration process that produces a refrigeration effect with mechanical equipment. A mechanical compression refrigeration system consists of a compressor, refrigerant, condenser, expansion device, evaporator, refrigerant lines, and accessories. **See Figure 12-1.**

A *compressor* is a mechanical device that compresses refrigerant or other fluid. A compressor increases the temperature of and pressure on refrigerant vapor and produces the high pressure side of the system. A *refrigerant* is a fluid that is used for transferring heat in a refrigeration system. Refrigerants have a low boiling (vaporization) point, which allows refrigerants to boil and vaporize at room temperature.

A *condenser* is a heat exchanger that removes heat from high-pressure refrigerant vapor. High-pressure refrigerant vapor flows through the condenser and the condensing medium passes across the outside of the condenser. Heat flows from the hot refrigerant vapor to the cold condensing medium. A *condensing medium* is the fluid in the condenser of a refrigeration system that carries heat away from the refrigerant. Air and water are condensing mediums used in refrigeration systems. As refrigerant vapor gives up heat to the condensing medium, the vapor condenses to a liquid.

An *expansion device* is a valve or mechanical device that reduces the pressure on a liquid refrigerant by allowing the refrigerant to expand. As the pressure on the liquid refrigerant decreases, some of the liquid refrigerant vaporizes because it has a lower boiling point. The refrigerant absorbs heat as it vaporizes, which cools the remainder of the refrigerant. The refrigerant then flows as a liquid-vapor mixture to the evaporator.

An *evaporator* is a heat exchanger that adds heat to low-pressure refrigerant liquid. Low-pressure refrigerant liquid flows through the evaporator and an evaporating medium passes across the outside of the evaporator. Heat flows from the hot evaporating medium to the cold refrigerant. An *evaporating medium* is a fluid that is cooled when heat is transferred in the evaporator from the evaporating medium to cold refrigerant. An evaporating medium adds heat to a refrigerant because it has a higher temperature than the refrigerant. As the liquid refrigerant absorbs heat from the evaporating medium, the refrigerant boils and vaporizes.

Refrigerant lines carry refrigerant and connect the components of a mechanical compression refrigeration system. Accessories monitor the system to ensure proper operation.

Pressure-Temperature Relationships

Pressure is the force per unit of area that is exerted by an object or a fluid. Pressure is expressed in pounds per square inch (psi). *Atmospheric pressure* is the force exerted by the weight of the atmosphere on the surface of the Earth. Atmospheric pressure is measured at sea level and is normally expressed in pounds per square inch absolute (14.7 psia). Atmospheric pressure is measured precisely with a mercury barometer. A *mercury barometer* is an instrument used to measure atmospheric pressure and is calibrated in inches of mercury absolute (29.92 in. Hg abs.). A mercury barometer consists of a glass tube that is closed on one end and filled completely with mercury. The tube is inverted in a dish of mercury. A vacuum is created at the top of the tube as the mercury tries to run out of the tube. *Vacuum* is a pressure lower than atmospheric pressure. Vacuum is expressed in inches of mercury (in. Hg). The pressure of the atmosphere on the mercury in the open dish prevents the mercury in the tube from running out of the tube. The height of the mercury in the tube corresponds to the pressure of the atmosphere on the mercury in the open dish. Minute pressure changes can be expressed in inches of water column (in. WC). **See Figure 12-2.**

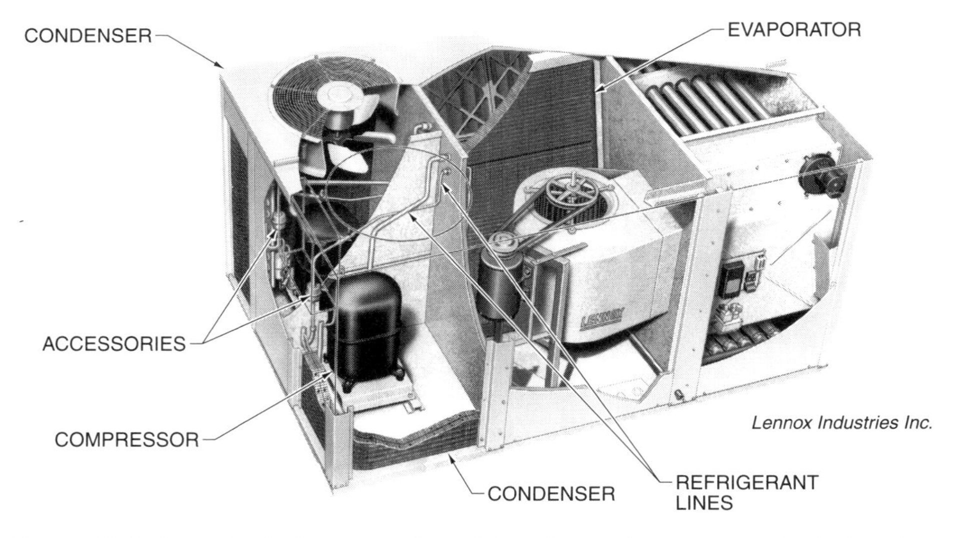

Figure 12-1. A mechanical compression refrigeration system uses mechanical equipment to produce a refrigeration effect.

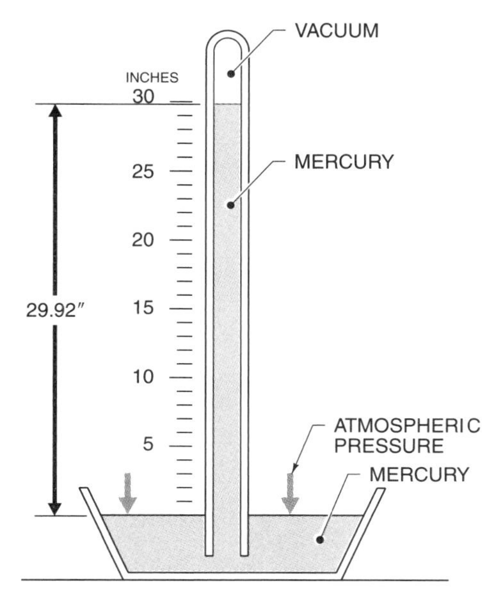

	PSI	PSIA	Hg ABS*	Hg*	WC*
2 Atmosphere	14.7	29.4	59.84	–	817
	12	26.7	54	–	738
	6	20.7	42	–	574
	1	15.7	32	–	438
1 Atmosphere	0	14.7	29.92 (30)	0	409
	–	9.8	20	10	273
	–	2.9	6	24	82
	–	1	2.035	28	28
	–	.491	1	29	14
	–	.036	.0732	29.85	1
Perfect Vacuum	–	0	0	29.92 (30)	0

* in in.

Figure 12-2. A mercury barometer is used to measure atmospheric pressure.

Gauge pressure is pressure above atmospheric pressure that is used to express pressures inside a closed system. Gauge pressure is expressed in pounds per square inch gauge (psig or psi). *Absolute pressure* is any pressure above a perfect vacuum (0 psia). Absolute pressure is the sum of gauge pressure plus atmospheric pressure. Absolute pressure is expressed in pounds per square inch absolute (psia).

Pressure outside a closed system such as normal air pressure is expressed in pounds per square inch absolute. The difference between gauge pressure and absolute pressure is the pressure of the atmosphere at sea level with standard conditions (14.7 psia).

A pressure gauge reads 0 psi at normal atmospheric pressure. To find absolute pressure when gauge pressure is known, add the atmospheric pressure of 14.7 to the gauge pressure. Absolute pressure is found by applying the formula:

$psia = psig + 14.7$

where

$psia$ = pounds per square inch absolute

$psig$ = pounds per square inch gauge

14.7 = constant

Example: Finding Absolute Pressure

A gauge reads 68 psi on the low-pressure side of an operating refrigeration system. Find the absolute pressure.

$psia = psig + 14.7$

$psia = 68 + 14.7$

$psia =$ **82.7 psia**

Boiling point is the temperature at which a liquid vaporizes. Boiling point of a liquid is directly related to the pressure on the liquid. If the pressure on a liquid increases, the boiling point increases. If the pressure on a liquid decreases, the boiling point of the liquid decreases. For example, at 14.7 psia the boiling point of water is 212°F. If the pressure on the water increases, the boiling point of the water increases. At 29.8 psia, the boiling point of water is 250°F. If the pressure on the water decreases, the boiling point of water decreases. At 11.0 in. Hg or 9.3 psia (vacuum), the boiling point of water is 190°F.

Condensing point is the temperature at which a vapor condenses to a liquid. If the pressure on a vapor decreases, the temperature at which the liquid condenses decreases. If the pressure on a vapor increases, the temperature at which the vapor condenses increases. All substances follow this pressure-temperature relationship.

Heat Transfer

In a mechanical compression refrigeration system, heat transfer occurs in the condenser and the evaporator. In the condenser, heat is transferred from the refrigerant that flows through the condenser to the condensing medium that passes across the outside of the condenser. The refrigerant condenses as it rejects heat to the condensing medium. This process heats the condensing medium. **See Figure 12-3.**

In the evaporator, heat is transferred from the evaporating medium that passes across the outside of the evaporator to the refrigerant that flows through the evaporator. The refrigerant vaporizes as it absorbs heat from the evaporating medium. This process cools the evaporating medium.

Pressure Control

Mechanical compression refrigeration systems have a high-pressure side and a low-pressure side. Pressure is controlled in the low-pressure side by reduced refrigerant flow, reduced pressure, and relatively low compressor suction. **See Figure 12-4.**

The expansion device meters the flow of refrigerant in a refrigeration system. An expansion device is located just ahead of the evaporator in the liquid line. The *liquid line* is the refrigerant line that connects the condenser and the expansion device.

As liquid refrigerant flows through the expansion device, the pressure on the refrigerant decreases. The decreased pressure causes some of the refrigerant to vaporize. This vaporization takes heat from and decreases the temperature of the rest of the liquid refrigerant. The liquid-vapor mixture flows from the expansion device directly to the evaporator. The pressure of the refrigerant remains the same through the low-pressure side of the system except for decreases in pressure caused by the evaporator, lines, and fittings. **See Figure 12-5.**

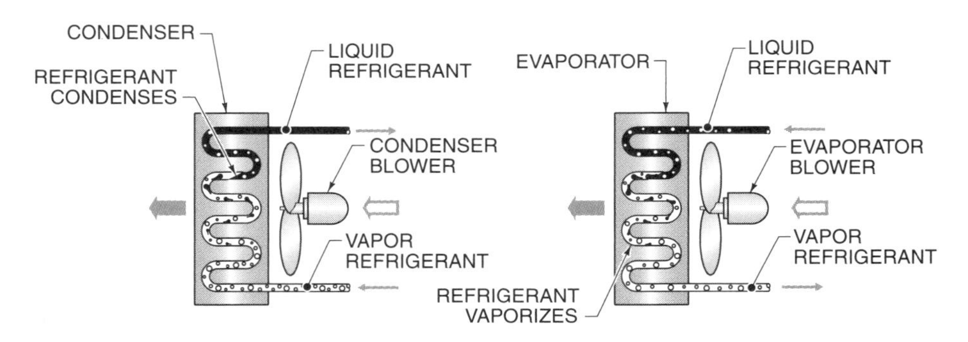

Figure 12-3. In a mechanical compression refrigeration system, heat transfer occurs in the condenser and the evaporator.

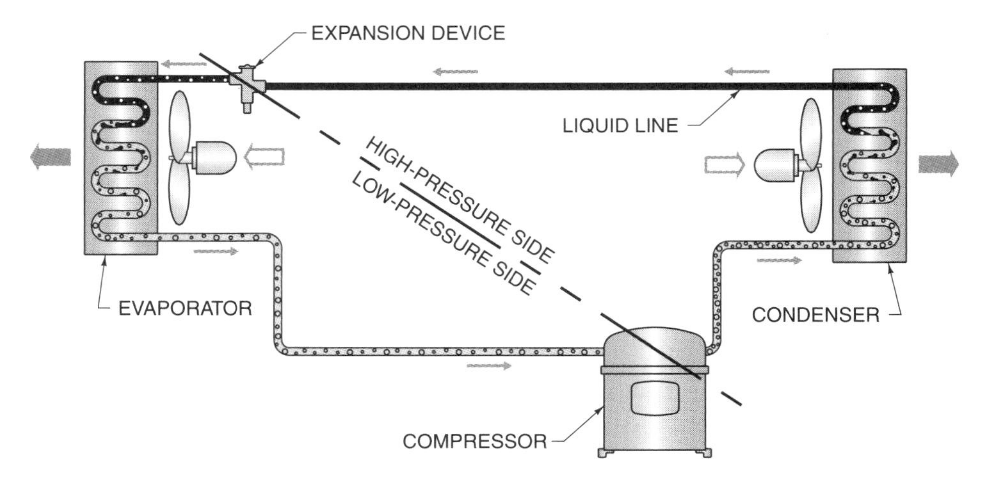

Figure 12-4. Mechanical compression refrigeration systems have a high-pressure side and a low-pressure side.

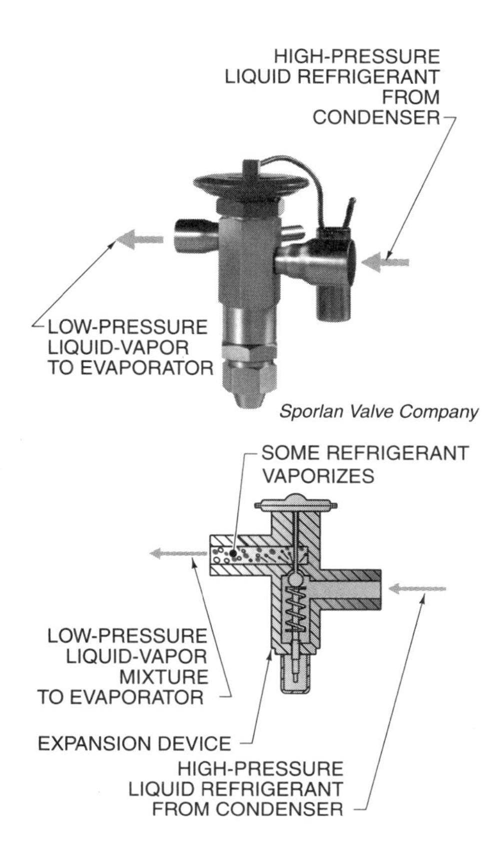

Figure 12-5. As liquid refrigerant flows through an expansion device, the decrease in pressure causes some of the refrigerant to vaporize.

Compressor suction pressure is the pressure created by the compressor when refrigerant is drawn into the compressor. The compressor suction maintains low pressure in the low-pressure side of the system. Because refrigerant is drawn out of the evaporator to the inlet of the compressor (suction port) as fast as it is introduced through the expansion device, the pressure on the low-pressure side of the system remains constant.

Compressor discharge flow maintains the high pressure in the high-pressure side of a refrigeration system. *Compressor discharge pressure* is the pressure created by the resistance to flow of the compressor when refrigerant is discharged from the compressor. The compressor compresses and circulates the refrigerant. Refrigerant leaves the evaporator as a vapor and flows to the compressor. The refrigerant vapor is compressed in the compressor and leaves the compressor at a higher pressure. **See Figure 12-6.** Because the refrigerant absorbs heat from the compression process and possibly the compressor motor windings, the refrigerant that leaves the compressor is hotter than the refrigerant in the rest of the refrigeration system. The hot refrigerant vapor discharged from the compressor flows to the condenser. The pressure in the high-pressure side of the system is constant except for minor pressure decreases created by the friction of fittings and refrigerant lines.

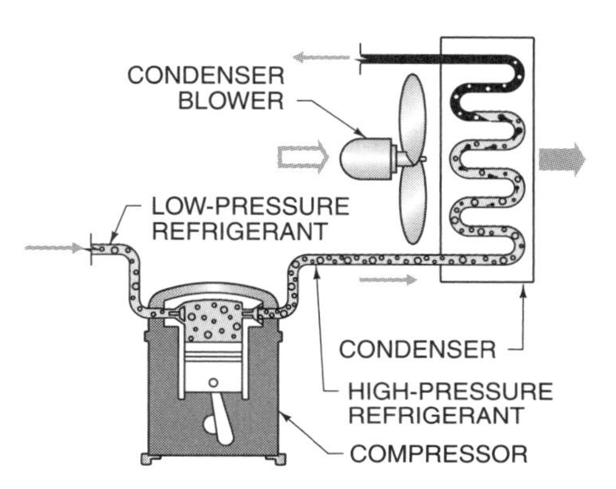

Figure 12-6. Refrigerant vapor is compressed in the compressor and leaves the compressor at a higher pressure and temperature.

Because of the low pressure in the evaporator and compressor suction port, the temperature of the refrigerant is decreased to a temperature lower than the temperature of the evaporating medium. The temperature difference causes heat to flow from the evaporating medium to the refrigerant. Because of the high pressure in the compressor discharge port and condenser, the temperature of the refrigerant is increased to a temperature higher than the temperature of the condensing medium. This temperature difference causes heat to flow from the refrigerant to the condensing medium.

An example of a mechanical compression refrigeration system is a household refrigerator. Inside a refrigerator, refrigerant circulates through the evaporator and absorbs heat from food. This heat vaporizes the refrigerant. The refrigerant vapor is drawn out of the evaporator by the compressor suction. The compressor compresses the refrigerant, which raises the pressure and temperature of the refrigerant. The refrigerant circulates to the condenser, which is located on the rear of the refrigerator. The heat absorbed from the food inside the refrigerator is released to the air outside the refrigerator by the condenser. **See Figure 12-7.**

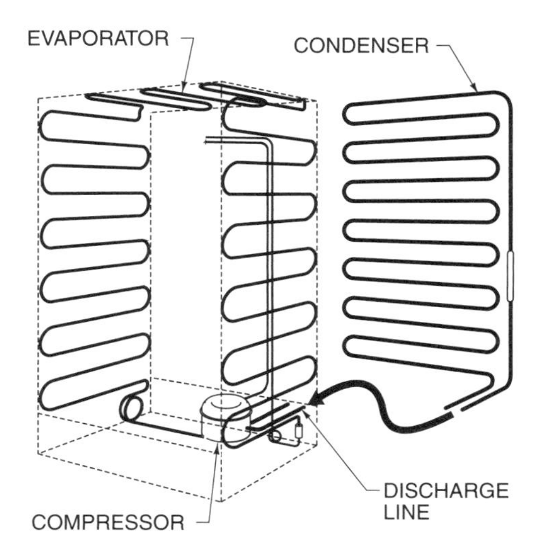

Figure 12-7. A household refrigerator removes heat from food and releases the heat to the air outside the refrigerator through the condenser.

Refrigerants

The first substance to be used as a refrigerant was probably ice, which was used for cooling food and drinks. As ice melts, it changes state from a solid to liquid. The ice absorbs heat from the air or liquid that surrounds it. The removal of heat cools the air or liquid. Water in the liquid state is also used as a refrigerant. In dry climates, water-soaked canvas bags are used to cool water. As the water evaporates, it absorbs heat from the water inside the bag. The evaporation provides a cooling effect on the water inside the bag.

In the early 1900s, mechanical refrigeration systems that used air or chemicals as refrigerants became common. Some of the chemicals used were ammonia, carbon dioxide, sulfur dioxide, methyl chloride, and hydrocarbons. Most of these chemicals have properties that make using them impractical. Some of the chemicals require very high pressures, requiring heavy equipment, and consume large amounts of power. Some of these refrigerants are toxic, flammable, and/or corrosive.

Halocarbon refrigerants were first developed by the General Motors Corporation. The refrigerants were later made jointly with the Du Pont Co. Du Pont produced refrigerants in their Freon® division. Freon® is a registered trademark of the Du Pont Co. and has been accepted as the common name for all halocarbon refrigerants.

Refrigerants used in systems today are derived from methane (CH_4) or ethane (C_2H_6). The base molecule of methane or ethane is halogenated by a fluorine atom. The important factor is that the molecule does not contain a chlorine atom. These refrigerants are known as hydrofluorocarbons (HFCs) and are considered safe to the environment. The most popular HFCs are R-134a, R-152a, R-32, R-143a, and R-125. HFC refrigerants operate at pressures easily attained in a refrigeration system, and are nontoxic, nonflammable, and relatively safe to handle.

Refrigerants are identified by numbers. The numbers are assigned according to the physical and chemical composition of a refrigerant. Numbers less than 170 are assigned to refrigerants based on the chemical composition of the refrigerant. Numbers above 170 are assigned arbitrarily.

Refrigerants derived from methane are assigned two-digit numbers. The first digit in two-digit numbers equals one more than the number of hydrogen atoms in each molecule. The second digit is the number of fluorine atoms in each molecule. Refrigerants derived from ethane are assigned three-digit numbers under 170. The first digit in three-digit numbers is arbitrarily assigned. The second digit in three-digit numbers equals one more than the number of hydrogen atoms in each molecule. The third digit is the number of fluorine atoms in each molecule. Atoms that are not accounted for are chlorine atoms. For example, the chemical formula for refrigerant 32 (R-32) is CH_2F_2 and the chemical name is difluoromethane. R-32 is a methane-base refrigerant that contains two atoms of fluorine and two atoms of hydrogen, which leaves room for two atoms of chlorine. The carbon atom is from the original ethane. The number of a refrigerant identifies a particular refrigerant regardless of the manufacturer. **See Figure 12-8.**

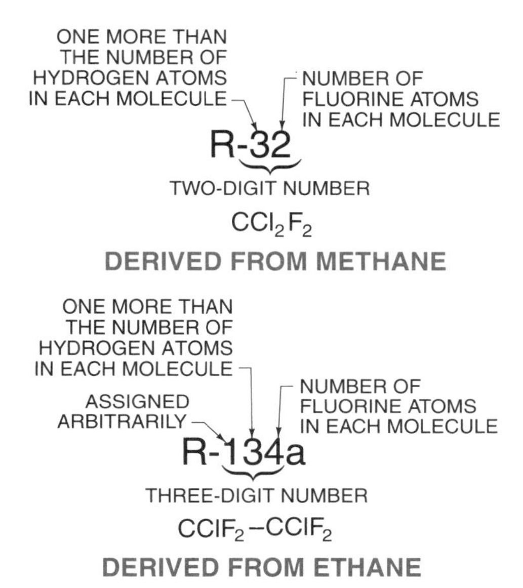

Figure 12-8. Refrigerants derived from methane are assigned two-digit numbers. Refrigerants derived from ethane are assigned three-digit numbers under 170.

There is concern about the effect of refrigerants on the atmosphere of the Earth. Refrigerants that use chlorofluorocarbons (CFCs) and fully halogenated chlorofluorocarbons (HCFCs) such as R-11, R-12, and R-22 are believed to be responsible for the depletion of the Earth's ozone layer and may cause global warming. As a result, venting CFCs from equipment was banned, with the CFC refrigerants being phased out of production in 1995 to comply with the provisions of the Clean Air Act of 1992. Contractors and HVAC technicians must be certified in refrigerant recovery and must document refrigeration system maintenance as required by federal, state, and local clean air regulations. When a refrigeration system is serviced, the refrigerant must be recovered and may be reclaimed or recycled if not CFC.

When used refrigerants are recycled, they are recovered, cleaned, and replaced into the refrigeration system. Recycling a refrigerant increases the risks of poor equipment performance. When used refrigerants are reclaimed, they are recovered and shipped to the manufacturer for purification. The refrigeration system is then refilled with new refrigerant, which guarantees proper system performance.

As more CFC and HCFC refrigerants are phased out and supplies disappear, companies will have to convert to alternative refrigerants that have improved environmental characteristics. Du Pont Co. is currently producing new SUVA® alternative refrigerants. SUVA® is the Du Pont Co. registered trademark for the new alternative refrigerants. SUVA® refrigerants are currently available for new and retrofitted systems. *Retrofitting* is the process of furnishing a system with new parts that were not available at the time the system was manufactured. Refrigeration systems that currently use CFC refrigerants can be retrofitted with new equipment to make the system compatible with the new, safer refrigerants. SUVA® refrigerants provide similar performance to CFCs with reduced potential for ozone depletion and global warming.

Properties. Thermodynamic and physical properties are the most important properties of a refrigerant. Properties of refrigerants determine efficiency, cooling, and flow rates. Properties are the amount of heat involved in a change of state, the temperature at which a change of state occurs, and the volume of the refrigerant in liquid and vapor states. Refrigerant property tables contain information about thermodynamic and physical properties of refrigerants.

A *refrigerant property table* is a table that contains values for and information about the properties of a refrigerant at saturation conditions or at other pressures and temperatures. **See Figure 12-9.** *Saturation conditions* are the temperature and pressure at which a refrigerant changes state. The properties listed on a refrigerant property table are volume, density, enthalpy, and entropy.

Determining the efficiency of the heat transfer in the evaporator and condenser is done by comparing the temperatures of the refrigerant in the evaporator and condenser to the temperatures of the evaporating and condensing mediums. The volume and quantity of refrigerant needed to provide a certain amount of cooling in a system determines the size of the components needed to hold the refrigerant. Refrigerants are selected for a specific application based on refrigerant properties. Desirable refrigerant properties include low boiling point, low freezing point, nontoxic, high critical point, low specific volume, high density, high latent heat, low compression ratio, nonflammable, noncorrosive, stable, and miscible with oil.

A refrigerant must have a boiling point that is lower than the desired temperature of the air that leaves the evaporator. The actual boiling point is controlled by the pressure in the low-pressure side of the system. The boiling point should be low enough that the temperature can be attained without special equipment. The pressure in the low-pressure side should be above atmospheric pressure. Pressure above atmospheric pressure prevents air from leaking into the system if an opening occurs.

The refrigerant in an air conditioning system must have a boiling point that is at least 20°F lower than the temperature of the air that leaves the evaporator. Because the temperature of the air that leaves the evaporator is about 60°F, the temperature of the refrigerant must be about 40°F or lower. R-134A is often used in air conditioning systems. The boiling point of R-134A is about –15°F at atmospheric pressure.

Refrigerants with lower boiling points are required for lower-temperature refrigeration systems. R-32 has a boiling point of –61.24°F at normal atmospheric pressure and is used in medium-temperature applications. **See Figure 12-10.** R-13, R-14, or R-503 were used for lower-temperature applications as low as –130°F. These refrigerants are used in two-stage compression or cascade systems. A *two-stage compression system* is a compression system that uses more than one compressor to raise the pressure of a refrigerant. The system raises the pressure above the pressure that can be achieved with a single compressor. A *cascade system* is a compression system that uses one refrigeration system to cool the refrigerant in another system.

Tech Facts

A refrigerant is a fluid that picks up heat by evaporating at a low temperature and pressure and gives up heat by condensing at a higher temperature and pressure.

R-134A PROPERTIES OF SATURATED LIQUID AND SATURATED VAPOR

Temp. °F	Pressure psia	Volume§	Density#	Enthalpy**		Entropy§§	
		Vapor	Liquid	Liquid	Vapor	Liquid	Vapor
–150	0.073	454.5455	98.86	-31.2	80.9	-0.0859	0.2761
–140	0.130	256.4103	97.91	-28.4	82.3	-0.0771	0.2693
–130	0.222	156.2500	96.96	-25.6	83.8	-0.0686	0.2633
–120	0.366	97.0874	96.01	-22.9	85.2	-0.0604	0.2579
–110	0.584	62.5000	95.06	-20.1	86.7	-0.0523	0.2531
–100	0.903	41.6667	94.11	-17.3	88.2	-0.0444	0.2488
–90	1.358	28.4091	93.15	-14.5	89.6	-0.0367	0.2450
–85	1.649	23.6407	92.68	-13.1	90.4	-0.0329	0.2433
–80	1.991	19.8413	92.20	-11.6	91.1	-0.0291	0.2416
–75	2.389	16.7224	91.72	-10.2	91.9	-0.0254	0.2401
–70	2.850	14.1844	91.24	-8.8	92.7	-0.0217	0.2386
–65	3.384	12.0773	90.76	-7.3	93.4	-0.0180	[illegible]

Figure 12-9. Refrigerant property tables contain values for and information about the properties of a refrigerant at saturation condition or at other pressures and temperatures.

A refrigerant should have a low freezing point. A refrigerant should be in the liquid state or the vapor state at all times and should never be in the solid state in a refrigeration system. If a refrigerant changes to a solid at normal operating pressures, the equipment will not operate. A refrigerant with a low freezing point does not change from a liquid to a solid at the temperatures in a typical refrigeration system.

A refrigerant should be nontoxic. Toxic materials can cause illness or death if inhaled or absorbed through the skin. A refrigerant should be nontoxic because refrigerant could escape when a system is charged or serviced and because refrigerant leaks could occur during operation.

A refrigerant should have a high critical point. *Critical point* is the pressure and temperature above which a material does not change state regardless of the absorption or rejection of heat. All refrigerants have critical points. If a system operates with a high-pressure side close to the critical point of the refrigerant, the cost of operation will be high because the compressor must circulate the refrigerant against the higher pressure. For efficiency, a system should operate with the high-pressure side far below the critical point of the refrigerant.

Specific volume is the volume of a substance per unit of the substance. The specific volume of a refrigerant is the volume (in cubic feet) that 1 lb of the refrigerant occupies at a given pressure. In a system at operating pressures, a refrigerant that has a high specific volume occupies more space than a refrigerant that has a low specific volume. Because circulating a large volume of refrigerant requires more energy, operating a system that uses a refrigerant with a high specific volume costs more than operating a system that uses a refrigerant with a low specific volume.

The volume of refrigerant that a compressor moves at required pressures determines the size of the compressor. A refrigeration system that has a refrigerant with a high specific volume requires a larger compressor than a refrigeration system that has a refrigerant with a low specific volume.

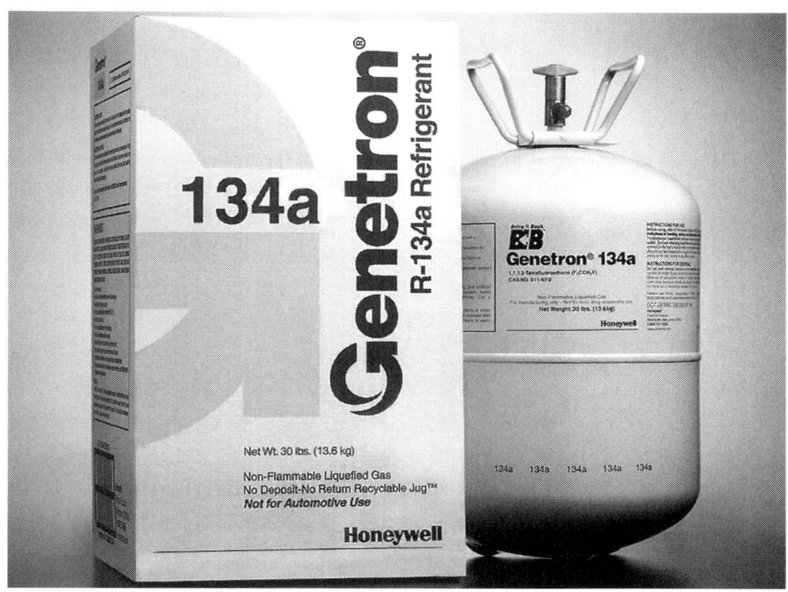

Honeywell Chemicals

Refrigerant containers must be labeled, tagged, and marked with the appropriate hazard warning per OSHA 29 CFR 1910.1200(f)—Labels and Other Forms of Warning.

Density is the weight of a substance per unit of volume. The density of a refrigerant is the weight (in pounds) of 1 cu ft of the refrigerant. Density is the reciprocal of the specific volume at a given pressure. If a refrigerant has a low specific volume, it will have a high density. A system that circulates a high-density refrigerant operates more efficiently than a system that circulates a low-density refrigerant. A centrifugal compressor, however, is designed to move large volumes of refrigerant at relatively low pressures. If a refrigerant with a low density is used in a system with a centrifugal compressor, the system will operate inefficiently.

R-32 REFRIGERANT			
Chemical Name	**Chemical Formula**	**Molecular Mass**	**Boiling Point***
Difluoromethane	CH2F2	52.02	-61.24
Freezing Point*	**Critical Temperature***	**Critical Pressure†**	**Critical Volume‡**
-212.8	173.12	846	.0821

* in °F
† in psi
‡ in cu ft/lb

Figure 12-10. Refrigerants with lower boiling points are required for lower temperature refrigeration systems.

A refrigerant should have high latent heat content. Latent heat is heat identified by a change of state and no temperature change. Heat transfer in a refrigeration system causes a refrigerant to change state from a liquid to a vapor in the evaporator and from a vapor to a liquid in the condenser. The refrigeration effect of a system is the amount of latent heat involved in the change in the evaporator.

A refrigerant should be nonflammable and noncorrosive. A flammable refrigerant could ignite when work such as brazing and soldering is done during installation or repair of a refrigeration system.

A refrigerant should be noncorrosive. A corrosive substance reacts chemically with materials and causes corrosion of surfaces and components. A refrigerant must not react chemically with any of the materials in a refrigeration system. A corrosive refrigerant breaks down parts of the system such as the refrigerant lines, the compressor, the motor in a hermetic compressor, or materials that may be used for brazing, soldering, or repairing a leak. A *hermetic compressor* is a compressor in which the motor and compressor are sealed in the same housing. If acid in a system causes severe corrosion of the refrigerant lines, the compressor could form a copper plating. Corrosion or plating of mechanical parts causes system failure.

A refrigerant should be stable at the pressures and temperatures in a refrigeration system. When exposed to temperatures and/or pressures above the critical point, a nonstable refrigerant can break down chemically into base compounds or recombine into new compounds. If a refrigerant is exposed to high temperatures and pressures, the properties of the refrigerant will change. A refrigerant must be stable at all pressures and temperatures within a system including high pressures and temperatures.

A refrigerant should be miscible with oil. *Miscibility* is the ability of a substance to mix with other substances. The oil that lubricates a compressor circulates with the refrigerant in a refrigeration system. A refrigerant must be able to mix with oil so that the refrigerant can carry the oil through the compressor. If the oil separates from the refrigerant and remains at any other point in the system, the oil can block heat transfer in the evaporator and condenser, resulting in a reduced refrigeration effect.

CFCs, HCFCs, and HFCs. The three most common classifications of refrigerants are chlorofluorocarbons (CFCs), hydrochlorofluorocarbons (HCFCs), and hydrofluorocarbons (HFCs). **See Figure 12-11.** CFCs contain chlorine, fluorine, and carbon. They were phased out in 1995 because they contain chlorine and they decompose in the Earth's stratosphere by reacting with the sun's ultraviolet radiation. Although CFCs are no longer manufactured, technicians must deal with them in older refrigeration equipment that has not been converted to the newer refrigerants. CFC refrigerants include R-11, R-12, R-113, R-114, and R-115.

STANDARD CFC, HCFC, AND HFC REFRIGERANTS		
R-11	Trichlorofluoromethane	CFC
R-12	Dichlorodifluoromethane	CFC
R-13	Chlorotrifluoromethane	CFC
R-13B1	Bromotrifluoromethane	CFC
R-22	Chlorodifluoromethane	HCFC
R-23	Trifluoromethane	HFC
R-32	Difluoromethane	HFC
R-113	Trichlorotrifluoroethane	CFC
R-114	Dichlorotetrafluoroethane	CFC
R-123	Dichlorotrifluoroethane	HCFC
R-124	Chlorotetrafluoroethane	HCFC
R-125	Pentafluoroethane	HFC
R-134a	Tetrafluoroethane	HFC
R-401a	Blend	HCFC
R-401b	Blend	HCFC
R-404a	Blend	HCFC
R-407a	Blend	HFC
R-407c	Blend	HFC
R-410a	Blend	HFC
R-500	Blend	CFC
R-502	Blend	CFC
R-507a	Blend	HFC
5-717	Ammonia	Inorganic

Figure 12-11. Refrigerants are classified as chlorofluorocarbons (CFCs), hydrochlorofluorocarbons (HCFCs), and hydrofluorocarbons (HFCs).

HCFCs contain hydrogen, chlorine, fluorine, and carbon. HCFCs are less stable in the atmosphere and break down more easily, releasing most of the chlorine before they reach the stratosphere. That is because of the added hydrogen atom. But because some chlorine does still reach the stratosphere, they can affect the environment. HCFCs are scheduled to be phased out in 2030. They include R-22, R-123, and R-124. R-22 is important because it is used in building air conditioning systems and in centrifugal chillers.

HFCs contain hydrogen, fluorine, and carbon. Because HFC refrigerants do not contain chlorine they are considered safe for the environment. Many are the replacements for CFCs and HCFCs. For example, R-134a can replace R-12 if some modifications to the system are done. The oil used with R-12 is not compatible with R-134a so a complete oil change is necessary. Some R-12 systems contain gaskets and seals that are not compatible with the R-134a refrigerants, requiring that the gaskets and seals be changed. All of the old systems that contained CFCs and HCFCs need to be modified in some manner before a new refrigerant can be used.

Mechanical Compression System Operation

A mechanical compression refrigeration system produces a refrigeration effect with mechanical equipment. For example, in a mechanical compression refrigeration system, R-407C enters the evaporator of a refrigeration system with a pressure of 68.5 psig, a temperature of 40°F, and a heat content of approximately 34.4 Btu/lb. **See Figure 12-12.** The boiling point of R-407C at 68.5 psig is 40°F. The temperature of the refrigerant remains at about 40°F as it moves through the evaporator. Air that is about 80°F passes across the outside of the evaporator and is warmer than the refrigerant in the evaporator. The refrigerant absorbs the heat and vaporizes.

The evaporator is large enough to allow the refrigerant to completely vaporize before it leaves the evaporator. The refrigerant absorbs superheat in the last few rows of coils in the evaporator. *Superheat* is heat added to a material after it has changed state. The refrigerant leaves the evaporator at a temperature higher than the saturated temperature for its pressure. The refrigerant has more heat than if it is saturated because of the superheat absorbed by the refrigerant in the evaporator.

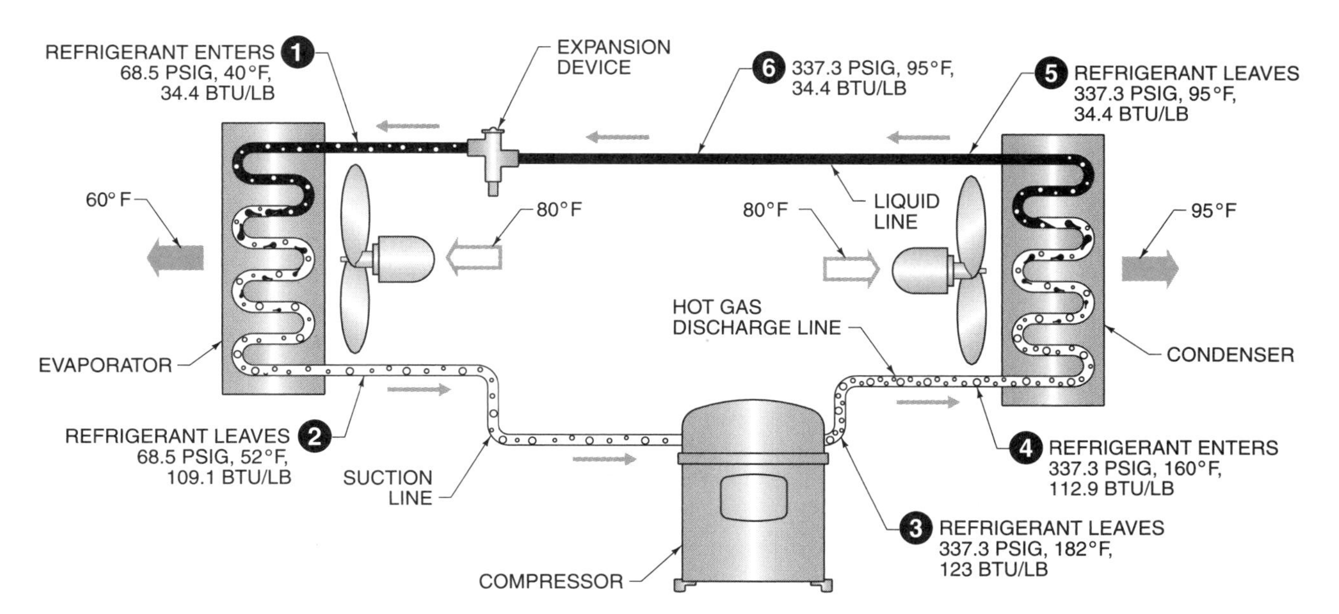

Figure 12-12. Mechanical compression refrigeration produces a refrigeration effect with mechanical equipment.

The refrigerant vapor then leaves the evaporator with a pressure of 68.5 psig, a temperature of 52°F, and a heat content of approximately 109.1 Btu/lb. The refrigerant absorbs heat from the evaporating medium, which raises the temperature and heat content of the refrigerant. The refrigerant leaves the evaporator through the suction line.

In the compressor the pressure of the refrigerant rises to 337.3 psig. The temperature of the refrigerant rises because of the heat added by compression and, in certain compressors, from cooling the compressor motor windings.

The refrigerant vapor then leaves the compressor with a pressure of 337.3 psig, a temperature of 182°F, and a heat content of 123 Btu/lb of refrigerant. The saturated temperature of the refrigerant at 337.3 psig is 140°F, but the actual temperature of refrigerant as it leaves the compressor is about 182°F because the refrigerant is superheated. The refrigerant leaves the compressor through the hot gas discharge line and moves to the condenser.

The *hot gas discharge line* is the line that connects the compressor to the condenser. The hot gas discharge line contains the hot gas (refrigerant vapor) that is cooled in the condenser. While the refrigerant moves through the hot gas discharge line, it loses some of the superheat it absorbed in the compressor. By the time the refrigerant reaches the condenser, it is close to saturated temperature.

The refrigerant then enters the condenser from the hot gas discharge line with a pressure of 337.3 psig, a temperature of 140°F, and a heat content of 112.9 Btu/lb. Heat flows from the refrigerant to the condensing medium because the temperature of the condensing medium is lower than the temperature of the refrigerant in the condenser. The amount of heat rejected by the refrigerant in the condenser is the same as the amount of heat absorbed by the refrigerant in the evaporator and compressor. As the refrigerant rejects heat in the condenser, the refrigerant changes state from a vapor back to a liquid.

The R-407C refrigerant leaves the condenser at the same pressure, 337.3 psig, but it is now a liquid. Most condensers have extra capacity so the refrigerant completely condenses to a liquid and extra cooling (subcooling) takes place in the liquid state. *Subcooling* is the cooling of a material such as a refrigerant to a temperature that is lower than the saturated temperature of the material for a particular pressure. Because of subcooling, the refrigerant leaves the condenser with a temperature of 95°F and a heat content of approximately 34.4 Btu/lb.

The refrigerant leaves the condenser through the liquid line and moves either directly to the expansion device or to a receiver tank. In either case the refrigerant enters the expansion device at the same pressure and temperature that it had when it left the condenser.

The refrigerant next flows through the expansion device. The restriction in the expansion device causes a pressure decrease. The pressure decrease is the difference between the high-pressure side and low-pressure side of the system. The decreased pressure allows 15% of the refrigerant to vaporize, which causes a temperature decrease from 95°F to 40°F. The boiling point of the refrigerant on the low-pressure side is 40°F. From the expansion device, the refrigerant enters the evaporator at 68.5 psig as a liquid-vapor mixture and the cycle begins again.

CHILLERS

A *chiller* is a component in a hydronic air conditioning system that cools water, which cools the air. The chilled water is circulated to the cooling coils of a building at about 45°F. It increases about 10°F through a cooling coil and is returned to the chiller at about 55°F to be cooled again. The three basic types of chillers are high-pressure chillers, low-pressure chillers, and absorption chilled-water systems. **See Figure 12-13.**

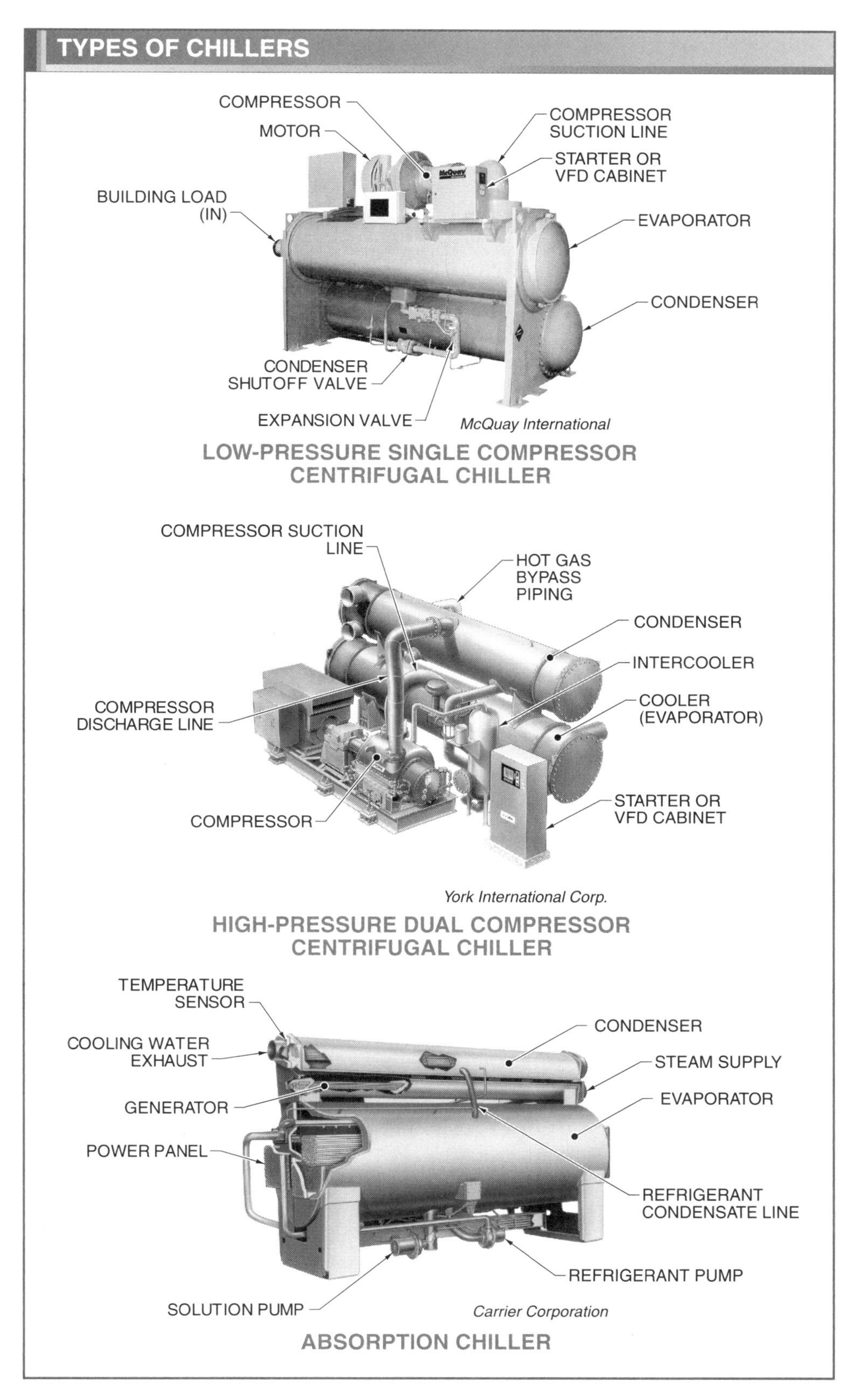

Figure 12-13. The three types of chillers used today in commercial buildings are high-pressure chillers, low-pressure chillers, and absorption chilled-water systems.

High- and Low-Pressure Chillers

A high-pressure chiller operates at 15 psi and a low-pressure chiller operates at 15 psi or lower. High-pressure and low-pressure chillers use a compressor in the compression cycle to create the pressure differences inside the chiller to vaporize and condense the refrigerant. Compression cycle chillers have components similar to basic refrigeration systems, such as a compressor, a condenser, metering devices, and evaporator. Low-pressure chillers also have a purge unit to remove "non-condensables" from the system.

Compressors. The compressors commonly used in high-pressure chillers are reciprocating, scroll, and screw. Low-pressure chillers use centrifugal compressors. The compressor is the pumping component that controls the pressure in the evaporator and condenser. This pressure corresponds to a design evaporating temperature of 38°F and condensing temperature of 105°F. For R-407C the evaporator pressure is about 35 psi and the condenser pressure is about 127 psi, which would be a high-pressure chiller. A low-pressure chiller might use R-123. R-123 has an evaporator pressure of about 7.9 psi and a condenser pressure of about 11 psi.

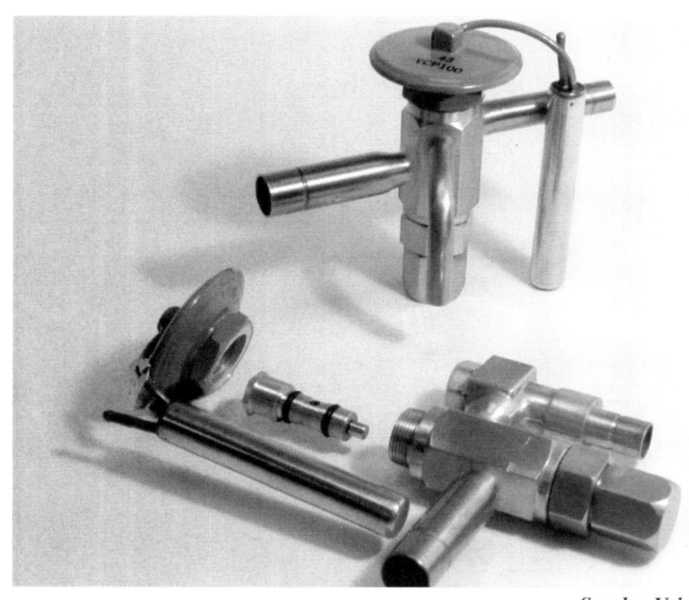

Sporlan Valve Company

Thermostatic expansion valves use the refrigerant discharge temperature from an evaporator to control the amount of refrigerant entering the evaporator.

Reciprocating compressors used for chillers are similar to those used in other air conditioning and refrigeration systems. Most reciprocating compressors have multiple stages to allow higher system pressures. This is accomplished by having two or more cylinders on one, two, or more compressors. For example, a chiller may have four stages of compression utilizing two reciprocating compressors.

Scroll compressors are also positive-displacement compressors. When used in chillers they normally are in the 10 ton to 15 ton size and operate the same as the small units. A ton of cooling is the amount of heat required to melt a ton of ice in a 24-hour period. The capacity control of the chiller is maintained by cycling the compressors ON and OFF as needed. For example, a 25-ton chiller may have two scroll compressors; one 10-ton unit and the other a 15-ton unit. The chiller has a capacity control of 10, 15, or 25 tons of cooling. Two advantages a scroll compressor has over a reciprocating compressor are that scroll compressors run quieter and are able to handle small amounts of liquid refrigerant.

Rotary screw compressors used in chillers are large capacity, positive-displacement compressors that have few moving parts. These compressors are reliable, trouble-free units that can handle slugs of liquid refrigerant. Capacity control is maintained by a slide valve that modulates open and closed to control the amount of refrigerant admitted to the compressor as determined by the cooling load.

Centrifugal compressors used in chillers utilize the centrifugal force applied to the refrigerant by a fast-spinning impeller. The motor of the compressor is directly connected to a transmission that can have gear ratios of up to 9 to 1. When the speed of the motor is 3450 rpm, the impeller on some high-pressure, single-stage compressors may approach 30,000 rpm. Centrifugal compressors do not have a great deal of force, but they can handle large volumes of refrigerant. If a greater pressure differential is required than one impeller can produce, multiple impellers (stages) are operated in series. The discharge of one impeller enters the inlet (eye) of the next impeller. Centrifugal chillers are available in units with capacity ratings of 200 tons and up.

Condensers. A *condenser* is a heat exchanger that removes heat from high-pressure refrigerant vapor. In a water-cooled system, the condenser is usually a two-pass, tube-and-shell heat exchanger. The refrigerant is in the tubes and the water surrounds the tubes. The refrigerant in the condenser transfers heat to the water, raising the water temperature to about 95°F. The heated water leaves the condenser and is circulated to a cooling tower where the water transfers its heat to the surrounding air. The water is circulated back to the condenser, at a lower design temperature of 85°F, to start the process again.

Metering Devices. A *metering device* is a valve or orifice in a refrigeration system that controls the flow of refrigerant into the evaporator to maintain the correct evaporating temperature. For large chillers there are four types of metering devices used: thermostatic expansion valve, orifice, high-side or low-side float, and electronic expansion valve. Low-pressure chillers normally use an orifice-type or a float-type metering device.

A thermostatic expansion valve (TXV) metering device used on chillers is the same type used in other refrigeration and air conditioning systems, except that larger ones are used to accommodate high refrigerant flows. A thermostatic expansion valve (TXV) senses the temperature of the refrigerant discharged from the evaporator and controls the liquid refrigerant flowing into the evaporator. The TXV maintains a constant superheat, usually of 8°F to 12°F. TXVs are used on chillers under 150 tons of cooling capacity.

An *orifice-type metering device* is a small, fixed opening which is used as a restriction in the liquid line between the condenser and the evaporator of a refrigeration system. The flow of refrigerant through the orifice is determined by the pressure differential between the high side and low side. As the cooling load increases, the condenser pressure increases, which causes an increase in the pressure differential, creating a higher refrigerant flow rate through the orifice. The amount of refrigerant allowed to enter a chiller using an orifice metering device is critical because overcharging can cause liquid refrigerant to enter the compressor.

High-side or low-side float metering devices are used on chillers with flooded evaporators. The high-side float is located on the liquid line to the evaporator. As the cooling load increases, more refrigerant is boiled off in the evaporator. That also means that the condenser is condensing more refrigerant, so the liquid level in the float chamber increases. The float ball rises with the liquid level, allowing more liquid refrigerant to flow into the evaporator.

A chiller that utilizes a low-side float metering device has the float located at the normal refrigerant level in the evaporator. As more refrigerant is boiled off on an increase in load, the liquid refrigerant level decreases. A drop in the level of the refrigerant causes the float ball to drop, opening a valve to allow more liquid refrigerant into the evaporator from the condenser.

Electronic expansion valve (EXV) metering devices are becoming more popular in large chillers. These devices are similar to EXVs used in other refrigeration and air conditioning equipment. The sensor for the EXV is a thermistor mounted in the liquid line from the evaporator to monitor the temperature of the refrigerant vapor. A signal is sent to the EXV to maintain a given superheat temperature. EXVs are capable of higher flow rates and are able to operate with lower condenser pressures than TXVs. The better capabilities of EXVs allow them to be used in systems with a wider variation in load conditions.

Evaporators. An *evaporator* is a heat exchanger that adds heat to low-pressure refrigerant liquid. As the heat is absorbed, the refrigerant boils, creating a vapor that is carried to the compressor. This exchange of heat takes place in tube-and-shell evaporators that are normally of a two-pass design. The water returning to the chiller from the building is cooled and circulated back to the cooling coils in the building. The water returning to the chiller is normally at a temperature of about 55°F, which is cooled to about 45°F before the water is sent back into the building.

The evaporators used for chillers are either direct expansion evaporators or flooded evaporators. A direct expansion evaporator has a specific superheat and normally uses a TXV to control the superheat. Direct expansion evaporators are used on small chillers because of the limitations of the TXV. Flooded evaporators are more popular for large chillers but require significantly more refrigerant than direct expansion chillers. The main advantage of flooded evaporator chillers is that there is a better exchange of heat from the water to the liquid refrigerant, making the chiller more efficient to operate.

Purge Units. When low-pressure chillers are running, the evaporator is in a vacuum. A leak in the refrigerant system allows air to enter the system. Air contains oxygen and moisture that mixes with the refrigerant, creating a mild acidic condition. The acid can break down motor windings over time and cause the windings to short out. The air can also collect in the condenser and cause an increase in condenser pressure. If enough air is present in the condenser, the increased pressure will shut down the chiller because of high head pressure. Air can be removed from the system and problems can be avoided by using a purge unit to collect the air from the top of the condenser.

A *purge unit* is a device used to maintain a system free of air and moisture. A purge unit is a small condensing unit that takes a sample of the gasses from the top of the condenser, compresses the sample using a compressor, and sends the sample into the condenser of the purge unit. If the sample does condense it is a refrigerant gas and is returned to the evaporator. If it does not condense it is "non-condensable" (air) and is released to the atmosphere.

Because of the high cost of the refrigerants and the environmental concerns, old purge units are being replaced by new, high-efficiency purge units. The older units allow a large percentage of the gasses that were released to the atmosphere to be refrigerant. Such practices are not allowed today.

ABSORPTION CHILLED WATER SYSTEMS

An absorption refrigeration system produces a refrigeration effect when a refrigerant absorbs heat as it vaporizes and rejects heat as it condenses. The components of the system control the flow of the refrigerant, which produces the refrigeration effect. **See Figure 12-14.**

An absorption refrigeration system works because the refrigerant is a liquid under pressure, and as the pressure decreases, it expands rapidly, vaporizes, and produces a cooling effect. An example of this cooling effect is a carbon dioxide cartridge that is pierced. Frost forms on the end of the cartridge as the carbon dioxide escapes because, as the carbon dioxide evaporates, it absorbs heat from the end of the cartridge and the air that surrounds the cartridge. In an absorption refrigeration system, the refrigerant pressure is increases by adding heat. The high-pressure liquid refrigerant vaporizes and is sent to a low-pressure area where heat is absorbed from an evaporating medium such as water.

Absorption refrigeration is used for air conditioning systems in large buildings or for small refrigeration systems in campers and motor homes. Large absorption air conditioning systems can have a capacity of several hundred tons of cooling.

The four components of an absorption refrigeration system are the absorber, generator, condenser, and evaporator. An absorption refrigeration system includes a circulating pump and an orifice which produce a refrigeration effect. **See Figure 12-15.**

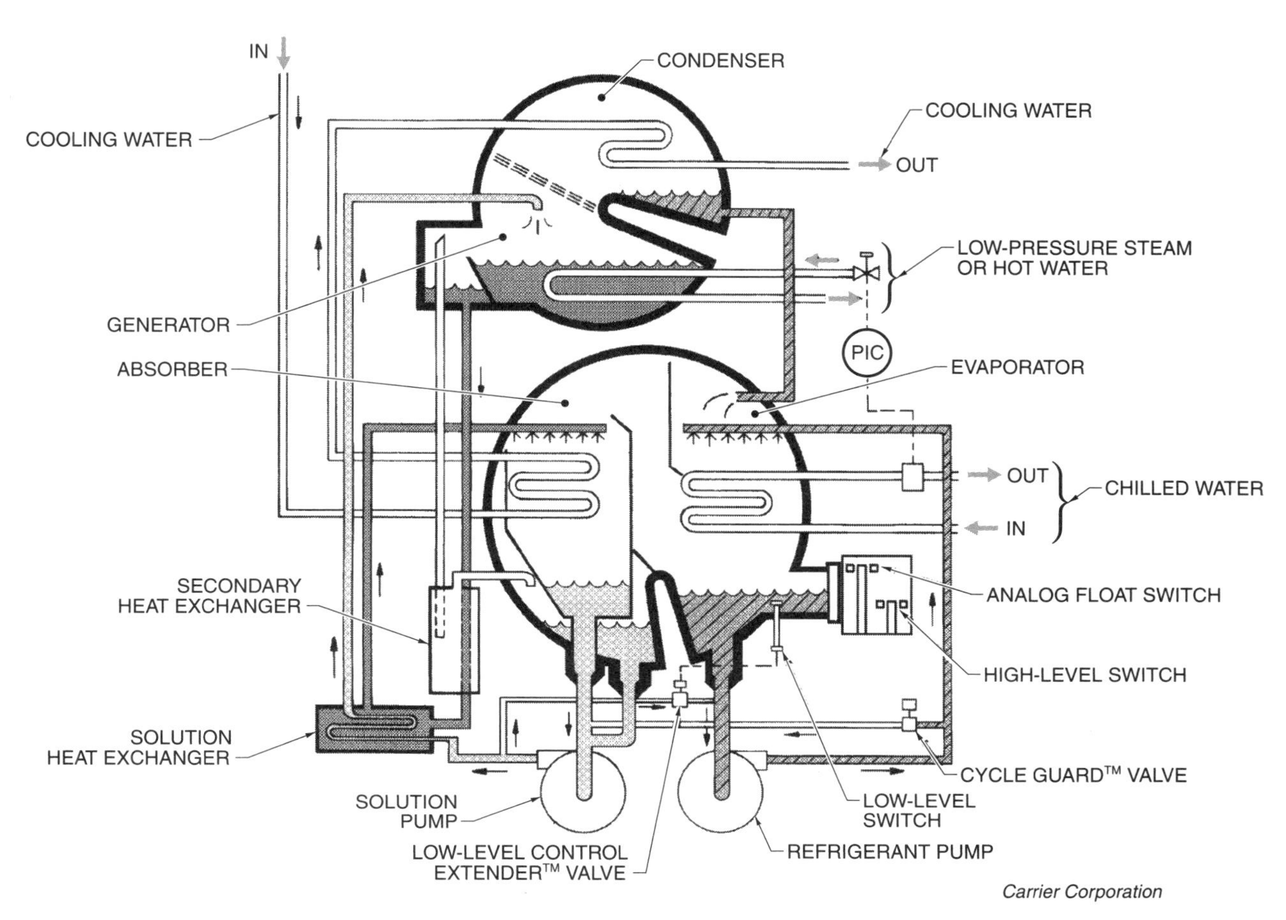

Figure 12-14. An absorption refrigeration system produces a refrigeration effect when a refrigerant absorbs heat as it vaporizes and rejects heat as it condenses.

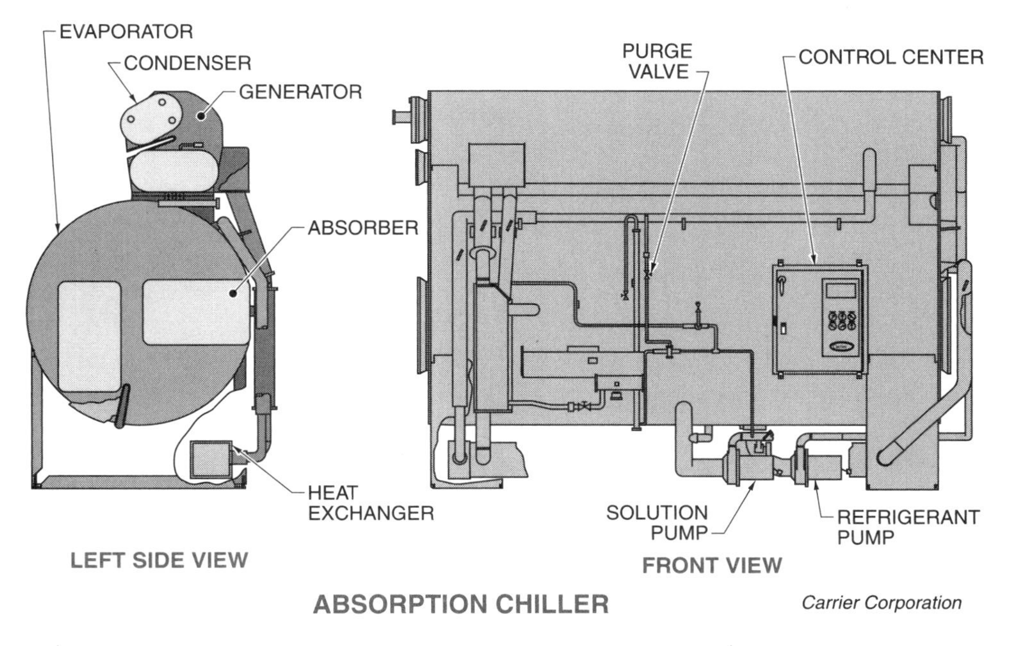

Figure 12-15. The four components of an absorption refrigeration system are the absorber, generator, condenser, and evaporator.

An *absorber* is the device in an absorption system in which the refrigerant is absorbed by an absorbant substance. The *absorbant* is a fluid that has a strong attraction for another fluid. As absorption occurs, the combined volume of the refrigerant-absorbant solution decreases, which causes a lower pressure in the absorber. Different combinations of refrigerants and absorbants are used. Ammonia and water or water and lithium bromide are the two most common combinations for refrigerant-absorbant solutions. In both cases the absorbant substance has a strong attraction for the refrigerant and absorbs it in large quantities. In the absorber, most of the refrigerant vapor from the evaporator mixes with the absorbant. The solution is then pumped to the generator by a circulating pump.

A *generator* is a device in the chiller that adds heat to the refrigerant-absorbant solution to vaporize the refrigerant, raise the pressure of the refrigerant, and separate the refrigerant from the absorbant. Once the refrigerant and absorbant are separated, the refrigerant vapor moves to the condenser and the absorbant returns to the absorber.

In the condenser, the high-pressure refrigerant vapor condenses to a liquid as heat is removed from it by the condenser heat exchanger. The condensing medium that runs through the condenser carries the heat away so it can be exhausted or used in other systems.

In the evaporator, the low-pressure, low-temperature refrigerant evaporates as it absorbs heat from the evaporating medium. The evaporating medium flows through the heat exchanger where it transfers heat to the refrigerant. The evaporating medium is then used for cooling.

A circulating pump is located in the refrigerant-absorbant line between the absorber and the generator. The circulating pump moves the refrigerant-absorbant solution from the absorber to the generator. The pump maintains a low pressure in the low-pressure side of the system by removing the refrigerant-absorbant solution from the low-pressure side as it rapidly enters through the orifice. Not all absorption refrigeration systems have circulating pumps.

An orifice is a restriction in the refrigerant line between the condenser and the evaporator. The orifice causes a pressure decrease in the refrigerant as it flows through the line. The refrigerant is at a high pressure before the orifice, and is at a low pressure after the orifice.

High- and Low-Pressure Sides

An absorption refrigeration system has a low-pressure side and a high-pressure side. The low-pressure side includes the orifice between the condenser and the evaporator, the evaporator, the absorber, and the refrigerant-absorbant solution line that leads to the inlet of the circulating pump. The high-pressure side includes the circulating pump (solution pump) discharge line, the generator, the condenser, and the refrigerant line that leads to the inlet of the orifice. **See Figure 12-16.**

On the low-pressure side of an absorption system, the refrigerant expands as it absorbs heat from the evaporating medium in the refrigeration system. The pressure decreases as the refrigerant recombines with the absorbant in the absorber.

The high-pressure side of an absorption system contains the components that remove heat from the refrigerant. The pressure on the refrigerant-absorbant solution increases at the circulating pump discharge. The pressure increases more when the temperature increases because of heat added in the generator. As the pressure increases, the temperature of the refrigerant increases. The temperature of the refrigerant that enters the condenser is higher than the temperature of the condensing medium, allowing heat to flow from the refrigerant to the condensing medium. The pressure of the refrigerant controls the temperature of the refrigerant.

The pressure in the high-pressure side of an absorption refrigeration system is maintained by introducing liquid refrigerant-absorbant solution through the circulating pump as rapidly as the vaporized refrigerant and absorbant flow through the orifice and bypass back to the absorber. The absorbant is separated from the refrigerant in the generator and returns to the absorber for reuse.

Heat is added to the refrigerant-absorbant solution by the generator. The heat separates the refrigerant from the absorbant and increases the pressure in that part of the system. In small systems that do not have a circulating pump, the generator provides the power that circulates the refrigerant through the system and increases the pressure of the refrigerant.

The refrigeration effect is produced in the evaporator. The refrigerant vaporizes in the evaporator because the pressure in the evaporator is low. The pressure is low because of the low pressure in the absorber. The low pressure in the absorber is caused by the orifice.

Heat Transfer

In an absorption system, heat is transferred from the evaporator medium to the refrigerant in the evaporator. The heat transfer cools the evaporator medium, which is used for air conditioning in a building. Heat is transferred from the refrigerant to the condensing medium in the condenser, where the heat is exhausted or used as auxiliary heat. Heat is added to the refrigerant-absorbant solution by the generator.

Tech Facts

Heat transfer in a refrigeration system occurs by moving heat from an area and exhausting it to another area. Liquid refrigerant is used to absorb the heat and transfer it to another area. To optimize heat transfer in water-filled coils, the entrapped air must be eliminated and efficient air must be created.

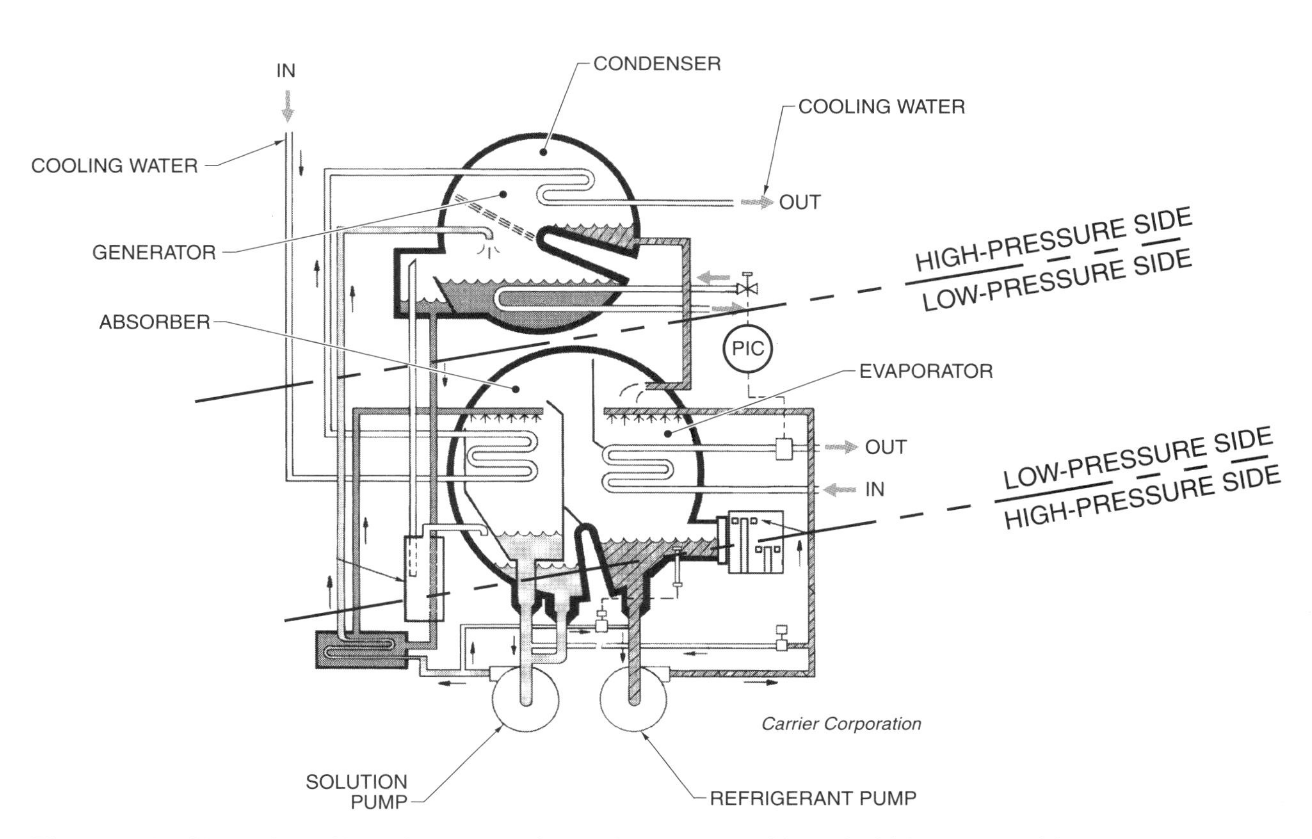

Figure 12-16. Absorption refrigeration systems have a low pressure side and a high pressure side.

Heat transfer occurs in the evaporator of an absorption system when liquid refrigerant flows from the condenser to the evaporator. The liquid refrigerant passes through the orifice, causing a pressure decrease. The refrigerant vaporizes because the pressure in the evaporator is lower than the pressure before the orifice. The evaporating medium flows through the evaporator heat exchanger, giving up its heat to the refrigerant, guaranteeing the refrigerant has totally vaporized. The transfer of heat cools the evaporating medium. The evaporating medium is then used for cooling building spaces. **See Figure 12-17.**

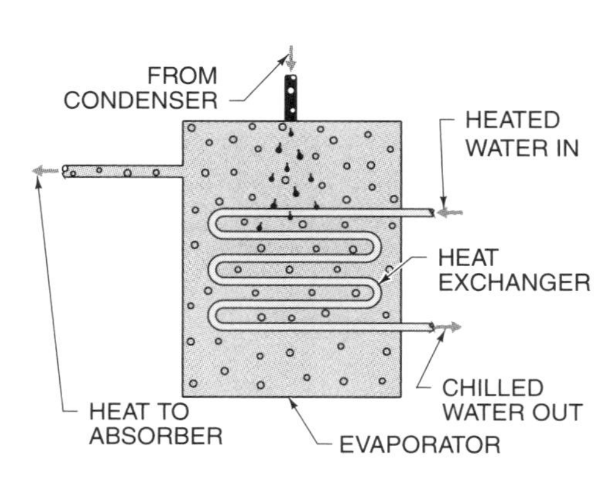

Figure 12-17. In an absorption refrigeration system, heat is transferred in the evaporator when liquid refrigerant flows from the condenser and is vaporized in the evaporator.

The *refrigeration effect* is the amount of heat in Btu/lb that the refrigerant absorbs from the evaporator medium. The amount of heat absorbed is equal to the amount of cooling the refrigeration system produces. The refrigeration effect of the system is found by applying the formula:

$RE = h_l - h_e$

where

RE = refrigeration effect (in Btu/lb)

h_l = enthalpy of refrigerant leaving evaporator (in Btu/lb)

h_e = enthalpy of refrigerant entering evaporator (in Btu/lb)

Example: Finding Refrigeration Effect — Absorption Refrigeration System

A refrigerant enters the evaporator of an absorption refrigeration system with an enthalpy of 77.9 Btu/lb and leaves the evaporator with an enthalpy of 1080.0 Btu/lb. Find the refrigeration effect.

$RE = h_l - h_e$

$RE = 1080.0 - 77.9$

$RE =$ **1002.1 Btu/lb**

In the condenser of an absorption refrigeration system, heat is rejected from the high-pressure refrigerant vapor to the condensing medium. The *heat of rejection* is the amount of heat in Btu/lb rejected by the refrigerant in the condenser. As the refrigerant flows through the condenser, heat is rejected to the condensing medium that flows through a heat exchanger in the condenser. As heat is rejected from the refrigerant vapor, the vapor condenses back to a liquid. **See Figure 12-18.**

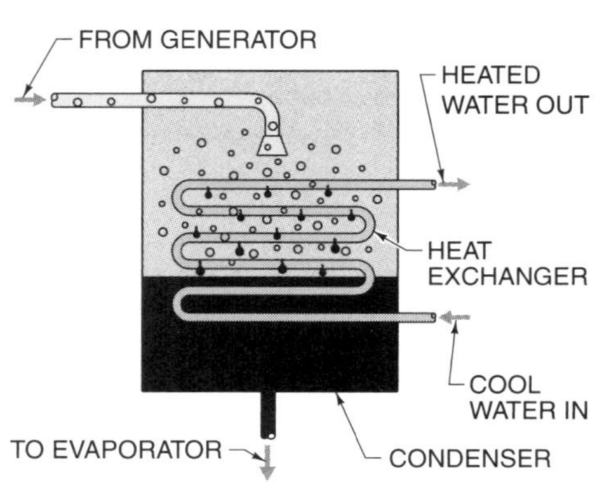

Figure 12-18. In an absorption refrigeration system, heat is rejected in the condenser as high-pressure refrigerant vapor from the generator condenses.

The heat of rejection in the condenser is found by applying the formula:

$HR = h_e - h_l$

where

HR = heat of rejection (in Btu/lb)

h_e = enthalpy of refrigerant entering condenser (in Btu/lb)

h_l = enthalpy of refrigerant leaving condenser (in Btu/lb)

Example: Finding Heat of Rejection — Absorption Refrigeration System

A refrigerant enters the condenser in an absorption refrigeration system with an enthalpy of 1080.0 Btu/lb. The refrigerant leaves the condenser as a liquid with an enthalpy of 77.9 Btu/lb. Find the heat of rejection.

$HR = h_e - h_l$

$HR = 1080.0 - 77.9$

$HR = \mathbf{1002.1\ Btu/lb}$

Because the refrigeration effect equals the heat of rejection, it appears that the system operates with no energy cost. However, heating and cooling the refrigerant and absorbant in the heat exchanger must be considered in a complete analysis of operation.

Heat is added to the refrigerant-absorbant solution in the generator of an absorption refrigeration system. The heat vaporizes the refrigerant and separates it from the absorbant. The source of the heat may be a flame, an electric heater, or a hot water or steam coil. The absorbant that remains after the separation returns to the absorber, and is used again. **See Figure 12-19.**

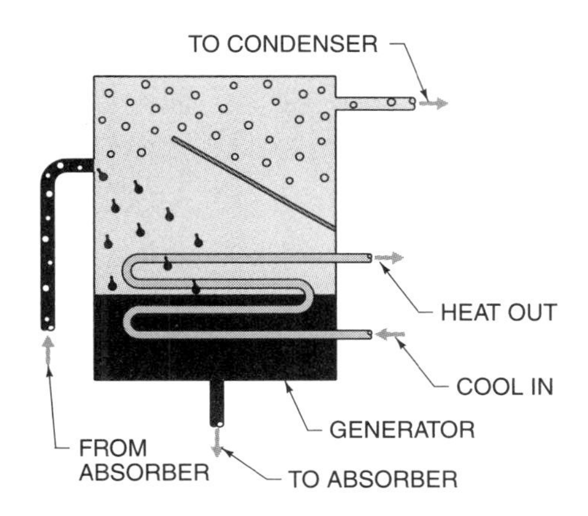

Figure 12-19. Heat vaporizes the refrigerant-absorbant solution in the generator. Vaporization separates the refrigerant from the absorbant.

Pressure Control

The temperature of the refrigerant in an absorption refrigeration system is controlled by the pressure of the refrigerant in the evaporator and the condenser. The pressure on the refrigerant decreases as the refrigerant passes through the orifice in the refrigerant line to the evaporator. The decreased pressure vaporizes some of the refrigerant after the orifice. The vaporization draws heat out of the remaining refrigerant, which decreases the temperature of the remaining refrigerant. The rest of the refrigerant vaporizes as it absorbs heat from the evaporating medium.

This added heat gives the refrigerant in the condenser a higher pressure and temperature than the refrigerant in the rest of the system. The refrigerant in the condenser is warmer than the condensing medium. Heat flows from the refrigerant to the condensing medium in the condenser heat exchanger because of the temperature difference.

Refrigerants

The refrigerants used in absorption refrigeration systems are solutions that consist of two chemical compounds. One of the two compounds must have a strong attraction for the other compound. Two commonly used chemical compounds are water and lithium bromide or ammonia and water. A water and lithium bromide solution uses water as the refrigerant and lithium bromide as the absorbant. An ammonia and water solution uses ammonia as the refrigerant and water as the absorbant.

Operation

Water and lithium bromide are often used as refrigerant and absorbant in small to medium-size absorption refrigeration systems. The water and lithium bromide are brought together in the absorber where they readily combine. **See Figure 12-20.** As the substances combine, the lithium bromide absorbs the water and forms a strong-bonding refrigerant-absorbant solution. Some heat is produced during absorption. The absorber may contain a heat exchanger with cool water flowing through it to help control the temperature of the solution.

In an absorption refrigeration system, the refrigerant-absorbant solution leaves the absorber with a temperature of 100°F and a heat content of 47.2 Btu/lb of refrigerant. Most systems have a heat exchanger between the absorber and the generator. Heat flows from the absorbant (going back to the absorber) to the refrigerant-absorbant solution (going to the generator). The temperature of the refrigerant- absorbant solution rises to 170°F with a heat content of 79 Btu/lb.

The generator heat exchanger adds heat to the solution to separate the refrigerant from the absorbant. The refrigerant leaves the generator as a vapor at a temperature of 200°F with a heat content of 1150 Btu/lb. The refrigerant moves to the condenser.

The absorbant that has separated from the refrigerant leaves the generator at a temperature of 210°F with a heat content of 107 Btu/lb. The absorbant returns to the absorber.

The refrigerant vapor in the condenser condenses as heat is removed by the condenser heat exchanger.

The refrigerant flows from the condenser to the evaporator through the orifice. The pressure decreases as the refrigerant flows through the orifice, decreasing the temperature of the refrigerant to 110°F with a heat content to 77.9 Btu/lb.

As the refrigerant flows through the evaporator, it vaporizes and absorbs heat from the evaporating medium. Most of the refrigerant leaves the evaporator as a vapor.

When the refrigerant leaves the evaporator, it is at a temperature of 41°F with a heat content of 1080 Btu/lb. The refrigerant returns to the absorber where the cycle begins again.

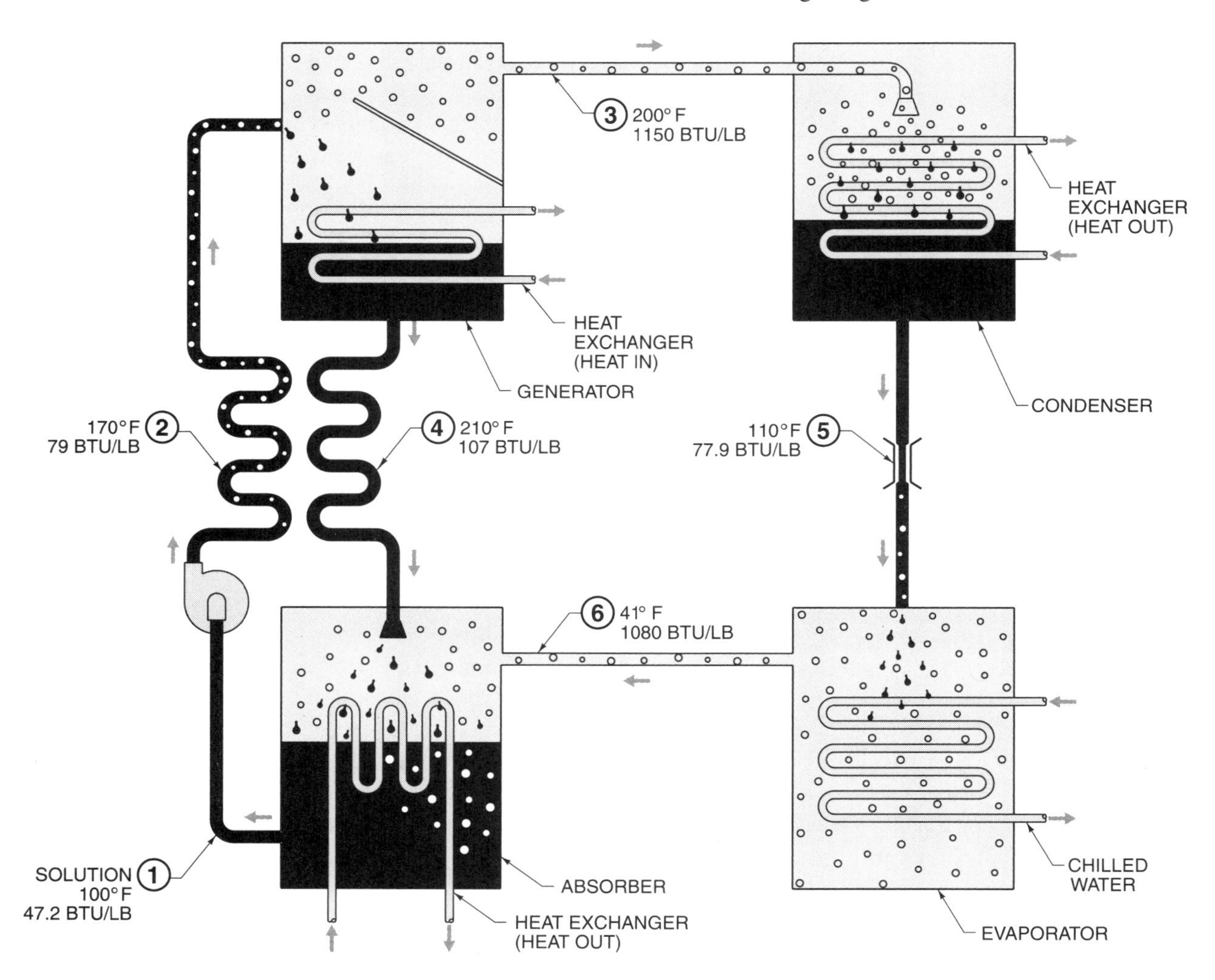

Figure 12-20. An absorption refrigeration system produces a refrigeration effect when one chemical is absorbed by another and heat is then removed.

An absorption refrigeration system always works with a coefficient of performance of less than 1. *Coefficient of performance* (*COP*) is the cooling capacity produced in a refrigeration system from the heat input. COP is found by dividing the cooling capacity produced from a system by the energy used in the system. COP is found by applying the formula:

$$COP = \frac{CC}{EU}$$

where
COP = coefficient of performance
CC = cooling capacity (in Btu/hr)
EU = equivalent of energy used (in Btu/hr)

Example: Finding Coefficient of Performance

A refrigeration system produces 48,500 Btu/hr of cooling while using the equivalent of 70,000 Btu/hr of energy. Find the COP of the system.

$$COP = \frac{CC}{EU}$$

$$COP = \frac{48{,}000}{70{,}000}$$

$$COP = \mathbf{.69\ Btu/hr}$$

Review Questions

1. Describe the refrigeration effect.
2. List and describe the function of the main parts of a mechanical compression refrigeration system.
3. What causes the temperature of the refrigerant to decrease as it leaves the expansion device and enters the evaporator?
4. Explain the relationship between the boiling point of a liquid and the pressure on the liquid.
5. Define *superheat.* Explain the function of superheat in a refrigeration system.
6. What two devices in a mechanical compression refrigeration system separate the low-pressure side from the high-pressure side?
7. Describe how water provides a refrigeration effect when it vaporizes.
8. Describe how the number of a refrigerant identifies the chemical composition of the refrigerant.
9. How has the Clean Air Act of 1992 impacted the duties of contractors and HVAC technicians when working with refrigerants?
10. List the desirable refrigerant properties.
11. Why is boiling point an important property of a refrigerant?
12. Define *critical point.* Why is it important?
13. Define *miscibility.* Why is it important that a refrigerant be miscible?
14. Compare the three classifications of refrigerants most commonly used today.
15. List and describe the function of the main parts of a chiller.
16. List and describe the function of each component of an absorption refrigeration system.
17. Define *heat rejection.* Explain how it is found.
18. What happens to the refrigerant-absorbent solution in the generator of an absorption refrigeration system?

19. How is heat removed from refrigerant in the condenser of an absorption refrigeration system?
20. How does the evaporator of an absorption refrigeration system cool the evaporating medium?
21. What device is located before the evaporator in an absorption refrigeration system? What does it do?
22. How is the coefficient of performance (COP) found?

Air conditioning systems produce a refrigeration effect to maintain comfort within building spaces. Forced-air air conditioning and hydronic air conditioning are the two kinds of air conditioning systems. Forced-air air conditioning systems include an air conditioner, blower, supply and return ductwork, registers, grills, and controls. Hydronic air conditioning systems include a chiller, circulating pump, supply and return piping, terminal devices, and controls.

13 Air Conditioning Systems

AIR CONDITIONING

Air conditioning is the process of cooling the air in a building to provide a comfortable temperature. An *air conditioner* is the component in a forced-air air conditioning system that cools the air. A *chiller* is the component in a hydronic air conditioning system that cools water, which cools the air. An air conditioning system is the equipment that produces a refrigeration effect and distributes cool air or water to building spaces. Air conditioning systems are classified by evaporating medium, condensing medium, physical arrangement, and cooling capacity.

Evaporating Mediums

An *evaporating medium* is the fluid that is cooled when heat is transferred in the evaporator from the evaporating medium to cold refrigerant. Air and water are evaporating mediums.

Air. Air is used as the evaporating medium when an air conditioning system is located close to building spaces. Cool air is distributed to building spaces through ductwork. Air is readily available and inexpensive to use. Blower operation is the main expense when using air as an evaporating medium. Exhausting air after it has been used is not difficult. Moving air over a long distance or moving a large amount of air requires a large amount of energy.

A relatively large quantity of air is required to carry heat. At standard conditions, air has a density of .0753 lb/cu ft and a specific heat of .24 Btu/lb. *Specific heat* is the amount of heat that is required to raise the temperature of 1 lb of a substance 1°F. Raising the temperature of 1 lb of air 1°F requires .24 Btu. Large ductwork is required to move large quantities of air. Large quantities of air are required to carry large amounts of heat.

Water. Water is used as an evaporating medium when an air conditioning system is located at a distance from building spaces. Cool water circulates from a cooler to terminal devices. At the terminal devices, the water cools the air and lowers the temperature of the air in the building spaces.

At standard conditions, water has a density of 62.32 lb/cu ft and a specific heat of 1.0 Btu/lb. Raising 1 lb of water 1°F requires 1 Btu. Because water is denser than air, water holds heat better. The water in a small pipe holds the same amount of heat as the air in a large duct.

Condensing Mediums

A *condensing medium* is the fluid in the condenser of a refrigeration system that carries heat away from the refrigerant. Air and water are condensing mediums. An air-cooled condenser in an air conditioning system cools air. A water-cooled condenser in a chiller cools water. An evaporative condenser uses both air and water as the condensing medium.

Air. Air is used as the condensing medium for an air conditioning system when the condenser can be located outdoors. Airflow to the condenser cannot be restricted. The normal ambient temperature of outdoor air should be cooler than the condensing temperature of the refrigerant in the air conditioning system. When the temperature of the air is lower than the temperature of the refrigerant, the air can carry heat away from the condenser. An air-cooled condenser has a large capacity, which allows a large amount of air to circulate through the condenser.

Water. Water is used as a condensing medium when the condenser cannot be located outdoors. The initial cost of the equipment for a water-cooled condenser is high. Eliminating the wastewater from a water-cooled condenser is also expensive. Water may be supplied to the condenser directly from a public or private water supply system or may be recirculated from a cooling tower. Because supply water temperature does not vary greatly, a water-cooled condenser works well regardless of changing outdoor temperatures. Wastewater from a water-cooled condenser is piped to a cooling tower for cooling and reuse or dumped into a storm sewer. Three kinds of cooling towers are forced draft, induced draft, and hyperbolic. **See Figure 13-1.**

A *cooling tower* is an evaporative heat exchanger that removes heat from water. A cooling tower is a large structure that contains louvered panels for airflow. Natural or forced convection causes airflow in the cooling tower. A reservoir or tank in the tower holds warm water from the condenser. The water is sprayed over a fill material inside the tower. The fill material breaks the streams of water into small, cascading droplets. The water droplets are cooled by evaporation as air passes through the louvers in the tower. The water that flows out of the bottom of the tower is much cooler than the water sprayed in at the top.

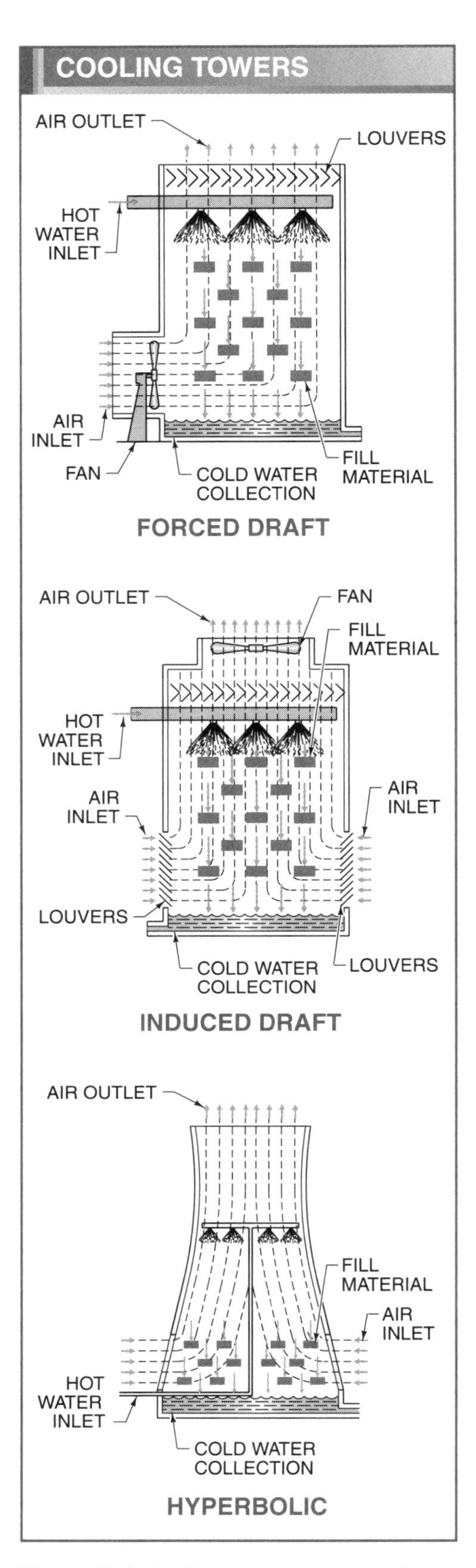

Figure 13-1. Cooling towers cool water for reuse in a water-cooled air conditioning system.

A *forced draft cooling tower* is a cooling tower that has a fan located at the bottom of the tower that forces a draft through the tower. An *induced draft cooling tower* is a cooling tower that has a fan located at the top of the tower that induces a draft by pulling the air through the tower. A *hyperbolic cooling tower* is a cooling tower that has no fan. Natural draft moves air through a hyperbolic cooling tower.

Air and Water. An evaporative condenser uses both air and water as condensing mediums. In an evaporative condenser, air passes over the condenser coil and water is sprayed on the coil. The refrigerant flowing through the coil rejects heat because of the air passing over the coil, the water flowing over the coil, and the water evaporating from the surface of the coil. Evaporative condensers remove heat more efficiently than air-cooled or water-cooled condensers. Evaporation makes evaporative condensers more efficient. **See Figure 13-2.**

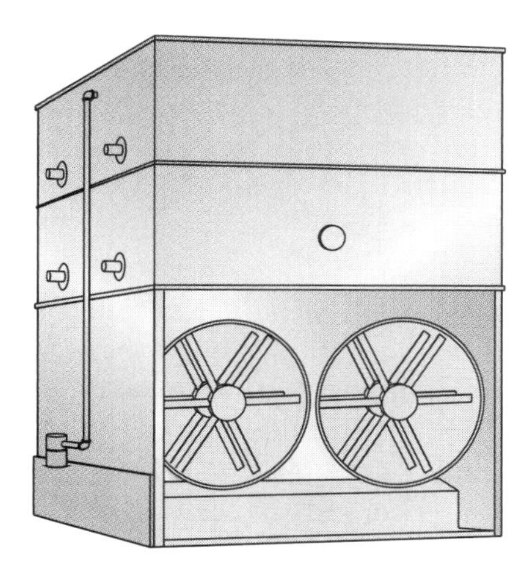

Figure 13-2. An evaporative condenser uses evaporation to remove heat from refrigerant.

By combining the two evaporating mediums and the two condensing mediums, four different refrigeration systems are possible. Air conditioners and chillers can have either an air-cooled condenser or a water-cooled condenser. An air conditioner uses air as the evaporating medium. A chiller uses water as the evaporating medium. These four systems can be arranged to fit almost any application.

Physical Arrangement

Air conditioning systems have different physical arrangements depending on the size and layout of a building. Split systems, package units, and combination units are examples of different physical arrangements.

Split Systems. A *split system* is an air conditioning system that has separate cabinets for the evaporator and the condenser. The cabinets are connected by refrigerant lines, plumbing, and electrical conductors. The evaporator of a split system is located inside a building, and the condenser is located outdoors. If the evaporator is located inside the furnace cabinet, the same blower can be used to move cool air and warm air. Split air conditioning systems have the evaporator in one cabinet and the condenser and compressor in another cabinet. Some split air conditioning systems have three parts, which are the compressor, the evaporator, and the condenser. **See Figure 13-3.**

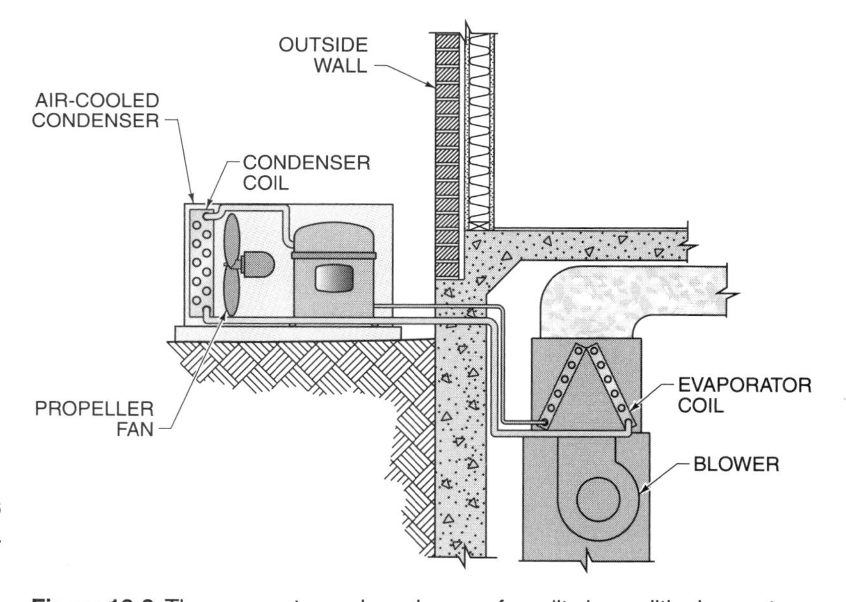

Figure 13-3. The evaporator and condenser of a split air conditioning system are enclosed in separate cabinets.

The major components of a split system must be compatible. Each component of a split system is designed to have the same capacity as the other components in the system. Equipment specification sheets describe the equipment and list combinations of equipment that are compatible.

Refrigerant lines connect the components of smaller split systems. Electrical conductors are field installed and must meet NEC® requirements.

Packaged Units. A *packaged unit* is a self-contained air conditioner that has all of the components contained in one sheet metal cabinet. A packaged unit requires electrical power and control connections and must be connected to an air distribution system.

Controls for a packaged unit are installed by the manufacturer. Controls operate the components of a packaged unit. The thermostat and other remote controls must be installed and connected on-site when the unit is installed.

Air-cooled packaged units use air as the condensing medium and are normally installed outdoors. Air-cooled packaged units take in ambient air from outdoors and exhaust the air outdoors. An air distribution system that consists of ductwork runs from an air-cooled packaged unit to building spaces.

A *rooftop unit* is an air-cooled packaged unit that is located on the roof of a building. Rooftop units save building space. **See Figure 13-4.** If the ductwork in a building is installed under the floor or if access to the ductwork from the roof is difficult, a unit may be located on the ground near the building.

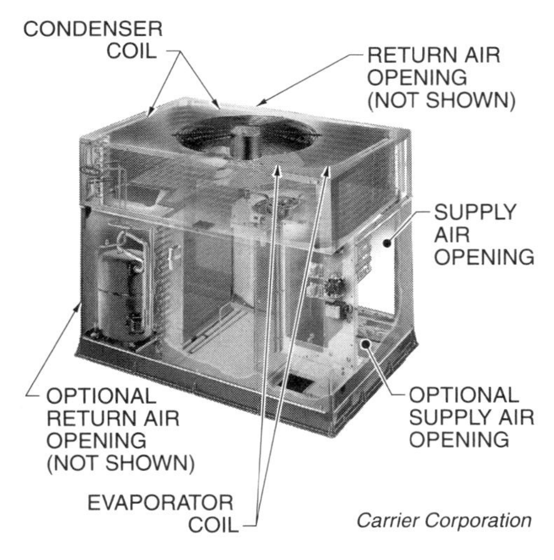

Figure 13-4. All of the components of a packaged unit are contained in one sheet metal cabinet.

Water-cooled packaged units use water as the condensing medium. Water-cooled packaged units can be installed wherever supply water and a wastewater outlet are available. Water-cooled packaged units are often located inside a building where using an air-cooled unit would be difficult. Water-cooled packaged units are often single-zone units, which cool a specific zone in a building. A *zone* is a specific section of a building that requires separate temperature control. Zones are also known as modules. When a water-cooled packaged unit is located in a zone, less piping is required than for an entire building. A large building with several zones may have several packaged water-cooled units. This method of cooling is known as modular cooling.

Combination Units. Buildings that require cooling in the summer and heating in the winter require combination units. A *combination unit* is an air conditioner that contains the components for cooling and heating in one sheet metal cabinet. Power lines, fuel line connections, and control connections are provided. Most combination units are air-cooled and are located outdoors.

The cabinet of a combination unit is divided into two sections. The heating unit, evaporator, and blower are in one section, and the condenser and compressor are in the other section. **See Figure 13-5.** Inlet and outlet air louvers are located in the condenser/compressor section. An intake air louver and an outlet for flue gas are also located on the combination unit cabinet.

Combination units are usually located on the roof of a building but sometimes on the ground near the building. The location of the unit depends on the location of the ductwork in the building. One combination unit may be used on a building or several combination units may be used for modular applications. Because a combination unit controls heating and cooling, air distribution systems for heating and cooling are used. Single duct or double duct systems are used with combination units.

A *single duct system* is an air distribution system that consists of a supply duct that carries both cool air and warm air and a return duct that returns the air. When the thermostat calls for cooling, the combination unit distributes cool air to the building spaces through the supply duct. When the thermostat calls for heating, the combination unit distributes warm air to the building spaces through the supply duct. The return duct returns the air to the heating and cooling unit.

A *double duct system* is an air distribution system that consists of a supply duct that carries cool air and a supply duct that carries heated air. Mixing boxes mix the air at take-offs for each zone or building space. A *mixing box* is a sheet metal box that is attached to the cool air duct and the warm air duct. Openings in the side of the box connect with the ducts. Dampers in the box are controlled to introduce both warm and cool air and mix them in the proper proportion to provide a certain temperature.

Cooling Capacity

Air conditioners produce a given cooling capacity, which is expressed in Btu per hour or in tons of cooling. A ton of cooling equals 12,000 Btu/hr. Air conditioners are rated in Btu per hour of cooling at certain standard conditions and are given a nominal size in tons of cooling. *Nominal size* is the cooling capacity of an air conditioner in Btu per hour rounded to the nearest ton or half-ton of cooling. **See Figure 13-6.**

Small air conditioners have cooling capacities of less than 1 ton of cooling to about 2 tons of cooling. Window and through-the-wall air conditioners are small package units. Small package units cool one room at a time and can be connected to the electrical system in most buildings.

Medium-size air conditioners have cooling capacities of 1.5 tons to 7.5 tons of cooling. Medium-size air conditioners are connected to ductwork and cool entire buildings such as homes or small commercial buildings. Medium-size air conditioners are either package or split systems.

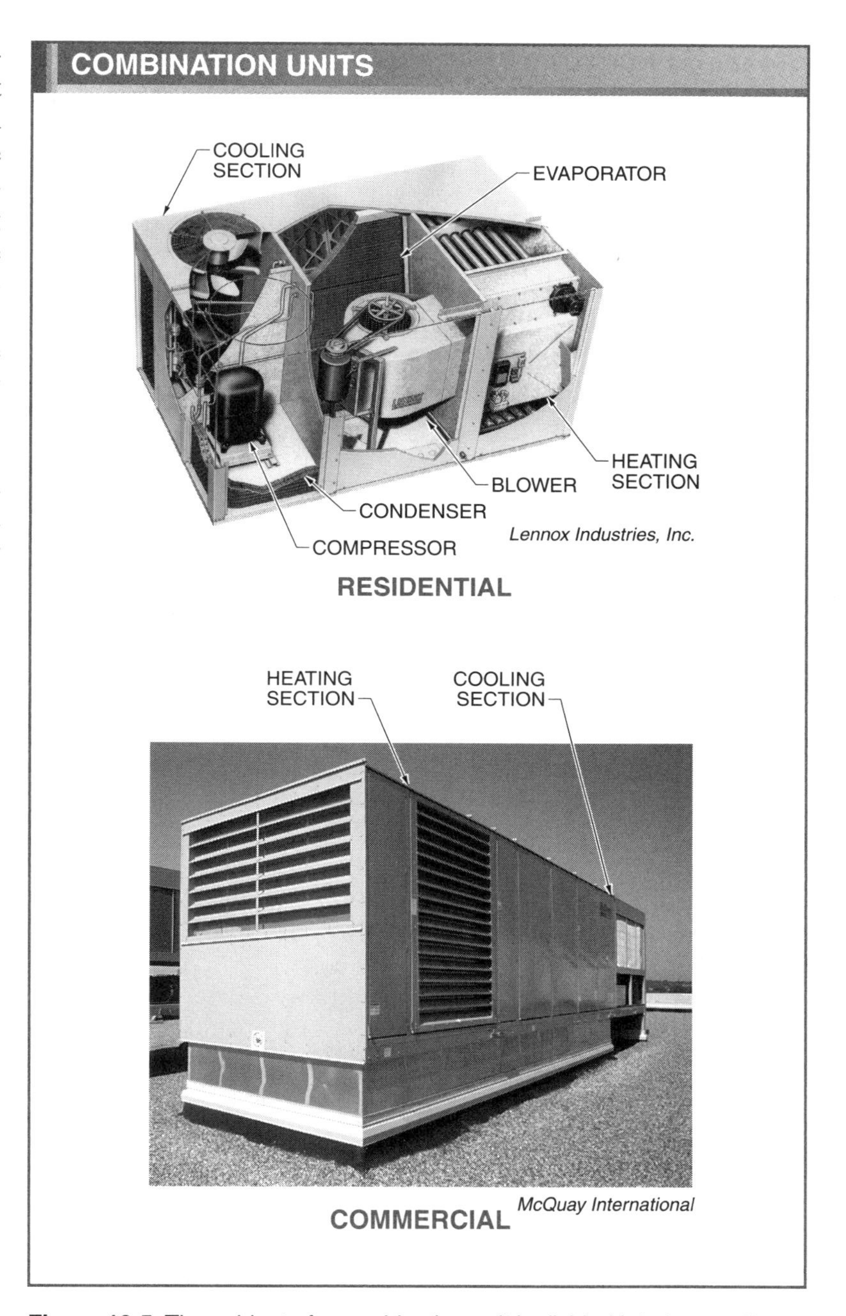

Figure 13-5. The cabinet of a combination unit is divided into two sections.

Tech Facts

Before "Btu per hour" was used to indicate cooling capacity, the cooling capacity of refrigeration units was indicated as "tons of refrigeration." A ton of refrigeration is equal to the amount of heat energy absorbed when a ton (2000 lb) of ice melts during a 24-hour period.

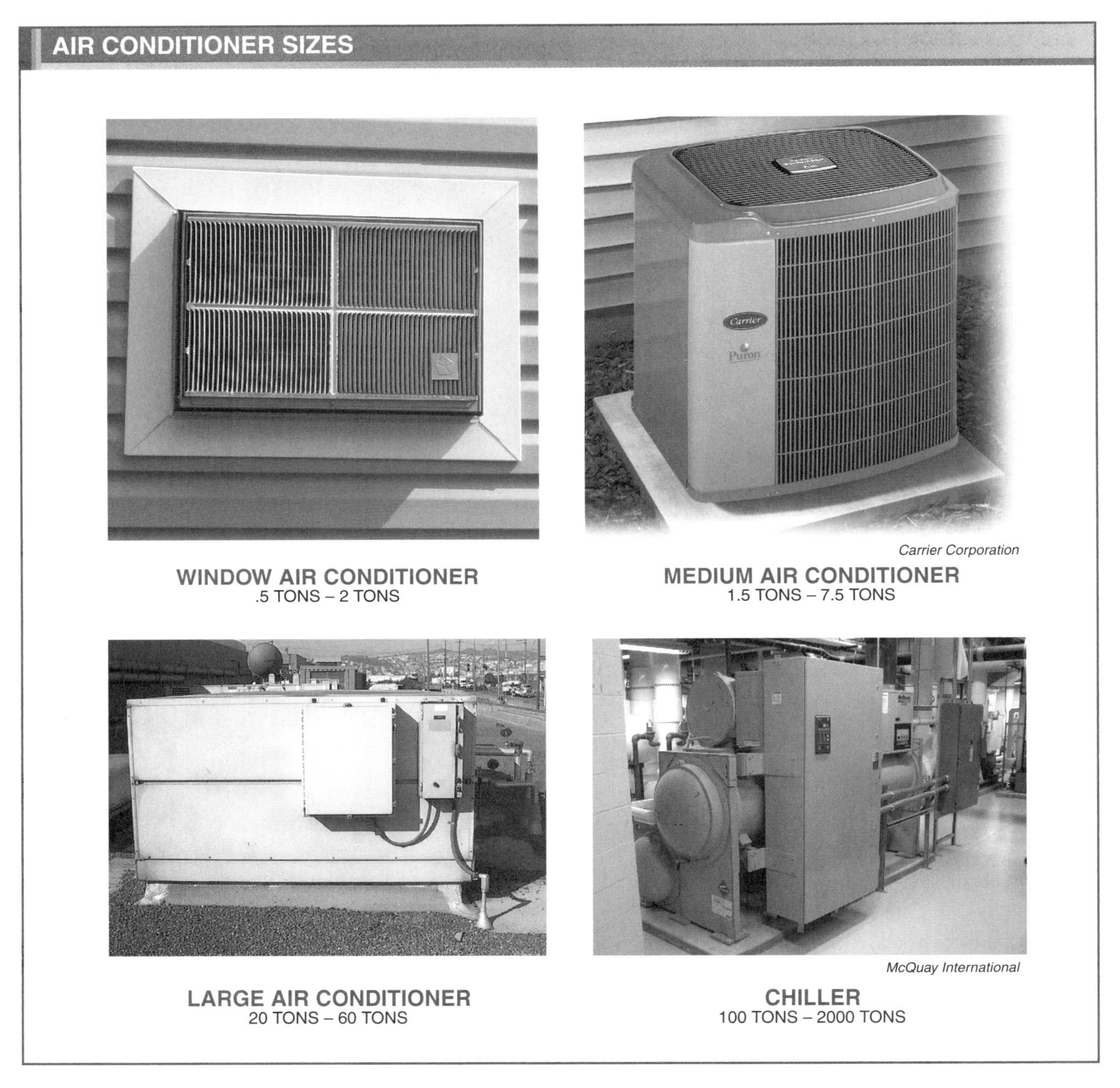

Figure 13-6. The cooling capacity of air conditioners ranges from less than .5 tons of cooling to 750 tons of cooling.

Large air conditioners are available with cooling capacities of 20 tons to 60 tons of cooling. Large air conditioners are available as package units and as components in built-up systems. Chillers have cooling capacities of 100 tons to 750 tons of cooling.

Refrigeration Process

Air conditioners use either the mechanical compression refrigeration process or absorption refrigeration process. Each refrigeration process removes heat from air or water and exhausts the heat where it is not objectionable.

Mechanical Compression Refrigeration. In the mechanical compression refrigeration process, a compressor increases the pressure and temperature of a refrigerant so the refrigerant can reject heat to the condensing medium as it condenses. A mechanical compression refrigeration system includes an expansion device, evaporator, compressor, condenser, blower or circulating pump, refrigerant lines, and controls.

The expansion device causes a pressure decrease in the refrigerant. The evaporator transfers heat from the evaporating medium to the refrigerant in the system. The compressor increases the pressure and temperature of the refrigerant. The condenser transfers heat from the refrigerant to the condensing medium. The blower or circulating pump moves the condensing medium through the system. Refrigerant lines connect the components, and controls regulate the operation of the components.

Absorption Refrigeration. In the absorption refrigeration process, an absorbant controls the pressure and temperature of a refrigerant. An absorption refrigeration system contains an absorber, generator, condenser, orifice, evaporator, and controls. An absorption system may also contain a heat exchanger and a circulating pump.

In the absorber, absorbant combines with refrigerant to form a strong refrigerant-absorbant solution. In the generator, the solution absorbs heat and separates into refrigerant and absorbant. In the condenser, the refrigerant rejects heat and condenses to a high-pressure liquid. In the orifice, the pressure decreases. In the evaporator, the low-pressure refrigerant evaporates. The refrigerant absorbs heat from the evaporating medium that flows through the evaporator. This cools the evaporating medium, which is used for cooling.

Controls

Air conditioning systems are designed to produce a refrigeration effect at a fixed rate. Air conditioning systems are selected to provide enough cooling output to compensate for the warmest outdoor temperature. Since the outdoor temperature is normally cooler than the expected maximum temperature, a typical air conditioning system produces more cooling than required most of the time. Controls cycle the components of an air conditioning system ON and OFF to produce the required refrigeration effect consistently, safely, and automatically.

Power Controls. Power controls control the flow of electricity to an air conditioning system. Power controls are located in the electrical circuit between the power source and the air conditioning system. Power controls should be installed by a licensed electrician per the National Electrical Code®. Power controls include disconnects, fuses, and circuit breakers.

At least two disconnects are located in the electrical power circuit to an air conditioning system. One disconnect is located in the electrical service panel where the circuit originates, and the other disconnect is located near the air conditioner. Manual disconnects are used on most air conditioners. A manual disconnect is a disconnect that is opened or closed manually by an operator. **See Figure 13-7.**

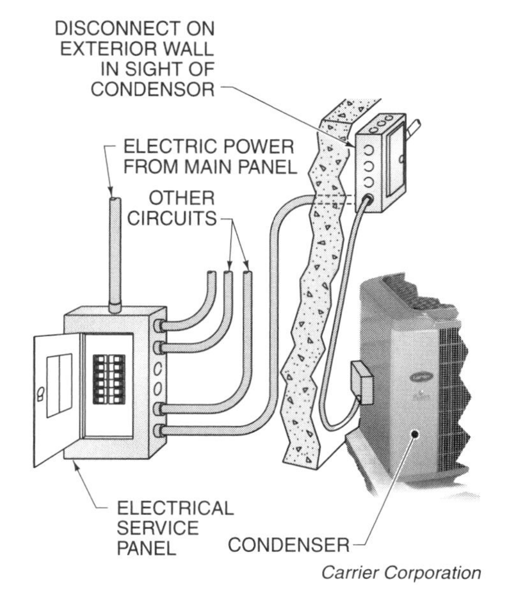

Figure 13-7. Power controls control the flow of electricity to an air conditioner.

Fuses or circuit breakers are electric overcurrent protection devices that protect the electrical conductors in a circuit from an overcurrent condition. If an overcurrent condition occurs, a fuse will burn out. The burned-out fuse breaks the circuit and shuts OFF the current flow before the circuit is damaged. Small air conditioners may have circuit breakers for combination disconnects and fusing. Large air conditioners have cartridge fuses for overcurrent protection.

Operating Controls. Operating controls on air conditioners and furnaces are similar. Operating controls regulate the operation of air conditioner components by cycling equipment ON or OFF. Each component of an air conditioner must function separately when combined with other components for proper performance. Operating controls include transformers, thermostats, blower controls, relays, contactors, magnetic starters, and solenoids. **See Figure 13-8.**

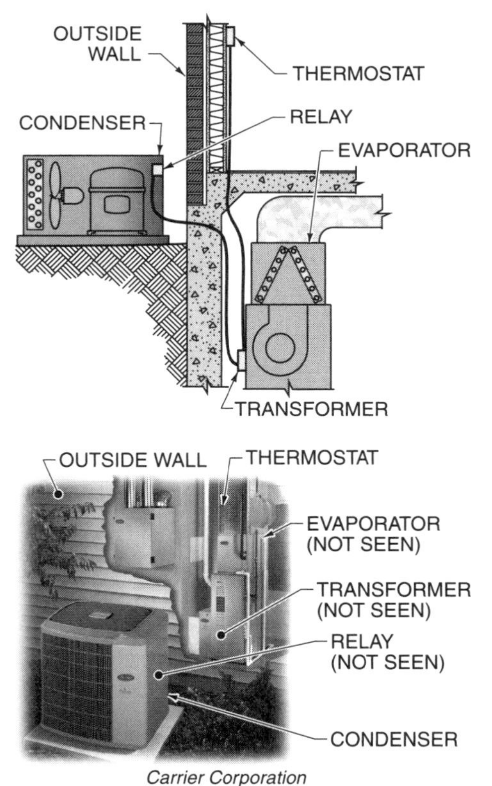

Figure 13-8. Operating controls regulate the operation of the components in an air conditioner.

A *transformer* is an electric device that uses electromagnetism to change (step-up or step-down) AC voltage from one level to another. The transformer on an air conditioner is wired on the primary side to the electrical power source and is wired on the secondary side to the control circuits. Most control circuit transformers are step-down transformers with a secondary voltage of 24 V.

A *thermostat* is a temperature-actuated electric switch that turns an air conditioner ON or OFF in response to temperature changes in building spaces. When the temperature at the cooling thermostat rises above a setpoint, the thermostat closes a switch to complete the electrical circuit. The system turns ON when the circuit is complete. When the temperature at the thermostat falls below a setpoint, the thermostat opens a switch to open the electrical circuit. The system turns OFF when the circuit is open. The setpoint is the temperature at which the switch in the thermostat opens and closes. The setpoint on most thermostats can be changed manually.

A *relay* is a device that controls the flow of electric current in one circuit (load circuit) with the electric current of another circuit (control circuit). Relays are used in control circuits where a low-voltage circuit is used to control a line voltage circuit. Electromechanical relays, contactors, and magnetic starters are used in air conditioners.

The condenser blower motor on small air conditioners is wired in parallel with the compressor motor so that the condenser blower and compressor motor run at the same time. The relay, contactor, or magnetic starter that controls the compressor also controls the condenser blower. In large air conditioners, the condenser blower may have a separate contactor or magnetic starter, but the control coil is wired in parallel with the control coil on the compressor starter. The compressor and condenser blower still run at the same time.

The evaporator blower is controlled by a separate relay, contactor, or magnetic starter. The relay is actuated by a signal from the thermostat. Most cooling thermostats are wired to the blower and are actuated whenever the thermostat calls for cooling or when the blower switch on the thermostat is set for constant operation. If the evaporator blower and compressor relay are wired in parallel, the evaporator blower relay may also be actuated whenever the compressor relay is actuated.

A solenoid is similar to a relay in that it controls one electrical circuit with another. In a solenoid, an electromagnet positions a movable core that opens or closes a set of contacts. The contacts are wired into a circuit so the circuit opens and closes as the switch is energized and de-energized. A solenoid coil is energized by the control circuit of an air conditioner. The contacts on a solenoid are wired into the load circuit.

Safety Controls. Safety controls on an air conditioner shut the unit OFF to prevent damage to the components in case one of the components malfunctions. Because the components of an air conditioner are interrelated, the failure of one component can cause damage to the others. The safety controls in an air conditioner include pressure switches.

A *pressure switch* is an electric switch that contains contacts and a spring-loaded lever arrangement. The lever opens and closes the contacts. The lever moves according to the amount of pressure acting on a diaphragm or bellows element. A change in the pressure of the refrigerant in the system moves the diaphragm or bellows element. This movement of the diaphragm or bellows is transmitted to a lever arm by mechanical linkage. The switch is wired in series with other devices in the electrical circuit that operates the compressor.

Pressure switches are automatic when the contacts in the pressure switch automatically close when the pressure in the system returns to normal, or the pressure switches are manually reset switches requiring a technician to close the contacts. On most pressure switches, the setpoint temperature and differential are adjusted separately.

Refrigerant pressure switches are actuated by refrigerant pressure in the system and may sense high or low pressures. Lubricating oil pressure switches are actuated by the pressure of the oil in equipment.

A high-pressure refrigerant switch is connected to the high-pressure side of an air conditioner near the hot gas discharge outlet of the compressor. The high-pressure refrigerant switch shuts the system OFF if the refrigerant pressure becomes too high. High pressure in an air conditioner can damage the compressor and burn out the compressor motor. **See Figure 13-9.**

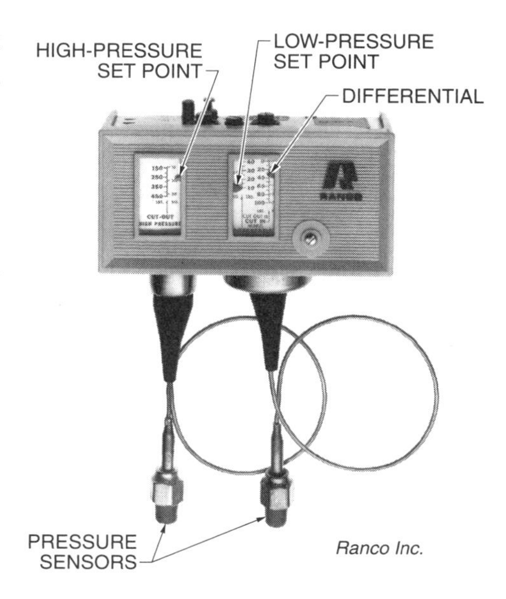

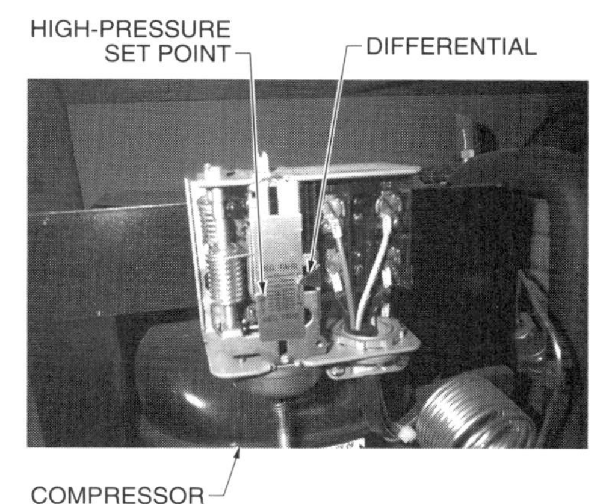

Figure 13-9. High- and low-pressure switches shut a compressor OFF when the refrigerant pressure becomes too high or too low.

Tech Facts

Pressure switches are a type of safety control consisting of electric switches that contain contacts and a spring-loaded lever arrangement. The spring-loaded lever opens and closes the contacts when there is a change in pressure.

A low-pressure refrigerant switch is connected into the low-pressure side of an air conditioner. The low-pressure refrigerant switch is connected to a refrigerant line near the suction inlet of the compressor or to the crankcase of the compressor. The low-pressure switch shuts the system OFF if the refrigerant pressure becomes too low. Low pressure can damage hermetic and semihermetic compressor motors. Low pressure in the system causes a low temperature. A low temperature in the evaporator coil can freeze and damage the coil.

An oil pressure switch is similar to a refrigerant pressure switch. An oil pressure switch ensures that there is lubricating oil in the system when the system is operating. An oil pressure switch is connected into the lubricating line of an air conditioner. Large air conditioners have positive-pressure lubrication systems. Positive-pressure lubrication systems contain an oil pump that distributes lubricating oil to compressor bearings. Failure of the lubrication system causes failure of the compressor.

The oil pressure switch sensors are connected to the oil line close to the oil pump discharge. The electric switch on an oil pressure switch shuts the compressor OFF and may actuate a bell or light to indicate oil pressure failure. **See Figure 13-10.**

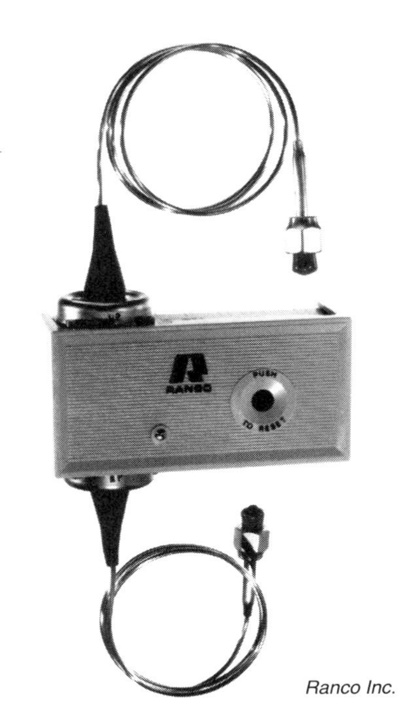

Ranco Inc.

Figure 13-10. An oil pressure switch ensures that lubrication oil is in the compressor when the compressor motor is operating.

FORCED-AIR AIR CONDITIONING SYSTEMS

Forced-air air conditioning systems use air as the evaporating medium. Air circulates through ductwork to building spaces. Forced-air air conditioning systems are used primarily where the air conditioning equipment is located close to building spaces and where only certain building spaces require cooling. Forced-air air conditioning systems are located close to building spaces because the required ductwork may be fairly large, and it is not practical to run large ductwork over long distances. In large commercial, multistory buildings, the conditioned space may be divided into smaller sections to accommodate the installation of individual forced-air systems for each section. The individual sections may encompass a single floor, two floors, three or more floors, or vertical sections through several floors.

In residential and small commercial buildings, a forced-air air conditioning system consists of an air conditioner, blower, supply and return ductwork, registers, grills, and controls. **See Figure 13-11.** In larger commercial buildings the air conditioner may be replaced with a cooling coil and an economizer. The cooling coil is considered a hydronic air conditioning system component. An economizer is a configuration of dampers that mix fresh outside air with building return air to be supplied to the building. The building return air is conditioned air, which is more economical to use in quantities as long as the mixed air supplied to the building has the required amount of fresh air.

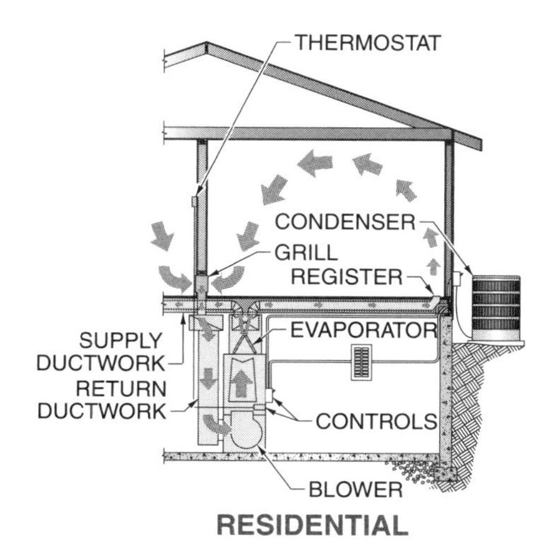

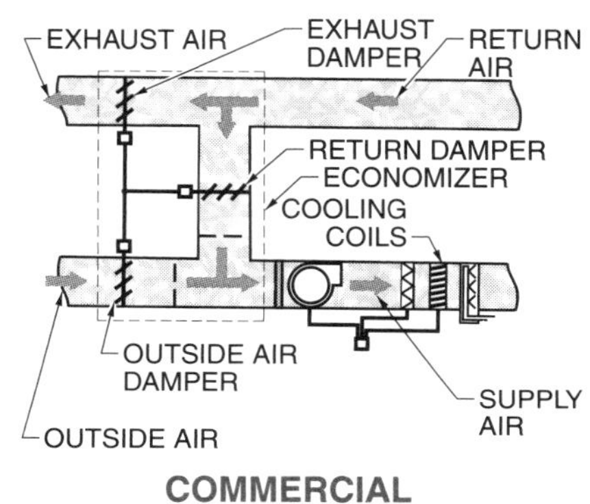

Figure 13-11. A forced air conditioning system consists of an air conditioner, blower, supply and return ductwork, registers, grills, and controls.

Air Conditioners

An *air conditioner* is the component in a forced-air air conditioning system that cools the air. An air conditioner uses either the mechanical compression or absorption refrigeration process to provide a refrigeration effect. The air conditioner contains a blower, filter, and controls. The supply and return air ductwork, registers, and grills have specifications for each application to provide the best air distribution for each building space.

Blowers

A blower moves air across the evaporator coil, circulates it through supply ductwork into the building spaces, and returns the air through return ductwork from the building spaces. Centrifugal blowers are used in air conditioning systems.

A centrifugal blower consists of a blower wheel mounted on a shaft inside a sheet metal scroll. Return air enters the wheel through openings on the sides of the sheet metal scroll. As an electric motor rotates the blower wheel, air is thrown out of the blower by centrifugal force from the tips of the vanes on the blower wheel. Air is drawn into the center of the wheel through the openings in the sides of the scroll because a low-pressure area is created inside the wheel.

Ductwork

Forced-air air conditioning systems include supply air ductwork and registers and return air ductwork and grills. The supply air ductwork runs from the blower in the air conditioner to building spaces. Registers are located on the supply branches and are sized and located to provide the proper amount of air required in each building space. **See Figure 13-12.**

Return air ductwork returns air from the building spaces to the air conditioner for reuse. Grills are sized and located to provide proper circulation of the air through the building.

HYDRONIC AIR CONDITIONING SYSTEMS

A hydronic air conditioning system uses water or a solution of water and antifreeze as the evaporating medium. Terminal devices use the evaporating medium to cool air in building spaces. The relatively high specific weight and high specific heat of water make it an excellent evaporating medium. The pipes that carry water in a hydronic air conditioning system are much smaller than ducts that carry air in a forced-air air conditioning system.

Hydronic air conditioning systems are used where the air conditioning equipment is located centrally and the building spaces are located remotely. Hydronic air conditioning systems are often located in one central plant that cools several other buildings, such as campus school buildings. Hydronic air conditioning systems are also used in multistory buildings where the cooling equipment is located on one floor and the cool water is distributed to terminal devices located on other floors. A typical hydronic air conditioning system contains a chiller, a piping system, circulating pumps, coils, terminal devices, and controls. **See Figure 13-13.**

Chillers

A *chiller* is the component in a hydronic air conditioning system that cools the water, which cools the air. A chiller uses either the mechanical compression or absorption refrigeration process to cool the water. Mechanical compression systems are used often for air conditioning applications. Absorption systems are used on large commercial or industrial applications. Chillers are available with almost any cooling capacity and can be used for almost any size air conditioning application.

> **Tech Facts**
>
> *Chillers are hydronic air conditioning components that are primarily used for large industrial cooling and commercial air conditioning. The types of refrigerants that are typically used with chillers are the HCFCs R-22 and R-123 and the HFC R-134a.*

Piping Systems

Piping systems distribute water from the chiller to the terminal devices. Piping systems are classified by the arrangement of the piping loop that carries the water. A typical piping system has a supply line that carries water from the chiller to the terminal devices and a return line that carries water back from the terminal devices to the chiller. Piping systems include one-, two-, three-, and four-pipe systems. Piping systems for hydronic air conditioning systems are similar to the piping systems on hydronic heating systems.

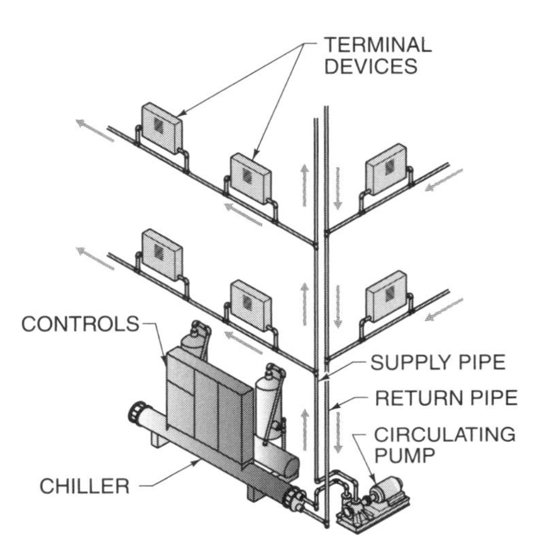

Figure 13-13. Hydronic air conditioning systems contain a chiller, piping system, circulating pump, terminal devices, and controls.

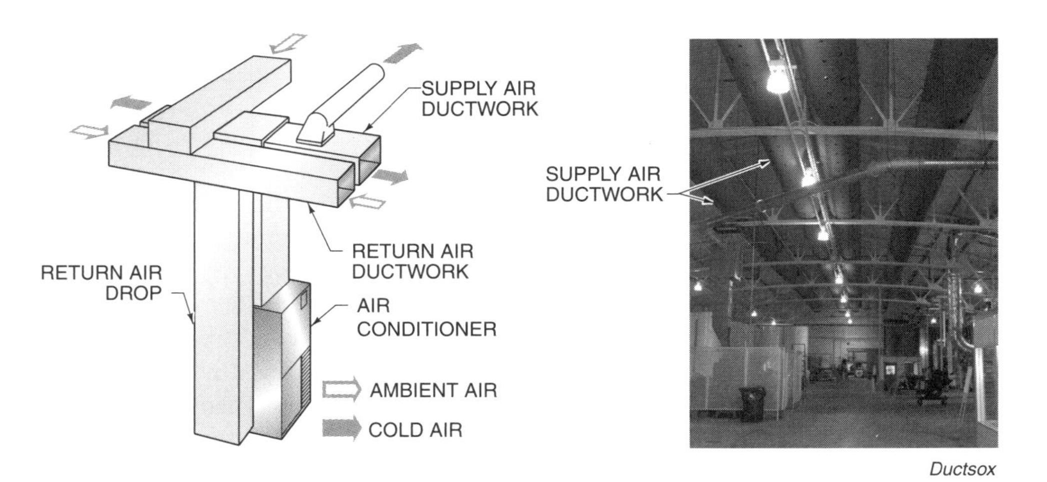

Ductsox

Figure 13-12. Forced air conditioning systems contain supply ductwork and registers and return ductwork and grills.

One-Pipe. Series and primary-secondary systems are one-pipe systems. In a one-pipe series system, the water in the system circulates through one pipe and flows through each terminal device in turn. Terminal devices are connected in series on the loop.

In a one-pipe primary-secondary system, the terminal devices are connected in parallel with the main supply pipe. Water flows from the supply pipe through the terminal devices and back into the supply pipe. The flow rate in each terminal device can be controlled within the loop. One-pipe systems are used in small- to medium-size applications.

Two-Pipe. A two-pipe system has a supply pipe and a return pipe that run to each terminal device. The terminal devices are arranged in parallel. Water flows from the supply pipe through each terminal device and back to the chiller through the return pipe. Direct-return and reverse-return systems are two-pipe systems.

In a two-pipe direct-return system, supply water flows into each terminal device from the supply pipe. The water then enters the return pipe and flows directly back to the chiller. The return water flows in the direction opposite to the water in the supply pipe.

In a two-pipe reverse-return system, supply water flows into each terminal device from the supply pipe. The water then enters the return pipe and flows in the same direction as the water in the supply pipe. The return water continues around the piping loop from one terminal device to the next.

Three-Pipe. A three-pipe system has two supply pipes and one return pipe. A three-pipe system is used for heating some building spaces and cooling other building spaces simultaneously. One supply pipe is connected to a boiler and the other is connected to a chiller. **See Figure 13-14.**

Both supply pipes are connected to each terminal device. Return water flows in a common pipe that is connected to the output of the terminal devices. Mixing valves that are controlled by system switches and thermostats regulate the flow of water from the supply pipes to the terminal devices. On a call for heat, water from the boiler flows through one supply pipe and into the terminal device. On a call for cooling, water from the chiller flows through the other supply pipe and into the terminal device.

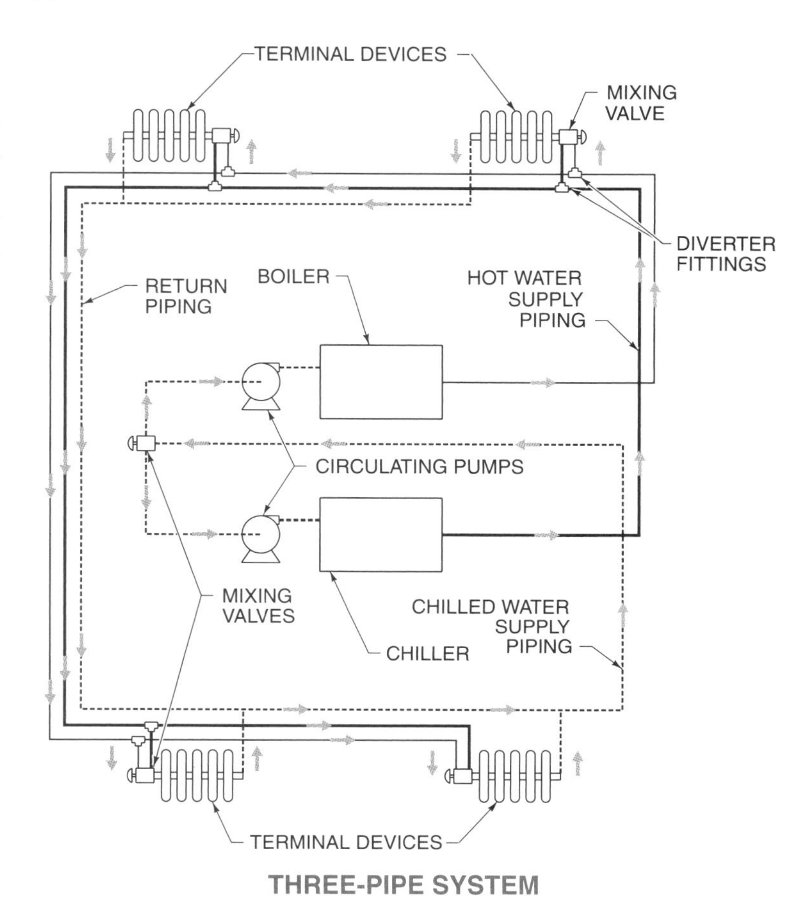

Figure 13-14. A three-pipe system has two supply pipes and one return pipe.

The Trane Company

Hydronic air conditioning piping systems distribute water from chillers to terminal devices and back to the chillers. Piping systems consist of at least one supply line and one return line connected to the terminal devices.

Three-pipe systems provide good control of air temperature but are more expensive to install than two-pipe systems. Three-pipe systems are also expensive to operate because the hot and cold return water is mixed.

Four-Pipe. A four-pipe system uses separate piping for heating and cooling. The terminal devices are connected to both heating and cooling pipes. Water flow to the terminal devices is controlled by mixing and diverting valves, which allow either hot or cold water to flow to any terminal device. Four-pipe systems are expensive to install but provide excellent control of air temperature. Four-pipe systems are more economical to operate than three-pipe systems. **See Figure 13-15.**

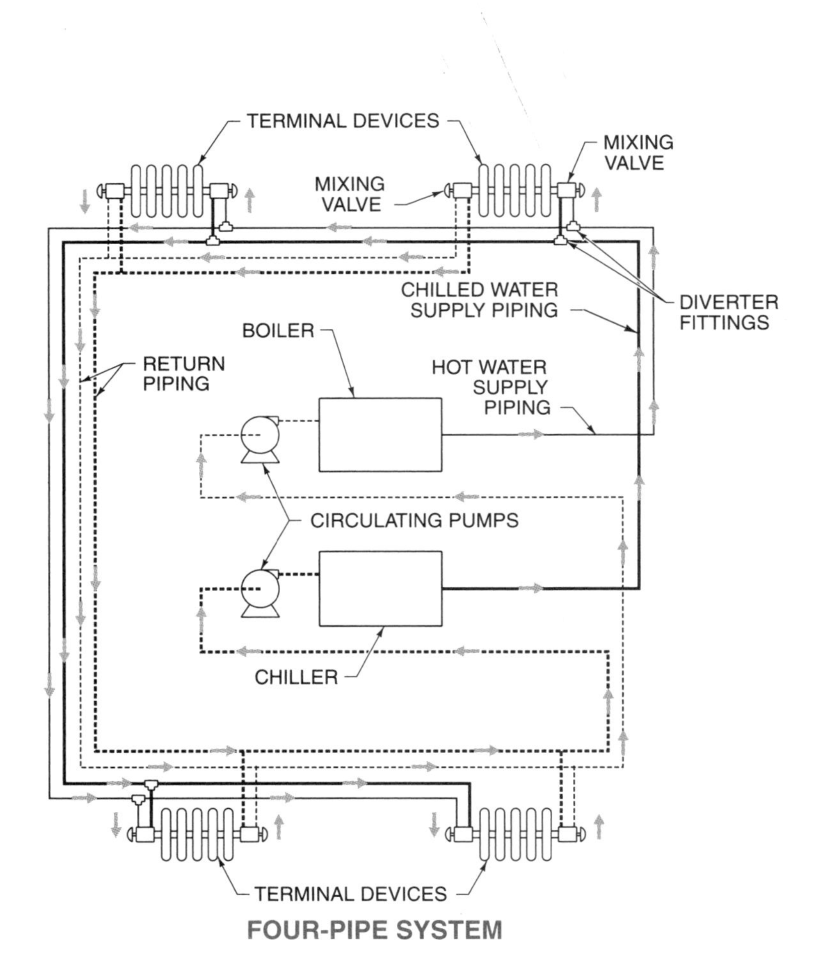

Figure 13-15. A four-pipe system has completely separate piping for heating and cooling.

Circulating Pumps

A circulating pump moves water from the chiller through the piping system to the terminal devices of a hydronic air conditioning system. A circulating pump may be connected to either the supply or return piping, but a circulating pump is usually connected to the return piping close to the chiller. The number and size of circulating pumps depends on the volume of water and resistance to flow in the system. If a hydronic air conditioning system has many piping loops feeding different zones in a building, a pump will be located on each loop.

Centrifugal pumps are often used on hydronic air conditioning systems. A *centrifugal pump* is a pump that has a rotating impeller inside a cast iron or steel housing. The housing has an opening parallel with the impeller shaft and another opening on the outer perimeter of the housing at right angles to the impeller shaft. The opening parallel with the impeller is the water inlet. The return water piping is connected at the water inlet. The opening on the outer perimeter of the housing is the water outlet. The supply water piping is connected at the water outlet. **See Figure 13-16.**

The impeller of a circulating pump rotates rapidly. As the impeller rotates, water is thrown out of the outlet opening by centrifugal force. This action creates a low-pressure area that draws water into the center of the impeller.

Terminal Devices

In a hydronic air conditioning system, a terminal device holds cold water from the chiller. Terminal devices receive a supply of cold water from the main supply pipe and return the water to the main return pipe. Unlike terminal devices for hydronic heating systems, terminal devices for hydronic air conditioning systems do not use thermal radiation for heat transfer. Blowers move air through the terminal device to cool the air. Terminal devices for hydronic air conditioning systems are cabinet air conditioners, unit ventilators, duct coils, and unit air conditioners.

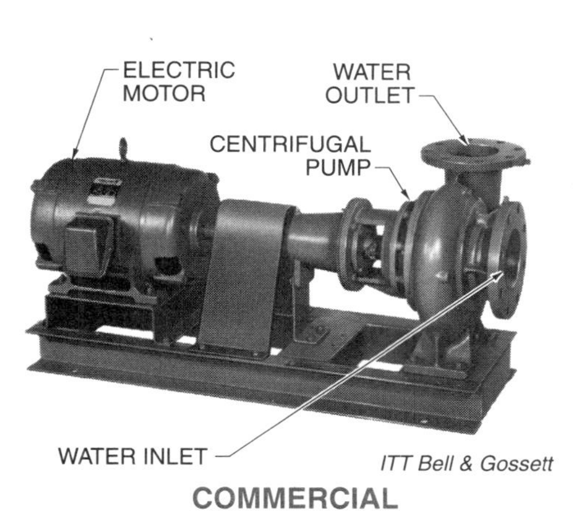

Figure 13-16. An impeller rotates inside the cast iron or steel housing of a centrifugal pump.

Cabinet Air Conditioners. Cabinet air conditioners are similar to cabinet heaters. A cabinet holds a water coil, blower, and filter. Cabinet air conditioners have louvers along the bottom of the cabinet for air intake and louvers along the top of the cabinet for exhaust. Cabinet air conditioners are installed on the floor along the outside walls of a room. **See Figure 13-17.**

Unit Ventilators. Unit ventilators are used for air conditioning applications and heating applications. Unit ventilators differ from cabinet air conditioners in that they have an opening to bring in air from outdoors. Unit ventilators can be used for cooling and ventilating a room simultaneously. Unit ventilators are often used in buildings such as schools and apartments that hold a large number of people and require constant ventilation. **See Figure 13-18.**

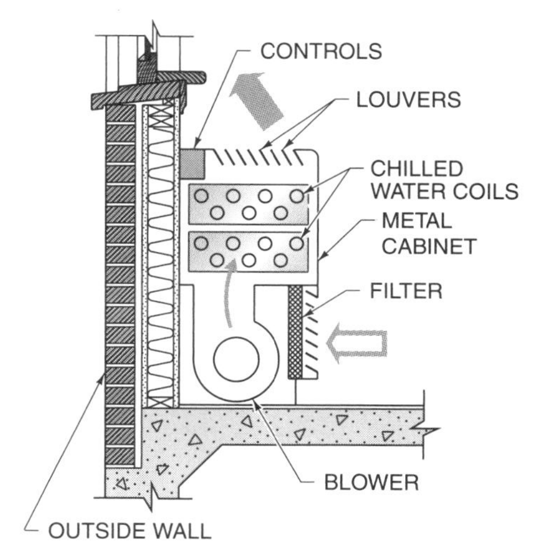

Figure 13-17. Cabinet air conditioners have a water coil, blower, and filter inside one cabinet.

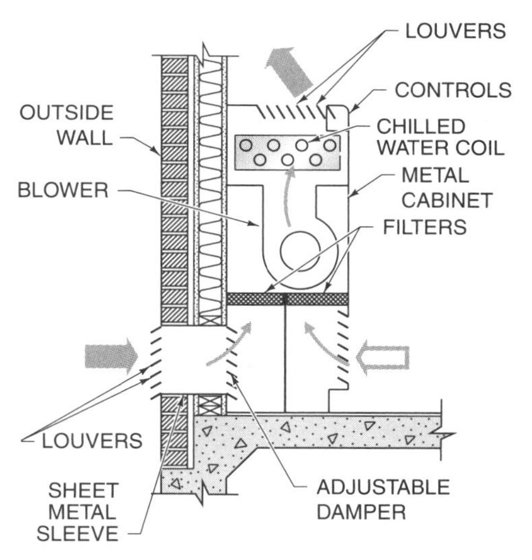

Figure 13-18. Unit ventilators have an opening to the outdoors that admits fresh air.

Duct Coils. A *duct coil* is a terminal device that is located in a duct. Air is supplied to the duct from a blower that may be remotely located. Duct coils are often used when a central chiller provides cooling to separate zones in a building. Each zone in the building has duct coils and a control system. **See Figure 13-19.**

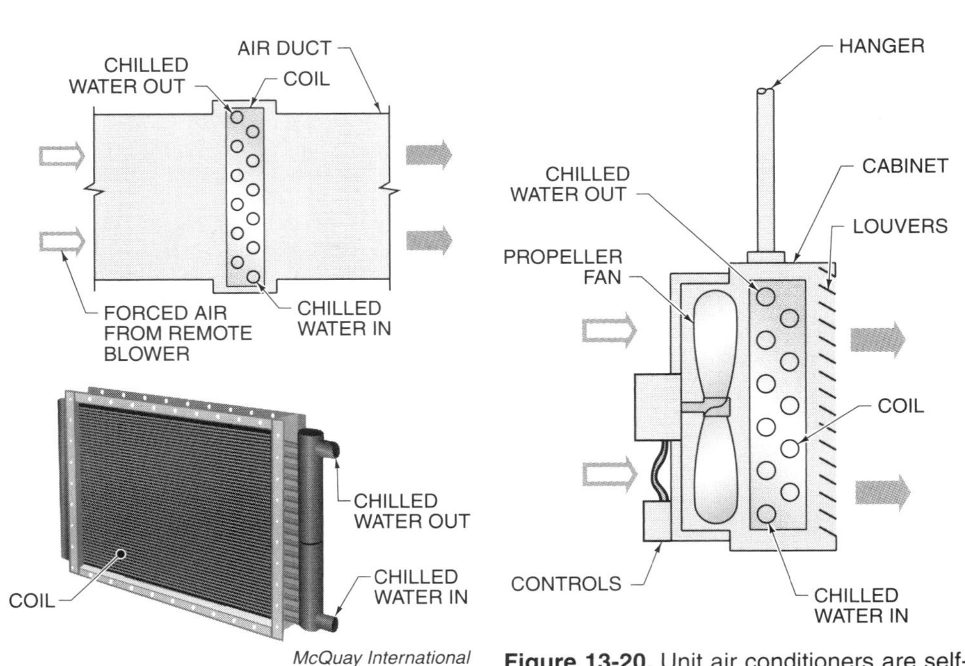

Figure 13-19. Duct coils are terminal devices installed in an air duct.

Figure 13-20. Unit air conditioners are self-contained air conditioners that contain a coil and a blower in one cabinet.

Unit Air Conditioners. A *unit air conditioner* is a self-contained air conditioner that contains a coil and a blower in one cabinet. Some unit air conditioners do not require ductwork. Unit air conditioners are similar to unit heaters. A propeller fan moves the air out of the outlet through louvers. **See Figure 13-20.** Unit air conditioners are often used in individual rooms or zones when controlling an individual space is important.

The Trane Company

Chillers are the main component in a hydronic air conditioning system. A chiller cools water that is circulated through the terminal devices, which cools the surrounding air.

TROUBLESHOOTING AND SERVICING AIR CONDITIONING SYSTEMS

To troubleshoot and service air conditioning systems, an individual must be familiar with the complete system. Many times a faulty component causes problems in other components in the system. Most of the major devices in an air conditioning system are made up of smaller components. For example, a centrifugal chiller includes an evaporator (cooler), a condenser, a compressor, a metering device, and controls. The controls include switches, relays, sensors, timers, and printed circuit boards.

Tools and Supplies

Using good quality test instruments that are accurate and dependable is important when troubleshooting air conditioning systems. The test instruments required must be able to take readings of voltage, amperage, resistance, temperature, water pressure, airflow, and static pressure. To check pressure signals, gauges or gauge sets are used.

To measure airflow out of diffusers or into grills, a balometer (capture hood) is used. To measure the airflow to or from diffusers or grills where there is not enough room to use a balometer, an anemometer is used with the appropriate accessories. **See Figure 13-21.**

Tech Facts

When servicing air conditioning systems, it is important to make sure that the system has the correct refrigerant charge. If the system is undercharged, the motor will operate continuously.

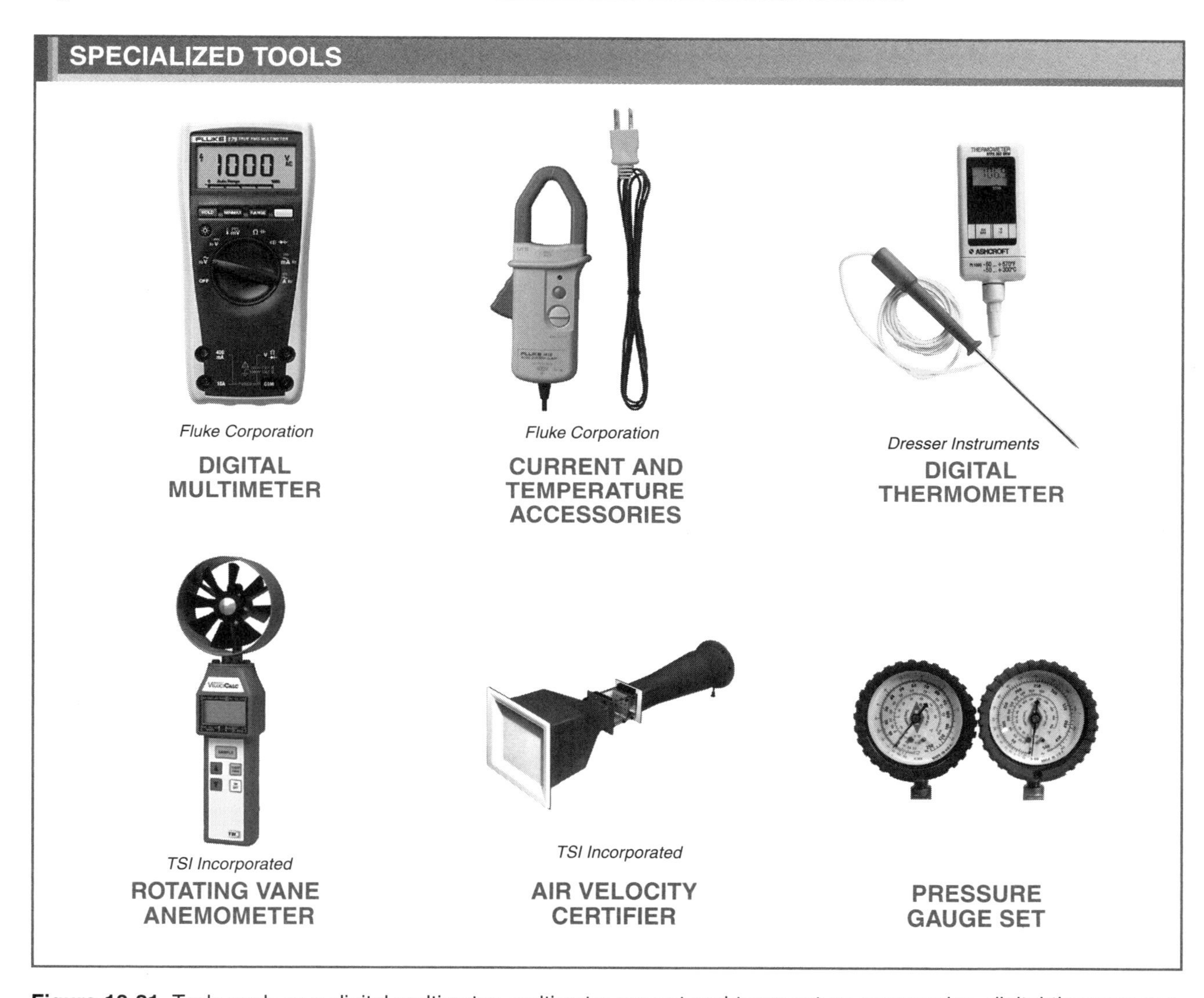

Figure 13-21. Tools such as a digital multimeter, multimeter current and temperature accessories, digital thermometer, anemometer, and pressure gauges are needed along with other specialized tools to service forced-air heating systems.

Review Questions

1. Define *air conditioning.* How does it differ from refrigeration?
2. List four ways in which air conditioning systems are classified.
3. Define *condensing medium.*
4. List and describe three kinds of condensers used in air conditioning systems.
5. Describe the difference between a split system and a package unit.
6. What is the main characteristic of a mechanical compression system?
7. List and describe the function of the six main parts of an absorption refrigeration system.
8. List and describe three classifications of controls found in a typical air conditioning system.
9. What is the main function of the operating controls on an air conditioning system?
10. What is the primary purpose of the safety controls in an air conditioning system?
11. Describe the operation of a forced-air air conditioning system. Where is this kind of system typically used?
12. What are the three main functions of the blower in an air conditioning system?
13. List the four parts of a forced-air air conditioning system that are not part of the air conditioner itself.
14. List common locations where hydronic air conditioning systems are typically used.
15. Why are pipes in hydronic air conditioning systems smaller than air ducts in forced air conditioning systems?
16. List and describe the four different kinds of piping systems for hydronic distribution systems.
17. How do the terminal devices in a hydronic air conditioning system cool the air in a building space?
18. What do balometers measure?

A heat pump is a mechanical compression refrigeration system that moves heat from one area to another area. When a heat pump is in the cooling mode, it moves heat from inside the building to outside the building. When a heat pump is in the heating mode, it moves heat form outside the building to the inside of the building. Changing the mode of the heat pump from cooling to heating reverses the flow of refrigerant through the system.

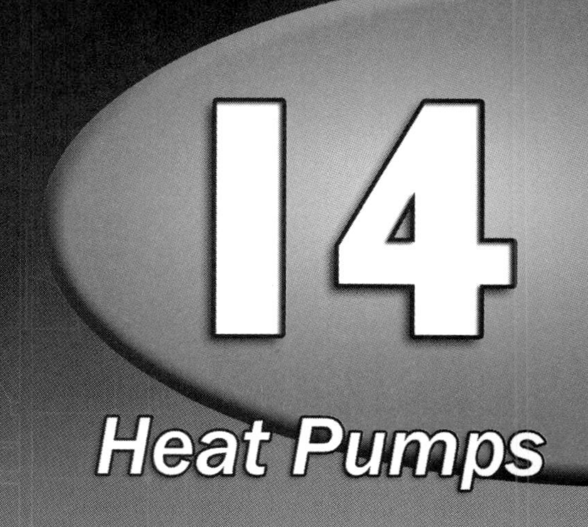

14 Heat Pumps

HEAT PUMPS

A *heat pump* is a mechanical compression refrigeration system that contains devices and controls that reverse the flow of refrigerant to move heat from one area to another area. Reversing the flow of refrigerant switches the relative position of the evaporator and condenser in a heat pump system. A reversing valve is used to reverse the flow of refrigerant through the system. A heat pump can absorb heat from inside a building and reject the heat outdoors, or the heat pump can absorb heat from the outdoors and reject the heat inside a building.

A heat pump in the cooling mode operates like an air conditioning system. Air conditioning systems or heat pumps in the cooling mode move heat from inside a building to the air outside a building. When a heat pump is in the cooling mode, the indoor unit is the evaporator and the outdoor unit is the condenser. A heat pump in the heating mode moves heat from the outside air to the air inside a building. When a heat pump is in the heating mode, the indoor unit is the condenser and the outdoor unit is the evaporator.

During heat pump operation, refrigerant flows from the compressor to a reversing valve. The refrigerant flows from the reversing valve to either the coil in the indoor unit or the coil in the outdoor unit. The direction of refrigerant flow depends on whether the system is in the cooling or heating mode.

Heat Sink

Unlike a furnace or boiler, a heat pump does not produce heat. A heat pump transfers heat from one area to another. A heat sink must be available to a heat pump. A *heat sink* is a substance with a relatively cold surface that is capable of absorbing heat. The heat of a heat sink is absorbed by a heat pump and is used for heating.

Any substance (material) that has a temperature above absolute zero (−460°F) contains heat. The higher the temperature of a material, the more heat energy it contains. As the heat energy contained in a material is increased, the better the material is as a heat sink (source of heat). Water and air are normally used as heat sink materials for most heat pump applications because they are readily available.

Air. Air is the most common material used as a heat sink for heat pumps. Because the temperature of outdoor air varies during different times of the year and during different times of the day, the amount of heat available from outdoor air varies. Air temperature above 15°F works well as a heat sink.

Water. Water used as a heat sink comes from lakes, streams, wells, and industrial holding ponds. When used as a heat sink, the temperature of the water should be as high as possible. Groundwater is also used as a heat sink for heat pumps. *Groundwater* is water that is found naturally in the ground. The average temperature of groundwater is 45°F to 55°F, which makes groundwater a better heat sink material for heat pumps than air.

Classifications

Heat pumps are classified according to the cooling medium and heat sink material. The four classifications of heat pumps are air-to-air, air-to-water, water-to-air, and water-to-water. The heat pump used for an application depends on the location of the heat pump and the availability of the cooling medium and heat sink material. The coil of a heat pump that uses air as a heat sink is located outdoors. Heat pumps that use air as a heat sink are split systems. The coil of a heat pump that uses water as a heat sink may be located outdoors or indoors.

Air-to-Air. An air-to-air heat pump system consists of a compressor, reversing valve, outdoor unit, expansion devices with bypass circuits, indoor unit, and refrigerant lines. **See Figure 14-1.** An *outdoor unit* is a package component that contains a coil heat exchanger and a fan. Depending on the direction of refrigerant flow, the outdoor unit acts as the evaporator or condenser of the heat pump.

A *bypass circuit* is a refrigerant line that contains a check valve on each side of an expansion device. The check valves in a bypass circuit are arranged to allow refrigerant to pass around the expansion device when the refrigerant flow is opposite to the direction of normal flow through the expansion device. Only capillary tube expansion devices produce a pressure decrease when refrigerant flows in either direction. An *indoor unit* is a package component that contains a coil heat exchanger and a blower. Depending on the direction of refrigerant flow, the indoor unit acts as the evaporator or condenser of the heat pump.

An air-to-air heat pump uses outdoor air as the heat sink material when the system operates in the heating mode. Return air from the building is blown across the indoor coil to absorb heat for heating the building spaces. Air-to-air heat pumps are available with cooling capacities from 1 ton to 5 tons of cooling. Air-to-air heat pumps are used in residences and small commercial buildings. Air-to-air heat pumps are available as split systems and as package systems.

Tech Facts

Air-to-air heat pumps are designed to heat or cool an area by using reversing valves to reverse the system operation.

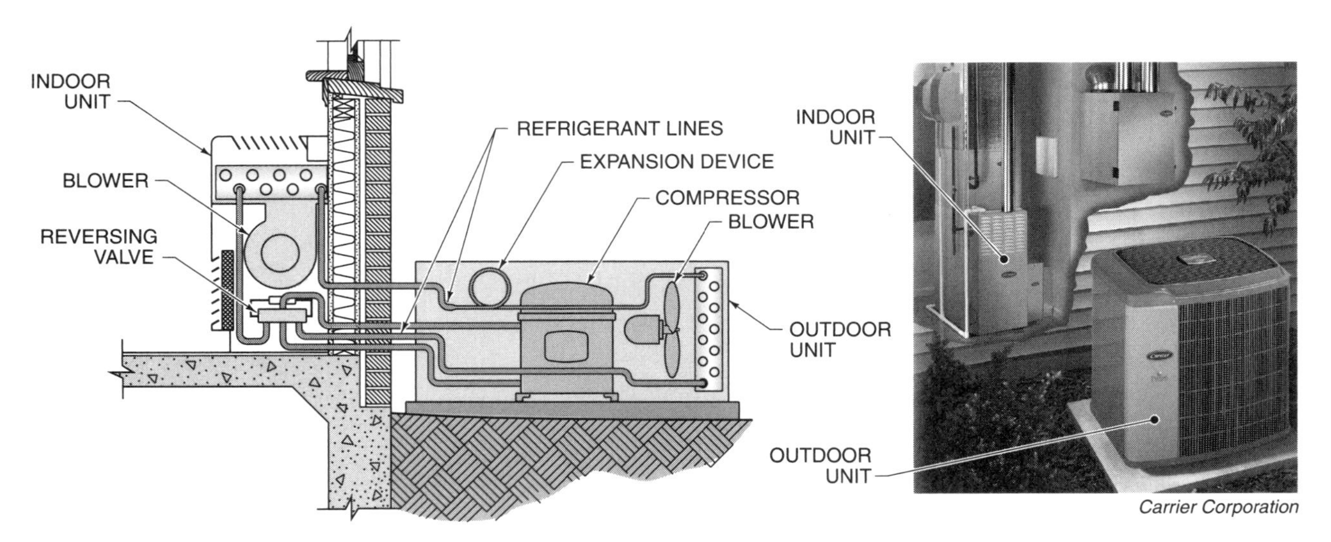

Figure 14-1. An air-to-air heat pump cools or heats the air and uses air as a heat sink.

Air-to-Water. An air-to-water heat pump system consists of a compressor, reversing valve, outdoor unit, expansion devices with bypass circuits, indoor unit, and refrigerant lines. The outdoor unit contains a coil heat exchanger. The indoor unit contains a coil heat exchanger and blower. An air-to-water heat pump uses outdoor air to cool or heat water that cools or heats the air in the building spaces. Outdoor air is used as the heat sink when a heat pump operates in the heating mode.

Air-to-water heat pumps are used to heat and cool large buildings that use hydronic heating and cooling systems. An air-to-water heat pump system is a built-up system. Built-up systems contain factory-built components that are assembled at the job site.

Water-to-Air. A water-to-air heat pump system consists of a compressor, reversing valve, outdoor unit, expansion devices with bypass circuits, indoor unit, and refrigerant lines. The outdoor unit contains a coil heat exchanger.

The indoor unit contains a finned-tube heat exchanger and blower. A water-to-air heat pump uses water as a heat sink and uses air to cool or heat building spaces. **See Figure 14-2.** Water-to-air heat pumps are often used as package units in applications where an outdoor unit cannot be located outdoors. Instead, the outdoor unit is located inside the building. The supply water is piped to what would be the outdoor unit inside the building. Small water-to-air heat pumps are often used to control the temperature of small spaces in high-rise buildings, such as hotel rooms.

Water-to-Water. A water-to-water heat pump system consists of a compressor, reversing valve, outdoor unit, expansion devices with bypass circuits, indoor unit, and refrigerant lines. The outdoor unit and the indoor unit each contain a coil heat exchanger. Neither unit contains a blower.

Carrier Corporation

Heat pumps contain a condenser and an evaporator that are used to transfer heat from one area to another.

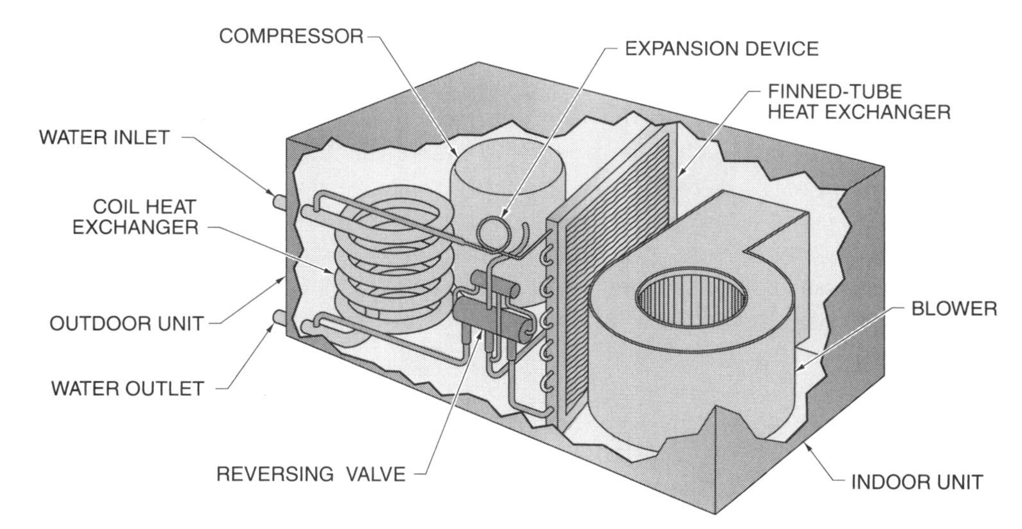

Figure 14-2. A water-to-air heat pump cools or heats water, which is used to cool or heat building spaces.

Water from outside the system is used as the heat sink for a water-to-water heat pump. The heat pump uses water to cool or heat water which is used to cool or heat building spaces. Water-to-water heat pump systems are used for cooling and heating large buildings. Most water-to-water systems are built-up systems which must be assembled at the job site. Centrifugal or screw compressors are often used in water-to-water heat pump systems.

COMPONENTS

Heat pumps contain a compressor, outdoor unit, indoor unit, expansion device, and refrigerant lines. Heat pumps also contain expansion devices with bypass circuits to allow the refrigerant to expand regardless of the direction of flow, and a reversing valve that reverses the flow of refrigerant in the refrigerant lines depending on whether the heat pump is in heating or cooling mode. The reversing valve is required because the compressor creates a flow of refrigerant in one direction only.

Reversing Valves

A *reversing valve* is a four-way directional valve that reverses the flow of refrigerant through a heat pump. A reversing valve consists of a piston and cylinder that have four refrigerant line connections. The piston moves from one end of the cylinder to the other, changing the direction of refrigerant flow by opening and closing outlets. **See Figure 14-3.**

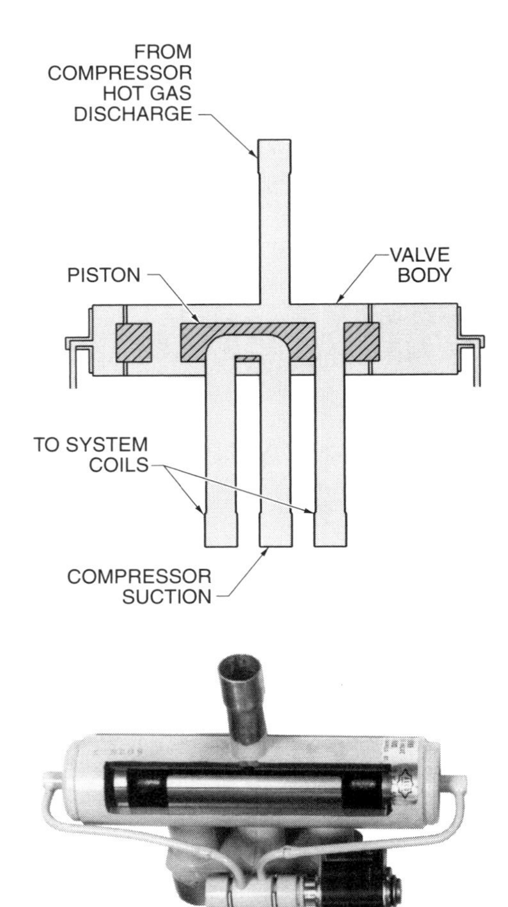

Figure 14-3. A reversing valve consists of a cylinder with a piston that moves from one end of the cylinder to the other.

When the piston is at one end of the cylinder, refrigerant flows into the valve through the hot gas discharge line and is directed out of one of the outlet ports to the opposing unit. When the piston moves to the other end of the cylinder, refrigerant flows into the valve through the same inlet port and is discharged to another outlet port. The refrigerant is then directed to the opposite unit of the heat pump. In either position, the suction line to the compressor returns the refrigerant from the reversing valve to the compressor. **See Figure 14-4.**

Reversing valves are controlled by a solenoid-operated pilot valve. A *solenoid-operated pilot valve* is a small valve connected to a refrigerant line that contains a solenoid and is used to control other valves. The solenoid-operated pilot valve is located near the reversing valve and is operated by a low-voltage electric signal from a thermostat or controller.

When an electric signal actuates the solenoid on the pilot valve, the solenoid moves a small piston back and forth in the pilot valve cylinder. A capillary tube runs from the pilot valve cylinder to the suction line on the reversing valve. Two other capillary tubes run from the pilot valve cylinder to each end of the reversing valve cylinder. **See Figure 14-5.**

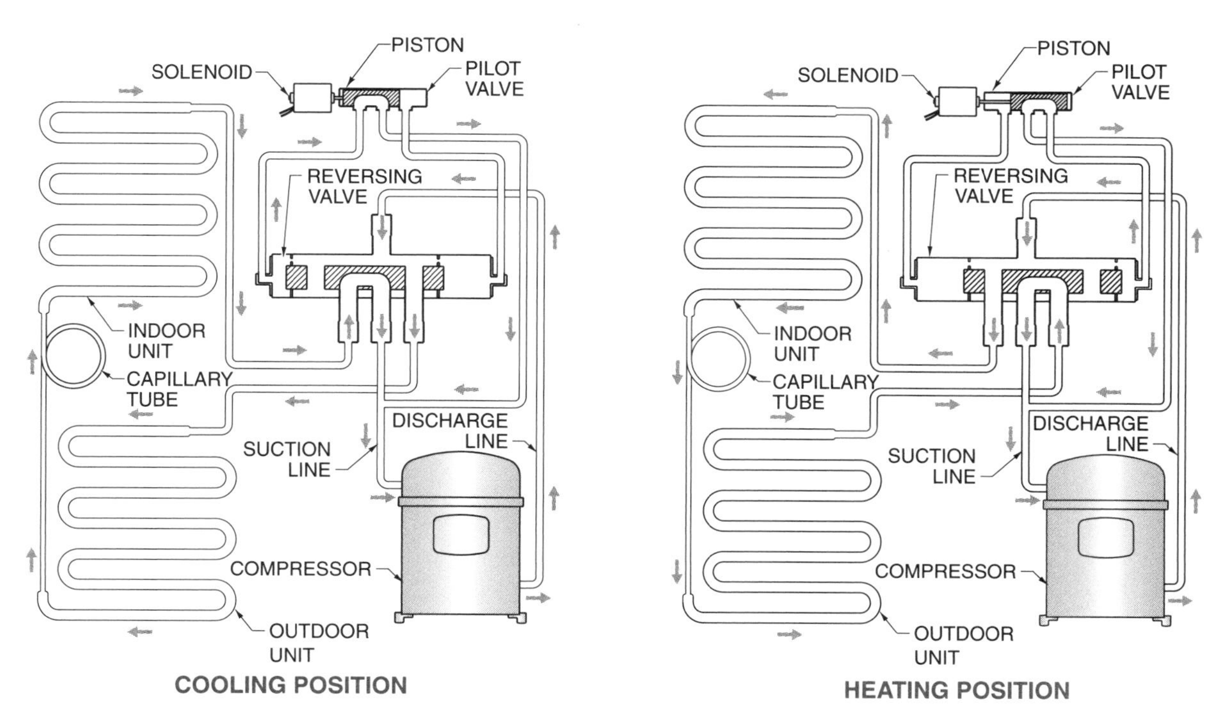

Figure 14-4. When a pilot valve receives refrigerant from a heat pump system, the pressure of the refrigerant is used to move the piston of the reversing valve.

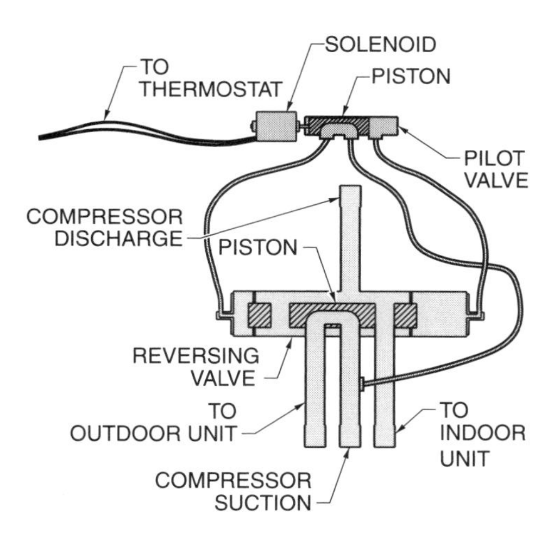

Figure 14-5. Capillary tubes connect the pilot valve to the reversing valve.

When the solenoid moves the pilot valve piston to the left side of the pilot valve cylinder, the capillary tube that runs to the left side of the reversing valve opens for refrigerant flow between the left side of the reversing valve and the pilot valve. The refrigerant flows from the pilot valve to the compressor suction line. The flow of refrigerant reduces the pressure in the left side of the reversing valve. The refrigerant pressure on the right side of the reversing valve also pushes the reversing valve piston to the left side of the cylinder at this time.

When the solenoid moves the pilot valve piston to the right side of the pilot valve cylinder, the capillary tube that runs to the right side of the reversing valve opens for refrigerant flow between the right side of the reversing valve and the pilot valve. The refrigerant flows from the pilot valve to the compressor suction line. The flow of refrigerant reduces the pressure in the right side of the reversing valve. The refrigerant pressure on the left side of the reversing valve also pushes the reversing valve piston to the right side of the cylinder at this time.

Expansion Devices

The coil in the indoor unit and the coil in the outdoor unit each require an expansion device because the refrigerant in a heat pump can flow in either direction. In small package heat pumps where the coils are close together, a capillary tube may be used as the expansion device. Some package heat pumps have refrigerant control valves, which are a combination check valve and orifice, as expansion devices. Large heat pumps, especially split systems, use thermostatic expansion valves as expansion devices.

Capillary Tubes. A capillary tube produces a pressure decrease regardless of the direction of refrigerant flow. Filters are located at each end of the capillary tube to remove small contaminant particles but allow liquid refrigerant to pass. When a capillary tube expansion device is used in a heat pump system, a bypass circuit is normally installed around the capillary tube because the refrigerant must be filtered. The capillary tube has a tube inside it to compensate for the higher pressures that occur in the heating mode. A capillary tube is used only on small package or split heat pump systems that have cooling capacities from 1 ton to 5 tons.

Refrigerant Control Valves. A *refrigerant control valve* is a combination expansion device and check valve. A refrigerant control valve has a piston inside the check valve that moves back and forth inside the valve housing. The piston moves in the same direction as the refrigerant flow. The piston partially closes the orifice in the end of the housing by moving to that end of the valve. One orifice is larger than the other to provide a pressure decrease for the refrigerant as refrigerant flows through the valve. **See Figure 14-6.**

Thermostatic Expansion Valves. Large heat pumps and split systems have thermostatic expansion valves as expansion devices. A *thermostatic expansion valve* is a valve that is controlled by pressure to control refrigerant flow.

In a thermostatic expansion valve, pressure from a remote bulb exerted against a diaphragm determines the size of the outlet. A pushrod connects the diaphragm in the top of the valve to a ball inside the valve body. If pressure on the diaphragm decreases, the ball closes the valve. If pressure on the diaphragm increases, the ball opens the valve. The pressure is exerted by a refrigerant-filled remote bulb. The bulb is attached by a capillary tube to the refrigerant line that leaves the outdoor coil. The temperature of the refrigerant leaving the outdoor coil is sensed by the refrigerant in the bulb. **See Figure 14-7.**

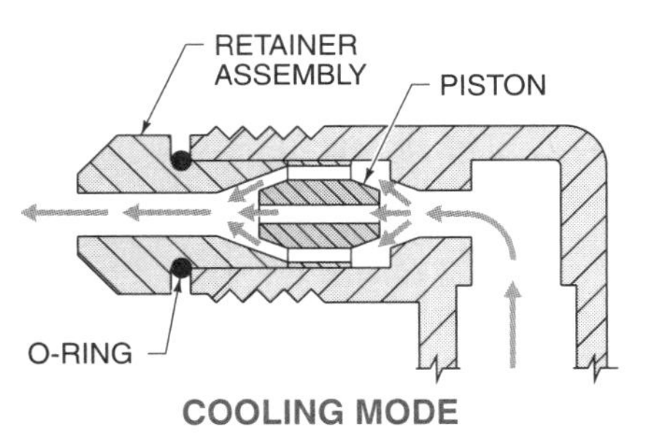

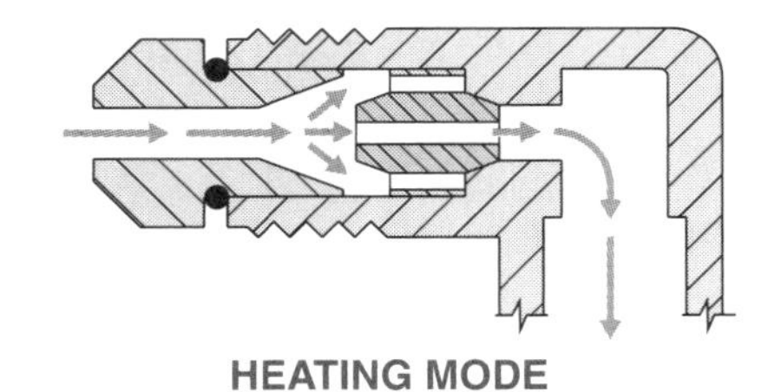

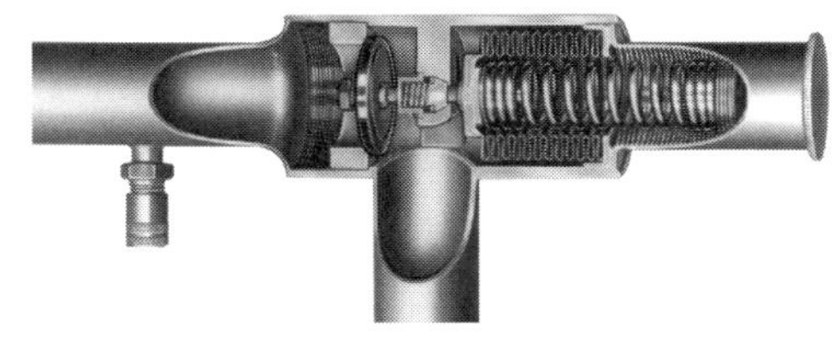

Sporlan Valve Company

Figure 14-6. A refrigerant control valve is an expansion device that controls the flow of refrigerant in either direction.

If the temperature of the refrigerant leaving the outdoor coil increases, the refrigerant in the bulb vaporizes and increases the pressure in the capillary tube. This pressure increase opens the thermostatic expansion valve, which increases the flow of refrigerant to the outdoor coil. If the temperature of the refrigerant leaving the outdoor coil decreases, the refrigerant in the bulb condenses and decreases the pressure in the capillary tube. This decrease in pressure closes the thermostatic expansion valve, which decreases the flow of refrigerant to the coil.

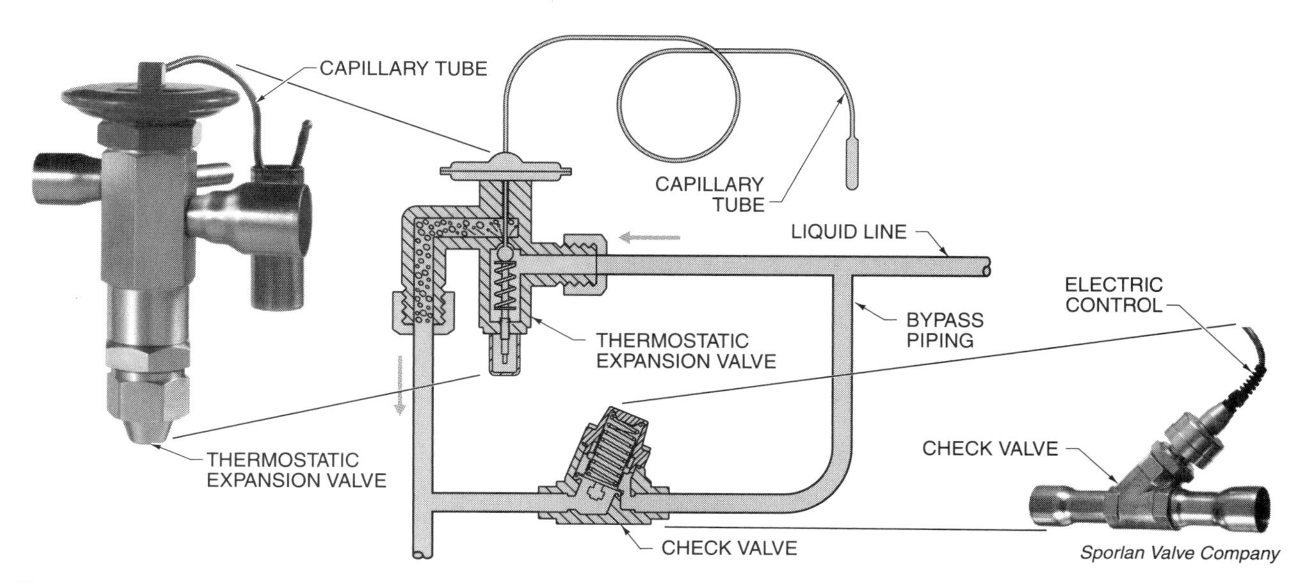

Figure 14-7. Large heat pumps and split systems have two thermostatic expansion valves. Bypass piping with a check valve allows refrigerant flow around the valve when flow is in the opposite direction.

In large heat pumps, a thermostatic expansion valve is located on each coil. The valves are necessary because the indoor coil is the evaporator in the cooling mode and the outdoor coil is the evaporator in the heating mode. A thermostatic expansion valve is located near each coil at the end of the liquid line.

Because refrigerant cannot be reversed in a thermostatic expansion valve, bypass piping is installed around each expansion device. The bypass piping has a check valve which prevents the liquid refrigerant from flowing back through the expansion valve by allowing the refrigerant to flow through the bypass. The thermostatic expansion valve of the outdoor coil is bypassed when the system operates in the cooling mode, and the expansion valve of the indoor coil is bypassed when the system operates in the heating mode.

AUXILIARY HEAT

An auxiliary heat source is used during cold weather because a heat pump that is sized for cooling a building in summer may not have a heat output that is adequate for heating the building in winter. The heat output of a heat pump is directly related to the amount of heat available from the heat sink.

Auxiliary Heat Source

In cold weather the temperature of air falls and the amount of heat needed to heat a building increases. If a heat pump is sized for the heat requirements of a building at the coldest outside temperatures, it is oversized for cooling. To prevent this problem, the heat pump is sized to provide the heat required to heat a building with an outdoor air temperature of about 35°F. An auxiliary heat source increases the heat output of the heat pump at outdoor temperatures below 35°F. Sources of auxiliary heat for heat pumps are resistance heating elements and gas fuel- or fuel oil-fired heaters.

Resistance Heating Elements. Resistance heating elements are the most common auxiliary heat source for a heat pump. Electricity is a convenient energy source for auxiliary heat because heat pumps are supplied with electricity. **See Figure 14-8.**

Tech Facts

Auxiliary heaters are needed when the outside air temperature drops to where the efficiency of air-to-air heat pumps significantly decreases. The auxiliary electric heaters use electric resistance heating to make up for the loss in efficiency.

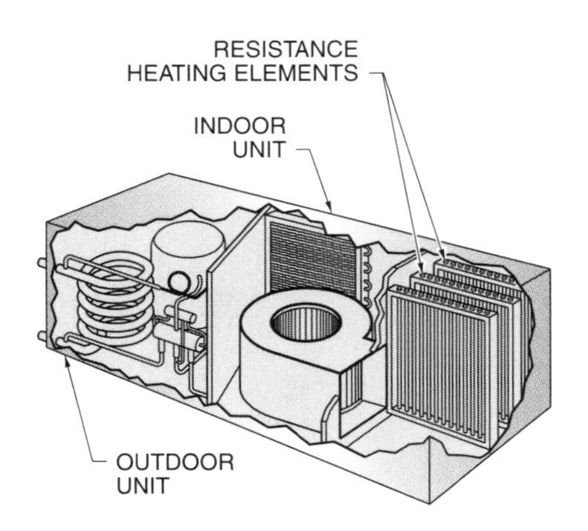

Figure 14-8. Resistance heating elements are often used as an auxiliary heat source for heat pump systems.

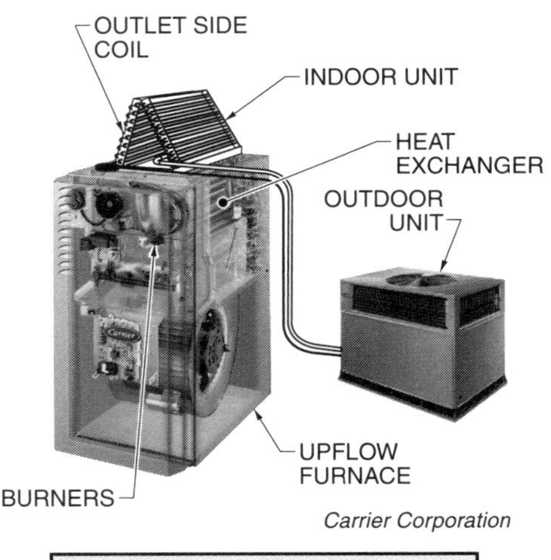

WARNING! OUTLET SIDE HEAT PUMP COIL
HEAT PUMP MUST NOT BE ON
WHEN FURNACE IS ON

Figure 14-9. Gas fuel- or fuel oil-fired heaters are used as auxiliary heating when a heat pump is added to an existing heating system.

The resistance heating elements used for auxiliary heat in heat pumps are similar to those used in electric furnaces. The elements consist of a grid of wire coils in porcelain insulators suspended in a frame. The coils are made of Nichrome® wire, which has a high resistance to the flow of electricity. When the coils are connected to an electrical circuit and the power is turned ON, the coils get extremely hot. Air is heated as it is blown across the coils. The auxiliary resistance heating elements of a heat pump should be installed after the indoor heating coil.

Gas Fuel- or Fuel Oil-Fired Heaters. Gas fuel- or fuel oil-fired heaters also provide auxiliary heat for heat pumps. Fuel fired heaters are normally used when a heat pump is added to an existing heating system, allowing the indoor coil to be placed on the supply side of the furnace. **See Figure 14-9.**

When a heat pump coil is used on the outlet side of an existing heating system, the controls must shut the heat pump down when the auxiliary heat is ON. The indoor coil of the heat pump must not be ON when the furnace is producing heat. The added heat from the furnace will cause excessive pressure in the heat pump system.

CONTROLS

Heat pumps have power, operating, and safety controls that are similar to the controls on a typical refrigeration system. Heat pumps include a heat pump thermostat, defrost control, and possibly an auxiliary heat source that has separate controls.

Thermostats

A *heat pump thermostat* is a component that incorporates a system switch, heating thermostat, and cooling thermostat. A heat pump thermostat switches the heat pump between the cooling mode and heating mode, and controls system operation in either mode. In the heating mode, the thermostat controls the reversing valve and compressor, and activates auxiliary heat source controls.

Different types of thermostats are used for controlling heat pumps. Some thermostats are switched manually from the cooling to heating mode while others switch automatically. Most heat pumps require a multibulb, multistage thermostat. A *multibulb thermostat* is a thermostat that contains more than one mercury bulb switch. A *multistage thermostat* is a thermostat that contains several mercury bulb switches that make and break contacts in stages. When a multibulb, multistage thermostat is connected to two different devices, the devices are energized at different times, depending on temperature changes.

For example, a thermostat used to control a heat pump may have two stages. In the cooling mode, the first-stage mercury bulb controls the reversing valve and the second-stage mercury bulb controls the compressor. When the temperature increases, the first-stage mercury bulb switches the system to the cooling mode, and the second-stage mercury bulb energizes the compressor to provide cooling.

When using the same heat pump system in the heating mode, the first-stage mercury bulb controls the compressor and the second-stage mercury bulb controls the auxiliary heat. As the temperature falls, the first-stage mercury bulb switches the compressor ON to provide heat. If the temperature continues to fall, the second-stage mercury bulb switches the auxiliary heat ON. **See Figure 14-10.**

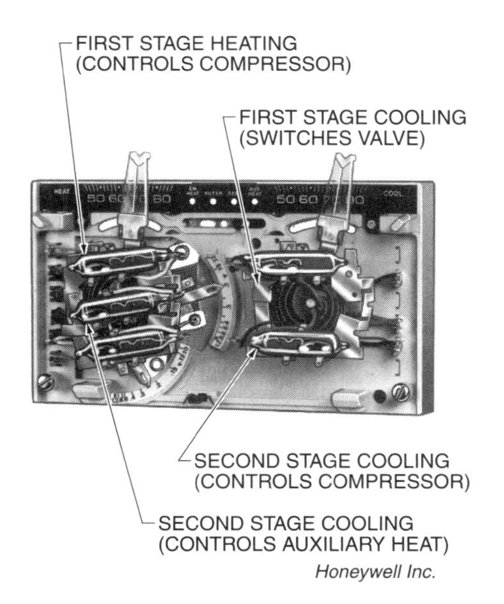

Figure 14-10. Most heat pump systems require a multibulb, multistage thermostat.

Defrost Controls

Defrost controls regulate the defrost cycles of a heat pump. A *defrost cycle* is a mechanical procedure that consists of reversing refrigerant flow in a heat pump to melt frost or ice that builds up on the outdoor coil. When an air-to-air heat pump operates in the heating mode, the outdoor coil functions as an evaporator. The temperature of the outdoor coil is usually below 32°F. At this temperature, moisture in the outdoor air freezes and forms frost or ice on the surface of the outdoor coil. The frozen moisture blocks air flow across the outdoor coil. **See Figure 14-11.**

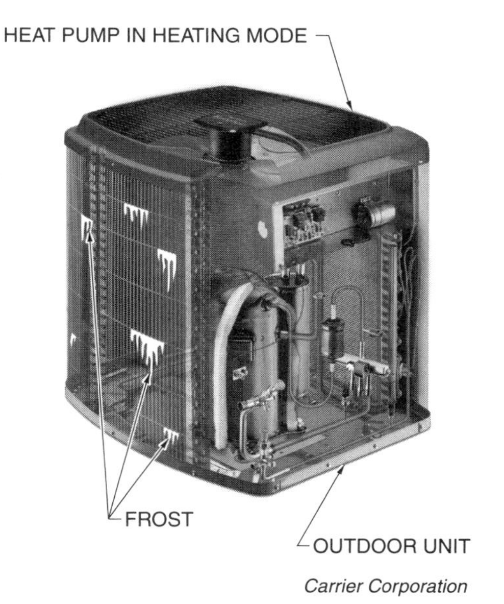

Figure 14-11. When the temperature of the outdoor coil is below 32°F, the outdoor air freezes and forms frost or ice on the surface of the coil.

When the flow of refrigerant in the heat pump system is reversed by temporarily switching the system to the cooling mode, hot refrigerant gas flows through the outdoor coil. The cooling mode is activated when the position of the reversing valve is changed. The outdoor fans are shut OFF, and the auxiliary heat is turned ON. This action raises the temperature of the hot gas, which flows through the outdoor coil and melts the frost or ice. The air blown into the building is cooled slightly during the defrost cycle.

Most defrost control systems are based on time cycles. A timer in the defrost control system calls for defrost at regular time intervals. If defrost is needed, an initiation sensor signals a need for defrost. Initiation sensors sense either the temperature of the air or temperature of the refrigerant leaving the outdoor coil to control the defrost cycle. Initiation sensors also use the pressure of the refrigerant leaving the outdoor coil to control the defrost cycle. If defrost is not needed when the timer signals, the initiation sensor prevents the system from going into a defrost cycle.

During a defrost cycle, a termination sensor senses the temperature of the outdoor coil either directly or through refrigerant pressure. When the temperature or pressure indicates that the frost or ice has melted, the cycle is terminated.

Time. Time is often used to initiate a defrost cycle. A timer in the defrost control system runs continuously. At intervals of 45, 60, or 90 minutes, the timer calls for a defrost cycle. The initiation sensor, which is located on the outdoor coil, determines if a defrost cycle is needed and activates the defrost cycle control system. **See Figure 14-12.**

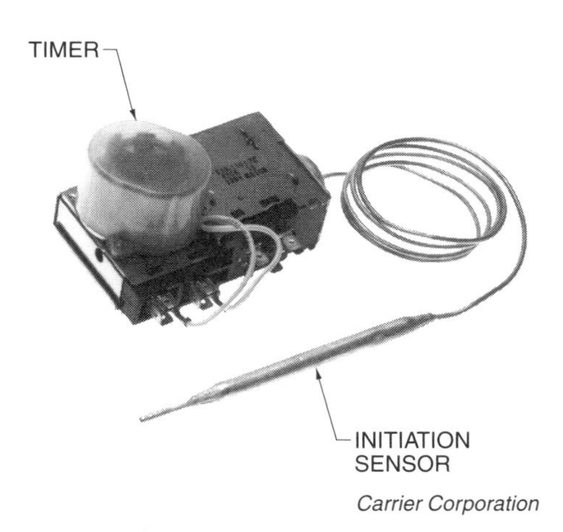

Figure 14-12. A timer is the most commonly used method of initializing a defrost cycle.

Temperature. The temperature of the air that enters and leaves the outdoor coil is also used to determine when a defrost cycle should begin and end. If frost or ice builds up on the outdoor coil, the temperature difference across the coil decreases. A temperature-actuated defrost control is mounted on the discharge side of the coil. A temperature-actuated defrost control consists of a remote bulb thermostat and an electric control switch. The switch is wired into the defrost control. The temperature-actuated defrost control sends a signal to the defrost control, which monitors temperature. If the temperature indicates a need for defrost, a defrost cycle is initiated. As the frost or ice on the coil is melted by the hot gas, the temperature of the refrigerant leaving the coil increases. This temperature increase is sensed by the thermostat, and the temperature-actuated defrost sensor sends a signal to the defrost control to terminate the defrost cycle. **See Figure 14-13.**

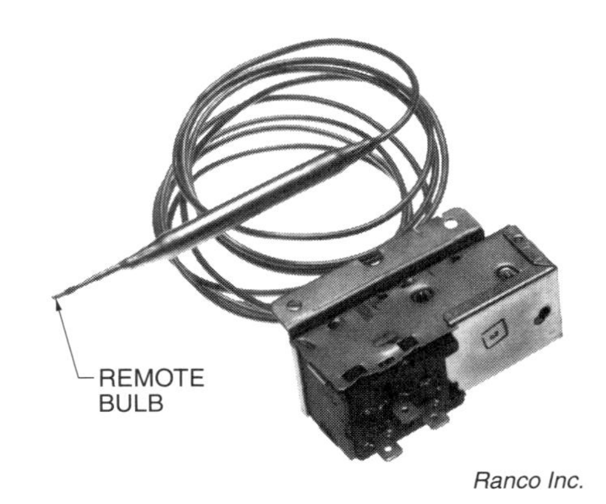

Figure 14-13. A temperature-actuated defrost control mounted on the coil of the outdoor unit senses the temperature of the coil.

Pressure. The pressure of the refrigerant in a refrigeration system can also be used to determine when a defrost cycle should begin and end. At saturation conditions, the refrigerant pressure in an outdoor coil represents the temperature of the refrigerant because of the pressure-temperature relationship.

The temperature of the outdoor coil is monitored by measuring the refrigerant pressure. A tapping valve is placed on one of the refrigerant lines in the outdoor coil, usually on a return bend. A *tapping valve* is a valve that is attached to and pierces a refrigerant line. **See Figure 14-14.** Tubing from the tapping valve leads to a pressure switch. When the coil begins to frost, the pressure of the refrigerant in the coil decreases. If the pressure decreases enough to indicate a freezing temperature, the pressure switch activates the defrost control.

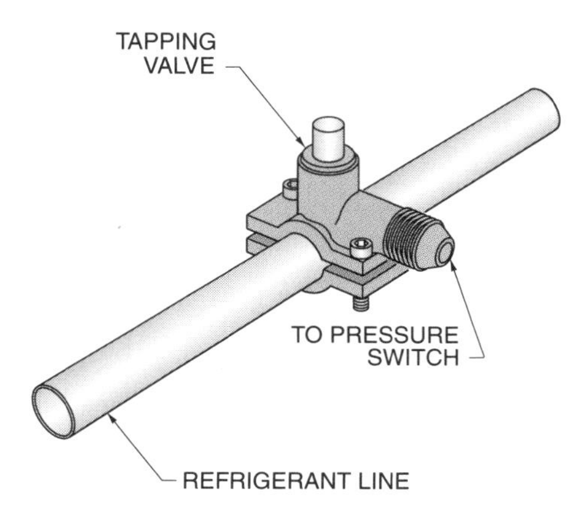

Figure 14-14. A tapping valve can provide the pressure of the refrigerant in a system to be used to begin and end a defrost cycle.

Auxiliary Heat Control

Auxiliary heat for a heat pump is controlled by the thermostat. On a call for heat, the reversing valve piston is in the heating mode position. The first-stage heating bulb of the thermostat operates the compressor, which provides heat from the heat pump. If heat pump output does not raise the temperature in the building to the thermostat setpoint in a specified time, the second-stage heating bulb calls for auxiliary heat. The auxiliary heat source controls are actuated by the heating relay when the second-stage bulb closes.

To ensure that the heat pump heats as much as possible and that auxiliary heat is used only when necessary, an outdoor thermostat is often used with the regular thermostat to control the amount of auxiliary heat produced. The outdoor thermostat has a set of contacts that close when the temperature falls below the outdoor temperature setpoint. If the outdoor temperature is above the setpoint temperature, the auxiliary heat will not turn ON.

The auxiliary heat is also controlled by the defrost control system. When the heat pump switches to a defrost cycle, the auxiliary heat is turned ON to heat the air that is blown into the building that normally would be cooled because of the refrigerant changing direction.

OPERATION

When a heat pump operates in the cooling mode, the refrigerant vapor is compressed in the compressor. The compressed refrigerant flows through the hot gas discharge line to the reversing valve. The refrigerant flows through the reversing valve to the outdoor coil. In the outdoor coil, the refrigerant condenses to a liquid as it rejects heat to the condensing medium (air). The refrigerant flows through the liquid line where the pressure of the refrigerant decreases as the liquid refrigerant flows through an expansion device. The refrigerant then flows to the indoor coil.

In the indoor coil, the refrigerant vaporizes as it absorbs heat from the evaporating medium. The refrigerant leaves the indoor coil through the compressor suction line and returns to the reversing valve. The refrigerant flows through the reversing valve into the compressor. **See Figure 14-15.**

A heat pump normally contains R-407c refrigerant. When operating in the cooling mode, the heat pump has a high-pressure-side pressure of 265 psig and a low-pressure-side pressure of 80 psig. In the cooling mode, the outdoor coil (condenser) is the high-pressure side and the indoor coil (evaporator) is the low-pressure side.

1. The refrigerant leaves the compressor with a pressure of 265 psig, a temperature of 168°F, and an enthalpy of 124 Btu/lb.
2. As the refrigerant enters the outdoor coil, it is still at a pressure of 265 psig. The temperature has decreased to 120°F, and the enthalpy is 112 Btu/lb. The temperature and enthalpy decreased because of sensible heat loss in the hot gas line between the compressor and the outdoor coil.

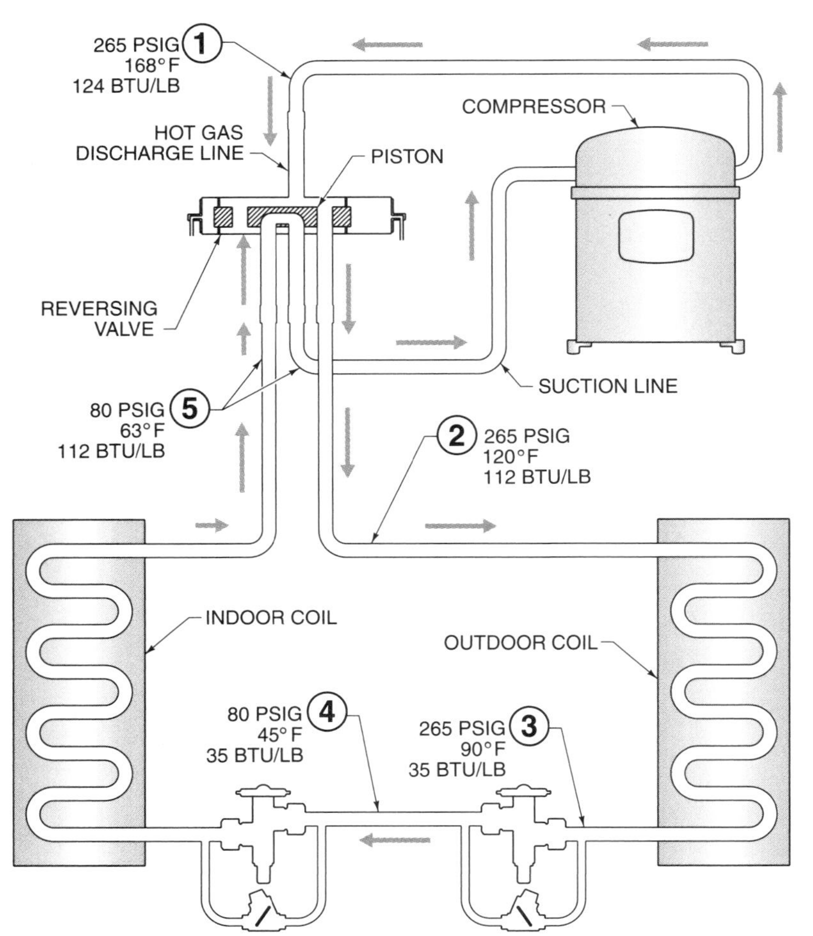

Figure 14-15. When a heat pump is in the cooling mode, the refrigerant vapor flows from the compressor, through the reversing valve, and into the outdoor coil.

3. The refrigerant leaves the outdoor coil with a pressure of 265 psig, a temperature of 90°F, and an enthalpy of 35 Btu/lb. The temperature decreases because the refrigerant is subcooled before it leaves the coil in the outdoor unit. The enthalpy decreases because the refrigerant rejects heat to the air as the refrigerant flows through the outdoor coil.

   The refrigerant leaves the outdoor coil and enters the expansion device with the same properties. The refrigerant has a pressure of 265 psig, temperature of 90°F, and enthalpy of 35 Btu/lb. After the refrigerant flows through the expansion device, the pressure has decreased to 80 psig.

4. The temperature now is 45°F, with an enthalpy of 35 Btu/lb. The pressure decrease is caused by the restriction in the expansion device. The temperature decreases because some of the refrigerant vaporizes at the lower pressure as the refrigerant enters the indoor coil.
5. The refrigerant leaves the indoor coil with a pressure of 80 psig, temperature of 63°F, and enthalpy of 112 Btu/lb. The temperature and enthalpy increase because the refrigerant in the coil of the indoor unit absorbs heat from the hot building air.

The refrigerant leaves the indoor coil and enters the reversing valve and compressor with the same properties. The pressure, temperature, and enthalpy increase as the refrigerant flows through the compressor. The temperature and enthalpy increase because of the heat of compression and possibly heat from the compressor motor.

When a heat pump operates in the heating mode, the refrigerant flows from the reversing valve to the indoor coil. The refrigerant flows through the indoor unit, liquid line, and expansion device. From the expansion device, the refrigerant flows through the outdoor unit and returns to the reversing valve.

For example, when a heat pump that contains R-407c refrigerant operates in the heating mode, the high-pressure-side pressure is 260 psig and the low-pressure-side pressure is 33 psig. In the heating mode, the indoor unit (condenser) is the high-pressure side and the outdoor unit (evaporator) is the low-pressure side. **See Figure 14-16.**

1. As the refrigerant leaves the compressor, it has a pressure of 260 psig, a temperature of 200°F, and an enthalpy of 130 Btu/lb of refrigerant.
2. The refrigerant enters the indoor coil with a pressure of 260 psig, a temperature of 120°F, and an enthalpy of 113 Btu/lb. The temperature and enthalpy decrease because of heat losses in the hot gas discharge line and the reversing valve.

3. The refrigerant leaves the indoor coil with a pressure of 260 psig, a temperature of 60°F, and an enthalpy of 27 Btu/lb. The temperature and enthalpy decrease because heat is rejected to the air at the indoor coil.
4. The refrigerant enters the expansion device at the outdoor coil with approximately the same conditions as the refrigerant that leaves the indoor coil. The refrigerant leaves the expansion device and enters the outdoor coil with a pressure of 33 psig, a temperature of 10°F, and an enthalpy of 27 Btu/lb. The pressure and temperature decrease because of the expansion device. The pressure decreases because of the restriction in the expansion device. The temperature decreases because some of the liquid refrigerant vaporizes due to the lower pressure.
5. The refrigerant leaves the outdoor coil at a pressure of 33 psig, a temperature of 30°F, and an enthalpy of 108 Btu/lb. The increase in temperature and enthalpy is due to heat absorbed from the heat sink at the outdoor coil.

The refrigerant enters the compressor at about the same conditions as it leaves the outdoor coil. As the refrigerant flows through the compressor, the pressure, temperature, and enthalpy increase. The temperature and enthalpy increase because of the heat of compression and possibly from the heat of the compressor motor.

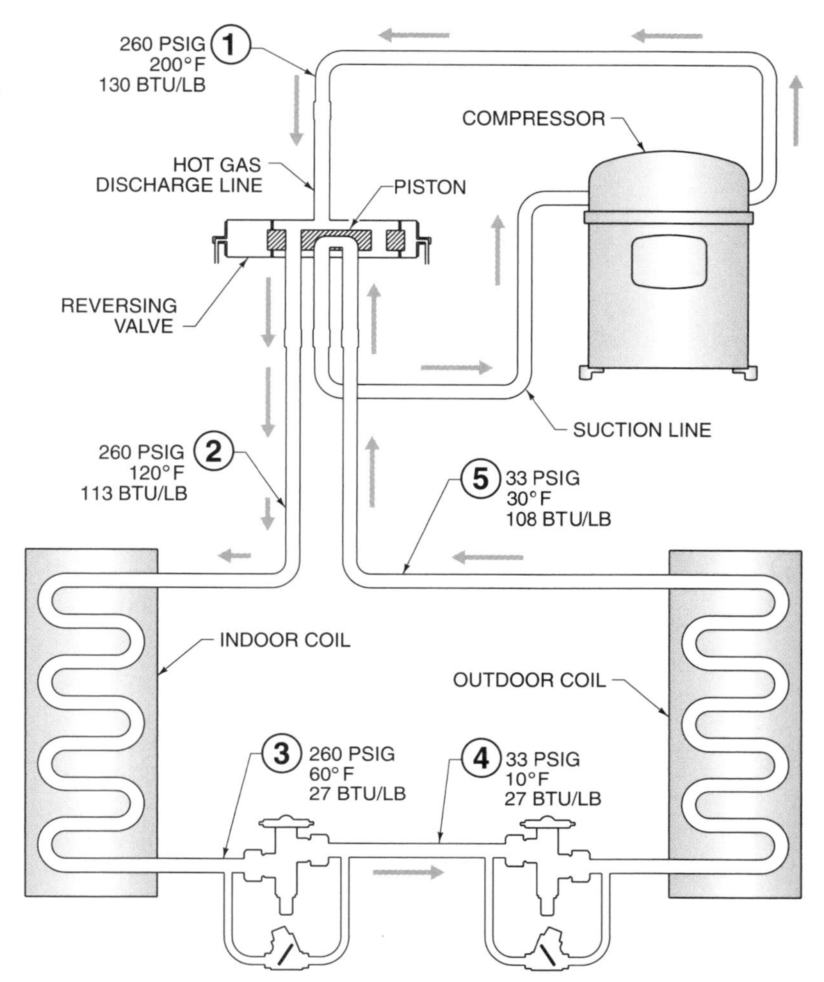

Figure 14-16. When a heat pump is in the heating mode, the refrigerant vapor flows from the compressor, through the reversing valve, and into the indoor coil.

HEAT PUMP SELECTION

A heat pump is selected for a given application based on the available source of heat, the type of application, and the cooling and heating capacity of the heat pump.

Sizing

A heat pump selected for cooling and heating a building located in a warm climate is sized to accommodate the cooling load of the building. Because of the warm climate, the heating capacity of a heat pump is great enough to handle the heating load.

A heat pump selected for cooling and heating a building located in a cold climate normally requires auxiliary heat to meet the required heating capacity. In this case, the heating capacity of the unit at the design temperature for the area should be checked against the heat loss of the building at the same design temperature to determine the amount of auxiliary heat required.

Balance Point Temperature. *Balance point temperature* is the temperature at which the output of a heat pump balances the heat loss of a building. The balance point temperature for a heat pump used to heat and cool a particular building space is found on a graph. The graph has the heat loss of the building space at different outdoor temperatures plotted against the heat output of the heat pump plotted against the same outdoor temperatures.

The heat loss of the building is plotted from the point that represents the total heat loss of the building at the outdoor design temperature. A line is drawn from this point to the base line of the graph at the outdoor air temperature at which the heat loss is 0 Btu/hr. The heat loss is usually about 5°F less than the indoor design temperature for the building.

After the building heat loss line is drawn on the graph, the heat output for a heat pump is drawn by using heat output data from the heat pump data sheet provided by the manufacturer. The data sheet shows that the heat output for a heat pump varies with outdoor temperature.

The heat output for a heat pump is found for several different outdoor temperatures, and the points are plotted on the graph. A line is drawn to connect the points. The line represents the heat output for the heat pump over a range of temperatures. Lines can be drawn for several sizes of heat pumps on the same graph. Some heat pump manufacturers provide completed output graphs on their heat pumps.

The point on the graph where the heat loss line for the building crosses the heat output line for the heat pump is the balance point temperature for the heat pump. At this temperature, the heat output from the heat pump balances the heat loss of the building. **See Figure 14-17.**

At temperatures above the balance point temperature, a heat pump heats the building without requiring auxiliary heat. At temperatures below the balance point temperature, auxiliary heat must be used with the heat pump to heat the building.

The amount of heat that must be provided by the auxiliary heat source is found on the scale at the left of the graph for any outdoor temperature by determining the difference between the two lines at that temperature (in Btu per hour).

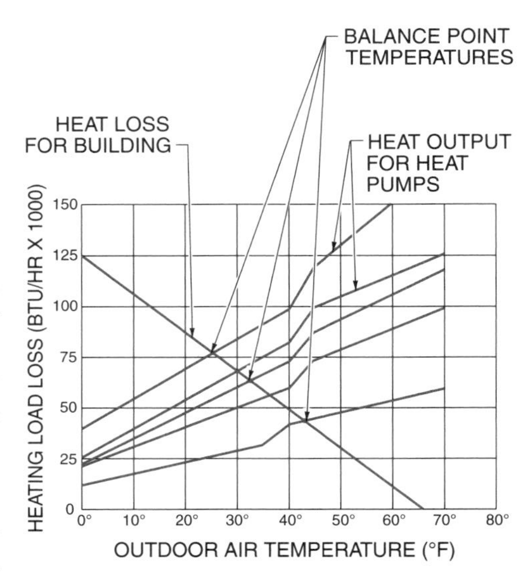

Figure 14-17. At the balance point temperature for the heat pump, the heat output from the heat pump balances the heat loss from the building.

TROUBLESHOOTING AND SERVICING HEAT PUMPS

To perform troubleshooting and servicing properly, individuals must have the proper tools, supplies, and materials. The individual must also understand the proper troubleshooting and service procedures for each type of heat pump.

Tools and Supplies

Tools should include a set of standard hand tools such as screwdrivers and wrenches. Specialized tools such as a digital multimeter, electronic thermometer, refrigerant recovery pump, refrigerant recovery tank, and refrigeration manifold gauges are needed as well. **See Figure 14-18.** Supplies needed may include refrigerant, filters, electrical tape, lubricant, and fasteners.

Written materials such as manufacturer information sheets, control diagrams, and written sequences of operation are needed. Any service and troubleshooting literature available from the manufacturer should be obtained and used. All applicable safety procedures must be followed, along with any specific recommendations from the manufacturer.

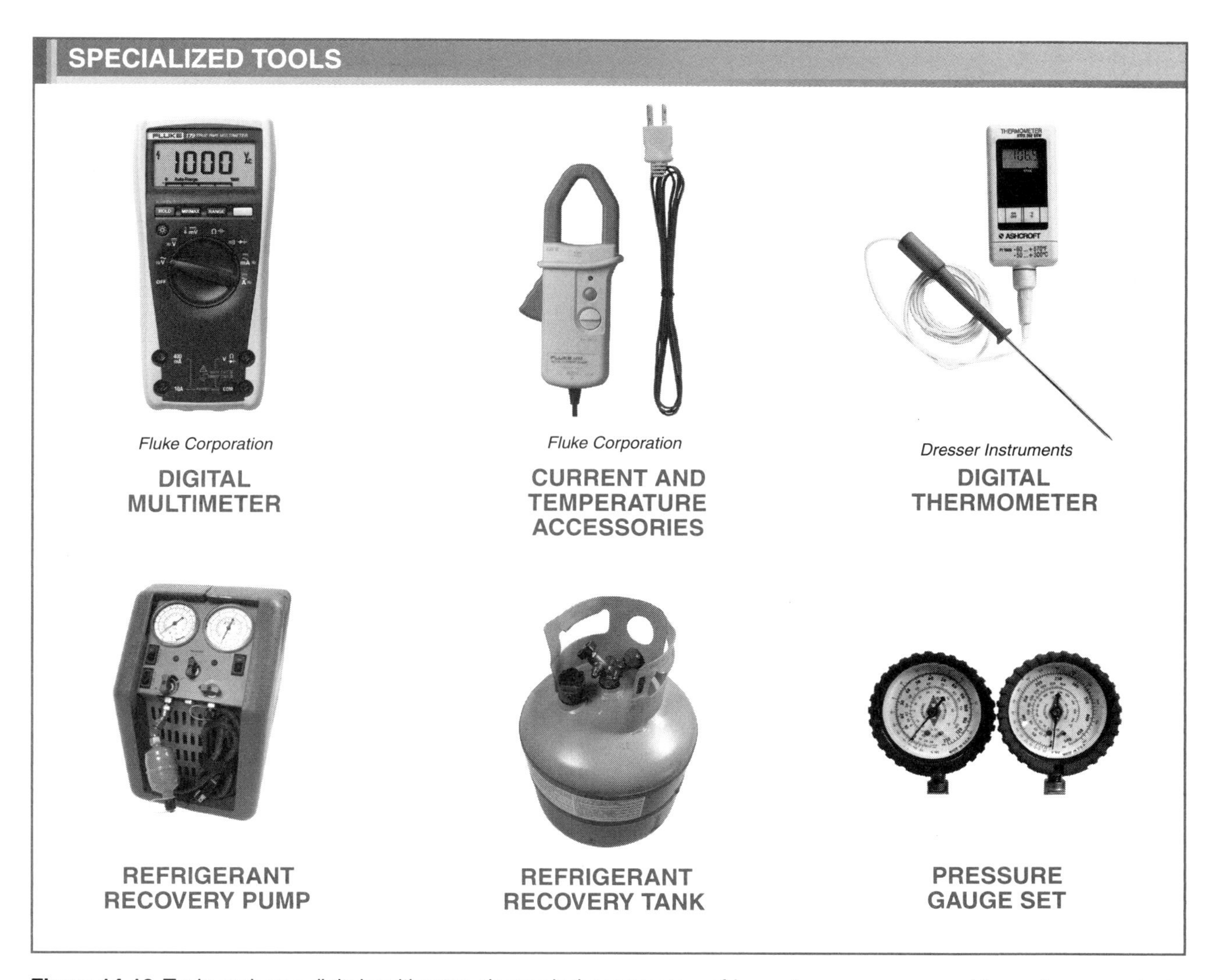

Figure 14-18. Tools such as a digital multimeter, electronic thermometer, refrigerant recovery pump, refrigerant recovery tank, and refrigeration manifold gauges may be needed along with other specialized tools to service heat pump systems.

The procedures for troubleshooting each type of heat pump and for equipment from various manufacturers may differ greatly. Specialized training from the equipment manufacturer may be needed to service some heat pumps.

Routine Service

Heat pumps need routine service to maintain optimum operating efficiency. Typically, the following service is performed based on a regular scheduled basis or the service is based on the heat pump run time.

- Lubricate the fan motors.
- Check the air filters. Change the filters as necessary.
- Verify that the system voltage and current are at specified values.
- Verify that the system operating pressures are at specified values.
- Verify that the system air flow rates are at specified values.

Review Questions

1. How does a heat pump system operate differently than an air conditioning system?
2. Define *heat sink*. How is a heat sink used with a heat pump system?
3. What two substances are used commonly as heat sinks?
4. List and describe the four classifications of heat pump systems.
5. In what types of buildings are air-to-air heat pump systems used?
6. In what types of buildings are water-to-water heat pump systems used?
7. Describe the function and operation of a reversing valve.
8. List and describe three types of expansion devices.
9. Why is the orifice at one end of a refrigerant control valve larger than the orifice at the other end of the valve?
10. Why is bypass piping required with thermostatic expansion valves on a heat pump system?
11. Why is auxiliary heat required when a heat pump is used to heat a building in cold weather climates?
12. Can a heat pump be used with auxiliary heat other than electric resistance heat? Explain.
13. Describe a heat pump thermostat. How does it control the heat during both the cooling and the heating modes?
14. What three main parts of a heat pump system does a heat pump thermostat control in heating mode?
15. What three components of a heat pump system are controlled directly during a defrost cycle?
16. Describe the operation of the timer in a defrost control.
17. Describe how temperature and pressure are used to determine when a defrost cycle should begin and end.
18. Describe the operation of a heat pump in cooling mode.
19. How does the pressure, temperature, and enthalpy of the refrigerant at different points in a heat pump system show how heat is transferred in the system?
20. What factors should be considered when selecting a heat pump?
21. Explain the process for determining balance point temperature.
22. List five routine service procedures that should be performed on a regular basis.

Building automation allows building systems to operate optimally with minimal intervention from building personnel. Each individual controller includes the logic to make decisions within its own system. Together, the controllers comprise a system that operates seamlessly while remaining adaptable to changing conditions and the needs of the occupants. Benefits include significant energy savings and improved control, comfort, security, and convenience.

15 Building Automation Control Systems

BUILDING AUTOMATION

Building automation is the control of the energy- and resource-using devices in a building for optimization of building system operations. Building automation uses sensors to measure variables in the building systems and provide the measured values as inputs to a controller. A controller makes decisions on how to optimize the system based on the input information. Output signals are then sent to change some aspect of the system, bringing it closer to the desired conditions. **See Figure 15-1.**

Building automation can be implemented in many different ways, with any combination of building systems, and for any type of building. The most common modern systems, though, are electronic control systems in commercial buildings, such as offices, schools, hospitals, warehouses, and retail establishments. These building automation systems make the biggest impact on improving energy efficiency, system control, and convenience.

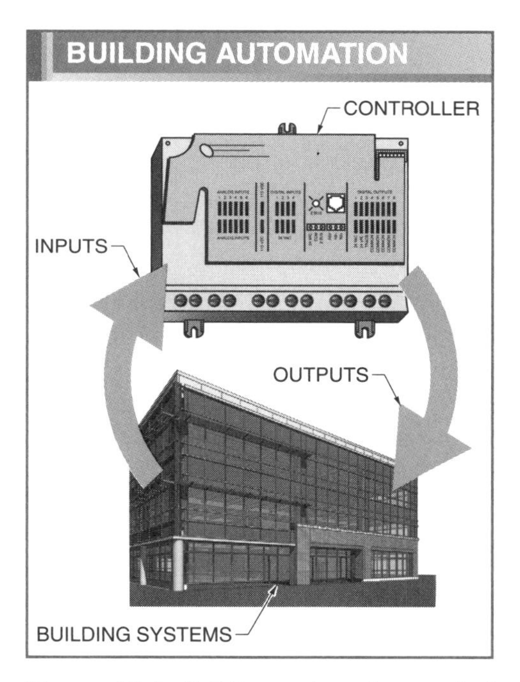

Figure 15-1. Building automation control involves controllers processing inputs and producing outputs that control a building system.

Tech Facts

The Leadership in Energy and Environmental Design (LEED®) Green Building Rating System™ is a method of quantifying the measures taken during the construction and operation of a building to save energy and use sustainable materials. Buildings are rated as Certified, Silver, Gold, or Platinum.

Building Automation Benefits

Building automation creates an intelligent building that improves occupant comfort and reduces energy use and maintenance. This can result in significant cost savings and improvements in safety and productivity. While building automation can be a significant initial investment, the return usually justifies the expense.

Energy Efficiency. Energy efficiency is probably the most significant benefit of a building automation system. The energy efficiency of a building is improved by controlling the electrical loads in such a way as to reduce their use without adversely affecting their purpose. For example, lighting is only used when rooms or areas are occupied, cutting down on electrical costs and increasing overall efficiency. However, with the automated controls, this reduction in lighting does not negatively impact the use of the space. As the largest consumers of energy in a building, the lighting and HVAC systems are the primary targets for improving energy efficiency through automation.

Building automation also helps conserve other resources, such as water and fuel. Plumbing controls manage water use and HVAC controls reduce fuel consumption by regulating the need for boiler or furnace use.

Improved Control. Most building system functions can be controlled manually, such as turning off lights and adjusting HVAC setpoints, but manual control is not as efficient as automated control. Automated controls can make control decisions must faster than a person and implement much smaller corrections to a system to optimize performance and efficiency. Automated controls work continuously and, if properly implemented, are not subject to common human errors such as inaccurate calculations, forgetting a control step, or missing an important input. This results in much more consistent and predictable control over the systems. For example, indoor air temperature remains consistent throughout the day because the automated controller is constantly monitoring all of the variables that affect this parameter, making small adjustments to the system to maintain the desired temperature. **See Figure 15-2.** Manual control of the same variables would not likely be as effective at providing such consistent results.

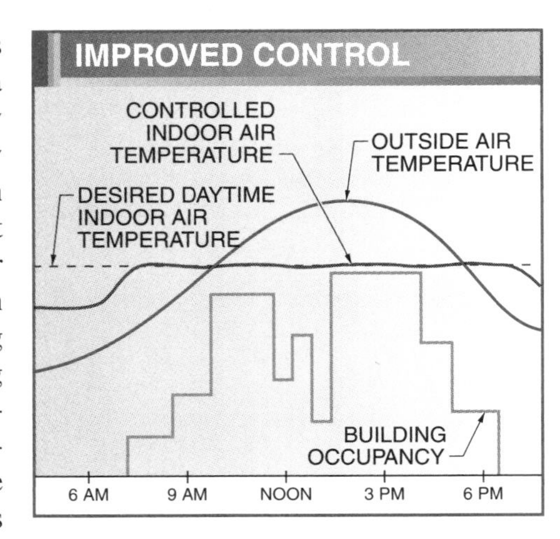

Figure 15-2. Building automation provides consistent and reliable control of building systems even while other parameters continuously change.

Automated controls also allow greater flexibility to implement control applications. For example, an output from one system can become an input to another, effectively integrating systems together in new and unique ways. The building systems may also be more secure from accidental or intentional tampering, since the automated controls can limit manual intervention.

Convenience. Building automation provides convenience for building owners, personnel, and occupants. After a system is commissioned and tested, it can operate with little human intervention. It reduces the workload of maintenance and security personnel by providing predictive maintenance information, automatically activating and deactivating some loads, and monitoring the building and its systems for problems. Building occupants find convenience in the automation of routine tasks, such as turning on lights when entering a room or unlocking doors when entering a secure area.

Building Automation Evolution

Early building system control was done manually. For example, an individual may have controlled the temperature in a living environment by adding fuel to a fire or allowing the fire to die down. As structures grew in size and contained a greater number of rooms, hot air from a central fire was regulated manually by opening or closing a diffuser (damper) with an adjusting pulley.

Control systems were developed in the early 1900s to provide automatic control of the building environment, primarily the HVAC system. First were pneumatic systems, but those were later replaced by electronic control systems, which provide better control and greater functionality. Currently, building automation systems are the most popular way to provide automatic control over several building systems.

Pneumatic Control Systems. Pneumatic control systems were the first automated control system for controlling the indoor environment in commercial buildings. A *pneumatic control system* is a control system in which compressed air is the medium for sharing control information and powering actuators. Pneumatic control systems can be separated into four main groups of components based on their function. These groups are the air compressor station, transmitters and controllers, auxiliary devices, and actuators. **See Figure 15-3.**

The air compressor station consists of the air compressor and other devices that ensure that the air supply is clean, dry, oil-free, and at the correct pressure. The compressed air supply is piped through the building to power control devices, mainly HVAC equipment. Transmitters and controllers are connected to the air supply and change the air pressure according to a measured variable, such as temperature or pressure.

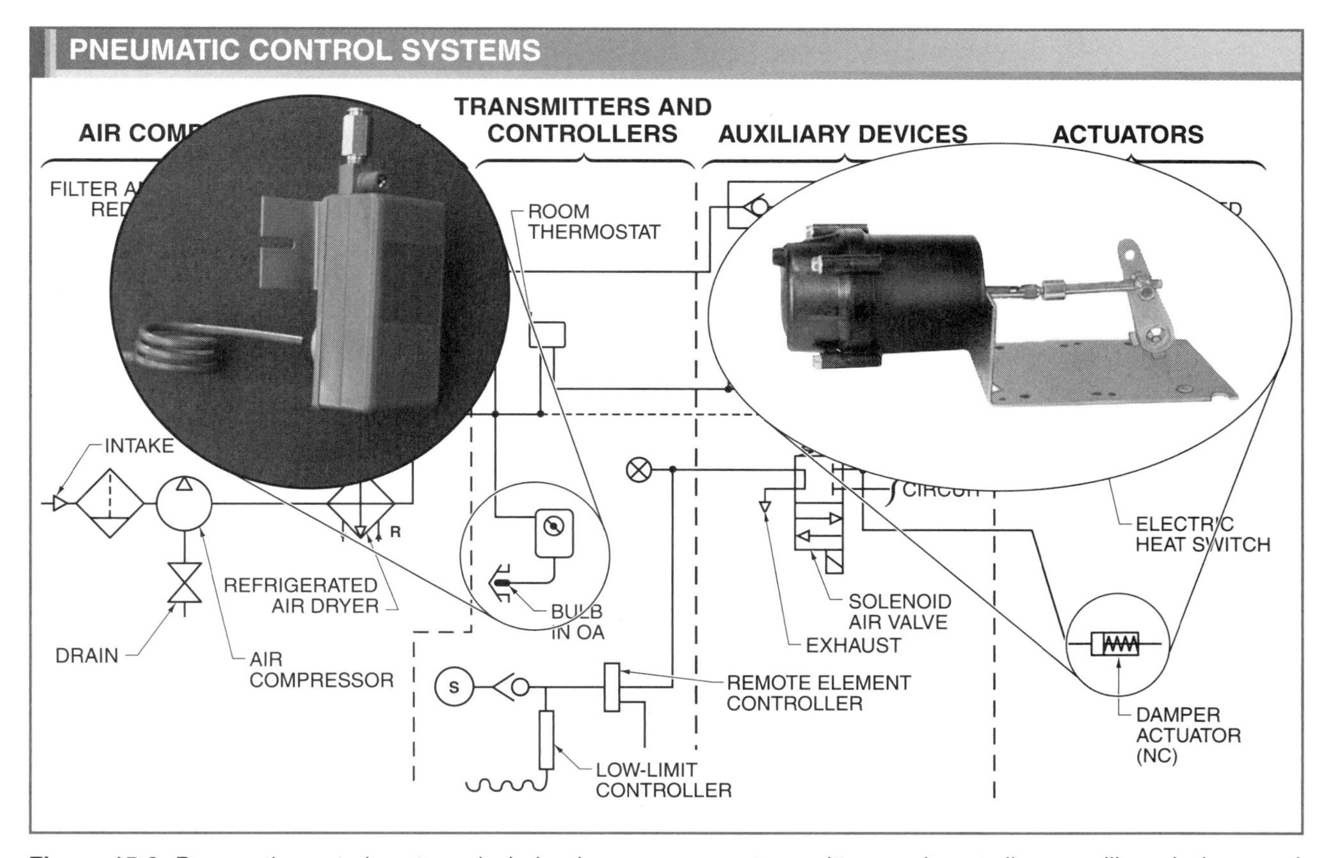

Figure 15-3. Pneumatic control systems include air compressors, transmitters and controllers, auxiliary devices, and actuators for managing an HVAC system.

Auxiliary devices are commonly located between a controller and a controlled device. Auxiliary devices change flow direction, change pressure, and interface between two devices. This modified air supply is routed to the controlled devices, such as dampers, valves, and switches. An actuator accepts an air pressure signal from a controller and causes a mechanical movement, which regulates some aspect of the controlled system, such as water or steam through a valve.

Pneumatic control systems have been primarily used in large commercial buildings such as hospitals and schools. They operate effectively and are very safe, but they may require a great deal of maintenance as well as specialized tools, calibration, and set-up procedures. With newer electronic controls available, pneumatic control systems are becoming increasingly rare.

Tech Facts

Technicians must understand building automation system instruments as they are installed in new and old buildings.

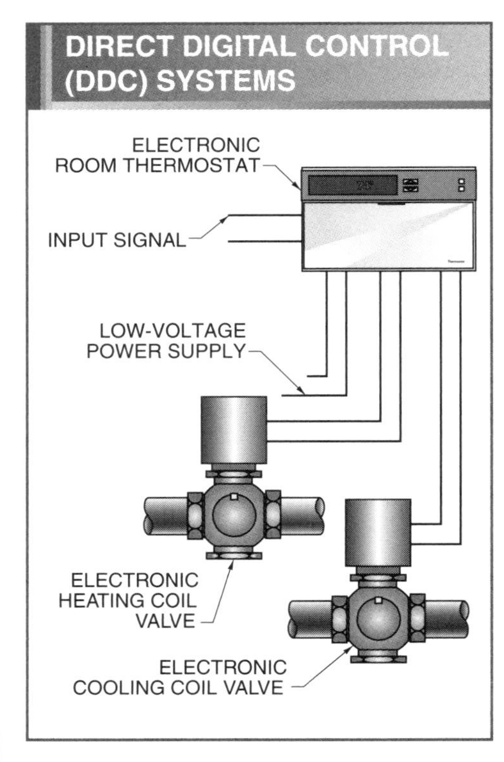

Figure 15-4. Direct digital control (DDC) systems use electronic controllers with their local sensors and actuators to control portions of a system.

Direct Digital Control (DDC) Systems. Improvements in electronics allowed for a significant improvement over pneumatic control systems. Electronics technology was adapted for use in building control systems, though still primarily for HVAC systems. A *direct digital control (DDC) system* is a control system in which electrical signals are used to measure and control system parameters.

Similar to pneumatic control systems, DDC systems include sensors, controllers, and actuators, though the electronic versions all operate from low-voltage DC power instead of air pressure. **See Figure 15-4.** Also, DDC controllers allow far greater functionality in making optimal control decisions because each includes a microprocessor to quickly and accurately calculate the necessary output signal based on the information from the input signals. The output of the controller is then sent to an electrically controlled actuator, such as a motor or solenoid.

DDC control systems are reliable, accurate, and relatively inexpensive. One disadvantage of electronic control systems is that they may require special diagnostic tools and procedures. DDC controllers are also limited to communicating only with the few sensors and actuators it is directly connected to.

Building Automation Systems. In order to share information between building systems, electronic controllers must have a way to communicate between themselves. A *building automation system* is a system that uses a distributed system of microprocessor-based controllers to automate any combination of building systems. Building automation systems can control almost any type of building system, including HVAC equipment, lighting, and security systems.

Building automation system controllers are similar to DDC controllers in that each includes a microprocessor, memory, and a control program. The controller is connected to local sensors and actuators and manages a part of a building system, such as a variable-air-volume (VAV) terminal box. The difference is that the building automation system devices can all communicate with each other on the same, shared network. **See Figure 15-5.** Information is shared in the form of structured network messages that contain details of many control parameters. This information is encoded into a series of digital signals and sent to any other device that requires that information. For example, if one device reads from a sensor measuring outside air temperature, it can share that temperature with any other device, in any building system, that may have use for that information.

Building automation systems are extremely accurate, offer sophisticated features such as data acquisition and remote control, can integrate a variety of building systems, and are very flexible. The disadvantages of building automation systems are that the design and programming can be more involved than with other controls requiring contractors with specialized knowledge of these systems.

Tech Facts

When integrating multiple building systems, all codes and standards required for stand-alone systems must be followed. With each additional system to be automated, more requirements are added.

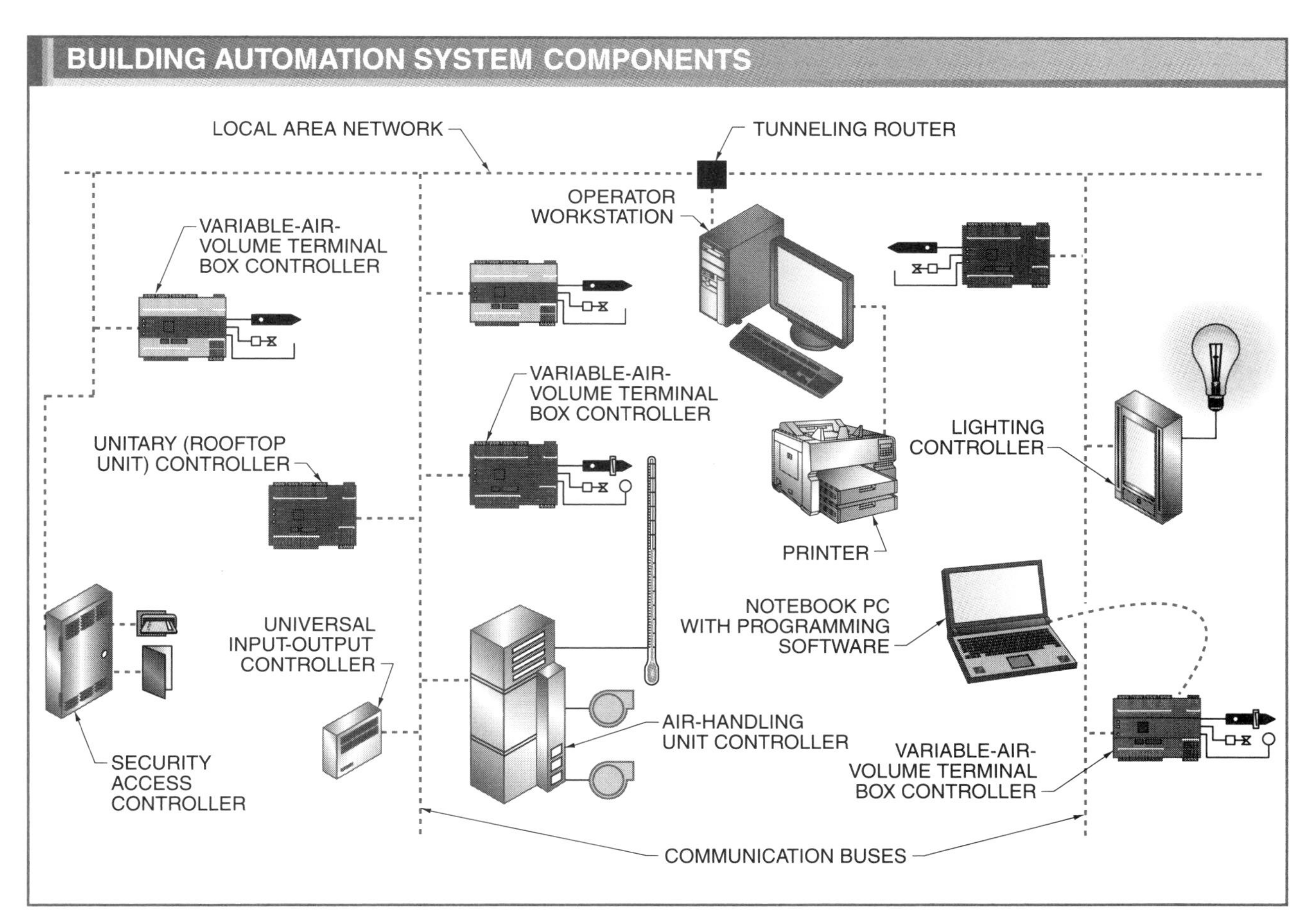

Figure 15-5. Building automation systems network electronic controllers together into a system that can share information between building systems.

Tech Facts

Building automation systems can be designed to communicate over a variety of network media, including twisted pair, fiber optic, existing power conductors, and even radio frequency. Multiple media can also be mixed within the same installation.

Automation Applications

Building automation systems can be implemented in any residential, commercial, or industrial facility. The basic principles of automation are the same, regardless of the building use, though different types of facilities will likely have different requirements for system automation. Device manufacturers have developed different ways to implement automation and often selectively target certain markets based on building use.

Residential Buildings. Residential buildings are the least automated as a group, but automation devices are becoming increasingly affordable for homeowners and landlords, while the installation and programming of the systems are becoming increasingly practical. One common automation technology for residential buildings is X10 technology.

X10 technology is a control protocol that uses powerline signals to communicate between devices. The devices overlay control signals on the 60 Hz AC sine wave, which can be read by any other X10 device connected to the same electrical power conductors. The control signals are used to energize, de-energize, dim, or monitor common electrical loads. This technology is particularly attractive for residential buildings because it uses the existing electrical infrastructure and requires no new wiring. The control devices simply plug into standard wall receptacles. The devices are also very simple to commission and program. **See Figure 15-6.**

For many of the same reasons, wireless devices are also becoming popular for residential control applications. No additional wiring is required and the risk of radio frequency interference in a residential environment is minimal. Also, residential control applications are not likely to be critical systems, so there is little risk to safety or property if there is a communication problem between control devices.

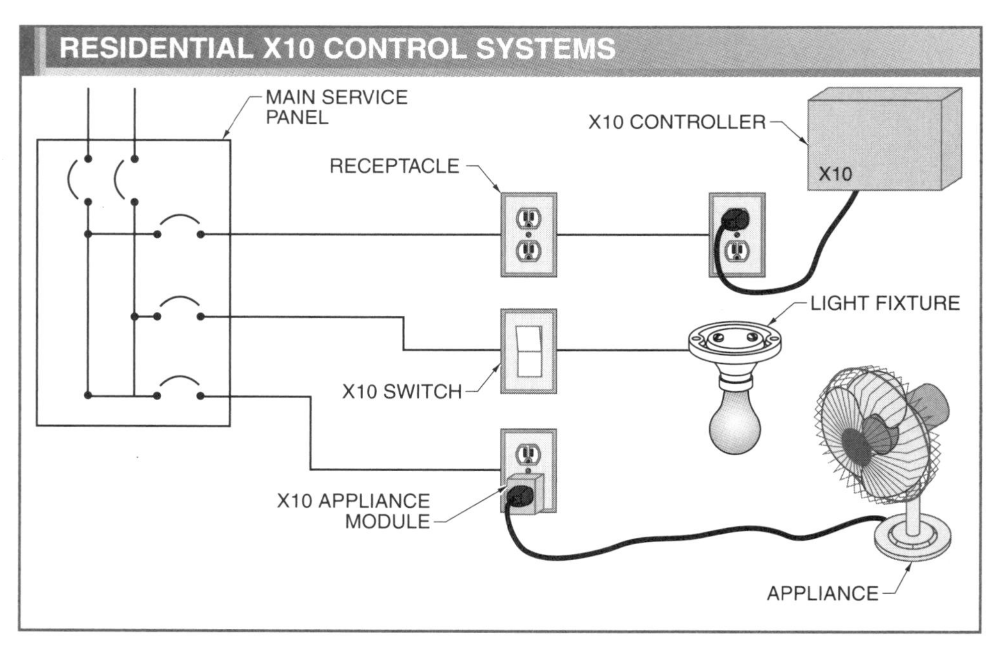

Figure 15-6. X10 is a common residential control system that uses the powerline as a medium for sharing control information between devices.

Commercial Buildings. Commercial buildings include offices, schools, hospitals, warehouses, and retail establishments. With growing emphasis on energy efficiency and "green" buildings, commercial buildings are probably the fastest growing market for implementing new automation systems. **See Figure 15-7.** New construction commercial buildings now commonly include some level of automation and many existing buildings are being retrofitted with automation systems.

Commercial building automation systems can control and integrate any or all of the building's systems, including electrical, lighting, HVAC, plumbing, fire protection, security, access control, VDV, and elevator systems. Building automation systems are inherently flexible, so building owners can choose almost any level of sophistication, from one specific application to all of the building systems. Upgrades or additional applications can be added and integrated into the system at any time in the future.

Figure 15-7. Commercial buildings are the most common applications of building automation systems.

Industrial/Process Facilities. Automation has a long history in industrial, manufacturing, and process facilities. In addition to the control of lighting, security, and sanitary water, industrial facilities typically automate manufacturing processes, though the building and manufacturing automation systems are typically separate. Manufacturing processes must be so tightly controlled for time, safety, productivity, equipment efficiency, and product quality that automation is essential. **See Figure 15-8.**

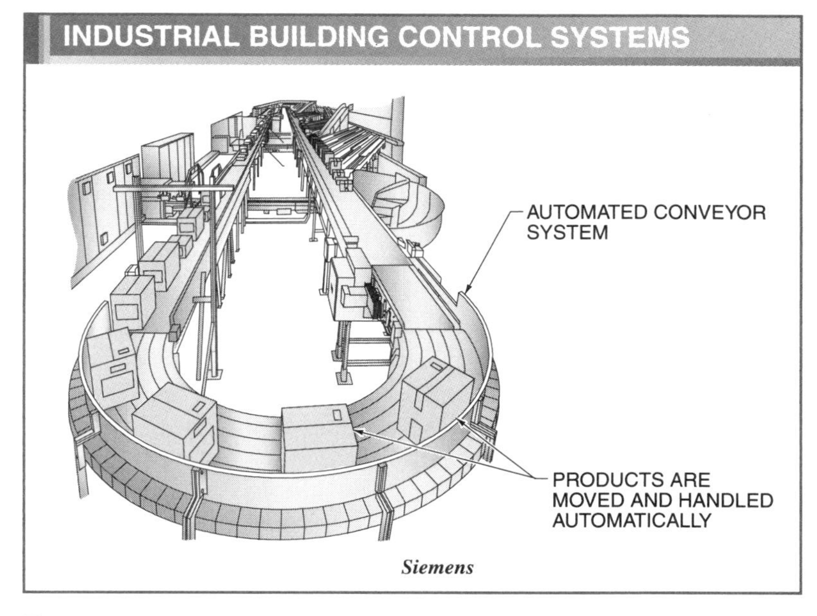

Figure 15-8. Industrial buildings include many control systems, but they are typically for manufacturing processes or material handling.

Industrial automation systems operate in much the same way as residential or commercial building automation, but for different purposes. The control devices, being in industrial environments, are more rugged. There may be more human interaction with the automation system in order to modify sequences for changing tasks. Industrial automation is commonly applied at the controller level rather than the device level. Primarily, the control sequences are focused on safety and product manufacturing.

Building Automation Industry

The building automation industry includes a wide range of individuals and groups that help take requirements, concerns, and ideas for a new or existing building and create a state-of-the-art automated building. The participants in the building automation industry are the building owners, consulting-specifying engineers, controls contractors, and authorities having jurisdiction.

Building Owners. Ultimately, it is the building owner's decision on which building automation strategies to implement based on the desired results. Owners may wish to build a fully automated showcase facility that is highly visible to the public, or they may wish to automate only the few building systems that are most important to the efficient operation of the building. The end use of the building may also affect the automation choices because the building's location, occupancy density, equipment and products, and special on-site processes dictate the automation priorities. For example, lighting and HVAC controls may be especially important for office buildings, while security and access control may be most important for warehouses full of valuable inventory.

The concerns of building owners also include initial costs, payback periods, maintenance, impact on occupants, system flexibility, system upgradeability, and environmental impacts. For example, automation systems typically return their extra costs within a few years, but still require a greater initial investment. These factors are even more important in retrofitting existing buildings for automation, as the implementation possibilities are typically more limited than in new construction.

Consulting-Specifying Engineers. For designing the most efficient system, the owner meets with consulting-specifying engineers to refine the automation plan. A *consulting-specifying engineer* is a building automation professional that designs the building automation system from the owner's list of desired features. This involves describing the necessary functionality of all the control devices and detailing the interactions between them. The consulting-specifying engineers work with the building owner, architect, and general contractor to make changes as needed to the planned automation system in order to maximize the benefits while working within any aesthetic, financial, or construction constraints.

Consulting-specifying engineers prepare contract specifications that include all of the control information. The *contract specification* is a document describing the desired performance of the purchased components and means and methods of installation. **See Figure 15-9.**

CONTRACT SPECIFICATIONS

1.1 ELECTRONIC DATA INPUTS AND OUTPUTS

A. Input/output sensors and devices matched to requirements of remote panel for accurate, responsive, noise free signal output/input. Control input to be highly sensitive and matched to loop gain requirements for precise and responsive control.
   1. In no case shall computer inputs be derived from pneumatic sensors.

B. Temperature sensors:
   1. Except as indicated below, all space temperature sensors shall be provided with single sliding setpoint adjustment. Scale on adjustment shall indicate temperature. The following are exceptions to this:
      a. The following locations shall have sensor without setpoint adjustment:
         1) All electrical and communication rooms.
         2) All mechanical rooms.
         3) All unit heaters.
         4) All public elevator lobbies and entrance vestibules.
         5) Elevator equipment rooms.
   2. Duct temperature sensors to be averaging type. Averaging sensors shall be of sufficient length (a maximum of 1.8 sqft of cross sectional area per 1 lineal foot of sensing element) to insure that the resistance represents an average

Figure 15-9. Contract specifications are produced by the consulting-specifying engineer and include the device requirements and control sequences for the building automation system.

Automation applications may be written out as detailed control sequence steps. These contract specifications are then made available to controls contractors to bid the project.

Controls Contractors. Controls engineering has traditionally been considered a subset of mechanical work, especially since HVAC systems are the most commonly automated building systems. However, controls technology has evolved into a highly technical field that has allowed contractors to specialize in automation. This has also given controls contractors the freedom to work with every building system so that they can be seamlessly integrated.

The controls contractor uses the contract specification to choose the specific components and infrastructure that can accomplish the requirements. For example, the specification may list a temperature sensor that must provide temperature measurements to an air-handling unit controller. The controls contractor chooses the manufacturer and model of the sensor that meets all the requirements for installation type, temperature range, environmental conditions, calibration needs, and signal type. The controls contractor submits a quote for the project based on the costs of the purchased equipment, installation labor, system commissioning, and tuning. If chosen, the controls contractor then performs this work.

Authorities Having Jurisdiction (AHJ). The *authority having jurisdiction (AHJ)* is the organization, office, or individual who is responsible for approving the equipment and materials used for building automation installation. This includes all agencies and organizations that regulate, legislate, or create rules that affect the building industry including building codes, National Electrical Code® (NEC®) regulations, Occupational Safety and Health Administration (OSHA) regulations, local and municipal codes, fire protection standards, waste disposal regulations, and all levels of government or other industry agencies. These groups may overlap. Building automation is most affected by policies related to energy efficiency and indoor air quality regulations.

Construction Documents

The planning and implementation of a building automation system requires detailed documentation. Controls contractors receive a set of construction documents from the consulting-specifying engineers, giving a comprehensive view of the project. For example, controls contractors must consider a number of trade areas in reviewing the project, including the sheet metal, plumbing, electrical, and piping drawings, as well as project schedules, scope, and other details.

Contract Documents. A *contract document* is a set of documents produced by the consulting-specifying engineer for use by a contractor to bid a project. Contract documents include all of the various design disciplines: architectural, civil, structural, mechanical, and electrical engineering.

The contract documents consist of construction drawings and a book of contract specifications. The contract specification describes the project requirements in words, rather than graphically. Contract documents are purposely written in a generic manner to accommodate a variety of available products and features. This allows the controls contractor the freedom to specify any manufacturers or models that fulfill the project requirements.

Shop Drawings. The controls contractor's project bid includes his or her shop drawings. A *shop drawing* is a document produced by the controls contractor with the details necessary for installation. The consulting-specifying engineer reviews the shop drawings to ensure that the intent of the contract documents has been met. Shop drawings show much more detail than the contract drawings because they are specific to a particular device manufacturer and model. For example, shop drawings (also known as cut sheets) include wiring details down to the individual device level. **See Figure 15-10.**

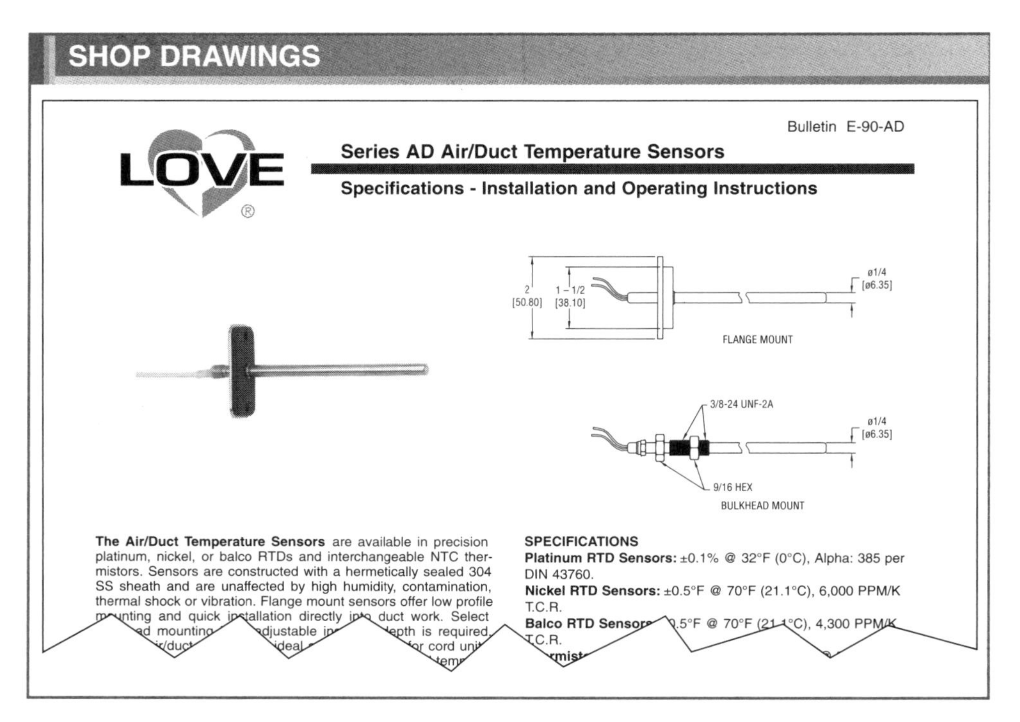

Figure 15-10. Shop drawings from a controls contractor include specific information on devices that satisfy the automation requirements.

In addition to playing a part in the construction process, the shop drawings are used by the owner's building maintenance staff for troubleshooting after the building has been turned over. Shop drawings include two basic categories: components and software/wiring.

The component portion of the shop drawing package includes information on every component and its size, configuration, and location. This allows the other trades workers to coordinate the installation of control components. For example, control valves are installed by pipe fitters long before the controls contractor connects signal wiring to them. Manufacturer data sheets are included to describe the features of the component in detail.

The component shop drawing package is often developed and reviewed by the consulting-specifying engineers first so that purchases with long lead times or decisions that impact other contractors can be made in a timely manner. For example, the controls contractor may choose to have the air terminal unit controllers mounted by the manufacturer at the factory to save installation time.

The software and wiring portion of the shop drawing package addresses all of the control interactions between the components. A building-level overview explains the control system architecture. Wiring interconnection and block diagrams illustrate both signal and power wiring for each panel and device, including information on power sources. Flow charts for each control sequence show interrelationships between inputs, calculations (control loops), and outputs. Sequences of operation describe all flow chart functions in words. A detailed explanation of color conventions clarifies drawing symbology.

Project Closeout Information. The *project closeout information* is a set of documents produced by the controls contractor for the owner's use while operating the building. Project closeout information includes operating and maintenance manuals for all equipment, owner instruction report, and a certified resolution of issues raised during the commissioning process.

CONTROL DEVICES

A *control device* is a building automation device for monitoring or changing system variables, making control decisions, or interfacing with other types of systems. A *variable* is some changing characteristic in a system. Control devices include sensors, controllers, actuators, and human-machine interfaces.

Sensors

A *sensor* is a device that measures the value of a variable and transmits a signal that conveys this information. The sensor output signal is commonly electrical but may also be conveyed by air pressure, optics, or radio frequency. Even though they output signals, sensors are considered input devices, which is relative to the point of view of the controllers that manage the systems.

As individual devices, sensors allow variables to be measured in any location or type of environment, while other control devices are installed in convenient and centralized places. For example, a controller may be connected to a number of individual temperature sensors that are installed throughout an HVAC unit. **See Figure 15-11.** Sensors may also be integral with a controller, as in a room thermostat.

Sensors must be carefully chosen to provide the appropriate information and in a way that is understandable by the receiving control devices. Factors to consider are sensor type, range, resolution, accuracy, calibration, signal type, power requirements, operating environment, mounting configuration, size, weight, construction materials, and agency listings.

Sensors that are compatible with modern automation systems are typically either electromechanical or electronic sensors. They are distinguished by the way in which the sensors monitor their environment. Electromechanical devices use mechanical means to measure a variable but include electrical components to convert motion into electrical signals and process and transmit the signals. Some variables that may require mechanical capabilities to be measured include pressure, position, and fluid level. Purely electronic devices are typically smaller and more reliable than electromechanical devices, since they have no moving parts. **See Figure 15-12.** The variable is measured by detecting the direct changes the variable has on the electrical properties of a solid-state component. Then other electronics process and transmit the information as a signal. Sensors are connected to terminals on a controller.

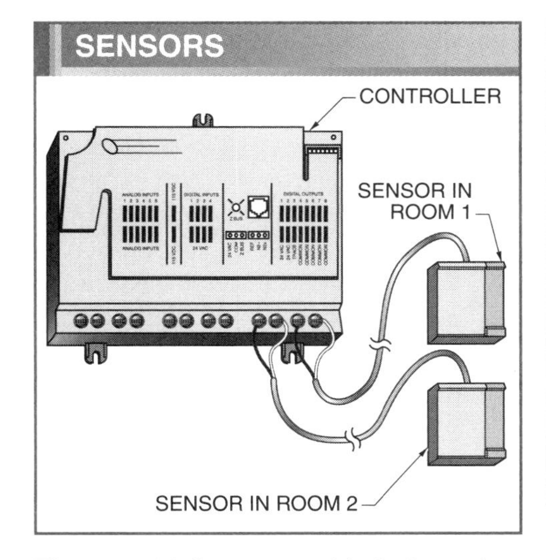

Figure 15-11. Sensors provide the inputs into the building automation controllers.

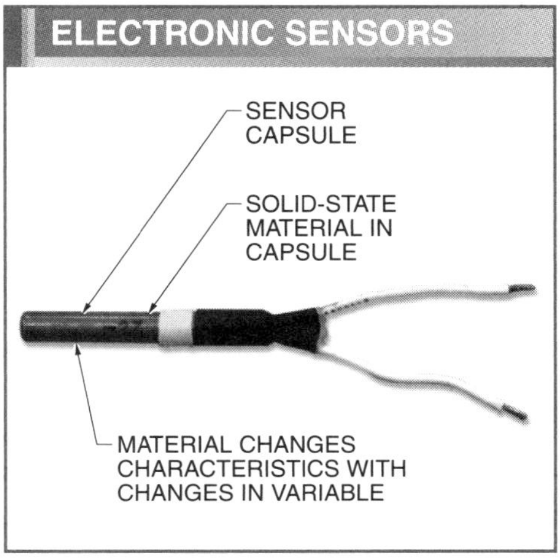

Figure 15-12. Electronic sensors detect changes in a variable from corresponding changes in the sensor's electrical properties.

Controllers

A *controller* is a device that makes decisions to change some aspect of a system based on sensor information and internal programming. The controller receives a signal from the sensor, compares it to a desired value to be maintained, and sends an appropriate output signal to an actuator.

Modern controllers are all electronic devices with processors, memory, and software that can execute control sequences very quickly and accurately. The electronics also allow relatively simple modification of the programming and settings to optimize building system operation. However, the electronics may be less rugged than some sensors.

An actuator node receives output information directed to its unique network address.

Some controllers are designed for specific control applications, such as controlling an air-handling unit. These provide many features and programs developed by the manufacturer to simplify installation and commissioning of the system, but may not allow much flexibility to change inputs, outputs, or decision making. Generic controllers, on the other hand, include only the electronics to read from standard input connections, run software, and write to standard output connections. Any type of device can be connected to the terminals and any type of decision can be made. These devices allow the greatest flexibility in applications, but the program must be custom-written and tested by the controls contractor to ensure correct operation and reliability.

Actuators

An *actuator* is a device that accepts a control signal and causes a mechanical motion. This motion may actuate a switch, rotate a valve stem, change a position, or cause some other change that affects one or more characteristics in a system. From the point of view of the controller, actuators are the output devices. For example, an actuator may open a valve to allow more steam into a heating coil, causing the temperature of the airflow across the coil to rise.

A large number of controlled devices may be used to change system characteristics, such as dampers for regulating airflow, valves for regulating water or steam flow, refrigeration compressors for delivering cooling, and gas valves and electric heating elements for regulating heating. Many of these are actually controlled by just a few different types of actuators. The most common types of actuators are relays, solenoids, and electric motors. **See Figure 15-13.**

A *relay* is an electrical switch that is actuated by a separate electrical circuit. It allows a low-voltage control circuit to open or close contacts in a higher-voltage load circuit. Many relays are electromechanical types, but some are completely solid-state, using no moving parts. Relays can be used with any device that is controlled by turning it ON or OFF. They are common for switching lighting circuits and HVAC package units, as well as sharing ON/OFF information between controllers.

A *solenoid* is a device that converts electrical energy into a linear mechanical force. Solenoids are typically used in quick-acting valves and for locking and unlocking doors. An *electric motor* is a device that converts electrical energy into rotating mechanical energy. Electric actuator motors can be used to turn something completely open or completely closed, but are particularly useful for actuating devices with multiple positions, such as dampers and valves.

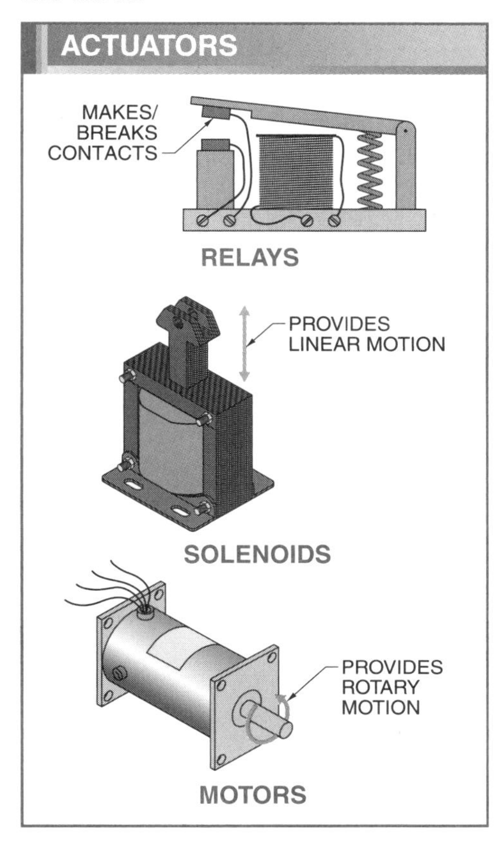

Figure 15-13. The most common types of actuators found in controlled devices are relays, solenoids, and motors.

Human-Machine Interfaces (HMIs)

A human-machine interface is connected into a building automation system to allow personnel to view and modify the information being shared between control devices. A *human-machine interface (HMI)* is an interface terminal that allows an individual to access and respond to building automation system information. Many HMIs show information graphically and include data over time so that current values and longer-term trends are easily visible. HMIs may also allow a user to interact with the system by manually inputting changes to values or device behavior.

Human-machine interfaces are either hardware-based or software-based. **See Figure 15-14.** Hardware-based HMIs are basically small computers that are specially designed to gather, process, and display system data. They are typically self-contained units with integral monitors, memory, connection terminals, and communication ports. If they are intended to be installed in extreme environments, they may be housed in special dust-resistant, moisture-resistant, rugged enclosures. If the HMI accepts inputs into the system from the user, it typically includes a keypad, touchscreen display, or an I/O port for connecting a separate keyboard or mouse.

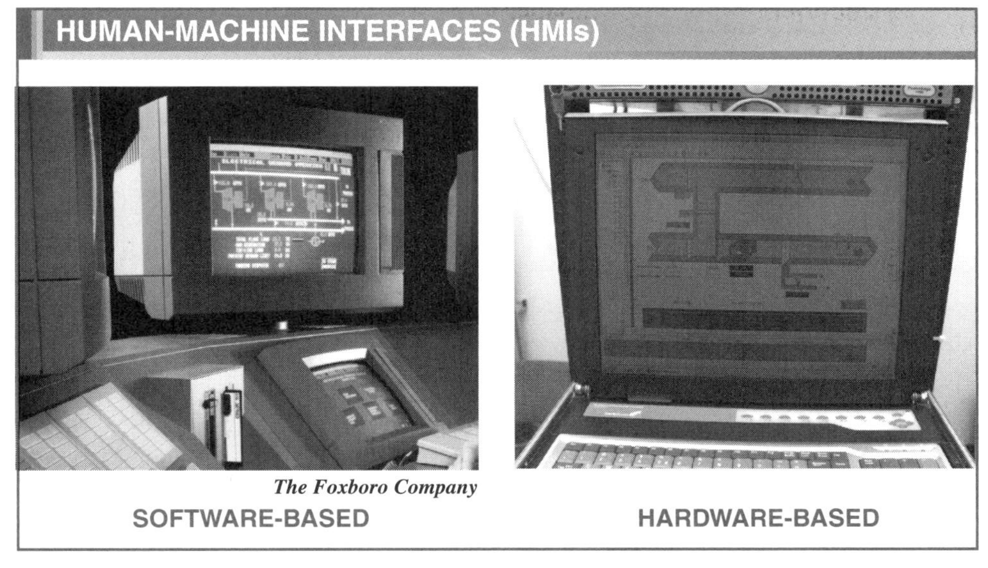

Figure 15-14. Human-machine interfaces (HMIs) may be software- or hardware-based.

Software-based HMIs may require a special piece of hardware to physically connect with the building automation system and retrieve data, but the interface portion relies on software running on separate personal computers. Some packages require special software (a "thick client" application) on the computer to communicate with the hardware. However, the current trend is to design the interface to be viewed with a standard web browser, which is common software on all computers and requires no additional installation (a "thin client" application). The hardware unit acts as a web server, delivering the building system information in a way that is similar to any other Internet site. These HMIs can be configured to provide this information to only computers within the building or to any computer anywhere in the world. Either type of software-based HMI is capable of being programmed to accept system inputs from the user. Security features ensure that only authorized personnel are able to make these changes.

CONTROL SIGNALS

A *control signal* is a changing characteristic used to communicate building automation information between control devices. Control signals are typically electronic signals but can also be transmitted by other media such as air pressure, light, and radio frequency. There are different types of control signals, depending on the type of information to be shared. The three common types of control signals are digital signals, analog signals, and structured network messages.

Digital Signals

A *digital signal* is a signal that has only two possible states. For this reason, digital signals may also be known as binary signals. Digital signals convey information as either one of two extremes, such as completely ON or completely OFF, or completely open or completely closed.

A digital signal is typically conveyed with a change in voltage. For example, 0 VDC can represent an OFF state and 5 VDC an ON state. **See Figure 15-15.** Alternatively, +12 VDC can represent an ON state and –12 VDC an OFF state. Any pair of two different voltages can be used to send digital signals, though they are commonly DC voltages of 24 VDC or less. Also, different digital signal voltage levels can be used within a system, as long as each device is compatible with the type of signal it is sending or receiving.

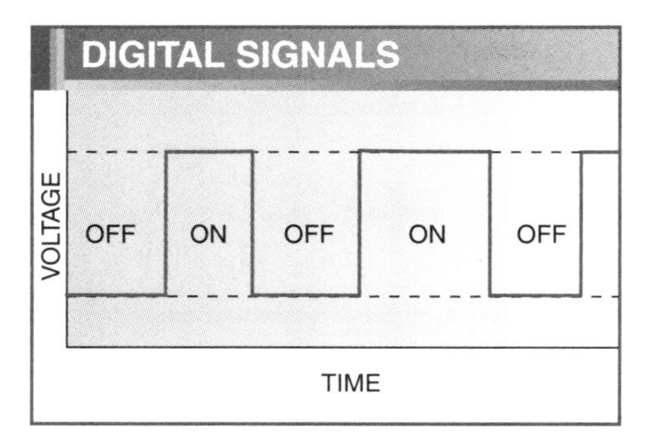

Figure 15-15. Digital signals are produced by a pair of voltage levels that represents either ON or OFF.

The devices also must agree on the voltage levels that define the two states. For example, due to a slight voltage drop in the signaling conductor, a 5 VDC (ON) signal may appear to the receiver as slightly less, such as 4.6 VDC. In order for this voltage to still be registered as an ON state, the two states may be more precisely defined as 0 V to 0.8 VDC for OFF and 2.0 VDC to 5.25 VDC for ON. Voltages between these two levels may be read as an erratically fluctuating ON and OFF, producing unpredictable results.

Digital signals can be generated by a device with a power supply by applying the necessary voltage to a signaling conductor. **See Figure 15-16.** Alternatively, digital signals can be generated by a set of contacts in an input device, which does not require an external power supply. The contact terminals are wired to a set of controller terminals with a pair of conductors. The controller senses the continuity or discontinuity through the terminals as closure or opening of the contacts in the input device.

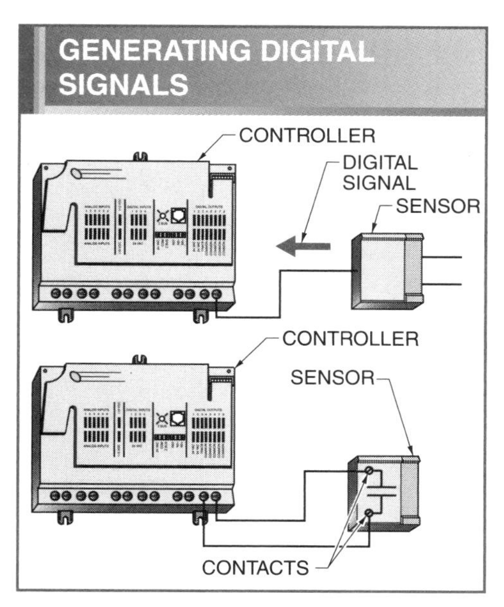

Figure 15-16. Digital signals can be generated by devices with their own power supply or by devices with switch contacts.

Digital signals are commonly used as an output to turn devices ON and OFF, typically through relays that switch the power needed to operate the device. For example, digital signals can turn electric motors, electric heating stages, and valves ON and OFF with digital signals. Digital signals can also be used to initiate different functions in package units. These units are energized by a separate, manual means, but a controller's digital signals may enable/disable the unit's primary function.

Analog Signals

An *analog signal* is a signal that has a continuous range of possible values between two points. **See Figure 15-17.** Analog signals can convey information that has units of measurement, such as degrees Fahrenheit, cubic feet per minute, meters per second, and inches of water column. They can also provide any value between 0% and 100% of some controllable characteristic. For example, analog signals can be used to control fan speed from 0% (stopped) to 100% of its full rated maximum speed.

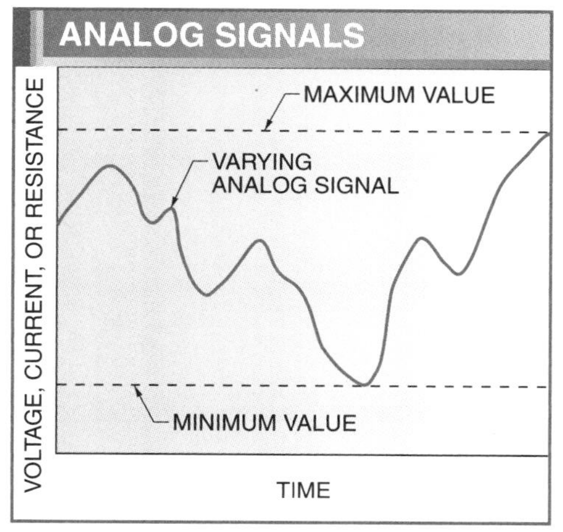

Figure 15-17. Analog signals can vary continuously between two points.

The most common electrical properties used to convey analog signals are voltage, current, and resistance. The most common analog signal ranges include 0 VDC to 10 VDC, 4 mA to 20 mA, and 0 kΩ to 10 kΩ. Devices may use other analog signal ranges, as long as they are compatible between the sender and receiver and represent the same values. For example, 0 VDC to 5 VDC, 1 mA to 10 mA, and 0 Ω to 1000 Ω ranges are common.

Analog signals differ from digital signals in that small fluctuations in the signal are meaningful. The range of possible signals is mapped to a range of possible values of the measured variable or the range of actuator positions. For example, an analog signal of 4 mA to 20 mA may be used to indicate a temperature between 32°F and 212°F. **See Figure 15-18.** Therefore, the range of 16 mA (20 mA – 4 mA = 16 mA) represents the range of 180°F (212°F – 32°F = 180°F). Each 1°F of change in temperature is indicated by a change of 0.089 mA. Units must be carefully noted in analog signal mapping, as many quantities can be mapped to other scales. For example, the same temperature analog signal also has the characteristic of 0.16 mA/°C (16 mA ÷ [100°C – 0°C] = 0.16 mA/°C).

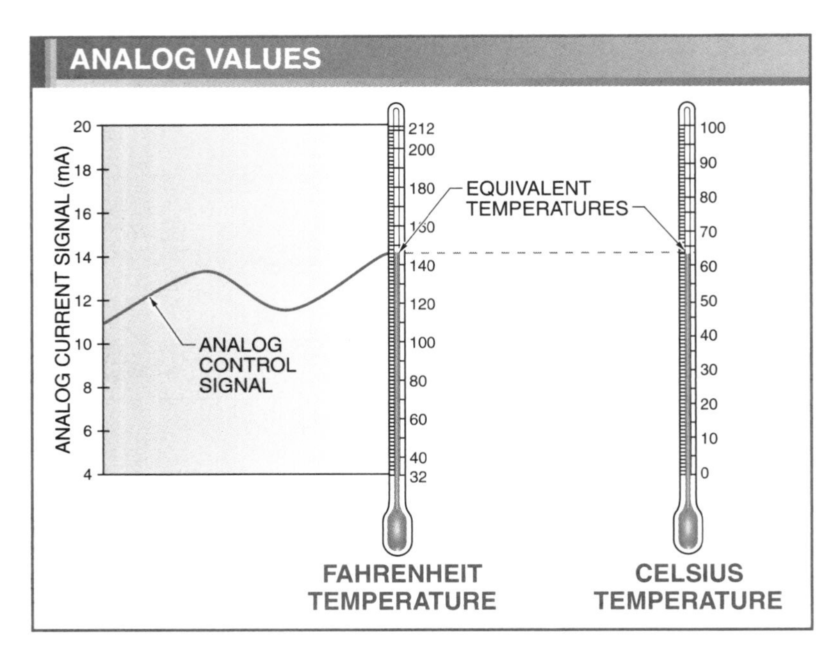

Figure 15-18. Analog signals are mapped to a certain range for the measured variable in order to determine the current value.

A disadvantage to analog signals is that electrical noise, which is normal to some degree in any electrical system, can affect the signals by changing their intended value slightly. While digital signals address this problem by providing ranges of values that correspond to the intended information, this cannot be done with analog signals, where every small change in value is potentially important. Depending on the system, noise can cause significant operational problems. Noise can be reduced to some degree by shielded conductors. Proper connections and careful design of the automated system, such as the length of conductor runs and proximity to noise-inducing electrical components, can help mitigate noise problems.

Tech Facts

Pulse width modulation (PWM) is a control technique in which a sequence of short pulses (digital signals) is used to communicate analog information. The amount of time the signal is ON versus OFF indicates any value between 0% and 100%.

Structured Network Messages

Some control devices can share much richer pieces of information by signaling in the form of structured network messages, which use protocols. A *protocol* is a set of rules and procedures for the exchange of information between two connected devices. The protocol is implemented in both the hardware and software of the devices. It defines the format, timing, and signals for reliable and repeatable data transmission between two devices.

Each structured network message includes information on the sending and receiving devices, the identification of the shared variable, its value, and any other necessary parameters. The information is encoded and transmitted via a series of digital signals, composing one complete message. **See Figure 15-19.** Almost any type of information can be shared in this way, as long as all the devices involved can work with the same protocol.

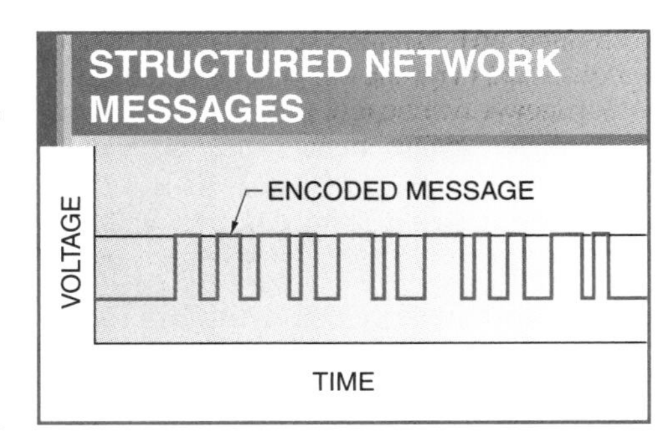

Figure 15-19. Structured network messages are several pieces of control information encoded into a series of digital signals.

Systems using digital and analog signals are typically connected point-to-point. That is, each device is connected with only the devices that need to receive its signal. However, systems using structured network messages are connected together in a similar way to a computer network. In this configuration, any device can communicate with any other device on the network. In fact, some systems can communicate over the same wiring and routing infrastructure as the building's computer local area network (LAN). There are also advantages to keeping the building automation system on a completely separate network, though this increases installation and maintenance costs.

CONTROL INFORMATION

Control devices are used in every type of control application to provide the required information for proper system operation. Hardware and software used to provide control information include control points, virtual points, setpoints, offsets, and deadband.

Control Points

The input information from sensors and output information to actuators creates control points in the building automation system. A *control point* is a variable in a control system. Control points change over time. **See Figure 15-20.** Controllers use input control points to make decisions about changes needed in the system and then determine the best value for the output control points.

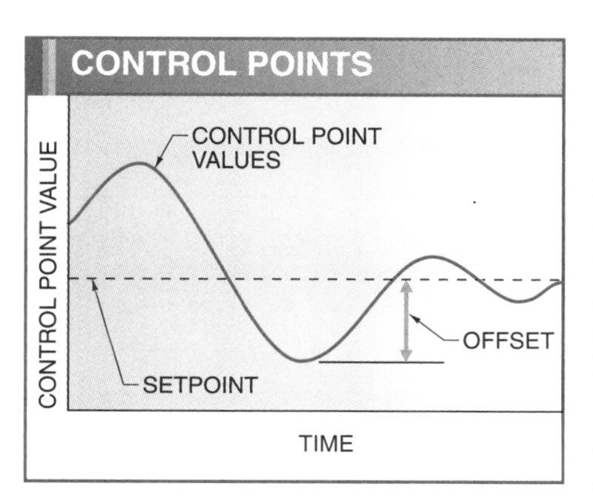

Figure 15-20. Controllers modify the building system until the control point variables equal the setpoint.

For example, a temperature sensor in a building space measures a temperature of 74°F, which is an input control point for the HVAC controllers. The information that the controllers generate and use to cause changes in the system is the output control point. For example, a damper actuator accepts a control point signal that corresponds to the position it must maintain.

Virtual Points

A *virtual point* is a control point that exists only in software. It is not a hard-wired point corresponding to a physical sensor or actuator. For example, programmed schedules produce virtual points that correspond to the status of the planned occupancy of a room. The resulting virtual point may have three states: OCCUPIED, UNOCCUPIED, and PRE-OCCUPANCY (meaning that the room systems are getting ready for OCCUPIED mode). This virtual point may be used by controllers to change the behavior of the HVAC and lighting systems, depending on whether people are expected to be using this room. Since they exist only in software, virtual points can only be monitored with some type of HMI.

Virtual points can also be used to hold snapshots of the values of other control points for different circumstances, such as a timestamp of an event, an extreme (high/low) value of a control point, or the extreme of multiple control points. For example, a series of three control points represent the current temperatures of three different rooms. A separate virtual point may represent the highest temperature any of the rooms reached during the last 24 hr period. Another virtual point may hold the timestamp for that event.

Lists of system inputs and outputs in the contract documents contain both hard-wired control points and virtual points, though they will be distinguished from each other in some way.

Tech Facts

Automated controls are often modified to collaborate with existing building systems in order to maximize the energy efficiency of the structure.

Setpoints

The purpose of the control system is to achieve and maintain certain conditions, which are quantified as setpoints. A *setpoint* is the desired value to be maintained by a system. Setpoints are matched to a corresponding control point variable, sharing the same variable types and units of measure, such as temperature in degrees Fahrenheit or flow rate in cubic feet per minute. For example, if it is desired to maintain a temperature of 72°F in a building space, the setpoint is 72°F.

There may be more than one setpoint associated with a control point. The currently active setpoint, as in the setpoint that the system is currently trying to achieve, can be changed according to schedules, occupancy, or any other parameter. For example, a cooling setpoint is raised from 74°F during the day to 85°F at night or when a building is unoccupied. This saves energy by reducing the system loads when a building is unoccupied, but still prevents the building conditions from getting excessively far from the occupied setpoint. Alternatively, setpoints can be reset based on other conditions. For example, a hydronic water heating setpoint is raised as the outside air temperature falls.

Offsets

Ideally, a control point is equal to its corresponding setpoint. However, there is often some difference, though it should be minimized as much as possible. The *offset* is the difference between the value of a control point and its corresponding setpoint. Offset values measure the accuracy of the control system. **See Figure 15-21.**

Expectations of control system accuracy have increased as the available technology has become more sophisticated. In the past, an accuracy of ±2°F was considered acceptable. With the widespread use of building automation systems, much tighter accuracies (such as ±0.5°F) are possible. Control system accuracy is a function of the accuracy of the controllers, the precision of the sensors and actuators, and the quality of the system design and tuning.

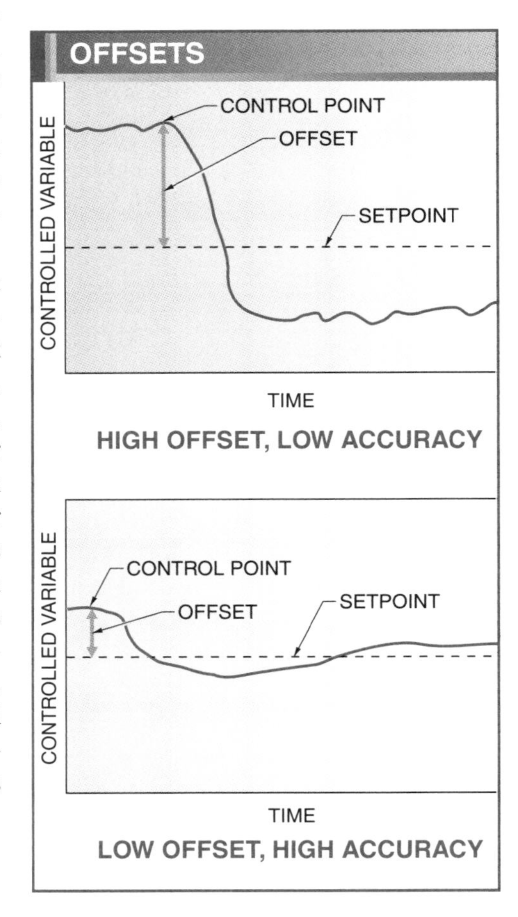

Figure 15-21. A control system with large offsets has low accuracy and a system with small offsets has high accuracy.

Deadband

Sometimes it is not necessary to maintain a condition at exactly one setpoint. Rather, the variable must only remain within a certain range. The range is defined by a pair of setpoints. **See Figure 15-22.** For example, the temperature within a space must be maintained between 70°F and 74°F. At 70°F, the heating system operates, and at 74°F, the cooling system operates, with a deadband in between. A *deadband* is the range between two setpoints in which no control action takes place.

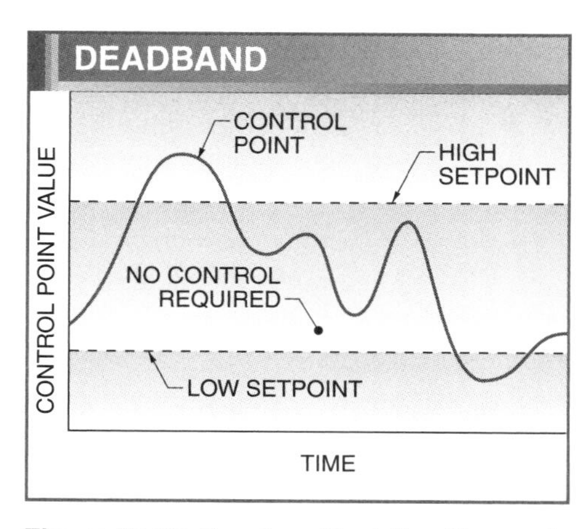

Figure 15-22. Deadband is defined by a pair of setpoints between which no control action is required.

As long as this does not adversely affect occupant comfort, then this deadband arrangement reduces energy use by the control system. It also reduces oscillation and frequent cycling of the system.

CONTROL LOGIC

The decisions that controllers make to change the operation of a building system involve control logic. *Control logic* is the portion of controller software that produces the necessary outputs based on the inputs. There are many different ways in which these decisions can be made, depending on the inputs used to make the decisions and the algorithms used to produce the results. An *algorithm* is a sequence of instructions for producing the optimal result to a problem. The decision-making process is described with a control loop. A *control loop* is the continuous repetition of the control logic decisions. Control loops are either open loops or closed loops.

Open-Loop Control

Open-loop control is the simpler of the two but has drawbacks. An *open-loop control system* is a control system in which decisions are made based only on the current state of the system and a model of how it should work. An example of an open-loop control system is a controller that turns a chilled water pump ON when the outside air temperature is above 65°F. The controller has no feedback to verify that the pump is actually ON. **See Figure 15-23.** Open-loop control requires perfect knowledge of the system, and assumes there are no disturbances to the system that would change the outcome otherwise. There is no connection between the controller's output and its input.

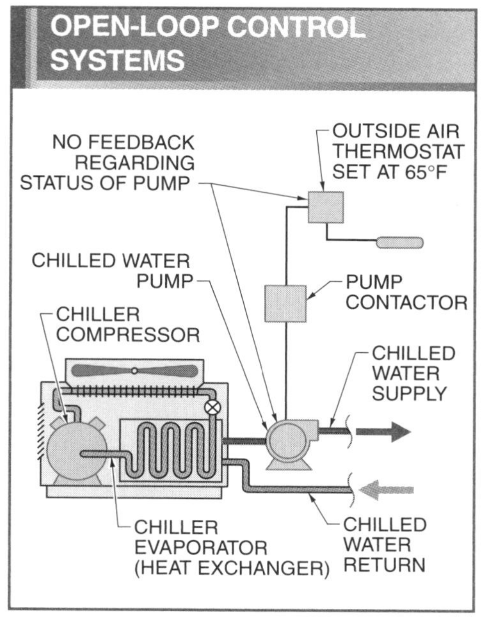

Figure 15-23. Open-loop control systems make changes to a system without verifying its effects.

The most common example of open-loop control in a building automation system is based on time schedules. Time-based control is a control strategy in which the time of day is used to determine the desired operation of a load. Time-based control turns a load ON or OFF at a specific time, without knowledge of any other factors that may affect the need for that load to operate. For example, open-loop time-based landscape irrigation control may activate the sprinklers based on a schedule. However, if it had recently rained, the system is overwatering the landscape and wasting water. Without a moisture sensor to provide an input based on the system output, the system has only open-loop control.

Closed-Loop Control

To address the limitation of open-loop control, most essential control loops include feedback. This makes them closed-loop. A *closed-loop control system* is a control system in which the result of an output is fed back into a controller as an input. For example, a thermostat controls the position of a valve in a hot water terminal device to maintain an air temperature setpoint. The thermostat in the building space provides the feedback of the air temperature that is used to continually adjust the hot water valve. **See Figure 15-24.**

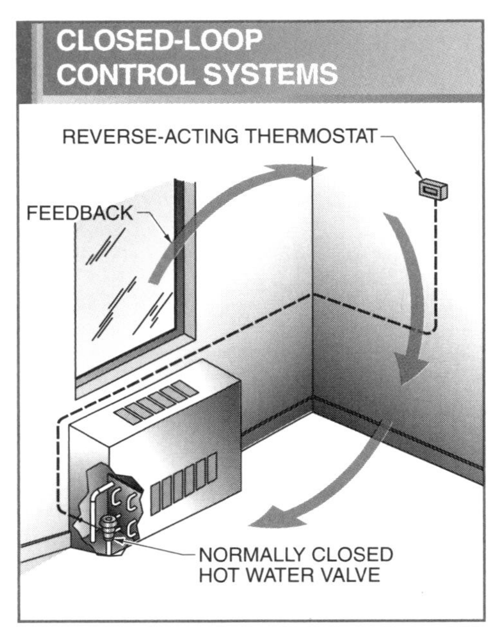

Figure 15-24. Closed-loop control systems provide feedback on system changes to the controller as an input.

Algorithms are used to calculate the output value based on the inputs. Different calculation algorithms can be used, each with different characteristics of accuracy, stability, and response time. The correct algorithm must be selected for the system. Common algorithms include proportional, integral, and derivative control algorithms.

Proportional Control Algorithms. Most modern commercial control applications use proportional control. *Proportional control* is a control algorithm in which the output is in direct response to the amount of offset in the system. For example, a 10% increase in room temperature results in a cooling control valve opening by 10%. Proportional controllers output an analog signal, which requires compatible actuators. **See Figure 15-25.**

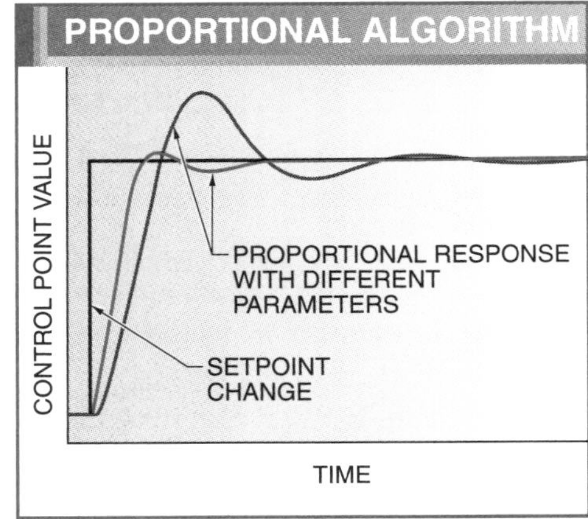

Figure 15-25. In order to bring a control point to a new setpoint, a proportional algorithm adjusts the output in direct response to the current offset.

Proportional control systems have a lower tendency to undershoot or overshoot. While proportional control is used successfully in most applications, it may be inaccurate if not set up properly. When the system reaches the setpoint, the controller outputs a default actuator position, typically a 50% setting. However, a load may require a different position when at the setpoint, resulting in increased offset and energy expenditure.

Tech Facts

"Building automation" is a broad term covering any automatic control of one or more building systems (such as an HVAC system). A "building automation system," however, specifically refers to a system of microprocessor-based devices that share information on a data network. Building automation systems also make the integration of multiple building systems practical.

Integral Control Algorithms. An *integral control algorithm* is a control algorithm in which the output is determined by the sum of the offset over time. *Integration* is a function that calculates the amount of offset over time as the area underneath a time-variable curve. **See Figure 15-26.** This offset area is then used to determine the output needed to eliminate the offset. Integral control algorithms move the system toward the setpoint faster than proportional algorithms.

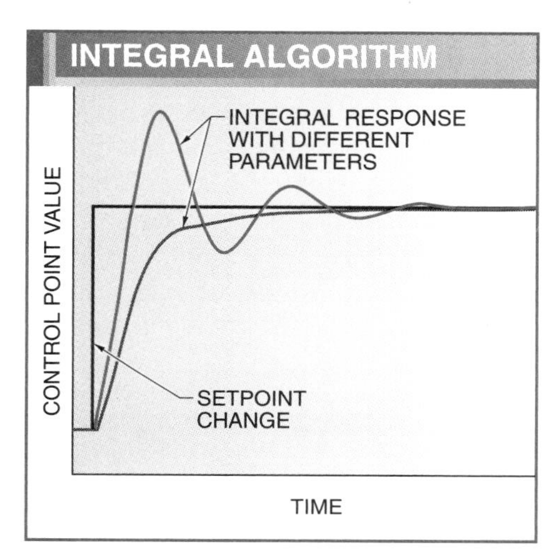

Figure 15-26. In order to bring a control point to a new setpoint, an integral algorithm adjusts the output according to the sum of the offsets in a preceding period.

However, since the integral is responding to accumulated errors from the past, it can cause the present value to overshoot the setpoint, crossing over the setpoint and creating an offset in the other direction. Proportional/integral (PI) control is the combination of proportional and integral control algorithms. These controllers can be more stable and accurate than integral-only controllers.

Derivative Control Algorithms. A *derivative control algorithm* is a control algorithm in which the output is determined by the instantaneous rate of change of a variable. **See Figure 15-27.** The rate of change is then used to determine the output needed to eliminate the offset. However, this calculation amplifies noise in the signal, which can cause the system to become unstable.

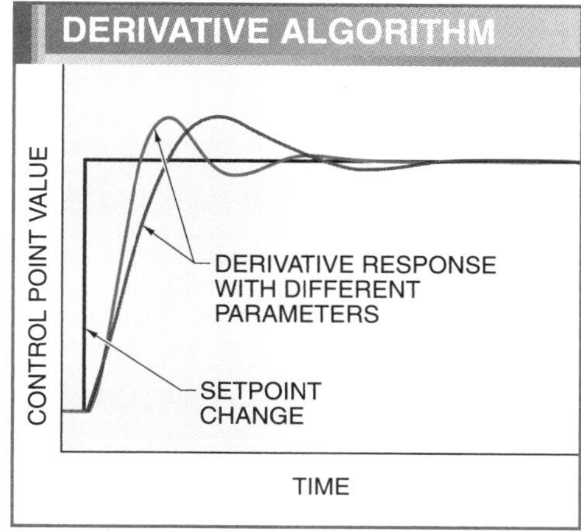

Figure 15-27. In order to bring a control point to a new setpoint, a derivative algorithm adjusts the output according to the rate of change of the control point.

Proportional/integral/derivative (PID) control is the combination of proportional, integral, and derivative algorithms. The derivative control algorithm slows the rate of change of the controller output, which is most noticeable close to the controller setpoint. Therefore, derivative control reduces the magnitude of the overshoot produced by the integral component and the combination improves stability. In building automation, only extremely sensitive control applications require PID control. Proportional/integral (PI) control is normally sufficient to achieve a setpoint.

PI and PID control systems must be carefully tuned to ensure accuracy and stability. *Tuning* is the adjustment of control parameters to the optimal values for the desired control response. **See Figure 15-28.** The parameters are the relative contributions of each algorithm to the final output decision and any parameters used in each algorithm calculation. There are several tuning methods by which experienced control professionals can determine the best system values.

EFFECTS OF CHANGING PID PARAMETERS				
Parameter	**Rise Time**	**Overshoot**	**Settling Time**	**Steady-State Error**
Proportional	Decrease	Increases	Small Change	Decreases
Integral	Decrease	Increases	Increases	Eliminates
Derivative	Small Change	Decreases	Decreases	None

Figure 15-28. Tuning involves adjusting the relative contributions of the proportional, integral, and derivative algorithms to quickly stabilize a control point at a setpoint.

BUILDING SYSTEMS

Building automation may involve any or all of the building systems. The level of sophistication of a building automation system is affected by the number of integrated building systems. Common building systems controlled by building automation include electrical, lighting, HVAC, plumbing, fire protection, security, access control, voice-data-video (VDV), and elevator systems.

Electrical Systems

An *electrical system* is a combination of electrical devices and components connected by conductors that distributes and controls the flow of electricity from its source to a point of use. The control of the electrical system involves ensuring a constant and reliable power supply to all building loads by managing building loads, uninterruptible power supplies (UPSs), and back-up power supplies. Sophisticated switching systems are used to connect and disconnect power supplies as needed to maintain building operation and occupant productivity. Electrical systems affect many other systems, such as lighting, security, and voice-data-video circuits. However, control of those systems is typically handled separately.

Lighting Systems

A *lighting system* is a building system that provides artificial light for indoor areas. Lighting is one of the single largest consumers of electricity in a commercial building. Therefore, improving energy efficiency and reducing electricity costs are the driving factors in lighting system control.

Lighting system control involves switching OFF or dimming lighting circuits as much as possible without adversely impacting the productivity and safety of the building occupants or the security of the building. Lighting systems are controlled based on schedules, occupancy, daylight harvesting, or timers. Specialty lighting control systems can also produce custom lighting scenes for special applications.

Tech Facts

A biometric reader analyzes a person's physical or behavioral characteristics for authentication. Fingerprint technology is among the most accurate biometric technologies and analyzes the patterns of the skin on the fingertip. The reader records an image of the fingerprint, which is then analyzed for a unique set of patterns. The patterns are compared to those stored in the access control database for that person. A sufficient match results in authentication.

HVAC Systems

A *heating, ventilating, and air conditioning (HVAC) system* is a building system that controls a building's indoor climate. Properly conditioned air improves the comfort and health of building occupants. HVAC systems are controlled to operate at optimum energy efficiency while maintaining desired environmental conditions.

HVAC systems are the most common building systems to automate but are also the most complicated. They include several parameters that must be controlled simultaneously and that are closely interrelated. Changes in one conditioned air parameter, such as temperature, can affect other parameters, such as humidity. Therefore, the systems use the most sophisticated control logic and must be carefully designed and tuned. A well-automated HVAC system, however, operates efficiently and with little manual input from occupants or maintenance personnel.

Plumbing Systems

A *plumbing system* is a system of pipes, fittings, and fixtures within a building that conveys a water supply and removes wastewater and waterborne waste. Plumbing systems are designed to have few active components. That is, most plumbing fixtures operate on water pressure and manual activation alone. However, there are a few applications in which certain parts of a commercial building's plumbing system can be actively controlled, such as water temperature, water pressure, and the supply of water for certain uses. These systems aim to use water the most effectively and provide a water supply with the optimal characteristics.

Fire Protection Systems

A *fire protection system* is a building system for protecting the safety of building occupants during a fire. Fire protection systems include both fire alarm systems and fire suppression systems. Fire protection systems automatically sense fire hazards, such as smoke and heat, and alert occupants to the dangers via strobes, sirens, and other devices. They also monitor their own devices for any wiring or device problems that may impair the system's proper operation during an emergency.

Fire protection systems are highly regulated due to their role as a life safety system. Fire alarm systems can be integrated with other building systems, but only through the fire alarm control panel (FACP). The fire-sensing devices and output devices may only be connected to the FACP. However, many FACPs include special features that enable integration with other automated systems. For example, some include special output connections that can be used to share fire alarm signals with other systems that have special functions during fire alarms.

Tech Facts

Bi-level switching is a technique that controls general light levels by switching individual lamps or groups of lamps in a fixture separately. This technique requires specially wired multi-lamp fixtures so that lamps must be ballasted separately.

Security Systems

A *security system* is a building system that protects against intruders, theft, and vandalism. Security systems are similar to fire protection systems in that they are typically implemented as separately designed package systems and allow connection with other building systems only at the security control panel. The control panel provides special output connections that can be used to initiate special control sequences in other systems, such as unoccupied modes, access, and security monitoring with surveillance systems.

Access Control Systems

An *access control system* is a system used to deny those without proper credentials access to a specific building, area, or room. Authorized personnel use keycards, access codes, or other means to verify their right to enter the restricted area.

Access control systems, through their control panels, can be integrated with other systems in much the same way as fire protection systems and security systems. In fact, due to their similarity, access control and security systems are often highly integrated. When a person's credentials have been verified by the access control system, the control panel can initiate the subsequent response of other systems, such as directing the surveillance system to monitor the person's entry.

Voice-Data-Video (VDV) Systems

A *voice-data-video (VDV) system* is a building system used for the transmission of information. Voice systems include telephone and intercom systems. Data systems include computer and control device networks, but can also carry voice or video information that has been encoded into compatible data streams. Video systems include closed-circuit television (CCTV) systems.

Building automation integration typically involves the VDV systems being controlled by other building systems. For example, a telephone line may be seized by the fire alarm system in order to alert the local authorities to a fire in the building. Alternatively, the CCTV system that is used for surveillance may be controlled by the security or access control systems to monitor certain areas where people have been detected.

Elevator Systems

An *elevator system* is a conveying system for transporting people and/or materials vertically between floors in a building. Elevator systems operate effectively largely on their own and with their own control devices but can accept inputs from other systems to modify their operating modes. A prime example of this is the integration of the fire alarm system with the elevator system. Since elevators can be dangerous to use during a fire, the elevator cars must be removed from service if the fire alarm is activated. An output on the fire alarm control panel (FACP) is connected to the elevator controller. In the event of a fire alarm signal, the elevator controller switches over to a fire service mode and disables the elevator car.

Review Questions

1. Briefly describe the three primary benefits of a building automation system.
2. What is the primary difference between a direct digital control (DDC) system and a building automation system?
3. How do the roles of the consulting-specifying engineer and the controls contractor differ with respect to designing the building automation system?
4. What is the difference between contract specifications and the component portion of the shop drawings?
5. Briefly describe the four primary categories of control devices.
6. What is the difference in the types of information shared between digital signals and analog signals?
7. Why might multiple setpoints be associated with one control point?
8. What is the difference between open-loop control and closed-loop control?
9. What is involved in tuning control loop algorithms for optimal control response?
10. Why are some building systems integrated only through their respective control panels?

Heating and cooling loads are the heat lost or gained by a building. The loads occur because of a difference between the indoor temperature and outdoor temperature, infiltration or ventilation in the building, and internal loads. When the air is cold outdoors, heat must be added to a building to offset the building heating and cooling loads to maintain a comfortable temperature indoors.

16

Heating and Cooling Loads

HEATING AND COOLING LOADS

Heating load is the amount of heat lost by a building. *Cooling load* is the amount of heat gained by a building. Heating and cooling loads occur because of a difference between the indoor temperature and outdoor temperature, infiltration or ventilation in the building, and internal loads. Heating and cooling loads are adjusted for internal loads, duct losses, and any other losses or gains not directly related to temperature difference.

Heating and cooling loads are calculated either manually or by a computer. The size of the heating and cooling equipment required in a building depends on the heating and cooling loads of the building. Heating and cooling loads are also necessary for designing a distribution system. Heating and cooling loads are found by using variables and factors. Variables and factors are arranged in a format that allows easy mathematical calculations.

Variables

Variables are data that are unique to a building. Variables relate to the specific location of the building and the specifications of the particular building. Variables include design temperature difference, areas of exposed building surfaces, infiltration or ventilation rates, people, lights, and appliances.

Design Temperature. *Design temperature* is the temperature of the air at a predetermined set of conditions. *Indoor design temperature* is the temperature selected for the inside of a building. The indoor design temperature used for buildings occupied by people is between 70°F and 75°F. *Outdoor design temperature* is the expected outdoor temperature that a heating or cooling load must balance.

The outdoor design temperature for an area is found on tables of outdoor design temperatures. *Outdoor design temperature tables* are tables of data developed from records of temperatures that have occurred in an area over many years. Outdoor design temperature tables are available from many sources. Outdoor design temperature tables include winter dry bulb temperature, summer dry bulb and wet bulb temperatures, daily range, and latitude degree. **See Figure 16-1.**

Winter dry bulb temperature is the coldest temperature expected to occur in an area while disregarding the lowest temperatures that occur in 1% to 2½% of the total hours in the three coldest months of the year. The coldest temperatures are disregarded because they occur for a short period of time during the night, when some temperature swing is acceptable. *Temperature swing* is the difference between the setpoint temperature and the actual temperature. Temperature swing slightly lowers the temperature in a building. A heating load for winter conditions is calculated at the winter dry bulb temperature.

OUTDOOR DESIGN TEMPERATURE					
State and City	Lat.*	Winter	Summer		
		DB§	DB§	DR#	WB§
ALABAMA					
Alexander City	32	18	96	21	79
Auburn	32	18	96	21	79
Birmingham	33	17	96	21	78
Mobile	30	25	95	18	80
Montgomery	32	22	96	21	79
Talladega	33	18	97	21	79
Tuscalossa	33	20	98	22	79
ALASKA					
Anchorage	61	-23	15	71	60
Fairbanks	64	-51	24	82	64
Juneau	58	- 4	15	74	61
Kodiak	57	10	10	69	60
Nome	64	-31	10	66	58

* in degrees
§ in °F
\# daily range in °F

Air Conditioning Contractors of America

Figure 16-1. Tables of outdoor design temperatures are developed based on temperature records over a period of many years.

Summer dry bulb temperature is the warmest dry bulb temperature expected to occur in an area while disregarding the highest temperature that occurs in 1% to 5% of the total hours in the three hottest months of the year. *Summer wet bulb temperature* is the wet bulb temperature that occurs concurrently with the summer dry bulb temperature. Wet bulb temperature is considered in cooling loads because of the effect of humidity on comfort at higher temperatures and because of the effect of humidity on cooling equipment capacity. A cooling load is calculated for summer conditions at the summer dry bulb temperature. *Design temperature difference* is the difference between the desired indoor temperature and the outdoor temperature for a particular season.

Area of Exposed Surfaces. To calculate heat loss and gain, the area of all exposed surfaces must be found. *Exposed surfaces* are building surfaces that are exposed to outdoor temperatures. Areas of the exposed surfaces of a building are calculated using building dimensions. The area of each wall, roof, or other exposed surface is calculated separately. For cooling loads, the areas should also be separated according to their exposure. *Exposure* is the geographic direction a wall faces. Exposure is important because of the solar gain on the surfaces. The solar gain on a surface depends on the exposure of the exterior walls, windows, and doors. Factors for solar gain, which are included in heating and cooling loads, are given for the different exposures. The area of each flat surface is found by multiplying the length of the surface by the height.

Gross wall area is the total area of a wall including windows, doors, and other openings. Gross wall area is calculated separately for each exposed wall of a building. Gross wall area is found by applying the formula:

$$A_g = w \times h$$

where

A_g = gross wall area (in sq ft)
w = width (in ft)
h = height (in ft)

Net wall area is the area of a wall after the area of windows, doors, and other openings have been subtracted. Net wall area is used for calculating heating and cooling loads. Net wall area is found by applying the formula:

$$A_n = A_g - A_o$$

where

A_n = Net wall area (in sq ft)
A_g = gross wall area (in sq ft)
A_o = area of opening (in sq ft)

Example: Finding Net Wall Area

The south wall of a building is 30′ wide and 8′ high. The wall has a door 3′ wide and 7′ high. Find the net wall area.

1. Find gross wall area.

$A_g = w \times h$
$A_g = 30 \times 8$
$A_g = \mathbf{240\ sq\ ft}$

2. Find the area of the opening.

$A_o = w \times h$
$A_o = 3 \times 7$
$A_o = \mathbf{21\ sq\ ft}$

3. Find net wall area.

$A_n = A_g - A_o$
$A_n = 240 - 21$
$A_n = \mathbf{219\ sq\ ft}$

The area of each window on each exposure of the building is calculated separately. The area of each door is also calculated separately. The values are categorized by exposure. The total area of windows and doors in each exposure are used for calculating window and door gain or loss and net wall area.

The area of the ceiling is calculated using inside dimensions. Overhangs that do not affect the transmission of heat are not included. For sloping ceilings such as cathedral ceilings, the exposed area, not the area of the horizontal plane, is used for calculating the area of the ceiling. **See Figure 16-2.**

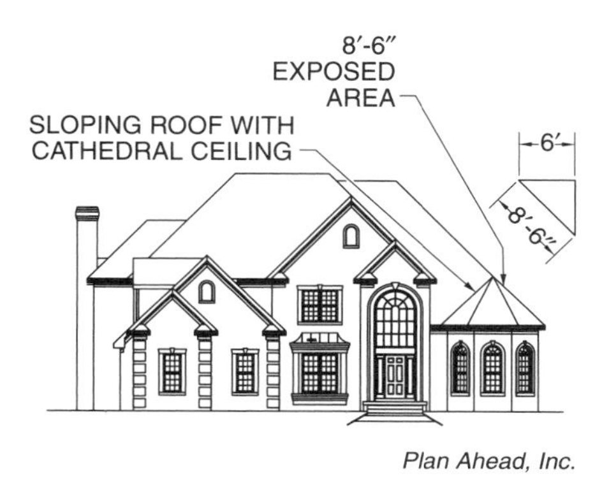

Figure 16-2. For sloping ceilings, the exposed area is used for calculating the area of the ceiling.

The area of a floor that is exposed to a space that is not conditioned is used to calculate heat loss or gain. In a building where a concrete slab floor is located on the ground, most of the heat loss from the floor occurs around the edge. Heat loss is calculated as the distance around the perimeter of the exposed edge of the floor. A concrete slab floor has no heat gain.

Number of People. Estimating the number of people occupying a residential building and estimating the number of people occupying a commercial building are done with different methods. When estimating the number of people that occupy a residential building, the living habits of the individuals must be considered. If a family entertains frequently, the number of people used for calculating loads should include family and guests. For family only, the estimate is based on the number of bedrooms in the building. The number is calculated based on two people in the main bedroom and one person for each additional bedroom. To include guests in the number of people, use an estimate of the actual number of guests.

In a commercial building, the estimate of the number of occupants is based on the area (in square feet) of floor space required per person and the activity level of the person. Tables of occupancy estimates show the area required per person for different activities. Care should be taken to ensure that only space that a person can actually occupy is used. Approximately one-half of the floor space is usually occupied by furniture, appliances, or displays.

Number of Lights and Appliances. Any device that produces heat inside a building must be compensated for by a cooling load. Electric lights produce heat as well as light. Appliances are powered by electricity or fuel. In either case, much of the energy that operates the appliance is converted to heat. The number of lights and appliances must be calculated to find the amount of heat given off by the devices.

Factors

Factors are numerical values that represent the heat produced or transferred under some specific condition. Factors are found in tables that have been prepared for typical applications. Factors include heat loss or gain from conduction, infiltration, ventilation, people, electricity used for lights, electricity or fuel used for operating appliances, and solar energy. Heat gain from people, lights, appliances, and solar energy normally only apply to cooling loads.

Conduction. According to the second law of thermodynamics, when the temperature on two sides of a material is different, heat flows from the warmer side to the cooler side of the material. The outside surfaces of a building, such as walls, windows, doors, floors, and ceilings are composed of materials across which temperature differences exist. When the temperature of the air outside a building is cooler than the temperature inside, heat flows from the indoors to the outdoors. When the temperature of the air outside a building is warmer, heat flows from the outdoors to the indoors. A *building component* is a main part of a building structure such as the exterior walls. **See Figure 16-3.**

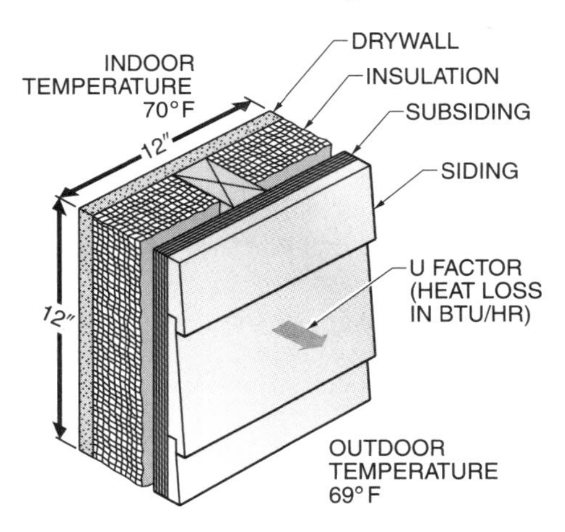

Figure 16-3. Conduction factors identify the amount of heat that flows through a building component because of a temperature difference.

A *conduction factor* (U) is a number that represents the amount of heat that flows through a building component because of a temperature difference. Conduction factors are expressed in Btu per hour per square foot of material per degree Fahrenheit temperature difference through the material. Factors that identify heat carried by air are based on the sensible heat of air, the amount of air moved, and the difference between the indoor temperature and outdoor temperature.

Heat transfer rates have been calculated for most building components used in the construction of residential and commercial buildings. Heat conduction factors are found in tables of conduction factors through various building materials.

To simplify calculations, conduction factors and either outdoor design temperatures or design temperature differences are combined in a table of heat transfer factors. A *heat transfer factor* is a conduction factor multiplied by a design temperature difference. A table of heat transfer factors is used for calculating cooling or heating loads for residential or small commercial buildings. **See Appendix.**

Infiltration. *Infiltration* is the process that occurs when outdoor air leaks into a building. *Infiltration air* is air that flows into a building when exterior doors are open or when air leaks in through cracks around doors, windows, or other openings.

Factors for calculating heat loss and gain from the exchange of air at different temperatures is found in tables of infiltration factors. **See Figure 16-4.** When a table of infiltration factors is used for calculating heat loss or gain for air, the volumetric flow rate (cfm) of air involved is multiplied by the infiltration factor.

The amount of heat transferred by infiltration is represented by factors derived from the specific heat of air. Most factors are expressed in Btu per hour per cubic foot of air per minute and the difference between the indoor and outdoor temperature. Some infiltration tables are based on the heat loss per square foot of exposed building surface.

To find the volumetric flow rate of the air involved, the volume of the building (in cu ft) or the volume of the building spaces involved in the calculation is multiplied by an air change factor. An *air change factor* is a value that represents the number of times per hour that the air in the building is completely replaced by outdoor air. Air change factors are found in tables for selected buildings. **See Figure 16-5.**

Ventilation. *Ventilation* is the process that occurs when outdoor air is brought into a building. *Ventilation air* is air that is brought into a building to keep building air fresh. The air must be heated if it is cooler than the indoor air or cooled if it is warmer than the indoor air.

Tech Facts

Typically, outdoor air that is brought into a building space for ventilation purposes is not filtered.

INFILTRATION FACTORS						
Outdoor design temperature*	85	90	95	100	105	110
Infiltration factors†	0.7	1.1	1.5	1.9	2.2	2.6

* in °F
† in Btu/hr/sq ft
Note: Infiltration factors are for an indoor design temperature of 75°F.

Air Conditioning Contractors of America

Figure 16-4. Tables of infiltration factors contain information for calculating heat loss and gain from air that seeps into a building space.

The volumetric flow rate of ventilation air required for a building is based on volumetric flow rate of air required per person inside the building. The flow rate of air required per person is found in the Uniform Building Code (UBC), or in local adaptations to the UBC. To find the required volumetric flow rate of ventilation air, the volumetric flow rate per person is multiplied by the number of people that occupy the building. The total volume of ventilation air required for a building is found by applying the formula:

$Q_t = p \times Q_p$

where

Q_t = total volumetric flow rate of ventilation air required (in cfm)

p = number of people

Q_p = required volumetric flow rate of air per person (in cfm)

Example: Finding Required Total Volumetric Flow Rate—Ventilation Air

A building is occupied by 70 people. The Uniform Building Code calls for 30 cfm of ventilation air per person. Find the required total volumetric flow rate of ventilation air.

$Q_t = p \times Q_p$

$Q_t = 70 \times 30$

$Q_t = \mathbf{2100\ cfm}$

To find the heat loss from the total volumetric flow rate of ventilation air, the flow rate is multiplied by 1.08 and the difference between the indoor and outdoor temperature. The heat loss from ventilation air is found by applying the formula:

$H_l = Q_t \times 1.08 \times \Delta T$

where

H_l = heat loss from ventilation air (in Btu/hr)

Q_t = total volumetric flow rate of ventilation

1.08 = constant

ΔT = temperature difference (in °F)

NUMBER OF AIR CHANGES PER HOUR			
Building Type	Construction Type		
	Loose	Medium	Tight
Residential	1.5	1.0	.5
Commerical	3.0	1.5	.5
Industrial	5.0	3.0	1.0

Air Conditioning Contractors of America

Figure 16-5. The number of air changes per hour in a building depends on the type of building and the construction of the building.

Example: Finding Heat Loss—Ventilation Air

A building is occupied by 35 people. The Uniform Building Code calls for 30 cfm of ventilation air per person. The difference between the indoor and outdoor temperature is 40°F. Find the heat loss from ventilation air.

1. Find the total volumetric flow rate of ventilation air required.

   $Q_t = p \times Q_p$

   $Q_t = 35 \times 30$

   $Q_t = \mathbf{1050\ cfm}$

2. Find the heat loss from ventilation air.

   $H_l = Q_t \times 1.08 \times \Delta T$

   $H_l = 1050 \times 1.08 \times 40$

   $H_l = \mathbf{45{,}360\ Btu/hr}$

The heat gain for a cooling load is calculated with the same procedure, except the humidity in the building must be considered. The most accurate way to calculate heat gain from ventilation air is by use of a psychrometric chart.

People. The human body produces heat and moisture. Heat is produced during digestion and is given off from the skin. Moisture is expelled from the body with every breath and during perspiration. The amount of heat given off varies with the amount of activity. A vigorous activity produces more heat than a passive activity. A table of heat gain from occupants is used to identify the heat produced by people when engaged in various activities. **See Figure 16-6.**

Lights. Heat gain from lights is a factor in a cooling load calculation. The factor for converting electricity to thermal energy is 3.41 Btu/W (Btu per watt). To calculate the heat gain from lights, the total wattage of the lights is multiplied by 3.41. If fluorescent lights are used, the conversion factor from electrical energy to thermal energy is 4.25 Btu/W. Fluorescent lights have a larger factor because larger quantities of heat are given off by the ballast in a fluorescent light.

Appliances. Heat gain from appliances is calculated by multiplying the wattage of the appliance by 3.41. Calculating heat gain for appliances that use fuel instead of electricity is based on the rate of fuel used and the heat output for the fuel. Heat gain values are also found on tables of heat gain from appliances. **See Figure 16-7.**

Solar Gain. *Solar gain* is heat gain caused by radiant energy from the sun that strikes opaque objects. Sunlight, which shines through the windows and strikes the interior of a building, adds heat to the building. This heat is solar gain. Solar gain is only used when calculating cooling loads. A cooling system must compensate for extra heat. In some commercial buildings, solar gain has a positive effect on heating loads.

Equivalent temperature difference is the design temperature difference that is adjusted for solar gain. Equivalent temperature differences are used when calculating heat gain through walls or ceilings of buildings. Equivalent temperature difference is used only for calculating cooling loads. Equivalent temperature differences are used in most cooling load calculations for commercial buildings.

When the sun shines on an exterior surface of a building or opaque objects inside a building, the solar energy becomes thermal energy. The thermal energy is then absorbed into the building. For the cooling load for a commercial building, the heat from the solar energy is added to the cooling load of the building because the system must compensate for the additional heat.

HEAT GAIN FROM OCCUPANTS

Degree of Activity	Typical Application	Total Heat*	Sensible Heat*	Latent Heat*
Seated, at rest	Theater, grade school classroom	330	225	105
Seated, very light work	Office, hotel, apartment, high school classroom	400	245	155
Moderately active office work	Office, hotel, apartment, college classroom	450	250	200
Standing, light work	Drug store, bank	500	250	250
Sedentary work	Restaurant	550	275	275
Light bench work	Factory	750	275	475
Moderate dancing	Dance hall	850	305	545
Walking, moderately heavy work	Factory	1000	375	625
Bowling, heavy work	Bowling alley, factory	1450	580	870

* in Btu/hr
Note: The above values are based on 75°F room dry bulb temperature.
For 80°F room dry bulb temperature, the total heat gain remains the same.

Air Conditioning Contractors of America

Figure 16-6. Heat gain from people is calculated by multiplying a factor from a table for heat gain from occupants.

HEAT GAIN FROM APPLIANCES								
Type of Appliance	Electric				Gas			
	Without Hood			Hood	Without Hood			Hood
	Sensible*	Latent*	Total*	All Sensible*	Sensible*	Latent*	Total*	All Sensible*
Broiler-griddle 31″ x 20″ x 18″					11,700	6300	18,000	3600
Coffee brewer/warmer					1750	750	2500	500
per burner	770	230	1000	340				
per warmer	230	70	300	90				
Coffee urn								
3 gal.	2550	850	3400	1000	3500	1500	5000	1000
5 gal.	3850	1250	5100	1600	5250	2250	7500	1500
8 gal.	5200	1600	6800	2100	7000	3000	10,000	2000
Deep fat fryer								
15 lb fat	2800	6600	9400	3000	7500	7500	15,000	3000
21 lb fat	4100	9600	13,700	4300				
Dry food warmer per sq ft top	320	80	400	130	560	140	700	140
Griddle, frying per sq ft top	3000	1600	4600	1500	4900	2600	7500	1500
Hot plate					5300	3600	8900	2800
Short order stove per burner					3200	1800	5000	1000
Toaster								
Continuous								
360 slices per hour	1960	1740	3700	1200	3600	2400	6000	1200
720 slices per hour	2700	2400	5100	1600	600	4000	10,000	2000
Pop-up (4 slice)	2230	1970	4200	1300				
Waffle iron 18″ x 20″ x 13″	1680	1120	2800	900				
Hair dryer								
Blower type	2300	400	2700					
Lab burners								
Bunsen					1680	420	2100	
Meeker	60				3360	840	4200	
Neon sign per foot of tube			60					
Sterilizer	650	1200	1850					
Vending machines								
Hot drink			1200					
Cold drink			625					

* Heat in Btu/hr created in 1 hour.

Air Conditioning Contractors of America

Figure 16-7. Tables for heat gain from appliances have factors that are included when calculating loads for commercial buildings.

Using an equivalent temperature difference is one way to include this increase in the cooling load. Tables of equivalent temperature differences show the equivalent temperature difference for sunlit walls and roofs. **See Figure 16-8.**

Tables of glass heat transfer factors include the Btu per hour gain per square foot of glass surface for typical windows. These tables also provide data for modifying the factor value for different types of glass such as regular, double-glazed, and heat-absorbing glass. **See Figure 16-9.**

DOE/NREL, Warren Gretz

Windows allow sunlight to enter a building, which adds heat to the building.

EQUIVALENT TEMPERATURE DIFFERENCES*								
Design Temperature§	85	90	95		100		105	110
Daily Temperature Range	L or M	L or M	M	H	M	H	H	H
Walls and Doors								
Wood frame walls and doors	13.6	18.6	23.6	18.6	28.6	23.6	28.6	33.6
Solid masonry, block, or brick walls	6.3	11.3	16.3	11.3	21.3	16.3	21.3	26.3
Partitions	5.0	10.0	15.0	10.0	20.0	15.0	20.0	25.0
Ceilings and Roofs								
Dark exterior	34.0	39.0	44.0	39.0	49.0	44.0	49.0	54.0
Light exterior	26.0	31.0	36.0	31.0	41.0	36.0	41.0	46.0
Floors								
Over unconditioned, vented, or open space	5.0	10.0	15.0	10.0	20.0	15.0	20.0	25.0
Over conditioned space or on or below grade	0	0	0	0	0	0	0	0

* Equivalent temperature differences are for an indoor design temperature of 75° F.
§ in °F

Air Conditioning Contractors of America

Figure 16-8. Equivalent temperature difference is the design temperature difference adjusted for solar gain.

COOLING HEAT TRANSFER FACTORS — WINDOWS AND DOORS*									
Exposure	Single Glass Temperature Diff.			Double Glass Temperature Diff.			Triple Glass Temperature Diff.		
	15°	20°	25°	15°	20°	25°	15°	20°	25°
N	18	22	26	14	16	18	11	12	13
NE & NW	37	41	46	31	33	35	26	27	28
E & W	52	56	60	44	46	48	38	39	40
SE & SW	45	49	53	39	41	43	33	34	35
S	28	32	36	23	25	27	19	20	21
Doors									
Wood	8.6	10.9	13.2	8.6	10.9	13.2	8.6	10.9	13.2
Metal	3.5	4.5	5.4	3.5	4.5	5.4	3.5	4.5	5.4

* Inside shading by venetian blinds or draperies.

Air Conditioning Contractors of America

Figure 16-9. Tables of glass heat transfer factors (cooling) contain the Btu per hour gain per square foot of glass surface for different types of windows.

Solar gain on walls and ceilings is combined with conduction gain in some factor tables. In other tables, solar gain is included with an equivalent temperature difference for the heating and cooling loads. When using a particular set of tables for calculating heat gain, the method for including solar gain through the building surfaces must be determined.

Tech Facts

When calculating the cooling load of a building, heat gain from windows must be factored into the cooling load.

LOAD FORMS

Load forms are documents that are used by designers for arranging the heating and cooling load variables and factors. Load forms simplify the calculation of heating and cooling loads. Two basic forms are prepared forms and columnar forms.

Prepared Forms

A *prepared form* is a preprinted form consisting of columns and rows that identify required information. The proper values for the variables and factors for an application are inserted in the spaces. The calculations are performed by working across the lines on the form.

Many different prepared forms are available. A prepared form has space at the top for the job name, date of the calculations, name of the person who prepared the form, and design temperatures. **See Figure 16-10.** The left-hand column has spaces for factors such as building components or other elements of the loads. The other columns are for information relating to individual rooms or zones in a building. Headings across the top of the form identify each room or zone column. Factors are entered in the factors column for each building component.

Prepared forms have multiple columns that are used to calculate loads on individual rooms or zones. Heat loss or gain is calculated for each room or zone so the hydronic piping or forced-air ductwork system can provide the correct amount of heat to each room or zone.

Small rooms or adjoining rooms without partitions may be combined on the form. A room or zone is identified at the top of each column. The area of each building component in the room or zone is entered in a space in the column. Factors are entered in the factors column. The areas in each column are multiplied by the factors on the same line. Factors are arranged to be entered once regardless of the number of rooms or zones.

Columnar Forms

A *columnar form* is a blank table that is divided into columns and rows by vertical and horizontal lines. Spaces at the top of a columnar form identify the job with the job name, date of the calculations, name of the person who prepared the form, and design temperatures.

A columnar form should be arranged with the procedure used for a prepared form. Headings are entered at the top of the columns and on the left-hand column on the form to identify information that is entered in the spaces. When a columnar form is used for calculating heating and cooling loads, the columns are used for room or zone data and the rows are used for variables and factors. Heating and cooling loads are calculated on two separate sections of the form. The data relating to rooms or zones is placed in individual columns. Factors are placed in the rows. **See Figure 16-11.**

A columnar form organizes the required data for heating and cooling loads so that the mathematical calculations are easy to perform. The load calculation data is readily available and the columns are easily added for sizing components of the distribution system and for balancing a system. A designer using a columnar form for calculating heating or cooling loads should be familiar with the method of calculating loads with a prepared form and the load calculation process.

Contractor: ____________ Date: ____________
Name of Job: ____________ By: ____________
Address: ____________

Winter: Indoor Design Temp. ____ Outdoor Design Temp. ____ Design Temperature Difference ____
Summer: Outdoor Design Temp. ____ Indoor Design Temp. ____ Design Temperature Difference ____

			Factors		1			4			5			Building Component Subtotals	
1	Name of Room				1			4			5				
2	Running Feet of Exposed Wall														
3	Room Dimensions														
4	Ceiling Height	Exposure													
	Types of Exposure		H	C	Area	Btu/hr H	Btu/hr C	Area	Btu/hr H	Btu/hr C	Area	Btu/hr H	Btu/hr C	Btu/hr H	Btu/hr C

Air Conditioning Contractors of America

Figure 16-10. A prepared form has headings for variables and factors used when calculating heating and cooling loads.

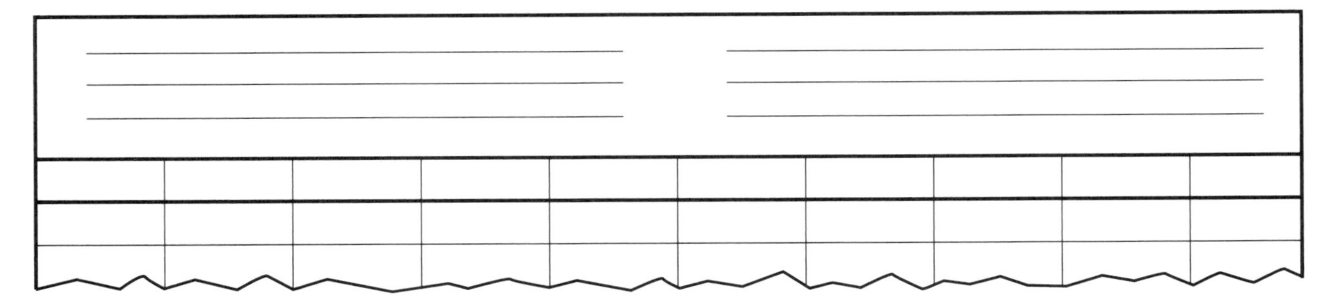

Figure 16-11. A columnar form is a blank grid. A design technician fills in headings for the columns and lines to fit each application.

Review Questions

1. List three elements that may be involved in calculating heating and cooling loads.
2. What two methods are used for calculating heating and cooling loads?
3. What two categories identify data used for calculating heating and cooling loads?
4. Define variables in relation to heating and cooling load data.
5. List and describe six variables used when calculating heating and cooling loads.
6. Why are the areas of exposed building surfaces important in calculating heating and cooling loads?
7. Define exposure in relation to calculating heating and cooling loads.
8. Explain how gross wall area is calculated differently than net wall area.
9. Why is net wall area of a building important?
10. List and explain the two ways window and door areas are used in calculating heating and cooling loads.
11. Explain how the number of people in a building is used when calculating heating and cooling loads.
12. Define factor in relation to heating and cooling load calculations.
13. Define conduction factor in relation to heat loss and heat gain.
14. What is the difference between a heat conduction factor and a heat transfer factor?
15. Define infiltration.
16. How are air change factors used in heating and cooling load calculations?
17. What is the difference between infiltration and ventilation?
18. How are design temperature difference figures adjusted to find the equivalent temperature difference?
19. What is the difference between a prepared load calculation form and a columnar load calculation form?
20. Explain the advantage of using a columnar form instead of a prepared form.

Heating and cooling load calculations determine the amount of heat or cooling that is required to maintain a constant temperature in a building. Heating and cooling loads are calculated by a conventional method or with a computer.

17 Load Calculations

LOAD CALCULATIONS—CONVENTIONAL METHOD

Conventional load calculations are performed manually. Two load forms are the prepared form and columnar form. The information for the cooling load is entered on the form first because more data is required for calculating cooling loads than for heating loads. The heating load information is taken from the cooling load information.

Once a load form is selected, the information about the job is entered in the spaces at the top of the form. This information includes the name of the contractor, name and address of job, date, and the name of the person who will prepare the form. **See Figure 17-1.**

Variables and factors are entered on a load calculation form and are used to calculate heat losses and gains for a specific building. Additional losses or gains that occur through the ductwork or piping are added to the building losses and gains to determine the total heating and cooling loads.

Variables

Variables for calculating heating and cooling loads include design temperature differences, area of building components, data for calculating infiltration or ventilation loss or gain, number of people occupying the building, and number of lights and appliances in the building. For most cooling applications, solar gain must also be considered. Solar gain through windows is usually calculated directly. Solar gain through walls and the roof is either included in the factors used for heat gain through walls and roof or is included as an equivalent temperature difference.

Design Temperature. *Design temperature* is the temperature of the air at a predetermined set of conditions. Indoor design temperature is the temperature selected for the inside of a building. A winter indoor design temperature of 70°F is usually selected to provide comfort for a normally clothed person during winter. A summer indoor design temperature of 75°F is usually selected to provide comfort for a normally clothed person during summer. Outdoor design temperatures are the coldest winter temperatures expected to occur in an area during 97.5% of the winter, and the warmest summer temperatures expected in an area during 95% of the summer.

Tech Facts

Load calculations are typically performed manually. The prepared form and columnar form are two load forms. Since more data is required for calculating cooling loads than for heating loads, the information for the cooling load is entered on the form first.

Contractor: City Heating & Cooling

Name of Job: Jones Residence

Address: 1875 W. Grant St.

Chicago, IL 60632

Date: 1/14/07

By: J. P.

Winter: Indoor Design Temp. 75°F Outdoor Design Temp. -9°F Design Temperature Difference 84°F

Summer: Outdoor Design Temp. 90°F Indoor Design Temp. 75°F Design Temperature Difference 15°F

														Building Component Subtotals	
1	Name of Room				1			4			5				
2	Running Feet of Exposed Wall														
3	Room Dimensions														
4	Ceiling Height	Exposure													
	Types		Factors		Area	Btu/hr	Btu/hr	Area	Btu/hr	Btu/hr	Area	Btu/hr	Btu/hr	Btu/hr	Btu/hr
	Exposure		H	C		H	C		H	C		H	C	H	C
5	Gross	N													
	Exposed	S													
	Wall	E													
	Area	W													
6	Windows and Glass														
	Doors (H)														
7	Windows	N													
	and Glass	E & W													
	Doors (C)	S													
8	Other Doors														
9	Net Exposed Walls														
10	Ceilings														
11	Floors														
12	People @ 300 Appliances @ 1200														
														Totals	
13	Subtotal Btu/hr Loss														
14	Sensible Btu/hr Gain														
15	Subtotal Btu/hr Gain (1.3)														

Figure 17-1. Specific information must be entered at the top of a load form.

Outdoor design temperatures are found in tables of outdoor design temperatures. Tables of outdoor design temperatures contain specific temperatures for many different areas. The outdoor design temperature for a building is found on a table by finding the area where the building is located. **See Figure 17-2. See Appendix.**

Design temperature difference is the difference between the indoor design temperature and the outdoor design temperature. Winter design temperature difference is found by subtracting the winter outdoor design temperature from the indoor design temperature. This data is entered on the form in the spaces at the top of the form. Winter design temperature difference is found by applying the formula:

$$\Delta T_W = T_i - T_o$$

where

ΔT_W = winter design temperature difference (in °F)

T_i = winter indoor design temperature (in °F)

T_o = winter outdoor design temperature (in °F)

Summer design temperature difference is found by subtracting the summer indoor design temperature from the summer outdoor design temperature for the area in which the building is located. This data is entered on the form in the corresponding spaces. Summer design temperature difference is found by applying the formula:

$\Delta T_S = T_o - T_i$

where

ΔT_S = summer design temperature difference (in °F)

T_o = summer outdoor design temperature (in °F)

T_i = summer indoor design temperature (in °F)

Example: Finding Design Temperature Difference

Find the winter and summer design temperature difference for a residence located in Chicago, Illinois. Winter design temperature difference and summer design temperature difference are found by applying the procedure:

1. Enter 75°F as the winter and summer indoor design temperatures on the load form.
2. On a table of outdoor design temperatures, locate the winter dry bulb and summer dry bulb temperatures for Chicago, Illinois. Enter the temperatures on the load form.
3. Find the winter design temperature difference.

   $\Delta T_W = T_i - T_o$

   $\Delta T_W = 75 - (-9)$

   $\Delta T_W =$ **84°F**
4. Find summer design temperature difference.

   $\Delta T_S = T_o - T_i$

   $\Delta T_S = 90 - 75$

   $\Delta T_S =$ **15°F**

Building Data. *Building data* is the name of each room, running feet of exposed wall, dimensions, ceiling height, and exposure of each room. Building data is entered at the top of the load calculation form. For a building in the planning or construction stage, the building data is found on the plans and specifications for the building. *Plans* are drawings of a building that show dimensions, construction materials, location, and arrangement of the spaces within the building. *Specifications* are written supplements to plans that describe the materials used for a building. Plans and specifications should be available from the building contractor.

OUTDOOR DESIGN TEMPERATURE

State and City	Lat.*	Winter	Summer		
		DB§	DB§	DR#	WB§
ILLINOIS					
Aurora	41	- 6	93	20	79
Champaign	40	- 3	95	21	78
Chicago	41	- 9	90	15	79
Galesburg	40	- 7	93	22	78
Joliet	41	- 5	93	20	78
Kankakee	41	- 4	93	21	78
Macomb	40	- 5	95	22	79
WASHINGTON					
Aberdeen	46	25	80	16	65
Olympia	46	16	87	32	67
Seattle-Tacoma	47	17	84	22	66
Spokane	47	- 6	93	28	65
Walla Walla	46	0	97	27	69

* in degrees
§ in °F
\# daily range in °F

Air Conditioning Contractors of America

Figure 17-2. Tables of outdoor design temperatures contain design temperatures for many different areas.

When plans and specifications are not available for a building that is already built, the floor plan of the building is sketched, and notes about the materials and construction methods are taken. The building data for the load form is taken from the sketches and notes and entered in the appropriate spaces on the form. **See Figure 17-3.**

Dimensions for walls, ceilings, and floors may be rounded to the nearest 6″. Dimensions for windows and doors may be rounded to the nearest 3″. Dimensions for rooms or zones should be taken from the outside of the building to the center of partitions between adjacent rooms or zones. Hallways, closets, stairways, and other spaces in the building not part of a room or zone are included in the adjacent rooms or zones. The dining room and kitchen of the Chicago residence are combined because they are not divided by a partition. The areas of the hallways are combined with the living room and bath.

Gross wall area is the total area of a wall including windows, doors, and other openings. Gross wall area is calculated separately for each exposed wall of the building. The gross exposed wall area for each room is entered on the load form according to the exposure of the wall. Exposure is the geographic direction a wall faces. Gross wall dimensions are taken from the outdoor dimensions of the building, which are found on sketches or plans of the building.

The information for calculating the area of windows and glass doors is found on sketches or plans of the building. For heating load calculations, the area of windows and glass doors is calculated and entered on the form. For cooling load calculations, the area of windows and glass doors is calculated and entered on the form according to exposure. The area of doors that are not glass is also entered on the form.

The net area of exposed walls is found by subtracting the area of windows and doors from the gross exposed wall area for each room. The area of the ceiling and floor is found and entered on the form. The area of the ceiling and floor is found by multiplying the dimensions of the rooms.

Three different methods are used to calculate the rate of infiltration of air into a building. In the first method, an infiltration factor is included with either the wall factors or the window factors taken from heat transfer factor tables. When calculating loads, factor tables should be checked to ensure that infiltration factors are included. For the Chicago residence, the infiltration factors are included in the heat transfer factors table.

The second method is the crackage method. *Crackage* is the opening around windows, doors, or other openings in the building such as between the foundation and framework of a building. In the crackage method, the actual length of crackage in a building is added up. A factor for calculating the flow rate or actual loss or gain for each foot of crackage is found on some calculation factor tables.

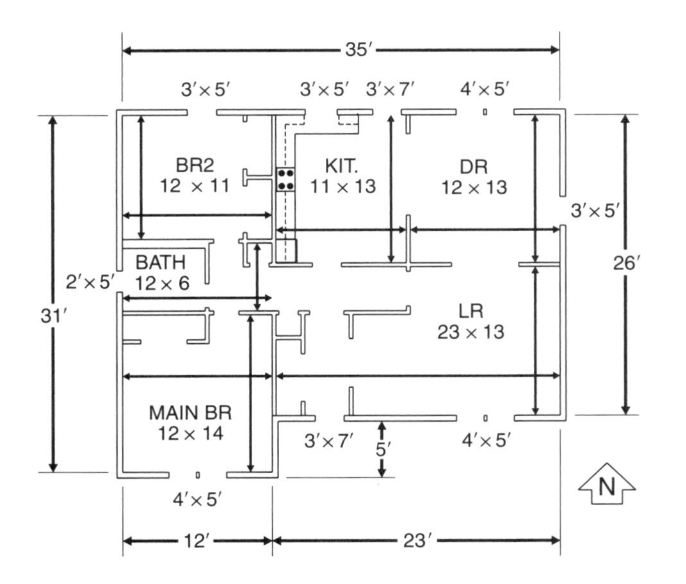

NOTES:
LOCATION – CHICAGO, ILLINOIS
WINDOWS – DOUBLE GLASS WITH STORM SASH
DOORS – WOOD WITH STORM DOOR
WALLS – WOOD FRAME WITH R-11 INSULATION
CEILING – R-44 INSULATION UNDER UNCONDITIONED SPACE
FLOOR – WOOD OVER VENTED SPACE WITH R-11 INSULATION

CHICAGO RESIDENCE

Figure 17-3. Information for load calculations is taken from sketches and notes describing a specific building.

The third method is the air change method. The air change method is based on the assumption that infiltrated air completely changes the air in a building. In the air change method, the volume of the building is found by multiplying the length times the width times the ceiling height of the building. The volume of the building is multiplied by a factor related to the number of times air change occurs per hour.

The number of people that occupy a building must also be determined. For a residence, the number of people that occupy a building is determined by the number of bedrooms in the building. The Chicago residence has two bedrooms. The main bedroom represents two people and the other bedroom represents one person. The number of people in the residence for load calculations is three. Enter this figure on the load form in the rooms that the people occupy during the day when cooling is required. This information is entered in the living room and kitchen columns of the building on the load calculation form.

The lighting load is usually not calculated for a residential building because the lights are not on during the day when the cooling load occurs. For residential applications, 1200 Btu/hr of heat gain is used for appliances and cooking processes. This amount is added to the kitchen gain for the cooling load. **See Figure 17-4.**

Contractor: City Heating & Cooling

Name of Job: Jones Residence

Address: 1875 W. Grant St.

Chicago, IL 60632

Date: 1/14/07

By: J.P.

Winter: Indoor Design Temp. ________ Outdoor Design Temp. ________ Design Temperature Difference ________

Summer: Outdoor Design Temp. ________ Indoor Design Temp. ________ Design Temperature Difference ________

1	Name of Room				1 LR			4 BR2			5 KIT./DINE			Building Component Subtotals	
2	Running Feet of Exposed Wall				36			23			36				
3	Room Dimensions				23′ x 13′			12′ x 11′			23′ x 13′				
4	Ceiling Height	Exposure			8′	S-E		8′	N-W		8′	N-E			
	Types		Factors		Area	Btu/hr	Btu/hr	Area	Btu/hr	Btu/hr	Area	Btu/hr	Btu/hr	Btu/hr	Btu/hr
	Exposure		H	C		H	C		H	C		H	C	H	C
5	Gross Exposed Wall Area	N						96			184				
		S			184										
		E			104						104				
		W						88							
6	Windows and Glass Doors (H)				20			15			50				
7	Windows and Glass Doors (C)	N						15			35				
		E & W									15				
		S			20										
8	Other Doors				21						21				
9	Net Exposed Walls				247			169			217				
10	Ceilings				299			132			299				
11	Floors				299			132			299				
12	People @ 300 Appliances @ 1200				2P		600				P+K		1500		

Figure 17-4. Variables used in calculating heating and cooling loads include design temperatures differences, area of building components, data related to infiltration, number of people occupying the building, and number of lights and appliances in a building.

Factors

Factors used in calculating heating and cooling loads include conduction factors, solar gain, infiltration factors, and factors for heat produced by people, lights, and appliances. Factors are taken from tables that have been developed for heating and cooling load calculations.

Heat transfer factors are used for calculating heat flow through individual components of a building. The heat transfer is due to a temperature difference between the two sides of a material. Tables of heat transfer factors are divided into separate sections based on building components.

Heating load factors are selected from a table of heat transfer factors. **See Appendix.** Factors for heat transfer through doors, windows, glass doors, walls, ceilings, and floors are entered on the load form according to the type of building material and design temperature difference.

Cooling load factors for walls, ceilings, and floors are selected from a table of cooling heat transfer factors and are entered on the load form. Cooling load factors for heat transfer through windows and doors are selected from a table of cooling heat transfer factors for doors and windows according to the exposure of the doors or windows and the difference between the indoor and outdoor temperatures. **See Appendix.** Infiltration factors are included with the wall, window, and door factors. Some factor tables include infiltration. Other tables require that infiltration be calculated separately. **See Figure 17-5.**

When using a table of infiltration factors that is based on crackage, the total amount of crackage (in feet) is multiplied by the appropriate factor. When using a table of infiltration factors that is based on flow rate of air, multiply the infiltration airflow rate, a factor of 1.08, and the winter design temperature difference to find heat loss or gain. The flow rate of infiltration air is the cooling load due to infiltration, but the actual cooling load is found on a psychrometric chart. The psychrometric chart gives a cooling load that includes the latent heat load for moisture in the air. Latent heat load for moisture in the air does not have to be considered in the heating load, but it must be considered in the cooling load. To find the heat loss for infiltration apply the formula:

$H = Q \times \Delta T \times 1.08$

where

H = heat transferred (in Btu/hr)

Q = flow rate (in cfm)

ΔT = temperature difference (in °F)

1.08 = constant

1	Name of Room			
2	Running Feet of Exposed Wall			
3	Room Dimensions			
4	Ceiling Height	Exposure		
	Types		Factors	
	Exposure		H	C
5	Gross	N		
	Exposed	S		
	Wall	E		
	Area	W		
6	Windows and Glass		75	
	Doors (H)			
7	Windows	N		14
	and Glass	E & W		44
	Doors (C)	S		23
8	Other Doors		205	8.6
9	Net Exposed Walls		6	1.7
10	Ceilings		2	.9
11	Floors		6	.8
12	People @ 300 Appliances @ 1200			

Figure 17-5. Heat transfer factors are used for calculating heat flow through individual components of a building due to temperature differences.

Example: Finding Heat Loss—Infiltration

Outdoor air at a dry bulb temperature of 45°F and a flow rate of 150 cfm is introduced into the heating system of a building for ventilation. The dry bulb temperature of the indoor air is 75°F. Find the heat loss for the air.

$H = Q \times \Delta T \times 1.08$

$H = 150 \times (75 - 45) \times 1.08$

$H = 150 \times 30 \times 1.08$

$H =$ **4860 Btu/hr**

The heat gain from people is not usually calculated in a heating load. Heat gain from people is usually disregarded in residential load calculations because relatively few people are included. For residential cooling applications, heat gain from people is 300 Btu/hr of sensible heat per person with a latent heat allowance added. This value is multiplied by the number of people and the resulting amount is entered on the load form in the corresponding spaces. The heat gain factor from appliances is 1200 Btu/hr and is added to the kitchen gains.

CALCULATIONS

After the variables and factors have been entered on a load form, calculations must be performed. The variables in the columns are multiplied by the factors in the factor column. The heating losses must be kept separate from the cooling gains. The results are the heating and cooling loads for each room or zone broken down into building components. Building components are the main part of a building structure such as the exterior walls, windows, doors, ceiling, and floor.

Subtotals

To find the subtotals for each room or zone, all of the losses or gains for the building components in a column are added together. The sum in the heating loss column is the heat loss for the room or zone. The sum in the cooling gain column is the sensible heat gain for each room or zone.

For a residence, the latent heat gain is 30% of the sensible heat gain. The total heat gain for each room or zone is the sensible heat gain plus the latent heat gain. To simplify the mathematical process of adding 30% to each sensible load, the load is multiplied by a factor of 1.3. *Note:* The component gains are for sensible heat gain only. To find the total component gain, multiply the sensible heat gains by 1.3. **See Figure 17-6.**

Totals

Total heat loss or gain for a building is the sum of all the room or zone subtotals. The sum is shown on the far right bottom of the load form. The total of all room or zone subtotals must equal the total of all building component subtotals.

ADJUSTMENTS

To obtain the actual heating and cooling loads for a building, adjustments have to be made to the totals. Adjustments for duct loss or gain are made to the totals. Adjustments for the actual heating or refrigeration effect of HVAC equipment are also made to the totals.

Ductwork or Piping

Heat lost or gained through the ductwork or piping that distributes air or water from the heating and/or cooling system is part of the total load for the building. If any part of the ductwork or piping system runs through an unconditioned space, heat will be lost in the ductwork or piping. This loss is added to the total heating load of the building.

Heat gain through ductwork or piping is usually not added in the calculation. Heat gain through ductwork or piping is normally not significant because of the lower temperature difference between the air or water and the material around it. Ductwork and piping in cooling systems should always be insulated to reduce heat gain. The load adjustment for loss through ductwork or piping is made before the heating equipment is selected.

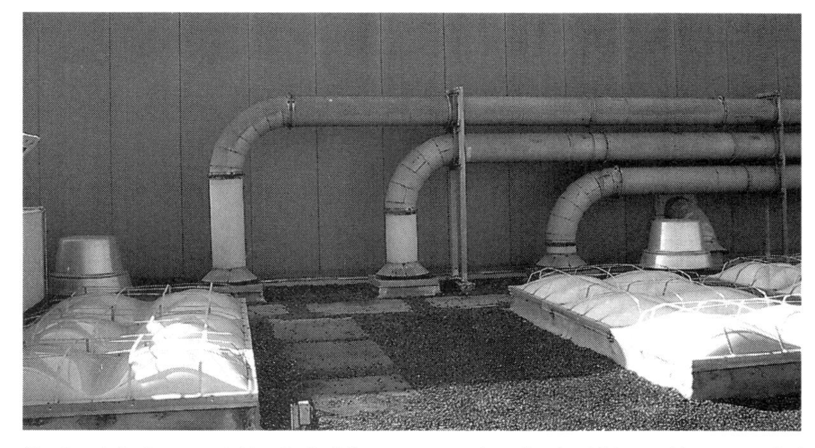

Ductwork that runs outside of a building space can lose heat, which must be accounted for when calculating the heating load.

Contractor: City Heating & Cooling

Name of Job: Jones Residence

Address: 1875 W. Grant St.
Chicago, IL 60632

Date: 1/14/07

By: J. P.

Winter: Indoor Design Temp. 75°F — Outdoor Design Temp. -9°F — Design Temperature Difference 84°F

Summer: Outdoor Design Temp. 90°F — Indoor Design Temp. 75°F — Design Temperature Difference 15°F

1	Name of Room				1 LR			2 MAIN BR			3 BATH			4 BR2			5 KIT./DR			Building Component Subtotals	
2	Running Feet of Exposed Wall				36			31			6			23			36				
3	Room Dimensions				23′ x 13′			12′ x 14′			12′ x 6′			12′ x 11′			23′ x 13′				
4	Ceiling Height	Exposure			8′	S-E		8′	W-S-E		8′	W		8′	N-E		8′	N-E			
	Types		Factors		Area	Btu/hr	Btu/hr	Area	Btu/hr	Btu/hr	Area	Btu/hr	Btu/hr	Area	Btu/hr	Btu/hr	Area	Btu/hr	Btu/hr	Btu/hr	Btu/hr
	Exposure		H	C		H	C		H	C		H	C		H	C		H	C	H	C
5	Gross Exposed Wall Area	N												96			184				
		S			184			96													
		E			104			40									104				
		W						112			48			88							
6	Windows and Glass Doors (H)		75		20	1500		20	1500		10	750		15	1125		50	3750		8625	
7	Windows and Glass Doors (C)	N		14										15		210	35		490		700
		E & W		44							10		440				15		660		1100
		S		23	20		460	20		460											920
8	Other Doors		205	8.6	21	4305	181		1368								21	4305	181	8610	362
9	Net Exposed Walls		6	1.7	247	1482	249	228	336	388	38	228	65	169	1014	287	217	1302	369	5394	1358
10	Ceilings		2	.9	299	598	269	168	1008	151	72	1444	65	132	264	119	299	598	269	1940	873
11	Floors		6	.8	299	1794	239	168		134	72	432	58	132	792	106	299	1794	239	5820	776
12	People @ 300 Appliances @ 1200				2P		600										P+K		1500		2100
																				Totals	
13	Subtotal Btu/hr Loss					9679			4212			1554			3195			11,749		30,389	
14	Sensible Btu/hr Gain						1998			1133			628			722			3708		8189
15	Subtotal Btu/hr Gain (1.3)						2597			1473			816			939			4820		10,646

Figure 17-6. Total heat loss or gain for a building is the sum of all the room or zone subtotals.

Load adjustment for ductwork or piping loss is found by using a table of duct heat loss multipliers that approximates the loss through the surfaces of the ductwork. A table of duct loss multipliers provides factors for finding ductwork heat loss.

A table of duct heat loss multipliers shows the increase in loss or gain as a percentage of the calculated loss or gain. **See Appendix.** To use the table, select the location and insulation of the ductwork and the winter outdoor design temperature. The ductwork heat loss multiplier value is multiplied by the building heating load. The result is an adjusted total that considers heat loss due to ductwork. Loss due to ductwork is found by applying the formula:

$$L_{DW} = TL \times DL$$

where

L_{DW} = loss due to ductwork (in Btu/hr)

TL = total loss (in Btu/hr)

DL = ductwork heat loss multiplier

Example: Finding Adjusted Total Due to Duct Heat Loss

A building has a total heat loss of 57,194 Btu/hr. The ductwork in the building runs the full length of the building in an unheated crawlspace. The ductwork is insulated with R-4 insulation. The outdoor design temperature is below 15°F. Find the adjusted heat loss due to ductwork.

On a table of duct loss multipliers, the duct loss multiplier is 1.10. **See Appendix.**

$$L_{DW} = TL \times DL$$

$$L_{DW} = 57{,}194 \times 1.10$$

$$L_{DW} = \mathbf{62{,}913.4\ Btu/hr}$$

Equipment Capacity

Heating and cooling equipment must have enough output capacity to satisfy the heating and cooling load on a building. The distribution system for a job must be sized for the equipment output capacity and not for the calculated load.

Since furnaces and air conditioners are manufactured in a limited selection of sizes, the equipment chosen for a job is usually oversized for the loads. Regardless of the capacity, the equipment is designed by the manufacturer to operate efficiently with a given supply of air or water circulating through it. If the heating and/or cooling equipment is oversized, the components of the distribution system must be oversized to fit the equipment.

To oversize the system components, an equipment ratio factor is used. The equipment ratio factor is found by dividing the equipment capacity by the calculated load. The ratio is then multiplied by the heating and cooling load for each room to adjust for the output capacity of the equipment. This allows the distribution system to be designed for the output capacity of the equipment. The sheet containing the equipment adjustment ratios becomes part of the permanent record of the calculations for the job. Heating equipment ratio is found by applying the formula:

$$R_{HE} = \frac{O_F}{T_L}$$

where

R_{HE} = heating equipment ratio

O_F = furnace output capacity (in Btu/hr)

TL = total load (in Btu/hr)

Example: Finding Heating Equipment Ratio

The total heating load for a building is 62,913 Btu/hr. The furnace with an output closest to and above 62,913 Btu/hr is a 75,000 Btu/hr model. Find the heating equipment ratio.

$$R_{HE} = \frac{O_F}{T_L}$$

$$R_{HE} = \frac{75{,}000}{62{,}913.4}$$

$$R_{HE} = \mathbf{1.19}$$

After the air conditioning equipment has been selected for a job, the actual refrigeration effect of the equipment is compared with the calculated heat gain. The cooling capacity of the air conditioner is divided by the total calculated heat gain to find the cooling equipment ratio. The room loads are multiplied by this ratio to find the air conditioning equipment adjustment factor.

Cooling equipment ratio is found by applying the formula:

$$R_{CE} = \frac{O_{AC}}{TG}$$

where

R_{CE} = cooling equipment ratio

O_{AC} = air conditioner output capacity (in Btu/hr)

TG = total heat gain (in Btu/hr)

Example: Finding Cooling Equipment Ratio

The total heat gain for a building is 17,559 Btu/hr. The air conditioner with an output closest to and above 17,559 Btu/hr is a 24,000 Btu/hr model. Find the cooling equipment ratio.

$$R_{CE} = \frac{O_{AC}}{TG}$$

$$R_{HE} = \frac{24{,}000}{17{,}559}$$

$$R_{HE} = \mathbf{1.37}$$

Example: Finding Heating and Cooling Load—Conventional Method

A sketch of a residence contains the dimensions required for calculating the heating and cooling load of the building. **See Figure 17-7.** Notes on the sketch describe construction features that affect the loads. The residence is located in Seattle, Washington. The dining room and kitchen are combined because they are not divided by a partition. The baths and utility room are combined because they are so small.

1. Enter job data at the top of the load form.
2. Enter indoor design temperatures, outdoor design temperatures, and design temperature differences on the load form. The outdoor design temperature is taken from an outdoor design temperature table under the listing for Seattle, Washington. The winter design temperature is 17°F db and the summer design temperature is 84°F db. **See Appendix.**
3. Enter names of rooms, running feet of exposed wall, room dimensions, ceiling height, and the exposure of the individual rooms in the appropriate areas on the load form. **See Figure 17-8.**
4. Find the gross exposed wall area for each room.
5. Find the area of windows and glass doors. Enter the area on the form for the heating load. Enter the area on the form according to exposure for the cooling load. Calculate area of other doors and enter on the form.
6. Find net exposed wall area for the individual rooms by subtracting the area of the windows from the gross exposed wall area. Enter the net exposed wall area on the form.
7. Find area of ceilings and floors. Enter the area on the form.
8. Determine the number of people occupying the building based on the number of bedrooms. A heat gain factor of 300 Btu/hr per person is added to the dining room and living room.

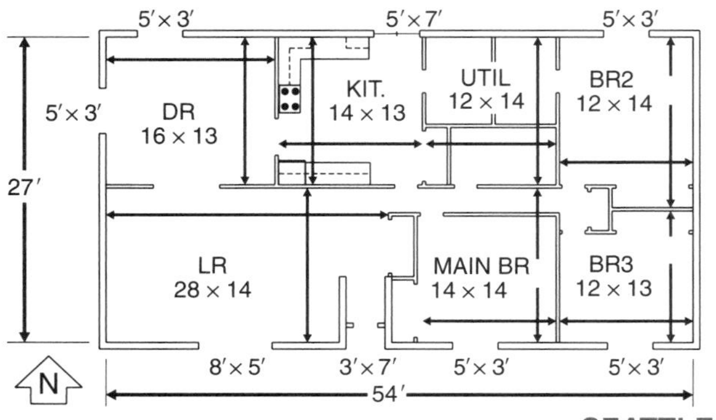

NOTES:
LOCATION – SEATTLE, WASHINGTON
WINDOWS – HORIZONTAL SLIDE, DOUBLE GLASS
DOORS – WOOD WITH STORM DOOR IN LIVING ROOM
DOUBLE SLIDING GLASS DOOR IN KITCHEN
WALLS – WOOD FRAME WITH SHEATHING AND SIDING WITH R-11 INSULATION
CEILING – UNDER UNCONDITIONED SPACE WITH R-38 INSULATION
FLOOR – WOOD OVER VENTED SPACE WITH R-11 INSULATION

SEATTLE RESIDENCE

Figure 17-7. A sketch of a residence shows the dimensions that are needed for heating and cooling load calculations. Notes on the sketch describe construction features that affect the loads.

Contractor:	Best Heating & Cooling		
Name of Job:	Patrick Residence	Date:	1/6/07
Address:	1541 N. Johnson	By:	G.H.
	Seattle, WA. 98031		

Winter: Indoor Design Temp. 75°F	Outdoor Design Temp. 17°F	Design Temperature Difference 58°F
Summer: Outdoor Design Temp. 84°F	Indoor Design Temp. 75°F	Design Temperature Difference 9°F

1	Name of Room				1 LR			4 BR3			5 MAIN BR			Building Component Subtotals	
2	Running Feet of Exposed Wall				42			25			14				
3	Room Dimensions				28′ x 14′			12′ x 13′			14′ x 14′				
4	Ceiling Height	Exposure			8′	S-W		8′	S-E		8′	S			
	Types Exposure		Factors H	Factors C	Area	Btu/hr H	Btu/hr C	Area	Btu/hr H	Btu/hr C	Area	Btu/hr H	Btu/hr C	Btu/hr H	Btu/hr C
5	Gross Exposed Wall Area	N													
		S			224			96			112				
		E						104							
		W			112										
6	Windows and Glass Doors (H)				40			15			15				
7	Windows and Glass Doors (C)	N													
		E & W													
		S			40			15			15				
8	Other Doors				21										
9	Net Exposed Walls				275			185			97				
10	Ceilings				392			156			196				
11	Floors				392			156			196				
12	People @ 300 Appliances @ 1200				2P		600								

Figure 17-8. The areas of each of the rooms, windows and glass doors, other doors, net exposed wall area, ceilings, and floors in the building are entered on the load form.

9. Determine the heat gain from appliances. A heat gain factor of 1200 Btu/hr is added to the kitchen of the residence.
10. Find factors for the heating load from a table of heat transfer factors. Find factors for cooling load from table of cooling heat transfer factors. **See Appendix. See Figure 17-9.**
11. Find heating and cooling load for each room by multiplying factors by variables for each room.
12. Find building component subtotals by adding the heating and cooling loads for all rooms in the building. **See Figure 17-10.**
13. Find the individual room loss or gain subtotals by adding the building component loss or gains for each room.
14. Make adjustments for ductwork and piping.
15. Find equipment adjustment ratios.

The required heating capacity is 25,293 Btu/hr. The heating unit with the nearest output is 30,000 Btu/hr. The heating adjustment ratio is 1.19. This value is multiplied by the subtotal heat loss for each room for designing the heating distribution system.

The cooling capacity required is 14,026 Btu/hr. The air conditioning unit with the nearest output is 24,000 Btu/hr.

The cooling equipment ratio is 1.71. This value is multiplied by the subtotal heat gain for each room for designing the air conditioning distribution system.

1	Name of Room			
2	Running Feet of Exposed Wall			
3	Room Dimensions			
4	Ceiling Height	Exposure		
	Types		Factors	
	Exposure		H	C
5	Gross	N		
	Exposed	S		
	Wall	E		
	Area	W		
6	Windows and Glass		60	
	Doors (H)			
7	Windows	N		14
	and Glass	E & W		44
	Doors (C)	S		23
8	Other Doors		145	8.6
9	Net Exposed Walls		4	1.7
10	Ceilings		2	1
11	Floors		4	.8
12	People @ 300 Appliances @ 1200			

Figure 17-9. The factors for the heating load are found on a table of heat transfer factors. The factors for the cooling load are found on a table of cooling heat transfer factors.

LOAD CALCULATIONS—COMPUTER-AIDED

Computer-aided load calculations are heating and cooling load calculations that are performed by a computer. A load calculation program is used to calculate heating and cooling loads for a building.

A *load calculation software program* is a series of commands that requests data from the operator and manipulates the data to determine the heating and cooling loads. The process is similar to a conventional load calculation, except that the computer does the calculations. The data about a building is entered and recorded on a disk. The data is converted to a format that is used for the calculations of the loads by the computer system.

A load calculation program requests the same variables and factors that are used for a conventional load calculation. The speed of the computer greatly reduces the amount of time required to perform the calculations. A person who operates load calculation programs should be familiar with conventional load calculation procedures.

Load calculation programs are available from engineering and technical associations, manufacturers of heating and air conditioning equipment, and other companies. An instruction manual is included in each program package. The instruction manual describes preparations that must be made to run the program and how to run the program.

Most load calculation program instructions include a job information sheet. A job information sheet provides spaces for all information that is required for the calculations. The information includes the job name, date, design temperatures, dimensions, building materials, and other information. A job information sheet should be completed before running the program. Data from the sheet is then input to the computer.

When all of the information requested by the program has been inputted, the computer calculates the loads. When the calculations are complete, the results are shown on the monitor. Some computer load calculation programs print the results automatically.

In most load calculation programs, questions usually begin by requesting general information such as the job, geographical location, and design temperatures. The next series of questions request the sizes of the building, rooms, or zones. Some programs have factors stored on the CD-ROM. Questions relating to factors are about construction methods and materials. Programs that do not have factors on the CD-ROM ask for many factors. To accurately calculate a load, the factors are obtained from charts and tables.

As a load calculation program runs, the operator inputs responses to questions that appear on the monitor. Explanations about procedures appear on the monitor, and many programs have options that give help when requested. A program instruction manual should answer any questions that the operator may have about the program. An instruction sheet that shows step-by-step procedures for running a program is usually included with a program.

Contractor: Best Heating & Cooling

Name of Job: Patrick Residence

Address: 1541 N. Johnson
Seattle, WA. 98031

Date: 1/6/07

By: G.H.

Winter: Indoor Design Temp. 75°F — Outdoor Design Temp. 17°F — Design Temperature Difference 58°F

Summer: Outdoor Design Temp. 84°F — Indoor Design Temp. 75°F — Design Temperature Difference 9°F

1	Name of Room				1 LR			2 DR/KIT.			3 UTIL			4 BR2			5 BR3			6 MAIN BR			Building Component Subtotals	
2	Running Feet of Exposed Wall				42			43			12			26			25			14				
3	Room Dimensions				28′ x 14′			30′ x 13′			12′ x 13′			12′ x 14′			12′ x 13′			14′ x 14′				
4	Ceiling Height	Exposure			8′	S-W		8′	N-W		8′	N		8′	N-E		8′	S-E		8′	S			
	Types		Factors		Area	Btu/hr	Btu/hr	Area	Btu/hr	Btu/hr	Area	Btu/hr	Btu/hr	Area	Btu/hr	Btu/hr	Area	Btu/hr	Btu/hr	Area	Btu/hr	Btu/hr	Btu/hr	Btu/hr
	Exposure		H	C		H	C		H	C		H	C		H	C		H	C		H	C	H	C
5	Gross	N						240			96			96										
	Exposed	S			224												96			112				
	Wall	E												112			104							
	Area	W			112			104																
6	Windows and Glass		60		40	2400		65	3900					15	900		15	900		15	900		9000	
	Doors (H)																							
7	Windows	N		14				50		700				15		210								910
	and Glass	E & W		44				15		660														660
	Doors (C)	S		23	40		920										15		345	15		345		1610
8	Other Doors		145	8.6	21	3045	181																3045	181
9	Net Exposed Walls		4	1.7	275	1100	468	279	1116	474	96	384	653	193	772	328	185	740	315	97	388	165	4500	2403
10	Ceilings		2	1	392	784	392	390	780	390	156	312	156	168	336	168	156	312	156	196	392	196	2916	1458
11	Floors		4	.8	392	1568	314	390	1560	312	156	624	125	168	672	134	156	624	125	196	784	157	5832	1167
12	People @ 300 Appliances @ 1200				2P		600	2P+K		1800					2									2400
																							Totals	
13	Subtotal Btu/hr Loss					8897			7356			1320			2680			2576			2464		25,293	
14	Sensible Btu/hr Gain						2875			4336			934			840			941			863		10,789
15	Subtotal Btu/hr Gain (1.3)						3738			5637			1214			1092			1223			1122		14,026

Figure 17-10. Total heat loss or gain for a building is the sum of all the room or zone subtotals.

After loads have been calculated, the results are shown on the monitor. The display includes a breakdown that shows subtotals for the components and for individual rooms or zones. Some programs also display information relating to the job variables, such as areas by exposure, infiltration data, and people and equipment input data. This information is used for checking the accuracy of the loads. A printout of the loads is made at this point. The printout may be used as a worksheet for checking the loads and is a permanent file copy.

The results of most load calculation programs are saved on a CD-ROM. Saving the job data is important when results need to be checked or modified. Several small jobs may be saved on one CD-ROM. Large jobs should be saved on individual CD-ROMs.

Analysis

After load calculations for a building have been run, the results of the calculations should be checked for accuracy. Accuracy is important because the calculations are used to design the entire heating or cooling system of a building. Errors can occur if information is input incorrectly. By comparing the results of the computer load calculations with estimated loads, accuracy is assured.

TSI Incorporated

Measurements taken to check system balance should confirm load calculations.

Total Loads. An analysis of computed loads usually starts with a check of the total loads. The total loads from the computer program can be compared with a standard for the type and location of the building or with a load calculated with the conventional method for a part of the building. Comparisons are made by preparing a sheet showing the results from the computer calculations and the results from a standard or conventional calculation method. **See Figure 17-11.**

Because different factors are used for each load calculation method, there is usually some difference between computer-aided load figures and conventional load figures. The totals of a computer-aided load calculation should be within 5% of the totals taken from a reliable standard. If the totals are not within 5%, the calculations should be checked further to find the reason for the difference.

Subtotals. Most commercial computer programs are written so a printout of the results shows subtotals of losses and gains for the rooms or zones in a building. These subtotals should be checked against a reliable standard to see if any errors appear. A positive error in one component can be offset by a negative error in another. If there is a noticeable difference between the computer subtotals and the standard, the component factors and variables used to get the subtotal should be checked. The best way to compare subtotals is by listing them on a comparison sheet.

Component Totals. If the computer printout shows component losses and gains, the component totals can be checked to see if any noticeable differences appear. A quick check of the component loads may indicate a higher or lower load on one component in a particular part of the building.

Variables and Factors. Make sure all data on the job information sheet is correct, and verify the data on the job information sheet against the data inputted in the computer. The computer program calculates component subtotals the same way they are calculated in a conventional calculation. Variables are multiplied by factors and the results are added.

Components		Heat Gain			Heat Loss		
		Conventional	Computer		Conventional	Computer	
Walls		1371	1128		3652	4361	
Windows	Includes solar gain and infiltration	7155	6021	Includes solar gain and infiltration	10,500	6191	
Doors		158	756		3045	2923	
Ceiling		1054	874		2108	3376	
Floor		528	611		4216	4725	
Infiltration			840			3240	
People & Appl.		2400	1200 + 871				
Subtotals		12,666	12,301				
		× 1.3	× 1.3				
Totals		16,466	15,991		23,521	24,816	

Figure 17-11. Comparisons for checking the computer-aided load calculations are made by preparing a worksheet that shows the results from the computer calculations and the results from a standard or a conventional calculation method.

Some load calculation programs include heat loss for ductwork and some do not. For many programs, adjustments must be made to the calculated loss data for the actual capacity of the equipment. This step must be done for a conventional calculation.

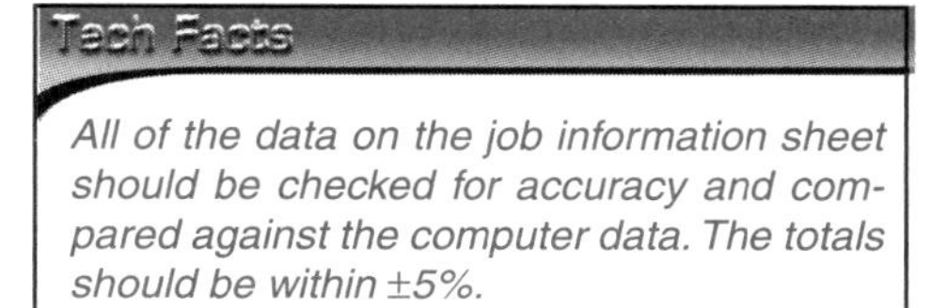

All of the data on the job information sheet should be checked for accuracy and compared against the computer data. The totals should be within ±5%.

Review Questions

1. List two ways that heating and cooling load calculations can be performed.
2. List two different load forms that are used for calculating heating and cooling loads.
3. Why should a cooling load form be filled in first when preparing forms for calculating loads?
4. List at least five of the six variables that are considered in a load calculation.
5. Which variable is not included in heating loads, but must be considered when calculating a cooling load?
6. Define design temperature.
7. Where are design temperatures found?
8. How is information that is needed for calculating heating and cooling loads obtained when plans and specifications for a building are not available?
9. Explain the difference between gross wall area and net wall area.

10. How are hallways, closets, and stairways included in load calculations?
11. List and describe three ways that infiltration loads are calculated.
12. What does air change mean in relation to infiltration loads?
13. How many people would be included in the calculations for the heating and cooling loads for a three-bedroom residence?
14. A heat transfer factor combines the difference between indoor and outdoor temperatures with what other element?
15. Why is the heating load for people not usually included in a residential heating load?
16. Where are factors included on a heating and cooling load form?
17. What is a building component?
18. How is sensible heat gain used to account for latent heat gain?
19. List and explain two adjustments that may have to be made to a heating or cooling load after the loads are calculated.
20. Why is the heating or cooling equipment usually oversized when it is chosen for a job?
21. Why do the heating and cooling loads have to be adjusted for the equipment size before a system is designed?
22. Describe how an equipment ratio is found.
23. What is a computer-aided load calculation?
24. What program is used for calculating heating and cooling loads?
25. How can errors occur when running a computer program?

A forced-air system uses air to carry heat. Air is circulated to building spaces through a distribution system from a furnace or an air conditioner. All of the parts of a distribution system must be designed to circulate the required amount of air to building spaces. The air offsets the heat gain or cooling load in each building space. Forced-air distribution systems for heating and cooling applications are similar.

18 Forced-Air System Design

AIR DISTRIBUTION SYSTEMS

Forced-air distribution systems consist of a blower, supply ductwork, registers, return ductwork, and grills. The distribution system circulates warm air from a furnace or cool air from an air conditioner to building spaces. The ductwork consists of supply and return ducts that are made of sheet metal, fiberglass, or fabric. **See Figure 18-1.**

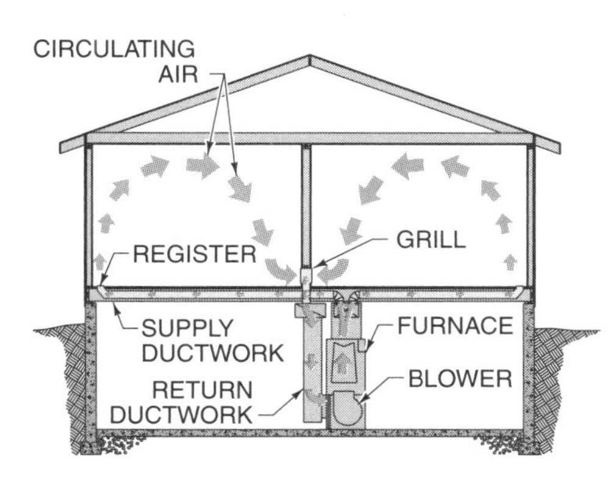

Figure 18-1. A forced-air distribution system consists of a blower, supply and return ductwork, registers, and grills.

Principles of Air Flow

The rate of air flow through a duct is a function of the air velocity, size of the duct, and friction loss. Air flow is expressed in cubic feet per minute. Duct dimensions are expressed in inches. Friction loss is expressed in inches of water column.

Air Flow Rate. Air flow rate is the volumetric flow rate of the air in the ductwork. Volumetric flow rate of air is the quantity of air moved per unit of time. The total air flow rate required for a forced-air system is a function of the capacity of the heating and cooling equipment. The air flow rate required for a heating system is determined by the temperature difference of the air through the heat exchanger of a furnace. The air flow rate required for an air conditioning system is determined by the temperature difference of the air through the evaporator coil of an air conditioner.

Ductsox

Fabric supply ductwork is used to distribute warm or cool air from a furnace or air conditioner to various building spaces.

A forced-air distribution system is designed to move air at the air flow rate that provides comfort to each building space. The air flow rate in each building space should be proportional to the heating and/or cooling load in each building space.

The heating and cooling load calculations for a building are used to find the air flow rate for each building space. The heating and cooling loads should include duct loss and duct gain and should be adjusted for the size of equipment. For a combination heating and cooling system, the larger value is used for duct sizing. The air flow rate for a heating load is found by applying the formula:

$$Q = \frac{L_h}{\Delta T \times 1.08}$$

where
Q = air flow rate (in cfm)
L_h = heating load (in Btu/hr)
ΔT = temperature difference (in °F)
1.08 = constant

Example: Finding Air Flow Rate—Heating Load

A room has an adjusted heating load of 13,300 Btu/hr. A temperature difference of 80°F is desired. Find the air flow rate required to provide comfort in the room.

$$Q = \frac{L_h}{\Delta T \times 1.08}$$

$$Q = \frac{13{,}300}{80 \times 1.08}$$

$$Q = \frac{13{,}300}{86.4}$$

Q = **153.94 cfm**

To find the air flow rate required to cool a room, divide the heat gain (cooling load) for the room by 30. Air conditioners are designed to have 400 cfm of air flowing across the evaporator coil per ton of cooling. With 12,000 Btu/hr per ton of cooling and 400 cfm of air per ton of cooling, 1 cfm of air equals 30 Btu/hr (12,000 ÷ 400 = 30). Dividing the heat gain for a room by 30 gives the air flow rate required to provide comfort in the room. The air flow rate required to cool a room is found by applying the formula:

$$Q = \frac{L_c}{30}$$

where
Q = air flow rate (in cfm)
L_c = cooling load (in Btu/hr)
30 = constant

Example: Finding Air Flow Rate—Cooling Load

A room has an adjusted heat gain of 7135 Btu/hr. Find the air flow rate required to provide comfort in the room.

$$Q = \frac{L_c}{30}$$

$$Q = \frac{7135}{30}$$

Q = **237.83 cfm**

Air Velocity. *Air velocity* is the speed at which air moves from one point to another. Proper circulation in building spaces depends on the correct air velocity. Proper circulation prevents temperature stratification and drafts. Temperature stratification is the variation of air temperature in a building space that occurs when warm air rises to the ceiling and cold air drops to the floor.

When temperature stratification occurs, individuals become uncomfortably warm due to insufficient air circulation. Drafts occur when the air velocity is above 40 fpm. Individuals become uncomfortably cool due to the high rate of evaporation of perspiration from the skin. The air velocity in a duct is a function of the air flow rate and duct size. Air velocity is found by applying the formula:

$$V = \frac{Q}{A}$$

where
V = air velocity (in fpm)
Q = air flow rate (in cfm)
A = area of duct (in sq ft)

Example: Finding Air Velocity

The blower in a forced-air distribution system moves 1200 cfm of air through a 24″ × 12″ duct. Find the air velocity.

$$V = \frac{Q}{A}$$

$$V = \frac{1200}{2}$$

$$V = \mathbf{600\ fpm}$$

Duct Size. Ducts carry the required air flow rate to building spaces to provide comfort. *Duct size* is the size of a duct, which is expressed in inches of diameter for round ducts and in inches of width and height for rectangular ducts. In dimensions for rectangular ducts, the width should always be given first. For example, a 20″ × 8″ duct is a duct 20″ wide and 8″ high.

Friction Loss. *Friction loss* is the decrease in air pressure due to the friction of the air moving through a duct. Friction loss occurs as air scrubs against the internal surfaces of a duct. Friction loss is due to the turbulence caused by a film of air moving along the surface of the duct. Friction loss is indicated by static pressure drop.

Static Pressure. *Static pressure* is air pressure in a duct measured at right angles to the direction of air flow. Static pressure is the pressure that has a tendency to burst a duct. *Static pressure drop* is the decrease in air pressure caused by friction between the air moving through a duct and the internal surfaces of the duct. **See Figure 18-2.**

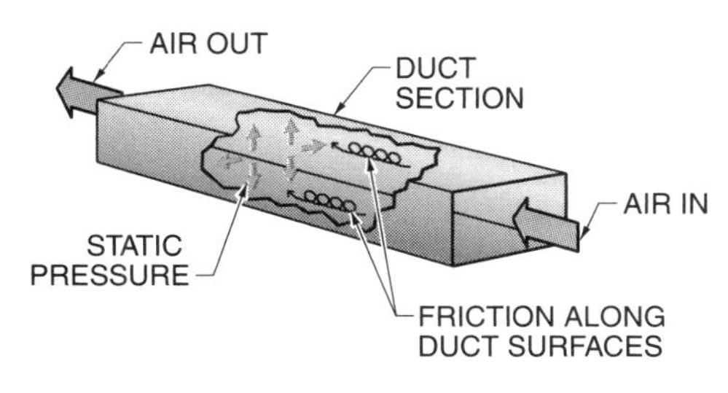

Figure 18-2. Friction loss occurs as air scrubs against the internal surfaces of a duct.

Static pressure is measured with a manometer. A *manometer* is a device that measures the pressures of vapors and gases. The two types of manometers commonly used to measure air pressure in ducts are U-tube and inclined manometers. A *U-tube manometer* is a U-shaped section of glass or plastic tubing that is partially filled with water or mercury.

The liquid in the two legs of a U-tube manometer remains level due to gravity when the pressure on the two legs is equal. If more pressure is exerted on the air in one leg, the liquid in that leg is pushed down. As the liquid in one leg goes down, the liquid in the other leg rises. The difference in the liquid levels is a measurement of air pressure. **See Figure 18-3.**

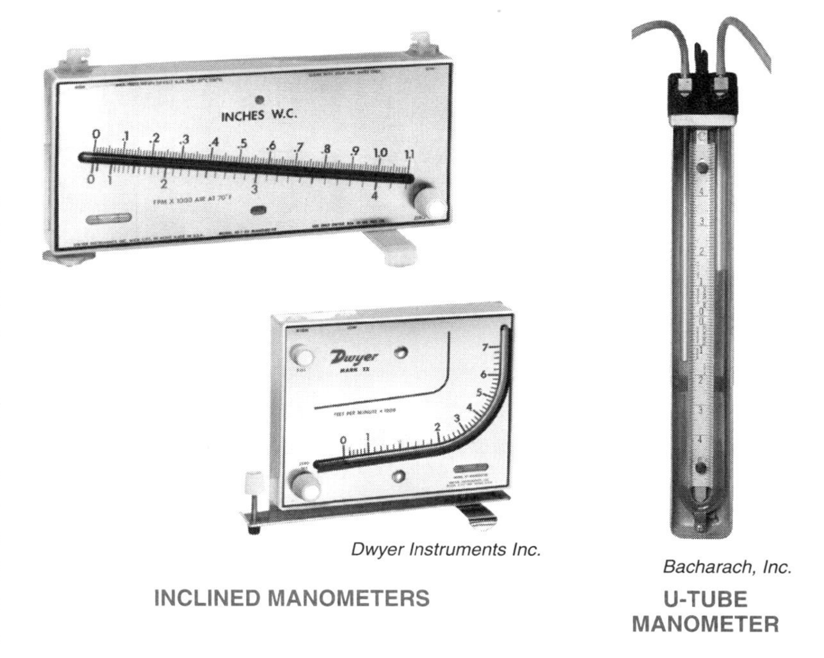

Figure 18-3. The difference between the liquid levels in the two legs of a manometer is a measurement of air pressure.

When using a manometer, air pressure is measured in inches of water column. Water column (WC) is the pressure exerted by a square inch of a column of water. Inches of water column are used to express small pressures above and below atmospheric pressure. One inch of water column equals .036 psi.

An *inclined manometer* is a U-tube manometer designed so the bottom of the "U" is a long inclined section of glass or plastic tubing. The liquid in an inclined manometer fills part of the inclined tube and part of one leg.

A change in air pressure in one leg of an inclined manometer causes the liquid to move along the inclined tube. The inclined tube is calibrated along the long slope of the incline. A small change in pressure moves the liquid a long distance along the incline. The calibration of an inclined manometer makes it possible to read pressure differentials on an incline manometer more accurately than on a U-tube manometer.

Static pressure drop in a duct reduces the pressure of the air as the air moves along the duct. To measure static pressure in a duct, a small hole is drilled or punched in the side of the duct, and a flexible tube is run from the hole to one leg of a manometer. The difference between the static pressure readings at two points in a duct is the friction loss that has occurred in the duct between the two points.

Static pressure drop is the difference between the static pressure at the beginning of a duct section and the static pressure at the end of the duct section. Static pressure drop in a duct section is found by dividing the length of the duct section by 100. The length of the duct section is divided by 100 because the design static pressure drop for the distribution system is per 100′ of duct. The *design static pressure drop* is the pressure drop per unit length of duct for a given size of duct at a given air flow rate. The result is multiplied by the design static pressure drop. The duct section pressure drop is found by applying the formula:

$$P_d = \frac{L}{100} \times p_d$$

where

P_d = static pressure drop (in in. WC)

L = length of duct section (in ft)

100 = constant

p_d = design static pressure drop (in in. WC)

Example: Finding Duct Static Pressure Drop

The air flow rate in a section of duct 140′ long is 1500 cfm. The duct section has a design static pressure drop of .08″ WC. The dimensions of the duct are 20″ × 12″. Find the static pressure drop of the duct section.

$$P_d = \frac{L}{100} \times p_d$$

$$P_d = \frac{140}{100} \times .08$$

$$P_d = 1.4 \times .08$$

$$P_d = \mathbf{.112'' \ WC}$$

The friction loss in a duct is directly related to the surface area exposed to air flow and the air flow rate. An *equal friction chart* is a chart that shows the relationship between the air flow rate, static pressure drop, duct size, and air velocity. **See Figure 18-4.**

An equal friction chart is used to size a duct when two values are given. To find the duct size required to provide an air flow rate at a given static pressure drop, apply the procedure:

1. Locate the air flow rate on the bottom of the equal friction chart and draw vertical line from that point.
2. Locate the static pressure drop on the left side or top of the chart and draw a horizontal line from that point.
3. Locate the size of the duct and the air velocity in the duct at the intersection of the two lines on the graph.

An equal friction chart shows the sizes of round ducts. A conversion table must be used to find the size of square or rectangular ducts that provide the same static pressure drop. **See Appendix.**

Example: Using Equal Friction Chart

The air flow rate in a section of duct is 1200 cfm. The static pressure drop is .10″ WC. Find the size of the duct using an equal friction chart.

1. Locate 1200 cfm on the scale at the bottom of the chart. Draw a vertical line from this air flow rate. **See Figure 18-5.**
2. Locate .10″ WC on the scale on the left side of the chart. Draw a horizontal line across the chart from this static pressure drop.
3. From the point where the lines intersect, draw a line parallel with the diagonal duct diameter lines to the scale at the top of the chart. The size of a round duct is approximately 13″.

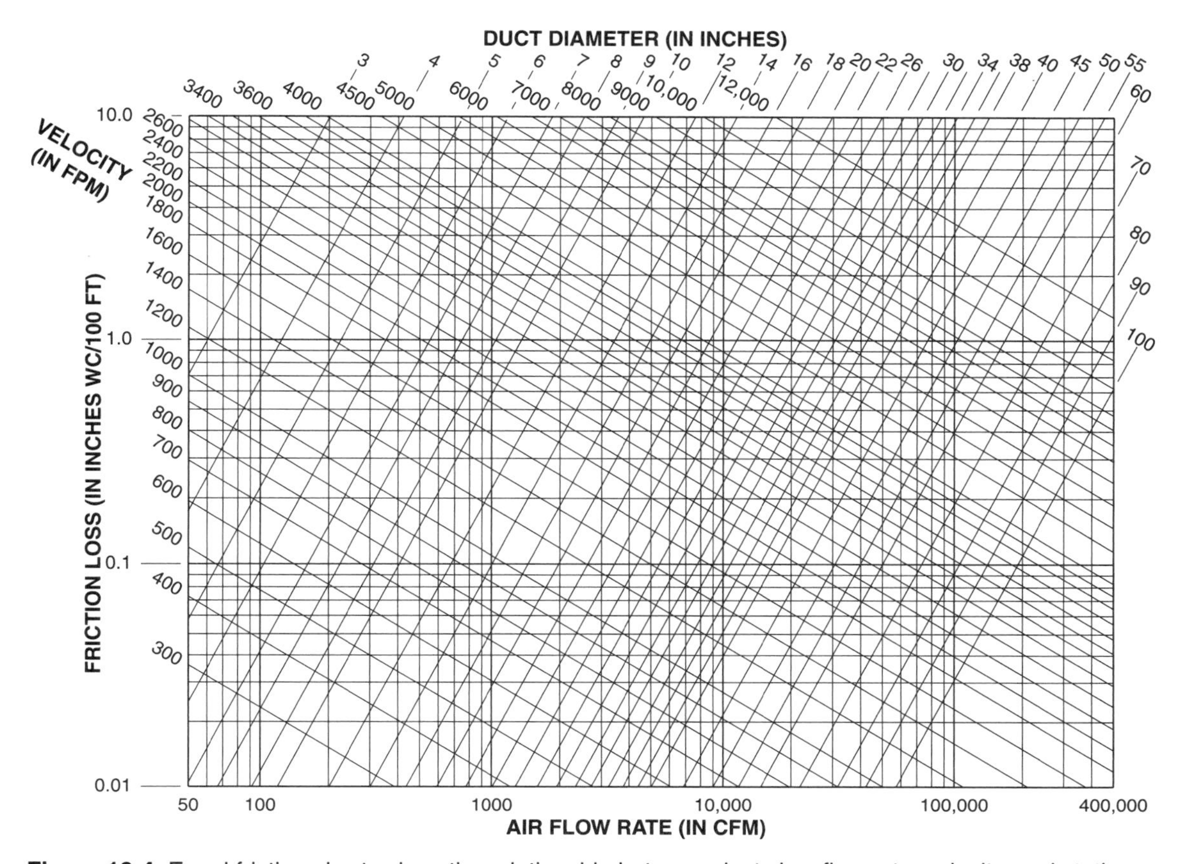

Figure 18-4. Equal friction charts show the relationship between duct size, flow rate, velocity, and static pressure drop.

Slide rules (ductulators), digital calculators, and computer programs can also be used for sizing ductwork. Each of these devices shows the same relationships as an equal friction chart.

System Pressure Drop. *System pressure drop* is the total static pressure drop in a system. Total static pressure drop in a forced-air distribution system is the sum of the static pressure drop through all duct sections, fittings, transitions, and accessories such as filters or coils, which add resistance to air flow in the distribution system.

Many forced-air distribution systems have several major duct sections. A *major duct section* is an independent part of a forced-air distribution system through which all or part of the air supply from the blower flows. A major duct section must be considered when system pressure drop is calculated. An *individual duct section* is part of a distribution system between fittings in which the air flow, direction, or velocity changes due to the configuration of the duct. **See Figure 18-6.**

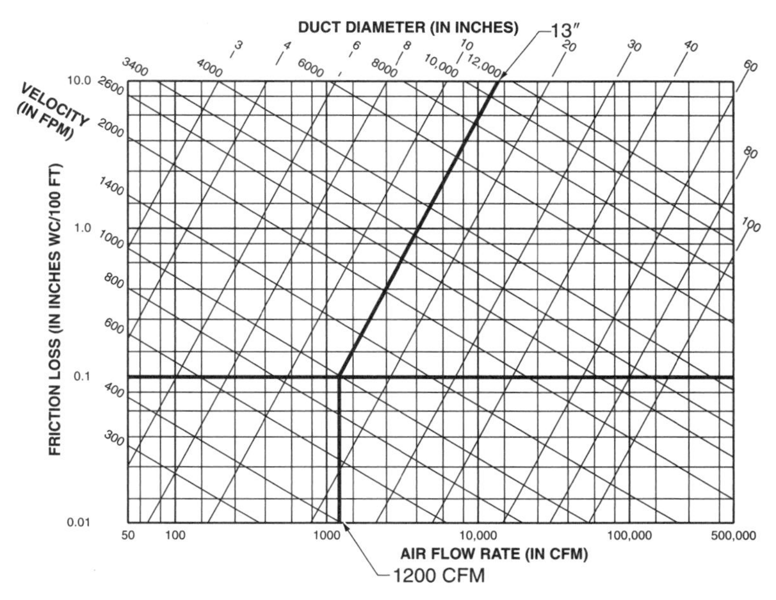

Figure 18-5. The size of a duct is found at the intersection of the air flow rate and the static pressure drop lines on the equal friction chart.

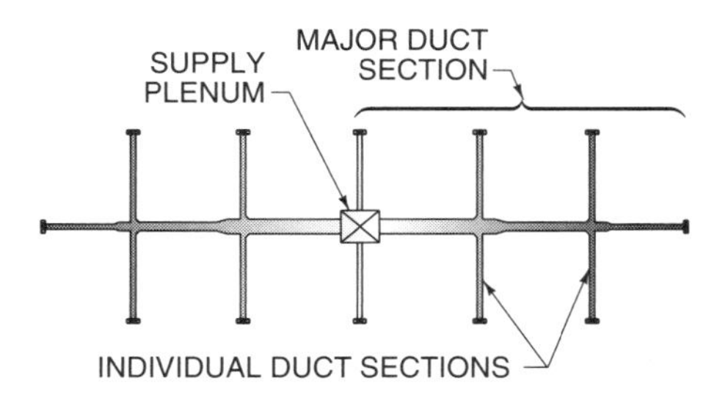

Figure 18-6. A duct section is part of a distribution system between fittings where the air flow, direction, or velocity changes due to the configuration of the duct.

Dynamic Pressure Drop. *Dynamic pressure drop* is the pressure drop in a duct fitting or transition caused by air turbulence as the air flows through the fitting or transition. Turbulence occurs wherever the air flow pattern is disturbed by a change in the air flow rate, direction of air flow, or size of the duct. Turbulence causes a drop in velocity pressure in the duct at the point where turbulence occurs.

Static pressure drop through a fitting or transition is calculated with a table of equivalent lengths or a table of fitting loss coefficients. A *table of equivalent lengths* is a table used to convert dynamic pressure drop to friction loss in feet of duct. A table of equivalent lengths contains the length of duct that gives the same static pressure drop as a particular metal fitting or transition. **See Figure 18-7.**

Velocity pressure is the air pressure in a duct that is measured parallel to the direction of air flow. Measuring velocity pressure without measuring static pressure is difficult. Velocity pressure is usually found by measuring total pressure and static pressure and subtracting the static pressure from the total pressure. The velocity pressure of the air entering a fitting is found by applying the formula:

$$P_v = \left(\frac{V}{4005}\right)^2$$

where

P_v = velocity pressure (in in. WC)

V = air velocity (in fpm)

4005 = constant

Example: Finding Velocity Pressure

A duct fitting handles 1200 cfm of air. The air velocity is 1500 fpm. Find the velocity pressure of the air entering the fitting.

$$P_v = \left(\frac{V}{4005}\right)^2$$

$$P_v = \left(\frac{1500}{4005}\right)^2$$

$$P_v = .3745^2$$

$$P_v = .3745 \times .3745$$

$$P_v = \mathbf{.1403'' \ WC}$$

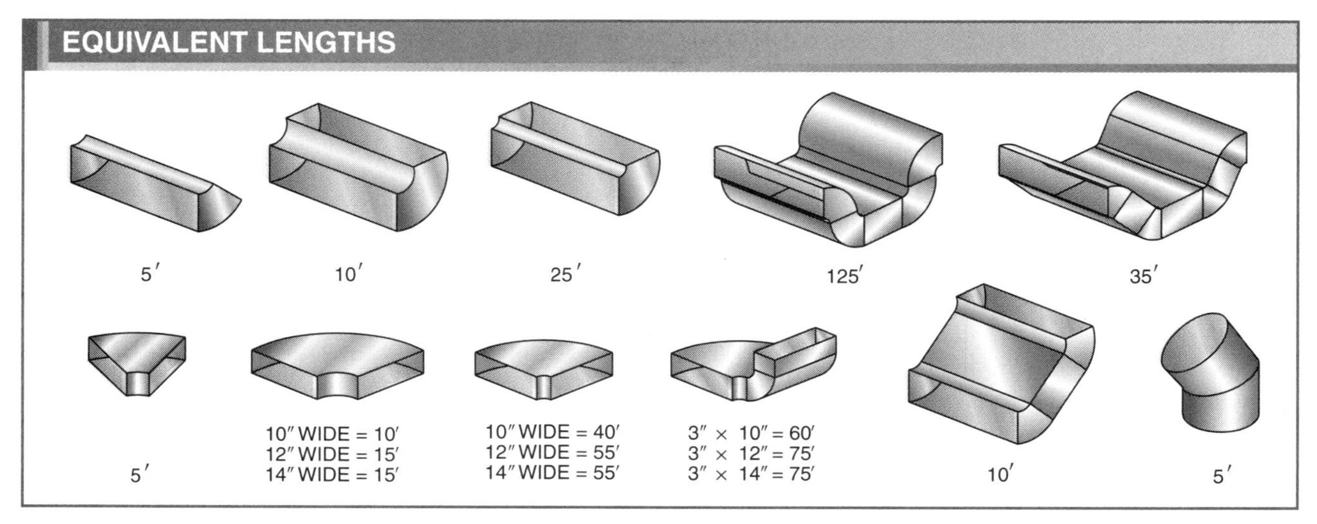

Figure 18-7. A table of equivalent lengths shows the length of a duct section that gives the same static pressure drop as a particular metal fitting or transition.

Fitting loss coefficients (*coefficient C*) are values that represent the ratio between the total static pressure loss through a fitting and the dynamic pressure at the fitting. To find the static pressure drop through a fitting, the coefficient C for the fitting is multiplied by the velocity pressure of the air entering the fitting. The static pressure drop through a fitting using coefficient C is found by applying the formula:

$$P_d = P_v \times C$$

where

P_d = static pressure drop (in in. WC)

P_v = velocity pressure (in in. WC)

C = fitting loss coefficient

Example: Finding Static Pressure Drop Through Fitting

An elbow with a 12″ radius has a velocity pressure of 0.1403″ WC. The fitting is a 18″ × 12″ smooth-radius elbow. Find the pressure drop through the fitting.

1. Find smooth-radius elbow on a fitting loss coefficient table and calculate coefficient C.
2. Multiply velocity pressure by the coefficient C.

$$P_d = P_v \times C$$

$$P_d = .1403 \times .19$$

$$P_d = \mathbf{0.0267'' \ WC}$$

Ductwork Design

A forced-air heating or air conditioning system is designed using a floor plan or sketch of the building. Each floor that has ductwork, registers, or grills requires a floor plan. To design the distribution system, the heating and cooling loads are calculated, and the ductwork is sized.

Layout Drawings. *Layout drawings* are drawings of the floor plan of a building that show walls, partitions, windows, doors, fixtures, and other details. These details may affect the location of the ductwork, registers, or grills. Layout drawings show the heating and cooling loads and required air flow rate for each room or zone. Because forced-air distribution systems are sized for a required air flow rate, the heating and cooling loads for a building must be converted to air flow rate for building spaces. **See Figure 18-8.**

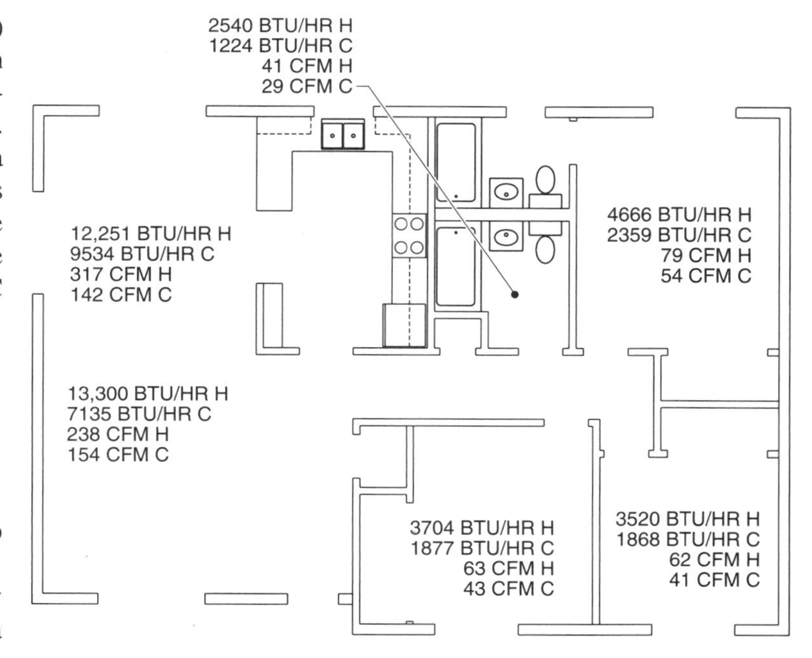

Figure 18-8. Layout drawings show the heating and cooling load for each room or zone.

Layout drawings include several overlays that show the placement of ductwork, registers, and grills on each floor of a building. Tentative locations of the registers and grills are shown on the overlay for each floor of the building.

Registers are located based on the air flow pattern desired out of each register. Grills are located so all of the air supplied by the registers is returned to the system blowers. Grills must be located where they are not covered by furniture or fixtures. Air must get back to the blowers through the return ductwork. Therefore, grills must be located so they are open at all times. Ceilings or walls are the best locations for grills because floor grills are easily covered with furniture or carpeting and collect dirt.

Grills should not be located close to registers. If grills are located close to registers, air from the registers will flow directly into the grills. This air flow upsets the distribution pattern of the registers. Grills should be located centrally in the building space. One or more grills may be used in a zone or building space to guarantee sufficient circulation. If some building spaces are isolated by partitions or doors, grills should be located in each building space. Grills may be connected to the same return ductwork but must be located so they return air to the system blowers from all parts of the building.

The distribution system is drawn on a separate overlay. The distribution system overlay contains the location of the registers and grills, which shows the duct work in relation to the registers and grills. Each register is connected to the supply ductwork, and each grill is connected to the return ductwork. The actual layout of a distribution system depends on the type of system chosen. On the drawings, ductwork must be located where it does not run through beams, columns, stairwells, elevator shafts, or where plumbing or electrical devices may be located. **See Figure 18-9.**

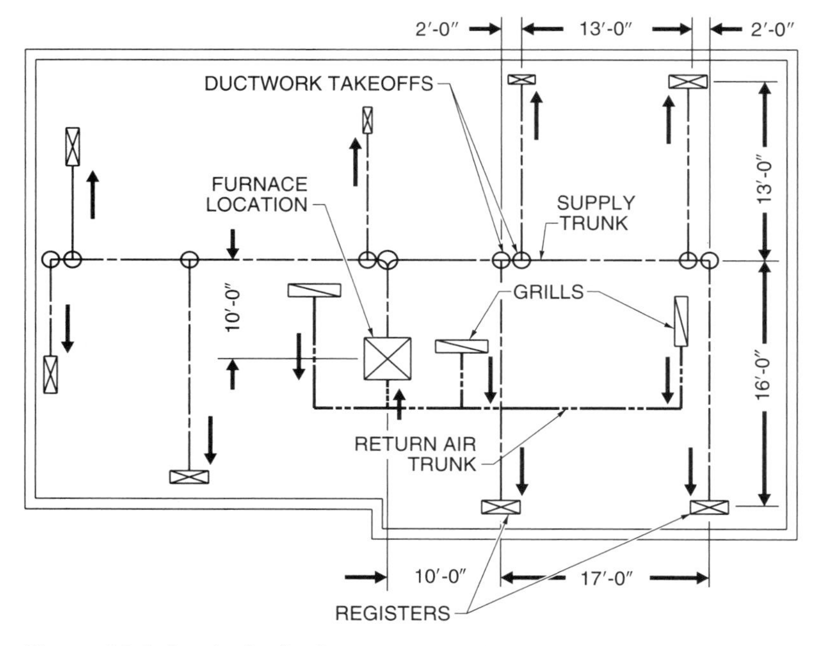

Figure 18-9. An air distribution system overlay shows the ductwork layout in relation to the placement of registers and grills.

After layout drawings and overlays have been prepared, the distribution system is chosen based on the temperature control and sophistication level required for each building space. After the distribution system has been chosen, the registers and grills are chosen based on the heating and cooling loads of each building space.

If the design of part of the distribution system is especially complicated, large-scale plans may be necessary to identify the actual location of each duct section. A ductwork layout drawing should show all fittings and transitions in the distribution system. Fittings and transitions are duct sections that change air flow direction or duct size. Fittings and transitions are shown on a single line drawing by an "X" marked on the duct drawing. A *single line drawing* is a drawing in which the walls and partitions are shown by a single line with no attempt to show actual wall thickness. Fittings and transitions are shown on a double line drawing in plan view as they actually appear. A *double line drawing* is a drawing in which walls and partitions are shown with double lines. Wall thickness is shown at actual scale on the drawing.

Section Identification. Identifying each duct section is necessary for keeping track of the proper size ducts, dampers, and fittings. A *duct section* is a section of ductwork between two fittings where the air flow rate changes. Duct sections occur between takeoffs where air enters or leaves the duct. Duct sections are identified using letters or numbers. The two most common methods of duct section identification are marking the duct sections or marking the fittings at the end of each section.

Letters on fittings are used to identify duct sections between the fittings. All of the fittings are marked with a different letter. The first fitting, which is at the furnace or air conditioner, is identified as A. The fitting at the end of the first duct section is identified as B. The duct section between fittings A and B is identified as section A-B. The letters (A-B) are shown on the distribution system overlay for that duct. **See Figure 18-10.**

Duct Sizing

A duct is sized to allow the required amount of air for proper heating or cooling to flow to each building space. The air flow rate is controlled by the size of the duct, and the duct is sized to carry the air at a given pressure loss.

Forced-air distribution systems often consist of major duct sections and individual duct sections. The major duct sections are connected to the supply plenum and individual duct sections run off of the major duct sections. For sizing and calculating pressure drop through a distribution system, each major section can be treated separately.

The most common methods for sizing HVAC ductwork are equal friction, velocity reduction, and static regain methods. Small distribution systems used in heating and air conditioning systems as well as small ventilation and exhaust systems are sized by either the equal friction method or velocity reduction method. Large, sophisticated distribution systems are sized using the static regain method. The equal velocity method is used for sizing exhaust systems that carry particulate matter.

In the past, duct sizes were calculated manually or with duct sizing charts, graphs, or ductulators. Today, computer programs are used for sizing ducts. Small jobs are easily sized by conventional methods, but large jobs are more easily sized using a computer duct-sizing program.

The *equal friction method* is duct sizing that considers that the static pressure is approximately equal at each branch takeoff in the distribution system. An equal friction chart is used to size the ductwork for a given static pressure drop using the equal friction method.

To use an equal friction chart for sizing ductwork, the static pressure drop is determined, and the chart is used to find the size of the duct. The static pressure drop can be selected arbitrarily. For larger systems, the static pressure drop is found by selecting an arbitrary air velocity and duct size for the first section of duct off the furnace or air conditioner. This data is used on an equal friction chart to find the static pressure drop. This static pressure drop is used for sizing the other ducts in the distribution system.

Small distribution systems are designed for about .06″ WC to .10″ WC static pressure drop per 100′ of duct. In larger distribution systems, the static pressure drop per 100′ of duct may be greater. To size ductwork using the equal friction method, apply the procedure:

1. Prepare layout drawings and overlays that show the heating and cooling load and airflow rate for each building space.
2. Select and locate registers and grills on the first overlay. In some applications, more than one register is required in a room. When more than one register is required in a room, more than one duct branch runs to the room. The air flow rate for the room should be divided between the duct branches supplying the room.
3. Lay out the distribution system on the second overlay.
4. Enter the duct sections in the first column of a columnar form.
5. Enter the air flow rate for each duct section in the second column.

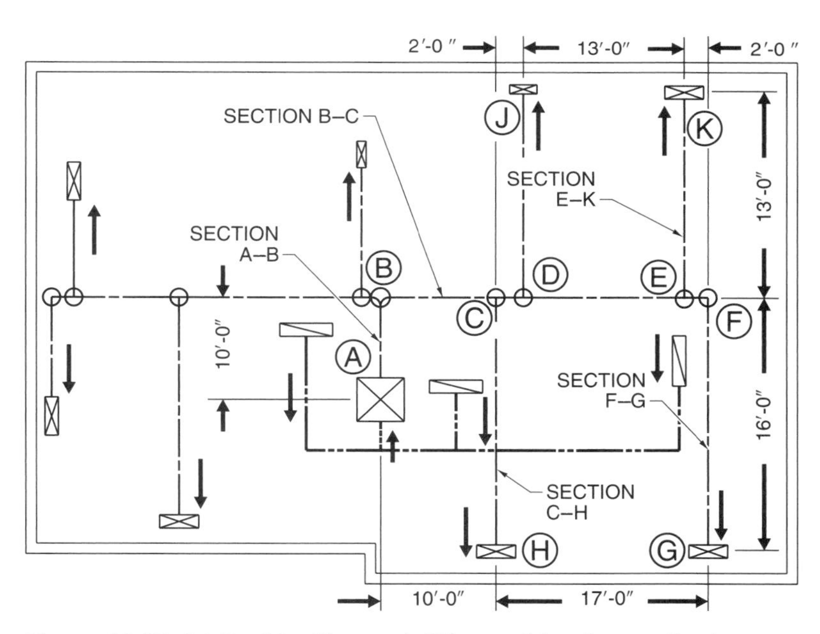

Figure 18-10. A letter identifies each fitting and the duct section between fittings is identified by the fitting letters.

Tech Facts

The volume of air to be delivered by the forced air system must be known to determine the size of the ducts. The volume is based on the amount of heat the forced air system must deliver and remove.

6. Select a design static pressure drop, which is the pressure drop that occurs per 100′ of duct in the actual distribution system design.

   The design static pressure drop is determined by selecting an air velocity, picking a duct size for that section of duct, and finding the design static pressure drop on an equal friction chart. The air velocity is for the first duct section off the blower to ensure quiet operation. This design static pressure drop is used for sizing the system ductwork.
7. Find the size of each duct section using an equal friction chart. Enter the values in the third column of a columnar form.
8. Convert sizes of round ducts to equivalent rectangular sizes. Enter the values in the fourth column of the columnar form. These values are taken from a round-to-rectangular duct conversion chart. **See Appendix.**

Example: Duct Sizing—Equal Friction Method

The blower of a furnace used to heat a three bedroom residence moves 800 cfm of air. Size the ductwork using the equal friction method.

1. Prepare layout drawings and overlays that show the heating and cooling load and air flow rate for each building space. **See Figure 18-11.**
2. Locate registers and grills on the first overlay.
3. Lay out the ductwork on the second overlay.
4. Enter the identifying letters for the duct sections in the first column of a columnar form.
5. Enter air flow rate for each duct section in the second column. **See Figure 18-12.**
6. Select a design static pressure drop of .1″ WC for the forced-air distribution system.
7. Find the size for each duct section using an equal friction chart. Enter the values in the third column of the columnar form.
8. For the duct sections that should be rectangular, convert the round duct sizes to equivalent rectangular sizes. Enter the values in the fourth column of the columnar form.

Rectangular ducts are used in distribution systems where head room is limited. The width of a rectangular duct can be greater than the height. This allows the bottom of the duct to be higher from the floor of a building space. Rectangular ducts are also used as trunks.

To function most efficiently, the aspect ratio of a rectangular duct should not be more than 3 to 4 times the height. *Aspect ratio* is the ratio between the height and width of a rectangular duct.

When duct sizes are taken from an equal friction chart or ductulator, the actual sizes are recorded on the columnar form. For practical use however, nominal sizes of ducts that are readily available or more economical to manufacture are shown on the drawing. If the actual size of a duct is not the same as the nominal size of an available duct, the next larger nominal size is used.

The *velocity reduction method* is duct sizing that considers that the air velocity is reduced at each branch takeoff. The air velocity in the first duct section off the furnace or air conditioner and the size of the first duct section are selected arbitrarily. The other duct sections are sized so the air velocity is reduced in each section according to some arbitrary scale. Tables show recommended air velocities in different duct sections. When a distribution system is sized using the velocity reduction method, the design static pressure drop may be different in each duct section. **See Figure 18-13.**

The *static regain method* is duct sizing that considers that each duct section is sized so that the static pressure increase at each takeoff offsets the friction loss in the preceding duct section. The static regain method is used for duct sizing in long distribution systems with many takeoffs or with registers located on the duct itself.

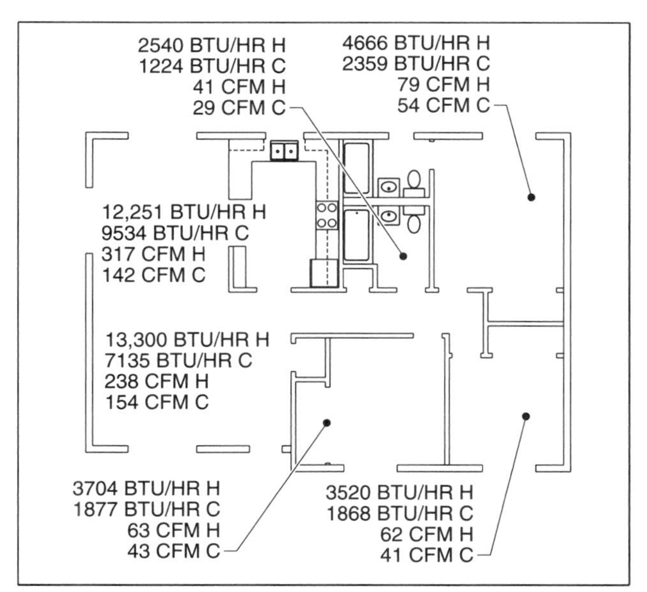

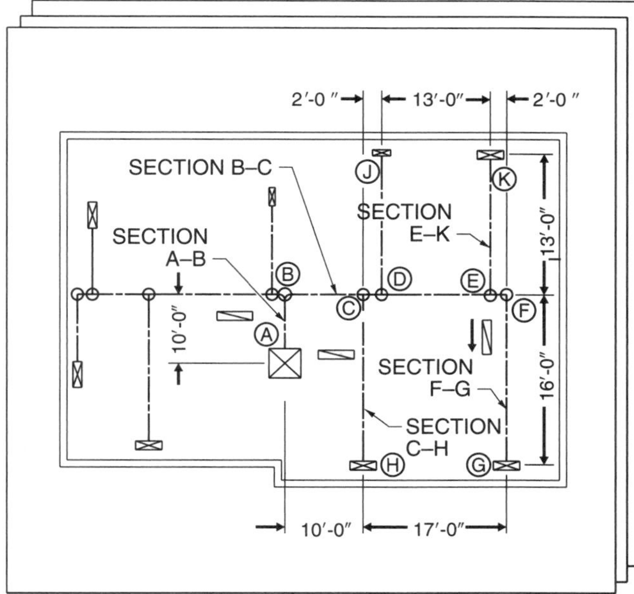

Figure 18-11. Layout drawings show the air flow rate for each building space. Overlays show the location of the ductwork system, registers, and grills.

3 bedroom residence
.1" WC

Section	Flow Rate*	Size§	Size#
A–B	800	12.7	20 x 8
B–C	245	8.1	12 x 8
C–D	182	7.3	12 x 8
D–E	141	6.6	12 x 8
E–F	62	4.8	8 x 8
F–G	62	4.8	
C–H	63	4.9	
D–J	41	4.1	
E–K	79	5.3	

* in cfm
§ in inches diameter
\# in inches

Figure 18-12. Duct sections, air flow rate, and duct size are recorded on a columnar form.

The static regain method causes approximately the same static pressure at each branch takeoff. The static regain method can result in large ducts with low air velocity at the ends of long runs. The static regain method is used for larger distribution systems and requires many calculations for each duct section and for each fitting.

The *equal velocity method* is duct sizing that considers that each duct section has the same air velocity. The ducts are sized with a ductulator and a velocity scale. The sizing may be done mathematically using the air velocity and the cross-sectional area of the duct. The equal velocity method is more applicable to sizing exhaust systems or material collection systems than it is for sizing distribution ducts for a heating or air conditioning system.

Blower Sizing

When designing a forced-air distribution system, the blower must have the capacity to move air at the required air flow rate against the resistance in the distribution system. When sizing a blower for a distribution system, the required air flow rate and the total allowable static pressure drop for the system are found on a blower performance table. The velocity of the blower wheel and the motor horsepower required to turn the blower wheel to produce the required air flow rate are determined from the blower performance table.

The blower in a forced-air distribution system must move the air at the air flow rate required for the system against the resistance to air flow of the system. The air flow rate required is determined when the heating or air conditioning equipment for the system is chosen. Resistance to air flow in a distribution system is due to friction in the ducts and to dynamic pressure losses through fittings and transitions in the ducts. Resistance to air flow occurs in both the supply and return ducts of a distribution system. A blower must be large enough to move the required air flow rate in the major section of supply and return ductwork having the greatest resistance to flow. The total resistance to air flow in a distribution system is calculated as static pressure drop in the system.

If a blower circulates air through more than one major section of supply ductwork and one major section of return ductwork in a system, the blower will not have to overcome the sum of the static pressure drops in all of the sections. The blower must overcome the static pressure drops in the supply and return sections having the largest static pressure drops.

When sizing a blower for a forced-air distribution system, it is not necessary to calculate the pressure drop through all the sections of a distribution system that have more than one major duct section running to or from the blower. It is necessary to find the supply duct section with the highest static pressure drop and the return duct section with the highest static pressure drop. The two static pressure drop values are added together to find the total static pressure drop against which the blower has to work.

To calculate the total static pressure drop through a major duct section that has been designed by the equal friction method, the actual length of the duct section that has the greatest resistance to air flow is found. The length is taken from the layout drawing. The equivalent length of each fitting or transition in the section is found from a table of equivalent lengths. The equivalent length of each fitting is added to the actual length of duct in the section. The result is the total equivalent length of that section of the distribution system.

The total equivalent length of the duct section is divided by 100′ and multiplied by the design static pressure drop that was used to size the duct. The result is the static pressure loss for that duct section.

To find the total static pressure drop through one section of a forced-air distribution system, start with the columnar form used for sizing the duct. Add a column for the actual length of each duct section and fitting equivalent lengths. **See Figure 18-14.**

Tech Facts

Blowers must be properly sized so that they have the capacity to move air at the required airflow rate against the resistance of the ducts and fittings in the distribution system. Smaller ducts are used to reduce duct size and save space. The smaller ducts require approximately twice the normal air velocity to operate properly.

RECOMMENDED DUCT AIR VELOCITIES*

Use		Residential Buildings		Commerical Buildings		Industrial Buildings	
		Recommended	Maximum	Recommended	Maximum	Recommended	Maximum
Total Face Area Velocities	Filters	250	300	300	350	350	350
	Coils	450	500	500	600	600	700
	Intakes	500	800	500	900	500	1200
	Air washers	650	650	650	650	650	650
Net Free Area Velocities	Ducts						
	Main	800	1000	1500	1600	1500	2200
	Branch	600	850	750	1300	900	1800
	Risers	500	800	650	1200	800	1300
	Outlets	1300	1700	1650	2200	2000	2800

* in fpm

Figure 18-13. Recommended duct air velocities are used when sizing ductwork by the velocity reduction method.

3 bedroom residence
.1" WC

Section	Flow Rate*	Size§	Size#	DAL**	FEL§§	TEL##			
A–B	800	12.7	20 x 8	10	35				
B–C	245	8.1	12 x 8	10	30				
C–D	182	7.3	12 x 8	2	25				
D–E	141	6.6	12 x 8	13	25				
E–F	62	4.8	8 x 8	2	25				
F–G	62	4.8		16	30				
C–H	63	4.9		16	30		$\left(\frac{355}{100}\right)$ x .1	= .355" WC	
D–J	41	4.1		13	30				
E–K	79	5.3		13	30				
		Total Actual Length		95 x	260 =	355'			

* in cfm
§ in inches diameter
# in inches
** duct actual length in ft
§§ fitting equivalent length in ft
total equivalent length in ft

Figure 18-14. The sum of the lengths of the duct sections and the fitting equivalent lengths is the total actual length of duct in the section.

Enter the actual duct length of each section from the overlays. In the next column, enter the equivalent lengths for the fittings in each duct section. Add the lengths of the duct sections, which gives the total actual length of duct in the section. Add the equivalent lengths for the fittings, which gives the total equivalent length for fittings. Add the duct actual length total and the fitting equivalent length total to get the total equivalent length of the major duct section.

Divide the total equivalent length by 100 and multiply by the design static pressure drop to get the static pressure drop through the major duct section. Repeat this step for the section of return ductwork that appears to have the greatest pressure drop. The sum of the pressure drop for the supply ductwork section and the return ductwork section is the total pressure drop for the ductwork in the system.

The static pressure drop for filters, coils, dampers, or other devices that add resistance to air flow in the distribution system should be added to the duct pressure drop. The pressure drop through an accessory is found in specifications sheets. The total static pressure drop for a major duct section is the duct pressure drop plus the pressure drop through accessories in that section.

If a distribution system has been designed by a method other than the equal friction method, the static pressure drop in each duct section between fittings may be different. To find the total static pressure drop in a major duct section, the static pressure drop in each individual duct section must be found. The individual section pressure drops are then added together. The pressure drops through fittings and transitions are calculated by using the equivalent length method or the coefficient C. The result is added to the sum of the section pressure drops. Pressure drop through accessories is added to the duct section and the fitting pressure drop to find the total major section pressure drop.

To find the static pressure drop through an individual duct section, the size of the duct and the air flow rate through the section are used. To find the static pressure drop through an individual duct section apply the procedure:

1. Using an equal friction chart or a ductulator, find the design static pressure drop.
2. Calculate the actual length of the duct section from the duct layout drawing. This is the length of the duct measured from centerline of the fitting at the start of the duct section to the centerline of the fitting at the end of the duct section.
3. Determine the equivalent length for the fittings in the section. Equivalent lengths are taken from an equivalent length table.

4. Add the equivalent lengths for fittings to the actual lengths for the section. The sum is the total equivalent length for the section.
5. Divide the total equivalent length of the section by 100 and multiply by the design static pressure drop for the system.

The result is the static pressure drop for the section of the distribution system. The pressure drop of accessories in the section, which add resistance to air flow, must be added to the duct section pressure drop to determine the total static pressure drop in the section.

The same procedure is followed for each individual duct section through a major section. The sum of the individual section pressure drops is the total static pressure drop for the major section.

Tech Facts

Typically, duct systems have several openings that forced air is diffused from. The forced air travels through the openings, which provide the path of least resistance.

Example: Sizing Blower—Velocity Reduction Method

In a forced-air distribution system, the design static pressure drop is .10″ WC per 100′. Size the blower using the velocity reduction method.

1. Enter the length of each individual duct section on a columnar form. **See Figure 18-15.**
2. Enter the static pressure drop for each duct section on the form.

   The static pressure drop in the duct section is found by dividing the length of the duct by 100 and multiplying by the design static pressure drop.
3. Enter a description of the fittings in each duct section on the next column of the form. Describe the fitting so it can be identified for calculating the static pressure loss.
4. List the equivalent lengths for the fittings identified for each duct section on the form.
5. Enter the static pressure drop for the fittings in each section. This is calculated with the method used for the duct sections. The equivalent length for fittings is divided by 100 and multiplied by the system design static pressure drop.
6. Add the static pressure drop for each duct section to the static pressure drop for the fittings in the section. Add across the form. The subtotal static pressure drops for the individual sections are added together to find the total pressure drop for the entire major duct section.

3 bedroom residence
.1" WC

Section	Flow Rate*	Size§	Size#	Length**	SP Drop§§	Fitting	EL##	SP Drop***	Sub-T§§§
A–B	800	12.7	20 x 8	10	.01	plenum T.O.	35	.035	.045
B–C	245	8.1	12 x 8	10	.01	splitter	30	.030	.040
C–D	182	7.3	12 x 8	2	.002	through T.O.	25	.025	.027
D–E	141	6.6	12 x 8	13	.013	through T.O.	25	.025	.038
E–F	62	4.8	8 x 8	2	.002	through T.O.	25	.025	.027
F–G	62	4.8		16	.016	BR T.O. & Boot	30	.030	.046
C–H	63	4.9		16	.016	BR T.O. & Boot	30	.030	.046
D–J	41	4.1		13	.013	BR T.O. & Boot	30	.030	.043
E–K	79	5.3		13	.013	BR T.O. & Boot	30	.030	.043
							Total Pressure Drop		.355

* in cfm
§ in inches diameter
# in inches
** duct length in ft
§§ static pressure drop in inches WC
equivalent length in ft
*** static pressure drop for fitting in inches WC
§§§ subtotal pressure drop in inches WC

Figure 18-15. The static pressure drop in a duct section is found by dividing the length of the duct by 100 and multiplying by the design static pressure drop.

The total pressure drop for the duct section is shown on the bottom right-hand corner of the columnar form. Before this figure is used for sizing a blower, the static pressure drop through the return system and the additional losses through registers, grills, or other accessories in the system should be found and added together.

Register Sizing

Registers and grills are selected and sized to allow the proper amount of air to flow into each building space to offset the heat loss or gain. Registers and grills are located to efficiently distribute the air to the building spaces.

The size of a register is determined by the air flow rate required from the register and the required air velocity in a building space. The air flow rate from a register is a function of the size of the free area of the register. *Free area* of a register is the face area of the register minus the area blocked by the frame or vanes. The air velocity at a register is a function of the free area of the register and the air flow rate. The air velocity at the register determines the distance the air is thrown from the register.

Registers are designed to deliver air in different patterns. Air patterns are defined by throw and spread. Throw is the distance that air travels directly away from a register. Throw is measured to the point where the air velocity drops below 50 fpm. Spread is the distance measured across the envelope of air that spreads out from a register. Spread is the distance on each side of the air envelope where the air velocity drops below 50 fpm. The pattern of the air flow out of the register is controlled by the vanes on the panel.

To size a register, the general type of register is chosen. The air flow rate required from the register, throw and spread characteristics, and the air velocity required to achieve those characteristics are determined. The register that produces these characteristics is found in a catalog. One or more types and sizes of supply registers may be used in a building space. The registers should be chosen for air pattern and sized to provide complete coverage of the room with conditioned air.

Registers are sized so the face velocity is not high enough to produce noise. Catalogs are marked to indicate the air delivery rate at which the registers produce noise.

Grill Sizing

Grills are sized according to velocity of the air at the grill face. Each grill must return the air from the building spaces at a low face velocity to prevent noise. A face velocity of 500 fpm to 750 fpm is allowable when sizing grills.

To size a grill, the air flow rate through the grill is divided by the face velocity desired (500 fpm to 750 fpm) and is multiplied by a factor of 1.3. A factor of 1.3 allows for 30% of the grill face area to be taken up by vanes and frame. The result is the free area of the face of the grill in feet. Grills are sized in inches. To find the size of the grill in inches, the free grill area in square feet must be converted to square inches. The constant 144 is used to convert square foot to square inches. Free grill area is found by applying the formula:

$$A_g = \frac{Q}{V} \times 1.3 \times 144$$

A_g = free grill area (in sq in.)
Q = air flow rate (in cfm)
V = air velocity (in fpm)
1.3 = constant
144 = constant

Example: Finding Free Area—Grill

A grill in a forced-air distribution system must handle 1500 cfm of air at a face velocity of 600 fpm. Find the required free area.

$$A_g = \frac{Q}{V} \times 1.3 \times 144$$

$$A_g = \frac{1500}{600} \times 1.3 \times 144$$

$$A_g = 2.5 \times 1.3 \times 144$$

$$A_g = 3.25 \times 144$$

$$A_g = \mathbf{468\ sq\ in.}$$

If a square grill is required, the size of the grill is found by taking the square root of the free area of the grill. The size of a square grill is found by applying the formula:

$$G_s = \sqrt{A_g}$$

where

G_s = size of square grill (in inches)

A_g = free grill face area (in sq in.)

Example: Finding Size of Square Grill

A grill has a free area of 468 sq in. A square grill is desired. Find the size of the grill.

$$G_s = \sqrt{A_g}$$

$$G_s = \sqrt{468}$$

$$G_s = \mathbf{21.63''}$$

The nominal grill size is 22″ × 22″.

If a rectangular grill is required, the area in square inches should be divided by the dimension of one side of the grill, and the result is the dimension of the other side of the grill. The size of a rectangular grill is found by applying the formula:

$$w = \frac{A_g}{h}$$

where

w = width (in inches)

A_g = free grill area (in sq in.)

h = height (in inches)

Example: Finding Size of Rectangular Grill

A grill in a distribution system must handle 1000 cfm of air at a face velocity of 500 fpm. The height must be 14″. Find the width of the grill.

1. Find the free area of the grill.

$$A_g = \frac{Q}{V} \times 1.3 \times 144$$

$$A_g = \frac{1000}{5000} \times 1.3 \times 144$$

$$A_g = 2 \times 1.3 \times 144$$

$$A_g = 2.6 \times 144$$

$$A_g = \mathbf{374.4\ sq\ in.}$$

2. Find the width of the grill.

$$w = \frac{A_g}{h}$$

$$w = \frac{374.4}{14}$$

$$w = \mathbf{27''}$$

After the duct, register, and grill sizing for a building is complete, the sizes are shown on the distribution system plans. The size of the duct in each section and the size of all fittings and transitions are shown on the plan.

Review Questions

1. List and describe the five main parts of a forced-air distribution system.
2. Why are the heating and cooling loads for a building required when finding the air flow rate?
3. Write the formula for finding the air flow rate when the temperature difference through the heating or cooling equipment and heating or cooling capacity is known.
4. Why is a factor of 30 used when finding the air flow rate for the cooling load for a building space? Explain.
5. The air velocity in a duct is a function of what two factors?
6. What causes friction loss in a duct section?
7. Name four variables of air flow found on an equal friction chart.
8. Explain static pressure drop, system pressure drop, and dynamic pressure drop.
9. Describe the process used to find the velocity pressure of air in a duct section.

10. Describe the first two steps when designing a duct system.
11. What data relative to a duct layout is shown on the layout drawing and overlays?
12. Why is it important that registers and grills not be located too close to each other?
13. What defines the two ends of a duct section?
14. Describe the difference between the three main methods used to size ductwork.
15. Name the two factors that must be considered when determining the design static pressure drop.
16. What is the maximum air velocity recommended for use in the main ducts in a public building?
17. Name the two variables of air flow that are used in sizing a blower for a duct system.
18. Name the two most important factors to consider when sizing registers.
19. What is the acceptable range of air velocity used for sizing grills?
20. Define aspect ratio.

Hydronic systems use water as the heat transfer medium. Heat is transformed from a boiler to the water in a hydronic heating system. The water is circulated to the terminal devices where it heats the air. Heat is transferred to a chiller from the water in the piping of a hydronic air conditioning system. The water is circulated to the terminal devices where it cools the air. The parts of a hydronic system are selected and sized to work together for efficient heat transfer.

19 Hydronic System Design

WATER DISTRIBUTION SYSTEMS

The distribution system for the water in a hydronic system is the piping system. The piping system distributes water from the boiler or chiller to the terminal devices. The piping system consists of iron or steel pipe or copper tubing, fittings, and valves. The piping system is installed in the frame of a building. The piping system must be designed to distribute the required quantity of water to each building space or zone to offset the heating or cooling loads. **See Figure 19-1.**

Principles of Water Flow

The rate of water flow through a pipe or an orifice is a function of the pressure exerted on the water and the size of the pipe. Water flow is expressed in gallons per minute. Water pressure is expressed in feet of head or pounds per square inch. Pipe size is indicated by the nominal outside diameter of the pipe and is expressed in inches. *Outside diameter* is the nominal distance from outside edge to outside edge of a pipe, or the actual distance from outside edge to outside edge of tubing. This measurement includes the thickness of the pipe or tubing walls.

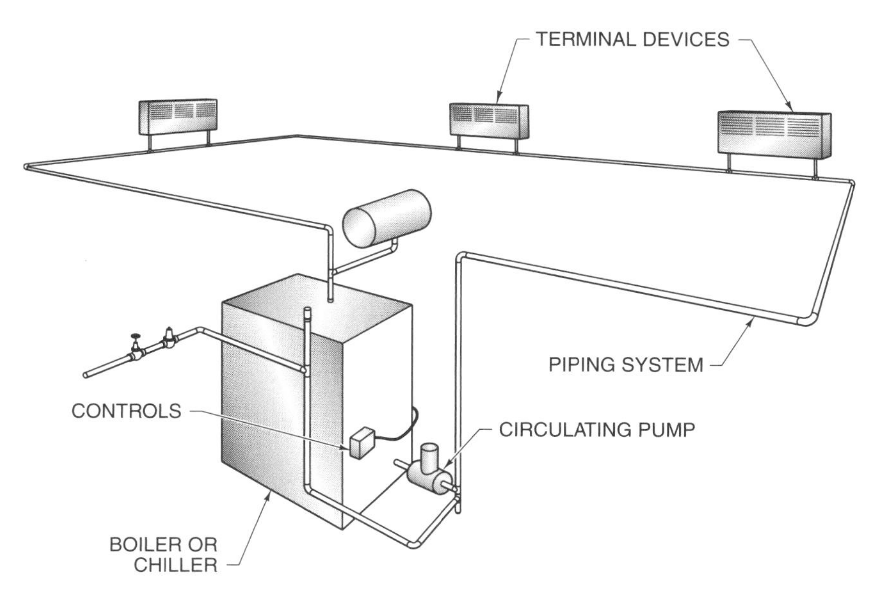

Figure 19-1. In a hydronic system, the piping circulates water form a boiler or chiller to the terminal devices.

Water Flow Rate. *Water flow rate* is the volumetric flow rate of the water as it moves through a given pipe section. Water flow rate is expressed in gallons per minute. Depending on the type of piping system, the flow rate may vary in different pipe sections within a piping loop.

If a piping system is sized for heating and cooling, the larger required flow rate is used for sizing the components for the system. Depending on the climate in which the loads are calculated, either the heating or the cooling load may require the larger flow rate.

Commercial water suppliers measure water flow rate with a meter. For design purposes, water flow rate is calculated mathematically. The water flow rate in a hydronic system is based on temperature difference of water at two points in the system and the amount of heat transferred during the change of temperature. The water flow rate is found by applying the formula:

$$Q = \frac{H}{\Delta T \times 500}$$

where

Q = water flow rate (in gpm)

H = heat transferred (in Btu/hr)

ΔT = temperature difference (in °F)

500 = constant

Example: Finding Water Flow Rate—Heating

A heating system has a total adjusted heat loss of 350,000 Btu/hr. A temperature difference of 20°F is used. Find the water flow rate.

$$Q = \frac{H}{\Delta T \times 500}$$

$$Q = \frac{350{,}000}{20 \times 500}$$

$$Q = \frac{350{,}000}{10{,}000}$$

$$Q = \mathbf{35\ gpm}$$

Water Pressure. Pressure is force exerted per unit of area. Water pressure is expressed in pounds per square inch or in feet of head. *Feet of head* is a unit of measure that expresses the height of a column of water that would be supported by a given pressure. For example, a foot of head is the pressure equal to the weight of a column of water 1′ high. Feet of head can be measured using a special manometer, but it is usually a calculated value.

System Pressure Drop. *Pressure drop* is a drop in water pressure caused by friction between water and the inside surface of a pipe as the water moves through the pipe. *System pressure drop* is the total static pressure drop in a system. System pressure drop is the difference in pressure between the point where water enters the system and the point where water leaves the system. System pressure drop is the resistance against which the circulating pump must move the water. System pressure drop is expressed in psi or in feet of head.

For water to flow through a pipe, pressure must be exerted on the water at the inlet end of the pipe. The pressure must be great enough to raise the water upward in vertical columns of pipe. This pressure must also overcome the system pressure drop. The sum of the friction head and static head forces is the total head of the system.

Friction head is the effect of friction in a pipe. Friction occurs between the water moving through the pipe and the interior surfaces of the pipe. In a straight pipe section, the friction is a function of the size of the pipe, amount of water flow, and the roughness of the surface of the pipe. Friction head can be calculated mathematically or found on tables that give pipe sizes relative to water flow rate and pressure drop.

Static head is the weight of water in a vertical column above a datum line. A *datum line* is the point at which the pressure exerted by water in a vertical column is zero. Pressure exerted by water in a vertical column represented by the weight of the water above the datum line is the static head of the system. Static head is a function of the weight of water and the vertical height of the water in the system above the pump. *Total head* is the sum of friction head and static head. Total head is the system pressure drop.

The pressure drop through a pipe fitting is found by converting the fitting pressure drop to the equivalent length of pipe that would produce the same pressure drop. Tables of elbow equivalents are used. **See Figure 19-2.**

The elbow equivalent value found on the table is multiplied by the equivalent length of pipe for the 90° elbow based on the size of the elbow and the velocity of the water. To find the equivalent length of pipe that has the same pressure drop as a fitting, apply the procedure:

1. Find the equivalent length of pipe based on the size of the elbow and the velocity of the water in the pipe on a table of equivalent pipe lengths for 90° elbows.

   If the water flow rate in a 3″ pipe is 6 fps (feet per second), each 90° elbow equals 8.9′ of pipe.

2. Find the number of 90° elbows that is equivalent to the fitting.

   If the fitting is an open globe valve, it will be equivalent to 12 elbows. **See Figure 19-3.**

ELBOW EQUIVALENTS		
Fitting	Iron Pipe	Copper Tubing
Elbow, 90°	1.0	1.0
Elbow, 45°	0.7	0.7
Elbow, 90°, long turn	0.5	0.5
Elbow, welded, 90°	0.5	0.5
Reduced coupling	0.4	0.4
Open return bend	1.0	1.0
Angle radiator valve	2.0	3.0
Radiator or convector	3.0	4.0
Boiler or heater	3.0	4.0
Open gate valve	0.5	0.7
Open globe valve	12.0	17.0

ASHRAE Handbook — Fundamentals

Figure 19-3. Tables of elbow equivalents give fitting pressure drop in number of 90° elbows.

Tech Facts

When designing a hydronic system, pressure drops through pipes and fittings must be considered. As water moves through a pipe, the water pressure tends to drop the further the water travels because there is friction between the water and the inside surface of the pipe.

EQUIVALENT LENGTH OF PIPE FOR 90° ELBOWS*															
Velocity §	Pipe Size														
	½	¾	1	1¼	1½	2	2½	3	3½	4	5	6	8	10	12
1	1.2	1.7	2.2	3.0	3.5	4.5	5.4	6.7	7.7	8.6	10.5	12.2	15.4	18.7	22.2
2	1.4	1.9	2.5	3.3	3.9	5.1	6.0	7.5	8.6	9.5	11.7	13.7	17.3	20.8	24.8
3	1.5	2.0	2.7	3.6	4.2	5.4	6.4	8.0	9.2	10.2	12.5	14.6	18.4	22.3	26.5
4	1.5	2.1	2.8	3.7	4.4	5.6	6.7	8.3	9.6	10.6	13.1	15.2	19.2	23.2	27.6
5	1.6	2.2	2.9	3.9	4.5	5.9	7.0	8.7	10.0	11.1	13.6	15.8	19.8	24.2	28.8
6	1.7	2.3	3.0	4.0	4.7	6.0	7.2	8.9	10.3	11.4	14.0	16.3	20.5	24.9	29.6
7	1.7	2.3	3.0	4.1	4.8	6.2	7.4	9.1	10.5	11.7	14.3	16.7	21.0	25.5	30.3
8	1.7	2.4	3.1	4.2	4.9	6.3	7.5	9.3	10.8	11.9	14.6	17.1	21.5	26.1	31.0
9	1.8	2.4	3.2	4.3	5.0	6.4	7.7	9.5	11.0	12.2	14.9	17.4	21.9	26.6	31.6
10	1.8	2.5	3.2	4.3	5.1	6.5	7.8	9.7	11.2	12.4	15.2	17.7	22.2	27.0	32.0

* in ft
§ in fps

ASHRAE Handbook — Fundamentals

Figure 19-2. Tables of equivalent lengths of straight pipe for 90° elbows convert the pressure drop in elbows to pressure drop per length of the pipe.

3. Calculate the equivalent length of pipe by multiplying the equivalent length of pipe for a 90° elbow by the number of elbows. The equivalent length of pipe is found by applying the formula:

   $L = E \times N$

   where

   L = equivalent length of pipe for fitting (in feet)

   E = equivalent length of pipe for a 90° elbow (in feet)

   N = equivalent number of elbows

If each 90° elbow is equal to 8.9′ of pipe and the fitting is equal to 12 elbows, find the equivalent length of pipe.

$L = E \times N$

$L = 8.9 \times 12$

$L =$ **106.8′**

Example: Finding Equivalent Length of Pipe

The copper tubing in a hydronic piping system contains an open gate valve. The copper tubing is 1″ in diameter and the velocity of the water in the valve is 2 fps. Find the equivalent length of pipe that will give the same pressure drop.

1. Find the equivalent length of pipe for 90° elbows.

   On the table for equivalent lengths, the equivalent length of pipe is 2.5′.

2. Find the number of 90° elbows.

   On the table for elbow equivalents, the gate valve is equal to .7 elbows.

3. Calculate the equivalent length of pipe.

   $L = E \times N$

   $L = 2.5 \times .7$

   $L =$ **1.75′**

Design Static Pressure Drop. *Hydronic design static pressure drop* is the pressure drop per unit length of pipe for a given size of pipe at a given water flow rate. Design static pressure drop is the pressure drop in feet of head per 100′ of pipe or mils per foot of pipe. A *mil* is a unit of measure equal to 1⁄1000 of an inch.

Pipe-sizing tables and charts show the size of pipe to use for a given flow rate at design static pressure drops. The charts also show the velocity of the water in feet per minute. Such tables and charts are available for iron, steel, copper, and other kinds of pipe. Sizing tables and charts show the pipe in nominal sizes. All pipes are nominally sized (a number that represents pipe size).

A specific design static pressure drop must be used when using a table or chart for sizing pipe. For most small- to medium-size applications, a design static pressure drop of about 2.5′ of head or 300 mpf (mils per foot) is used. These values give a total pressure drop through a typical piping system within the range of nominal pipe sizes and circulating pumps.

To use the tubing sizing chart, find the 300 mpf row in the friction loss column, which is on the left side of the chart. Follow the row across the chart to the required flow rate. Values for flow rate are listed in the columns under tubing sizes. The heading on the column with the required flow rate is the correct pipe size. The correct pipe size allows the proper water flow at 300 mpf pressure drop. A design pressure drop of 300 mpf of pipe is equivalent to 2.5′ of head. **See Figure 19-4.**

It is not always possible to find a pipe size that gives the exact flow rate. When this is the case, select a nominal pipe size that provides a flow rate equal to or greater than that needed.

If an arbitrary design pressure drop is not used, the design static pressure drop can be found by applying the procedure:

1. Determine the water flow rate in the first pipe section in the supply side of the system.
2. Choose a velocity that ensures quiet operation for the water in the first pipe section.
3. Select a size of pipe for the first pipe section that provides a reasonable water flow rate for the application.
4. Using the above data, select the static pressure drop from a pipe sizing table. This pressure drop is used as the design static pressure drop for the entire system.

FLOW OF WATER IN TYPE "L" COPPER TUBING*									
Friction Loss§	Actual Tubing Size#								Friction Loss**
	3/8	1/2	5/8	3/4	1	1 1/4	1 1/2	2	
100	—	.53	.96	1.44	3.1	5.3	8.5	18.2	.83
125	—	.59	1.05	1.63	3.5	6.0	9.6	20.5	1.04
150	—	.65	1.15	1.79	3.8	6.6	10.6	22.6	1.25
175	—	.71	1.26	1.95	4.2	7.3	11.5	24.6	1.46
200	.41	.76	1.35	2.10	4.5	7.8	12.4	26.5	1.67
225	.44	.81	1.44	2.24	4.7	8.3	13.2	28.3	1.88
250	.46	.86	1.53	2.36	5.0	8.7	14.0	30.0	2.08
275	.48	.91	1.61	2.49	5.3	9.2	14.7	31.5	2.29
300	.51	.95	1.68	2.61	5.6	9.6	15.3	33.0	2.50
325	.53	.99	1.75	2.70	5.8	10.0	16.0	34.5	2.71
350	.56	1.03	1.82	2.82	6.1	10.5	16.7	35.8	2.92
375	.58	1.07	1.90	2.93	6.3	10.8	17.3	37.0	3.13
400	.59	1.11	1.96	3.05	6.5	11.1	18.0	38.3	3.33
425	.61	1.15	2.03	3.15	6.7	11.5	18.5	39.6	3.54
450	.63	1.18	2.10	3.24	6.9	11.9	19.1	40.9	3.75

* in gpm
§ in mpf
\# in inches
** in ft/100′

ASHRAE Handbook — Fundamentals

Figure 19-4. Tubing sizing charts show the size of tubing to use for a given water flow rate at design pressure drops.

Temperature Difference. *Temperature difference* is the difference between the initial and final temperature of a material through which heat has been transferred. When designing a piping system, the temperature difference is the difference in temperature of the water at the beginning and end of a heating or cooling process. The temperature difference through each pipe section and terminal device is important because temperature difference represents heat loss or gain. In some piping systems, the temperature difference through each pipe section and terminal device is different.

Total system temperature difference for a heating application is the difference between the temperature of the water entering the piping system at the boiler and the water returning to the boiler. For a cooling application, total system temperature difference is the difference between the temperature of the water entering the piping system at the chiller and the water returning to the chiller. For a typical small- to medium-size application, a temperature difference of about 20°F may be used for both heating and cooling applications. Typically, the temperature of water leaving a boiler is from 160°F to 200°F, and the temperature of water returning to the boiler is from 140°F to 180°F. The temperature of water leaving a chiller is around 45°F and the temperature of the water returning is 65°F.

Piping System Design

A floor plan or sketch of the building is required when designing a piping system for a hydronic system. Each floor that has piping or terminal devices requires a floor plan. The heating and cooling loads must be calculated. When the plans have been made and the loads have been calculated, the pipe sizes can be chosen.

Layout Drawings. *Layout drawings* are drawings of the floor plan of a building that show walls, partitions, windows, doors, fixtures, and other details. These details may affect the location of the piping and terminal devices. Layout drawings also show the heating and cooling load for each room or zone and the required water flow rate. Piping systems are sized for water flow rates in gallons per minute. Heating and cooling loads for each building must be converted to water flow rate for each section. **See Figure 19-5.**

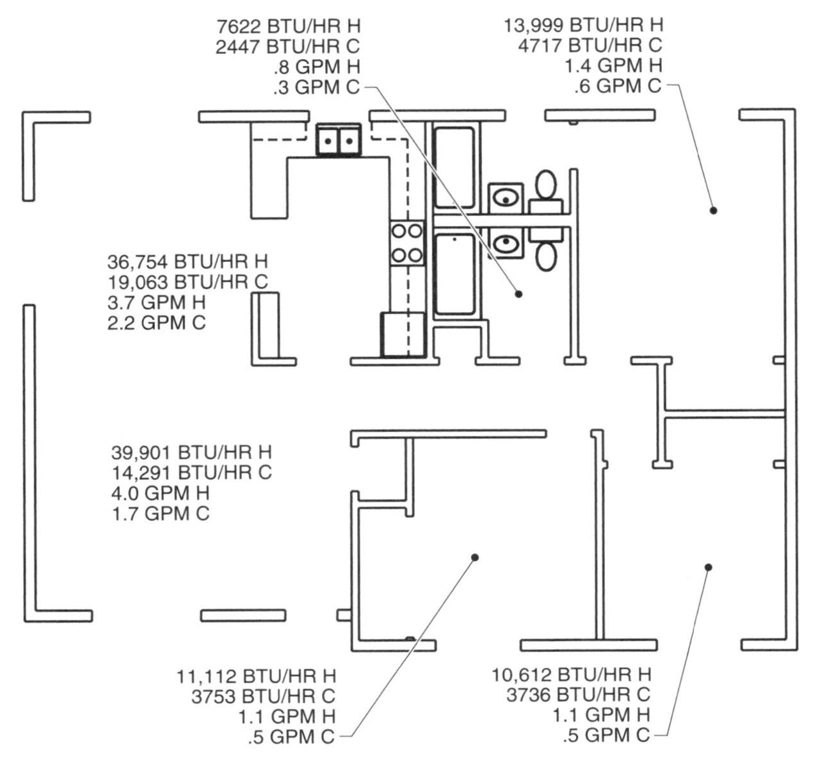

Figure 19-5. Layout drawings are drawings of the floor plan of a building, which contain the heating and cooling load for each room or zone.

A valve that is actuated by a solenoid is used to isolate and control terminal devices in a hydronic system.

Layout drawings contain several overlays, which show the placement of the terminal devices and the piping system. The overlays show the floors in the building that contain terminal devices and the places where pipes are run. Locations for terminal devices are on the overlay for each floor of the building.

The piping system is drawn on a separate overlay, which contains the location of the terminal devices. With this arrangement, the piping layout can be seen in relation to the terminal devices. On the piping layout, each terminal device is connected to the supply pipe and the return pipe. The actual layout of the piping system depends on the type of piping system chosen.

On the drawings and in a building, pipes must not run through beams, columns, stairwells, elevator shafts, or areas where plumbing or electrical devices may be located. When the layout drawings and overlays are ready, the piping system is chosen based on the temperature control and sophistication level required in each building space. After the piping system has been chosen, the terminal devices for each building space are chosen based on the heating and cooling requirements for each building space. **See Figure 19-6.**

Section Identification. Identifying each pipe section is necessary for keeping track of the proper size pipe, valves, and fittings for each pipe section. A *pipe section* is a length of pipe that runs from one fitting to the next fitting. The length of a pipe section is measured from the outlet of one fitting to the outlet of the previous fitting. Pipe sections are identified by letters or numbers, which are marked on the pipe sections themselves or are marked on the fittings at the ends of each pipe section.

When letters are used to identify pipe sections, the pipe section between fittings is identified by the letters marked on the fittings. For example, all of the fittings are marked with a different letter. The first fitting in a system, which is on the boiler or chiller, is identified as A. The fitting at the other end of this first pipe section, which is at the beginning of the second pipe section, is identified as B. The pipe section between fittings A and B is identified as section A-B. The letters that identify the sections and the fittings are shown on the piping overlay. **See Figure 19-7.**

Pipe Sizing

Pipe sizing depends on the material of the pipe used. The pipe made from some materials is rougher on the inside than other materials. The roughness of the inside of a pipe affects friction loss. Because one element of pipe sizing is friction loss, the type of pipe used affects the pipe size. Pipes used in hydronic systems are wrought iron or galvanized Schedule 40 pipe, or copper tubing. Copper tubing is used for small applications, and pipe is used for large applications.

The procedure for sizing a piping system differs depending on the type of piping system used. In a one-pipe series system, the pipe size remains the same throughout the system because the flow rate remains the same. In a one-pipe primary-secondary system, both the supply and return pipe sizes change between each terminal device because the flow rate changes. In a two-pipe direct-return system, the supply pipe and return pipes are usually the same size in each pipe section between terminal devices. The return pipes get progressively smaller as terminal devices are added. In a two-pipe reverse-return system, the supply pipe gets progressively smaller as it goes out from the boiler or chiller, and the return pipe gets progressively larger as it comes back to the boiler or chiller from the terminal device at the end of the piping loop. To design a piping system for a building, apply the procedure:

1. Prepare layout drawings and overlays, which show the heating and cooling load and water flow rate for each building space.
2. Select and locate the required terminal devices on the first overlay.
3. Lay out the piping system on the second overlay.
4. Enter the names of the piping sections in the first column of a columnar form.
5. Enter the water flow rate for each pipe section in the second column.
6. Select a design static pressure drop for the system. This is the pressure drop that occurs per 100′ of pipe in the final pipe design.

   The design static pressure drop should be approximately 300 mils per foot (mpf) or 2.5′ of head per 100′. Pipe sized with these pressure drops gives a system average-size pipe within the range of available circulating pumps.
7. Size each pipe section with a pipe sizing chart.

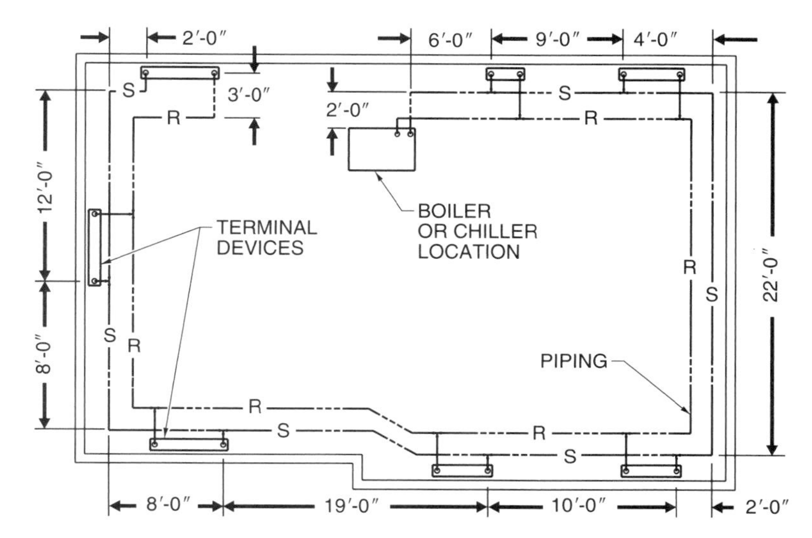

Figure 19-6. A piping system overlay includes the location of piping and terminal devices.

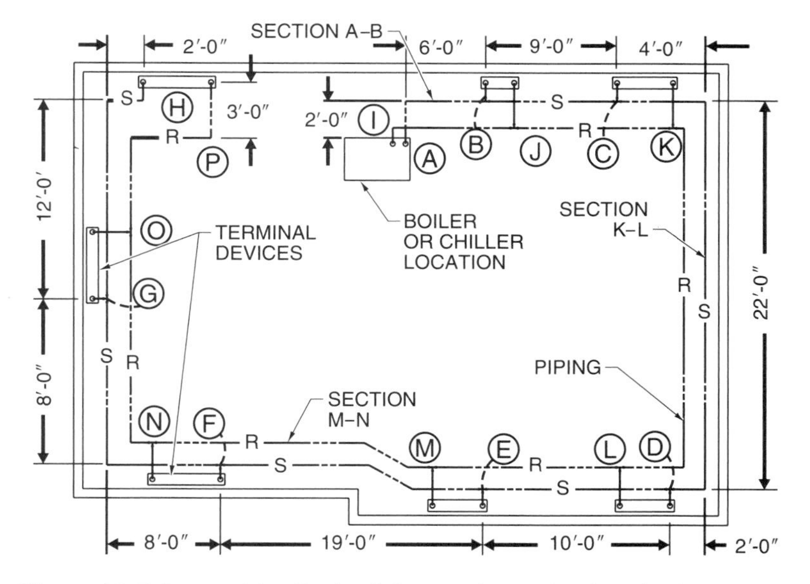

Figure 19-7. Letters identify the fittings at the ends of each pipe section. Pipe sections are identified by letters on the fittings.

Example: Sizing a Piping System

A small residence requires a hydronic system. The system is a two-pipe reverse-return system made of copper tubing. A 20°F temperature difference is used for the heating flow rate and a 17°F temperature difference is used for the cooling flow rate. Find the pipe sizes for the system.

1. Prepare layout drawings and overlays. Determine the heating and cooling load and water flow rate for each building space. **See Figure 19-8.**
2. Select and locate the required terminal devices on the first overlay.
3. Lay out the piping system on the second overlay.
4. Enter the names of the piping sections in the first column of a columnar form.
5. Enter the water flow rate for each pipe section in the second column. **See Figure 19-9.**
6. Select a design static pressure drop.
7. Size each pipe section with a pipe sizing chart.

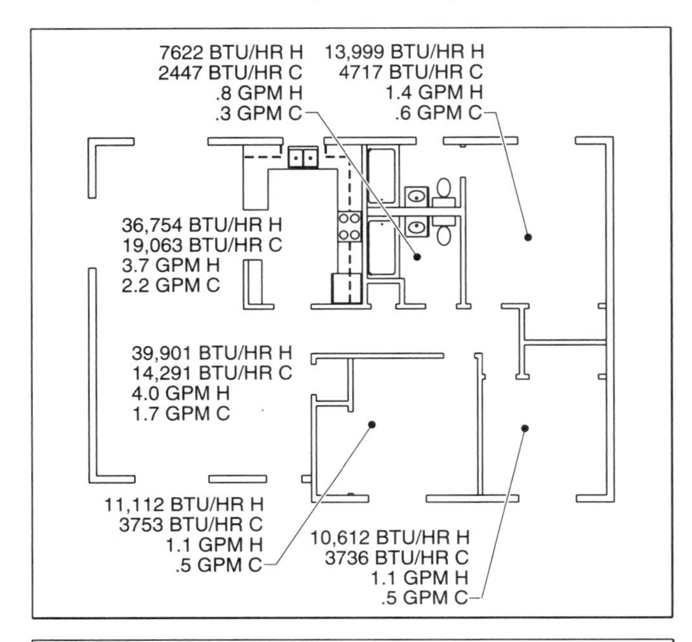

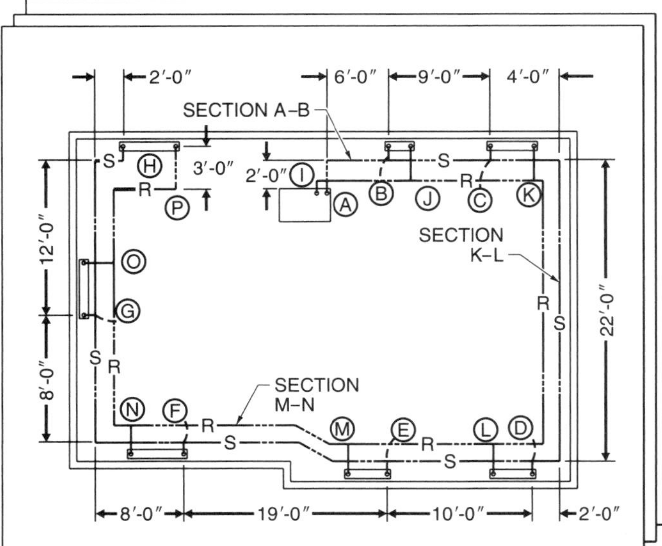

Figure 19-8. Layout drawings and overlays show the heating and cooling load, water flow rate, terminal device layout, and piping system layout.

3 bedroom residence
300 mpf pressure drop

Section	Flow Rate*	Size§
A–B	12.0	1.5
B–C	11.2	1.5
C–D	9.8	1.25
D–E	8.7	1.25
E–F	7.6	1.25
F–G	5.6	1
G–H	3.7	1
B–J	.8	.5
C–K	1.4	.625
D–L	1.1	.625
E–M	1.1	.625
F–N	2.0	.75
G–O	2.0	.75
H–P	3.7	1

* in gpm
§ in inches

Figure 19-9. The name of the pipe section, the water flow rate, and the pipe size are entered on a columnar form.

Terminal Devices

The type of terminal device used in a hydronic heating or air conditioning system depends on the degree of temperature control required in the building. Radiators or convectors are used as terminal devices with one-pipe series piping systems. In a one-pipe series piping system, all of the sections of the system, including each terminal device, have the same water flow rate. Each terminal device selected must be sized for that flow rate.

In a one-pipe primary-secondary system, the terminal devices are located on secondary loops that are in parallel with the main system loop. In this type of system, each of the terminal devices can be a different size. The piping to each terminal device must be sized for the flow rate required for the load handled by that terminal device only. The main piping loop is sized for the total load on the main loop. This load usually includes several terminal devices. In a two-pipe system, each terminal device may have a different flow rate.

Selecting Terminal Devices. Heating and cooling capacities of terminal devices are shown on specifications sheets. To choose the correct size, the heating and cooling load on each terminal device must be known. A terminal device specifications sheet that describes the terminal device is also required. The model number of each terminal device is recorded on the overlay that shows the terminal devices. **See Figure 19-10.**

Tech Facts

When selecting terminal devices, the heating and cooling load on each terminal device must be known so that the correct size terminal device can be used.

Circulating Pump Sizing

There should be one circulating pump on each piping loop in the system. Small piping systems usually have only one pump for the whole system. Larger systems may have several pumps. Circulating pumps are selected and sized to circulate the required amount of water against the total resistance of the piping system.

Selecting Circulating Pumps. Graphs of performance data show the capacity of different pumps. The graphs of performance data show the pumping capacity in gallons per minute for different pressure drops. **See Figure 19-11.**

The total pressure drop in a piping system is the sum of the friction head and the static head in the supply side and return side of a piping system. The friction head in a piping loop is found by multiplying the total equivalent length of the loop by the design pressure drop for the system. The static head is the sum of the lengths of the vertical risers in the loop.

The columnar form is used to find the total head in the system. A column for the actual length of each pipe section is added to the form. The pipe section lengths are taken from the layout drawing of the system. Enter the lengths on the columnar form. The fittings in each pipe section are described in the fifth column of the form. **See Figure 19-12.**

CAPACITY RATINGS FOR TERMINAL DEVICES IN HYDRONIC SYSTEMS*

			Hydronic Air Conditioning			Hydronic Heating		
	Flow Rate			45°F Water 80°F db, 67°F wb Air			140°F Water	
Model	Air#	Water**	Pressure Drop§	Total*	Sensible*	Pressure Drop§	50°F Air Total*	70°F Air Total*
Q–100	200	2	1.00	6500	5500	1.4	13,500	10,000
Q–160	250	2.5	1.60	8200	5900	1.6	15,500	12,000
Q–200	275	3	2.00	9500	6400	1.9	18,000	14,000
H–250	300	4.5	2.70	10,000	7600	2.2	21,000	16,000
H–750	400	4	2.80	13,500	10,200	3.9	27,000	21,000
H–1000	600	5	3.90	21,000	16,500	7.1	43,000	34,000

* heat in Btu/hr
§ in ft of water
# in cfm
** in gpm

Figure 19-10. When the type of terminal device has been selected, a model and size is chosen for each room or building space from specifications sheets, which are printed by manufacturers.

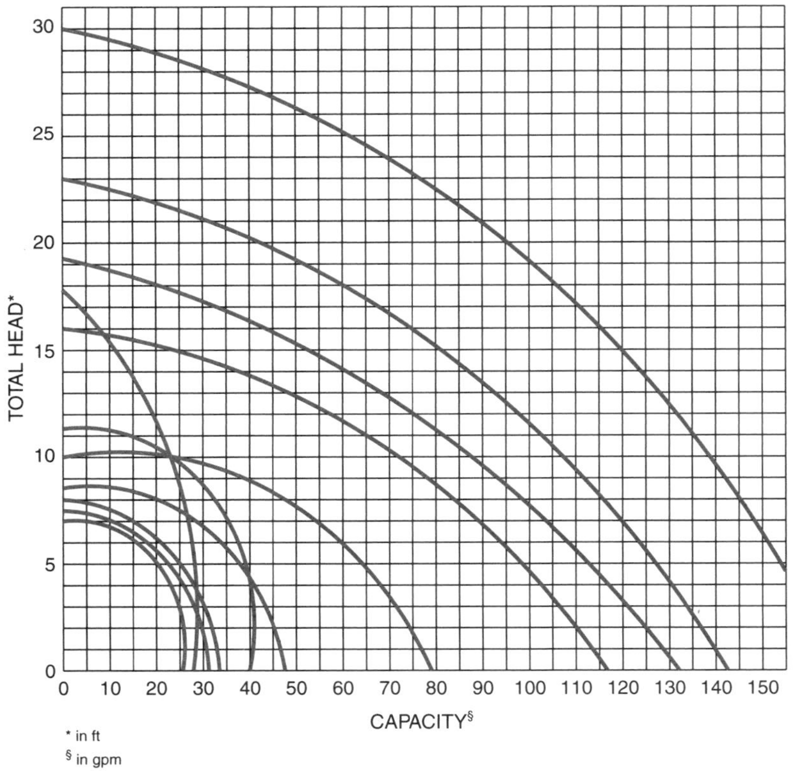

Figure 19-11. Performance data tables show the capacity of circulating pumps.

The equivalent lengths of the fittings in each pipe section are entered in the sixth column on the form. The final column on the form is used to record the equivalent length for each pipe section. The length of each pipe section is found by adding the actual length and equivalent lengths for the fittings horizontally across each row on the sheet. By adding the figures in the last column, the total equivalent length of the system is found. The total equivalent length is used to find the total pressure drop of the system.

The total pressure drop for the system in mils is found by multiplying the total equivalent length by 300. The total pressure drop for the system in feet of head is found by dividing the total equivalent length by 100′ and multiplying by 2.5. The result is the total pressure drop in the pipe and fittings in the system.

Tech Facts

For hydronic systems to operate properly, there should be at least one circulating pump on each piping loop in a system.

3 bedroom residence

Section	Flow Rate*	Size§	Length#	Fittings	FEL**	SEL§§			
A–B	12.0	1.5	8	1 Valve 1 El	1.7	9.7			
B–C	11.2	1.5	9	1 Tee	.5	9.5			
C–D	9.8	1.25	28	2 El 1 Tee	2.5	30.5			
D–E	8.7	1.25	10	1 Tee	.5	10.5			
E–F	7.6	1.25	19	2 45°El 1 Tee	1.9	20.9			
F–G	5.6	1	16	1 El 1 Tee	1.5	17.5			
G–H	3.7	1	14	1 Tee 2 El	2.5	16.5			
B–J	.8	.5	2	1 Tee 1 Valve	1.2	3.2			
C–K	1.4	.625	2	1 Tee 1 Valve	1.2	3.2			
D–L	1.1	.625	2	1 Tee 1 Valve	1.2	3.2			
E–M	1.1	.625	2	1 Tee 1 Valve	1.2	3.2			
F–N	2.0	.75	2	1 Tee 1 Valve	1.2	3.2			
G–O	2.0	.75	2	1 Tee 1 Valve	1.2	3.2			
H–P	3.7	1	2	1 El 1 Valve	1.7	3.7			
				Total Equivalent Length		138			

* in gpm
§ in inches
\# in ft
** fitting equivalent length in ft
§§ section equivalent length in ft

Figure 19-12. The length of the pipe sections and the length of the fittings are added to the columnar form.

The pressure drop through the terminal devices, boiler or chiller, and any other device that would add additional resistance to water flow must be added to the piping pressure drop to find the total system pressure drop. Total system pressure drop is found by applying the formula:

$$PD_t = PD_p + PD_{td} + PD_{hu}$$

where

PD_t = total system pressure drop (in ft of head)

PD_p = piping pressure drop (in ft of head)

PD_{td} = terminal device pressure drop (in ft of head)

PD_{hu} = heating unit pressure drop (in ft of head)

Example: Finding Total System Pressure Drop—Piping

The total system pressure drop through a piping system is 3.5′ of head. The water must flow through six terminal devices with a pressure drop of .65′ of head per device. The water also flows through a boiler with a pressure drop of 1.25′ of head. Find the total pressure drop through the system.

$$PD_t = PD_p + PD_{td} + PD_{hu}$$

$$PD_t = 3.5 + (.65 \times 6) + 1.25$$

$$PD_t = 3.5 + 3.9 + 1.25$$

$$PD_t = \mathbf{8.65'}$$

The size of the circulating pump required for each piping loop is found by using the total flow rate for the piping loop, the total pressure drop in the loop, and a sizing table for the type of pump desired. The circulating pump is selected based on the result.

Fittings and Valves

Pipe fittings and valves control the flow of water. Shutoff valves are located on each side of all major parts of a system, such as the boiler and terminal devices. Balancing valves are located in each branch line off the main supply line and in lines going directly to terminal devices.

Balancing valves are used for balancing water flow through a terminal device to provide the proper heating or cooling effect. Data relating to valves and fittings, especially diverting and balancing valves, should be obtained from manufacturers.

Some piping systems have modulating flow valves, which are located on the supply or return connections to terminal devices. Data and information for using and controlling these valves should be obtained from manufacturers.

Tech Facts

Pipe fittings and valves control the flow of water in a hydronic system.

Review Questions

1. Define hydronic as related to a heating or air conditioning system.
2. List and describe three parts of a piping system used with a hydronic heating or cooling system.
3. List and describe the two factors used when calculating water flow rate through a pipe.
4. Define foot of head as related to pipe sizing.
5. What causes pressure drop in a pipe?
6. How is design static pressure drop used in sizing pipe?
7. Why is the temperature drop through a piping system important?
8. Define friction head, static head, and total head as related to a hydronic system.

9. Why can water flow rate be different in pipes of the same size when the pipes are made of different material?
10. List the seven steps required for designing a hydronic heating or air conditioning system.
11. What is the difference between a one-pipe series piping system and a one-pipe primary-secondary system?
12. What difference must be considered when selecting terminal devices for a one-pipe series system as compared to a one-pipe primary-secondary system?
13. What two variables are used when selecting a circulating pump for a hydronic heating or air conditioning system?
14. What is the difference between the equivalent lengths of the fittings in a hydronic piping system and the total equivalent length of the system?
15. What are the balancing valves used for in a hydronic piping system?
16. What are shutoff valves used for in a hydronic piping system?
17. What variables are used to find the total pressure drop through a piping system?
18. Why are layout drawings important when designing a hydronic system?

A thorough knowledge of electricity and the operation of electrical systems reduces the chances of electrical shock. Proper grounding and the use of overload protection devices are required for safe operation of an electrical system. The proper tools must be used for the applications for which they are designed. Tools must be used correctly for safe and efficient electrical work. Individuals must follow all OSHA and NFPA 70E safety codes when performing electrical work.

20 Electrical Principles, Tools, and Safety

ELECTRICITY

Electricity is the movement of electrons from atom to atom. An understanding of the electrical properties of matter is required because the movement of electrons from atom to atom produces electrical energy.

Matter

Matter is anything that has mass and occupies space. All objects consist of matter. Matter can exist in the state of a solid, liquid, or gas. A *solid* is a state of matter that has a definite volume and shape. A *liquid* is a state of matter that has a definite volume but not a definite shape. A *gas* is a state of matter that is fluid, has a relatively low density, and is highly compressible.

All matter has electrical properties. The electrical behavior of matter varies according to the physical makeup of the matter. Some types of matter, such as copper, allow electricity to easily move through them and can act as conductors. A *conductor* is material that has very little resistance and permits electrons to move through it easily.

Other types of matter, such as rubber, do not allow electricity to easily move through them and can act as insulators. An *insulator* is any material that has a very high resistance and resists the flow of electrons. The properties of different matter must be understood when designing electrical components and circuits, working around electrical equipment, and troubleshooting electrical circuits.

Atoms

An *atom* is the smallest particle that an element can be reduced to and still maintain the properties of that element. The three principle parts of an atom are the electron, neutron, and proton.

An *electron* is a negatively charged particle in an atom. A *neutron* is a neutral particle, with a mass approximately the same as a proton, that exists in the nucleus of the atom. A *proton* is a positively charged particle that also exists in the nucleus of the atom. Every atom has a definite number of protons, neutrons, and electrons. **See Figure 20-1.** The number of protons in an atom determines the atom's weight (mass) and its atomic number. For example, a hydrogen (H) atom has the fewest protons of all atoms, has the least amount of mass, and is assigned the atomic number of one.

A *positive charge* is produced when there are fewer electrons than normal. A *negative charge* is produced when there are more electrons than normal. A proton of an atom has a positive (+) charge, the electron has a negative (–) charge, and the neutron has no charge. The neutrons and protons combine to form the nucleus of an atom. Since the neutron has no charge, the nucleus has a positive (+) charge.

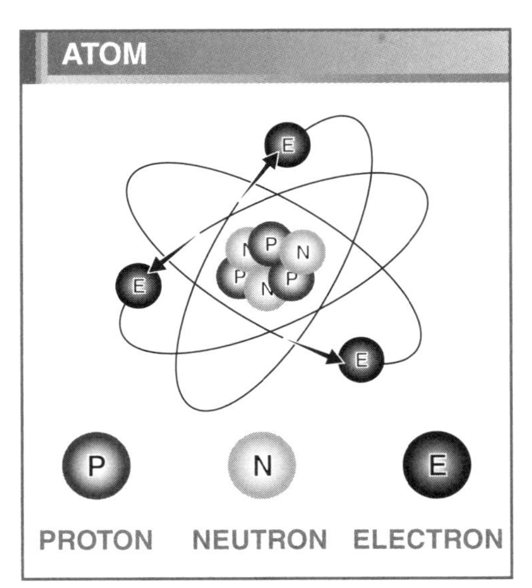

Figure 20-1. An atom is the smallest particle that an element can be reduced to and still maintain the properties of that element.

The electrons orbit the nucleus of an atom, completing billions of trips around the nucleus each millionth of a second. In all atoms the electrons are arranged in shells at various distances from the nucleus according to the amount of energy they have. The total number of shells an atom has varies from one to seven shells. The shells are numbered innermost to outermost 1, 2, 3, 4, 5, 6, and 7 or are lettered K, L, M, N, O, P, and Q. **See Figure 20-2.**

Each shell can hold only a specific number of electrons. The innermost shell can hold two electrons, the second shell can hold eight electrons, the third shell can hold 18 electrons, the fourth shell can hold 32 electrons, the fifth shell can hold 50 electrons, the sixth shell can hold 72 electrons, and the seventh shell can hold 98 electrons. The shells are filled starting with the inner shells and working outward, so that when the inner shell is filled with as many electrons as it can hold, the next shell is started. Most materials used in the electrical/electronic field contain four shells or less.

Valence Electrons

A *valence shell* is the outermost shell of an atom and contains the electrons that form new compounds. For example, two hydrogen atoms combine with the electrons in the outer shell of an oxygen atom to form water. The electrons in the valence shell are important because they can be used to produce an electric current flow.

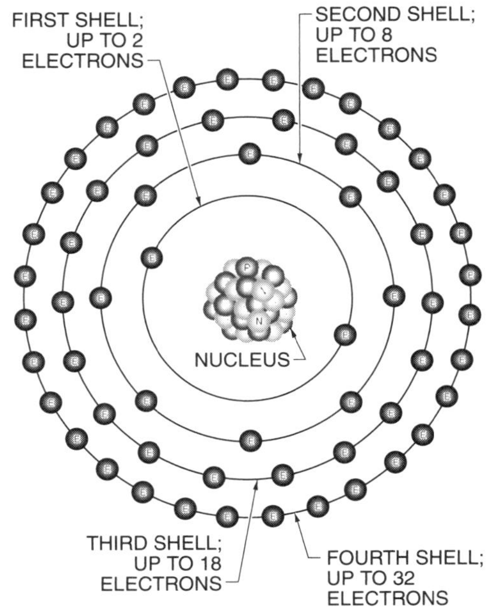

ELECTRON CONFIGURATION

Shell Number	Shell Letter	Maximum Number of Electrons
1	K	2
2	L	8
3	M	18
4	N	32
5	O	50
6	P	72
7	Q	98

Figure 20-2. The electrons in an atom are arranged in shells at various distances from the nucleus according to the amount of energy they have.

Most elements do not have a completed valence shell containing the maximum allowable number of electrons. The number of electrons in the valence shell determines whether an element allows electrons to easily move from atom to atom or resists the flow of electrons. A material that allows electrons to easily move from atom to atom can be used as a conductor. For example, copper is a good conductor because copper atoms allow their valence electrons to move from atom to atom. **See Figure 20-3.**

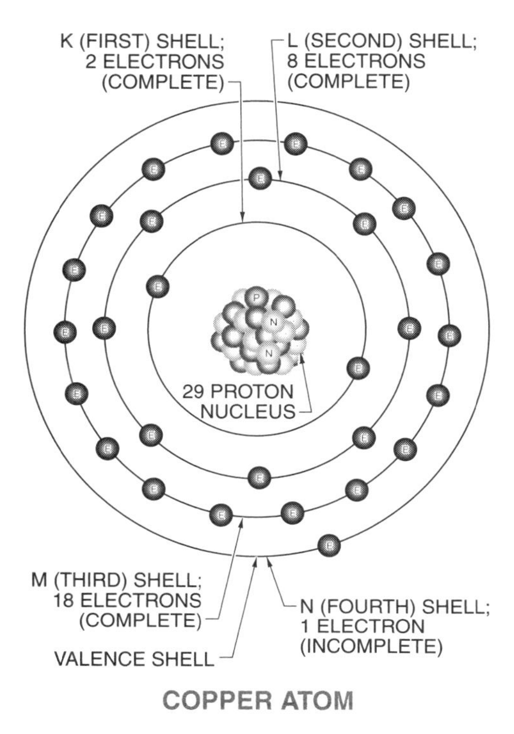

Figure 20-3. Copper is a good conductor because copper atoms allow their valence electrons to move from atom to atom easily.

A material that does not allow electrons to easily move from atom to atom can be used as an insulator. Most insulators used in the electrical field are made from compounds such as glass, rubber, plastic, or paper. Insulators have five or more valence electrons.

Conductors have three or less valence electrons. Most conductors have only one or two valence electrons. Most metals are good conductors. Silver is the best conductor, followed by copper, gold, aluminum, and iron.

A *semiconductor* is an electronic device that has electrical conductivity between that of a conductor (high conductivity) and that of an insulator (low conductivity). Semiconductors (carbon, germanium, and silicon) are made from materials that have exactly four valence electrons. Semiconductor materials are not conductors and not insulators. Semiconductor materials do not conduct electricity easily and are not good insulators.

In a conductor atom, an outside force can be applied to force the atom to lose or gain valence electrons. Electrons can be forced to move by chemical reaction, friction, pressure, heat, light, or a magnetic field. These forces can occur naturally, as with lightning and static electricity, or they can be produced, as in a battery or generator. **See Figure 20-4.**

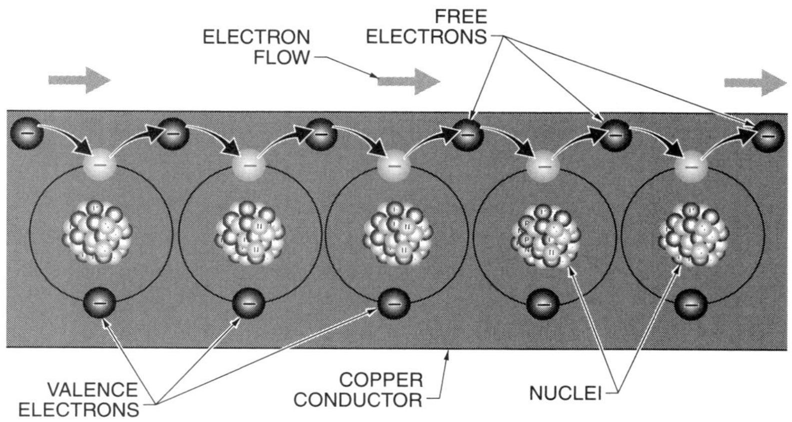

Figure 20-4. In a conductor atom, an outside force can be applied to force the atom to lose or gain valence electrons.

Static Electricity

Electricity may be static or generated electricity. *Static electricity* is an electrical charge at rest. The electrical shock a human encounters by touching an object is an example of the transfer of a static charge by contact. Lightning is an example of the transfer of a static charge by a spark. Static electricity has some limited practical uses, such as in electrostatic air filters and electrostatic spray-painting.

Generated Electricity

The other form of electricity is electricity that is generated. *Generated electricity* is the alternating current (AC) created by power plant generators. Power plant generators create three-phase alternating current electricity. Generators produce electricity when magnetic lines of force are cut by a rotating wire coil (armature). **See Figure 20-5.** The stronger the magnetic field and/or the faster the rotation, the higher the voltage produced.

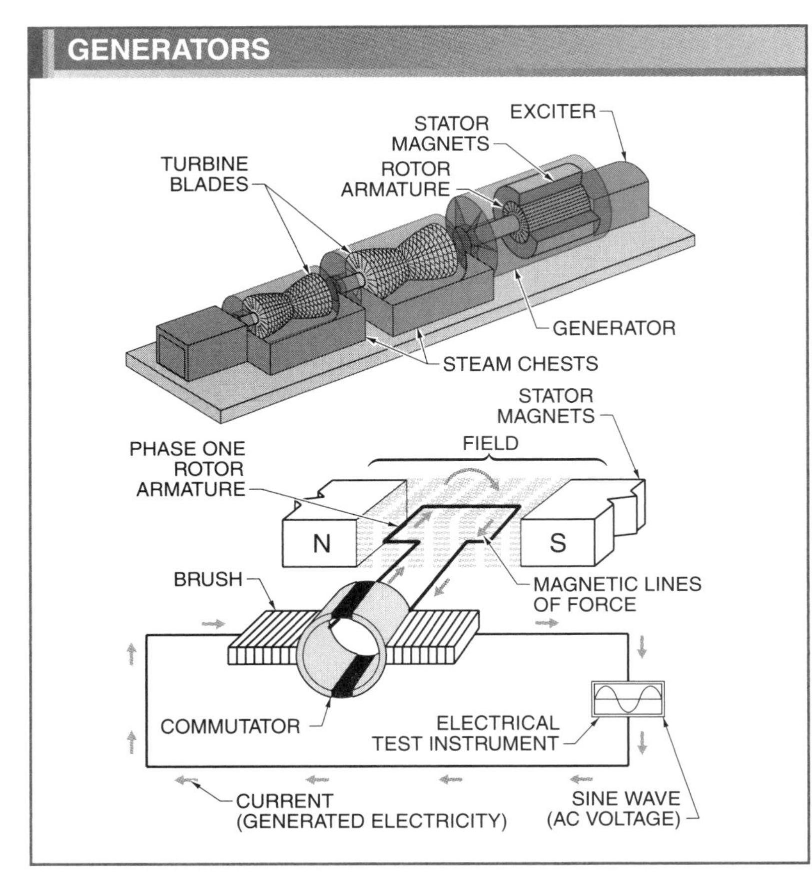

Figure 20-5. Generators produce electricity when magnetic lines of force are cut by a rotating wire coil (armature).

The voltage output of commercial generators is several thousand volts (typically 22,000 V). The AC electrical output of commercial (power plant) generators travels to transformers for voltage and current control prior to distribution. A *transformer* is an electric device that uses electromagnetism to change (step-up or step-down) AC voltage from one level to another. **See Figure 20-6.** Electricity is used in circuits specifically designed to carry voltage and current through controlled paths (conductors) to operate specific loads.

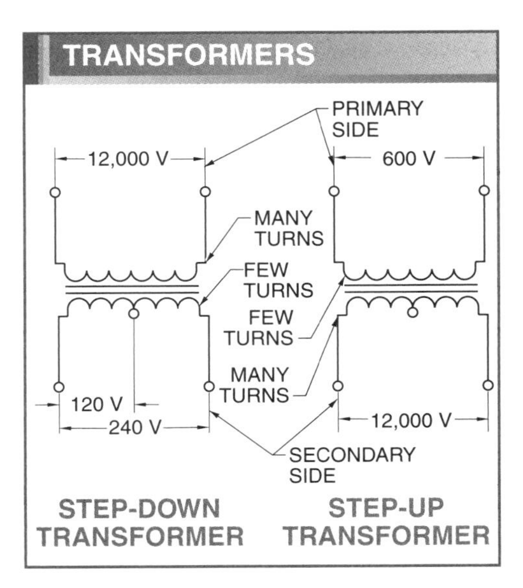

Figure 20-6. A transformer uses electromagnetism to change (step-up or step-down) AC voltage from one level to another.

> **Tech Facts**
>
> *Alternating current (AC) is created by power plant generators. Power plant generators create three-phase alternating current electricity.*

Electricity Distribution

The transformers at a power plant step-up the voltage from the generators (22,000 V) to approximately 240,000 V. **See Figure 20-7.** The high-voltage electricity is distributed across high-voltage power lines to step-down transformer substations. The substations step-down the electrical voltage from 240,000 V to 12,000 V for local distribution. The local voltage distribution level of 12,000 V is step-downed further by transformers to voltages of 600 V, 480 V, 240 V, or 120 V for consumer use.

Local underground vaults and pole transformers step-down electricity to 120V/240V to service apartments, condominiums, townhomes, and homes. **See Figure 20-8.** Service entrances (overhead or underground) provide the connection for the distributed power to the power company meter of the residence. Electric meters record the amount of electricity used.

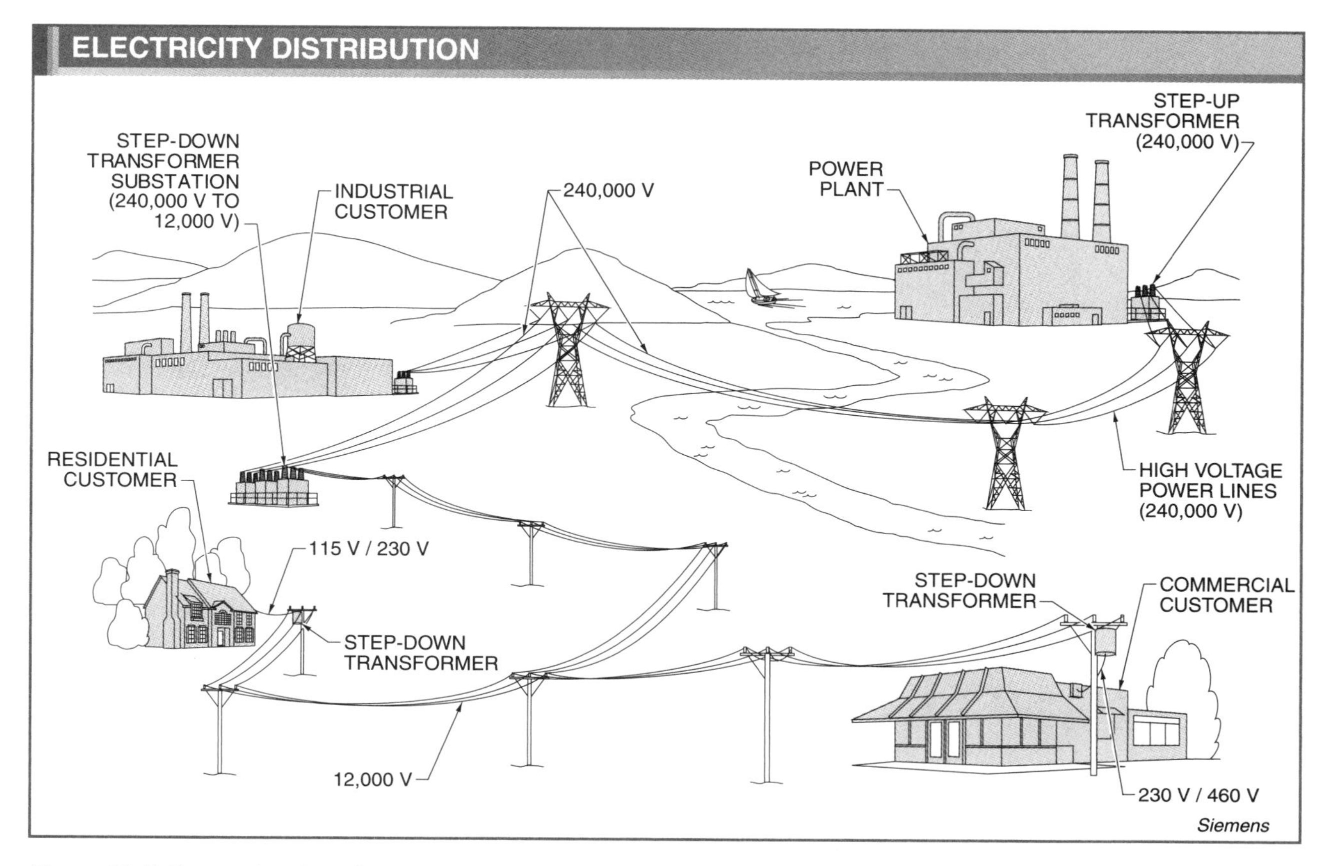

Figure 20-7. Power plant transformers step-up voltage to 240,000 V for the efficient distribution of electricity.

Service Panels

A *service panel* is an electrical device containing fuses or circuit breakers for protecting the individual circuits of a residence and serves as a means of disconnecting the entire residence from the distribution system. **See Figure 20-9.** Service panels are wall-mounted in basements, utility rooms, and attached garages. Wherever a service panel is mounted, the panel must be easily accessible to the occupants of the residence.

Tech Facts

Manufacturers of electrical equipment, such as appliances, tools, wires, and fuses, must meet minimum required codes and standards to ensure their products are built for safe operation. Exceeding minimum requirements helps establish the reputation of a company for producing reliable quality products and also reduces product liability issues that may arise from electrical shock, fire, or injury.

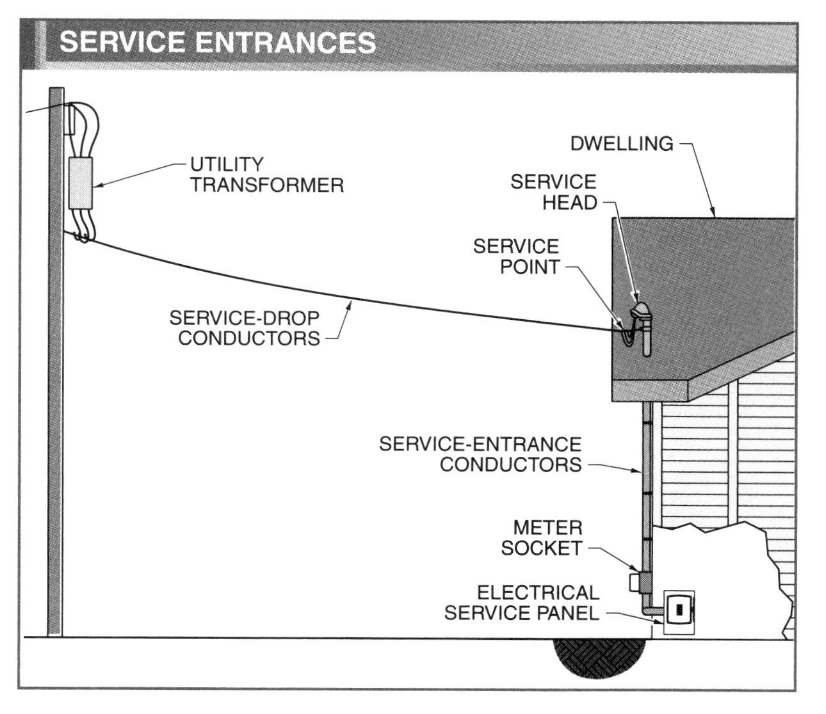

Figure 20-8. Clearance requirements protect service-drop conductors from physical damage and protect personnel from contact with the conductors.

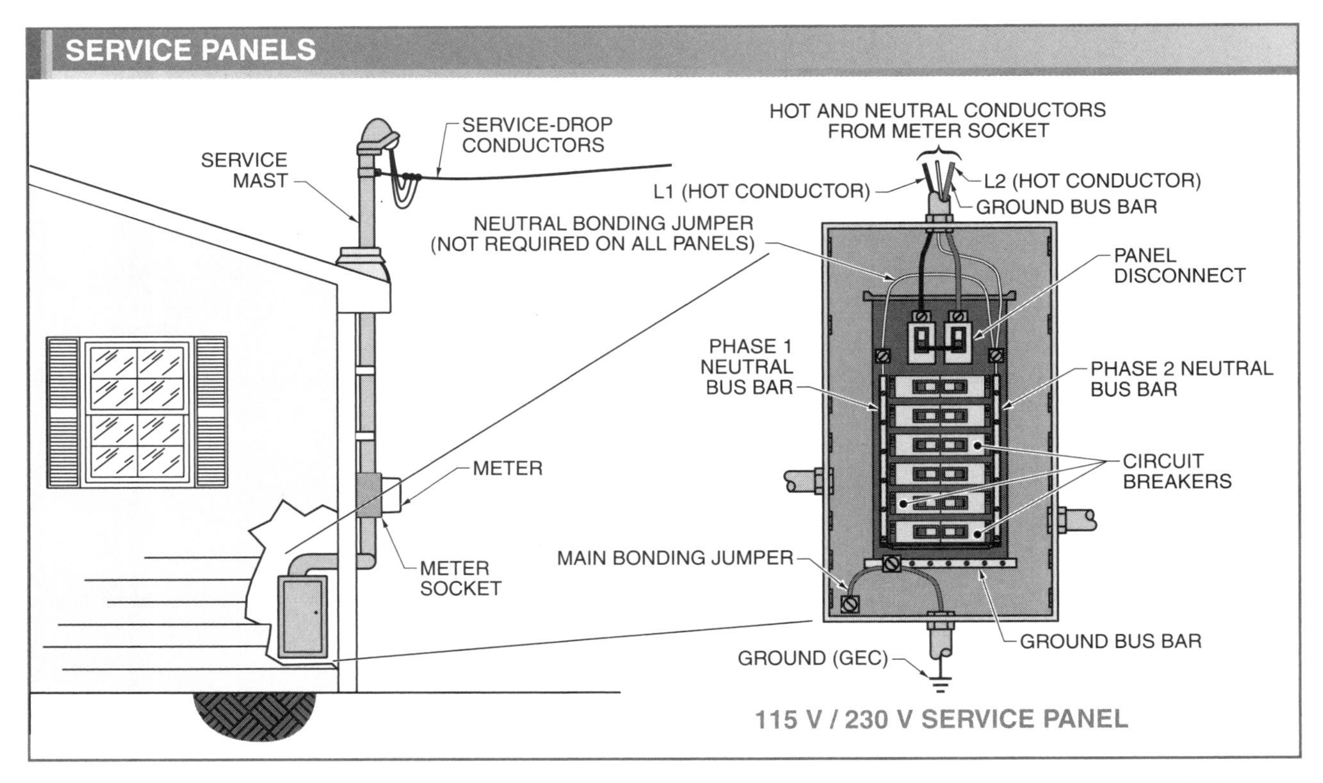

Figure 20-9. Service panels provide the disconnect from the utility company, system ground, circuit breakers for hot conductors, and neutral bus bars for returning common conductors.

RESIDENTIAL ELECTRICAL CIRCUITS

Residential electrical circuits include lighting circuits, general-purpose receptacle (small appliance) circuits, special-purpose receptacle (ranges, ovens, and GFCIs) circuits, and special hard-wired loads such as room heaters, water heaters, and air conditioner circuits. A *circuit* is a complete path (when ON) for current to take that includes electrical control devices, circuit protection, conductors, and load(s). A complete path is a grouping of electrical devices and wires that create a path for current to take from the power source (service panel), through controls (switches), to the load (light fixtures and receptacles), and back to the power source. The service panel also provides circuit protection by fuses or circuit breakers.

Grounding

Grounding is an integral part of any properly operating electrical system. In residences, grounding protects the occupants by providing a safe pathway for unwanted electricity that might otherwise create a hazard. Electricity always takes the easiest flow path to ground.

Grounding is typically established at two levels: system grounding and equipment grounding. A *system ground* is a special circuit designed to protect the entire distribution system of a residence. An *equipment ground* is a circuit designed to protect individual components connected to an electrical system.

The primary function of system grounding is to protect the service entrance wiring and connections from lightning and high voltage surges. A system ground provides protection by safely routing lightning and high voltage surges away from the service entrance through the use of a highly conductive pathway to Earth. The two methods of grounding an electrical system are metal water pipe grounding and alternative electrode grounding. **See Figure 20-10.**

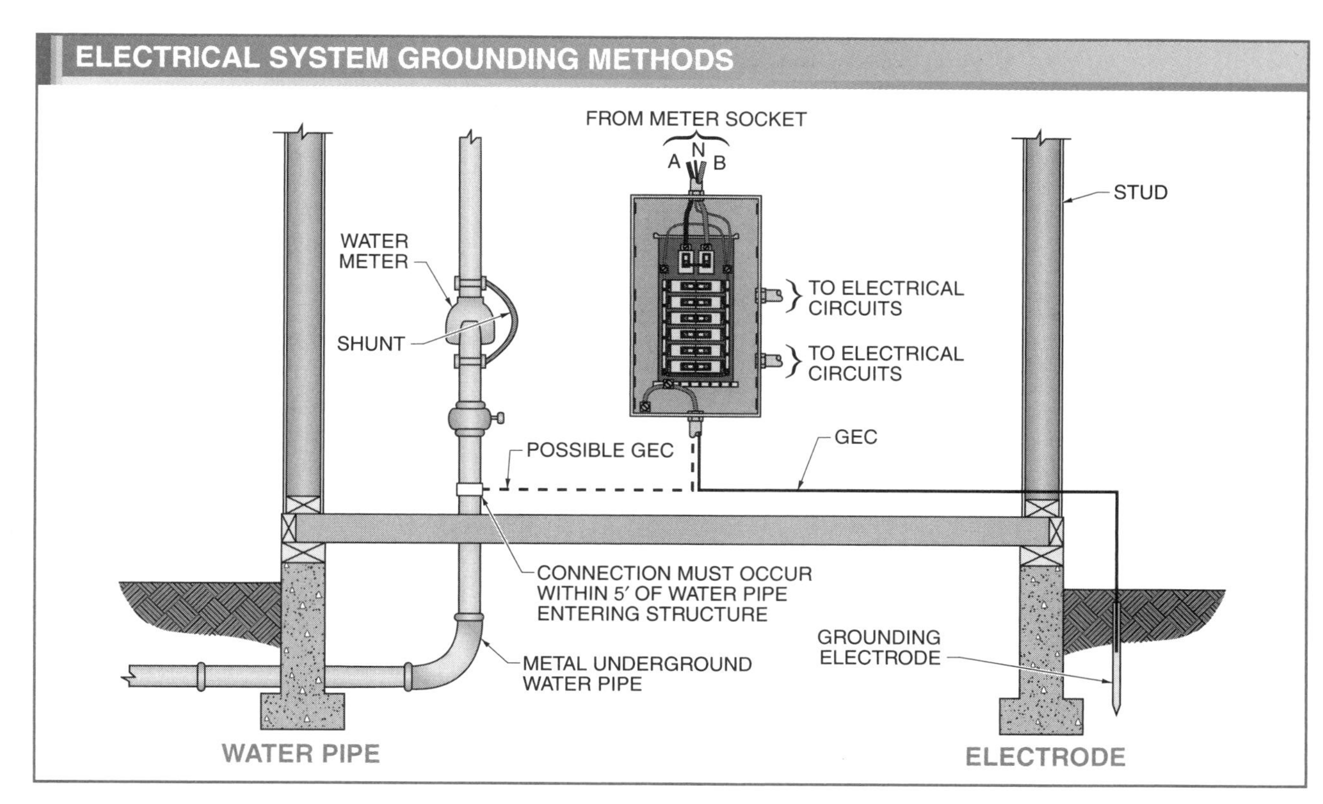

Figure 20-10. Water pipe grounding and electrode grounding are the two main types of grounding for residences.

Water Pipe Grounding. A water pipe ground is a continuous underground metallic pipe that supplies a residence with water and is typically the best electrical ground for a residential electrical system. Water pipes work well as grounds because a large surface area of the pipe is in contact with the earth. The large surface area reduces resistance and allows any unwanted electricity to easily pass through the pipe to the earth.

When a water pipe is used for grounding, the water pipe run must never be interrupted by a plastic fitting or have an open section of plumbing. Water meters are a source of open circuits when removed. To provide protection when a water meter is removed, a shunt, or meter bonding wire, must be permanently installed. A *shunt* is a permanent conductor placed across a water meter to provide a continuous flow path for ground current.

Electrode Grounding. An *electrode* is a long metal rod used to make contact with the earth for grounding purposes. When no satisfactory grounding electrode is readily available, the common practice is to drive one or more metal rods (connected in parallel) into the ground. The electrode and circuit must provide a flow path to earth with less than 25 Ω of resistance. When local soils are extremely salty, acidic, or alkaline, the local building inspector or the NEC® should be consulted to determine the correct type of electrode to use.

Equipment Grounding. The primary function of equipment grounding is to protect individual electrical devices. *Equipment grounding* safely grounds any devices attached to a system or plugged into receptacles inside a home. **See Figure 20-11.** When an appliance has not been properly grounded, electrical current caused by a short will seek the easiest path to earth. Unfortunately, the human body is an electrical conductor and allows current to reach earth by traveling through the body (electric shock). Properly grounded devices protect the body by harmlessly conducting unwanted electricity to ground.

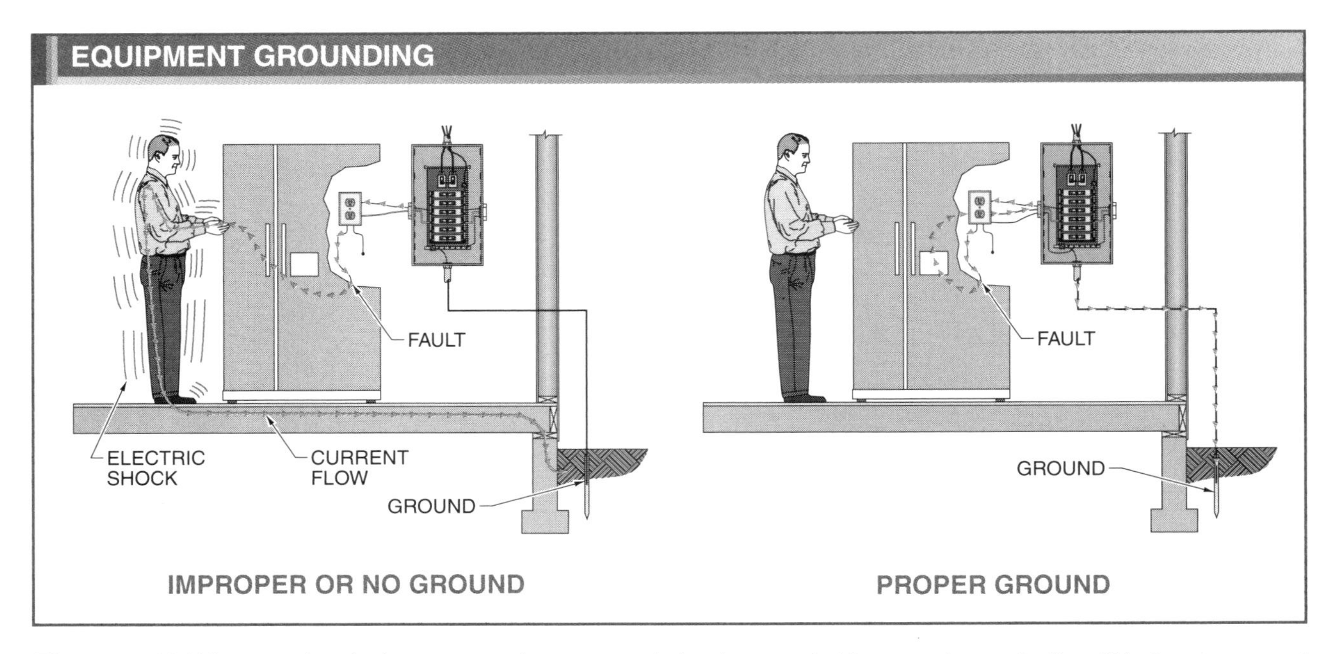

Figure 20-11. When an electrical component is not grounded or is grounded improperly, any faults within the component can create a potentially hazardous situation.

Small appliances are easily incorporated into a grounded system. Most small electrical appliances are designed with three-prong grounded plugs that match a standard three-prong grounded receptacle.

The U-shaped blade of the plug and the U-shaped hole in the receptacle are the ground connections. The U-shaped blade of a plug is longer than the current-carrying blades. The added length insures a strong ground connection while the plug is being inserted or removed from a receptacle.

Electrical circuits typically include a component (load), such as a ceiling fan.

Fuses

A *fuse* is an electric overcurrent protection device used to limit the rate of current flow in a circuit. Fuses typically contain a piece of soft metal that melts, opening the circuit when the fuse is overloaded.

The three broad categories of fuses are plug, cartridge, and blade. The typical homeowner generally requires 15 A or 20 A plug fuses and 30 A or 40 A cartridge fuses. Occasionally a service panel blade fuse may need to be replaced. Blade fuses typically have a 60 A or 100 A rating and are found in the main disconnect of the service panel. For each category of fuse, several specific types of fuses can be found for specific applications. **See Figure 20-12.** Time-delay fuses are one specific type of fuse that is quite popular.

Standard Plug Fuses. A *standard plug fuse* is a screw-in type electrical safety device that contains a metal conducting element designed to melt when the current through the fuse exceeds the rated value. During a serious overload condition, the melting of the conductor strip is almost instantaneous.

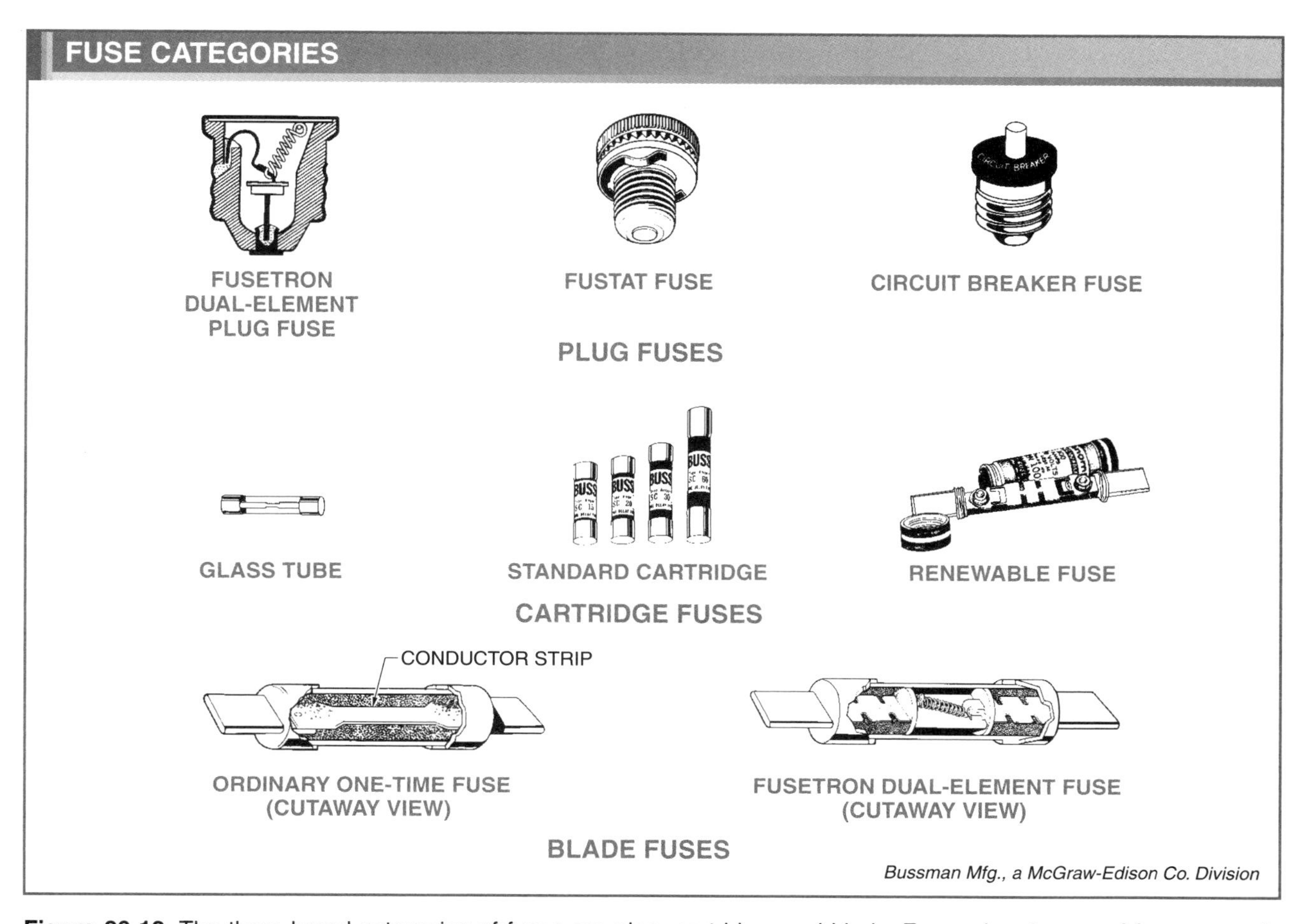

Figure 20-12. The three broad categories of fuses are plug, cartridge, and blade. For each category of fuse, several specific types of fuses can be found for specific applications.

Time-Delay Plug Fuses. A *time-delay plug fuse* is a screw-in type electrical safety device with a dual element. The first element provides the protection of a standard plug fuse for short circuits, and the second element protects against heating due to light overloads (the second element does not instantaneously melt). The second element is useful in preventing nuisance tripping caused by motor-driven appliances such as refrigerators and air conditioners starting up. Manufacturers of time-delay plug fuses have created plugs with various responses to shorts and overloads.

Type S Plug Fuses. A *Type S plug fuse* is a screw-in type electrical safety device that has all the operating characteristics of a time-delay plug fuse plus the added advantage of being non-tamperable. A Type S fuse is considered non-tamperable because the fuse cannot be installed into a base unless the fuse matches the size of the base. Each Type S base adapter is sized for a particular size fuse. Due to the non-tamperable design, a 20 A Type S fuse will not fit into a 15 A Type S base.

Cartridge Fuses. A *cartridge fuse* is a snap-in type electrical safety device that contains a wire that is designed to carry a specific amount of current. However, instead of a screw base, cartridge fuses are secured by clips. Many cartridge fuses have dual elements that absorb temporary overloads (such as with motor startup) without blowing, while protecting against dangerous overloads and short circuits. Cartridge fuses should be removed and installed using fuse pullers.

Cartridge fuses can be purchased as one-time-use fuses or as renewable element fuses. Renewable element cartridge fuses have replaceable elements and are typically used for industrial applications.

Blade Cartridge Fuses. A blade cartridge fuse is another snap-in type electrical safety device that also operates on the heating effect of the element. Blade cartridge fuses are typically used for high amperage rated applications found in industry. Blade fuses are secured by clips and may be one-time use or renewable.

Circuit Breaker Fuses. A *circuit breaker fuse* is a screw-in electrical safety device that has the operating characteristics of a circuit breaker. The advantage of a circuit breaker fuse is that the fuse can be reset after an overload and reused.

Circuit Breakers

Circuit breakers are similar to fuses in function. A *circuit breaker* is an overcurrent protection device with a mechanical mechanism that may manually or automatically open a circuit when an overload condition or short circuit occurs. Unlike most fuses, circuit breakers can be reset. They are the most popular overcurrent safety device. **See Figure 20-13.** The trip lever handle of a circuit breaker represents the position of the contacts when the circuit breaker is ON and when the breaker is tripped OFF. A circuit breaker may be reset by moving the trip lever handle to the full OFF position and then returning the handle to the ON position. Individuals must ensure the source of an overload is repaired before attempting to reset a breaker. The three types of circuit breakers are thermal, magnetic, and thermal-magnetic.

Thermal Circuit Breakers. A *thermal circuit breaker* is an electrical safety device that operates with a bimetallic strip that warps when overheated. The bimetallic strip is two pieces of metal made up of dissimilar materials that are permanently joined together. Because metals expand and contract at different rates, heating the bimetallic strip causes the strip to warp or curve. The warping effect of the bimetallic strip is utilized as the tripping mechanism for the thermal breaker.

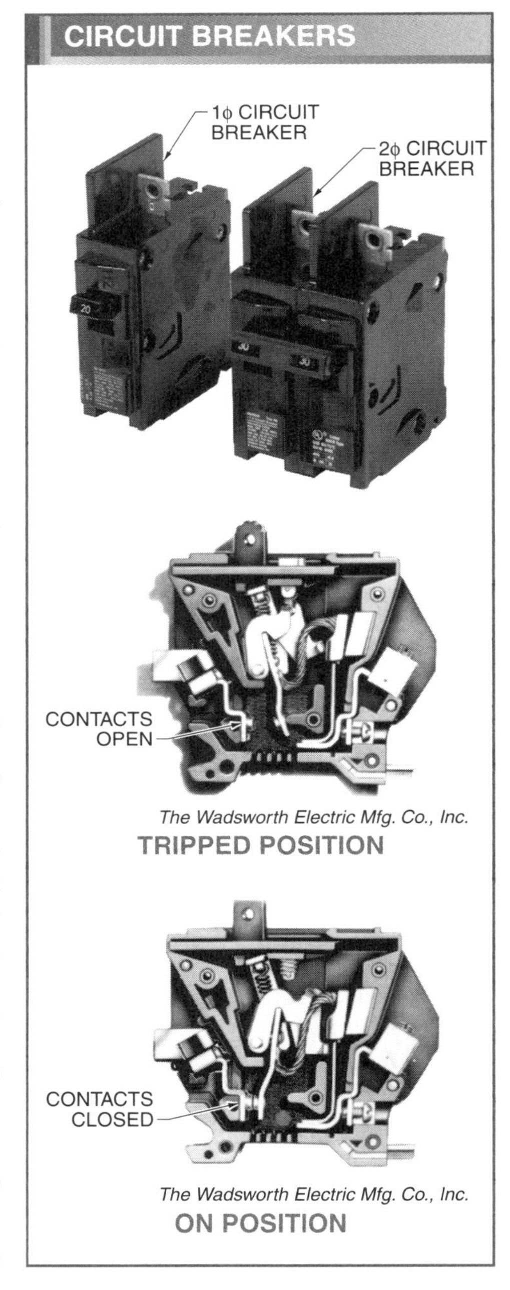

Figure 20-13. Circuit breakers have standard voltage ratings with varying amperages, similar to fuses, but can be reset after encountering an overload.

Once a thermal breaker has tripped, the bimetallic strip cools allowing the strip to re-shape and allowing the breaker to be reset. Individuals must reset the trip lever handle to reactivate the circuit breaker and restore power to the de-energized circuit. As with all circuit breakers, the source of an overload must be removed before attempting to reset the breaker.

Magnetic Circuit Breakers. A *magnetic circuit breaker* is an electrical safety device that operates with miniature electromagnets. Electromagnets are created by passing current through a coil of wire. The greater the current, the stronger the magnetic field created by the coil. When the current through the coil exceeds the rated value of the breaker, the magnetic attraction becomes strong enough to activate the trip bar and open the circuit. Once the overload is removed, the trip bar can be reset to the original position reactivating the circuit.

Thermal-Magnetic Circuit Breakers. A *thermal-magnetic circuit breaker* is an electrical safety device that combines the heating effect of a bimetallic strip with the pulling strength of a magnet to move the trip bar. The magnetic portion consists of a permanent magnet or electromagnet in series with the bimetallic strip. Thermal-magnetic breakers have the fastest response times to serious overloads of any type of circuit breaker.

Circuit breakers are available in a variety of amperages, but the voltage is typically rated as 115 V for single pole residential breakers or 230 V for double pole residential breakers.

Ground Fault Circuit Interrupters

A *ground fault circuit interrupter (GFCI)* is a fast-acting receptacle that detects low levels of leakage current to ground and opens the circuit in response to the leakage. **See Figure 20-14.** GFCIs limit electric shock by opening circuits before an individual receives a serious injury. GFCIs do not protect against line-to-line contact.

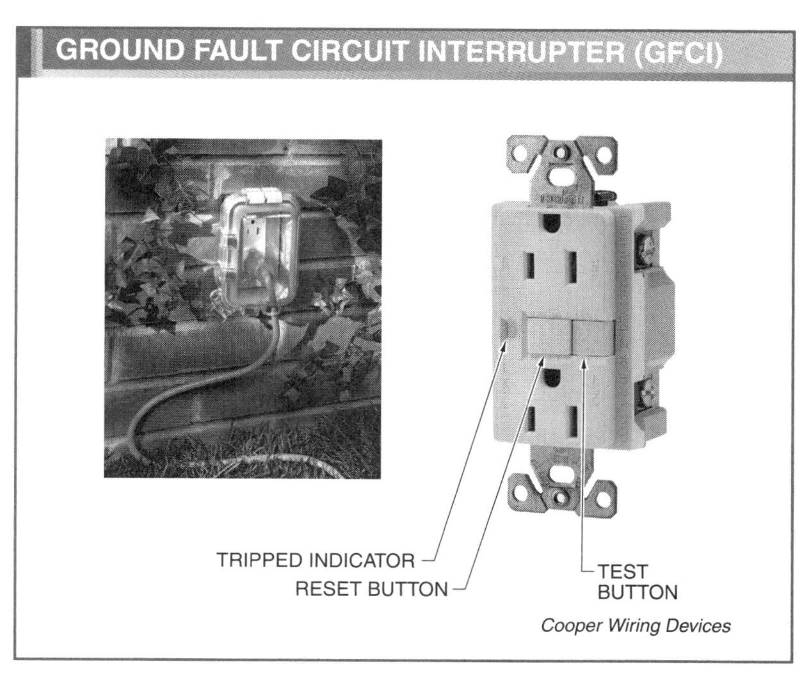

Figure 20-14. GFCIs are used for outdoor receptacle applications and anywhere extra protection is required.

Ground fault circuit interrupters were created because of the limitations of commonly used circuit breakers. Individuals have been electrocuted by equipment in electrical systems where the fault current was not great enough to trip a standard circuit breaker. The situation exists because standard circuit breakers are designed to trip only when large currents are present (short or overload). Currents present from deteriorating insulation and minor equipment damage do not produce enough current flow to open a standard circuit breaker. As an alternative, GFCIs are specifically designed with leakage currents in mind. A GFCI can react to a current as small as .005 A in a fraction of a second.

GFCI Operation. A sequence of events takes place when a fault current is detected by a GFCI. Typically, the load (electric shaver, drill, or garden tool) has the same amount of current flowing to the load (black or red wire) as flowing away from the load (white wire). However, in the event of a ground fault, some of the current, which normally returns to the power source through the white wire (neutral), is diverted to ground.

Because a current imbalance exists between the hot and neutral wires, the sensing device detects the current difference and signals an amplifier. When the amplified signal is large enough, the amplifier activates the interrupting device. Once the interrupting device is activated, the circuit is opened and current to the load is shut off. After the ground fault has been discovered and repaired, the GFCI is reset and the process begins again. GFCIs have a test circuit built into the unit so that the GFCI can be tested on a monthly basis without a ground fault condition.

The National Electrical Code (NEC®) requires ground fault circuit interrupters for protection in the following areas:

- Residential
- Outdoor receptacles
- Bathrooms
- Residential garages
- Kitchens (countertop areas)
- Unfinished basements and crawl spaces
- Construction Sites
- Mobile Homes and Mobile Home Parks
- Outdoor receptacles
- Bathrooms
- Swimming Pools and Fountains
- Receptacles near pools
- Lighting fixtures and lighting outlets near pools
- Underwater lighting fixtures over 15 V
- Electrical equipment used with storable pools
- Fountain equipment operating at over 15 V
- Cord-connected and plug-connected equipment for fountains

GFCIs are installed in standard receptacle boxes indoors and in weatherproof boxes outdoors. GFCI receptacles are typically installed individually. GFCIs can also be installed to protect several standard receptacles in one circuit. GFCIs are considered easy to install and most manufacturers supply an installation kit to aid in installation. In addition to receptacle types, GFCIs are available as circuit breakers to protect entire circuits. Kits are also available to match GFCIs to existing circuit breaker panels.

TOOLS

Electrical work requires quality tools. Experience has shown that individuals are more productive when they use quality tools and instruments. Quality tools last longer and more consistently provide quality results.

Organized Tool Systems

Because electrical tools are expensive to replace, individuals must protect their investment. To be effective, electrical tools must be available when needed, and protected from damage caused by the environment and daily use. An organized tool system must provide both a central location for retrieval of electrical tools and a means of protecting the tools during storage.

Electrical tools are organized in several ways depending upon where and how frequently the tools are used. When tools are used at a repair bench or at home for hobbies, a pegboard may be appropriate. When electrical tools are used at a construction site, a leather pouch is typically used to carry the most commonly used tools. When tools are to be transported to another job site or home, a toolbox is typically used.

Pegboard is a thin constructed board typically available in 4′ x 8′ sheets that is perforated with equally spaced holes for accepting hooks. Pegboard is constructed of heavy-duty tempered material that is best suited for the weight of tools. When pegboard is mounted, outlines of tools can be made at the hooks for inventory purposes.

An electrical tool pouch is a small, open tool container (pouch) for storing commonly used electrical tools. An electrical tool pouch is typically made of heavy-duty leather, and is designed to be used with a belt that holds the pouch in place.

Tool pouches can be simple in design and hold only a few tools, or they can be relatively large and designed to hold a wide selection of tools. The type of pouch chosen depends on the type of work performed.

Most individuals prefer to store tools in a quality toolbox. A well-designed toolbox can be locked and will keep tools clean and dry. In addition, toolboxes, whether fixed or portable, provide a place where all tools can be collected. To ensure a complete inventory of tools at the end of the day or after a job, a list of all tools owned can be kept in the toolbox as a checklist. Whichever system is chosen, organization is a must. An organized tool system ensures that tools are kept clean, dry, and at the fingertips when needed.

TOOL SAFETY

All tools can be dangerous when left in the wrong place. Many accidents are caused by tools falling off ladders, shelves, and scaffolds that are being used or moved.

Every tool should have a designated place in a toolbox. Tools should not be carried in clothes pockets unless the pocket is designed for the tool. Keep pencils in a pocket designed for pencils.

Keep tools with sharp points away from the edges of benches or work areas, because individuals brushing against a tool can cause the tool to fall and injure a leg or foot. When carrying edged and sharply pointed tools, carry the tool with the cutting edge or the point down and outward from the body. An individual setting a tool down should place the tool back in a toolbox, or at least in a safe location. Electrical tools may be classified as hand tools, power tools, and test instruments.

Hand Tools

Tool safety requires tool knowledge. Tool safety not only requires that the proper type of hand tool be chosen, but also that the correct size of tool be chosen. Individuals should use quality hand tools, and use the tools only in the manner in which they were designed to be used. **See Figure 20-15.**

Individuals must understand how to correctly use each tool owned. Tools must not be forced into a use beyond normal function. Questions must be asked when the proper and safe operation of a tool is not understood. The cost of a tool and the time taken to purchase a tool will prove far less costly than a serious accident caused by using a tool incorrectly.

Periodic checks of hand tools will aid in keeping tools in good condition. Always inspect a tool before using the tool. Do not use a tool that is in poor condition. Tool handles must be free of cracks and splinters and be fastened securely to the working part of the tool. Damaged tools are not only dangerous to the user but are also less productive than tools in good working condition. When inspection indicates a dangerous condition, repair or replace the tool immediately.

Cutting tools must be sharp and clean, because dull tools are dangerous. The extra force exerted when using dull tools often results in losing control of the tool. Dirt or oil on a tool can cause the tool to slip on the work and cause injury.

Power Tools

Do not use any power tools without knowing the operation, methods of use, and safety precautions of the power tool. **See Figure 20-16.** Authorization from a supervisor can be required before using a power tool.

All power tools are grounded except for power tools that are approved double-insulated. Non-double insulated power tools must have a three-conductor cord. A three-prong plug connects to a grounded receptacle (outlet). OSHA and local codes must be consulted for the proper grounding specifications of power tools. A proper ground ensures that a faulty power tool causing an electrical short will trip a circuit breaker or blow a fuse in the circuit.

Double-insulated tools have two prongs and have a notation on the specification plate that the tool is double-insulated. Double-insulated tools are relatively safe but grounded tools are typically used on job sites. Electrical parts in the motor of a double-insulated tool are surrounded by extra insulation to help prevent shock; therefore, the tool does not have to be grounded. Both the interior and exterior of double-insulated tools must be kept clean of grease, dirt, and water that might conduct electricity.

HAND TOOLS

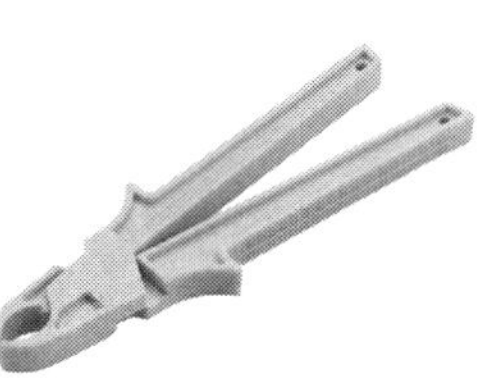

Ideal Industries, Inc.

FUSE PULLER

Safely removes cartridge-type fuses

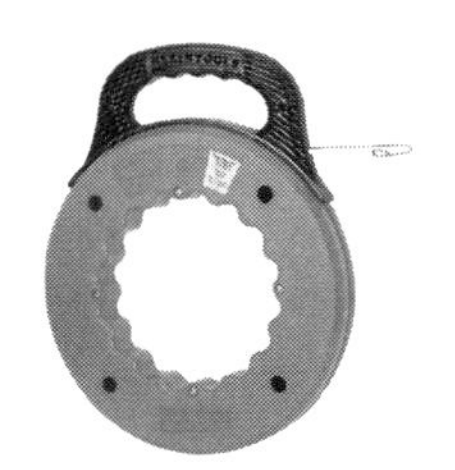

Klein Tools, Inc.

FISH TAPE

Used to pull wire through conduit and wires around obstructions in walls

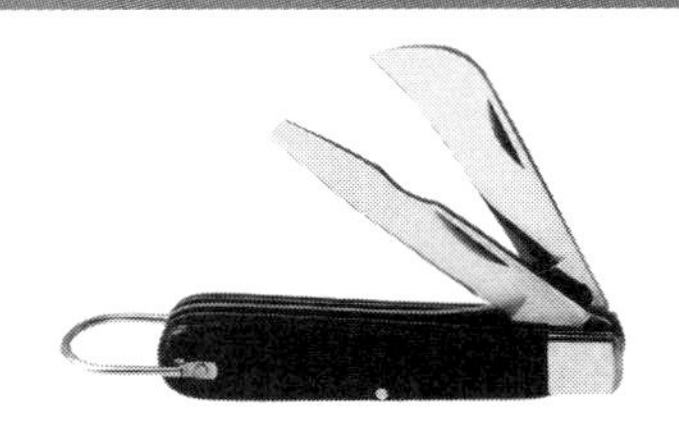

Klein Tools, Inc.

ELECTRICIAN'S KNIFE

Removes insulation from nonmetallic cable and service conductors

Greenlee Textron Inc.

WIRE STRIPPER

Removes insulation from small-diameter wire

Klein Tools, Inc.

CRIMPING PLIERS

Used to crimp die marked insulated and noninsulated solderless terminals

The Stanley Works

DIAGONAL CUTTING PLIERS

Useful for cutting wire and cables that are difficult to reach

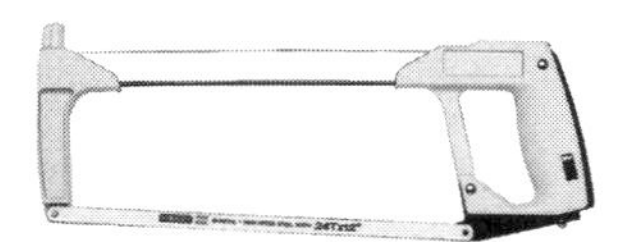

The Stanley Works

HACKSAW

Cuts heavy cable, pipe, and conduit

Klein Tools, Inc.

LONG-NOSE PLIERS

Useful for bending wire and positioning small components

The Stanley Works

HAMMER

Used to mount electrical boxes with nails, and determine height of receptacle boxes

Klein Tools, Inc.

ADJUSTABLE WRENCH

Tightens thick items such as hex head bolts and nuts, conduit couplings, and lag bolts

The Stanley Works

SIDE-CUTTING PLIERS

Used to cut cable, remove knockouts, twist wire, and debur conduit

Klein Tools, Inc.

POCKET LEVEL

Useful in leveling thermostats and conduit bends

Klein Tools, Inc.

UTILITY KNIFE

Cuts and scores drywall for drywall operations

Klein Tools, Inc.

TONGUE-AND-GROOVE PLIERS

Used to tighten box connectors, locknuts, and conduit couplings

The Stanley Works

FLATHEAD SCREWDRIVER

Used to install and remove threaded fasteners

Figure 20-15. Individuals should use quality tools and instruments, and choose the correct type and size of hand tool for the task.

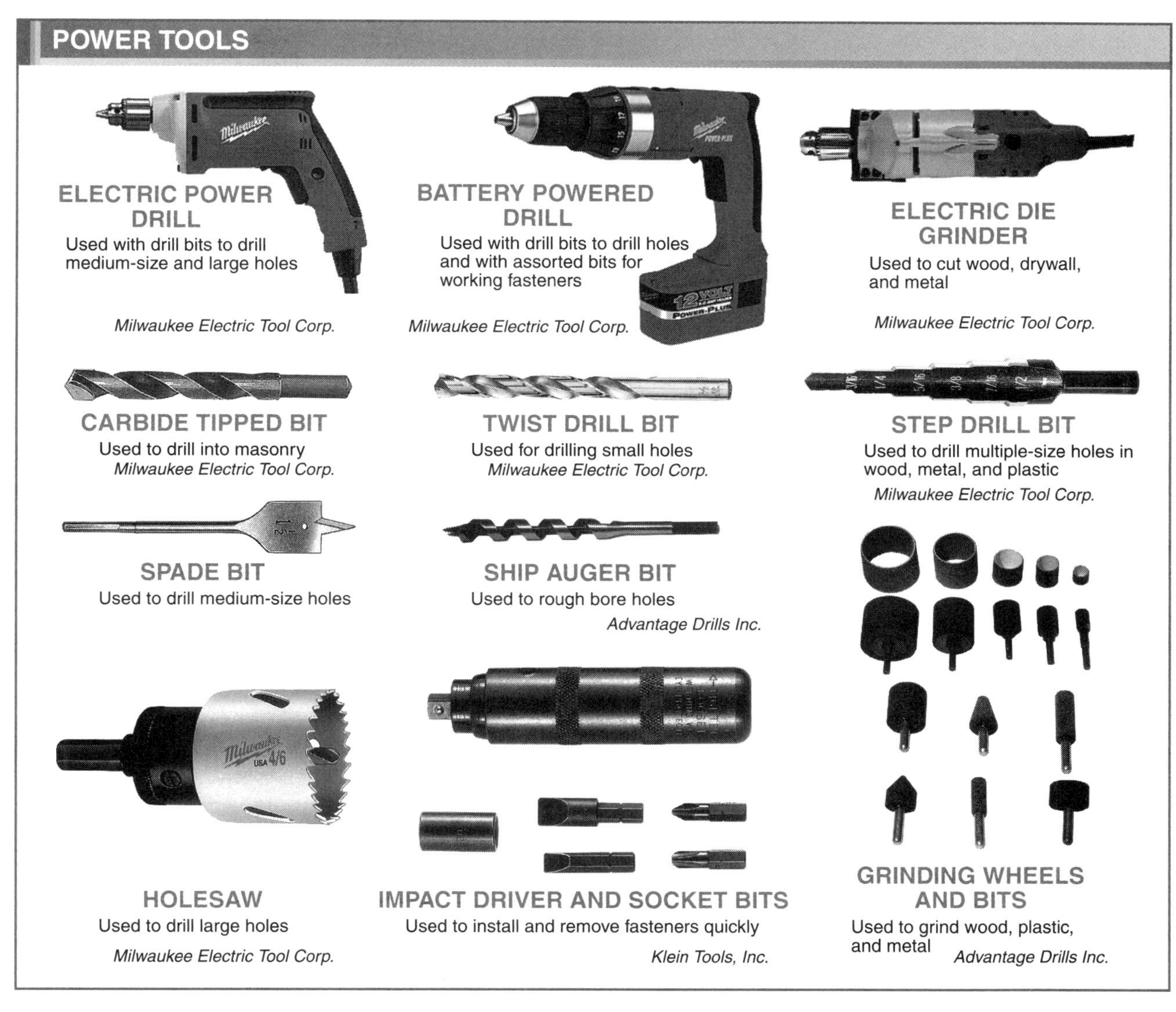

Figure 20-16. Power tools must only be used with a proper understanding of tool operation and the safety precautions associated with the tool.

Power Tool Safety Rules. When handling power tools, the following safety rules must be followed:

- Power tools must be inspected and serviced by qualified repair personnel at regular intervals as specified by the manufacturer or OSHA.
- Know and understand all safety recommendations of the manufacturer.
- Inspect electrical cords to verify that the cords are in good condition.
- Verify that all safety guards are in place and in proper working order. Do not remove guards or any other safety devices.
- Make all adjustments, blade changes, and inspections with the power OFF (cord disconnected).
- Before connecting a tool to a power source, ensure that the on/off switch is in the OFF position.
- Wear safety goggles and a dust mask when using power tools.
- Ensure that the material to be worked is free of obstructions and securely clamped.

- Be attentive (focused on the work) at all times.
- Investigate any change in sound immediately. A change in the sound of a power tool during operation normally indicates a problem.
- Shut the power OFF when work is completed. Wait until a portable tool stops before leaving or laying down the tool.
- Remove a defective power tool from service. Alert other personnel to the situation.
- Avoid operating power tools in locations where sparks could ignite flammable materials.
- Unplug power tools by pulling on the plugs. Never pull on the cord.

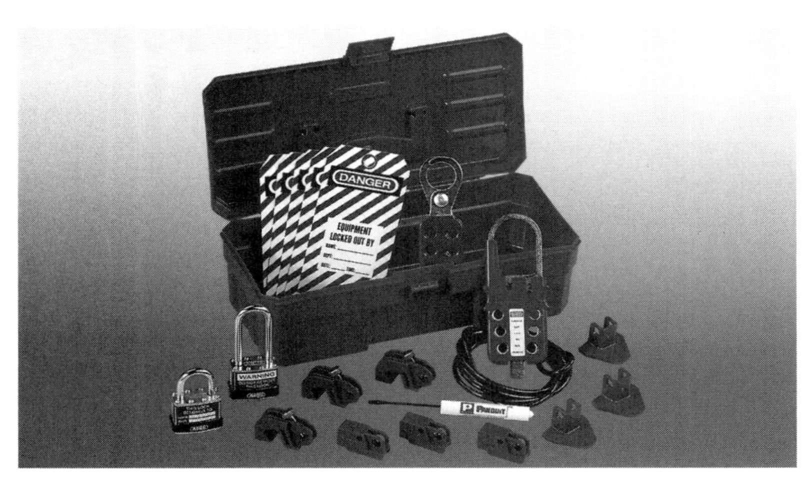

Lockout/tagout devices are available for locking out and tagging out any type of electrical equipment.

ELECTRICAL SAFETY

An individual must understand electricity and be able to apply the principles of safety to all tasks performed. State and local safety laws, NFPA 70E, and OSHA safety rules are in place to create a safe working environment. Individuals using common sense, safeguards, proper equipment, and safety practices are safe when working on electrical equipment. Whenever an individual has a doubt about electricity, the individual must ask for assistance from an electrician. All unsafe conditions, equipment, tools, or work practices must be reported to appropriate persons as soon as possible.

Electric Shock

Electric shock is the condition that occurs when an individual comes in contact with two conductors of a circuit or when the body of an individual becomes part of an electrical circuit. In either case, a severe shock can cause the heart and the lungs to stop functioning. Also, due to the heat produced by current flow, severe burns can occur where the electricity (current) enters and exits the body. **See Figure 20-17.**

Individuals using portable power tools must ensure that the power tool is safe and in proper operating condition. Portable power tools that are not double insulated must have a third wire on the plug for grounding in case a short occurs. Theoretically, when electric power tools are grounded, fault current flows through the third wire to ground instead of through the body to ground.

Out of Service Protection

Before any repair is performed on a piece of electrical equipment, an individual must be absolutely certain the source of electricity is locked out and tagged out (out of service) by a lockout/tagout device. Whenever an individual leaves a job for any reason, upon returning to the job the individual must ensure that the source of electricity is still locked out and tagged out. **See Figure 20-18.**

Safety Color Codes and Safety Labels

Federal law (OSHA) has established specific colors to designate cautions and dangers that are encountered when performing electrical tasks. **See Figure 20-19.** Individuals must understand the significance of colors, signal words, and symbols as related to electrical hazards.

ELECTRICAL CODES

The National Electrical Code® (NEC®) is a book of electrical standards that indicate how electrical systems must be installed and how work must be performed. The NEC® states the standards for all residential, commercial, and industrial electrical work. The NEC® is amended every three years to stay current with new safety issues, electrical products, and procedures.

In most areas, the NEC® and local code requirements are the same, but the NEC®sets only minimum requirements. Local codes, particularly in large cities, may be stricter than those of the NEC®. In the case of stricter local codes, the local code must be followed.

Notes explaining some of the NEC® requirements for general lighting and appliance circuits can be shown on electrical plans. Numbers and notes on electrical plans refer to specific articles and tables in the NEC® code. When more specific requirements (codes) are established by local city councils, the local codes are obtained from the local building inspector.

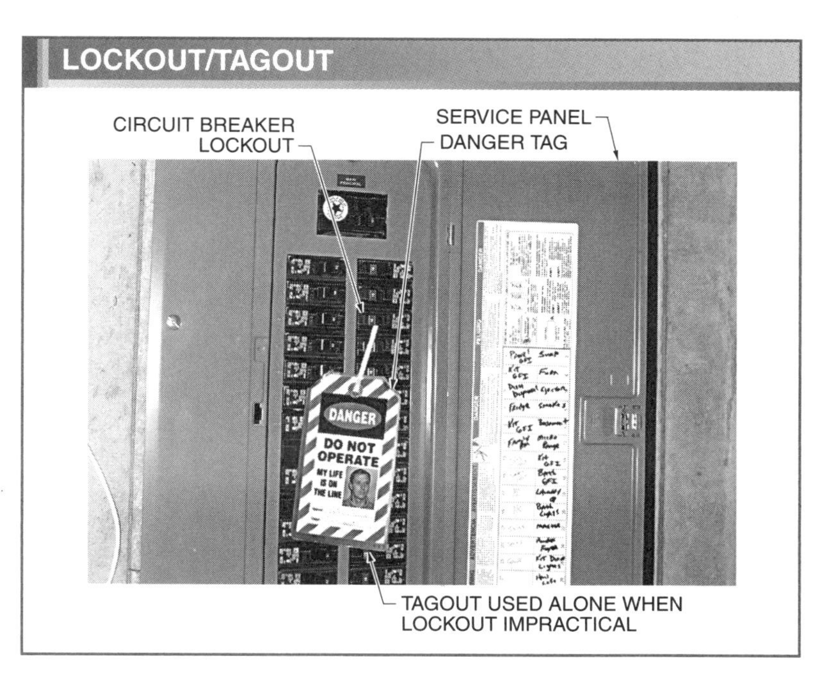

Figure 20-18. Lockout/tagout devices are available for locking out and tagging out any type of electrical equipment.

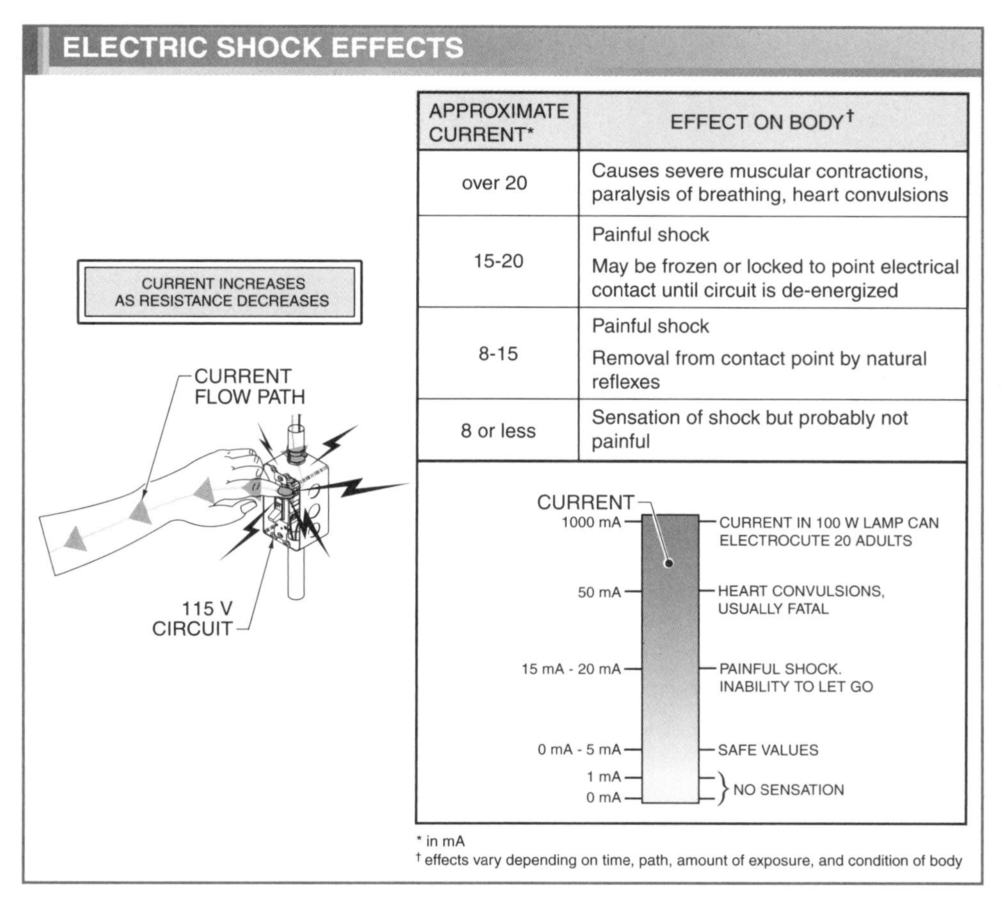

APPROXIMATE CURRENT*	EFFECT ON BODY†
over 20	Causes severe muscular contractions, paralysis of breathing, heart convulsions
15-20	Painful shock May be frozen or locked to point electrical contact until circuit is de-energized
8-15	Painful shock Removal from contact point by natural reflexes
8 or less	Sensation of shock but probably not painful

* in mA
† effects vary depending on time, path, amount of exposure, and condition of body

Figure 20-17. Possible effects of electric shock include the heart and lungs ceasing to function, and/or severe burns where the electricity (current) enters and exits the body.

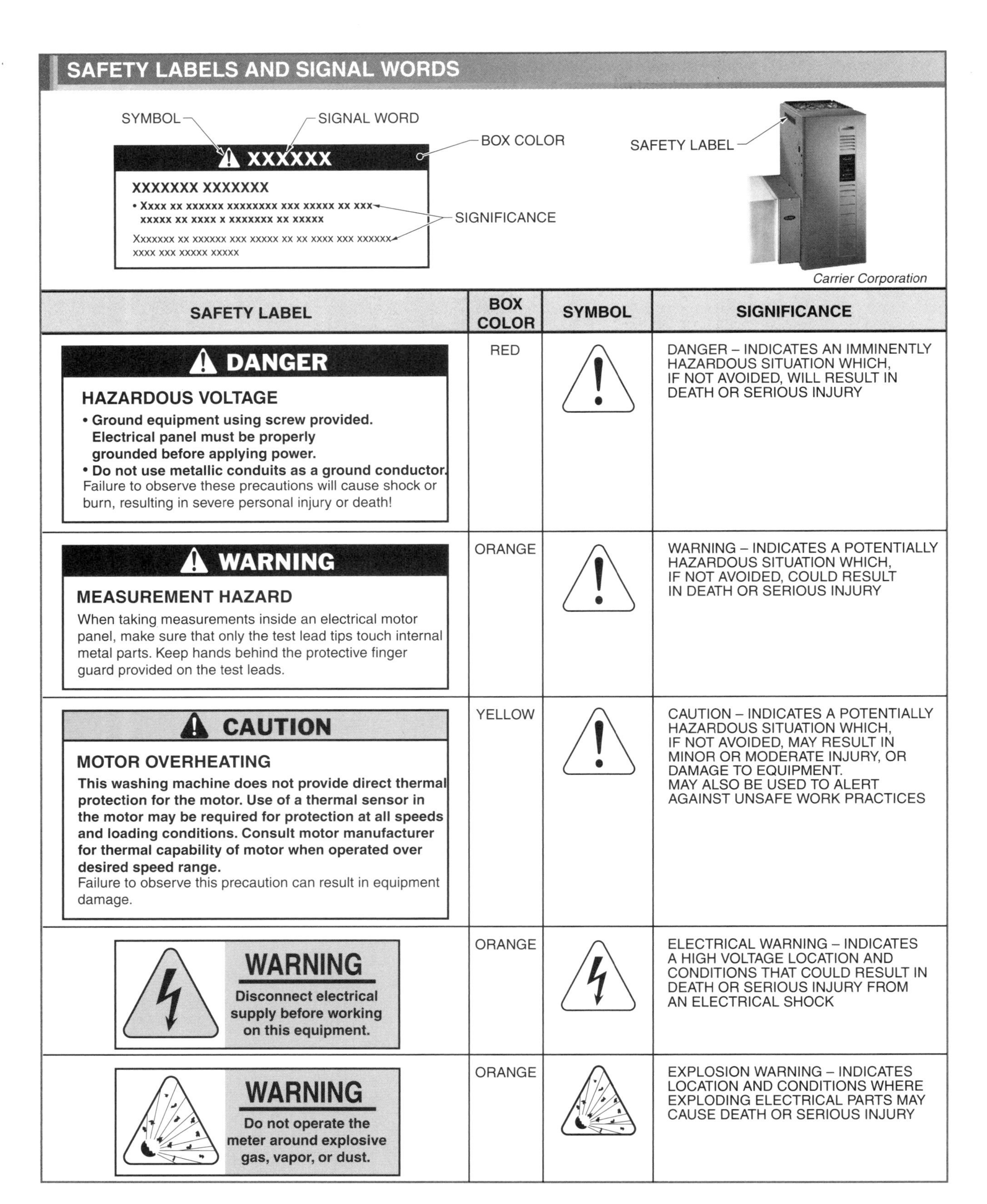

SAFETY LABEL	BOX COLOR	SYMBOL	SIGNIFICANCE
DANGER HAZARDOUS VOLTAGE • Ground equipment using screw provided. Electrical panel must be properly grounded before applying power. • Do not use metallic conduits as a ground conductor. Failure to observe these precautions will cause shock or burn, resulting in severe personal injury or death!	RED		DANGER – INDICATES AN IMMINENTLY HAZARDOUS SITUATION WHICH, IF NOT AVOIDED, WILL RESULT IN DEATH OR SERIOUS INJURY
WARNING MEASUREMENT HAZARD When taking measurements inside an electrical motor panel, make sure that only the test lead tips touch internal metal parts. Keep hands behind the protective finger guard provided on the test leads.	ORANGE		WARNING – INDICATES A POTENTIALLY HAZARDOUS SITUATION WHICH, IF NOT AVOIDED, COULD RESULT IN DEATH OR SERIOUS INJURY
CAUTION MOTOR OVERHEATING This washing machine does not provide direct thermal protection for the motor. Use of a thermal sensor in the motor may be required for protection at all speeds and loading conditions. Consult motor manufacturer for thermal capability of motor when operated over desired speed range. Failure to observe this precaution can result in equipment damage.	YELLOW		CAUTION – INDICATES A POTENTIALLY HAZARDOUS SITUATION WHICH, IF NOT AVOIDED, MAY RESULT IN MINOR OR MODERATE INJURY, OR DAMAGE TO EQUIPMENT. MAY ALSO BE USED TO ALERT AGAINST UNSAFE WORK PRACTICES
WARNING Disconnect electrical supply before working on this equipment.	ORANGE		ELECTRICAL WARNING – INDICATES A HIGH VOLTAGE LOCATION AND CONDITIONS THAT COULD RESULT IN DEATH OR SERIOUS INJURY FROM AN ELECTRICAL SHOCK
WARNING Do not operate the meter around explosive gas, vapor, or dust.	ORANGE		EXPLOSION WARNING – INDICATES LOCATION AND CONDITIONS WHERE EXPLODING ELECTRICAL PARTS MAY CAUSE DEATH OR SERIOUS INJURY

Figure 20-19. Federal law (OSHA) has established specific colors to designate cautions and dangers that individuals must understand.

PERSONAL PROTECTIVE EQUIPMENT

Personal protective equipment (PPE) is clothing and/or equipment worn by individuals to reduce the possibility of injury in the work area. The use of personal protective equipment is required whenever work occurs on or near energized exposed electrical circuits. The National Fire Protection Association standard NFPA 70E, Standard for Electrical Safety in the Workplace, addresses electrical safety requirements for employee workplaces that are necessary for safeguarding employees in pursuit of gainful employment.

NFPA 70E

For maximum safety, personal protective equipment and safety requirements for test instrument procedures must be followed as specified in NFPA 70E, OSHA Standard Part 1910 Subpart 1—Personal Equipment (1910.132 through 1910.138), and other applicable safety mandates. Personal protective equipment includes protective clothing, head protection, eye protection, ear protection, hand and foot protection, back protection, knee protection, and rubber insulated matting. **See Figure 20-20.**

Personal Protective Equipment Safety Rules

The following personal protective equipment safety rules must be followed when performing tasks on electrical systems:

- Wear protective clothing for the task being performed.
- Wear safety glasses, goggles, or an arc face shield, depending on the task being performed.
- Wear thick-soled rubber work shoes for protection against sharp objects and insulation against electrocution. Wear work shoes with safety toes when required for the task.
- Wear rubber boots in damp locations.
- Wear an approved safety helmet (hard hat). Confine long hair and be careful to avoid having long hair near powered tools.
- Wear electrical gloves and cover gloves when taking measurements on energized circuits. Electrical gloves provide protection against electrocution and cover gloves protect electrical gloves from damage.

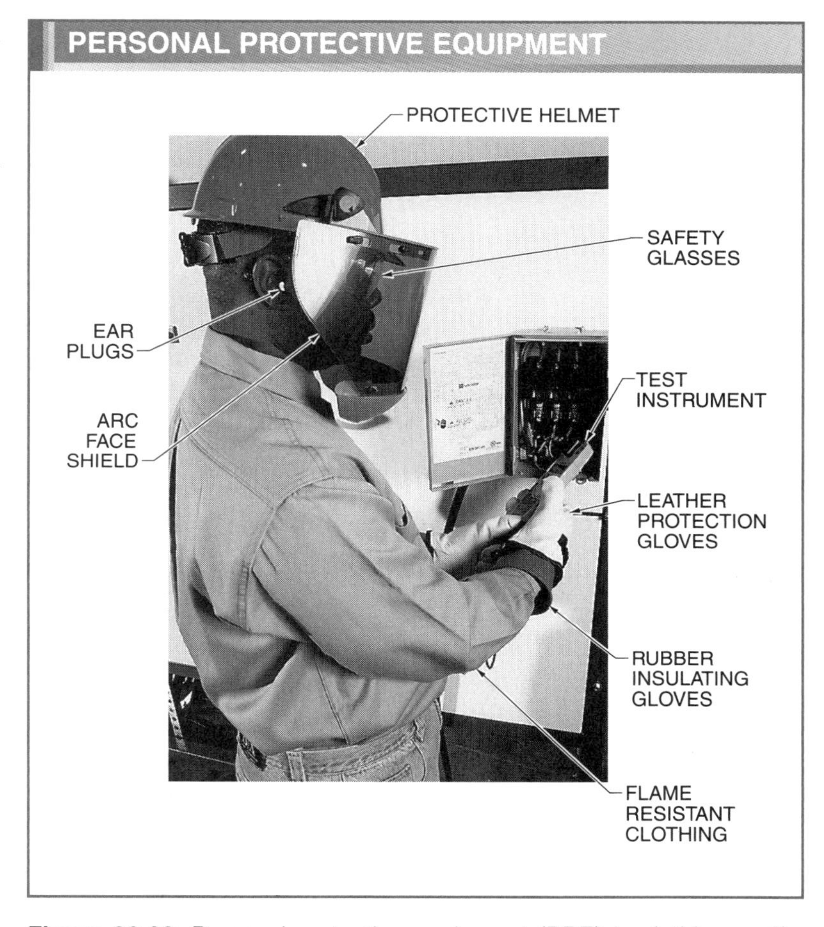

Figure 20-20. Personal protective equipment (PPE) is clothing and/or equipment worn by individuals to reduce the possibility of injury in the work area and is specified by the National Fire Protection Association Standard NFPA 70E, Standard for Electrical Safety in the Workplace.

FIRE SAFETY

The chance of fire is greatly reduced by good housekeeping. Keep debris in designated containers and away from buildings. If a fire should occur, the first thing an individual must do is provide an alarm. The fire department must be called and all personnel on the job must be alerted.

During the time before the fire department arrives, a reasonable effort should be made to contain the fire. In the case of a small fire, portable fire extinguishers (available around the site) are used to extinguish the fire. The stream from fire extinguishers must be directed to the base of fires. Each type of fire is designated by a class letter. Not all fire extinguishers can be used on all types of fires. The five common classes of fires are Class A, Class B, Class C, Class D, and Class K. **See Figure 20-21.**

Class A fires occur in wood, clothing, paper, rubbish, and other such items. Class A fires are typically controlled with water. (Symbol: green triangle.)

Class B fires occur in flammable liquids such as gasoline, fuel oil, grease, thinner, and paints. The agents required for extinguishing Class B fires dilute or eliminate the air to the fire by blanketing the surface of the fire. Chemicals such as foam, CO_2, dry chemical, and Halon are used on Class B fires. (Symbol: red square.)

Class C fires occur in facilities near or in electrical equipment. The extinguishing agent for Class C fires must be a nonconductor of electricity and provide a smothering effect. CO_2, dry chemical, and Halon extinguishers are typically used. (Symbol: blue circle.)

Class D fires occur in combustible metals such as magnesium, potassium, powdered aluminum, zinc, sodium, titanium, zirconium, and lithium. The extinguishing agent for Class D fires is a dry powdered compound. The powdered compound creates a smothering effect that is somewhat effective. (Symbol: yellow star.)

Class K fires occur in kitchens with grease. The extinguishing agent for Class K fires is a wet chemical (potassium acetate).

Tech Facts

Individuals must follow all OSHA and NFPA 70E safety rules and codes when performing electrical work. Electrical fires must be extinguished by a nonconductor of electricity and provide a smothering effect.

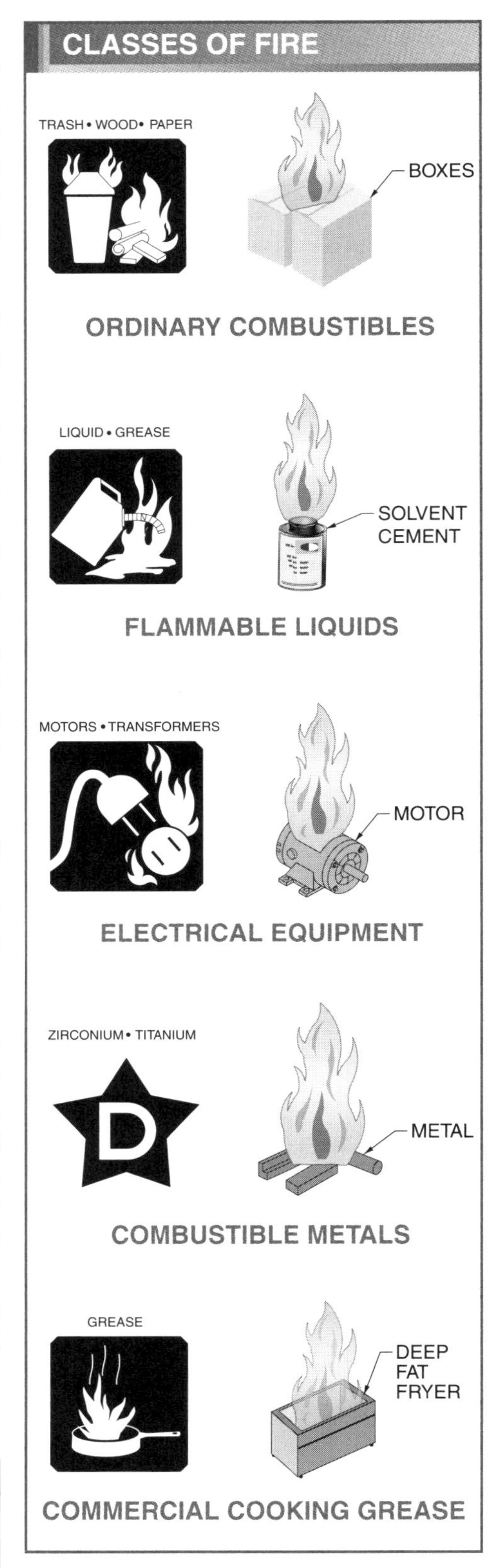

Figure 20-21. The five common classes of fires are Class A, Class B, Class C, Class D, and Class K.

Review Questions

1. Explain the difference between the three states of matter.
2. List and explain the three principal parts of an atom.
3. Define valence shell.
4. Discuss the difference between static electricity and generated electricity.
5. What is a transformer and how is it used?
6. What is grounding?
7. Identify and compare the three equipment grounding methods.
8. List the three categories of fuses and describe two specific fuse types in detail.
9. List the three types of circuit breakers.
10. Explain the purpose of a GFCI.
11. Name four important power tool safety rules.
12. What is the NEC®?
13. List three pieces of personal protective equipment and explain the function of each.
14. List the five common classes of fires and where they occur.

When potential energy is converted into electrical energy, it can be used to produce work. Basic electrical characteristics include voltage, current, power, resistance, capacitance, and inductance and can be quantified. Several formulas that show the relationship between the various quantities are used to understand electricity and the energy electricity can produce.

21 Basic Quantities

ENERGY

Energy is the capacity to do work. The two forms of energy are potential energy and kinetic energy. *Potential energy* is stored energy a body has due to its position, chemical state, or condition. For example, water behind a dam has potential energy because of the position of the water. A compressed spring has potential energy because of its condition. Gasoline has potential energy based on its chemical state. **See Figure 21-1.**

Kinetic energy is the energy of motion. Examples of kinetic energy include falling water, a released spring, a speeding automobile, etc. Kinetic energy is released potential energy. Energy released when water falls through a dam is used to generate electricity for residential, commercial, and industrial use. Energy released by a compressed spring is used to apply a braking force on a motor shaft. Energy released by the burning of gasoline is used to propel a vehicle forward and drive the alternator that produces the electrical power.

The major sources of energy are coal, oil, gas, wood, and nuclear power. Solar, wind, water, and thermal sources also provide energy. These energy forms are used to produce work when converted to electricity, steam, heat, and mechanical force. Some energy sources, such as coal, oil, and gas, cannot be replaced. They are consumed when used. Other energy sources, such as solar power, wind, and water, are not consumed when used.

Electricity is produced by converting potential energy directly or indirectly into electricity. Solar cells convert solar power directly into electricity. The vast majority of all electricity is produced by converting potential energy into some type of force that turns a generator. The major forms of energy used to produce electricity through a generator include fossil fuels (coal, natural gas, and oil), nuclear power, and hydroelectric power.

Voltage

All electrical circuits must have a source of power to produce work. The source of power used depends on the application and the amount of power required. All sources of power produce a set voltage level or voltage range.

Voltage (E) is the amount of electrical pressure in a circuit. Voltage is measured in volts (V). Voltage is also known as electromotive force (EMF) or potential difference. Voltage may be produced by electromagnetism (generators), chemicals (batteries), light (photocells), heat (thermocouples), pressure (piezoelectricity), or friction (static electricity). **See Figure 21-2.**

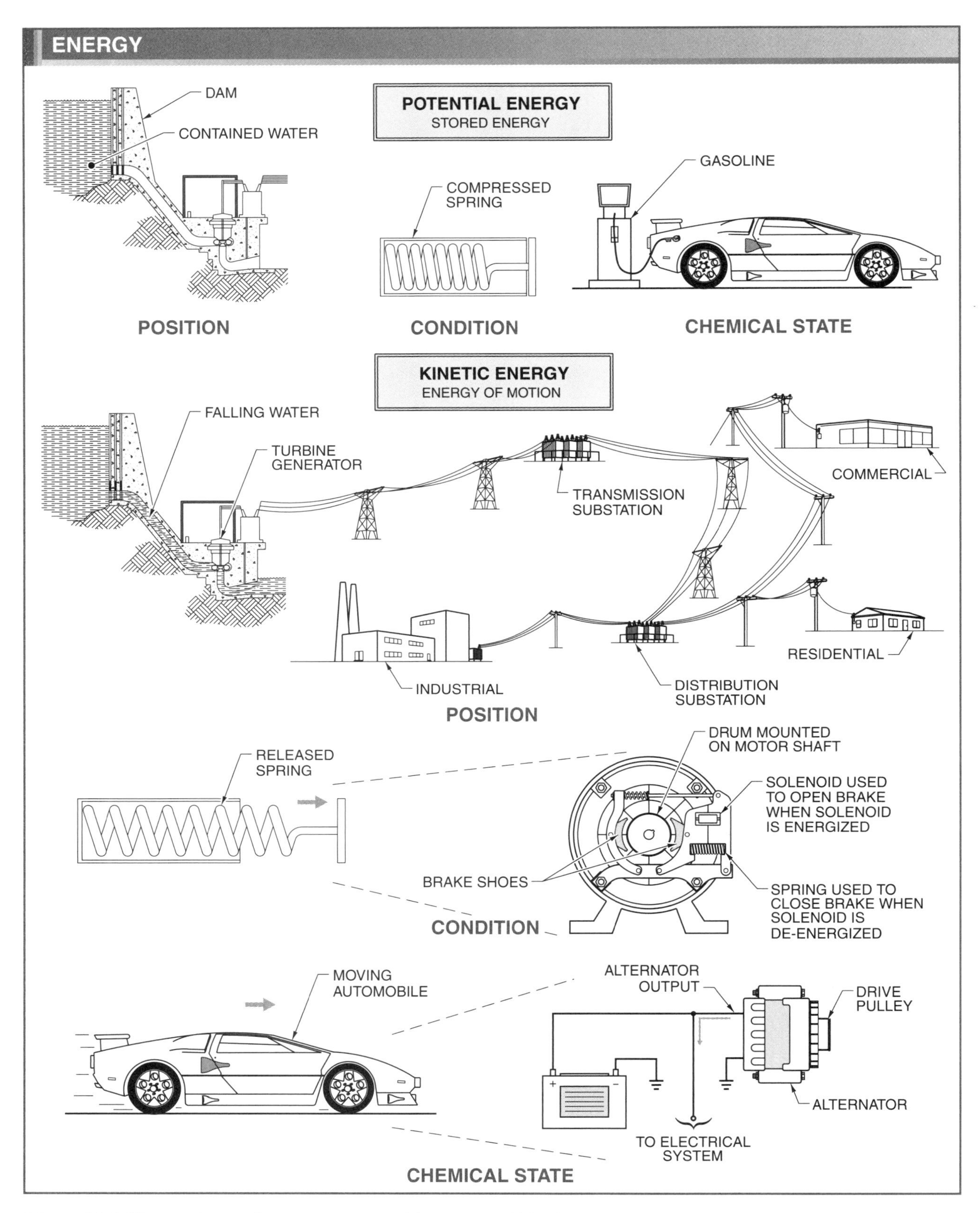

Figure 21-1. The two forms of energy are potential energy and kinetic energy.

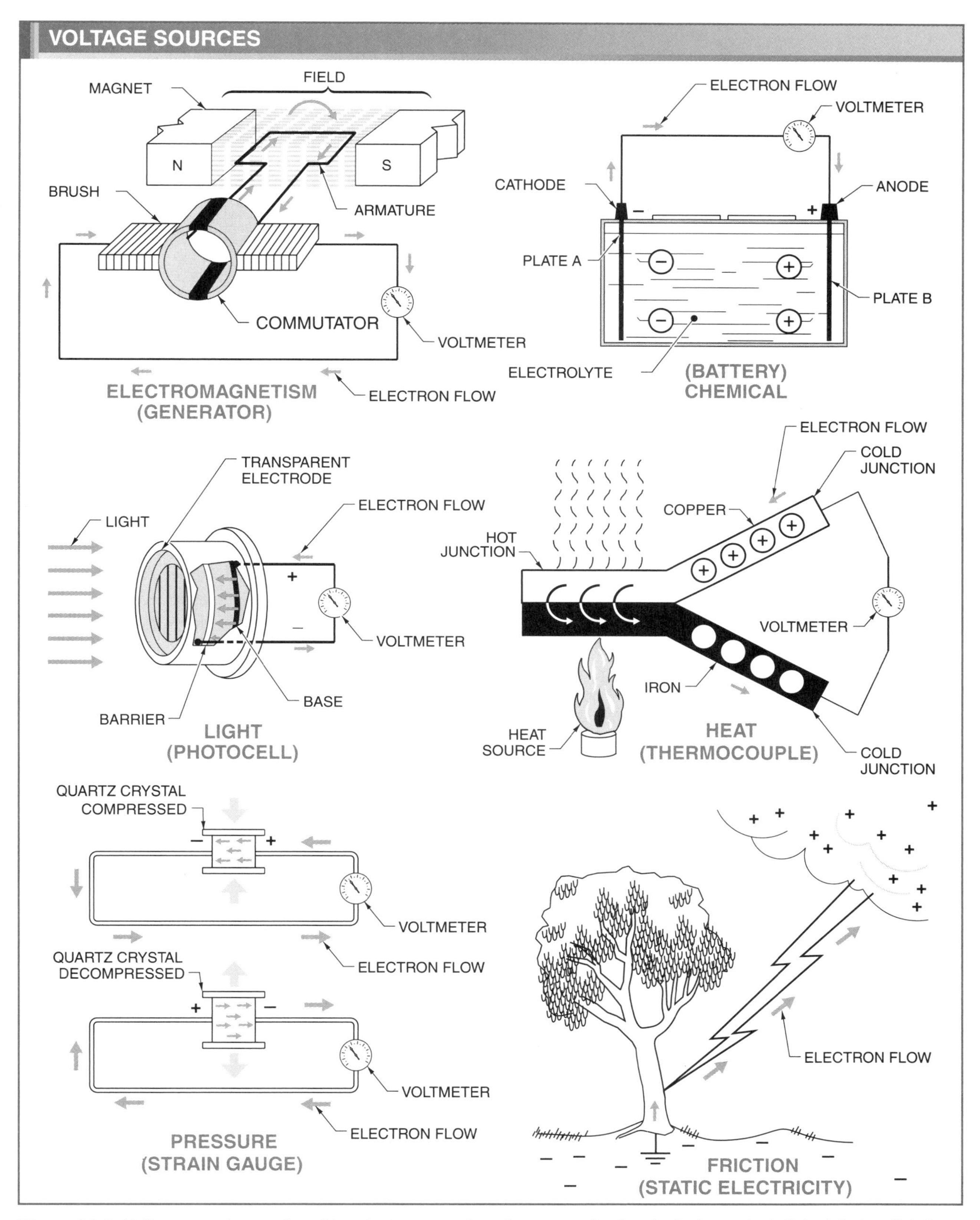

Figure 21-2. Voltage may be produced by electromagnetism (generators), chemicals (batteries), light (photocells), heat (thermocouples), pressure (piezoelectricity), or friction (static electricity).

Voltage is produced any time there is an excess of electrons at one terminal of a voltage source and a deficiency of electrons at the other terminal. The greater the difference in electrons between the terminals, the higher the voltage. The amount of voltage in a circuit depends on the application. For example, powering a flashlight requires a low voltage level. However, producing a picture on a TV screen requires an extremely high voltage level. **See Figure 21-3.**

Voltage is either direct (DC) or alternating (AC). *DC voltage* is voltage that flows in one direction only. *AC voltage* is voltage that reverses its direction of flow at regular intervals. DC voltage is used in almost all portable equipment (automobiles, golf carts, flashlights, cameras, etc.). AC voltage is used in residential, commercial, and industrial lighting and power distribution systems.

DC Voltage. All DC voltage sources have a positive and a negative terminal. The positive and negative terminals establish polarity in a circuit. *Polarity* is the positive (+) or negative (–) state of an object. All points in a DC circuit have polarity. The most common power sources that directly produce DC voltage are batteries and photocells.

In addition to obtaining DC voltage directly from batteries and photocells, DC voltage is also obtained from a rectified AC voltage supply. DC voltage is obtained any time an AC voltage is passed through a rectifier. A *rectifier* is a device that converts AC voltage to DC voltage by allowing the voltage and current to move in only one direction. DC voltage obtained from a rectified AC voltage supply varies from almost pure DC voltage to half-wave DC voltage. Common DC voltage levels include 1.5 V, 3 V, 6 V, 9 V, 12 V, 24 V, 36 V, 125 V, 250 V, 600 V, 1200 V, 1500 V, and 3000 V. **See Figure 21-4.**

AC Voltage. AC voltage is the most common type of voltage used to produce work. AC voltage is produced by generators which produce AC sine waves as they rotate. An *AC sine wave* is a symmetrical waveform that contains 360 electrical degrees. The wave reaches its peak positive value at 90°, returns to 0 V at 180°, increases to its peak negative value at 270°, and returns to 0 V at 360°.

A *cycle* is one complete positive and negative alternation of a wave form. An *alternation* is half of a cycle. A sine wave has one positive alternation and one negative alternation per cycle. **See Figure 21-5.**

COMMON VOLTAGE LEVELS	
Device	**Level***
Flashlight battery (AAA, AA, C, D)	1.5
Automobile battery	12
Golf cart	36
Refrigerator, TV, VCR	115
Central air conditioner	230
Industrial motor	460
TV picture tube	25,000
High-tension power line	up to 500,000

* in V

Figure 21-3. The amount of voltage in a circuit depends on the application.

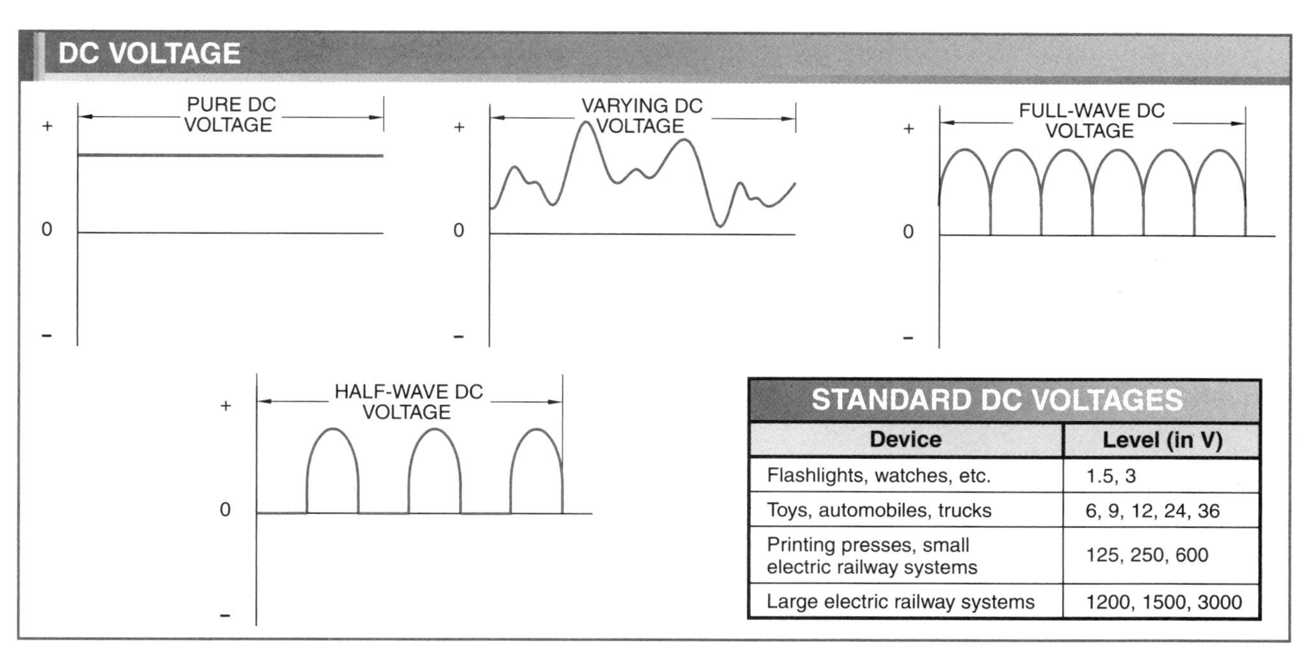

STANDARD DC VOLTAGES	
Device	**Level (in V)**
Flashlights, watches, etc.	1.5, 3
Toys, automobiles, trucks	6, 9, 12, 24, 36
Printing presses, small electric railway systems	125, 250, 600
Large electric railway systems	1200, 1500, 3000

Figure 21-4. Common DC voltage levels include 1.5 V, 3 V, 6 V, 9 V, 12 V, 24 V, 36 V, 125 V, 250 V, 600 V, 1200 V, 1500 V, and 3000 V.

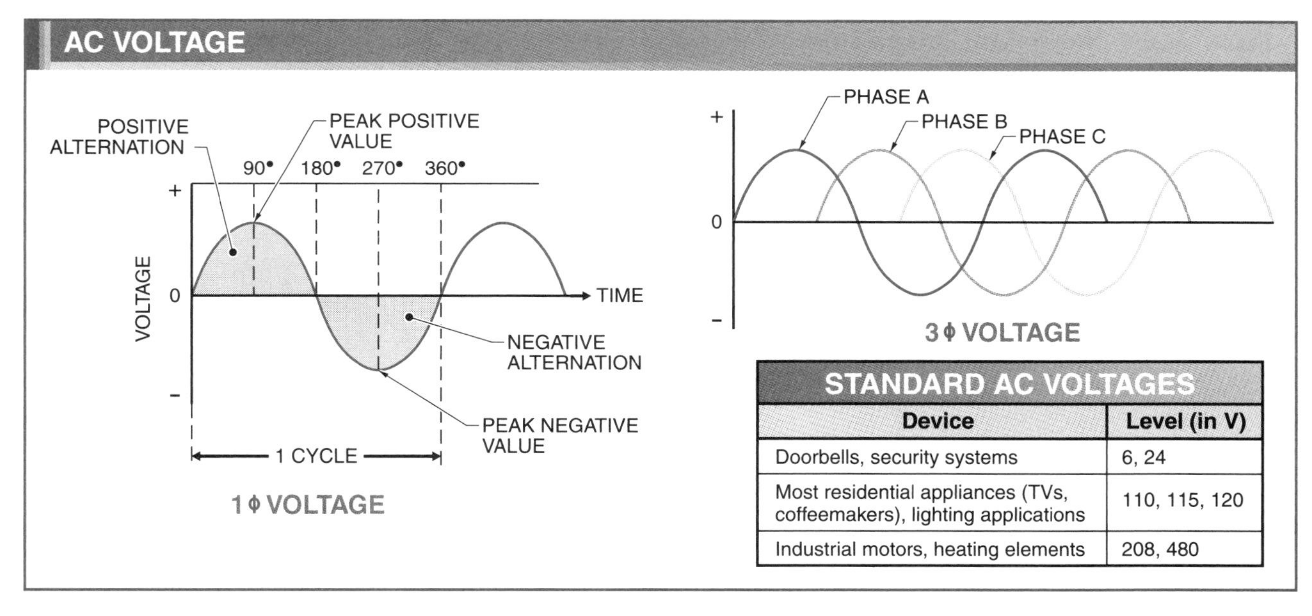

STANDARD AC VOLTAGES	
Device	**Level (in V)**
Doorbells, security systems	6, 24
Most residential appliances (TVs, coffeemakers), lighting applications	110, 115, 120
Industrial motors, heating elements	208, 480

Figure 21-5. A sine wave has one positive alternation and one negative alternation per cycle.

AC voltage is either single-phase (1ϕ) or three-phase (3ϕ). Single-phase AC voltage contains only one alternating voltage waveform. Three-phase AC voltage is a combination of three alternating voltage waveforms, each displaced 120 electrical degrees (one-third of a cycle) apart. Three-phase voltage is produced when three coils are simultaneously rotated in a generator.

Almost any level of AC voltage is available. Low AC voltages (6 V to 24 V) are used for doorbells and security systems. Medium AC voltages (110 V to 120 V) are used in residential applications for lighting, heating, cooling, cooking, running motors, etc.

High AC voltages (208 V to 480 V) are used in commercial applications for large-scale cooking, lighting, heating, cooling, etc., as required by restaurants, office buildings, shopping malls, schools, etc. High AC voltages are also used in industrial applications to convert raw materials into usable products, in addition to providing lighting, heating, and cooling for plant personnel.

AC voltage is stated and measured as peak, average, or rms values. The *peak value* (V_{max}) of a sine wave is the maximum value of either the positive or negative alternation. The positive and negative alternation are equal in a sine wave. The *peak-to-peak value* ($V_{p\text{-}p}$) is the value measured from the maximum positive alternation to the maximum negative alternation. **See Figure 21-6.** To calculate peak-to-peak value, apply the formula:

$$V_{p\text{-}p} = 2 \times V_{max}$$

where

$V_{p\text{-}p}$ = peak-to-peak value (in V)

2 = constant (to double peak value)

V_{max} = peak value (in V)

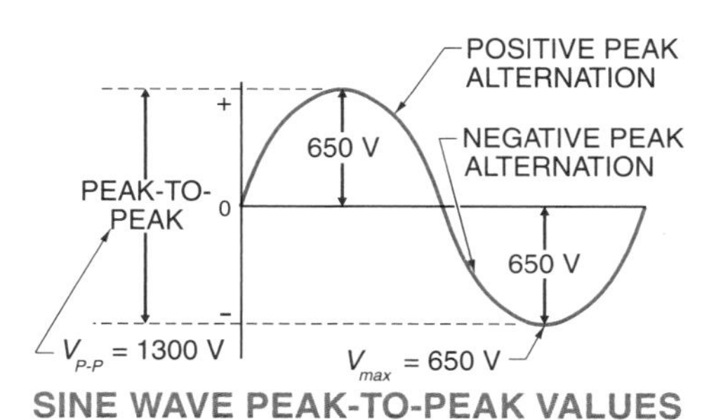

Figure 21-6. The peak-to-peak value (Vp-p) is the value measured from the maximum positive alternation to the maximum negative alternation.

Example: Calculating Peak-to-Peak Value

An AC voltage source with a positive peak alternation of 650 V equals what peak-to-peak value?

$$V_{p\text{-}p} = 2 \times V_{max}$$
$$V_{p\text{-}p} = 2 \times 650$$
$$V_{p\text{-}p} = \mathbf{1300\ V}$$

The *average value* (V_{avg}) of a sine wave is the mathematical mean of all instantaneous voltage values in the sine wave. The average value is equal to .637 of the peak value of a standard sine wave. **See Figure 21-7.** To calculate average value, apply the formula:

$$V_{avg} = V_{max} \times .637$$

where

V_{avg} = average value (in V)

V_{max} = peak value (in V)

.637 = constant (mean of instantaneous values)

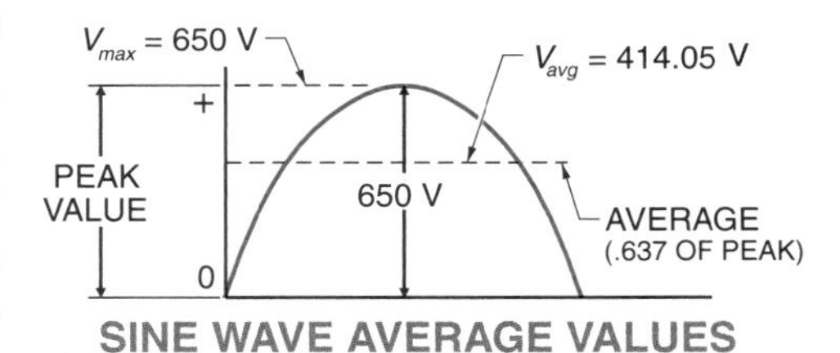

Figure 21-7. The average value is equal to .637 of the peak value of a standard sine wave.

Example: Calculating Average Value

An AC voltage source with a positive peak alternation of 650 V equals what average value?

$$V_{avg} = V_{max} \times .637$$
$$V_{avg} = 650 \times .637$$
$$V_{avg} = \mathbf{414.05\ V}$$

The *root-mean-square (effective) value* (V_{rms}) of a sine wave is the value that produces the same amount of heat in a pure resistive circuit as a DC current of the same value. The rms value is equal to .707 of the peak value in a sine wave. **See Figure 21-8.** To calculate rms value, apply the formula:

$$V_{rms} = V_{max} \times .707$$

where

V_{rms} = rms value (in V)

V_{max} = peak value (in V)

.707 = constant ($1 \div \sqrt{2}$)

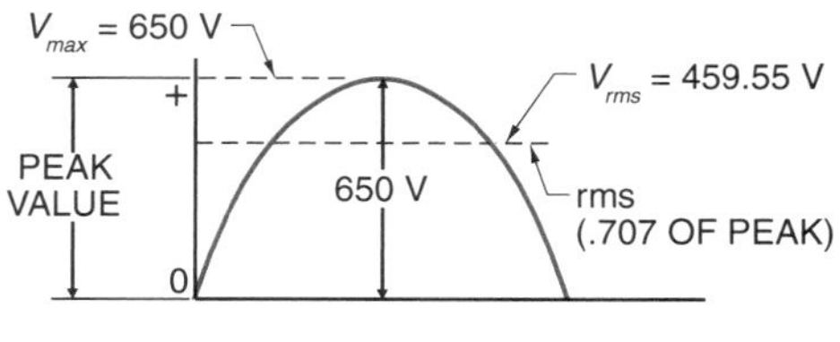

Figure 21-8. The rms value is equal to .707 of the peak value in a sine wave.

Example: Calculating rms Value

An AC voltage source with a positive peak alternation of 650 V equals what rms value?

$V_{rms} = V_{max} \times .707$

$V_{rms} = 650 \times .707$

$V_{rms} = \mathbf{459.55\ V}$

Current

Current flows through a circuit when a source of power is connected to a device that uses electricity. *Current (I)* is the amount of electrons flowing through an electrical circuit. Current is measured in amperes (A). An *ampere* is the number of electrons passing a given point in one second. The more power a load requires, the larger the amount of current flow. **See Figure 21-9.**

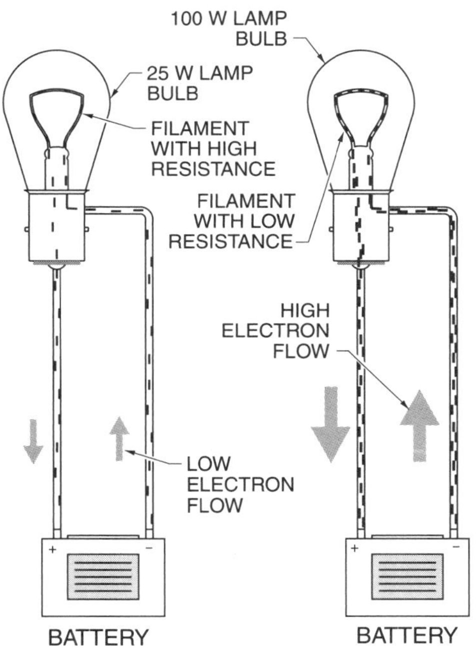

Figure 21-9. The more power a load requires, the larger the amount of current flow.

Tech Facts

According to Ohm's Law, current is equal to voltage divided by resistance. This means that as voltage increases, current will increase. However, as resistance increases, current will decrease.

Different voltage sources produce different amounts of current. For example, standard AA, C, and D size batteries all produce 1.5 V, but each size is capable of delivering different amounts of current. Size AA batteries are capable of delivering the least amount of current, and size D batteries are capable of delivering the most amount of current. For this reason, a load connected to a size D battery lasts longer than a load connected to a size A battery. Current may be direct or alternating. *Direct current (DC)* is current that flows in only one direction. Direct current flows in any circuit connected to a power supply producing a DC voltage. *Alternating current (AC)* is current that reverses its direction of flow at regular intervals. Alternating current flows in any circuit connected to a power supply producing an AC voltage.

Current Flow. Early scientists believed that electrons flowed from positive (+) to negative (–). Later, when atomic structure was studied, electron flow from negative to positive was introduced. *Conventional current flow* is current flow from positive to negative. *Electron current flow* is current flow from negative to positive.

Power

Electrical energy is converted into another form of energy any time current flows in a circuit. Electrical energy is converted into sound (speakers), rotary motion (motors), light (lamps), linear motion (solenoids), and heat (heating elements). *Power (P)* is the rate of doing work or using energy. Power may be expressed as either true power or apparent power. *True power (P_T)* is the actual power used in an electrical circuit. True power is expressed in watts (W). *Apparent power (P_A)* is the product of the voltage and current in a circuit calculated without considering the phase shift that may be present between the voltage and current in the circuit. Apparent power is expressed in volt amps (VA). True power is always less than apparent power in any circuit in which there is a phase shift between voltage and current.

Resistive Circuits

A *resistive circuit* is a circuit that contains only resistive components, such as heating elements and incandescent lamps. Alternating voltage and current are in-phase in resistive circuits. *In-phase* is the state when voltage and current reach their maximum amplitude and zero level simultaneously. **See Figure 21-10.**

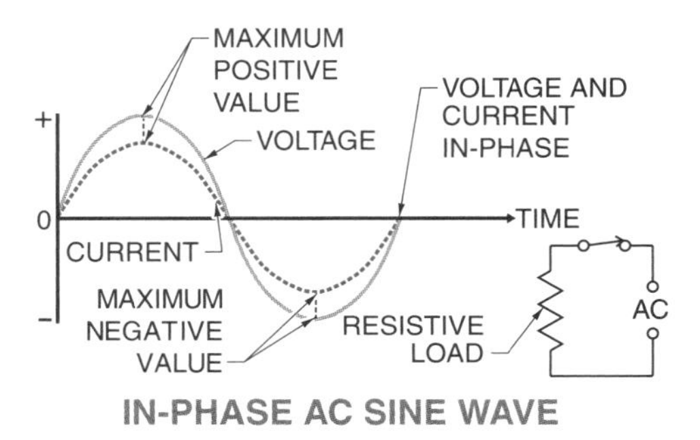

Figure 21-10. In-phase is the state when voltage and current reach their maximum amplitude and zero level simultaneously.

National Wood Flooring Association

A resistive circuit can be composed of recessed lights that use incandescent lamps.

Inductive Circuits

Inductance is the property of a circuit that causes it to oppose a change in current due to energy stored in a magnetic field. All coils (motor windings, transformers, solenoids, etc.) create inductance in an electrical circuit. A phase shift exists between alternating voltage and current in an inductive circuit. An inductive circuit is a circuit in which current lags voltage. The greater the inductance in a circuit, the larger the phase shift. **See Figure 21-11.**

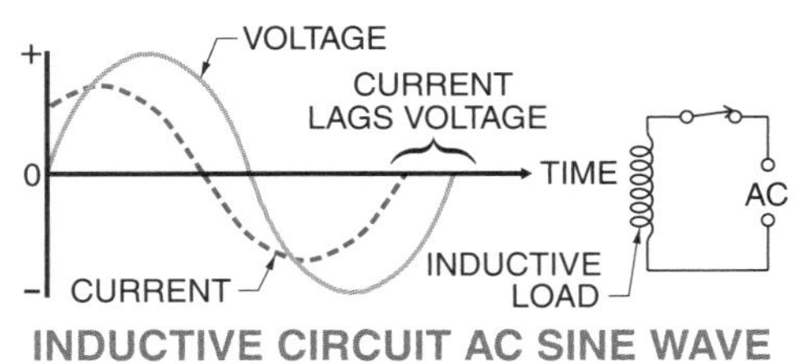

Figure 21-11. The greater the inductance in a circuit, the larger the phase shift.

Capacitive Circuits

Capacitance is the ability of a component or circuit to store energy in the form of an electrical charge. Capacitors create capacitance in an electrical circuit. A phase shift exists between voltage and current in a capacitive circuit. A capacitive circuit is a circuit in which current leads voltage. The greater the capacitance in a circuit, the larger the phase shift. **See Figure 21-12.**

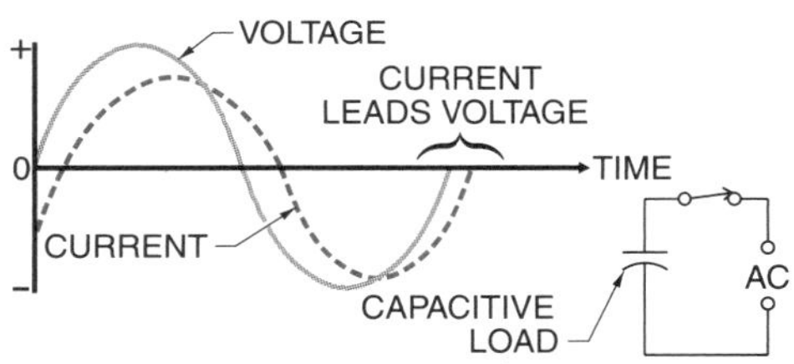

Figure 21-12. The greater the capacitance in a circuit, the larger the phase shift.

Power Factor

Power factor (PF) is the ratio of true power used in an AC circuit to apparent power delivered to the circuit. Power factor is expressed as a percentage. True power equals apparent power when the power factor is 100%. When the power factor is less than 100%, the circuit is less efficient and has higher operating cost. To calculate power factor, apply the formula:

$$PT = \frac{P_T}{P_A} \times 100$$

where
PF = power factor (percentage)
P_T = true power (in W)
P_A = apparent power (in VA)
100 = constant (to convert decimal to percent)

Example: Calculating Power Factor
What is the power factor when the true power of an electrical circuit is 1500 W and the apparent power is 1700 VA?

$$PF = \frac{P_T}{P_A} \times 100$$

$$PF = \frac{1500}{1700} \times 100$$

$$PF = .882 \times 100$$

$$PF = \mathbf{88.2\%}$$

Conductors and Insulators

All materials have some amount of resistance to current flow. A *conductor* is a material that has very little resistance and permits electrons to move through it easily. Most metals are good conductors. Copper is the most commonly used conductor. Silver is a better conductor than copper, but is too expensive for most applications. Aluminum is not as good of a conductor as copper, but costs less and is lighter. For this reason, aluminum is also commonly used as a conductor.

An *insulator* is a material that has a very high resistance and resists the flow of electrons. Common insulators include rubber, plastic, air, glass, and paper. Electrical insulations are classified by their temperature ratings as 60°C (140°F), 75°C (167°F), and 90°C (194°F).

Heat

Heat is thermal energy. Anything that gives off heat is a heat source. Electrical energy may be converted into heat. Electricity may be used to heat almost any gas, liquid, or solid. Heat produced from electricity is used in residential and commercial applications to heat rooms and water, cook food, dry clothes, etc. Most industrial processes require the heating of a variety of liquids, solids, and gases. Heat is used to separate metals from their ores, refine crude oil, process foods, remove moisture, shape metals and plastics, join metals and plastics, harden metals, and to cure finishes such as paint, enamels, and varnishes.

Heat is measured in British thermal units (Btu) or calories (cal). British thermal units are used to measure the amount of heat in the U.S. system of measurements.

A *Btu* is the amount of heat required to raise 1 lb of water 1°F. Calories are used to measure the amount of heat in the Metric system of measurements. A *calorie* is the amount of heat required to raise 1 g of water 1°C. One Btu is equivalent to 252 calories or .252 kilocalories.

Most electric heating elements are rated in watts and/or Btu. The amount of electrical energy used by a heating element depends on the power rating of the heating element (in watts) and the amount of time (in hours) the heating element is ON. To calculate the amount of electrical energy used by a heating element, apply the formula:

$$E = P \times T$$

where
E = energy (in Wh)
P = power (in W)
T = time (in hr)

Example: Calculating Electrical Energy
A heater rated at 3000 W uses how much energy in 1 hr, and how much energy in 10 hr?

One Hour

$$E = P \times T$$

$$E = 3000 \times 1$$

E = **3000 Wh (3 kWh)**

Ten Hours

$$E = P \times T$$

$$E = 3000 \times 10$$

E = **30,000 Wh (30 kWh)**

Temperature

All heating elements produce heat. Temperature increases when heat is produced. *Temperature* is the measurement of the intensity of heat. The amount of increase in temperature depends on the amount of heat produced, the mass of the body being heated, and the material of which the heated body is made.

The higher the amount of heat produced, the faster the temperature increases in the body being heated. The larger the body to be heated, the slower the temperature rise of the body. The better conductor of heat the material being heated is, the faster the heat transfer. The rate a material conducts heat depends on the thermal conductivity rating of the material.

Thermal conductivity is the property of a material to conduct heat in the form of thermal energy. The higher the thermal conductivity rating, the faster the material conducts (transfers) heat. The lower the thermal conductivity rating, the slower the material conducts (transfers) heat. The thermal conductivity rating (number) assigned to a given material is based on that material's ability to transfer heat through 1 sq ft of the surface area for a given thickness in Btu/hr, per 1°F temperature difference.

Gases and liquids such as air and water have poor thermal conductivity. Solids such as aluminum and copper have good thermal conductivity. The thermal conductivity number assigned to a given material is based on the ability of the material to transfer heat. The number is based on the amount of heat transferred through 1 sq ft of the surface area for a given thickness in Btu per hour, per 1°F difference through the material. Aluminum (thermal conductivity = 128) is a much better thermal conductor than steel (thermal conductivity = 26.2). Since aluminum is a much better conductor of heat, it is used in most heat sink applications. A *heat sink* is a device that conducts and dissipates heat away from a component. **See Figure 21-13.**

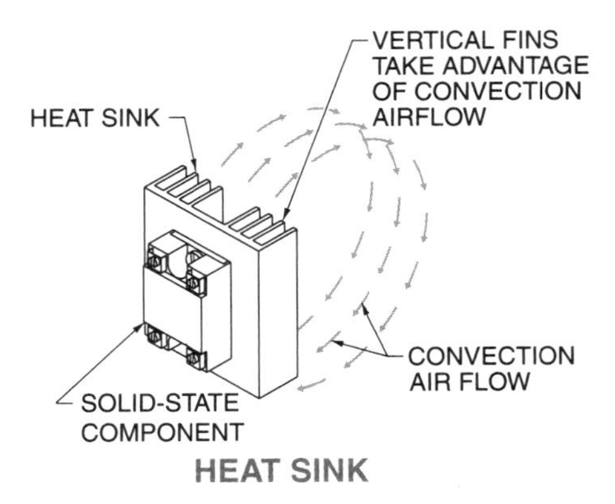

Figure 21-13. A heat sink is a device that conducts and dissipates heat away from a component.

Temperature Conversion. Temperature is normally measured in degrees Fahrenheit (°F) or degrees Celsius (°C). Converting one unit to the other is required because both Fahrenheit and Celsius are commonly used in the electrical field.

To convert a Fahrenheit temperature reading to Celsius, subtract 32 from the Fahrenheit reading and divide by 1.8. To convert Fahrenheit to Celsius, apply the formula:

$$°C = \frac{°F - 32}{1.8}$$

where

$°C$ = degrees Celsius

$°F$ = degrees Fahrenheit

32 = difference between bases

1.8 = ratio between bases

Example: Converting Fahrenheit to Celsius

A temperature reading of 90°F equals what °C?

$$°C = \frac{°F - 32}{1.8}$$

$$°C = \frac{90 - 32}{1.8}$$

$$°C = \frac{58}{1.8}$$

$$°C = \mathbf{32.2\bar{2}\ C}$$

To convert a Celsius temperature reading to Fahrenheit, multiply 1.8 by the Celsius reading and add 32. To convert Celsius to Fahrenheit, apply the formula:

$$°F = (1.8 \times °C) + 32$$

where

$°F$ = degrees Fahrenheit

1.8 = ratio between bases

$°C$ = degrees Celsius

32 = difference between bases

Example: Converting Celsius to Fahrenheit

A temperature reading of 150°C equals what °F?

$°F = (1.8 \times °C) + 32$

$°F = (1.8 \times 150) + 32$

$°F = 270 + 32$

$°F = $ **302°F**

Light

Light is that portion of the electromagnetic spectrum which produces radiant energy. The electromagnetic spectrum ranges from cosmic rays with extremely short wavelengths to electric power frequencies with extremely long wavelengths. **See Figure 21-14.**

Light can be in the form of visible light or invisible light. *Visible light* is the portion of the electromagnetic spectrum to which the human eye responds. Visible light includes the part of the electromagnetic spectrum that ranges from violet to red light. *Invisible light* is the portion of the electromagnetic spectrum on either side of the visible light spectrum. Invisible light includes ultraviolet and infrared light.

The color of light is determined by its wavelength. Visible light with the shortest wavelengths produce the color violet. Visible light with the longest wavelengths produce red. Wavelengths between violet and red produce blue, green, yellow, and orange. The combination of colored light produces white light when a light source, such as the sun, produces energy over the entire visible spectrum in approximately equal quantities. A non-white light is produced when a light source, such as a low-pressure sodium lamp, produces energy that mostly lies in a narrow band of the spectrum (yellow-orange). **See Figure 21-15.**

Color distortion of light-colored objects viewed under the lamp may be extreme because the low-pressure sodium lamp produces most of its energy in the yellow-orange area of the light spectrum. However, since the low-pressure sodium lamp produces more light per watt of power than any other lamp type, it is used in applications such as street lighting in which color distortion may be tolerated.

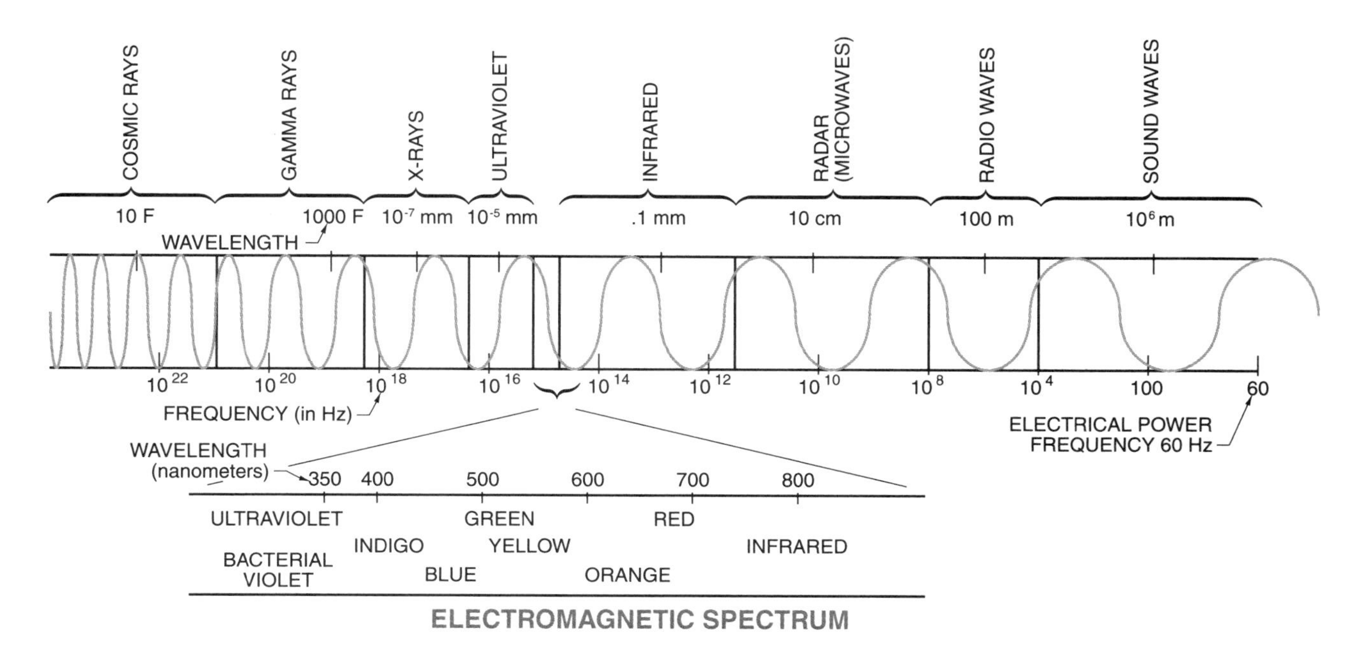

Figure 21-14. The electromagnetic spectrum ranges from cosmic rays with extremely short wavelengths to electric power frequencies with extremely long wavelengths.

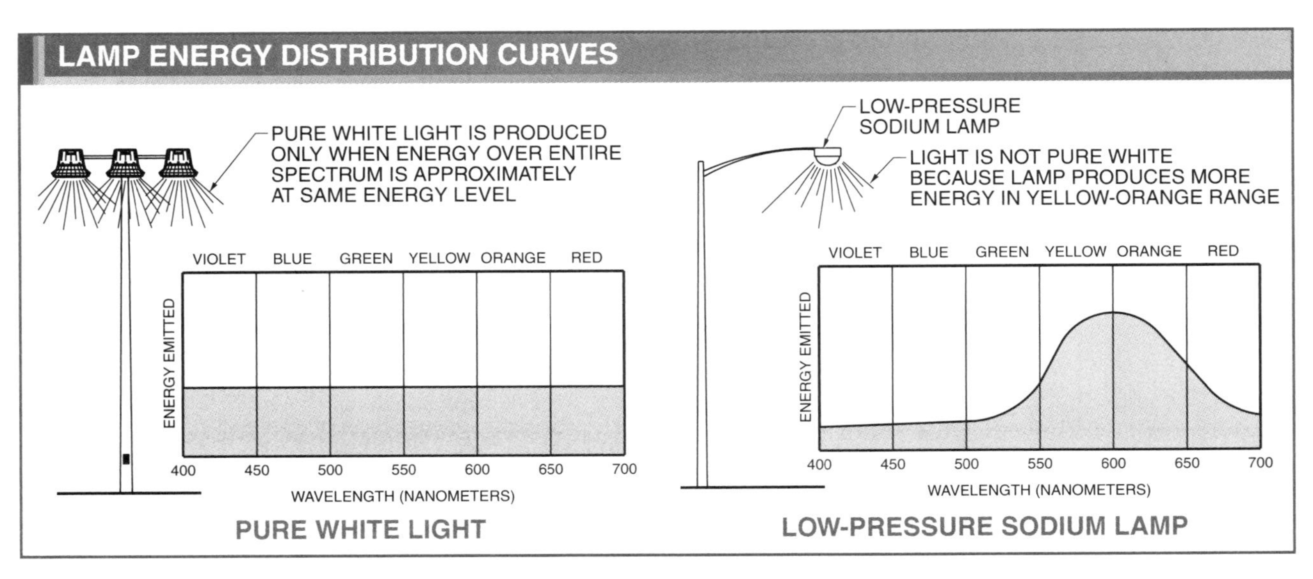

Figure 21-15. The color of light is determined by its wavelength.

The *ultraviolet region* is the region of the spectrum with wavelengths just short of the color violet. The infrared region is the region of the spectrum with wavelengths slightly longer than the color red. Both ultraviolet and infrared light are invisible to the human eye. However, ultraviolet and infrared light is used in special applications. Depending on the wavelength, ultraviolet light is used in germicidal (bacteria killing) lamps, photochemical lamps, black lights, and sun lamps. Infrared light is used in heat lamps.

A *lamp* is an output device that converts electrical energy into light. The amount of light a lamp produces is expressed in lumens. A *lumen (lm)* is the unit used to measure the total amount of light produced by a light source. For example, a standard 40 W incandescent lamp produces about 480 lm, and a standard 40 W fluorescent lamp produces about 3100 lm. Manufacturers rate lamps (light bulbs) in the total amount of light (lumen) produced by the lamps. Since the lumen is the total amount of light, it is comparable to the amount of current (amperes) in an electrical circuit or the amount of flow (gpm) in a hydraulic system.

The light produced by a light source causes illumination. *Illumination* is the effect that occurs when light falls on a surface. The unit of measure of illumination is the footcandle. A *footcandle (fc)* is the amount of light produced by a lamp (lumens) divided by the area that is illuminated.

Light spreads as it travels farther from the light-producing source. The relationship between the amount of light produced at the source and the amount of illumination at different distances from the light source is expressed by the inverse square law. The *inverse square law* states that the amount of illumination on a surface varies inversely with the square of the distance from the light source. **See Figure 21-16.**

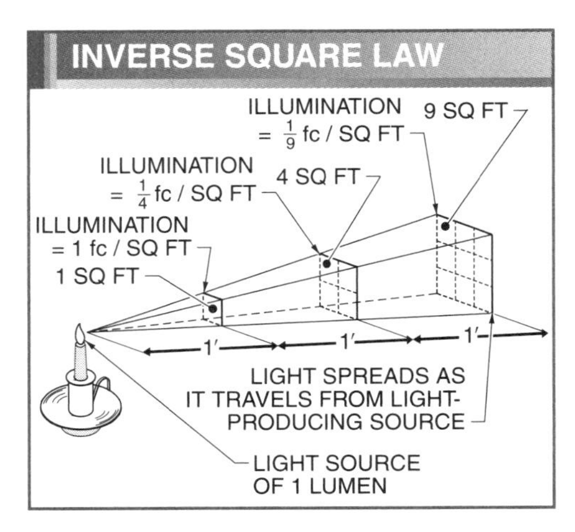

Figure 21-16. The inverse square law states that the amount of illumination on a surface varies inversely with the square of the distance from the light source.

One lm produces an illumination of 1 fc on a 1 sq ft area. Thus, if the amount of light falling on a surface 1′ away from the light source produces 1 fc, then the light source produces ¼ fc at a 2′ distance, and ⅑ fc at a 3′ distance. This is because the light produced at the source covers four times the area at a 2′ distance and nine times the area at a 3′ distance, than it does at a 1′ distance. **See Figure 21-17.**

The amount of light produced on a surface is dependent upon the amount of lumens produced by the light source and the distance the surface is from the light source. The light level required on a surface varies widely. For example, an operating table in a hospital requires much more light than a teller's station in a bank, and an expressway requires much less light than a baseball infield.

RECOMMENDED LIGHT LEVELS					
Interior Lighting		**Exterior Lighting**		**Sports Lighting**	
Area	**fc**	**Area**	**fc**	**Area**	**fc**
Assembly Rough, easy seeing Medium Fine	 30 100 500	**Airports** Terminal Apron – Loading	 2	**Baseball** Outfield Infield	 100 150
Auditorium Exhibitions	 30	**Building Construction** General Excavation	 10 2	**Basketball** College and Professional Recreational	 50 10
Banks Lobby, general Waiting areas Teller station	 50 70 150	**Buildings** Light surface Dark surface	 15 50	**Billiards** Recreational Tournament	 30 50
Clothing Manufacturing Pattern making Shops	 50 100	**Loading areas**	20	**Boxing** Professional Championship	 200 500
Hospital/Medical Lobby Dental chair Operating table	 30 1000 2500	**Parking areas** Industrial Shopping	 2 5	**Golf** Tee Fairway Green Miniature	 5 2 5 10
Machine Shop Rough bench Medium bench	 50 100	**Piers** Freight/Passenger	 20	**Racing** Auto, Horse Dog	 20 30
Offices Regular office work Accounting Detailed work	 100 150 200	**Railroad Yards** Switch points	 2	**Ski Slope**	1
Printing Proofreading Color inspecting	 150 200	**Service Station** Pump island area Service areas	 25 5	**Volleyball** Recreational Tournament	 10 20
Warehouses Inactive Active	 5 30	**Streets** Local Expressway	 .9 1.4	**Tennis Courts** Recreational Tournament	 15 30

Figure 21-17. One lm produces an illumination of 1 fc on a 1 sq ft area.

Review Questions

1. Define kinetic energy and explain its relationship to potential energy.
2. List seven sources of energy.
3. Explain the difference between DC voltage and AC voltage.
4. Which voltage (DC or AC) would you consider to be more versatile, and why?
5. Describe the relationship between power and current.
6. Discuss the possible benefits of using true power measurements rather than apparent power.
7. Compare and rate three metal conductors based on effectiveness and cost.
8. What is a Btu?
9. What three factors influence the increase of temperature?
10. Give examples of materials that have good thermal conductivity and materials that have poor thermal conductivity.
11. How is white light produced?
12. What relationship is explained by the inverse square law?

Test instruments are devices that measure quantities such as voltage, resistance, temperature, and speed. Test instrument abbreviations and symbols are used on switches, scales, displays, and in manuals to convey information such as settings and units of measure.

22 Test Instruments, Abbreviations, and Measurements

TEST INSTRUMENTS

Test instruments measure quantities such as voltage, current, resistance, temperature, speed, etc. Test instruments are either single-function or multifunction. A *single-function test instrument* is an instrument capable of measuring and displaying only one quantity. A *multifunction test instrument* is a test instrument that is capable of measuring two or more quantities. Quantities measured by test instruments are either electrical quantities or nonelectrical quantities. The test instrument used depends on the quantity measured and the application.

Test Instrument Abbreviations

Test instrument switches, scales, displays, and manuals use abbreviations and symbols to convey information, such as settings and units of measure. An *abbreviation* is a letter or combination of letters that represents a word. Abbreviations depend on a particular language. All test instruments use standard abbreviations to represent a quantity or term. Abbreviations can be used individually (100 V) or in combinations (100 kV). **See Figure 22-1.**

Test Instrument Symbols

A *symbol* is a graphic element that represents a quantity or unit. Symbols are independent of language because a symbol can be recognized regardless of the language a person speaks. Most test instruments use standard symbols to represent an electrical component (battery, etc.), term (ground, etc.), or message to the user (warning, etc.). **See Figure 22-2.**

Analog Displays

An *analog display* is an electromechanical device that indicates readings by the mechanical motion of a pointer. Analog displays use scales to display measured values. Analog scales may be linear or nonlinear. A *linear scale* is a scale that is divided into equally spaced segments. A *nonlinear scale* is a scale that is divided into unequally spaced segments. **See Figure 22-3.**

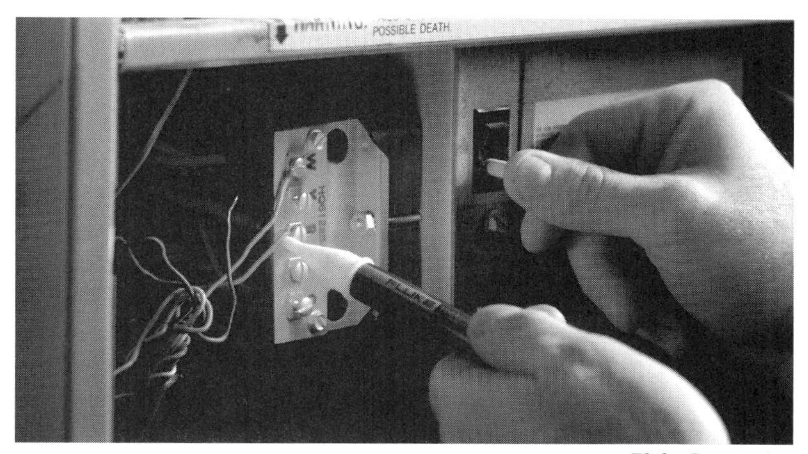

Fluke Corporation

A voltage indicator is a pocket test instrument that is easy to use. It provides a quick visual indication of the presence of voltage when its tip glows red.

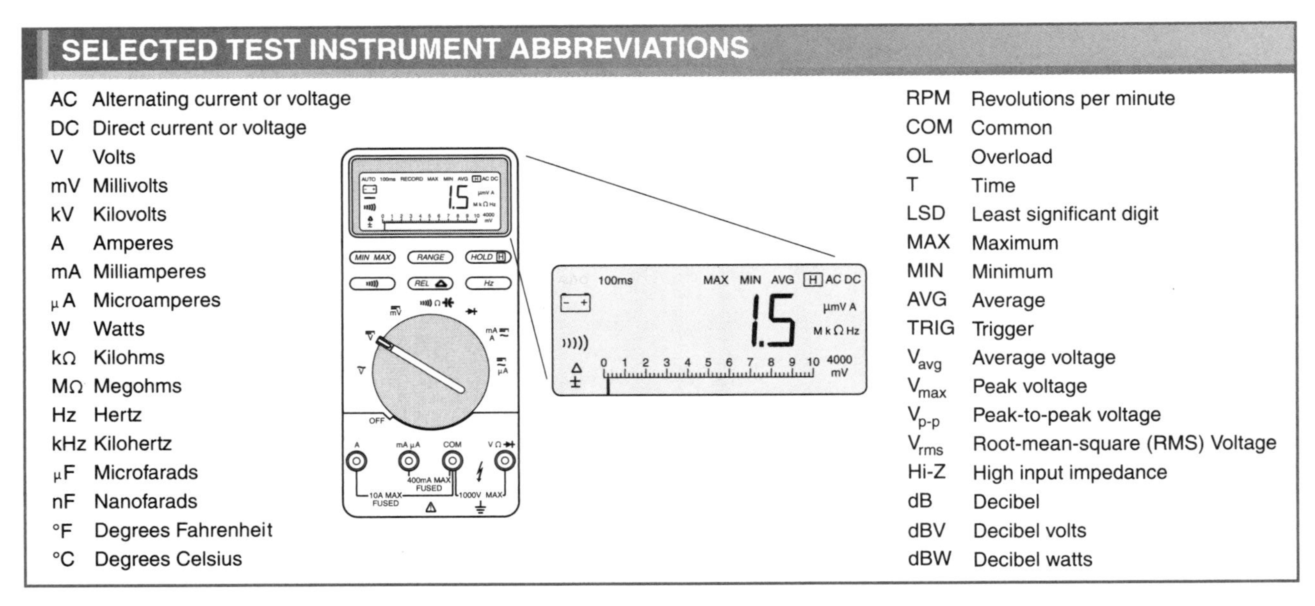

Figure 22-1. All test instruments use standard abbreviations to represent a quantity or term.

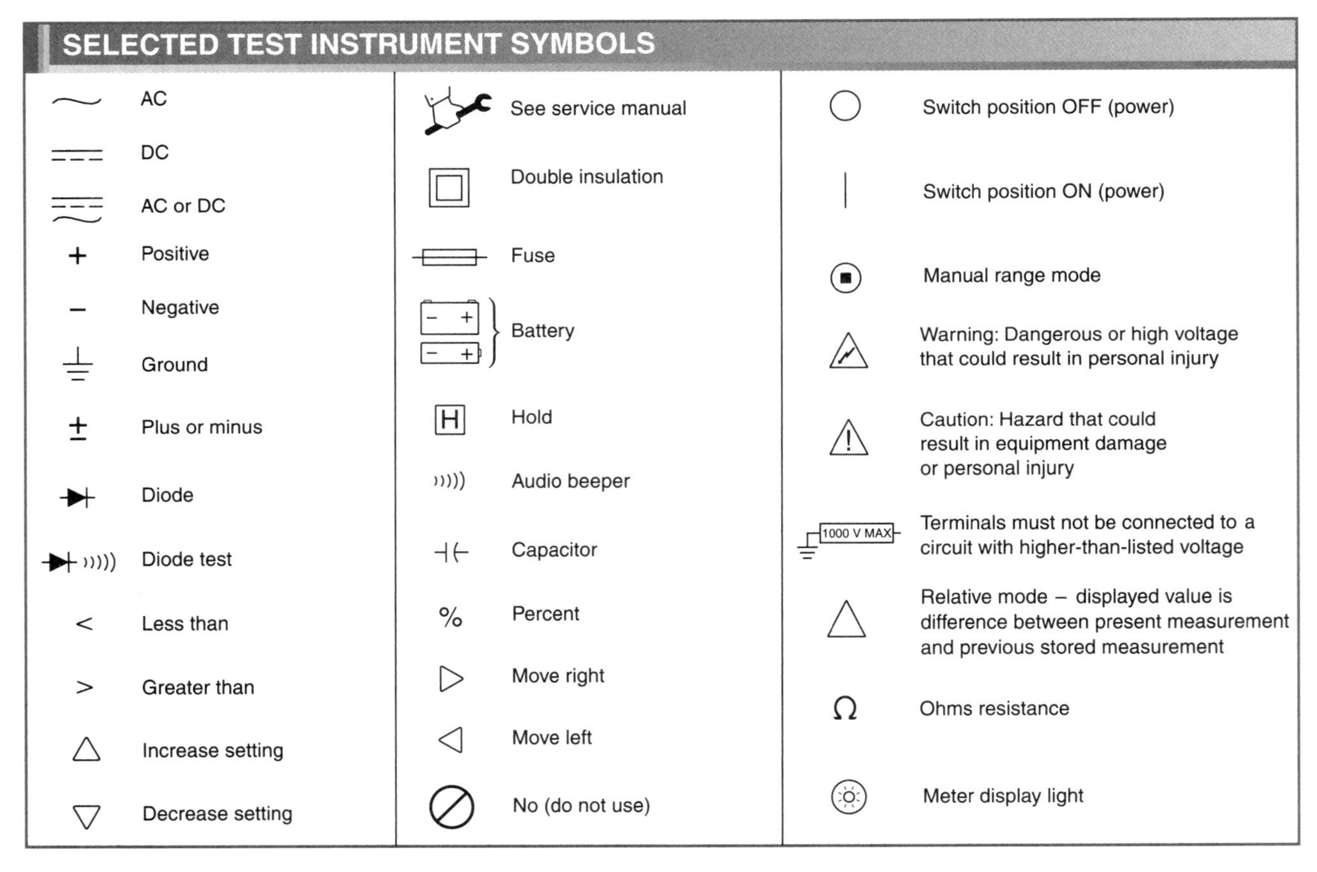

Figure 22-2. Most test instruments use standard symbols to represent an electrical component (battery, etc.), term (ground, etc.), or message to the user (warning, etc.).

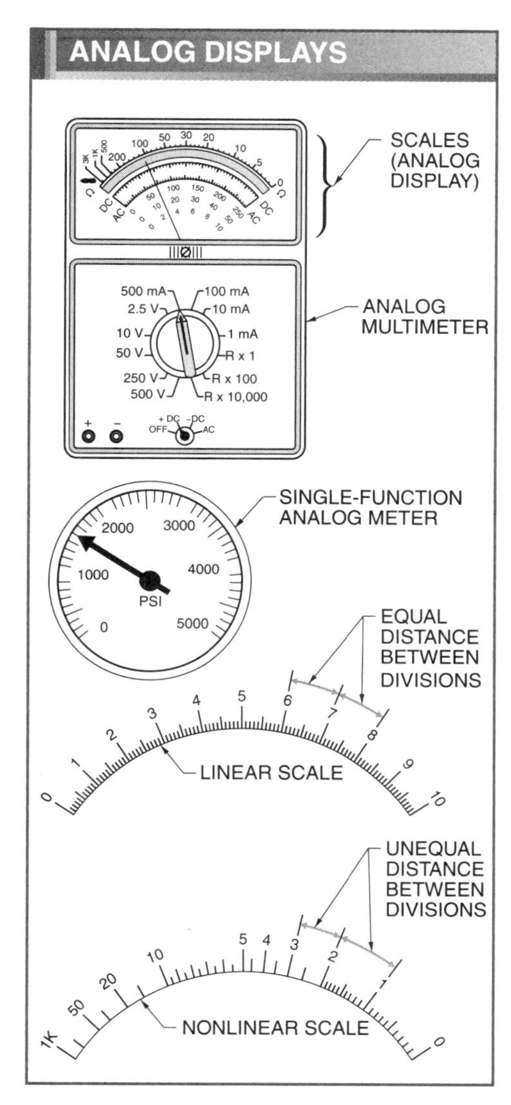

Figure 22-3. Analog displays use scales to display measured values.

Analog scales are divided using primary divisions, secondary divisions, and subdivisions. A *primary division* is a division with a listed value. A *secondary division* is a division that divides primary divisions in halves, thirds, fourths, fifths, etc. A *subdivision* is a division that divides secondary divisions in halves, thirds, fourths, fifths, etc. Secondary divisions and subdivisions do not have listed numerical values. When reading an analog scale, add the primary, secondary, and subdivision readings. **See Figure 22-4.** To read an analog scale, apply the following procedure:

1. Read the primary division.
2. Read the secondary division if the pointer moves past a secondary division. *Note:* This may not occur with very low readings.
3. Read the subdivision if the pointer is not directly on a primary or secondary division. Round the reading to the nearest subdivision if the pointer is not directly on a subdivision. Round the reading to the next highest subdivision if rounding to the nearest subdivision is unclear.
4. Add the primary division, secondary division, and subdivision readings to obtain the analog reading.

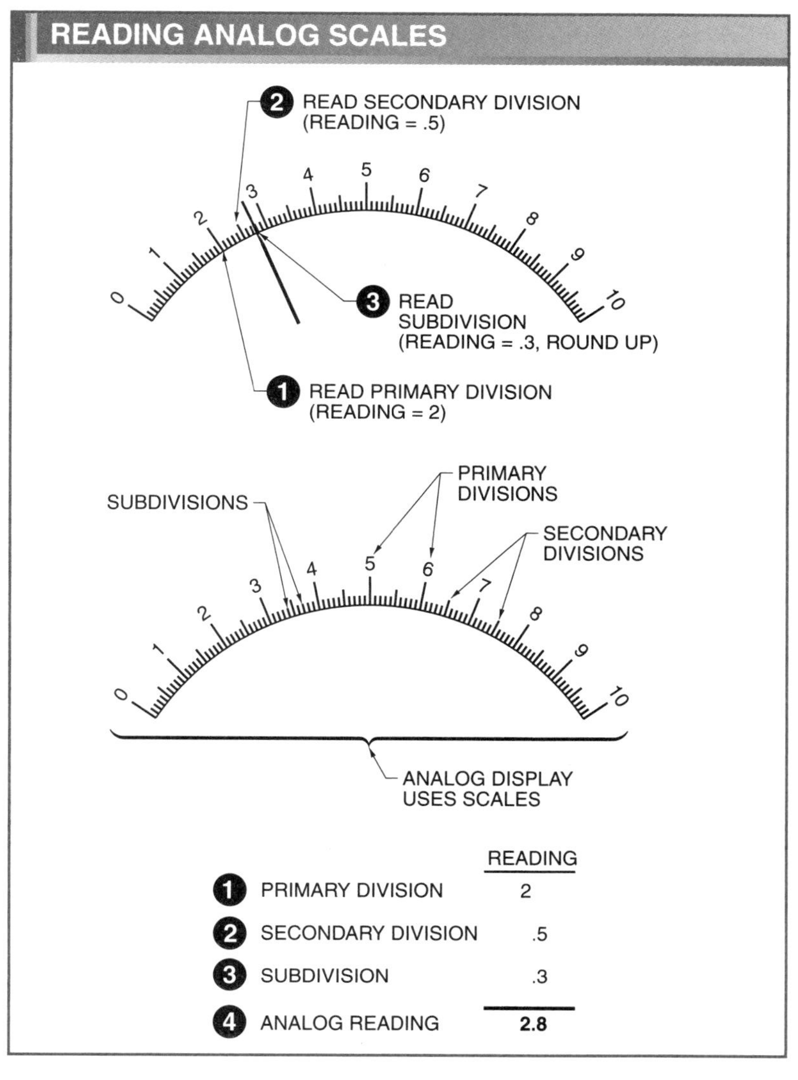

Figure 22-4. When reading an analog scale, add the primary, secondary, and subdivision readings.

Digital Displays

A *digital display* is an electronic device that displays readings as numerical values. Digital displays help eliminate human error when taking readings by displaying exact values measured. Errors occur when reading a digital display if the displayed prefixes, symbols, and decimal points are not properly applied.

The exact value on a digital display is determined from the numbers displayed and the position of the decimal point. A range switch determines the placement of the decimal point. Accurate readings are obtained by using the range that gives the best resolution without overloading.

Bar Graphs

Most digital displays include a bar graph to show changes and trends in a circuit. A *bar graph* is a graph composed of segments that function as an analog pointer. The displayed bar graph segments increase as the measured value increases and decrease as the measured value decreases. Reverse the polarity of test leads if a negative sign is displayed at the beginning of the bar graph. **See Figure 22-5.** A *wrap-around bar graph* is a bar graph that displays a fraction of the full range on the graph. The pointer wraps around and starts over when the limit of the bar graph is reached.

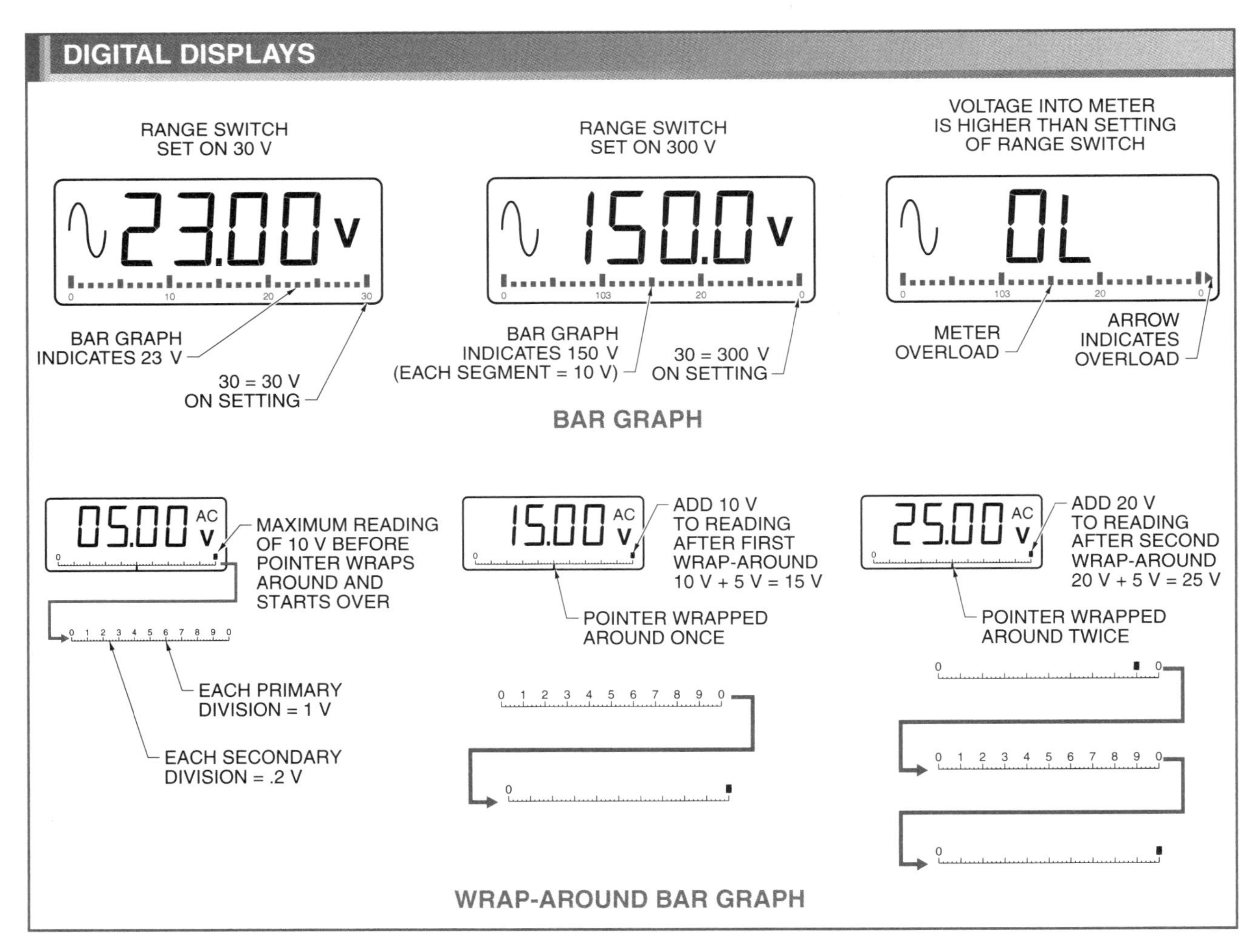

Figure 22-5. The displayed bar graph segments increase as the measured value increases and decrease as the measured value decreases.

A bar graph reading is updated 30 times per second. A numerical reading is updated 4 times per second. The bar graph is used when quickly changing signals cause the numerical display to flash or when there is a change in the circuit that is too rapid for the numerical display to detect.

Ghost Voltages

A test instrument set to measure voltage may display a reading before the instrument is connected to a powered circuit. The displayed voltage is a ghost voltage that appears as changing numbers on a digital display or as a vibrating analog display. A *ghost voltage* is a voltage that appears on a test instrument that is not connected to a circuit.

Ghost voltages are produced by the magnetic fields generated by current-carrying conductors, fluorescent lighting, and operating electrical equipment. Ghost voltages enter a test instrument through the test leads because test leads not connected to a circuit act as antennae for stray voltages. **See Figure 22-6.**

Ghost voltages do not damage a test instrument. Ghost voltages may be misread as circuit voltages when a test instrument is connected to a circuit that is believed to be powered. A circuit that is not powered can also act as an antenna for stray voltages. To ensure true circuit voltage readings, connect a test instrument to a circuit long enough so that the instrument displays a constant reading.

TEST INSTRUMENT MEASUREMENT PROCEDURES

Proper equipment and procedures are required when taking measurements with test instruments. To take a measurement, the test instrument must be set to the correct measuring position and properly connected into the circuit to be tested. The displayed measurement is then read and interpreted. Test instruments not properly connected to a circuit increase the likelihood of an improper measurement and may create an unsafe condition.

In most electrical and electronic circuits, measurements are taken when there is a problem (troubleshooting), or to display the presence and level of a measurable quantity, such as voltage, current, speed, or temperature during normal operation. Taking measurements during troubleshooting gives a clear picture of any problems in a circuit, circuit operation, and possible future problems. Test instruments include voltage indicators, test lights, continuity testers, voltage testers, digital multimeters, megohmmeters, thermometers, tachometers, and scopes.

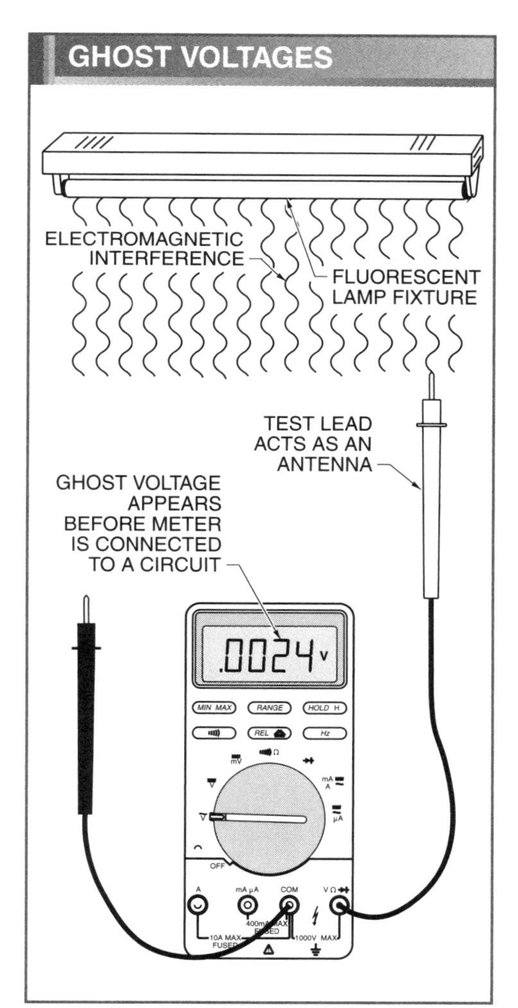

Figure 22-6. Ghost voltages enter a test instrument through the test leads because test leads not connected to a circuit act as antennae for stray voltages.

Voltage Indicators

Voltage measurements are normally taken to establish that there is voltage at a given point in a circuit and to determine if the voltage is at the proper level. Most electrical devices, circuits, and power sources operate at a rated voltage level (1.5 V, 9 V, 24 V, 115 V, 230 V, etc.). The exact voltage level in a circuit may vary. For most electrical circuits, the voltage level may vary ±10% from the rated voltage.

A *voltage indicator* is a test instrument that indicates the presence of voltage when the test tip touches, or is near, an energized hot conductor or energized metal part. The tip glows and/or the device creates a sound when voltage is present at the test point.

Voltage indicators are used to test receptacles, fuses, circuit breakers, cables, and other devices in which the presence of voltage must be detected. The hot side (short slot or black wire) of a standard receptacle is typically used to test a voltage indicator for proper operation. Voltage indicators are available in various voltage ranges (a few volts to hundreds of volts) and in the different voltage types (AC, DC, AC/DC) for testing various types of circuits.

Voltage Indicator Measurement Procedure. Before taking any measurements using a voltage indicator, ensure the indicator is designed to take measurements on the circuit being tested. Refer to the operating manual of the test instrument for all measuring precautions, limitations, and procedures. Always wear required personal protective equipment and follow all safety rules when taking a measurement. **See Figure 22-7.** To test for voltage using a voltage indicator, apply the following procedure:

1. Verify that the voltage indicator has a voltage rating higher than the highest potential voltage in the circuit being tested. When circuit voltage is unknown, slowly bring the voltage indicator near the conductor or slot being tested. The voltage indicator will glow and/or sound when voltage is present. The brighter a voltage indicator glows, the higher the voltage or the closer the voltage indicator is to the voltage source.
2. Place the tip of the voltage indicator on or near the wire or device being tested. When testing an extension cord for a break, test several points along the wire.
3. Remove the voltage indicator from the test area.

When a voltage indicator does not indicate the presence of voltage by glowing or making a sound, do not assume that there is no voltage and start working on exposed components of a circuit. Always take a second test instrument (voltmeter or multimeter) and measure for the presence of voltage before working around or on exposed wires and electrical components.

Test Lights

A *test light (neon tester)* is a test instrument that is designed to illuminate in the presence of 115 V and 230 V circuits. A soft glow indicates a 115 V circuit. A brighter glow indicates a 230 V circuit.

Test lights are primarily used to determine when voltage is present in a circuit (the circuit is energized), such as when testing receptacles. When testing a receptacle, the test light bulb illuminates when the receptacle is properly wired and energized.

When a receptacle is properly wired, a test light bulb will illuminate when the test light leads are connected between the neutral slot and the hot slot. A test light bulb will also illuminate when the leads are connected between the ground slot and the hot slot. If the test light illuminates when the leads are connected from the neutral slot to the ground slot, the hot (black) and neutral (white) wires are reverse wired.

If a test light illuminates when the test leads are connected to the neutral slot and hot slot but does not light when connected to the ground slot and hot slot, the receptacle is not grounded. When a test light illuminates, but is dimmer than when connected between the neutral slot and hot slot, the receptacle has an improper ground (having high resistance). Improper grounds are also a safety hazard and must be corrected.

Test Light Measurement Procedure. Before using a test light, always check the test light on a known energized circuit that is within the test light's rating to ensure that the test light is operating correctly. Always use a test light that is designed for use on the circuit being tested. Follow all electrical safety practices and all manufacturer recommendations and procedures. Check and wear personal protective equipment (PPE) for the procedure being performed. Refer to the operating manual of the test instrument for all measuring procedures, limitations, and precautions. **See Figure 22-8.** To test for voltage using a test light, apply the following procedure:

1. Verify that the test light has a voltage rating higher than the highest potential voltage in the circuit. Care must be taken to guarantee that the exposed metal tips of the test light leads do not touch fingers or any metal parts not being tested.
2. Connect one test lead of the test light to the neutral side of the circuit or ground. When testing a circuit that has a neutral or ground, connect to the neutral or ground side of the circuit first.
3. Connect the other test lead of the test light to the other side (hot side) of the circuit. Voltage is present when the test light bulb illuminates. Voltage is less than the rating of the test light when the test light is dimly lit, and is higher than the rating of the test light when the test light glows brighter than normal. Voltage is not present in a circuit or present at a very low level when a test light does not illuminate.
4. Remove the test light from the circuit.

Tech Facts

Voltage indicators such as test lights should be tested on a known energized circuit if the test light does not illuminate near the circuit being tested. This will verify that the voltage indicator is operating properly. Properly working test equipment is essential for safety and taking accurate measurements.

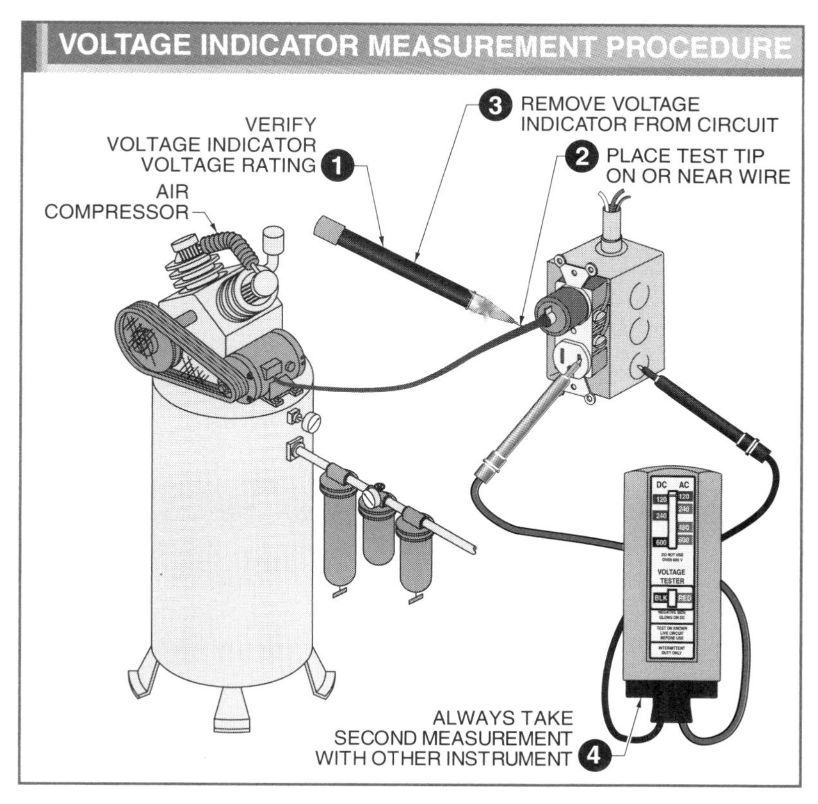

Figure 22-7. Always verify that a voltage indicator or any test instrument has a voltage rating higher than the highest potential voltage in the circuit being tested.

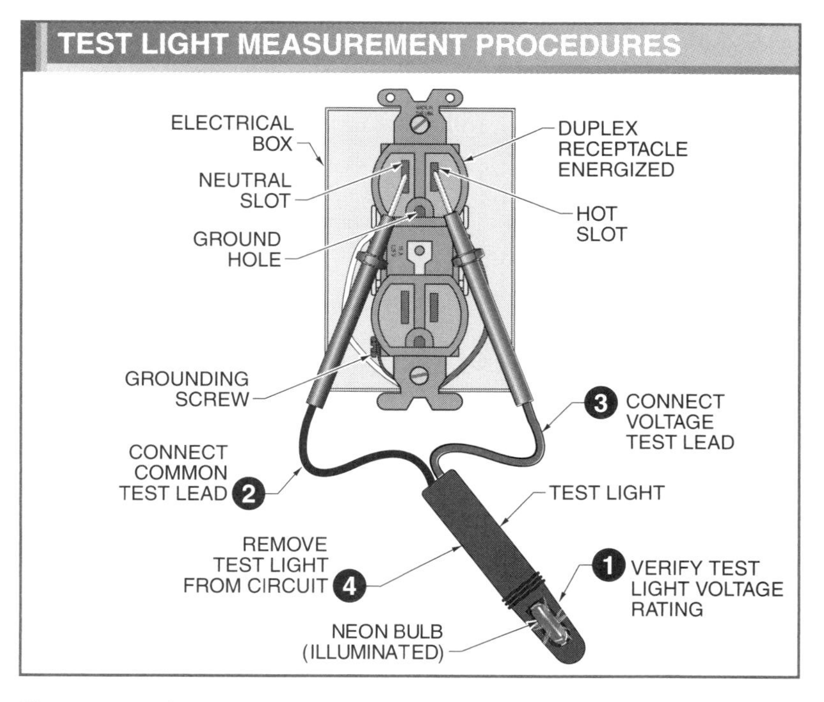

Figure 22-8. Care must be taken to ensure that the exposed metal tips of the test light leads or any test instrument do not touch fingers or any metal parts not being tested.

Continuity Testers

A *continuity tester* is a test instrument that is used to test a circuit for a complete path for current to flow. A closed switch that is operating correctly has continuity while an open switch does not have continuity. A continuity tester or test instrument with continuity test mode indicates when an electrical device or circuit has a complete path by emitting an audible sound (beep). Indication of a complete path is used to determine the condition of a device, such as a switch, as open or closed.

Continuity Tester Measurement Procedure. Continuity is tested with a test instrument set to the continuity test mode. Before taking any continuity measurements using a continuity tester, ensure the meter is designed to take measurements on the circuit being tested. Follow all electrical safety practices and all manufacturer recommendations and procedures. Check and wear personal protective equipment (PPE) for the procedure being performed. Refer to the operating manual of the test instrument for all measuring procedures, limitations, and precautions. **See Figure 22-9.** To take continuity measurements with a continuity tester, apply the following procedure:

1. Set the continuity tester function switch to continuity test mode. Most test instruments have the continuity test mode and resistance mode sharing the same function switch position.
2. With the circuit de-energized, connect the test leads across the component being tested. The position of the test leads is arbitrary.
3. When there is a complete path (continuity), the continuity tester beeps. When there is no continuity (open circuit), the continuity tester does not beep.
4. After completing all continuity tests, remove the continuity tester from the circuit or component being tested and turn the instrument OFF to prevent battery drain.

Voltage Testers

A *voltage tester* is an electrical test instrument that indicates the approximate amount of voltage and the type of voltage (AC or DC) in a circuit by the movement of a pointer (and vibration, on some models). When a voltage tester includes a solenoid, the solenoid vibrates when the tester is connected to AC voltage. Some voltage testers include a colored plunger or other indicator such as a light that indicates the polarity of the test leads as positive or negative when measuring a DC circuit. Voltage testers are used to take measurements any time an exact voltage measurement is not required.

Voltage Tester Measurement Procedure. Before using a voltage tester, always check the voltage tester on a known energized circuit that is within the voltage rating of the voltage tester to verify proper operation. Follow all electrical safety practices and all manufacturer recommendations and procedures. Check and wear personal protective equipment (PPE) for the procedure being performed. Refer to the operating manual of the test instrument for all measuring procedures, limitations, and precautions. **See Figure 22-10.** To take a voltage measurement with a voltage tester, apply the following procedure:

1. Verify that the voltage tester has a voltage rating higher than the highest potential voltage in the circuit being tested.
2. Connect the common test lead to the point of testing (neutral or ground).
3. Connect the voltage test lead to the point of testing (ungrounded conductor). The pointer of the voltage tester indicates a voltage reading and vibrates when the current in the circuit is AC. The indicator shows a voltage reading and does not vibrate when the current in the circuit is DC.
4. Observe the voltage measurement displayed.
5. Remove the voltage tester from the circuit.

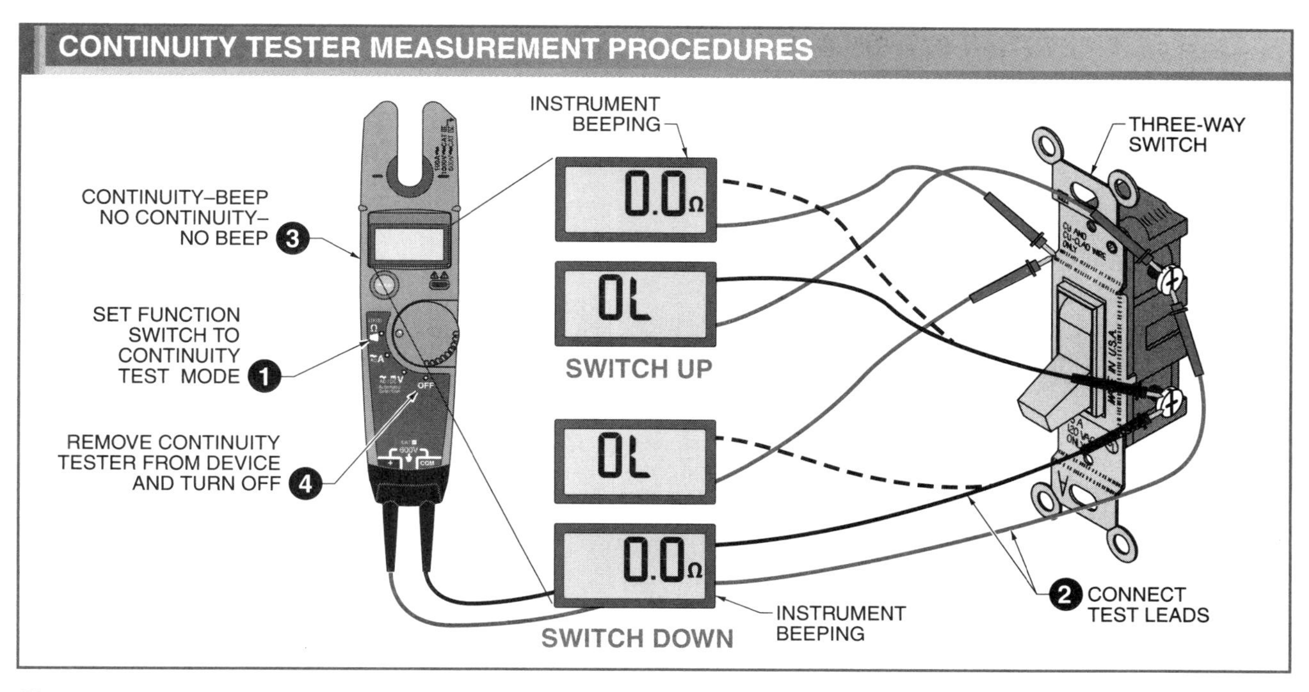

Figure 22-9. A continuity tester indicates a complete electrical path by emitting an audible sound (beep).

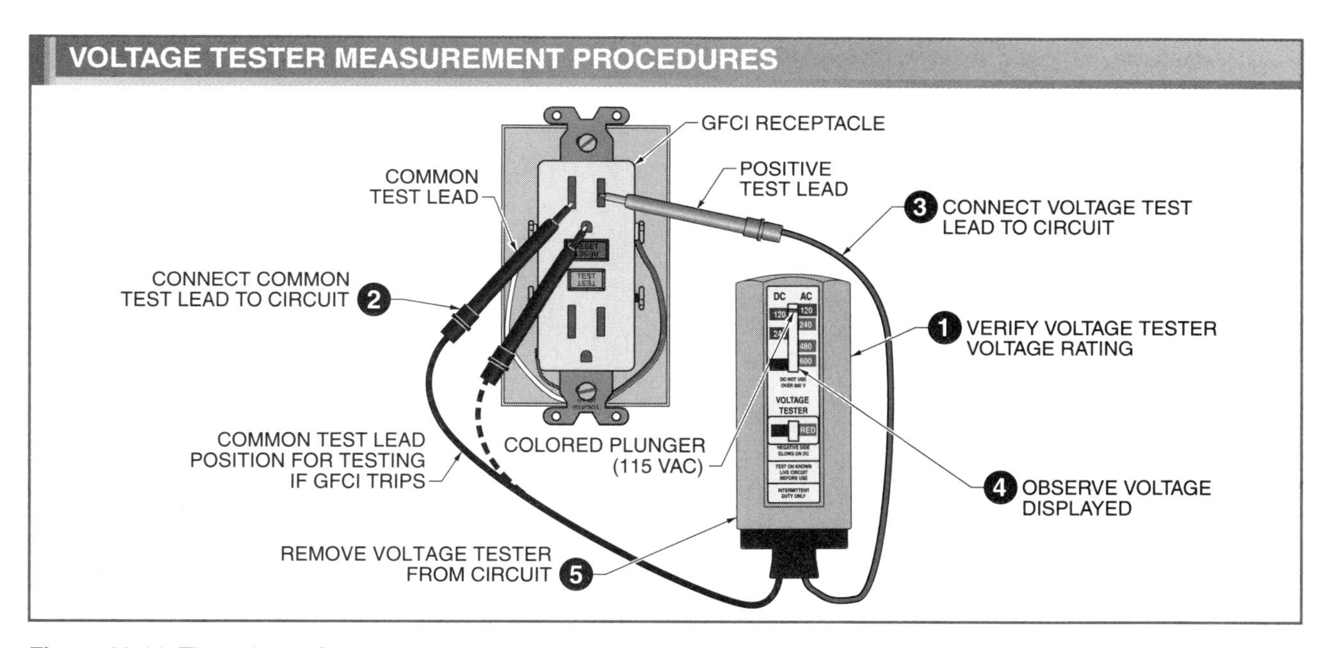

Figure 22-10. The pointer of a voltage tester indicates the voltage reading and vibrates when the current in a circuit is AC.

Digital Multimeters

A *digital multimeter (DMM)* is an electrical test instrument that can measure two or more electrical properties and display the measured properties as numerical values. Basic DMMs measure resistance (continuity) and AC or DC voltage and current. Advanced DMMs include special functions such as measuring frequency, capacitance, and temperature. DMMs are used to take measurements any time the voltage, current, or resistance of the circuit being tested is within the rating of the meter and an exact measurement is expected.

DMM AC Voltage Measurement Procedure. Before taking any voltage measurements using a DMM, ensure the meter is designed to take measurements on the circuit being tested. Follow all electrical safety practices and all manufacturer recommendations and procedures. Check and wear personal protective equipment (PPE) for the procedure being performed. Refer to the operating manual of the test instrument for all measuring procedures, limitations, and precautions. **Warning:** Ensure that no body part contacts any part of the live circuit, including the metal contact points at the tip of the test leads. **See Figure 22-11.** To take an AC voltage measurement with a DMM, apply the following procedure:

1. Set the function switch to AC voltage. Set the test instrument on the highest voltage setting if the voltage in the circuit is unknown.
2. Plug the black test lead into the common jack.
3. Plug the red test lead into the voltage jack.
4. Connect the test leads to the circuit. The position of the test leads is arbitrary. Common industrial practice is to connect the black test lead to the grounded (neutral) side of the AC voltage.
5. Read the voltage displayed on the test instrument.

DMM DC Voltage Measurement Procedure. Safety precautions must be taken to ensure that a DMM is properly used because voltage measurements are taken when a circuit is powered. Exercise caution when measuring DC voltages over 60 V. **Warning:** Ensure that no body part contacts any part of a live circuit, including the metal contact points at the tip of the test leads. **See Figure 22-12.** To measure DC voltage with a DMM, apply the following procedure:

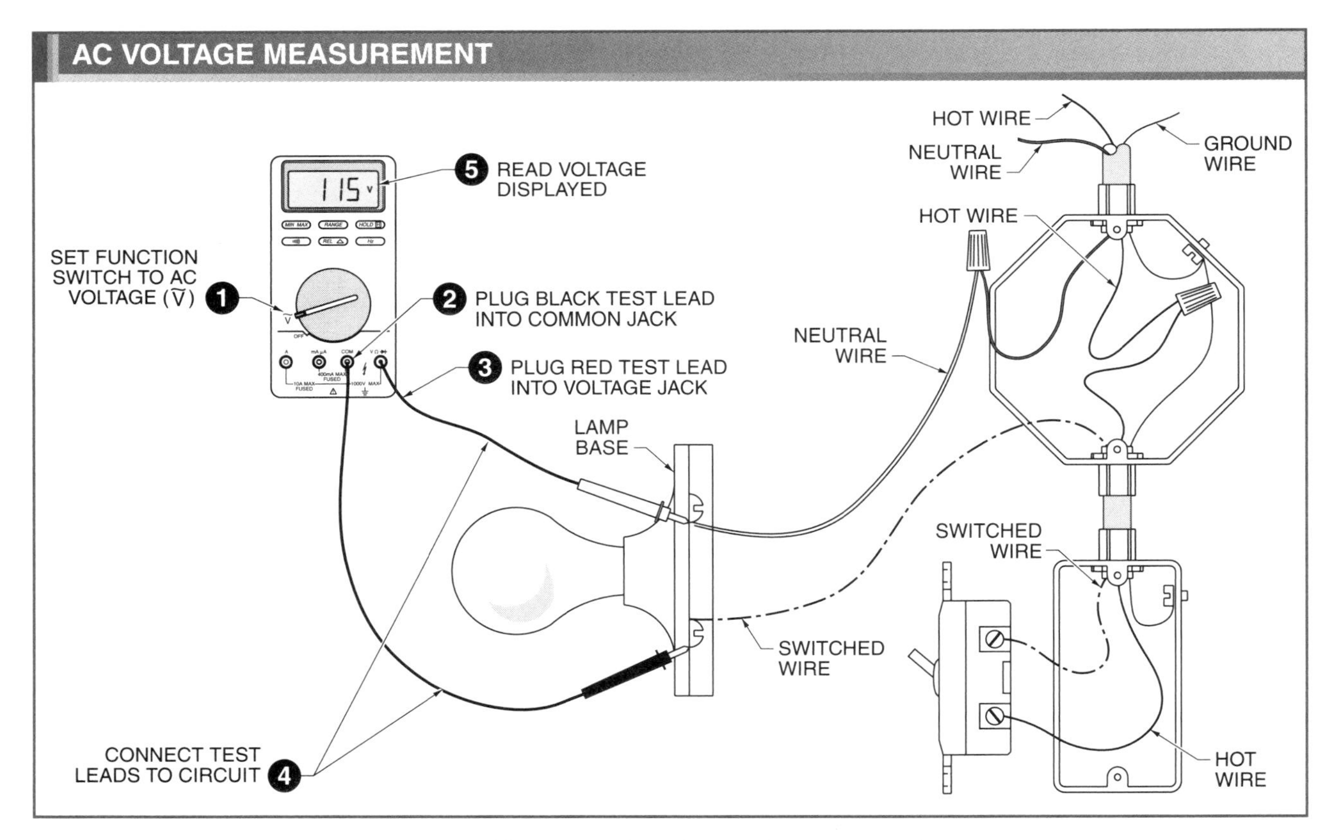

Figure 22-11. The function switch is set to VAC when a DMM is used to measure AC voltage.

1. Set the function switch to DC voltage. Select a setting high enough to measure the highest possible circuit voltage if the test instrument has more than one voltage position or if the circuit voltage is unknown.
2. Plug the black test lead into the common jack.
3. Plug the red test lead into the voltage jack.
4. Connect the test instrument test leads in the circuit. Connect the black test lead to circuit ground and the red test lead to the point at which the voltage is under test. Reverse the black and red test leads if a negative sign appears in front of the reading on a digital instrument.
5. Read the voltage displayed on the test instrument.

In-Line Ammeter DC Measurement Procedure. The amount of current in a circuit can be measured using an ammeter. An *in-line ammeter* is a meter that measures current in a circuit by inserting the test instrument in series with the component(s) under test. In-line ammeter readings require the circuit to be opened so the ammeter can be inserted. Ammeters are used to measure the amount of current flowing in a circuit that is powered. Safety precautions must be taken to ensure that no part of the body comes in contact with a live part of the circuit.

In-line ammeters are used to measure very small amounts of current (milli and microamperes) and are normally limited to measuring currents of less than 10 A. In-line ammeters are normally used to measure DC currents. Always check the type (AC and/or DC) and limits (maximum measurable amount) of the test instrument being used.

Care must be taken to protect the test instrument, circuit, and person using the test instrument when measuring DC with an in-line ammeter. Always apply the following rules when using an in-line ammeter:

- Check that the power to the test circuit is OFF before connecting and disconnecting test leads.
- Do not change the position of any switches or settings on the test instrument while the circuit under test is energized.
- Turn the power to the test instrument and circuit OFF before any settings are changed.
- Connect the ammeter in series with the component(s) to be tested.
- Do not take current readings from any circuit in which the current may exceed the limit of the test instrument.

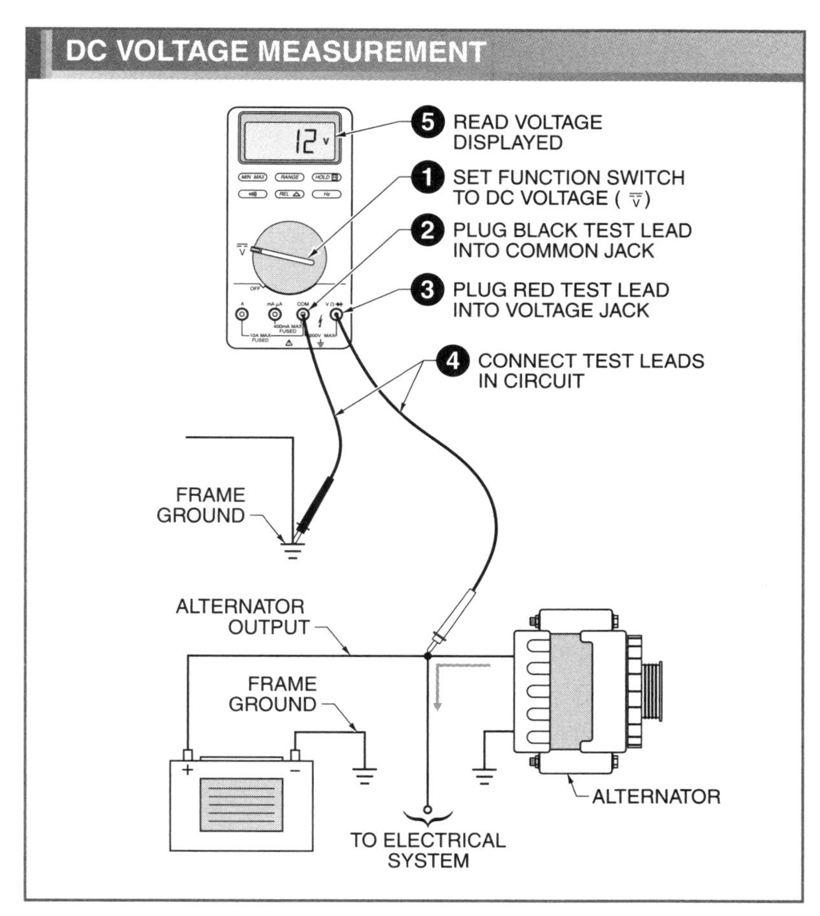

Figure 22-12. When using a DMM to measure DC voltage, the black test lead is connected to circuit ground and the red test lead is connected to the point at which the voltage is under test. Reverse the black and red test leads if a negative sign appears in front of the reading on a digital meter.

Tech Facts

Digital multimeters record measurements and display values so that they can be easily read.

Many test instruments include a fuse in the low-ampere range to prevent damage caused by excessive current. Before using a test instrument, check to see if it is fused on the current range being used. The test instrument is marked as fused or not fused at the test lead terminals. To protect the test instrument, the fuse rating should not exceed the current range of the instrument. **Warning:** Ensure that no body parts contact any part of the live circuit, including the metal contact points at the tip of the test leads. **See Figure 22-13.** To measure DC using an in-line ammeter, apply the following procedure:

1. Set the selector switch to DC. Select a setting high enough to measure the highest possible circuit current if the test instrument has more than one DC position.
2. Plug the black test lead into the common jack.
3. Plug the red test lead into the current jack.
4. Turn the power to the circuit or device under test OFF and discharge all capacitors if possible.
5. Open the circuit and connect the test leads to each side of the opening. The black (negative) test lead is connected to the negative side of the opening and the red (positive) test lead is connected to the positive side of the opening. Reverse the black and red test leads if a negative sign appears in front of the displayed reading.
6. Turn the power to the circuit under test ON.
7. Read the current displayed on the test instrument.
8. Turn the power OFF and remove the test instrument from the circuit.

The same procedure is used to measure AC with an in-line ammeter, except that the selector switch is set on AC current.

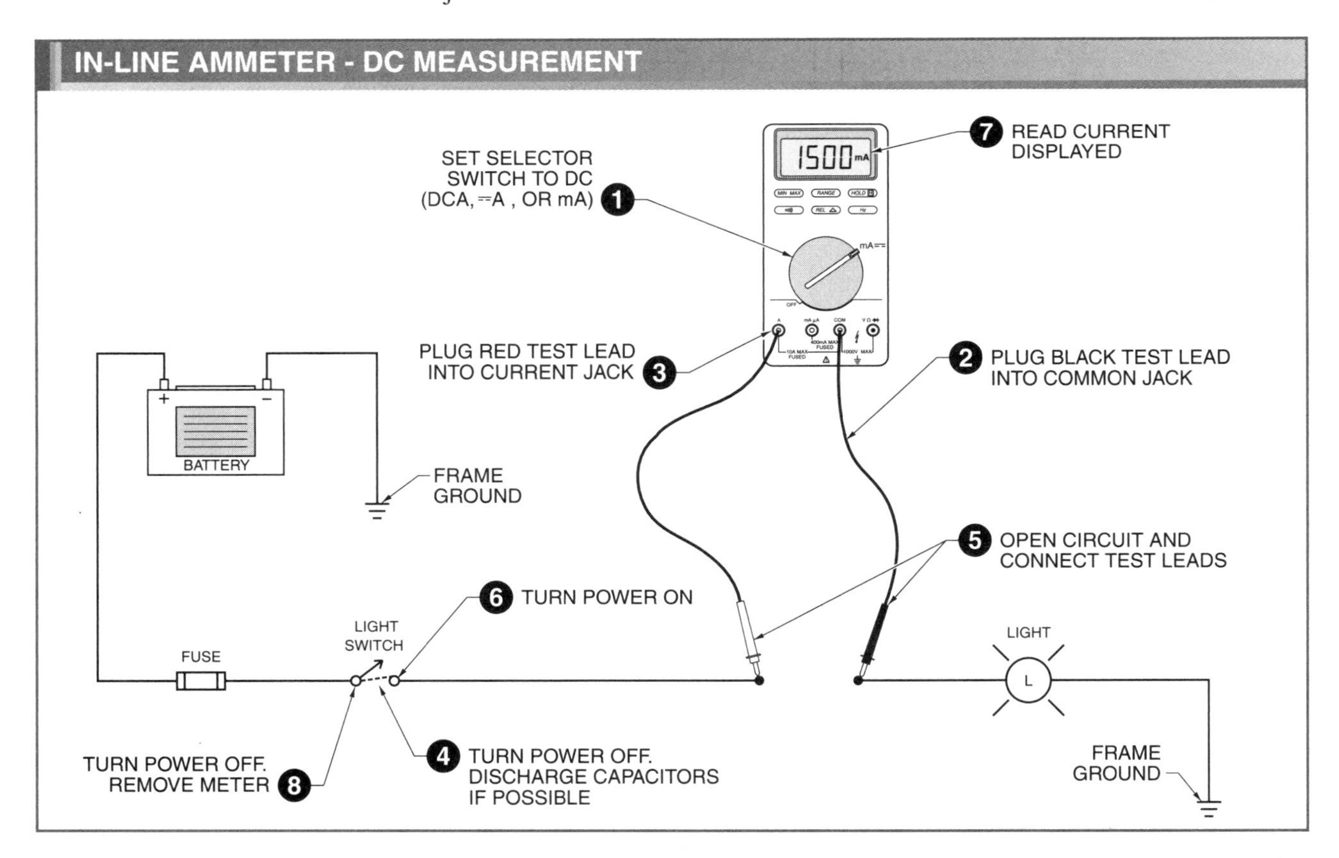

Figure 22-13. Care must be taken to protect the test instrument, circuit, and person using the instrument when measuring DC with an in-line ammeter.

Clamp-On Ammeter AC Measurement Procedure. A *clamp-on ammeter* is a meter that measures current in a circuit by measuring the strength of the magnetic field around a conductor. Clamp-on ammeters take current readings without opening the circuit. Clamp-on ammeters are normally used to measure AC currents from .01 A or less to 1000 A or more. **Warning:** Care must be taken to ensure that the test instrument does not pick up stray magnetic fields. Whenever possible, separate the conductors under test from other surrounding conductors by a few inches. **See Figure 22-14.** To measure AC using a clamp-on ammeter, apply the following procedure:

1. Set the function switch to AC current. Select the proper setting to measure the highest possible circuit current if the test instrument has more than one current position or if the circuit current is unknown.
2. Plug the current probe accessory into the test instrument when using a DMM that requires a current probe.
3. Open the jaws by pressing against the trigger.
4. Enclose one conductor in the jaws. Ensure that the jaws are completely closed before taking readings.
5. Read the current displayed on the test instrument.

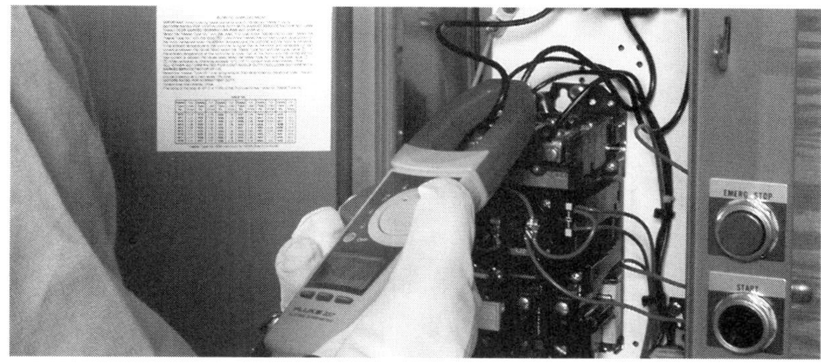

AEMC® Instruments

Clamp-on ammeters are used to measure the current flowing through an isolated conductor by measuring the strength of the magnetic field around the conductor.

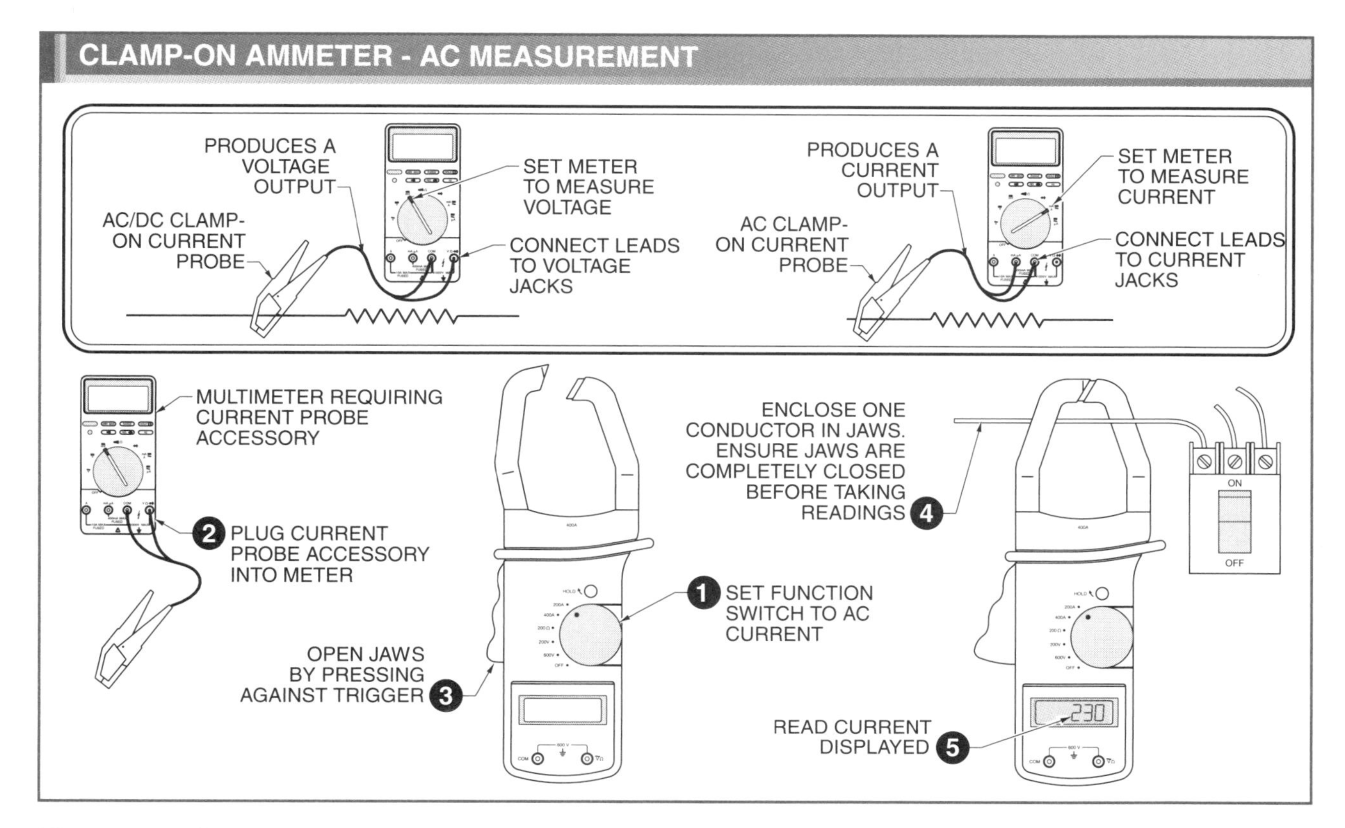

Figure 22-14. Clamp-on ammeters measure the current in a circuit by measuring the strength of the magnetic field around a conductor.

Ohmmeter Resistance Measurement Procedure. An *ohmmeter* is a device that is used to measure the amount of resistance in a component (or circuit) that is not powered. Resistance measurements are normally taken to indicate the condition of a component or circuit. The higher the resistance, the lower the current flow. Likewise, the lower the resistance, the higher the current flow.

Low voltage applied to a test instrument set to measure resistance causes inaccurate readings. High voltage applied to a test instrument set to measure resistance causes instrument damage. Check for voltage using a voltmeter. **Warning:** Ensure that no voltage is present in the circuit or component under test before taking resistance measurements. **See Figure 22-15.** To measure resistance using an ohmmeter, apply the following procedure:

1. Check to ensure that all power is OFF in the circuit or component under test and disconnect the component from the circuit.
2. Set the function switch to the resistance position.
3. Plug the black test lead into the common jack.
4. Plug the red test lead into the resistance jack.
5. Ensure that the test instrument batteries are in good condition. The battery symbol is displayed when the batteries are low.
6. Connect the test instrument test leads across the component under test. Ensure that contact between the test leads and the circuit is good. Dirt, solder flux, oil, and other foreign substances greatly affect resistance readings.
7. Read the resistance displayed on the test instrument. Check the circuit schematic for parallel paths. Parallel paths with the resistance under test cause reading errors. Do not touch exposed metal parts of the test leads during the test. The resistance of a person's body can cause reading errors.
8. Turn the test instrument OFF after measurements are taken to save battery life.

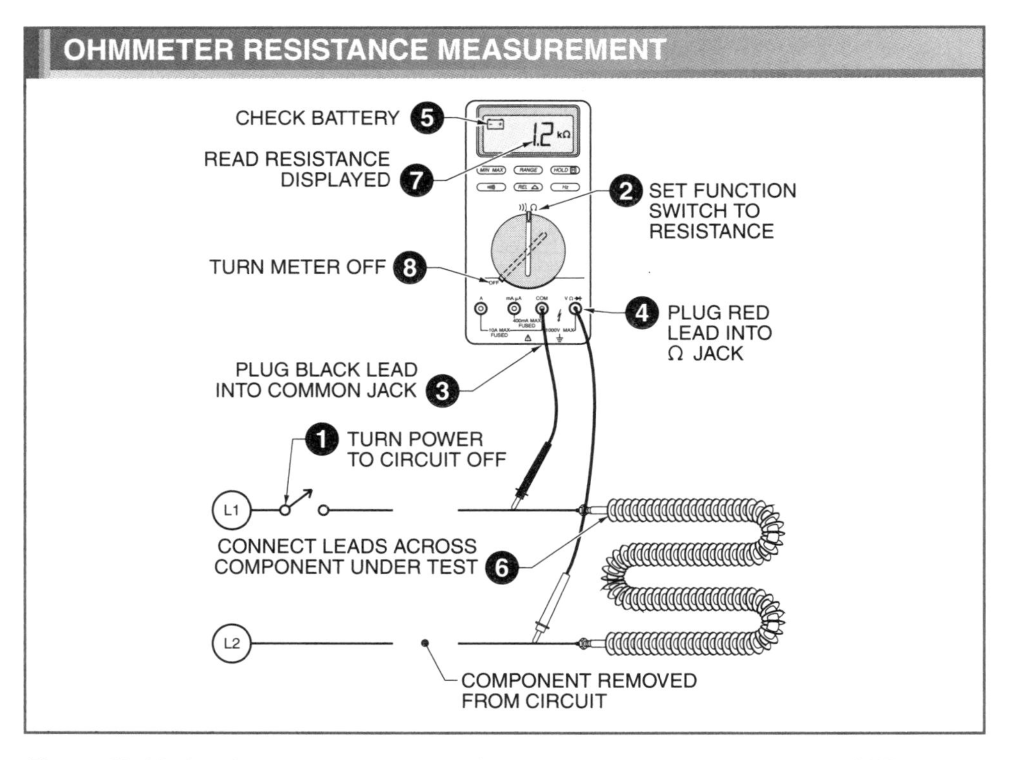

Figure 22-15. An ohmmeter measures resistance with all power to the circuit OFF.

Megohmmeter Resistance Measurement Procedure. A *megohmmeter* is a device that detects insulation deterioration by measuring high resistance values under high test voltage conditions. Megohmmeters deliver a high voltage to the circuit under test. Megohmmeter test voltages range from 50 V to 5000 V. **Warning:** Ensure that no voltage is present in a circuit or component under test before taking any resistance measurements. Ensure that no body part contacts the high voltage. **See Figure 22-16.** To measure resistance using a megohmmeter, apply the following procedure:

1. Ensure that all power is OFF in the circuit or component under test. Test for voltage using a voltmeter if uncertain.
2. Set the selector switch to the voltage at which the circuit is to be tested. The test voltage should be as high or higher than the highest voltage to which the circuit under test is exposed.
3. Plug the black test lead into the negative (earth) jack.
4. Plug the red test lead into the positive (line) jack.
5. Ensure that the batteries are in good condition. The megohmmeter contains no batteries if it includes a crank. The meter contains no batteries or crank if the meter plugs into a standard outlet.
6. Connect the line test lead to the conductor under test.
7. Connect the earth test lead to a second conductor in the circuit or earth ground.
8. Press the test button or turn the crank and read the resistance displayed. Change the resistance or voltage range if required.
9. Consult the equipment manufacturer or megohmmeter manufacturer for the minimum recommended resistance values. The insulation is good if the meter reading is equal to or higher than the minimum value.

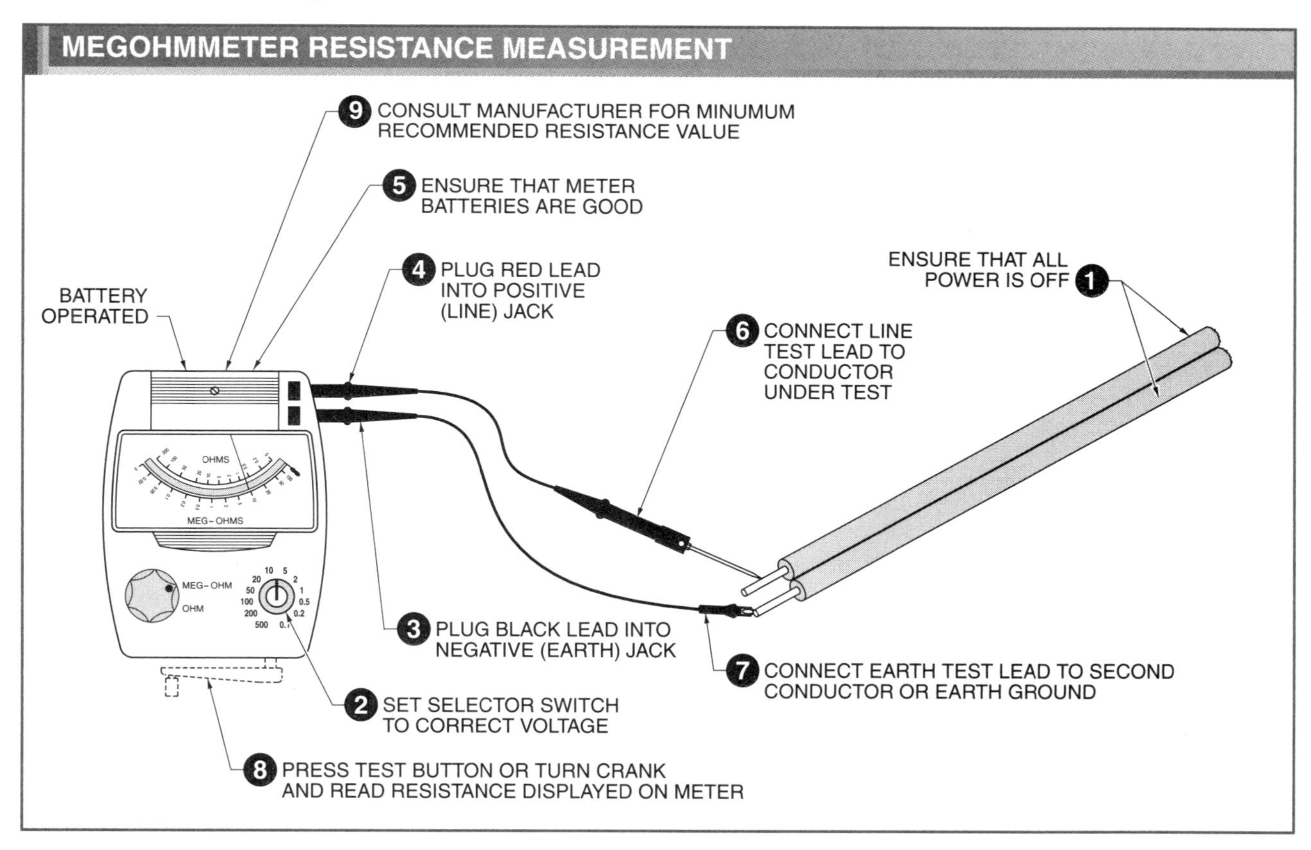

Figure 22-16. Megohmmeters deliver a high voltage to the circuit under test.

Contact Thermometer Temperature Measurement. *Temperature* is the measurement of the intensity of heat. Temperature is normally measured in degrees Fahrenheit (°F) or degrees Celsius (°C). All materials emit infrared energy in proportion to their temperature. A *contact thermometer* is an instrument that measures temperature at a single point. Use a temperature probe that is rated higher than the highest possible temperature and always wear safety glasses. **Warning:** Avoid contact with any material that can cause burns. **See Figure 22-17.** To measure temperature using a contact thermometer, apply the following procedure:

1. Select a temperature probe. A probe must have a higher temperature rating than the highest temperature it may contact and should have a shape that allows good contact with the device under test.
2. Connect the temperature probe to the test instrument.
3. Set the test instrument to the correct temperature range. Select the highest range if the temperature is unknown.
4. Place the temperature probe tip on the object or in the area to be measured.
5. Read the temperature displayed.
6. Remove the temperature probe from the object or area under test.

Infrared Meter Temperature Measurement Procedure. An *infrared meter* is a meter that measures heat energy by measuring the infrared energy that a material emits. An infrared meter displays an image on a screen that shows different temperatures indicated by different colors or gives a direct digital temperature readout. **Warning:** Avoid contact with any material that can cause burns. **See Figure 22-18.** To measure temperature using an infrared meter, apply the following procedure:

1. Aim meter at area to be measured. Focus meter based on the distance the object is from the meter.
2. Take the temperature reading of any areas suspected to have temperatures above ambient temperature.

To determine temperature differential between ambient temperature and the area of increased temperature, subtract the ambient temperature reading from the reading obtained for the area of increased temperature.

3. Take the ambient temperature reading for reference.

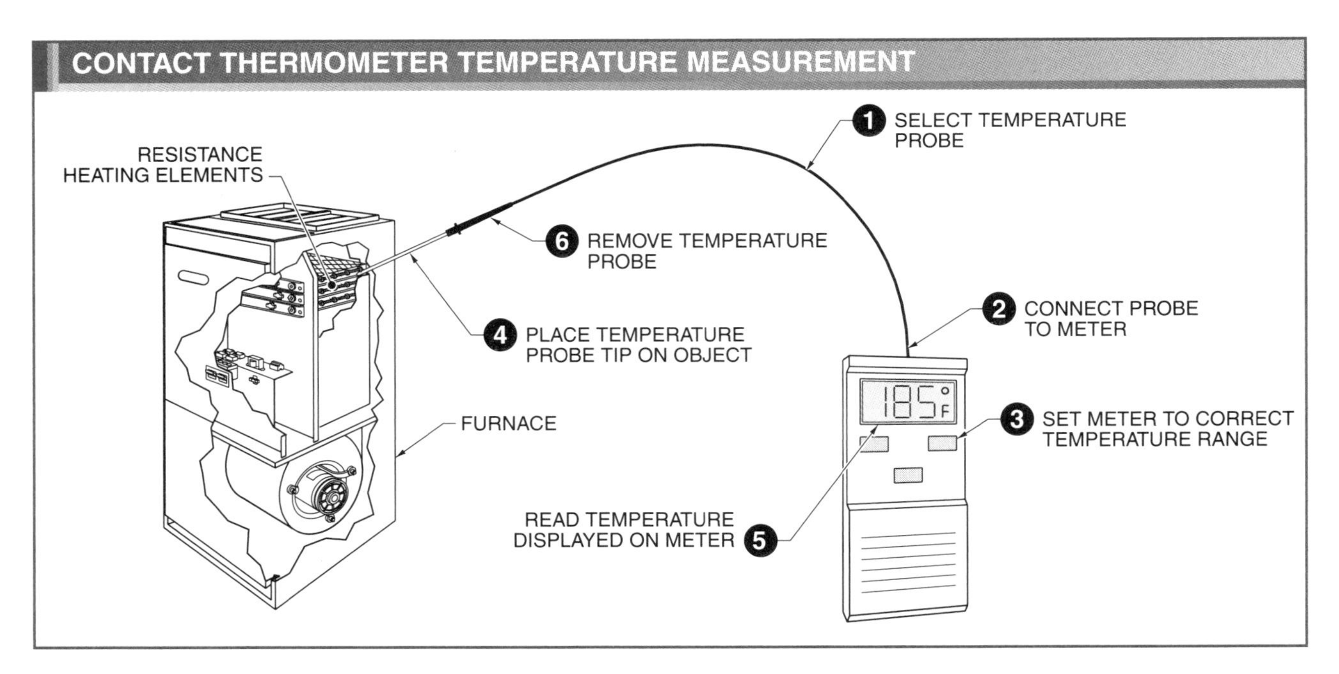

Figure 22-17. A contact thermometer measures temperature at a single point.

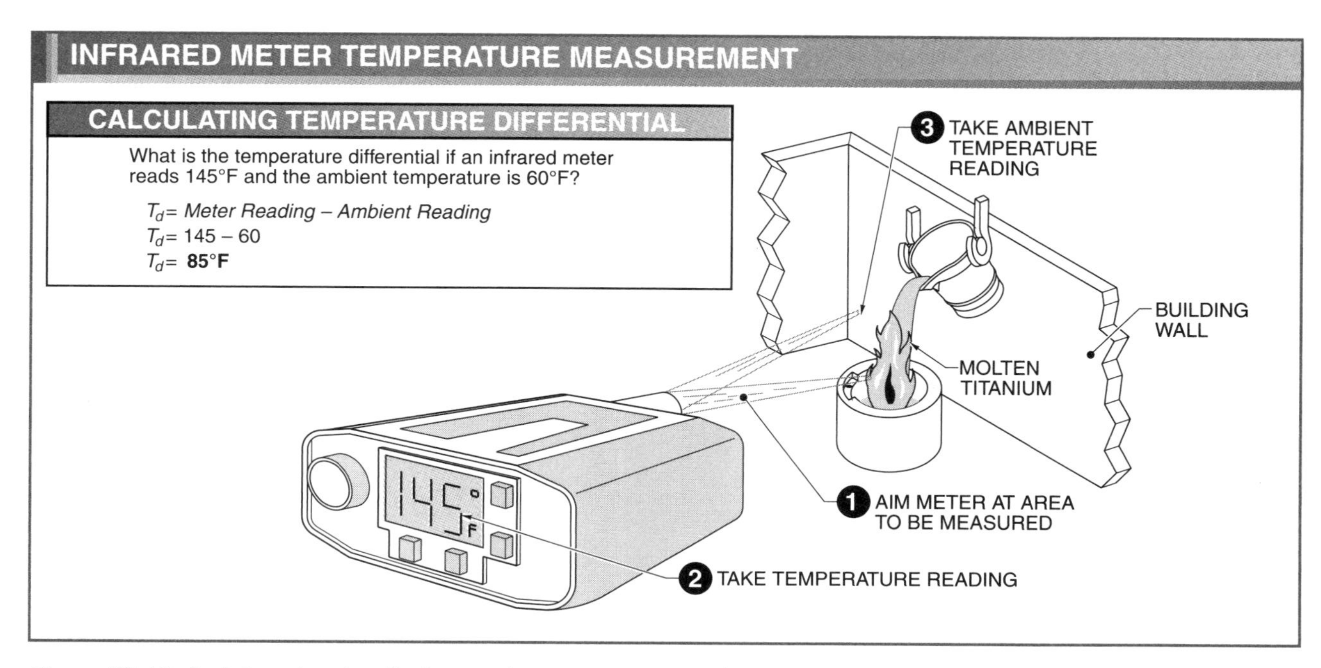

Figure 22-18. An infrared meter displays an image on a screen that shows different temperatures indicated by different colors or gives a direct digital temperature readout.

Contact Tachometer Speed Measurement Procedure. *Speed* is the rate at which an object is moving. Speed of a rotating object is measured in revolutions per minute (rpm). A *contact tachometer* is a device that measures the rotational speed of an object through direct contact of the tachometer tip with the object to be measured. A contact tachometer measures speeds from .1 rpm to 25,000 rpm. **Warning:** Exercise caution when working around moving objects. **See Figure 22-19.** To measure speed with a contact tachometer, apply the following procedure:

1. Place the tip of the tachometer in direct contact with the moving object.
2. Read the speed displayed on the instrument.

Strobe Tachometer Speed Measurement Procedure. A *strobe tachometer* is a device that uses a flashing light to measure the speed of a moving object. Strobe tachometers measure speeds from 20 rpm to 1,000,000 rpm. A strobe tachometer measures speed by synchronizing its light's flash rate with the speed of the moving object. **Warning:** Exercise caution when working around moving objects. **See Figure 22-20.** To measure speed with a strobe tachometer, apply the following procedure:

1. Set the test instrument for the best speed range for the application.
2. Turn the tachometer ON and align the visible light beam with the object to be measured.
3. Read the speed displayed.

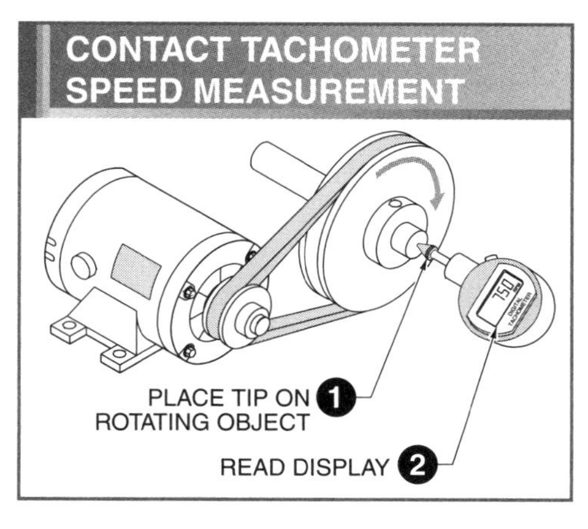

Figure 22-19. A contact tachometer must be placed in contact with the shaft of the rotating object.

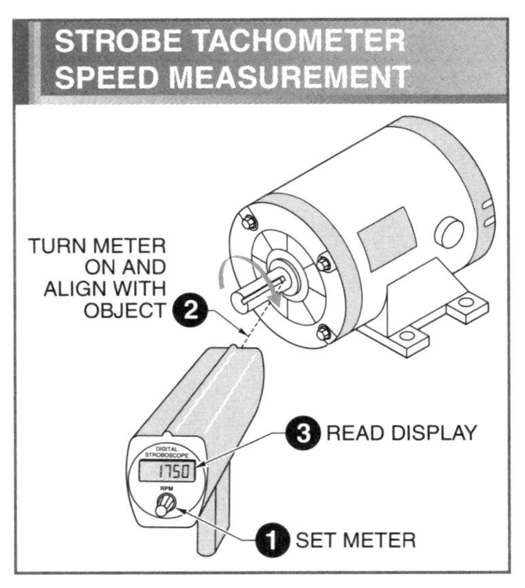

Figure 22-20. A strobe tachometer measures speed by synchronizing its light's flash rate with the speed of the moving object.

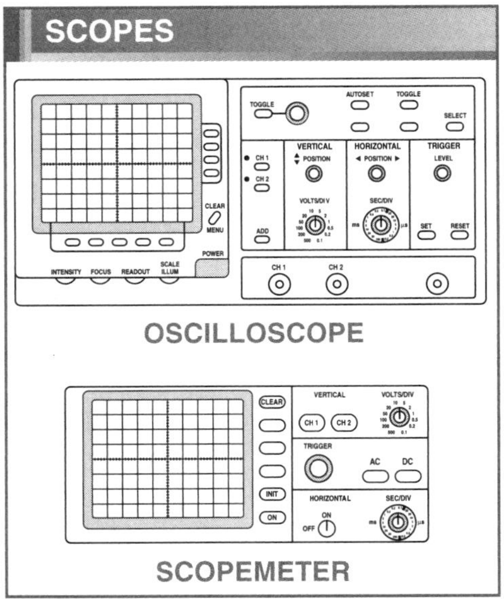

SCOPES

OSCILLOSCOPE

SCOPEMETER

Figure 22-21. The two basic types of scopes used in troubleshooting are oscilloscopes and scopemeters.

Scopes

A graphical display of circuit voltage has more meaning than a numerical value in some troubleshooting situations. Circuits that include rapidly fluctuating signals, stray signals, and phase shifts are easily detected on a graphical display. A *scope* is a device that gives a visual display of voltages. A scope shows the shape of a circuit's voltage and allows the voltage level, frequency, and phase to be measured. The two basic types of scopes used in troubleshooting are oscilloscopes and scopemeters. **See Figure 22-21.**

Oscilloscopes. An *oscilloscope* is an instrument that displays an instantaneous voltage. An oscilloscope is used to display the shape of a voltage waveform when bench testing electronic circuits. Oscilloscopes are used to troubleshoot digital circuits, communication circuits, TVs, VCRs, and computers.

Scopemeters. A *scopemeter* is a combination oscilloscope and digital multimeter. A scopemeter is used to display the shape of a voltage waveform when troubleshooting circuits in the field. A scopemeter is portable and can be used as a multimeter and a scope. A scopemeter does not have all the features of an oscilloscope.

Scope Displays. A scope displays the voltage under test on the scope screen. The scope screen contains horizontal and vertical axes. The horizontal (x) axis represents time. The vertical (y) axis represents the amplitude of the voltage waveform. The horizontal and vertical lines divide the screen into equal divisions. The divisions help to measure the voltage level and frequency of the displayed waveforms. **See Figure 22-22.**

Scope Trace. A trace is established on the screen before a circuit under test is connected. A *trace* is a reference point/line that is visually displayed on the face of the scope screen. The trace is normally positioned over the horizontal center line on the screen.

The starting point of the trace is located near the left side of the screen. *Sweep* is the movement of the displayed trace across the scope screen. The sweep of the scope trace is from left to right. **See Figure 22-23.**

Manually-operated controls are adjusted to view a waveform. Typical manually-operated scope adjustment controls include intensity, focus, horizontal positioning, vertical positioning, volts/division, and time/division.

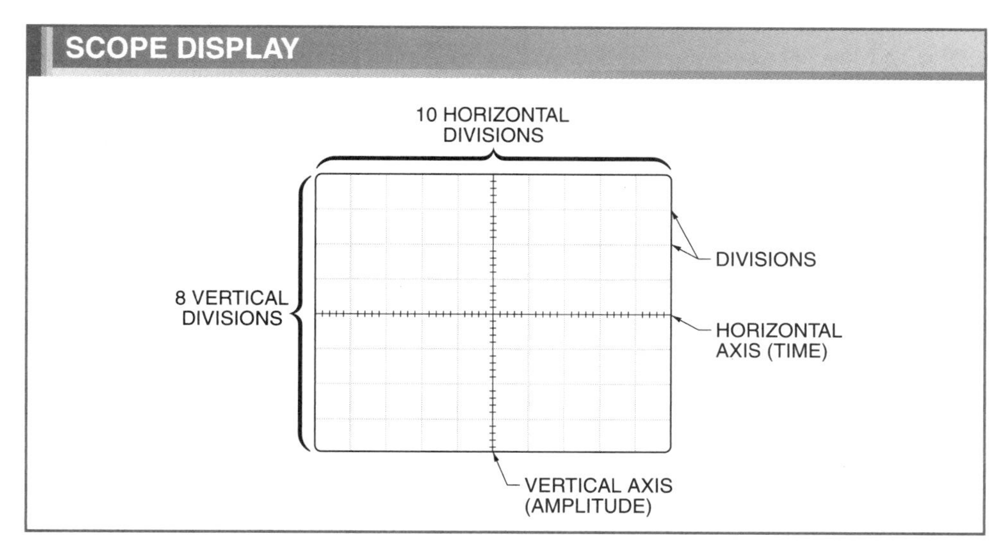

Figure 22-22. The scope screen contains horizontal and vertical axes.

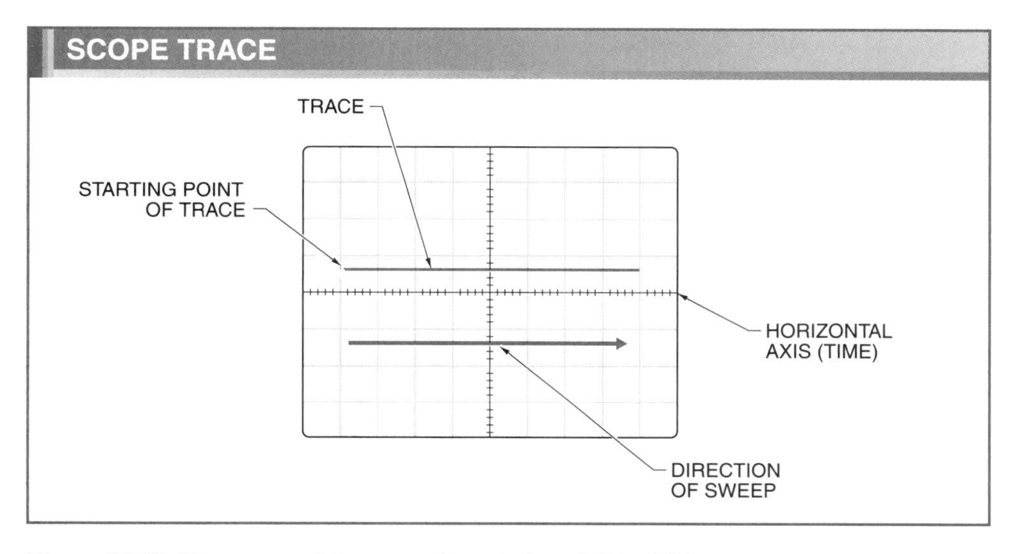

Figure 22-23. The sweep of the scope trace is from left to right.

Intensity is the level of brightness. The intensity control sets the level of brightness on the displayed voltage trace. The intensity level is kept as low as possible to keep the trace in focus. The focus control adjusts the sharpness of the displayed voltage trace.

The horizontal control adjusts the left and right positions of the displayed voltage trace. The horizontal control sets the starting point of the trace. The vertical control adjusts the up and down positions of the displayed voltage trace.

The volts/division (volts per division) control selects the height of the displayed waveform. The setting determines the number of volts each horizontal screen division represents. For example, if a waveform occupies 4 divisions and the volts/division control is set at 20, the peak-to-peak voltage (V_{p-p}) equals 80 V ($4 \times 20 = 80$ V). Eighty volts peak-to-peak equals 40 V peak (V_{max}) ($80 \div 2 = 40$ V). Forty volts peak equals 28.28 V_{rms} ($40 \times .707 = 28.28$ V). **See Figure 22-24.**

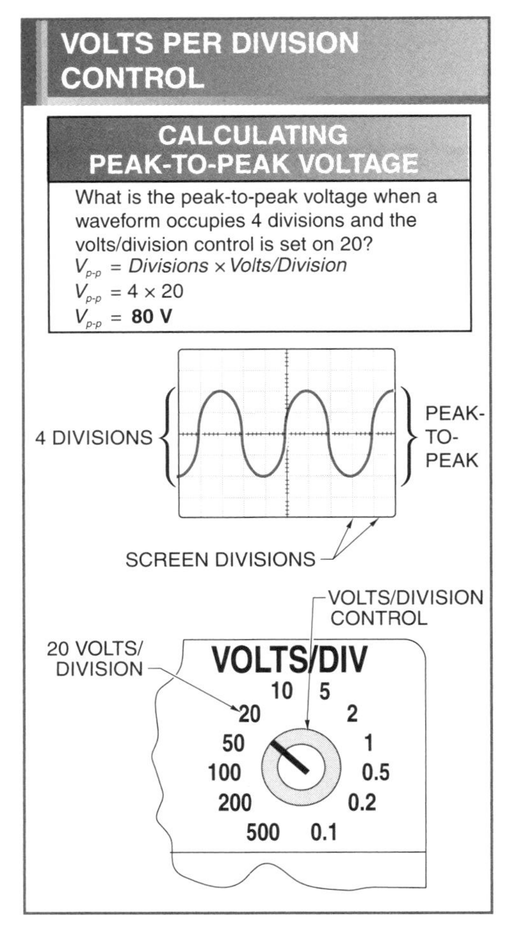

Figure 22-24. The volts/division (volts per division) control selects the height of the displayed waveform.

The time/division (time per division) control selects the width of the displayed waveform. The setting determines the length of time each cycle takes to move across the screen. For example, if the time/division control is set at 10, each vertical screen division equals 10 ms (milliseconds). If one cycle of a waveform equals 4 divisions, the displayed time equals 40 ms (4 × 10 = 40 ms). **See Figure 22-25.**

To determine the frequency of the displayed waveform, first calculate the period of the waveform and then calculate the frequency. To calculate the frequency of a waveform, apply the following procedure:

1. Calculate time period. To calculate the time period, multiply the number of divisions by the time/division setting.
2. Calculate frequency. To calculate frequency, apply the formula:

$$f = \frac{1}{T}$$

where

f = frequency (in Hz)

1 = constant (for reciprocal relation between f and T)

T = time period (in sec)

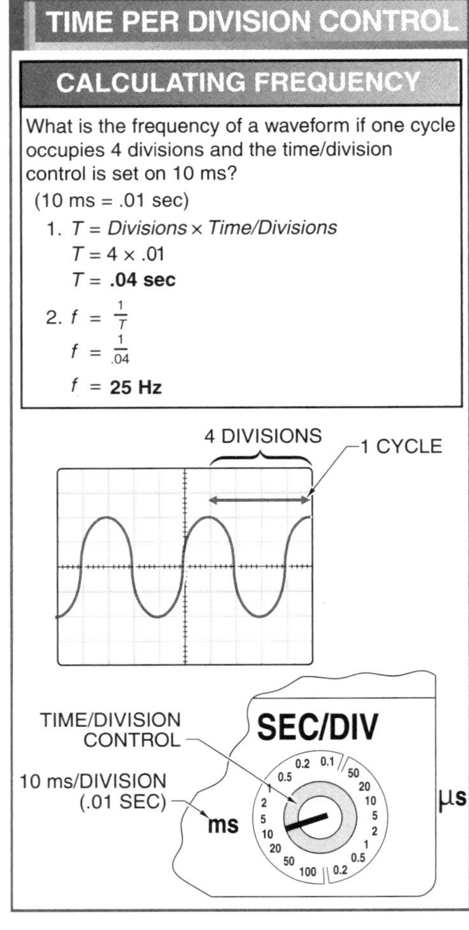

Figure 22-25. The time/division (time per division) control selects the width of the displayed waveform.

Tech Facts

Oscilloscopes are test instruments that provide a visual display of voltages. Oscilloscopes display the shape of a voltage waveform. Oscilloscopes may include high-capacity hard discs, removable storage, and nonvolatile memory.

Example: Calculating Frequency

What is the frequency of a waveform if one cycle of the waveform occupies 4 divisions and the time/division control is set on 10 ms (10 ms = .01 sec).

1. Calculate time period.

   $T = 4 \times .01$

   $T = .04$ sec

2. Calculate frequency.

   $f = \frac{1}{T}$

   $f = \frac{1}{.0001}$

   $f = \mathbf{10{,}000\ Hz}$

Scope AC Voltage Measurement Procedure. A scope is connected in parallel with a circuit or component under test. The scope is connected by a probe on the end of a test lead. A 1X probe (1 to 1) is used to connect the input of the scope to the circuit under test when the test voltage is lower than the voltage limit of the scope.

A 10X probe (10 to 1) is used to divide the input voltage by 10. The scope voltage limit equals 10 times the normal rated voltage when a 10X probe is used. The amount of measured voltage displayed on the scope screen must be multiplied by 10 to obtain the actual circuit voltage when using a 10X probe. For example, if the measured scope voltage is 25 V when using a 10X probe, the actual circuit voltage is 250 V (25 V × 10 = 250 V). **See Figure 22-26.** To use a scope to measure AC voltage, apply the following procedure:

1. Turn the power switch ON and adjust the trace brightness on the screen.
2. Set the AC/DC control switch to AC.
3. Set the volts/division control to display the voltage level under test. Set the control to the highest value if the voltage level is unknown.
4. Connect the scope probe to the AC voltage under test.
5. Adjust the volts/division control to display the full waveform of the voltage under test.
6. Set the time/division control to display several cycles of the voltage under test.
7. Adjust the waveform by using the vertical control to set the lower edge of the waveform on one of the lower lines.
8. Measure the vertical amplitude of the waveform by counting the number of divisions displayed (V_{p-p}).

To calculate V_{rms}, first calculate V_{p-p}. V_{p-p} is calculated by multiplying the number of divisions by the volts/division setting. For example, if a waveform occupies 4 divisions and the volts/division setting is 10, V_{p-p} equals 40 V (4 × 10 = 40 V). V_{max} equals 20 V (40 ÷ 2 = 20 V). V_{max} is multiplied by .707 to find V_{rms}. For example, if V_{max} is 20 V, V_{rms} equals 14.14 V (20 V × .707 = 14.14 V). V_{rms} is the value of the voltage under test as measured on a voltmeter.

Scope Frequency Measurement Procedure. In AC applications such as variable frequency motor drives, it may be necessary to measure the frequency in the circuit. A frequency meter should be used in applications where high accuracy is required. A scope gives an adequate reading in most frequency measurement applications. A scope also shows any distortion present in the circuit under test. To measure frequency, the scope probes are connected in parallel with the circuit or component under test. **See Figure 22-27.** To use a scope to measure frequency, apply the following procedure:

1. Turn the power switch ON and adjust the trace brightness.
2. Set the AC/DC control switch to AC.
3. Set the volts/division control to display the voltage level under test. Set the control to the highest value if the voltage level is unknown.
4. Connect the scope probe to the AC voltage under test.
5. Adjust the volts/division control to display the vertical amplitude of the waveform under test.
6. Set the time/division control to display approximately two cycles of the waveform under test.
7. Set the vertical control so that the center of the waveform is on the centerline of the scope screen.

8. Set the horizontal control so that the start of one cycle of the waveform begins at the vertical centerline on the scope screen.
9. Measure the number of divisions between the start point and end point of one cycle.

Tech Facts

Before taking frequency measurements, ensure that the meter is designed to take measurements on the circuit being tested.

To determine frequency, multiply the number of divisions by the time/division setting. This value is the time period for one cycle. To determine the frequency of the waveform, divide the time period by 1.

Example: Calculating Frequency

What is the frequency if one cycle of a waveform occupies 5 divisions and the time/division setting is 20 ms (microseconds).

1. Calculate time period.

$T = 5 \times .00002$

$T = .0001$ sec

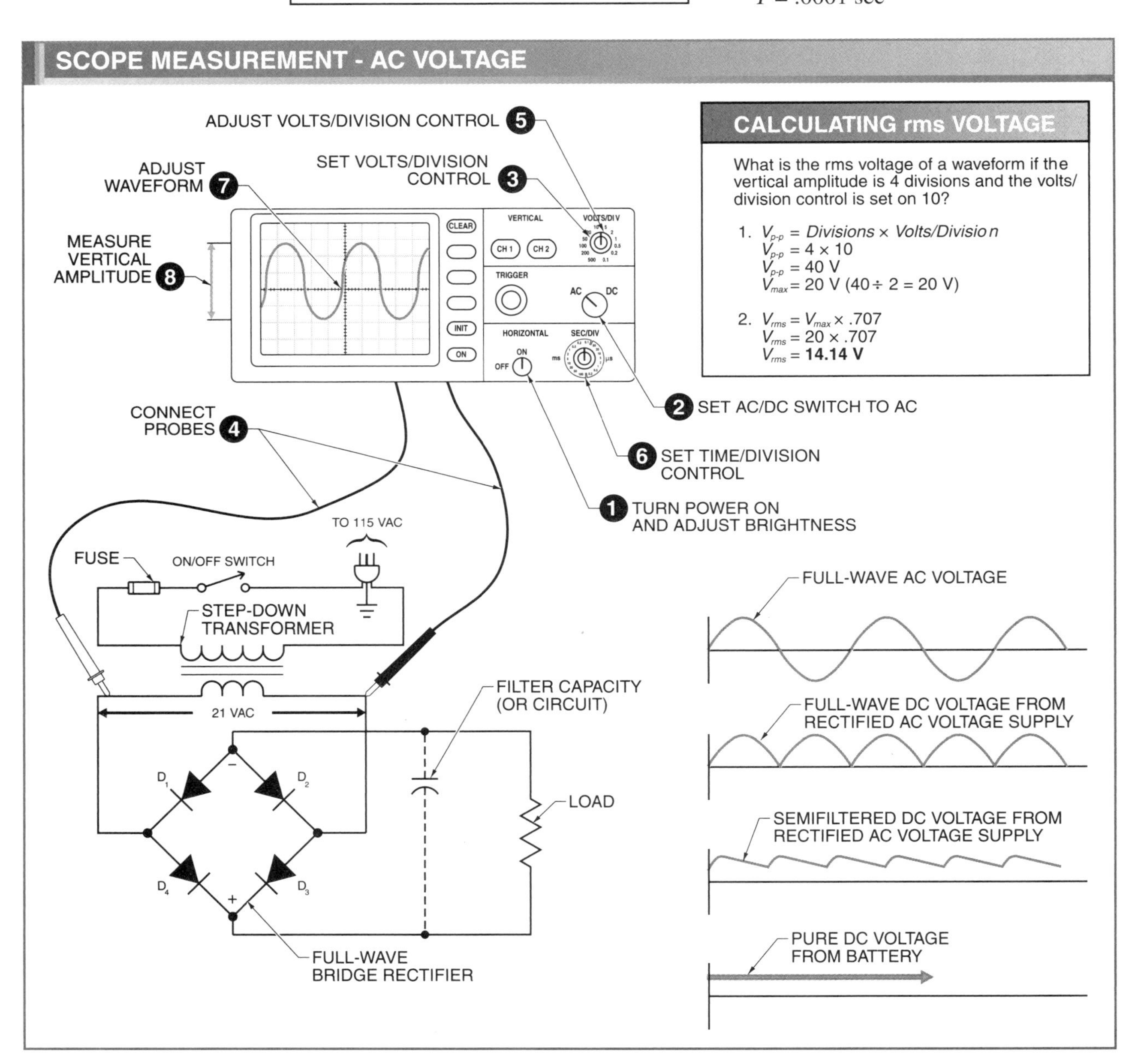

Figure 22-26. A scope is connected in parallel with a circuit or component under test.

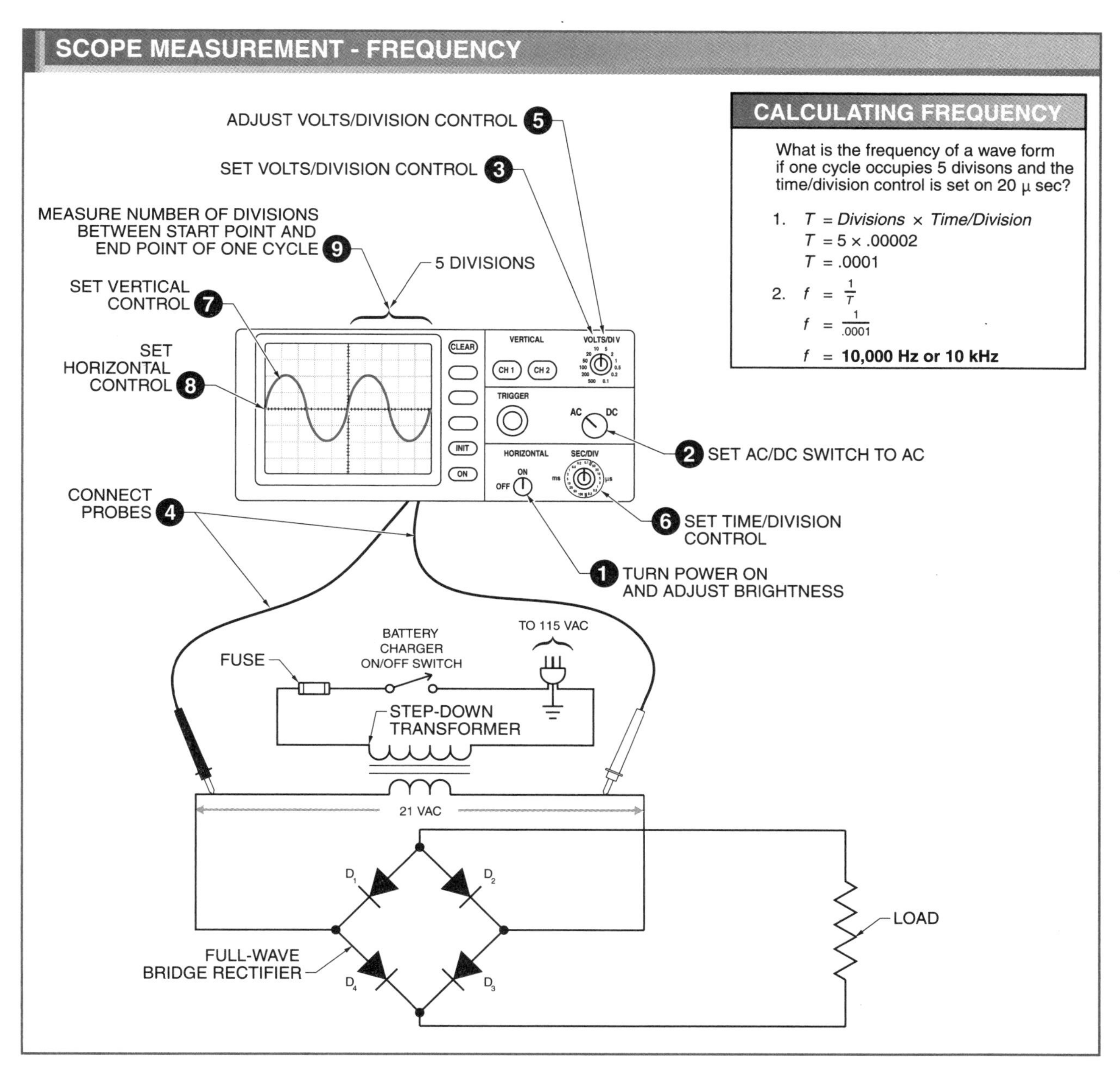

Figure 22-27. To measure frequency, the scope probes are connected in parallel with the circuit or component under test.

2. Calculate frequency.

$$f = \frac{1}{T}$$

$$f = \frac{1}{.0001}$$

$$f = \mathbf{10{,}000\ Hz}$$

Scope DC Voltage Measurement Procedure. To measure DC voltage using a scope, a test probe is connected to the point in the circuit where the DC voltage is to be measured. The ground lead of the scope is connected to the ground of the circuit. The voltage is positive if the trace moves above the centerline. The voltage is negative if the trace moves below the centerline. **See Figure 22-28.** To use a scope to measure DC voltage, apply the following procedure:

1. Turn the power switch ON and adjust the trace brightness.
2. Set the AC/DC control switch to DC.

3. Set the volts/division control to display the voltage level under test. Set the control to the highest value possible if the voltage level under test is unknown.
4. Connect the scope probe to the ground point of the circuit under test.
5. Set the vertical control so that the displayed line is in the center of the screen. The displayed line represents 0 VDC.
6. Remove the scope probe from ground point and connect it to the DC voltage under test. The displayed voltage moves above or below the scope centerline depending on the polarity of the DC voltage under test.
7. Measure the vertical amplitude of the voltage from the centerline by counting the number of divisions from the centerline.

The number of displayed divisions is multiplied by the volts/division setting to determine the DC voltage under test. For example, if a waveform is located at 3 divisions above the centerline and the volts/division control is set at 5 V, the voltage equals 15 VDC (3 × 5 = 15 V).

Test Instrument Safety Rules

A variety of test instruments are available that measure electrical quantities such as voltage, current, resistance, etc. Each instrument includes an instruction manual. The instruction manual shows and explains individual instrument specifications and features, proper operating procedures and safety, and specific warnings and applications.

Always read the manufacturer's instruction manual before using any test equipment and refer to it for information concerning specific use, safety, and instrument capacity. Safety precautions are required when using some test instruments because of the high voltage and currents in some circuits. When using test instruments, the following safety rules must be followed:

- Choose the correct test instrument for the application.
- Do not use test instruments that are damaged (cracked or with missing parts or worn insulation).
- When taking a measurement, keep fingers behind the finger guards on the test leads.

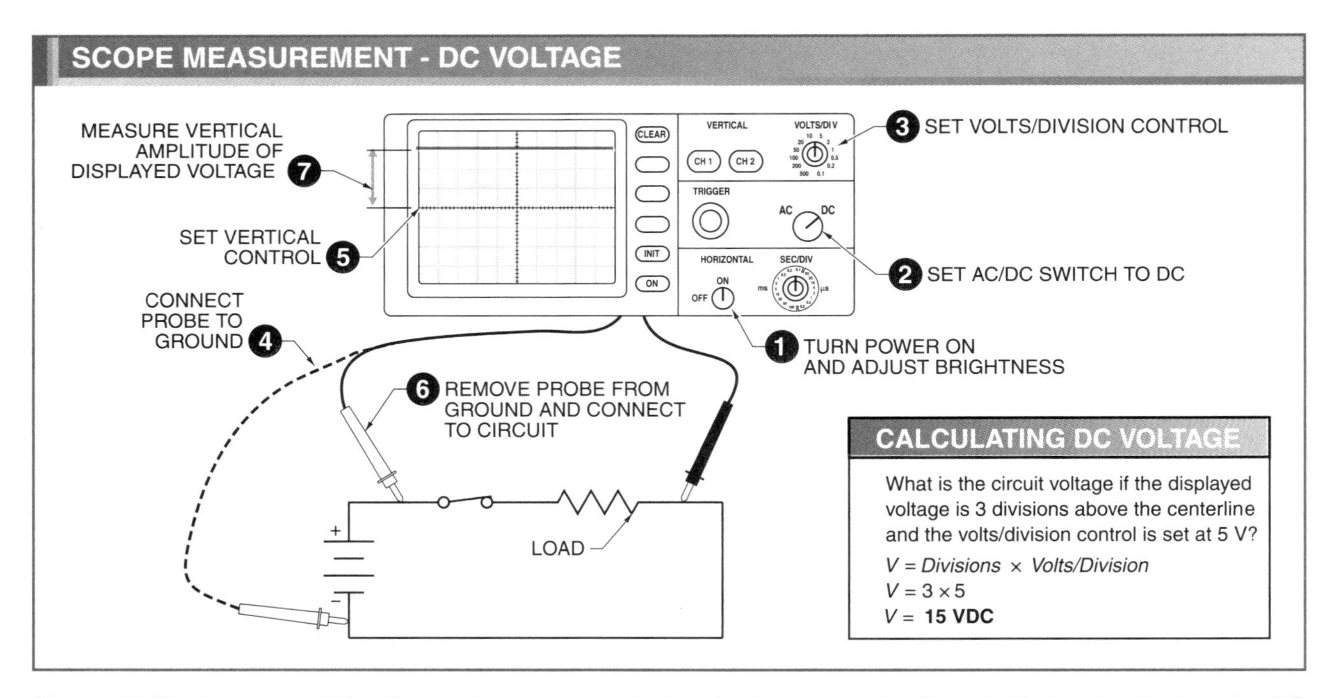

Figure 22-28. To measure DC voltage using a scope, a test probe is connected to the point in the circuit where the DC voltage is to be measured.

- Do not use test instruments in hazardous environments (explosive gas, vapor, or dust) unless specifically designed for hazardous environments.
- Before use, verify that the test leads of a test instrument are in the proper jacks — the jacks that correspond to the setting of the function switch — before taking any measurements.
- Before taking a measurement, verify that the function switch of a test instrument matches the desired measurement and the connections of the test leads to the circuit.

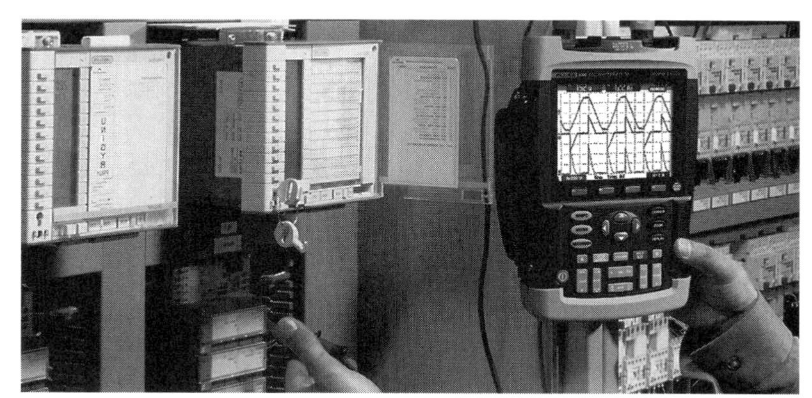

Fluke Corporation

Handheld scope meters are used when signal measurements must be taken in the field.

Review Questions

1. What is the difference between an analog display and a digital display?
2. What causes ghost voltages?
3. Why might a voltage indicator be used?
4. Describe the procedure for using a voltage tester.
5. Why should personal protective equipment be worn when using a DMM?
6. How can an in-line ammeter be used to measure current within a circuit?
7. What can be gained by measuring resistance in a circuit?
8. Define speed.
9. Why is a scope sometimes more useful than a DMM when examining voltage?
10. List four safety rules for using test instruments.

Ohm's law states that current in a circuit is equal to voltage divided by the resistance. The power formula states that power in a circuit equals voltage multiplied by current. A series circuit has two or more components connected with only one path for the current to flow through the circuit. A parallel circuit has two or more components connected with more than one path for the current to flow.

23 Ohm's Law, the Power Formula, and Series and Parallel Circuits

OHM'S LAW

Ohm's law is the relationship between voltage, current, and resistance in a circuit. Ohm's law states that current in a circuit is proportional to the voltage and inversely proportional to the resistance. Any value in this relationship can be found when the other two are known. The relationship between voltage, current, and resistance may be visualized by presenting Ohm's law in pie chart form. **See Figure 23-1.**

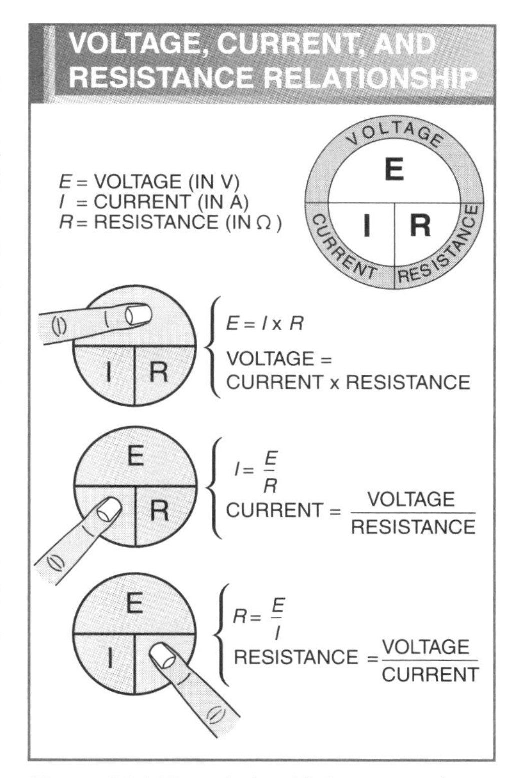

Figure 23-1. The relationship between voltage, current, and resistance may be visualized by presenting Ohm's law in pie chart form.

Calculating Voltage Using Ohm's Law

Ohm's law states that voltage (E) in a circuit is equal to resistance (R) times current (I). To calculate voltage using Ohm's law, apply the formula:

$E = I \times R$

where

E = voltage (in V)

I = current (in A)

R = resistance (in Ω)

Tech Facts

Ohm's Law is named after Georg Simon Ohm, a German physicist who, in 1827, discovered through experimentation the relationship between voltage (E), current (I), and resistance (R) and created the formula E = I × R to mathematically express the relationship.

Example: Calculating Voltage Using Ohm's Law

What is the voltage in a circuit that includes a 3 Ω heating element that draws 4 A?

$E = I \times R$

$E = 4 \times 3$

$\mathbf{E = 12\ V}$

Calculating Current Using Ohm's Law

Ohm's law states that current (I) in a circuit is equal to voltage (E) divided by resistance (R). To calculate current using Ohm's law, apply the formula:

$$I = \frac{E}{R}$$

where
I = current (in A)
E = voltage (in V)
R = resistance (in Ω)

Example: Calculating Current Using Ohm's Law

What is the current in a circuit with a 3 Ω heating element connected to a 12 V supply?

$$I = \frac{E}{R}$$

$$I = \frac{12}{3}$$

$$I = \mathbf{4\ A}$$

Calculating Resistance Using Ohm's Law

Ohm's law states that the resistance (R) in a circuit is equal to voltage (E) divided by current (I). To calculate resistance using Ohm's law, apply the formula:

$$R = \frac{E}{I}$$

where
R = resistance (in Ω)
E = voltage (in V)
I = current (in A)

Example: Calculating Resistance Using Ohm's Law

What is the resistance of a circuit in which a load that draws 4 A is connected to a 12 V supply?

$$R = \frac{E}{I}$$

$$R = \frac{12}{4}$$

$$R = \mathbf{3\ \Omega}$$

Voltage/Current Relationship

Ohm's law states that if the resistance in a circuit remains constant, a change in current is directly proportional to a change in voltage. This is shown in an application that uses a variable power supply and has a fixed resistance load. An example is a heat shrink sealing gun connected to a variable power supply. **See Figure 23-2.**

In this circuit, a fixed load resistance of 4 Ω is connected to a variable power supply which may be varied from 0 V to 24 V. The current in the circuit may be found for any given voltage by applying Ohm's law. For example, if the voltage in the circuit is set at 8 V, the current equals 2 A (8 ÷ 4 = 2 A).

The graph shows the direct proportional relationship between voltage and current in the circuit. The voltage values are marked on the horizontal axis and the current values are marked on the vertical axis. A linear resistance line is developed when the given values for voltage and current are plotted on the graph. The plotted resistance values show how any change in voltage results in a proportional change in current.

Current/Resistance Relationship

Ohm's law states that if the voltage in a circuit remains constant, a change in resistance produces an inversely proportional change in current. The current in a circuit decreases with an increase in resistance, and the current in the circuit increases with a decrease in resistance. This is shown in any application that uses a variable resistance load. **See Figure 23-3.**

In this circuit, a variable resistance of between 2 Ω and 24 Ω is connected to a fixed power supply. The current in the circuit may be found for any given resistance by applying Ohm's law. For example, if the resistance in the circuit is set at 24 Ω, the current equals 5 A (120 ÷ 24 = 5 A).

Tech Facts

When voltage in a circuit remains constant, a change in resistance produces an inversely proportional change in current.

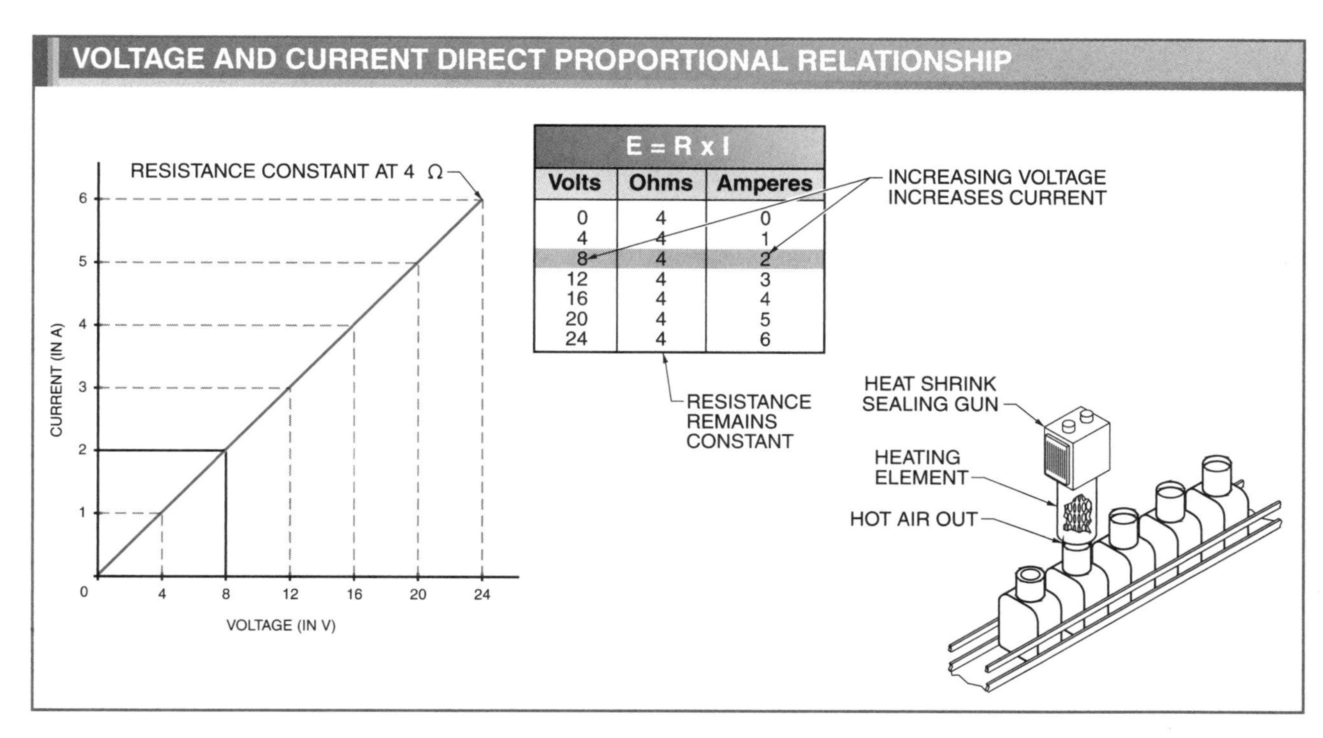

Figure 23-2. Ohm's law states that if the resistance in a circuit remains constant, a change in current is directly proportional to a change in voltage.

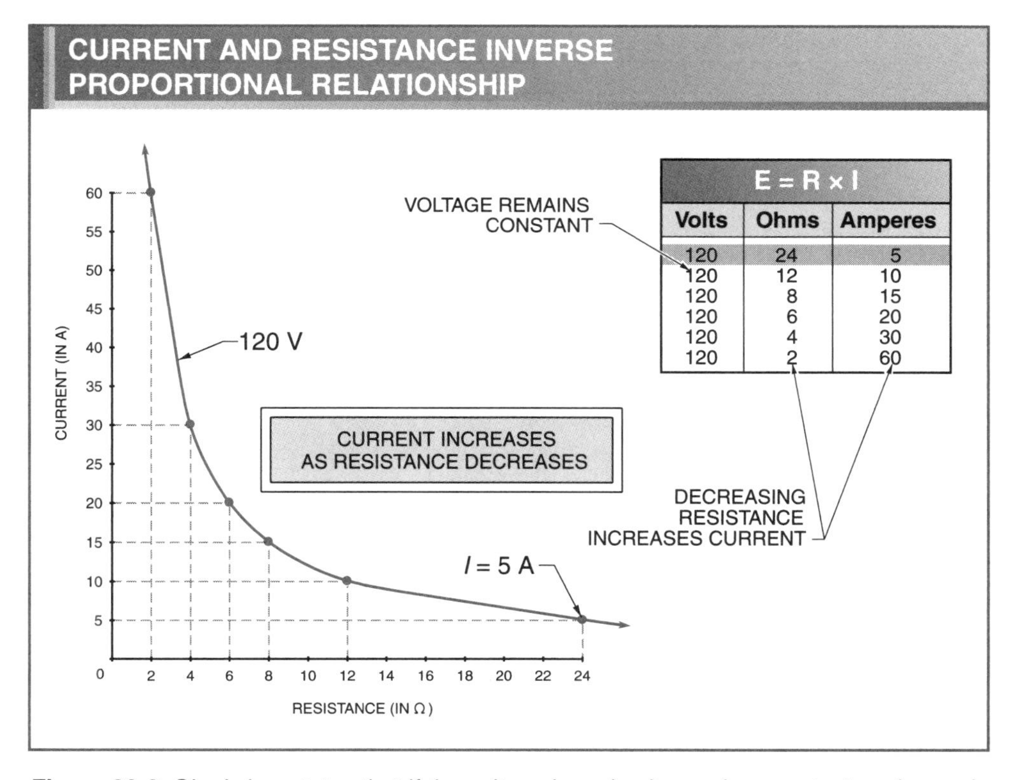

Figure 23-3. Ohm's law states that if the voltage in a circuit remains constant, a change in resistance produces an inversely proportional change in current.

The graph shows the inverse proportional relationship between current and resistance in the circuit. The changing resistance values are marked on the horizontal axis. The changing current values are marked on the vertical axis. The voltage is plotted on the graph. The graph shows that current in a circuit is inversely proportional to the resistance.

Applications Using Ohm's Law

Ohm's law is used in design applications and troubleshooting applications. In design applications, Ohm's law is used to solve for the proper values of voltage, current, or resistance during circuit design and to predict circuit characteristics before power is applied to a circuit when two of the three electrical values are known. Calculating these values helps determine the correct size of conductors and components and the voltage level that must be used for proper circuit operation. In troubleshooting applications, Ohm's law is used to determine how a circuit should operate and how it is operating under power. When troubleshooting a circuit, the circuit is already designed. To determine the problem, voltage and current measurements are taken. These measured values are used to calculate the circuit's resistance and help locate the problem.

Using Ohm's Law when Designing Circuits

Changing a circuit's resistance changes the amount of current flowing though the circuit. This principle is applied in any circuit that uses variable resistor (potentiometer or rheostat) to control a load. For example, variable resistors are used to control the sound output of speakers (volume controls), to control the brightness level of lamps (dimmer switches), and to control the amount of heat output of heating elements (temperature controls).

When designing a circuit like the automobile lighting circuit, the design engineer uses Ohm's law to determine the resistance range required to control the lamps through their full range of brightness. The design engineer determines the required resistance range based on the known electrical values. In this circuit, the known electrical values are the battery voltage and the current draw of the lamps selected.

For example, if the lamps are rated at .25 A per bulb and four bulbs are used, the total lamp current is 1 A ($.25 \times 4 = 1$ A). The resistance of the circuit at full lamp brightness equals 12 Ω ($12 \div 1 = 12\ \Omega$).

Knowing that the total resistance is 12 Ω and the total current is 1 A, the design engineer can determine the required variable resistor range to control the lamps. For example, if the midrange of the dimmer switch is to produce a 50% lamp brightness, the current has to be reduced by 50%. A 50% current value of .5 A ($1 \times .5 = .5$ A) is used to find the total circuit resistance required. The total required circuit resistance equals 24 Ω ($12 \div .5 = 24\ \Omega$). Fifty percent lamp brightness resistance equals 24 Ω.

Thus, 24 Ω total resistance is required to reduce the brightness of the lamps by 50%. Since the four lamps produce 12 Ω of resistance, an additional 12 Ω of resistance must be produced by the dimmer control. In this application, a variable resistor that produced 12 Ω of resistance at midrange is required. By using Ohm's law, the design engineer has determined that a variable resistor that has a full range of 0 Ω to 24 Ω (using 12 Ω as the midrange) could be used as the panel lamp dimmer control.

Using Ohm's Law when Troubleshooting

Ohm's law is used during troubleshooting to determine circuit conditions. Voltage and current measurements are taken because resistance cannot be measured on a circuit that is powered. After voltage and current measurements are taken, Ohm's law is applied to determine the resistance of the circuit. **See Figure 23-4.**

In this circuit, the resistance of an electric heater may be determined by measuring the circuit's voltage and current. Using Ohm's law, the normal circuit's resistance is equal to 60 Ω ($240 \div 4 = 60\ \Omega$).

A troubleshooter can use the 60 Ω resistance value to help determine the condition of the circuit. For example, if the ammeter measured 3 A instead of 4 A, the troubleshooter knows that the circuit's resistance has increased (240 ÷ 3 = 80 Ω). An additional 20 Ω of circuit resistance is present (80 Ω – 60 Ω = 20 Ω). The troubleshooter knows that the additional 20 Ω could come from a loose (or dirty) connection or open coil sections. Open coil sections increase the total circuit resistance.

Likewise, if the ammeter measured 6 A instead of 4 A, the troubleshooter knows that the circuit's resistance has decreased (240/6 = 40 Ω). Because the circuit's resistance has decreased by 20 Ω, the troubleshooter knows that the decreased resistance could be caused by a partially shorted coil or insulation breakdown.

POWER FORMULA

The *power formula* is the relationship between power (P), voltage (E), and current (I) in an electrical circuit. Any value in this relationship may be found using the power formula when the other two are known. Using the formula listed in a chart form, the unknown value is converted. The relationship between power, current, and voltage may be visualized by presenting the power formula in pie chart form. **See Figure 23-5.**

Tech Facts

The power formula is the relationship between power, voltage, and current. The power formula states that power (P) in a circuit equals voltage (E) times current (I).

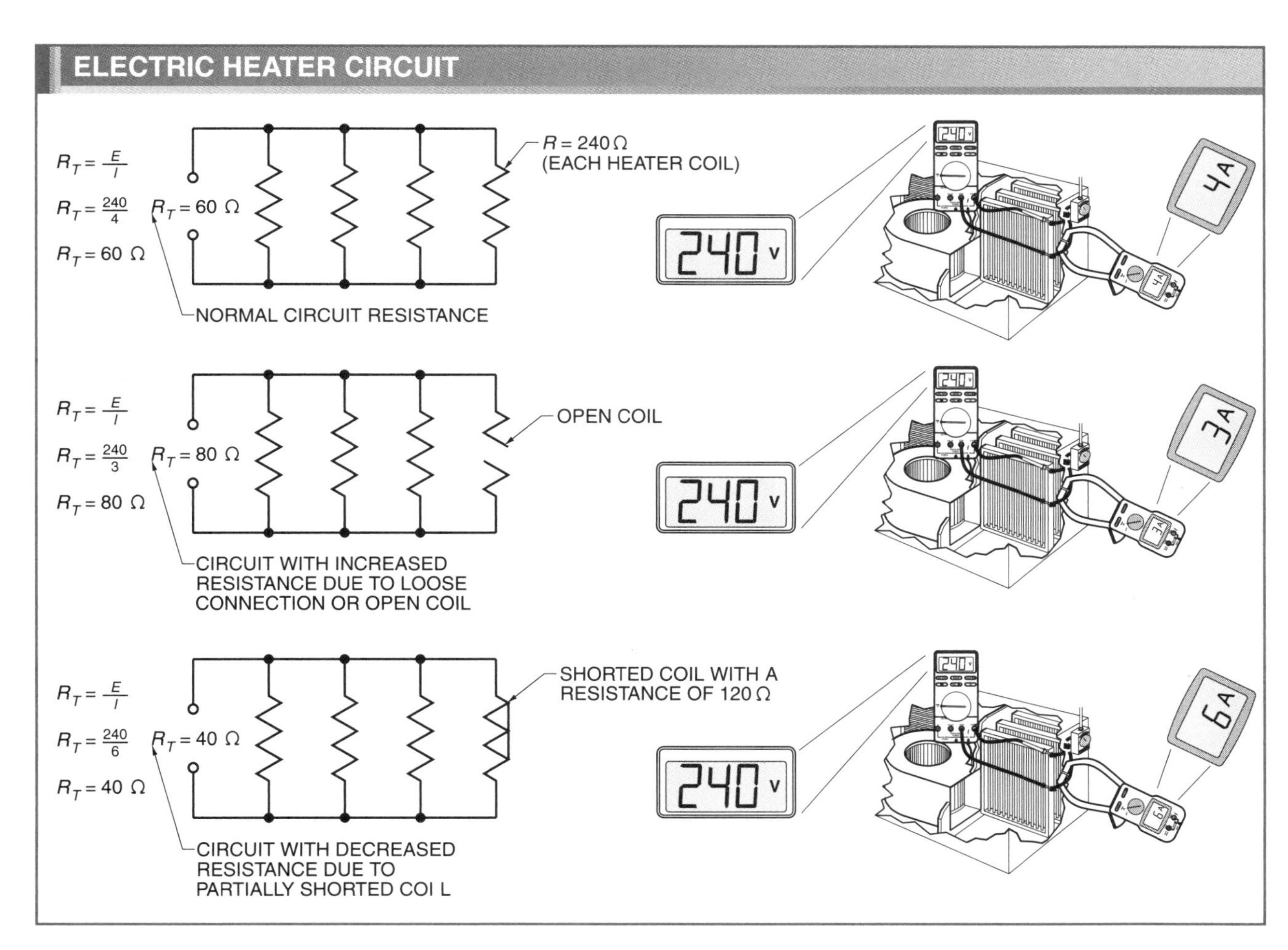

Figure 23-4. Ohm's law is applied to determine the resistance of the circuit.

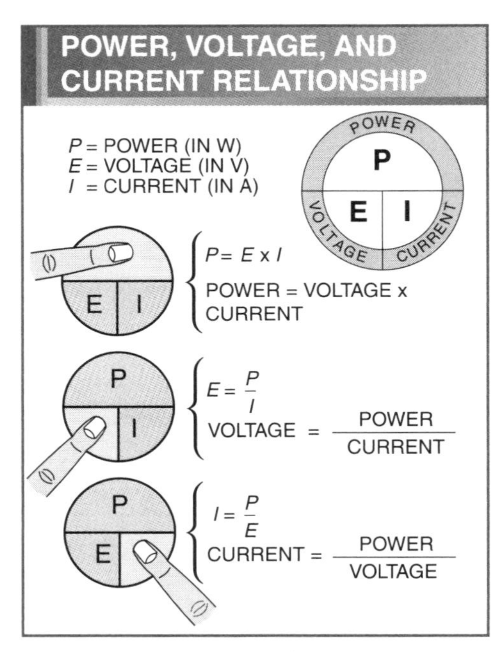

Figure 23-5. The relationship between power, current, and voltage may be visualized by presenting the power formula in pie chart form.

Calculating Power Using Power Formula

The power formula states that power (P) in a circuit is equal to voltage (E) times current (I). To calculate power using the power formula, apply the formula:

$$P = E \times I$$

where

P = power (in W)

E = voltage (in V)

I = current (in A)

Example: Calculating Power Using Power Formula

What is the power of a load that draws 0.5 A when connected to a 120 V supply?

$$P = E \times I$$

$$P = 120 \times 0.5$$

$$P = \mathbf{60\ W}$$

Calculating Voltage Using Power Formula

The power formula states that voltage (E) in a circuit is equal to power (P) divided by current (I). To calculate voltage using the power formula, apply the formula:

$$E = \frac{P}{I}$$

where

E = voltage (in V)

P = power (in W)

I = current (in A)

Example: Calculating Voltage Using Power Formula

What is the voltage in a circuit in which a 60 W load draws 0.5 A?

$$E = \frac{P}{I}$$

$$E = \frac{60}{.05}$$

$$E = \mathbf{120\ V}$$

Calculating Current Using Power Formula

The power formula states that current (I) in a circuit is equal to power (P) divided by voltage (E). To calculate current using the power formula, apply the formula:

$$I = \frac{P}{E}$$

where

I = current (in A)

P = power (in W)

E = voltage (in V)

Example: Calculating Current Using Power Formula

What is the current in a circuit in which a 60 W load is connected to a 120 V supply?

$$I = \frac{P}{E}$$

$$I = \frac{60}{120}$$

$$I = \mathbf{.5\ A}$$

Power/Current Relationship

The power formula states that if the voltage in a circuit remains constant and the power required from the circuit changes, the current in the circuit also changes. The power required from a circuit changes any time loads are added (power increase) or removed (power decrease). This principle is shown in any circuit when loads are added. **See Figure 23-6.**

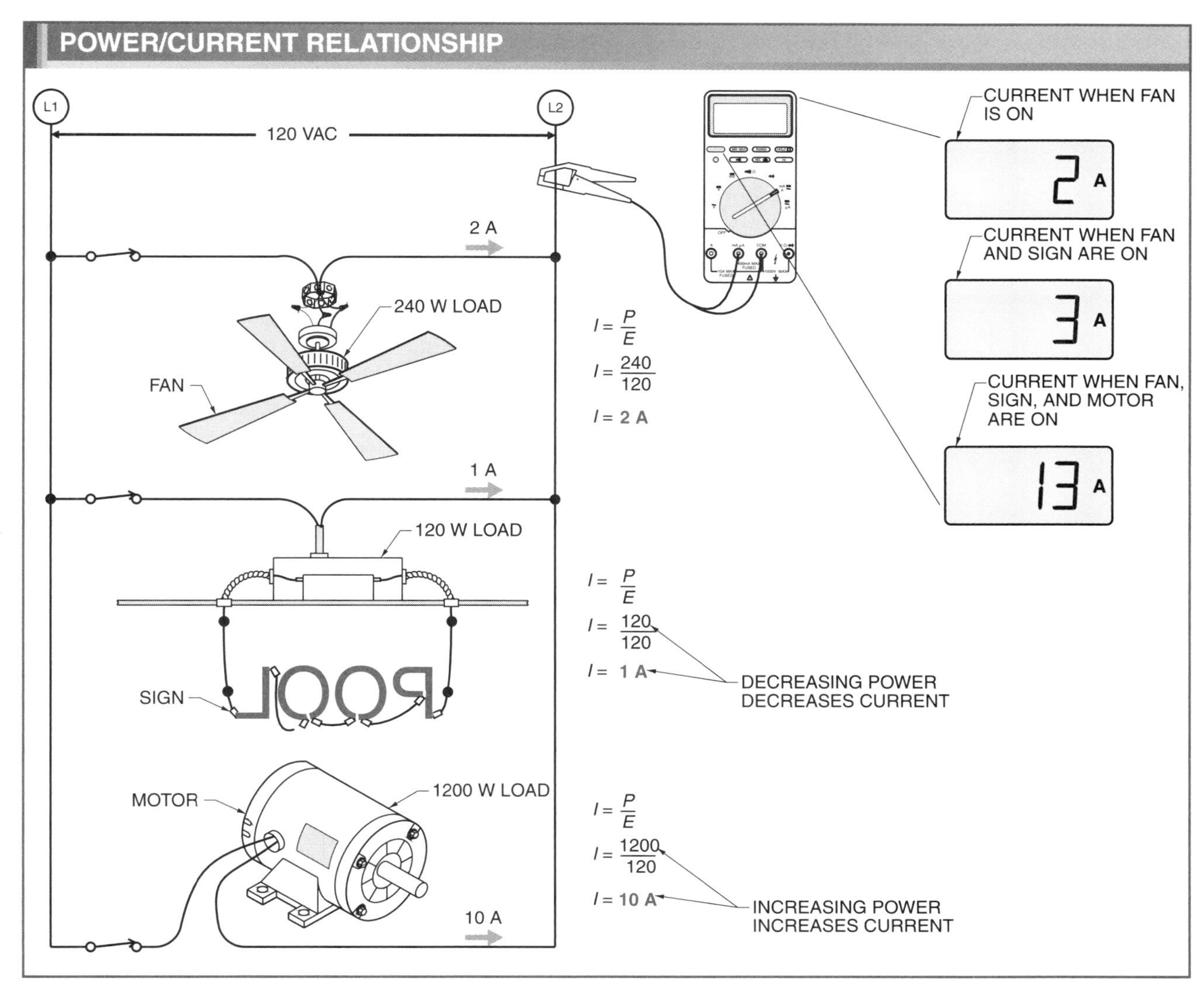

Figure 23-6. The power formula states that if the voltage in a circuit remains constant and the power required from the circuit changes, the current in the circuit also changes.

In this circuit, a 240 W fan, a 120 W sign, and a 1200 W motor are connected to a 120 V supply. The total circuit current may be found using the power formula. For example, when the 240 W fan is turned ON, the total circuit current equals 2 A (240 ÷ 120 = 2 A). The current of the fan equals 2 A. If, in addition to the fan being ON, the sign is also turned ON, the circuit's current increases by 1 A (120 ÷ 120 = 1 A). The current of the sign equals 1 A. The total current with both the fan and the sign ON equals 3 A (2 + 1 = 3 A).

If, in addition to the fan and sign being ON, the motor is also turned ON, the circuit's current increases by 10 A (1200 ÷ 120 = 10 A). The current of the motor equals 10 A. The total circuit current with all three loads ON equals 13 A (2 + 1 + 10 = 13 A).

Applications Using Power Formula

Like Ohm's law, the power formula is used when troubleshooting and to predict circuit characteristics before power is applied. Any electrical value (P, E, or I) can be calculated when the other two values are known or measured. The power formula is useful when determining expected current values because most electrical equipment lists a voltage and power rating. The listed power rating is given in watts for most appliances and heating elements or in horsepower for motors.

Combining Ohm's Law and Power Formula

Ohm's law and the power formula may be combined mathematically and written as any combination of voltage (E), current (I), resistance (R), or power (P). This combination lists six basic formulas and six rearranged formulas. **See Figure 23-7.**

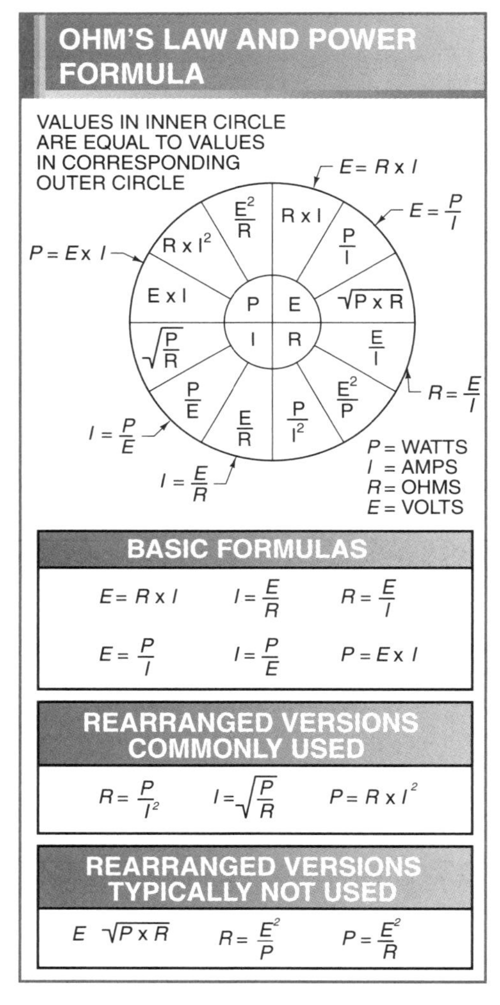

Figure 23-7. Ohm's law and the power formula may be combined mathematically and written as any combination of voltage (E), current (I), resistance (R), or power (P).

Ohm's Law and Impedance

Ohm's law and the power formula are limited to circuits in which electrical resistance is the only significant opposition to the flow of current. This limitation includes all direct current (DC) circuits and any alternating current (AC) circuits that do not contain a significant amount of inductance and/or capacitance. AC circuits that do not include inductance and/or capacitance include such devices as heating elements and incandescent lamps. AC circuits that include inductance are any circuits that include a coil as the load. Motors, transformers, and solenoids all include coils. AC circuits that include capacitance are any circuits that include a capacitor(s).

In DC circuits and AC circuits that do not contain a significant amount of inductance and/or capacitance, the opposition to the flow of current is resistance (R). In circuits that contain inductance (X_L) or capacitance (X_C), the opposition to the flow of current is reactance (X). In circuits that contain resistance (R) and reactance (X), the combined opposition to the flow of current is impedance (Z). Impedance is stated in ohms.

Ohm's law is used in circuits that contain impedance, however, Z is substituted for R in the formula. Z represents the total resistive force (resistance and reactance) opposing current flow. The relationship between voltage (E), current (I), and impedance (Z) may be visualized by presenting the relationship in pie chart form. **See Figure 23-8.**

SERIES CIRCUITS

Switches, loads, meters, fuses, circuit breakers, and other electric components can be connected in series. A *series connection* has two or more components connected so there is only one path for current flow. Opening the circuit at any point stops the flow of current. Current stops flowing any time a fuse blows, a circuit breaker trips, or a switch or load opens.

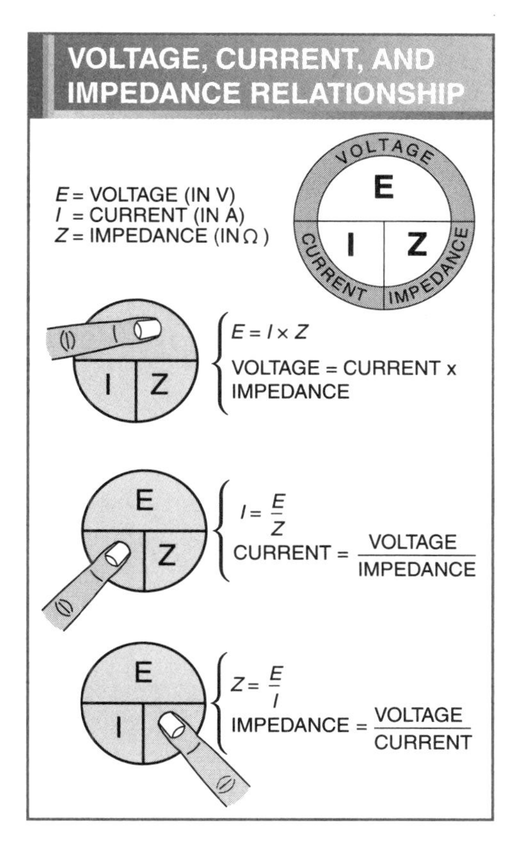

Figure 23-8. The relationship between voltage (E), current (I), and impedance (Z) may be visualized by presenting the relationship in pie chart form.

A fuse blows or circuit breaker trips when the circuit is overloaded. A switch is opened any time the circuit is turned OFF. The opening of a switch may be manual (toggle switches, etc.), mechanical (limit switches), or automatic (temperature switches). A load opens any time the load's current-consuming component (lamp filament, motor winding, etc.) burns or breaks open. **See Figure 23-9.**

Recognizing series-connected components and circuits enables a troubleshooter to take proper measurements, make circuit modifications, and troubleshoot the circuit. A schematic or line diagram is normally used when illustrating a series-connected circuit. A schematic or line diagram shows the circuit in its simplest form.

Series Switch Connections

All switches connected in series must be closed before current flows in the circuit. Opening any one or more of the switches stops current flow. Series-connected switches include any manual, mechanical, or automatic switch. Switches are usually connected in series to build safety into a circuit. For example, a microwave oven includes a limit switch in the door. The limit switch is connected in series with the oven ON switch. The oven cannot be turned ON unless the limit switch is closed and the oven ON switch is closed.

Manually-operated switches are often connected in series with automatic switches. The manual switch provides a point in the circuit for removing power when switched to the OFF position. For example, a manually-operated system ON/OFF switch is included with most thermostats. The manual switch can be used to turn OFF the automatically-operated heating and air conditioning system. **See Figure 23-10.**

Resistance in Series Circuits

The total resistance in a circuit containing series-connected loads equals the sum of the resistances of all loads. The resistance in the circuit increases if loads are added in series and decreases if loads are removed. To calculate total resistance of a series circuit, apply the formula:

$$R_T = R_1 + R_2 + R_3 + ...$$

where

R_T = total resistance (in Ω)

R_1 = resistance 1 (in Ω)

R_2 = resistance 2 (in Ω)

R_3 = resistance 3 (in Ω)

Example: Calculating Total Resistance in Series Circuits

What is the total resistance of a circuit containing resistors of 25 Ω, 50 Ω, and 75 Ω connected in series?

$R_T = R_1 + R_2 + R_3$

$R_T = 25 + 50 + 75$

$R_T = \mathbf{150\ \Omega}$

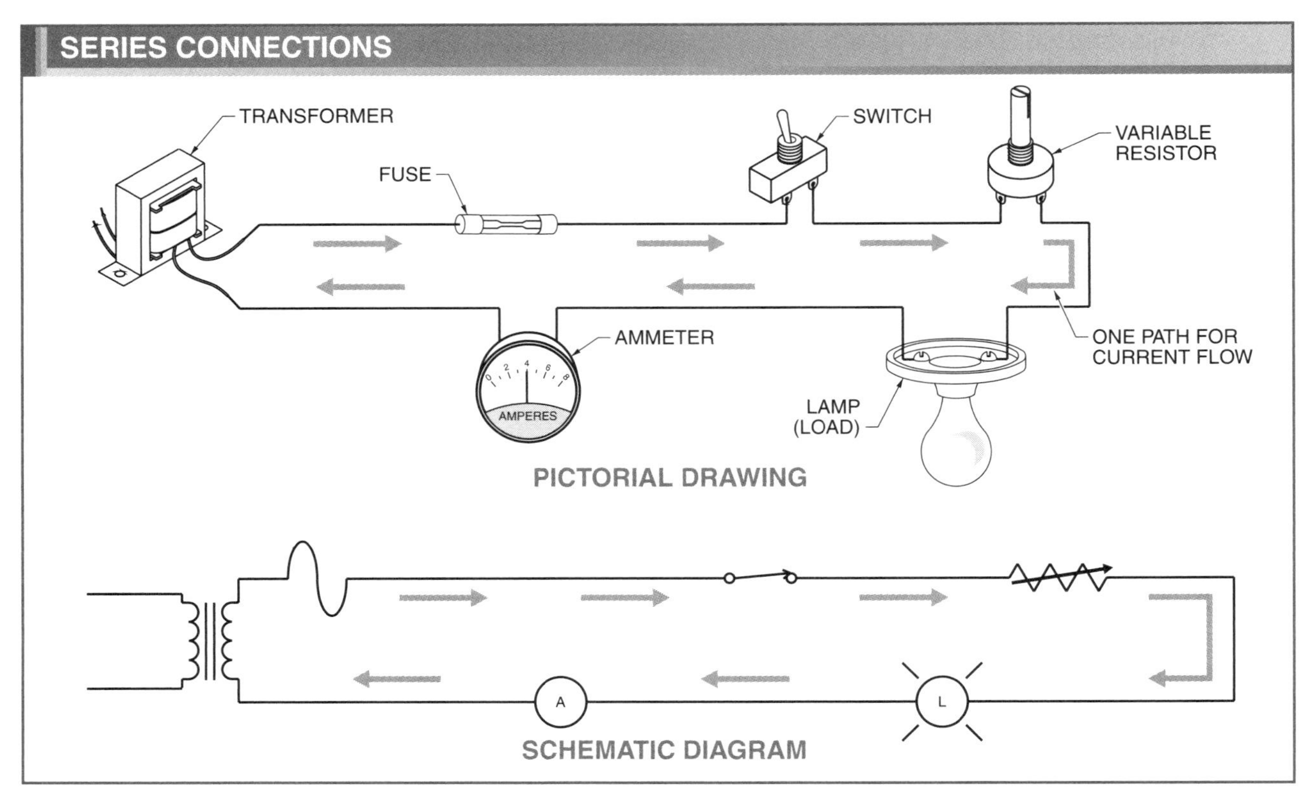

Figure 23-9. Switches, loads, meters, fuses, circuit breakers, and other electric components can be connected in series.

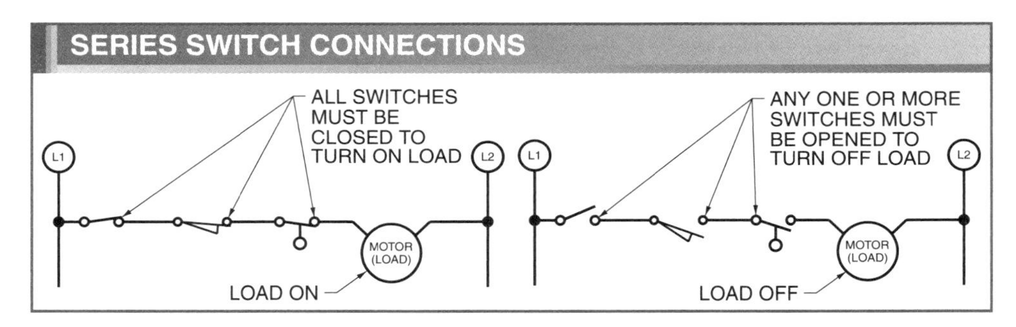

Figure 23-10. Series-connected switches include any manual, mechanical, or automatic switch.

Voltage in Series Circuits

The voltage applied across loads connected in series is divided across the loads. Each load drops a set percentage of the applied voltage. The exact voltage drop across each load depends on the resistance of each load. The voltage drop across each load is the same if the resistance values are the same. The load with the lowest resistance has the lowest voltage drop and the load with the highest resistance has the highest voltage drop. **See Figure 23-11.**

In an ideal circuit, the load(s) should be the only devices in the circuit that have resistance. For this reason, all of the power supply voltage should be dropped across them. However, any other part of an electrical circuit may also have a voltage drop across it. For example, mechanical switches that are burnt, corroded, or not making good contact have a voltage drop across them. The higher the voltage drop, the more damaged the switch. Likewise, the conductors (wires, etc.) that connect the electrical circuit together also have a voltage drop across them. In theory, the voltage drop across conductors should be zero. In fact, all conductors have some voltage drop across them.

Small conductors have a higher resistance than large conductors made of the same material. Thus, undersized conductors have a higher voltage drop across them. Conductors should be sized large enough to prevent no more than a 3% voltage drop from the power source to the load. The voltage drop across conductors and switches can be measured using a voltmeter.

All loads have some resistance value. The higher the power rating of a load, the lower the resistance value. The lower the power rating of a load, the higher the resistance value. Loads can be lamps, heating elements, solenoids, motors, bells, horns, or any device that uses electrical energy.

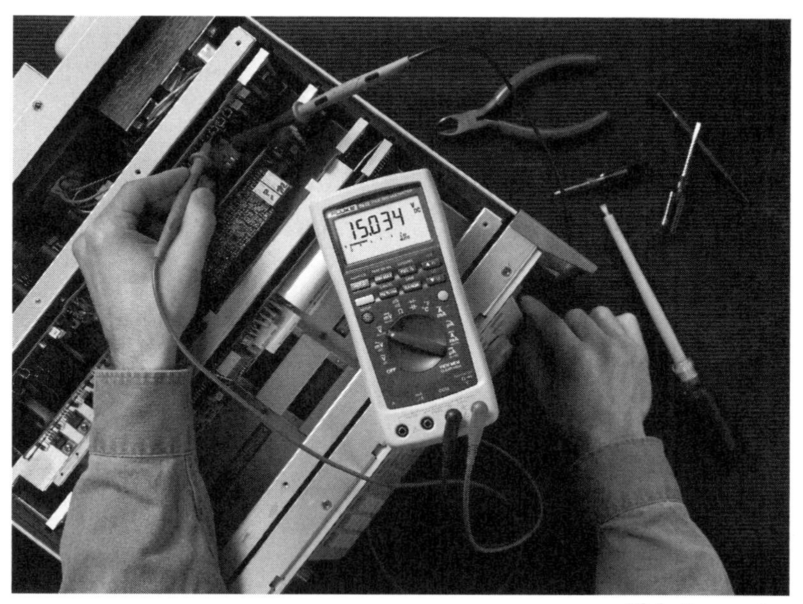

Fluke Corporation

Voltage across several loads in a series circuit should drop or decrease across each load.

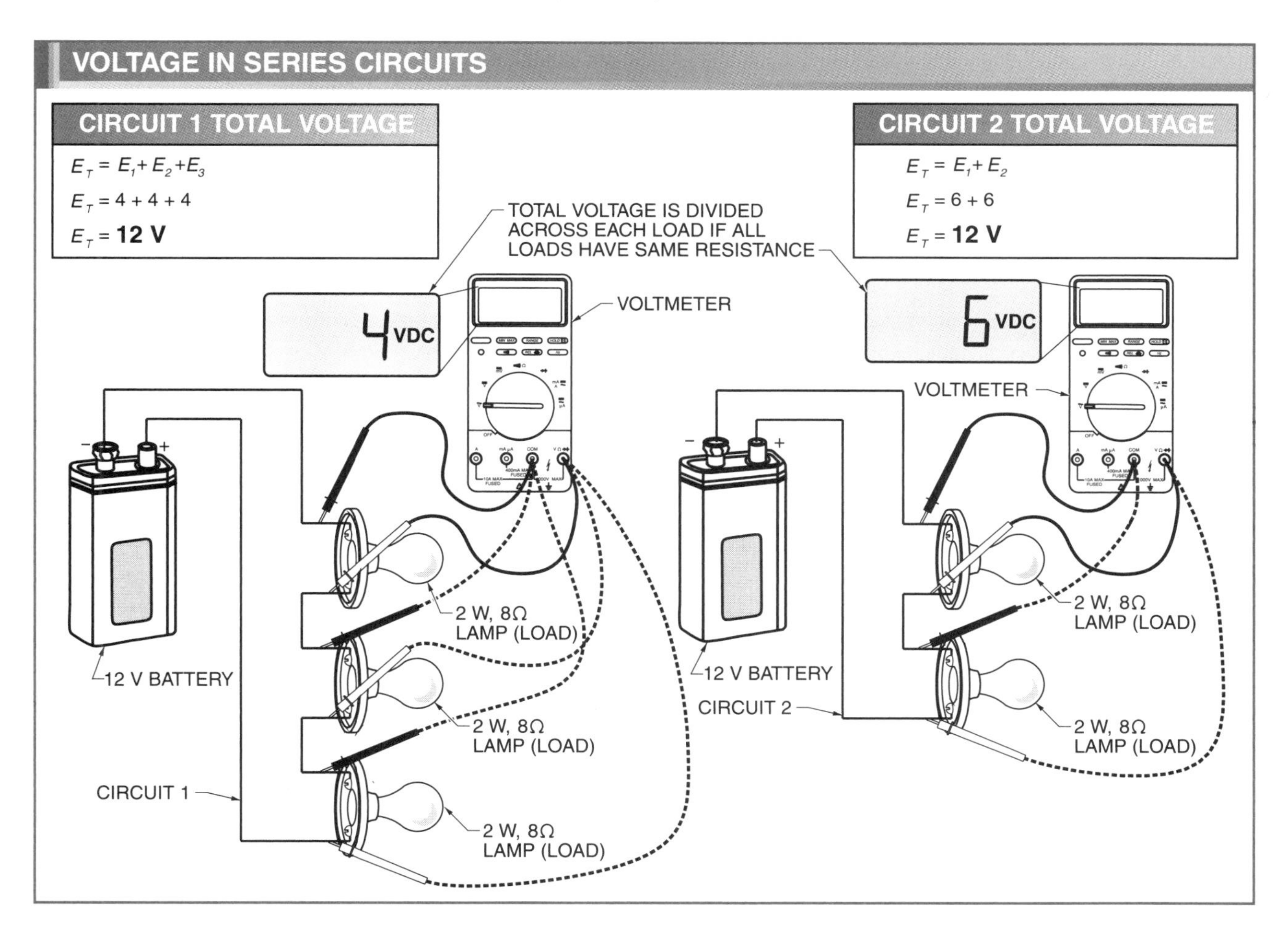

Figure 23-11. The voltage applied across loads connected in series is divided across the loads.

Loads are normally represented in a schematic diagram by the resistor symbol. The sum of all the voltage drops in a series circuit is equal to the applied voltage because the voltage in a series circuit is divided across all the loads in the circuit. To calculate total voltage in a series circuit when the voltage across each load is known or measured, apply the formula:

$E_T = E_1 + E_2 + E_3 + ...$

where

E_T = total applied voltage (in V)

E_1 = voltage drop across load 1 (in V)

E_2 = voltage drop across load 2 (in V)

E_3 = voltage drop across load 3 (in V)

Example: Calculating Voltage in Series Circuits

What is the total applied voltage of a circuit containing 2 V, 4 V, and 6 V drops across three loads?

$E_T = E_1 + E_2 + E_3 + ...$

$E_T = 2 + 4 + 6$

$E_T = \mathbf{12\ V}$

Fluke Corporation

Current in a series circuit should remain constant throughout the entire circuit.

Current in Series Circuits

The current in a circuit containing series-connected loads is the same throughout the circuit. The current in the circuit decreases if the circuit's resistance increases and the current increases if the circuit's resistance decreases. **See Figure 23-12.** To calculate total current of a series circuit, apply the formula:

$I_T = I_1 = I_2 = I_3$

where

I_T = total circuit current (in A)

I_1 = current through component 1 (in A)

I_2 = current through component 2 (in A)

I_3 = current through component 3 (in A)

Example: Calculating Current in Series Circuits

What is the total current through a series circuit if the current measured at each of six different loads is .25 A (250 mA)?

$I_T = I_1 = I_2 = I_3 = I_4 = I_5 = I_6$

$I_T = .25 = .25 = .25 = .25 = .25 = .25$

$I_T = \mathbf{.25\ A}$

Each component must be sized to safely pass the current of the loads because the current through the switches, fuses, and other components equals the current through each load.

Power in Series Circuits

Power is produced when voltage is applied to a load and current flows through the load. The power produced is used to produce light (lamps), heat (heating elements), rotary motion (motors), or linear motion (solenoids).

The lower the load's resistance or higher the applied voltage, the more power produced. The higher the load's resistance or lower the applied voltage, the less power produced. The amount of power produced is measured in watts (W).

The amount of power produced by a load is equal to the voltage drop across the load times the current through the load. The total power in a circuit is equal to the sum of the individual power produced by each load. **See Figure 23-13.**

To calculate total power in a series circuit when the power across each load is known or measured, apply the formula:

$P_T = P_1 + P_2 + P_3 ...$

where

P_T = total circuit power (in W)

P_1 = power of load 1 (in W)

P_2 = power of load 2 (in W)

P_3 = power of load 3 (in W)

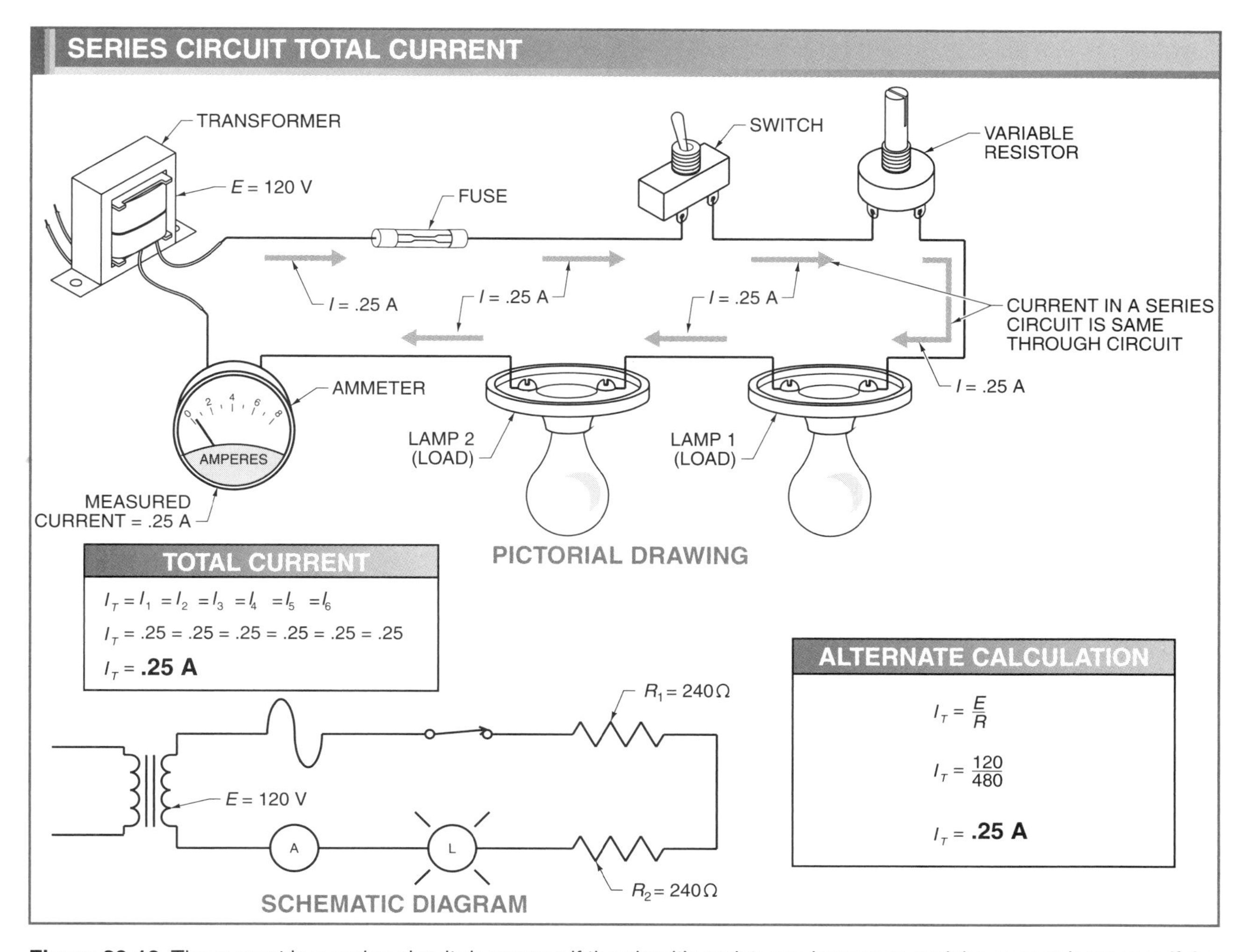

Figure 23-12. The current in a series circuit decreases if the circuit's resistance increases and the current increases if the circuit's resistance decreases.

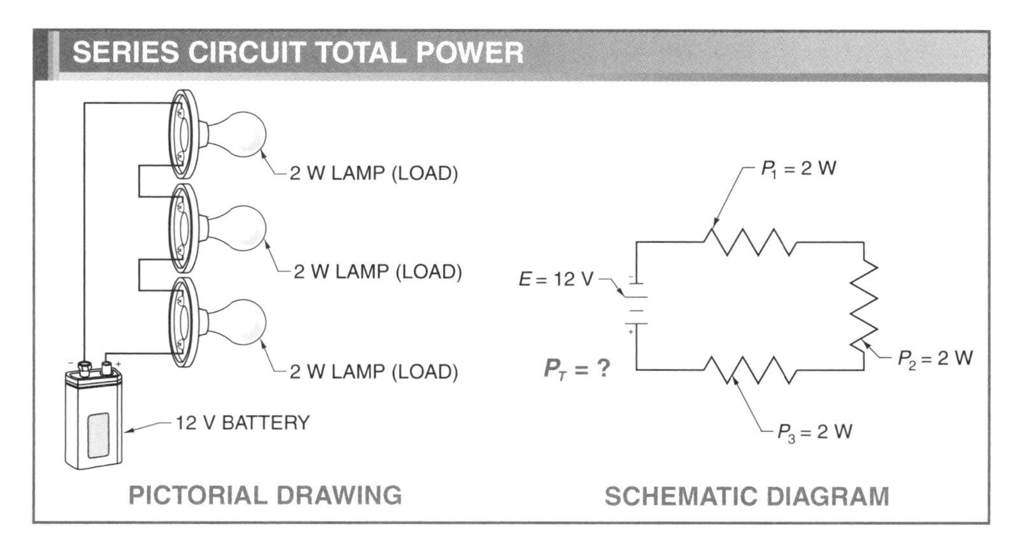

Figure 23-13. The total power in a series circuit is equal to the sum of the individual power produced by each load.

Example: Calculating Total Power in Series Circuits

What is the total power in a series circuit if three loads are connected in series and each load produces 2 W?

$P_T = P_1 + P_2 + P_3 + ...$

$P_T = 2 + 2 + 2$

$P_T = \mathbf{6\ W}$

Series Circuit Application

The individual concepts and formulas for series circuits can be applied for a total understanding of circuit operation when designing and troubleshooting circuits. For example, a typical coffeemaker applies the concepts of a basic series circuit to brew and keep coffee warm. **See Figure 23-14.**

The coffeemaker includes a brew heating element, a warm heating element, an ON/OFF switch, and a temperature switch. The brew heating element heats the water and forces it over the coffee grounds. The warm heating element keeps the coffee warm after brewing.

Since the brew heating element must produce more power than the warm heating element, the brew element has a lower resistance value than the warm heating element.

The ON/OFF switch starts and stops the process. The temperature switch changes the circuit's total resistance after the coffee is brewed.

The total resistance of the circuit is equal to the resistance of the brew heating element when the coffee brew cycle is started. The warm heating element is not a part of the electrical circuit at this time because the normally closed temperature switch short circuits the warm heating element during brewing.

Tech Facts

Series circuits have the same current flowing through the entire circuit. The total resistance of a series circuit can be calculated by adding the resistances of the individual loads in the circuit.

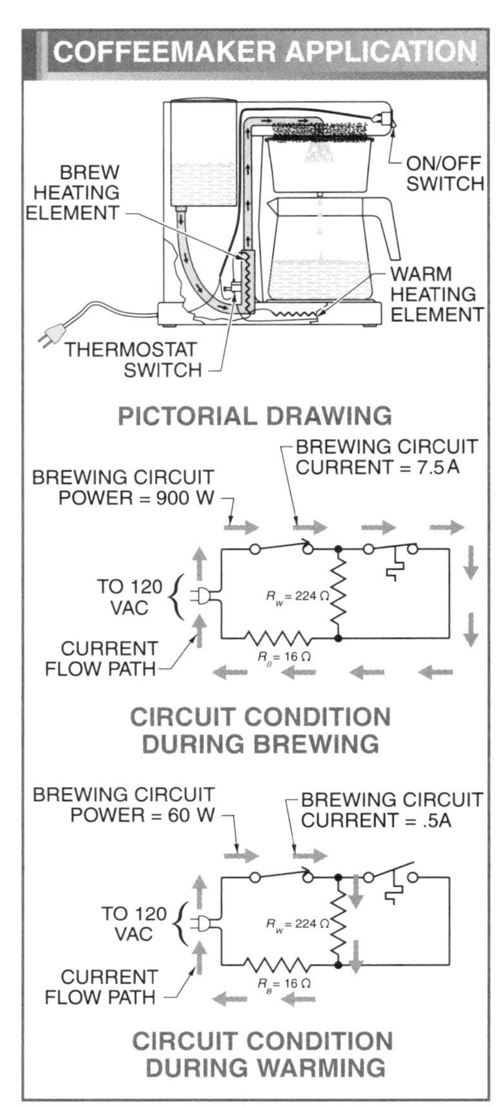

Figure 23-14. A typical coffeemaker applies the concepts of a basic series circuit to brew and keep coffee warm.

The current in the circuit during brewing is found by applying Ohm's law. To calculate circuit current during brewing, apply the formula:

$$I = \frac{E}{R}$$

where

I = current during brewing (in A)

E = applied voltage (in V)

R = brew heating element resistance (in Ω)

Example: Calculating Brewing Current
What is the current in the circuit during brewing when the applied voltage is 120 V and the brew heating element resistance is 16 Ω?

$I = \frac{E}{R}$

$I = \frac{120}{16}$

$I = \mathbf{7.5\ A}$

The power in the circuit during brewing can be found by applying the power formula. To calculate power during brewing, apply the formula:

$P = E \times I$

where

P = power during brewing (in W)

E = applied voltage (in V)

I = current during brewing (in A)

Example: Calculating Brewing Power
What is the power in the circuit during brewing when the applied voltage is 120 V and the circuit current is 7.5 A?

$P = E \times I$

$P = 120 \times 7.5$

$P = \mathbf{900\ W}$

The temperature increases after all the water is forced over the brew heating element. The increased temperature opens the temperature switch. The warm heating element and the brew heating element are connected in series when the temperature switch opens. The total resistance of the series circuit is found by adding the resistances of each series-connected component. To calculate the total resistance during warming, apply the formula:

$R_T = R_1 + R_2$

where

R_T = total circuit resistance (in Ω)

R_1 = resistance of warm heating element (in Ω)

R_2 = resistance of brew heating element (in Ω)

Example: Calculating Total Resistance During Warming
What is the total resistance in a circuit containing resistances of 224 Ω and 16 Ω?

$R_T = R_1 + R_2$

$R_T = 224 + 16$

$R_T = \mathbf{240\ \Omega}$

The current in the circuit during warming is found by applying Ohm's law.

Example: Calculating Warming Current
What is the current in the circuit during warming when the applied voltage is 120 V and the total circuit resistance is 240 Ω?

$I = \frac{E}{R}$

$I = \frac{120}{240}$

$I = \mathbf{.5\ A}$

The power in the circuit during warming is found by applying the power formula.

Example: Calculating Warming Power
What is the power in the circuit during warming when the applied voltage is 120 V and the circuit current is .5 A?

$P = E \times I$

$P = 120 \times .5$

$P = \mathbf{60\ W}$

PARALLEL CIRCUITS

Switches, loads, meters, and other components can be connected in parallel. A *parallel connection* has two or more components connected so that there is more than one path for current flow. Care must be taken when working with parallel circuits because current can be flowing in one part of the circuit even though another part of the circuit is turned OFF. Understanding and recognizing parallel-connected components and circuits enables a technician or troubleshooter to take proper measurements, make circuit modifications, and troubleshoot the circuit. **See Figure 23-15.**

Parallel Switch Connections

Parallel-connected switches include any manual, mechanical, or automatic switches. One or more parallel-connected switches must be closed to start current flow. All switches must be opened to stop current flow. Switches are usually connected in parallel to provide additional turn-ON points. For example, several pushbuttons may be connected in parallel to turn ON a doorbell. The switches may be located at the front, back, and side doors. **See Figure 23-16.**

Tech Facts

Unlike series circuits, parallel circuits have different currents flowing through each device. However, the voltage in a parallel circuit is the same across each component.

Manually-operated switches are often connected in parallel with mechanical and automatic switches. A manual switch is used to override the mechanical or automatic switch. For example, the dome light in an automobile is automatically turned ON when a door is opened and turned OFF when the door is closed by a limit switch built into the door. A manual switch (usually built into the headlight control switch) is connected in parallel with the door limit switch so that the dome light can be manually turned ON even if all the doors are closed.

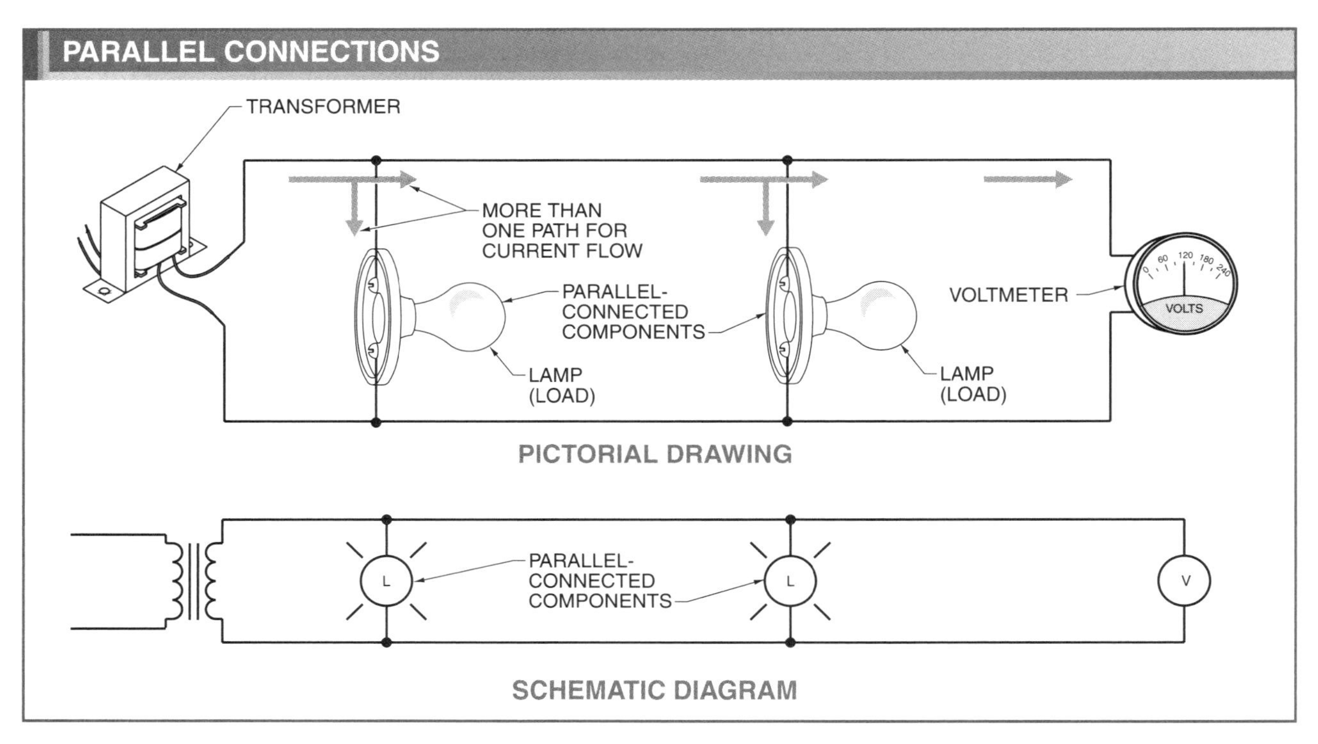

Figure 23-15. Understanding and recognizing parallel-connected components and circuits enables a technician or troubleshooter to take proper measurements, make circuit modifications, and troubleshoot the circuit.

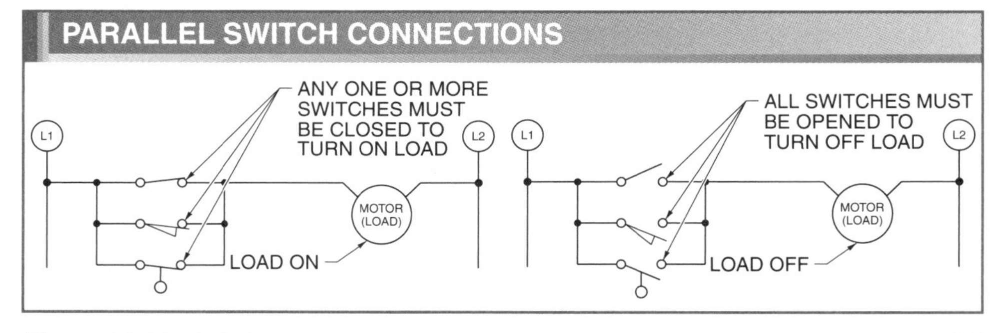

Figure 23-16. Switches are usually connected in parallel to provide additional turn-ON points.

Resistance in Parallel Circuits

Resistance is the opposition to current flow. The total resistance in a circuit containing parallel-connected loads is less than the smallest resistance value. The total resistance decreases if loads are added in parallel and increases if loads are removed. To calculate total resistance in a parallel circuit containing two resistors, apply the formula:

$$R_T = \frac{R_1 \times R_2}{R_1 + R_2}$$

where
R_T = total resistance (in Ω)
R_1 = resistance 1 (in Ω)
R_2 = resistance 2 (in Ω)

Example: Calculating Total Resistance—Two Resistors Connected in Parallel

What is the total resistance in a circuit containing resistors of 300 Ω and 600 Ω connected in parallel?

$$R_T = \frac{R_1 \times R_2}{R_1 + R_2}$$

$$R_T = \frac{300 \times 600}{300 + 600}$$

$$R_T = \frac{180,000}{900}$$

$$R_T = \mathbf{200\ \Omega}$$

To calculate total resistance in a parallel circuit with three or more resistors, apply the formula:

$$R_T = \frac{1}{\frac{1}{R_1} + \frac{1}{R_2} + \frac{1}{R_3}}$$

where
R_T = total resistance (in Ω)
R_1 = resistance 1 (in Ω)
R_2 = resistance 2 (in Ω)
R_3 = resistance 3 (in Ω)

Example: Calculating Total Resistance—Three Resistors Connected in Parallel

What is the total resistance in a circuit containing three resistors of 5000 Ω, 2000 Ω, and 10,000 Ω connected in parallel?

$$R_T = \frac{1}{\frac{1}{R_1} + \frac{1}{R_2} + \frac{1}{R_3}}$$

$$R_T = \frac{1}{\frac{1}{5000} + \frac{1}{2000} + \frac{1}{10,000}}$$

$$R_T = \frac{1}{.0002 + .0005 + .0001}$$

$$R_T = \mathbf{1250\ \Omega}$$

Voltage in Parallel Circuits

The voltage across each load is the same when loads are connected in parallel. The voltage across each load remains the same if parallel loads are added or removed. **See Figure 23-17.** Connecting loads in parallel is the most common method used to connect loads. For example, loads such as lamps, small appliances, fans, TVs, etc., are all connected in parallel when connected to a branch circuit.

To calculate total voltage in a parallel circuit when the voltage across a load is known or measured, apply the formula:

$$E_T = E_1 = E_2 = \ldots$$

where
E_T = total applied voltage (in V)
E_1 = voltage across load 1 (in V)
E_2 = voltage across load 2 (in V)

Example: Calculating Voltage in Parallel Circuits

What is the total voltage if the voltage across each load is measured at 112 VAC with a voltmeter?

$$E_T = E_1 = E_2$$

$$E_T = 112 = 112$$

$$E_T = 112\text{ VAC}$$

Tech Facts

The voltage across several loads in a circuit connected in parallel is the same for each of the loads. Parallel circuits are typically used in residential and commercial buildings to maintain the same voltage across multiple loads on the same circuit. This allows devices to operate on the same voltage, but different current.

Current in Parallel Circuits

Total current in a circuit containing parallel-connected loads equals the sum of the current through all the loads. Total current increases if loads are added in parallel and decreases if loads are removed. **See Figure 23-18.** To calculate total current in a parallel circuit, apply the formula:

$I_T = I_1 + I_2 + I_3 + \ldots$

where

I_T = total circuit current (in A)

I_1 = current through load 1 (in A)

I_2 = current through load 2 (in A)

I_3 = current through load 3 (in A)

Example: Calculating Current in Parallel Circuits with Several Loads

What is the total current in a circuit containing four loads connected in parallel if the current through the four loads is 0.833 A, 6.25 A, 8.33 A, and 1.2 A.

$I_T = I_1 + I_2 + I_3 + I_4$

$I_T = 0.833 + 6.25 + 8.33 + 1.2$

$I_T =$ **16.613 A**

The current through each load and the total circuit current must be known when designing and troubleshooting a circuit. When designing a circuit, the size of wire used is based on the amount of expected current. The greater the current flow, the larger the required wire size. The lesser the current flow, the smaller the required wire size, which reduces circuit cost. When troubleshooting a circuit, an overloaded circuit has a higher than-normal current reading. A circuit with a lower-than-expected current reading may have a loose connection. A loose connection increases the circuit resistance and decreases the circuit current.

Power in Parallel Circuits

Power is produced when voltage is applied to a load and current flows through the load. The power produced is used to produce light, heat, rotary motion, or linear motion.

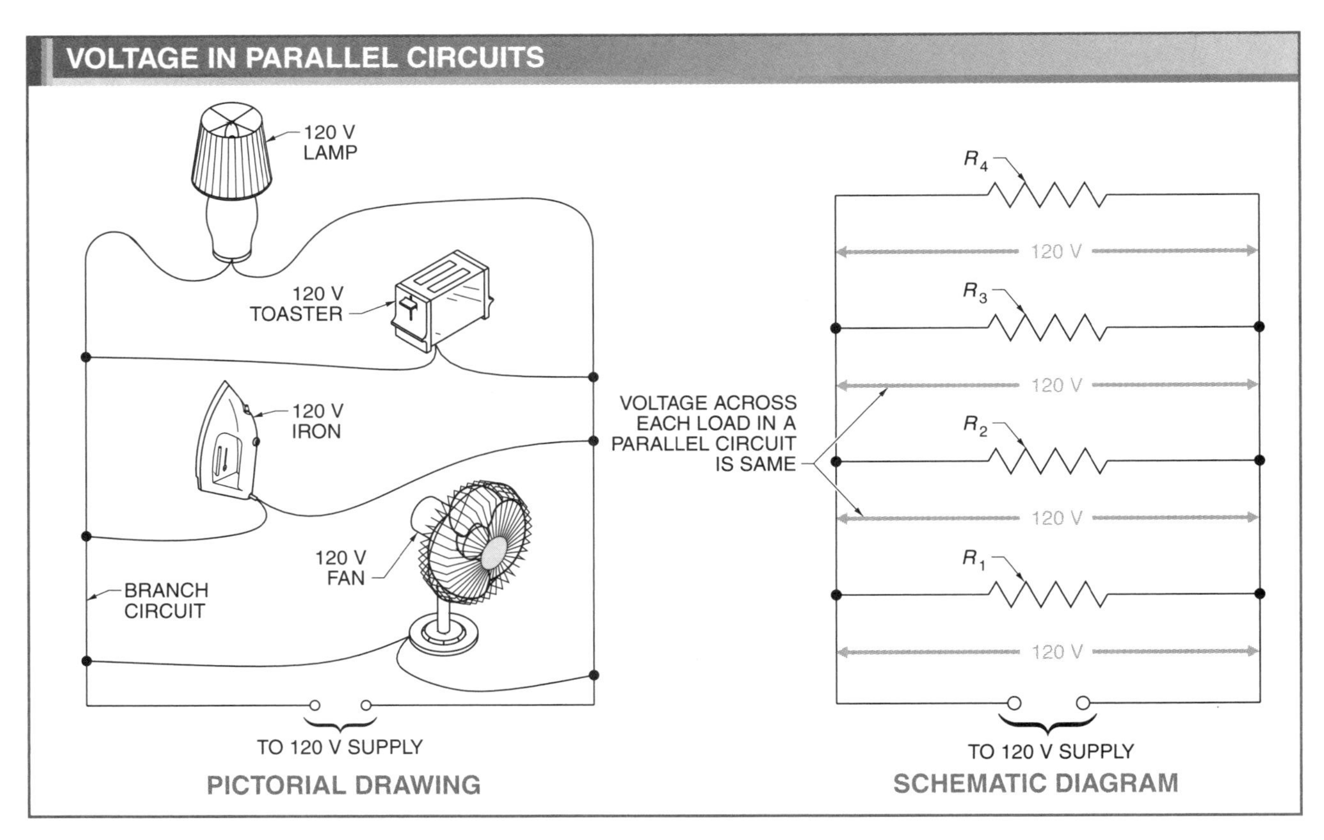

Figure 23-17. The voltage across each load is the same when loads are connected in parallel.

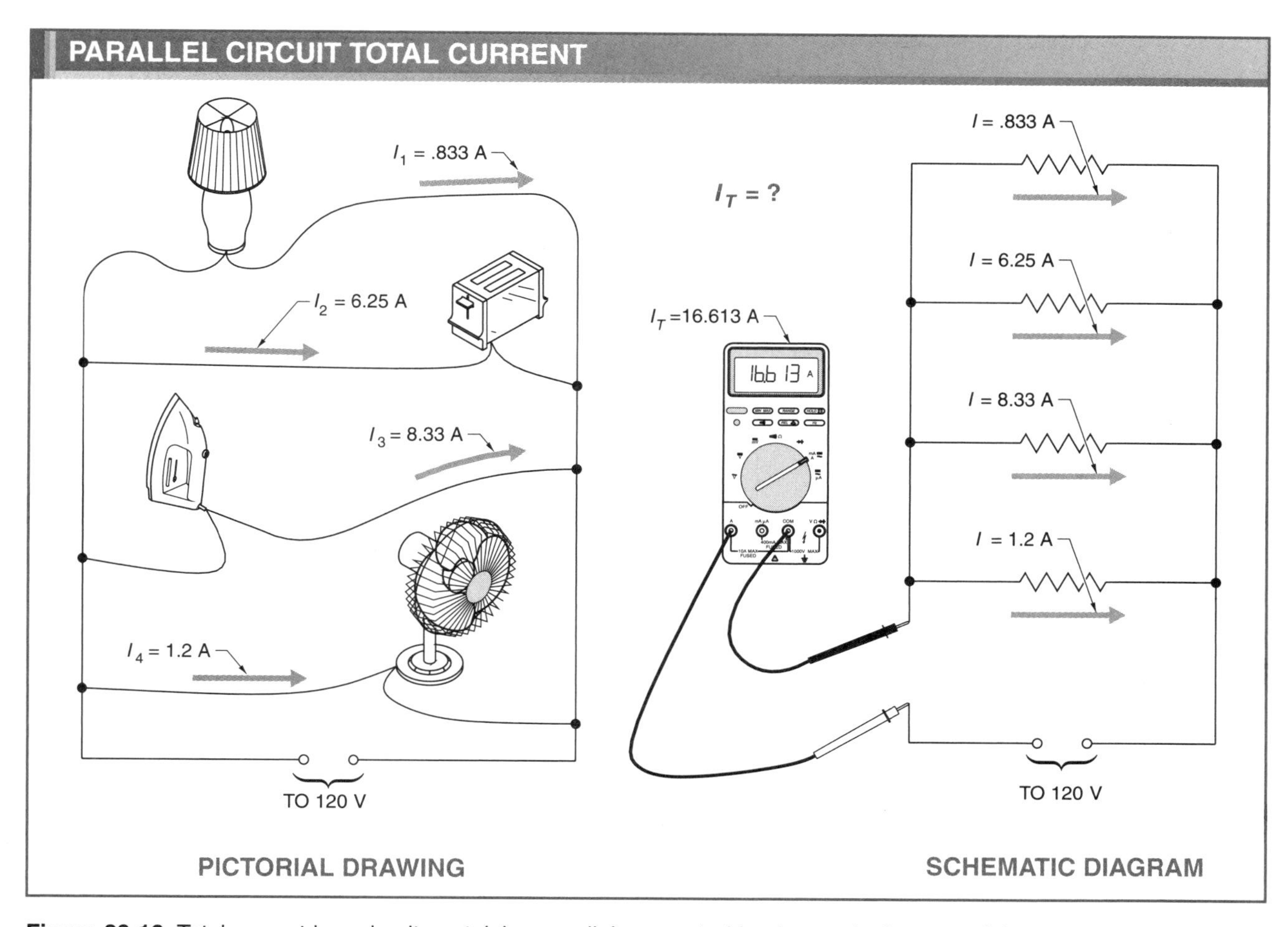

Figure 23-18. Total current in a circuit containing parallel-connected loads equals the sum of the current through all the loads.

The lower the load's resistance or higher the applied voltage, the more power produced. The higher the load's resistance or lower the applied voltage, the less power produced.

Power produced by a load is equal to the voltage drop across the load times the current through the load. Total power is equal to the sum of the power produced by each load. **See Figure 23-19.** To calculate total power in a parallel circuit when the power across each load is known or measured, apply the formula:

$P_T = P_1 + P_2 + P_3 + \ldots$

where

P_T = total circuit power (in W)

P_1 = power of load 1 (in W)

P_2 = power of load 2 (in W)

P_3 = power of load 3 (in W)

Example: Calculating Power in Parallel Circuits with Several Loads

What is the total circuit power if four loads are connected in parallel and the loads produce 100 W, 750 W, 1000 W, and 150 W?

$P_T = P_1 + P_2 + P_3 + P_4$

$P_T = 100 + 750 + 1000 + 150$

$P_T = \mathbf{2000\ W}$

Parallel Circuit Application

The individual concepts and formulas for parallel circuits can be applied for a total understanding of circuit operation when designing and troubleshooting circuits. For example, a waffle iron applies the concepts of a basic parallel circuit to heat the upper and lower heating elements. **See Figure 23-20.**

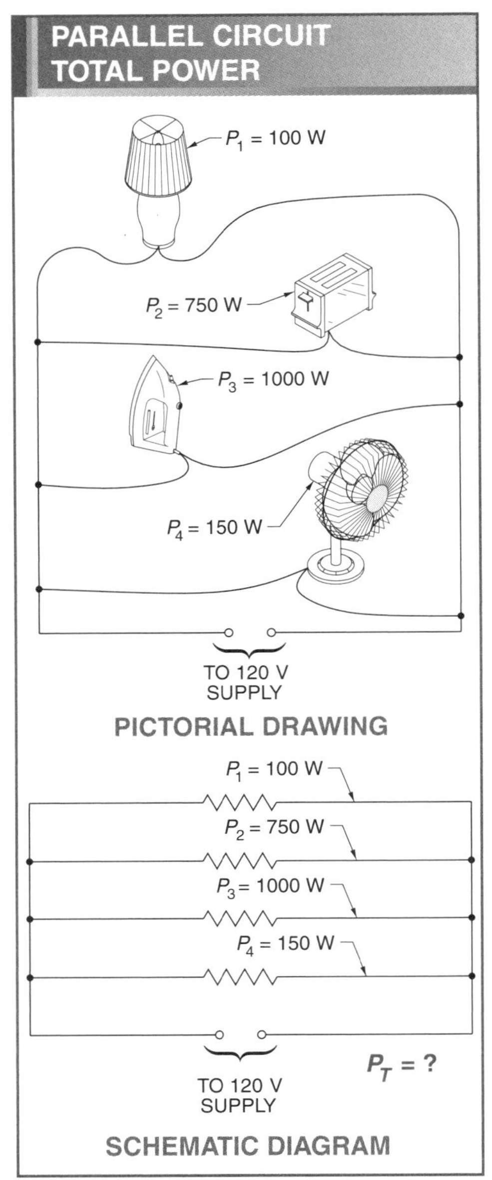

Figure 23-19. Total power is equal to the sum of the power produced by each load.

The waffle iron includes a heating element that heats the upper section and a heating element that heats the lower section. The two heating elements are connected in parallel. An ON/OFF temperature control switch is used to turn ON the heating elements and control the temperature. A temperature switch is wired in series with the ON/OFF temperature control switch to turn power OFF if overheating occurs in the circuit. To calculate the total power of the waffle iron, apply the formula:

$$P_T = P_1 + P_2$$

where

P_T = waffle iron total power (in W)

P_1 = power of upper heating element (in W)

P_2 = power of lower heating element (in W)

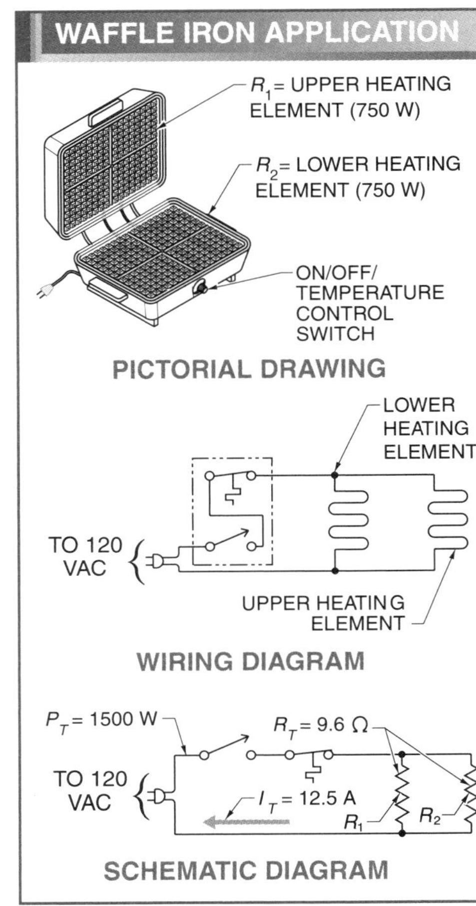

Figure 23-20. A waffle iron applies the concepts of a basic parallel circuit to heat the upper and lower heating elements.

Example: Calculating Waffle Iron Total Power

What is the total power in the waffle iron circuit containing two heating elements with power ratings of 750 W?

$$P_T = P_1 + P_2$$

$$P_T = 750 + 750$$

$$P_T = \mathbf{1500\ W}$$

To calculate the total current used by both waffle iron heating elements, apply the formula:

$$I_T = \frac{P}{E}$$

where

I_T = total current of both heating elements (in A)

P = total power of waffle iron (in W)

E = total voltage (in V)

Example: Calculating Waffle Iron Total Current

What is the total current in the waffle iron if the total power is 1500 W and the total voltage is 120 V?

$$I_T = \frac{P}{E}$$

$$I_T = \frac{1500}{120}$$

$$I_T = \mathbf{12.5\ A}$$

To calculate total resistance of the waffle iron, apply the formula:

$$R_T = \frac{E}{I}$$

where

R_T = waffle iron total resistance (in Ω)

E = waffle iron total voltage (in V)

I = waffle iron total current (in A)

Example: Calculating Waffle Iron Total Resistance

What is the total resistance of the waffle iron if the total voltage is 120 V and the total current is 12.5 A?

$$R_T = \frac{E}{I}$$

$$R_T = \frac{120}{12.5}$$

$$R_T = \mathbf{9.6\ \Omega}$$

Review Questions

1. State Ohm's law.
2. Compare the voltage/current relationship with the current/resistance relationship.
3. What are some of the applications of Ohm's law?
4. How is current calculated using the power formula?
5. Which variable of the power formula would be easiest to calculate for electrical equipment?
6. Define impedance.
7. List several items that can have a voltage drop across them.
8. Explain the relationship between resistance, applied voltage, and power produced.
9. How does a parallel connection differ from a series connection?
10. Which connection method is most commonly used when connecting loads?

An electrical plan is a drawing and list that indicates the electrical devices to be used and the electrical wiring methods. A good electrical plan also indicates the location of all major electrical devices and connections. Electrical connections must be mechanically and electrically secure. Electrical connections require supplies such as solder, tape, wire nuts, and wire markers.

24 Electrical Plans and Connections

ELECTRICAL PRINTS

All trades have a specific language that must be understood in order to transfer information efficiently. This language may include symbols, drawings or diagrams, schematics, words, phrases, and abbreviations. Electrical prints are used to convey electrical information. Electrical prints include pictorial drawings, electrical layouts, schematic diagrams, and line diagrams.

Pictorial Drawings

A *pictorial drawing* is a drawing that shows the length, height, and depth of an object in one view. Pictorial drawings indicate physical details such as holes and shoulders of an object as seen by the eye. **See Figure 24-1.**

Electrical Layouts

An *electrical layout* is a drawing that indicates the connections of all devices and components in a residential electrical system. Electrical layouts represent as closely as possible the actual location of each part of an electrical circuit. Electrical layouts often include details of the type of wire and the kind of hardware used to fasten wires and cables to terminals and boxes. **See Figure 24-2.**

An electrical layout typically indicates the devices and components of a system with rectangles or circles. The location or layout of all system devices and components is accurate on the electrical diagram for the particular dwelling. All connecting wires are shown connected from one part to another. Individuals should use electrical layouts when constructing a dwelling and when performing maintenance.

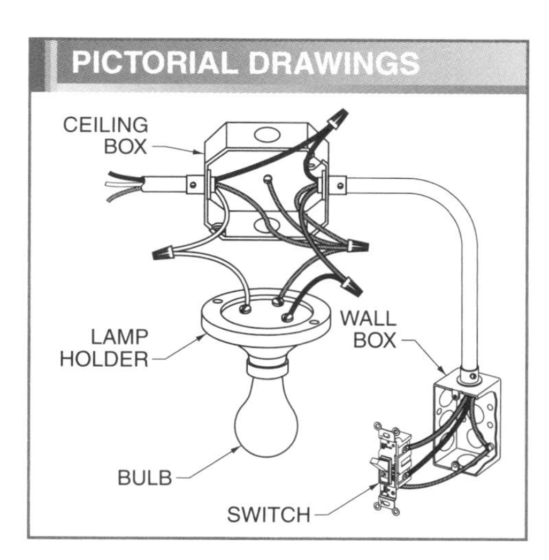

Figure 24-1. Pictorial drawings show the physical details of electrical components as seen by the eye.

RESIDENTIAL ELECTRICAL CIRCUITS

An *electrical circuit* is an assembly of conductors (wires), electrical devices (switches and receptacles), and components (lights and motors) through which current flows.

An electrical circuit is complete (a closed circuit) when current flows from the power source (service panel) to the load (lights and motors), and back to the power source. A circuit is not complete (an open circuit) when current does not flow. A broken wire, a loose connection, or a switch in the OFF position stops current from flowing in an electrical circuit.

Receptacles, housed in electrical boxes, are indicated on electrical prints by the use of symbols.

All industrial and residential electrical circuits include five basic parts. The five parts include a component (load) that converts electrical energy into some other usable form of energy such as light, heat, or motion; a source of electricity; conductors to connect the individual devices and components; a method of controlling the flow of electricity (switch); and protection devices (fuses or circuit breakers) to ensure that the circuit operates safely within electrical limits. **See Figure 24-3.**

All residential circuits use the service panel as the power source for the dwelling. Conductors are the wires pulled through conduit or metallic and nonmetallic cables. The most common control device for a residence is a single-pole switch. Control devices can be single-pole switches, three-way switches, motion detectors, timers, or thermostats. Loads found in residential circuits include lamps, motors, televisions, radios, and appliances. The typical overcurrent device used to protect a residential electrical circuit is a circuit breaker.

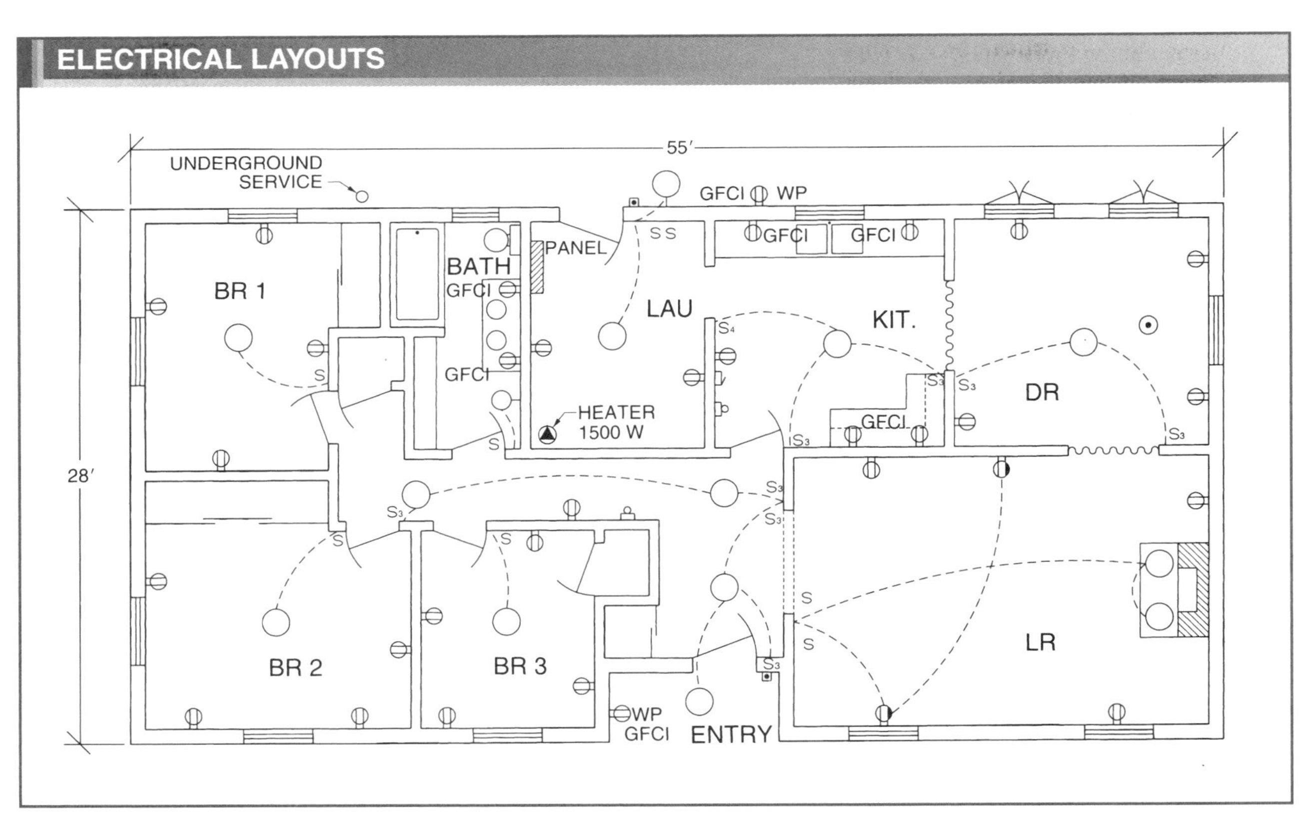

Figure 24-2. Electrical layouts show as accurately as possible the actual location of each component of an electrical circuit.

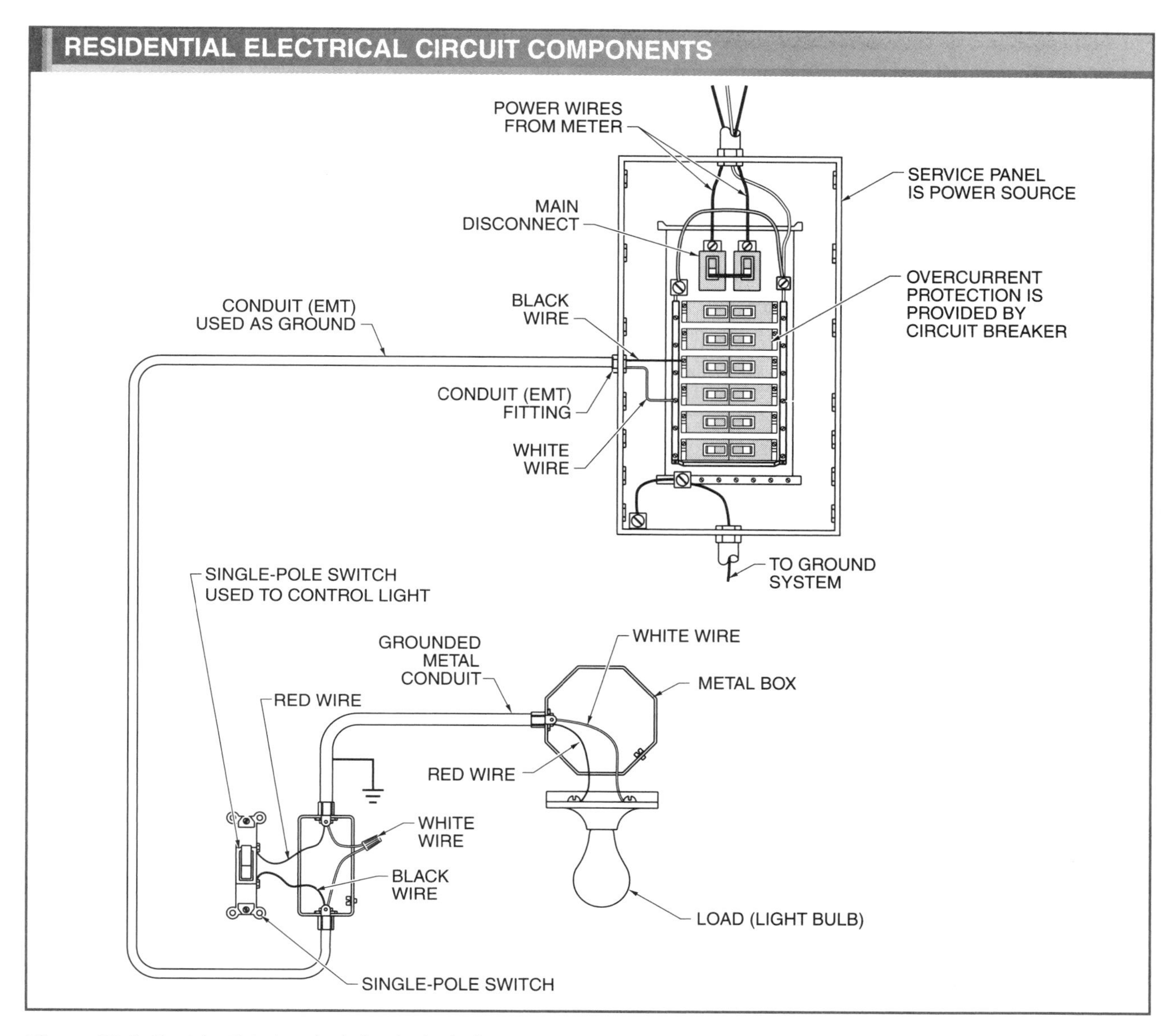

Figure 24-3. Residential electrical circuits include a power source (service panel), conductors (wiring), a load (such as a light), control devices (switches), and overload protection (fuses or circuit breakers).

Manually and Automatically Controlled Circuits

A *manually controlled circuit* is any circuit that requires a person to initiate an action for the circuit to operate. An example of a manually controlled circuits is a typical lighting circuit controlled by a switch. Circuit voltages in a residence are typically 115 VAC, but can be 12 VAC, 18 VAC, 24 VAC, or 230 VAC.

Flipping a single-pole switch to the ON position allows current to pass from the hot conductor, through the closed contacts of the wall switch, through the lamp, and on through the common conductor, forming a complete path (circuit) for current to flow and allow the light to illuminate. Flipping the switch to the OFF position opens the contacts of the switch, stopping the flow of current to the lamp and turning the light OFF.

Automatically controlled devices have replaced many manually controlled devices. Any manually controlled circuit can be converted to automatic operation. For example, an electric motor on a sump pump can be turned ON and OFF automatically by adding an automatic control device such as a float or pressure switch. **See Figure 24-4.** Float and pressure switch control circuits are used to turn sump pumps OFF and ON to prevent flooding automatically. An automatic control circuit turns on a pump when water reaches a predetermined level, which removes the water. The float or pressure switch senses the change in water level and automatically stops the pump.

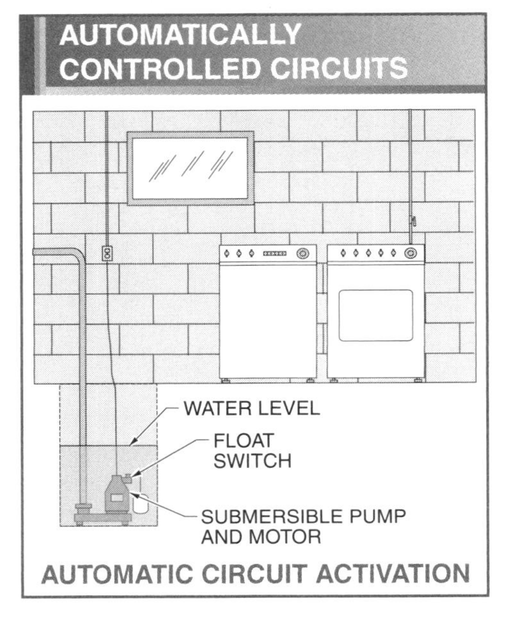

Figure 24-4. Automatically controlled circuits do not require the interaction of a person to control a function, such as turning a pump ON.

When a sump pump uses an automatic float switch circuit, float switch contacts determine when current passes through the circuit to start the motor. Current passes through float switch contacts and the motor when the float switch contacts are closed. To start the pump motor, current must flow from the power source, through the load, and back to the power source, making a complete circuit. The pump moves water until the water level drops enough to open the float switch contacts and shut the pump motor OFF. A power failure or the manual opening of the contacts prevents the pump from automatically moving water even after the water reaches the predetermined level.

Another example of automatic control is a photocell controller. **See Figure 24-5.** Photocell controllers automatically detect the absence of light and respond by sending commands to loads (lights). A photocell controller responds to the absence of light by turning lights inside or outside a dwelling ON.

Photocell controllers can also transmit OFF commands in response to approaching daylight. A photocell controller responds to light by turning connected lighting loads inside or outside the dwelling OFF. Photocell controllers are mounted to standard octagon electrical boxes or directly to controlled lighting.

A ceiling fan typically uses a remote control to manually control the electrical circuit that operates the fan motor and lights.

ELECTRICAL PLANS

An *electrical plan* is a drawing and list that indicates the devices to be used, the location of the electrical devices, and wiring methods. **See Figure 24-6.** Electrical plans indicate the use of switches, lights, and receptacles with symbols. Symbols always represent real-world devices. Because electrical symbols are easy to draw, symbols are used in place of actual pictorial drawings of electrical devices. Electrical plans are easier to draw and understand when the plans are drawn using standard symbols.

Understanding electrical device symbols is the key to understanding electrical plans.

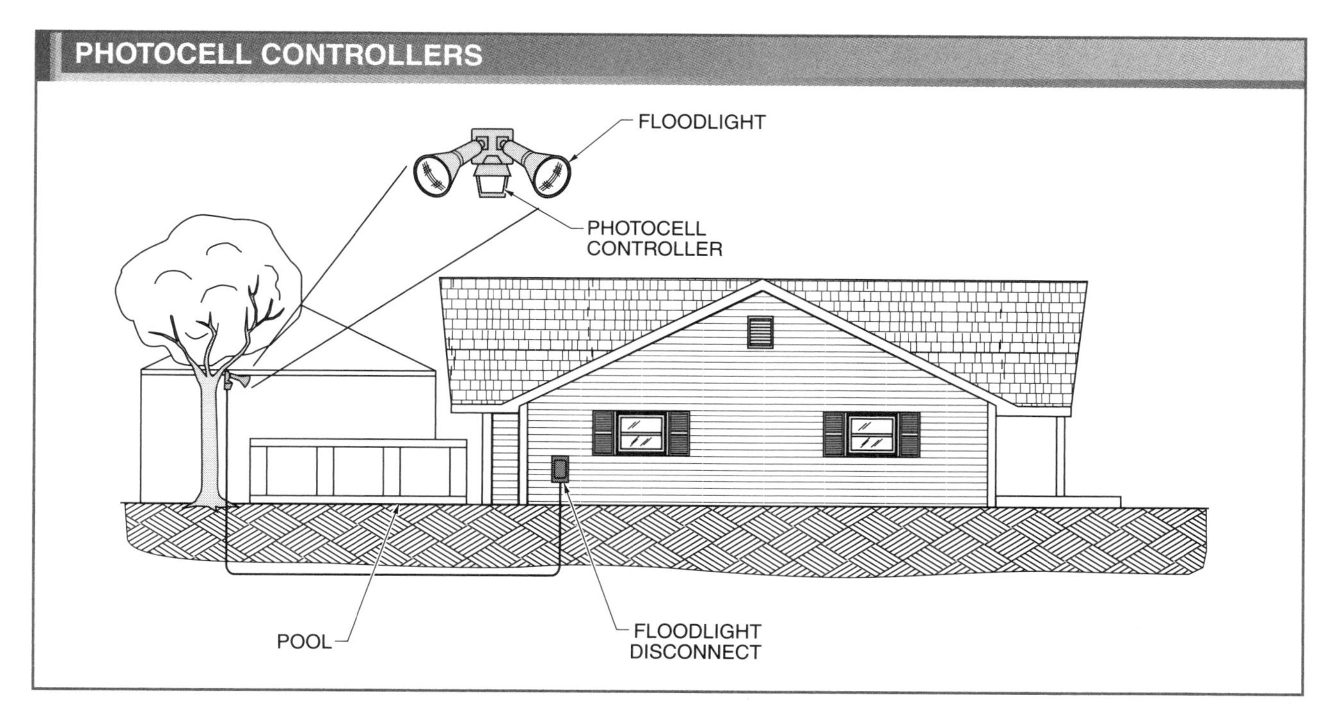

Figure 24-5. Photocell controllers respond to the absence of light by turning connected lighting loads inside or outside a dwelling ON, and respond to light by turning lights OFF.

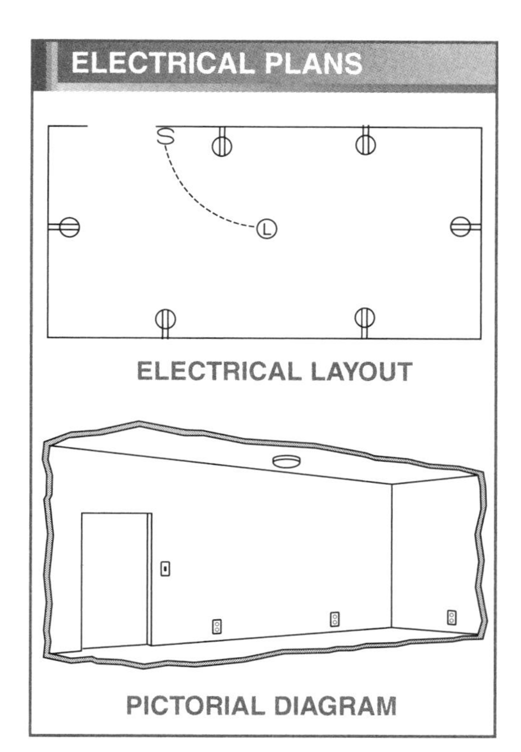

Figure 24-6. Basic electrical plans provide information about the type and location of device used in a room. Pictorial diagrams can be used to show the room after wire and drywall installation.

Architects use standard symbols when creating electrical plans. Electrical device symbols and the operation of the devices must be understood to properly wire circuits from electrical plans. When interpreting electrical plans, unfamiliar symbols must be identified before any work is started. **See Figure 24-7.**

Electrical Plan Information

A good electrical plan indicates the location of all major electrical devices and connections. Electrical plans are also used to determine the number of openings in walls and ceilings and to estimate the amount of wire and conduit that is required for the job. Electrical plans must provide enough information to indicate the best and shortest way to complete a job.

Developing Electrical Plans

A complete electrical plan or set of plans is developed in two main parts: component plan(s) and wiring plan(s).

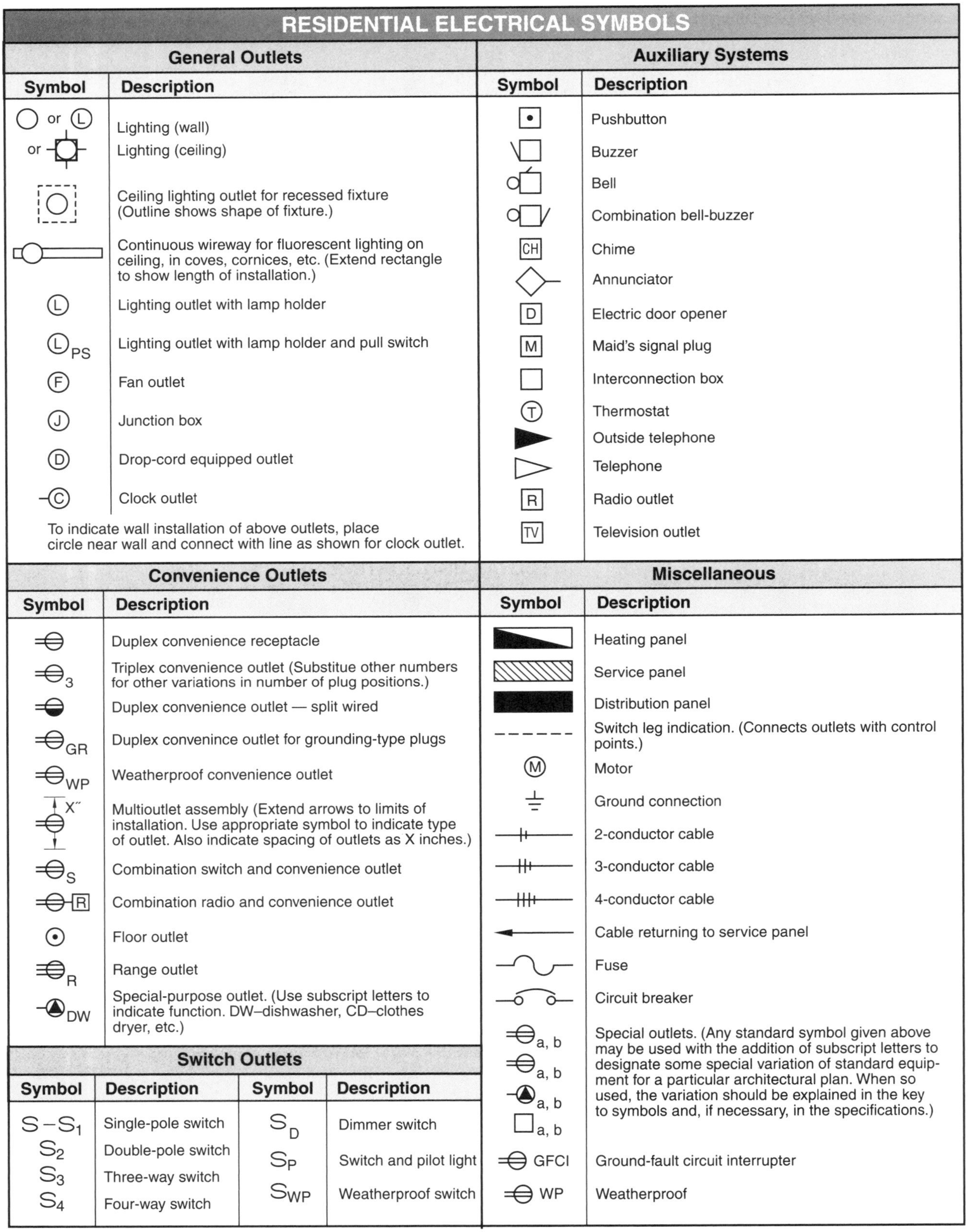

RESIDENTIAL ELECTRICAL SYMBOLS

General Outlets

Symbol	Description
or L	Lighting (wall)
or	Lighting (ceiling)
	Ceiling lighting outlet for recessed fixture (Outline shows shape of fixture.)
	Continuous wireway for fluorescent lighting on ceiling, in coves, cornices, etc. (Extend rectangle to show length of installation.)
L	Lighting outlet with lamp holder
L PS	Lighting outlet with lamp holder and pull switch
F	Fan outlet
J	Junction box
D	Drop-cord equipped outlet
C	Clock outlet

To indicate wall installation of above outlets, place circle near wall and connect with line as shown for clock outlet.

Auxiliary Systems

Symbol	Description
	Pushbutton
	Buzzer
	Bell
	Combination bell-buzzer
CH	Chime
	Annunciator
D	Electric door opener
M	Maid's signal plug
	Interconnection box
T	Thermostat
	Outside telephone
	Telephone
R	Radio outlet
TV	Television outlet

Convenience Outlets

Symbol	Description
	Duplex convenience receptacle
3	Triplex convenience outlet (Substitue other numbers for other variations in number of plug positions.)
	Duplex convenience outlet — split wired
GR	Duplex convenince outlet for grounding-type plugs
WP	Weatherproof convenience outlet
X″	Multioutlet assembly (Extend arrows to limits of installation. Use appropriate symbol to indicate type of outlet. Also indicate spacing of outlets as X inches.)
S	Combination switch and convenience outlet
R	Combination radio and convenience outlet
	Floor outlet
R	Range outlet
DW	Special-purpose outlet. (Use subscript letters to indicate function. DW–dishwasher, CD–clothes dryer, etc.)

Switch Outlets

Symbol	Description	Symbol	Description
$S - S_1$	Single-pole switch	S_D	Dimmer switch
S_2	Double-pole switch	S_P	Switch and pilot light
S_3	Three-way switch	S_{WP}	Weatherproof switch
S_4	Four-way switch		

Miscellaneous

Symbol	Description
	Heating panel
	Service panel
	Distribution panel
- - - - - -	Switch leg indication. (Connects outlets with control points.)
M	Motor
	Ground connection
	2-conductor cable
	3-conductor cable
	4-conductor cable
	Cable returning to service panel
	Fuse
	Circuit breaker
a, b / a, b / a, b / a, b	Special outlets. (Any standard symbol given above may be used with the addition of subscript letters to designate some special variation of standard equipment for a particular architectural plan. When so used, the variation should be explained in the key to symbols and, if necessary, in the specifications.)
GFCI	Ground-fault circuit interrupter
WP	Weatherproof

Figure 24-7. Understanding electrical device symbols is the key to understanding electrical plans.

Component Plans. A *component plan* is a group of schedules that state the required locations for receptacles, lights, and switches according to the NEC® and local codes. To be effective, a component plan must include devices that satisfy all the electrical requirements for light, heat, and power to a given area.

To aid in the determination of the electrical devices required for a system, electrical component lists are created with the component plan for each area of a residence such as the living room, kitchen, bedrooms, and bathrooms. A *component list* is a list of electrical equipment indicating manufacturer, specifications, and the number of each electrical component required for a room or area. A partial component list is a component schedule. **See Figure 24-8.** Careful study of an electrical component list or schedule aids in developing a checklist for work to be performed.

Wiring Plans. Once the locations of receptacles, lights, and switches have been determined and indicated on a component plan, a wiring plan can be drawn. **See Figure 24-9.** A *wiring plan* is a drawing that indicates the placement of all electrical devices and components and the wiring required to connect all the equipment into circuits. Wiring plans are the basis for all electrical work performed at a residence. The purpose of a wiring plan is to group individual electrical devices into specific circuits. A wiring plan is essential when determining the number of components per circuit (load) and the best route for wire. Time spent studying a wiring plan aids in anticipating problems and reducing the amount of materials required. Slash marks indicate the number of conducting wires in cables or conduit and an arrow symbolizes the circuit connection to the service panel.

COMPONENT PLAN SCHEDULES

Area	Convenience Receptacles	Special-Purpose Outlets	General Lighting	Major Appliances	General Switching for All Areas
Bedrooms	No space along a wall should be more than 6′ from a receptacle outlet. Any wall space 2′ or larger should have a minimum of 1 receptacle. NEC® 210.52 (A).	TV outlet, intercom, speakers (music), and telephone jack	Ceiling, wall or valence light, lamp switched at receptacle	Room air conditioner, electric baseboard heat	Switches are typically placed opposite the hinged side of door. When there are two or more entrances to a room, multiple switching should be used. Door switch can be used for closet light so that the switch turns the light ON and OFF as the door opens and closes. Switches with pilot lights should be used on lights, fans, and other electrical devices which are in locations not readily observable.
Living Room	No space along a wall should be more than 6′ from a receptacle outlet. Any wall space 2′ or larger should have a minimum of 1 receptacle. NEC® 210.52 (A).	TV outlet, intercom, speakers (music)	Ceiling fixture, recessed lighting, valence light, lamp switched at receptacle, possible dimmer	Room air conditioner, built-in stereo system, electric baseboard heating	
Family Room	No space along a wall should be more than 6′ from a receptacle outlet. Any wall space 2′ or larger should have a minimum of 1 receptacle. NEC® 210.52 (A).	TV outlet, intercom, speakers (music), bar area (ice maker, blenders, small refrigerator, hot plate), telephone jack, thermostat	Ceiling fixture, recessed lighting, valence lights, studio spot lights, lamp switched at receptacle, fluorescent light, possible dimmer	Room air conditioner, built-in stereo system, electric baseboard heating	
Dining Room	No space along a wall should be more than 6′ from a receptacle outlet. Any wall space 2′ or larger should have a minimum of 1 receptacle. NEC® 210.52 (A).	Elevated receptacles (48″) for buffet tables, speakers (music)	Ceiling chandelier, recessed lighting, valence lighting for china hutch	Room air conditioner, electric baseboard heating	

Figure 24-8. Component lists and schedules are created for each area of a residence such as the living room, kitchen, bedrooms, and bathrooms.

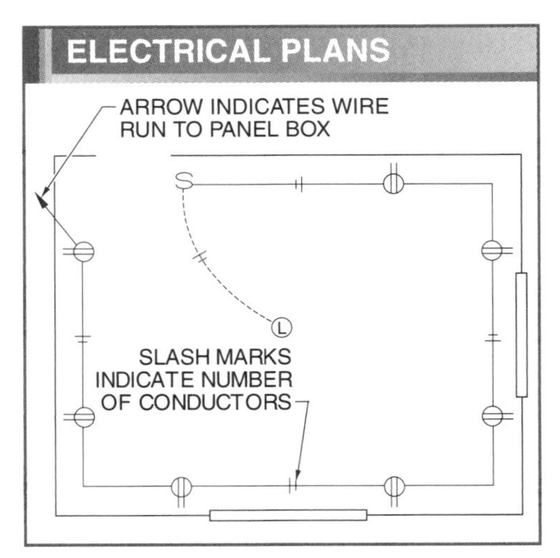

Figure 24-9. A wiring plan provides a detailed layout of cable or wire (in conduit), routes, and number of conductors.

Manufacturer Specifications. When an individual installs an electrical appliance such as washing machine, dryer, or electric oven, the individual must follow the directions or specifications provided by the manufacturer. Manufacturer specifications indicate the size of wire that must be used and the amount of overload protection that must be provided. The specifications set by manufacturers must agree with or surpass the standards of the NEC®.

ROOM LAYOUTS

All the knowledge of electrical components and installation is expressed in the finished electrical system. The use of standard plans and private plans result in a full-scale house plan and electrical system that meets the needs of the occupants.

Component Layout

Electrical components are installed in locations typically shown on the floor plans. An isometric drawing (construction layout) can help to clarify the approximate location of all switches, receptacles, and light fixtures. **See Figure 24-10 through Figure 24-19.** Component layouts are directly coordinated to the actual construction layout which includes the stud walls and electrical boxes.

Electrical Layouts

An electrical layout is a drawing that indicates how the component parts of a circuit will be connected to one another and where the wires will be run. Careful planning of the electrical layout results in substantial savings by eliminating long runs of wire. Electrical layouts are used by individuals to determine where to place electrical boxes and how many boxes are required. Electrical layouts are helpful in determining a bill of material and approximate costs of an electrical project.

Wire runs are laid out on a wiring layout in a smooth and definite pattern to make the drawing easier to follow. In many cases, wire runs shown at right angles on the wiring layout could have been run diagonally in the room to conserve wire. When any cable runs are routed on a job site, shortening the wire runs always results in lower installation costs.

Exploded Views

Exploded views or component layouts provide key points for the construction of an electrical circuit by providing detailed representations of how each component is actually wired into the system. When component wiring is duplicated, the component can be given a letter designation to indicate the device is wired exactly the same way.

Wiring Methods

Although it is assumed that conduit is being used in most construction layouts, all circuits can be wired with metallic sheathed cable or nonmetallic cable.

Use of Room Designs

Wiring layouts provide enough information to successfully wire each room. Wiring layouts must be studied carefully to determine how each component will be connected. All residential electrical systems vary, but all systems have items in common. The design of an electrical circuit for one room may require only minor changes to be used for another room.

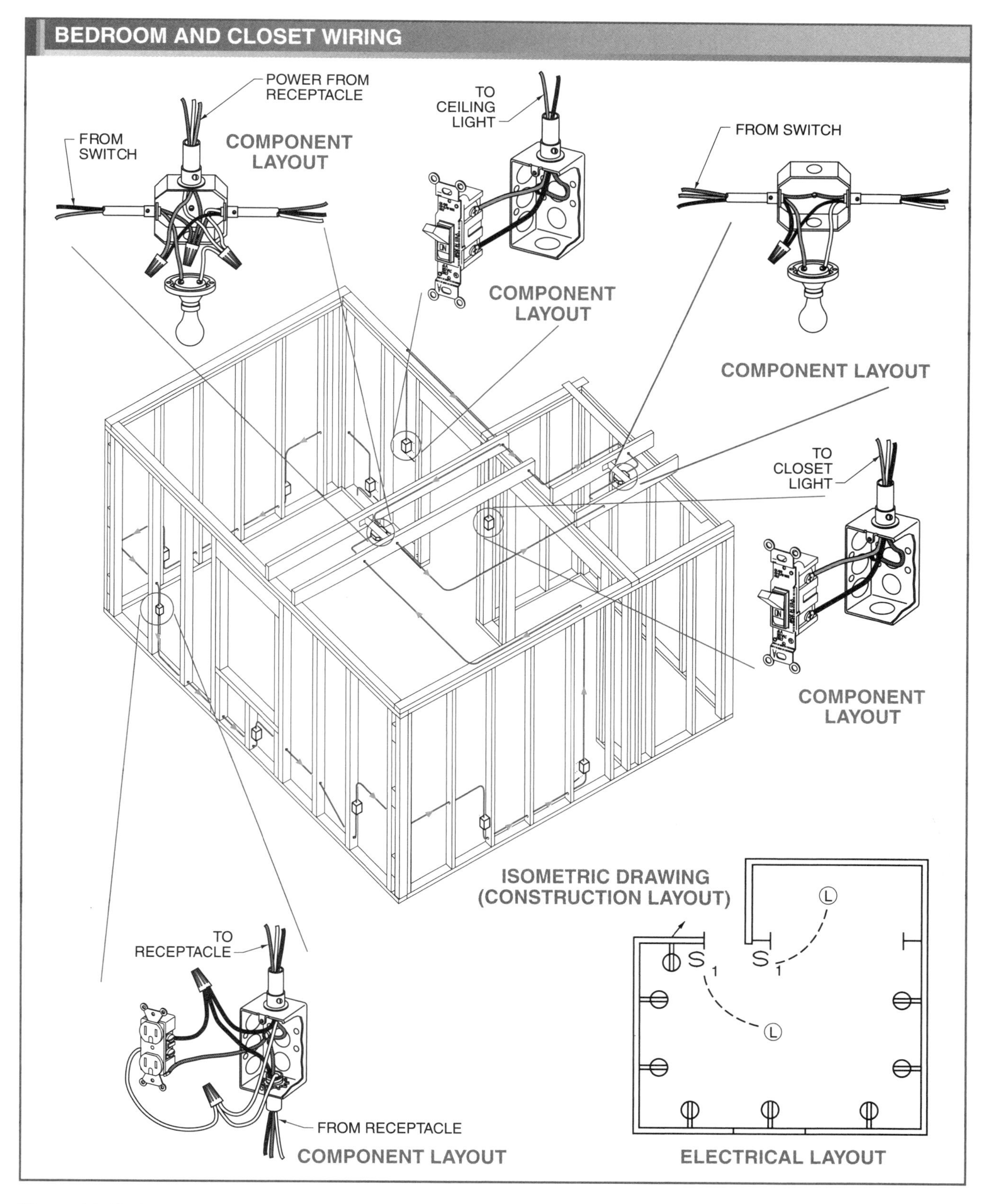

Figure 24-10. A typical bedroom with closet wiring layout shows receptacles, two ceiling lights, and switches for the two ceiling lights.

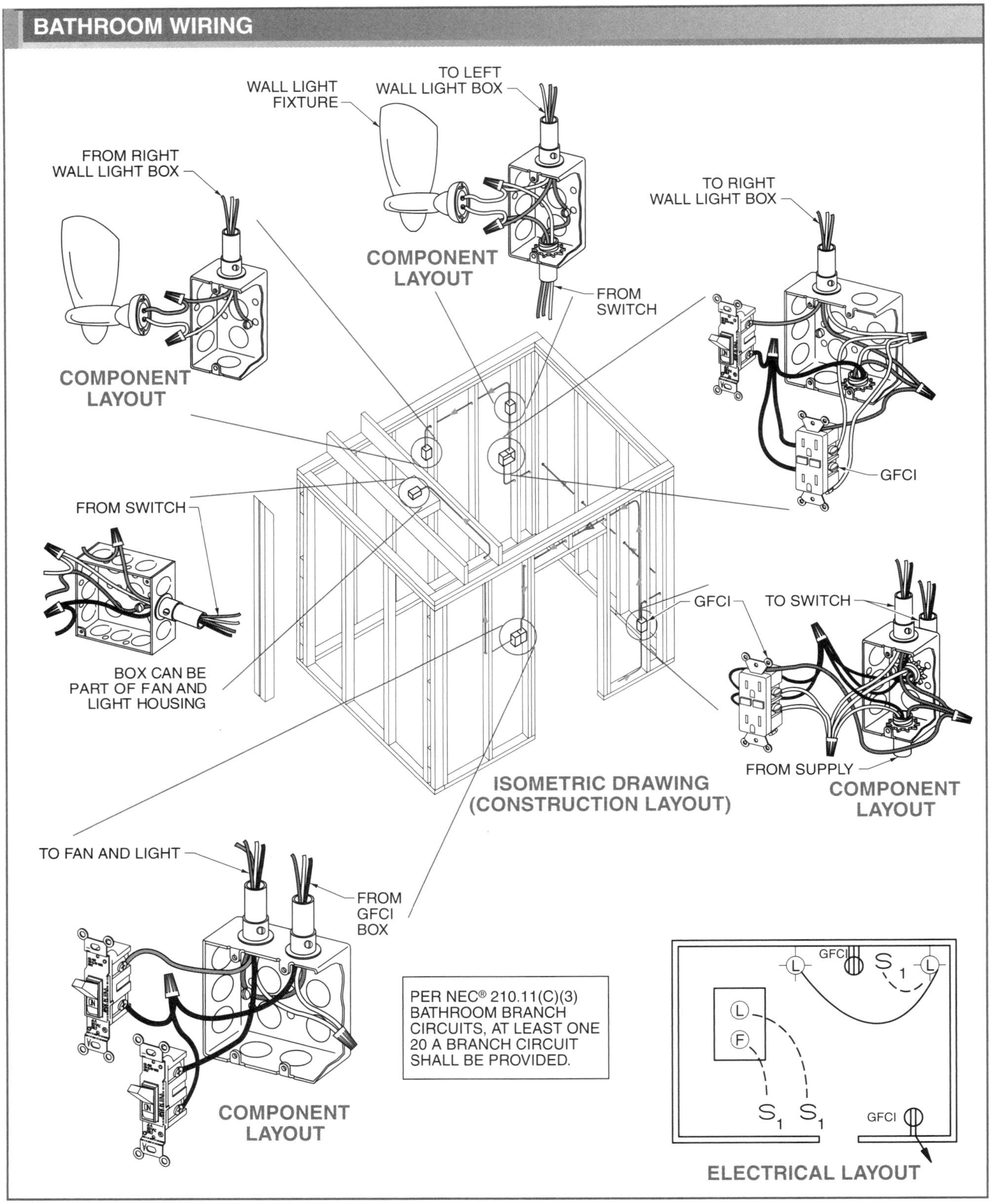

BATHROOM WIRING

PER NEC® 210.11(C)(3) BATHROOM BRANCH CIRCUITS, AT LEAST ONE 20 A BRANCH CIRCUIT SHALL BE PROVIDED.

Figure 24-11. A typical bathroom wiring layout shows two lighting fixtures, a combined fan and ceiling light, and two GFCI receptacles.

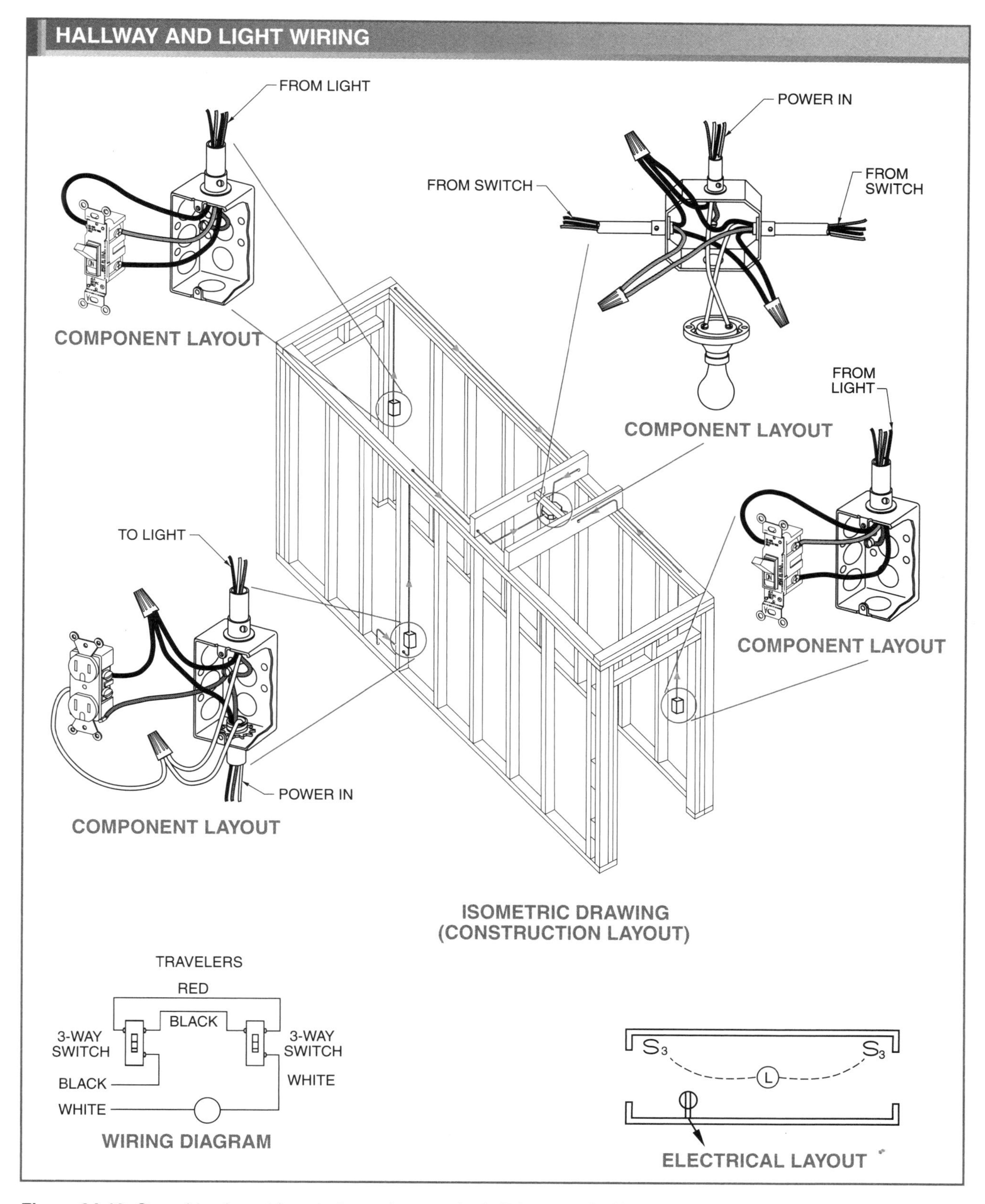

Figure 24-12. One wiring layout for a hallway shows a single light controlled from two locations.

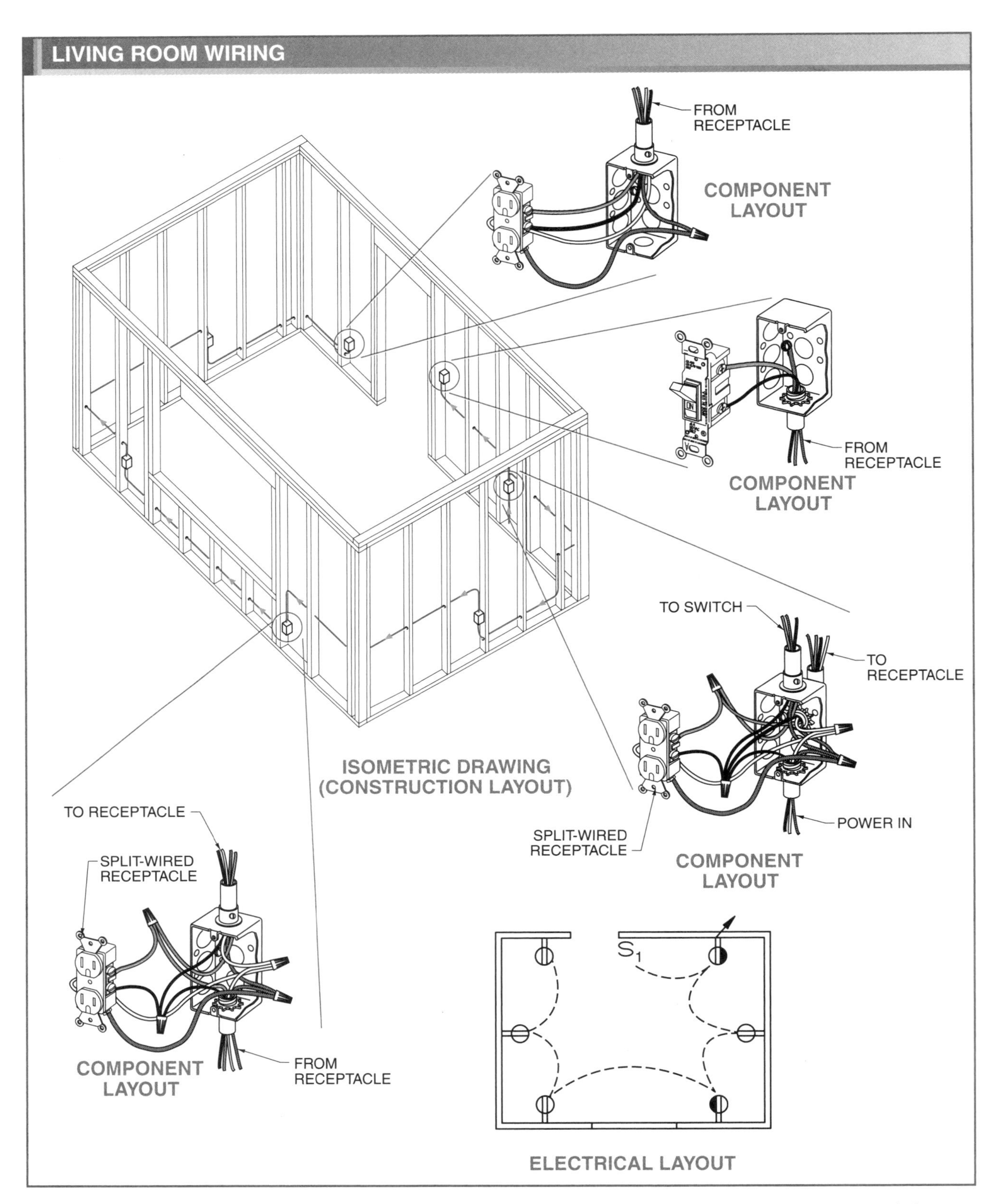

Figure 24-13. A living room wiring layout with split-wired receptacles controlled from one location uses one switch.

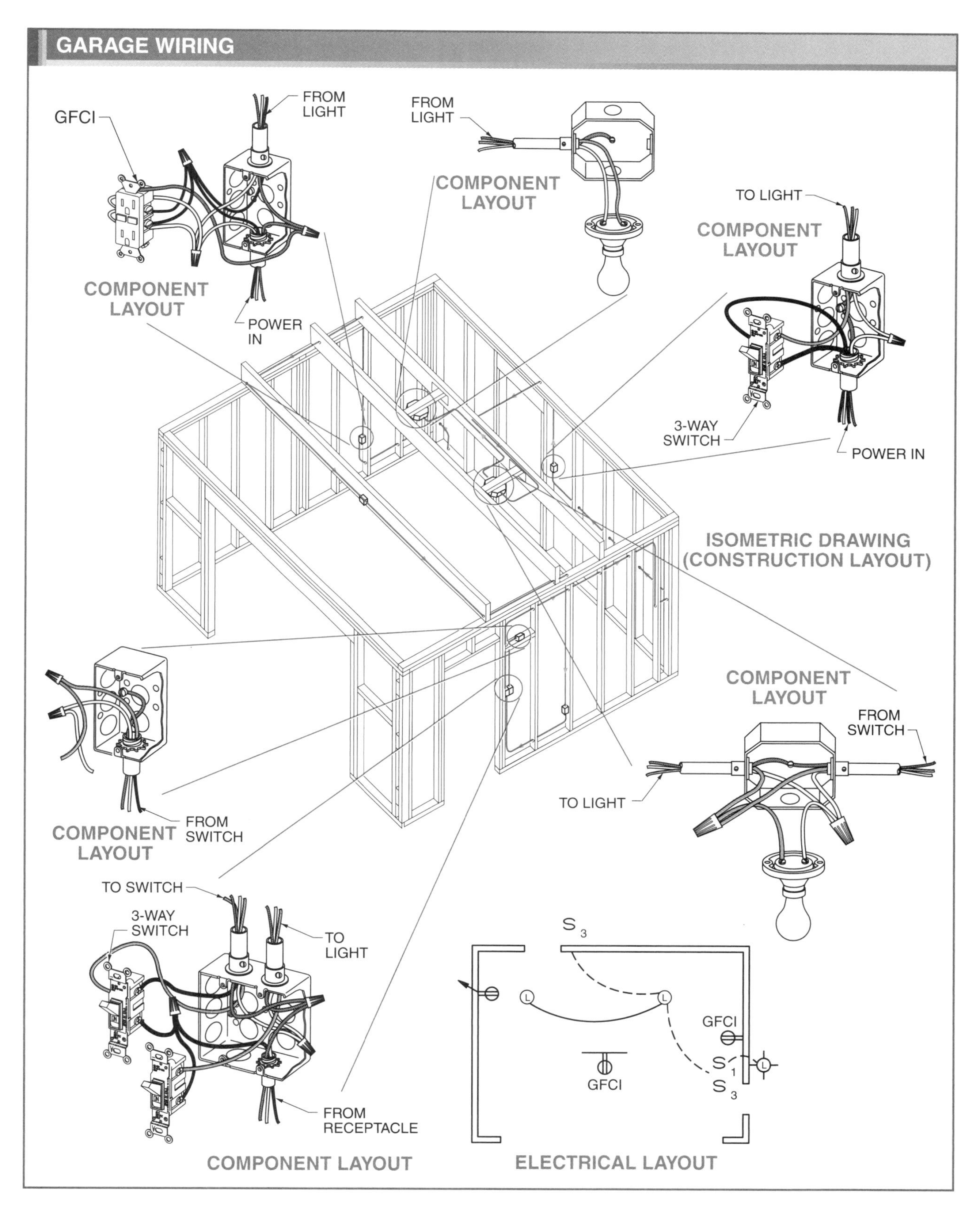

Figure 24-14. A garage wiring layout has two lights and uses one 1-way switch and two 3-way switches.

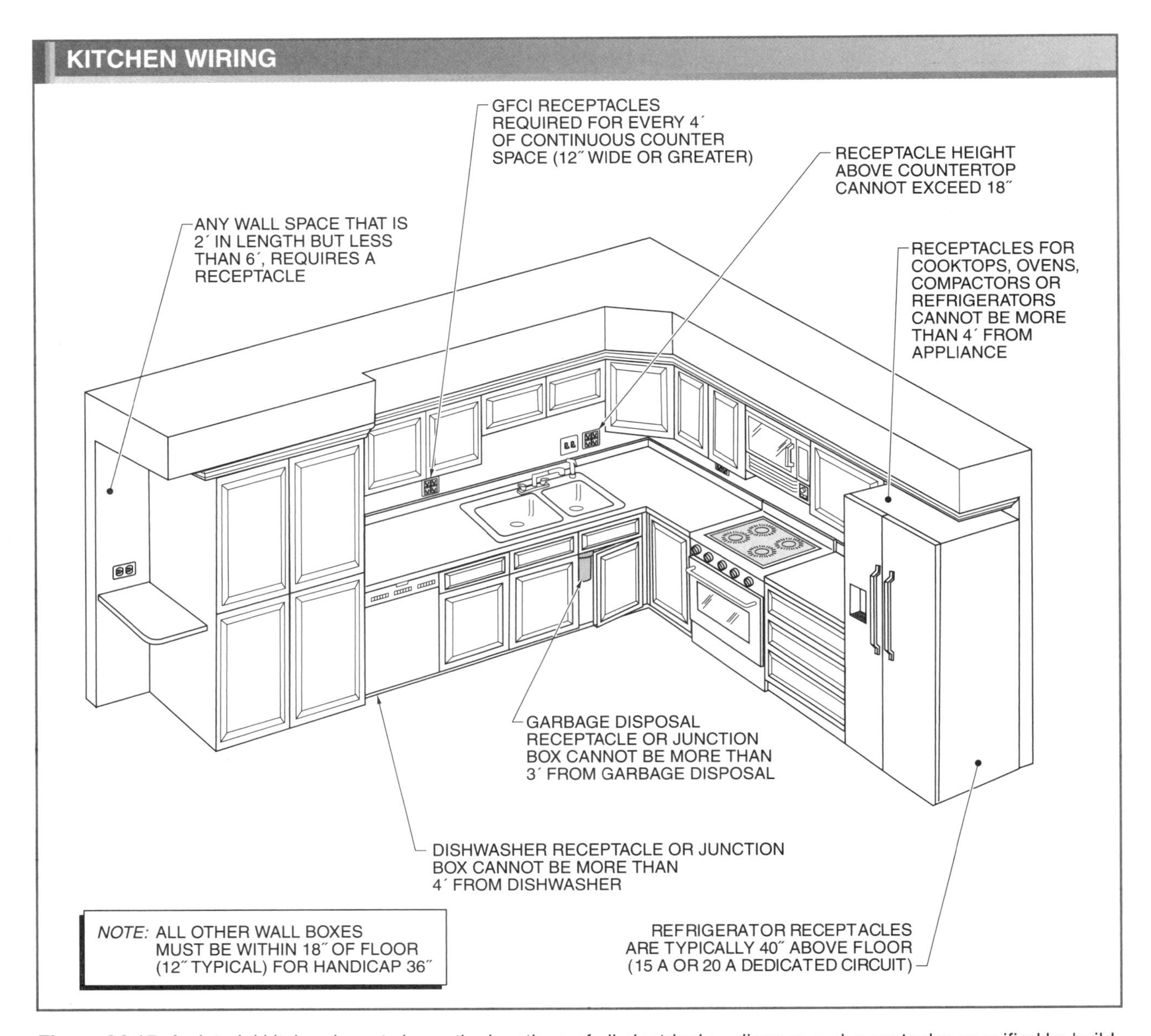

Figure 24-15. A pictorial kitchen layout shows the locations of all electrical appliances and receptacles specified by building and electrical codes.

Leviton Manufacturing Co., Inc.

GFCI receptacles are available in a variety of colors to help coordinate with kitchen and bathroom decor.

Tech Facts

When installing GFCI receptacles near a kitchen countertop, consideration must be given to the height of the backsplash, which is installed to prevent water from getting between the countertop and the wall. Typical backsplash height is 4″ but can vary. Always verify dimensions with kitchen elevation drawings, electrical layout, or the builder before installing receptacles near a kitchen counter.

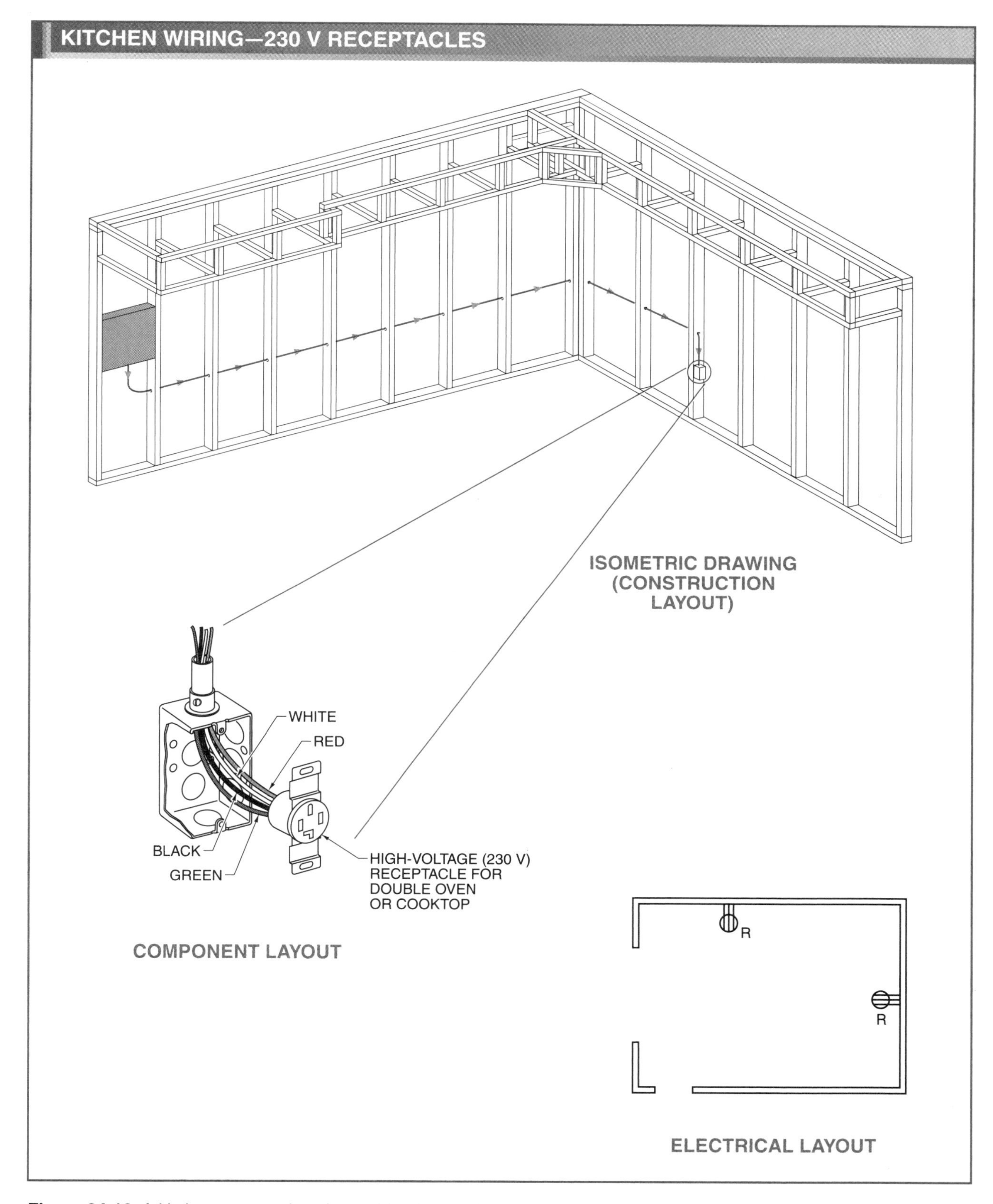

Figure 24-16. A kitchen oven and cooktop wiring layout shows the locations of 230 V receptacles.

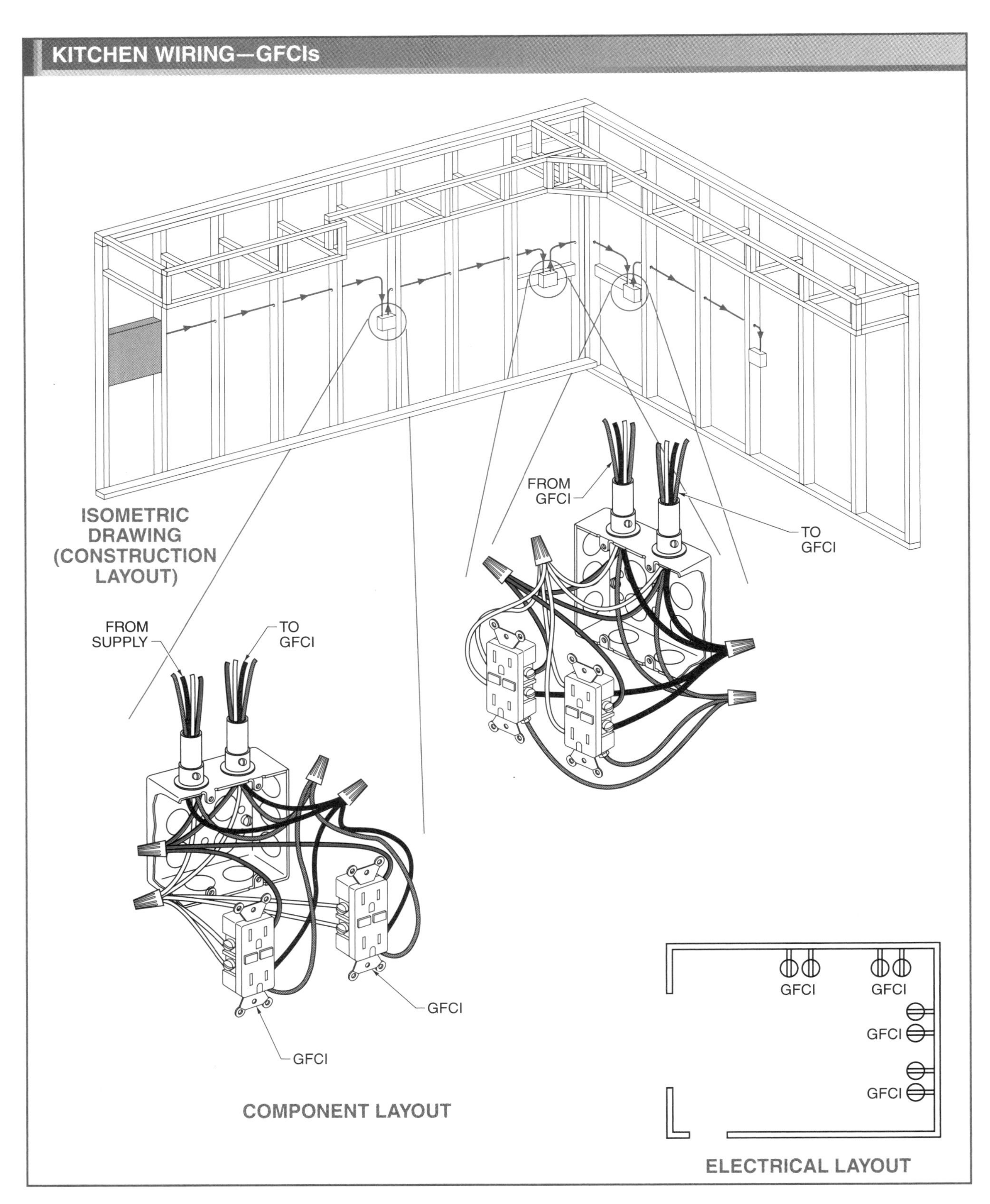

Figure 24-17. A kitchen wiring layout shows the ground fault circuit interrupters (GFCIs).

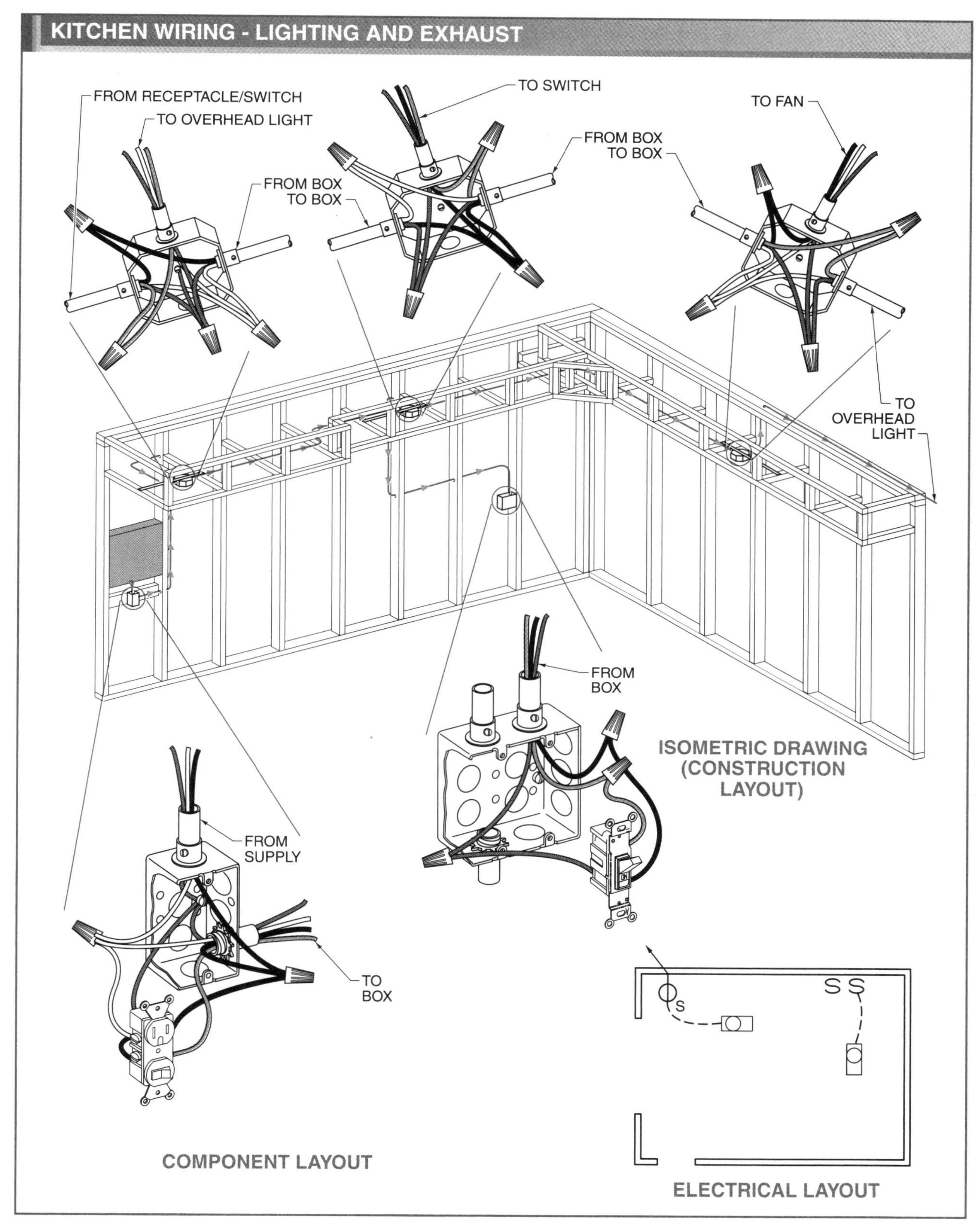

Figure 24-18. A kitchen wiring layout shows lighting and exhaust fan wiring.

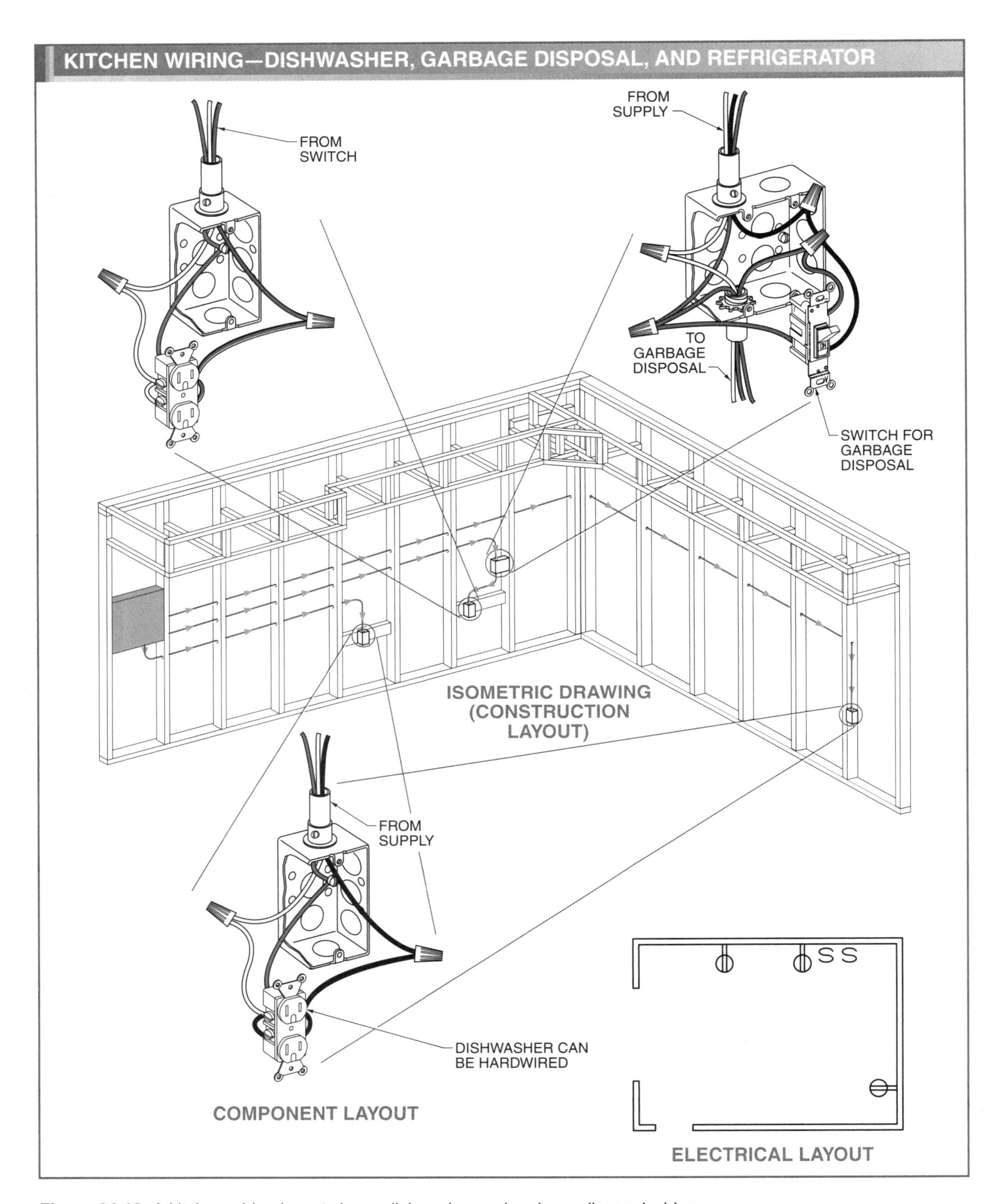

Figure 24-19. A kitchen wiring layout shows dishwasher and garbage disposal wiring.

ELECTRICAL CONNECTIONS

Electrical connections must be mechanically and electrically secure. Electrical connections require supplies such as solder, tape, wire nuts, and wire markers.

Solder

Solder commonly used in electrical work is an alloy of 60% tin and 40% lead. The amount of tin determines the strength of the solder and the amount of lead determines the melting point of the solder. The 60/40 mixture provides dependable strength at a reasonable melting temperature. Flux is added to solder to clean metal surfaces and ensure the solder adheres to the metal surface. Flux removes oxides and other small impurities from metal surfaces to improve the connection.

The two most common types of solder are rosin core and acid core. Rosin core solder is preferred on small electrical applications because the flux is contained within the solder. **See Figure 24-20.** Acid core solder must never be used on electrical connections because acid core solder corrodes electrical connections.

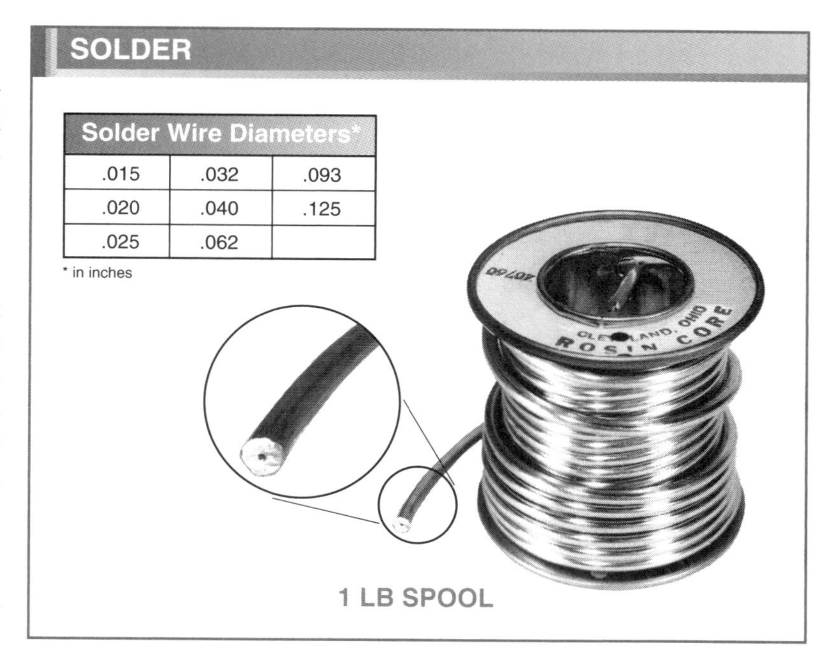

Solder Wire Diameters*		
.015	.032	.093
.020	.040	.125
.025	.062	

* in inches

Figure 24-20. Solder is available as rosin core (wire) and acid core (wire) and in various sizes.

Tape

Although many types of tape can be used as insulation on electrical wiring, plastic tape, which insulates against voltages up to 600 V per wrap, is typically used for residential wiring.

Wire Nuts

Wire nuts are designed to hold several electrical wires firmly together and to provide an insulating cover for the connection. Wire nuts are available in several sizes. The size of a wire nut is determined by the number and size of wires (conductors) to be connected. Manufacturers of wire nuts use color coding to indicate the maximum number of conductors allowed per connection. Color codes may vary by manufacturer. **See Figure 24-21.**

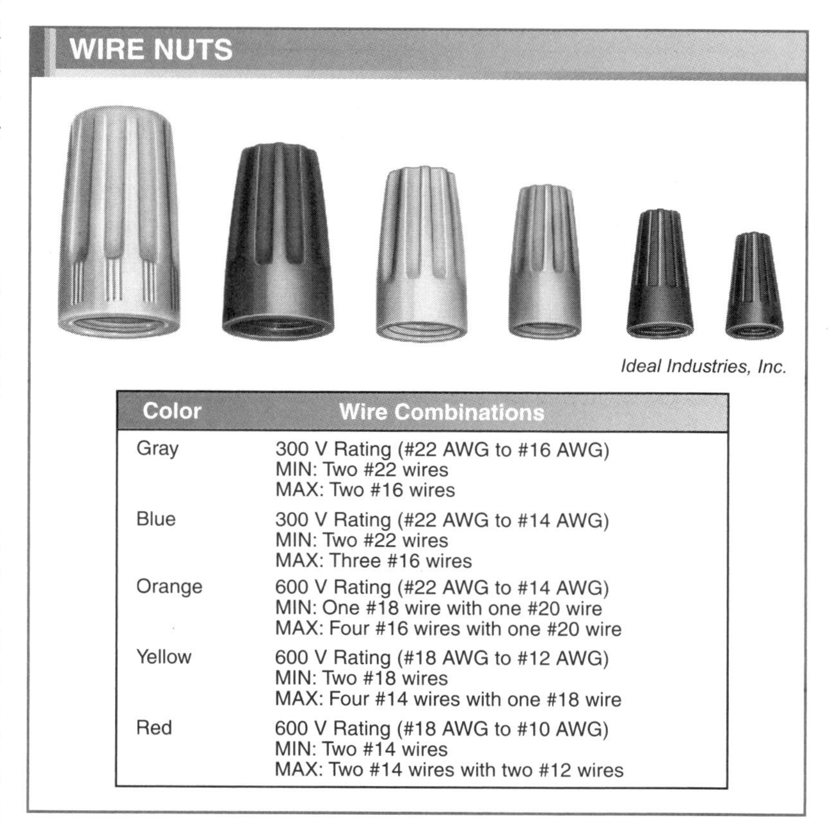

Color	Wire Combinations
Gray	300 V Rating (#22 AWG to #16 AWG) MIN: Two #22 wires MAX: Two #16 wires
Blue	300 V Rating (#22 AWG to #14 AWG) MIN: Two #22 wires MAX: Three #16 wires
Orange	600 V Rating (#22 AWG to #14 AWG) MIN: One #18 wire with one #20 wire MAX: Four #16 wires with one #20 wire
Yellow	600 V Rating (#18 AWG to #12 AWG) MIN: Two #18 wires MAX: Four #14 wires with one #18 wire
Red	600 V Rating (#18 AWG to #10 AWG) MIN: Two #14 wires MAX: Two #14 wires with two #12 wires

Figure 24-21. Wire nuts are used extensively for splices in outlet and junction boxes.

Wire Markers

A *wire marker* is a preprinted peel-off sticker designed to adhere to insulation when wrapped around a conductor. Wire markers resist moisture, dirt, and oil and are used to identify conductors that have the same color but different uses. For example, the two hot black conductors (L1 and L2) of a home can each be marked with a different lettered and numbered wire marker. Wire markers are still used even when different-colored conductors are used. Using wire markers in addition to color coding further clarifies the uses of all conductors.

SPLICES

A *splice* is the joining of two or more electrical conductors by mechanically twisting the conductors together or by using a special splicing device. Splices must be able to withstand any reasonable mechanical strain that might be placed on the connection.

Electrical splices must also allow electricity to pass through the connection as if the connection were one wire.

Wire amperage is a measurement of the amount of current a wire can carry safely. Ampacity varies according to the size of the wire. When wiring a circuit, wire that has amperage ratings that exceed circuit demands must be installed. For dedicated appliance circuits, the wattage rating of the appliance should be checked to ensure the appliance does not exceed the maximum wattage rating of the circuit. **See Figure 24-22.** Many types of wire splices can be found in a residence. Typical residential wire splices are pigtail, Western Union, T-tap, and portable cord splices.

Tech Facts

Splices are used to join two or more electrical conductors together by twisting them or by using a special splicing device.

CONDUCTOR AMPERAGE AND WATTAGE RATINGS

	Gauge	Amperage	Maximum Watt Load
For Branch Circuits			
	8	40	7680 W (240 V)
	10	30	2880 W (120 V) 5760 W (240 V)
	12	25	1920 W (120 V) 3840 W (240 V)
	14	20	1440 W (120 V)
For Bells, Thermostats, and Cords			
	16	11	---
	18	8	---
	20	6	---

Figure 24-22. The amperage and wattage ratings of conductors vary with the AWG size of the wire.

Pigtail Splices

A pigtail splice is the most commonly used electrical splice. A pigtail splice is made by twisting two ends of wire together. **See Figure 24-23.** When a pigtail splice is taped, the ends must be bent over so the sharp wire points do not penetrate the tape. When wire nuts are used instead of tape, the ends of the conductors are cut off.

When more than two wires are joined by a pigtail splice, all wires must be twisted together securely before the wire nut is installed. Twisting the wires together insures that all the wires are properly fastened before installing a wire nut.

Western Union Splices

A *Western Union splice* is a splice that is used when the connection must be strong enough to support long lengths of heavy wire. **See Figure 24-24.** When a Western Union splice is to be taped, care must be taken to eliminate any sharp edges from the wire ends. In the 1800s, Western Union splices were used to repair telegraph wires.

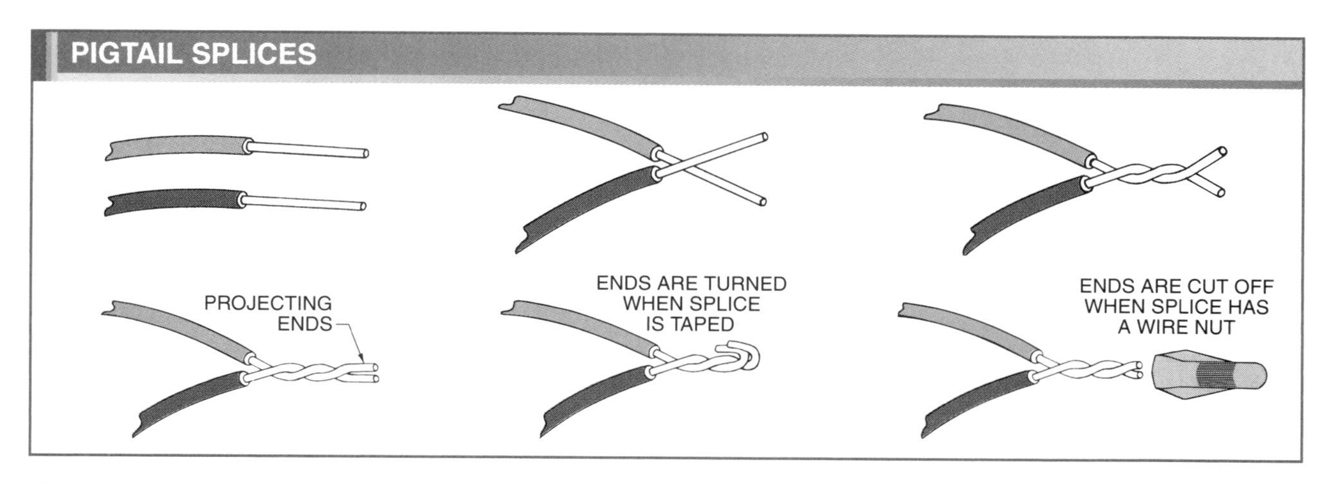

Figure 24-23. Pigtail splices are the most common splices used in residential wiring.

T-Tap Splices

A *T-tap splice* is a splice that allows a connection to be made without cutting the main wire. **See Figure 24-25.** A T-tap splice is one of the most difficult splices to perform correctly. A certain amount of practice is required to make a T-tap connection look neat. Good technique must be used to ensure proper T-tap splicing.

Portable Cord Splices

Portable cord splices are a weak splice because there is no connector to hold the conductors together. **See Figure 24-26.** Portable cords with stranded wires or solid wires can be spliced if the conductors are 14 AWG or larger. Electrified tape cannot be used so the splice must retain the insulation and outer covering properties of the portable cord.

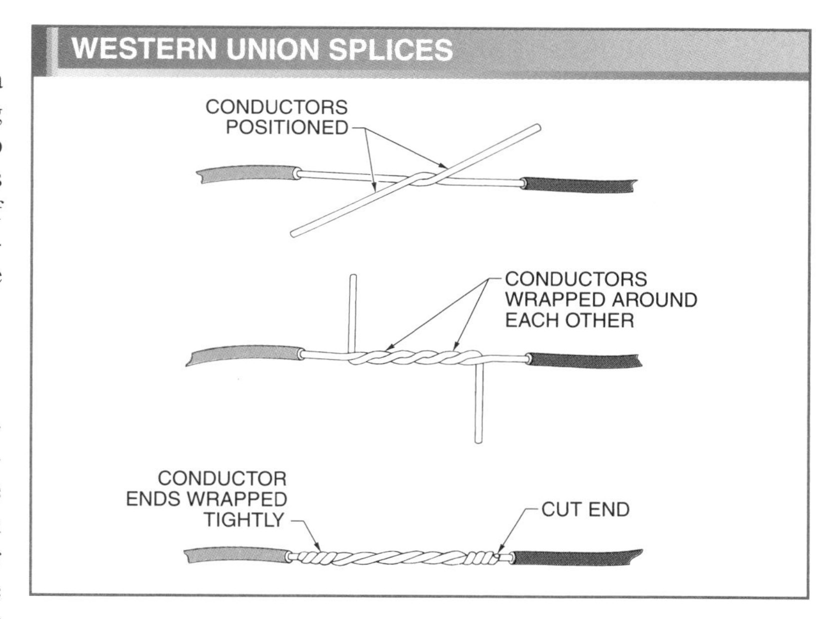

Figure 24-24. Western Union splices are used where considerable strain is placed on the connection.

Taping Splices

Taping is required to protect splices from oxidation (formation of rust) and to insulate individuals against electrical shock. Taping must provide at least as much insulative and mechanical protection for the splice as the original insulation. Although one wrap of tape (plastic or vinyl) provides insulative protection up to 600 V, several wraps are required to provide a strong mechanical protection. When plastic tape is used, the tape is stretched as it is applied. Stretching secures the tape more firmly.

Soldering Splices

Once the decision to solder an electrical splice has been made and the insulation has been stripped off the wire, the splice should be soldered as soon as possible. The longer a metal conductor is exposed to dirt and air, the greater the oxidation present on the wire, and the less chance an individual has of achieving a properly soldered connection. Clean metal surfaces are required to allow molten solder to flow freely around the connection.

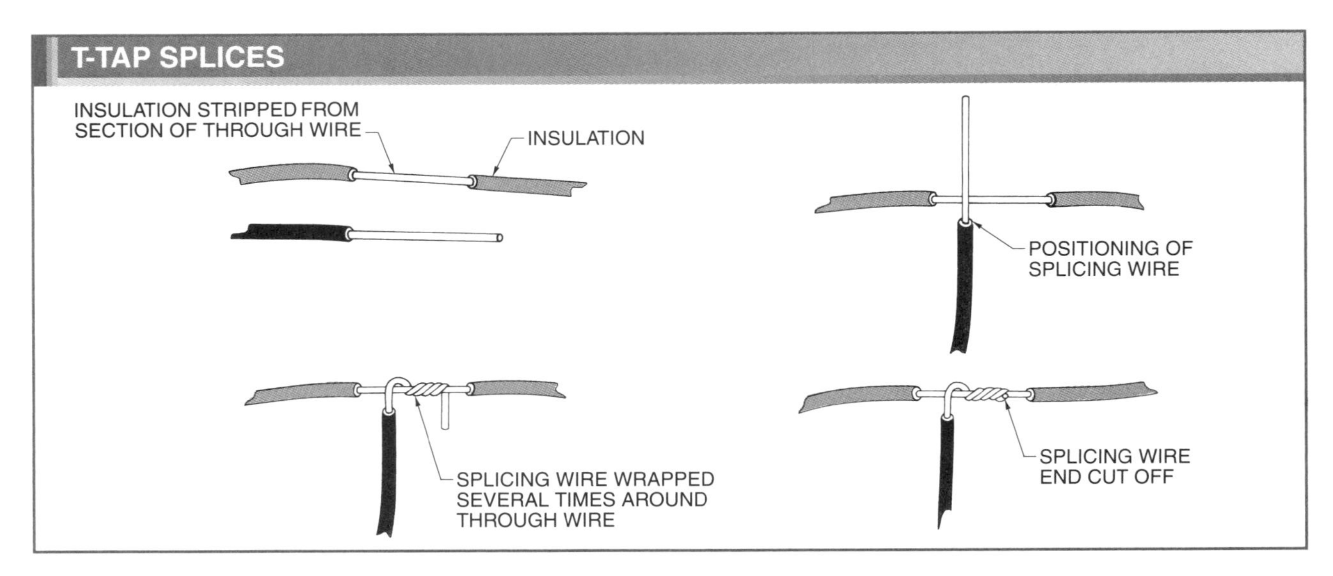

Figure 24-25. T-tap splices are used to splice a conductor to a through wire.

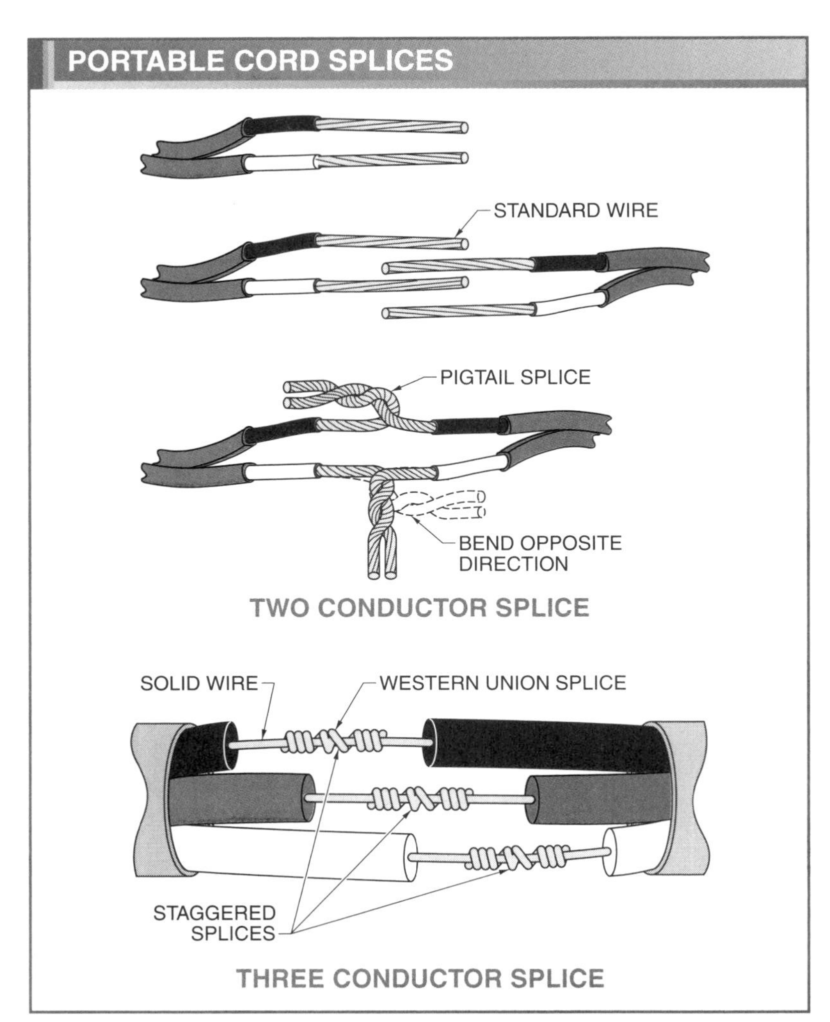

Figure 24-26. Staggering the splices of a large portable cord reduces the bump created.

The metal surfaces of the conductors are cleaned with light sandpaper, emery cloth, or by applying flux to the surfaces. Solder typically is in wire form and is melted with heat from a soldering device such as a soldering iron, soldering gun, or propane torch.

Electric soldering irons and soldering guns are used when electricity is available. A propane torch is used to solder large wires or when there is no electricity at the job site. Whatever heating method is used, individuals must apply solder to the splice on the side opposite the point where heat is being applied. **See Figure 24-27.** Melting solder flows toward the source of heat. Thus, if the top of the wire is hot enough to melt the solder, the bottom of the wire closest to the heat source will draw the solder down through all the wires. The splice must be allowed to cool naturally without movement. Once cooled, the splice is cleaned of any excess flux with a damp rag and taped.

SOLDERLESS CONNECTORS

A *solderless connector* is a device used to join wires firmly without the use of solder. Because solderless connectors are convenient and save time, several types have been developed. Some of the most commonly used solderless connectors approved by Underwriters Laboratories Inc.® are split-bolt connectors, screw terminals, back-wired connectors, wire nuts, and crimp connectors.

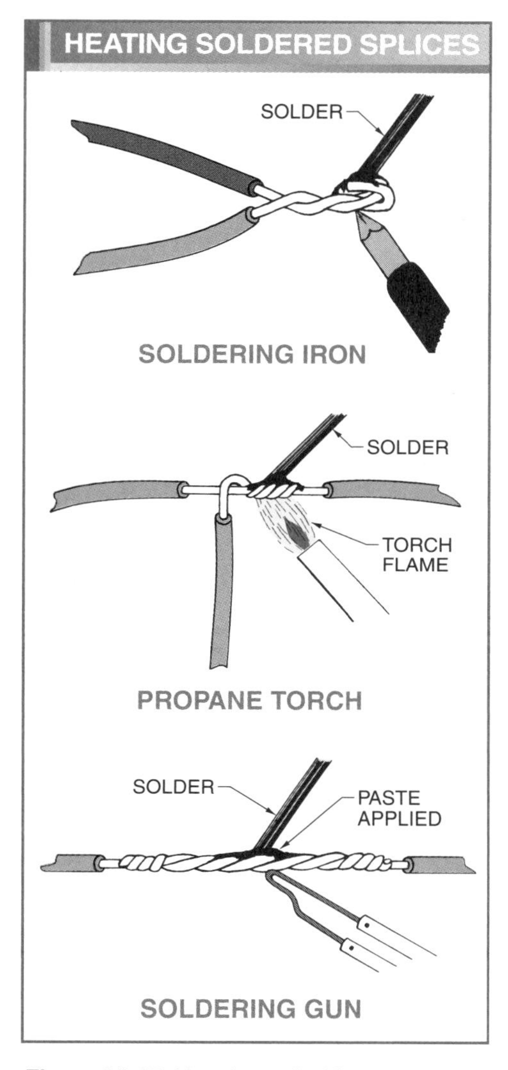

Figure 24-27. Heat is applied from beneath a splice while solder is applied from above.

Split-Bolt Connectors

A *split-bolt connector* is a solderless mechanical connection used for joining large cables. A split bolt is slipped over the wires to be connected so that a nut can be attached. Split-bolt connectors must be made of the same material as the conductors to prevent corrosion. **See Figure 24-28.**

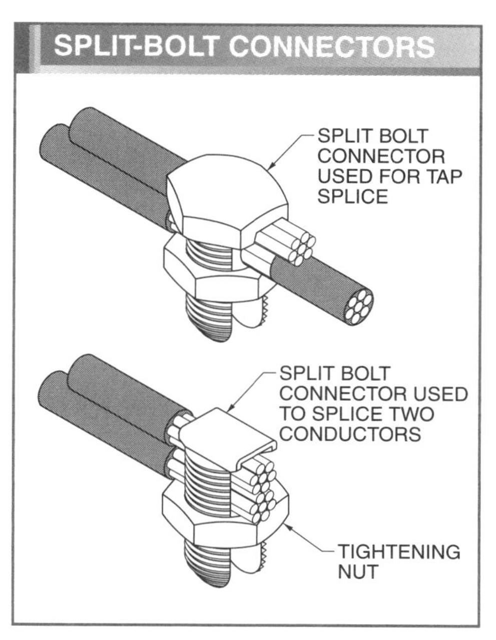

Figure 24-28. Split-bolt connectors are used to splice large cables.

Screw Terminals

Screw terminals provide a mechanically and electrically secure connection. Since wiring is always attached to electrical equipment with right-hand screws, the wire should be bent around a screw in a clockwise direction. **See Figure 24-29.** When the terminal screw method of wire connections is used, the screw draws the wire tight without pushing the wire away from the terminal.

Back-Wired (Quick) Connectors

A *quick connector* is a mechanical connection method used to secure wires to the backs of switches and receptacles. The wires are held in place by either spring or screw tension, with screw tension being the most secure. **See Figure 24-30.**

To remove a wire from a back-wired (quick) connector, a screwdriver or a stiff piece of wire is inserted into the spring opening next to the connection. Pressing down the spring through the opening releases the wire. Screw-type connectors release a wire as the screw is loosened.

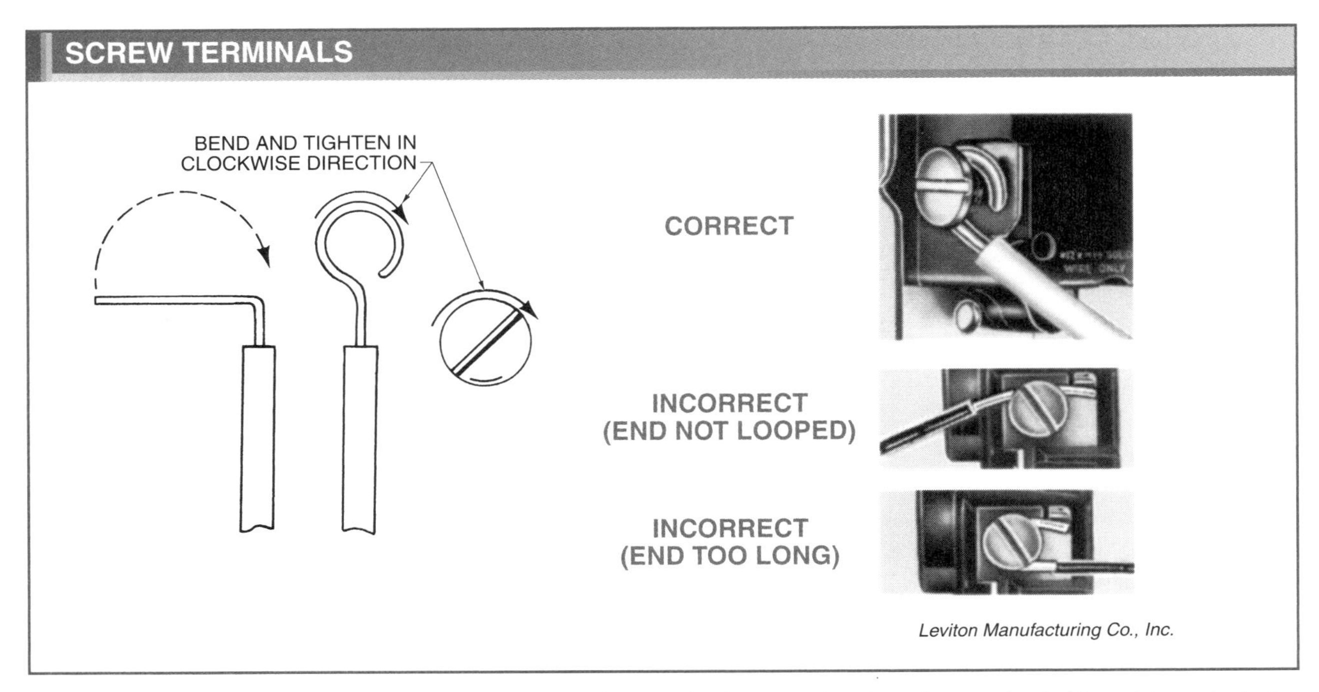

Figure 24-29. Conductors attached to screw terminals must be bent so that the wire can be tightened in a clockwise direction.

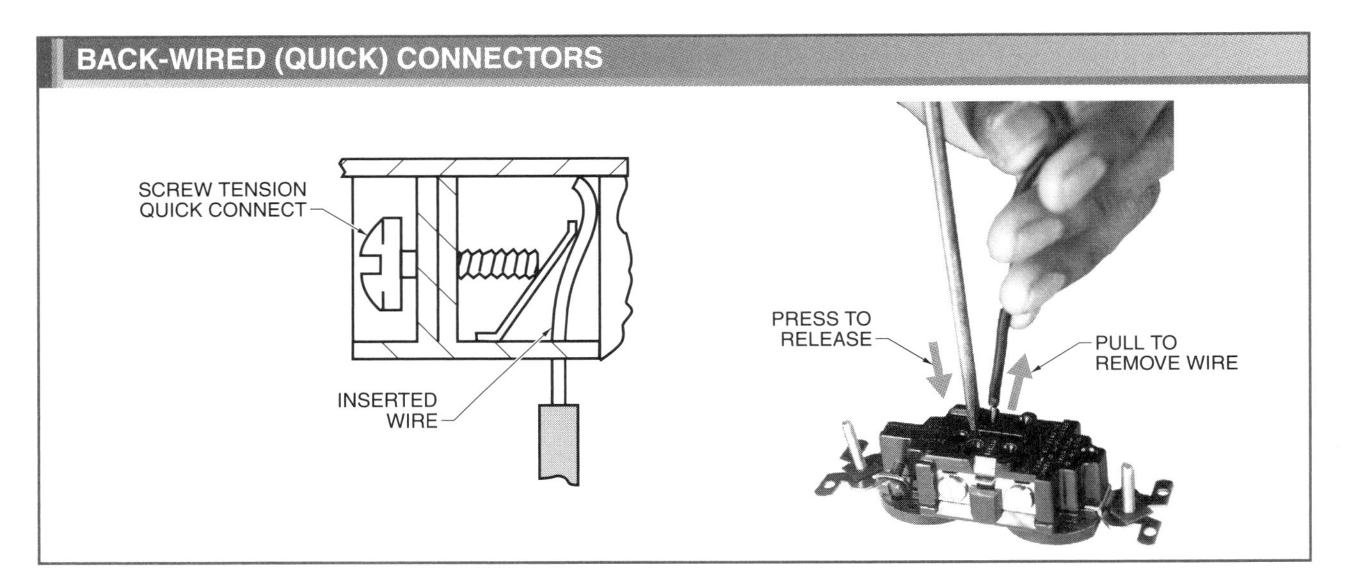

Figure 24-30. A conductor attached to a back-wired (quick) connector is held in place by spring tension or screw tension.

Wire Nut Connectors

Wire nuts have almost eliminated the need for soldering and taping. Wire nuts are manufactured in a variety of sizes and shapes. For the most effective use of wire nuts, the wires are twisted together clockwise before the wire nut is twisted on and taped. **See Figure 24-31.** When solid and stranded wires are joined together, the solid wire is bent back over the stranded wire.

The size of wire nuts is determined by the size and number of wires to be connected. To ensure safe connections, every wire nut is rated for minimum and maximum wire capacity. Wire nuts are used to connect both conducting wires and ground wires. Green wire nuts are used only for ground wires.

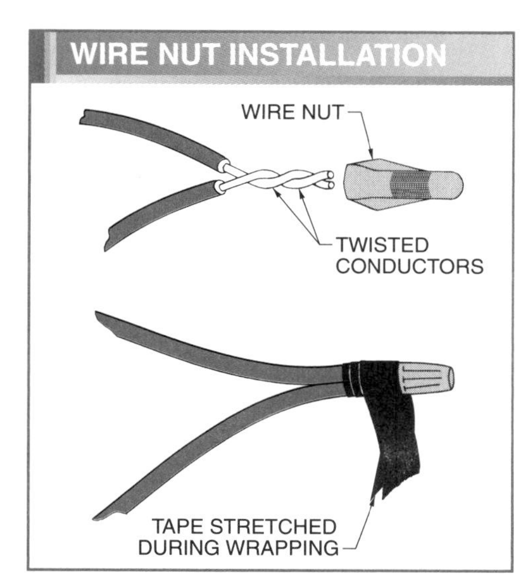

Figure 24-31. To ensure a proper connection, conductors are twisted together before the wire nut and tape are applied.

Crimp Connectors

A *crimp connector* is an electrical device that is used to join wires together or to serve as terminal ends for screw connections. **See Figure 24-32.** Crimp connectors are manufactured as insulated or noninsulated. Noninsulated crimp connectors are less expensive and are used where there is no danger of shorting the connector to a metal surface. When working with crimp connectors, identification can be a problem. To avoid confusion, a wire marker is used to identify each conductor. Crimp connectors are secured into place with a multipurpose crimp tool.

SOLDER CONNECTIONS

Soldering is the process of joining metals by using filler metal and heat to make a strong electrical and mechanical connection. The parts to be soldered must be clean, fluxed, and hot enough to melt solder for the solder to properly adhere.

In addition to surface dirt, oil, and corrosion, the metal surfaces must be cleaned of all surface oxides. Oxide is formed on metal surfaces by oxygen in the air. For example, when copper is exposed to air long enough, the oxide appears as a green tarnish. Flux is used with solder to remove surface oxides. Flux removes the oxide by making the oxide soluble and evaporating the oxide as the flux boils off during heating. A soldered connection must be mechanically strong before soldering. Heat is applied until the materials are hot, allowing solder to be applied.

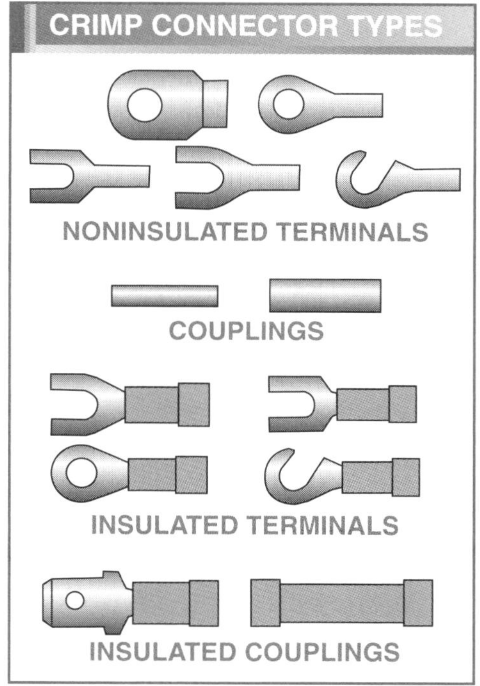

Figure 24-32. Crimp connectors are used to join wires together or to serve as terminal ends for screw connections.

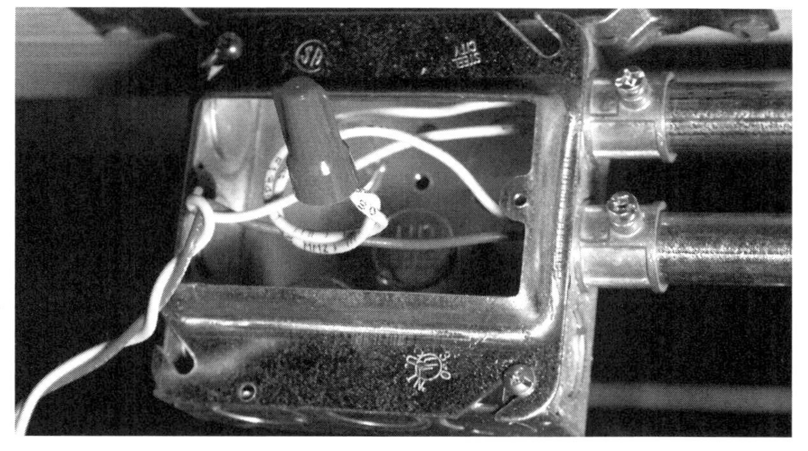

Wire nuts are used to join two wires together by twisting the wire nut over a pair of conductors.

Only a small amount of solder should be used. The connection should appear smooth and shiny. When the solder appears dull and crackly, the connection is a cold solder joint. A *cold solder joint* is a soldered joint with poor electrical and mechanical properties. Cold solder joints are typically caused by insufficient heat during soldering or the parts being moved after the solder is applied but before cooling.

Care must be taken not to damage surrounding parts by overheating them. Semiconductor components, such as transistors and ICs, are very sensitive to heat. A heat sink, such as an alligator clip, is used to help prevent heat damage. The heat sink is placed between the soldered connection point and component that requires protection. The heat sink absorbs the heat produced during soldering.

Review Questions

1. Name some of the elements that may be included in a trade-specific language.
2. List the five basic parts of any industrial or residential electrical circuit.
3. What can be done to a manually controlled circuit to make it automatic?
4. Why are electrical plans necessary?
5. How does a component plan differ from a wiring plan?
6. How can an electrical layout result in cost savings?
7. Define splice.
8. Describe three types of splices and explain the differences between the types.
9. Why would a solderless connector be used?
10. What should be done to parts before they are soldered?

Switches and receptacles are necessary parts of a residential wiring system. Originally, switches were used to open and close circuits by physically switching the circuit ON (closed) or OFF (open). With the help of modern technology, switches can be activated by light, heat, water, smoke, and radio waves. Switches are classified as simple or complex. A receptacle is a contact device installed for the connection of plugs and flexible cords to supply current to portable electronic equipment. Receptacles are often called convenience outlets.

25 Switches and Receptacles

SWITCHES

In most residential electrical circuits, switches are used as control devices. A *switch* is an electrical device used to stop, start, or redirect the flow of current in an electrical circuit.

Switches were originally used to open and close circuits by physically changing the switch to either the ON (closed) or OFF (open) condition. Modern technology and the demand for convenience have greatly expanded the role of switches.

Today, switches can be activated by light, heat, water, smoke, and radio waves. Switches can turn lights ON and OFF or vary the intensity (brightness) of lights. Switches are classified as simple or complex depending on the requirements of the user.

Switch Markings

A great amount of information can be found on the switch. The information provided can be in the form of color-coding or abbreviations. **See Figure 25-1.** The information determines where and how a switch can be used.

Color Coding. *Color coding* is a technique used to identify switch terminal screws through the use of various colors. Switches use black, steel, and green colors to identify certain screw terminals. Black is used on three-way switches to indicate the screw that is the common terminal for the switch. Power will always come into a switch at the common and leave the switch by the traveler terminal. The common screw terminal is typically black or darker than the other screws. Green always indicates a ground terminal screw. Switches are constructed with a green ground screw attached to the metallic strap of the switch for grounding.

Inscribed Information. In addition to color coding, abbreviated words and symbols are inscribed on a switch to provide information such as wire to be used, voltage and current ratings, etc.

UL Label. A *UL label* is a stamped or printed icon that indicates that a device or material has been approved for consumer use by Underwriters Laboratories Inc®. Underwriters Laboratories (UL) was created by the National Board of Fire Underwriters to test electrical materials and devices. Manufacturers, who wish to obtain UL approval, must submit samples to a UL testing laboratory in New York, Chicago, or Santa Clara, California. When a sample passes an extensive testing program, the sample is listed in a UL category as having met the minimum safety requirements.

A UL label indicates the minimum safety requirements and is not meant to serve as a quality comparison between manufacturers. Two switches, both with UL approval, can differ greatly in quality and performance.

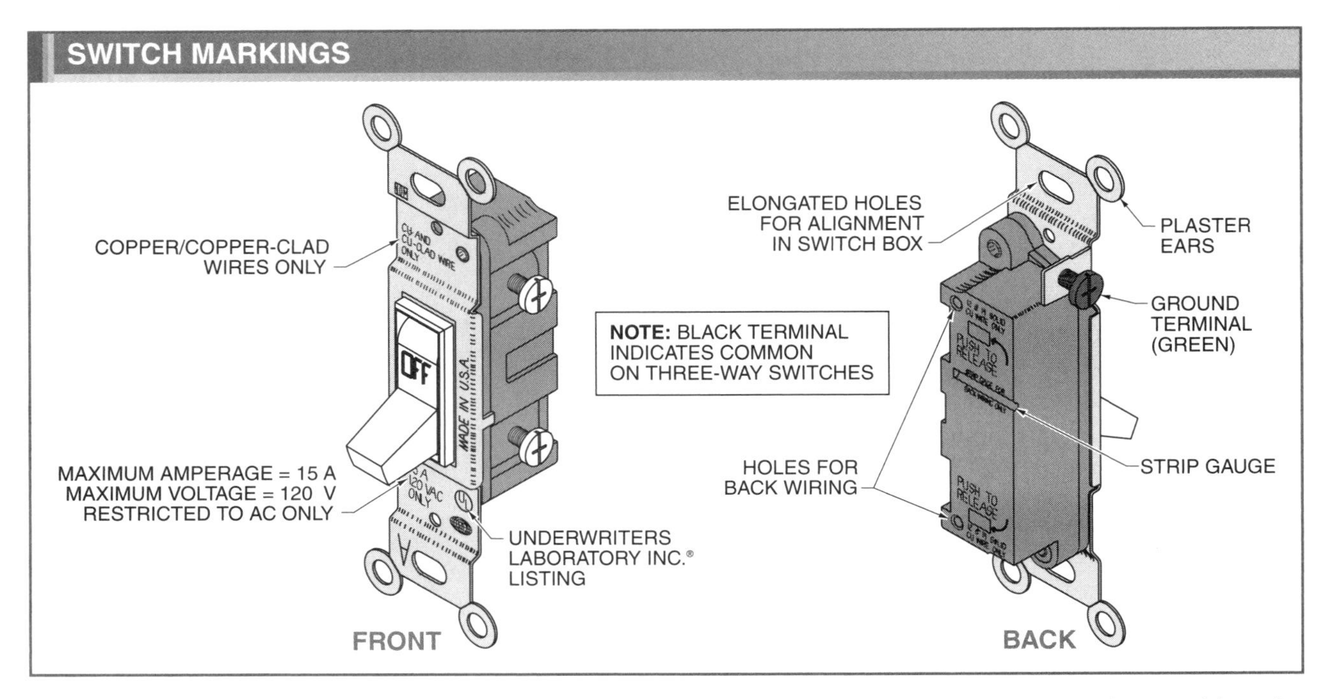

Figure 25-1. Switches control loads by opening and closing circuits and contain information as to where and how the switch can be used.

CSA Label. A *Canadian Standard Association (CSA) label* is a marking that indicates that extensive tests have been conducted on a device by the Canadian Standards Association. Both CSA and UL labels are found on many devices to indicate that the device is accepted in both Canada and the United States.

Rockwell Automation, Allen-Bradley Company, Inc.

The UL label indicates that a device has been approved for consumers by Underwriters Laboratories Inc.®, and the CSA label indicates that extensive tests have been conducted on a device by the Canadian Standards Association.

Conductor Symbols. A *conductor symbol* is an electrical symbol that represents copper and aluminum respectively. Certain electrical devices are made to work with copper only and some devices with aluminum or copper. When a device is specified CU/AL, the device can be used with copper or aluminum wire.

When aluminum wire is used with copper connectors, the typical result is overheating, but electrical fires due to the chemical reaction between the dissimilar metals are also possible.

Current and Voltage Ratings. Current and voltage ratings are always provided on a switch. The maximum current rating of a switch for residential use is 15 A or 20 A. The maximum voltage rating is typically 120 V. In addition, some switches are specified for alternating current (AC) use only. Failure to follow current and voltage ratings along with other specifications reduces the useful life of a switch.

T Ratings. Switches used to control loads with a tungsten filament (such as standard incandescent lamps) must be marked with the letter T. A *T rating* is special switch information that indicates a switch is capable of handling the severe overloading created by a tungsten load as the switch is closed. Tungsten has a very low resistance when cold and increases in resistance as heated. At the moment a switch is closed, the low resistance of a tungsten load causes the current draw to be 8 to 10 times the normal operating current. Once the switch is closed, the tungsten filament (load) heats up and the current flow drops immediately. The process of heating up a tungsten filament takes about 1/240 of a second. Failure to observe the T rating of a switch will reduce the life expectancy of the switch.

Strip Gauges. Strip gauges are found on the rear portion of a back-wired switch. **See Figure 25-2.** A *strip gauge* is a short groove that indicates the amount of insulation that must be removed from a wire so the wire can be properly inserted into a switch. When a wire has not had enough insulation removed, a proper connection is not made. When too much insulation is removed, bare wire is left exposed allowing a short to possibly occur.

STRIP GAUGE

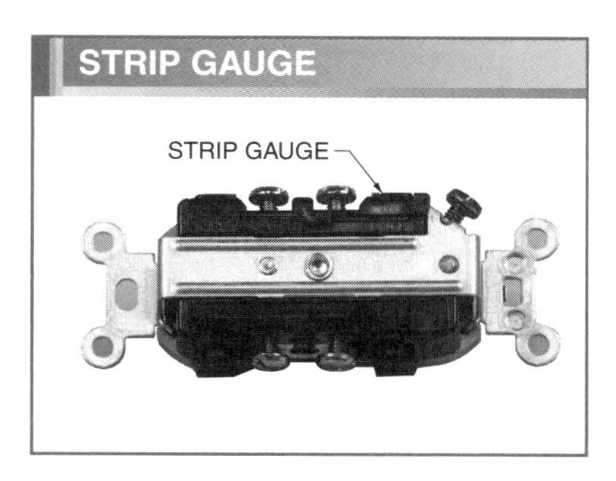

Figure 25-2. A strip gauge is a marker on the back of a switch that identifies the amount of insulation that must be removed from a wire so the wire can be correctly attached to a switch.

Tech Facts

Switches may have more than two poles or two throws. A single- or double-pole switch can have up to five or more throw positions.

Ruud Lighting, Inc.

Low-voltage track lighting systems are available with a photocontrol/timer, where the photocontrol turns the lights on at dusk and after a set period, the timer turns them off.

Types of Switches

Switches are designed for a variety of applications. **See Figure 25-3.** Common switches used in residences are the single-pole, three-way, four-way, and double-pole switches. Some modern residential-use switches are the timer, programmable, and motion sensor switches. A timer switch is set to the times that the switch is turned ON and OFF. A programmable switch allows a sequence of events to be established, such as turning lights ON when no one is home. Motion sensor switches are used as security devices or energy savers. As a security device, motion sensor switches turn ON lights when someone enters an area. As an energy saver, motion sensor switches turn lights OFF when no one is detected in an area.

Single-Pole Switches. A single-pole switch is a control device used to turn lights or appliances ON and OFF from a single location. Switches are connected into a circuit in the positive or hot side (leg) of a circuit.

SWITCH TYPES . . .

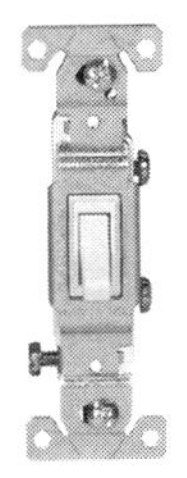

Cooper Wiring Devices

SINGLE-POLE SWITCH

Leviton Manufacturing Co., Inc.

THREE-WAY SWITCH

Standard ON/OFF switch provides control from one location; a three-way switch used with another three-way switch provides control from two locations (has no ON/OFF markings)

Leviton Manufacturing Co., Inc.

FOUR-WAY SWITCH

Used with two three-way switches to provide control from three or more locations

Cooper Wiring Devices

DOUBLE-POLE SINGLE-THROW SWITCH

Equivalent to two single-pole switches. May open and close two circuits or wires at the same time; Looks identical to a four-way switch except double-pole switch is marked with an ON/OFF

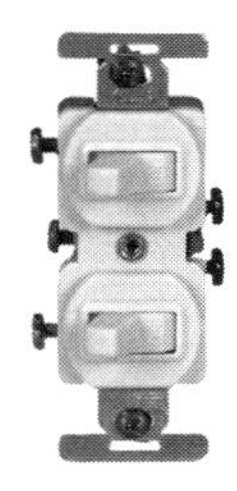

Cooper Wiring Devices

DUPLEX

Useful when space is a problem; duplex switch allows two single pole switches to be placed in a standard switch box

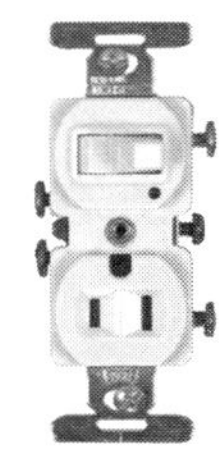

Cooper Wiring Devices

SWITCH/RECEPTACLE COMBINATION

Switch may be wired to control receptacle or receptacle can be hot while the switch controls another circuit

Cooper Wiring Devices

DIMMER

Provides light adjustment from low levels to full brilliancy; Available in single-pole or three-way operation

Leviton Manufacturing Co., Inc.

KEY

Used where access to the switch is to be restricted

Cooper Wiring Devices

AC PUSHBUTTON

Replaces old and loud snap action switches; Quiet switches are also available with toggle operation

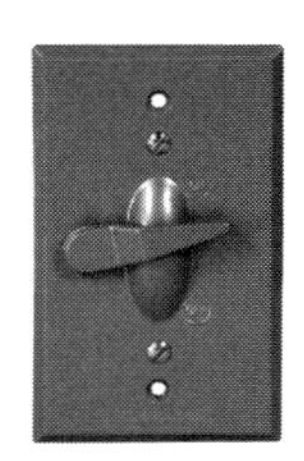

Copper Wiring Devices

WEATHERPROOF

Suitable for outside use

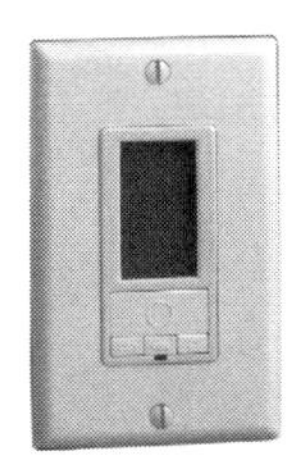

Leviton Manufacturing Co., Inc.

TIMER

Used to turn ON and OFF lights or appliances at predetermined times

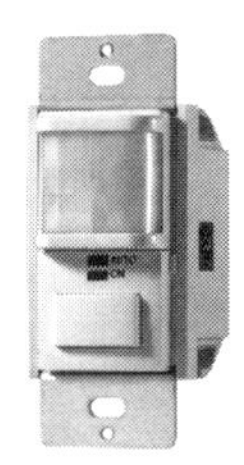

Cooper Wiring Devices

MOTION SENSOR

Allows a circuit to be turned ON and OFF according to any movement within a 400 to 1600 sq ft distance

Figure 25-3. . . .

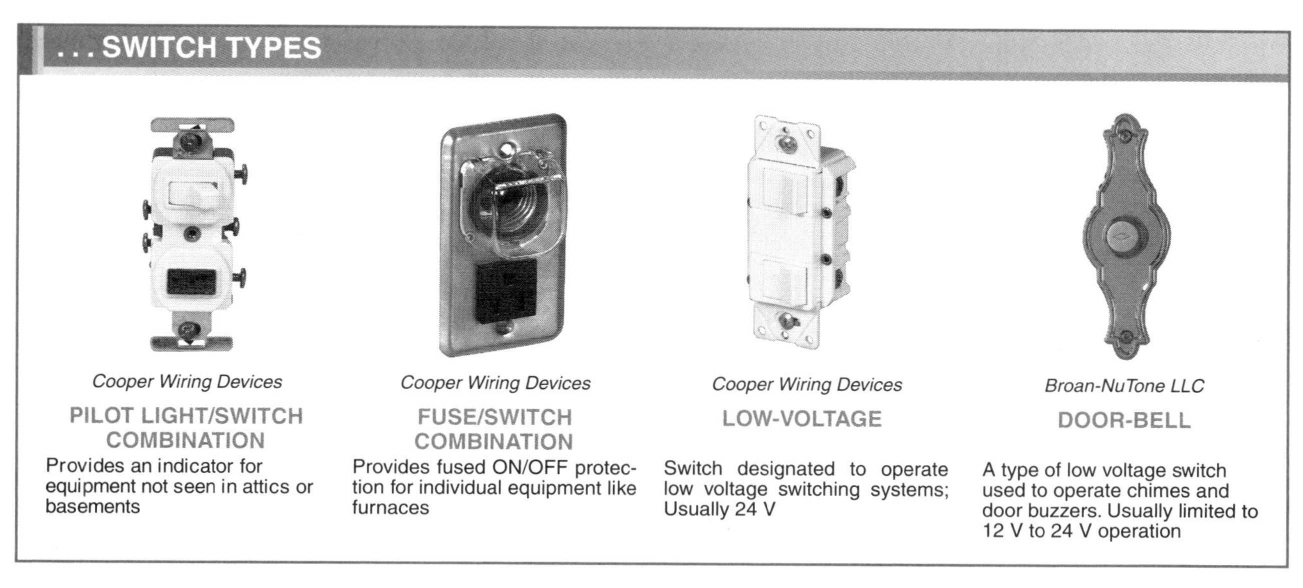

. . . **Figure 25-3.** Common switches used in residences are the single-pole, three-way, four-way, and double-pole switches.

As a general rule, neutral conductors (white wires) are not switched or used as a leg in a switch loop. Not switching the neutral conductor is an easy rule to follow for conduit systems but does not apply to armored and nonmetallic-sheathed cable. When armored or nonmetallic cable is used, an identified white conductor can be used as the feeder conductor from a switch. **See Figure 25-4.** The NEC® allows this exception. Some applications require that a load (lamp) be controlled from two separate locations.

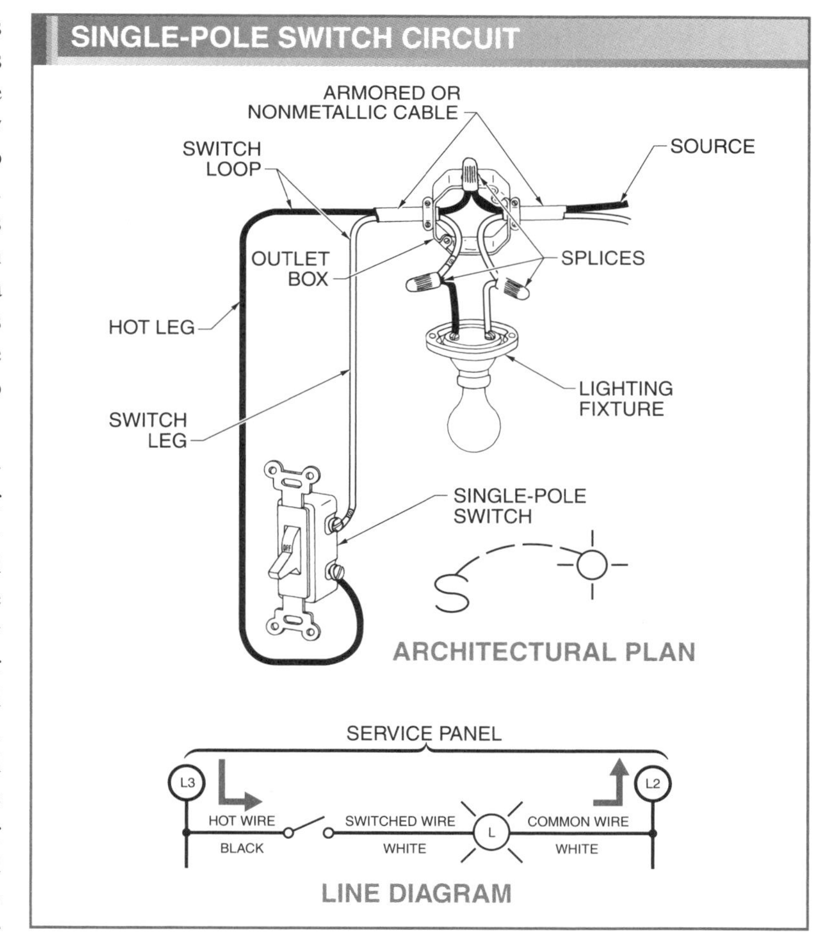

Figure 25-4. The color of wire can vary when wiring with armored or nonmetallic cables.

Three-Way Switches. Three-way switches are used in pairs to control a light or load from two locations. The term "three-way" is the name given to the switch and in no way describes the operation of the switch. The terminals of a three-way switch are identified as common, traveler A, and traveler B. The single terminal at one end of the switch is the common. The common terminal is easily identified because the terminal is darker than the other terminals or is black. The positive or hot wire of a circuit is always connected to the darker terminal. The remaining two terminals are the connecting points for the two traveler wires that connect paired switches together.

A three-way switch operates by moving a pole between two positions. A three-way switch is really 2 two-position switches. When the handle is down, contact is made with terminal 1. When the handle is up, contact is made with terminal 2. Three-way switches have no ON or OFF markings because the handle indicates the position of the contacts from one three-way switch in relation to another three-way switch. The pole position of each switch determines whether or not the circuit is complete (load ON).

When the same respective traveler of each switch is connected, a conducting path is completed, which allows a load (light) to turn ON. When either switch contact is moved, the light will turn OFF because the conducting path is broken. Because either switch can break the conducting path, the load can be turned OFF from two locations. When the light must be turned back ON, either switch can be moved to allow a flow path to be completed. Two three-way switches are typically used for hallways, stairways, and rooms with two entrances.

A lamp that is used to light a staircase (six steps or more) must have a three-way switch at the top and bottom of the staircase. Rooms that have two entrances (or exits) require two three-way switches to control the light(s) from each location. Two three-way switches must be used to control a lamp from two locations. **See Figure 25-5.**

National Wood Flooring Asociation

Three-way switches are installed both at the top and bottom of a staircase to control the lamp used to light the staircase.

Tech Facts

Three-way switches are used in pairs to control a light or load from two locations. Three-way switches are wired using a common conductor and two travelers. Rooms with two entrances often have a three-way switch at each entrance.

Three-way switch circuits have the common terminal of one three-way switch connected to then hot power supply at all times. The common point of the other three-way switch is connected directly to the load (lamp) at all times. The traveler terminals of both three-way switches are connected together. A black wire is used to connect the hot power line to the first three-way switch. Red wires are used for the travelers. The load wire connecting the second three-way switch to the lamp is typically red or black with colored tape or marker used to distinguish the load wire from others.

Residential circuits typically use a green or bare wire for the ground when nonmetallic cable is used. A ground wire is required to ground all non-current carrying metal to earth. Metal conduit may be used to maintain a solid ground connection when metal conduit is used.

Four-Way Switches. A four-way switch is used in combination with two three-way switches to allow control of a load from three locations. One or more four-way switches are used with two three-way switches to provide control of a load from three or more locations. The through-wired type of a four-way switch is the most popular. Any number of four-way switches can be connected into a circuit, but all four-way switches must be connected between two three-way switches. The positions of the switch contacts in relation to each other determine whether a load (lamp or receptacle) is ON or OFF.

For example, a room that contains three entrances requires a switch to turn a lamp ON from any of the three entrances. A four-way switch is placed between two three-way switches to control the lamp from three locations. **See Figure 25-6.**

Two three-way switches and two four-way switches can be used to control a light from four locations. The two four-way switches must always be wired between two three-way switches. Any number of four-way switches can be added to a lamp control circuit to increase the number of control locations.

Tech Facts

Commercial-grade switches should be used for extra protection or longer service. Commercial-grade switches have an increased voltage capability from 120 V to 277 V, horsepower capability from ½ HP to 2 HP, and amperage capability from 15 A to 20 A or 30 A.

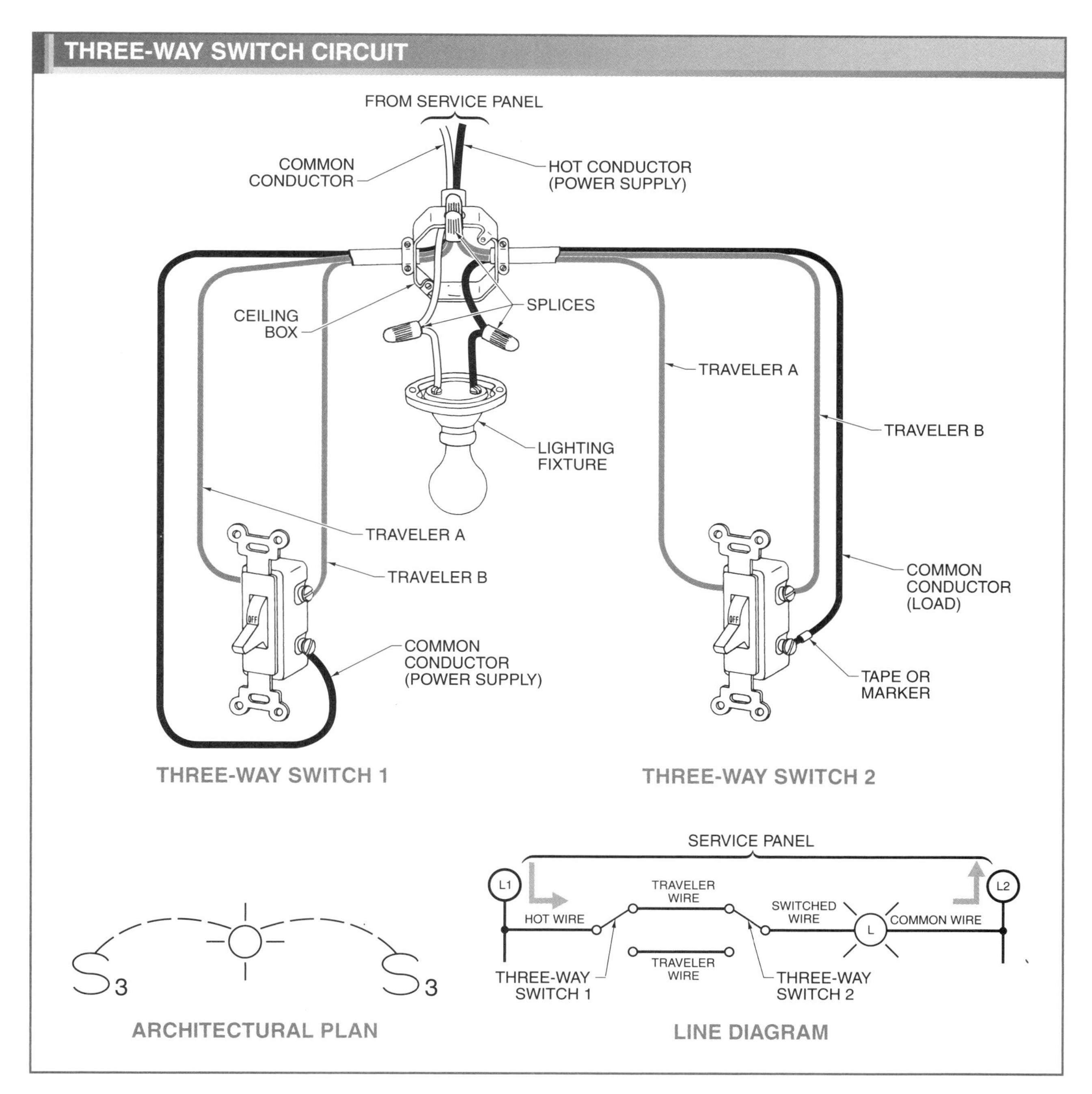

Figure 25-5. Two three-way switches are used to control a lamp from two locations.

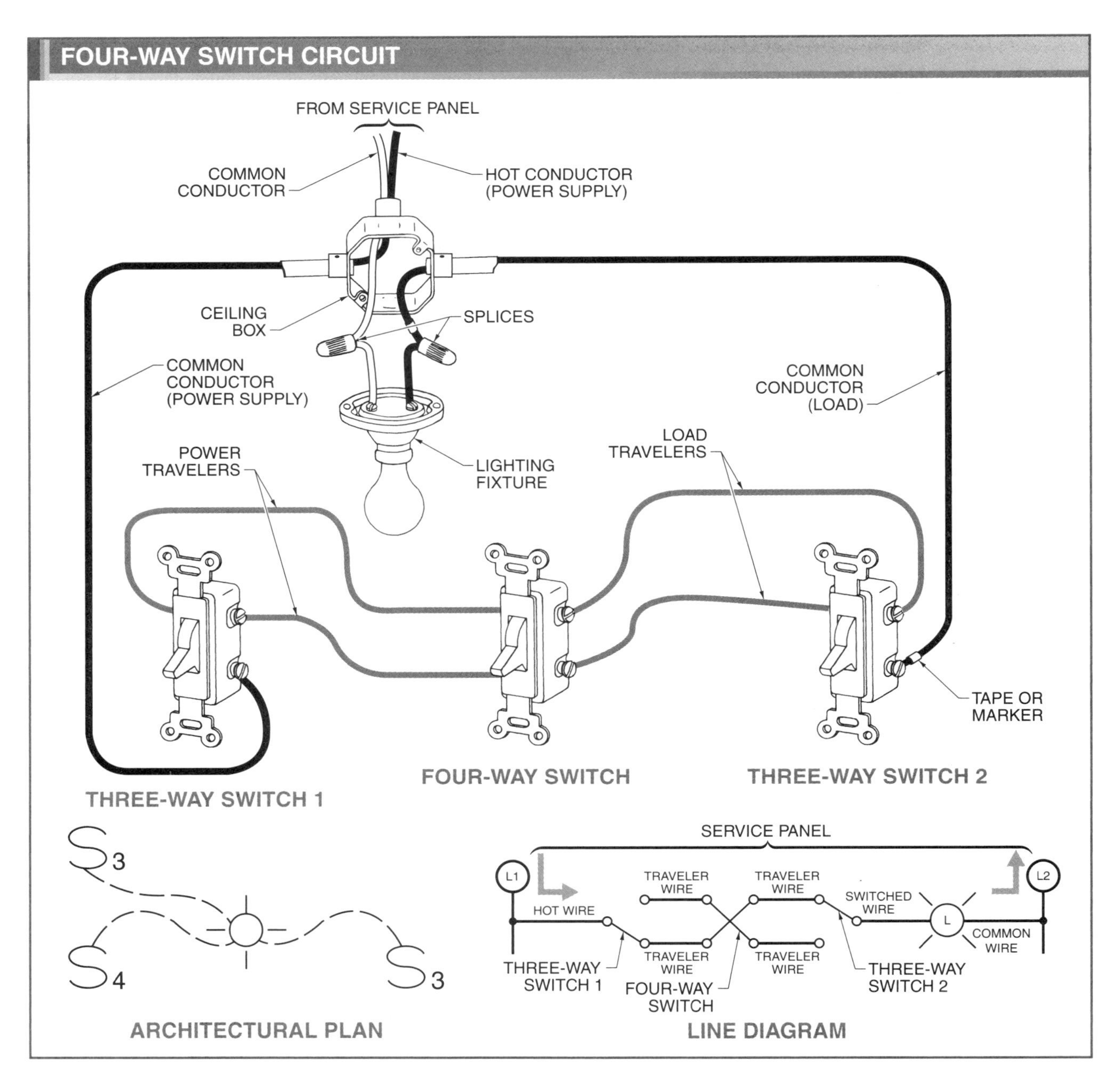

Figure 25-6. Four-way switches add as many control locations to a circuit as required: 1 four-way equals three locations, 2 four-ways equal four locations, 3 four-ways equal five locations, etc.

Double-Pole Switches. A double-pole switch is a device that consists of two switches in one package for controlling two separate loads. Double-pole switches are designed to connect or disconnect (open or close) at the same time. **See Figure 25-7.** Double-pole switches are common for 230 V circuits where both conductors are hot conductors. Double-pole switches open or close the dual path to the 230 V load, turning the load ON or OFF. Double-pole switches look similar to four-way switches, however double-pole switches are distinctly marked with an ON and OFF position.

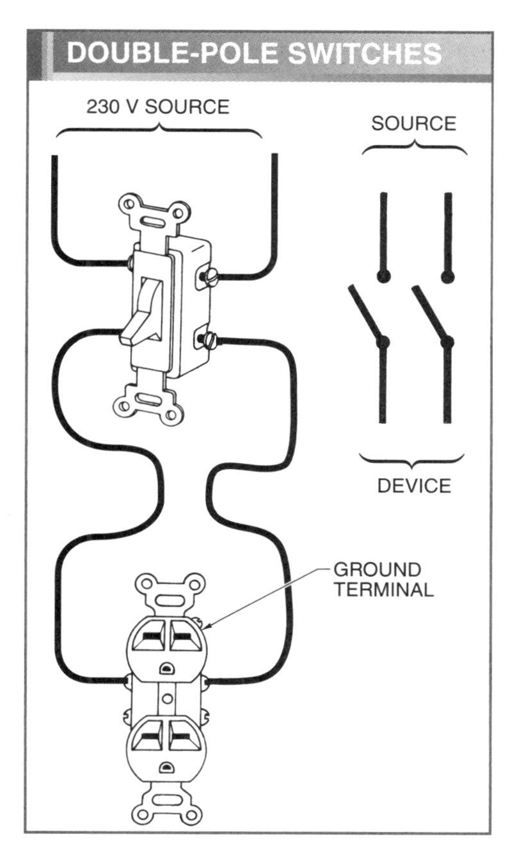

Figure 25-7. Double-pole switches open two lines or circuits simultaneously.

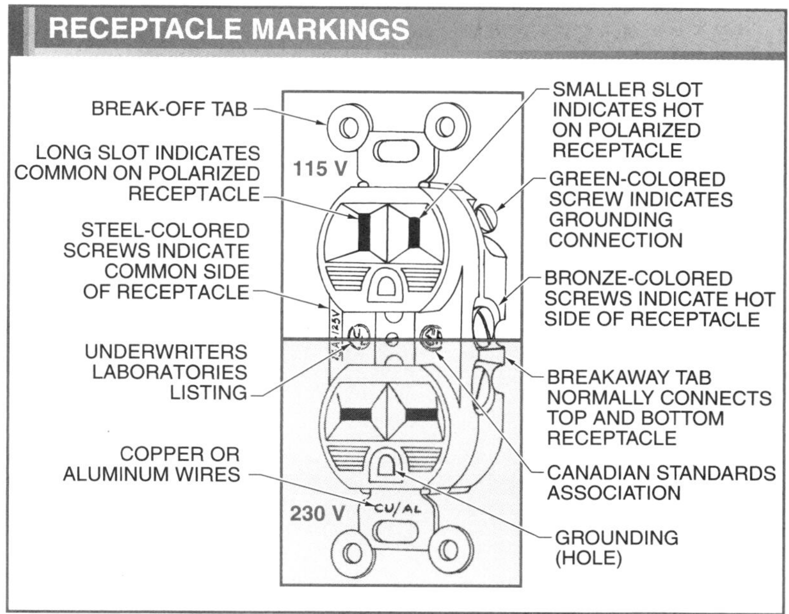

Figure 25-8. Receptacles are installed for the connection of plugs and flexible cords to supply current to portable electrical equipment.

RECEPTACLES

Receptacles are often called convenience outlets. A *receptacle* is a contact device installed for the connection of plugs and flexible cords to supply current to portable electrical equipment.

Receptacle Markings

Receptacles, like switches, convey important information by colors and symbols. In addition, the shapes and positions of the openings (slots) in a receptacle also determine proper use. **See Figure 25-8.**

Shape and Position. The shape and position of the connection slots are used to differentiate between a 15 A, 125 V receptacle and a 15 A, 250 V receptacle. A 125 V receptacle has connection slots that are vertical and typically different lengths. A 250 V receptacle has connection slots that are horizontal and the same size.

A 125 V receptacle with different size connection slots is a polarized receptacle. A *polarized receptacle* is a receptacle in which the size of the connection slots determines the plug connection. The short connection slot is the hot connection and the long connection slot is the common (neutral) connection. Because the slots are designated and a plug can only be connected one way, the receptacle is a polarized receptacle.

Ratings determine the number of contacts and the configuration in which the slots are positioned in the receptacle. **See Figure 25-9.** There are a large number of configurations and diagrams for receptacles encountered in residential construction work.

Receptacles are designed for straight blade or locking type plugs. Locking type receptacles are more common in commercial and industrial applications than in residential work.

Color Coding. Steel, bronze, and green colors are typically used on receptacle terminal screws. A silver-colored screw indicates the neutral terminal, a bronze-colored screw indicates the hot terminal, and a green colored screw indicates the ground terminal.

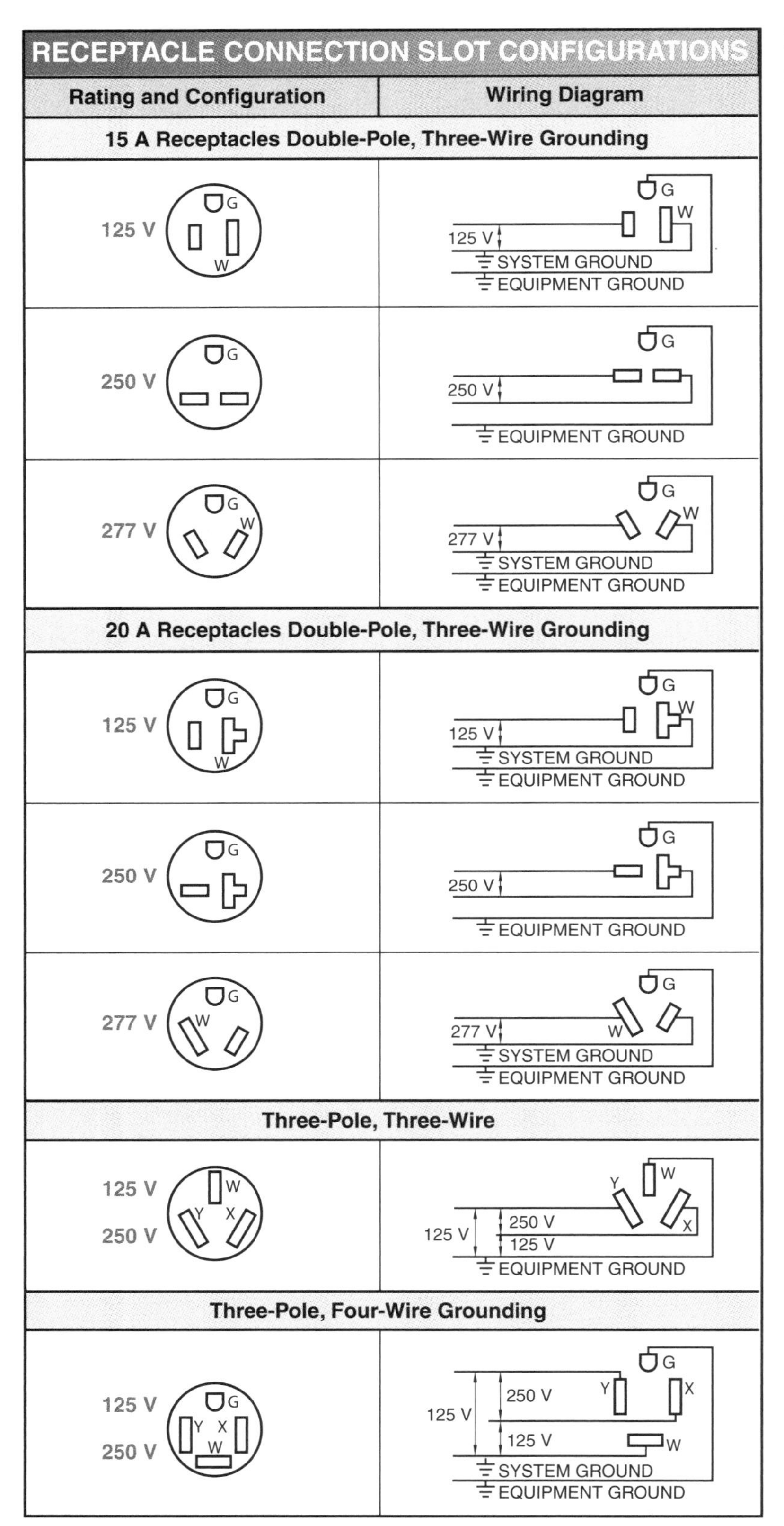

Figure 25-9. Receptacle amperage and voltage ratings determine the number of connection slots and the configuration in which the slots are positioned.

Inscribed Information. Receptacles are marked with "UL", "CSA", "CU/AL", "15 A", and "125 V". Standard receptacles are marked 15 A, 125 or 20 A, 125 V (maximum values). Receptacles marked 20 A, 125 V are typically required for laundry and kitchen circuits. Other 125 V receptacles found in a residence are 15 A.

Receptacles marked CU are used with solid copper wire only. Receptacles marked CU-CLAD are used with copper-coated aluminum wire. Receptacles marked CO/ALR are used with solid aluminum wire.

Types of Receptacles

Receptacles are found in a variety of shapes and sizes depending on the application. Typical receptacles found in residences are standard, isolated ground, split-wired, and GFCI. **See Figure 25-10.**

Standard Receptacles. Standard receptacles include a long neutral slot, a short hot slot, and a U-shaped ground hole. Wires are attached to the receptacle at screw terminals or push-in fittings. A connecting tab between the two hot and two neutral screw terminals provides an electrical connection between the terminals. The electrical connection allows for both terminals to be powered when one wire is connected to either terminal screw.

When wiring a 120 V duplex receptacle, the black or red (hot) wire is connected to the brass-colored screw, the white wire is connected to the steel-colored (common) screw, and the green or bare wire is connected to the green screw. **See Figure 25-11.**

Split-Wired Receptacles. A *split-wired receptacle* is a standard receptacle that has had the tab between the two brass-colored (hot) terminal screws removed. The tab between the two steel-colored (neutral) terminals has not been removed. Split-wired receptacles are used to provide a standard and switched circuit or two separate circuits at the same duplex outlet.

RECEPTACLE TYPES . . .

STANDARD DUPLEX

A grounded receptacle commonly used for 15 A, 120 V appliances in the home

STANDARD DUPLEX

A grounded receptacle commonly used for 20 A, 120 V appliances in the home

COMBINATION HIGH/LOW VOLTAGE

Used with two three-way switches to provide control from three or more locations.

SINGLE GROUNDED 15 AMPERE

Maximum load is 15 A

30 AMPERE

Used on clothes dryers for 120/240 V operation; maximum load is 30 A

TWIST LOCK

Used with twist lock plugs in areas where positive connection is to be maintained

SINGLE GROUNDED 20 AMPERE

Maximum load is 20 A

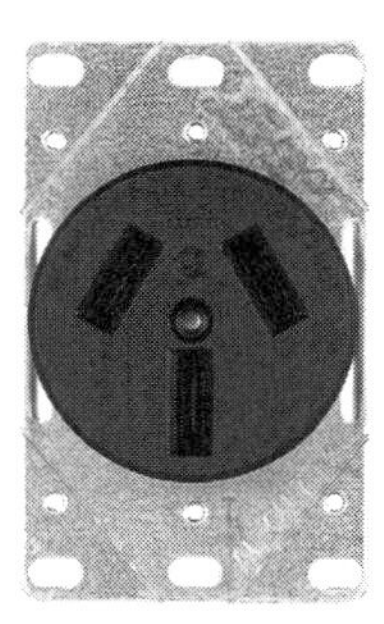

50 AMPERE

Used on electric ranges to supply 120/240 V operation; maximum load is 50 A

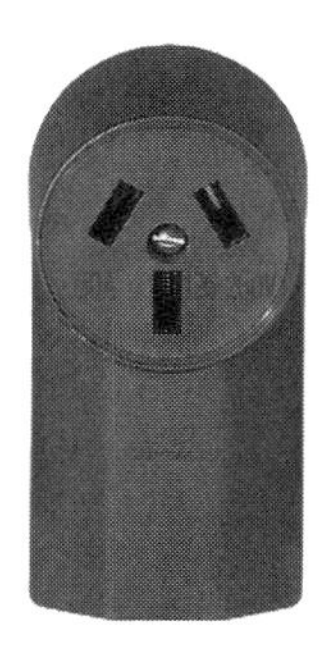

50 AMPERE FLOOR MOUNT

Used on electric devices to supply 120/240 V operation; maximum load is 50 A

Cooper Wiring Devices

Figure 25-10. . . .

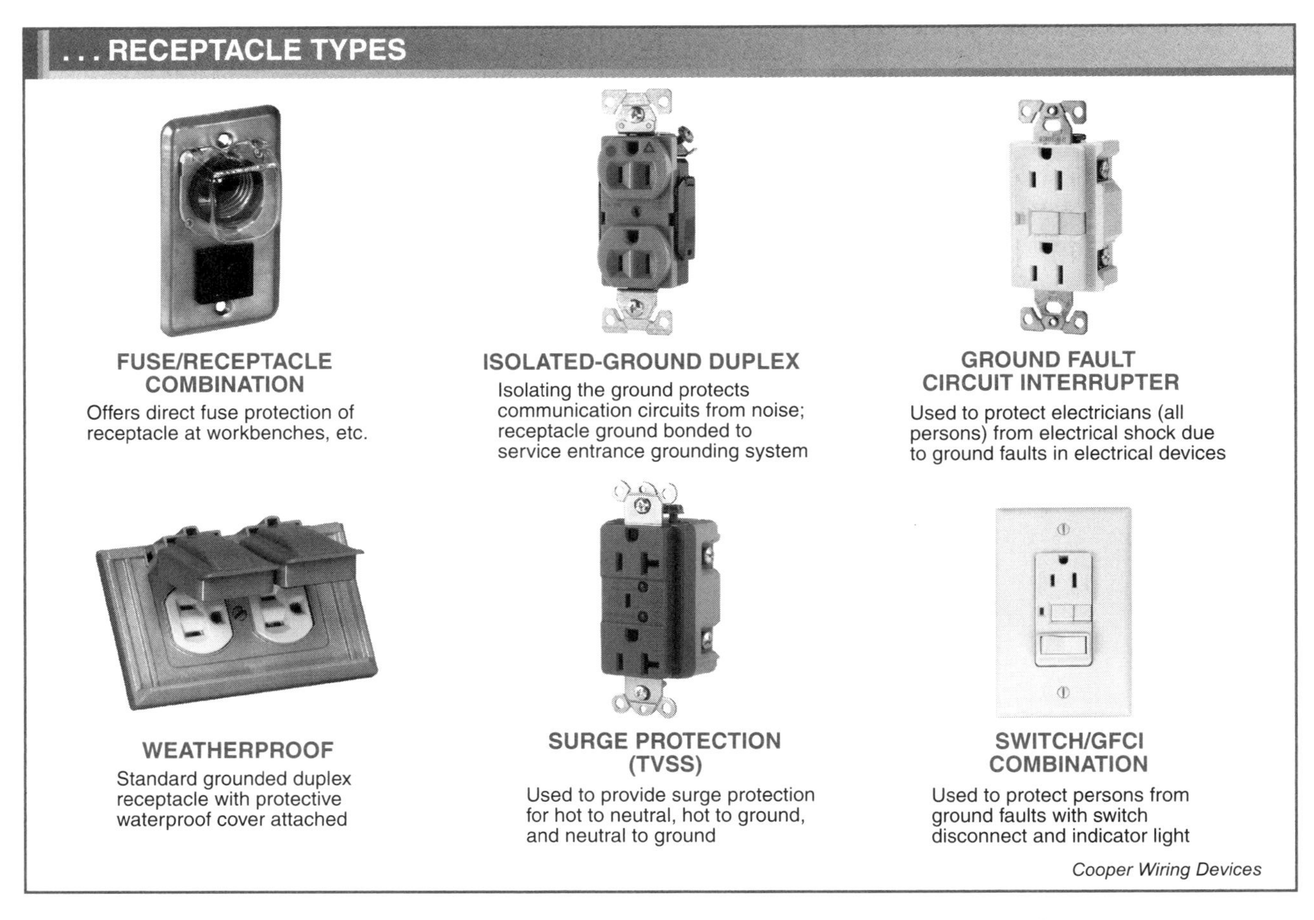

. . . **Figure 25-10.** Receptacles are found in a variety of shapes and sizes depending on the application.

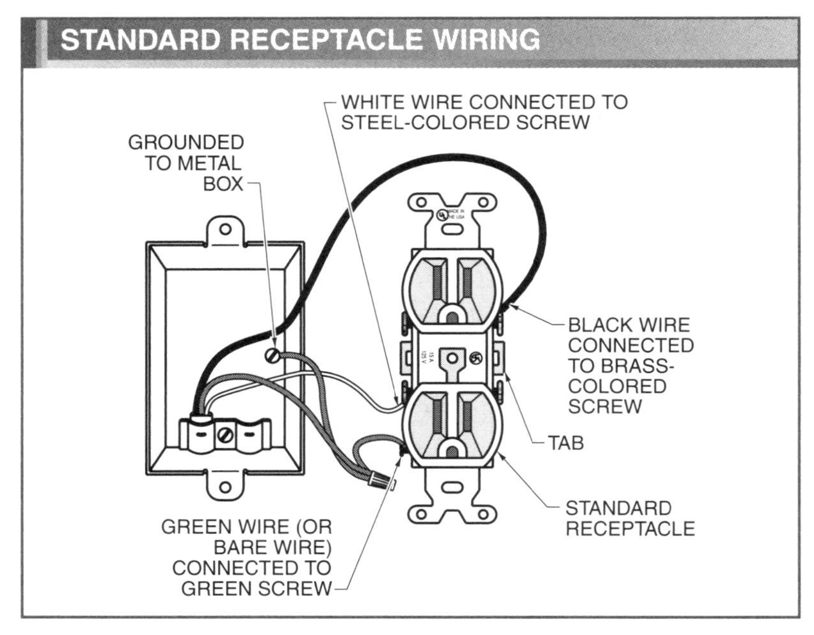

Figure 25-11. Most local codes require that a standard receptacle be installed as a polarized receptacle.

Isolated-Ground Receptacles. In a standard receptacle, the receptacle ground is connected to the common grounding system when the receptacle is installed in a metal outlet box. A common grounding system normally includes all metal wiring, boxes, conduit, water pipes, and the non-current-carrying metal parts of most electrical equipment. The receptacle ground becomes part of the larger grounding system when a piece of electrical equipment is plugged into the receptacle.

The common grounding system acts as a large antenna and conducts electrical noise. This electrical noise causes interference in computers, medical, security, military, and communication equipment.

An isolated-ground receptacle is used to minimize problems in sensitive applications or areas of high electrical noise.

An *isolated-ground receptacle* is a special receptacle that minimizes electrical noise by providing a separate grounding path for each connected device. Isolated-ground receptacles are identified by an orange color. A separate ground conductor is run with the circuit conductors in an isolated grounding system.

GFCI Receptacles. A *ground fault circuit interrupter (GFCI)* is a fast-acting receptacle that detects low levels of leakage current to ground and opens the circuit in response to the leakage (ground fault). A *ground fault* is any current above the level that is required for a dangerous shock. GFCIs provide greater protection than standard or isolated-ground receptacles.

SWITCH AND RECEPTACLE COVERS

To properly trim out a switch or receptacle, the device must be properly covered. **See Figure 25-12.** Switches can be combined with receptacles and trimmed out by using the correct cover plate configuration. Switches and receptacles can also be located outdoors. For outdoor applications, weatherproof covers are required.

Tech Facts

Per 210.8(A) of the NEC®, dwelling units must have GFCI protection in bathrooms, garages, outdoor areas, crawl spaces, unfinished basements, countertops in kitchens, and within 6′ of wet bar sinks.

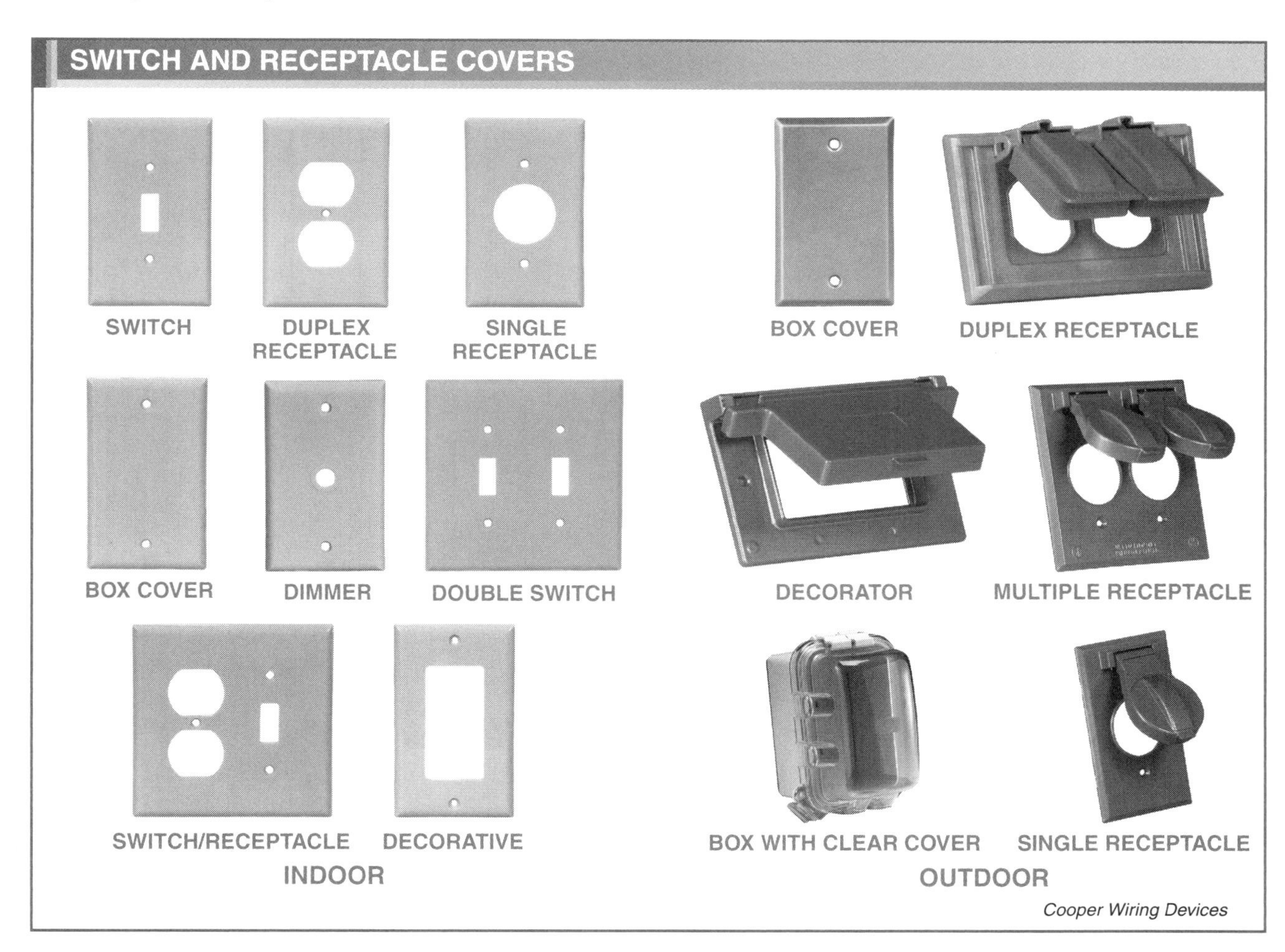

Figure 25-12. Switches and receptacles must be properly covered according to NEC® requirements.

Review Questions

1. Explain the purpose of a switch.
2. List several pieces of information that are included on a switch.
3. What types of switches are commonly used in residences?
4. What is signified by the word “three-way” in the term “three-way switch”?
5. How is a four-way switch used?
6. Describe the design of a double-pole switch.
7. What is a polarized receptacle?
8. What is determined by the ratings on a receptacle?
9. How is a split-wired receptacle different from a standard receptacle?

Nonmetallic-sheathed cable is a type of cable that consists of three or more insulated electrical conductors held together and protected by a strong plastic jacket. The plastic jacket can contain any number of conductors. Metallic-sheathed cable is a type of cable that consists of two or more individually insulated wires protected by a flexible metal outer jacket and is often referred to as armored cable. Conduit is a rugged protective tube, typically metal, through which insulated conductors are pulled.

26 Nonmetallic-Sheathed Cable, Metallic-Sheathed Cable, and Conduit

NONMETALLIC-SHEATHED CABLE

Nonmetallic-sheathed cable is a type of cable that consists of three insulated electrical conductors held together and protected by a strong plastic jacket. The plastic jacket can contain any number of conductors; however, the typical cable contains two or three insulated conductors and a separate bare ground wire. Nonmetallic-sheathed cable is very popular in residential wiring because it is inexpensive and easy to install. **See Figure 26-1.**

Information is printed on the outside jacket of nonmetallic-sheathed cable. The printed information typically includes size of conductors (wire), number of conductors (with or without ground), and type of jacket. The numbers provide information about the size of the wire and the number of conductors in the cable, such as cable that is marked 14/2 or 12/3. The first number (in this case 14 or 12) indicates the size of the wire. The smaller the number, the larger the wire size. The second number (in this case 2 or 3) refers to the number of current-carrying conductors in the cable, not counting the ground wire. To determine the load carrying capacity of any wire, the National Electrical Code® (NEC®) should be referred to for wire size and ampacity.

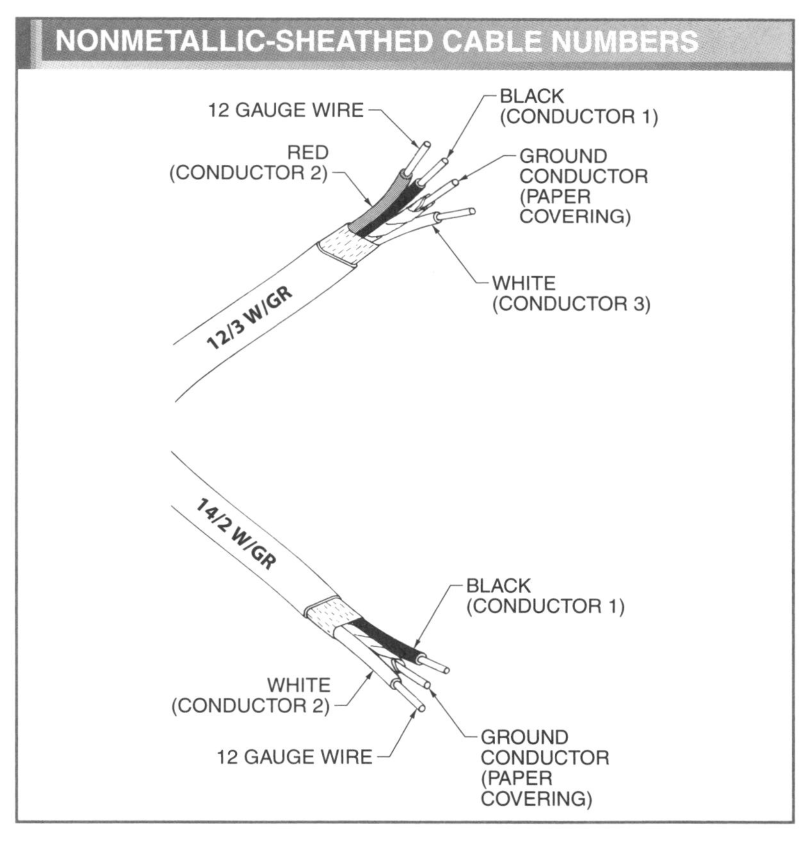

Figure 26-1. Nonmetallic-sheathed cable is manufactured in various wire sizes and with a specified number of conductors.

The letters NM and NMC indicate specific types of nonmetallic cable and the characteristics of the cable. *Type NM cable* is a nonmetallic-sheathed cable that has the conductors enclosed within a nonmetallic jacket and is the type typically used for dry interior wiring. *Type NMC cable* is a nonmetallic-sheathed cable that has the conductors enclosed within a corrosion-resistant, nonmetallic jacket. Article 334 of the NEC® contains detailed information on NM and NMC cable.

The letters WITH GR (with ground) indicate that a separate ground conductor is present. When a nonmetallic-sheathed cable does not clearly specify WITH GR, the cable does not contain a ground wire.

Nonmetallic-sheathed cable is typically sold in boxed rolls of 250′. For small jobs, smaller quantities are available at electrical supply stores. When purchasing nonmetallic-sheathed cable, electricians must specify the length and type of cable required. For example, when an application requires 100′ of cable that has two conductors of 14 gauge wire with a separate ground wire, the cable is specified as 100′ of 14/2 WITH GR Type NM or NMC.

Preparing Nonmetallic Sheathed Cable

Nonmetallic-sheathed cable must be prepared. Typically the cable must be cut to length, have the outer jacket removed to a distance from the end, and have a specified amount of insulation removed from the individual conductors.

Nonmetallic-sheathed cable is easy to install and requires few tools to perform a quality job. Lineman side cutting pliers are capable of cutting NM cable quickly to length. When NM cable has been cut to length, the ends of the plastic jacket must be removed before the cable is inserted into outlet or junction boxes. To open the plastic jacket of an NM cable, a knife is used to cut through the center of the jacket where the ground wire is located. The bare ground wire will guide the path of the knife blade. At least 8″ of the jacket should be removed from each end of the cable.

When great lengths of cable must be prepared, cable rippers are used to save time. Cable rippers have a small, razor-sharp blade designed to penetrate a short distance into the cable to remove only the outer plastic cover. When the outer jacket is cut away, insulation from the individual conductors (wires) can be removed. Various multipurpose wire-stripping tools are used to remove insulation from individual conductors.

Rough-in is the placement of electrical boxes and wires before wall coverings and ceilings are installed. Roughing-in must be performed in such a way that the entire electrical system (wires and boxes) can be easily traced after walls and ceilings are in place. All rough-in work must be carefully checked by the Authority having Jurisdiction (AHJ). Errors in the roughing-in process are compounded once the wall coverings are in place.

Receptacle Boxes

A *receptacle box* is an electrical device designed to house electrical components and protect wiring connections. Depending on the specific application, number of wires, and size of wires, receptacle boxes can have various shapes, sizes, and standard accessories.

Receptacle boxes and junction boxes must be installed at every point in the electrical system where NM cable is spliced or terminated. Receptacle boxes are installed so that every point in the system is accessible for future repairs or additions. Boxes must never be completely covered to the point that a box becomes inaccessible.

Typically three shapes of receptacle boxes are used in residential wiring. The three boxes are square, octagonal, and rectangular. **See Figure 26-2.** Square, octagonal, and rectangular boxes come in various widths and depths, and with various knockout arrangements. Octagonal boxes are typically used in ceilings for lighting fixtures. Square and rectangular boxes are typically used for switches and receptacles. Any of the three are used for junction boxes.

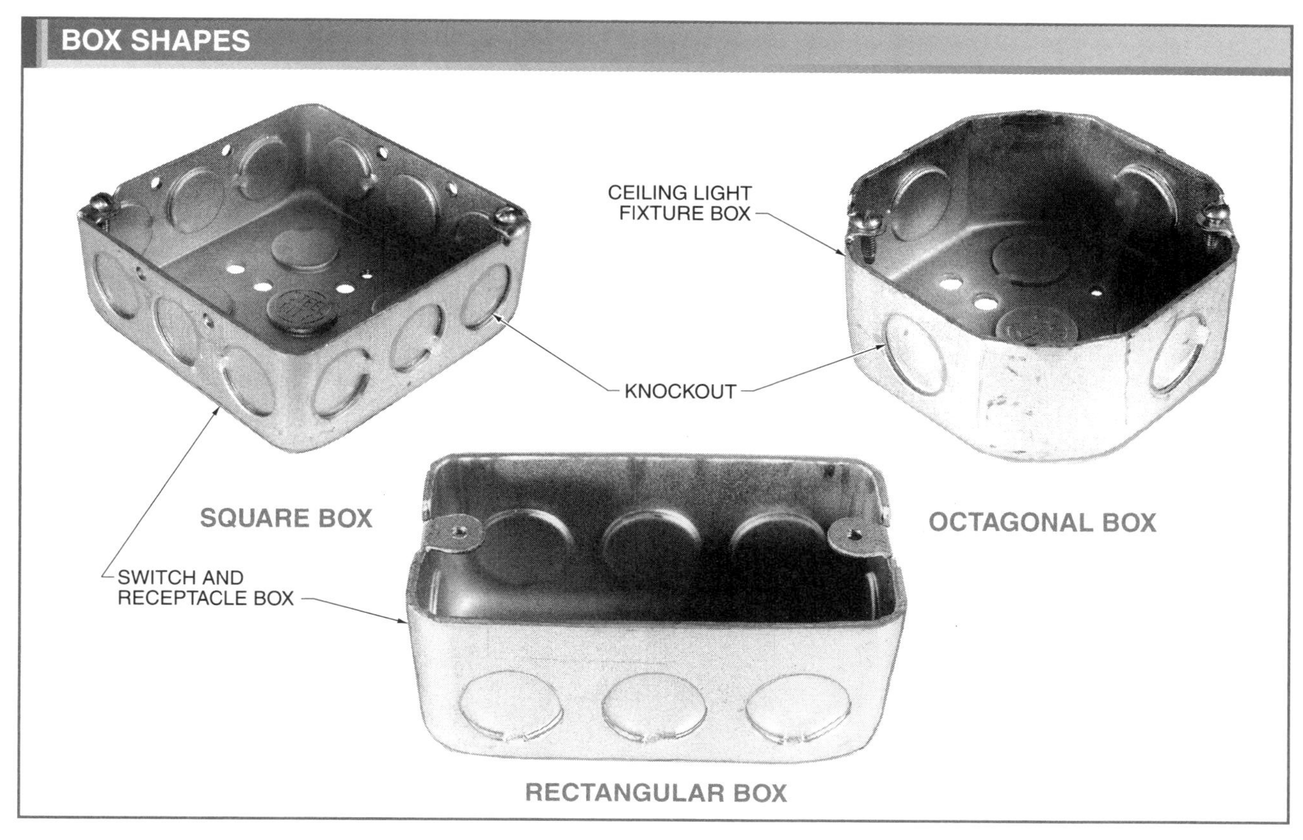

Figure 26-2. Typical boxes found in residential wiring are square and rectangular (for switches and receptacles) and octagonal (for ceiling fixtures).

The size of an outlet box is based on the number and the size of the wires entering the box. More wires and larger size wires require a larger or deeper box. Manufacturer specification sheets must be consulted when ordering boxes to find the appropriate volume (cubic inch capacity) box. Additional information can be obtained from the NEC®.

A *knockout* is a round indentation punched into the metal of a box and held in place by unpunched narrow strips of metal. Knockouts (abbreviated KO) are located in the sides and bottoms of outlet boxes and provide ready access from any direction. Knockouts must be removed to attach fittings for wiring.

Boxes are typically secured by nails. The nails may be part of the box or driven through straps in the box. **See Figure 26-3.** When a box must be mounted between studs and joists, hangers are used. Almost all ceiling outlet boxes are mounted with bar hangers.

When mounting plastic boxes, the box should be positioned against the stud so that the front of the box will be flush with the finished surface of the wall. For example, if the finished wallboard is ½″ drywall, the box should be positioned ½″ past the face of the stud. When the wallboard will be covered by tile, the thickness of the tile must be taken into account.

Tech Facts

Never use a nonmetallic receptacle box for hanging lights or ceiling fans. A metal ceiling-mount receptacle box has large attachment screws and will be more secure than a nonmetallic receptacle box.

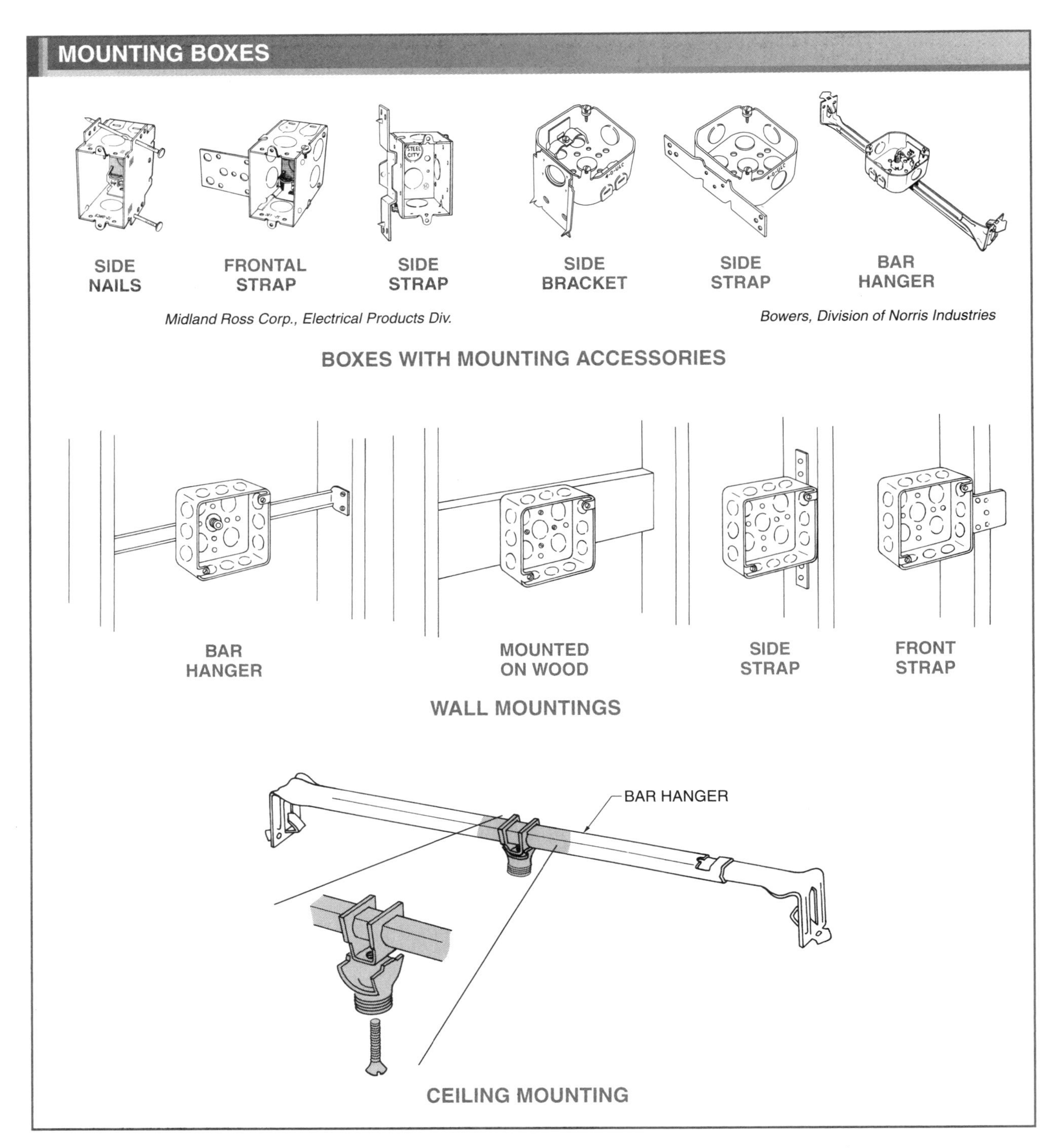

Figure 26-3. A variety of box mounting accessories and methods are used for mounting (securing) boxes.

In each plastic box there is an area where knockout holes can be created. The knockouts can be removed before or after a box is mounted. **See Figure 26-4.** Any sharp edges that might penetrate the plastic sheathing that surrounds the NM cable should be removed. Burrs or sharp edges can typically be removed by rotating a screwdriver around the knockout hole.

When securing cables to a box, at least 8″ of wire must extend beyond the bottom of the box to allow enough wire for making connections to receptacles and switches. When securing nonmetallic cables to studs and joists, a wide range of staples are available. **See Figure 26-5.**

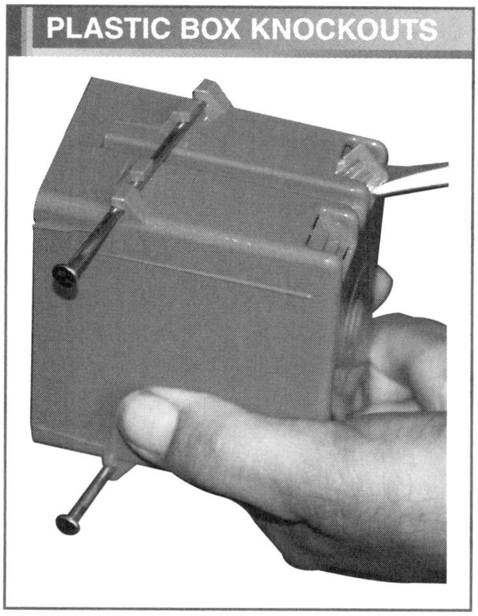

PLASTIC BOX KNOCKOUTS

Figure 26-4. Plastic box knockouts are typically removed with a screwdriver.

Installing Cable Runs

When all holes have been properly drilled, NM cable can be installed quickly and easily. To obtain the proper length of cable required, the cable should be pulled into position before cutting. Positioning the cable ensures that enough cable is pulled.

Cable and conductors should be stripped before being installed in a box. Stripping the wire before installing the wire into a box saves time later when outlets and switches are being installed. Stripping cable is much easier to perform from outside a box than from inside a box. When stripping the cable, leave at least 8″ of wire for rough-in purposes. It is easier to cut off excess wire later than to try to add wire.

Tech Facts

Boxes must have enough volume to account for the receptacle and incoming and outgoing conductors. Nonmetallic boxes are available in sizes of 16 cu in., 18 cu in., 20.3 cu in., and 22.5 cu in. and have the volume stamped on the inside of the box. Having too many conductors in a box is known as a "cable fill violation." Check the manufacturer's catalog for the volume of metal boxes.

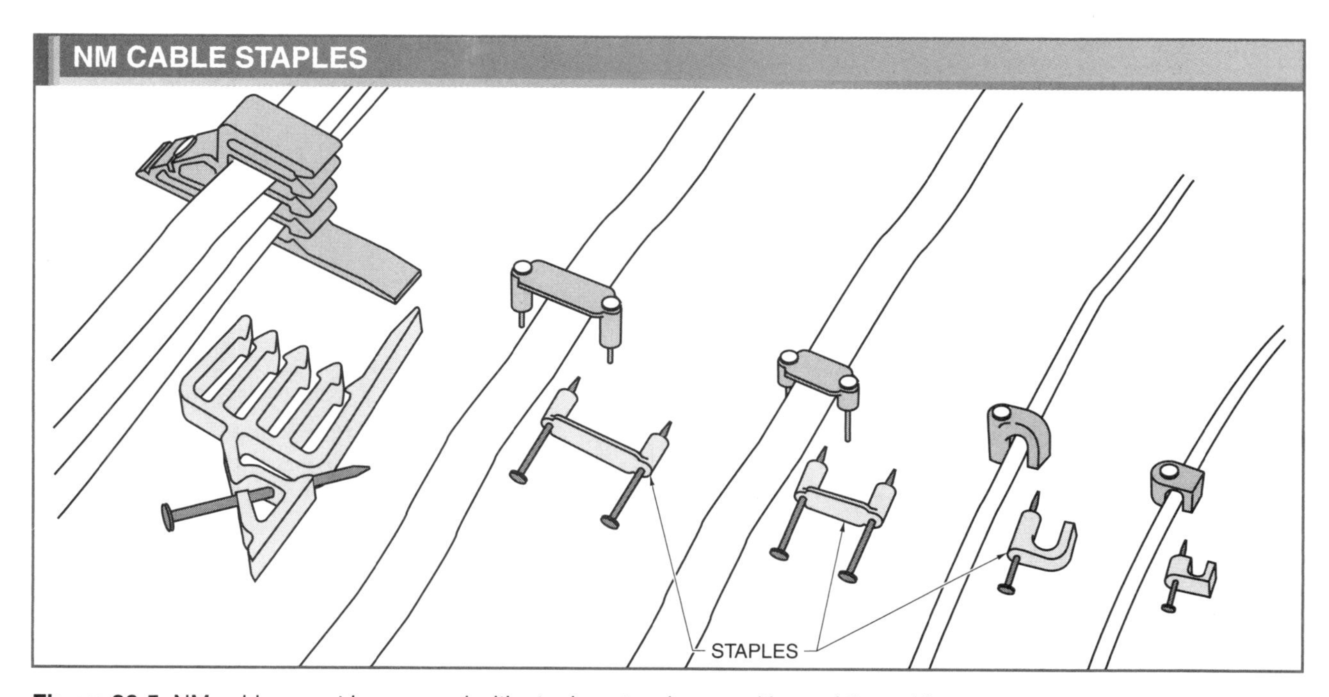

Figure 26-5. NM cables must be secured with staples at various positions of the cable run.

Running nonmetallic cable through stud walls and through floor joists is the most popular way of routing. **See Figure 26-6.** Typically, holes are drilled through the center of the studs and joists. This practice reduces the possibility of a nail penetrating deep enough to puncture the cable. When cable is run through drilled holes, no additional support is required. In some instances, studs may have to be notched instead of drilled. When notches are substituted for holes, the cable run must be protected from nails by a steel plate at least 1⁄16″ thick.

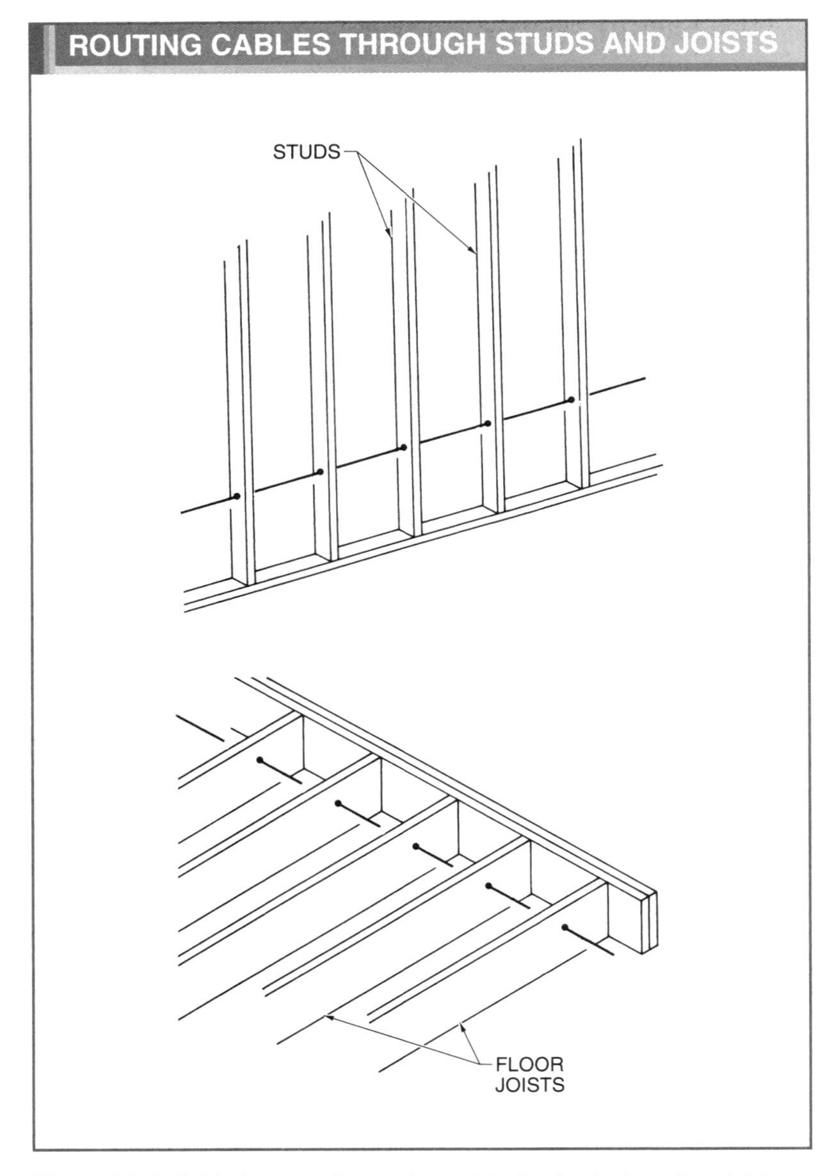

Figure 26-6. Cable is normally run through holes in studs or floor joists.

Cables routed through solid corners by using various methods. One method is to drill holes from each side at an angle to accommodate a cable run. **See Figure 26-7.** Drilling holes from each side will save wire but may be time-consuming if the holes do not line up properly. Another method is to notch the corner studs, using steel plates for protection.

Routing over or under a corner is another method of routing cable. At first glance, the looping under method appears to be the most economical method, but looping under requires extra time. The extra time is due to someone being required to go into the basement or crawl space to send the wire back up through the floor. The over method is accomplished rapidly by one electrician without leaving the room.

Type NM cable may be fished through masonry walls if no moisture is present and the wall is above grade. **See Figure 26-8.** When moisture is present, type NMC cable must be used. Neither NM nor NMC cable can be embedded in poured cement, concrete, or aggregate.

Securing Cable

NM and NMC cable must be supported or secured (stapled) every 4½′ of cable run and within 12″ of a box. NM and NMC cables can also be secured using guard strips. When cable has been pulled into position, the cable must be securely fastened with special staples. Nonmetallic cable staples typically have a plastic strap that reduces the possibility of damage to the cable. The NEC® requires that all cables be secured near an outlet box.

When cable is not supported by holes or notches, but rests along the sides of joists, studs, or rafters, the cable must be stapled every 4½′. Cables must also be stapled under floor joists in such a way that the cables do not present problems later when the basement ceiling is being installed.

Securing Ground Wires. When a cable is secured, a good mechanical and electrical ground must be established to provide electrical protection. The three widely accepted methods of proper residential grounding are component, pigtail, and clip. **See Figure 26-9.**

Component grounding is a grounding method where the ground wire is attached directly to an electrical component such as a receptacle. Component grounding requires that the ground wire be attached before the electrical component is permanently mounted.

Pigtail grounding is a grounding method where two ground wires are used to connect an electrical device to a ground screw in the box and then to system ground. The box ground wire is secured to a threaded hole in the bottom of the box. Once secured, the cable ground wire is pigtailed to the box ground wire.

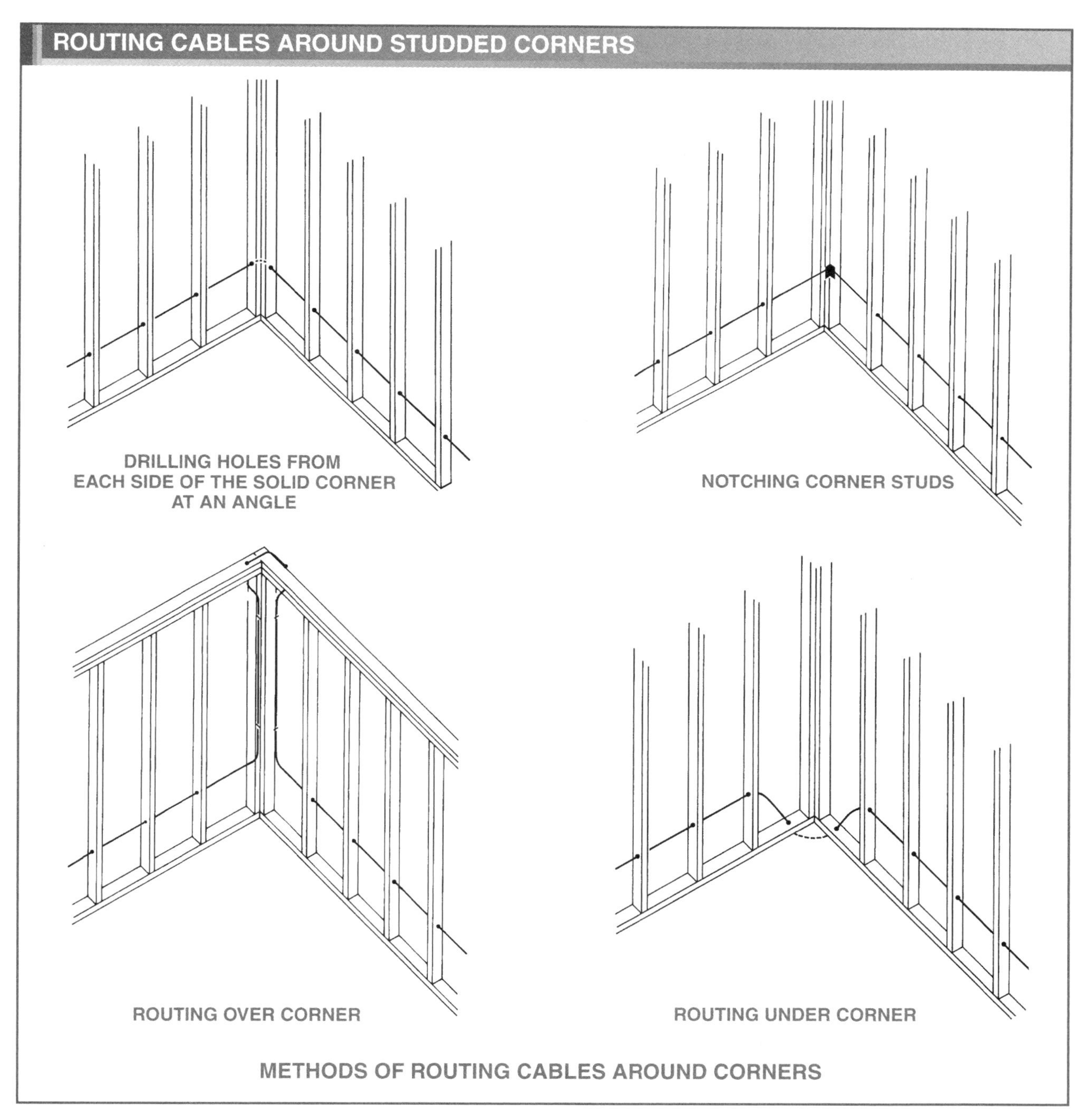

Figure 26-7. Cables can be run through holes or notches in corner studs, or run over or under the corner.

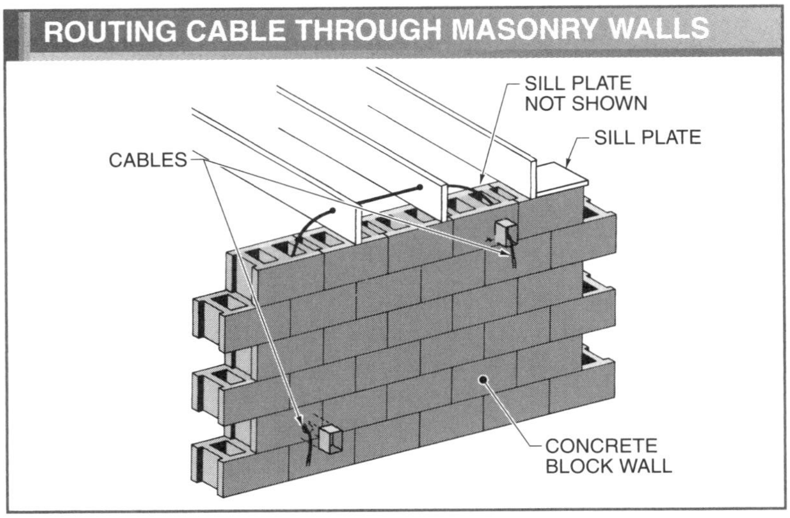

ROUTING CABLE THROUGH MASONRY WALLS

SILL PLATE NOT SHOWN
SILL PLATE
CABLES
CONCRETE BLOCK WALL

Figure 26-8. Type NM cable may be fished in masonry walls when no moisture is present and the cable run is above grade; when moisture is present, type NMC must be used.

Clip grounding, like pigtail grounding, can be secured as the cable is put in place. *Clip grounding* is a grounding method where a ground clip is slipped over the ground wire from the electrical device. The ground wire and ground clip are secured with pressure using a screwdriver.

Securing Cables in Outlet Boxes. Cables are secured to outlet boxes by clamps. The three types of clamps typically available for nonmetallic cables are saddle, straight, and cable connectors. **See Figure 26-10.** Saddle clamps and straight clamps are part of an outlet box when manufactured. Cable connectors are installed by electricians. Locknuts are installed so that the points of the nut point inward to dig firmly into the metal box.

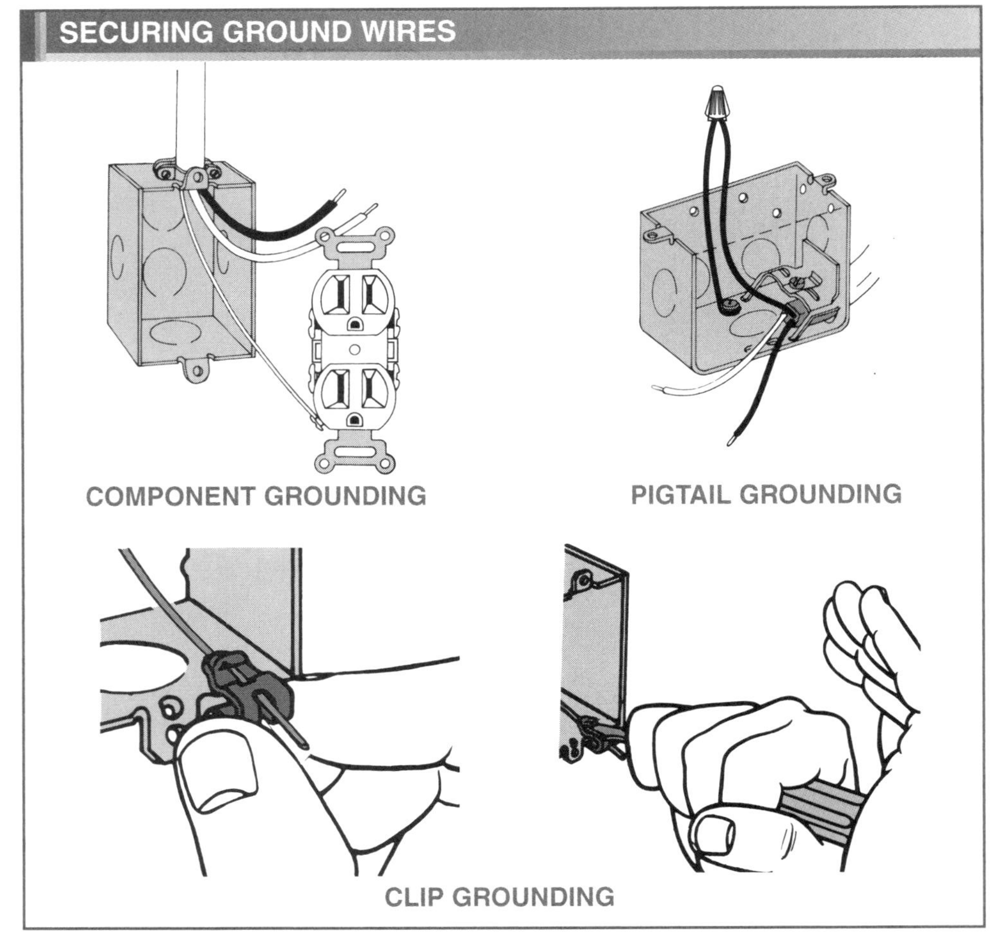

Figure 26-9. The three widely accepted methods of proper residential grounding are component grounding, pigtail grounding, and clip grounding.

When inserting cable through clamps, care must be taken not to damage any insulation by scraping the cable on the rough surfaces and edges. Care must also be taken when tightening cable clamps. Too much pressure can cause the clamp to penetrate the insulation and cause a short circuit.

METALLIC-SHEATHED CABLE

Metallic-sheathed cable is a type of cable that consists of two or more individually insulated wires protected by a flexible metal outer jacket. Metallic-sheathed cable is often referred to as armored cable or BX. Armored cable provides additional protection to wires by using a metal outer jacket. The trade name BX was used to denote the armored cable produced by one specific manufacturer. The BX trade name became so popular that even armored cable produced by other manufacturers was called BX.

Armored cable has a typical construction. In addition to insulation, each current-carrying conductor has a paper wrapping. The paper provides additional protection within the cable but is removed from the conductors when the cable is being installed into boxes. **See Figure 26-11.**

A *bonding wire* is an uninsulated conductor in armored cable that is used for grounding. The bonding wire is in contact with the flexible metal outer jacket to assure a proper conducting (ground) path along the entire length of an armored cable. Bonding wires are typically made of aluminum or copper.

As with nonmetallic cable, armored cable is purchased with a specific number of current-carrying conductors. Armored cable is typically sold in coils of 250′. When armored cable is ordered, the cable is typically specified as 12/3 or 14/2 to indicate the wire size (8, 10, 12, or 14) and the number of current carrying conductors (2, 3, 4, or 5) required. When designating armored cable, the phrase "with ground" is not required because the outer metal jacket and bonding wire automatically establish a conductive grounding path.

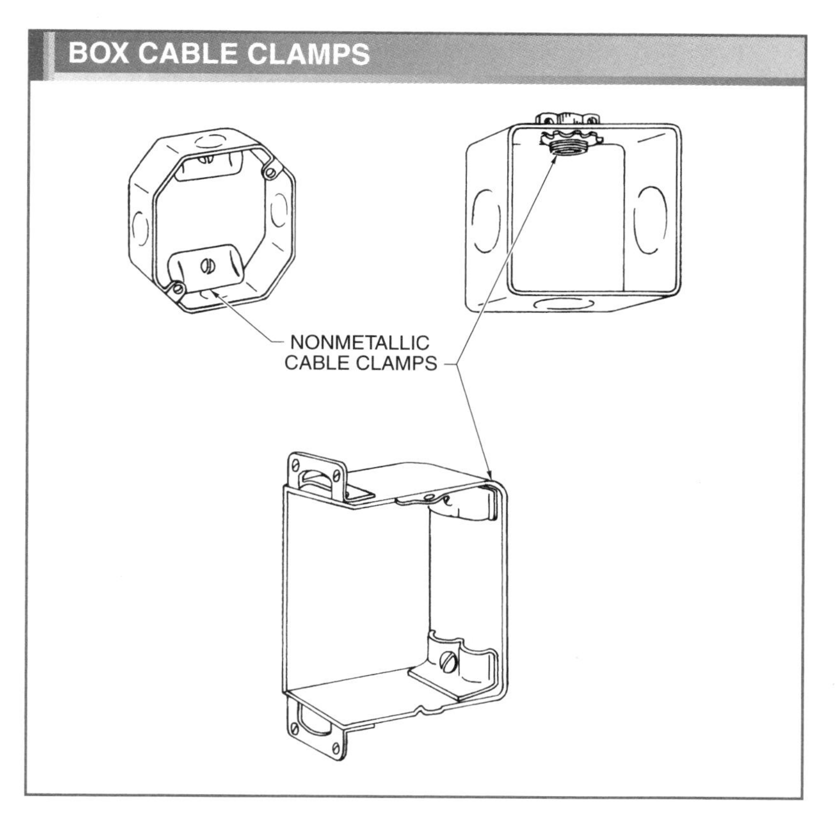

Figure 26-10. Various types of cable clamps are used to secure nonmetallic cable to outlet and junction boxes.

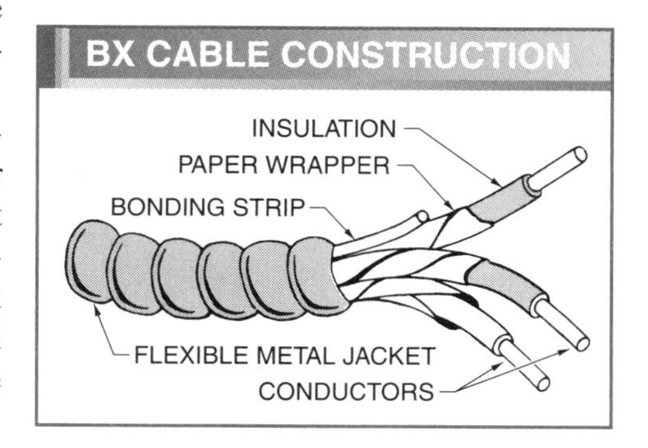

Figure 26-11. BX cable has a bonding strip, and each insulated conductor of the cable is wrapped with paper for extra protection.

The NEC® recognizes three types of armored cable for residential work: AC, ACT, and ACL. AC and ACT armored cables are used in dry locations. Both AC and ACT may be fished through the air voids of masonry walls when the walls are not exposed to excessive moisture.

Both AC and ACT may be used for underplaster electrical extensions. ACT armored cable is used for either exposed work or concealed work.

ACL armored cable is embedded in concrete or masonry, run underground, or used where gasoline or oil is present. ACL armored cable is lead-covered to provide additional protection for masonry and underground applications. Article 332 of the NEC® contains detailed information on ACL armored cabling.

Preparing Armored Cable

Installing armored cable requires many of the same techniques and equipment used when working with nonmetallic cable. However, armored cable requires more attention when cutting and splicing because of the sharp metal edges.

Most electricians use a hacksaw to cut through the metal jacket of armored cable. The first step is to cut through the outer armor (one of the convolutions) at a 45° angle about 6″ to 8″ from an end. To avoid damage to the insulation of conductors, care must be taken not to cut too deeply into the jacket of the cable. When the armored cable is cut, the cable is separated by twisting the two sections apart. The armored cable must be carefully flexed until the cable breaks.

Twisting the cable can open the convolutions enough so that tin snips or cable cutters can be inserted. Armored cable cutters are used to avoid damaging the insulation on the conductors. The use of protective gloves helps protect hands from cuts and scratches. When the outer armor and conductors are cut, a second trimming cut may be required to remove any of the metal outer jacket that is bending outward. A center cut is used to separate the armored cable while side cuts prepare each end for mounting the cable to electrical boxes.

Whenever armored cable is cut, an unavoidable sharp edge remains on the ends of the metal armor. To avoid any damage to the insulation of the conductors, anti-short bushings must be installed. **See Figure 26-12.**

An *anti-short bushing* is a plastic or heavy fiber paper device used to protect the conductors of armored cable. The bushing covers the sharp edges at the ends of the armor to reduce the possibility of damage to conductor insulation.

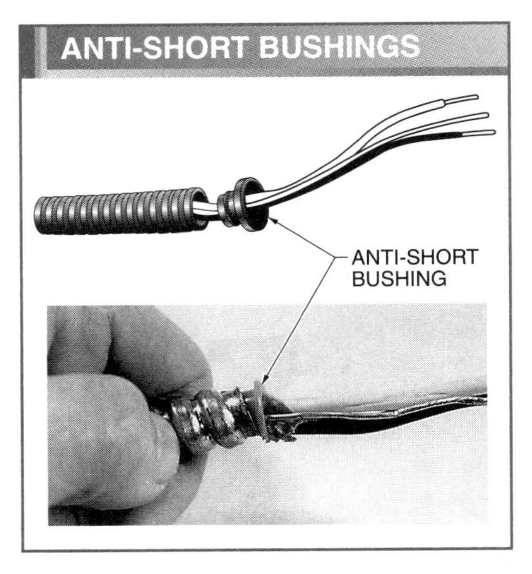

Figure 26-12. To protect conductors from the sharp edges of cut armored cable, plastic anti-short bushings must be used.

To rough-in armored cable, the cable must be pulled, cut, secured, and grounded properly. When roughing-in armored cable into studded walls, the cable must be pulled into position first and then cut to length. Pulling the armored cable into position, then cutting to length, reduces waste. To avoid twisting and kinking the cable, it should be unwound from the center of the coil. After the cable has been pulled into position, three cuts are made in the armor about 6″ to 8″ apart. The entire cable is cut through with a center cut.

The two side cuts are cut through the metal armor (sheath) only; the conductors and insulation are left intact. One end of the cable is used to terminate the run previously started, and the other end is used to start the next run. When wires are concealed in walls, cable runs can be shortened by allowing the cable to run wild, thus saving an amount of cable.

See Figure 26-13. To avoid breaking armored cable, sharp bends should not be made. Typically a bend in armored cable must have a minimum radius at least 5 times the diameter of the cable.

Armored cable is secured in place with cable clamps or cable connectors. **See Figure 26-14.** Cable clamps are typically part of a box, while cable connectors are installed before a box is wired. A right angle connector (90° connector) is used where the radius bend of the armored cable would be so tight that the cable would snap if forced into the bend.

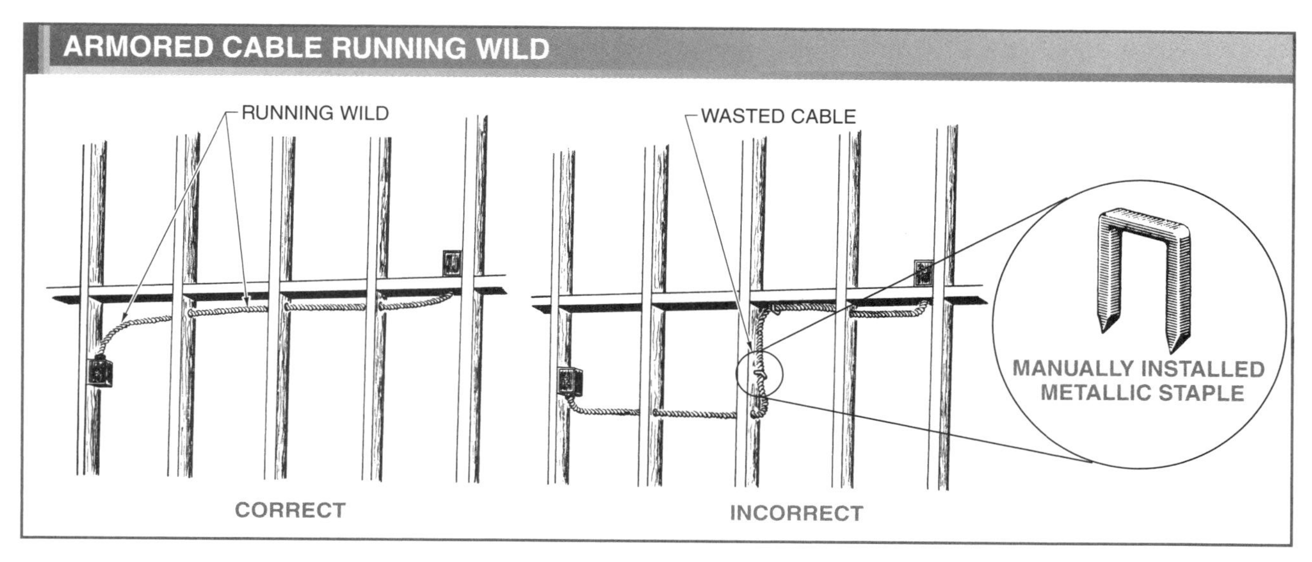

Figure 26-13. The length of an armored cable that is concealed in a wall can be shortened by allowing the cable to run wild.

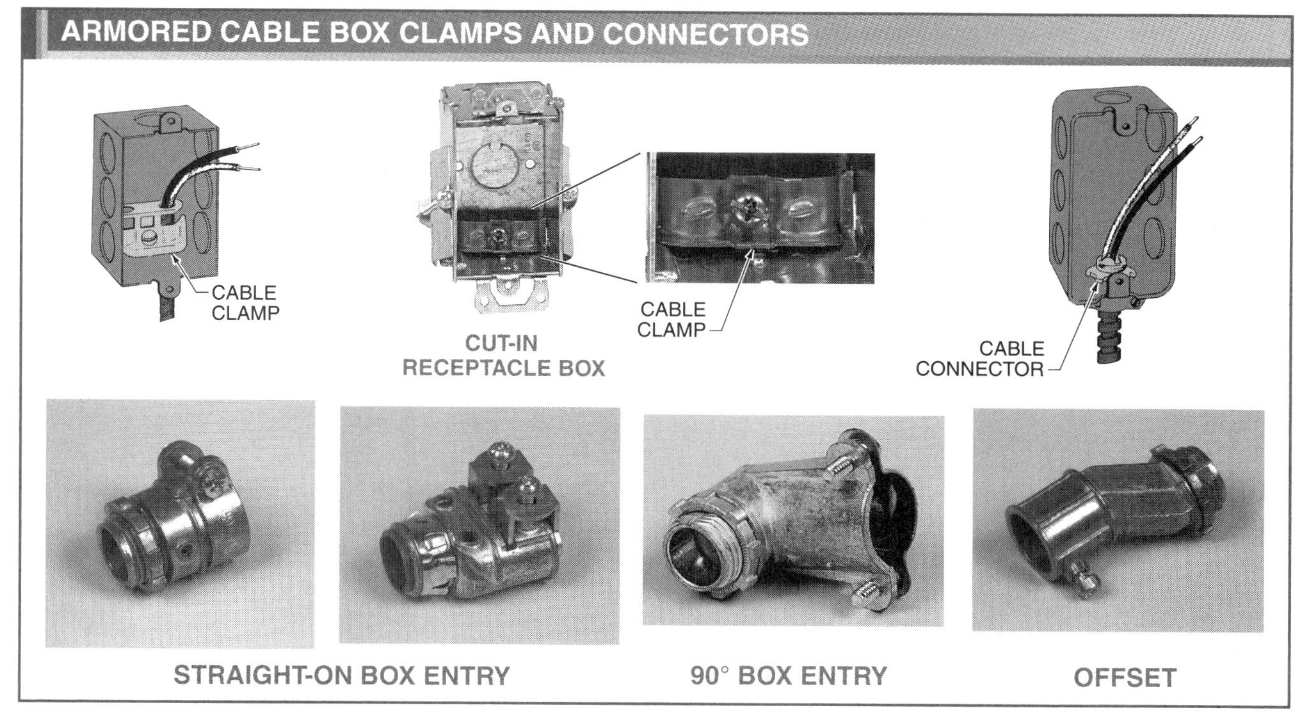

Figure 26-14. Armored cables can be secured to a box with cable clamps, which are part of the box, or cable connectors, which are installed at wiring.

Grounding Armored Cables

Grounding armored cable requires that the metal jacket and bonding wire be firmly secured. **See Figure 26-15.** A typical method used to ground armored cable is to secure the bonding wire of the cable between the connector and box. Other methods of grounding armored cable are the same as with grounding nonmetallic cable.

CONDUIT

Conduit is a rugged protective tube (typically metal) through which insulated conductors are pulled. Although several types of conduit are used in the electrical industry, only three types are commonly found in residential installations. The three most common types of conduit are electrical metallic tubing (EMT), rigid conduit, and flexible metal conduit. **See Figure 26-16.**

Electrical Metallic Tubing (EMT)

Electrical metallic tubing (EMT) is a light-gauge electrical pipe often referred to as thin-wall conduit. EMT has a wall thickness that is about 40% the thickness of rigid conduit. Because EMT is lighter, easier to bend, and requires no threading, EMT is typically used for a complete residential layout. Additional information on the installation of EMT can be found in Article 358 of the NEC®.

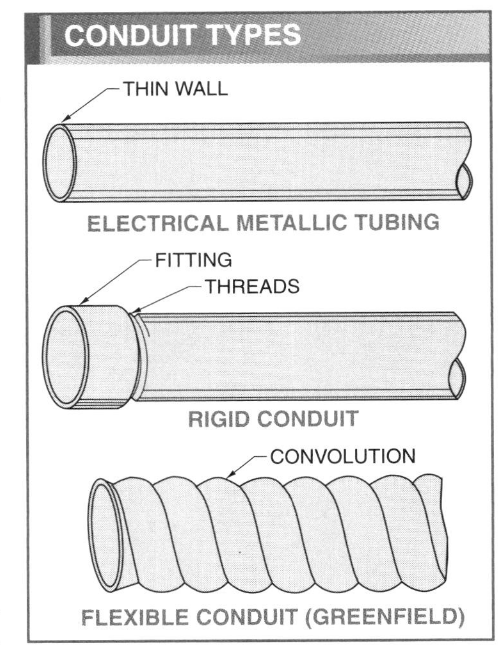

Figure 26-16. Electrical metallic tubing (EMT) is the most common conduit used in residential wiring.

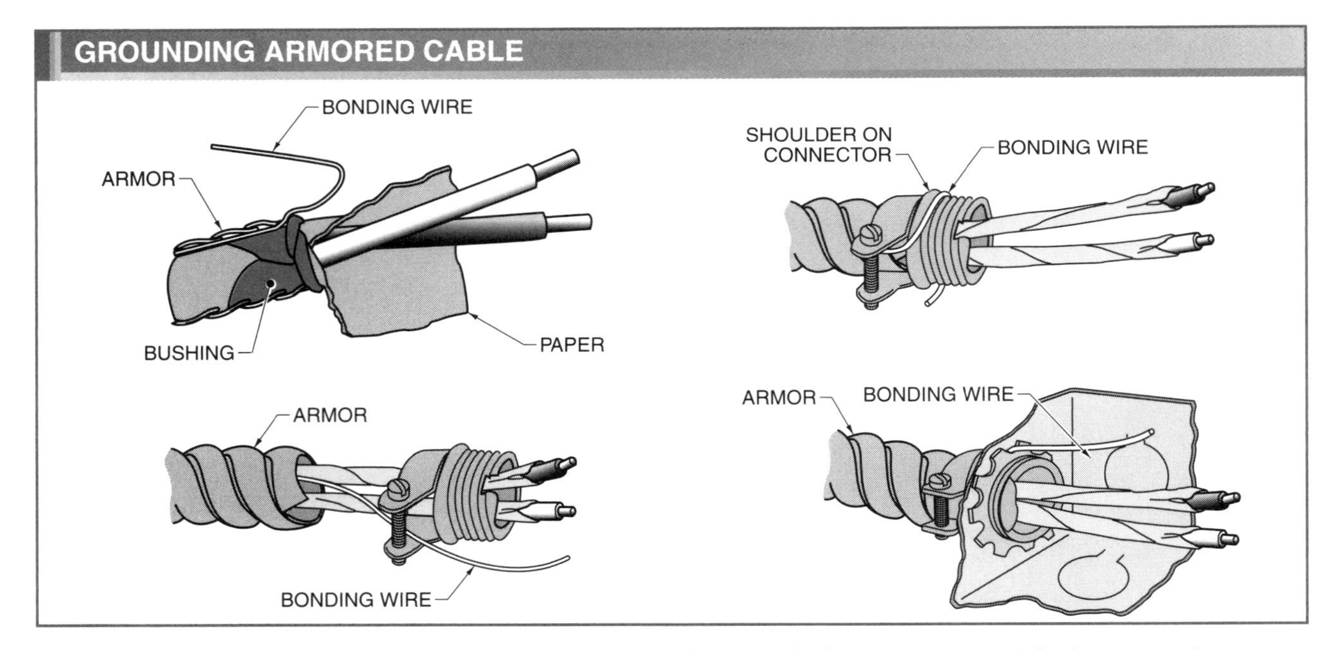

Figure 26-15. The bonding wire of armored cable is secured between the box connector and the box to create a proper grounding path.

Rigid Metal Conduit (RMC)

Rigid metal conduit (RMC) is a heavy-duty pipe that is threaded on the ends much like standard plumbing pipe. For residential use, rigid conduit is limited mainly to risers for service entrances. Rigid conduit can be cut with a standard hacksaw and must be threaded with a standard cutting die (plumbing die) providing 3/4″ per foot taper. Many electrical supply houses carry a number of conduit sizes in various lengths that are prethreaded and ready for use. Additional information on the installation of rigid metal conduit can be found in Article 344 of the NEC®.

Flexible Metal Conduit (FMC)

Flexible metal conduit (FMC) is a conduit that has no wires, and can be bent by hand. The conductors (wires) are not installed until the system is complete. Generally, flexible metal conduit is used where some type of movement or vibration is present. For example, flexible conduit is suited for wiring to a motor. Flexible conduit (Greenfield) is also used where other conduits might be difficult to bend. Additional information on the installation of flexible metal conduit can be found in Article 348 of the NEC®.

Installing Electrical Metallic Tubing (EMT)

EMT is typically purchased in bundles of 100′. The bundles contain ten lengths of conduit, each 10′ long. Most electrical supply houses warehouse conduit; therefore, individual lengths of 10′ can typically be purchased. The trade sizes for conduit range from 1/2″ to 4″. **See Figure 26-17.**

All sizes of conduit have limitations on the size and number of conductors that are allowed in the conduit for an application such as residential wiring. Further information on the allowable number of conductors per size of conduit can be obtained from Annex C of the NEC®. **See Figure 26-18.**

Tech Facts

Conduit fittings, such as connectors and elbows, are the weak points of a raceway system. Fittings may loosen or corrode and destroy the grounding path.

EMT CONDUIT SPECIFICATIONS

Inside Diameter Nearest 1/16″	Inside Diameter	Outside Diameter	Trade Size in Inches
5/8	0.622	0.706	1/2
13/16	0.824	0.922	3/4
1 1/16	1.049	1.163	1
1 3/8	1.380	1.508	1 1/4
1 5/8	1.610	1.738	1 1/2
2 1/16	2.067	2.195	2
2 3/4	2.731	2.875	2 1/2
3 3/8	3.356	3.500	3
3 13/16	3.834	4.000	3 1/2
4 5/16	4.334	4.500	4

Figure 26-17. EMT conduit is typically specified by nominal size.

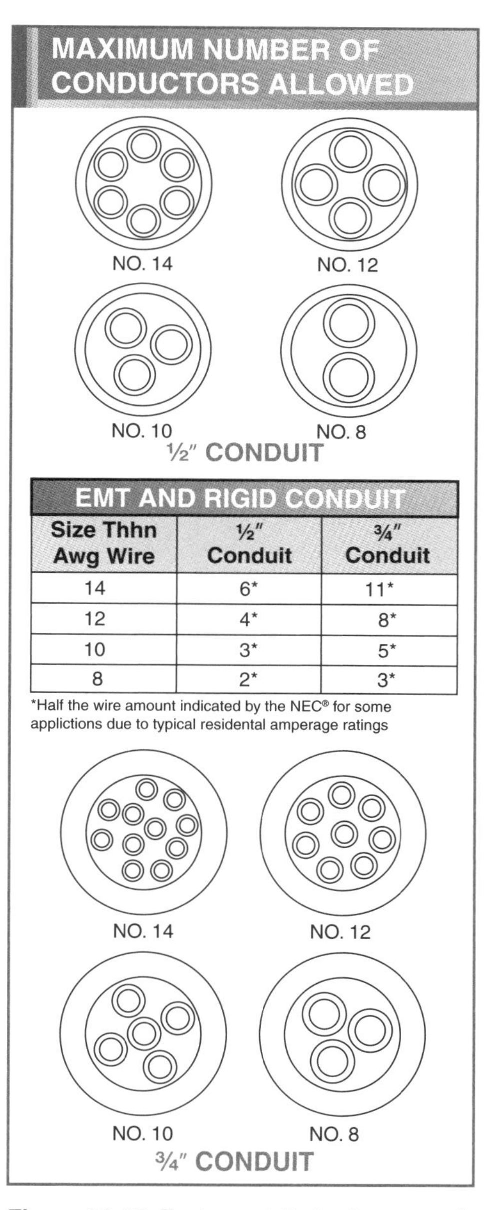

EMT AND RIGID CONDUIT

Size Thhn Awg Wire	1/2″ Conduit	3/4″ Conduit
14	6*	11*
12	4*	8*
10	3*	5*
8	2*	3*

*Half the wire amount indicated by the NEC® for some applictions due to typical residental amperage ratings

Figure 26-18. Each conduit size has a maximum number of wires allowed.

Conduit is typically cut to length using a hacksaw. However, all the rough edges from cutting must be removed before wire can be pulled through conduit. Removing the rough edges is called "deburring" and is accomplished with a conduit deburring tool designed for this purpose. Conduit can also be deburred with a file or a special deburring tool called a reamer.

Bending EMT conduit properly requires the aid of a hand conduit bender. A hand rigid or heavywall conduit bender is sometimes referred to in the trade as a "hickey." A hand conduit bender has high supporting sidewalls to prevent flattening or kinking of the EMT conduit.

Conduit Bends. To create a 45° angle (bend), a hand bender is placed on EMT conduit. The handle of the bender is raised until the handle is in the vertical position. The handle is pulled until the angle required (45°) is completed. The hand bender is removed, releasing the EMT conduit with a smooth-flowing 45° bend. Hand benders also have a long arc that permits making 90° bends in a single sweep without moving the bender to a new position. **See Figure 26-19.**

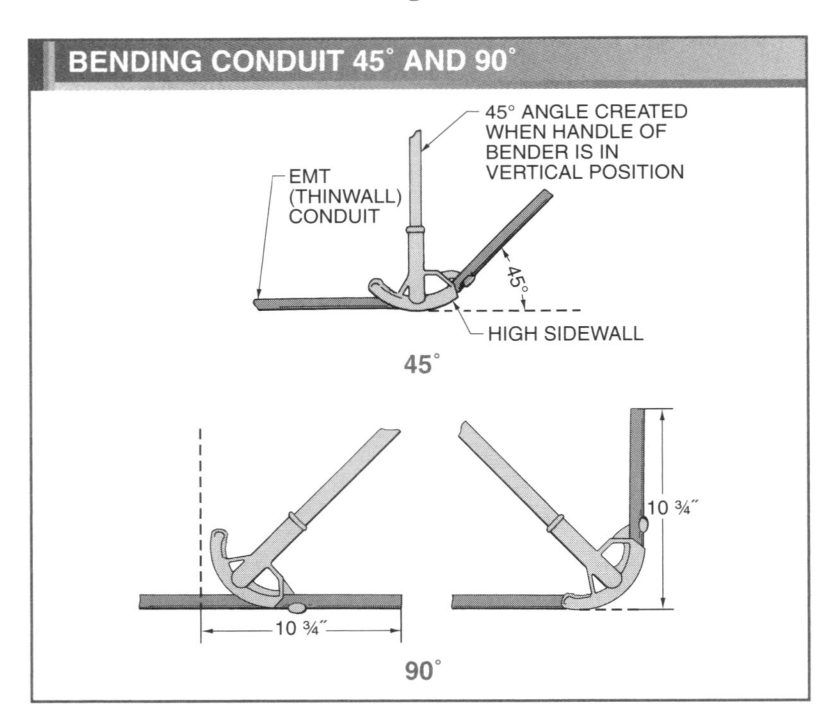

Figure 26-19. Hand conduit benders are specifically designed to bend EMT conduit.

Ninety degree bends are used to turn corners or to form a predetermined length for use as a stub up through a floor that will be added onto later. The typical method for laying out a 90° bend requires that the conduit be marked to a predetermined measurement.

When running conduit from one outlet box to another, the run can require that two 90° bends be made to one piece of conduit. The use of two 90° bends is called back-to-back bending. To make back-to-back bends, the first bend is made as a regular 90° bend. **See Figure 26-20.** Then the distance between the boxes is laid out on the conduit as distance *D*. The direction of the hand bender is reversed and point A of the bender is placed over the mark at distance *D*. Raising the hand bender to form a normal 90° bend finishes the second right angle. When creating back-to-back bends, individuals must ensure that both bends line up with each other (are on the same plane).

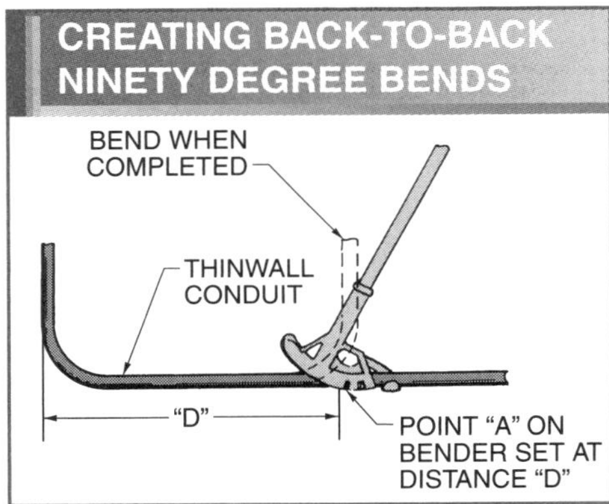

Figure 26-20. When creating back-to-back bends, electricians must ensure that both bends line up with each other (are on the same plane).

Conduit runs are often interrupted by obstructions requiring changes in direction that cannot be accomplished by using 45° and 90° bends. An *offset* is a compound bend in conduit used to bypass many types of obstructions. **See Figure 26-21.** An offset is laid out with the aid of parallel lines drawn on a floor or other smooth surface.

The hand bender must be repositioned on the conduit to create both offset bends. A simple method for checking the accuracy of an offset bend uses a pocket level.

A *double offset,* or saddle, is a common complex bend made in conduit to bypass obstructions. Standard procedures are used in creating a double offset. **See Figure 26-22.** Parallel lines are used to provide a frame of reference for a double offset. Parallel lines on each side of the conduit are used to help eliminate any deviation from center.

Per Article 358.24 of the NEC®, *Bends – How Made,* bends in tubing shall be made so that the tubing will not be damaged and that the internal diameter of the tubing will not be effectively reduced. In addition, per Article 358.26 of the NEC®, a run between pull points (between outlet and outlet, between fitting and fitting, or between outlet and fitting) shall not contain more than the equivalent of four 90° bends (360° total), including bends located immediately at the outlet box or fitting. The NEC® table Annex – Table 2, Radius of Conduit and Tubing Bends, indicates the minimum acceptable radius for various sizes of conduit.

The NEC® requires that EMT be installed as a complete system and be securely fastened in place at least every 10′ and within 3′ of each outlet box, junction box, cabinet, or fitting. Conduit must be installed into smooth-flowing pathways through which wires can be easily pulled. Conduit is secured in place and attached to electrical boxes and other conduit using straps, connectors, and couplings.

Conduit Connectors. EMT is firmly secured to electrical boxes using compression, indenter, and set screw connectors. **See Figure 26-23.** Each type of connector has a distinctly different technique for holding the conduit securely.

A *compression connector* is a box fitting that firmly secures conduit to a box by utilizing a nut that compresses a tapered metal ring (ferrule) into the conduit. As the compression nut is tightened, the nut forces the ferrule into the conduit, locking the conduit in position. Compression connectors can be loosened to remove the conduit and reused multiple times to attach the same piece of conduit.

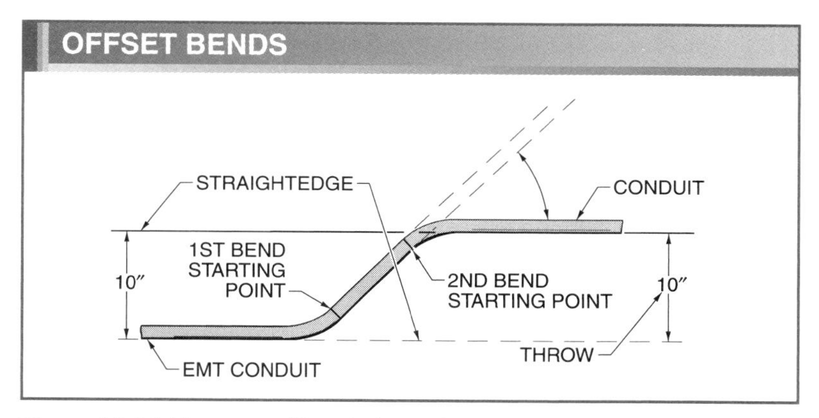

Figure 26-21. The second bend of an offset is readily determined by using parallel lines drawn on the floor.

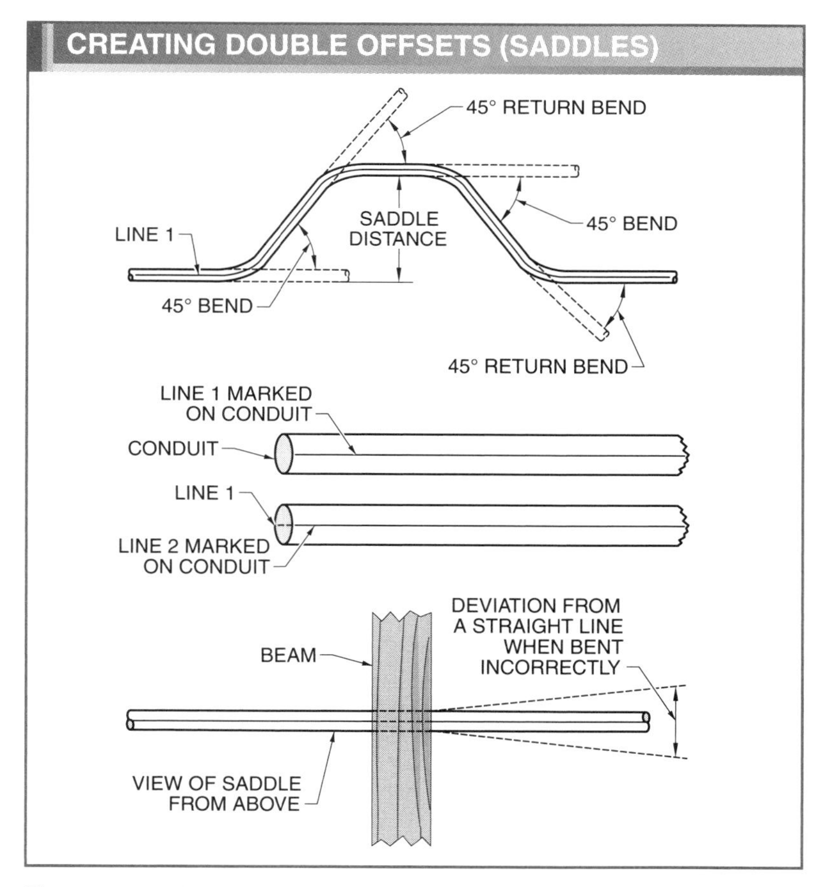

Figure 26-22. Double offsets or saddles are used to clear obstructions.

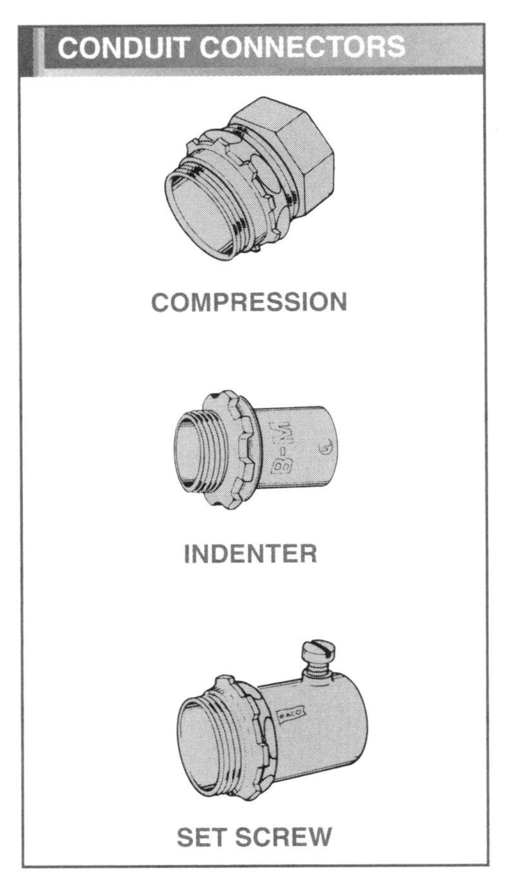

Figure 26-23. Conduit connectors hold conduit securely to boxes.

An *indenter connector* is a box fitting that secures conduit to a box with the use of a special indenting tool. Indenter coupling connectors are not reusable and must be cut off with a hacksaw to be removed. A *set screw connector* is a box fitting that relies on the pressure of a screw against the conduit to hold the conduit in place. Set screw connectors are also reusable.

Conduit Couplings. Unlike metallic and nonmetallic cable, conduit must be joined together using conduit couplings when a run exceeds 10′. A *conduit coupling* is a fitting used to join one length of conduit to another length of conduit and still maintain a smooth inner surface. **See Figure 26-24.** Conduit couplings are similar in design to back-to-back conduit connectors and operate on the same principles of compression, indentation, and pressure. Applications that require conduit to be run where water is present require that special liquid-tight conduit and watertight conduit couplings be used.

Installing Rigid Metal Conduit (RMC)

Bending rigid metal conduit requires many of the skills used to bend EMT. The primary difference is that rigid metal conduit is heavier and much more difficult to bend, requiring use of heavy-duty hydraulic machinery. Due to the difficulty in bending, rigid metal conduit is typically threaded. The threads allow threaded fittings such as elbows to be used instead of bending the conduit.

Fortunately, very little rigid metal conduit is used in residential construction. For residences, rigid metal conduit is typically only used for overhead service risers and for underground services. For service installations, rigid metal conduit can be purchased precut and prethreaded. Electrical supply houses or hardware stores typically stock a variety of lengths and sizes of rigid metal conduit ready for use. Electrical supply houses or stores also carry stock fittings such as 45° and 90° bends. When special lengths are required, rigid metal conduit can be cut and threaded by a supply house for an additional fee.

Most electrical metallic tubing is easily bent using a hand bender.

Installing Flexible Metal Conduit (FMC)

Flexible metal conduit (Greenfield) uses the same techniques for installation as metallic sheathed cable. Flexible metallic conduit is a metal conduit, with no wires, that can be bent by hand. Flexible metal conduit differs from metallic sheathed cable in that wires must be pulled through flexible metal conduit after installation.

Fishing is the term used for the process of pulling wires through conduit. A *fish tape* is a device used to pull wires through conduit. The fish tape is extended and pushed through the conduit until the tape reaches an opening. At the opening the wires being installed are firmly secured to the fish tape. The fish tape is retrieved by pulling the tape out of the conduit with the wires attached. The fish tape must be pulled evenly and smoothly to allow the wires to move easily through the conduit. Typically, one electrician operates the fish tape while another electrician feeds the wires into the conduit.

For the most part, installing switches and receptacles in a conduit system is the same as in any other system. In a properly installed metal conduit system, switches and receptacles are grounded through the conduit. However, an additional green wire pulled through the system and attached to the green grounding screw provides additional grounding security.

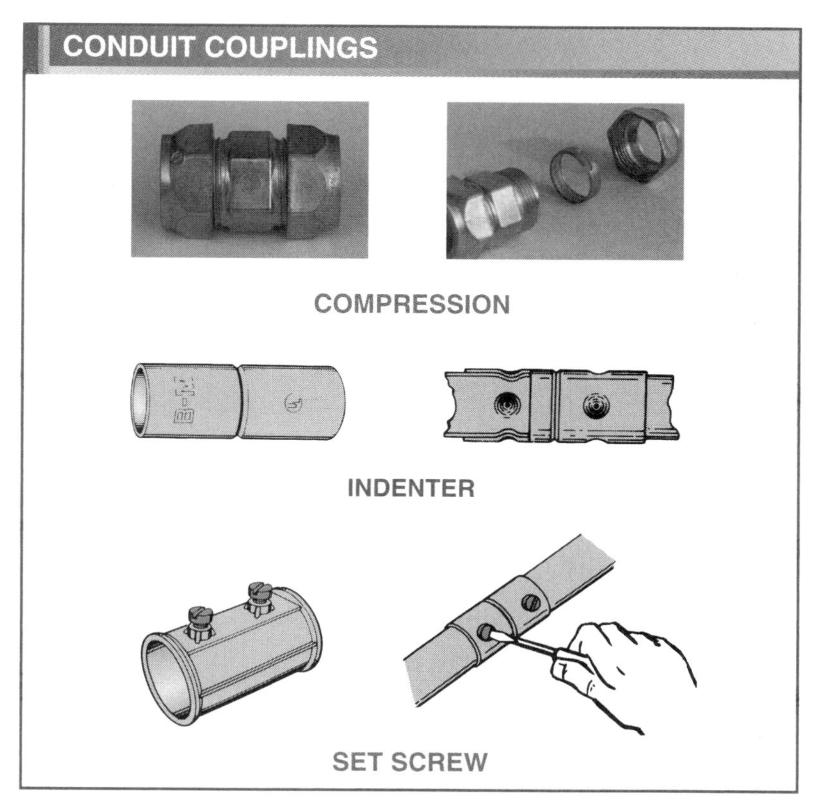

Figure 26-24. Conduit couplings use the same techniques as conduit connectors (compression, indenting, and set screw) to join two lengths of conduit.

Review Questions

1. State the purpose of nonmetallic-sheathed cable.
2. What do the two numbers marked on nonmetallic-sheathed cable indicate?
3. How is nonmetallic-sheathed cable prepared?
4. Where and how are receptacle boxes for NM cable installed?
5. Why is cable stripped before being installed in a box?
6. Compare the three methods of residential grounding.
7. Why would metallic-sheathed cable be used rather than nonmetallic-sheathed cable?
8. List and define the three most common types of conduit.
9. What is deburring?

10. List the three types of conduit connectors.
11. How is rigid metal conduit used in residential construction?
12. What does the term “fishing” refer to and when is it performed?

Service entrances are the link between a building and the power company. Service entrances provide all the equipment necessary to obtain electricity and distribute it throughout a residence. Cost and convenience are both factors when determining the placement of a service entrance.

27 Service Entrances

RESIDENTIAL SERVICE ENTRANCES

Installing a service entrance is typically the first priority when wiring a new home. A service entrance is the link between a residence and the power company. Service entrances provide all the equipment necessary to obtain electricity and distribute it throughout a residence.

Electrical service to a residence is provided by wires that run overhead from a utility pole to the residence, or by wires that are buried in the ground. Electrical service may be provided to a residence through a service drop or underground (service lateral). A service drop has the electrical service wires run from a utility pole to a service head on or above the residence. An underground (service lateral) has the electrical service wires to a residence buried in the ground. **See Figure 27-1.**

Placement of Service Entrances

Placement of a service entrance must take into consideration both cost and convenience. Electricians must try to make the service panel of the residence as accessible as possible using the minimum of materials. The three tips that are useful in determining the placement of a service entrance are the following:

- Keep the service wires as short as possible.
- Locate the service panel as close to the kitchen as possible to avoid costly wire runs to major appliances.
- Avoid having the meter socket placed on the outside of a bedroom wall.

Sizing Service Entrances

Service entrances for single family dwellings require a minimum 100 A capability. For larger residences and structures using electric heat, a 200 A or larger service entrance can be required. Residential service entrances are typically rated in amperes because voltage is typically a standard (115/230 V). Further information concerning service entrances can be obtained from Article 230 of the NEC® or the local building inspector.

Clearance requirements are provided to protect the service-drop conductors from physical damage and to protect personnel from contact with the conductors.

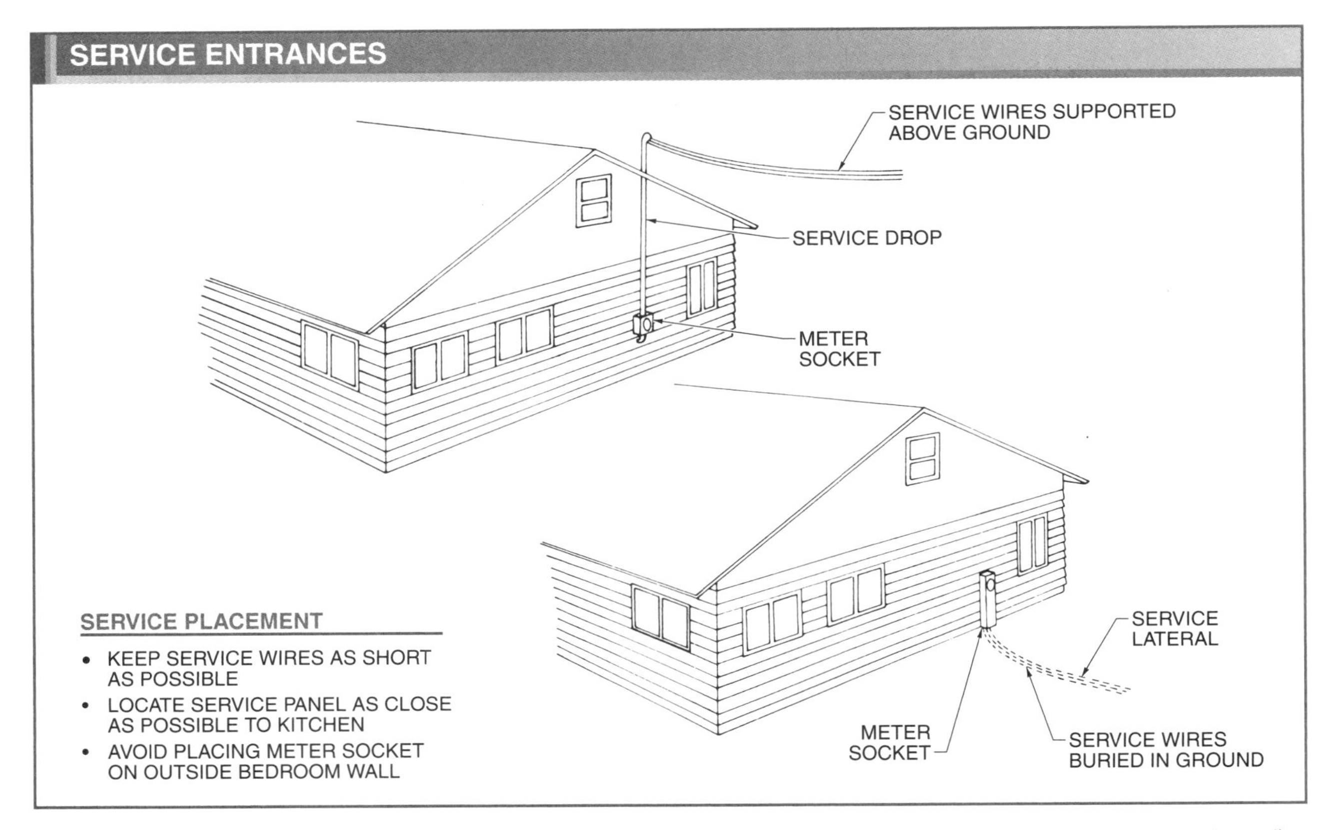

Figure 27-1. Electrical service may be provided to a residence through a service drop or underground (service lateral).

Types of Service Entrances

Not all electrical services are installed in the same manner. Services differ because of the application or because of the location in which the service is located. Service entrances may be overhead riser, lateral (conduit), lateral (cabinet), and mobile home service entrances.

The service line distributes power to the service panel, and individual lines distribute power throughout a building.

Overhead Riser Service Entrances. Many established neighborhoods and urban areas use overhead riser service entrances. **See Figure 27-2.** An overhead riser service entrance has all service wires running from a utility pole to a service head, where the meter socket and riser are firmly secured to the exterior of a dwelling, while the service panel is placed inside. An appropriate-size conduit typically connects the meter socket and meter to the service panel.

Lateral Service Entrances (Conduit). In most new developments it is customary to bury all service wires underground. **See Figure 27-3.** A lateral service entrance (conduit) has all service wires buried underground, creating a condition where the wires are subjected to less damage from the environment (weather) and allowing unsightly poles and service wires to be removed from streets and alleys. Typically, underground services are built using conduit.

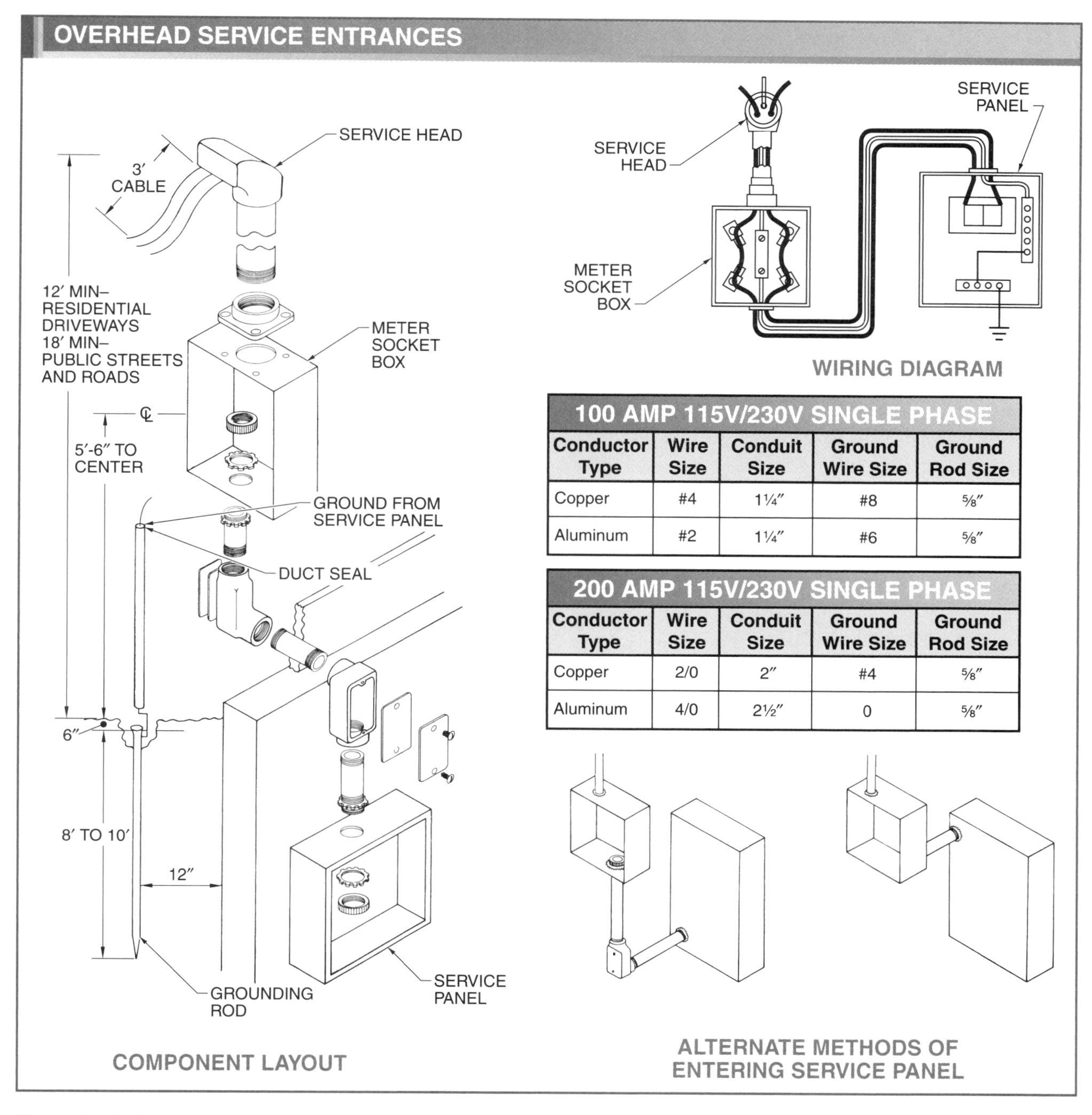

100 AMP 115V/230V SINGLE PHASE				
Conductor Type	Wire Size	Conduit Size	Ground Wire Size	Ground Rod Size
Copper	#4	1¼″	#8	⅝″
Aluminum	#2	1¼″	#6	⅝″

200 AMP 115V/230V SINGLE PHASE				
Conductor Type	Wire Size	Conduit Size	Ground Wire Size	Ground Rod Size
Copper	2/0	2″	#4	⅝″
Aluminum	4/0	2½″	0	⅝″

Figure 27-2. An overhead riser service entrance has the meter socket and riser firmly secured to the exterior of a dwelling, while the service panel is located inside.

Lateral Service Entrances (Cabinet). A lateral service entrance (cabinet) utilizes a cabinet that encloses the service wires, meter socket, and meter. **See Figure 27-4.** The entire enclosure is then covered with a metal cover that is typically locked. A cabinet lateral service enables easy wire routing.

Securing Service Equipment

Typically, the service wires are not connected to the power company until the service and inside of the residence are properly wired. All boxes, panels, conduit, and wires must be securely held in place to avoid problems from wind, rain, snow, hail, and other environmental concerns.

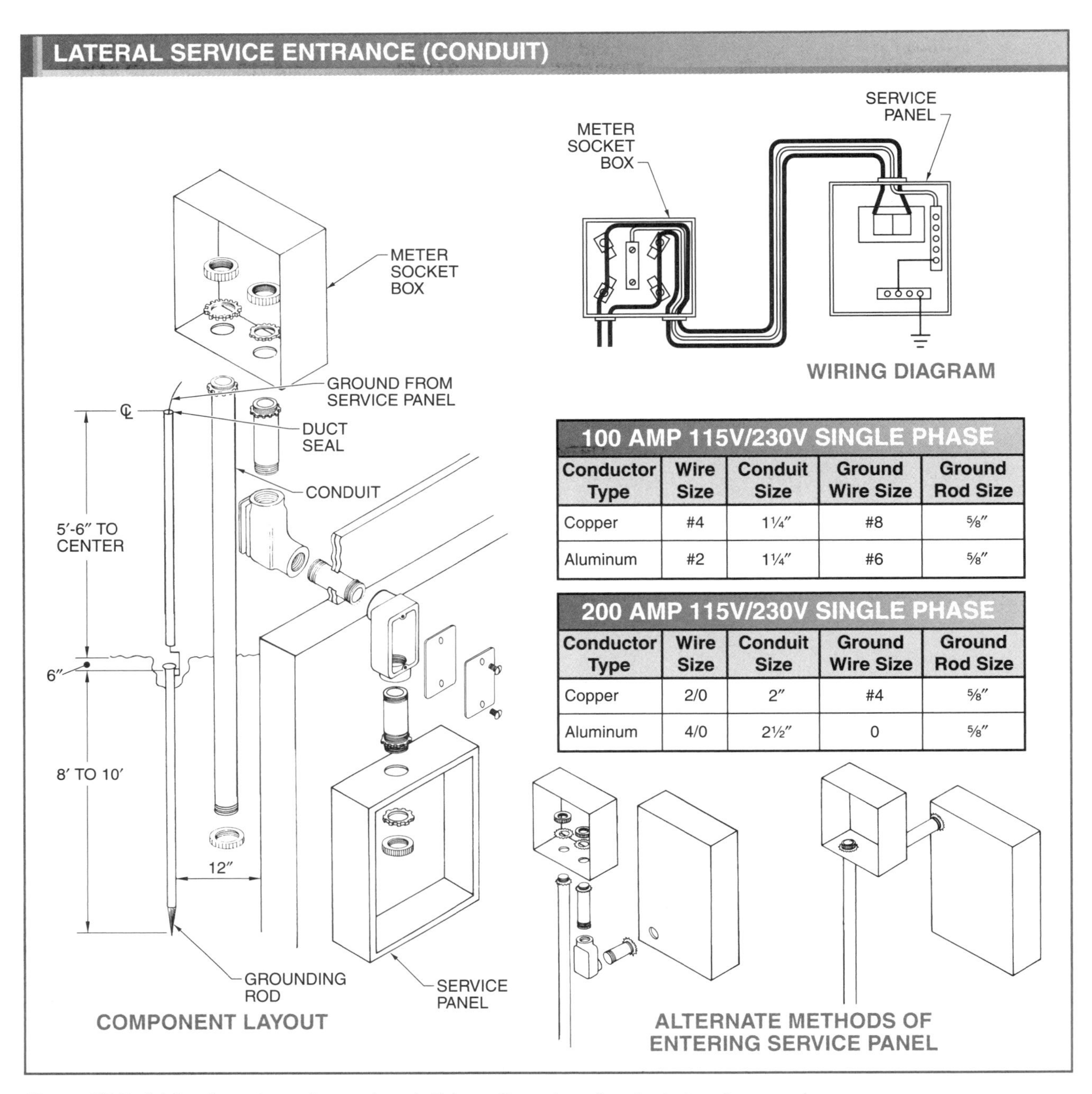

100 AMP 115V/230V SINGLE PHASE				
Conductor Type	Wire Size	Conduit Size	Ground Wire Size	Ground Rod Size
Copper	#4	1¼″	#8	⅝″
Aluminum	#2	1¼″	#6	⅝″

200 AMP 115V/230V SINGLE PHASE				
Conductor Type	Wire Size	Conduit Size	Ground Wire Size	Ground Rod Size
Copper	2/0	2″	#4	⅝″
Aluminum	4/0	2½″	0	⅝″

Figure 27-3. A lateral service entrance (conduit) has all service wires buried underground.

During the installation of a service entrance, large and small box knockouts must be removed from the service panel. Whenever possible, large knockouts should be removed before the service panel is mounted. Removing the knockouts before mounting reduces any chance of loosening the panel. Box knockouts are removed from the service panel using standard procedures.

Meter sockets and service panels must be firmly secured before any wires (cables) are pulled or conduit attached. When service equipment is mounted on wood, wood screws or lag bolts are used to hold the equipment in place. When service equipment is mounted to masonry, drills and special bolts with anchors are used. Plastic anchors support a considerable amount of weight when properly installed.

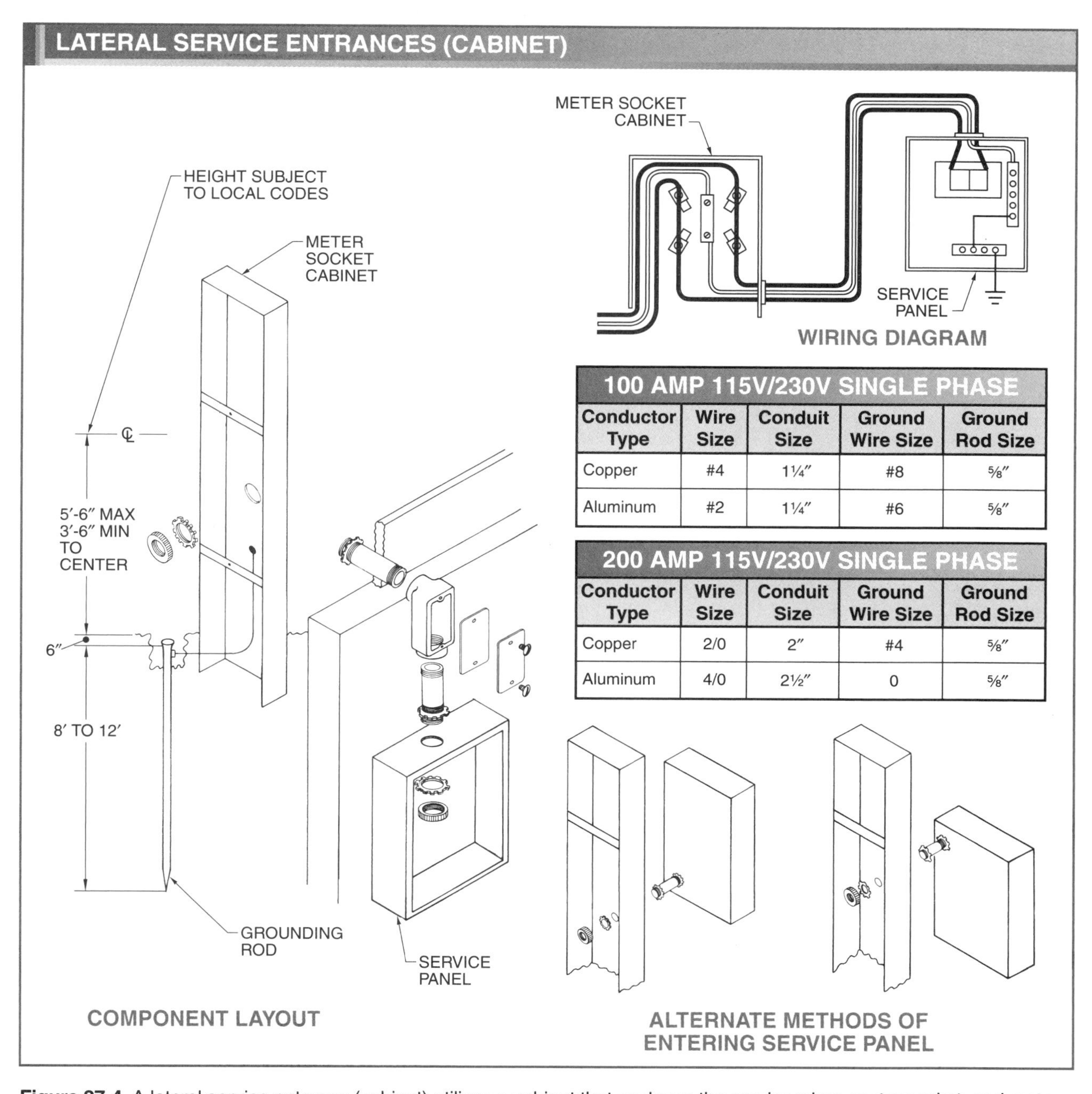

100 AMP 115V/230V SINGLE PHASE				
Conductor Type	**Wire Size**	**Conduit Size**	**Ground Wire Size**	**Ground Rod Size**
Copper	#4	1¼″	#8	⅝″
Aluminum	#2	1¼″	#6	⅝″

200 AMP 115V/230V SINGLE PHASE				
Conductor Type	**Wire Size**	**Conduit Size**	**Ground Wire Size**	**Ground Rod Size**
Copper	2/0	2″	#4	⅝″
Aluminum	4/0	2½″	0	⅝″

Figure 27-4. A lateral service entrance (cabinet) utilizes a cabinet that encloses the service wires, meter socket, and meter.

SERVICE PANELS

The electric power brought to a residence is typically a three-wire 115 V/230 V, 1ϕ system. Red, black, and white wires for the service panel enter the residence through a service head. **See Figure 27-5.** The red and black wires represent the hot or live wires, and the white wire represents the neutral. As with most residential services, 230 V can be obtained between the red and black hot wires. One hundred and fifteen volts can also be obtained between either of the hot wires and neutral. The 115 V and 230 V voltages are constant throughout the meter socket and into the service panel.

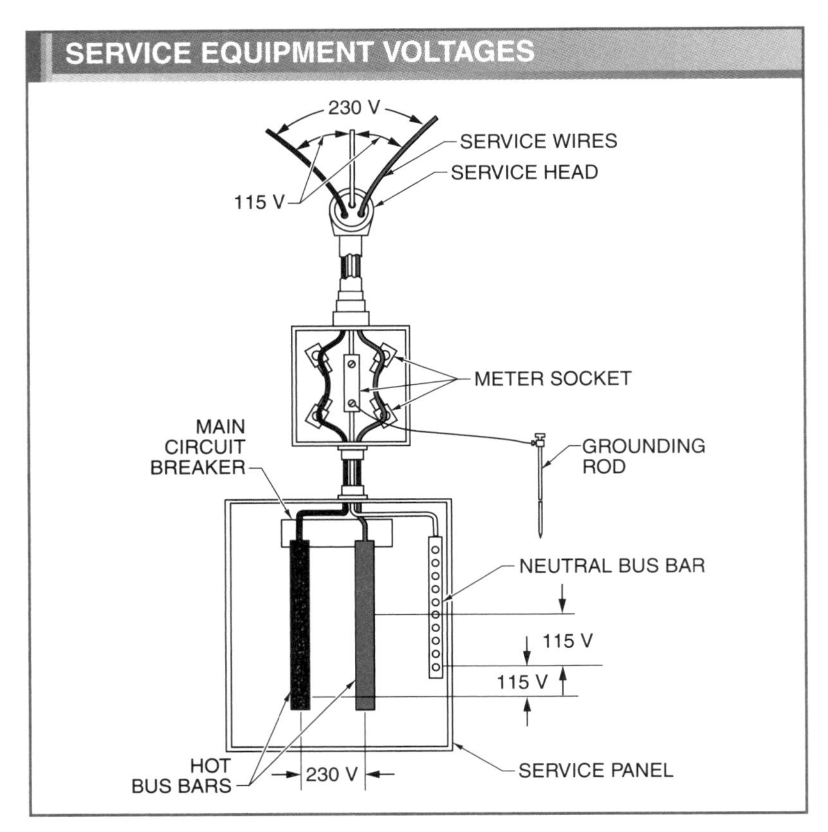

Figure 27-5. A standard 115 V/230 V system consists of two incoming hot lines (red and black) and a neutral (white).

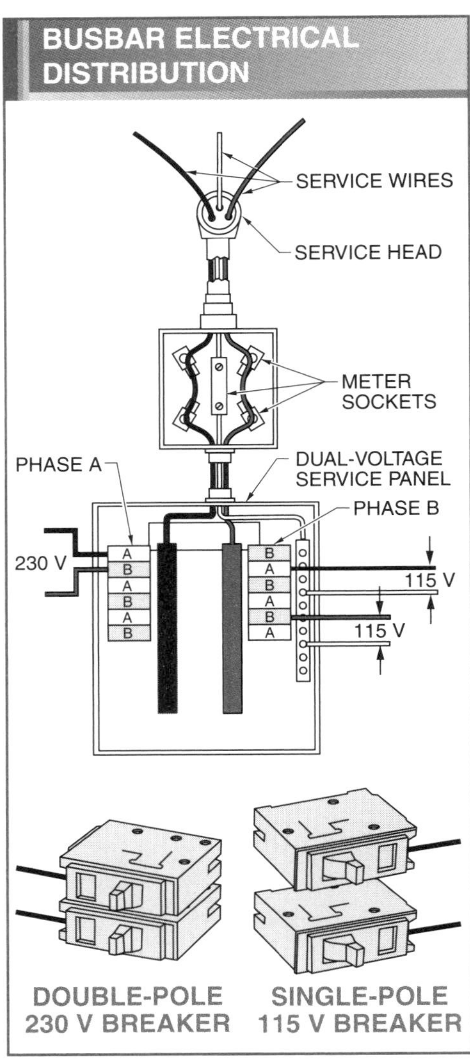

Figure 27-6. Busbars alternately supply circuit breakers with phase 1 and phase 2 to form 115 V and 230 V circuits.

Inside the service panel, where the wires are terminated, metal conductors called bus bars continue the system. Bus bars distribute the power to the circuit breakers in an organized manner. **See Figure 27-6.** Service panel bus bars convey power to single-pole and double-pole circuit breakers to form 115 V and 230 V circuits.

Any single circuit breaker can only furnish power to one 115 V circuit. Two consecutive circuit breakers are used to provide power for a 230 V circuit. The circuit breakers must be consecutive so that the breaker handles can be fastened together as one unit. Fastening the handles of two circuit breakers together ensures that, in the event of a 230 V overload, both hot lines are disconnected simultaneously. Circuit breakers are snapped into position when being installed to the busbars of a service panel.

Service Entrance Cables

Instead of individual wires being pulled through conduit, quite often service entrance cables are used. Service entrance cables have a bare conductor that is wound around the insulated conductors. The spiral winding prevents tampering with the conductors ahead of the meter and also serves as the neutral conductor.

Service entrance cables are accepted in many areas and require no special tools to install. Service entrance caps are shaped to accept oval service cables. The sill plate covers the cable and the cable opening where the service enters the dwelling. Sealing compound or caulking is applied to make the sill plate watertight. Service cable connectors are available as standard box connectors or as watertight connectors when used on top of the meter socket. Mounting straps are used to secure service cables in place. **See Figure 27-7.**

SERVICE DROPS

A *service drop* consists of overhead wires and devices that connect the power company power lines to a residence. Service drops are owned and maintained by the power company. The National Electrical Code® (NEC®) provides guidelines for the installation of service drops in relation to platforms, windows, service heads, and structure elevations.

To comply with NEC® standards, a service head must be at least 10′ above and 3′ to the side of any platform. The NEC® has standards governing service drops in relation to windows. Service conductors must be at least 3′ from the bottom and sides of a window. Conductors that run above a window are considered out of reach from the window.

NEC® standards governing the elevation of service conductors specify a 3′ clearance over adjacent buildings. Sidewalks require a 10′ minimum clearance. Private drives require a 12′ minimum clearance. Alleys and streets require an 18′ minimum clearance.

The NEC® has standards relating to the use of service heads. Very explicit specifications are in place to ensure the performance and safety of service head installations. Careful consideration of each NEC® standard during installation ensures a permanent weatherproof connection out of the reach of potentially destructive obstacles. **See Figure 27-8.**

SERVICE LATERALS

A *service lateral* is any service to a residence that is achieved by burying the wires underground. The four types of service laterals used for residences are the following:

- from utility manhole in street to service panel disconnect.
- from utility sidewalk handhole to service panel disconnect.
- from pole riser to service panel disconnect.
- from transformer pad to service panel disconnect.

When service originates in a manhole, the work to be performed must be arranged with the utility company for joint installation of all conduit and wires. The manhole requirement is necessary because a manhole is a potentially dangerous space and is restricted to utility workers.

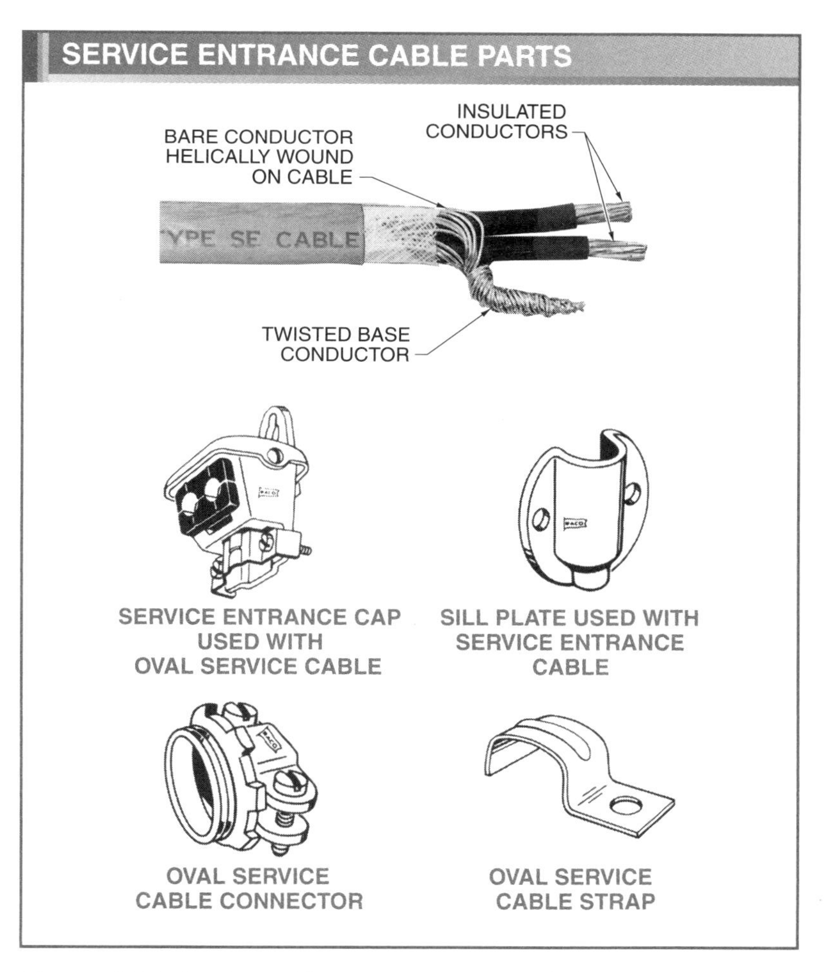

Figure 27-7. Service cable connectors are available as standard box connectors or as watertight connectors when used on top of the meter socket.

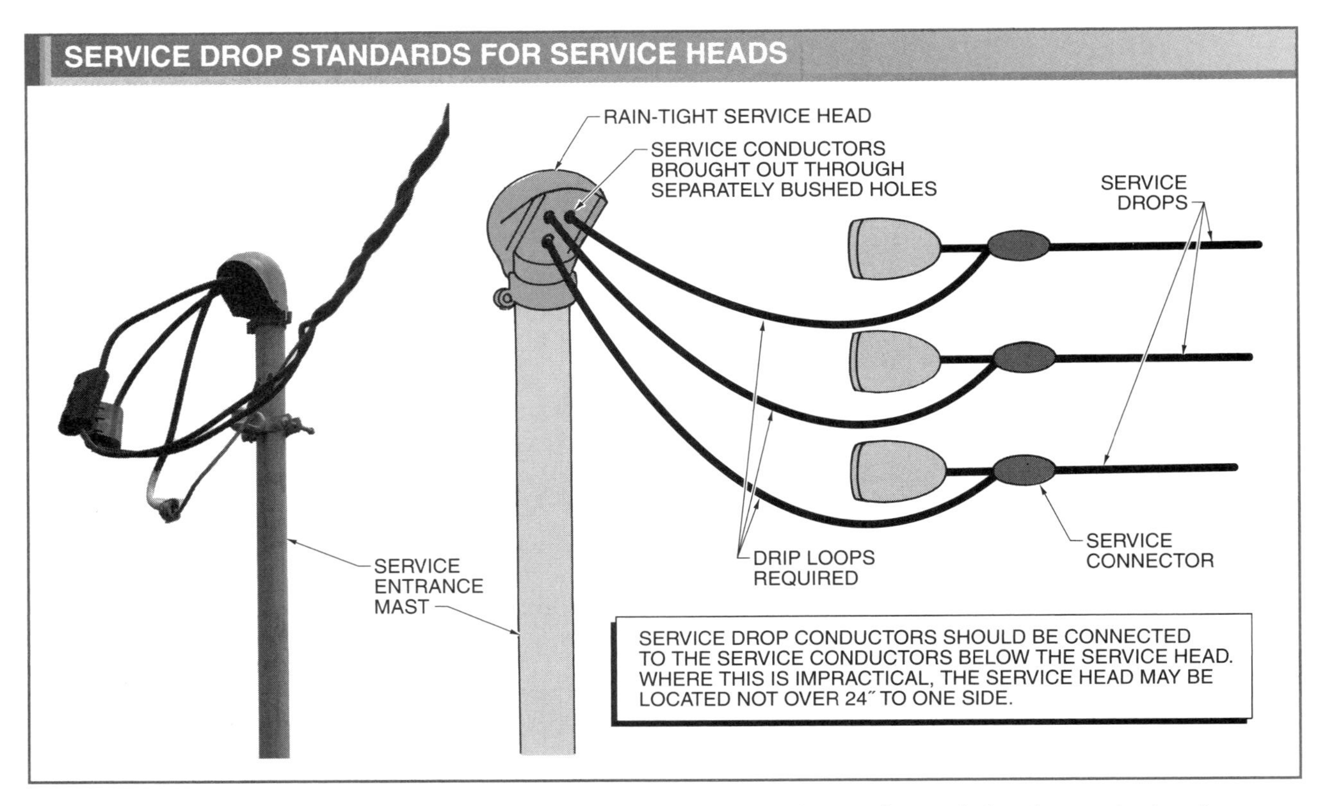

Figure 27-8. Service drop conductors are typically connected to the service conductors below the service head.

When service originates in a sidewalk handhole, electricians are typically permitted to run conduit and wires into the handhole. **See Figure 27-9.** The utility company will make the appropriate connections in the handhole.

When a pole riser is used, the service installation is shared by the homeowner and utility company. In many instances, electricians simply install conduit to a point at least 8′ above grade and pull enough wire to reach the crossarm at the top of the pole. The utility company extends the protection for the drop cables with approved molding to the crossarm upon final connection. In some cases a short piece of conduit with a service head is used to terminate the run at the crossarm. When the utility company extends protection, the homeowner furnishes the conduit and service head. In other cases, homeowners must furnish conduit all the way with a service head at the end.

When a service entrance originates from a transformer pad, special direct-burial cable is used to make the installation. **See Figure 27-10.** Transformer pad service entrances require the homeowner to furnish a transformer pad in addition to the regular electrical service equipment. The utility company typically creates the trench from the transformer pad to the dwelling, buries the cable, and makes all final connections.

TRIMMING OUT A SERVICE ENTRANCE

Trimming out or finishing a service entrance requires the installation of internal service conductors, circuit breakers, and branch circuit wiring. Wires must be bent to shape to fit some service heads. Similar bends are required when wiring the meter socket and service panel.

Wire (cable) coming from branch circuits must also be bent to fit the configuration of the service panel. Electricians should always take care and provide a professional appearance and organization to the wires to each circuit breaker.

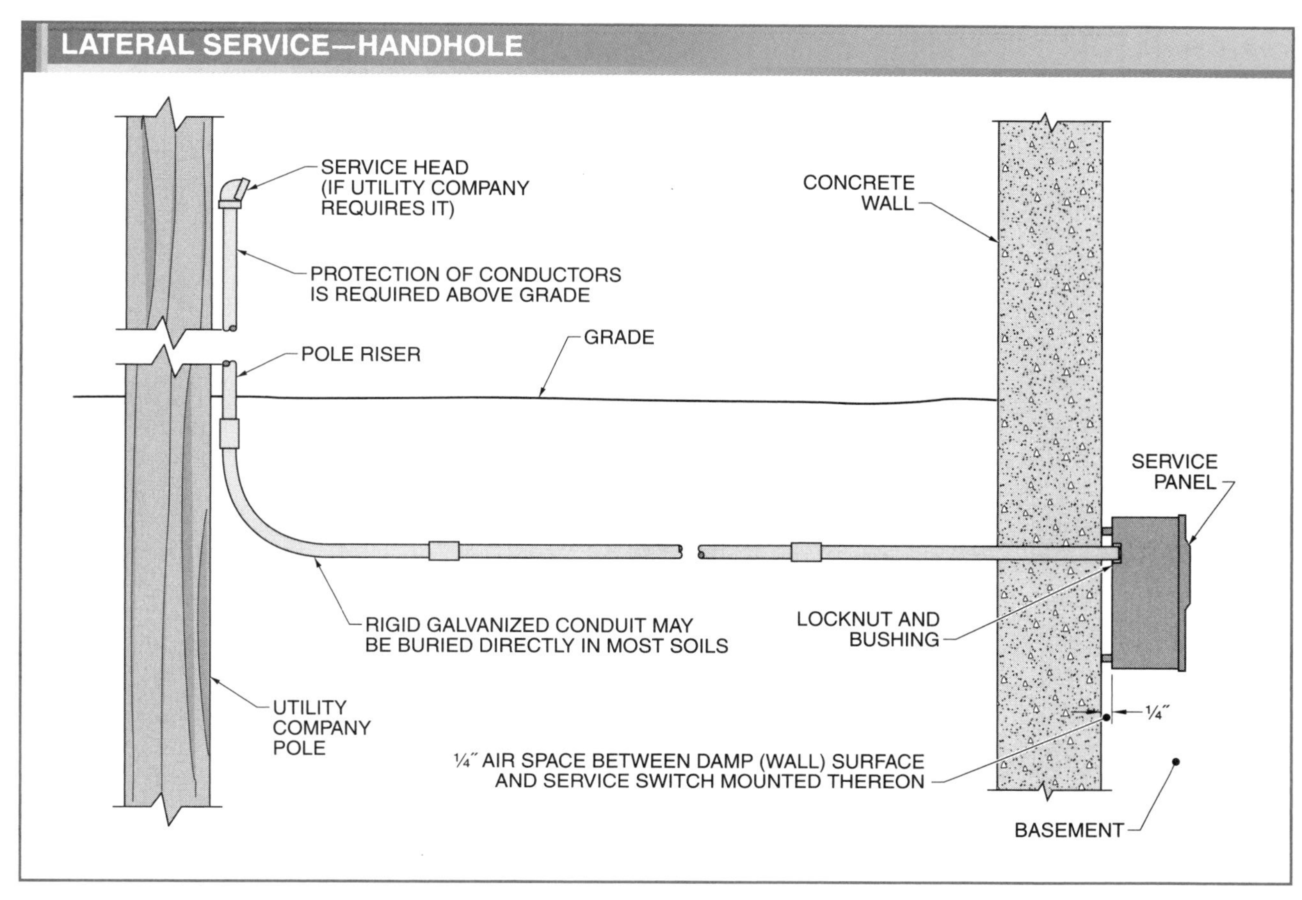

Figure 27-9. Underground service can be provided to a dwelling by running wires through buried conduit from a sidewalk handhole.

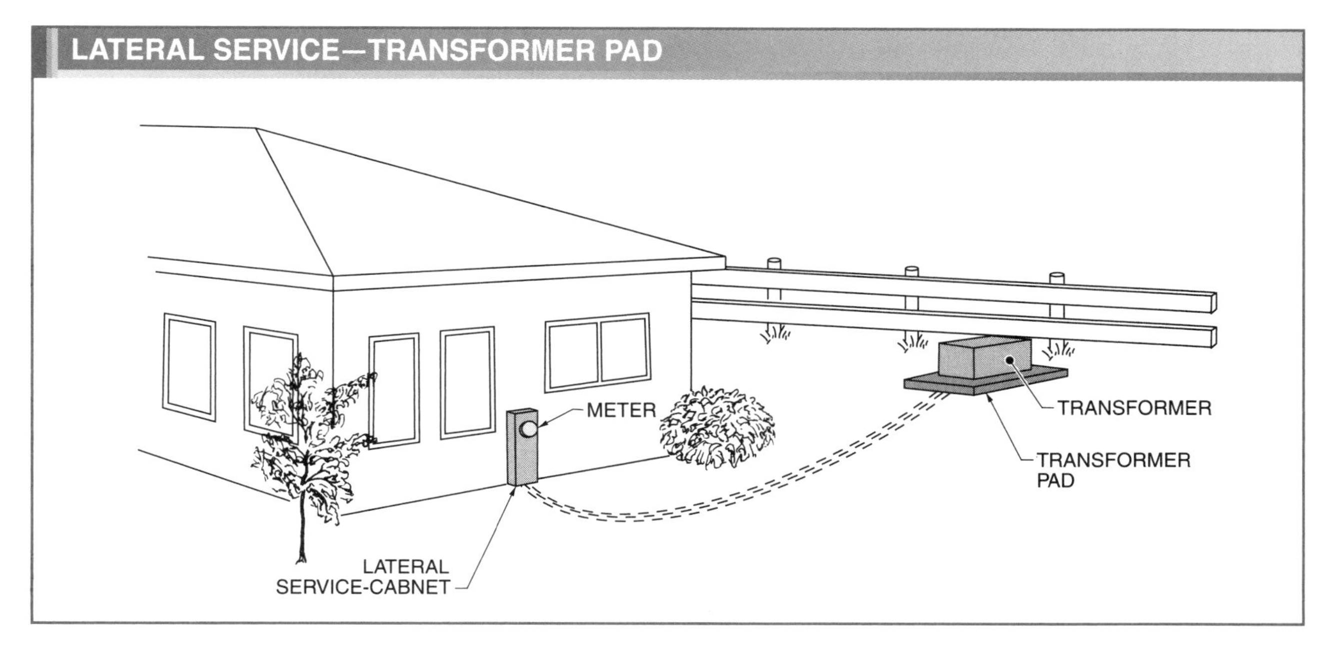

Figure 27-10. Transformer pad service entrances require the homeowner to furnish a transformer pad and service equipment.

After all the circuit breakers have been properly installed and each circuit tested, electricians must prepare the trim cover to the service panel. Proper technique must be used to remove the breaker openings from the service panel trim cover. Remove only the openings necessary for the circuit breakers installed. The service panel cover is installed to finish trimming out a service panel. The cover panel has a directory on the inside of the door. The directory is filled out to indicate the location and use of each circuit breaker. **See Figure 27-11.**

Figure 27-11. A service panel is finished when the cover is installed and each breaker is labeled according to position and operation (circuit).

Tech Facts

A service head or gooseneck should be installed at a point that is above the point of attachment for the service-drop conductors. Installing a service head above the point of attachment ensures that water does not enter the service raceway or service-entrance cable.

AC GENERATORS

A *generator* is an electromechanical device that converts mechanical energy into electrical energy by means of electromagnetic induction. AC generators (alternators) convert mechanical energy into AC voltage and current. AC generators consist of field windings, an armature, slip rings, and brushes. **See Figure 27-12.**

Field windings are magnets used to produce the magnetic field in a generator. The magnetic field in a generator can be produced by permanent magnets or electromagnets. Most generators use electromagnets, which must be supplied with current. An *armature* is the movable coil of wire in a generator that rotates through the magnetic field. The armature may consist of many coils. The ends of the coils are connected to slip rings.

Slip rings are metallic rings connected to the ends of the armature and are used to connect the induced voltage to the generator brushes. When the armature is rotated in the magnetic field, a voltage is generated in each half of the armature coil. A *brush* is the sliding contact that rides against the slip rings and is used to connect the armature to the external circuit (power grid).

AC generators are similar in construction and operation to DC generators. The major difference between AC and DC generators is that DC generators contain a commutator that reverses the connections to the brushes every half cycle. The commutator maintains a constant polarity of voltage to the load. AC generators use slip rings to connect the armature to the external circuit (load). The slip rings do not reverse the polarity of the output voltage produced by the generator. The result is an alternating sine wave output. **See Figure 27-13.**

As the armature is rotated, each half cuts across the magnetic lines of force at the same speed. The strength of the voltage induced in one side of the armature is always the same as the strength of the voltage induced in the other side of the armature. However, since the two halves of the coil are connected in a closed loop, the voltages add to each other. The result is that the total voltage of a full rotation of the armature is twice the voltage of each coil half. The total voltage is obtained at the brushes connected to the slip rings, and is applied to an external circuit (grid).

Home Emergency Standby Generators

Standby generators are installed during new construction or retrofitted into existing residences and businesses. Home generators are installed outside similarly to the condenser of a home air conditioning unit. Standby generators are typically placed on a solid surface as with air conditioning condensers. **See Figure 27-14.** Standby generators are powered by gasoline, natural gas, or LPG. Typical home emergency standby generators range in capacity from 6500 W to 40,000 W. The size of a generator is determined by the number of preselected lighting and appliance circuits that must be powered during a power outage.

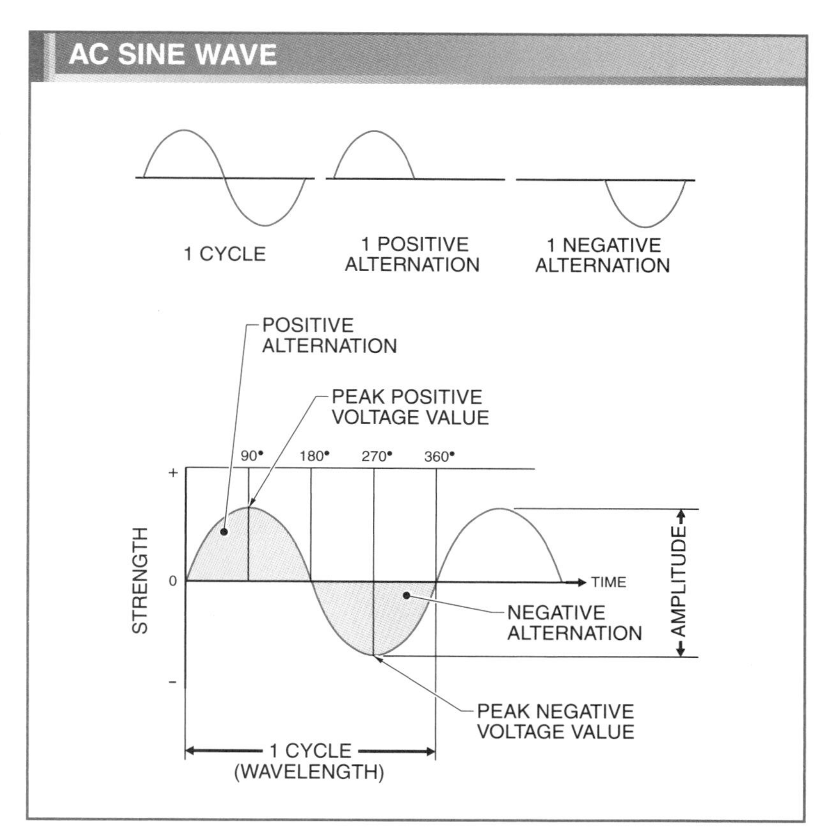

Figure 27-13. The armature of a generator induces a varying voltage (AC) output that is positive then negative.

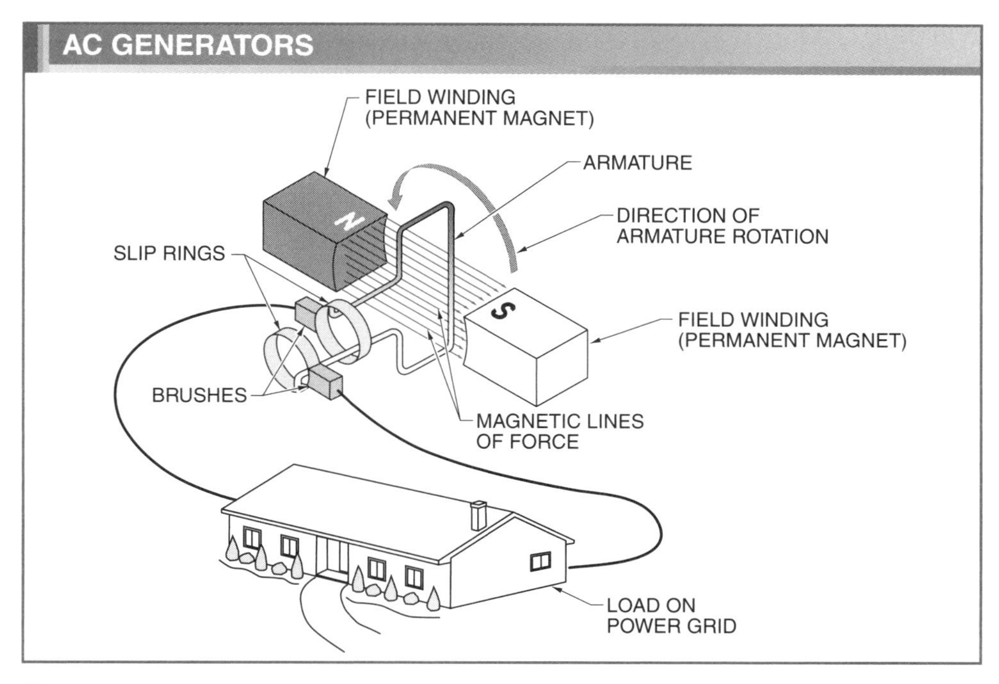

Figure 27-12. AC generators consist of field windings, a coil (armature), slip rings, and brushes.

Figure 27-14. Residential emergency standby generators typically range in capacity from 6500 W to 40,000 W.

APPLIANCE STARTING AND RUNNING WATTAGES

Appliance	Wattage (Starting)*	Wattage (Running)*
Refrigerator	2800	700
Freezer	2500	500
Well Pump-2 HP	6000	2000
Well Pump-3 HP	9000	3000
Gas Furnace Fan	500 to 2350	300 to 875

* in watts

Figure 27-15. Some appliances draw more power when starting than when running. Ensure startup wattages are known for use in circuit load calculations.

Because generators are chosen based on capacity in watts, it is important to understand the relationship between volts (V), amperage (A), and watts (W). The basic formula for calculating wattage when the voltage (V) and amperage (A) are known is watts equal volts times amps ($W = V \times A$). When an appliance is plugged into a 120 V outlet and draws 5 A, the watts are 600 W ($120\ V \times 5\ A = 600\ W$). In most cases an appliance has a nameplate stating the start and running wattages.

When computing the total wattage required from a generator, the calculation must account for some appliances that require more power (watts) to start than to run. **See Figure 27-15.** All appliances, HVAC systems, lighting, and receptacles of residences have wattage ratings for determining the total load created. It is also important to understand which circuits (appliances) are considered critical circuits and must have power supplied during a power outage. For example, in winter it is critical that the furnace of a residence operate, but in summer it is more important to have air conditioning and a freezer operating.

Reading Electric Meters

An electric meter has four dials. Each dial has a hand. Electric meters have two dials that read clockwise and two dials that read counterclockwise. To read a meter, always use the number that the hand has just passed (CW or CCW) on each of the dials. **See Figure 27-16.**

Calculating Appliance Operating Costs. To calculate the cost to operate an appliance for a month, the nameplate on the appliance must be checked for amperage and wattage specifications. When amperage is provided, the power formula converts amperage and voltage into watts. The formula states that $1\ A \times 115\ V = 115\ W$. Therefore, a 10 A iron consumes 1150 W ($10\ A \times 115\ V = 1150\ W$). When the wattage of an appliance is known, operating cost per hour can be determined. **See Figure 27-17.**

Automatic Transfer Switches

An *automatic transfer switch* is an electrical device that transfers the load of a residence from public utility circuits to the output of a standby generator during a power failure. An automatic transfer switch (ATS) has electronic circuitry that can sense an interruption in power. When a power failure is detected, an ATS turns on a standby generator. The transfer switch also disconnects all the dedicated circuits in the service panel from the public utility source and transfers the dedicated circuits to the output of the emergency generator. When power is transferred, an ATS prevents electricity from flowing into the public service lines and accidentally injuring electric company workers who are working on the system to restore power.

Automatic transfer switches come in various sizes depending upon the power rating required. An ATS is installed in close proximity to the electrical distribution panel and should be installed by a licensed electrician.

A checklist to use when installing a home emergency generator is as follows:

- Generator should be Underwriters Laboratories listed (UL listed).
- Generator should be properly sized for the calculated load.
- Generator should be properly positioned so the noise generated is acceptable.
- Generator should be properly housed for outdoor environmental exposure, such as to temperature, snow, and rain.
- Generator should operate with the available fuel, whether gasoline, natural gas, or LPG.

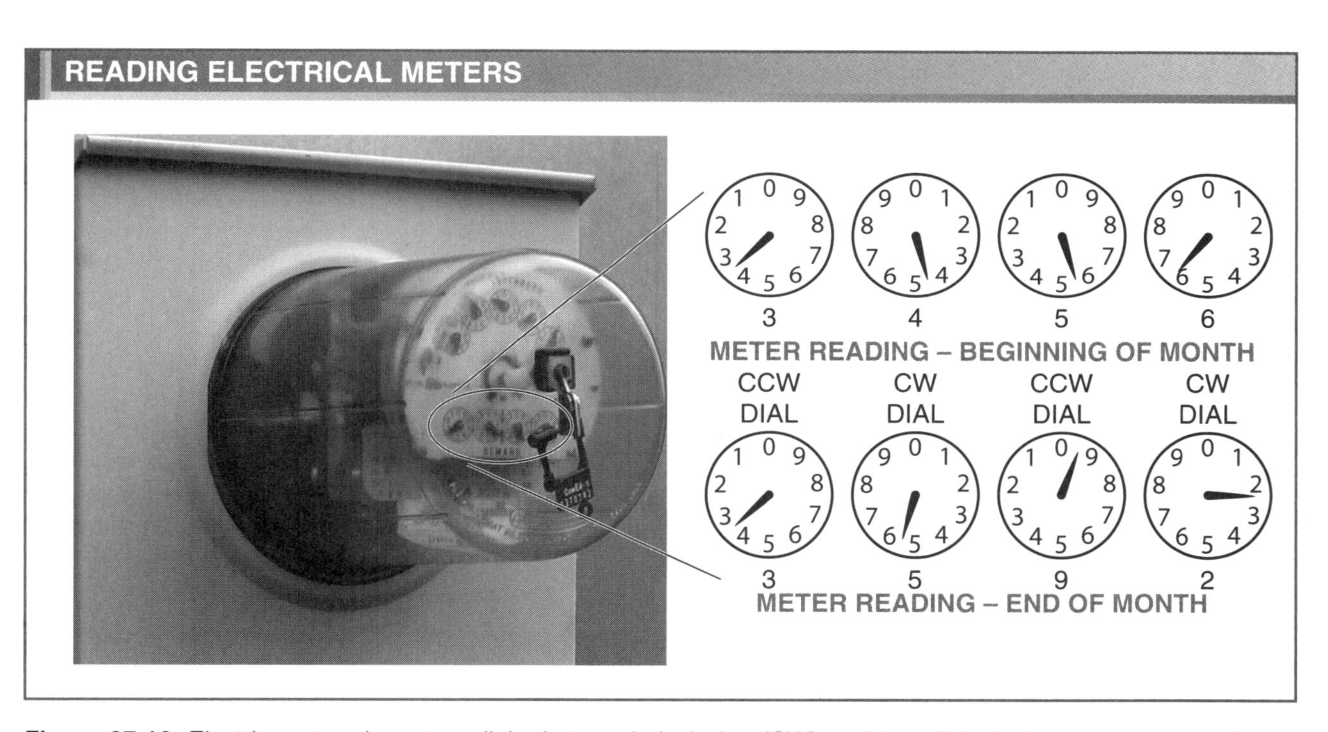

Figure 27-16. Electric meters have two dials that read clockwise (CW) and two dials that read counter-clockwise (CCW).

SMALL APPLIANCE HOURLY OPERATING COSTS				
Wattage Consumed by Appliance	**.03 kWh**	**.04 kWh**	**.05 kWh**	**.06 kWh**
100	10 hr for 3¢	7.5 hr for 3¢	6 hr for 3¢	5 hr for 3¢
300	3 hr for 2.7¢	2,5 hr for 3¢	2 hr for 3¢	1.66 hr for 3¢
500	2 hr for 3¢	1 hr for 2¢	1.2 hr for 3¢	1 hr for 3¢
700	1.4 hr for 3¢	1.1 hr for 3¢	1 hr for 3.5¢	1 hr for 4.2¢
900	1.1 hr for 3¢	1 hr for 3.6¢	1 hr for 4.5¢	1 hr for 5.4¢
1000	1 hr for 3¢	1 hr for 4¢	1 hr for 5¢	1 hr for 6¢

Figure 27-17. To calculate the cost of operating an appliance for a month, the nameplate on the appliance must be checked for amperage and wattage specifications.

- Follow any special considerations due to the type of fuel.
- Generator battery should be charged automatically.
- Generator should have automatic safety shut downs to protect engine and generator in event of low oil pressure, high temperature, or overspeed.

Voltage Changes

AC generators are designed to produce a specific amount of output current at a rated voltage. In addition, all electrical and electronic equipment is rated for operation at a specific voltage. *Rated voltage* is a voltage range that is typically within ±10% of ideal voltage. Today, however, with many components derated to save energy and operating cost, the range is typically +5% to –10%. A voltage range is used because an overvoltage is more damaging than an undervoltage. Equipment manufacturers, utility companies, and regulating agencies must routinely compensate for changes in system voltage.

Back-up generators are used to compensate for voltage changes and are powered by diesel, gasoline, natural gas, or propane engines connected to the generator. When there is time between the loss of main utility power and when the generator starts providing power, the generator is classified as a standby (emergency) power supply. Voltage changes in an electrical system are categorized as momentary, temporary, or sustained.

Momentary Power Interruptions. A *momentary power interruption* is a decrease to 0 V on one or more power lines lasting from 0.5 cycles up to 3 sec. All power distribution systems have momentary power interruptions during normal operation. Momentary power interruptions can be caused when lightning strikes nearby, by utility grid switching during a problem situation (a short on one line), or during open circuit transition switching. *Open circuit transition switching* is a process in which power is momentarily disconnected when switching a circuit from one voltage supply or level to another.

Temporary Power Interruptions. A *temporary power interruption* is a decrease to 0 V on one or more power lines lasting between 3 sec and 1 min. Automatic circuit breakers and other circuit protection equipment protect all power distribution systems.

Circuit protection equipment is designed to remove faults and restore power. An automatic circuit breaker takes from 20 cycles to about 5 sec to close. When power is restored, the power interruption is only temporary. If power is not restored, a temporary power interruption becomes a sustained power interruption. A temporary power interruption can also be caused by a time gap between a power interruption and when a back-up power supply (generator) takes over, or if someone accidentally opens the circuit by switching the wrong circuit breaker switch.

Sustained Power Interruptions. A *sustained power interruption* is a decrease to 0 V on all power lines for a period of more than 1 min. All power distribution systems have a complete loss of power at some time. Sustained power interruptions (outages) are commonly the result of storms, tripped circuit breakers, blown fuses, and/or damaged equipment.

The effect of a power interruption on a load depends on the load and the application. When a power interruption causes equipment, production, and/or security problems that are not acceptable, an uninterruptible power system is used. An *uninterruptible power system (UPS)* is a power supply that provides constant on-line power when the primary power supply is interrupted. For long-term power interruption protection, a generator/UPS is used. For short-term power interruptions, a static UPS is used.

Transients

A *transient* is a temporary, unwanted voltage in an electrical circuit. Transient voltages are typically large erratic voltages, or spikes that have a short duration and a short rise time. Computers, electronic circuits, and specialized electrical equipment require protection against transient voltages. Protection methods commonly include proper wiring, grounding, shielding of the power lines, and use of surge suppressors. A *surge suppressor* is a receptacle that provides protection from high-level transients by limiting the level of voltage allowed downstream from the surge suppressor. Surge suppressors are installed at service entrance panels and at individual loads.

Tech Facts

Transients or temporary unwanted voltage in an electrical circuit can damage sensitive electronic equipment, including computers. A surge suppressor provides protection from high-level transients by limiting the level of voltage allowed downstream from the surge suppressor.

Review Questions

1. What is a service entrance?
2. What are the benefits of using lateral service entrances as opposed to overhead riser service entrances?
3. What colors of wires are used for electrical wiring in a residence and what is indicated by each color?
4. Define service drop.
5. List the types of service laterals used for residences.
6. Define the four main components of an AC generator.
7. How is wattage calculated?
8. When are automatic transfer switches used?
9. What is rated voltage?
10. What are the three types of power interruptions?

Transformers are electrical devices that use electromagnetism to change voltage from one level to another or to isolate one voltage from another. They are primarily used to step up or step down voltage. Transformers operate on the principle of mutual induction.

28 Transformers

MAGNETISM

Magnetism is a force that acts at a distance and is caused by a magnetic field. *Ferromagnetic materials* are materials, such as soft iron, that are easily magnetized. Magnetism is used to produce most of the electricity consumed, develop rotary motion in motors, and develop linear motion in solenoids.

A *magnet* is a device that attracts iron and steel because of the molecular alignment of its material. All magnets or magnetized material have a north (N) and south (S) pole. The basic law of magnetism states that unlike magnetic poles (N and S) attract each other and like magnetic poles (N and N or S and S) repel each other. The force of attraction between two magnets increases as the distance between the magnets decreases. Likewise, the force of attraction between two magnets decreases as the distance between the magnets increases. **See Figure 28-1.**

Magnetic flux is the invisible lines of force that make up the magnetic field. The more dense the flux, the stronger the magnetic force. Flux is most dense at the ends of a magnet. For this reason, the magnetic force is strongest at the ends of a magnet. The lines of flux leave the north pole and enter the south pole of a magnet or magnetic field.

Magnets may be permanent or temporary. *Permanent magnets* are magnets that hold their magnetism for a long period of time. *Temporary magnets* are magnets that lose their magnetism as soon as the magnetizing force is removed. Permanent magnets are used in electrical applications, such as in permanent magnet DC motors and reed switches. Temporary magnets are used in most electrical applications, such as motors, transformers, and solenoids.

Electromagnetism

A magnetic field is produced any time electricity passes through a conductor (wire). *Electromagnetism* is the magnetic field produced when electricity passes through a conductor. Electromagnetism is a temporary magnetic force because the magnetic field is present only as long as current flows. The magnetic field is reduced to zero when the current flow stops.

The magnetic field around a straight conductor is not strong and is of little practical use. The strength of the magnetic field is increased by wrapping the conductor into a coil, increasing the amount of current flowing through the conductor, or wrapping the conductor around an iron core. A strong, concentrated magnetic field is developed when a conductor is wrapped into a coil. The strength of a magnetic field is directly proportional to the number of turns in the coil and the amount of current flowing through the conductor. An iron core increases the strength of the magnetic field by concentrating the field. **See Figure 28-2.**

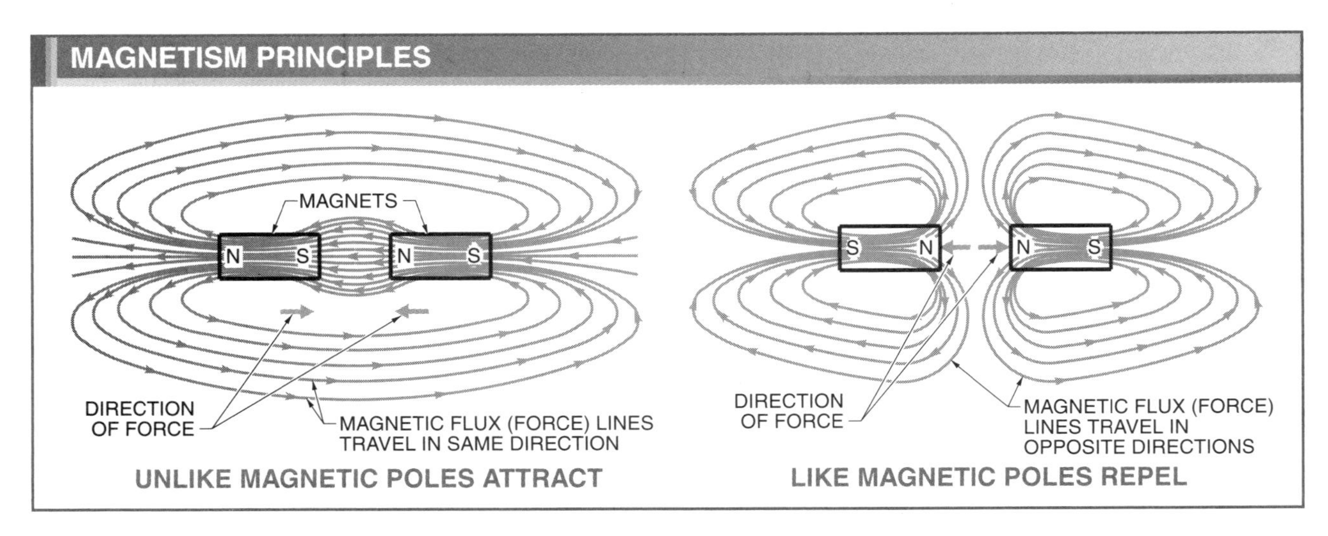

Figure 28-1. The basic law of magnetism states that unlike magnetic poles (N and S) attract each other and like magnetic poles (N and N or S and S) repel each other.

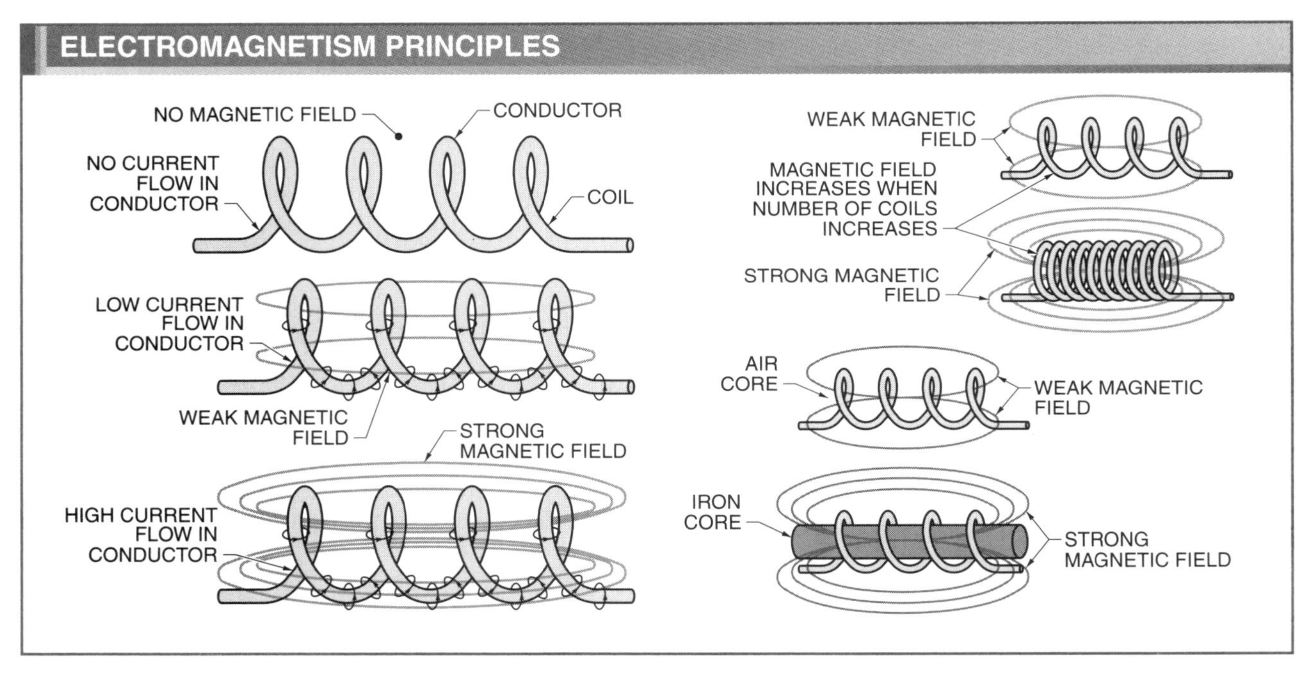

Figure 28-2. The strength of the magnetic field is increased by wrapping the conductor into a coil, increasing the amount of current flowing through the conductor, or wrapping the conductor around an iron core.

Direct current applied to a conductor starts at zero and goes to its maximum value almost instantly. The magnetic field around the conductor also starts at zero and goes to its maximum strength almost instantly. The current and strength of the magnetic field remain at their maximum value as long as the load resistance does not change. The current and the strength of the magnetic field increase if the resistance of the circuit decreases. The current and the magnetic field drop to zero when the direct current is removed.

Alternating current applied to a conductor causes the current to continuously vary in magnitude and the magnetic field to continuously vary in strength. Current flow and magnetic field strength are at their maximum value at the positive and negative peak of the AC sine wave. The current is zero and no magnetic field is produced at the zero points of the AC sine wave. The direction of current flow and polarity of the magnetic field change every time the current passes the zero point of the AC sine wave. On standard 60 Hz (cycle) power frequencies, the current passes the zero point 120 times per second.

SOLENOIDS

Electromagnetism is used in solenoids. A *solenoid* is a device that converts electrical energy into a linear, mechanical force. Solenoids produce a linear, mechanical force when electricity is applied to the coil. The coil produces an electromagnetic force when current passes through the coil windings.

The coil windings are wound around an iron frame to increase the magnetic force produced by the coil. An iron core placed near the energized coil causes the magnetic force to draw the iron core into the coil. The magnetic field is not produced when power is removed from the coil. The iron core is removed from the coil by a spring. **See Figure 28-3.**

The amount of linear force a solenoid develops depends on the number of turns of wire in the coil and the amount of applied current. The more coil turns a solenoid has, the greater the linear force. However, as the number of coil turns is increased, the overall size of the solenoid increases. The higher the current flow through the coil, the greater the linear force. The amount of current drawn by the solenoid depends on the applied voltage. For any given size solenoid, the higher the applied voltage, the higher the current. However, if the applied voltage is increased beyond the voltage rating of the solenoid, the high current causes the solenoid to overheat and burn out.

Solenoids develop linear force in applications that use short strokes at low force. Solenoids are used in many electrical applications. Common solenoid applications include the control the flow of water or gas in washing machines, dishwashers, dryers, furnaces, and automatic sprinkling systems. In commercial and industrial application, solenoids control door locks, open and close valves dispensing product, control movement of parts, stamp information on products, clamp parts, and dispense coins.

TRANSFORMERS

A *transformer* is an electric device that uses electromagnetism to change voltage from one level to another or to isolate one voltage from another. Transformers are primarily used to step up or step down voltage. Transformers operate on the electromagnetic mutual induction principle. *Mutual inductance* is the effect of one coil inducing a voltage into another coil.

The principle of electromagnetic mutual induction states that when the magnetic flux lines from one expanding and contracting magnetic field cut the windings of a second coil, a voltage is induced in the second coil. The amount of voltage induced in the second coil depends on the relative position of the two coils and the number of turns in each coil.

The induced voltage in the second coil depends on the relative position of the two coils if the number of turns in each coil are equal. The highest mutual inductance occurs when all the magnetic flux lines from each coil cut through all the turns of wire in the opposite coil. Likewise, no mutual inductance occurs when no magnetic flux lines from one coil cut through any of the turns of wire in the other coil.

Transformers consist of two or more coils of insulated wire wound on a laminated steel core. The steel core is magnetized when a voltage is applied to one coil. The magnetized steel core induces a voltage into the secondary coil. The *primary coil (input side)* of a transformer is the coil to which the voltage is connected. The *secondary coil (output side)* of a transformer is the coil in which the voltage is induced.

The laminated steel core is used so that the magnetic induction between the two coils is as high as possible. **See Figure 28-4.**

The amount of change in voltage between the primary coil and secondary coil depends on the turns ratio (voltage ratio) of the two coils. A *step-up transformer* is a transformer in which the secondary coil has more turns of wire than the primary coil. A step-up transformer produces a higher voltage on the secondary coil than the voltage applied to the primary. A *step-down transformer* is a transformer in which the secondary coil has fewer turns of wire than the primary coil. A step-down transformer produces a lower voltage on the secondary coil than the voltage applied to the primary. To calculate the relationship between the number of turns and the voltage, apply the formula:

$$\frac{N_P}{N_S} = \frac{E_P}{E_S}$$

where

N_P = number of turns in primary coil

N_S = number of turns in secondary coil

E_P = voltage applied to primary coil (in V)

E_S = voltage induced in secondary coil (in V)

This formula may be rearranged to calculate any one value when the other three are known.

Tech Facts

Ferromagnetic materials are materials that can be magnetized easily. Paramagnetic materials are materials that cannot be magnetized easily. Diamagnetic materials are materials that cannot be magnetized under normal conditions.

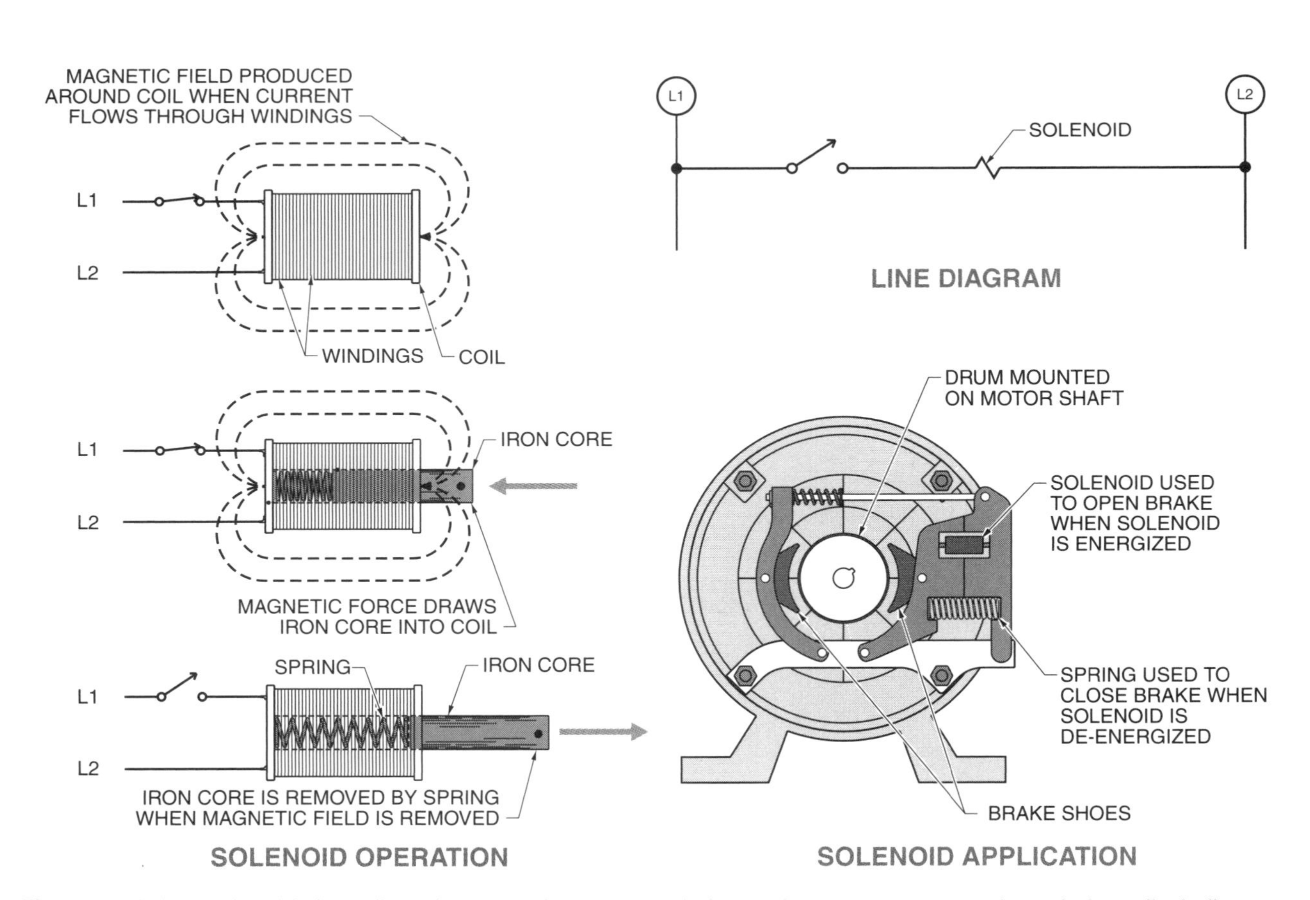

Figure 28-3. In a solenoid, the coil produces an electromagnetic force when current passes through the coil windings.

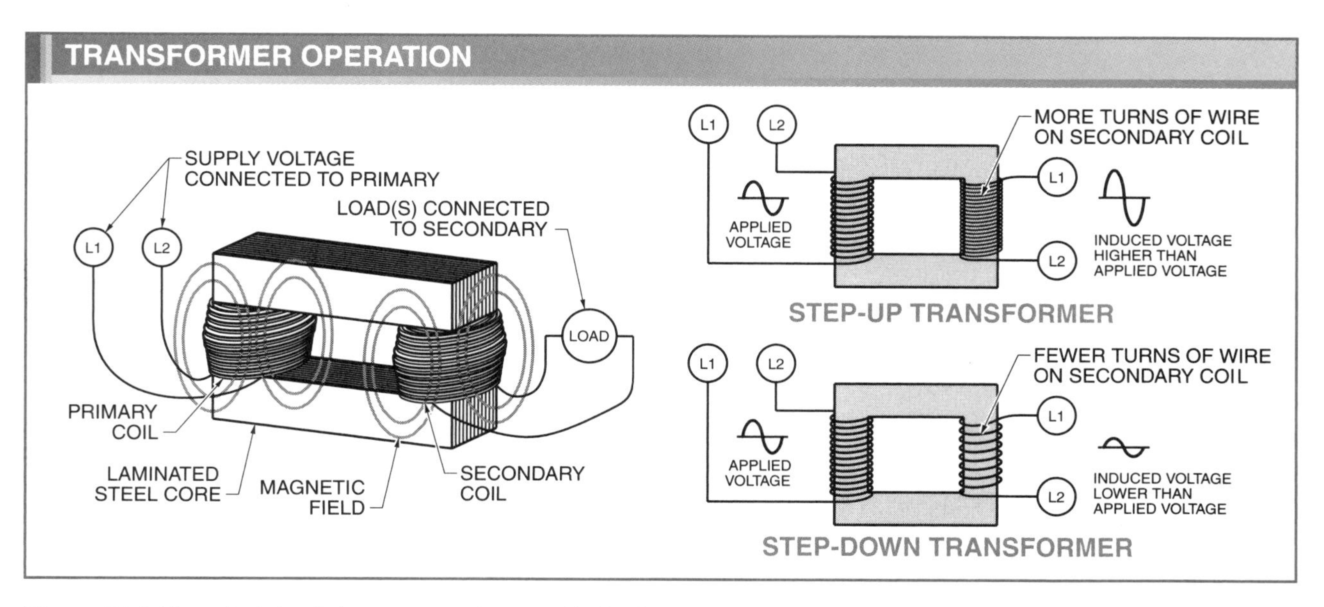

Figure 28-4. The principle of electromagnetic mutual induction states that when the magnetic flux lines from one expanding and contracting magnetic field cut the windings of a second coil, a voltage is induced in the second coil.

Example: Calculating Secondary Induced Voltage

What is the secondary induced voltage of a transformer that has a secondary coil containing 100 turns, a primary coil containing 500 turns, and 25 V applied to the primary coil?

$$E_S = E_P \times \frac{N_S}{N_P}$$

$$E_S = 25 \times \frac{100}{500}$$

$$E_S = \mathbf{5\ V}$$

This transformer is a step-down transformer because the voltage is reduced from 25 V to 5 V. The same formula applies to a step-up transformer. For example, 25 V applied to a transformer primary coil containing 500 turns induces 100 V in the transformer secondary coil containing 2000 turns (25 × (2000 ÷ 500) = 100).

In the process of changing voltage from one level to another, transformers also change the current to a lower or higher level. The ratio of primary current to secondary current is inversely proportional to the voltage ratio. For example, if the voltage ratio is 10:1, the current ratio is 1:10. To calculate the current and voltage relationship, apply the formula:

$$\frac{I_P}{I_S} = \frac{E_S}{E_P}$$

where

I_P = current in primary coil (in A)

I_S = current in secondary coil (in A)

E_S = voltage induced in secondary coil (in V)

E_P = voltage applied to primary coil (in V)

This formula may be rearranged to calculate any one value when the other three are known.

Example: Calculating Transformer Primary Current

What is the current in the primary winding of a transformer when the primary voltage is 240 V, the secondary voltage is 120 V, and a 10 A load is connected to the secondary?

$$I_P = I_S \times \frac{E_S}{E_P}$$

$$I_P = 10 \times \frac{120}{240}$$

$$I_P = 10 \times .5$$

$$I_P = \mathbf{5\ A}$$

The current potential of a transformer is stepped down any time a transformer steps up the voltage. Likewise, the current potential is stepped up any time a transformer steps down the voltage.

Transformer Ratings

Transformers are designed to transform power at one voltage level to power at another voltage level. In an ideal transformer, there is no loss or gain. Energy is simply transferred from the primary circuit to the secondary circuit. For example, if the secondary of a transformer requires 500 W of power to operate the loads connected to it, the primary must deliver 500 W.

It is standard practice with transformers to use voltage and current ratings and not wattage ratings. It is also standard practice to rate a transformer for its output capabilities because it is the output of the transformer to which the loads are connected. Thus, transformers are rated by their volt-ampere (VA), or kilovolt-ampere (kVA) output. Small transformers are rated in either VA or kVA. Large transformers are rated in kVA. For example, a 50 VA transformer may be rated as a 50 VA or 0.05 kVA transformer. A 5000 VA transformer is rated as a 5 kVA transformer.

In an ideal transformer, energy is transferred from the primary circuit to the secondary circuit and there is no power loss. However, all transformers have some power loss. Even though most transformers operate with little power loss (normally 0.5% to 8%), there is always a power loss in a transformer. The less the power loss, the more efficient the transformer. The efficiency of a transformer is expressed as a percentage. To calculate the efficiency of a transformer, apply the formula:

$$Eff = \frac{P_S}{P_P} \times 100$$

where

Eff = efficiency (in %)

P_S = power of secondary circuit (in W)

P_P = power of primary circuit (in W)

Example: Calculating Transformer Efficiency

What is the efficiency of a transformer that uses 1200 W of primary power to deliver 1110 W of secondary power?

$$Eff = \frac{P_S}{P_P} \times 100$$

$$Eff = \frac{1110}{1200} \times 100$$

$$Eff = .925 \times 100$$

$$Eff = \mathbf{92.5\%}$$

Power loss in a transformer is caused by hysteresis loss, eddy-current loss, and copper loss. The total amount of power loss is a combination of the three.

Transformers are used on AC circuits. In all AC circuits, the voltage (and magnetic field) are constantly changing. *Hysteresis loss* is loss caused by magnetism that remains (lags) in a material after them magnetizing force has been removed. Hysteresis loss occurs every half-cycle of AC when the current reverses direction and some magnetism remains in the iron core.

Eddy-current loss is loss caused by the induced currents that are produced in metal parts that are being magnetized. In a transformer, eddy-current loss occurs in the iron core because the iron core is magnetized. The induced currents in the iron core cause heat in the iron core, primary windings, and secondary windings. Eddy currents are reduced by using a laminated steel core.

Copper is a good conductor of electricity. However, copper has some resistance. *Copper loss* is loss caused by the resistance of the copper wire to the flow of current. The higher the resistance, the greater the loss.

Hysteresis loss, eddy-current loss, and copper loss all produce losses in the transformer due to the heat they develop in the metal parts. These losses may be reduced by proper design and installation. Hysteresis loss is reduced by using a silicon-steel core instead of an iron core. Eddy-current loss is reduced by using a laminated core instead of a solid core. Copper loss is reduced by using as large a conductor as possible.

Transformer Types

All transformers are used to either step up, step down, or isolate voltage. However, when transformers are used for specific applications, they are normally referred to by the application name, such as a distribution transformer, control transformer, instrument transformer, etc. The basic types of transformers include appliance/equipment, control, bell/chime, instrument (current), distribution, isolation, neon sign, and power transformers. **See Figure 28-5.**

Appliance/equipment transformers are specifically designed for the appliance/equipment in which they are used. Electrical and electronic devices, such as TVs, garage door openers, and computers require different voltages to operate their different circuits. The transformer normally has many secondary voltage output levels that can be used to operate several different circuits. Appliance/equipment transformers are rated from 2 VA (or less) to several thousand VA and deliver secondary voltages from 5 V (or less) to several thousand volts.

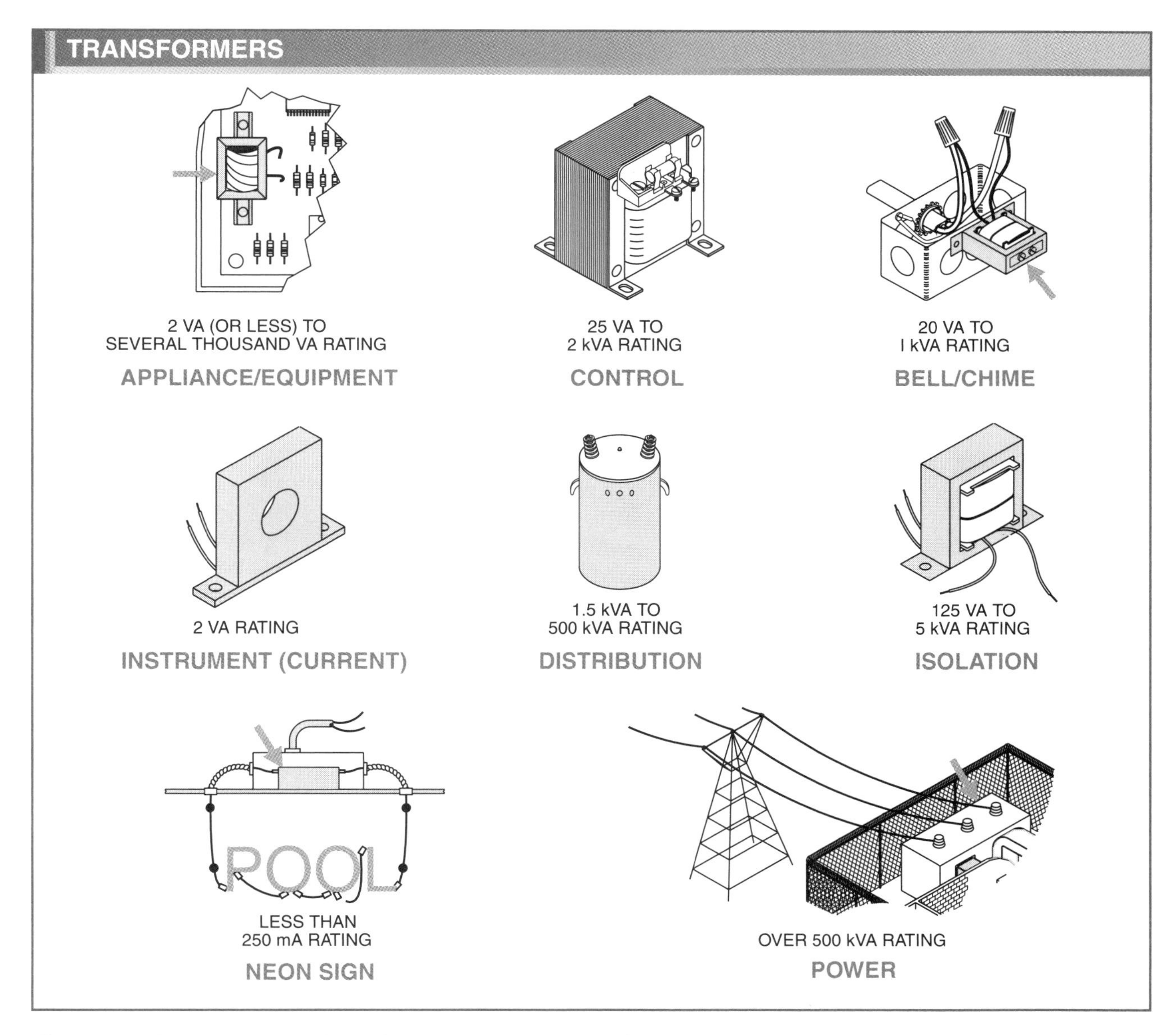

Figure 28-5. The basic types of transformers include appliance/equipment, control, bell/chime, instrument (current), distribution, isolation, neon sign, and power transformers.

Control transformers are step-down transformers that are used to lower the voltage to the control circuit in which the control switches, contactors, and motor starters are connected. Industrial motors and heating elements normally operate on 460/480 V. However, the control circuits used to control the 460/480 V loads normally operate on 115/120 V. Control transformers are normally rated from 25 VA (.025 kVA) to 2000 VA (2 kVA) and deliver a secondary voltage of 115/120 V. Control transformers are mounted inside control cabinets and motor control centers.

Bells, chimes, buzzers, fire systems, and security systems normally operate at voltage levels from 8 V to 24 V. Bell/chime transformers are specifically designed to step down 115/120 V to 8 V, 12 V, 16 V, or 24 V. Bell/chime transformers are normally rated from 20 VA to 1000 VA (1 kVA). The size (VA rating) of the transformer is based on the number and power rating of the loads. The larger the number of loads, the higher the required power rating. Likewise, the higher the power rating of the loads connected to the transformer, the higher the required power rating.

Instrument transformers are transformers specifically designed to step down the voltage or current of a circuit to a lower value that can safely be used to measure the voltage, current, or power in the circuit with a meter (instrument). Instrument transformers are rated according to their primary to secondary ratio, such as 10:1, 50:1, 400:5, etc. For example, a current transformer that has a 10:1 ratio delivers 1 A output on the secondary for every 10 A flowing through the primary. There is 2.5 A flowing through the secondary if there is 25 A flowing through the primary. The ammeter (instrument) that is measuring the circuit current (25 A) is connected to the secondary so that only 2.5 A is flowing into the meter. The meter is calibrated to display 25 A whenever there is a 2.5 A input. Instruments and meters may use an external instrument transformer or include an instrument transformer as an integral part of the meter.

Distribution transformers are step-down transformers that reduce high transmitted voltage down to usable residential, commercial, and industrial voltage levels. Although power is efficiently transmitted at high voltages, high voltage is not safe for use in practical applications. Distribution transformers are normally rated from 1.5 kVA to 500 kVA and deliver a secondary voltage of 115 V, 120 V, 208 V, 230 V, 240 V, 460 V, or 480 V.

Isolation transformers are transformers that are designed to isolate the load from the power source. Isolation transformers have a 1:1 turns (and voltage) ratio. Therefore, the primary voltage is equal to the secondary voltage. The transformer provides isolation between the two sections of the circuit even though the primary and secondary are operating at the same voltage.

Neon sign transformers are step-up transformers designed to step up the secondary voltage to several thousand volts when connected to a primary voltage of 115/230 V. Neon sign transformers are rated according to their secondary voltage. Secondary voltages normally range from 2000 V to 15,000 V. Although the secondary voltage is high, the current rating is relatively low. Current ratings normally range less than 250 mA.

Power transformers are large transformers used by power companies in generating plants to step up voltages for power transmission and in substations to step down voltages. Generated voltage must be stepped up to a high level for efficient transmittal over long distances. The voltage is stepped up to a high level to step the current down to a low level. Low current levels require small wire sizes, reducing cost. Power transformers are normally rated over 500 kVA (500,000 VA) and used on power lines that operate on voltages over 67 kV (67,000 V).

Distribution transformers are the same as power transformers but are smaller and serve a different purpose in the power distribution system. Small distribution transformers are mounted on poles or the ground. Large distribution transformers are mounted on the ground.

Transformer Overloading

All transformers have a power (VA or kVA) rating. The power rating of a transformer indicates the amount of power the transformer can safely deliver. However, like most electrical devices, this rating is not an absolute value. For example, a 100 VA rated transformer is not destroyed if required to deliver 110 VA for a short period of time.

The heat produced by the power destroys the transformer. The heat destroys the transformer by breaking down the insulation, causing short circuits within the windings of the transformer. Thus, temperature is the limiting factor in transformer loading. The more power the transformer must deliver, the higher the temperature produced at the transformer.

Transformers are used to deliver power to a set number of loads. For example, a transformer is used to deliver power to a school. As loads in the school are switched ON and OFF, the power delivered by the transformer changes. At certain times (night), the power output required from the transformer may be low. At other times (during school hours), the power output required from the transformer may be high. *Peak load* is the maximum output required of a transformer. **See Figure 28-6.**

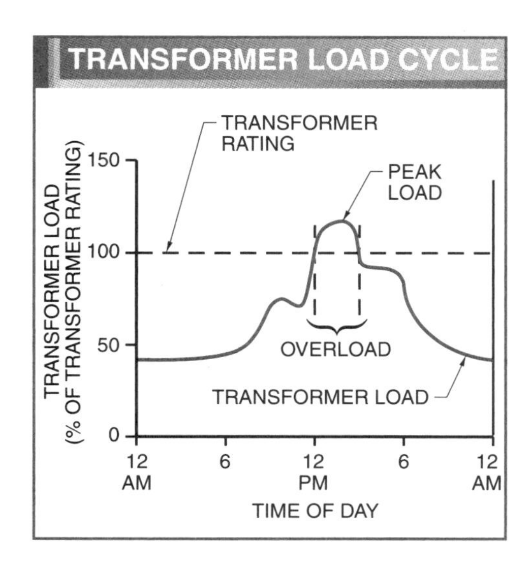

Figure 28-6. A transformer is overloaded when it is required to deliver more power than its rating.

A transformer is overloaded when it is required to deliver more power than its rating. A transformer is not damaged when overloaded for a short time period. This is because the heat storage capacity of a transformer ensures a relatively slow increase in internal transformer temperature. Transformer manufacturers list the length of time a transformer may be safely overloaded at a given peak level.

Transformer Cooling Methods

The methods used to dissipate heat in a transformer include self-air cooled, forced-air cooled, liquid-immersed/ self-air cooled, and liquid-immersed/forced-air cooled.

Self-air cooled transformers are transformers that dissipate heat through the air surrounding the transformer. Heat produced in the windings and core is dissipated into the surrounding air by convection. Convection heat transfer is increased by adding radiating fins to the transformer.

Forced-air cooled transformers are transformers that use a fan to move air over the transformer. Using a fan to speed the convection process increases the power that the transformer can deliver by about 30% over the power that can be delivered without a fan.

Multiple high-velocity fans are used in some applications to increase the transformer power output by more than 30%. The fans may be designed to remain ON at all times or may be automatically turned ON when the transformer reaches a set temperature.

Liquid-immersed/self-air cooled transformers are transformers that use refined oil or synthetic oil to help cool the transformer windings. The transformer coils and core are enclosed in a metal tank which is immersed in the oil. The oil is used to conduct heat from the windings and core to the outer surface of the transformer. The oil helps slow the heating process by increasing the heat storage capacity of the transformer. This is useful when the transformer is temporarily overloaded during peak usage times.

Liquid-immersed/forced-air cooled transformers are transformers that use refined oil or synthetic oil and fans to cool the transformer. The oil conducts the heat to the outer surface of the transformer and the fans dissipate the heat to the surrounding air.

Sizing Single-Phase Transformers

A transformer is used any time power is delivered to a residential, commercial, industrial, construction, or other site. The size of the transformer is based on the amount of expected power required. This amount normally takes into consideration present and future needs, peak loading, ambient temperature, load types, and other factors that affect transformer operation and temperature. **See Figure 28-7.** To size a 1ϕ transformer, apply the procedure:

1. Determine the total voltage required by the loads if more than one load is connected. The secondary side of the transformer must have a rating equal to the voltage of the loads.
2. Determine the amperage rating or kVA capacity required by the load(s). Add all loads that are (or may be) ON concurrently.
3. Check load(s) frequency on the nameplate. The frequency of the supply voltage and the electrical load(s) must be the same.
4. Check the supply voltage to the primary side of the transformer. The primary side of a transformer must have a rating equal to the supply voltage. Consider each voltage if there is more than one source voltage available. Use a transformer that has primary taps if there is a variation in the supply voltage. The transformer must have a kVA capacity of at least 10% greater than that required by the loads.

Tech Facts

Bar magnets should be stored in parallel with each other so their unlike poles touch. Horseshoe magnets should be stored with a soft iron bar placed across the pole ends.

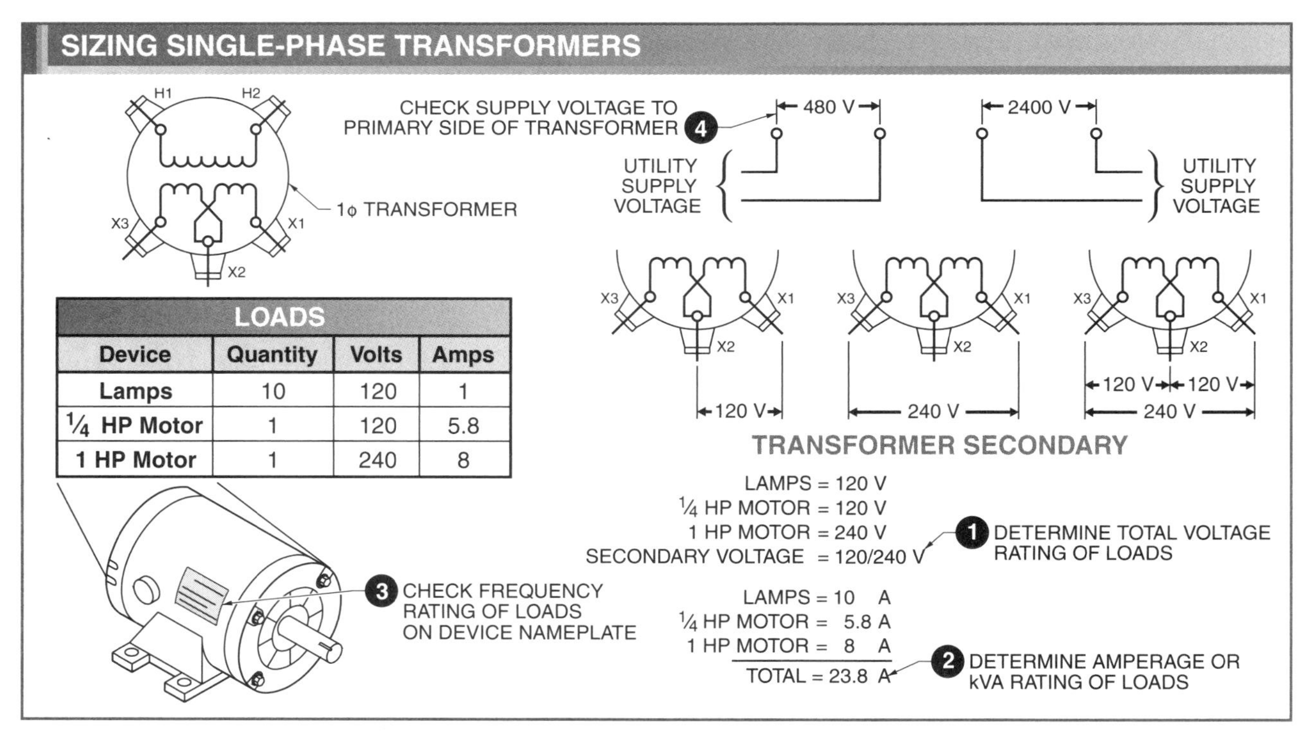

Device	Quantity	Volts	Amps
Lamps	10	120	1
¼ HP Motor	1	120	5.8
1 HP Motor	1	240	8

Figure 28-7. The size of the transformer is based on the amount of expected power required.

A single-phase full-load currents conversion table may be used to determine proper kVA capacity when the load rating is given in amperes. **See Figure 28-8.** To calculate kVA capacity of a 1ϕ transformer when voltage and current are known, apply the formula:

$$kVA_{CAP} = E \times \frac{I}{1000}$$

where

kVA_{CAP} = transformer capacity (in kVA)

E = voltage (in V)

I = current (in A)

1000 = constant

Example: Calculating 1ϕ Transformer Capacity

What is the kVA capacity of a 1ϕ transformer used in a 240 V circuit that has loads of 5 A, 3 A, and 8 A connected to the secondary?

$$kVA_{CAP} = E \times \frac{I}{1000}$$

$$kVA_{CAP} = 240 \times \frac{16}{1000}$$

$$kVA_{CAP} = \frac{3840}{1000}$$

$$kVA_{CAP} = \mathbf{3.84\ kVA}$$

A 5 kVA transformer is selected for the application because it is the next size available with a capacity of at least 10% greater than that required by the loads.

Sizing Three-Phase Transformers

The size of a 3ϕ transformer is based on the amount of expected power required in the circuit. **See Figure 28-9.** To size a 3ϕ transformer, apply the procedure:

1. Determine the total voltage required by the loads if more than one load is connected. The secondary side of the transformer must have a rating equal to the voltage of the loads.
2. Determine the amperage rating or kVA capacity required by the load(s). Add all loads that are (or may be) ON concurrently.
3. Check the frequency of the load(s) on the nameplate. The frequency of the supply voltage and the electrical load(s) must be the same.
4. Determine the type of 3ϕ voltage available. This includes three-wire no ground or three-wire with ground (four-wire).
5. Check the supply voltage to the primary side of the transformer.

SINGLE-PHASE FULL-LOAD CURRENTS*						
kVA	**120 V**	**208 V**	**240 V**	**277 V**	**380 V**	**480 V**
.050	.4	.2	.2	.2	.1	.1
.100	.8	.5	.4	.3	.2	.2
.150	1.2	.7	.6	.5	.4	.3
.250	2.0	1.2	1	.9	.6	.5
.500	4.2	2.4	2.1	1.8	1.3	1
.750	6.3	3.6	3.1	2.7	2	1.6
1	8.3	4.8	4.2	3.6	2.6	2.1
1½	12.5	7.2	6.2	5.4	3.9	3.1
2	16.7	9.6	8.3	7.2	5.2	4.2
3	25	14.4	12.5	10.8	7.9	62
5	41	24	20.8	18	13.1	10.4
7½	62	36	31	27	19.7	15.6
10	83	48	41	36	26	20.8
15	125	72	62	54	39	31

* in A

Figure 28-8. A single-phase full-load currents conversion table may be used to determine proper kVA capacity when the load rating is given in amperes.

The primary side of a transformer must have a rating equal to the supply voltage. Consider each voltage when there is more than one source of voltage available. Use a transformer that has primary taps when there is a variation in the supply voltage. The transformer must have a kVA capacity of at least 10% greater than that required by the loads. A three-phase full-load current conversion table is used to determine the kVA capacity of 3ϕ circuits when the load rating is given in amperes. **See Figure 28-10.** To calculate the kVA capacity of a 3ϕ transformer when voltage and current are known, apply the formula:

$$kVA_{CAP} = E \times 1.732 \times \frac{I}{1000}$$

where

kVA_{CAP} = transformer capacity (in kVA)

E = voltage (in V)

1.732 = constant (for 3ϕ power)

I = current (in A)

1000 = constant

Transformers are often placed on concrete pads. Barriers are then put in place to protect the transformer from damage.

THREE-PHASE FULL-LOAD CURRENTS*

kVA	208 V	240 V	480 V	600 V
3	8.3	7.2	3.6	2.9
4	12.5	10.8	5.4	4.3
6	16.6	14.4	7.2	5.8
9	25	21.6	10.8	8.6
15	41	36	18	14.4
22	62	54	27	21.6
30	83	72	36	28
45	124	108	54	43

* in A

Figure 28-10. A three-phase full-load current conversion table is used to determine the kVA capacity of 3ϕ circuits when the load rating is given in amperes.

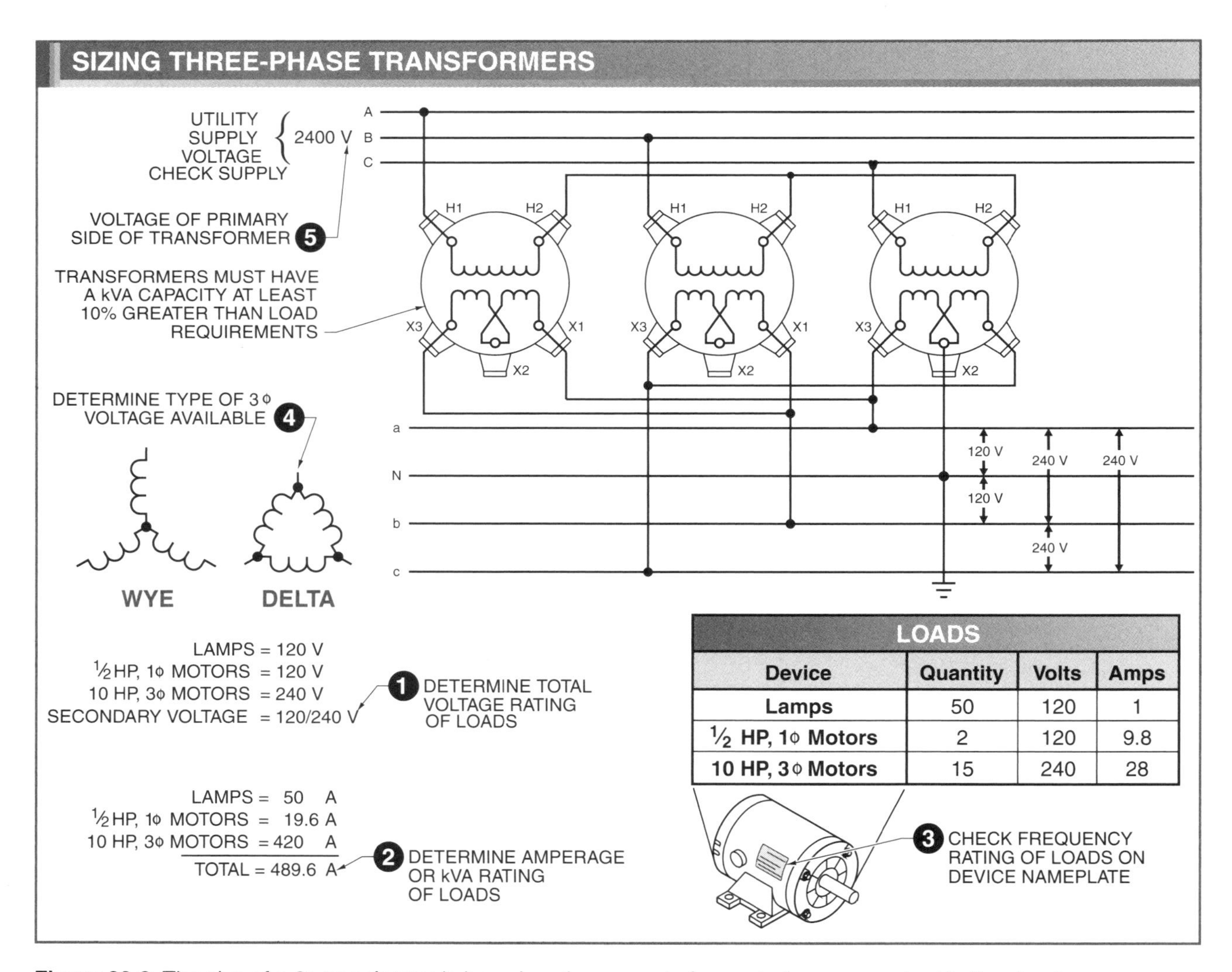

Device	Quantity	Volts	Amps
Lamps	50	120	1
½ HP, 1ϕ Motors	2	120	9.8
10 HP, 3ϕ Motors	15	240	28

Figure 28-9. The size of a 3ϕ transformer is based on the amount of expected power required in the circuit.

Example: Calculating 3ϕ Transformer kVA Capacity

What is the kVA capacity of a 3ϕ transformer used in a 240 V circuit that has loads of 25 A, 30 A, and 8 A connected to its secondary?

$$kVA_{CAP} = E \times 1.732 \times \frac{I}{1000}$$

$$kVA_{CAP} = 240 \times 1.732 \times \frac{63}{1000}$$

$$kVA_{CAP} = \frac{26{,}188}{1000}$$

$$kVA_{CAP} = \mathbf{26.188\ kVA}$$

A 30 kVA transformer is selected for the application because it is the next size available with a capacity of at least 10% greater than that required by the loads.

Tech Facts

Some transformers are provided with a tap changer to provide a uniform secondary voltage when the primary voltage varies due to line voltage drop.

Determining Single-Phase Transformer Current Draw

A transformer has a voltage ready to be used. However, no current flows and no power is used if no load is connected to the transformer. Current flows and electrical power is consumed when loads are connected to the available voltage source. The amount of current flow depends on the power rating of the load. The higher the power rating, the larger the amount of current flow. Likewise, the lower the power rating, the smaller the amount of current flow. Knowing the amount of current flowing in a circuit is necessary because the conductor size is based on the amount of expected current flow. To calculate transformer current draw when kVA capacity and voltage are known, apply the formula:

$$I = kVA_{CAP} \times \frac{1000}{E}$$

where

I = current draw (in A)

kVA_{CAP} = transformer capacity (in kVA)

1000 = constant (to convert VA to kVA)

E = voltage (in V)

Example: Calculating 1ϕ Transformer Current Draw

What is the current draw of a 1ϕ, 41 kVA, 120 V rated transformer when fully loaded?

$$I = kVA_{CAP} \times \frac{1000}{E}$$

$$I = 41 \times \frac{1000}{120}$$

$$I = \frac{41{,}000}{120}$$

$$I = \mathbf{341\ A}$$

Determining Three-Phase Transformer Current Draw

The current draw of a 3ϕ transformer is calculated similarly to the current draw of a 1ϕ transformer. The only difference is the addition of the constant (1.732) for 3ϕ power. To calculate current draw of a 3ϕ transformer when kVA capacity and voltage are known, apply the formula:

$$I = kVA_{CAP} \times \frac{1000}{E \times 1.732}$$

where

I = current (in A)

kVA_{CAP} = transformer capacity (in kVA)

1000 = constant

E = voltage (in V)

1.732 = ($\sqrt{3}$)

Example: Calculating 3ϕ Transformer Current Draw

What is the current draw of a 3ϕ, 30 kVA, 480 V rated transformer when fully loaded?

$$I = kVA_{CAP} \times \frac{1000}{E \times 1.732}$$

$$I = 30 \times \frac{1000}{480 \times 1.732}$$

$$I = \frac{30{,}000}{831}$$

$$I = \mathbf{36.1\ A}$$

Transformer Standard Ambient Temperature Compensation

Temperature rise in a transformer is the temperature of the windings above the existing ambient temperature. Transformer nameplates list their maximum temperature rise. Transformer normal ambient temperature is 40°C.

A transformer must be derated if the ambient temperature exceeds 40°C. Transformer derating charts are used to derate transformers in high ambient temperatures. **See Figure 28-11.** To calculate the derated kVA capacity of a transformer operating at a higher than- normal ambient temperature condition, apply the formula:

kVA = *rated kVA* × *maximum load*

where

kVA = derated transformer capacity (in kVA)

rated kVA = manufacturer transformer rating (in kVA)

maximum load = maximum transformer loading (in %)

TRANSFORMER DERATINGS	
Maximum Ambient Temperature (°C)	**Maximum Transformer Loading (%)**
40	100
45	96
50	92
55	88
60	81
65	80
70	76

Figure 28-11. Transformer derating charts are used to derate transformers in high ambient temperatures.

Example: Calculating Standard Transformer Derating

What is the derated kVA value of a 30 kVA rated transformer installed in an ambient temperature of 50°C?

kVA = *rated kVA* × *maximum load*

kVA = 30 × 0.92

kVA = **27.6 kVA**

Transformer Special Ambient Temperature Compensation

Standard ambient temperature is the average temperature of the air that cools a transformer over a 24-hour period. Standard ambient temperature assumes that maximum temperature does not exceed 10°C above average ambient temperature.

A transformer is derated above the standard values when the maximum temperature exceeds the average temperature by more than 10°C. A transformer is derated by 1½% for each 1°C above 40°C when the maximum ambient temperature exceeds 10°C above the average temperature.

For example, a 30 kVA rated transformer installed in an ambient temperature of 50°C (maximum temperature exceeds ambient temperature by 10°C) is derated to 25.5 kVA. The transformer is derated by 1½% for each degree above 40°C. A 50°C ambient temperature is 10°C above 40°C so the transformer is derated 15% (10°C × 1½% = 15%).

kVA = *rated kVA* × *maximum load*

kVA = 30 × 0.85 (15% derating)

kVA = **25.5 kVA**

Single-Phase Residential Transformer Connections

Electricity is used in residential applications (one-family, two-family, and multifamily dwellings) to provide energy for lighting, heating, cooling, cooking, running motors, etc. The electrical service to dwellings is normally 1ϕ, 120/240 V. The low voltage (120 V) is used for general-purpose receptacles and general lighting. The high voltage (240 V) is used for heating, cooling, cooking, etc.

Three-Phase Transformer Connections

Three 1ϕ transformers may be connected to develop 3ϕ voltage. The three transformers may be connected in a wye or delta configuration. A *wye configuration* is a transformer connection that has one end of each transformer coil connected together. The remaining end of each coil is connected to the incoming power lines (primary side) or used to supply power to the load(s) (secondary side). A *delta configuration* is a transformer connection that has each transformer coil connected end-to-end to form a closed loop. Each connecting point is connected to the incoming power lines or used to supply power to the load(s). The voltage output and type available for the load(s) is determined by whether the transformer is connected in a wye or delta configuration. **See Figure 28-12.**

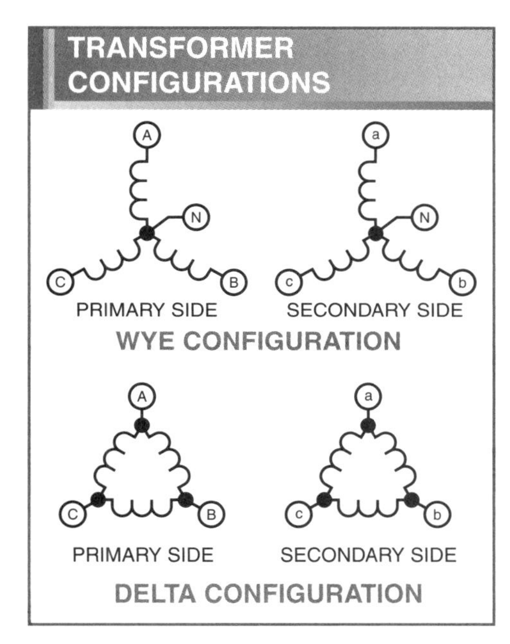

Figure 28-12. The voltage output and type available for the load(s) is determined by whether the transformer is connected in a wye or delta configuration.

Three-Phase, Delta-to-Delta Connections

Three transformers may be connected in a delta-to-delta connection. A delta-to-delta transformer connection is used to supply 3ϕ voltage on the secondary. In a delta-to-delta connection, each transformer is connected end-to-end. **See Figure 28-13.**

The advantage of a delta-to-delta connection is that if one transformer is disabled, the other two may be used in an open-delta connection for emergency power. The rating of the open-delta bank is 57.7% of the original three transformer bank, but 3ϕ power is available until repairs are made.

One of the delta transformers is center-tapped to supply both 3ϕ voltage and 1ϕ voltage. Single-phase voltage at 120/240 V is available when the transformer is center-tapped. However, because only one transformer is tapped, the transformer that is tapped carries all of the 1ϕ, 120/240 V load and $\frac{1}{3}$ of the 3ϕ, 240 V load. The other two transformers each carry $\frac{1}{3}$ of the 3ϕ, 240 V load. For this reason, this connection should be used in applications that require a large amount of 3ϕ power and a small amount of 1ϕ power.

Three-Phase, Wye-to-Wye Connections

Three transformers may be connected in a wye-to-wye connection. A wye-to-wye transformer connection is used to supply both 1ϕ and 3ϕ voltage. In a wye-to-wye transformer connection, the ends of each transformer are connected together. **See Figure 28-14.**

The advantage of a wye-connected secondary is that the 1ϕ power draw may be divided equally over the three transformers. Each transformer carries $\frac{1}{3}$ of the 1ϕ and 3ϕ power if the loads are divided equally. A disadvantage of a wye-to-wye connection is that interference with telephone circuits may result.

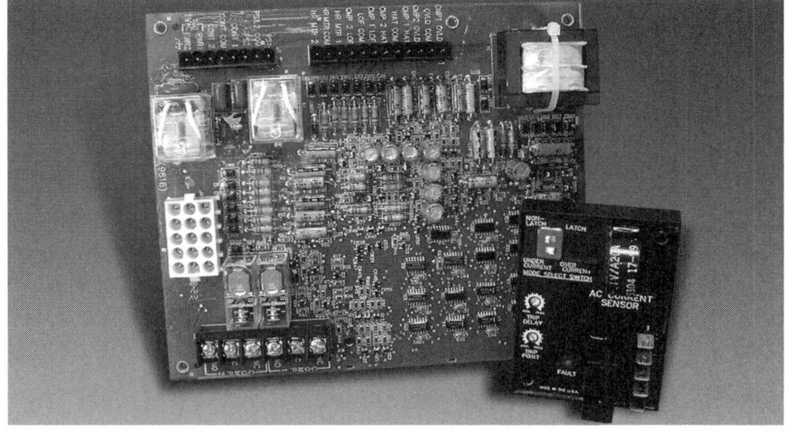

Ametek®

Small transformers are often used in circuit boards to raise or lower the voltage or for circuit isolation.

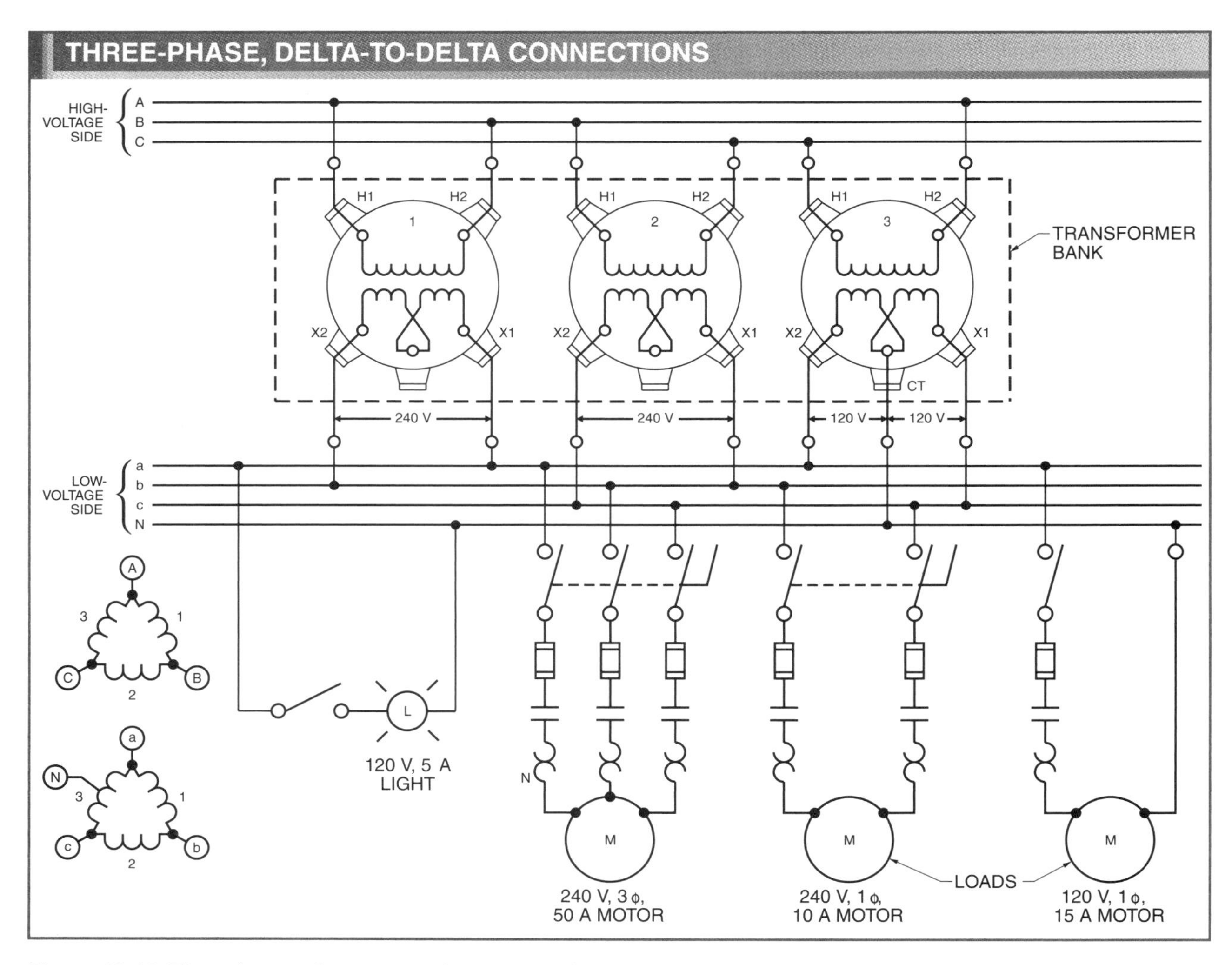

Figure 28-13. Three 1ϕ transformers may be connected to develop 3ϕ voltage.

Bushings provide electrical resistance, which insulates power lines from transformer enclosures.

Delta-to-Wye/Wye-to-Delta Connections

Transformers may also be connected in a delta-to-wye or wye-to-delta connection. The connection used depends on the incoming supply voltage, the requirements of the loads, and the practice of the local power company. A delta-to-wye transformer connection delivers the same voltage output as the wye-to-wye transformer connection. The difference is the primary is supplied from a delta system. A wye-to-delta transformer connection delivers the same voltage output as the delta-to-delta transformer connection. The difference is that the primary is supplied from a wye system.

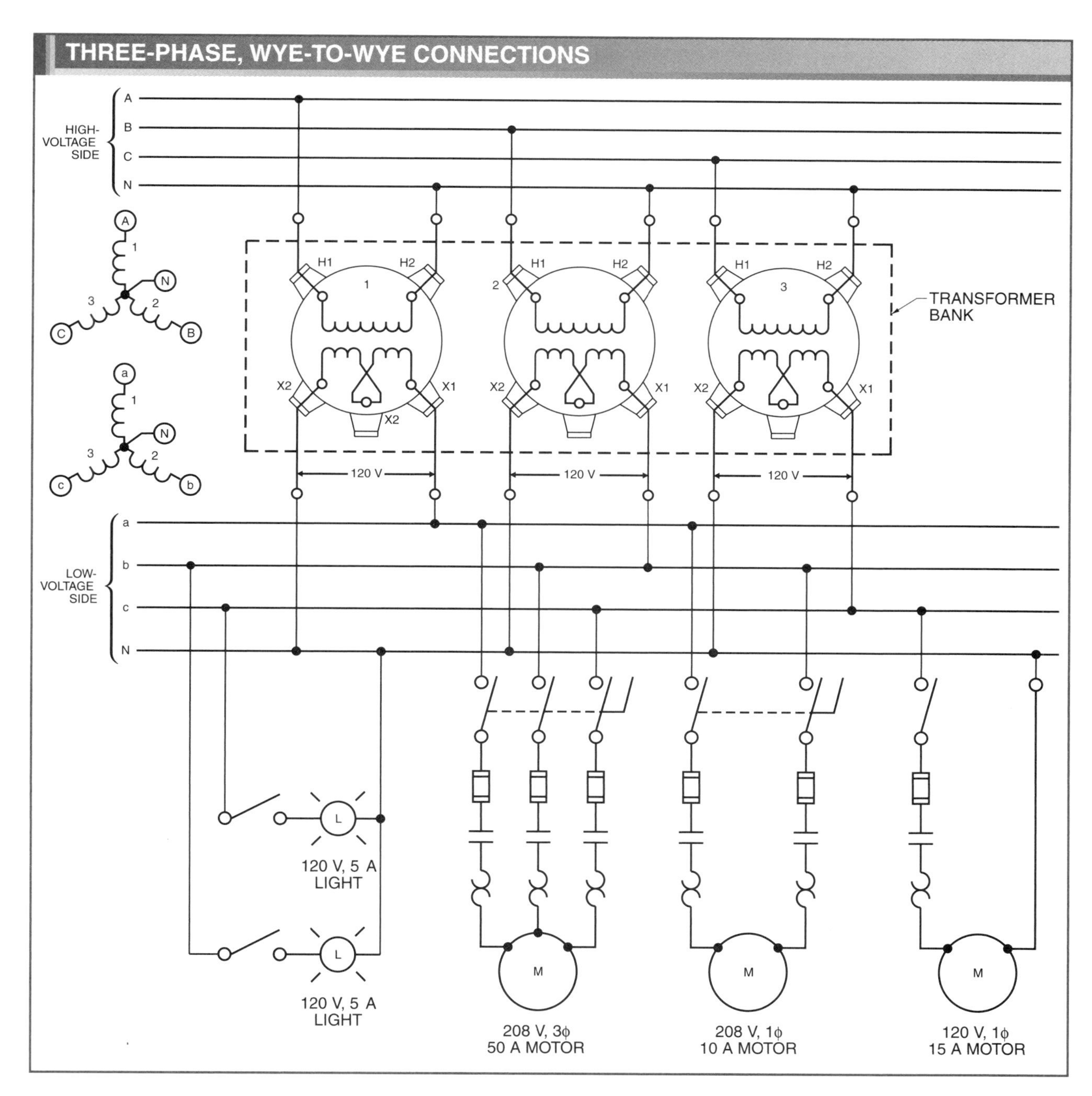

Figure 28-14. A wye-to-wye transformer connection is used to supply both 1ϕ and 3ϕ voltage.

Transformer Load Balancing

The loads connected to a transformer should be connected so that the transformer is as electrically balanced as possible. Electrical balance occurs when loads on a transformer are placed so that each coil of the transformer carries the same amount of current for the various loads, such as for several motors. **See Figure 28-15.**

Transformer Tap Connections

Transformer taps are connecting points that are provided along the transformer coil. Taps are available on some transformers to correct for excessively high or low voltage conditions. The taps are located on the primary side of the transformer. Standard taps are provided for 2% and 5% of rated primary voltage. **See Figure 28-16.**

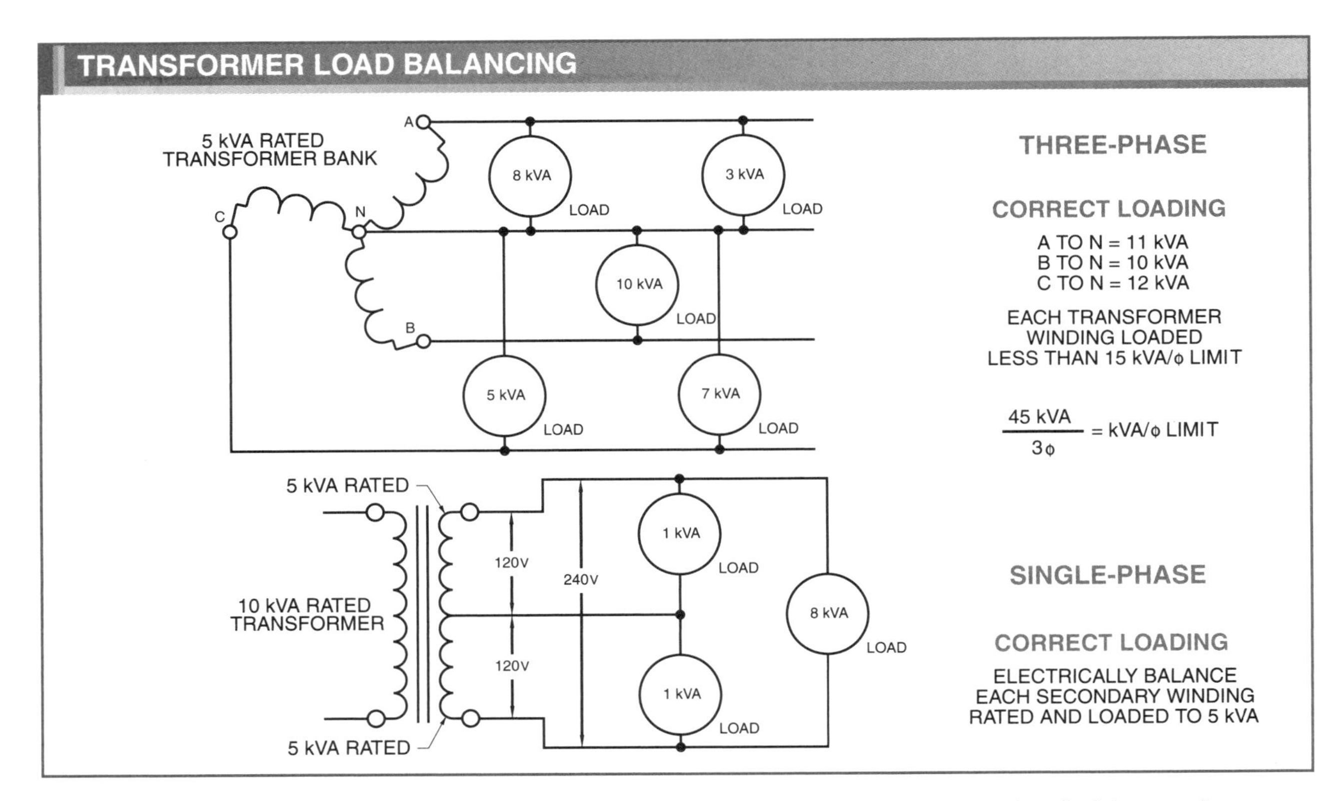

Figure 28-15. Electrical balance occurs when loads on a transformer are placed so that each coil of the transformer carries the same amount of current.

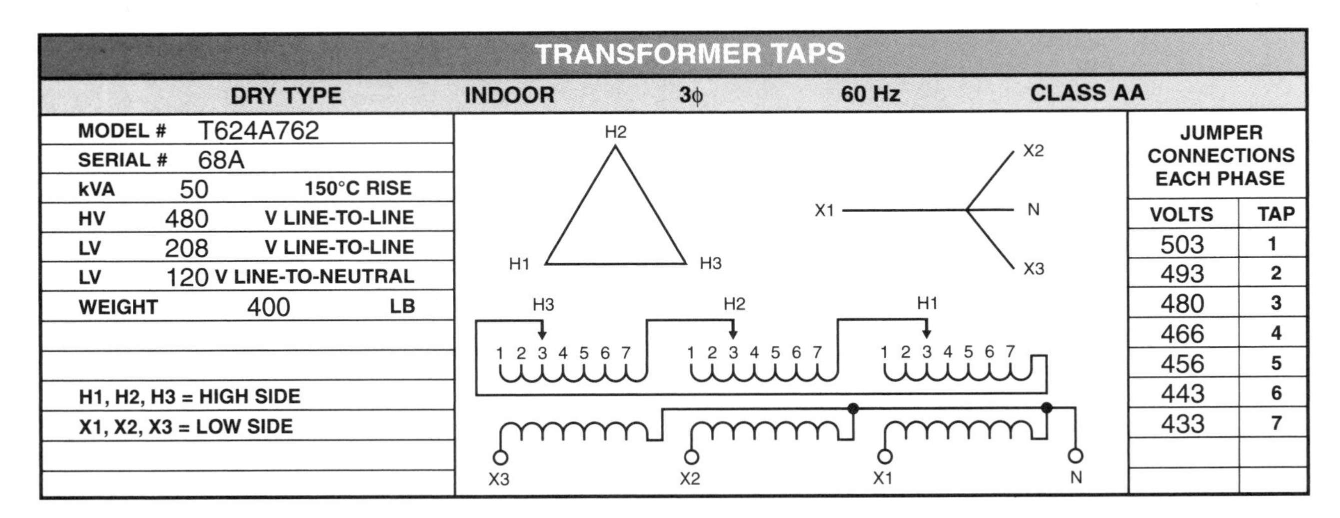

Figure 28-16. Transformer taps are connecting points that are provided along the transformer coil.

For example, if a transformer has a 480 V primary rating and the available supply voltage is 504 V, the primary should be connected to the 5% above-normal tap. This ensures that the secondary voltage is correct even when the primary voltage is high.

Single-Phase Transformer Parallel Connections

Additional power is required when the capacity of a transformer is insufficient for the power requirements of the load(s). Additional power may be obtained by changing the overloaded transformer to a larger size (higher kVA rating) or adding a second transformer in parallel with the overloaded transformer. The best and most efficient method is to replace the overloaded transformer with a larger one. However, in some applications, it is easier to add a second transformer in parallel. These include systems where extra power is needed only temporarily or a larger transformer is not available.

Single-phase transformers may be connected in parallel as long as certain conditions are met. These conditions include:

- Primary and secondary voltage ratings are identical.
- Frequencies are the same.
- Tap settings are identical.
- Impedance of either transformer is within ±7% (93% to 107%) of the other. **See Figure 28-17.**

The total power rating of two compatible, 1ϕ transformers connected in parallel is equal to the sum of the individual power ratings. To calculate the total power rating of two 1ϕ transformers connected in parallel, apply the formula:

$kVA_T = kVA_1 + kVA_2$

where

kVA_T = Total rating of transformer combination (in kVA)

kVA_1 = Rating of transformer 1 (in kVA)

kVA_2 = Rating of transformer 2 (in kVA)

Example: Calculating Total Power Rating—Parallel-Connected Transformers

What is the total output rating of two compatible 1ϕ, 5 kVA transformers connected in parallel?

$kVA_T = kVA_1 + kVA_2$

$kVA_T = 5 + 5$

$kVA_T = \mathbf{10\ kVA}$

Three-Phase Transformer Parallel Connections

Like 1ϕ transformers, 3ϕ transformers may also be connected in parallel. The conditions that must be met to connect 3ϕ transformers in parallel include:

- Primary and secondary voltage ratings are identical.
- Frequencies are the same.
- Tap settings are identical.
- Impedance of either transformer is within ±7% (93% to 107%) of the other.
- Angular displacement of transformer banks is the same. For example, both banks must have a 0°, 30°, or 180° angular displacement. Standard angular displacements are 0° for wye-to-wye or delta-to-delta connected banks and 30° for wye-to-delta or delta-to-wye connected banks. **See Figure 28-18.**

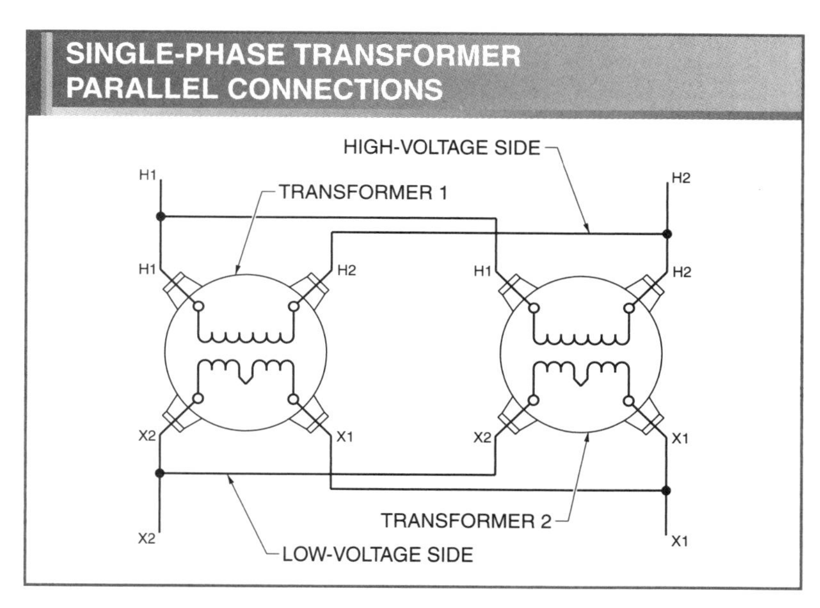

Figure 28-17. Additional power may be obtained by adding a second transformer in parallel with the overloaded transformer.

Calculating the total power rating of two compatible, 3ϕ transformers connected in parallel is similar to calculating the power rating of two compatible 1ϕ transformers connected in parallel. The total power rating of two compatible 3ϕ transformers connected in parallel equals the sum of the individual power ratings (kVA).

Tech Facts

It is dangerous and illegal to improperly dispose of transformer oil containing polychlorinated biphenyls (PCBs). Check local and state codes for proper transformer-oil handling procedures.

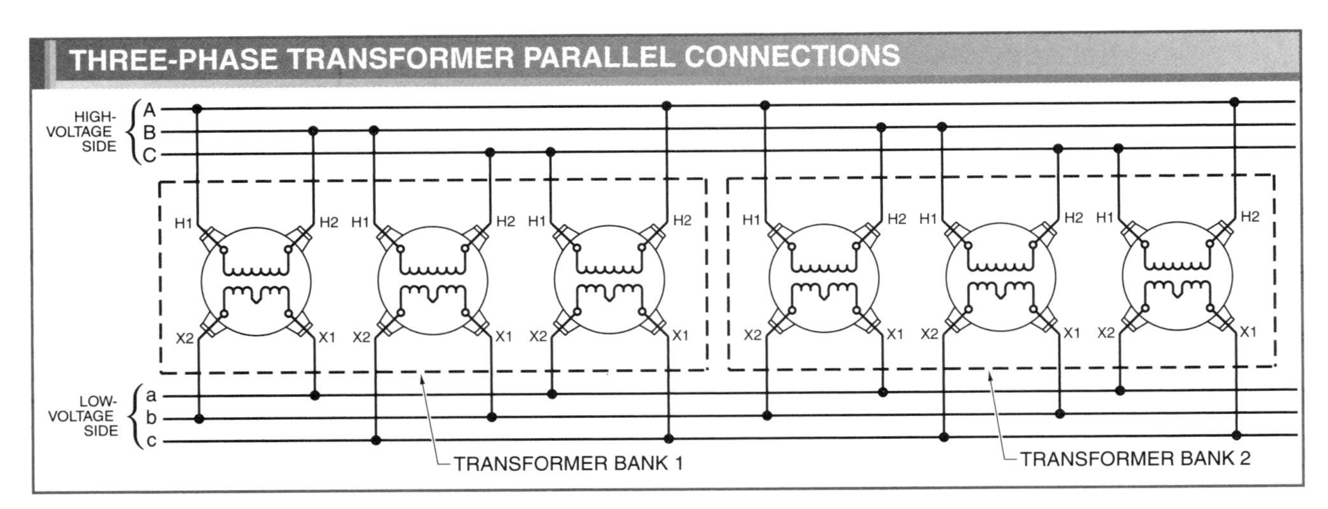

Figure 28-18. The total power rating of two compatible 3ϕ transformers connected in parallel equals the sum of the individual power ratings (kVA).

Review Questions

1. What are some of the ways that magnetism is utilized?
2. Where is magnetic flux the strongest and why?
3. What is a solenoid and how is it used?
4. Which electromagnetic principle provides the basis for transformer operation?
5. Explain the difference between a step-up transformer and a step-down transformer.
6. Describe the three types of power loss that occur in transformers.
7. Compare two types of step-down transformers.
8. When does a transformer become overloaded?
9. Describe a delta configuration.
10. What conditions must be met before a single-phase transformer can be connected in parallel?

Resistance, inductance, and capacitance are all attributes of an electric circuit. A resistive circuit is a circuit that contains only resistance. Inductance is the property of a circuit that causes it to oppose a change in current due to energy stored in a magnetic field. Capacitance is the ability of a component to store energy in the form of an electrical charge.

29 Resistance, Inductance, and Capacitance

BASIC CIRCUIT ELEMENTS

Electrical circuits range from simple to complex. Simple electrical circuits may include only a power supply, switch, and load. The power supply may be a DC, AC, DC to AC, or AC to DC power supply. The switch starts or stops the flow of current in the circuit. Switches may be mechanical or solid-state. A *load* is any device that converts electrical energy to motion, heat, light, or sound. Common loads include heating coils (electrical energy to heat), lamps (electrical energy to light), speakers (electrical energy to sound), and motors (electrical energy to motion).

A flashlight is an example of a simple electrical circuit. A flashlight includes a power supply (batteries), switch, and a load (lamp). A digital clock is an example of a more complex electrical circuit. A digital clock that plugs into 115 VAC includes a step-down transformer, rectifier (AC to DC) circuit, filter circuits, digital logic circuits, LED displays, and manual input controls (time setting switches).

In a flashlight circuit, the only opposition to current flow is the lamp. A high-wattage lamp offers less resistance (allowing a higher current flow) and a low-wattage lamp offers more resistance (allowing a smaller current flow). Ohm's law and the power formula may be directly applied for determining voltage (E), current (I), resistance (R), and power (P), because the lamp is a resistive load and the power supply (batteries) is pure DC.

Ohm's law and the power formula cannot always be directly applied to more complex DC circuits and most AC circuits. For example, the digital clock circuit includes inductance (step-down transformer) and capacitance (capacitors) in addition to resistance (resistors). Inductance (L) and capacitance (C) affect electrical circuits differently than only resistance (R). Thus, resistance, inductance, and capacitance must all be considered when analyzing a circuit that includes more than just a resistive load. **See Figure 29-1.**

A flashlight contains a simple electrical circuit that consists of a power supply, switch, and load.

Resistive Circuits

A *resistive circuit* is a circuit that contains only resistive components, such as heating elements and incandescent lamps. Resistive circuits are the simplest circuits. In a resistive circuit, the electrical characteristics of the circuit are the same regardless of whether an AC or DC voltage source is used. **See Figure 29-2.**

In a resistive circuit, voltage, current, resistance, and power are easily calculated. Ohm's law and the power formula may be applied if any two electrical quantities are known (or measured). All electrical circuits include some resistance, because all conductors, switch contacts, connections, loads, etc. have resistance.

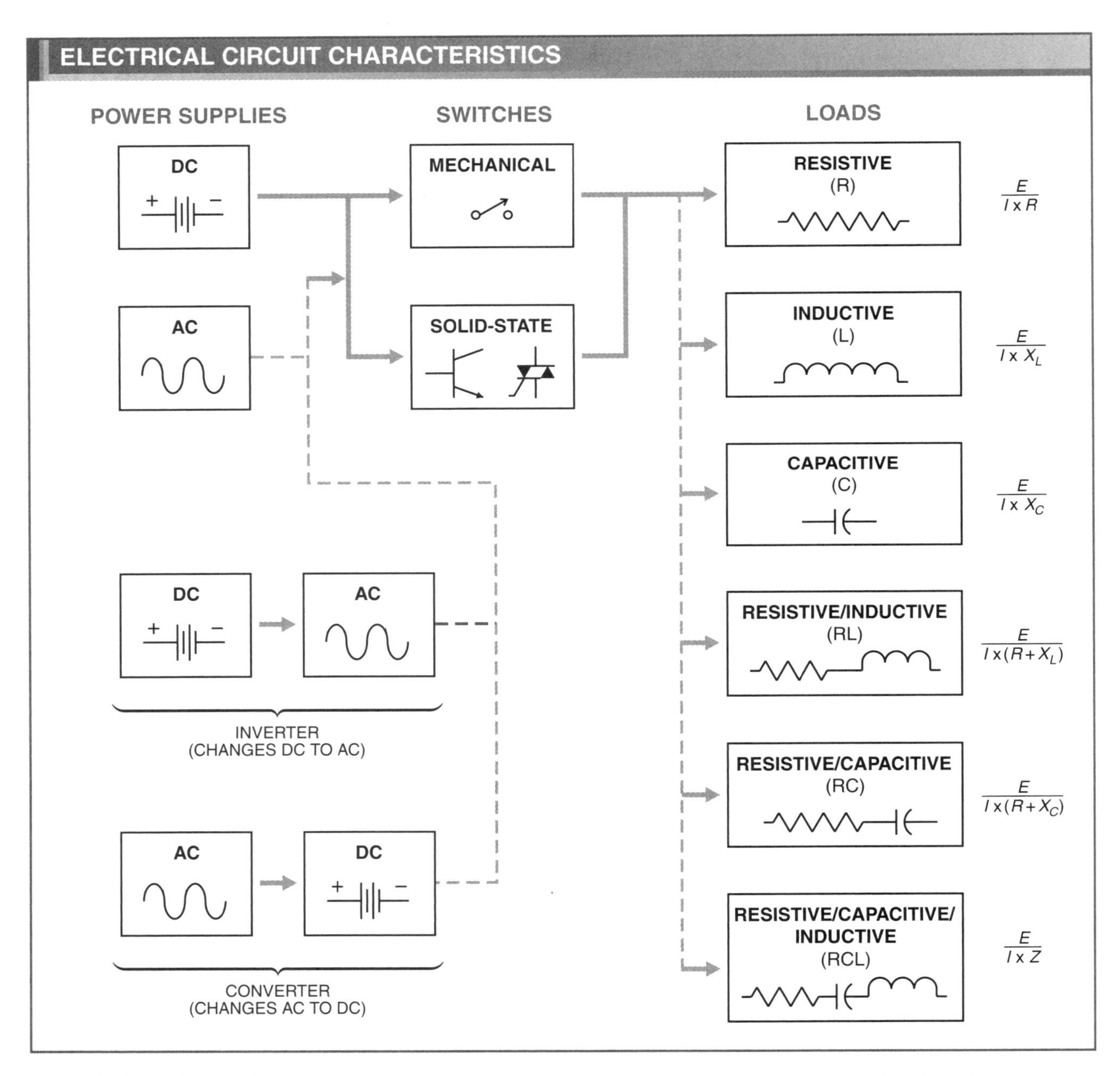

Figure 29-1. Resistance, inductance, and capacitance must all be considered when analyzing a circuit that includes more than just a resistive load.

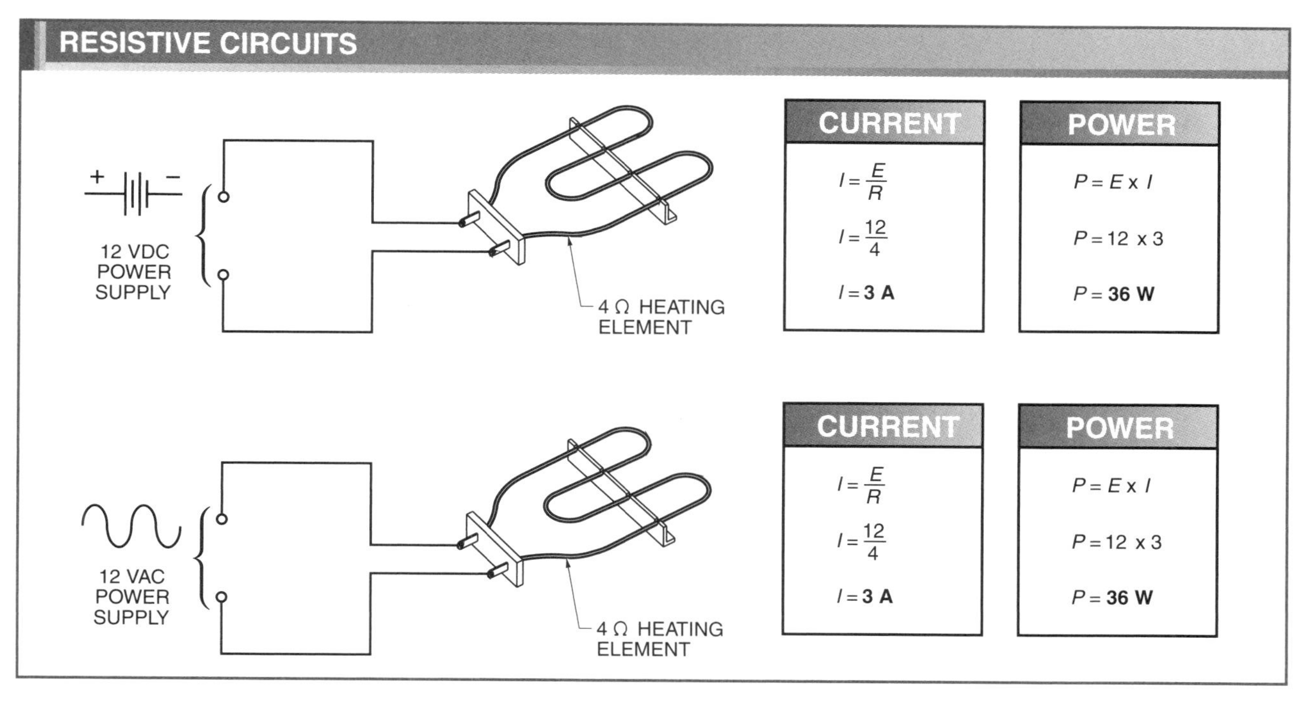

Figure 29-2. In a resistive circuit, the electrical characteristics of the circuit are the same regardless of whether an AC or DC voltage source is used.

Conductor Resistance. All electrical circuits use conductors to carry electricity to different parts of the circuit. A *conductor* is a material that has very little resistance and permits electrons to move through it easily. All conductors have some resistance. Conductor resistance should be kept to a value low enough that the conductor has little or no effect on circuit operation. Conductor resistance is kept to a minimum by limiting the temperature in the circuit and by using the correct size, length, and material. The resistance of any conductor (or material) with a uniform cross-sectional area is based on:

- Material Temperature—For metals, the higher the temperature, the greater the resistance.
- Conductor Size (cross-sectional area)—The smaller the conductor, the greater the resistance.
- Conductor Length—The longer the wire, the greater the resistance.
- Material—Copper has less resistance than aluminum. Aluminum has less resistance than iron.

As the area of the conductor increases, its resistance decreases. As the length of the conductor (current path) increases, its resistance increases. Thus, the resistance of a conductor is directly proportional to its length and inversely proportional to its cross-sectional area.

Resistivity is the resistance of a conductor having a specific length and cross-sectional area. Resistivity is used to compare the resistance characteristics of different materials. Materials with high resistivity are poor conductors (good insulators) and materials with low resistivity are good conductors (poor insulators).

As the temperature of a conductor increases, the free electrons between the atoms move faster and their vibrations increase. For this reason, more collisions occur between the free electrons drifting along the conductor and the atoms of the material. As the number of collisions increases, the flow of current decreases. Thus, an increase in temperature affects the resistance of the conductor and the circuit. For example, the resistance of copper at 100°C is about 43% greater than its resistance at 0°C. In general, resistance increases proportionally with an increase in temperature for most conductor materials. However, in most nonmetals, resistance decreases with an increase in temperature.

Although some circuits include only resistance, most circuits include other electrical characteristics in addition to resistance. For example, any circuit that includes a motor, solenoid, bell, alarm, speaker, fluorescent lamp, HID (high-intensity discharge) lamp, transformer, mechanical relay, motor starter, contactor, or an electronic circuit, includes inductance and/or capacitance in addition to resistance. Inductance and capacitance, like resistance, affect how a circuit operates and the actual amount of power the circuit uses. In some circuits, inductance and/or capacitance has more effect on current flow than resistance.

Inductance

Inductance (L) is the property of a circuit that causes it to oppose a change in current due to energy stored in a magnetic field. The opposition to a change in current is a result of the energy stored in the magnetic field of a coil. The unit of inductance is the henry (H). Inductance is normally stated in henrys (H), millihenrys (mH), or microhenrys (μH).

Any time current flows through a conductor, a magnetic field is produced around the conductor. A conductor wrapped into a coil produces a strong magnetic field around the coil any time current flows through the coil. Direct current flow produces a constant magnetic field around the coil. Alternating current flow produces an alternating magnetic field around the coil. **See Figure 29-3.**

In a DC circuit, once the magnetic field is built, it remains at its maximum potential until the circuit (switch) is opened. Once the circuit is opened, the magnetic field collapses. In an AC circuit, the magnetic field is continuously building and collapsing until the circuit is opened. In an AC circuit, the magnetic field also changes direction with each change in sine wave alternation.

Magnetic Field Strength. The amount of inductance produced by a coil depends on the strength of the magnetic field produced by the coil. The strength of the magnetic field depends on the number of turns of wire, spacing between turns, core used, and wire size (gauge).

Anything that increases the strength of an inductor's magnetic field increases inductance. Factors that determine a coil's inductance include:

- Number of turns in the coil—The greater the number of turns, the higher the inductance. The inductance is greater because more counterelectromotive force (reverse voltage produced by the magnetic lines of force) is induced. Counterelectromotive force (CEMF) pushes back against the supply voltage and attempts to keep the circuit current flow at zero.
- Length and spacing of the coil—The more area (length) a coil has and/or the closer the spacing between turns, the higher the inductance.
- Core material and the relative permeability of the core material—Ferrous (containing iron) materials are the best material to use for magnetic fields. A copper core weakens the magnetic field and reduces inductance. A ferrous core increases inductance.
- Size of the wire used to make the coil—Large wire sizes allow more current to flow, which produces a stronger magnetic field. The larger the wire (and higher current), the higher the inductance.

The number of turns, size of the coil, and core material are fixed by the manufacturer. Thus, the only variable in most coils is the amount of current flowing through the coil. The amount of current flowing through the coil is determined by the total circuit resistance. The higher the total resistance, the lower the circuit current. Likewise, the lower the total resistance, the higher the circuit current.

Tech Facts

The inductance in a coil is 1 H when the change in current is 1 A/sec and the induced voltage is 1 V.

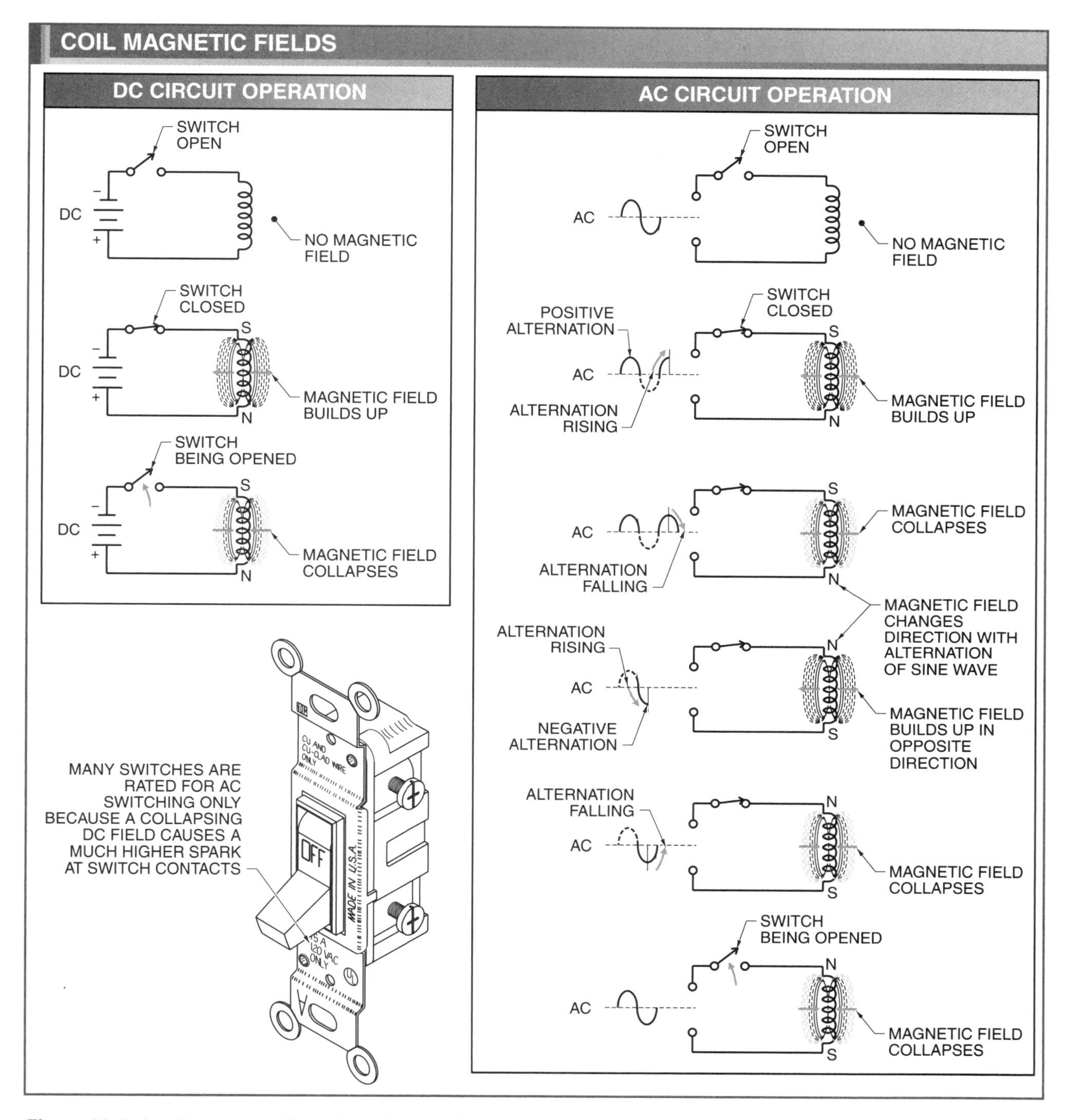

Figure 29-3. Any time current flows through a conductor, a magnetic field is produced around the conductor.

A component may have inductance by design or due to individual components used to make it operate. Coils designed to have specific values of inductance are referred to as inductors, coils, or chokes. These components are intentionally added into a circuit to take advantage of the properties of inductance that they produce. Inductors are generally added into a circuit to control the amount of current flowing in a circuit or a branch of the circuit. For example, a ballast (coil) is added to fluorescent and HID lamp circuits to provide a high-voltage starting surge and limit current flow. **See Figure 29-4.**

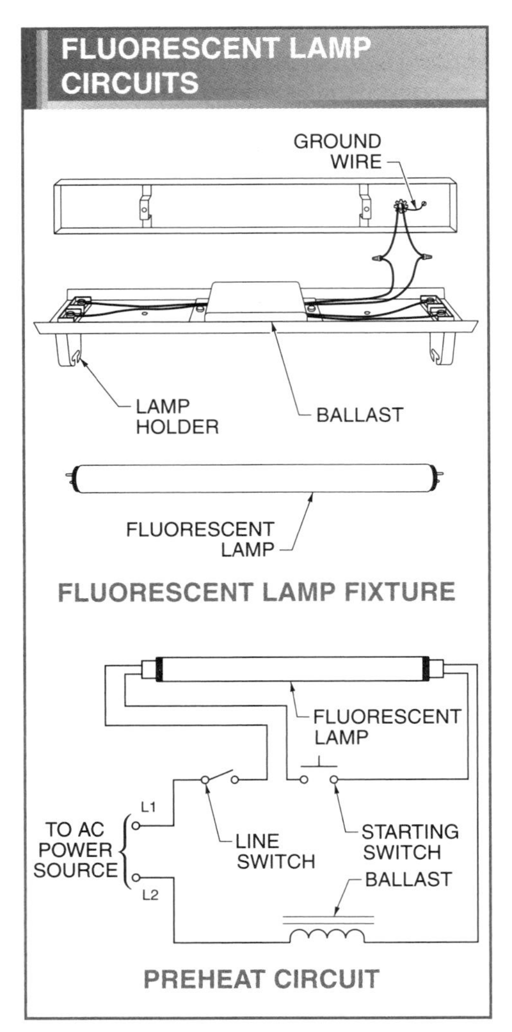

Figure 29-4. A ballast (coil) is added to fluorescent and HID lamp circuits to provide a high-voltage starting surge and limit current flow.

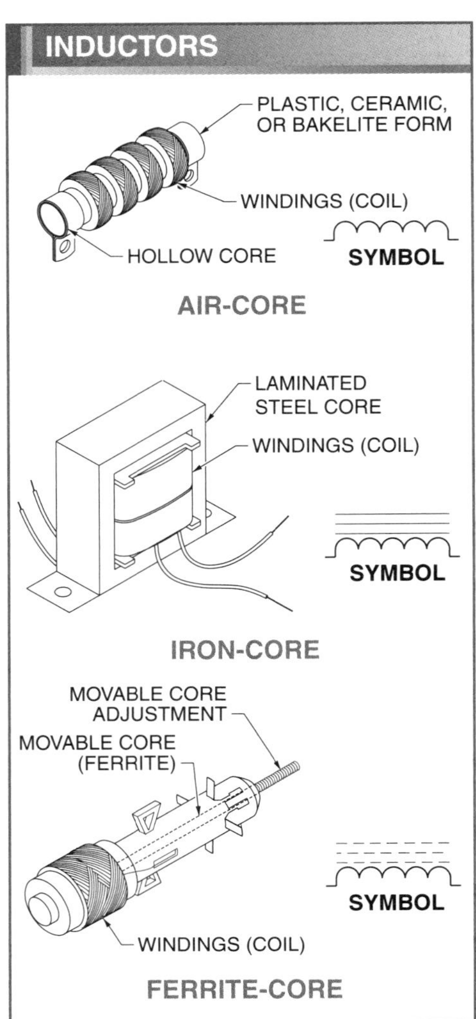

Figure 29-5. The three basic inductors include air-core, iron-core, and ferrite-core.

Other components are not designed specifically to have inductance, but include inductance anyway. For example, all solenoids, motors, generators, transformers, and any component including a coil (mechanical relay, etc.) have inductance.

Inductors. Inductors operate on the same basic electrical principles of magnetism as transformers. The difference between an inductor and a transformer is the number of coils used. Inductors use one coil, transformers use two or more coils. The three basic inductors include air-core, iron-core, and ferrite-core. **See Figure 29-5.**

Air-core inductors consist of a coil (copper wire) wrapped around a form (plastic, ceramic, or Bakelite) with no material in the middle. Air-core inductors do not change their inductance value (as do iron-core and ferrite-core) with a change in current. Air-core inductors are used in high frequency equipment circuits, such as radios (AM and FM), TVs, and other transmitter/receiver circuits.

Iron-core inductors consist of a coil (copper wire) wound around a laminated steel core. Iron-core inductors are commonly used in circuits that operate at standard AC power frequencies (50 Hz and 60 Hz) and at audio frequencies (20 Hz to 20 kHz).

Ferrite-core inductors consist of a core made up of ceramic material (ferrite). *Ferrite* is a chemical compound consisting of powdered iron oxide and ceramic. Ferrite-core inductors are available with a movable ferrite core that can be adjusted for different circuit tuning applications. As the core is moved out of the coil, the air gap reduces the strength of the magnetic field and the amount of inductance. Likewise, as the core is moved into the coil, the air gap is replaced by the core and inductance is increased. Ferrite-core inductors are used to tune many electronic circuits.

AC Circuit Inductive Reactance. *Inductive reactance (X_L)* is an inductor's opposition to alternating current. Like resistance, inductive reactance is measured in ohms. The amount of inductive reactance in a circuit depends on:

- The amount of inductance (in henrys) of the coil (inductor). Inductance is normally a fixed amount.
- The frequency of the current. Frequency may be a fixed or variable amount.

To calculate inductive reactance for AC circuits, apply the formula:

$$X_L = 2\pi f L$$

where

X_L = inductive reactance (in Ω)

2π = 6.28 (indicates circular motion that produces the AC sine wave)

f = applied frequency (in Hz)

L = inductance (in H)

The formula indicates that the higher the frequency or greater the inductance, the higher the inductive reactance. Likewise, the lower the frequency or lesser the inductance, the lower the inductive reactance. **See Figure 29-6.**

Example: Calculating Inductive Reactance

What is the inductive reactance of an 8 mH coil connected to a 50 kHz circuit?

$$X_L = 2\pi f L$$

$$X_L = 6.28 \times 50{,}000 \times 0.008$$

$$X_L = \mathbf{2512\ \Omega}$$

Inductive reactance can take the place of resistance when using Ohm's law because inductive reactance is, in effect, AC resistance. This can be done if voltage and current measurements are taken on a circuit in which frequency is known (or measured). Inductive reactance can be found when the voltage across and current through a coil is measured. To calculate inductive reactance when voltage across a coil (E_L) and current through a coil (I_L) are known (or measured), apply the formula:

$$X_L = \frac{E_L}{I_L}$$

where

X_L = inductive reactance (in Ω)

E_L = voltage across coil (in V)

I_L = current through coil (in A)

Example: Calculating Inductance Reactance—Voltage and Current Measured

What is the inductive reactance of a circuit that contains a coil connected to a 120 VAC power source and has 80 mA of current flow?

$$X_L = \frac{E_L}{I_L}$$

$$X_L = \frac{120}{.08}$$

$$X_L = \mathbf{1500\ \Omega}$$

Inductive reactance may be calculated when the frequency and inductance are known. Also, the third variable may be found if any two factors (inductive reactance, frequency, or inductance) are known. To calculate any one of the unknown quantities, apply the formulas:

$$X_L = 2\pi f L,\ L = \frac{X_L}{2\pi f},\ or\ f = \frac{X_L}{2\pi L}$$

where

X_L = inductive reactance (in Ω)

2π = 6.28 (indicates circular motion that produces the AC sine wave)

f = applied frequency (in Hz)

L = inductance (in H)

A voltmeter and ammeter may be connected into a circuit to measure the circuit's applied voltage and current flow. These values may be used to calculate the inductance of the coil.

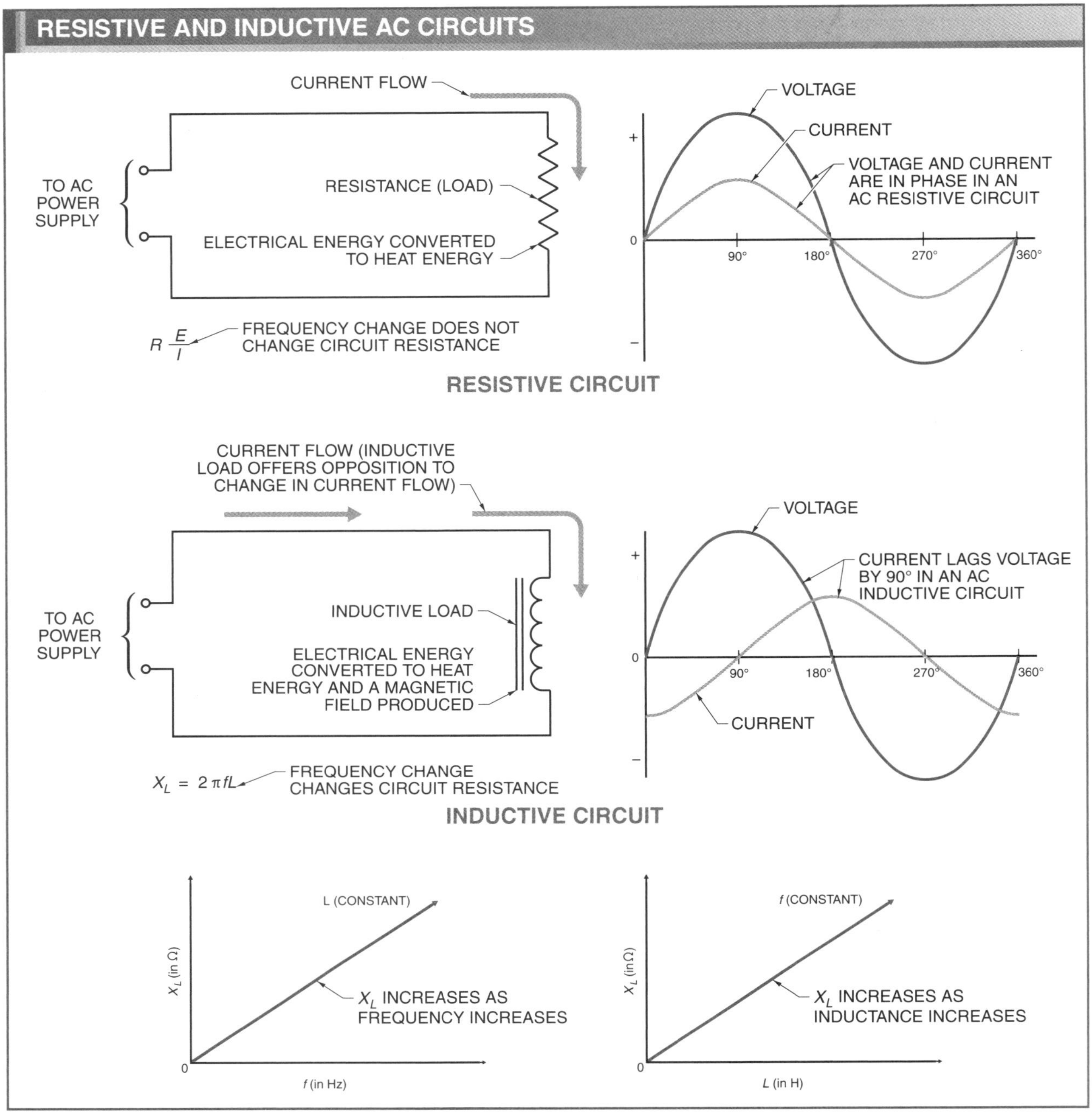

Figure 29-6. Inductive reactance (X_L) is an inductor's opposition to alternating current.

> **Tech Facts**
>
> *Inductive reactance is measured in ohms and can be determined when the values of inductance and frequency are known. The formula for determining inductive reactance is $X_L = 2\pi fL$.*

Example: Calculating Inductance—Inductive Reactance and Frequency Known

What is the inductance of a coil when the voltage in the circuit is 24 VAC, the current is 60 mA, and the frequency is 60 Hz?

1. Calculate inductive reactance using measured voltage and current.

$$X_L = \frac{E_L}{I_L}$$
$$X_L = \frac{24}{.06}$$
$$X_L = \mathbf{400\ \Omega}$$

2. Calculate inductance based on inductive reactance and frequency.

$$L = \frac{X_L}{2\pi f}$$
$$L = \frac{400}{6.28 \times 60}$$
$$L = \frac{400}{376.8}$$
$$L = \mathbf{1.06\ H}$$

Note: Line frequency may be changed (or adjusted) intentionally as it is in most electronic communication equipment or it may be changed when a piece of equipment is used in a different country. For example, the standard line frequency in the U.S. is 60 Hz. The standard line frequency in Europe is 50 Hz. Such circuit changes must be considered when designing (or using) equipment overseas.

DC Circuit Inductive Reactance. At zero frequency (DC circuits), there is no opposition due to inductance. In DC circuits, a coil has low resistance and no inductive reactance. Current flow through a coil connected to DC is limited only by the resistance of the wire. The inductor has no resistance (X_L) except the small resistance of the wire. Thus, for a steady direct current without any change in current, inductive reactance is normally considered to be zero. The amount of resistance is less with a large wire or less turns. The amount of resistance is greater with a small wire or more turns.

Current flow through a coil connected to AC is limited by the wire resistance and the inductive reactance. In AC circuits, a coil has low resistance and high inductive reactance. This reduces current flow in the circuit and produces a voltage drop across the coil.

There are many circuit applications that use inductors. For example, circuits use inductors where it is desired to have high resistance (X_L) to alternating current flow but little resistance (X_L) to direct current flow. Inductors are also used in circuits to produce more opposition to a high frequency alternating current compared with low frequency alternating current.

Series Inductive Reactance. The inductive reactance of a circuit containing more than one coil is found by first determining the total inductance of the circuit and then calculating the inductive reactance. The inductive reactance of three coils connected in series is calculated by adding the inductance of each coil to find the total inductance and then calculating the inductive reactance.

Example: Calculating Inductive Reactance—Coils Connected in Series

What is the inductive reactance of a circuit containing coils of 4 mH, 6 mH, and 20 mH connected to a 1000 Hz AC supply?

1. Calculate total inductance of coils connected in series.

$$L_T = L_1 + L_2 + L_3 + \ldots$$
$$L_T = 4 + 6 + 20$$
$$L_T = \mathbf{30\ mH}$$

2. Calculate inductive reactance based on total inductance.

$$X_L = 2\pi fL$$
$$X_L = 6.28 \times 1000 \times 0.03$$
$$X_L = \mathbf{188.4\ \Omega}$$

Milwaukee Electric Tool Corporation

A cordless drill is an example of a DC circuit being used to power a motor.

Parallel Inductive Reactance. The total inductance in a circuit containing parallel-connected coils is less than the smallest coil value. The total inductance of two coils connected in parallel is found by applying the same formula for calculating the resistance of two loads connected in parallel. The inductive reactance in a circuit containing two parallel-connected coils is found by calculating the total inductance of the two coils and then calculating the inductive reactance.

Example: Calculating Inductive Reactance—Two Coils Connected in Parallel
What is the inductive reactance of a circuit containing a 4 mH coil and a 6 mH coil connected in parallel to a 1000 Hz AC supply?

1. Calculate total inductance of two coils connected in parallel.

$$L_T = \frac{L_1 \times L_2}{L_1 + L_2}$$

$$L_T = \frac{4 \times 6}{4 + 6}$$

$$L_T = \frac{24}{10}$$

$$L_T = \mathbf{2.4\ mH}$$

2. Calculate inductive reactance based on total reactance.

$$X_L = 2\pi fL$$

$$X_L = 6.28 \times 1000 \times 0.0024$$

$$X_L = \mathbf{15.072\ \Omega}$$

The total inductance of three coils connected in parallel is found by applying the same formula for calculating the resistance of three loads connected in parallel. The total circuit inductance decreases if coils are added in parallel and increases if coils are removed. The inductive reactance in a circuit containing three parallel-connected coils is found by calculating the total inductance of the three coils and then calculating the inductive reactance.

Example: Calculating Inductive Reactance—Three Coils Connected in Parallel
What is the inductive reactance of a circuit containing three parallel-connected coils with inductances of 5 mH, 10 mH, and 20 mH connected to a 1000 Hz supply?

1. Calculate total inductance of three coils connected in parallel.

$$L_T = \frac{1}{\frac{1}{L_1} + \frac{1}{L_2} + \frac{1}{L_3}}$$

$$L_T = \frac{1}{\frac{1}{5} + \frac{1}{10} + \frac{1}{20}}$$

$$L_T = \frac{1}{.20 + .10 + .05}$$

$$L_T = \frac{1}{.35}$$

$$L_T = \mathbf{2.86\ mH}$$

2. Calculate inductive reactance based on circuit total inductance.

$$X_L = 2\pi fL$$

$$X_L = 6.28 \times 1000 \times 0.00286$$

$$X_L = \mathbf{17.9608\ \Omega}$$

Capacitance

Capacitance (C) is the ability of a component or circuit to store energy in the form of an electrical charge. A *capacitor* is an electric device specifically designed to store a charge of energy. The unit of capacitance is the farad (F). The farad is a unit that is too large for most electrical/electronic applications. Thus, capacitance (and capacitor values) are normally stated in microfarads (μF) or picofarads (pF).

A capacitor consists of two conductors (plates) separated by an insulator (dielectric), which allows an electrostatic charge to be developed. The electrostatic charge becomes a source of stored energy. The strength of the charge depends on the applied voltage, size of the conductors, and quality of the insulation. The dielectric may be air, paper, oil, ceramic, mica, or any other nonconducting material. **See Figure 29-7.**

The closer the two plates are placed together, the greater the charge on them. A capacitor is charged when its leads are connected to a DC voltage source. The positive terminal of the voltage source attracts electrons from the plate connected to it and the negative terminal of the voltage source repels an equal number of electrons into the negative plate. Charging (current flow) continues until the voltage across the charged plates is equal to the applied voltage. At this time the capacitor is fully charged.

Once charged, the voltage source may be disconnected from the capacitor. The stored energy remains between the two capacitor plates. Thus, a capacitor has the ability to store the amount of charge equal to the applied charging voltage. The capacitor holds this charge until something is connected between the two leads. At this time the capacitor discharges its stored energy. Once discharged, the capacitor must be recharged.

Capacitors are rated for voltage and capacitance. The voltage rating indicates the maximum voltage that can safely be applied to the capacitor leads. The amount of capacitance a capacitor has is determined by:

- The area of the plates. The greater the plate area, the higher the capacitance value.
- The spacing between the plates. The closer the plates, the higher the capacitance.
- The dielectric used. The better the dielectric material, the higher the capacitance value.
- Connection arrangement. Capacitors are connected in parallel to increase total circuit capacitance.

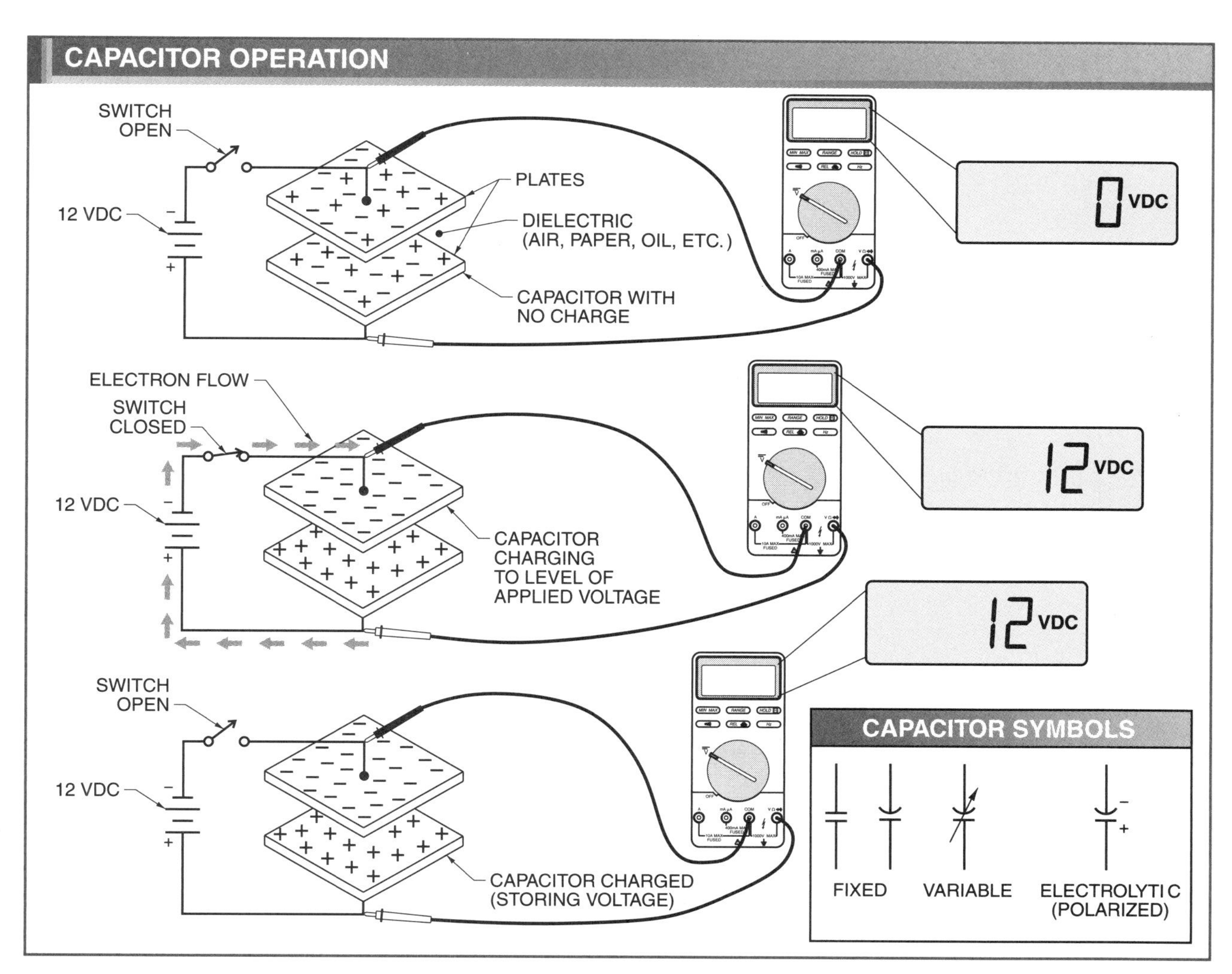

Figure 29-7. A capacitor consists of two conductors (plates) separated by an insulator (dielectric), which allows an electrostatic charge to be developed.

Capacitors. Capacitors may be fixed or variable. A *fixed capacitor* is a capacitor that has one value of capacitance. A *variable capacitor* is a capacitor that varies in capacitance value. Fixed capacitors are classified according to the dielectric used. Fixed capacitors include mica, paper, ceramic, and electrolytic capacitors. Mica, paper, and ceramic capacitors are not polarized. These capacitors may be connected in a circuit with either terminal lead connected to the positive side of the circuit. Electrolytic capacitors are polarized. They must be connected in a circuit with the positive terminal lead connected to the positive side of the circuit. **See Figure 29-8.**

Mica capacitors use mica as the dielectric. Mica capacitors are made by stacking mica sheets between a metal foil (tinfoil or aluminum sheets). The strips of tinfoil are connected to create one plate and the mica sheets are connected to create the other plate.

Tech Facts

Current in a capacitor can only flow during the time period when the capacitor is either being charged or discharged. A capacitor is created by separating strips of metal with a dielectric or insulating material.

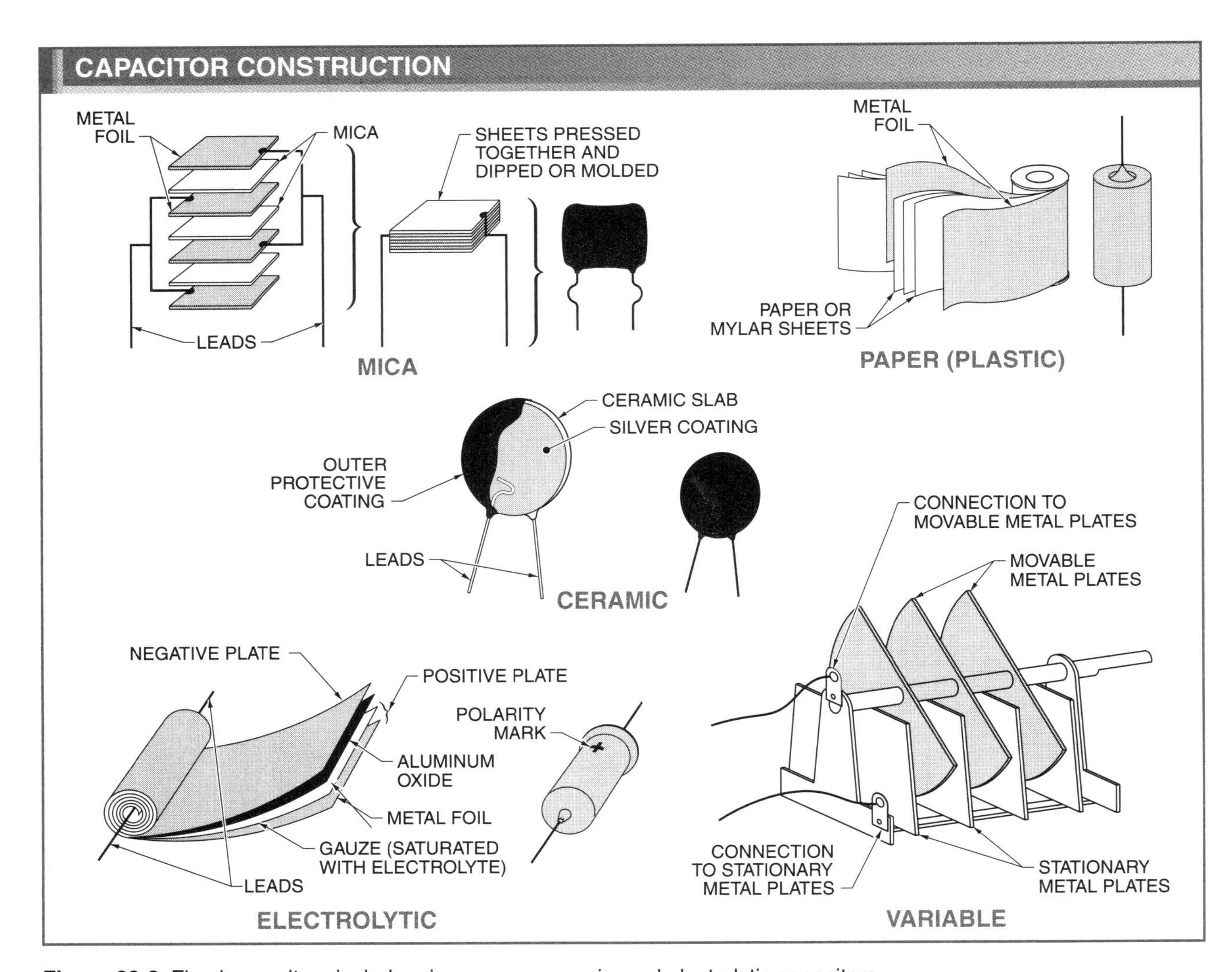

Figure 29-8. Fixed capacitors include mica, paper, ceramic, and electrolytic capacitors.

Mica capacitors are small in size and have low capacitance ratings. Although mica capacitors have low capacitance values, they have high-voltage ratings. For this reason, mica capacitors are often used in high-voltage circuits. Mica capacitors commonly range from 1 pF to about 0.1 μF.

Paper capacitors use wax paper or plastic film as the dielectric. In newer capacitors, mylar has replaced the paper film. Paper capacitors are made of flat strips of metal foil separated by a dielectric rolled into a compact cylinder. To prevent moisture problems, they are normally sealed in wax. Paper capacitors have low capacitance ratings from 0.001 μF to 1 μF. Their voltage ratings are normally less than 600 V.

Ceramic capacitors use ceramic as the dielectric. The advantage of using ceramic is that it can produce the same effect as a paper dielectric, but in less space. Ceramic capacitors commonly range from 500 pF to 0.01 μF.

Electrolytic capacitors use a paste as the dielectric. Electrolytic capacitors provide more capacitance for their size than any other type of capacitor. For this reason, electrolytic capacitors are used when a very high capacitance is required. Electrolytic capacitors may have 100,000 μF of capacitance or more.

Variable capacitors normally use air or mylar as the dielectric and include movable and stationary metal plates. The capacitance is maximum when the movable plates are fully meshed (but not touching) with the stationary plates. Moving the plates apart lowers the capacitance. Variable capacitors are used as tuning capacitors in radio receivers. The capacitance is varied to tune in different stations.

Capacitive Reactance. The charges on the plates of a capacitor reverse with each change in the applied voltage polarity when a capacitor is connected to an AC power supply. The plates are alternately charged and discharged. Capacitors offer opposition to the flow of current in a circuit when connected to an AC power supply. *Capacitive reactance* (X_C) is the opposition to current flow by a capacitor. Like inductive reactance (X_L), capacitive reactance is expressed in ohms. **See Figure 29-9.** To calculate capacitive reactance, apply the formula:

$$X_C = \frac{1}{2\pi fC}$$

where

X_C = capacitive reactance (in Ω)

2π = 6.28 (indicates circular motion that produces the AC sine wave)

f = applied frequency (in Hz)

C = capacitance (in F)

The formula for capacitive reactance shows that the higher the frequency or the greater the capacitance, the lower the capacitive reactance. Likewise, the lower the frequency or the lower the capacitance, the higher the capacitive reactance.

Example: Calculating Capacitive Reactance

What is the capacitive reactance of a 50 μF capacitor connected to a 60 Hz supply?

$$X_C = \frac{1}{2\pi fC}$$

$$X_C = \frac{1}{6.28 \times 60 \times .00005}$$

$$X_C = \frac{1}{.01884}$$

$$X_C = \mathbf{53.08\ \Omega}$$

Capacitive reactance, like inductive reactance is, in effect, AC resistance. For this reason, it may take the place of resistance when using Ohm's law. This may be done if voltage and current measurements are taken on a circuit in which frequency is known (or measured). To calculate capacitive reactance when voltage across the capacitor (E_C) and current through the capacitor (I_C) are known, apply the formula:

$$X_C = \frac{E_C}{I_C}$$

where

X_C = capacitive reactance (in Ω)

E_C = voltage across capacitor (in V)

I_C = current through capacitor (in A)

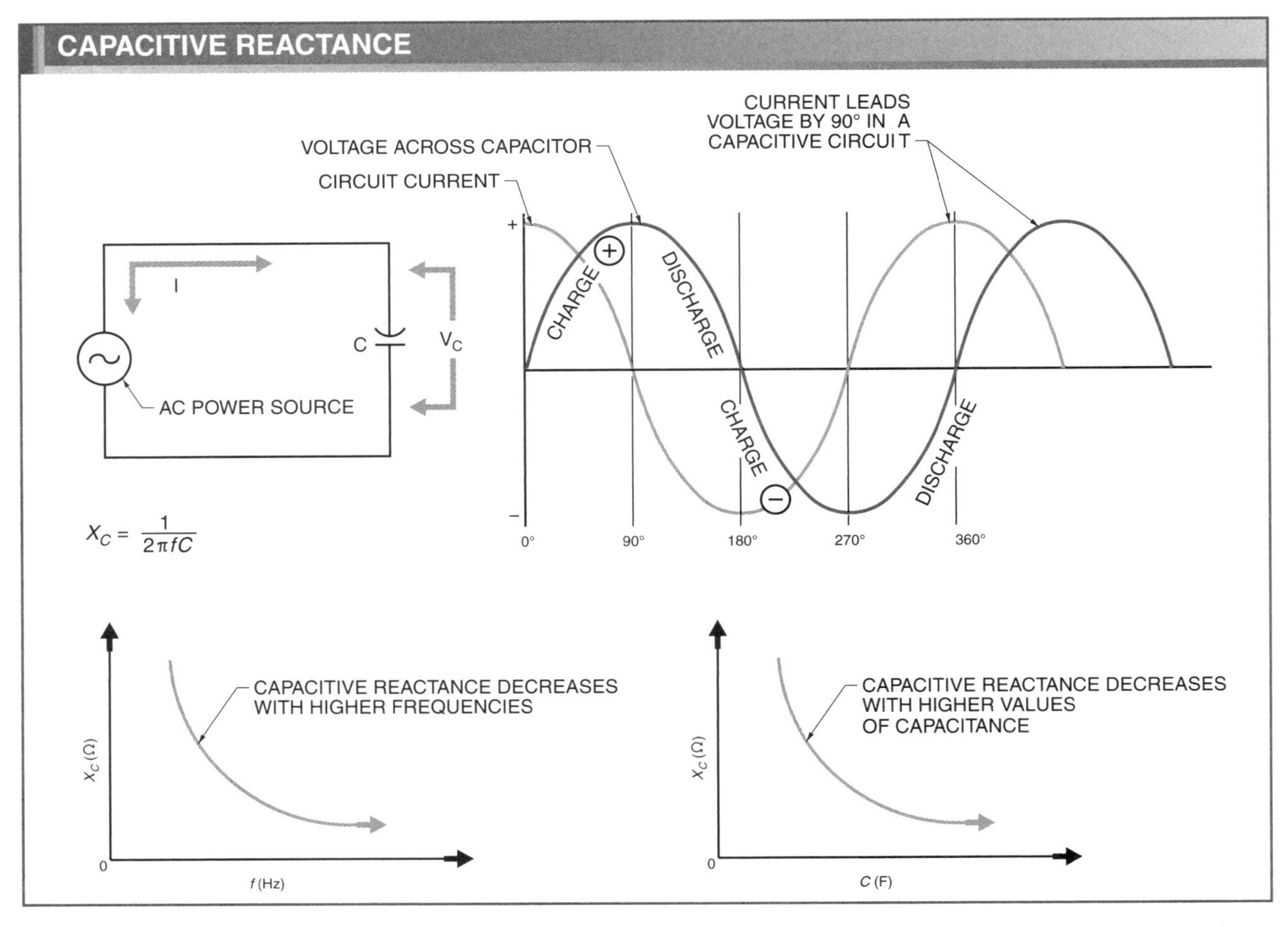

Figure 29-9. Capacitive reactance (X_C) is the opposition to current flow by a capacitor.

Example: Calculating Capacitive Reactance—Voltage and Current Known

What is the capacitive reactance of a 120 VAC circuit with 20 mA of current flow?

$$X_C = \frac{E_C}{I_C}$$

$$X_C = \frac{120}{.02}$$

$$X_C = \mathbf{6000\ \Omega}$$

Series Capacitive Reactance. The capacitive reactance of a circuit containing capacitors connected in series is found by determining the total capacitance of the circuit and then calculating the capacitive reactance based on the total capacitance. The total capacitance of two capacitors connected in series is determined by using the same formula used to calculate total resistance of two resistors connected in parallel.

Example: Calculating Capacitive Reactance—Two Capacitors Connected in Series

What is the capacitive reactance of a 60 Hz circuit containing a 60 μF and a 90 μF capacitor connected in series?

1. Calculate total capacitance of two capacitors connected in series.

$$C_T = \frac{C_1 \times C_2}{C_1 + C_2}$$

$$C_T = \frac{60 \times 90}{60 + 90}$$

$$C_T = \frac{5400}{150}$$

$$C_T = \mathbf{36\ \mu F}$$

2. Calculate capacitive reactance based on circuit total capacitance.

$$X_C = \frac{1}{2\pi fC}$$

$$X_C = \frac{1}{6.28 \times 60 \times .000036}$$

$$X_C = \frac{1}{.01356}$$

$$X_C = \mathbf{73.75\ \Omega}$$

The capacitive reactance of three capacitors connected in series is determined by calculating the total capacitance of the three capacitors and using this value to determine capacitive reactance. The total capacitance of three (or more) capacitors connected in series is determined by using the same formula used to calculate total resistance of three (or more) resistors connected in parallel.

Example: Calculating Capacitive Reactance—Three Capacitors Connected in Series

What is the capacitive reactance of a 60 Hz circuit containing a 200 μF, 200 μF, and 400 μF capacitor connected in series?

1. Calculate total capacitance of three capacitors connected in series.

$$C_T = \frac{1}{\frac{1}{C_1} + \frac{1}{C_2} + \frac{1}{C_3}}$$

$$C_T = \frac{1}{\frac{1}{200} + \frac{1}{200} + \frac{1}{400}}$$

$$C_T = \frac{1}{.005 + .005 + .0025}$$

$$C_T = \frac{1}{.0125}$$

$$C_T = \mathbf{80\ \mu H}$$

2. Calculate capacitive reactance based on circuit total capacitance.

$$X_C = \frac{1}{2\pi fC}$$

$$X_C = \frac{1}{6.28 \times 60 \times .00008}$$

$$X_C = \frac{1}{.03}$$

$$X_C = \mathbf{33.33\ \Omega}$$

Parallel Capacitive Reactance. The capacitive reactance of capacitors connected in parallel is determined by calculating the total capacitance of the parallel-connected capacitors and using this value to calculate capacitive reactance. The total capacitance of a parallel circuit is determined by adding the value of the individual capacitors.

Example: Calculating Capacitive Reactance—Capacitors Connected in Parallel

What is the total capacitive reactance of a 60 Hz circuit containing a 50 μF, 75 μF, and a 25 μF capacitor connected in parallel?

1. Calculate total capacitance of capacitors connected in parallel.

$$C_T = C_1 + C_2 + C_3 + \ldots$$

$$C_T = 50 + 75 + 25$$

$$C_T = \mathbf{150\ \mu F}$$

2. Calculate capacitive reactance based on circuit total capacitance.

$$X_C = \frac{1}{2\pi fC}$$

$$X_C = \frac{1}{6.28 \times 60 \times .000015}$$

$$X_C = \frac{1}{.0565}$$

$$X_C = \mathbf{17.7\ \Omega}$$

Impedance

Inductors and capacitors are electrical energy-storing devices that are used in AC and DC applications such as timing circuits, surge protection, filtering circuits, fluorescent and HID lamp circuits, 1φ motors, motor control circuits, etc. Inductors are used to store a current charge and capacitors are used to store a voltage charge. Inductors and capacitors, when used together, each contribute to the circuit operation. For example, a capacitor and inductor may be used in a filter circuit to smooth a pulsating direct current.

In AC circuits, resistance, inductive reactance, and capacitive reactance all limit current flow. Most AC circuits contain all three oppositions to current flow. The exact behavior of current in an AC circuit depends on the amount of resistance, inductive reactance, and capacitive reactance.

Impedance (Z) is the total opposition to the flow of alternating current, consisting of any combination of resistance, inductive reactance, and capacitive reactance. Impedance, like inductive reactance and capacitive reactance, is measured in ohms. **See Figure 29-10.**

Tech Facts

Impedance, the total opposition to the flow of alternating current that consists of any combination of resistance, inductive reactance, and capacitive reactance, is measured in ohms and may be substituted for resistance in Ohm's law.

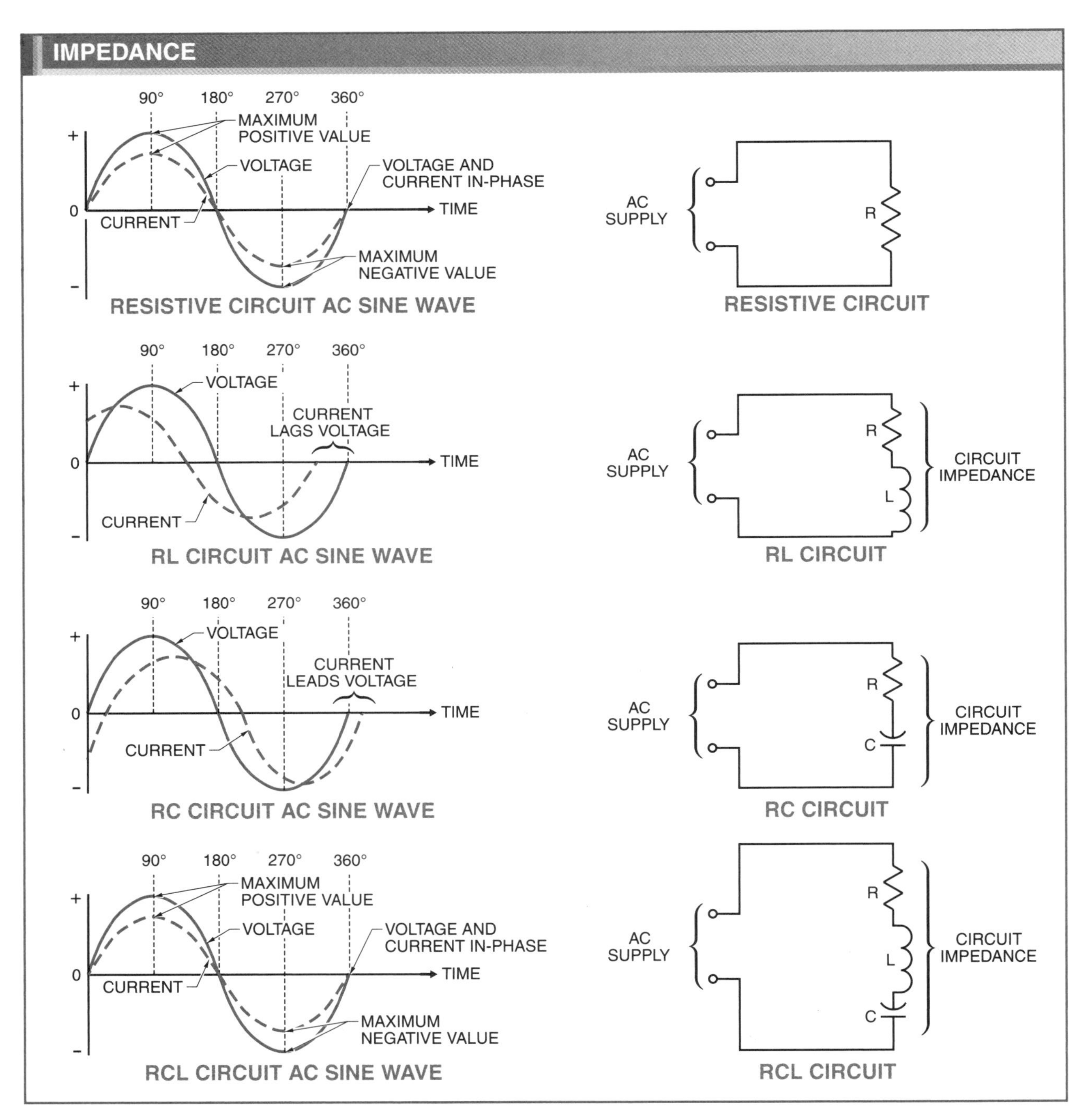

Figure 29-10. Impedance (*Z*) is the total opposition of any combination of resistance, inductive reactance, and capacitive reactance offered to the flow of alternating current.

In circuits containing only resistance, current and voltage are normally in-phase. The phase angle between voltage and current is zero (or close to zero).

In circuits containing resistance and inductive reactance (RL), resistance and inductive reactance oppose current flow. RL circuit impedance is the combined effect of the resistance and inductive reactance. For this reason, current lags voltage by a phase angle between 0° and 90°. The more inductive reactance in the circuit, the greater the phase shift (closer to 90°).

In circuits containing resistance and capacitive reactance (RC), resistance and capacitive reactance oppose current flow. RC circuit impedance is the combined effect of the resistance and capacitive reactance. For this reason, current leads voltage by a phase angle between 0° and 90°. The more capacitive the circuit, the greater the phase shift (closer to 90°).

In circuits containing resistance, inductive reactance, and capacitive reactance (RCL), resistance, inductive reactance, and capacitive reactance oppose current flow. RCL circuit impedance is the combined effects of the resistance, inductive reactance, and capacitive reactance. For this reason, the lagging current caused by inductance tends to cancel the leading current caused by capacitance because they are in direct opposition to each other. The total current flow in a circuit that includes resistance, inductive reactance, and capacitive reactance is determined by the applied voltage and total circuit impedance.

Impedance may be substituted for resistance in Ohm's law because impedance represents all of the opposition to current flow in a circuit. Ohm's law may be used to determine any one value when the other two are known. To determine impedance, current, or voltage, apply the formulas:

$$Z = \frac{E}{I}, I\frac{E}{Z}, \text{or } E = I \times Z$$

where

Z = impedance (in Ω)

E = voltage (in V)

I = current (in A)

Example: Calculating Impedance—Voltage and Current Known

What is the impedance of a 120 VAC circuit that has 5 mA of current flow?

$$Z = \frac{E}{I}$$

$$Z = \frac{120}{.005}$$

$$Z = \mathbf{24,000\ \Omega}$$

Review Questions

1. Define load and provide four examples of loads.
2. How is conductor resistance kept to a minimum?
3. What is inductance?
4. List four factors that determine the inductance of a coil.
5. What are the three basic inductor types?
6. Provide two different formulas for calculating inductive reactance.
7. Describe the composition of a capacitor.
8. Name five types of material used as dielectric.
9. What factors determine the amount of capacitance a capacitor has?
10. Explain the difference between a fixed capacitor and a variable capacitor.
11. What three factors affect current flow in an AC circuit?
12. Define impedance.

Commercial circuits are high-power circuits used for lighting, motors, heating elements, and other loads found in a commercial building or area. Commercial buildings typically require both residential- and commercial-type power systems.

30 Commercial Circuits

COMMERCIAL CIRCUITS

Commercial buildings such as retail shops, restaurants, apartment buildings, hotels, motels, gas stations, bakeries, sports complexes, etc. require high levels of power for lighting, motors, heating elements, and other loads. This high energy demand requires a greater number of individual circuits than residential buildings. Different size power distribution systems are used depending on the amount of power required and the type of loads to be controlled. Relatively low voltage levels (208 V–240 V) are delivered for small commercial applications (gas stations, restaurants, etc.) by the local utility. Medium commercial voltage levels (460 V–480 V) are delivered for mid-sized commercial applications (small office buildings, etc.).

Large commercial applications (colleges, small businesses, hotels, etc.) require voltages of over 1000 V to be delivered and distributed to different transformer banks throughout the building. The high voltage reduces the conductor size required to deliver high amounts of power.

The high voltage is reduced at a substation. A *substation* is an assemblage of equipment installed for switching, changing, or regulating the voltage of electricity. The substations contain transformers and secondary switches that distribute the low-voltage levels (208 V–480 V) to feeder panels, other secondary transformers, and branch-circuit panels. **See Figure 30-1.**

NEC® Branch Circuit Voltage Limits

The exact voltage level used to operate different loads in commercial applications depends on the load, the given application, and the NEC®. For safety reasons, the NEC® places restrictions on the voltage level of some locations. For example, subsection 210.6(A) of the NEC® limits the voltage level to 120 V between conductors that supply lighting fixtures, cords, and plugs that connect loads up to 1440 VA (volt-amperes), and motors up to ¼ HP in dwelling units and guest rooms of hotels, motels, and similar rooms of occupancies. For this reason, most lighting is standard 120 V incandescent or fluorescent fixtures. Less chance exists for an individual to accidentally contact a fatal voltage when the voltage is limited to a low level.

Stadium lights are considered commercial loads because of the large amount of power used.

Fluke Corporation

Commercial transformers require that loads be balanced across the three phases.

However, it is not practical to limit all circuits to a 120 V level. This is because the lower the voltage, the higher the current for any given power (wattage) rating. Increasing the voltage level to reduce the current level allows for a more practical installation because conductor size, conduit size, and switching equipment size is determined by the amount of current. Thus, higher voltage levels are allowed in commercial areas in which electrical circuits are serviced by qualified maintenance persons. For example, subsection 210.6(C) of the NEC® allows voltage levels up to 277 V for incandescent lamps (equipped with mogul-base screw shell lampholders), fluorescent lamps, and HID lamps used in commercial applications. Such lighting fixtures should not be serviced by nonqualified persons.

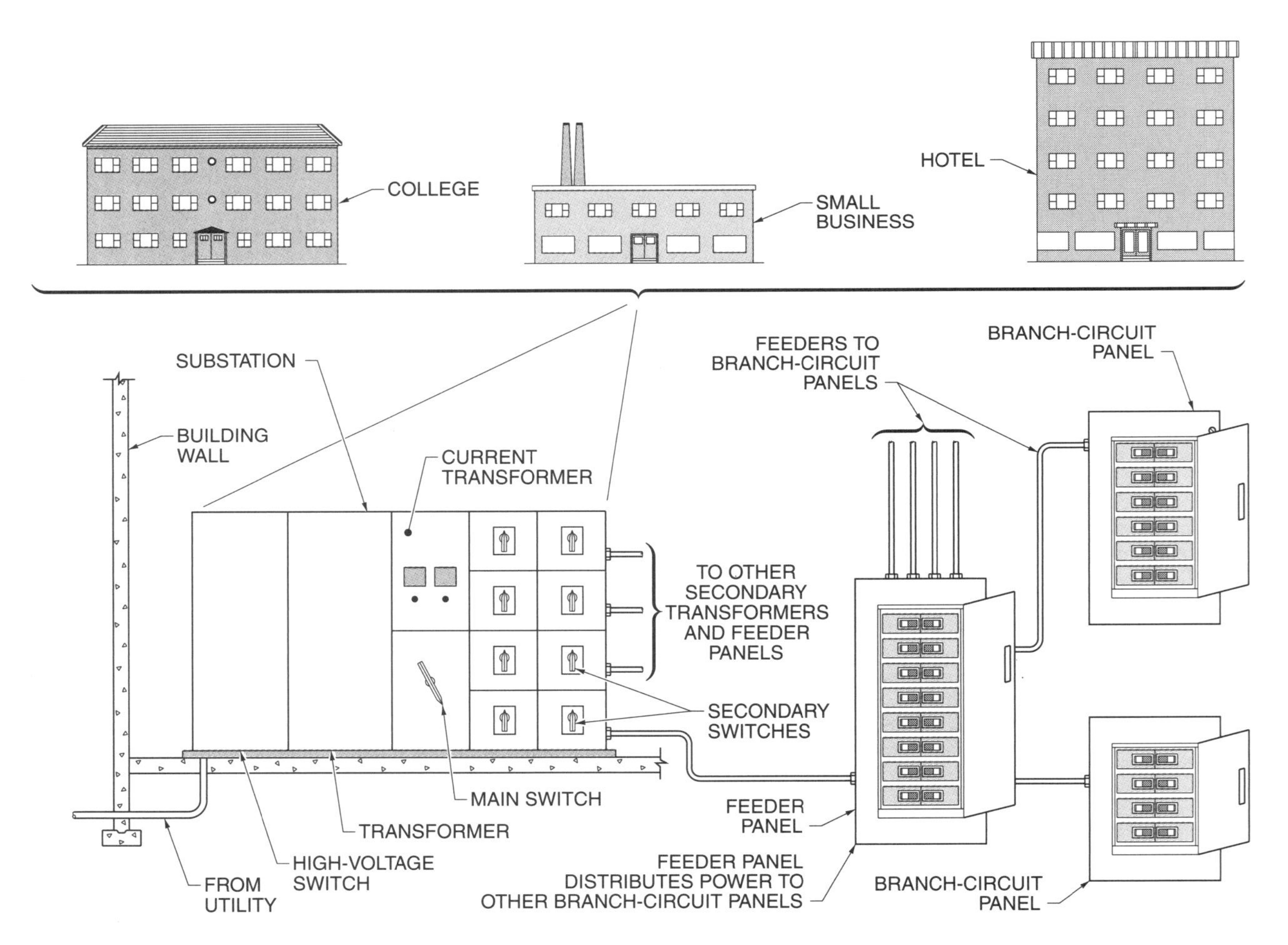

Figure 30-1. Substations contain transformers and secondary switches that distribute voltage to feeder panels, other secondary transformers, and branch-circuit panels.

Distribution Systems

Electrical power companies generate and distribute high-voltage 3ϕ power. However, commercial customers use low-voltage 3ϕ and 1ϕ power. For this reason, power companies use transformers to deliver power to customers through different distribution systems. Power companies primarily use 1ϕ transformers connected in different combinations to deliver the correct power to customers. Three transformers (except in an open-delta system) are normally interconnected for delivering 3ϕ voltage to commercial customers. The exact voltage level is determined by customer needs and the utility company's existing system.

Commercial power needs include the same needs as residential (1ϕ) and industrial (3ϕ) systems. Apartment buildings and offices require lighting and low-voltage 1ϕ small appliance circuits that are similar to residential buildings except on a much larger scale. Other commercial buildings, such as bakeries and restaurants, require about the same lighting and high-voltage 1ϕ and 3ϕ large appliance circuits as some industrial systems except on a smaller scale. Still other commercial systems, such as outdoor sports stadiums, require a great number of large scale lighting circuits.

The distribution system used varies based on the application, but a few common systems are used for most commercial applications. Common commercial distribution systems include 120/240 V, 1ϕ, 3-wire; 120/208 V, 3ϕ, 4-wire; 120/240 V, 3ϕ, 4-wire; and 277/480 V, 3ϕ, 4-wire services.

120/240 V, 1ϕ, 3-Wire Service. A 120/240 V, 1ϕ, 3-wire service is used to supply power to customers that require 120 V and 240 V, 1ϕ power. This service provides 120 V, 1ϕ; 240 V, 1ϕ; and 120/240 V, 1ϕ circuits. The neutral wire is grounded, therefore it should not be fused or switched at any point. **See Figure 30-2.**

A 120/240 V, 1ϕ, 3-wire service is commonly used for interior wiring for lighting and small appliance use. For this reason, 120/240 V, 1ϕ, 3-wire service is the primary service used to supply most residential buildings. This service is also used for small commercial applications, although a large power panel is used (or there are additional panels).

In commercial wiring, small appliance usage includes more than typical residential small appliance usage. This is because commercial appliance circuits include high-power devices such as copying machines, commercial refrigerators, washers, etc., and different appliances such as hot tubs and saunas in hotels, automatic ice melting machines, heating elements embedded in the concrete of entranceways, etc. Other commercial small appliance usage includes office computers, printers, motors less than 5 HP, cooking equipment, security equipment, large entertainment systems, hospital equipment, etc. In addition to different appliances, commercial circuits often require duplication because there may be hundreds of rooms or offices at one location.

Tech Facts

All ungrounded (hot) conductors read a voltage between the conductor and a ground point (grounded conduit, green ground wire, etc.).

In commercial power distribution, each load requires a separate disconnect.

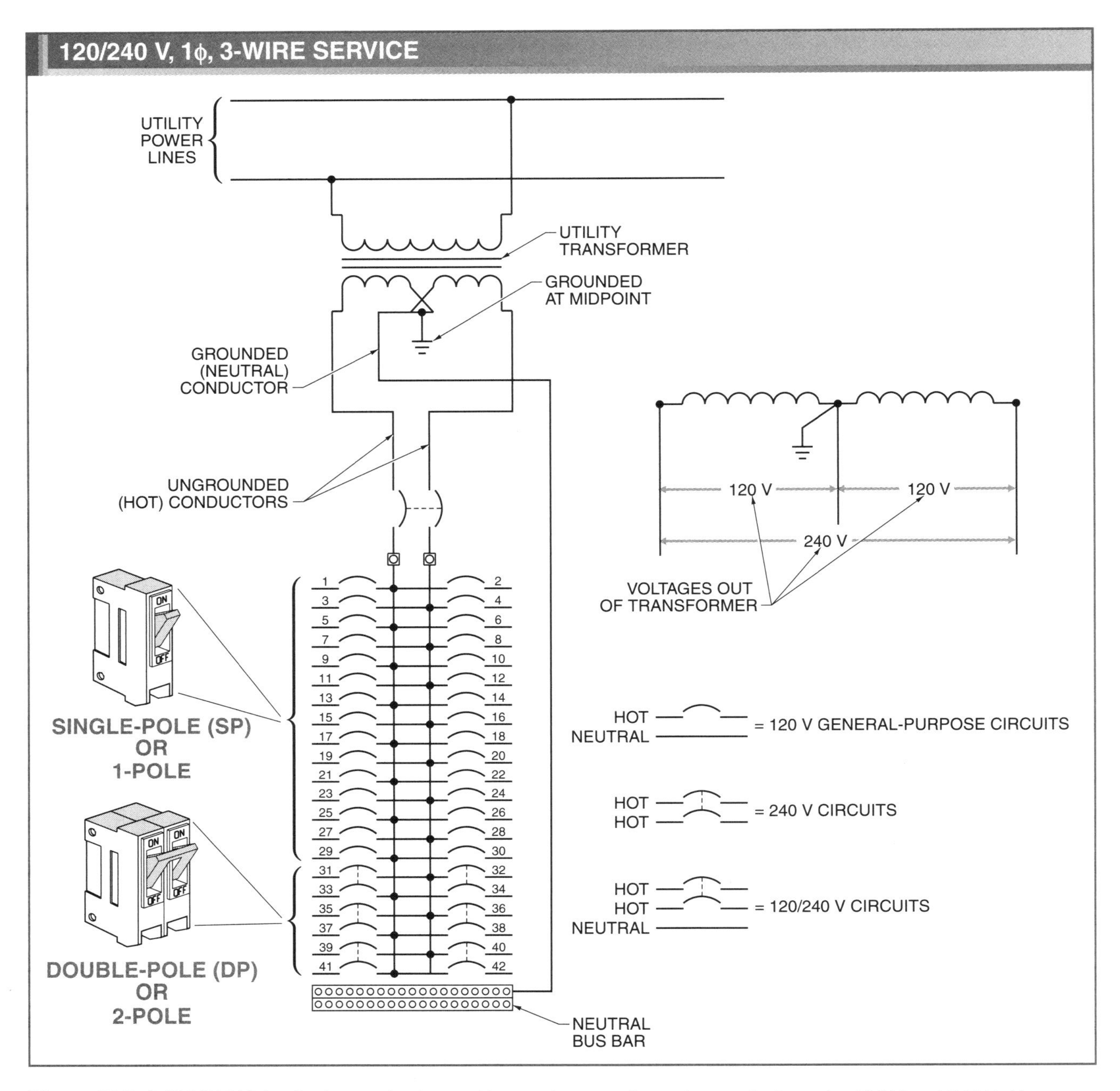

Figure 30-2. A 120/240 V, 1ϕ, 3-wire service is used to supply power to customers that require 120 V and 240 V, 1ϕ power.

Three-phase circuits include three individual ungrounded (hot) power lines. These power lines are referred to as phases A (L1), B (L2), and C (L3). Phases A, B, and C are connected to a switchboard or panelboard according to NEC® subsection 408.3(E). The phases must be arranged A, B, C from front to back, top to bottom, and left to right, as viewed from the front of the switchboard or panelboard. **See Figure 30-3.**

The high leg must be phase B and colored orange (or clearly marked) when the switchboard or panelboard is fed from a 120/240 V, 3ϕ, 4-wire service. This is because in this system, there are 195 V between phase B and the neutral (grounded conductor). This is considered an unreliable source of power because 195 V is too high for standard 115 V or 120 V loads and too low for standard 230 V or 240 V loads.

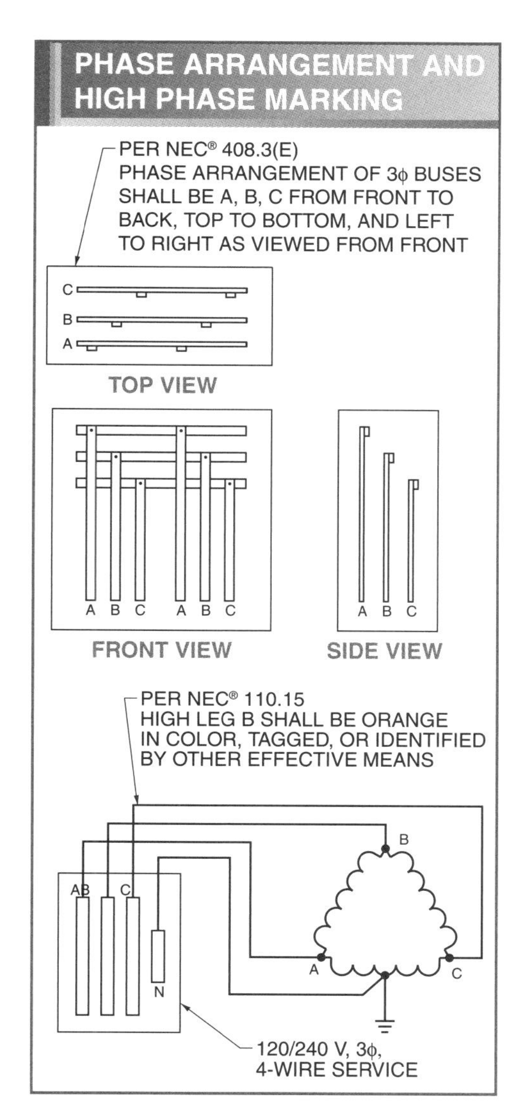

Figure 30-3. Phases A, B, and C are connected to a switchboard or panelboard according to NEC® subsection 408.3(E).

120/208 V, 3ϕ, 4-Wire Service. A 120/208 V, 3ϕ, 4-wire service is used to supply customers that require a large amount of 120 V, 1ϕ power; 208 V, 1ϕ power; and low-voltage 3ϕ power. This service includes three ungrounded (hot) lines and one grounded (neutral) line. Each hot line has 120 V to ground when connected to the neutral line. **See Figure 30-4.**

A 120/208 V, 3ϕ, 4-wire service is used to provide large amounts of low voltage (120 V, 1ϕ) power. The 120 V circuits should be balanced to equally distribute the power among the three hot lines. This is accomplished by alternately connecting the 120 V circuits to the power panel so each phase (A to N, B to N, C to N) is divided among the loads (lamps, receptacles, etc.). Likewise, 208 V, 1ϕ loads such as 208 V lamps and heating appliances should also be balanced between phases (A to B, B to C, C to A).

Three-phase loads, such as heating elements designed to be connected to 3ϕ power, can be connected to phases A, B, and C. Three-phase motors may also be connected to the 3ϕ power lines. However, the applied 3ϕ voltage to the motor is only 208 V and may present a problem when using some 3ϕ motors. This is because 208 V is at the lowest end of the 3ϕ voltage range. Any voltage variation on an AC induction motor affects motor operation.

120/240 V, 3ϕ, 4-Wire Service. A 120/240 V, 3ϕ, 4-wire service is used to supply customers that require large amounts of 3ϕ power with some 120 V and 240 V, 1ϕ power. This service supplies 1ϕ power delivered by one of the three transformers and 3ϕ power delivered by using all three transformers. The 1ϕ power is provided by center tapping one of the transformers. **See Figure 30-5.**

Because only one transformer delivers all of the 1ϕ power, this service is used in applications that require mostly 3ϕ power or 240 V, 1ϕ power and some 120 V, 1ϕ power. However, large power-consuming electric loads (including most motors over 1 HP) are connected to 3ϕ power. Thus, in many commercial applications, the total amount of 1ϕ power used is small when compared to the total amount of 3ϕ power used. Each transformer may be center tapped if large amounts of 1ϕ power are required.

Tech Facts

The voltage on the secondary side of a transformer begins to drop if the transformer is overloaded. The higher the overload, the greater the voltage drop.

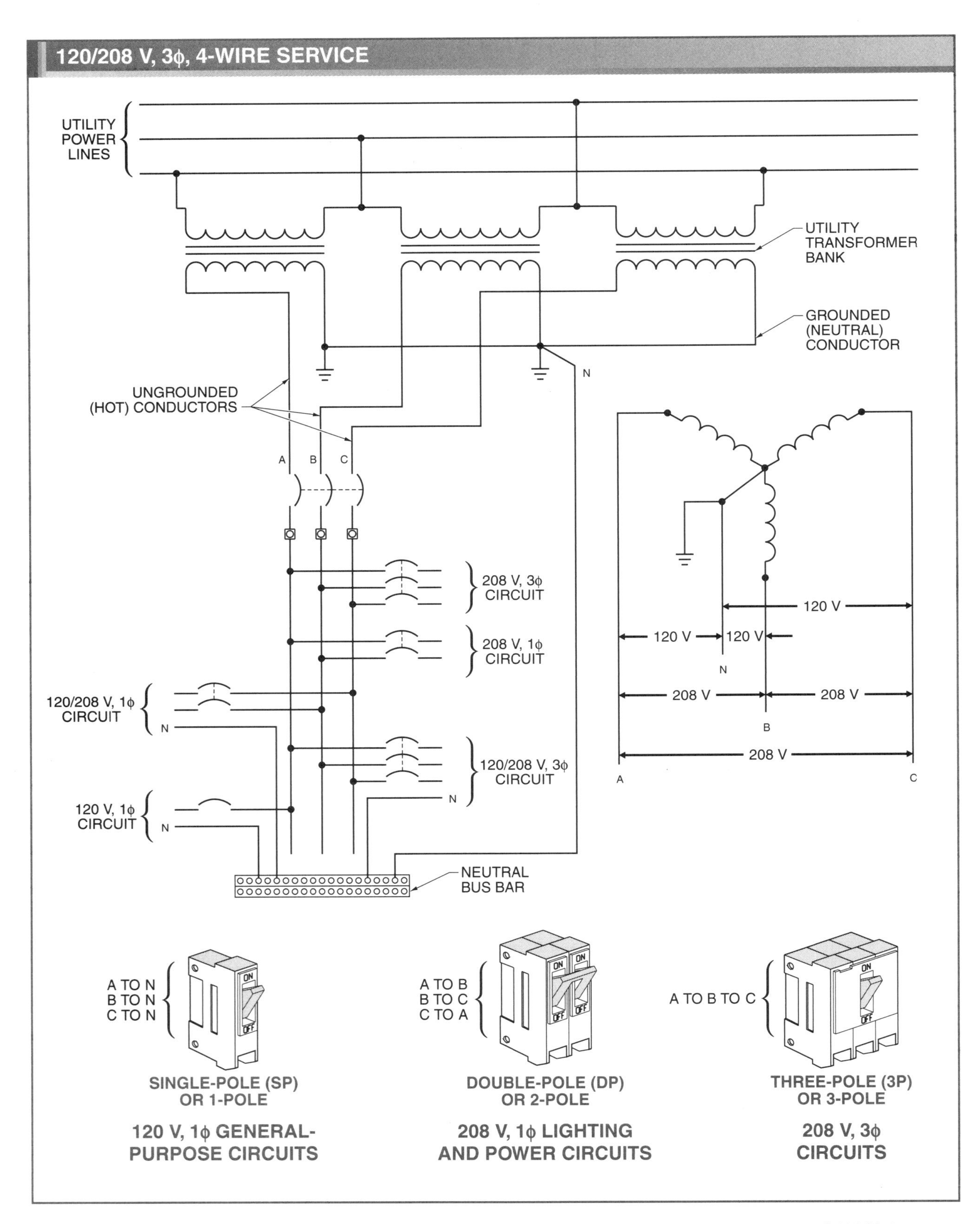

Figure 30-4. A 120/208 V, 3ϕ, 4-wire service is used to supply customers that require a large amount of 120 V, 1ϕ power; 208 V, 1ϕ power; and low-voltage 3ϕ power.

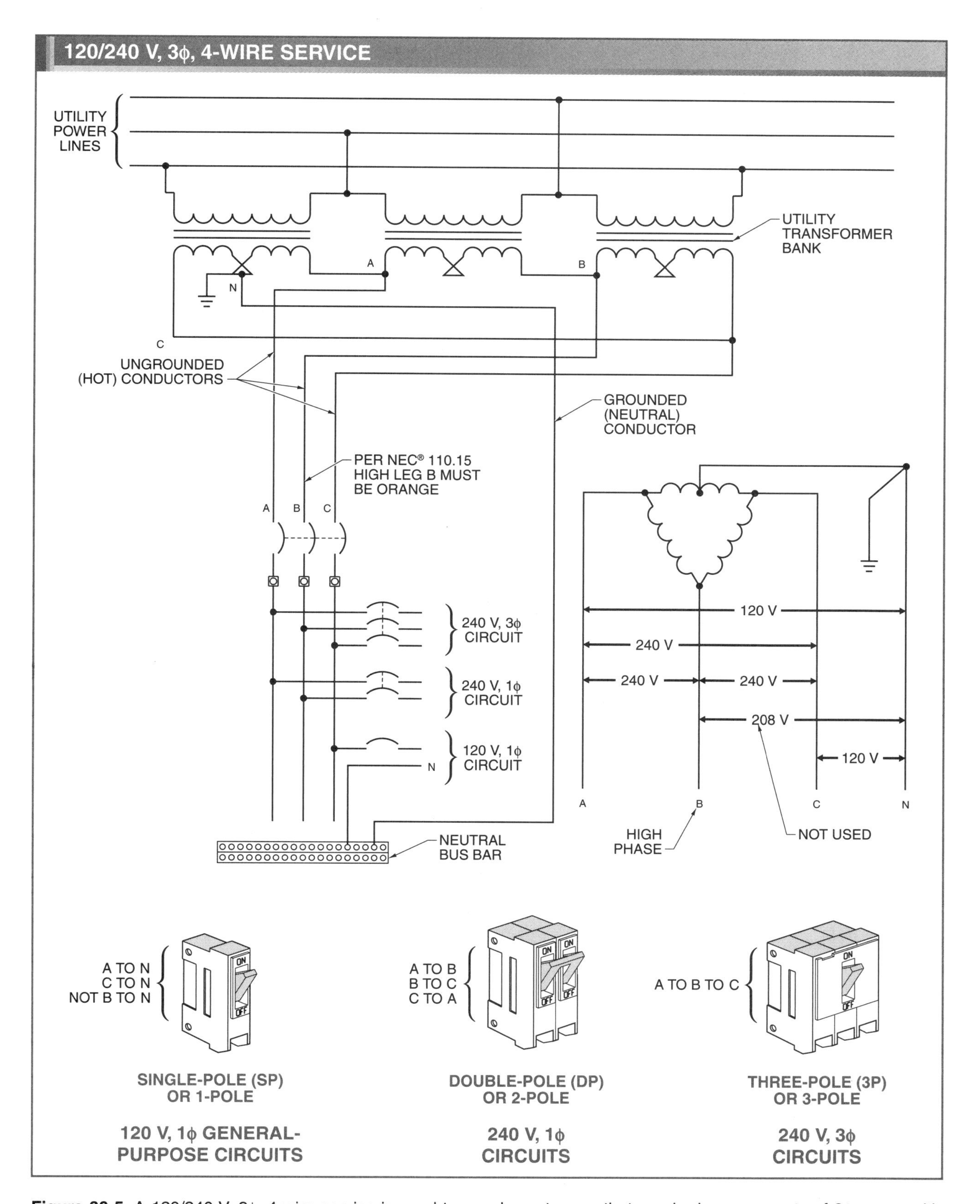

Figure 30-5. A 120/240 V, 3ϕ, 4-wire service is used to supply customers that require large amounts of 3ϕ power with some 120 V and 240 V, 1ϕ power.

277/480 V, 3ϕ, 4-Wire Service. The 277/480 V, 3ϕ, 4-wire service is the same as the 120/208 V, 3ϕ, 4-wire service except the voltage levels are higher. This service includes three ungrounded (hot) lines and one grounded (neutral) line. Each hot line has 277 V to ground when connected to the neutral or 480 V when connected between any two hot (A to B, B to C, or C to A) lines. **See Figure 30-6.**

This service provides 277 V, 1ϕ or 480 V, 1ϕ power, but not 120 V, 1ϕ power. For this reason, 277/480 V, 3ϕ, 4-wire service is not used to supply 120 V, 1ϕ general lighting and appliance circuits. However, this service can be used to supply 277 V and 480 V, 1ϕ lighting circuits. Such high-voltage lighting circuits are used in commercial fluorescent and HID (high-intensity discharge) lighting circuits.

A system that cannot deliver 120 V, 1ϕ power appears to have limited use. However, in many commercial applications (sport complexes, schools, offices, parking lots, etc.), lighting is a major part of the electrical system. Because large commercial applications include several sets of transformer banks, 120 V, 1ϕ power is available through other transformers. Additional transformers can also be connected to the 277/480 V, 3ϕ, 4-wire service to reduce the voltage to 120 V, 1ϕ.

Voltage Variation Effects on AC Motors

All motors are designed to operate at a given (nameplate rated) voltage. A motor operates if the voltage varies below or above the rated voltage, but there is a change in the motor's performance. Normally, the applied voltage should be within ±10% of the motor's rated voltage. **See Figure 30-7.**

A motor delivers less torque and operates at a higher temperature when connected to a voltage source lower than the motor's rated voltage. However, this does not adversely effect the motor unless the motor is operating in an application that requires it to deliver full-rated power. In such applications, the motor size and/or the delivered voltage must be increased.

Voltage Variation Effects on Heating Elements

Voltage is applied to heating elements to produce heat. Single-phase voltage is used for small heating applications and 3ϕ voltage is used for large commercial applications.

Any variation in applied voltage to a heating element increases or decreases the amount of heat produced. However, unlike motors, varying the voltage to heating elements is often done intentionally. The heating element output (amount of heat) can be reduced, and thus controlled, by reducing the applied voltage. Reducing the applied voltage to heating elements does not harm the heating elements. An SCR (silicon controlled rectifier) circuit may be used to set and maintain the amount of heat output.

Voltage Variation Effects on Lamps

The effects of voltage variations on a lamp depend on the lamp used. The voltage on incandescent lamps can vary from 0 V to 10% higher than the lamp's rated voltage. As the voltage is reduced, the lamp's output is reduced (lamp dims). Adjusting the voltage applied to an incandescent lamp is the basic principle of incandescent lamp dimmers.

Unlike incandescent lamps, HID lamps must be operated within a closer voltage range. Low-wattage bulbs (up to 200 W) should have an applied voltage within ±5% of the lamp fixture's rating. Lamps over 200 W should have an applied voltage within ±10% of the lamp fixture's rating.

Fluorescent lamps are available in many different styles. In general, fluorescent lamps should have an applied voltage that is within ±5% of the lamp fixture's rating. Check manufacturer's specifications for voltage variations greater or less than 5%.

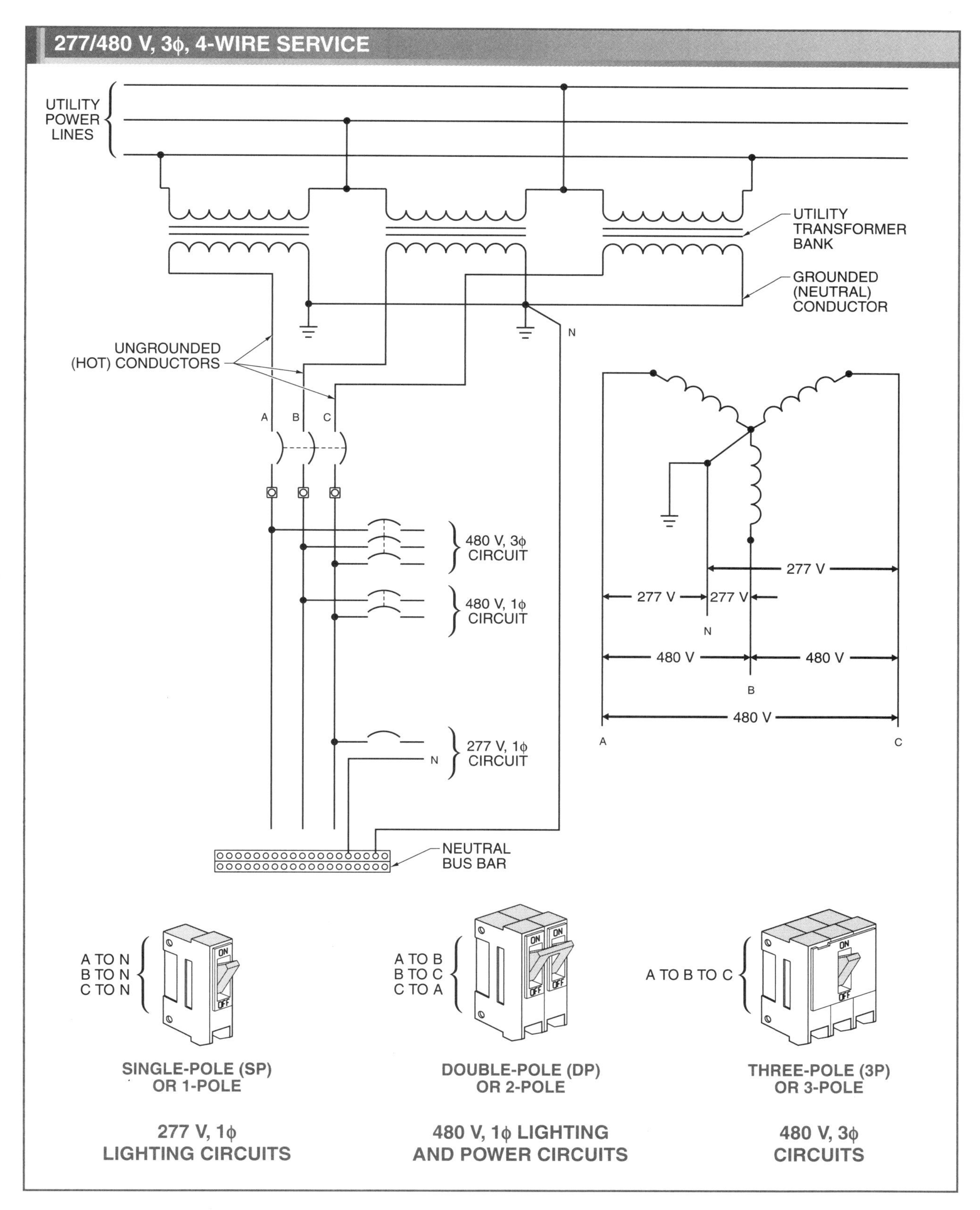

Figure 30-6. In a 277/480 V, 3ϕ, 4-wire service, each hot line has 277 V to ground when connected to the neutral or 480 V when connected between any two hot (A to B, B to C, or C to A) lines.

INDUCTION MOTOR PERFORMANCE CHANGES*

Characteristic	Change in Design Voltage	
	110%	90%
Torque Starting and maximum running	Increase 21%	Decrease 19%
Speed† Synchronous Full load Percent slip	No change Increase 1% Decrease 17%	No change Decrease 1.5% Increase 23%
Efficiency Full load ¾ load ½ load	Increase .5 to 1 Little change Decrease 1 to 2	Decrease 2 Little change Increase 1 to 2
Power factor Full load ¾ load ½ load	Decrease 3 Decrease 4 Decrease 5 to 6	Increase 1 Increase 2 to 3 Increase 4 to 5
Current Starting Full load	Increase 10% to 12% Decrease 7%	Decrease 10% to 12% Increase 11%
Temperature Rise	Decrease 3°C to 4°C	Increase 6°C to 7°C
Maximum Overload Capacity	Increase 21%	Decrease 19%
Magnetic Noise	Slight increase	Slight decrease

* 2-, 4-, 6-, and 8-pole motors
† speed of AC induction motors varies directly with frequency

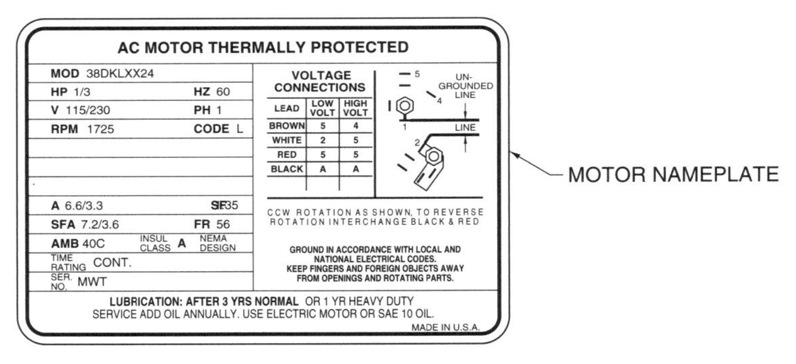

Figure 30-7. A motor operates if the voltage varies below or above the rated voltage, but there is a change in the motor's performance.

Conductor Identification—Color Coding

Conductors (wires) are covered with an insulating material that is available in different colors. The advantage of using different colors on conductors is that the function of each conductor can be easily determined. Some colors have definite meaning. For example, the color green always indicates a conductor used for grounding. Other colors may have more than one meaning depending on the circuit. For example, a red conductor may be used to indicate a hot wire in a 230 V circuit or switched wire in a 115 V circuit. Conductor color coding makes balancing loads among the different phases easier and aids when troubleshooting. Always use standard colors (green, white, black, etc.) where applicable.

Tech Facts

Never assume that conductors are properly marked or color-coded. Take voltage measurements to ensure that voltage is OFF before working on a circuit.

Green or green with a yellow stripe is the standard color for a grounding conductor in the electrical field. A solid green conductor is the most common color used for grounding conductors. Green is used to indicate a grounding conductor regardless of the voltage level (115 V, 230 V, 460 V, etc.) or circuit (1ϕ or 3ϕ). A *grounding conductor* is a conductor that does not normally carry current, except during a fault (short circuit).

The color white or natural gray is used for the neutral (grounded circuit) conductor. A *neutral conductor* is a current-carrying conductor that is intentionally grounded. Neutral conductors carry current from one side of a load (lamp, heating element, motor, etc.) to ground. Neutral conductors are connected directly to loads and never connected through fuses, circuit breakers, or switches. The conductor colors green (or green with yellow stripe) and white (or natural gray) are required per NEC® Article 210.5.

Electrical circuits include ungrounded (hot) conductors in addition to grounding and neutral conductors. An *ungrounded conductor* is a current-carrying conductor that is connected to loads through fuses, circuit breakers, and switches. Ungrounded conductors can be any color other than white, natural gray, green, or green with yellow stripe. Black is the most common color used for ungrounded conductors. Red, blue, orange, and yellow are also used for ungrounded conductors. Although such colors as red, blue, orange, and yellow are used to indicate a hot conductor, the exact color used to indicate different hot conductors (A [L1], B [L2], and C [L3] in a 3ϕ system) may vary. The one exception to this is listed in NEC® Article 215.8. This section states that in a 4-wire delta-connected secondary system, the higher voltage phase should be colored orange (or clearly marked) because it is too high for low voltage 1ϕ power and too low for 1ϕ low voltage. **See Figure 30-8.**

Conductor Identification–Wire Markers

Hot conductors in a 3ϕ system may be identified with other types of markings in addition to color coding. For example, a 3ϕ system may use three black conductors for each of the three hot (A, B, and C) conductors and identification markings for each conductor.

One band of colored (red, yellow, etc.) electrician's tape may be placed around the line one (L1) black conductor when marking 3ϕ hot conductors. Likewise, two bands of colored electrician's tape may be placed around the line two (L2) black conductor, and three bands of colored electrician's tape may be placed around the line three (L3) black conductor.

Wire markers may be used in place of tape bands. A *wire marker* is a preprinted peel-off marker designed to adhere when wrapped around a conductor. Wire markers resist moisture, dirt, oil, etc. and can be used to identify conductors of the same color that have different meanings. For example, the three hot black conductors (L1, L2, and L3) in a 3ϕ system can each be marked with a different numbered wire marker. Wire markers can still be used even when different colored conductors are used. Using wire markers in addition to color coding further clarifies the meanings of the conductors.

Commercial Lighting Systems

Outdoor commercial lighting is required for parking lots, entranceways, sports complexes, airports, buildings, piers, yards, storage areas, etc. Outdoor commercial lighting systems may use low- or high-pressure sodium, metal-halide, or mercury-vapor lamps in voltages of 120 V, 208 V, 240 V, 277 V, or 480 V.

Power ratings range from 35 W to above 1500 W. Indoor commercial lighting is required for all general rooms of occupancy, halls, restrooms, work areas, etc. Indoor commercial lighting systems may use incandescent, halogen, or fluorescent lamps in voltages of 120 V, 208 V, 220 V, 240 V, or 277 V.

Power ratings range from almost any size up to 300 W for incandescent and 15 W to 215 W for fluorescent lamps. Commercial lighting also includes lighting for information (signs and advertising), effect (stage lighting, black lights, etc.), and function (heat lamps, germicidal lamps, etc.). Informational commercial lighting systems may use incandescent or fluorescent lamps in voltages of 120 V and 277 V. Power ratings vary based on size of the sign.

Lamps are used to produce light. A *lamp* is an output device that converts electrical energy into light. The amount of light a lamp produces is expressed in lumens. A *lumen (lm)* is the unit used to measure the total amount of light produced by a light source. The more lumens produced by a light source per watt of power used, the more efficient the light source is in producing light. For example, a standard 40 W incandescent lamp produces about 480 lm (or 12 lm/W), and a standard 40 W fluorescent lamp produces about 3100 lm (or 77.5 lm/W). Thus, a 40 W fluorescent lamp source is more light-efficient than a 40 W incandescent lamp. The amount of lumens produced per watt varies depending on the wattage rating of the lamp. Manufacturer's specification sheets are used when determining exact energy efficiency for a given lamp.

The light produced by a light source causes illumination. *Illumination* is the effect that occurs when light falls on a surface. The unit of measure of illumination is the footcandle. A *footcandle (fc)* is the amount of light produced by a lamp (lumens) divided by the area that is illuminated. The amount of footcandles required for any given area varies depending on the application. **See Figure 30-9.**

CONDUCTOR COLOR CODING COMBINATIONS

Voltage*	Circuit	Conductor Colors
120	1ϕ, 2-wire with ground	One black (hot wire), one white (neutral wire), and one green (ground wire)
120/240	1ϕ, 3-wire with ground	One black (one hot wire), one red (other hot wire), one white (neutral wire), and one green (ground wire)
120/208	3ϕ, 4-wire wye with ground	One black (phase 1 hot wire), one red (phase 2 hot wire), one blue (phase 3 hot wire), one white (neutral wire), and one green (ground wire)
240	3ϕ, 3-wire delta with ground	One black (phase 1 hot wire), one red (phase 2 hot wire), one blue (phase 3 hot wire), and one green (ground wire)
120/240	3ϕ, 4-wire delta with ground	One black (first low phase hot wire), one red (second low phase hot wire), one orange (high phase leg wire), one white (neutral wire), and one green (ground wire)
277/480	3ϕ, 4-wire wye with ground	One brown (phase 1 hot wire), one orange (phase 2 hot wire), one yellow (phase 3 hot wire), and one green (ground wire)
480	3ϕ, 3-wire delta with ground	One brown (phase 1 hot wire), one orange (phase 2 hot wire), one yellow (phase 3 hot wire), and one green (ground wire)

* in V

Figure 30-8. Conductor color coding makes balancing loads among the different phases easier and aids when troubleshooting.

RECOMMENDED LIGHT LEVELS

ILLUMINATION

Illumination Category	Ranges of Illuminances*			Activity
	Low	Average	High	
A	2	3	5	Public spaces with dark surroundings
B	5	7.5	10	Simple orientation for short temporary visits
C	10	15	20	Working spaces where visual tasks are only occasionally performed
D	20	30	50	Performance of visual tasks of high contrast or large size
E	50	75	100	Performance of visual tasks of medium contrast or small size
F	100	150	200	Performance of visual tasks of low contrast or very small size
G	200	300	500	Performance of visual tasks of low contrast and very small size over a prolonged period
H	500	750	1000	Performance of prolonged and exacting visual tasks
I	1000	1500	2000	Performance of special visual tasks of extremely low contrast and small size

* in fc

ILLUMINANCE CORRECTION FACTORS – CATEGORIES A THROUGH C

Characteristics	Correction Factor		
	Low Value	Average Value	High Value
Occupant's Age	Under 40	40 to 55	Over 55
Surface Reflectances	Reflective	Average	Absorbent

ILLUMINANCE CORRECTION FACTORS – CATEGORIES D THROUGH I

Characteristics	Correction Factor		
	Low Value	Average Value	High Value
Worker's Age	Under 40	40 to 55	Over 55
Speed and/or Accuracy	Less important	Important	More important

Figure 30-9. Illumination is the effect that occurs when light falls on a surface.

Additional factors may be considered when determining the required amount of footcandles for a given application. For example, the size of the object to be viewed, the speed of a moving object, the reflectiveness of the surrounding surfaces, and the age of the viewer all affect the amount of total light required. Small objects are harder to see and require more light than large objects. The eye requires time to adjust and view an object when the object is moving. Fast moving objects require more light than slow moving objects.

In general, older people need more light for proper viewing. The surface on which an object is viewed must also be considered when determining the required amount of light. For example, the words on a textbook page are placed on a white paper background which has a reflectance of approximately 80%.

Brightness is the perceived amount of light reflecting from an object. Brightness depends on the amount of light falling on an object and the reflecting ability of the object. The object appears to have a uniform brightness if it is uniform and equally illuminated. However, most objects are not uniform nor equally illuminated so they appear to have varying degrees of brightness (shadows). Brightness is described as "perceived" because the eye is constantly adjusting to the amount of brightness. For example, when walking into a dark room, many objects cannot be seen. However, as the eyes adjust to the darkness, objects start appearing in more detail.

Likewise, glare takes place when there is too much brightness. Glare is reduced when the brightness is reduced and/or the eye reduces (squinting) the amount of light received. Proper lamp spacing, number of lamps, and size (wattage) determines the evenness of the brightness. *Contrast* is the ratio of brightness between different objects. For example, the black-on-white letters on a printed page have a high contrast. However, black-on-gray letters have a low contrast. High contrast is helpful when distinguishing outlines and small objects. Contrast between surfaces is sharpened by greater brightness and/or contrasting surfaces. **See Figure 30-10.**

Incandescent Lamps

Incandescent lamps are the most widely used lamps in the world. An *incandescent lamp* is an electric lamp that produces light by the flow of current through a tungsten filament inside a sealed glass bulb sometimes filled with a gas. Incandescent lamps have a low initial cost and are simple to install and service. These lamps have a lower electrical efficiency and shorter life than other lamps.

An incandescent lamp operates as current flows through the filament. A *filament* is a conductor with a resistance high enough to cause the conductor to heat. The filament glows white-hot and produces light. The filament is made of tungsten which limits the current to a safe operating level.

Inrush current is higher than operating current because the filament has a low resistance when cold. This high inrush current, when first switched ON, is the major cause of most incandescent lamp failures. The air inside the bulb is removed before the bulb is sealed. This prevents oxidation of the filament. The filament burns out quickly if oxygen is present. A gas mixture of nitrogen and argon is placed inside most incandescent bulbs to increase the life of the lamp.

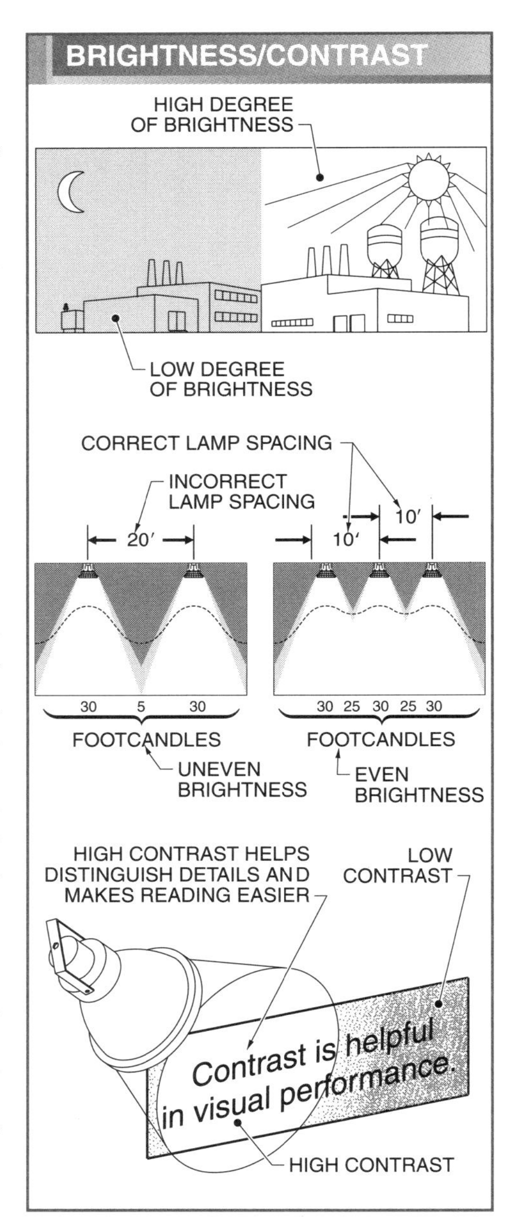

Figure 30-10. Contrast between surfaces is sharpened by greater brightness and/or contrasting surfaces.

Incandescent Lamp Bases. The base of an incandescent lamp holds the lamp firmly in the socket and connects electricity from the outside circuit to the filament. The ends of the filament are brought out to the base. The base simplifies the replacement of an incandescent lamp. The base of a lamp is used to hold the lamp in place and to connect the bulb to power. The size of the bulb and its current rating normally determine the type of base used. The two basic base configurations are bayonet and screw bases.

Small lamps are fitted with bayonet bases. A *bayonet base* is a bulb base that has two pins located on opposite sides. The pins slide into slots located in the bulb's socket. The advantage of a bayonet base is that it firmly holds the bulb. Bayonet base bulbs are used in applications that have high vibration, such as sewing machines and automobile lighting.

Screw bases are used in low-vibration applications. The medium screw and mogul screw are the most common bases used for general lighting service lamps. The medium screw base is normally used for bulbs up to 300 W. The mogul screw base is normally used for bulbs over 300 W. The miniature and candelabra bases are used for low-wattage bulbs, particularly the decorative types.

Incandescent Lamp Bulb Shapes. Bulb sizes and shapes are determined by the end use of the lamp. For any given wattage rating, the larger the bulb, the greater the cost. Large bulbs produce more light over time because the large bulb produces less blacking of the glass. This is because there is a large area over which the vaporized tungsten burning off the filament can dissipate. Bulb shape and size is determined by economics, esthetics, and performance.

Incandescent lamp bulbs are available in a variety of shapes and sizes. **See Figure 30-11.** Bulb shapes are designated by letters. The most common bulb shape is the "A" bulb. The "A" bulb is used for most residential and commercial indoor lighting applications. The "A" bulb is normally available in sizes from 15 W to 200 W.

Two common variations of the "A" bulb are the "PS" and the "P" bulbs. The "PS" bulb has a longer neck than the "A" bulb. The "PS" bulb is available in sizes from 150 W to 2000 W and is used for commercial and industrial lighting applications. The "P" bulb has a shorter neck than the "A" bulb and is used in indoor lighting applications that require a smaller bulb than the "A" bulb.

Other bulbs are used for decorative or special-purpose applications. The "C" bulb is designed to withstand moderate vibration and is used in sewing, washing, and other machines. The "B," "F," and "G" bulbs are used for decorative applications and are normally of low wattage. They are used for lighting applications in which the bulb is clearly seen, such as chandeliers, lamps, and mirrors in washrooms.

The "PAR" bulb is normally used as an outdoor spotlight. Floodlights normally use a reflector "R" bulb (indoors), or "PAR" bulb (outdoors). The "PAR" bulb is preferred outdoors because its shape allows for a more watertight seal. The "R" bulb is also used as a decorative light. Its high wattage is useful for applications such as showroom displays and advertising boards.

The "S" bulb is a straight bulb that is normally used in sign applications. The "T" bulb is a tubular decorative light that is used on display cabinets. "T" bulbs containing bayonet bases are used in high vibration applications such as automobile lighting.

Incandescent Lamp Sizes and Ratings. The size of an incandescent lamp is determined by the diameter of the lamp's bulb. The diameter of a bulb is expressed in eighths of an inch (1⁄8″). For example, an A-21 bulb is 21-eighths (21⁄8″) or 25⁄8″ in diameter at the bulb's maximum dimension. A scale calibrated in eighths is used to measure the size of a lamp. The scale is placed at the largest diameter of the bulb.

The power rating of incandescent lamp bulbs is given in watts. A *watt (W)* is a unit of power equal to the power produced by a current of 1 A across a potential difference of 1 V. To calculate power, apply the formula:

$P = E \times I$

where

P = power (in W)

E = voltage (in V)

I = current (in A)

Example: Calculating Power

How many watts are produced by a 120 V bulb pulling .5 A?

$P = E \times I$

$P = 120 \times .5$

$P = \mathbf{60\ W}$

The higher the wattage rating, the larger the amount (lumens) of light produced. However, the amount of light produced by an incandescent lamp is not directly proportional to the wattage rating. The higher the wattage rating, the more lumens per watt output delivered by the bulb.

Tech Facts

Fiberglass-reinforced lamp fixtures should be used in environments that contain corrosive materials (salt water, oil, sulfuric acid, steam, sodium hydroxide, hydrochloric acid, etc.).

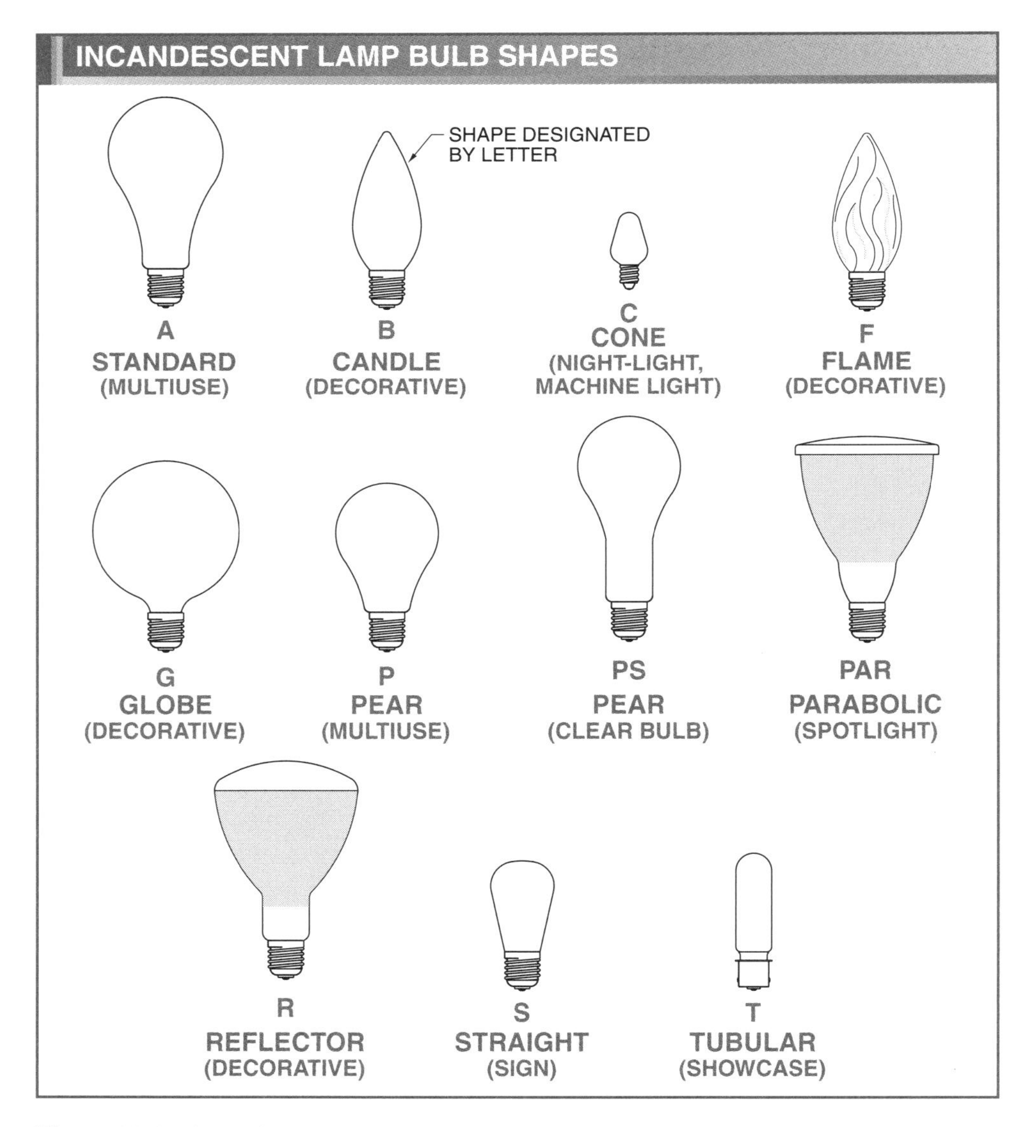

Figure 30-11. Incandescent lamp bulbs are available in a variety of shapes and sizes.

Tungsten-Halogen Lamps

A *tungsten-halogen lamp* is an incandescent lamp filled with a halogen gas (iodine or bromine). The gas combines with tungsten evaporated from the filament as the lamp burns. Tungsten-halogen lamps are used for display lighting, outdoor lighting, and in photocopy machines because they produce a large amount of light instantly. A tungsten-halogen lamp lasts about twice as long as a standard incandescent lamp.

Tungsten-halogen lamp wattages range from 15 W to 1500 W. However, because they are not as widely used as the standard incandescent lamp, they are not available in as wide a selection of bulb sizes and wattages.

Care must be taken when replacing a burned-out tungsten-halogen lamp. Ensure that all power is OFF because a high temperature is produced when the lamp is turned ON. The bulb is made of quartz to withstand the high temperature of the lamp. High-wattage tungsten-halogen lamps are available in "T" or "PAR" bulbs. The replacement cost is about three times the cost of an incandescent lamp. **See Figure 30-12.**

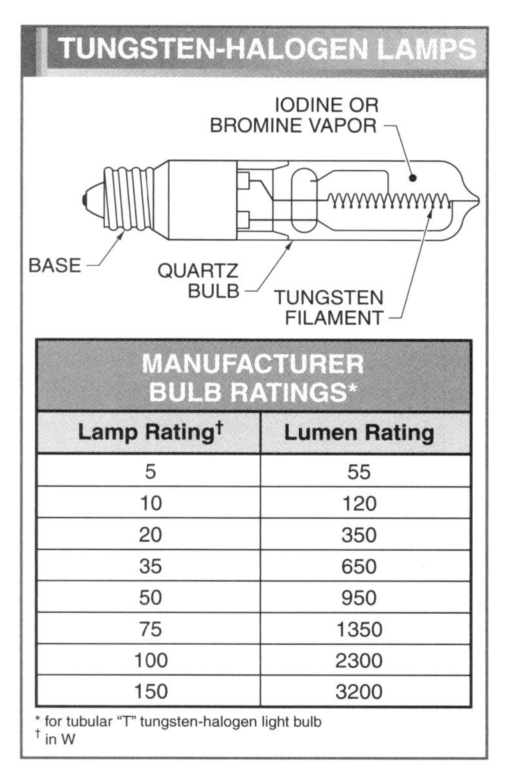

MANUFACTURER BULB RATINGS*

Lamp Rating†	Lumen Rating
5	55
10	120
20	350
35	650
50	950
75	1350
100	2300
150	3200

* for tubular "T" tungsten-halogen light bulb
† in W

Figure 30-12. Tungsten-halogen lamps are incandescent lamps filled with a halogen gas (iodine or bromine) that are used for display lighting, outdoor lighting, and in photocopy machines because they produce a large amount of light instantly.

Fluorescent Lamps

A *fluorescent lamp* is a low-pressure discharge lamp in which ionization of mercury vapor transforms ultraviolet energy generated by the discharge into light. The bulb contains a mixture of inert gas (normally argon and mercury vapor) which is bombarded by electrons from the cathode. This provides ultraviolet light. The ultraviolet light causes the fluorescent material on the inner surface of the bulb to emit visible light.

A fluorescent lamp consists of a cylindrical glass tube (bulb) sealed at both ends. Each end (base) includes a cathode that supplies the current to start and maintain the arc. A *cathode* is a tungsten coil coated with electron-emissive material which releases electrons when heated. The cathode in each end is constructed to allow conduction in either direction. Conduction in either direction allows the lamp to be operated on alternating current.

Standard fluorescent lamps vary in diameter from ⅝″ to 2⅛″ and in length from 6″ to 96″. The overall diameter of the lamp is designated by a number that indicates eighths of an inch. The most common size is the 40 W lamp, which is designated T-12 and has a length of 48″. The T-12 is 1½″ in diameter (12 × ⅛″ = 1½″).

Fluorescent lamp holders are designed for various types of lamp bases. The base is used to support the lamp and provide the required electrical connections. Fluorescent lamps that use preheat or rapid-start cathodes require four electrical contacts (pins). The four connections are made by using a bi-pin base at either end. Three standard types of bi-pin bases are miniature bi-pin, medium bi-pin, and mogul bi-pin. Instant-start fluorescent lamps require a single pin at each end of the lamp. **See Figure 30-13.**

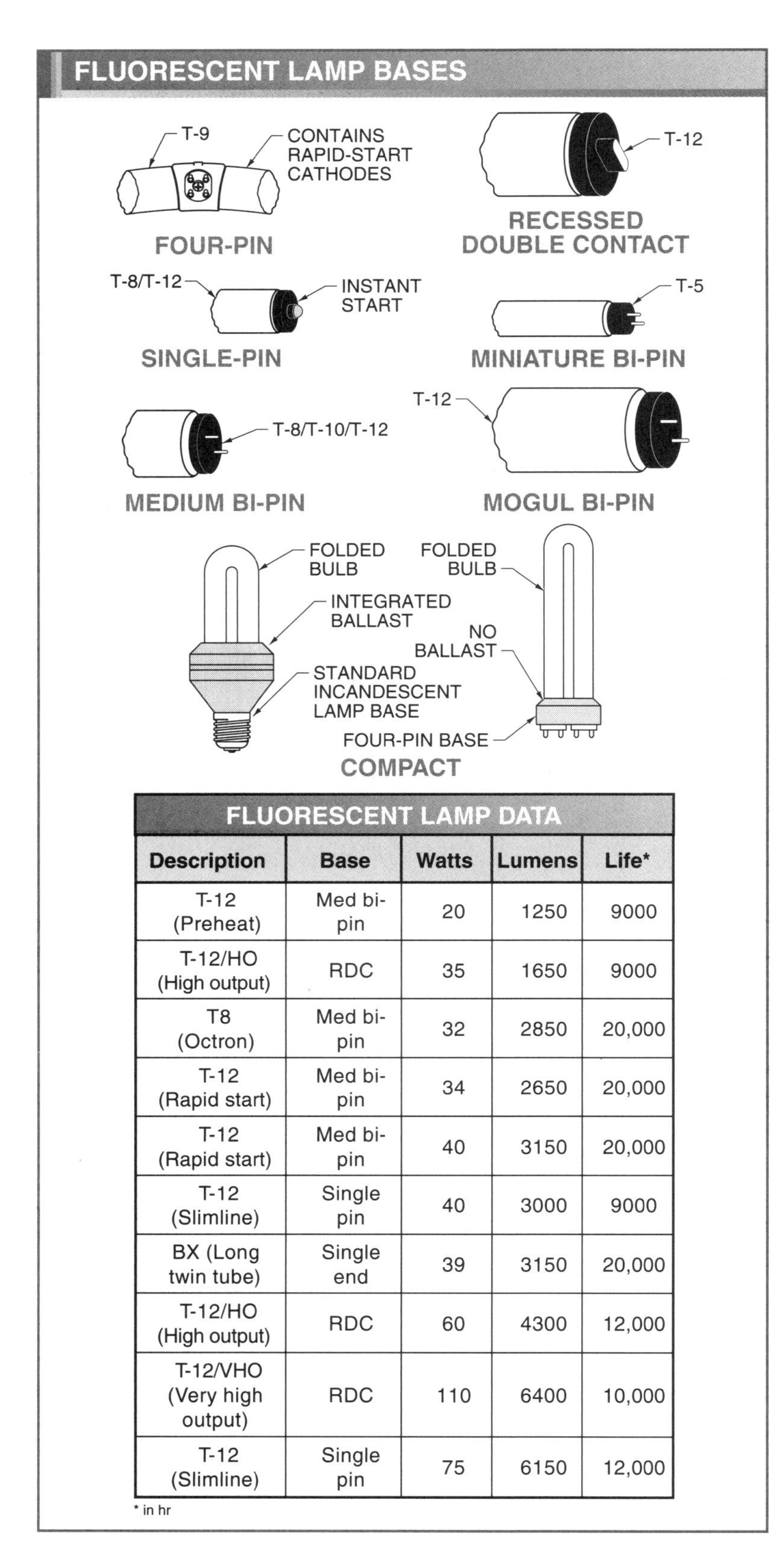

FLUORESCENT LAMP DATA

Description	Base	Watts	Lumens	Life*
T-12 (Preheat)	Med bi-pin	20	1250	9000
T-12/HO (High output)	RDC	35	1650	9000
T8 (Octron)	Med bi-pin	32	2850	20,000
T-12 (Rapid start)	Med bi-pin	34	2650	20,000
T-12 (Rapid start)	Med bi-pin	40	3150	20,000
T-12 (Slimline)	Single pin	40	3000	9000
BX (Long twin tube)	Single end	39	3150	20,000
T-12/HO (High output)	RDC	60	4300	12,000
T-12/VHO (Very high output)	RDC	110	6400	10,000
T-12 (Slimline)	Single pin	75	6150	12,000

* in hr

Figure 30-13. Fluorescent lamp holders are designed for various types of lamp bases.

A *compact fluorescent lamp* is a fluorescent lamp that has a smaller diameter than a conventional fluorescent lamp and a folded bulb configuration. Compact fluorescent lamps provide high light output in much smaller sizes than conventional fluorescent lamps.

Compact fluorescent lamps are available with no ballast or with an integrated ballast. Compact fluorescent lamps containing no ballasts have standard fluorescent lamp bases and are connected to standard fluorescent lamp circuits. Compact fluorescent lamps with an integrated ballast are available with standard incandescent lamp bases and are connected to standard incandescent lamp circuits.

Compact fluorescent lamps are available in a wide variety of sizes and output ratings. Compact fluorescent lamps provide increased energy savings compared to incandescent lamps of comparable light output. Compact fluorescent lamps have an average rated life of 10,000 hr. For this reason, they require less replacement than incandescent lamps. Compact fluorescent lamps are more expensive than incandescent lamps, but, due to their long life, they can provide the same lumens as incandescent lamps at nearly one-fourth the cost.

Fluorescent bulbs do not start well in cold climate conditions because the electrons released by the heated cathode are thermally released. The colder the bulb, the longer the time required to heat the cathode and release electrons into the mercury vapor. Even after the bulb is turned ON, it delivers less light in cold climate conditions as low temperatures affect the mercury-vapor pressure inside the bulb. The colder the bulb, the less the light output. Standard indoor fluorescent lamps are designed to deliver a peak light output under approximately 75°F ambient temperature conditions.

The 75°F ambient temperature is considered to be room temperature for most fluorescent lamp installations. However, if some temperature other than the standard room temperature is used to rate the lamp, the manufacturer normally lists a luminaire temperature. *Luminaire temperature* is the temperature at which a lamp delivers its peak light output. Moderate changes in ambient temperature (50°F to 105°F) have little effect (less than 10%) on light output. Temperatures lower than 50°F or higher than 105°F have a greater effect on light output.

To help hold the heat, outdoor rated fluorescent lamps include an outer glass jacket. The jacket helps hold in heat which shifts the peak output to a lower ambient temperature point. Jacketed lamps are available with different peak output points. Jacketed lamps are recommended for many outdoor applications and certain indoor applications such as freezer warehouses, subways, and tunnels where cold and windy conditions exist.

Fluorescent lamp bulbs that produce different shades of colors are available. The shade of color produced by a fluorescent bulb helps produce a certain environment. Fluorescent bulbs are available for cool, moderate, and warm environments.

Cool Environments. Cool white or light white fluorescent bulbs produce a pale blue-green whiteness. They are used to enhance blues and greens. This bulb is suggested for high lighting areas, such as industrial and commercial work spaces. It is also used in marketing displays to make people and merchandise look more attractive and natural.

Moderate Environments. White fluorescent bulbs produce a very pale yellow-whiteness. White fluorescent bulbs may also be labeled as natural or daylight. They are used to produce a light environment midway between the blueness of cool white and the yellowness of warm white. White fluorescent bulbs are used in areas where a more neutral effect is desired. For example, white fluorescent bulbs are used in supermarkets, libraries, classrooms, and where fluorescent lighting is mixed with incandescent lighting.

Warm Environments. Warm, white fluorescent bulbs produce a more yellow-whiteness. They are used to give a warm feeling and enhance brighter colors. This bulb type is suggested for low lighting levels. For example, restaurants may use warm, white fluorescent bulbs to help set a particular mood they wish to create.

Preheat Circuits. A *preheat circuit* is a fluorescent lamp-starting circuit that heats the cathode before an arc is created. Preheat circuits are used in some low-wattage lamps (4 W to 20 W) and are common in desk lamps. A preheat circuit uses a separate starting switch that is connected in series with the ballast. The electrodes in a preheat circuit require a few seconds to reach the proper temperature to start the lamp after the starting switch is pressed. **See Figure 30-14.**

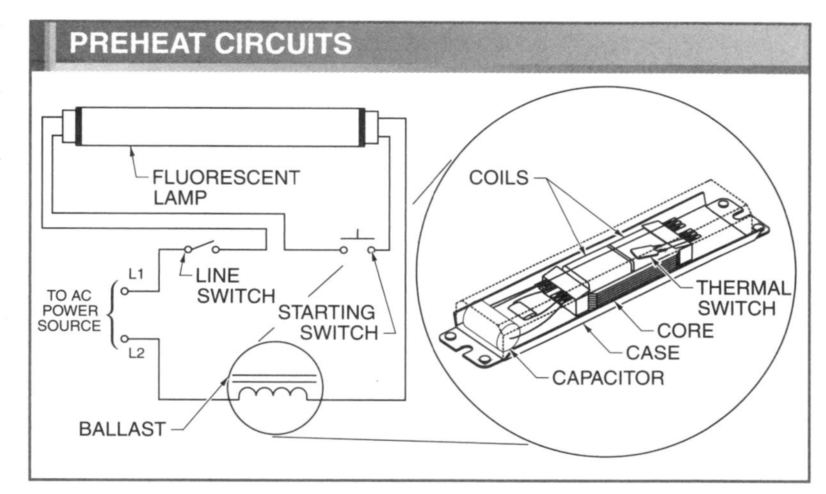

Figure 30-14. A preheat circuit is a fluorescent lamp-starting circuit that heats the cathode before an arc is created.

Industrial motor drives use three-phase power to supply electric motors.

The opening of the starting switch breaks the path of current flow through the starting switch. This leaves the gas in the lamp as the only path for current to travel. A high-voltage surge is produced by the collapsing magnetic field as the starting switch is no longer connected to power. This high-voltage surge starts the flow of current through the gas in the lamp. Once started, the ballast limits the flow of current through the lamp.

Instant-Start Circuits. An *instant-start circuit* is a fluorescent lamp-starting circuit that provides sufficient voltage to strike an arc instantly. The instant-start circuit was developed to eliminate the starting switch and overcome the starting delay of preheat circuits. An arc strikes without preheating the cathodes when a high enough voltage is applied across a fluorescent lamp. **See Figure 30-15.**

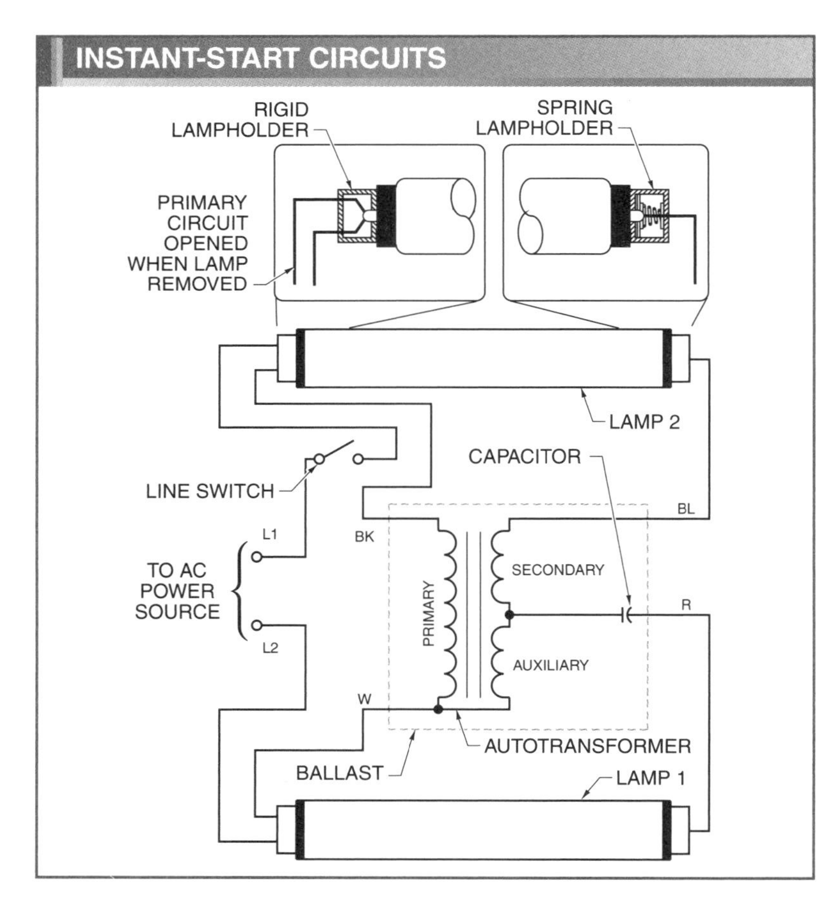

Figure 30-15. An instant-start circuit is a fluorescent lamp-starting circuit that provides sufficient voltage to strike an arc instantly.

Tech Facts

Passive infrared wall switch sensors may be used in commercial rooms to automatically turn OFF lights when no one is present and automatically turn ON lights when a person enters the room.

The high initial voltage requires a large autotransformer as an integral part of the ballast. The autotransformer delivers an instant voltage of 270 V to 600 V, depending on the bulb size and voltage rating. The larger the bulb, the higher the voltage. Instant-start lamps require only one pin at each base because no preheating of the electrode is required.

A safety circuit is used with an instant-start lamp to prevent electrical shock due to the high starting voltage. The base pin acts as a switch, interrupting the circuit to the ballast when the lamp is removed.

Rapid-Start Circuits. A *rapid-start circuit* is a fluorescent lamp-starting circuit that has separate windings to provide continuous heating voltage on the lamp cathodes. The lamp starting time is reduced because the cathodes are continuously heated. The rapid-start circuit brings the lamp to full brightness in about two seconds. A rapid-start circuit is the most common circuit used in fluorescent lighting.

A rapid-start circuit uses lamps that have short, low-voltage electrodes which are automatically preheated by the lamp ballast. The rapid-start ballast preheats the cathodes by means of a heater winding. The heater winding continues to provide current to the lamp after ignition. **See Figure 30-16.**

High-Intensity Discharge Lamps

A *high-intensity discharge (HID) lamp* is a lamp that produces light from an arc tube. An *arc tube* is the light-producing element of an HID lamp. The arc tube contains metallic and gaseous vapors and electrodes. An arc is produced in the tube between the electrodes. The arc tube is enclosed in a bulb which may contain phosphor or a diffusing coating that improves color rendering, increases light output, and reduces surface brightness. HID lamps include low-pressure sodium, mercury-vapor, metal-halide, and high-pressure sodium lamps. **See Figure 30-17.**

All HID lamps are electric discharge lamps. An *electric discharge lamp* is a lamp that produces light by an arc discharged between two electrodes. High vapor pressure is used to convert a large percentage of the energy produced into visible light. Arc tube pressure for most HID lamps is normally from one to eight atmospheres. HID lamps provide an efficient, long-lasting source of light and are used for street, parking lot, and general lighting applications.

HID Lamp Color Rendering. *Color rendering* is the appearance of a color when illuminated by a light source. For example, a red color may be rendered light, dark, pinkish, or yellowish depending on the light source under which it is viewed. Color rendering of HID lamps varies depending on the lamp used.

Low-pressure sodium lamps produce yellow to yellow-orange light. The yellow light is produced by the sodium in the low-pressure sodium lamp. Low-pressure sodium lamps are normally not used where the appearance of people and colors are important. The yellow light produces severe color distortion on most light-colored objects.

Mercury-vapor lamps produce a light yellow to white light. Mercury-vapor lamps with clear bulbs have poor rendering of reds. Blue colors appear purplish with most other colors appearing normal.

Metal-halide lamps produce a light yellow to white light. Metal-halide lamps produce good overall color rendering. Red colors appear slightly muted with some pinkish overtones.

High-pressure sodium lamps produce a golden white light. High-pressure sodium lamps have good color rendering with all colors being clearly distinguishable. However, reds, greens, blues, and violets are muted.

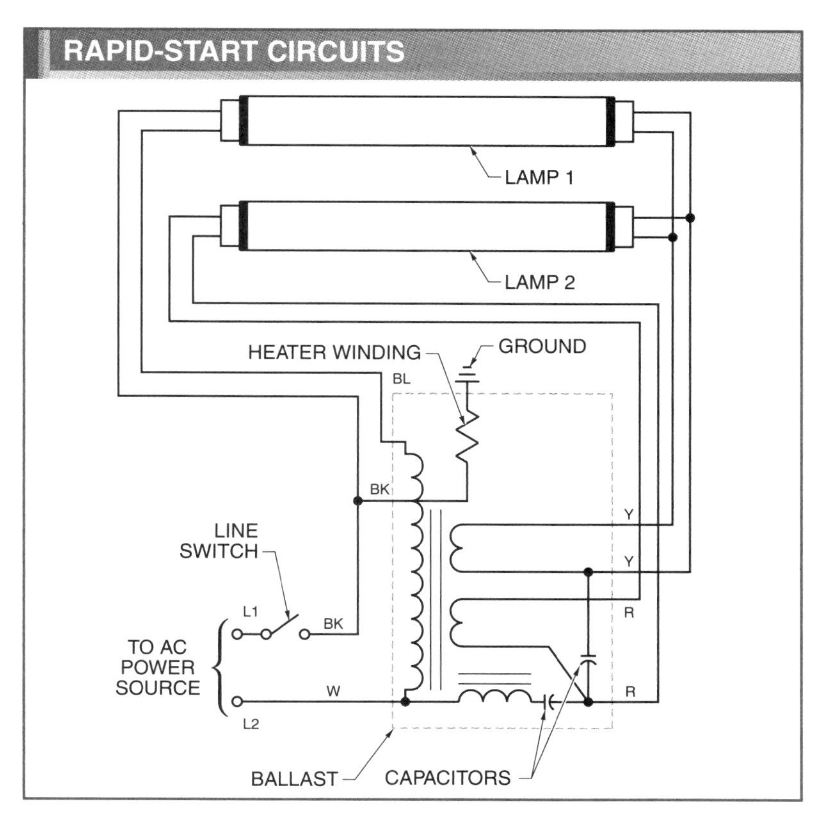

Figure 30-16. A rapid-start circuit is a fluorescent lamp-starting circuit that has separate windings to provide continuous heating voltage on the lamp cathodes.

HID Lamp Selection. Low-pressure sodium lamps are good for outdoor lighting installations. The yellow-orange color is acceptable for street, highway, parking lot, and floodlight applications. The long start-up time is not a problem for outdoor lighting because the lights are normally turned ON by a photoelectric cell at dusk. Photoelectric cell switches may be adjusted to turn ON at different light levels. Low-pressure sodium lamps are also used for some indoor applications such as warehouse lighting and other areas where color distortion is not critical to an operation.

Metal-halide lamps may be used when the color rendering of low-pressure sodium lamps is not acceptable for the application. Metal-halide lamps are used for sport, street, highway, parking lot, and floodlight applications. Metal-halide lamps are normally specified for most sport, indoor, and outdoor lighting applications.

HID LAMPS

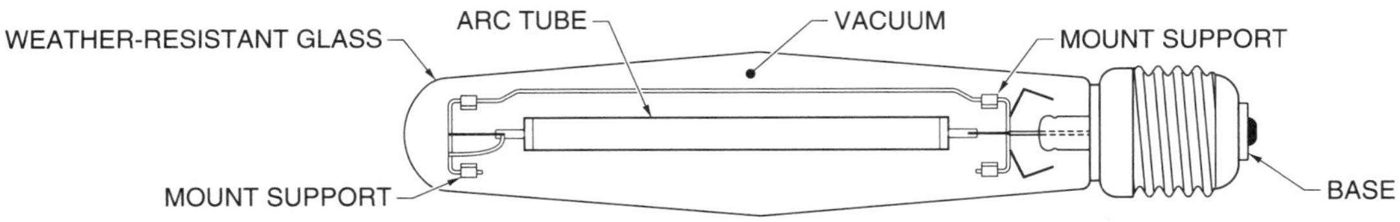

Low-Pressure Sodium	Mercury-Vapor	Metal-Halide	High-Presure Sodium
• Uses sodium vapor under low pressure to produce light	• Uses mercury vapor to produce light	• Uses mercury vapor with metal halides to produce light	• Uses sodium vapor under high pressure to produce light
• 6 min to 12 min start time	• 5 min to 6 min start time	• 2 min to 5 min start time	• 3 min to 4 min start time
• 4 sec to 12 sec restart time	• 3 min to 5 min restart time	• 10 min to 15 min restart time	• 30 sec to 60 sec restart time
• 190 lm/W to 200 lm/W	• 50 lm/W to 60 lm/W	• 80 lm/W to 125 lm/W	• 65 lm/W to 115 lm/W
• Produces yellow to yellow-orange light	• Produces white light with blue colors appearing purplish and yellow colors having a greenish overtone	• Produces white light with red colors slightly muted	• Produces golden white light
• 1800 hr rated bulb life*	• 16,000 hr to 24,000 hr rated bulb life*	• 3000 hr to 20,000 hr rated bulb life*	• 7500 hr to 14,000 hr rated bulb life*

* rated bulb life depends on usage, bulb size, and cycle time

HID LAMP DATA

Lamp	Description	Base	Watts	Lumens	Life*
Low-pressure sodium	T-16 clear	Bayonet	18	1570	16,000
	T-16 clear	Bayonet	35	4000	16,000
	T-16 clear	Bayonet	55	6655	16,000
	T-21 clear	Bayonet	90	11,095	16,000
	T-21 clear	Bayonet	135	19,140	16,000
Mercury-vapor	E-23½ white	Mogul	100	4200	24,000
	E-28 white	Mogul	175	8600	24,000
	E-28 white	Mogul	250	12,100	24,000
	E-37 or BT-37 white	Mogul	400	22,500	24,000
	BT-56 white	Mogul	1000	63,000	24,000
High-pressure sodium	B-17 clear	Medium	35	2250	24,000
	E-23½ clear	Mogul	50	4000	24,000
	E-23½ clear	Mogul	70	6400	24,000
	B-17 clear	Medium	100	9500	24,000
	B-17 clear	Medium	150	16,000	24,000
	E-23½ clear	Mogul	150	16,000	24,000
	E-18 clear	Mogul	250	30,000	24,000
	E-25 clear	Mogul	1000	140,000	24,000
Metal-halide	ED-17 coated	Medium	70	4800	10,000
	ED-17 coated	Medium	150	12,200	10,000
	E-28/BT-28 coated	Mogul	175	14,000	10,000
	E-28/BT-28 coated	Mogul	250	20,500	10,000
	E-37/BT-37 coated	Mogul	400	34,000	20,000
	BT-56 coated	Mogul	1000	110,000	12,000

* in hr

Figure 30-17. A high-intensity discharge (HID) lamp is a lamp that produces light from an arc tube.

Mercury-vapor lamps may be used when the initial installation cost is of major importance. Mercury-vapor lamps are used in all types of applications. Metal-halide lamps may be used when true color appearance is important such as in a car dealer's lot.

High-pressure sodium lamps may be used when lower operating cost is important and some color distortion is acceptable. High-pressure sodium lamps may be used in parking lots, street lighting, shopping centers, exterior buildings, and most storage areas.

Low-Pressure Sodium Lamps. A *low-pressure sodium lamp* is an HID lamp that operates at a low vapor pressure and uses sodium as the vapor. A low-pressure sodium lamp has a U-shaped arc tube. The arc tube has both electrodes located at the same end. The arc tube is placed inside a glass bulb and contains a mixture of neon, argon, and sodium-metal. On start-up, an arc is discharged through the neon, argon, and sodium-metal. As the sodium-metal heats and vaporizes, the amber color of sodium is produced.

A low-pressure sodium lamp is named for its use of sodium inside the arc tube. This lamp has the highest efficiency rating of any lamp. Some low-pressure sodium lamps deliver up to 200 lm/W of power. This is 10 times the output of an incandescent lamp.

Low-pressure sodium lamps must be operated on a ballast that is designed to meet the lamp's starting and running requirements. Low-pressure sodium lamps do not have a starting electrode or ignitor. The ballast must provide an open-circuit voltage of approximately three to seven times the lamp's rated voltage to start and sustain the arc. **See Figure 30-18.**

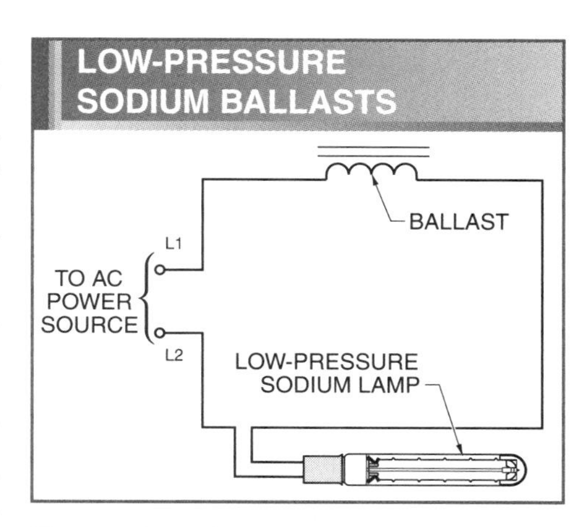

Figure 30-18. Low-pressure sodium lamps must be operated on a ballast that is designed to meet the lamp's starting and running requirements.

Mercury-Vapor Lamps. A *mercury-vapor lamp* is an HID lamp that produces light by an electrical discharge through mercury vapor. Mercury-vapor lamps are used for general lighting applications. Phosphor coating is added to the inside of the bulb to improve color-rendering characteristics.

A mercury-vapor lamp contains a starting electrode and two main electrodes. An electrical field is set up between the starting electrode and one main electrode when power is first applied to the lamp. The electrical field causes current to flow and an arc to strike. Current flows between the two main electrodes as the heat vaporizes the mercury.

Mercury-vapor ballasts include reactor, high-reactance autotransformer, constant-wattage autotransformer, and two-winding, constant-wattage ballasts. The ballast used normally depends on economics. **See Figure 30-19.** A *reactor ballast* is a ballast that connects a coil (reactor) in series with the power line leading to the lamp. Reactor ballasts are used when the incoming supply voltage meets the starting voltage requirements of the lamp. This is common when the incoming supply voltage is 240 V or 277 V. Both 240 V and 277 V mercury-vapor lamps are standard. A capacitor is added to some reactor ballasts to improve the power factor.

Reactor ballasts cost the least of all mercury-vapor ballasts but have poor lamp-wattage regulation. They should only be used when line voltage regulation is good because a 5% change in line voltage produces a 10% change in lamp wattage in a reactor ballast.

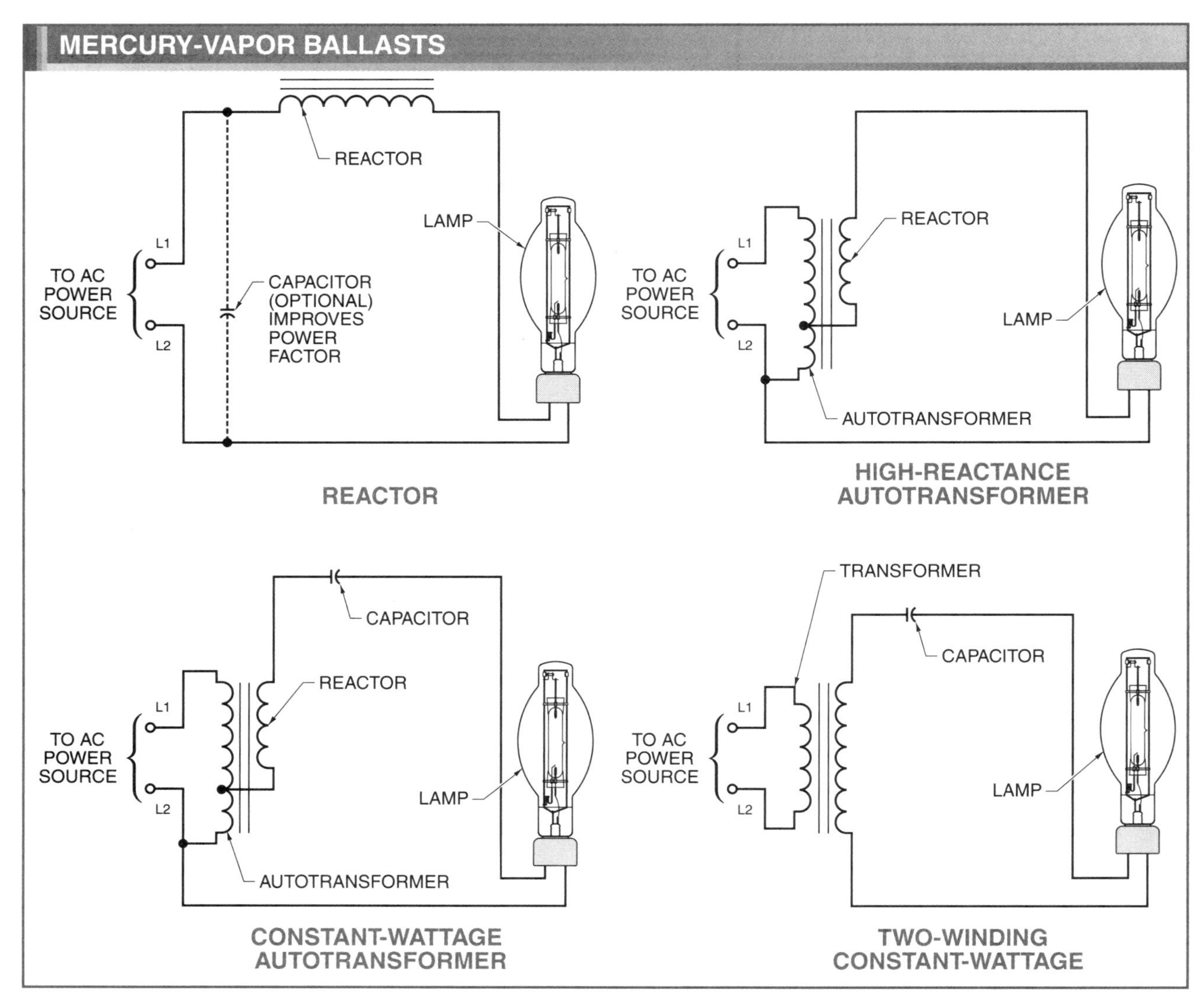

Figure 30-19. Mercury-vapor ballasts include reactor, high-reactance autotransformer, constant-wattage autotransformer, and two-winding, constant-wattage ballasts.

A *high-reactance autotransformer ballast* is a high-reactance ballast that uses two coils (primary and secondary) to regulate both voltage and current. High-reactance autotransformer ballasts are used when the incoming supply voltage does not meet the starting requirements of the lamp. Incoming voltages of 115 V, 208 V, and 460 V require a voltage change to the lamp. The voltage is regulated to within ±5% of the bulb's rating because a small percentage change in voltage results in a large percentage change in light output.

A *constant-wattage autotransformer ballast* is a high-reactance autotransformer ballast with a capacitor added to the circuit. The capacitor improves the power factor. A constant-wattage autotransformer ballast is the most commonly used ballast.

A *two-winding constant-wattage ballast* is a ballast that uses a transformer which provides isolation between the primary and secondary circuits. A two-winding constant-wattage ballast has excellent lamp-wattage regulation. A 13% change in line voltage produces only a 2% to 3% change in lamp wattage.

Metal-Halide Lamps. A *metal-halide lamp* is an HID lamp that produces light by an electrical discharge through mercury vapor and metal halide in the arc tube. A *metal halide* is an element (normally sodium and scandium iodide) which is added to the mercury in small amounts. A metal-halide lamp produces more lumens per watt than a mercury-vapor lamp.

The light produced by a metal-halide lamp does not produce as much color distortion as a mercury-vapor lamp. A metal-halide lamp is an efficient source of white light. It has a shorter bulb life than the other HID lamps.

A metal-halide ballast uses the same basic circuit as the constant-wattage autotransformer mercury-vapor ballast. The ballast is modified to provide high starting voltage required by metal-halide lamps. **See Figure 30-20.**

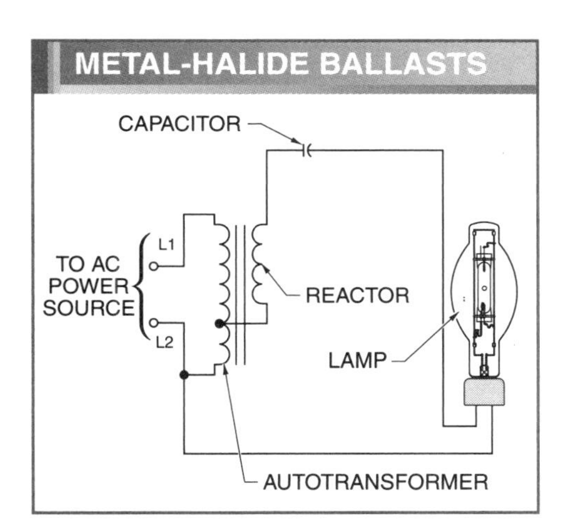

Figure 30-20. A metal-halide ballast uses the same basic circuit as the constant-wattage autotransformer mercury-vapor ballast.

High-Pressure Sodium Lamps. A *high-pressure sodium lamp* is an HID lamp that produces light when current flows through sodium vapor under high pressure and high temperature. A high-pressure sodium lamp is a more efficient lamp than a mercury-vapor or metal-halide lamp. The light produced from a high-pressure sodium lamp appears as a golden-white color.

A high-pressure sodium lamp is constructed with a bulb and an arc tube. The arc tube is made of ceramic to withstand high temperature. The bulb is made of weather-resistant glass to prevent heat loss and protect the arc tube.

High-pressure sodium lamps do not have a starting electrode. The ballast must deliver a voltage pulse high enough to start and maintain the arc. This voltage pulse must be delivered every cycle and must be 4000 V to 6000 V for 1000 W lamps and 2500 V to 4000 V for smaller lamps. The starter (ignitor) is the device inside the ballast that produces the high starting voltage. **See Figure 30-21.**

A high-pressure sodium ballast is similar to a mercury-vapor reactor ballast. The main difference is the added starter. The reactor ballast is used where the input voltage meets the lamp's requirement. A transformer or autotransformer is added to the ballast circuit when the incoming voltage does not meet the lamp's requirements.

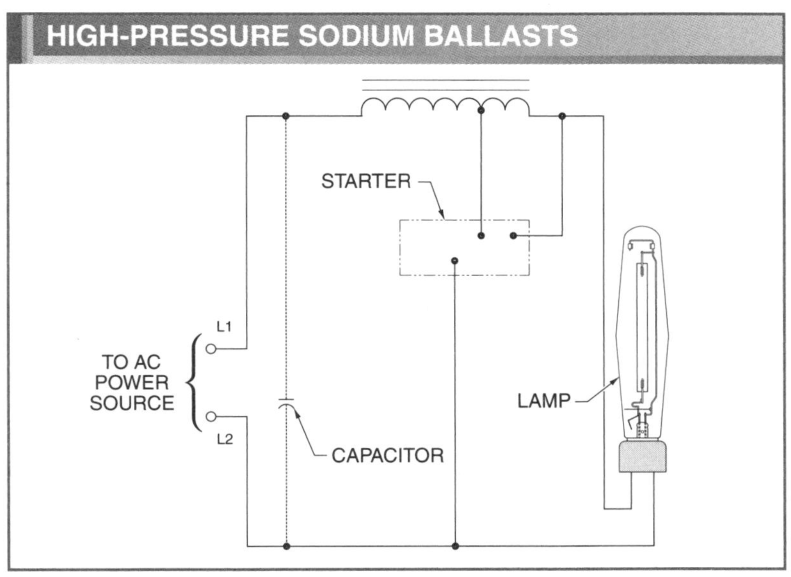

Figure 30-21. The starter (ignitor) is the device inside the ballast that produces the high starting voltage in a high-pressure sodium ballast.

Review Questions

1. What is the purpose of a substation?
2. What kind of transformer (single-phase or three-phase) do power companies normally use to distribute power to customers?
3. How high can the voltage on an incandescent lamp become?
4. What is the standard color for a grounding conductor?
5. What kind of conductor uses a white-colored wire?
6. What is a footcandle?
7. What type of lamp is most widely used?
8. Describe how power is calculated.
9. What are two common applications for high-intensity discharge lamps?
10. What is a reactor ballast?

Appendix

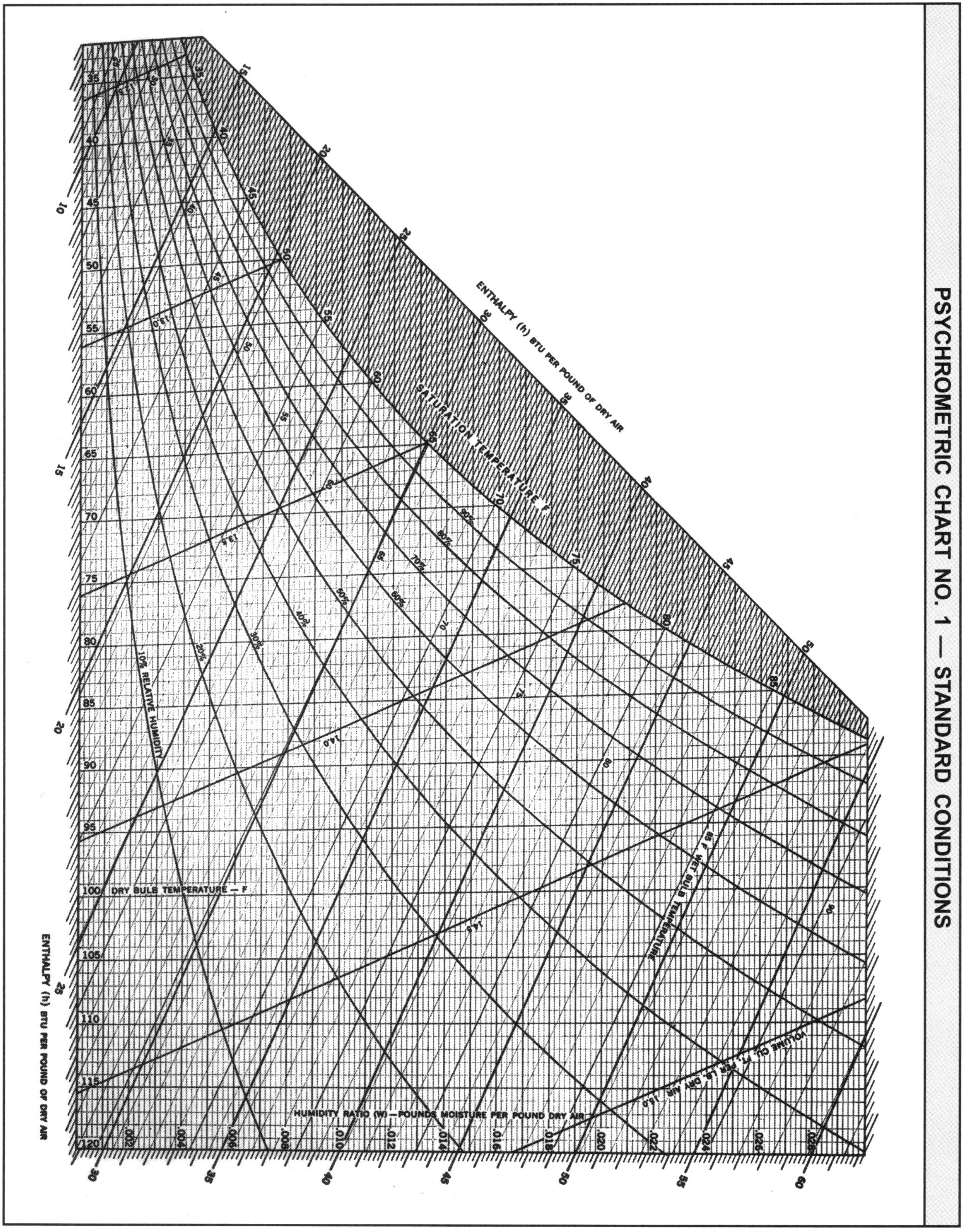

ASHRAE Handbook — Fundamentals

OUTDOOR DESIGN TEMPERATURE					
State and City	**Lat.§**	**Winter**	**Summer**		
		DB*	**DB***	**Daily Range***	**WB***
ALABAMA					
Alexander City	32	18	96	21	79
Auburn	32	18	96	21	79
Birmingham	33	17	96	21	78
Huntsville	34	11	95	23	78
Mobile	30	25	95	18	80
Montgomery	32	22	96	21	79
Talladega	33	18	97	21	79
Tuscaloosa	33	20	98	22	79
ALASKA					
Anchorage	61	–23	71	15	60
Barrow	71	–45	57	12	54
Fairbanks	64	–51	82	24	64
Juneau	58	– 4	74	15	61
Kodiak	57	10	69	10	60
Nome	64	–31	66	10	58
ARIZONA					
Flagstaff	35	– 2	84	31	61
Phoenix	33	31	109	27	76
Tucson	32	28	104	26	72
Yuma	32	36	111	27	79
ARKANSAS					
Fort Smith	35	12	101	24	80
Hot Springs	34	17	101	22	80
Little Rock	34	15	99	22	80
Pine Bluff	34	16	100	22	81
CALIFORNIA					
Bakersfield	35	30	104	32	73
Burbank	34	37	95	25	71
Fresno	36	28	102	34	72
Laguna Beach	33	41	83	18	70
Long Beach	33	41	83	22	70
Los Angeles	33	41	83	15	70
Monterey	36	35	75	20	64
Napa	38	30	100	30	71
Oakland	37	34	85	19	66
Oceanside	33	41	83	13	70
Palm Springs	33	33	112	35	76
Pasadena	34	32	98	29	73
Sacramento	38	30	101	36	72
San Diego	32	42	83	12	71
San Fernando	34	37	95	38	71
San Francisco	37	38	74	14	64
San Jose	37	34	85	26	68
Santa Barbara	34	34	81	24	68
Santa Cruz	36	35	75	28	64
Santa Monica	34	41	83	16	70
Stockton	37	28	100	37	71
COLORADO					
Boulder	40	2	93	27	64

OUTDOOR DESIGN TEMPERATURE					
State and City	**Lat.§**	**Winter**	**Summer**		
		DB*	**DB***	**Daily Range***	**WB***
Colorado Spgs.	38	– 3	91	30	63
Denver	39	– 5	93	28	64
Pueblo	38	– 7	97	31	67
CONNECTICUT					
Bridgeport	41	6	86	18	75
Hartford	41	3	91	22	77
New Haven	41	3	88	17	76
Waterbury	41	– 4	88	21	75
DELAWARE					
Dover	39	11	92	18	79
Wilmington	39	10	92	20	77
DISTRICT OF COLUMBIA					
Washington	38	14	93	18	78
FLORIDA					
Cape Kennedy	28	35	90	15	80
Daytona Beach	29	32	92	15	80
Fort Lauderdale	26	42	92	15	80
Key West	24	55	90	9	80
Miami	25	44	91	15	79
Orlando	28	35	94	17	79
Pensacola	30	25	94	14	80
St. Petersburg	27	36	92	16	79
Sarasota	27	39	93	17	79
Tallahassee	30	27	94	19	79
Tampa	27	36	92	17	79
GEORGIA					
Athens	33	18	94	21	78
Atlanta	33	17	94	19	77
Augusta	33	20	97	19	80
Griffin	33	18	93	21	78
Macon	32	21	96	22	79
Savannah	32	24	96	20	80
HAWAII					
Honolulu	21	62	87	12	76
Kaneohe Bay	21	65	85	12	76
IDAHO					
Boise	43	3	96	31	68
Idaho Falls	43	–11	89	38	65
Lewiston	46	– 1	96	32	67
Twin Falls	42	– 3	99	34	64
ILLINOIS					
Aurora	41	– 6	93	20	79
Bloomington	40	– 6	92	21	78
Carbondale	37	2	95	21	80
Champaign	40	– 3	95	21	78
Chicago	41	– 9	90	15	79
Galesburg	40	– 7	93	22	78
Joliet	41	– 5	93	20	78
Kankakee	41	– 4	93	21	78
Macomb	40	– 5	95	22	79

* in °F   § in degrees

continued

continued

OUTDOOR DESIGN TEMPERATURE

State and City	Lat.§	Winter	Summer		
		DB*	DB*	Daily Range*	WB*
Peoria	40	– 8	91	22	78
Rantoul	40	– 4	94	21	78
Rockford	42	– 9	91	24	77
Springfield	39	– 3	94	21	79
Waukegan	42	– 6	92	21	78
INDIANA					
Fort Wayne	41	– 4	92	24	77
Hobart	41	– 4	91	21	77
Indianapolis	39	– 2	92	22	78
Kokomo	40	– 4	91	22	77
Lafayette	40	– 3	94	22	78
Muncie	40	– 3	92	22	76
South Bend	41	– 3	91	22	77
Terre Haute	39	– 2	95	22	79
Valparaiso	41	– 3	93	22	78
IOWA					
Ames	42	–11	93	23	78
Burlington	40	– 7	94	22	78
Cedar Rapids	41	–10	91	23	78
Des Moines	41	–10	94	23	78
Dubuque	42	–12	90	22	77
Iowa City	41	–11	92	22	80
Keokuk	40	– 5	95	22	79
Sioux City	42	–11	95	24	78
KANSAS					
Garden City	37	– 1	99	28	74
Liberal	37	2	99	28	73
Russell	38	0	101	29	78
Topeka	39	0	99	24	79
Wichita	37	3	101	23	77
KENTUCKY					
Bowling Green	35	4	94	21	79
Lexington	38	3	93	22	77
Louisville	38	5	95	23	79
LOUISIANA					
Alexandria	31	23	95	20	80
Baton Rouge	30	25	95	19	80
Lafayette	30	26	95	18	81
Monroe	32	20	99	20	79
New Orleans	29	29	93	16	81
Shreveport	32	20	99	20	79
MAINE					
Augusta	44	– 7	88	22	74
Bangor	44	–11	86	22	73
Caribou	46	–18	84	21	71
Portland	43	– 6	87	22	74
MARYLAND					
Baltimore	39	10	94	21	78
Frederick	39	8	94	22	78
Salisbury	38	12	93	18	79

OUTDOOR DESIGN TEMPERATURE

State and City	Lat.§	Winter	Summer		
		DB*	DB*	Daily Range*	WB*
MASSACHUSETTS					
Boston	42	6	91	16	75
Clinton	42	– 2	90	17	75
Lawrence	42	– 6	90	22	76
Lowell	42	– 4	91	21	76
New Bedford	41	5	85	19	74
Pittsfield	42	– 8	87	23	73
Worcester	42	0	87	18	73
MICHIGAN					
Battle Creek	42	1	92	23	76
Benton Harbor	42	1	91	20	75
Detroit	42	3	91	20	76
Flint	42	– 4	90	25	76
Grand Rapids	42	1	91	24	75
Holland	42	2	88	22	75
Kalamazoo	42	1	92	23	76
Lansing	42	– 3	90	24	75
Marquette	46	–12	84	18	72
Pontiac	42	0	90	21	76
Port Huron	42	0	90	21	76
Saginaw	43	0	91	23	76
Sault Ste. Marie	46	–12	84	23	72
MINNESOTA					
Alexandria	45	–22	91	24	76
Duluth	46	–21	85	22	72
International Falls	48	–29	85	26	71
Minneapolis	44	–16	92	22	77
Rochester	43	–17	90	24	77
MISSISSIPPI					
Biloxi	30	28	94	16	82
Clarksdale	34	14	96	21	80
Jackson	32	21	97	21	79
Laurel	31	24	96	21	81
Natchez	31	23	96	21	81
Vicksburg	32	22	97	21	81
MISSOURI					
Columbia	38	– 1	97	22	78
Hannibal	39	– 2	96	22	80
Jefferson City	38	2	98	23	78
Kansas City	39	2	99	20	78
Kirksville	40	– 5	96	24	78
Moberly	39	– 2	97	23	78
St. Joseph	39	– 3	96	23	81
St. Louis	38	2	97	21	78
Springfield	37	3	96	23	78
MONTANA					
Billings	45	–15	94	31	67
Butte	45	–24	86	35	60
Great Falls	47	–21	91	28	64
Lewiston	47	–22	90	30	65

* in °F   § in degrees

continued

continued

OUTDOOR DESIGN TEMPERATURE					
State and City	Lat.§	Winter	Summer		
		DB*	DB*	Daily Range*	WB*
Missoula	46	–13	92	36	65
NEBRASKA					
Columbus	41	– 6	98	25	77
Fremont	41	– 6	98	22	78
Grand Island	40	– 8	97	28	75
Lincoln	40	– 5	99	24	78
Norfolk	41	– 8	97	30	78
North Platte	41	– 8	97	28	74
Omaha	41	– 8	94	22	78
NEVADA					
Carson City	39	4	94	42	63
Las Vegas	36	25	108	30	71
Reno	39	6	96	45	64
NEW HAMPSHIRE					
Claremont	43	– 9	89	24	74
Concord	43	– 8	90	26	74
Manchester	42	– 8	91	24	75
Portsmouth	43	– 2	89	22	75
NEW JERSEY					
Atlantic City	39	10	92	18	78
Long Branch	40	10	93	18	78
Newark	40	10	94	20	77
New Brunswick	40	6	92	19	77
Trenton	40	11	91	19	78
NEW MEXICO					
Albuquerque	35	12	96	27	66
Carlsbad	32	13	103	28	72
Gallup	35	0	90	32	64
Los Alamos	35	5	89	32	62
Santa Fe	35	6	90	28	63
Silver City	32	5	95	30	66
NEW YORK					
Albany	42	– 6	91	23	75
Batavia	43	1	90	22	75
Buffalo	42	2	88	21	74
Geneva	42	– 3	90	22	75
Glens Falls	43	–11	88	23	74
Ithaca	42	– 5	88	24	74
Kingston	41	– 3	91	22	76
Lockport	43	4	89	21	76
New York City	40	11	92	17	76
Niagara Falls	43	4	89	20	76
Rochester	43	1	91	22	75
Syracuse	43	– 3	90	20	75
Utica	43	–12	88	22	75
NORTH CAROLINA					
Charlotte	35	18	95	20	77
Durham	35	16	94	20	78
Greensboro	36	14	93	21	77
Jacksonville	34	20	92	18	80

* in °F § in degrees

OUTDOOR DESIGN TEMPERATURE					
State and City	Lat.§	Winter	Summer		
		DB*	DB*	Daily Range*	WB*
Wilmington	34	23	93	18	81
Winston-Salem	36	16	94	20	76
NORTH DAKOTA					
Bismark	46	–23	95	27	73
Fargo	46	–22	92	25	76
Grand Forks	47	–26	91	25	74
Williston	48	–25	91	25	72
OHIO					
Akron-Canton	40	1	89	21	75
Athens	39	0	95	22	78
Bowling Green	41	– 2	92	23	76
Cambridge	40	1	93	23	78
Cincinnati	39	1	92	21	77
Cleveland	41	1	91	22	76
Columbus	40	0	92	24	77
Dayton	39	– 1	91	20	76
Fremont	41	– 3	90	24	76
Marion	40	0	93	23	77
Newark	40	– 1	94	23	77
Portsmouth	38	5	95	22	78
Toledo	41	– 3	90	25	76
Warren	41	0	89	23	74
OKLAHOMA					
Bartlesville	36	6	101	23	77
Chickasha	35	10	101	24	78
Lawton	34	12	101	24	78
McAlester	34	14	99	23	77
Norman	35	9	99	24	77
Oklahoma City	35	9	100	23	78
Seminole	35	11	99	23	77
Stillwater	36	8	100	24	77
Tulsa	36	8	101	22	79
Woodward	36	6	100	26	78
OREGON					
Albany	44	18	92	31	69
Astoria	46	25	75	16	65
Baker	44	– 1	92	30	65
Eugene	44	17	92	31	69
Grants Pass	42	20	99	33	71
Klamath Falls	42	4	90	36	63
Medford	42	19	98	35	70
Portland	45	17	89	23	69
Salem	44	18	92	31	69
PENNSYLVANIA					
Allentown	40	4	92	22	76
Altoona	40	0	90	23	74
Butler	40	1	90	22	75
Erie	42	4	88	18	75
Harrisburg	40	7	94	21	77
New Castle	41	2	91	23	75

continued

continued

OUTDOOR DESIGN TEMPERATURE

State and City	Lat.§	Winter	Summer		
		DB*	DB*	Daily Range*	WB*
Philadelphia	39	10	93	21	77
Pittsburgh	40	1	89	22	74
Reading	40	9	92	19	76
West Chester	39	9	92	20	77
Williamsport	41	2	92	23	75
RHODE ISLAND					
Newport	41	5	88	16	76
Providence	41	5	89	19	75
SOUTH CAROLINA					
Charleston	32	25	94	13	81
Columbia	33	20	97	22	79
Florence	34	22	94	21	80
Sumter	33	22	95	21	79
SOUTH DAKOTA					
Aberdeen	45	−19	94	27	77
Brookings	44	−17	95	25	77
Huron	44	−18	96	28	77
Rapid City	44	−11	95	28	71
Sioux Falls	43	−15	94	24	76
TENNESSEE					
Athens	35	13	95	22	77
Chattanooga	35	13	96	22	78
Dyersburg	36	10	96	21	81
Knoxville	35	13	94	21	77
Memphis	35	13	98	21	80
Murfreesboro	34	9	97	22	78
Nashville	36	9	97	21	78
TEXAS					
Abilene	32	15	101	22	75
Alice	27	31	100	20	82
Amarillo	35	6	98	26	71
Austin	30	24	100	22	78
Beaumont	29	27	95	19	81
Big Spring	32	16	100	26	74
Brownsville	25	35	94	18	80
Corpus Christi	27	31	95	19	80
Dallas	32	18	102	20	78
El Paso	31	20	100	27	69
Forth Worth	32	17	101	22	78
Galveston	29	31	90	10	81
Houston	29	27	96	18	80
Huntsville	30	22	100	20	78
Laredo	27	32	102	23	78
Lubbock	33	10	98	26	73
Mcallen	26	35	97	21	80
Midland	31	16	100	26	73
Pecos	31	16	100	27	73
San Antonio	29	25	99	19	77
Temple	31	22	100	22	78
Tyler	32	19	99	21	80

OUTDOOR DESIGN TEMPERATURE

State and City	Lat.§	Winter	Summer		
		DB*	DB*	Daily Range*	WB*
Victoria	28	29	98	18	82
Waco	31	21	101	22	78
Wichita Falls	33	14	103	24	77
UTAH					
Cedar City	37	− 2	93	32	65
Logan	41	− 3	93	33	65
Provo	40	1	98	32	66
Salt Lake City	40	3	97	32	66
VERMONT					
Barre	44	−16	84	23	73
Burlington	44	−12	88	23	74
Rutland	43	−13	87	23	74
VIRGINIA					
Charlottesville	38	14	94	23	77
Fredericksburg	38	10	96	21	78
Harrisonburg	38	12	93	23	75
Lynchburg	37	12	93	21	77
Norfolk	36	20	93	18	79
Petersburg	37	14	95	20	79
Richmond	37	14	95	21	79
Roanoke	37	12	93	23	75
Winchester	39	6	93	21	77
WASHINGTON					
Aberdeen	46	25	80	16	65
Bellingham	48	10	81	19	68
Olympia	46	16	87	32	67
Seattle-Tacoma	47	17	84	22	66
Spokane	47	− 6	93	28	65
Walla Walla	46	0	97	27	69
WEST VIRGINIA					
Charleston	38	7	92	20	76
Clarksburg	39	6	92	21	76
Parkersburg	39	7	93	21	77
Wheeling	40	1	89	21	74
WISCONSIN					
Beloit	42	− 7	92	24	78
Fon Du Lac	43	−12	89	23	76
Green Bay	44	−13	88	23	76
La Crosse	43	−13	91	22	77
Madison	43	−11	91	22	77
Milwaukee	42	− 8	90	21	76
Racine	42	− 6	91	21	77
Sheboygan	43	−10	89	20	77
Wausau	44	−16	91	23	76
WYOMING					
Casper	42	−11	92	31	63
Cheyenne	41	− 9	89	30	63
Laramie	41	−14	84	28	61
Rock Springs	41	− 9	86	32	59
Sheridan	44	−14	94	32	66

* in °F § in degrees

Air Conditioning Contractors of America

HEAT TRANSFER FACTORS															
Item	**Design Temperature Difference***														
	30	**35**	**40**	**45**	**50**	**55**	**60**	**65**	**70**	**75**	**80**	**85**	**90**	**95**	**100**
Windows§ — Wood or Metal Frame															
Double-hung, Horizontal-slide, Casement, or Awning															
Single glass	45	50	60	65	75	80	90	95	105	110	120	125	135	140	150
With double glass or insulating glass	30	35	40	45	50	55	60	65	70	75	80	80	85	90	95
With storm sash	25	30	35	40	45	50	55	60	60	65	70	75	80	85	90
Fixed or Picture															
Single glass	40	50	55	60	70	75	85	90	95	105	110	115	125	130	140
Double glass or with storm sash	25	30	35	40	45	45	50	55	60	65	70	75	75	80	85
Jalousie															
Single glass	225	265	300	340	375	415	450	490	525	565	600	640	675	715	750
With storm sash	65	75	90	100	110	120	135	145	155	165	175	190	200	210	220
Doors§															
Sliding Glass Doors															
Single glass	75	85	100	115	125	140	150	165	175	190	200	210	225	240	250
Double glass	60	70	80	90	100	110	120	130	140	150	160	170	180	190	200
Other Doors															
Weatherstripped and with storm door	40	45	55	60	65	75	80	85	90	100	105	110	120	125	130
Weatherstripped or with storm door	70	85	95	110	120	135	145	155	170	180	195	205	215	230	240
No weatherstripping or storm door	135	160	180	200	225	250	270	290	315	340	360	380	405	430	450
Walls and Partitions§ — Wood Frame with Sheathing and Siding															
No insulation	8	9	10	11	13	14	15	16	18	19	20	21	23	24	25
R-5 polystyrene sheathing	3	4	4	5	6	6	7	7	8	8	9	9	10	10	11
R-7 batt insulation (2″ – 2¾″)	3	4	4	5	5	6	6	7	7	8	8	9	9	10	10
R-11 batt insulation (3″ – 3½″)	2	2	3	3	4	4	4	5	5	5	6	6	6	7	7
Partition Between Conditioned and Unconditioned Spaces															
Finished one side only, no insulation	17	19	22	25	28	30	33	36	39	41	44	47	49	52	55
Finished both sides, no insulation	9	11	12	14	16	17	19	20	22	23	25	26	28	29	31
Partition with 1″ polystyrene board R-5	4	4	5	5	6	7	7	8	8	9	10	10	11	11	12
R-7 insulation finished both sides	3	4	4	5	5	6	6	7	7	8	8	9	9	10	10
R-11 insulation finished both sides	2	3	3	4	4	4	5	5	6	6	6	7	7	8	8
Solid Masonry, Block, or Brick															
Plastered or plain	14	16	18	20	22	25	27	29	32	34	36	38	40	43	45
Furred, no insulation	9	10	12	13	14	16	17	19	20	22	23	25	26	28	29
Furred, with R-5 insulation	4	5	5	6	6	7	8	8	9	10	10	11	12	12	13
Basement or Crawl Space															
Above grade, no insulation	15	18	20	23	26	28	31	33	36	38	41	43	46	48	51
R-3.57 insulation (molded bead bd.)	5	6	7	8	9	10	11	12	13	14	14	15	16	17	18
R-5 insulation (ext. polystrene bd.)	4	5	6	6	7	8	9	9	10	11	12	12	13	14	14

* in °F

§ in Btu/hr per sq ft (Factors include heat loss for transmission and infiltration.)

Note: R values on this chart refer to thermal resistance value.

continued

continued

HEAT TRANSFER FACTORS

Item	Design Temperature Difference*														
	30	35	40	45	50	55	60	65	70	75	80	85	90	95	100
Basement or Crawl Space															
Wall of crawl space used as supply plenum, R-3.57 insulation	11	12	13	14	15	16	16	17	18	19	20	21	22	23	24
Wall of crawl space used as supply plenum, R-5 insulation	9	10	10	11	12	12	13	14	15	15	16	17	17	18	19
Below grade wall	2	2	2	3	3	3	4	4	4	5	5	5	5	6	6
Ceilings and Roofs[§] — Ceiling Under Unconditioned Space or Vented Roof															
No insulation	18	21	24	27	30	33	36	39	42	45	48	51	54	57	60
R-11 insulation (3″ – 3¼″)	2	3	3	4	4	4	5	5	6	6	6	7	7	8	8
R-19 insulation (5¼″ – 6½″)	2	2	2	2	2	3	3	3	4	4	4	4	4	5	5
R-22 insulation (6″ – 7″)	1	1	2	2	2	2	2	3	3	3	3	3	4	4	4
R-30 insulation	1	1	1	2	2	2	2	2	2	3	3	3	3	3	3
R-38 insulation	1	1	1	1	1	1	2	2	2	2	2	2	2	3	3
R-44 insulation	1	1	1	1	1	1	1	2	2	2	2	2	2	2	2
Roof on Exposed Beams or Rafters															
Roofing on 1½ ″ wood decking no ins.	10	12	14	15	17	19	20	22	24	26	27	29	31	32	34
Roofing on 1½ ″ wood decking 1″ insulation between roofing and decking	5	6	7	8	8	9	10	11	12	13	13	14	15	16	17
Roofing on 1½ ″ wood decking 1½″ insulation between roofing and decking	4	5	5	6	7	7	8	9	10	10	11	12	12	13	14
Roofing on 2″ wood plank	6	7	8	9	10	11	12	14	15	16	17	18	19	20	21
Roofing on 3″ wood plank	5	5	6	7	8	8	9	10	11	11	12	13	14	14	15
Roofing on 1½ ″ fiberboard decking	6	7	8	9	10	10	11	12	13	14	15	16	17	18	19
Roofing on 2″ fiberboard decking	4	5	6	7	8	8	9	10	11	11	12	13	14	14	15
Roofing on 3″ firberboard decking	3	4	4	5	6	6	7	7	8	8	9	9	10	10	11
Roofing-Ceiling Combination															
No insulation	9	11	12	14	16	17	19	20	22	23	25	26	28	29	31
R-11 insulation	2	2	3	3	4	4	4	5	5	5	6	6	6	7	7
R-19 insulation	1	2	2	2	2	2	3	3	3	3	4	4	4	4	5
R-22 insulation (6″ – 7″)	1	1	2	2	2	2	2	3	3	3	3	3	4	4	4
Floors[§] — Floors Over Unconditioned Space															
Over unconditioned room	4	5	6	6	7	8	8	9	10	11	11	12	13	13	14
No insulation	8	10	11	13	14	15	17	18	20	21	22	24	25	27	28
R-7 insulation (2″ – 2¾″)	3	3	4	4	5	5	6	6	7	7	8	8	9	9	9
R-11 insulation (3″ – 3½″)	2	2	3	3	4	4	4	5	5	5	6	6	6	7	7
R-19 insulation (5¼″ – 6½″)	1	2	2	2	2	2	3	3	3	3	4	4	4	4	4
Floor of Room Over Heated Crawl Space															
Less than 18″ below grade	35	40	40	45	45	50	50	55	55	60	60	65	65	70	75
18″ or more below grade	15	20	20	25	25	30	30	35	35	40	40	45	45	50	50

* in °F
§ in Btu/hr per sq ft (Factors include heat loss for transmission and infiltration.)

Air Conditioning Contractors of America

COOLING HEAT TRANSFER FACTORS			
Type of Construction	Cooling Factor*		
	15°	20°	25°
Walls			
Wood Frame with Sheeting, Siding, and Veneer or Other Finish			
No insulation, ½″ gypsum board	5.0	6.4	7.8
R-11 cavity insulation + ½″ gypsum board	1.7	2.1	2.6
R-13 cavity insulation + ½″ gypsum board	1.5	1.9	2.3
R-13 cavity insulation + ¾″ bead board (R-2.7)	1.3	1.7	2.0
R-19 cavity insulation + ½″ gypsum board	1.1	1.4	1.7
R-19 cavity insulation + ¾″ extruded poly	0.9	1.2	1.4
Masonry			
Above grade – No insulation	5.8	8.3	10.9
Above grade + R-5	1.6	2.3	3.1
Above grade + R-11	0.9	1.3	1.6
Below grade – No insulation	0.0	0.0	0.0
Below grade + R-5	0.0	0.0	0.0
Below grade + R-11	0.0	0.0	0.0
Ceilings			
No insulation	17.0	19.2	21.4
2″ – 2½″ insulation R-7	4.4	4.9	5.5
3″ – 3½″ insulation R-11	3.2	3.7	4.1
5¼″ – 6½″ insulation R-19	2.1	2.3	2.6
6″ – 7″ insulation R-22	1.9	2.1	2.4
10″ – 12″ insulation R-38	1.0	1.1	1.3
12″ – 13″ insulation R-44	0.9	1.0	1.1
Cathedral type (roof/ceiling combination)			
No insulation	11.2	12.6	14.1
R-11	2.8	3.2	3.5
R-19	1.9	2.2	2.4
R-22	1.8	2.0	2.2
Floors			
Over Unconditioned Space			
Over basement or enclosed crawl space (not vented)	0.0	0.0	0.0
Over vented space or garage	3.9	5.8	7.7
Over vented space or garage + R-11 insulation	0.8	1.3	1.7
Over vented space or garage + R-19 insulation	0.5	0.8	1.1
Basement Concrete Slab Floor Unheated			
No edge insulation	0.0	0.0	0.0
1″ edge insulation R-5	0.0	0.0	0.0
2″ edge insulation R-9	0.0	0.0	0.0
Basement Concrete Slab Floor Duct in Slab			
No edge insulation	0.0	0.0	0.0
1″ edge insulation R-5	0.0	0.0	0.0
2″ edge insulation R-9	0.0	0.0	0.0

* in °F

continued

Note: R values on this chart refer to thermal resistance value.

COOLING HEAT TRANSFER FACTORS — WINDOWS AND DOORS*									
	Single Glass			Double Glass			Triple Glass		
	Temperature Difference			Temperature Difference			Temperature Difference		
Exposure	15°	20°	25°	15°	20°	25°	15°	20°	25°
N	18	22	26	14	16	18	11	12	13
NE & NW	37	41	46	31	33	35	26	27	28
E & W	52	56	60	44	46	48	38	39	40
SE & SW	45	49	53	39	41	43	33	34	35
S	28	32	36	23	25	27	19	20	21
Wood	8.6	10.9	13.2	8.6	10.9	13.2	8.6	10.9	13.2
Metal	3.5	4.5	5.4	3.5	4.5	5.4	3.5	4.5	5.4

* Inside shading by venetian blinds or draperies.

Air Conditioning Contractors of America

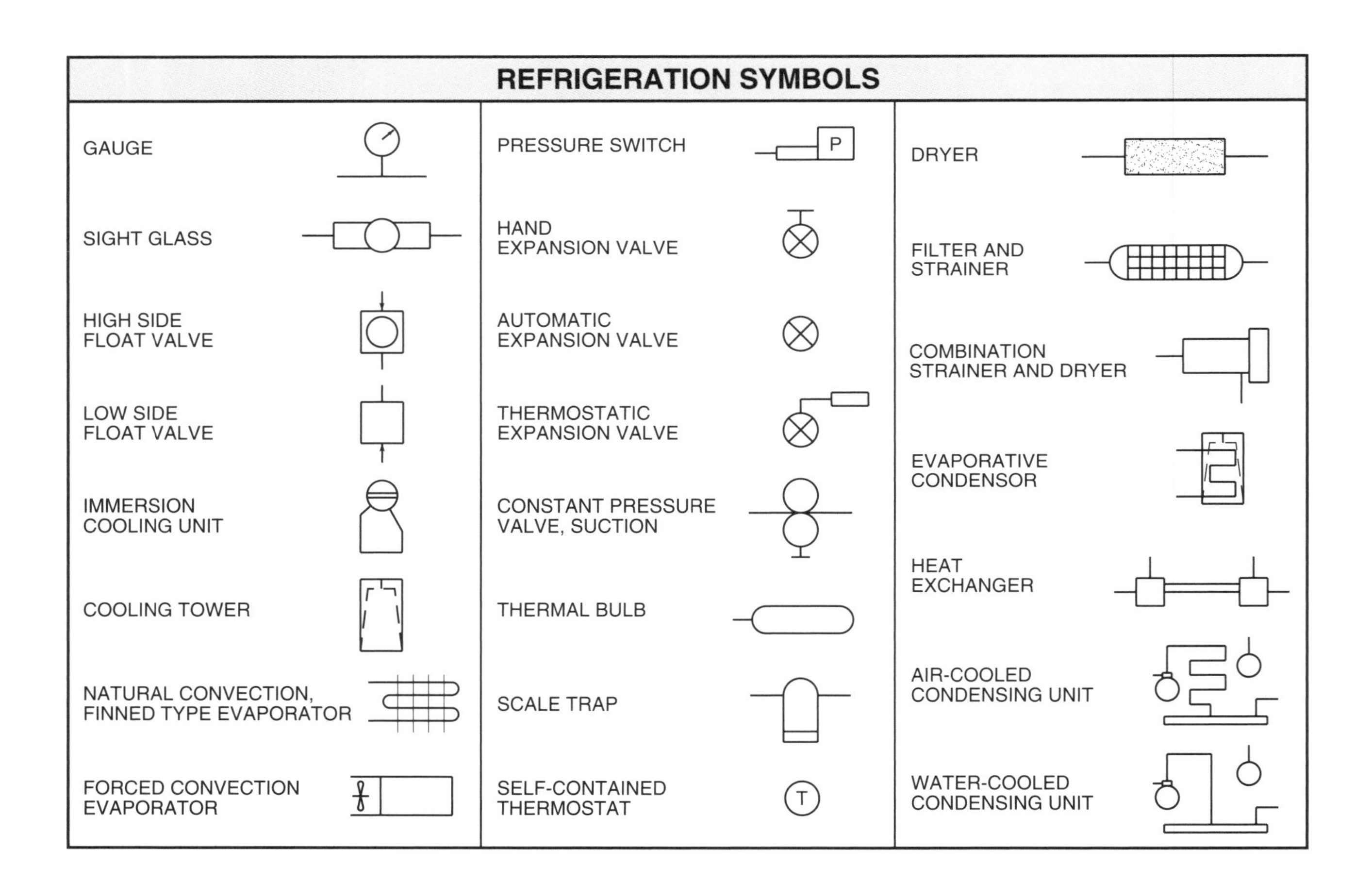

DUCT HEAT LOSS MULTIPLIERS		
Duct Location and Insulation Value	**Duct Loss Multipliers**	
Exposed to Outdoor Ambient Air — Attic, Garage, Exterior Wall, Open Crawl Space	**Winter Design Below 15° F**	**Winter Design Above 15° F**
None	1.30	1.25
R-2	1.20	1.15
R-4	1.15	1.10
R-6	1.10	1.05
Enclosed in Unheated Space — Vented or Unvented Crawl Space or Basement		
None	1.20	1.15
R-2	1.15	1.10
R-4	1.10	1.05
R-6	1.05	1.00
Duct Buried in or Under Concrete Slab — Edge Insulation		
None	1.25	1.20
R value = 3 to 4	1.15	1.10
R value = 5 to 7	1.10	1.05
R value = 7 to 9	1.05	1.00

Air Conditioning Contractors of America

DUCT HEAT GAIN MULTIPLIERS	
Duct Location and Insulation Value Exposed to Outdoor Ambient Air — Attic, Garage, Exterior Wall, Open Crawl Space	**Duct Gain Multiplier**
None	1.30
R-2	1.20
R-4	1.15
R-6	1.10
Enclosed in Unconditioned Space — Vented or Unvented Crawl Space or Basement	
None	1.15
R-2	1.10
R-4	1.05
R-6	1.00
Duct Buried in or Under Concrete Slab — Edge Insulation	
None	1.10
R value = 3 to 4	1.05
R value = 5 to 7	1.00
R value = 7 to 9	1.00

Air Conditioning Contractors of America

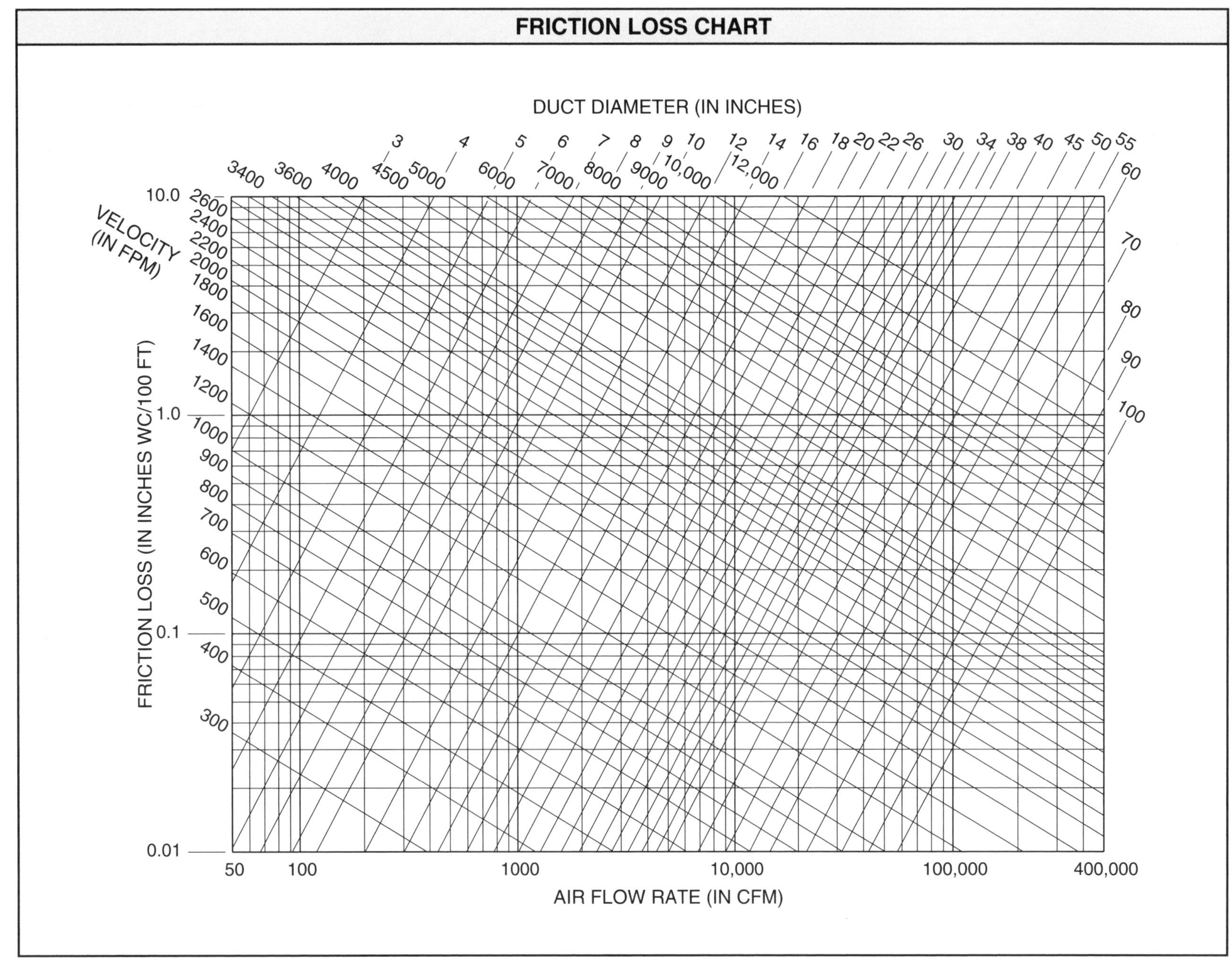

ASHRAE Handbook — Fundamentals

ROUND-TO-RECTANGULAR DUCT CONVERSION CHART

Lgth. Adj.	Length of One Side of Rectangular Duct*																
	4.0	**4.5**	**5.0**	**5.5**	**6.0**	**6.5**	**7.0**	**7.5**	**8.0**	**9.0**	**10.0**	**11.0**	**12.0**	**13.0**	**14.0**	**15.0**	**16.0**
3.0	3.8	4.0	4.2	4.4	4.6	4.7	4.9	5.1	5.2	5.5	5.7	6.0	6.2	6.4	6.6	6.8	7.0
3.5	4.1	4.3	4.6	4.8	5.0	5.2	5.3	5.5	5.7	6.0	6.3	6.5	6.8	7.0	7.2	7.5	7.7
4.0	4.4	4.6	4.9	5.1	5.3	5.5	5.7	5.9	6.1	6.4	6.7	7.0	7.3	7.6	7.8	8.0	8.3
4.5	4.6	4.9	5.2	5.4	5.7	5.9	6.1	6.3	6.5	6.9	7.2	7.5	7.8	8.1	8.4	8.6	8.8
5.0	4.9	5.2	5.5	5.7	6.0	6.2	6.4	6.7	6.9	7.3	7.6	8.0	8.3	8.6	8.9	9.1	9.4
5.5	5.1	5.4	5.7	6.0	6.3	6.5	6.8	7.0	7.2	7.6	8.0	8.4	8.7	9.0	9.3	9.6	9.9

Lgth. Adj.	Length of One Side of Rectangular Duct*																				Lgth. Adj.
	6	**7**	**8**	**9**	**10**	**11**	**12**	**13**	**14**	**15**	**16**	**17**	**18**	**19**	**20**	**22**	**24**	**26**	**28**	**30**	
6	6.6																				6
7	7.1	7.7																			7
8	7.6	8.2	8.7																		8
9	8.0	8.7	9.3	9.8																	9
10	8.4	9.1	9.8	10.4	10.9																10
11	8.8	9.5	10.2	10.9	11.5	12.0															11
12	9.1	9.9	10.7	11.3	12.0	12.6	13.1														12
13	9.5	10.3	11.1	11.8	12.4	13.1	13.7	14.2													13
14	9.8	10.7	11.5	12.2	12.9	13.5	14.2	14.7	15.3												14
15	10.1	11.0	11.8	12.6	13.3	14.0	14.6	15.3	15.8	16.4											15
16	10.4	11.3	12.2	13.0	13.7	14.4	15.1	15.7	16.4	16.9	17.5										16
17	10.7	11.6	12.5	13.4	14.1	14.9	15.6	16.2	16.8	17.4	18.0	18.6									17
18	11.0	11.9	12.9	13.7	14.5	15.3	16.0	16.7	17.3	17.9	18.5	19.1	19.7								18
19	11.2	12.2	13.2	14.1	14.9	15.7	16.4	17.1	17.8	18.4	19.0	19.6	20.2	20.8							19
20	11.5	12.5	13.5	14.4	15.2	16.0	16.8	17.5	18.2	18.9	19.5	20.1	20.7	21.3	21.9						20
22	12.0	13.0	14.1	15.0	15.9	16.8	17.6	18.3	19.1	19.8	20.4	21.1	21.7	22.3	22.9	24.0					22
24	12.4	13.5	14.6	15.6	16.5	17.4	18.3	19.1	19.9	20.6	21.3	22.0	22.7	23.3	23.9	25.1	26.2				24
26	12.8	14.0	15.1	16.2	17.1	18.1	19.0	19.8	20.6	21.4	22.1	22.9	23.5	24.2	24.9	26.1	27.3	28.4			26
28	13.2	14.5	15.6	16.7	17.7	18.7	19.6	20.5	21.3	22.1	22.9	23.7	24.4	25.1	25.8	27.1	28.3	29.5	30.6		28
30	13.6	14.9	16.1	17.2	18.3	19.3	20.2	21.1	22.0	22.9	23.7	24.4	25.2	25.9	26.6	28.0	29.3	30.5	31.7	32.8	30
32	14.0	15.3	16.5	17.7	18.8	19.8	20.8	21.8	22.7	23.5	24.4	25.2	26.0	26.7	27.5	28.9	30.2	31.5	32.7	33.9	32
34	14.4	15.7	17.0	18.2	19.3	20.4	21.4	22.4	23.3	24.2	25.1	25.9	26.7	27.5	28.3	29.7	31.0	32.4	33.7	34.9	34
36	14.7	16.1	17.4	18.6	19.8	20.9	21.9	22.9	23.9	24.8	25.7	26.6	27.4	28.2	29.0	30.5	32.0	33.3	34.6	35.9	36
38	15.0	16.5	17.8	19.0	20.2	21.4	22.4	23.5	24.5	25.4	26.4	27.2	28.1	28.9	29.8	31.3	32.8	34.2	35.6	36.8	38
40	15.3	16.8	18.2	19.5	20.7	21.8	22.9	24.0	25.0	26.0	27.0	27.9	28.8	29.6	30.5	32.1	33.6	35.1	36.4	37.8	40
42	15.6	17.1	18.5	19.9	21.1	22.3	23.4	24.5	25.6	26.6	27.6	28.5	29.4	30.3	31.2	32.8	34.4	35.9	37.3	38.7	42
44	15.9	17.5	18.9	20.3	21.5	22.7	23.9	25.0	26.1	27.1	28.1	29.1	30.0	30.9	31.8	33.5	35.1	36.7	38.1	39.5	44
46	16.2	17.8	19.3	20.6	21.9	23.2	24.4	25.5	26.6	27.7	28.7	29.7	30.6	31.6	32.5	34.2	35.9	37.4	38.9	40.4	46
48	16.5	18.1	19.6	21.0	22.3	23.6	24.8	26.0	27.1	28.2	29.2	30.2	31.2	32.2	33.1	34.9	36.6	38.2	39.7	41.2	48
50	16.8	18.4	19.9	21.4	22.7	24.0	25.2	26.4	27.6	28.7	29.8	30.8	31.8	32.8	33.7	35.5	37.2	38.9	40.5	42.0	50
52	17.1	18.7	20.2	21.7	23.1	24.4	25.7	26.9	28.0	29.2	30.3	31.3	32.3	33.3	34.3	36.2	37.9	39.6	41.2	42.8	52
54	17.3	19.0	20.6	22.0	23.5	24.8	26.1	27.3	28.5	29.7	30.8	31.8	32.9	33.9	34.9	36.8	38.6	40.3	41.9	43.5	54
56	17.6	19.3	20.9	22.4	23.8	25.2	26.5	27.7	28.9	30.1	31.2	32.3	33.4	34.4	35.4	37.4	39.2	41.0	42.7	44.3	56
58	17.8	19.5	21.2	22.7	24.2	25.5	26.9	28.2	29.4	30.6	31.7	32.8	33.9	35.0	36.0	38.0	39.8	41.6	43.3	45.0	58
60	18.1	19.8	21.5	23.0	24.5	25.9	27.3	28.6	29.8	31.0	32.2	33.3	34.4	35.5	36.5	38.5	40.4	42.3	44.0	45.7	60

* in inches

ASHRAE Handbook — Fundamentals

HVAC SYMBOLS

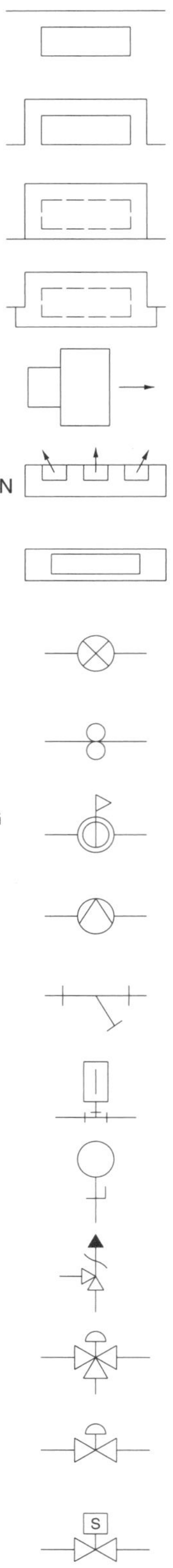

EQUIPMENT SYMBOLS

- EXPOSED RADIATOR
- RECESSED RADIATOR
- FLUSH ENCLOSED RADIATOR
- PROJECTING ENCLOSED RADIATOR
- UNIT HEATER (PROPELLER) – PLAN
- UNIT HEATER (CENTRIFUGAL) – PLAN
- UNIT VENTILATOR – PLAN
- STEAM
- DUPLEX STRAINER
- PRESSURE-REDUCING VALVE
- AIR LINE VALVE
- STRAINER
- THERMOMETER
- PRESSURE GAUGE AND COCK
- RELIEF VALVE
- AUTOMATIC 3-WAY VALVE
- AUTOMATIC 2-WAY VALVE
- SOLENOID VALVE (S)

DUCTWORK

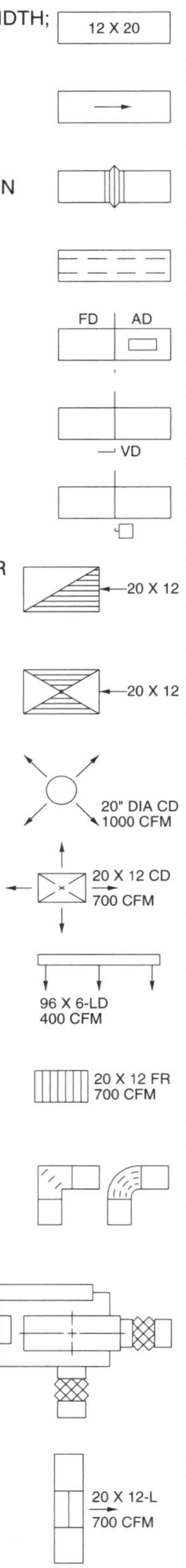

Item	Symbol label
DUCT (1ST FIGURE, WIDTH; 2ND FIGURE, DEPTH)	12 X 20
DIRECTION OF FLOW	
FLEXIBLE CONNECTION	
DUCTWORK WITH ACOUSTICAL LINING	
FIRE DAMPER WITH ACCESS DOOR	FD AD
MANUAL VOLUME DAMPER	VD
AUTOMATIC VOLUME DAMPER	
EXHAUST, RETURN OR OUTSIDE AIR DUCT – SECTION	20 X 12
SUPPLY DUCT – SECTION	20 X 12
CEILING DIFFUSER SUPPLY OUTLET	20" DIA CD 1000 CFM
CEILING DIFFUSER SUPPLY OUTLET	20 X 12 CD 700 CFM
LINEAR DIFFUSER	96 X 6-LD 400 CFM
FLOOR REGISTER	20 X 12 FR 700 CFM
TURNING VANES	
FAN AND MOTOR WITH BELT GUARD	
LOUVER OPENING	20 X 12-L 700 CFM

HEATING PIPING

HIGH-PRESSURE STEAM	—— HPS ——
MEDIUM-PRESSURE STEAM	—— MPS ——
LOW-PRESSURE STEAM	—— LPS ——
HIGH-PRESSURE RETURN	—— HPR ——
MEDIUM-PRESSURE RETURN	—— MPR ——
LOW-PRESSURE RETURN	—— LPR ——
BOILER BLOW OFF	—— BD ——
CONDENSATE OR VACUUM PUMP DISCHARGE	—— VPD ——
FEEDWATER PUMP DISCHARGE	—— PPD ——
MAKEUP WATER	—— MU ——
AIR RELIEF LINE	—— V ——
FUEL OIL SUCTION	—— FOS ——
FUEL OIL RETURN	—— FOR ——
FUEL OIL VENT	—— FOV ——
COMPRESSED AIR	—— A ——
HOT WATER HEATING SUPPLY	—— HW ——
HOT WATER HEATING RETURN	—— HWR ——

AIR CONDITIONING PIPING

REFRIGERANT LIQUID	—— RL ——
REFRIGERANT DISCHARGE	—— RD ——
REFRIGERANT SUCTION	—— RS ——
CONDENSER WATER SUPPLY	—— CWS ——
CONDENSER WATER RETURN	—— CWR ——
CHILLED WATER SUPPLY	—— CHWS ——
CHILLED WATER RETURN	—— CHWR ——
MAKEUP WATER	—— MU ——
HUMIDIFICATION LINE	—— H ——
DRAIN	—— D ——

SELECTED PLUMBING STANDARDS*. . .	
TITLE	**ARTICLE**
Eye and Face Protection.	OSHA 29 CFR 1926.102
Occupational and Educational Eye and Face Protection	ANSI Z87.1
Personal Protection Protective Footwear	ANSI Z41
Ladders	OSHA 29 CFR 1926.1053
Ladders—Portable Wood, Safety Requirements for	ANSI A14.1
Ladders—Fixed-Safety Requirements	ANSI A14.3
Ladders—Portable Reinforced Plastic-Safety Requirements	ANSI A14.5
Aerial Lifts	OSHA 29 CFR 1926.1053
Powered Platforms for Building Maintenance	OSHA 29 CFR 1910.66
Vehicle-Mounted Elevating and Rotating Aerial Devices	ANSI A92.2
Scaffolds	OSHA 29 CFR 1926 Subpart L
Scaffold General Requirements	OSHA 29 CFR 1926.451
Safety Requirements for Scaffolding	ANSI A10.8
Cranes, Derricks, Hoists, and Conveyors	OSHA 29 CFR 1926 Subpart N
Hoisting and Rigging	DOE Standard 1090-99
Overhead Hoists (Underhung)	ANSI/ASME B30.16
Powered Industrial Trucks	ANSI B56.1/OSHA 29 CFR 1910.178
Motor Vehicles	OSHA 29 CFR 1926.601
Material Handling Equipment	OSHA 29 CFR 1926.602
Hazard Communication	OSHA 29 CFR 1910.1200
Hazardous Industrial Chemicals Material Safety Data Sheet Preparation	ANSI Z400.1
Rigging Equipment for Material Handling	OSHA 29 CFR 1926.251
Asbestos	OSHA 29 CFR 1926.1101
Standard Practice for Safe Handling of Solvent Cements, Primers, and Cleaners Used for Joining Thermoplastic Pipe and Fittings	ASTM F402
Bloodborne Pathogens Compliance Directive	OSHA 29 CFR 1910.1030
Standard Specification for Acrylonitrile Butadiene-Styrene (ABS) Schedule 40 Plastic Drain, Waste, and Vent Pipe with a Cellular Core	ASTM F628
Standard Specification for Acrylonitrile Butadiene-Styrene (ABS) Schedule 40 Plastic Drain, Waste, and Vent Pipe Fittings	ASTM D2661
Standard Specification for Poly(Vinyl Chloride) (PVC) Plastic Drain, Waste, and Vent Pipe and Fittings	ASTM D2665
Standard Specification for Poly(Vinyl Chloride) (PVC) Plastic Pipe Fittings, Schedule 40	ASTM D2466
Standard Specification for Socket-Type Poly(Vinyl Chloride) (PVC) Plastic Pipe Fittings, Schedule 80	ASTM D2467
Standard Specification for Threaded Poly(Vinyl Chloride) (PVC) Plastic Pipe Fittings, Schedule 80	ASTM D2464
Standard Specification for Chlorinated Poly(Vinyl Chloride) (CPVC) Plastic Hot- and Cold-Water Distribution Systems	ASTM D2846
Standard Specification for Cross-Linked Polyethylene (PEX) Tubing	ASTM F876
Standard Specification for Cross-Linked Polyethylene (PEX) Plastic Cold- and Hot-Water Distribution Systems	ASTM F877
Cast Copper Alloy Solder Joint Pressure Fittings	ASME B16.18
Wrought Copper and Copper Alloy Solder Joint Pressure Fittings	ASME B16.22
Cast Copper Alloy Solder Joint Drainage Fittings-DWV	ASME B16.23
Wrought Copper and Wrought Copper Alloy Solder Joint Drainage Fittings-DWV	ASME B16.29
Standard Specification for Performance of Gasketed Mechanical Couplings for Use in Piping Applications	ASTM F1476
Standard Specification for the Performance of Fittings for Use with Gasketed Mechanical Couplings Used in Piping Applications	ASTM F1548

. . . SELECTED PLUMBING STANDARDS*. . .	
TITLE	**ARTICLE**
Cast Copper Alloy Fittings for Flared Copper Tube	ASME B16.26
Standard Specification for Hubless Cast Iron Soil Pipe and Fittings for Sanitary, Storm Drain, Waste, and Vent Piping Applications	ASTM A888
Standard Specification for Shielded Couplings Joining Hubless Cast Iron Soil Pipe and Fittings	ASTM C1277
Standard Specification for Rubber Gaskets for Cast Iron Soil Pipe and Fittings	ASTM C564
Standard Specification for Cast Iron Soil Pipe and Fittings	ASTM A74
Malleable Iron from Threaded Fittings	ASME B16.3
Gate Valves for Water and Sewage Systems	ANSI/AWWA C500
Standard Specifications for Envelope Dimensions for Bronze Globe Valves NPS ¼ to 2	ASTM F885
Performance Required for Hose Connection Backflow Preventers	ANSI/ASSE 1052
Face-to-Face Dimensions of Flangeless Control Valves (ANSI Classes 150, 300, and 600)	ANSI/ISA S75.04
Standard Specifications for Envelope Dimensions for Butterfly Valves—NPS 2 to 24	ASTM F1098
Backwater Valves	ASME A112.14.1
Water Pressure-Reducing Valves for Domestic Water Supply Systems	ANSI/ASSE 1003
Relief Valves for Hot Water Supply Systems	ANSI Z21.22
Cold Water Meters-Displacements Type Bronze Main Case	ANSI/AWWA C700
Cold Water Meters—Turbine Type for Customer Service	ANSI/AWWA C701
Cold Water Meters—Compound Type	ANSI/AWWA C702
Standard Specification for Solvent Cement for Acrylonitrile-Butadiene-Styrene (ABS) Plastic Pipe and Fittings	ASTM D2235
Standard Specification for Solvent Cements for Poly (Vinyl Chloride) (PVC) Plastic Piping Systems	ASTM D2564
Standard Specification for Solvent Cements for Chlorinated Poly (Vinyl Chloride) (CPVC) Plastic Pipe and Fittings	ASTM F493
Standard Practice for Safe Handling of Solvent Cements, Primers, and Cleaners Used for Joining Thermoplastic Pipe and Fittings	ASTM F402
Safety of Public Water Systems	SDWA 42 CFR
Standard Specification for Solder Metal	ASTM B32
Standard Specification for Liquid and Paste Fluxes for Soldering of Copper and Copper Alloy Tube	ASTM B813
Specification for Filler Metals for Brazing and Braze Welding	AWS A5.8
Specifications for Fluxes for Brazing and Braze Welding	AWS A5.31
Standard Specification for Rubber Gaskets for Cast Iron Soil Pipe and Fittings	ASTM C564
Pipe Threads, General Purpose (Inch)	ASME B1.20.1
Pipe Hangers and Supports—Materials, Design and Manufacture	MSS-SP58
Pipe Hangers and Supports—Selection and Application	MSS-SP69
Plumbing Fixture Fittings	ASME A112.12.12.1M
Trap Seal Primer Valves	ANSI/ASSE 1018
Water Pressure-Reducing Valves	ASSE 1003
Air Gaps in Plumbing Systems	ASME 112.1.2
Pipe Applied Atmospheric Type Vacuum Breakers	ASSE 1001
Hose Connection Vacuum Breakers	ASSE 1011
Performance Requirements for Hose Connection Backflow Preventers	ASSE 1052
Diverters for Plumbing Faucets with Hose Spray, Anti-siphon Type, Residential Applications	ASSE 1025
Pressure Vacuum Breaker Assemblies	ASSE 1020
Performance Requirements for Backflow Preventer with Intermediate Atmospheric Vent	ASSE 1012
Double Check Valve Backflow Prevention Assembly	AWWA C510

. . . SELECTED PLUMBING STANDARDS*	
TITLE	**ARTICLE**
Dual Check Valve Type Backflow Preventers	ASSE 1024
Performance Requirements for Backflow Preventers on Carbonated Beverage Machines	ASSE 1022
Performance Requirements for Reduced Pressure Principle Backflow Preventers and Reduced Pressure Principle Fire Protection Backflow Preventers	ASSE 1013
Plumbing Fixture Fittings	ASME A112.18.1M
Enameled Cast Iron Plumbing Fixtures	ASME A112.19.1M
Vitreous China Plumbing Fixtures	ASME A112.19.2M
Stainless Steel Plumbing Fixtures	ASME A112.19.3M
Porcelain Enameled Formed Steel Plumbing Fixtures	ASME A112.19.4M
Hydraulic Performance Requirements for Water Closets and Urinals	ASME A112.19.6
Anti-siphon Fill Valves (Ball Cocks) for Gravity Water Closet Flush Tanks	ANSI/ASSE 1002
Pressurized Flushing Devices (Flushometers) for Plumbing Fixtures	ANSI/ASSE 1037
Plastic Toilet (Water Closet) Seats	ANSI Z124.5
Wall-Mounted and Pedestal-Mounted, Adjustable and Pivoting Lavatory and Sink Carrier Systems	ASME A112.19.12
Individual Thermostatic Pressure-Balancing, and Combination Pressure-Balancing and Thermostatic Control Valves for Individual Fixture Fittings	ANSI/ASSE 1016
Plumbing Requirements for Household Dishwashers	ANSI/ASSE 1006
Floor Drains	ASME A112.21.1M
Gas Appliance Thermostats	ANSI Z21.23
Performance Requirements for Water Hammer Arrestors	ANSI/ASSE 1010
Relief Valves for Hot Water Supply Systems	ANSI Z21.22

* The standards listed in this table are selected from several standards organizations that apply to the plumbing trade and do not encompass all standards for the trade.

Glossary

A

abbreviation: Letter or combination of letters that represents a word.

absolute pressure: Any pressure above a perfect vacuum (0 psia).

absorbant: Fluid that has a strong attraction for another fluid.

absorber: Device in an absorption system in which the refrigerant is absorbed.

access control system: A system used to deny those without proper credentials access to a specific building, area, or room.

accessories: Components not directly attached to the boiler.

acrylonitrile-butadiene-styrene (ABS) pipe and fittings: Black plastic pipe and fittings used for sanitary drainage and vent piping and aboveground and underground storm water drainage.

AC sine wave: Symmetrical waveform that contains 360 electrical degrees.

actuator: A device that accepts a control signal and causes a mechanical motion.

AC voltage: Voltage that reverses its direction of flow at regular intervals.

air change factor: Value that represents the number of times per hour that the air in a building is completely replaced by outdoor air.

air conditioner: Component in forced-air air conditioning system that cools the air.

air conditioning: Process of cooling the air in a building to provide a comfortable temperature.

air velocity: Speed at which air moves form one point to another.

air vent: Automatic or manual device that vents air from a hydronic heating system without allowing water to escape or air the enter the system.

air test: Plumbing system test in which inlets and outlets to the system are sealed and air is forced into the system until a uniform air pressure of 5 psi is reached and maintained for 15 min without additional air being added to the system.

alternating current (AC): Current that reverses its direction of flow at regular intervals.

alternation: Half of a waveform cycle.

ampere: Number of electrons passing a given point in one second.

analog display: Electromechanical device that indicates readings by the mechanical motion of a pointer.

analog signal: A signal that has a continuous range of possible values between two points.

angle valve: Globe valve in which the inlet and outlet are at 90° to each other.

annealed copper tube: Drawn copper tube that is heated to a specific temperature and cooled at a predetermined rate to impart desired strength and hardness characteristics.

anti-short bushing: Plastic or heavy fiber paper device used to protect the conductors of armored cable.

apparent power: Product of the voltage and current in a circuit calculated without considering the phase shift that may be present between the voltage and current in the circuit.

appliance: Plumbing fixture that performs a special function and is controlled and/or energized by motors, heating elements, or pressure- or temperature-sensing elements.

arc tube: Light-producing element of a high-intensity discharge (HID) lamp.

area of influence: Area from the front of the register to a point where the air velocity drops below 50 fpm (feet per minute).

armature: Movable coil of wire in a generator that rotates through the magnetic field.

aspect ratio: Ratio between the height and width of a rectangular duct.

atmospheric air: Mixture of dry air, moisture, and particles.

atmospheric burner: Burner that uses ambient air supplied at normal atmospheric air pressure for combustion air.

atmospheric pressure: Force exerted by the weight of the atmosphere on the surface of the Earth.

atom: Smallest particle that an element can be reduced to and still maintain the properties of that element.

authority having jurisdiction (AHJ): The organization, office, or individual responsible for approving the equipment and materials used for building automation installation.

automatic flow control valve: Check valve that opens or closes automatically.

automatic transfer switch: Electrical device that transfers the load of a residence from public utility circuits to the output of a standby generator during a power failure.

automatic zone valve: Valve that is opened or closed automatically by a valve motor, electric solenoid coil, or pneumatic actuator.

average value: Mathematical mean of all instantaneous voltage values in a sine wave.

axial flow blower: Blower that contains a blower wheel, which works like a turbine wheel.

B

backwater valve: Check valve used to prevent the backflow of sewage into a building.

balanced draft: Mechanical draft that uses a combination of forced draft and induced draft.

balance point temperature: Temperature at which the output of a heat pump balances the heat loss of a building.

balancing: Adjustment of the resistance to the flow of water through the piping loops in a system.

ball-type backwater valve: Backwater valve in which backflow is prevented through the used of a ball enclosed within the valve body.

ball valve: Valve in which fluid flow is controlled by a ball that fits tightly against a resilient (pliable) seat in the valve body.

bar graph: Graph composed of segments that function as an analog pointer.

barometric damper: Metal plate positioned in an opening in the flue so that atmospheric pressure can control the airflow.

baseboard radiator: Section of horizontal tubes that are connected with headers at each end and that can be concealed.

basket strainer: Drain fitting installed in a kitchen sink that consists of a strainer body fitted with a fixed strainer and a removable basket with a rubber stopper.

bathtub: Plumbing fixture used to bathe the entire body.

bayonet base: Bulb base that has two pins located on opposite sides.

belt drive system: Motor-to-wheel connection that has a blower motor mounted on the sroll.

bidet: Plumbing fixture used to bathe the external genitals and posterior parts of the body and also to provide relief of certain health conditions.

bimetal element: Sensor that consists of two different kinds of metal that are bonded together into a strip or a coil.

bimetal overload relay: Relay that contains a set of contacts that are actuated by a bimetal element.

black pipe: Steel pipe that is coated with varnish to protect it against corrosion.

blower: Mechanical device that consists of moving blades or vanes that force air through a venturi.

blower drive: Connection from an electric motor to a blower wheel.

blower wheel: Sheet metal cylinder with curved vanes along its perimeter.

blowout urinal: Wall-hung urinal with a nonsiphonic passageway at the rear of the bowl and an integral flush rim and jet.

blowout water closet: Water closet with a nonsiphonic passageway at the rear of the bowl and an integral flush rim and jet.

boiler: Pressure vessel that safely and efficiently transfers heat to water.

boiler combustion safety control: Control that is similar to the combustion safety controls used on forced-air furnaces.

boiler drain: Valve with hose threads that is installed on a tank, such as a water heater, to drain and/or flush the tank.

boiling point: Temperature at which a liquid vaporizes.

bonding wire: Uninsulated conductor in armored cable that is used for grounding.

bottom blowdown valve: Valve located on the blowdown line located at the lowest point of the waterside of a boiler; used to release water from the bottom of the boiler.

branch interval (BI): Vertical length of stack at least 8′ high within which the horizontal branches from one story or floor of the building are connected to the stack.

brightness: Perceived amount of light reflecting from an object.

British thermal unit (Btu): Amount of heat required to raise the temperature of 1 lb of water 1°F.

brush: Sliding contact that rides against the slip rings and is used to connect the armature to the external circuit (power grid).

building automation: The control of the energy- and resource-using devices in a building for optimization of building system operations.

building automation system: A system that uses a distributed system of microprocessor-based controllers to automate any combination of building systems.

building component: Main part of a building structure such as the exterior walls.

building data: Name of each room, running feet of exposed wall, dimensions, ceiling height, and exposure of each room.

burner: Heat-producing component of a combustion furnace.

burner tube: Tube that has an opening on one end and burner ports located along the top.

burner vestibule: Area where burner(s) and controls are located.

butterfly valve: Valve used to control fluid flow consisting of a rotating disk that seats against a resilient material within the valve body.

bypass circuit: Refrigerant line that contains a check valve on each side of an expansion device.

C

cabinet: Sheet metal enclosure that completely covers and provides support for the components of a furnace.

cabinet automatic water softener: Water softener that has a mineral tank within the brine tank.

cabinet convector: Convection heater that is enclosed in a cabinet.

cabinet heater: Forced convection heater that has a blower, hot water coils, filter, and controls in one cabinet.

calorie: Amount of heat required to raise 1 g of water 1°C.

Canadian Standards Association label: Marking that indicates that extensive tests have been conducted on a device.

capacitance: Ability of a component or circuit to store energy in the form of an electrical charge.

capacitive reactance: Opposition to current flow by a capacitor.

capacitor: Electric device specifically designed to store a charge of energy.

cartridge fuse: Snap-in type electrical safety device that contains a wire that is designed to carry a specific amount of current.

cascade system: Compression system that uses one refrigeration system to cool the refrigerant in another system.

cast iron (sectional) boilers: Boilers made of hollow sections that are bolted together.

cathode: Tungsten coil coated with electron-emissive material that releases electrons when heated.

centerset faucet: Combination lavatory fitting that consists of two faucet handles, spout, and pop-up waste fitting lift rod mounted on a raised base.

central-forced air heating system: Forced-air heating system that uses a centrally located furnace to produce heat for a building.

centrifugal blower: Blower that consists of a scroll, blower wheel, shaft, and inlet vanes.

centrifugal feedwater pump: Feedwater pump that converts centrifugal force to flow for transporting feedwater.

centrifugal force: Force that pulls a body outward when it is spinning around a center.

centrifugal pump: Pump that has a rotating impeller inside a cast iron or steel housing.

change in direction: Various turns that may be required in drainage piping.

check valve: Valve that allows flow in only one direction.

chiller: Component in a hydronic air conditioning system that cools water, which cools the air.

chlorinated polyvinyl chloride (CPVC) pipe and fittings: Cream-colored thermoplastic material specially formulated to withstand higher temperatures than other plastics; used in potable water distribution, corrosive industrial fluid handling, and fire suppression systems.

circuit: Complete path (when ON) for current to take that includes electrical control devices, circuit protection, conductors, and loads.

circuit breaker: Overcurrent protection device with a mechanical mechanism that may manually or automatically open a circuit when an overload condition or short circuit occurs.

circuit breaker fuse: Screw-in electrical safety device that has the operating characteristics of circuit breaker.

circulating pump: Pump that moves water from a boiler through the piping system, through terminal devices, and then through return piping of a hydronic heating system.

circulation: The movement of air.

clamp-on ammeter: Meter that measures current in a circuit by measuring the strength of the magnetic field around a conductor.

clamshell heat exchanger: Heat exchanger that has multiple clam-shaped sections.

clip grounding: Grounding method where a ground clip is slipped over the ground wire from the electrical device.

closed-loop control system: A control system in which the result of an output is fed back into a controller as an input.

close nipple: Nipple that is threaded its entire length.

clothes washer outlet box: Plastic enclosure that accommodates water supply and waste connections for a clothes waster.

coefficient of performance: Cooling capacity produced in a refrigeration system from the heat input.

cold solder joint: Soldered joint with poor electrical and mechanical properties.

color coding: Technique used to identify switch terminal screws through the use of various colors.

color rendering: Appearance of a color when illuminated by a light source.

columnar form: Blank table that is divided into columns and rows by vertical and horizontal lines.

combination unit: Air conditioner that contains the components for cooling and heating in one sheet metal cabinet.

combination waste and overflow fitting: Bathtub drain fitting that is an outlet for bathtub waste and allows excess water to drain from the fixture so that it does not overflow onto the bathroom floor.

combustion air blower: Blower used to provide combustion air at a positive pressure at the burner face.

combustion chamber: Area in a heating unit where combustion takes place.

combustion safety controls: Controls that shut down a burner(s) if a malfunction occurs.

comfort: Condition that occurs when a person cannot sense a difference between themselves and the surrounding air.

compact fluorescent lamp: Fluorescent lamp that has a smaller diameter than a conventional fluorescent lamp and a folded bulb configuration.

component grounding: Grounding method where the ground wire is attached directly to an electrical component such as a receptacle.

component list: List of electrical equipment indicating manufacturer, specifications, and the number of each electrical component required for a room or area.

component plan: Group of schedules that state the required locations for receptacles, lights, and switches according to the NEC® and local codes.

compression connector: Box fitting that firmly secures conduit to a box by utilizing a nut that compresses a tapered metal ring (ferrule) into the conduit.

compressive force: Force that squeezes air together.

compressor: Mechanical device that compresses refrigerant or other fluid.

compressor discharge pressure: Pressure created by the resistance to flow of the compressor when refrigerant is discharged from the compressor.

compressor suction pressure: Pressure created by the compressor when refrigerant is drawn into the compressor.

compound gauge: Pressure gauge that indicates vacuum in inches of mercury (Hg) and pressure in pounds per square inch (psi).

compound water meter: Water meter that combines a disc and turbine meter; used in buildings in which there is a large fluctuation of water flow such as an office building, which has large water usage during business hours and little water usage during evening hours and weekends.

compression faucet: Faucet in which the flow of water is shut off by means of a washer that is forced down (compressed) onto its seat, as in a globe valve.

concealed faucet: Combination lavatory fitting that consists of one or two faucet handles and spout mounted above the lavatory or countertop and pop-up waste fitting lift rod, with the faucet bodies below the fixture.

condensate return system: System used to return condensate back to the boiler after all the useful heat has been removed.

condensate return tank: Steam boiler accessory that collects condensate returned from heating units.

condensation: Formation of liquid (condensate) as a gas or vapor cools below its dew point.

condenser: Heat exchanger that removes heat from high-pressure refrigerant vapor.

condensing heat exchanger: Heat exchanger that reduces the temperature of the flue gas below the dew point temperature of the heat exchanger.

condensing medium: Fluid in the condenser of a refrigeration system that carries heat away from the refrigerant.

condensing point: Temperature at which a vapor condenses to a liquid.

conduction factor: Number that represents the amount of heat the flows through a building component because of a temperature difference.

conductor symbol: Electrical symbol that represents copper and aluminum respectively.

conductor: Material that has very little resistance and permits electrons to move through it easily.

conduit: Rugged protective tube (typically metal) through which wires are pulled.

conduit coupling: Fitting used to join one length of conduit to another length of conduit and still maintain a smooth inner surface.

constant-wattage autotransformer ballast: High-reactance autotransformer ballast with a capacitor added to the circuit.

consulting-specifying engineer: A building automation professional that designs the building automation system from the owner's list of desired features.

contactor: Heavy-duty relay that can be rebuilt.

contact tachometer: Device that measures the rotational speed of an object through direct contact of the tachometer tip with the object to be measured.

contact thermometer: Instrument that measures temperature at a single point.

continuity tester: Test instrument that is used to test a circuit for a complete path for current to flow.

continuous waste fitting: Drainage fitting that consists of a section of horizontal drainage pipe and sanitary tee; used to convey waste from a kitchen sink drain to a common P-trap.

contract document: A set of documents produced by the consulting-specifying engineer for use by a contractor to bid a project.

contract specification: A document describing the desired performance of the purchased components and means and methods of installation.

contrast: Ratio of brightness between different objects.

control device: A building automation device for monitoring or changing system variables, making control decisions, or interfacing with other types of systems.

controller: A device that makes decisions to change some aspect of a system based on sensor information and internal programming.

control logic: The portion of controller software that produces the necessary outputs based on the inputs.

control loop: The continuous repetition of the control logic decisions.

control point: A variable in a control system.

control signal: A changing characteristic used to communicate building automation information between control devices.

control valve: Valve designed to control fluid flow rate by partially opening and/or closing the valve.

conventional current flow: Current flow from positive to negative.

convection heating: Heat transfer that occurs when currents circulate between warm and cool regions of a fluid.

cooling load: Amount of heat gained by a building.

cooling tower: Evaporative heat exchanger that removes heat from water.

copper loss: Loss caused by the resistance of the copper wire to the flow of current.

core cock: Valve through which water or gas flow is controlled by a circular core or plug that fits closely in a machined seat.

corporation cock: Core cock placed on the water main to which the water service of the building is connected.

countertop lavatory: Lavatory installed in an opening of a bathroom or rest room cabinet or countertop, or resting on a cabinet frame.

crackage: Opening around windows, doors, or other openings in the building, such as between the foundation and framework of a building.

crimp connector: Electrical device that is used to join wires together or to serve as terminal ends for screw connections.

critical point: Pressure and temperature above which a material does not change state regardless of the absorption or rejection of heat.

cross-linked polyethylene (PEX): Thermosetting plastic made from medium- or high-density cross-linkable polyethylene; used for water service piping and cold and hot water distribution piping.

curb cock: Core cock installed on the water service to turn on or off the potable water flow to a building.

current: Flow of electrons through an electrical circuit.

cycle: One complete positive and negative alternation of a wave form.

D

datum line: Point at which the pressure exerted by water in a vertical column is zero

damper: Device that controls airflow.

DC voltage: Voltage that flows in one direction only.

deadband: **1.** Setpoint differential when no mechanical heating or cooling is allowed because the temperature is between the two setpoints. **2.** The range between two setpoints in which no control action takes place.

defrost cycle: Mechanical procedure that consists of reversing refrigerant flow in a heat pump to melt frost or ice that builds up on the outdoor coil.

dehumidifier: Device that removes moisture from air by causing moisture to condense.

delta configuration: Transformer connection that has each transformer coil connected end-to-end to form a closed loop.

density: Weight of a substance per unit of volume.

design static pressure drop: Pressure drop per unit length of duct for a given size of duct at a given air flow rate.

design temperature: Temperature of the air at a predetermined set of conditions.

design temperature difference: Difference between the desired indoor temperature and the outdoor temperature for a particular season.

derivative control algorithm: A control algorithm in which the output is determined by the instantaneous rate of change of a variable.

dew point: Temperature below which moisture begins to condense in the air.

diaphragm flushometer valve: Flushometer valve in which a segmented diaphragm within the valve body controls the flushing water by equalizing pressure on both sides of the diaphragm.

differential: Difference between the temperature at which the switch in the thermostat turns to burner(s) ON and the temperature at which the thermostat turns the burner(s) OFF.

digital display: Electronic device that displays readings as numerical values.

digital multimeter: Electrical test instrument that can measure two or more electrical properties and display the measured properties as numerical values.

digital signal: A signal that has only two possible states.

digital valve: Two-position (ON/OFF) valve.

dilution air: Atmospheric air that mixes with, dilutes, and cools the products of combustion.

dimensional change hygrometer: Hygrometer that operates on the principle that some materials absorb moisture and change size and shape depending on the amount of moisture in the air.

direct current (DC): Current that flows in only one direction.

direct digital control (DDC) system: A control system in which electrical signals are used to measure and control system parameters.

direct drive system: Motor-to-wheel connection that has a blower wheel mounted directly on the motor shaft.

direct-fired heater: Unit heater that does not have a heat exchanger.

disc water meter: Water meter used to measure water flow through small water services.

discomfort: The condition that occurs when a person can sense a difference between themselves and the surrounding air.

dishwasher: Electric plumbing appliance used to wash dishes.

diverter fitting: Tee that meters water flow.

diverter valve: Three-way valve that has one inlet with two outlets to divert the water flow to a boiler and/or chiller, depending on building needs.

double duct system: Air distribution that consists of a supply duct that carries cool air and a supply duct that carries heated air.

double line drawing: Drawing in which the walls and partitions are shown with double lines.

double offset: Saddle is a common complex bend made in conduit to bypass obstructions.

downflow furnace: Furnace in which heated air flows downward as it leaves the furnace.

draft: Movement of air across a fire and through a heat exchanger.

draft diverter: Box made of sheet metal that runs the width of the heat exchanger.

draft inducer: Blower installed in the flue pipe to provide positive pressure in the flue.

draft system: Boiler system that regulates the flow of air into and from the boiler burner.

drainage fixture unit (dfu): Measure of the probable discharge of wastewater and waterborne waste into the drainage system.

drawn-copper tube: Copper tube that is pulled through a single die or series of dies to achieve a desired diameter.

drinking fountain: Wall-hung plumbing fixture that delivers a stream of drinking water through a nozzle at an upward angle to permit the user of the fixture to conveniently drink from the fountain.

drop-in bathtub: Bathtub that is installed in an enclosure that supports the fixture.

drum heat exchanger: Round drum or tube that is located on a combustion chamber to make the products of combustion flow through it.

dry air: The elements that make up atmospheric air with the moisture and particles removed.

dry-bulb temperature: Measurement of sensible heat.

dry-bulb thermometer: Thermometer that measures dry bulb temperature.

duct chase: Special space provided in a building for installing ductwork.

duct coil: Terminal device that is located in a duct.

duct heater: Unit heater that is installed in a duct and supplied with air from a remote blower.

duct section: Section of ductwork between two fittings where the air flow rate changes.

duct size: Size of a duct, which is expressed in inches of diameter for round ducts and in inches of width and height for rectangular ducts.

dynamic pressure drop: Pressure drop in duct fitting or transition caused by air turbulence as the air flows through the fitting or transition.

E

eddy-current loss: Loss caused by the induced currents that are produced in metal parts that are being magnetized.

efficiency rating: Comparison of the furnace input rating with the output rating.

electrical circuit: Assembly of conductors (wires), electrical devices (switches and receptacles), and components (lights and motors) through which current flows.

electrical impedance hygrometer: Hygrometer based on the principle that the electrical conductivity of a substance changes as the amount of moisture in the air changes.

electrical layout: Drawing that indicates the connections of all devices and components in a residential electrical system.

electrical metallic tubing (EMT): Light-gauge electrical pipe often referred to as thin-wall conduit.

electrical plan: Drawing and list that indicates the devices to be used, the location of the electrical devices, and wiring methods.

electrical system: A combination of electrical devices and components, connected by conductors, that distributes and controls the flow of electricity from its source to a point of use.

electric boiler: Boiler that uses heat produced by resistance heating elements to produce steam or heat water; classified separately from combustion boilers.

electric discharge lamp: Lamp that produces light by an arc discharged between two electrodes.

electric motor: A device that converts electrical energy into rotating mechanical energy.

electric spark igniter: Device that produces an electric spark.

electric water heater: Water heater that utilizes heat produced by the flow of electricity through a resistance wire contained in the heating elements to heat cold water contained within the storage tank.

electricity: Movement of electrons from atom to atom.

electrode: Long metal rod used to make contact with the earth for grounding purposes.

electromagnetism: Magnetic field produced when electricity passes through a conductor.

electromechanical relay: Electric device that uses a magnetic coil to open or close one or more sets of contacts.

electron: Negatively charged particle in an atom.

electron current flow: Current flow from negative to positive.

electrostatic filters: Devices that clean the air as the air passes through electrically charged plates and collector cells.

energy: Capacity to do work.

elevator: A conveying system for transporting people and/or materials vertically between floors in a building.

Engel process: PEX manufacturing process in which peroxides (heat-activated chemicals) release molecules for cross-linking.

enthalpy: Total heat contained in a substance measured from a baseline of 32°F; it is the sum of sensible heat and latent heat.

equal friction chart: Chart that shows the relationship between the air flow rate, static pressure drop, duct size, and air velocity.

equal friction method: Duct sizing that considers that the static pressure is approximately equal at each branch takeoff in the distribution system.

equal velocity method: Duct sizing that considers that each duct section has the same air velocity.

equipment ground: Circuit designed to protect individual components connected to an electrical system.

equipment grounding: Type of grounding that safely grounds any devices attached to a system or plugged into receptacles inside a home.

equivalent temperature difference: Design temperature difference that is adjusted for solar gain.

evaporating medium: Fluid that is cooled when heat is transferred in the evaporator from the evaporating medium to cold refrigerant.

evaporator: Heat exchanger that adds heat to low-pressure refrigerant liquid.

expansion device: Valve or mechanical device that reduces the pressure on a liquid refrigerant by allowing the refrigerant to expand.

expansion loop: Loop in plastic tubing; provides an area for the tubing to expand and contract without stressing.

expansion tank: Vessel that allows the water in a hydronic heating system to expand without raising the system water pressure to dangerous levels.

exposure: Geographic direction a wall faces.

exposed surfaces: Building surfaces that are exposed to outdoor temperatures.

F

factors: Numerical values that represent the heat produced or transferred under some specific condition.

feedwater heater: Steam boiler accessory that heats feedwater before the feedwater enters the boiler.

feedwater pump: Steam boiler accessory used to pump feedwater into a boiler.

feedwater regulator: Steam boiler accessory that maintains the normal operating water level (NOWL) in the boiler.

feedwater system: System that supplies the proper amount of water to a boiler.

feedwater valve: Valve that controls the flow of makeup water into a boiler to compensate for losses.

feet of head: Unit of measure that expresses the height of a column of water that would be supported by a given pressure.

ferrite: Chemical compound consisting of powdered iron oxide and ceramic.

ferromagnetic materials: Materials, such as soft iron, that are easily magnetized.

filament: Conductor with a resistance high enough to cause the conductor to heat.

filter medium: Any porous material that removes particles from a moving fluid.

filtration: Process of removing particles and contaminants from air that circulates through an air distribution system.

final air test: Test of plumbing fixtures and their connections to a sanitary drainage system.

final condition: Point on a psychrometric chart that represents the properties of air after it goes through a process.

fire protection system: A building system for protecting the safety of building occupants during a fire.

firetube boiler: Boiler that heats water that surrounds the fire tubes as the hot gases of combustion pass through the tubes.

fishing: Process of pulling wires through conduit.

fish tape: Device used to pull wires through conduit.

fitting: Device fastened to the ends of pipes to make connections between individual pipes.

fittings: Components directly attached to the boiler and boiler devices that are required for the operation of the boiler.

fixed capacitor: Capacitor that has one value of capacitance.

fixture trim: Water supply and drainage fittings installed on a fixture or appliance to control water flow into a fixture and wastewater flow from the fixture to the sanitary drainage system.

flexible metal conduit: Conduit that has no wires and can be bent by hand.

floor drain: Cast iron or plastic plumbing fixture set flush with the finished floor and used to receive water drained from the floor and convey it to the drainage system.

floor-set water closet: Water closet installed directly on the floor; common in residential construction.

floor sink: Floor drain installed in commercial kitchens and food markets to indirectly receive waste from food preparation and storage equipment and fixtures.

flow control valve: Valve that regulates the flow (in gallons per minute) of water in a hydronic heating system.

flow pressure: Water pressure in the water supply pipe near an outlet, such as a faucet, and is measured while the outlet is wide open and flowing.

flow rate: 1. Volume of water used by a fixture in a given amount of time. **2.** Rate at which a furnace burns fuel.

flushometer valve: Flush device actuated by direct water pressure to supply a fixed quantity of water for flushing purposes.

fluorescent lamp: Low-pressure discharge lamp in which ionization of mercury vapor transforms ultraviolet energy generated by the discharge into light.

flush tank: Reservoir that retains a supply of water used to flush one water closet.

food waste disposer: Electric appliance supplied with water from the kitchen sink faucet that grinds food waste into pulp and discharges the pulp into the drainage system.

footcandle (fc): Amount of light produced by a lamp (lumens) divided by the area that is illuminated.

forced-air cooled transformers: Transformers that use a fan to move air over the transformer.

forced-air heating system: Heating system that uses air to carry heat.

forced draft: Mechanical draft created by air pushed through the firebox by blowers located on the front of a boiler.

forced draft cooling tower: Cooling tower that has a fan located at the bottom of the tower that forces a draft through the tower.

free area: Face area of a register minus the area blocked by the frame or vanes.

freestanding bathtub: Bathtub supported by legs that is not permanently attached to the bathroom walls or floor.

friction head: Effect of friction in a pipe.

friction loss: Decrease in air pressure due to the friction of the air moving through a duct.

fuel oil burner: Boiler component that provides atomized fuel oil to the boiler.

fuel oil heater: Boiler component used to heat cold heavy fuel to assist in pumping and allow for efficient burning.

fuel oil power burner: Burner that atomizes fuel oil.

fuel oil strainer: Device that removes foreign matter from a fuel oil system.

fuel system: System that provides fuel for combustion to produce the necessary heat in a boiler.

full-way valve: Valve designed to be used in its fully open or fully closed position.

furnace: Self-contained heating unit that includes a blower, burner(s) and heat exchanger or electric heating elements, and controls.

fuse: Electric overcurrent protection device used to limit the rate of current flow in a circuit.

fusible plug: Warning device used to indicate extreme overheating of a boiler from a low water condition.

G

galvanized pipe: Steel pipe that is cleaned and dipped into a hot (870°F) molten zinc bath to create a protective coating.

gas: State of matter that is fluid, has a relatively low density, and is highly compressible.

gas butterfly valve: Metering valve that controls the volume of gas sent to a burner.

gas cock: Valve for controlling gas flow to a gas appliance.

gas fuel-fired atmospheric burner: Burner that mixes ambient air with a gas fuel to create a flame.

gas fuel power burner: Power burner that uses natural or liquefied petroleum (LP) gas and has a fan or blower on the outside of the combustion chamber.

gas fuel valve: A 100% shutoff safety valve that controls the flow of fuel to the main burner and the pilot burner.

gas pressure regulator: Device that reduces the gas pressure from the utility to the gas pressure required for the burner.

gas shutoff cock: Special valve used on natural gas fuel systems to manually isolate the fuel system or parts of the system.

gas vent valve: Valve used as a safety device on high-pressure gas systems that use two main gas valves and/or two pilot light valves.

gas water heater: Water heater that utilizes heat produced by the combustion of natural or liquefied petroleum (LP) gas to heat cold water contained within a storage tank.

gate valve: Valve that has an internal gate that slides over an opening through which water flows.

gauge pressure: Pressure above atmospheric pressure that is used to express pressures inside a closed system.

generated electricity: Alternating current (AC) created by power plant generators.

generator: 1. Device in the chiller that adds heat to the refrigerant-absorbant solution to vaporize the refrigerant, raise the pressure of the refrigerant, and separate the refrigerant from the absorbant. **2.** Electromechanical device that converts mechanical energy into electrical energy by means of electromagnetic induction.

ghost voltage: Voltage that appears on a test instrument that is not connected to a circuit.

globe valve: Valve that has a disk (globe) that rises or lowers over a seat through which water flows.

grade: Slope of a horizontal run of pipe, expressed as a fractional inch per foot length of pipe; for example, ¼″ per foot.

grill: Device that covers the opening of the return air ductwork.

grooved end pipe: Steel pipe that is used for grooved joints.

gross wall area: Total area of a wall including windows, doors, and other openings.

gross unit output: Heat output of a boiler when it is fired continuously.

ground fault: Current above the level that is required for a dangerous shock.

ground-fault circuit interrupter (GFCI): Fast-acting receptacle that detects low levels of leakage current to ground and opens the circuit in response to the leakage (ground fault).

grounding conductor: Conductor that does not normally carry current, except during a fault (short circuit).

groundwater: Water that is found naturally in the ground.

H

hard water: Potable water that contains excessive amounts of calcium and magnesium.

heat: Thermal energy.

heat anticipator: Small heating element that is located inside a thermostat.

heat exchanger: Device that transfers heat from one substance to another without allowing the substances to mix.

heating, ventilating, and air conditioning (HVAC) system: A building system that controls a building's indoor climate.

heating load: Amount of heat lost by a building.

heating unit: Heat exchanger, such as a radiator or coil, in which heat that was transported by steam is transferred to the air in a building space.

heat pump: Mechanical compression refrigeration system that contains devices and controls that reverse the flow of refrigerant to move heat from one area to another area.

heat pump thermostat: Component that incorporates a system switch, heating thermostat, and cooling thermostat.

heat of rejection: Amount of heat in British thermal unit/pound rejected by the refrigerant in the condenser.

heat sink: **1.** In heat pumps, a substance with a relatively cold surface that is capable of absorbing heat. **2.** In the electronic field, a device that conducts and dissipates heat away from a component.

heat transfer factor: Conduction factor multiplied by a design temperature difference.

hermetic compressor: Compressor in which the motor and compressor are sealed in the same housing.

high-efficiency filters: Filters that contain filter media made of large bags of filter paper.

high-intensity discharge (HID) lamp: Lamp that produces light from an arc tube.

high-pressure gas switch: Normally closed switch that remains closed and opens if the gas pressure is too high.

high-pressure gas system: System that mixes gas at a pressure of up the 5 psi with air.

high-reactance autotransformer ballast: Ballast that uses two coils (primary and secondary) to regulate both voltage and current.

high-pressure sodium lamp: High-intensity discharge (HID) lamp that produces light when current flows through sodium vapor under high pressure and high temperature

home run: Plumbing design in which centrally located manifolds distribute water to each fixture with dedicated hot and cold water lines.

horizontal branch drain: Drainage pipe extending horizontally from a soil or waste stack or building drain, with or without vertical sections or branches.

horizontal furnace: Furnace in which heated air flows horizontally as it leaves the furnace.

horizontal pipe: Pipe or fitting that makes an angle of less than 45° with the horizontal plane.

hot gas discharge line: Line that connects the compressor to the condenser.

hot surface igniter: Device that uses a small piece of silicon carbide that glows when electric current passes through it.

hot water supply boiler: Boiler having a volume exceeding 120 gal., a heat input exceeding 200,000 Btu/hr, or an operating temperature exceeding 200°F that provides hot water for uses other than heating.

human-machine interface (HMI): An interface terminal that allows an individual to access and respond to building automation system information.

humidifier: Device that adds moisture to air by causing water to evaporate into the air.

humidity: Amount of moisture (water vapor) in air.

humidity ratio: Ratio of the mass (weight) of the moisture in a quantity of air to the mass of the air and moisture together.

hydronic design static pressure drop: Pressure drop per unit length of pipe for a given size of pipe at a given water flow rate.

hydronic heating system: Heating system that uses water or other fluid to carry heat from the point of generation to the point of use.

hydronic radiant heater: Heater that has a radiant surface that is heated by hot water to a temperature high enough to radiate energy.

hygrometer: Any instrument used for measuring humidity.

hyperbolic cooling tower: Cooling tower that has no fan.

hysteresis loss: Loss caused by magnetism that remains (lags) in a material after them magnetizing force has been removed.

I

immersion element: Electric heating device that is inserted into the storage tank of an electric water heater and makes direct contact with the water to provide fast and efficient heat transfer to the water.

immersion heaters: Copper rods enclosed in insulated waterproof tubes that are installed so that water surrounds the heaters.

impedance: Total opposition to the flow of alternating current, consisting of any combination of resistance, inductive reactance, and capacitive reactance.

impeller: Plate with blades that radiate from a central hub (eye).

incandescent lamp: Electric lamp that produces light by the flow of current through a tungsten filament inside a sealed glass bulb sometimes filled with a gas.

inclined manometer: U-tube manometer designed so the bottom of the "U" is a long inclined section of glass or plastic tubing.

indenter connector: Box fitting that secures conduit to a box with the use of a special indenting tool.

indoor unit: Package component that contains a coil heat exchanger and a blower.

individual duct section: Part of a distribution system between fittings in which the air flow, direction, or velocity changes due to the configuration of the duct.

induced draft: Mechanical draft of air pulled through the boiler firebox by a blower located in the breaching after the boiler.

induced draft cooling tower: Cooling tower that has a fan located at the top of the tower that induces a draft by pulling air through the tower.

inductance: Property of a circuit that causes it to oppose a change in current due to energy stored in a magnetic field.

inductive reactance: Opposition of an inductor to alternating current.

infiltration: Process that occurs when outdoor air leaks into a building.

infiltration air: Air that flows into a building when exterior doors are open or when air leaks in through cracks around doors, windows, or other openings.

inflatable test plug: Inflatable rubber device inserted into the plumbing system to seal openings during an air or water test.

infrared meter: Meter that measures heat energy by measuring the infrared energy that a material emits.

infrared radiant heater: Heating unit that heats by radiation only.

initial condition: Point on a psychrometric chart that represents the properties of air before it goes through a process.

inlet vanes: Adjustable dampers that control the airflow to the blower.

in-line ammeter: Meter that measures current in a circuit by inserting the test instrument in series with the component(s) under test.

illumination: Effect that occurs when light falls on a surface.

in-phase: State when voltage and current reach their maximum amplitude and zero level simultaneously.

input rating: Heat produced in Btu/hr per unit of fuel burned.

insulator: Any material that has a very high resistance and resists the flow of electrons.

instant-start circuit: Fluorescent lamp-starting circuit that provides sufficient voltage to strike an arc instantly.

integral control algorithm: A control algorithm in which the output is determined by the sum of the offset over time.

integration: A function that calculates the amount of offset over time as the area underneath a time-variable curve.

intensity: Level of brightness.

internal gear pump: Positive-displacement pump that consists of a small drive gear mounted inside a large internal ring gear.

inverse square law: States that the amount of illumination on a surface varies inversely with the square of the distance from the light source.

invisible light: Portion of the electromagnetic spectrum on either side of the visible light spectrum.

ions: Individual atoms or groups of atoms that carry an electrical charge.

isolated-ground receptacle: Special receptacle that minimizes electrical noise by providing a separate grounding path for each connected device.

J

jet: Orifice (opening) at the base of the bowl that directs water into the passageway inlet to help create a siphonic flushing action.

K

kinetic energy: Energy of motion.

knockout: Round indentation punched into the metal of a box and held in place by unpunched narrow strips of metal.

L

lamp: Output device that converts electrical energy into light.

layout drawings: Drawings of the floor plan of a building that show walls, partitions, windows, doors, fixtures, and other details.

laundry tray: Plumbing fixture used for prewashing clothes that is commonly installed in a residential laundry room.

lavatory: Plumbing fixture used to wash the hands and face.

left-hand bathtub: Bathtub with the drain on the left end as a person faces the tub.

lift check valve: Check valve in which backflow is prevented through the use of a disk that move vertically within the valve body.

lift-and-turn waste fitting: Combination waste and overflow fitting consisting of a stopper with a raised knob at the top that is raised and rotated to allow the fitting to remain in the drain position.

lift waste fitting: Combination waste and overflow fitting in which a lifting mechanism within the overflow tube is connected by a lever to the stopper in the waste shoe (bathtub drain outlet).

light: Portion of the electromagnetic spectrum which produces radiant energy.

lighting system: A building system that provides artificial light for indoor areas.

liquid: State of matter that has a definite volume but not a definite shape.

liquid line: Refrigerant line that connects the condenser and the expansion device.

limit switch: Electric switch that shuts a furnace OFF if the furnace overheats.

linear scale: Scale that is divided into equally spaced segments.

liquid-immersed/forced-air cooled transformers: Transformers that use refined oil or synthetic oil and fans to cool the transformer.

liquid-immersed/self-air cooled transformers: Transformers that use refined oil or synthetic oil to help cool the transformer windings.

load: Device that converts electrical energy to motion, heat, light, or sound.

load calculation software program: Series of commands that requests data from the operator and manipulates the data to determine the heating and cooling loads.

load forms: Documents that are used by designers for arranging the heating and cooling load variables and factors.

low-efficiency filters: Filters that contain filter media made of fiberglass or other fibrous material.

low excess air burner: Burner that uses only the amount of air necessary for complete combustion.

low-pressure gas switch: Normally open switch that is held closed by gas pressure and will open if the gas pressure is too low.

low-pressure gas system: System that mixes gas with air at approximately atmospheric pressure.

low-pressure hot water heating boiler: Boiler in which water is heated for the purpose of supplying heat at pressures not to exceed 160 psi and temperatures not to exceed 250°F.

low-pressure sodium lamp: High-intensity discharge (HID) lamp that operates at a low vapor pressure and uses sodium as the vapor.

low-pressure steam heating boiler: Boiler operated at pressures not to exceed 15 psi of steam.

low-temperature limit control: Temperature-actuated electric switch that may energize a damper motor to shut the damper and open the water valve if the ventilation air temperature drops below a setpoint (35°F).

low water fuel cutoff valve: Steam boiler fitting that shuts the burner OFF in the event of a low water condition in the boiler.

lumen: Unit used to measure the total amount of light produced by a light source.

luminaire temperature: Temperature at which a lamp delivers its peak light output.

M

magnet: Device that attracts iron and steel because of the molecular alignment of its material.

magnetic circuit breaker: Electrical safety device that operates with miniature electromagnets.

magnetic flux: Invisible lines of force that make up a magnetic field.

magnetic starter: Contactor with an overload relay (contacts) added to it.

magnetism: Force that acts at a distance and is caused by a magnetic field.

major duct section: Independent part of a forced-air distribution system through which all or part of the air supply from the blower flows.

main gas valve: Electrically operated valve used to either allow gas to flow to the burner or to stop the flow of gas to the burner.

main steam stop valve: Valve used to place the boiler on-line or take the boiler off-line.

makeup air: Air that is used to replace air that is lost to exhaust.

manifold: Pipe that has outlets for connecting other pipes.

manometer: Device that measures the pressures of vapors and gases.

manual disconnect: Protective metal box that contains fuses or circuit breakers and the disconnect.

manual flow control valve: Globe valve that manually controls the flow of water in a hydronic heating system.

manually controlled circuit: Circuit that requires a person to initiate an action for the circuit to operate.

manual zone valve: Valve that is set by hand to regulate the flow of water in a piping loop.

matter: Anything that has mass and occupies space.

mechanical test plug: Test device inserted into the end of a pipe or other opening and secured in position by tightening a hex or wing nut.

mechanical compression refrigeration: Refrigeration process that produces a refrigeration effect with mechanical equipment.

medium-efficiency filters: Filters that contain filter media made of dense fibrous mats or filter paper.

megohmmeter: Device that detects insulation deterioration by measuring high resistance values under high test voltage conditions.

mercury barometer: Instrument used to measure atmospheric pressure; calibrated in inches of mercury absolute (29.92 in. Hg abs).

mercury-vapor lamp: High-intensity discharge (HID) lamp that produces light by an electrical discharge through mercury vapor.

metal halide: Element (normally sodium and scandium iodide) that is added to mercury in small amounts.

metal-halide lamp: High-intensity discharge (HID) lamp that produces light by an electrical discharge through mercury vapor and metal halide in the arc tube.

metallic-sheathed cable: Type of cable that consists of two or more individually insulated wires protected by a flexible metal outer jacket.

meter: 1. In plumbing systems, a device used to measure and indicate fluid flow, such as a gas or water meter. **2.** In electrical systems, a device used to measure an electrical property.

metering device: Valve or orifice in a refrigeration system that controls the flow of refrigerant into the evaporator to maintain the correct evaporating temperature.

mil: Unit of pressure equal to 1⁄1000 of an inch.

miscibility: Ability of a substance to mix with other substances.

mixing box: Sheet metal box that is attached to the cool air duct and the warm air duct in an HVAC system.

mixing valve: Three-way valve that has two inlets and one outlet and is used to mix two water supplies into one desired flow to the terminal device.

moist air: Mixture of dry air and moisture.

modular force-air heating system: Forced-air heating system that uses more than one heating unit to produce heat for a building.

modulating valve: Infinite position valve.

momentary power interruption: Decrease to 0 V on one or more power lines lasting from 0.5 cycles up to 3 sec.

mop basin: Floor-set basin used for cleaning and other maintenance tasks.

multibulb thermostat: Thermostat that contains more than one mercury bulb switch.

multifunction test instrument: Test instrument that is capable of measuring two or more quantities.

multistage thermostat: Thermostat that contains several mercury bulb switches that make and break contacts in stages.

mutual inductance: Effect of one coil inducing a voltage into another coil.

N

natural draft: Draft produced by natural action resulting from temperature and density differences between atmospheric air and the gases of combustion.

negative charge: Charge that is produced when there are fewer electrons than normal.

net unit output: Gross boiler output multiplied by a percentage of loss because of pickup.

neutral conductor: Current-carrying conductor that is intentionally grounded.

neutron: Neutral particle, with a mass approximately the same as a proton, that exits in the nucleus of an atom.

nipple: Short piece of pipe.

NM cable: Nonmetallic-sheathed cable that has the conductors enclosed within a nonmetallic jacket and is the type typically used for dry interior wiring.

NMC cable: Nonmetallic-sheathed cable that has the conductors enclosed within a corrosion-resistant, nonmetallic jacket.

nominal pipe size: Approximate inside diameter of steel pipe.

nominal size: Cooling capacity of an air conditioner; expressed in British thermal units per hour, or in tons of cooling.

nonlinear scale: Scale that is divided into unequally spaced segments.

nonmetallic-sheathed cable: Electrical conductor (cable) that has a set of insulated electrical conductors held together and protected by a strong plastic jacket.

non-rated globe valve: Globe valve in which the valve seat diameter is less than the stated size of the valve.

nonrising stem-inside screw gate valve: Gate valve in which neither the handwheel nor the stem rises when the valve is opened.

O

offset: **1.** In plumbing, the combination of elbows or bends that brings one section of the pipe out of line but into a line parallel with the other section. **2.** In conduit bending, a compound bend used to bypass many types of obstructions. **3.** In building automation, the difference between the value of a control point and its corresponding setpoint.

ohmmeter: Device that is used to measure the amount of resistance in a component (or circuit) that is not powered.

Ohm's law: Relationship between voltage, current, and resistance in a circuit.

one-pipe primary-secondary piping system: Piping system that circulates boiler water through a primary loop.

one-pipe series piping system: Piping system that circulates water through one pipe, looped through each terminal device.

open circuit transition switching: Process in which power is momentarily disconnected when switching a circuit from one voltage supply or level to another.

open-loop control system: A control system in which decisions are made based only on the current state of the system and a model of how it should work.

operating controls: Controls that cycle equipment ON or OFF.

orifice: Precisely sized hole through which gas fuel flows.

orifice-type metering device: Small, fixed opening that is used as a restriction in the liquid line between the condenser and the evaporator of a refrigeration system.

oscilloscope: Instrument that displays an instantaneous voltage.

os&y valve: Gate valve that indicates whether it is open or closed by the position of the stem.

outdoor design temperature: Expected outdoor temperature that a heating or cooling load must balance.

outdoor design temperature tables: Tables of data developed from records of temperatures that have occurred in an area over many years.

outdoor unit: Package component that contains a coil heat exchanger and a fan.

output rating: Actual heat output produced by a heater in Btu/hr after heat losses from draft.

outside diameter: Nominal distance from outside edge to outside edge of a pipe or the actual distance from outside edge to outside edge of tubing.

overload: Condition that occurs when a motor is connected to an excessive load.

overload relay: Electric switch that protects a motor against overheating and mechanical overloading.

overrim bathtub fitting: Bathtub water supply fitting that consists of a faucet assembly and mixing spout.

P

packaged unit: Self-contained air conditioner that has all of the components contained in one sheet metal cabinet.

parallel connection: Two or more components connected so that there is more than one path for current flow.

passageway: Channel that connects a water closet bowl to the outlet.

pedestal lavatory: Two-piece lavatory with the washbasin resting directly on a pedestal base.

peak load: Maximum output required of a transformer.

peak-to-peak value: Value measured from the maximum positive alternation to the maximum negative alternation of a sine wave.

peak value: Maximum value of either the positive or negative alternation of a sine wave.

perimeter loop system: System that consists of a single loop of ductwork with feeder branches that supply air to the loop.

physiological functions: Natural physical and chemical functions of an organism.

pickup: Additional heat that is needed to warm the water in a hydronic heating system once the period of off-time, such as overnight, has passed.

pictorial drawing: Drawing that shows the length, height, and depth of an object in one view.

pigtail grounding: Grounding method where two ground wires are used to connect an electrical device to a ground screw in the box and then to system ground.

pilot burner: Small burner located near the burner tubes.

pilot safety control: Safety control that determines if the pilot light is burning.

pipe: Cylindrical tube used for conveying potable water, wastewater, waterborne waste, and air from one location to another.

pipe cap: Test device installed on the outside of a pipe end.

pipe plug: Test device installed on the inside of a pipe end.

pipe section: Length of pipe that runs from one fitting to the next fitting.

plain-end pipe: Steel pipe that is not threaded or grooved on the ends.

plans: Drawings of a building that show dimensions, construction materials, location, and arrangement of the spaces within the building.

plastics: Family of synthetic materials manufactured from petroleum-based product and chemicals, including oil, natural gas, and coal.

plumbing fixture: Receptacle or device that is connected permanently or temporarily to the water distribution system, demands a supply of potable water, and discharges the waste directly or indirectly into the sanitary drainage system.

plumbing system: A system of pipes, fittings, and fixtures within a building that conveys a water supply and removes wastewater and waterborne waste.

pneumatic control system: A control system in which compressed air is the medium for sharing control information and powering actuators.

polarity: Positive or negative state of an object.

polarized receptacle: Receptacle in which the size of the connection slots determines the plug connection.

poles: Number of load circuits that contacts control at one time.

polyvinyl chloride (PVC) pipe and fittings: Plastic pipe and fittings used for sanitary drainage and vent piping, aboveground and underground storm water drainage, water mains, and water service lines.

potential energy: Stored energy a body has due to its position, chemical state, or condition.

potentiometer: Variable-resistance electric device that divides voltage proportionally between two circuits.

pop-up waste fitting: Lavatory drain fitting that consists of a brass waste outlet into which a sliding metal or plastic stopper is fitted.

port control faucet: Single-handle noncompression faucet that contains ports for the hot and cold water supply and a cartridge or ceramic disc that opens and closes the ports as the faucet handle is move or rotated.

positive charge: Charge that is produced when there are fewer electrons than normal.

positive pressure: Pressure greater than atmospheric pressure.

power: Rate of doing work or using energy.

power burner: Burner that uses a fan or blower to supply and control combustion air.

power controls: Controls that control the electrical current to a furnace.

power factor: Ratio of true power used in an AC circuit to apparent power delivered to the circuit.

power formula: Relationship between power (P), voltage (E), and current (I) in an electrical circuit.

power hot water (high-temperature) boiler: Boiler used for heating water or liquid to a pressure exceeding 160.psi or to a temperature exceeding 250°F.

power steam boiler: Boiler in which steam or other vapor is used at pressures exceeding 15 psi.

prefilters: Filters installed ahead of bag filters in the airstream to filter large particulate matter.

preheat circuit: Fluorescent lamp-starting circuit that heats the cathode before an arc is created.

prepared form: Preprinted form consisting of columns and rows that identify required information.

pressure: Force per unit of area that is exerted by an object or a fluid.

pressure-balancing valve: Integral temperature-control device that prevents surges of hot and cold water through the use of sensitive diaphragm within the device.

pressure control: Switch on a steam boiler that starts the burner, controls the firing rate, or shuts down the burner based on the steam pressure in the boiler.

pressure drop: Drop in water pressure caused by friction between water and the inside surface of a pipe as the water moves through the pipe.

pressure flush water closet: Water closet in which water is stored in a 1.6 gal. pressure tank before it is discharged to the fixture at a high velocity.

pressure loss due to friction: Pressure variation resulting from friction within the pipe between the street water main and the water supply outlet where the water is being used.

pressure-reducing valve: Automatic device used to convert high and/or fluctuating inlet water pressure to a lower or constant outlet pressure.

pressure-relief valve: Safety device used to automatically lower excessive pressure in a closed plumbing system.

pressure switch: Electric switch that contains contacts and a spring-loaded lever arrangement.

pressure-temperature gauge: Gauge that measures the pressure and temperature of the water at the point on the boiler where the gauge is located.

pressure vessel: Tank or container that operates at a pressure greater than atmospheric pressure.

primary coil: Transformer coil to which voltage is connected.

primary division: Division with a listed value.

programmable thermostat: Stand-alone thermostat that can be programmed by building occupants.

programmer: Control device that functions as the mastermind of the burner control system to control the firing cycle of a boiler.

project closeout information: A set of documents produced by the controls contractor for the owner's use while operating the building.

propeller fan: Mechanical device that consists of blades mounted on a central hub.

properties of air: Characteristics of air, which are temperature, humidity, enthalpy, and volume.

proportional control: A control algorithm in which the output is in direct response to the amount of offset in the system.

proportional thermostat: Thermostat that contains a potentiometer that sends out an electric signal that varies as the temperature varies.

protocol: A set of rules and procedures for the exchange of information between two connected devices.

proton: Positively charged particle that exists in the nucleus of the atom.

psychrometer: Instrument used for measuring humidity that consists of a dry-bulb thermometer and a wet-bulb thermometer mounted on a common base.

psychrometric chart: Chart that defines the condition of the air for any give property.

pulse burner: Low excess air burner that introduces the air-fuel mixture to the burner face in small amounts or pulses.

pulverized coal: Coal that is ground to a fine powder and blown into a boiler where it is burned in suspension.

purge unit: Device used to maintain a system free of air and moisture.

Q

quick connector: Mechanical connection method used to secure wires to the backs of switches and receptacles.

R

radial system: System that consists of branches that run out radially from the supply plenum of a furnace.

radiant heating: Heating that occurs when a surface is heated and the surface gives off heat in the form of radiant energy waves.

radiant panel: Factory-built panel with a radiant surface that is heated by a piping coil or grid built into it.

radiant surface: Heated surface from which radiant energy waves are generated.

radiation process: PEX manufacturing process in which polyethylene is subjected to high-energy electrons to form a cross-linked bond.

rapid-start circuit: Fluorescent lamp-starting circuit that has separate windings to provide continuous heating voltage on the lamp cathodes.

rated globe valve: Globe valve that has a full-size valve seat opening.

rated valve: Valve that meets or exceeds engineering criteria for the normal pressure range of the fluids.

rated voltage: Voltage range that is typically within ±10% of ideal voltage.

reactor ballast: Ballast that connects a coil (reactor) in series with the power line leading to the lamp.

receptacle: Contact device installed for the connection of plugs and flexible cords to supply current to portable electrical equipment.

receptacle box: Electrical device designed to house electrical components and protect wiring connections.

recessed bathtub: Bathtub permanently attached or built into the walls and floor of the bathroom.

rectifier: Device that converts AC voltage to DC voltage by allowing the voltage and current to move in only one direction.

reducing fitting: Pipe fitting in which the dimension of at least one opening is smaller than other openings, such as a ½″ × ¼″ 90° ell or ¾″ × ½″ tee.

refrigerant: Fluid that is used for transferring heat in a refrigeration system.

refrigerant control valve: Combination expansion device and check valve.

refrigeration: Process of moving heat from an area where it is undesirable to an area where the heat is not objectionable.

refrigeration effect: Amount of heat in British thermal unit/pound that a refrigerant absorbs from an evaporator medium.

refrigeration system: Closed system that controls the pressure and temperature of a refrigerant to regulate the absorption and rejection of heat by the refrigerant.

refrigerant property table: Table that contains values for and information about the properties of a refrigerant at saturation conditions or at other pressures and temperatures.

register: Device that covers the opening of the supply air ductwork.

reheat: Heat that is supplied at the point of use while air supply comes from a central location.

relative humidity: Amount of moisture in air compared to the amount of moisture the air would hold at the same temperature if it were saturated (full of water).

relay: An electrical switch that is actuated by a separate electrical circuit.

relief valve: Safety device that is activated to open when pressure and/or temperature in a closed plumbing system exceeds safe operating limits.

remote bulb: Sensor that consists of a small refrigerant-filled metal bulb connected to a thermostat by a thin tube.

resistance heating element: Electric heating element that consists of a grid of electrical resistance wires that are attached to a support frame with ceramic insulators.

resistive circuit: Circuit that contains only resistive components, such as heating elements and incandescent lamps.

resistivity: Resistance of a conductor having a specific length and cross-sectional area.

retrofitting: Process or furnishing a system with new parts that were not available at the time the system was manufactured.

reversing valve: Four-way directional valve that reverses the flow of refrigerant through a heat pump.

right-hand bathtub: Bathtub with the drain on the right end as a person faces the tub.

rigid metal conduit: Heavy-duty pipe that is threaded on the ends much like standard plumbing pipe.

rising stem-inside screw gate valve: Gate valve in which the unthreaded stem and handwheel rise as the valve is opened to indicate the position of the wedge disk.

rising stem-outside stem and yoke (os&y) gate valve: Gate valve in which the threaded stem of the valve rises and the valve is opened.

rooftop unit: Air-cooled package unit that is located on the roof a building.

root-mean-square (effective) value (V_{rms}): Value of a sine wave that produces the same amount of heat in a pure resistive circuit as a DC current of the same value.

rough in: **1.** In plumbing systems, the installation of all parts of a plumbing system that can be completed prior to the installation of the fixtures. **2.** In electrical systems, the placement of electrical boxes and wires before wall coverings and ceilings are installed.

S

safety controls: Controls that monitor the operation of a furnace.

safety relief valve: Valve on a hot water boiler that prevents excessive pressure form building up during a malfunction.

safety valve: Steam boiler fitting (valve) that prevents the boiler from exceeding a maximum allowable working pressure (MAWP).

saturation conditions: Temperature and pressure at which a refrigerant changes state.

saturation line: Curve where the wet-bulb temperature and dew point scales begin on a psychrometric chart.

scope: Device that gives a visual display of voltages.

scopemeter: Combination of an oscilloscope and digital multimeter.

scroll: Sheet metal enclosure that surrounds a blower wheel.

seamless pipe: Steel pipe made by piercing a solid cylindrical steel billet with a series of mandrels while passing the billet through rollers.

secondary coil: Transformer coil in which the voltage is induced.

secondary division: Division that divides primary divisions in halves, thirds, fourths, fifths, etc.

security system: A building system that protects against intruders, theft, and vandalism.

self-air cooled transformers: Transformers that dissipate heat through the air surrounding the transformer.

self-rimming kitchen sink: Kitchen sink in which the bowl is placed in an opening in the countertop and the edge of the fixture rests directly on top of the countertop.

self-rimming lavatory: Lavatory in which the bowl is placed in an opening in the countertop and the edge of the fixture rests directly on top of the countertop.

sensor: A device that measures the value of a variable and transmits a signal that conveys this information.

series connection: Two or more components connected so there is only one path for current flow.

service drop: Overhead wires and devices that connect the power company power lines to a residence.

service lateral: Service to a residence that is achieved by burying the wires underground.

service panel: Electrical device containing fuses or circuit breakers for protecting the individual circuits of a residence; serves as a means for disconnecting the entire residence from the distribution system.

service sink: High-back sink with a deep basin used for filling and emptying scrub pails, rinsing mops, and disposing of cleaning water.

setpoint: The desired value to be maintained by a system.

setpoint adjustor: Lever or dial that indicates the desired temperature on an exposed scale.

setpoint temperature: Temperature at which the switch in a thermostat opens and closes.

set screw connector: Box fitting that relies on the pressure of a screw against the conduit to hold the conduit in place.

sheave: Pulley, which is a grooved wheel.

shop drawing: A document produced by the controls contractor with the details necessary for installation.

shoulder nipple: Nipple that is threaded on the ends and has a short portion of unthreaded pipe in the middle.

shower: Plumbing fixture that discharges water from above a person who is bathing.

shunt: Permanent conductor placed across a water meter to provide a continuous flow path for ground current.

sillcock: Valve with integral external threads installed on the exterior of a building for attachment of a garden hose.

silane process: PEX manufacturing process in which silane molecules are bonded to polyethylene molecules during the extrusion process, resulting in greater manufacturing efficiency and productivity.

single duct system: Air distribution system that consists of a supply duct that carries both cool air and warm air and a return duct that returns the air.

single-function test instrument: Instrument capable of measuring and displaying only one quantity.

single line drawing: Drawing in which the walls and partitions are shown by a single line with no attempt to show actual wall thickness.

siphon jet urinal: Wall-hung urinal in which the trap seal is forced from the trap through a large opening at the trap inlet.

siphon jet water closet: Water closet with a siphonic passageway at the rear of the bowl and an integral flush rim and jet.

sling psychrometer: Instrument used for measuring humidity that consists of a wet-bulb and a dry-bulb thermometer mounted on a base.

slip rings: Metallic rings connected to the ends of the armature and are used to connect the induced voltage to the generator brushes.

small power boiler: Boiler with pressures exceeding 15 psi but not exceeding 100 psi and having less that 440,000 Btu/hr heat input.

smoke chamber: Device constructed of a short section of large-diameter pipe and two reducer fittings in which a smoke cartridge is placed.

specific heat: Amount of heat that is required to raise the temperature of 1 lb of a substance 1°F.

solar gain: Heat gain caused by radiant energy from the sun that strikes opaque objects.

soldering: Process of joining metals by using filler metal and heat to make a strong electrical and mechanical connection.

solderless connector: Device used to join wires firmly without the use of solder.

solenoid: Device that converts electrical energy into a linear, mechanical force.

solenoid-operated pilot valve: Small valve connected to a refrigerant line that contains a solenoid and is used to control other valves.

solid: State of matter that has a definite volume and shape.

solid-wedge disk gate valve: Gate valve in which a one-piece solid bronze wedge fits against the valve seat to restrict fluid flow.

specifications: Written supplements to plans that describe the materials used for a building.

speed: Rate at which an object is moving.

spot heating: Heating that provides heat for a specific building space.

specific volume: Volume of a substance per unit of the substance.

splice: Joining of two or more electrical conductors by mechanically twisting the conductors together or by using a special splicing device.

split-bolt connector: Solderless mechanical connection used for joining large cables.

split system: Air conditioning system that has separate cabinets for the evaporator and the condenser.

split-wedge disk gate valve: Gate valve in which a two-piece bronze wedge fits against the seat to restrict fluid flow.

split-wired receptacle: Standard receptacle that has had the tab between the two brass-colored (hot) terminal screws removed.

stack switch: Mechanical combustion safety control device that contains a bimetal element that senses flue-gas temperature and converts it to mechanical motion.

standard plug fuse: Screw-in type electrical safety device that contains a metal conducting element designed to melt when the current through the fuse exceeds the rated value.

static electricity: Electrical charge at rest.

stagnant air: Air that contains an excess of impurities such as carbon dioxide (CO_2) and lacks the oxygen required to provide comfort.

standard conditions: Values used as a reference for comparing properties of air at different elevations and pressures.

standing radiator: Section of vertical tubes that are connected with headers at each end.

static head: Weight of water in a vertical column above a datum point.

static pressure: 1. Pressure that acts through weight only with no motion. **2.** Air pressure in a duct measured at right angles to the direction of air flow.

static pressure drop: The decrease in air pressure caused by friction between the air moving through a duct and the internal surfaces of the duct.

static regain method: Duct sizing that considers that each duct section is sized so that the static pressure increase at each takeoff offsets the friction loss in the preceding duct section.

steam header: Distribution pipe that supplies steam to branch lines.

steam heating system: Heating system that uses steam to carry heat from the point of generation to the point of use.

steam pressure gauge: Boiler fitting (gauge) that displays the amount of pressure in pounds per square inch (psi) inside a steam boiler.

steam strainer: Steam system accessory that is located in steam lines before the steam trap to remove foreign matter from the steam that could cause a steam trap malfunction.

steam supply system: Piping and support systems that provide a path for steam flow from the boiler to the point of use.

steam system: System that collects, controls, and distributes the steam produced in a boiler.

steam trap: Steam boiler accessory that removes air and condensate from steam lines and heating units.

steel boilers: Boilers that are welded together to form a single unit that has a specific output rating.

step-down transformer: Transformer in which the secondary coil has fewer turns of wire than the primary coil.

step-up transformer: Transformer in which the secondary coil has more turns of wire than the primary coil.

stoker: Mechanical device for feeding coal to a burner at a constant rate.

stop valve: Valve that stops water flow.

stop-and-waste valve: Non-rated globe valve with a side port in the valve body, which is used to drain fluid from the outlet side of the valve.

straight fitting: Pipe fitting in which all openings are the same dimension.

strip gauge: Short groove that indicates the amount of insulation that must be removed from a wire so the wire can be properly inserted into a switch.

strobe tachometer: Device that uses a flashing light to measure the speed of a moving object.

subcooling: Cooling of a material such as a refrigerant to a temperature that is lower than the saturated temperature of the material for a particular pressure.

subdivision: Division that divides secondary divisions in halves, thirds, fourths, fifths, etc.

substation: Assemblage of equipment installed for switching, changing, or regulating the voltage of electricity.

summer dry-bulb temperature: Warmest dry-bulb temperature expected to occur in an area, disregarding the highest temperature that occurs in 1% to 5% of the total hours in the three hottest months of the year.

summer wet-bulb temperature: Wet-bulb temperature that occurs concurrently with the summer dry-bulb temperature.

superheat: Heat added to a material after it has a changed state.

superheated water: Water under pressure that is heated above 212°F without becoming steam.

supply plenum: Sealed sheet metal chamber that connects the furnace supply air opening to the supply ductwork.

surface blowdown valve: Valve located on the surface blowdown line used to release water and surface impurities from the steam boiler.

surface radiation system: Heating system that uses the interior surfaces of a room as radiant surfaces.

surge suppressor: Receptacle that provides protection from high-level transients by limiting the level of voltage allowed downstream from the surge suppressor.

sustained power interruption: Decrease to 0 V on all power lines for a period of more than 1 min.

static head: Weight of water in a vertical column above a datum line.

sweep: Movement of the displayed trace across the scope screen.

switch: Electrical device used to stop, start, or redirect the flow of current in an electrical circuit.

swing-check backwater valve: Backwater valve in which backflow is prevented through the use of hinged disk or flapper.

swing-check valve: Check valve in which backflow is prevented through the use of a hinged disk within the valve body.

symbol: Graphic element that represents a quantity or unit.

system ground: Special circuit designed to protect the entire distribution system of a residence.

system pressure drop: Total static pressure drop in a system.

T

table of equivalent lengths: Table used to convert dynamic pressure drop to friction loss in feet of duct.

tapping valve: Valve that is attached to and pierces a refrigerant line.

temperature: Measurement of the intensity of heat.

temperature difference: Difference between the initial and final temperature of a material through which heat has been transferred.

temperature and pressure (T&P) relief valve: Safety device used to protect against excessive temperature and/or pressure in a water heater.

temperature stratification: Variation of air temperature in a building space that occurs when warm air rises to the ceiling and cold air drops to the floor.

temperature swing: Difference between the setpoint temperature and the actual temperature.

temporary magnets: Magnets that lose their magnetism as soon as the magnetizing force is removed.

temporary power interruption: Decrease to 0 V on one or more power lines lasting between 3 sec and 1 min.

terminal device: Heating system component that transfers heat from the hot water in a hydronic heating system to the air in building spaces.

test cap: Reinforced rubber cap installed on the outside of an opening and secured in position with a stainless steel hose clamp.

test gauge assembly: Test device used to measure pressure within waste and vent, water, gas, air, or other piping systems to ensure that the system is maintaining the proper pressure.

test light: Test instrument that is designed to illuminate in the presence of 115 V and 230 V circuits.

thermal circuit breaker: Electrical safety device that operates with a bimetallic strip that warps when overheated.

thermal conductivity: Property of a material to conduct heat in the form of thermal energy.

thermal magnet circuit breaker: Electrical safety device that combines the heating effect of a bimetallic strip with the pulling strength of a magnet to move the trip bar.

thermal radiation: Heat transfer that occurs as radiant energy waves transport energy from one object to another.

thermistor: Electronic device that changes resistance in response to a temperature change.

thermocouple: Pair of electrical wires (usually constantan and copper) that have different current-carrying characteristics and that are welded together at one end.

thermosetting resin: Plastic resin that cannot be remelted after it is formed and cured in its final shape.

thermoplastic resin: Plastic resin that can be heated and reformed repeatedly with little or no degradation in physical characteristics.

thermostat: Temperature-actuated electric switch.

thermostat setback: Reduction in heating setpoint at night when occupants are asleep or a space is unoccupied.

thermostatic expansion valve: Valve that is controlled by pressure to control refrigerant flow.

thermostatic valve: Integral temperature-control device that regulates the flow of hot and cold water through the use of a thermostat within the device.

threaded and coupled (T and C) pipe: Steel pipe used for threaded joints that has threads on both ends of the pipe and coupling on one end.

thermostat: Temperature-actuated electric switch that turns an air conditioner ON or OFF in response to temperature changes in building spaces.

throws: Number of different closed contact positions per pole, which is the number of circuits that each individual pole controls.

time-delay plug fuse: Screw-in type electrical safety device with a dual element.

total head: Sum of friction head and static head.

ton of cooling: Amount of heat required to melt a ton of ice (2000 lb) over a 24-hr period.

trace: Reference point/line that is visually displayed on the face of the scope screen.

transformer: Electric device that uses electromagnetism to change (step-up or step-down) AC voltage from one level to another.

transient: Temporary unwanted voltage in an electrical circuit.

trap seal: Height of water in a water closet bowl, which prevents sewer gases from entering the living area.

transformer taps: Connecting points that are provided along the transformer coil.

T rating: Special switch information that indicates a switch is capable of handling the severe overloading created by a tungsten load as the switch is closed.

true power: Actual power used in an electrical circuit.

trunk: Main supply duct that extends from the supply plenum.

trunk and branch system: System that consists of one or more trunks that run out from the supply plenum of a furnace.

try cocks: Valves located on the water column that are used to determine the boiler water level if the gauge glass is not functional.

T-tap splice: Splice that allows a connection to be made without cutting the main wire.

tungsten-halogen lamp: Incandescent lamp filled with a halogen gas (iodine or bromine).

tuning: The adjustment of control parameters to the optimal values for the desired control response.

turbine water meter: Water meter used to measure large and constant volumes of water in buildings.

Type-S plug fuse: Screw-in safety device that has all the operating characteristics of a time-delay plug fuse plus the added advantage of being non-tamperable.

two-pipe direct-return piping system: Piping system that circulates the supply water in the opposite direction of the circulation of the return water.

two-pipe reverse-return piping system: Piping system that circulates return water in the same direction as supply water.

two-stage compression system: Compression system that uses more than one compressor to raise the pressure of a refrigerant.

two-tank automatic water softener: Water softener that has separate mineral (zeolite) and brink tanks.

two-winding constant-wattage ballast: Ballast that uses a transformer which provides isolation between the primary and secondary circuits.

U

UL label: Stamped or printed icon that indicates that a device or material has been approved for consumer use by Underwriters Laboratories Inc.®

ultrasonic leak detector: Sensitive microphone probe and amplifier that is used to locate leaks in water supply and distribution piping.

ultraviolet region: Region of the spectrum with wavelengths just short of the color violet.

unit heater: Self-contained heating unit that is not connected to ductwork.

undercounter lavatory: Lavatory attached to the underside of a countertop using rim clamps or other proprietary hardware.

ungrounded conductor: Current-carrying conductor that is connected to loads through fuses, circuit breakers, and switches.

unit air conditioner: Self-contained air conditioner that contains a coil and a blower in one cabinet.

uninterruptible power supply (UPS): Power supply that provides constant on-line power when the primary power supply is interrupted.

unit ventilator: Forced convection heater that has a blower, hot water coil, filters, controls, and a cabinet that has an opening for outdoor ventilation.

upflow furnace: Furnace in which heated air flows upward as it leaves the furnace.

urinal: Plumbing fixture that receives only liquid body waste and conveys the waste through a trap seal into a sanitary drainage system.

U-tube manometer: U-shaped section of glass or plastic tubing that is partially filled with water or mercury.

V

vacuum: Pressure lower than atmospheric pressure.

vacuum breaker: Backflow prevention device that consists of a spring-loaded check valve that seals against an atmospheric outlet when water is turned on.

valence shell: Outermost shell of an atom; contains the electrons that form new compounds.

valve: Fitting used to regulate fluid flow within a system.

vanity-top lavatory: One-piece washbasin and countertop installed on top of a bathroom or rest room cabinet or supported from wall framing members.

variables: Data that are unique to a building.

variable capacitor: Capacitor that varies in capacitance value.

velocity reduction method: Duct sizing that considers that the air velocity is reduced at each branch takeoff.

vented closet cross: Specially designed fittings for offsetting a water closet on a lower floor from the stack to properly vent the water closet.

vented closet tee: Specially designed fitting for offsetting a water closet on a lower floor from the stack to properly vent the water closet.

ventilation: Process that occurs when outdoor air is brought into a building.

ventilation air: Air that is brought into a building to keep building air fresh.

venturi: Restriction that causes increased pressure as air moves through it.

vent valve (boiler vent or air cock): Boiler fitting used to vent air from inside a boiler.

vertical pipe: Pipe or fitting that makes an angle of 45° or less with the vertical plane.

virtual point: A control point that only exists in software.

visible light: Portion of the electromagnetic spectrum to which the human eye responds.

vitreous china: Ceramic compound fired at high temperature to form a nonporous material with a ceramic glaze fused to the surface.

voice-data-video (VDV) system: A building system used for the transmission of information.

voltage: Amount of electrical pressure in a circuit.

voltage indicator: Test instrument that indicates the presence of voltage when the test tip touches, or is near, an energized hot conductor or energized metal part.

voltage tester: Electrical test instrument that indicates the approximate amount of voltage and the type of voltage (AC or DC) in a circuit by the movement of a pointer (and vibration, on some models).

volume: Amount of space occupied by a three-dimensional figure.

W

wall-hung lavatory: Lavatory supported by a stamped steel or cast iron bracket fastened to a backing board installed between wall framing members when the fixture is roughed in.

wall-hung water closet: Water closet suspended from the wall on a supporting chair carrier; used in commercial construction.

wall units: Packaged units that provide heat to a building space by using a fan and heat exchanger mounted in a metal enclosure that is mounted against an outside wall.

washout urinal: Urinal in which the liquid waste is washed from a trap rather than flushed.

water closet: Water-flushed plumbing fixture that receives human liquid and solid waste in a water-containing receptacle and, upon flushing, conveys the waste to a soil pipe.

water column: Steam boiler fitting that reduces the movement of boiler water in the gauge glass to provide an accurate water level.

water cooler: Wall-hung plumbing appliance that incorporates an electric cooling unit into a drinking fountain to provide cooled drinking water at a desired temperature.

water filter: Plumbing appliance that removes sand, sediment, chlorine, lead, and other undesirable elements from water and protects water heaters and other fixtures and appliances from collecting residue.

water flow rate: Volumetric flow rate of the water as it moves through a given pipe section.

water hammer: 1. Water supply system defect in which a loud noise is created when a quick-closing valve, such as a clothes washer valve or ball valve, is suddenly closed. **2.** Banging noise caused by water in steam lines moving rapidly and hitting obstructions such as elbows and valves.

water hammer arrestor: Device installed on water supply pipe near the fixture with the quick-closing valve to control the effect of water hammer.

water heater: Plumbing appliance used to heat water for purposes other than heating a structure.

water meter: Device used to measure, in cubic feet or gallons, the amount of water flowing through the water service.

water softener: Plumbing appliance that removes dissolved minerals, such as calcium and magnesium, from water by an ion-exchange process.

water spot: Surface area of the water in the water closet bowl when a flush is completed; usually measured in inches of width by length.

water strainer: Device in a hydronic heating system that removes particles from the water inside a piping system.

water supply fixture unit (wsfu): Measure of the estimated water demand of a plumbing fixture.

water supply fixture unit sizing method: Common method of properly sizing water supply piping for buildings that require a 2″ or smaller water service and in which the distribution piping does not exceed 2 ½″ in size.

water test: Plumbing system test in which pipe openings are sealed with plugs or caps and the pipe is filled with water to provide a specified amount of pressure to the plumbing system to determine the tightness of the system.

watertube boiler: Boiler that heats water as hot products of combustion pass over the outside surfaces of the water tubes.

watt: Equal to the power produced by a current of 1 A across a potential difference of 1 V.

welded pipe: Steel pipe manufactured by drawing flat steel strips through a die to form a cylindrical shape and then electric butt-welding the seam to create a leakproof joint.

Western Union splice: Splice that is used when the connection must be strong enough to support long lengths of heavy wire.

wet-bulb depression: Difference between the wet-bulb and dry-bulb temperature readings; directly related to the amount of moisture in the air.

wet-bulb temperature: Measurement of the amount of moisture in the air.

wet-bulb thermometer: Thermometer that has a small cotton sock placed over the bulb.

whirlpool bathtub: Plumbing fixture equipped with water and air circulation equipment and used to bathe and massage the entire body.

winter dry-bulb temperature: Coldest temperature expected to occur in an area, disregarding the lowest temperature that occurs in 1% of 2½% of the total hours in the three coldest months of the year.

wire marker: Preprinted peel-off sticker designed to adhere to insulation when wrapped around a conductor.

wiring plan: Drawing that indicates the placement of all electrical devices and components and the wiring required to connect all the equipment into circuits.

wrap-around bar graph: Bar graph that displays a fraction of the full range on the graph.

wrist blade handle: ADA-compliant faucet handle that does not require tight grasping by the hand or twisting of the wrist to properly operate the faucet.

wye configuration: Transformer connection that has one end of each transformer coil connected together.

Z

zeolite: Synthetic resin bead used as the ion-exchange medium in the ion-exchange process.

zone: Specific section of a building that requires separate temperature control.

zone valve: Valve that regulates the flow of water in a zone or terminal device of a building.

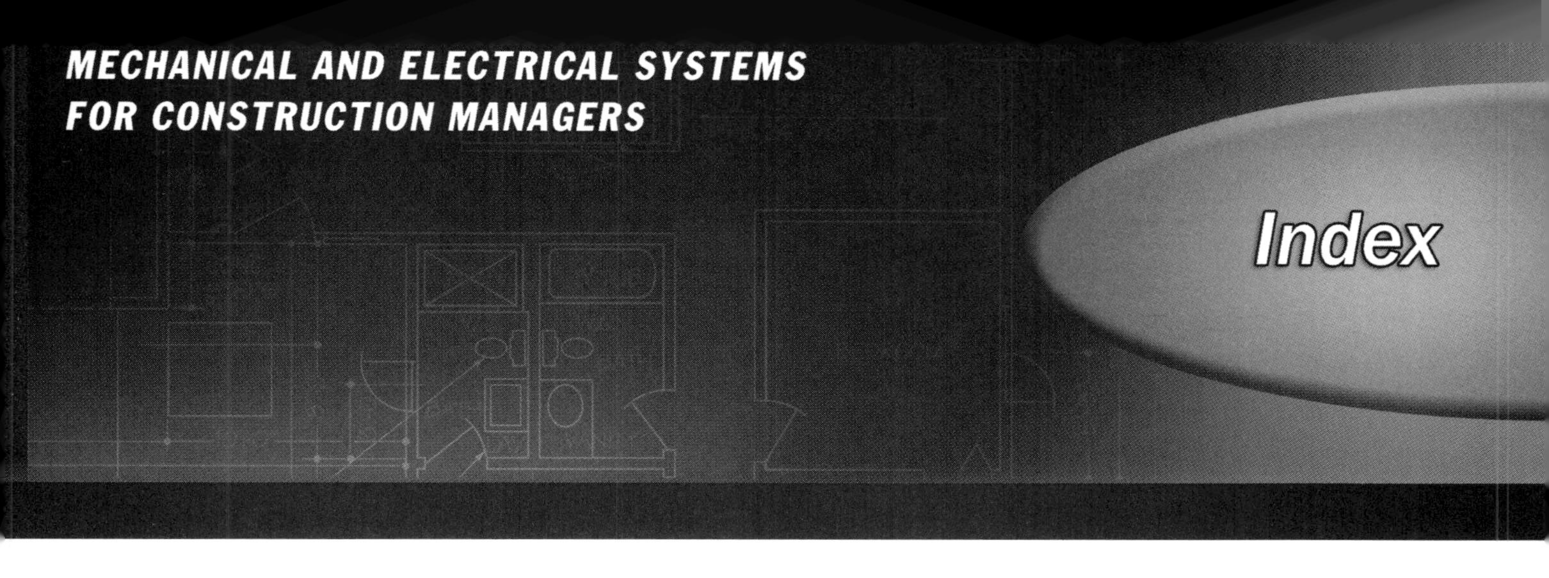
MECHANICAL AND ELECTRICAL SYSTEMS
FOR CONSTRUCTION MANAGERS
Index

G

Q

R

T

U